Recommended Dietary Allowances (RDA) and Adequate Intakes (AI) for Vitamins

Age (yr)	Thiamin RDA (mg/day)	Riboflavin RDA (mg/day)	Niacin RDA (mg/day)[a]	Biotin AI (μg/day)	Pantothenic acid AI (mg/day)	Vitamin B_6 RDA (mg/day)	Folate RDA (μg/day)[b]	Vitamin B_{12} RDA (μg/day)	Choline AI (mg/day)	Vitamin C RDA (mg/day)	Vitamin A RDA (μg/day)[c]	Vitamin D AI (μg/day)[d]	Vitamin E RDA (mg/day)[e]	Vitamin K AI (μg/day)
Infants														
0–0.5	0.2	0.3	2	5	1.7	0.1	65	0.4	125					2.0
0.5–1	0.3	0.4	4	6	1.8	0.3	80	0.5	150					2.5
Children														
1–3	0.5	0.5	6	8	2	0.5	150	0.9	200					30
4–8	0.6	0.6	8	12	3	0.6	200	1.2	250					55
Males														
9–13	0.9	0.9	12	20	4	1.0	300	1.8	375	45	600	5	11	60
14–18	1.2	1.3	16	25	5	1.3	400	2.4	550	75	900	5	15	75
19–30	1.2	1.3	16	30	5	1.3	400	2.4	550	90	900	5	15	120
31–50	1.2	1.3	16	30	5	1.3	400	2.4	550	90	900	5	15	120
51–70	1.2	1.3	16	30	5	1.7	400	2.4	550	90	900	10	15	120
>70	1.2	1.3	16	30	5	1.7	400	2.4	550	90	900	15	15	120
Females														
9–13	0.9	0.9	12	20	4	1.0	300	1.8	375	45	600	5	11	60
14–18	1.0	1.0	14	25	5	1.2	400	2.4	400	65	700	5	15	75
19–30	1.1	1.1	14	30	5	1.3	400	2.4	425	75	700	5	15	90
31–50	1.1	1.1	14	30	5	1.3	400	2.4	425	75	700	5	15	90
51–70	1.1	1.1	14	30	5	1.5	400	2.4	425	75	700	10	15	90
>70	1.1	1.1	14	30	5	1.5	400	2.4	425	75	700	15	15	90
Pregnancy														
≤18	1.4	1.4	18	30	6	1.9	600	2.6	450	80	750	5	15	75
19–30	1.4	1.4	18	30	6	1.9	600	2.6	450	85	770	5	15	90
31–50	1.4	1.4	18	30	6	1.9	600	2.6	450	85	770	5	15	90
Lactation														
≤18	1.4	1.6	17	35	7	2.0	500	2.8	550	115	1200	5	19	75
19–30	1.4	1.6	17	35	7	2.0	500	2.8	550	120	1300	5	19	90
31–50	1.4	1.6	17	35	7	2.0	500	2.8	550	120	1300	5	19	90

NOTE: For all nutrients, values for infants are AI.

[a] Niacin recommendations are expressed as niacin equivalents (NE), except for recommendations for infants younger than 6 months, which are expressed as preformed niacin.

[b] Folate recommendations are expressed as dietary folate equivalents (DFE).

[c] Vitamin A recommendations are expressed as retinol activity equivalents (RAE).

[d] Vitamin D recommendations are expressed as cholecalciferol and assume an absence of adequate exposure to sunlight.

[e] Vitamin E recommendations are expressed as α-tocopherol.

Recommended Dietary Allowances (RDA) and Adequate Intakes (AI) for Minerals

Age (yr)	Sodium AI (mg/day)	Chloride AI (mg/day)	Potassium AI (mg/day)	Calcium AI (mg/day)	Phosphorus RDA (mg/day)	Magnesium RDA (mg/day)	Iron RDA (mg/day)	Zinc RDA (mg/day)	Iodine RDA (μg/day)	Selenium RDA (μg/day)	Copper RDA (μg/day)	Manganese AI (mg/day)	Fluoride AI (mg/day)	Chromium AI (μg/day)	Molybdenum RDA (μg/day)
Infants															
0–0.5	120	180	400	210	100	30	0.27	2	110	15	200	0.003	0.01	0.2	2
0.5–1	370	570	700	270	275	75	11	3	130	20	220	0.6	0.5	5.5	3
Children															
1–3	1000	1500	3000	500	460	80	7	3	90	20	340	1.2	0.7	11	17
4–8	1200	1900	3800	800	500	130	10	5	90	30	440	1.5	1.0	15	22
Males															
9–13	1500	2300	4500	1300	1250	240	8	8	120	40	700	1.9	2	25	34
14–18	1500	2300	4700	1300	1250	410	11	11	150	55	890	2.2	3	35	43
19–30	1500	2300	4700	1000	700	400	8	11	150	55	900	2.3	4	35	45
31–50	1500	2300	4700	1000	700	420	8	11	150	55	900	2.3	4	35	45
51–70	1300	2000	4700	1200	700	420	8	11	150	55	900	2.3	4	30	45
>70	1200	1800	4700	1200	700	420	8	11	150	55	900	2.3	4	30	45
Females															
9–13	1500	2300	4500	1300	1250	240	8	8	120	40	700	1.6	2	21	34
14–18	1500	2300	4700	1300	1250	360	15	9	150	55	890	1.6	3	24	43
19–30	1500	2300	4700	1000	700	310	18	8	150	55	900	1.8	3	25	45
31–50	1500	2300	4700	1000	700	320	18	8	150	55	900	1.8	3	25	45
51–70	1300	2000	4700	1200	700	320	8	8	150	55	900	1.8	3	20	45
>70	1200	1800	4700	1200	700	320	8	8	150	55	900	1.8	3	20	45
Pregnancy															
≤18	1500	2300	4700	1300	1250	400	27	12	220	60	1000	2.0	3	29	50
19–30	1500	2300	4700	1000	700	350	27	11	220	60	1000	2.0	3	30	50
31–50	1500	2300	4700	1000	700	360	27	11	220	60	1000	2.0	3	30	50
Lactation															
≤18	1500	2300	5100	1300	1250	360	10	13	290	70	1300	2.6	3	44	50
19–30	1500	2300	5100	1000	700	310	9	12	290	70	1300	2.6	3	45	50
13–50	1500	2300	5100	1000	700	320	9	12	290	70	1300	2.6	3	45	50

Tolerable Upper Intake Levels (UL) for Vitamins

Age (yr)	Niacin (mg/day)[a]	Vitamin B$_6$ (mg/day)	Folate (μg/day)[a]	Choline (mg/day)	Vitamin C (mg/day)	Vitamin A (μg/day)[b]	Vitamin D (μg/day)	Vitamin E (mg/day)[c]
Infants								
0–0.5	—	—	—	—	—	600	25	—
0.5–1	—	—	—	—	—	600	25	—
Children								
1–3	10	30	300	1000	400	600	50	200
4–8	15	40	400	1000	650	900	50	300
9–13	20	60	600	2000	1200	1700	50	600
Adolescents								
14–18	30	80	800	3000	1800	2800	50	800
Adults								
19–70	35	100	1000	3500	2000	3000	50	1000
>70	35	100	1000	3500	2000	3000	50	1000
Pregnancy								
≤18	30	80	800	3000	1800	2800	50	800
19–50	35	100	1000	3500	2000	3000	50	1000
Lactation								
≤18	30	80	800	3000	1800	2800	50	800
19–50	35	100	1000	3500	2000	3000	50	1000

[a] The UL for niacin and folate apply to synthetic forms obtained from supplements, fortified foods, or a combination of the two.

[b] The UL for vitamin A applies to the preformed vitamin only.

[c] The UL for vitamin E applies to any form of supplemental α-tocopherol, fortified foods, or a combination of the two.

Tolerable Upper Intake Levels (UL) for Minerals

Age (yr)	Sodium (mg/day)	Chloride (mg/day)	Calcium (mg/day)	Phosphorus (mg/day)	Magnesium (mg/day)[d]	Iron (mg/day)[b]	Zinc (mg/day)	Iodine (μg/day)	Selenium (μg/day)	Copper (μg/day)	Manganese (mg/day)	Fluoride (mg/day)	Molybdenum (μg/day)	Boron (mg/day)	Nickel (mg/day)	Vanadium (mg/day)
Infants																
0–0.5	—[e]	—[e]	—	—	—	40	4	—	45	—	—	0.7	—	—	—	—
0.5–1	—[e]	—[e]	—	—	—	40	5	—	60	—	—	0.9	—	—	—	—
Children																
1–3	1500	2300	2500	3000	65	40	7	200	90	1000	2	1.3	300	3	0.2	—
4–8	1900	2900	2500	3000	110	40	12	300	150	3000	3	2.2	600	6	0.3	—
9–13	2200	3400	2500	4000	350	40	23	600	280	5000	6	10	1100	11	0.6	—
Adolescents																
14–18	2300	3600	2500	4000	350	45	34	900	400	8000	9	10	1700	17	1.0	—
Adults																
19–70	2300	3600	2500	4000	350	45	40	1100	400	10,000	11	10	2000	20	1.0	1.8
>70	2300	3600	2500	3000	350	45	40	1100	400	10,000	11	10	2000	20	1.0	1.8
Pregnancy																
≤18	2300	3600	2500	3500	350	45	34	900	400	8000	9	10	1700	17	1.0	—
19–50	2300	3600	2500	3500	350	45	40	1100	400	10,000	11	10	2000	20	1.0	—
Lactation																
≤18	2300	3600	2500	4000	350	45	34	900	400	8000	9	10	1700	17	1.0	—
19–50	2300	3600	2500	4000	350	45	40	1100	400	10,000	11	10	2000	20	1.0	—

[d] The UL for magnesium applies to synthetic forms obtained from supplements or drugs only.

[e] Source of intake should be from human milk (or formula) and food only.

NOTE: An Upper Limit was not established for vitamins and minerals not listed and for those age groups listed with a dash (—) because of a lack of data, not because these nutrients are safe to consume at any level of intake. All nutrients can have adverse effects when intakes are excessive.

SOURCE: Adapted with permission from the *Dietary Reference Intakes* series, National Academy Press. Copyright 1997, 1998, 2000, 2001, 2005 by the National Academy of Sciences. Courtesy of the National Academy Press, Washington, D.C.

SEVENTH EDITION

Nutrition and Diet Therapy

PRINCIPLES AND PRACTICE

SEVENTH EDITION

Nutrition and Diet Therapy

PRINCIPLES AND PRACTICE

Linda Kelly DeBruyne

Kathryn Pinna

Ellie Whitney

THOMSON

WADSWORTH

Australia • Canada • Mexico • Singapore • Spain • United Kingdom • United States

Dedication

To my adventurous son Zak and his beautiful bride Summer -- Love, health, and happiness always.
Linda

To my ever-young mother "tinine," who bravely leapt into cyberspace this year. With love.
Kathryn

To all of my grandchildren and stepgrandchildren: Max, Zoey, Emily, Rebecca, Sarah, Will, Toot-Toot, and Jacob.
Ellie

Acquisitions Editor: Peter Adams
Development Editor: Kate Franco
Assistant Editor: Elesha Feldman
Editorial Assistant: Elizabeth Downs
Technology Project Manager: Ericka Yeoman-Saler
Marketing Manager: Shemika Britt
Project Manager, Editorial Production: Belinda Krohmer
Creative Director: Rob Hugel
Art Director: John Walker
Print Buyer: Rebecca Cross

Permissions Editor: Timothy Sisler
Production Service: Pre-Press PMG
Text Designer: Randall Goodall, Seventeenth Street Studios
Photo Researcher: Pre-Press PMG
Copy Editor: Pre-Press PMG
Cover Designer: John Walker
Cover Image: © image 100/Corbis
Cover Printer: Courier Kendallville
Compositor: Pre-Press PMG
Printer: Courier Kendallville

Printed in the United States of America
1 2 3 4 5 6 7 11 10 09 08 07

For more information about our products, contact us at:
Thomson Learning Academic Resource Center
1-800-423-0563
For permission to use material from this text or product, submit a request online at http://www.thomsonrights.com. Any additional questions about permissions can be submitted by e-mail to thomsonrights@thomson.com.
Thomson Higher Education
10 Davis Drive Belmont, CA 94002-3098 USA

Library of Congress Control Number: 2007930095

ISBN-13: 978-0-495-11916-6
ISBN-10: 0-495-11916-4

Wadsworth/Thomson Learning
10 Davis Drive
Belmont, CA 94002-3098
USA

Asia
Thomson Learning
5 Shenton Way #01-01
UIC Building
Singapore 068808

Australia
Nelson Thomson Learning
102 Dodds Street
South Melbourne, Victoria 3205
Australia

Canada
Nelson Thomson Learning
1120 Birchmount Road
Toronto, Ontario M1K 5G4
Canada

Europe/Middle East/Africa
Thomson Learning
High Holborn House
50/51 Bedford Row
London WC1R 4LR
United Kingdom

Contents in Brief

Contents

CHAPTER FIVE

Digestion and Absorption 113

CHAPTER SIX

Metabolism, Energy Balance, and Body Composition 139

CHAPTER SEVEN

Weight Management: Overweight and Underweight 171

CHAPTER EIGHT

The Vitamins 201

CHAPTER NINE

Water and the Minerals 239

CHAPTER TEN

Fitness and Nutrition 273

PART TWO

Nutrition Throughout Life

CHAPTER ELEVEN

Nutrition through the Life Span: Pregnancy and Lactation 305

PART THREE

Nutrition in Health Care

How to Features Appear on the Following Pages

Case Studies Appear in Chapters of Parts Two, Three, and Four

Preface

Nutrition is a rapidly expanding science with many unanswered questions and new "facts" surfacing every day. This seventh edition of *Nutrition and Diet Therapy* reflects the many changes that have occurred in the field of nutrition since the last edition. The goal of the text is to provide basic nutrition information and advice on how to apply that information to everyday situations. A second mission is to help students evaluate and interpret new nutrition research.

The book is structured in a deliberate way. It first introduces the basics of nutrition and shows how nutrition supports health. It then describes how nutrient needs change throughout the life span. The second half of the book begins by looking at some ways that poor nutrition may lead to disease and then describes the potential impact of illnesses, and medications on nutrient needs and nutritional health. The remaining chapters of the book focus on medical nutrition therapy and its role in a variety of medical conditions.

Each chapter includes a wealth of nutrition tools to facilitate teaching and learning. Definitions of key terms appear at the bottom of each page, and notes in the margins clarify nutrition information, remind readers of previously defined terms, and provide cross-references. "How To" skill boxes help readers work through calculations or give practical suggestions for applying nutrition information. The "Self Check" feature at the end of each chapter helps readers review and test their understanding of the chapter material. "In Summary" statements at the end of each major chapter section help students assimilate the material and assess reading comprehension.

Special features help apply the chapter concepts and provide a rich learning experience for students. The applied and personal "Analysis: Your Diet" exercises in Chapters 1–10 (excluding Chapters 5 and 6) ask students to apply nutrition information from each chapter to their own diets. Case studies in the later chapters challenge readers to apply chapter information to clinical situations. "Clinical Applications" provide practice with mathematical calculations and help students understand the impact of nutrition-related issues on health care professionals and their clients. Later chapters also include "Nutrition Assessment Checklists," which summarize assessment parameters relevant to different stages of the life cycle or groups of disorders. The clinical chapters include "Diet-Drug Interaction" boxes, which point out interactions relevant to the medications described in each chapter.

A hallmark of the text is the "Nutrition in Practice" section located at the end of every chapter. These sections provide coverage of current research topics, advanced subjects, or specialty areas such as nutrition and dental health. This edition includes new Nutrition in Practice sections examining different kinds of fat and their health effects, food safety, fad diets, nutritional genomics and its implications for nutrition care, and the glycemic index in nutrition practice.

A robust set of appendixes supports the book. The appendixes include a wealth of information on the contents of foods and enteral formulas, U.S. nutrient intake recommendations and the exchange system, Canadian guidelines and food guides physical activity and energy requirements, additional information about nutrition assessment, and aids to calculations.

We hope that as you discover the many fascinating aspects of nutrition, you will enthusiastically apply the concepts in both your professional and your personal life. For nutrition updates and other resources, we invite you to visit our website: http://www.wadsworth.com/nutrition.

Acknowledgments

Among the most difficult words to write are those that express the depth of our gratitude to the many dedicated people whose efforts have made this book possible. A special note of appreciation to Sharon Rolfes for her numerous contributions to the chapters and Nutrition in Practice sections as well as to the Dietary Reference Intakes on the inside front cover and the appendices. Many thanks to Fran Webb for sharing her knowledge, ideas, and resources about the latest nutrition developments. Special thanks to Alexandra Rodriguez for help in manuscript preparation. We also wish to acknowledge the efforts of Donna Kelly and the folks at Axxya for their assistance in creating the food composition appendix. We are indebted to our editorial team, Peter Adams and Kate Franco, and our production team—Belinda Krohmer and Stephanie Kling—for seeing this project through from start to finish. We would also like to acknowledge Jennifer Somerville for her marketing efforts. To the many others involved in designing, indexing, typesetting, dummying, and marketing, we offer our thanks. We are especially grateful to our associates, family, and friends for their continued encouragement and support and to our reviewers who consistently offer excellent suggestions for improving the text.

Reviewers

Dan Benardot
Georgia State University

Lynnette K. Bourne
Mott Community College

Glenda Dupuis
Lamar State College, Port Arthur

Andrea Hutchins
University of Colorado, Colorado
     Springs

Fran Lukacik
Community College of Philadelphia

Myrtle R. McCulloch
Georgetown University

Mollie Smith
California State University, Fresno

Pamela Towery
Arkansas State University

Janet Westhoff
Mott Community College

About the Authors

Linda Kelly DeBruyne, M.S., R.D., received her B.S. in 1980 and her M.S. in 1982 in nutrition and food science at Florida State University. She is a founding member of Nutrition and Health Associates, an information resource center in Tallahassee, Florida, where her specialty areas are life cycle nutrition and fitness. Her other publications include the textbooks *Nutrition for Health and Health Care, Life Span Nutrition: Conception through Life, Health: Making Life Choices,* and a multimedia CD-ROM called *Nutrition Interactive.* As a consultant for a group of Tallahassee pediatricians, she teaches infant nutrition classes to parents. She maintains a professional membership in the American Dietetic Association.

Kathryn Pinna, Ph.D., R.D., received her M.S. and Ph.D. degrees in nutrition from the University of California at Berkeley. She has taught nutrition and food science courses in the San Francisco Bay Area for nearly 20 years and currently teaches nutrition at City College of San Francisco. She is coauthor of the textbooks *Understanding Normal and Clinical Nutrition* and *Nutrition for Health*

and Health Care. She is a Registered Dietitian and a member of the American Dietetic Association and the American Society for Nutrition.

Eleanor Noss Whitney, Ph.D., received her B.A. in biology from Radcliffe College in 1960 and her Ph.D. in biology from Washington University in St. Louis in 1970. Formerly on the faculty at Florida State University, and a dietitian registered with the American Dietetic Association she now devotes full time to research, writing, and consulting. Her earlier publications include articles in *Science, Genetics,* and other journals. Her textbooks include *Understanding Nutrition, Nutrition Concepts and Controversies, Life Span Nutrition: Conception through Life, Understanding Normal and Clinical Nutrition, Nutrition for Health and Health Care,* and *Essential Life Choices* for college students and *Making Life Choices* for high school students. Her most intense interests currently include energy conservation, solar energy uses, alternatively fueled vehicles, and ecosystem restoration.

Overview of Nutrition and Health

1

Every day, several times a day, you make choices that will either improve your **health** or harm it. Each choice may influence your health only a little, but when these choices are repeated over years and decades, their effects become significant.

The choices people make each day affect not only their physical health, but also their **wellness**—all the characteristics that make a person strong, confident, and able to function well with family, friends, and others. People who consistently make poor lifestyle choices, on a daily basis, increase their risks of developing diseases. Figure 1–1 shows how a person's health can fall anywhere along a continuum, from maximum wellness on the one end to total failure to function (death) on the other.

As health care professionals, when you take responsibility for your own health by making daily choices and practicing behaviors that enhance your well-being, you prepare yourself physically, mentally, and emotionally to meet the demands of your profession. As health care professionals, however, you have a responsibility to your clients ■ as well as to yourselves. You have unique opportunities to make your clients aware of the benefits of positive health choices and behaviors, to show them how to change their behaviors and make daily choices to enhance their own health, and to serve as role models for those behaviors.

This text focuses on how nutrition choices affect health and disease. The early chapters introduce the basics of nutrition to support good health. The later chapters emphasize diet therapy and its role in supporting health and treating diseases and symptoms.

■ Health care professionals generally use either client or patient when referring to an individual under their care. The first 13 chapters of this text emphasize the nutrition concerns of people in good health; therefore, the term client is used in these chapters.

FIGURE 1–1

The Health Line
No matter how well you maintain your health today, you may still be able to improve tomorrow. Likewise, a person who is well today can slip by failing to maintain health-promoting habits.

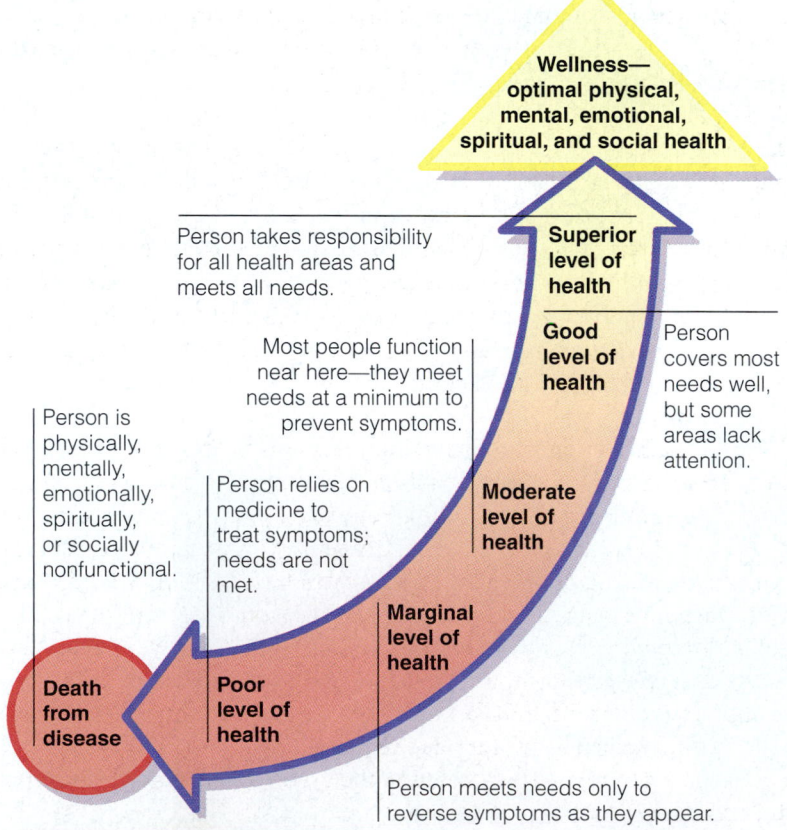

Wellness—optimal physical, mental, emotional, spiritual, and social health

Person takes responsibility for all health areas and meets all needs.

Superior level of health

Most people function near here—they meet needs at a minimum to prevent symptoms.

Good level of health

Person covers most needs well, but some areas lack attention.

Person is physically, mentally, emotionally, spiritually, or socially nonfunctional.

Person relies on medicine to treat symptoms; needs are not met.

Moderate level of health

Marginal level of health

Death from disease

Poor level of health

Person meets needs only to reverse symptoms as they appear.

Food Choices

Sound **nutrition** throughout life does not ensure good health and long life, but it can certainly help to tip the balance in their favor. Nevertheless, most people choose foods for reasons other than their nourishing value. Even people who claim to choose foods primarily for the sake of health or nutrition will admit that other factors also influence their food choices. Because food choices become an integral part of people's lifestyles, they sometimes find it difficult to change their eating habits. Health care professionals who help clients make diet changes must understand the dynamics of food choices because people will alter their eating habits only if their preferences are honored.

Preference Why do people like certain foods? One reason, of course, is their preference for certain tastes. Some tastes are widely liked, such as the sweetness of sugar and the zest of salt. Research suggests that genetics may influence people's taste preferences, a finding that may eventually have implications for clinical nutrition.[1] For example, sensitivity to bitter taste is an inheritable trait. People born with great sensitivity to bitter tastes tend to avoid foods with bitter flavors such as broccoli, cabbage, brussels sprouts, spinach, and grapefruit juice. These foods, as well as many other fruits and vegetables, contain compounds called **phytochemicals** that may reduce the risk of cancer. ■ Thus the role that genetics may play in food selection is gaining importance in cancer research.

Habit Sometimes habit dictates people's food choices. People eat a sandwich for lunch or drink orange juice at breakfast simply because they have always done so.

Associations People also like foods with happy associations—foods eaten in the midst of warm family gatherings on traditional holidays or given to them as children by someone who loved them. By the same token, people can attach intense and unalterable dislikes to foods that they ate when they were sick or that were forced on them when they were not hungry.

Ethnic Heritage and Tradition Every country, and every region of a country, has its own typical foods and ways of combining them into meals. The foodways of North America reflect the many different cultural and ethnic backgrounds of its inhabitants. Many foods with foreign origins are familiar items on North American menus: tacos, egg rolls, lasagna, and gyros, to name a few. Still others, such as spaghetti and croissants, are almost staples in the "American diet." North American regional cuisines such as Cajun and Tex-Mex blend the traditions of several cultures. Table 1–1 (p. 4) presents selected **ethnic diets** and food choices.

Values People's values, environmental ethics, religious beliefs, and political views also influence their food choices. By choosing to eat some foods or avoid others, people make statements that reflect their values. For example, people may select only foods that come in containers that can be reused or recycled. A political activist may boycott fruit or vegetables picked by migrant workers who have been exploited. Some people choose only brands of canned tuna fish that state "dolphin safe" on the label, meaning that dolphins were not killed when the tuna were netted. Labels on other foods carry similar statements or symbols that imply that the foods have been produced in ways that are considered environmentally favorable.[2] These are known as ecolabels. One of the most successful ecolabels is "Certified Organic," meaning the food was grown and processed according to regulations defining the use of synthetic fertilizers, herbicides, pesticides, and other chemical ingredients.

© Hill Street Studios/Jupiter Images

Nutrition is only one of the many factors that influence people's food choices.

■ Nutrition in Practice 8 addresses phytochemicals and their role in disease prevention.

health: a range of states with physical, mental, emotional, spiritual, and social components. At a minimum, health means freedom from physical disease, mental disturbances, emotional distress, spiritual discontent, social maladjustment, and other negative states. At a maximum, health means "wellness."

wellness: maximum well-being; the top range of health states; the goal of the person who strives toward realizing his or her full potential physically, mentally, emotionally, spiritually, and socially.

nutrition: the science of foods and the nutrients and other substances they contain and of their ingestion, digestion, absorption, transport, metabolism, interaction, storage, and excretion. A broader definition includes the study of the environment and of human behavior as it relates to these processes.

phytochemicals: nonnutrient compounds in plant-derived foods that have biological activity in the body.

ethnic diets: foodways and cuisines typical of national origins, races, cultural heritages, or geographic locations.

TABLE 1–1

Selected Ethnic Cuisines and Food Choices

	Grains	Vegetables	Fruits	Meats, Poultry, Fish, Legumes, Eggs, and Nuts	Milk
Asian[a] © Becky Luigart-Stayner/Corbis	Barley, buckwheat, millet, rice, wheat, rice or wheat noodles	Amaranth, baby corn, bamboo shoots, bok choy, cabbages, mung bean sprouts, scallions, seaweed, snow peas, straw mushrooms, water chestnuts, wild yam	Kumquats, loquats, lychee, mandarin oranges, melons, pears, persimmon, plums	Beef, pork, poultry, fish and other seafood, squid, soybeans, tofu, duck eggs, cashews, peanuts	Soy milk
Mediterranean © Photodisc, Inc.	Bulgur, couscous, focaccia, Italian bread, pastas, pita pocket bread, polenta, rice	Cucumbers, eggplant, grape leaves, onions, peppers, tomatoes	Dates, figs, grapes, lemons, melons, olives, raisins	Beef, gyros, lamb, pork, sausage, chicken, fish and other seafood, fava beans, lentils, almonds, walnuts	Feta, goat, mozzarella, parmesan, provolone, and ricotta cheeses; yogurt
Mexican © Photodisc, Inc.	Taco shells, tortillas (corn or flour), rice	Cactus, cassava, chayote, chilies, corn, jicama, onions, tomatoes, tomato salsa, yams	Avocado, bananas, guava, lemons, limes, mango, oranges, papaya, plantain	Beef, chorizo, chicken, fish, refried beans, eggs	Cheese, flan (caramel custard)
U.S. Deep South (West African Influence)[a] © Bonnie Kamin/Photoedit	Biscuits, corn bread, grits, macaroni, rice	Black-eyed peas, collards (other leafy greens), corn, hominy, okra, onions, pole beans, potatoes, snap beans, summer squash, sweet potatoes, tomatoes	Apples, bananas, berries, melons, peaches, pears	Beef, pork, sausages, poultry, fish, beans, peanuts	Buttermilk, milk, American cheese, cheddar cheese

[a] Traditional cuisines of China and of West African influence exclude fluid milk as a beverage for adults and use few or no milk products in cooking. Calcium and certain other nutrients of milk are supplied by other foods, such as small fish eaten with the bones or large servings of leafy green vegetables.

Religion also influences many people's food choices. Jewish law sets forth an extensive set of dietary rules. Many Christians forgo meat on Fridays during Lent, the period prior to Easter. In Islamic dietary laws, permitted or lawful foods are called *halal*. Other faiths prohibit some dietary practices and promote others. Diet planners can foster sound nutrition practices only if they respect and honor each person's values.

Social Interaction Social interaction is another powerful influence on people's food choices. Meals are social events, and the sharing of food is part of hospitality. It is often considered rude to refuse food or drink being shared by a group or offered by a host. Food brings people together for many different reasons: to celebrate a holiday or special event, to renew an old friendship, to make new friends, to conduct business, and many more. Sometimes food defines status. Lobster and caviar are often associated with wealth; peanut butter and jelly, with children. Sometimes food is

Ethnic meals and family gatherings nourish the spirit as well as the body.

used to influence or impress someone. For example, a business executive invites a prospective new client out to dinner in hope of edging out the competition. In each case, for whatever the purpose, food plays an integral part of the social interaction.

Emotional State People may eat in response to emotional stimuli—for example, to relieve boredom or depression or to calm anxiety. A lonely person may choose to eat rather than to call a friend and risk rejection. A person who has returned home from an exciting evening out may unwind with a late-night snack. Eating in response to emotions can easily lead to overeating and obesity but may be appropriate at times. For example, sharing food at times of bereavement serves both the giver's need to provide comfort and the receiver's need to be cared for and to interact with others.

In contrast, some people do not eat at all, or eat very little, in response to emotions. A depressed person may simply have no appetite for food. A person overcome with grief can easily lose all interest in food. As long as a person's appetite or interest in food returns quickly, little harm is done.

Availability, Convenience, and Economy The influence of these factors on people's food selections is clear. You cannot eat foods if they are not available, if you cannot get to the grocery store, if you do not have the time or skill to prepare them, or if you cannot afford them. Convenience plays a major role in many people's food selections today, but along with convenience, consumers want great taste and quality foods that are fast, delicious, and nutritious.[3] The demand for fully prepared ready-to-eat foods or foods that can be easily prepared in a microwave oven demonstrates this influence. Consumers today spend more than half their food budget on meals that require no preparation.[4] They frequently eat out, bring ready-to-eat meals home, or have food delivered. Consumer demand for microwavable foods that taste and smell as though they were prepared in a conventional oven has prompted manufacturers to expand the variety of microwavable foods available and to change the formulation and packaging of the foods.[5] Food processors are developing many more microwavable meals and snacks for use at home and at the office—rising crust pizzas, chicken and fish dishes, precooked entrées, and side dishes such as french fries and rice. Many of the new products are packaged in single-serve containers.

Age Age influences people's food choices. Infants, for example, depend on others to choose foods for them. Older children also rely on others but become more active in selecting foods that taste sweet and are familiar to them and rejecting those whose taste or texture they dislike. In contrast, the links between taste preferences and food choices in adults are less direct than in children. Adults often choose foods based on health concerns such as body weight. Indeed, adults may avoid sweet or familiar foods because of such concerns.

Occupation Some people have jobs that keep them away from home for days at a time, or require them to conduct business in restaurants or at conventions, or involve hectic schedules that allow little or no time for meals at home. For these people, the kinds of restaurants available to them and the cost of eating out so often may limit food choices.

Body Image Sometimes people select foods that they associate with ideals of body image. The fashion and movie industries, not the medical community, have defined what people believe to be the ideal body—sometimes an excessively thin body for women or an excessively muscular body for men. Both men and women seek "beautiful bodies," and in doing so, they select or avoid foods that they believe will improve or impair their physical appearance. Such intentions are rational when based on sound nutrition and fitness knowledge, but when based on faddism or carried to extremes, they undermine good health.

Medical Conditions Sometimes medical conditions and their treatments (including medications) limit the foods a person can select. For example, a person with heart disease might need to adopt a diet low in certain types of fats. The chemotherapy needed to treat cancer can interfere with a person's appetite or limit food choices by causing vomiting. Allergy to certain foods can also limit choices. The second half of this text discusses how diet can be modified to accommodate different medical conditions.

Health and Nutrition Consumers today cite nutrition as a primary concern in making food choices, yet the foods they choose do not always reflect this concern. Food manufacturers and restaurant chefs have responded to scientific findings linking health with nutrition by offering an abundant selection of health-promoting foods and beverages. In some cases, the health-promoting foods are as simple and familiar as oatmeal or tomatoes. In other cases, the foods have been processed or prepared in a way that provides health benefits, perhaps by reducing the sodium content. In still other cases, manufacturers have developed **functional foods**—products that may contain physiologically active ingredients that provide health benefits. ■ Examples of functional foods include margarine made with plant sterol and stanol esters that lower blood cholesterol and orange juice fortified with calcium to help build strong bones. More and more functional foods are being developed—a trend that will no doubt continue as consumer demand for such products grows.[6]

Consumers may be led to believe that functional foods are a new category of foods. All foods contain thousands of nonnutrient compounds that are biologically active in the body, however, so virtually all of them have some special value in supporting health.

■ Nutrition in Practice 8 offers more discussion of functional foods.

- A person selects foods for many different reasons.

- Food choices influence health—both positively and negatively. Individual food selections neither make nor break a diet's healthfulness, but the balance of foods selected over time can make an important difference to health.

- In the interest of health, people are wise to think "nutrition" when making their food choices.

The Nutrients

You are a collection of molecules that move. All these moving parts are arranged in patterns of extraordinary complexity and order—cells, tissues, and organs. Although the arrangement remains constant, the parts are continually changing, using **nutrients** and energy derived from nutrients.

Almost any food you eat is composed of dozens or even hundreds of different kinds of materials. Spinach, for example, is composed mostly of water (95 percent), and most of its solid materials are the compounds carbohydrates, fats (properly called lipids), and proteins. If you could remove these materials, you would find a tiny residue of minerals, vitamins, and other compounds. Water, carbohydrates, fats, proteins, vitamins, and minerals are the six classes of nutrients commonly found in spinach and other foods. ■ Some of the other materials in foods, such as the pigments and other phytochemicals, are not nutrients but may still be important to health. The body can make some nutrients for itself, at least in limited quantities, but it cannot make them all, and it makes some in insufficient quantities to meet its needs. Therefore, the body must obtain many nutrients from foods. The nutrients that foods must supply are called **essential nutrients.**

Carbohydrates, Fats, and Proteins Four of the six classes of nutrients (carbohydrates, fats, proteins, and vitamins) contain carbon, which is found in all living things. They are therefore **organic** (meaning, literally, "alive"). During metabolism, three of these four (carbohydrates, fats, and proteins) provide energy the body can use. ■ These **energy-yielding nutrients** continually replenish the energy you spend daily. Carbohydrates and fats meet most of the body's energy needs; proteins make a significant contribution only when other fuels are unavailable.

Vitamins, Minerals, and Water Vitamins are organic but do not provide energy to the body. They facilitate the release of energy from the three energy-yielding nutrients. In contrast, minerals and water are **inorganic** nutrients. Minerals yield no energy in the human body, but like vitamins, they help to regulate the release of energy, among their many other roles. As for water, it is the medium in which all of the body's processes take place.

kCalories: A Measure of Energy The amount of energy that carbohydrates, fats, and proteins release can be measured in **calories**—tiny units of energy so small that a single apple provides tens of thousands of them. To ease calculations, energy is expressed in 1,000-calorie metric units known as **kilocalories** (shortened to **kcalories**, but commonly called "calories"). When you read in popular books or magazines that an apple

■ The six classes of nutrients are water, carbohydrates, fats, proteins, vitamins, and minerals.

■ *Metabolism* is the set of processes by which nutrients are rearranged into body structures or broken down to yield energy.

functional foods: foods that may provide health benefits beyond their nutrient contributions. Functional foods may include whole foods, fortified foods, and modified foods.

nutrients: substances obtained from food and used in the body to provide energy and structural materials and to serve as regulating agents to promote growth, maintenance, and repair. Nutrients may also reduce the risks of some diseases.

essential nutrients: nutrients a person must obtain from food because the body cannot make them for itself in sufficient quantities to meet physiological needs.

organic: carbon containing. The four organic nutrients are carbohydrate, fat, protein, and vitamins.

energy-yielding nutrients: the nutrients that break down to yield energy the body can use. The three energy-yielding nutrients are carbohydrate, protein, and fat.

inorganic: not containing carbon or pertaining to living things.

calories: units by which energy is measured. Food energy is measured in **kilocalories** (1,000 calories equal 1 kilocalorie), abbreviated **kcalories** or kcal. One kcalorie is the amount of heat necessary to raise the temperature of 1 kilogram (kg) of water 1°C. The scientific use of the term *kcalorie* is the same as the popular use of the term *calorie.*

■ Food energy can also be measured in kilojoules (kJ). The kilojoule is the international unit of
energy. One kcalorie equals 4.2 kJ.

provides "100 calories," understand that it means 100 kcalories. This book uses the
term *kcalorie* and its abbreviation *kcal* throughout, as do other scientific books and
journals. ■ kCalories are not constituents of foods; they are a measure of the energy
foods provide. The energy a food provides depends on how much carbohydrate, fat,
and protein the food contains.

Carbohydrate yields 4 kcalories of energy from each gram, and so does protein. Fat
yields 9 kcalories per gram. Thus fat has a greater **energy density** than either carbohydrate or protein. If you know how many grams of each nutrient a food contains, you
can derive the number of kcalories potentially available from the food. Simply multiply
the carbohydrate grams times 4, the protein grams times 4, and the fat grams times 9,
and add the results together (the accompanying "How to" describes how to calculate
the energy a food provides).

Energy Nutrients in Foods Practically all foods contain mixtures of the energy-
yielding nutrients, although foods are sometimes classified by their predominant
nutrient. To speak of meat as "a protein" or of bread as "a carbohydrate," however, is
inaccurate. Each is rich in a particular nutrient, but a protein-rich food such as beef
contains a lot of fat along with the protein, and a carbohydrate-rich food such as
corn bread also contains fat (corn oil) and protein. Only a few foods are exceptions
to this rule, the common ones being sugar (which is pure carbohydrate) and oil
(which is pure fat).

Energy Storage in the Body The body first uses the energy-yielding nutrients to
build new compounds and fuel metabolic and physical activities. Excesses are then rearranged into storage compounds, primarily body fat, and put away for later use.
Thus, if you take in more energy than you expend, whether from carbohydrate, fat, or
protein, the result is usually a gain of body fat. Too much meat (a protein-rich food) is
just as fattening as too many potatoes (a carbohydrate-rich food).

Alcohol, Not a Nutrient One other substance contributes energy: alcohol. The body
derives energy from alcohol at the rate of 7 kcalories per gram. Alcohol is not a nutrient, however, because it cannot support the body's growth, maintenance, or repair.
Nutrition in Practice 20 discusses alcohol's effects on nutrition.

How To

CALCULATE THE ENERGY AVAILABLE FROM FOODS

To calculate the energy available from a food, multiply the number of grams of carbohydrate, protein, and fat by 4, 4, and 9, respectively. Then add the results together. For example, 1 slice of bread with 1 tablespoon of peanut butter on it contains 16 grams of carbohydrate, 7 grams of protein, and 9 grams of fat:

16 g carbohydrate × 4 kcal/g = 64 kcal.

7 g protein × 4 kcal/g = 28 kcal.

9 g fat × 9 kcal/g = 81 kcal.

Total = 173 kcal.

From this information, you can calculate the percentage of kcalories each of the energy nutrients contributes to the total. To determine the percentage of kcalories from fat, for example, divide the 81 fat kcalories by the total 173 kcalories:

81 fat kcal ÷ 173 total kcal = 0.468 (rounded to 0.47).

Then multiply by 100 to get the percentage:

0.47 × 100 = 47%

Dietary recommendations that urge people to limit fat intake to 20 to 35 percent of kcalories refer to the day's total energy intake, not to individual foods. Still, if the proportion of fat in each food choice throughout a day exceeds 35 percent of kcalories, then the day's total surely will, too. Knowing that this snack provides 47 percent of its kcalories from fat alerts a person to the need to make lower-fat selections at other times that day.

- Foods provide nutrients—substances that support the growth, maintenance, and repair of the body's tissues.

- The six classes of nutrients are water, carbohydrates, fats, proteins, vitamins, and minerals.

- Vitamins, minerals, and water facilitate a variety of activities in the body.

- Foods rich in the energy-yielding nutrients (carbohydrates, fats, and proteins) provide the major materials for building the body's tissues and yield energy the body can use or store.

- Energy is measured in kcalories.

Nutrient Recommendations

Nutrient recommendations are used as standards to measure healthy people's energy and nutrient intakes. ■ Nutrition experts use the recommendations to assess nutrient intakes and to guide people on amounts to consume. Individuals can use them to decide how much of a nutrient they need to consume.

Dietary Reference Intakes

Defining the amounts of energy, nutrients, and other dietary components that best support health is a huge task. Nutrition experts have produced a set of standards that define the amounts of energy, nutrients, other dietary components, and physical activity that best support health. These recommendations are called **Dietary Reference Intakes (DRI)** and reflect the collaborative efforts of scientists in both the United States and Canada.*[7] The DRI values are presented on the inside front covers of this text.

Setting Nutrient Recommendations: RDA and AI One advantage of the DRI is that they apply to the diets of individuals. The DRI committee offers two sets of values to be used as nutrient intake goals by individuals: a set called the **Recommended Dietary Allowances (RDA)** and a set called **Adequate Intakes (AI)**.

Based on solid experimental evidence and other reliable observations, the RDA are the foundation of the DRI. The AI values are also based on scientific findings as much as possible, but estimating their values requires some educated guesswork. The committee establishes an AI value whenever scientific evidence is insufficient to generate a RDA.[8] To see which nutrients have an AI and which have a RDA, turn to the inside front cover.

Abundant new research has linked nutrients in the diet with the promotion of health and the prevention of chronic diseases. An advantage of the DRI is that they take into account disease prevention where appropriate, as well as an adequate nutrient intake. For example, the AI for calcium is based on intakes thought to reduce the likelihood of osteoporosis-related fractures later in life.

To ensure that the vitamin and mineral recommendations meet the needs of as many people as possible, the recommendations are set near the top end of the range of the population's estimated average requirements (see Figure 1–2 on p. 10). Small amounts above the daily **requirement** do no harm, whereas amounts below the requirement

---

* The DRI reports are produced by the Food and Nutrition Board, Institute of Medicine of the National Academies, with active involvement of scientists from Canada.

■ Appendix B presents nutrient recommendations developed by two international groups, the Food and Agriculture Organization and the World Health Organization (the FAO/WHO recommendations).

energy density: a measure of the energy a food provides relative to the amount of food (kcalories per gram).

Dietary Reference Intakes (DRI): a set of values for the dietary nutrient intakes of healthy people in the United States and Canada. These values are used for planning and assessing diets.

Recommended Dietary Allowances (RDA): a set of values reflecting the average daily amounts of nutrients considered adequate to meet the known nutrient needs of practically all healthy people in a particular life stage and gender group; a goal for dietary intake by individuals.

Adequate Intakes (AI): a set of values that are used as guides for nutrient intakes when scientific evidence is insufficient to determine a RDA.

requirement: the lowest continuing intake of a nutrient that will maintain a specified criterion of adequacy.

FIGURE 1–2

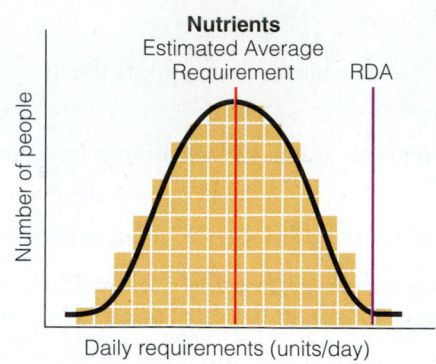

Nutrient Intake Recommendations
The nutrient intake recommendations are set high enough to cover nearly everyone's requirements (the boxes represent people).

may lead to health problems. When people's intakes are consistently **deficient**, their nutrient stores decline, and over time this decline leads to deficiency symptoms and poor health.

Facilitating Nutrition Research and Policy: EAR In addition to the RDA and AI, the DRI committee has established another set of values: **Estimated Average Requirements (EAR).** These values establish average requirements for given life stage and gender groups that researchers and nutrition policymakers use in their work. Nutrition scientists may use the EAR as standards in research. Public health officials may use them to assess nutrient intakes of populations and make recommendations. The EAR values form the scientific basis on which the RDA are set.

Establishing Safety Guidelines: UL The DRI committee also establishes upper limits of intake for nutrients posing a hazard when consumed in excess. These values, the **Tolerable Upper Intake Levels (UL),** are indispensable to consumers who take supplements. Consumers need to know how much of a nutrient is too much. The UL are also of value to public health officials who set allowances for nutrients that are added to foods and water. The UL values are listed on the inside front cover.

Using Nutrient Recommendations Each of the four DRI categories serves a unique purpose. For example, the EAR are most appropriately used to develop and evaluate nutrition programs for *groups* such as schoolchildren or military personnel. The RDA (or AI if a RDA is not available) can be used to set goals for *individuals*. The UL help to keep nutrient intakes below the amounts that increase the risk of toxicity. With these understandings, professionals can use the DRI for a variety of purposes.

In addition to understanding the unique purposes of the DRI, it is important to keep their uses in perspective. Consider the following:

- The values are recommendations for safe intakes, not minimum requirements; except for energy, they include a generous margin of safety. Figure 1–3 presents an accurate view of how a person's nutrient needs fall within a range, with marginal and danger zones both below and above the range.

- The values reflect daily intakes to be achieved on average, over time. They assume that intakes will vary from day to day, and they are set high enough to ensure that body nutrient stores will meet nutrient needs during periods of inadequate intakes lasting a day or two for some nutrients and up to a month or two for others.

- The values are chosen in reference to specific indicators of nutrient adequacy, such as blood nutrient concentrations, normal growth, and reduction of certain chronic diseases or other disorders when appropriate, rather than prevention of deficiency symptoms alone.

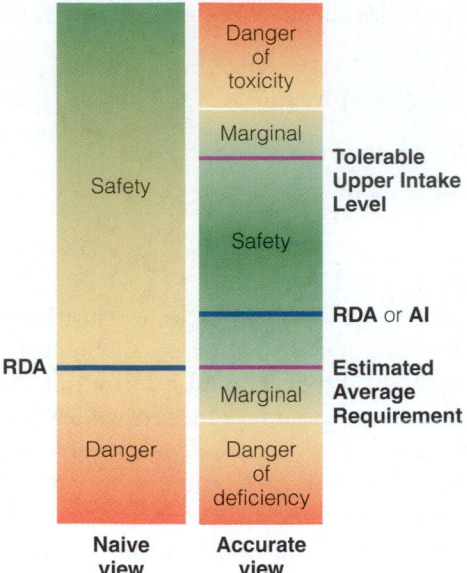

Naive view | Accurate view

FIGURE 1–3

Naive versus Accurate View of Nutrient Intakes
The RDA or AI for a given nutrient represents a point that lies within a range of appropriate and reasonable intakes between toxicity and deficiency. Both of these recommendations are high enough to provide reserves in times of short-term dietary inadequacies, but not so high as to approach toxicity. Nutrient intakes above or below this range may be equally harmful.

■ The recommendations are designed to meet the needs of most healthy people. Medical problems alter nutrient needs as later chapters describe.

■ The recommendations are specific for people of both genders as well as various ages and stages of life: infants, children, adolescents, men, women, pregnant women, and lactating women.

Setting Energy Recommendations   In contrast to the vitamin and mineral recommendations, the recommendation for energy, called the **Estimated Energy Requirement (EER)**, is not generous because excess energy cannot be excreted and is eventually stored as body fat. Rather, the key to the energy recommendation is balance. For a person who has a body weight, body composition, and physical activity level consistent with good health, energy intake from food should match energy expenditure, so the person achieves energy balance. Enough energy is needed to sustain a healthy, active life, but too much energy leads to obesity. The EER is therefore set at a level of energy intake predicted to maintain energy balance in a healthy adult of a defined age, gender, weight, height, and physical activity level.* Another difference between the requirements for other nutrients and those for energy is that each person has an obvious indicator of whether energy intake is inadequate, adequate, or excessive: body weight. Because any amount of energy in excess of need leads to weight gain, the DRI committee did not set a Tolerable Upper Intake Level.

Acceptable Macronutrient Distribution Ranges (AMDR)

As noted earlier, the DRI committee considers prevention of chronic disease as well as nutrient adequacy when establishing recommendations. To that end, the committee established healthy ranges of intakes for the energy-yielding nutrients—carbohydrate, fat, and protein—known as **Acceptable Macronutrient Distribution Ranges (AMDR)**. Each of these three energy-yielding nutrients contributes to a person's total energy (kcalorie) intake, and those contributions vary in relation to each other. The DRI committee has determined that a diet that provides the energy-yielding nutrients in the

* The EER for children, pregnant women, and lactating women includes energy needs associated with the deposition of tissue or the secretion of milk at rates consistent with good health.

deficient: in regard to nutrient intake, the amount below which almost all healthy people can be expected, over time, to experience deficiency symptoms.

Estimated Average Requirements (EAR): the average daily nutrient intake levels estimated to meet the requirements of half of the healthy individuals in a given age and gender group; used in nutrition research and policymaking and the basis on which RDA values are set.

Tolerable Upper Intake Levels (UL): a set of values reflecting the highest average daily nutrient intake levels that are likely to pose no risk of toxicity to almost all healthy individuals in a particular life stage and gender group. As intake increases above the UL, the potential risk of adverse health effects increases.

Estimated Energy Requirement (EER): the average dietary energy intake that maintains energy balance in a healthy person of a defined age, gender, and weight, and a level of physical activity that is consistent with good health.

Acceptable Macronutrient Distribution Ranges (AMDR): ranges of intakes for the energy-yielding nutrients that provide adequate energy and nutrients and reduce the risk of chronic disease.

following proportions provides adequate energy and nutrients and reduces the risk of chronic disease:

- 45 to 65 percent from carbohydrate

- 20 to 35 percent from fat

- 10 to 35 percent from protein

Nutrition Surveys

How do nutrition experts know whether people are meeting nutrient recommendations? The Dietary Reference Intakes and other major reports that examine the relationships between diet and health depend on information collected from nutrition surveys. Researchers use nutrition surveys to learn which foods they are eating and which supplements they are taking, to assess people's nutritional health, and to determine people's knowledge, attitudes, and behaviors about nutrition and how these relate to health.[9] The resulting wealth of information can be used for a variety of purposes. For example, Congress uses this information to establish public policy on nutrition education, assess food assistance programs, and regulate the food supply. Scientists use the information to establish research priorities. One of the first nutrition surveys, taken before World War II, suggested that up to a third of the U.S. population might be eating poorly. Programs to correct **malnutrition** have been evolving ever since.

Coordinating Nutrition Survey Data The National Nutrition Monitoring program coordinates the many nutrition-related activities of various federal agencies. All major reports that examine the contribution of diet and nutrition status to the health of the people of the United States depend on information collected and coordinated by this national program. One of its most recent projects is the integration of two major national surveys to provide comprehensive data efficiently.[10] One portion of the survey, previously known as the Continuing Survey of Food Intakes by Individuals (CSFII), collects data on the kinds and amounts of foods people eat. Researchers then calculate the energy and nutrients in the foods and compare the amounts consumed with standards such as the DRI. The other portion of the survey, the National Health and Nutrition Examination Survey (NHANES), examines the people themselves, using nutrition assessment methods. The new integrated survey is called What We Eat in America-NHANES and is conducted by the U.S. Department of Agriculture (USDA) and the U.S. Department of Health and Human Services (HHS). Data will be collected for What We Eat in America on a continuous yearly basis. Two days of data will be collected for all respondents. The data provide information on several nutrition-related conditions, including growth retardation, heart disease, and nutrient deficiencies. These data also provide the basis for developing and monitoring national health goals.

Healthy People Reports Healthy People, a program that sets goals for improving the nation's health, was initiated more than 20 years ago. At the start of each decade, the program sets goals for improving the nation's health during the following 10 years. Healthy People 2010 focuses on two overall goals:

- To help people of all ages increase life expectancy and improve their quality of life.

- To eliminate disparities in health that occur by gender, race, ethnicity, disability, income, education, and sexual orientation.

In short, Healthy People 2010 challenges individuals, communities, and professionals—indeed, all of us—to take specific steps to ensure that everyone will enjoy good health and long life.

Nutrition is one of many focus areas in Healthy People 2010; each has its own specific objectives. Table 1–2 lists the nutrition and overweight objectives for 2010.

TABLE 1–2

Healthy People 2010 Nutrition and Overweight Objectives

Increase the proportion of adults who are at a *healthy weight*.

Reduce the proportion of adults who are *obese*.

Reduce the proportion of children and adolescents who are *overweight* or *obese*.

Reduce *growth retardation* among low-income children under age 5 years.

Increase the proportion of persons aged 2 years and older who consume at least two daily servings of *fruit*.

Increase the proportion of persons aged 2 years and older who consume at least three daily servings of *vegetables,* with at least one-third being dark green or orange vegetables.

Increase the proportion of persons aged 2 years and older who consume at least six daily servings of *grain products,* with at least three being whole grains.

Increase the proportion of persons aged 2 years and older who consume less than 10 percent of kcalories from *saturated fat*.

Increase the proportion of persons aged 2 years and older who consume no more than 30 percent of kcalories from *total fat*.

Increase the proportion of persons aged 2 years and older who consume 2,400 milligrams or less of *sodium*.

Increase the proportion of persons aged 2 years and older who meet dietary recommendations for *calcium*.

Reduce *iron deficiency* among young children, females of childbearing age, and pregnant females.

Reduce *anemia* among low-income pregnant females in their third trimester.

Increase the proportion of children and adolescents aged 6 to 19 years whose intake of *meals and snacks at school* contributes to good overall dietary quality.

Increase the proportion of schools that teach all essential *nutrition education* topics in one course.

Increase the proportion of worksites that offer *nutrition or weight management classes or counseling*.

Increase the proportion of primary care providers who provide nutrition *assessment* when appropriate and who formulate a diet plan for those who need *intervention*.

Increase the proportion of physician office visits made by patients with a diagnosis of cardiovascular disease, diabetes, or hyperlipidemia that include *counseling or education related to diet and nutrition*.

Increase *food security* among U.S. households and in so doing reduce hunger.

Note: "Nutrition and Overweight" is one of 28 focus areas, each with numerous objectives. Several of the other focus areas have nutrition-related objectives.

Source: Healthy People 2010, **www.healthypeople.gov**

IN SUMMARY

- The Dietary Reference Intakes (DRI) are a set of nutrient intake values that can be used to plan and evaluate dietary intakes for healthy people.

- The Estimated Energy Requirement (EER) defines the energy intake level needed to maintain energy balance in a healthy person of a defined age, gender, weight, height, and physical activity level.

- The Acceptable Macronutrient Distribution Ranges (AMDR) define the proportions contributed by carbohydrate, fat, and protein to a healthy diet.

- Nutrition surveys measure people's food consumption and evaluate the nutrition status of populations.

- Information gathered from nutrition surveys serves as the basis for many major diet and nutrition reports, including Healthy People 2010.

malnutrition: any condition caused by deficient or excess energy or nutrient intake or by an imbalance of nutrients.

Dietary Guidelines and Food Guides

Today, government authorities are as much concerned about **overnutrition** as they once were about **undernutrition.** Research confirms that dietary excesses, especially of energy, certain fats, and alcohol, contribute to many **chronic diseases**, including heart disease, cancer, stroke, diabetes, and liver disease.[11] Only two common lifestyle habits have more influence on health than a person's choice of diet: smoking and other tobacco use and excessive drinking of alcohol. Table 1–3 lists the leading causes of death in the United States and shows that the top three are nutrition related (and related to tobacco use), while accidents are alcohol related. Note, however, that although diet is a powerful influence on these diseases, they cannot be prevented by a healthy diet alone; genetics, physical activity, and lifestyle also play a role. Within the range set by genetic inheritance, however, disease development is strongly influenced by the foods a person chooses to eat.

So how can health care professionals help people select foods to create a diet that supplies all the needed nutrients in amounts consistent with good health? The principle is simple enough: encourage clients to eat a variety of foods that supply all the nutrients the body needs. In practice, how do people do this? It helps to keep in mind that a nutritious diet achieves six basic ideals.

Dietary Ideals

A nutritious diet has six characteristics. The first is **adequacy.** The earlier discussion on the DRI already addressed the ideal of nutrient adequacy. An adequate diet has enough energy and enough of every nutrient (as well as fiber) to meet the needs of healthy people. Second is **balance:** the food choices do not overemphasize one nutrient or food type at the expense of another.

The essential minerals calcium and iron illustrate the importance of dietary balance. Meats, fish, and poultry are rich in iron but poor in calcium. Conversely, milk and milk products are rich in calcium but poor in iron. Use some meat or meat alternatives for iron; use some milk and milk products for calcium; and save some space for other foods, too, since a diet consisting of milk and meat alone would not be adequate. For other nutrients, people need grains, vegetables, and fruit.

TABLE 1–3

Leading Causes of Death in the United States

The four diseases in bold italics, are nutrition related. The fifth-ranked cause, motor vehicle and other accidents, represented in italics, is alcohol related.

1. *Heart disease*
2. *Cancers*
3. *Strokes*
4. Chronic lower respiratory diseases
5. *Motor vehicle and other accidents*
6. *Diabetes mellitus*
7. Alzheimer's disease
8. Pneumonia and influenza
9. Kidney disease
10. Infections of the blood

Source: National Center for Health Statistics, 2006.

The third characteristic is **kcalorie control:** the foods provide the amount of energy needed to maintain a healthy body weight—not more, not less. The key to kcalorie control is to select foods that deliver the most nutrients for the least food energy. This fourth characteristic is known as **nutrient density**. Consider calcium sources, for example. Ice cream and fat-free milk both supply calcium, but the milk is "denser" in calcium per kcalorie. A cup of rich ice cream contributes more than 250 kcalories, a cup of fat-free milk only 85—and with almost double the calcium.

The fifth characteristic of a nutritious diet is **moderation**. Foods rich in fat and sugar provide enjoyment and energy but relatively few nutrients. In addition, they promote weight gain when eaten in excess. A person who practices moderation eats such foods only on occasion and regularly selects foods low in fat and sugar, a practice that automatically improves nutrient density. Returning to the example of ice cream and fat-free milk, the milk not only offers more calcium for less energy, but it also contains far less fat than ice cream.

Finally, the sixth characteristic of a nutritious diet is **variety:** the foods chosen differ from one day to the next. A diet may have all the virtues just described and still lack variety, if a person eats the same foods day after day. People should select foods from each of the food groups daily and vary their choices within each food group from day to day for a couple of reasons. First, different foods within the same group contain different arrays of nutrients. Among the fruits, for example, strawberries are especially rich in vitamin C while apricots are rich in vitamin A. Second, no food is guaranteed entirely free of substances that, in excess, could be harmful. The strawberries might contain trace amounts of one contaminant, the apricots another. By alternating fruit choices, a person will ingest very little of either contaminant.

Dietary Guidelines for Americans

The *Dietary Guidelines for Americans 2005* originated from convincing scientific evidence that nutritious food, kcalorie control, and physical activity promote health and reduce the risk of chronic disease.[12] In general, the *Dietary Guidelines* answer the question: What should an individual eat to stay healthy? Table 1–4 (p. 16) presents the nine *Dietary Guidelines* topics with their key recommendations. The first three topics focus on choosing nutrient-rich foods within energy needs, maintaining a healthy body weight, and engaging in regular physical activity. The fourth topic, "Food Groups to Encourage," focuses on the selection of a variety of fruits and vegetables, whole grains, and milk. The next four topics advise people to choose sensibly in their use of fats, carbohydrates, salt, and alcoholic beverages for those who partake. Finally, consumers are reminded to keep food safe. Together, the *Dietary Guidelines* point the way toward better health. Table 1–5 (p. 17) presents Canada's *Guidelines for Healthy Eating*.

IN SUMMARY

- A well-planned diet delivers adequate nutrients, a balanced array of nutrients, and an appropriate amount of energy.

- A well-planned diet is based on nutrient-dense foods, moderate in substances that can be detrimental to health, and varied in its selections.

- The Dietary Guidelines apply these principles, offering practical advice on how to eat for good health.

overnutrition: overconsumption of food energy or nutrients sufficient to cause disease or increased susceptibility to disease; a form of malnutrition.

undernutrition: underconsumption of food energy or nutrients severe enough to cause disease or increased susceptibility to disease; a form of malnutrition.

chronic diseases: degenerative diseases characterized by deterioration of the body organs. Examples include heart disease, cancer, and diabetes.

adequacy: the characteristic of a diet that provides all the essential nutrients, fiber, and energy necessary to maintain health and body weight.

balance: the dietary characteristic of providing foods of a number of types in proportion to each other, such that foods rich in some nutrients do not crowd out foods that are rich in other nutrients.

kcalorie control: management of food energy intake.

nutrient density: a measure of the nutrients a food provides relative to the energy it provides; the more nutrients and the fewer kcalories, the higher the nutrient density.

moderation: providing enough, but not too much, of a substance.

variety: eating a wide selection of foods within and among the major food groups (the opposite of monotony).

TABLE 1-4

Key Recommendations of the *Dietary Guidelines for Americans 2005*

Adequate Nutrients within Energy Needs

Consume a variety of nutrient-dense foods and beverages within and among the basic food groups; limit intakes of saturated and *trans* fats, cholesterol, added sugars, salt, and alcohol.

Meet recommended intakes within energy needs by adopting a balanced eating pattern, such as the USDA Food Guide (Figure 1–4, pp. 18–19).

Weight Management

To maintain body weight in a healthy range, balance kcalories from foods and beverages with kcalories expended (see Chapter 6).

To prevent gradual weight gain over time, make small decreases in food and beverage kcalories and increase physical activity.

Physical Activity

Engage in regular physical activity and reduce sedentary activities to promote health, psychological well-being, and a healthy body weight.

To reduce the risk of chronic disease in adulthood, engage in at least 30 minutes of moderate-intensity physical activity, above usual activity, at home or work on most days of the week.

For most people, greater health benefits can be obtained by engaging in physical activity of more vigorous intensity or longer duration.

Achieve physical fitness by including cardiovascular conditioning, stretching exercises for flexibility, and resistance exercises or calisthenics for muscle strength and endurance.

Food Groups to Encourage

Consume a sufficient amount of fruits, vegetables, milk and milk products, and whole grains while staying within energy needs.

Select a variety of fruits and vegetables each day, including selections from all five vegetable subgroups (dark green, orange, legumes, starchy vegetables, and other vegetables) several times a week. Make at least half of the grain selections whole grains. Select fat-free or low-fat milk products.

Fats

Consume less than 10 percent of kcalories from saturated fats and less than 300 milligrams of cholesterol per day, and keep *trans* fat consumption as low as possible (see Chapter 3).

Keep total fat intake between 20 and 35 percent of kcalories; choose from mostly polyunsaturated and monounsaturated fat sources such as fish, nuts, and vegetable oils.

Select and prepare foods that are lean, low fat, or fat-free and low in saturated and/or *trans* fat.

Carbohydrates

Choose fiber-rich fruits, vegetables, and whole grains often.

Choose and prepare foods and beverages with little added sugars (see Chapter 2).

Reduce the incidence of dental caries by practicing good oral hygiene and consuming sugar- and starch-containing foods and beverages less frequently.

Sodium and Potassium

Choose and prepare foods with little salt (less than 2,300 milligrams sodium or approximately 1 teaspoon salt). At the same time, consume potassium-rich foods, such as fruits and vegetables (see Chapter 9).

Alcoholic Beverages

Those who choose to drink alcoholic beverages should do so sensibly and in moderation.

Some individuals should not consume alcoholic beverages (see Nutrition in Practice 20).

Food Safety

To avoid microbial foodborne illness, keep foods safe: clean hands, food contact surfaces, and fruits and vegetables; separate raw, cooked, and ready-to-eat foods; cook foods to a safe internal temperature; chill perishable food promptly; and defrost food properly.

Avoid unpasteurized milk and products made from it; raw or undercooked eggs, meat, poultry, fish, and shellfish; unpasteurized juices; and raw sprouts.

Note: These guidelines are intended for adults and healthy children ages 2 and older.

Source: The *Dietary Guidelines for Americans 2005,* available at **www.healthierus.gov/dietaryguidelines**.

TABLE 1–5

Canada's *Guidelines for Healthy Eating*

Enjoy a variety of foods.

Emphasize cereals, breads, other grain products, vegetables, and fruits.

Choose lower-fat dairy products, leaner meats, and foods prepared with little or no fat.

Achieve and maintain a healthy body weight by enjoying regular physical activity and healthy eating.

Limit salt, alcohol, and caffeine.

Source: These guidelines derive from *Action Towards Healthy Eating—Canada's Guidelines for Healthy Eating and Recommended Strategies for Implementation.*

The USDA Food Guide

To help people achieve the goals set forth by the *Dietary Guidelines for Americans 2005* (see Table 1–4), the USDA provides a **food group plan**—the **USDA Food Guide**—that builds a diet from categories of foods that are similar in vitamin and mineral content. ■ Thus each group provides a set of nutrients that differs somewhat from the nutrients supplied by the other groups. Selecting foods from each of the groups eases the task of creating an adequate and balanced diet.

Figure 1–4 (pp. 18–19) presents the USDA Food Guide. The USDA Food Guide assigns foods to five major food groups and recommends daily amounts of foods from each group to meet nutrient needs. ■ The judicious use of oils, solid fats, and added sugars is also described in the USDA Food Guide. In addition to presenting the food groups, the figure indicates the most notable nutrients of each group and lists some foods within each group sorted by nutrient density. ■

Recommended amounts for fruits, vegetables, and milk are given in cups, and those for grains and meats are given in ounces. Figure 1–4 provides equivalent measures for foods that are not readily measured in cups and ounces. For example, users learn that 1 ounce of grains is equivalent to 1 slice of bread or $1/2$ cup of cooked rice.

■ The DASH Eating Plan, presented in Chapter 22, is another dietary pattern that meets the goals of the Dietary Guidelines for Americans 2005.

■ Five food groups:
  ■ Fruits
  ■ Vegetables
  ■ Grains
  ■ Meat, poultry, fish, legumes, eggs, and nuts
  ■ Milk, yogurt, and cheese

■ Chapter 12 provides a food guide for young children, and Appendix B presents Canada's food group plan, *Eating Well with Canada's Food Guide.*

© Matthew Farruggio

A portion of grains is 1 ounce, yet most bagels today weigh 4 ounces or more—meaning that a single bagel can easily supply four or more portions of grains, not one as many people assume.

food group plan: a diet-planning tool that sorts foods into groups based on nutrient content and then specifies that people should eat certain amounts of food from each group.

USDA Food Guide: the USDA's food group plan for ensuring dietary adequacy that assigns foods to five major food groups.

FIGURE 1–4

USDA Food Guide, 2005

Key:
- ● Foods generally high in nutrient density (choose most often)
- ▲ Foods lower in nutrient density (limit selections)

GRAINS

© Polara Studios, Inc.

Make at least half of the grain selections whole grains.

These foods contribute folate, niacin, riboflavin, thiamin, iron, magnesium, selenium, and fiber.

> 1 oz grains is equivalent to 1 slice bread; ½ c cooked rice, pasta, or cereal; 1 oz dry pasta or rice; 1 c ready-to-eat cereal; 3 c popped popcorn.

- ● Whole grains (barley, brown rice, bulgur, millet, oats, rye, wheat) and whole-grain, low-fat breads, cereals, crackers, and pastas; popcorn.

- ● Enriched bagels, breads, cereals, pastas (couscous, macaroni, spaghetti), pretzels, rice, rolls, tortillas.

- ▲ Biscuits, cakes, cookies, corn bread, crackers, croissants, doughnuts, French toast, fried rice, granola, muffins, pancakes, pastries, pies, presweetened cereals, taco shells, waffles.

VEGETABLES

© Polara Studios, Inc.

Choose a variety of vegetables from all five subgroups several times a week.

These foods contribute folate, vitamin A, vitamin C, vitamin E, magnesium, potassium, and fiber.

> ½ c vegetables is equivalent to ½ c cut-up raw or cooked vegetables; ½ c cooked legumes; ½ c vegetable juice; 1 c raw, leafy greens.

- ● Dark green vegetables: Broccoli and leafy greens such as arugula, beet greens, bok choy, collard greens, kale, mustard greens, romaine lettuce, spinach, turnip greens.

- ● Orange and deep yellow vegetables: Carrots, carrot juice, pumpkin, sweet potatoes, winter squash (acorn, butternut).

- ● Legumes: Black beans, black-eyed peas, garbanzo beans (chickpeas), kidney beans, lentils, navy beans, pinto beans, soybeans and soy products such as tofu, split peas.

- ● Starchy vegetables: Cassava, corn, green peas, hominy, lima beans, potatoes.

- ● Other vegetables: Artichokes, asparagus, bamboo shoots, bean sprouts, beets, brussels sprouts, cabbages, cactus, cauliflower, celery, cucumbers, eggplant, green beans, iceberg lettuce, mushrooms, okra, onions, peppers, seaweed, snow peas, tomatoes, vegetable juices, zucchini.

- ▲ Baked beans, candied sweet potatoes, coleslaw, french fries, potato salad, refried beans, scalloped potatoes, tempura vegetables.

FRUITS

© Polara Studios, Inc.

Consume a variety of fruits and no more than one-third of the recommended intake as fruit juice.

These foods contribute folate, vitamin A, vitamin C, potassium, and fiber.

> ½ c fruit is equivalent to ½ c fresh, frozen, or canned fruit; 1 small fruit; ¼ c dried fruit; ½ c fruit juice.

- ● Apples, apricots, avocados, bananas, blueberries, cantaloupe, cherries, grapefruit, grapes, guava, kiwi, mango, oranges, papaya, peaches, pears, pineapples, plums, raspberries, strawberries, watermelon; dried fruit; unsweetened juices.

- ▲ Canned or frozen fruit in syrup; juices, punches, ades, and fruit drinks with added sugars; fried plantains.

FIGURE 1–4

(continued)

MILK, YOGURT, AND CHEESE

© Polara Studios, Inc.

Make fat-free or low-fat choices. Choose lactose-free products or other calcium-rich foods if you don't consume milk.

These foods contribute protein, riboflavin, vitamin B_{12}, calcium, magnesium, potassium, and, when fortified, vitamin A and vitamin D.

> 1 c milk is equivalent to 1 c fat-free milk or yogurt; $1\frac{1}{2}$ oz fat-free natural cheese; 2 oz fat-free processed cheese.

● Fat-free milk and fat-free milk products such as buttermilk, cheeses, cottage cheese, yogurt; fat-free fortified soy milk.

▲ 1% low-fat milk, 2% reduced-fat milk, and whole milk; low-fat, reduced-fat, and whole-milk products such as cheeses, cottage cheese, and yogurt; milk products with added sugars such as chocolate milk, custard, ice cream, ice milk, milk shakes, pudding, sherbet; fortified soy milk.

MEAT, POULTRY, FISH, LEGUMES, EGGS, AND NUTS

© Polara Studios, Inc.

Make lean or low-fat choices. Prepare them with little, or no, added fat.

Meat, poultry, fish, and eggs contribute protein, niacin, thiamin, vitamin B_6, vitamin B_{12}, iron, magnesium, potassium, and zinc; legumes and nuts are notable for their protein, folate, thiamin, vitamin E, iron, magnesium, potassium, zinc, and fiber.

> 1 oz meat is equivalent to 1 oz cooked lean meat, poultry, or fish; 1 egg; $\frac{1}{4}$ c cooked legumes or tofu; 1 tbs peanut butter; $\frac{1}{2}$ oz nuts or seeds.

● Poultry (no skin), fish, shellfish, legumes, eggs, lean meat (fat-trimmed beef, game, ham, lamb, pork); low-fat tofu, tempeh, peanut butter, nuts or seeds.

▲ Bacon; baked beans; fried meat, fish, poultry, eggs, or tofu; refried beans; ground beef; hot dogs; luncheon meats; marbled steaks; poultry with skin; sausages; spare ribs.

OILS

© Matthew Farruggio

Select the recommended amounts of oils from among these sources.

These foods contribute vitamin E and essential fatty acids (see Chapter 3), along with abundant kcalories.

> 1 tsp oil is equivalent to 1 tbs low-fat mayonnaise; 2 tbs light salad dressing; 1 tsp vegetable oil; 1 tsp soft margarine.

● Liquid vegetable oils such as canola, corn, flaxseed, nut, olive, peanut, safflower, sesame, soybean, and sunflower oils; mayonnaise, oil-based salad dressing, soft trans-free margarine.

● Unsaturated oils that occur naturally in foods such as avocados, fatty fish, nuts, olives, and shellfish.

SOLID FATS AND ADDED SUGARS

© Matthew Farruggio

Limit intakes of food and beverages with solid fats and added sugars.

Solid fats deliver saturated fat and *trans* fat, and intake should be kept low. Solid fats and added sugars contribute abundant kcalories but few nutrients, and intakes should not exceed the discretionary kcalorie allowance—kcalories to meet energy needs after all nutrient needs have been met with nutrient-dense foods. Alcohol also contributes abundant kcalories but few nutrients, and its kcalories are counted among discretionary kcalories. See Table 1–6 for some discretionary kcalorie allowances.

▲ Solid fats that occur in foods naturally such as milk fat and meat fat (see ▲ in previous lists).

▲ Solid fats that are often added to foods such as butter, cream cheese, hard margarine, lard, sour cream, and shortening.

▲ Added sugars such as brown sugar, candy, honey, jelly, molasses, soft drinks, sugar, and syrup.

▲ Alcoholic beverages include beer, wine, and liquor.

TABLE 1–6

Recommended Daily Amounts from Each Food Group

FOOD GROUP	1,600 kcal	1,800 kcal	2,000 kcal	2,200 kcal	2,400 kcal	2,600 kcal	2,800 kcal	3,000 kcal
Fruits	1½ c	1½ c	2 c	2 c	2 c	2 c	2½ c	2½ c
Vegetables	2 c	2½ c	2½ c	3 c	3 c	3½ c	3½ c	4 c
Grains	5 oz	6 oz	6 oz	7 oz	8 oz	9 oz	10 oz	10 oz
Meat and legumes	5 oz	5 oz	5½ oz	6 oz	6½ oz	6½ oz	7 oz	7 oz
Milk	3 c	3 c	3 c	3 c	3 c	3 c	3 c	3 c
Oils	5 tsp	5 tsp	6 tsp	6 tsp	7 tsp	8 tsp	8 tsp	10 tsp
Discretionary kcalorie allowance	132 kcal	195 kcal	267 kcal	290 kcal	362 kcal	410 kcal	426 kcal	512 kcal

■ Chapter 6 explains how to determine energy needs.

■ Reminder: *Phytochemicals* are the nonnutrient compounds found in plant-derived foods that have biological activity in the body.

Recommended Daily Food Amounts All food groups offer valuable nutrients, and people should make selections from each group daily. Table 1–6 specifies the amounts of food needed from each group daily to create a healthful diet for several energy (kcalorie) levels. ■ Estimated daily kcalorie needs for sedentary and active men and women are shown in Table 1–7. For example, a person needing 2,000 kcalories per day would select 2 cups of fruit; 2¹/₂ cups of vegetables (dispersed among the vegetable subgroups); 6 ounces of grain foods (with at least half coming from whole grains); 5¹/₂ ounces of meat, poultry, or fish, or the equivalents of **legumes**, eggs, seeds, or nuts; and 3 cups of milk or yogurt, or the equivalent of cheese or fortified soy products. Additionally, a small amount of unsaturated oil, such as vegetable oil, or the oils of nuts, olives, or fatty fish, is required to supply needed nutrients.

All vegetables provide an array of vitamins, fiber, and the mineral potassium, but some vegetables are especially good sources of certain nutrients and beneficial phytochemicals. ■ For this reason, the USDA Food Guide sorts the vegetable group into subgroups. The dark green vegetables deliver the B vitamin folate; the orange vegetables provide abundant vitamin A; legumes supply iron and protein; the starchy vegetables contribute abundant carbohydrate energy; and the other vegetables fill in the gaps and add more of these same nutrients.

TABLE 1–7

Estimated Daily Kcalorie Needs for Adults

	Sedentary[a]	Active[b]
Women		
19–30 yr	2,000	2,400
31–50 yr	1,800	2,200
51 + yr	1,600	2,100
Men		
19–30 yr	2,400	3,000
31–50 yr	2,200	2,900
51 + yr	2,000	2,600

Note: In addition to gender, age, and activity level, energy needs vary with height and weight (see Chapter 6).

[a] *Sedentary* describes a lifestyle that includes only the activities typical of day-to-day life.

[b] *Active* describes a lifestyle that includes physical activity equivalent to walking more than 3 miles per day at a rate of 3 to 4 miles per hour, in addition to the activities typical of day-to-day life. kCalorie values for active people reflect the midpoint of the range appropriate for age and gender, but within each group, older adults may need fewer kcalories and younger adults may need more.

TABLE 1–8

Recommended Weekly Amounts from the Vegetable Subgroups

Vegetable Subgroups	1,600 kcal	1,800 kcal	2,000 kcal	2,200 kcal	2,400 kcal	2,600 kcal	2,800 kcal	3,000 kcal
Dark green	2 c	3 c	3 c	3 c	3 c	3 c	3 c	3 c
Orange and deep yellow	1½ c	2 c	2 c	2 c	2 c	2½ c	2½ c	2½ c
Legumes	2½ c	3 c	3 c	3 c	3 c	3½ c	3½ c	3½ c
Starchy	2½ c	3 c	3 c	6 c	6 c	7 c	7 c	9 c
Other	5½ c	6½ c	6½ c	7 c	7 c	8½ c	8½ c	10 c

Table 1–6 specifies the recommended amounts of total vegetables per day. This table shows those amounts dispersed among five vegetable subgroups per week.

In a 2,000-kcalorie diet, the recommended 2½ cups of daily vegetables should be spread among the subgroups over one week's time, as shown in Table 1–8. In other words, eating 2½ cups of potatoes or even spinach every day for seven days would *not* meet the recommended vegetable intakes. Potatoes and spinach make excellent choices when consumed in balance with other vegetables from other subgroups. Intakes of vegetables are appropriately averaged over one week's time—it is not necessary to include every subgroup every day.

Notable Nutrients As Figure 1–4 notes, each food group contributes key nutrients. This feature provides flexibility in diet planning: a person can select any food from a food group and receive similar nutrients. For example, a person can choose milk, cheese, or yogurt and receive the same key nutrients the milk group offers. Legumes are included in the meat, fish, and poultry group because these foods are notable for their contributions of protein, iron, and zinc. Legumes are unique, however, because they are also grouped with the vegetables by virtue of their fiber and vitamins. Importantly, foods provide not only the nutrients for which each group is noted, but small amounts of other nutrients and phytochemicals as well.

The USDA Food Guide encourages greater consumption from certain food groups to provide the nutrients most often lacking in the diets of Americans. ■ In general, most people need to eat:

- *More* dark green vegetables, orange vegetables, legumes, fruits, whole grains, and low-fat milk and milk products

- *Less* refined grains, total fats (especially saturated fat, *trans* fat, and cholesterol), added sugars, and total kcalories

Nutrient Density The USDA Food Guide provides a foundation for a healthy diet by emphasizing nutrient-dense options within each food group. By consistently selecting nutrient-dense foods, a person can both obtain all the nutrients needed and keep kcalories under control. In contrast, eating foods that are low in nutrient density makes it difficult to get enough nutrients without exceeding energy needs and gaining weight. For this reason, consumers should select low-fat foods from each group and foods without added fats or sugars—for example, fat-free milk instead of whole milk, baked chicken without the skin instead of hot dogs, green beans instead of french fries, orange juice instead of fruit punch, and whole wheat bread instead of biscuits. Notice that Figure 1–4 provides a key indicating which foods *within each group* are high or low in nutrient density. Oil is a notable exception: even though oil is pure fat and therefore rich in kcalories, a small amount of oil from sources such as nuts, fish, or vegetable oils is necessary every day to provide nutrients lacking from other foods. Consequently these high-fat foods are listed among the nutrient-dense foods (see Nutrition in Practice 3 to learn why).

Discretionary kCalorie Allowance At each kcalorie level, people who consistently choose nutrient-dense foods may be able to meet their nutrient needs without

■ The USDA nutrients of concern are vitamins A, C, and E; the minerals calcium, magnesium, and potassium; and fiber.

© Polara Studios,

This cola and bunch of grapes illustrate nutrient density. Each provides about 150 kcalories (mainly from carbohydrate), but the grapes offer a trace of protein, some vitamins, minerals, and fiber along with the energy; the cola beverage offers only "empty" kcalories. Grapes, or any fruit, are more nutrient dense than cola beverages.

legumes (lay-GYOOMS, LEG-yooms): plants of the bean and pea family, with seeds that are rich in protein compared with other plant-derived foods.

FIGURE 1–5

Discretionary kCalorie Allowance in
a 2,000-kcalorie Diet

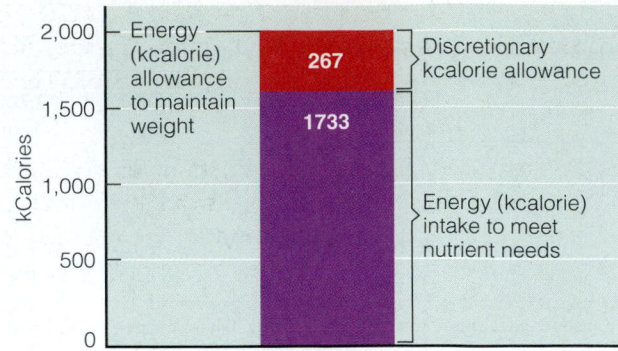

consuming their full allotment of kcalories. This difference between the kcalories needed to supply nutrients and those needed for energy—known as the **discretionary kcalorie allowance**—is illustrated in Figure 1–5.

Table 1–6 (p. 20) includes the discretionary kcalorie allowance for several kcalorie levels. A person with discretionary kcalories available might choose to:

■ Eat more nutrient-dense foods, such as an extra serving of skinless chicken or a second ear of corn.

■ Select a few foods with fats or added sugars, such as reduced-fat milk or sweetened cereal.

■ Add a little fat or sugar to foods, such as butter or jelly on toast.

■ Consume some alcohol. (Nutrition in Practice 20 explains why this may not be a good choice for some individuals.)

Alternatively, a person wanting to lose weight might choose to:

■ *Not* use the kcalories available from the discretionary kcalorie allowance.

Compared to physically active adults, sedentary adults have lower kcalorie needs and so have a lower discretionary kcalorie allowance. Physically active adults expend more energy and therefore have a higher energy need and a greater discretionary kcalorie allowance.

Added fats and sugars are always counted as discretionary kcalories. The kcalories from the fat in higher-fat milks and meats are also counted among discretionary kcalories. It helps to think of fat-free milk as "milk" and whole milk or reduced-fat milk as "milk with added fat." Similarly, "meats" should be the leanest; other cuts are "meats with added fat." Puddings and other desserts made from whole milk provide discretionary kcalories from both the sugar added to sweeten them and the naturally occurring fat in the whole milk they contain. Even fruits, vegetables, and grains can carry discretionary kcalories into the diet in the form of peaches canned in syrup, french fries, or high-fat crackers.

Discretionary kcalories must be counted separately from the kcalories of the nutrient-dense foods of which they may be a part. A fried chicken leg, for example, provides discretionary kcalories from two sources: the naturally occurring fat of the chicken skin and the added fat absorbed during frying. The kcalories of the skinless chicken underneath are not discretionary kcalories—they are necessary to provide the nutrients of chicken.

Portion Control To control kcalories, the diet planner must also learn to control food portions (USDA serving equivalents were listed in Figure 1–4). The trend in the United States has been toward consuming larger food portions, especially of foods rich in fat and sugar. At the same time, body weights have been steadily creeping upward, suggesting an increasing need to control portion sizes. In contrast to the random-size (and often large) helpings offered in restaurants, fast-food franchises, and elsewhere, the quantities recommended in the USDA Food Guide are specific, precise, and reliable for delivering certain amounts of key nutrients in food. The margin offers some tips for estimating portion sizes. ■

Mixtures of Foods Some foods—such as casseroles, soups, and sandwiches—fall into two or more food groups. With a little practice, users can learn to divide these foods into food groups. From the USDA Food Guide point of view, a taco represents four different food groups: the taco shell from the grains group; the onions, lettuce, and tomatoes from the "other vegetable" group; the ground beef from the meat group; and the cheese from the milk group.

Vegetarian Food Guide Vegetarian diets rely mainly on plant foods: grains, vegetables, legumes, fruits, seeds, and nuts. Some vegetarian diets include eggs, milk products, or both. People who do not eat meats or milk products can still use the USDA Food Guide to create an adequate diet.[13] The food groups are similar, and the recommended amounts of foods remain the same. Vegetarians select *meat alternates* from the meat group—foods such as legumes, seeds, nuts, tofu, and, for those who eat them, eggs. Legumes and at least one cup of dark green leafy vegetables help to supply the iron that meats usually provide. Vegetarians who do not drink cow's milk can use soy "milk"—a product made from soybeans that provides similar nutrients if it has been fortified with calcium, vitamin D, and vitamin B_{12}. Nutrition in Practice 4 presents a Food Guide for Vegetarians, defines vegetarian terms, and provides more information on vegetarian diet planning.

Ethnic Food Choices People can use the USDA Food Guide and still enjoy a diverse array of culinary styles by sorting ethnic foods into their appropriate food groups. For example, a person eating Mexican foods would find tortillas in the grains group, jicama in the vegetable group, and guava in the fruit group. Table 1–1 (p. 4) features ethnic food choices.

MyPyramid

The USDA created an educational tool called MyPyramid to illustrate the concepts presented in the *Dietary Guidelines* and the USDA Food Guide. Figure 1–6 (p. 24) presents a graphic image of MyPyramid, which was designed to encourage consumers to make healthy food and physical activity choices every day.

An abundance of material supporting MyPyramid is available to consumers who want to find the kinds and amounts of foods to eat each day (www.MyPyramid.gov). In addition to creating a personal plan, consumers can find tips to help them improve their diet and lifestyle by taking small steps each day.

IN SUMMARY

- Food group plans such as the USDA Food Guide serve as the basis for planning adequate, balanced, and varied diets.

- Each food group contributes key nutrients, a feature that provides flexibility in diet planning.

- The USDA Food Guide emphasizes nutrient-dense foods within each group.

- The discretionary kcalorie allowance is the difference between the kcalories needed to meet nutrient needs and those needed for energy.

- MyPyramid is an educational tool used to illustrate the concepts presented in the Dietary Guidelines and the USDA Food Guide.

discretionary kcalorie allowance: the kcalories remaining in a person's energy allowance after consuming enough nutrient-dense foods to meet all nutrient needs for one day.

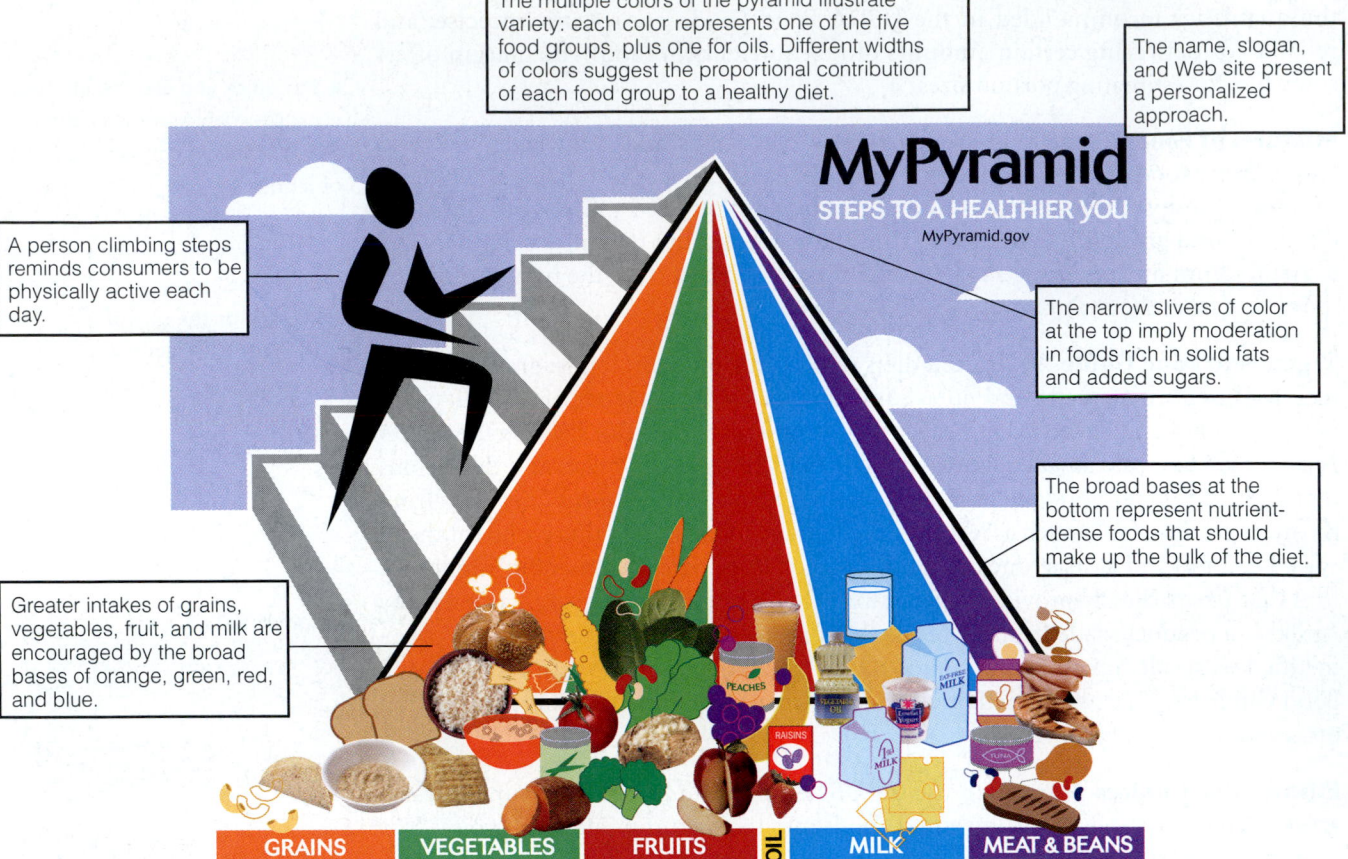

The multiple colors of the pyramid illustrate variety: each color represents one of the five food groups, plus one for oils. Different widths of colors suggest the proportional contribution of each food group to a healthy diet.

The name, slogan, and Web site present a personalized approach.

A person climbing steps reminds consumers to be physically active each day.

The narrow slivers of color at the top imply moderation in foods rich in solid fats and added sugars.

The broad bases at the bottom represent nutrient-dense foods that should make up the bulk of the diet.

Greater intakes of grains, vegetables, fruit, and milk are encouraged by the broad bases of orange, green, red, and blue.

MyPyramid
STEPS TO A HEALTHIER YOU
MyPyramid.gov

GRAINS VEGETABLES FRUITS OIL MILK MEAT & BEANS

FIGURE 1–6

MyPyramid: Steps to a Healthier You
Source: USDA, 2005.

Food Labels

Today, consumers know more about the links between diet and disease than they did in the past, and they are demanding still more information on disease prevention. Many people rely on food labels to help them select foods with less saturated fat, *trans* fat, cholesterol, and sodium and more vitamins, minerals, and dietary fiber. Figure 1–7 illustrates the requirements for label information. Most food labels must conform to all these requirements. Exceptions include plain coffee, tea, spices, and other foods contributing few nutrients; foods produced by small businesses; packages with fewer than 12 square inches of surface area; and those prepared and sold in the same establishment as long as the foods do not make nutrient or health claims.*

The Ingredient List All foods must list all ingredients on the label in descending order of predominance by weight. Knowing that the first ingredient predominates by weight, consumers can glean much information. Compare these products, for example:

■ A beverage powder that contains "sugar, citric acid, natural flavors . . ." versus a juice that contains "water, tomato concentrate, concentrated juices of carrots, celery . . ."

■ A cereal that contains "puffed milled corn, sugar, corn syrup, molasses, salt . . ." versus one that contains "100 percent rolled oats"

* For example, restaurants need not provide complete nutrition information unless they are making "heart-healthy" claims for menu items.

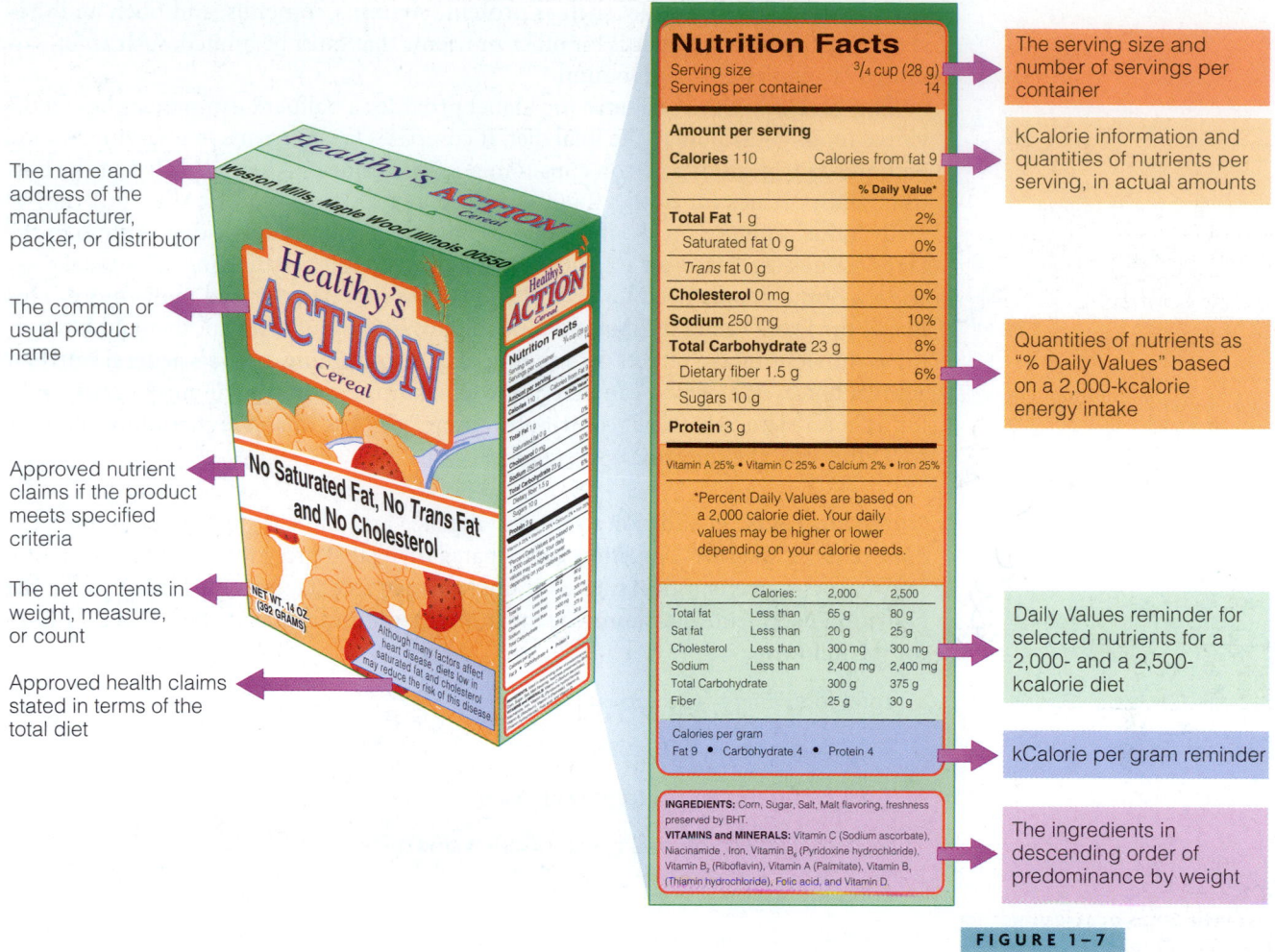

Nutrition Facts

Serving size	¾ cup (28 g)
Servings per container	14

Amount per serving

Calories 110	Calories from fat 9

	% Daily Value*
Total Fat 1 g	2%
Saturated fat 0 g	0%
Trans fat 0 g	
Cholesterol 0 mg	0%
Sodium 250 mg	10%
Total Carbohydrate 23 g	8%
Dietary fiber 1.5 g	6%
Sugars 10 g	
Protein 3 g	

Vitamin A 25% • Vitamin C 25% • Calcium 2% • Iron 25%

*Percent Daily Values are based on a 2,000 calorie diet. Your daily values may be higher or lower depending on your calorie needs.

		Calories:	2,000	2,500
Total fat	Less than		65 g	80 g
Sat fat	Less than		20 g	25 g
Cholesterol	Less than		300 mg	300 mg
Sodium	Less than		2,400 mg	2,400 mg
Total Carbohydrate			300 g	375 g
Fiber			25 g	30 g

Calories per gram
Fat 9 • Carbohydrate 4 • Protein 4

INGREDIENTS: Corn, Sugar, Salt, Malt flavoring, freshness preserved by BHT.
VITAMINS and MINERALS: Vitamin C (Sodium ascorbate), Niacinamide , Iron, Vitamin B₆ (Pyridoxine hydrochloride), Vitamin B₂ (Riboflavin), Vitamin A (Palmitate), Vitamin B₁ (Thiamin hydrochloride), Folic acid, and Vitamin D.

Callouts on the package (left):

- The name and address of the manufacturer, packer, or distributor
- The common or usual product name
- Approved nutrient claims if the product meets specified criteria
- The net contents in weight, measure, or count
- Approved health claims stated in terms of the total diet

Callouts on the label (right):

- The serving size and number of servings per container
- kCalorie information and quantities of nutrients per serving, in actual amounts
- Quantities of nutrients as "% Daily Values" based on a 2,000-kcalorie energy intake
- Daily Values reminder for selected nutrients for a 2,000- and a 2,500-kcalorie diet
- kCalorie per gram reminder
- The ingredients in descending order of predominance by weight

FIGURE 1–7

Example of a Food Label

In each comparison, consumers can tell that the second product is the more nutrient dense.

Serving Sizes Because labels present nutrient information per serving, they must identify the size of a serving. The Food and Drug Administration (FDA) has established specific serving sizes for various foods and requires that all labels for a given product use the same serving size. For example, the serving size for all ice creams is ½ cup and for all beverages, 8 fluid ounces. This facilitates comparison shopping. Consumers can see at a glance which brand has more or fewer kcalories or grams of fat, for example. Standard serving sizes are expressed in both common household measures, such as cups, and metric measures, such as milliliters, to accommodate users of both types of measures.

When examining the nutrition information on a food label, consumers need to consider how the serving size compares with the actual quantity eaten. If it is not the same, they will need to adjust the quantities accordingly. For example, if the serving size is four cookies and you only eat two, then you need to cut the nutrient and kcalorie values in half; similarly, if you eat eight cookies, then you need to double the values. Notice, too, that small bags or individually wrapped items, such as chips or candy bars, may contain more than a single serving. The number of servings per container is listed just below the serving size.

The Daily Values To help consumers evaluate the information found on labels, the FDA created a set of nutrient standards called the **Daily Values** specifically for use on food labels. The Daily Values do two things: they set adequacy standards for nutrients

Daily Values: reference values developed by the FDA specifically for use on food labels.

that are desirable in the diet such as protein, vitamins, minerals, and fiber, and they also set moderation standards for other nutrients that must be limited, such as fat, saturated fat, cholesterol, and sodium.

The "% Daily Value" column on a label provides a ballpark estimate of how individual foods contribute to the total diet. It compares key nutrients in a serving of food with the daily goals of a person consuming 2,000 kcalories. Most labels list, at the bottom, Daily Values for both a 2,000-kcalorie and a 2,500-kcalorie diet, but the "% Daily Value" column on all labels applies only to a 2,000-kcalorie diet. Although the Daily Values are based on a 2,000-kcalorie diet, people's actual energy intakes vary widely. Some people need fewer kcalories, and some people need many more. This makes the Daily Values most useful for comparing one food with another and less useful as nutrient intake targets for individuals. By examining a food's general nutrient profile, however, a person can determine whether the food contributes "a little" or "a lot" of a nutrient, whether it contributes "more" or "less" than another food, and how well it fits into the consumer's overall diet.

Nutrition Facts In addition to the serving size and the servings per container, the FDA requires that the "Nutrition Facts" panel on a label present nutrient information in two ways—in quantities (such as grams) and as percentages of the Daily Values. The Nutrition Facts panel must provide the nutrient amount, percent Daily Value, or both for the following:

- Total food energy (kcalories)

- Food energy from fat (kcalories)

- Total fat (grams and percent Daily Value)

- Saturated fat (grams and percent Daily Value)

- *Trans* fat (grams)

- Cholesterol (milligrams and percent Daily Value)

- Sodium (milligrams and percent Daily Value)

- Total carbohydrate, including starch, sugar, and fiber (grams and percent Daily Value)

- Dietary fiber (grams and percent Daily Value)

- Sugars (grams), including both those naturally present in and those added to the food

- Protein (grams)

The labels must also present nutrient content information as a percentage of the Daily Values for the following vitamins and minerals:

- Vitamin A

- Vitamin C

- Iron

- Calcium

The FDA developed the Daily Values for use on food labels because comparing nutrient amounts against a standard helps make them meaningful to consumers. A person might wonder, for example, whether 1 milligram of iron or calcium is a little or a lot. As Table 1–9 shows, the Daily Value for iron is 18 milligrams, so 1 milligram of iron is enough to take notice of: it is over 5 percent. But the Daily Value for calcium on food labels is 1,000 milligrams, so 1 milligram of calcium is essentially nothing.

TABLE 1–9

Daily Values for Food Labels

Food labels must present the "% Daily Value" for these nutrients.

Food Component	Daily Value	Calculation Factors
Fat	65 g	30% of kcalories
Saturated fat	20 g	10% of kcalories
Cholesterol	300 mg	—
Carbohydrate (total)	300 g	60% of kcalories
Fiber	25 g	11.5 g per 1,000 kcalories
Protein	50 g	10% of kcalories
Sodium	2,400 mg	—
Potassium	3,500 mg	—
Vitamin C	60 mg	—
Vitamin A	1,500 µg	—
Calcium	1,000 mg	—
Iron	18 mg	—

Note: Daily Values were established for adults and children over 4 years old. The values for energy-yielding nutrients are based on 2,000 kcalories per day. For fiber, the Daily Value was rounded up from 23.

Nutrient Claims The FDA defines the **nutrient claims** a label may use to describe the contents of a product (see Table 1–10). Definitions include the conditions under which each term can be used. For example, in addition to having less than 2 milligrams of cholesterol, a "cholesterol-free" product may not contain more than 2 grams of saturated fat and *trans* fat combined per serving.

TABLE 1–10

Terms Used on Food Labels

General Terms

free, without, no, zero none or a trivial amount. *Calorie free* means containing fewer than 5 kcalories per serving; *sugar free* or *fat free* means containing less than half a gram per serving.

fresh raw, unprocessed, or minimally processed with no added preservatives.

good source 10 to 19% of the Daily Value per serving.

healthy low in fat, saturated fat, *trans* fat, cholesterol, and sodium and containing at least 10% of the Daily Value for vitamin A, vitamin C, iron, calcium, protein, or fiber.

high in 20% or more of the Daily Value for a given nutrient per serving; synonyms include "rich in" or "excellent source."

less, fewer, reduced containing at least 25% less of a nutrient or kcalories than a reference food. This may occur naturally or as a result of altering the food. For example, pretzels, which are usually low in fat, can claim to provide less fat than potato chips, a comparable food.

light this descriptor has three meanings on labels:

1. A serving provides one-third fewer kcalories or half the fat of the regular product.
2. A serving of a low-kcalorie, low-fat food provides half the sodium normally present.
3. The product is light in color and texture, so long as the label makes this intent clear, as in "light brown sugar."

more, extra at least 10% more of the Daily Value than in a reference food. The nutrient may be added or may occur naturally.

continued

nutrient claims: statements that characterize the quantity of a nutrient in a food.

TABLE 1–10 (continued)

Energy Terms

low calorie 40 kcalories or fewer per serving.

reduced calorie at least 25% lower in kcalories than a "regular," or reference, food.

calorie free fewer than 5 kcalories per serving.

Fat Terms (meat and poultry products)

extra lean

less than 5 g of fat *and*

less than 2 g of saturated fat and *trans* fat combined, *and*

less than 95 mg of cholesterol per serving.

lean[a]

less than 10 g of fat *and*

less than 4.5 g of saturated fat and *trans* fat combined, *and*

less than 95 mg of cholesterol per serving.

Fat and Cholesterol Terms (all products)

cholesterol free

less than 2 mg of cholesterol *and*

2 g or less saturated fat and *trans* fat combined per serving.

fat free less than 0.5 g of fat per serving.

less saturated fat 25% or less saturated fat and *trans* fat combined than the comparison food.

low cholesterol

20 mg or less of cholesterol *and*

2 g or less saturated fat per serving.

low fat 3 g or less fat per serving.[b]

low saturated fat 1 g or less saturated fat and less than 0.5 g of *trans* fat per serving.

percent fat free may be used only if the product meets the definition of *low fat* or *fat free*. Requires disclosure of grams of fat per 100 g food.

Fat and Cholesterol Terms (all products)

reduced or **less cholesterol**

at least 25% less cholesterol than a reference food *and*

2 g or less saturated fat per serving.

reduced saturated fat

at least 25% less saturated fat *and*

reduced by more than 1 g saturated fat per serving compared with a reference food.

saturated fat free

less than 0.5 g of saturated fat *and*

less than 0.5 g of *trans* fat.

***trans* fat free**

less than 0.5 g of *trans* fat *and*

less than 0.5 g of saturated fat per serving.

Fiber Terms

high fiber 5 g or more per serving. (Foods making high-fiber claims must fit the definition of low fat, or the level of total fat must appear next to the high-fiber claim.)

good source of fiber 2.5 g to 4.9 g per serving.

more or added fiber at least 2.5 g more per serving than a reference food.

Sodium Terms

low sodium 140 mg or less sodium per serving.

reduced sodium at least 25% lower in sodium than the regular product.

sodium free less than 5 mg per serving.

very low sodium 35 mg or less sodium per serving.

[a] The word *lean* as part of the brand name (as in "Lean Supreme") indicates that the product contains fewer than 10 grams of fat per serving.
[b] Exceptions may be granted for foods with low levels of saturated and *trans* fats.

TABLE 1–11		
Food Label Health Claims—The "A" List		
Calcium and reduced risk of osteoporosis		
Sodium and reduced risk of hypertension		
Dietary saturated fat and cholesterol and reduced risk of coronary heart disease		
Dietary fat and reduced risk of cancer		
Fiber-containing grain products, fruits, and vegetables and reduced risk of cancer		
Fruits, vegetables, and grain products that contain fiber, particularly soluble fiber, and reduced risk of coronary heart disease		
Fruits and vegetables and reduced risk of cancer		
Folate and reduced risk of neural tube defects		
Sugar alcohols and reduced risk of tooth decay		
Soluble fiber from whole oats and from psyllium seed husk and reduced risk of heart disease		
Soy protein and reduced risk of heart disease		
Whole grains and reduced risk of heart disease and certain cancers		
Plant sterol and plant stanol esters and heart disease		
Potassium and reduced risk of hypertension and stroke		

Some descriptions imply that a food contains, or does not contain, a nutrient. Implied claims are prohibited unless they meet specified criteria. For example, a claim that a product "contains no oil" implies that the food contains no fat. If the product is truly fat-free, then it may make the no-oil claim, but if it contains another source of fat, such as butter, it may not.

Health Claims Until recently, the FDA held manufacturers to the highest standards of scientific evidence before allowing them to place **health claims** on food labels. When a label stated, "diets low in sodium may reduce the risk of high blood pressure," for example, consumers could be sure that the FDA had examined much scientific evidence and found substantial support for the claim. Such reliable health claims make up the FDA's "A" list and still appear on some food labels (see Table 1–11).

The FDA recently created a ranking system that includes three new categories of allowable health claims that are supported by evidence that is less conclusive.[14] The new system assigns each claim a letter grade reflecting the degree to which the claim is backed by science (see Table 1–12 on p. 30). The FDA can no longer demand that only health claims with the highest degree of scientific support appear on food labels.

Under the new system, the reliable health claims shown in Table 1–11 receive an "A" grade. Claims with a "B," "C," or "D" grade are called "qualified" health claims because they must bear a statement explaining the degree of scientific evidence backing them up. The FDA is relying on consumers to notice these printed disclaimers.

Structure-Function Claims Consumers need to be aware that a different kind of claim, known as a "structure-function" claim, may also appear on food or dietary supplement labels. **Structure-function claims** are statements about a food substance's effect on a structure or function of the body—for example, "antioxidants support heart health." Structure-function claims are not required to have FDA approval, but the claims may not refer to the reduction of disease risk. The claims must be carefully worded to avoid any mention of a specific disease. Typical claims are "slows aging," "improves memory," and "builds strong bones." To make a more specific claim such as "prevents osteoporosis," the manufacturer would have to submit to the rigorous requirements for health claims or meet the even stricter safety and

health claims: statements that characterize the relationship between a nutrient or other substance in food and a disease or health-related condition.

structure-function claims: statements that describe how a product may affect a structure or function of the body; for example, "calcium builds strong bones." Structure-function claims do not require FDA authorization.

TABLE 1–12

The FDA's Health Claims Report Card

Grade	Level of Confidence in Health Claim	Required Label Disclaimers
A	High: Significant scientific agreement	These health claims do not require disclaimers; see Table 1–11 for examples
B	Moderate: Evidence is supportive, but not conclusive	"[Health claim.] Although there is scientific evidence supporting this claim, the evidence is not conclusive."
C	Low: Evidence is limited and not conclusive	"Some scientific evidence suggests [health claim]. However, FDA has determined that this evidence is limited and not conclusive."
D	Very low: Little scientific evidence supporting this claim	"Very limited and preliminary scientific research suggests [health claim]. FDA concludes that there is little scientific evidence supporting this claim."

efficacy standards applied to drugs. These rules ensure that consumers can have confidence that when a claim names a specific disease, there is substantial scientific agreement that the food, in the context of a healthy diet, may help protect against that disease.

IN SUMMARY

- Food labels provide consumers with information they need to select foods that will help them meet their nutrition and health goals.

- Daily Values are a set of nutrient standards created by the FDA for use on food labels.

- Health claims that are graded "A" are backed by the highest standards of scientific evidence. Health claims with a "B," "C," or "D" grade are supported by less conclusive scientific evidence than those graded with an "A."

Analysis

YOUR DIET

The secret to making healthy food choices is learning to incorporate the 2005 *Dietary Guidelines* (Table 1–4 on p. 16) and the USDA Food Guide (Figure 1–4 on pp. 18–19) into your decision-making process. Before completing this assignment, you may want to review the section called "Mixtures of Foods" on p. 23 to see where foods such as casseroles or soups fall in the food group plan.

- Keep a record of the foods you eat for 24 hours. Record both the types and amounts of foods eaten.

- Turn to Table 1–7 on p. 20 to determine your estimated kcalorie needs (Chapter 6 offers a more individualized approach to estimate your kcalorie needs). What is your estimated daily kcalorie need? _____ kcalories.

- Compare the foods you ate with the USDA Food Guide recommendations for your energy needs (see Table 1–6 on p. 20) by listing them in the accompanying table (keep a copy of this for later reference). For example, if you need 2,400 kcalories a day, the recommended amount of fruits is

2 cups, while that of grains is 8 ounces. The USDA Food Guide on pp. 18–19 provides equivalent measures for foods that are not readily measured in cups or ounces. Thus, 1 slice of bread counts as 1 ounce of grains, while 1 small apple counts as $\frac{1}{2}$ cup of fruit.

■ Do your food choices include foods from each of the food groups?

■ Are some food groups over- or underrepresented? If so, what can you do to ensure a more balanced diet?

■ Do you eat a variety of foods within each group? If not, suggest ways to enhance the variety in your diet.

■ Did you choose at least some whole grain foods from the grain group, and fat-free or low-fat foods from the milk, yogurt, and cheese group and the meat group? In your list below (foods consumed), circle the foods you selected that are whole grain (grain group), fat-free, and low-fat (milk and meat groups)

Food Group	Recommended Amount	Foods Consumed	Your Diet: Total Amount Consumed
Grains		. . . .	_____oz
Vegetables		. . . .	_____cups
Fruits		. . . .	_____cups
Milk, yogurt, and cheese		. . . .	_____cups
Meat, poultry, fish, legumes, eggs, and nuts		. . . .	_____oz
Oils		. . . .	_____tsp

CLINICAL APPLICATIONS

1. Make a list of the foods and beverages you have consumed in the last two days. Look at each item on your list and consider why you chose the particular food or beverage you did. Did you eat cereal for breakfast because that is what you always eat (habit) or because it was the easiest, quickest food to prepare (convenience)? Did you put fat-free milk on the cereal because you want to control your energy intake (nutrition)? In going down your list, you may be surprised to discover exactly why you chose certain foods.

2. As a health care professional, you can uncover clues about a client's food choices by paying close attention. You may be surprised to discover why a client chooses certain foods, but you can then use this knowledge to serve the best interests of the client. For example, an elderly, undernourished widower may eat the same sandwich for lunch every day. In talking with the client, you discover that is what he and his wife fixed together each day. Consider ways you might be able to help the client learn to eat other foods and vary his choices.

Self Check

1. When people eat the foods typical of their families or geographic region, their choices are influenced by:
   a. occupation.
   b. nutrition.
   c. emotional state.
   d. ethnic heritage or tradition.

2. The energy-yielding nutrients are:
   a. fats, minerals, and water.
   b. minerals, proteins, and vitamins.
   c. carbohydrates, fats, and vitamins.
   d. carbohydrates, fats, and proteins.

3. The inorganic nutrients are:
   a. proteins and fats.
   b. vitamins and minerals.
   c. minerals and water.
   d. vitamins and proteins.

4. Alcohol is not a nutrient because:
   a. the body derives no energy from it.
   b. it is organic.
   c. it is converted to body fat.
   d. it does not contribute to the body's growth or repair.

5. DRI stands for:
   a. Daily Recommended Intakes.
   b. Dietary Requirements for Individuals.
   c. Dietary Reference Intakes.
   d. Daily Recommendations for Individuals.

6. Which of the following is consistent with the *Dietary Guidelines for Americans 2005*?
   a. Limit intakes of fruits, vegetables, and whole grains.
   b. Engage in regular physical activity and reduce sedentary activities to promote health, psychological well-being, and a healthy body weight.
   c. Choose a diet with plenty of whole-milk products.
   d. Eat an abundance of foods to ensure nutrient adequacy.

7. In a food group plan such as the USDA Food Guide, foods within a given food group are similar in their contents of:
   a. energy.
   b. proteins and fibers.
   c. vitamins and minerals.
   d. carbohydrates and fats.

8. A slice of apple pie supplies 350 kcalories with 3 grams of fiber; an apple provides 80 kcalories and the same 3 grams of fiber. This is an example of:
   a. kcalorie control.
   b. nutrient density.
   c. variety
   d. essential nutrients

9. Which of the following statements does *not* apply to a person's discretionary kcalorie allowance?
   a. Compared to physically active adults, sedentary adults have a lower discretionary kcalorie allowance.
   b. It is the difference between the kcalories needed to supply nutrients and those needed for energy.
   c. Compared to physically active adults, sedentary adults have a higher discretionary kcalorie allowance.
   d. A person with discretionary kcalories available can, if he or she chooses to, select a few foods with added fat or sugar.

10. Food labels list ingredients in:
    a. alphabetical order.
    b. ascending order of predominance by weight.
    c. descending order of predominance by weight.
    d. manufacturer's order of preference.

Answers to these questions appear in Appendix H.

Notes

1. A. Mennella, M. Y. Pepino, and D. R. Reed, Genetic and environmental determinants of bitter perception and sweet preferences, *Pediatrics* 115 (2005): e216.

2. W. Lockeretz and K. A. Merrigan, Selling to the eco-conscious food shopper, *Nutrition Today* 40 (2005): 45–49.

3. K. Bertrand, Microwavable foods satisfy need for speed and palatability, *Food Technology* 59 (2005): 30–34; J. E. Tillotson, Fast-casual dining: Our next eating passion? *Nutrition Today* 38 (2003): 91–94; J. E. Tillotson, Our ready-prepared ready-to-eat nation, *Nutrition Today* 37 (2002): 36–38.

4. Tillotson, 2002.

5. K. Bertrand, Microwavable foods satisfy need for speed and palatability, *Food Technology* 59 (2005): 30–34.

6. Position of the American Dietetic Association: Functional foods, *Journal of the American Dietetic Association* 104 (2005): 814–826.

7. Standing Committee on the Scientific Evaluation of Dietary Reference Intakes, Food and Nutrition Board, Institute of Medicine, *Dietary Reference Intakes for Water, Potassium, Sodium, Chloride, and Sulfate* (Washington, D.C.: National Academies Press, 2004); Standing Committee on the Scientific Evaluation of Dietary Reference Intakes, Food and Nutrition Board, Institute of Medicine, *Dietary Reference Intakes for Energy, Carbohydrate, Fiber, Fat, Fatty Acids, Cholesterol, Protein, and Amino Acids* (Washington, D.C.: National Academies Press, 2005); Standing Committee on the Scientific Evaluation of Dietary Reference Intakes, Food and Nutrition Board, Institute of Medicine, *Dietary Reference Intakes for Vitamin A, Vitamin K, Arsenic, Boron, Chromium, Copper, Iodine, Iron, Manganese, Molybdenum, Nickel, Silicon, Vanadium, and Zinc* (Washington, D.C.: National Academy Press, 2001); Standing Committee on the Scientific Evaluation of Dietary Reference Intakes, Food and Nutrition Board, Institute of Medicine, *Dietary Reference Intakes for Vitamin C, Vitamin E, Selenium, and Carotenoids* (Washington, D.C.: National Academy Press, 2000); Standing Committee on the Scientific Evaluation of Dietary Reference Intakes, Food and Nutrition Board, Institute of Medicine, *Dietary Reference Intakes for Thiamin, Riboflavin, Niacin, Vitamin B_6, Folate, Vitamin B_{12}, Pantothenic Acid, Biotin, and Choline* (Washington, D.C.: National Academy Press, 1998); Standing Committee on the Scientific Evaluation of Dietary Reference Intakes, Food and Nutrition Board, Institute of Medicine, *Dietary Reference Intakes for Calcium, Phosphorus, Magnesium, Vitamin D, and Fluoride* (Washington, D.C.: National Academy Press, 1997).

8. Subcommittee on Interpretation and Uses of Dietary Reference Intakes and the Standing Committee on the Scientific Evaluation of Dietary Reference Intakes, *Applications in Dietary Planning* (Washington, D.C.: National Academies Press, 2003).

9. J. Dwyer and coauthors, Collection of food and dietary supplement intake data: What we eat in America—NHANES, *Journal of Nutrition* 133 (2003): 590S–600S.

10. Dwyer and coauthors, 2003.

11. Centers for Disease Control, Physical activity and good nutrition: Essential elements to prevent chronic diseases and obesity, 2004, available at www.cdc.gov/nccdphp/dnpa; D. Yach and coauthors, The global burden of chronic diseases: Overcoming impediments to prevention and control, *Journal of the American Medical Association* 291 (2004): 2616–2622; K. T. B. Knoops and coauthors, Mediterranean diet, lifestyle factors, and 10-year mortality in elderly European men and women, *Journal of the American Medical Association* 292 (2004): 1433–1439; D. Lee, L. M. Steffen, and D. R. Jacobs, Association between serum γ-glutamyltransferase and dietary factors: The Coronary Artery Risk Development in Young Adults (CARDIA) Study, *American Journal of Clinical Nutrition* 79 (2004): 600–605.

12. U.S. Department of Agriculture and U.S. Department of Health and Human Services, *Nutrition and Your Health-Dietary Guidelines for Americans 2005*, 6th ed., Home and Garden Bulletin no. 232 (Washington, D.C.: 2005), available online at www.healthierus.gov/dietaryguidelines or call (888) 878-3256.

13. Position of the American Dietetic Association and Dietitians of Canada: Vegetarian diets, *Journal of the American Dietetic Association* 103 (2003): 748–765.

14. J. E. Tillotson, Health claims 2005: A new tower of Babel? *Nutrition Today* 40 (2005): 88–91; U. S. Food and Drug Administration, FDA to encourage science-based labeling and competition for healthier dietary choices, *FDA News*, available at www.fda.gov/bbs/topics/NEWS/2003/NEW00923.html.

FINDING THE TRUTH ABOUT NUTRITION

Nutrition and health receive so much attention on television, in the popular press, and on the Internet that it is easy to be overwhelmed with conflicting, confusing information. Determining whether nutrition information is accurate can be a challenging task. It is also an important task because nutrition affects a person both professionally and personally.

A person watches a nutrition report on television and then reads a conflicting report in the newspaper. Why do nutrition news reports and claims for nutrition products seem to contradict each other so often?

The problem of conflicting messages arises for several reasons:

- Popular media, often faced with tight deadlines and limited time or space to report new information, rush to present the latest "breakthrough" in a headline or a 60-second spot. They can hardly help omitting important facts about the study or studies on which the "breakthrough" is based.

- Despite tremendous advances in the last few decades, scientists still have much to learn about the human body and nutrition. Scientists themselves often disagree on their first tentative interpretations of new research findings, yet these are the very findings about which the public hears most.

- The popular media often broadcast preliminary findings in hope of grabbing attention and boosting readership or television ratings.

- Commercial promoters turn preliminary findings into advertisements for products or supplements long before the findings have been validated—or disproved. The scientific process requires many experiments or trials to confirm a new finding. Seldom do promoters wait as long as they should to make their claims.

- Promoters are aware that consumers like to try new products or treatments even though they probably will not withstand the tests of time and scientific scrutiny.

So how can a person tell what claims to believe?

The Food and Nutrition Science Alliance (FANSA), whose partners include the American Dietetic Association (ADA), the American Society for Nutritional Sciences (ASNS), and the Institute of Food Technologists (IFT), attempts to help consumers distinguish valid from misleading nutrition information. FANSA has created a list of ten red flags for detecting "junk science" (see Table NP1–1).[1]

TABLE NP1–1
FANSA's Red Flags of Junk Science

1. Promises of a quick and easy fix.
2. Dire warnings of danger from a single product or regimen.
3. Too good to be true. The claim says what most people want to hear.
4. Enticingly simple conclusions drawn from a complex study.
5. Recommendations based on a single study.
6. Dramatic statements that are refuted by reputable scientific organizations.
7. Lists of "good" and "bad" foods.
8. Claims are made to help sell a product.
9. Claims or recommendations based on studies published without peer review.
10. Recommendations from studies that ignore differences among individuals or groups.

Source: Adapted from Position of the American Dietetic Association: Food and nutrition misinformation, *'Journal of the American Dietetic Association',* (2006) 106: 601–607. Used by permission of Elsevier, Inc.

Because nutrition misinformation harms the health and economic status of consumers, the ADA works with health care professionals and educators to present sound nutrition information to the public and to actively confront nutrition misinformation.[2] Table NP1–2 offers a list of credible sources of nutrition information.

TABLE NP1–2
Credible Sources of Nutrition Information

Professional health organizations, government health agencies, volunteer health agencies, and consumer groups provide consumers with reliable health and nutrition information. Credible sources of nutrition information include:

Professional health organizations, especially the American Dietetic Association **www.eatright.org**, also the Society for Nutrition Education **www.sne.org** and the American Medical Association **www.ama-assn.org**

Government health agencies such as the Federal Trade Commission (FTC) **www.ftc.gov**, the U.S. Department of Health and Human Services (HHS) **www.os.dhhs.gov**, the Food and Drug Administration (FDA) **www.fda.gov**, and the U.S. Department of Agriculture (USDA) **www.usda.gov**

Volunteer health agencies such as the American Cancer Society **www.cancer.org**, the American Diabetes Association **www.diabetes.org**, and the American Heart Association **www.americanheart.org**

Reputable consumer groups such as the Better Business Bureau **www.bbb.org**, the Consumers Union **www.consumer-sunion.org**, the American Council on Science and Health **www.acsh.org**, and the National Council Against Health Fraud **www.ncahf.org**

What about nutrition and health information found on the Internet? How does a person know whether the Web sites are reliable?

With hundreds of millions of Web sites on the Internet, searching for nutrition and health information can be daunting. The Internet offers no guarantee of the accuracy of the information found there, and much of it is pure fiction. Web sites must be evaluated for their accuracy, just like every other source. Table NP1–3 provides clues to identifying reliable nutrition information sites and lists some credible sites.

The Federal Trade Commission (FTC), the Food and Drug Administration (FDA), and other law enforcement agencies have launched "Operation Cure.All" (see Table NP1–3) to take action against fraudulent marketing of supplements and health products on the Internet.[3] The latest actions target unscrupulous companies that use the Internet to promote products to the most vulnerable consumers—those with diseases such as AIDS, Alzheimer's, and cancer. Of greatest concern are those products that not only make false promises, but also are potentially dangerous. For example, herbal products touted as safe treatments for serious illnesses such as AIDS may interact with medications and impair the effectiveness of the medicines. The FTC advises consumers to be suspicious of:

- Claims that a product is "natural" or "nontoxic." "Natural" or "nontoxic" does not always mean safe.

- Claims that a product is a "scientific breakthrough," "miraculous cure," "secret ingredient," or "ancient remedy."

- Claims that a product cures a wide range of illnesses.

- Claims that use impressive-sounding medical terms.

- Claims of a "money-back" guarantee.

Everyone seems to be giving advice on nutrition. How can a person tell whom to listen to?

Registered dietitians (RDs) and nutrition professionals with advanced degrees (M.S., Ph.D.) are experts (see the Glossary, p. 36).[4] These professionals are probably in the best position to answer a person's nutrition questions. On the other hand, **nutritionists** may be experts or quacks, depending on the state where they practice. Some states require people who use this title to meet strict standards. In other states, a "nutritionist" may be any individual who claims a career connection with the nutrition field.

Other purveyors of nutrition information may also lack credentials. A health food store owner may be in the nutrition business simply because it is a lucrative market. The owner may have a background in business or sales and no education in nutrition at all. Such a person is not qualified to provide nutrition information to customers. For accurate nutrition information, seek out a trained professional with a college education in nutrition—an expert in the field of **dietetics**.

What about nurses and other health care professionals?

All members of the health care team share responsibility for helping each client to achieve optimal health, but the registered dietitian is usually the primary nutrition expert. Each of the other team members has a related specialty. Some physicians are specialists in clinical nutrition and are also experts in the field. Other physicians, nurses, and **dietetic technicians** often assist dietitians in providing nutrition information and may help to administer direct nutrition care. Nurses play central roles in client care management and client relationships. Visiting nurses and home health care nurses may become intimately involved in clients' nutrition care at home, teaching them both theory and cooking techniques. Physical therapists

---

TABLE NP1–3

Evaluating the Reliability of Web Sites

To judge whether an Internet site offers reliable nutrition information, answer the following questions.

Who is responsible for the site? Clues can be found in the three-letter "tag" that follows the dot in the site's name. For example, "gov" and "edu" indicate government and university sites, usually reliable sources of information.

Do the names and credentials of information providers appear? Is an editorial board identified? Many legitimate sources provide e-mail addresses or other ways to obtain more information about the site and the information providers behind it.

Are links with other reliable information sites provided? Reputable organizations almost always provide links with other similar sites because they want you to know of other experts in their area of knowledge. Caution is needed when you evaluate a site by its links, however. Anyone, even a quack, can link a Web page to a reputable site without the organization's permission. Doing so may give the quack's site the appearance of legitimacy, just the effect for which the quack is hoping.

Is the site updated regularly? Nutrition information changes rapidly, and sites should be updated often.

Is the site selling a product or service? Commercial sites may provide accurate information, but they also may not. Their profit motive increases the risk of bias.

Does the site charge a fee to gain access to it? Many academic and government sites offer the best information, usually for free. Some legitimate sites do charge fees, but before paying up, check the free sites. Chances are good you will find what you are looking for without paying.

Some credible Web sites include:

National Council Against Health Fraud
www.ncahf.org

Stephen Barrett's Quackwatch
www.quackwatch.com

Centers for Disease Control and Prevention's Current Health Related Hoaxes and Rumors
www.cdc.gov/hoax_rumors.htm

Federal Trade Commission's Operation Cure.All
www.ftc.gov/opa/2001/06/cureall.htm

GLOSSARY OF TERMS ASSOCIATED WITH NUTRITION EXPERTS

dietetic technicians: persons who have completed a two-year academic degree from an accredited college or university and an approved dietetic technician program. A **dietetic technician, registered (DTR)** has also passed a national examination and maintains registration through continuing professional education.

dietetics: the practical application of nutrition, including the assessment of nutrition status, recommendation of appropriate diets, nutrition education, and the planning and serving of meals.

nutritionists: persons who specialize in the study of nutrition. Some nutritionists are registered dietitians, but others are self-described experts whose training may be minimal or nonexistent. Some states make the term meaningful by allowing it to apply only to people who have master's (M.S.) or doctoral (Ph.D.) degrees from institutions accredited to offer such degrees in nutrition or related fields.

registered dietitians (RDs): dietitians who have graduated from a university or college after completing a program of dietetics that has been accredited by the American Dietetic Association (or Dietitians of Canada). The dietitian must serve in an approved internship or coordinated program to practice the necessary skills, pass the association registration examination, and maintain competency through continuing education. Many states require licensing for practicing dietitians. Licensed dietitians (LDs) have met all *state* requirements to offer nutrition advice.

can provide individualized exercise programs related to nutrition—for example, to help control obesity. Social workers may provide practical and emotional support.

What roles might these other health care professionals play in nutrition care?

Some of the responsibilities of the health care professional might be:

- Helping people understand why nutrition is important to them.

- Answering questions about food and diet.

- Explaining to clients how modified diets work.

- Collecting information about clients that may influence their nutritional health.

- Identifying clients at risk for poor nutrition status (see Chapter 14) and recommending or taking appropriate action.

- Recognizing when clients need extra help with nutrition problems (in such cases, the problems should be referred to a dietitian or physician).

Health care professionals may routinely perform these nutrition-related tasks:

- Obtaining diet histories.

- Taking weight and height measurements.

- Feeding clients who cannot feed themselves.

- Recording what clients eat or drink.

- Observing clients' responses and reactions to foods.

- Helping clients mark menus.

- Monitoring weight changes.

- Monitoring food and drug interactions.

- Encouraging clients to eat.

- Assisting clients at home in planning their diets and managing their kitchen chores.

- Alerting the physician or dietitian when nutrition problems are identified.

- Charting actions taken and communicating on these matters with other professionals as needed.

Thus, although the dietitian assumes the primary role as the nutrition expert on a health care team, other health care professionals play important roles in administering nutrition care.

Notes

1. Position of the American Dietetic Association: Food and nutrition misinformation, *Journal of the American Dietetic Association* 102 (2006): 601–607.

2. Position of the American Dietetic Association, 2006.

3. Federal Trade Commission, "Operation Cure.All" wages new battle in ongoing war against Internet health fraud, www.ftc. gov/opa/2001/06/cureall.htm, site visited April 30, 2005.

4. Position of the American Dietetic Association: The roles of registered dietitians and dietetic technicians, registered in health promotion and disease prevention, *Journal of the American Dietetic Association* 106 (2006): 1875–1884.

© James Darell/Digital Vision/Getty Images

2 Carbohydrates

Grains, vegetables, legumes, fruits, and milk offer ample carbohydrate.

■ Chapter 3 describes the roles of fats in health and disease.

M ost people would like to feel good all the time. Part of the secret of feeling well is replenishing the body's energy supply with food. That means choosing foods that contain the energy nutrients—carbohydrate and fat, primarily. But which to choose?

Carbohydrate is the preferred energy source for many of the body's functions. As long as carbohydrate is available, the human brain depends exclusively on it as an energy source. Athletes eat a "high-carb" diet to store as much muscle fuel as possible, and dietary recommendations urge people to eat carbohydrate-rich foods for better health. Many people, however, mistakenly think of carbohydrate-rich foods as "fattening" and avoid them. In truth, people who wish to lose fat and to maintain lean tissue and the health of the body can best do so by being physically active, paying close attention to portion sizes, and designing a diet based on foods that supply carbohydrate in balance with other energy nutrients.[1] Most unrefined plant foods—grains, vegetables, legumes, and fruits—provide ample carbohydrate and fiber with little or no fat. Milk is the only animal-derived food that contains significant amounts of carbohydrate.

Carbohydrate shares its fuel-providing responsibility with fat. Fat, however, normally is not used as fuel by the brain and central nervous system, and diets high in certain types of fat are associated with chronic diseases. ■ The other energy sources available to the body—protein and alcohol—offer no advantage as fuels. Protein is best left to serve its own diverse functions, as discussed in Chapter 4. Alcohol, of course, has well-known undesirable side effects when used in excess. Alcohol and its relationships with health and disease are the subject of Nutrition in Practice 20.

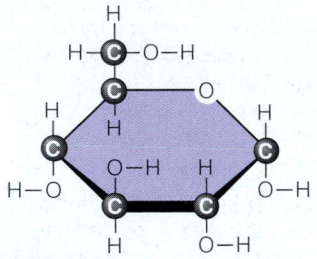

FIGURE 2–1

Chemical Structure of Glucose
On paper, the structure of glucose has to be drawn flat, but in nature the five carbons and oxygen are roughly in a plane, with the H, OH, and CH_2OH extending out above and below it.

■ The Chemist's View of Carbohydrates

The dietary **carbohydrates** include the **simple sugars** and starch and fiber. Chemists describe the simple sugars as:

- **Monosaccharides** (single sugars)
- **Disaccharides** (double sugars)

Starch and fiber are:*

- **Polysaccharides**—compounds composed of chains of monosaccharide units.

All of these carbohydrates are composed of the simple sugar **glucose** and other compounds that are much like glucose in composition and structure. Figure 2–1 shows the chemical structure of glucose.

Monosaccharides

Three monosaccharides are important in nutrition: glucose, **fructose**, and **galactose**. All three monosaccharides have the same number and kinds of atoms but in different arrangements.

Glucose Most cells depend on glucose for their fuel to some extent, and the cells of the brain and the rest of the nervous system depend almost exclusively on glucose for their energy. The body can obtain this glucose from carbohydrates. To function optimally, the body must maintain blood glucose within limits that allow the cells to nourish themselves. If blood glucose falls below normal, the person may become dizzy and

carbohydrates: energy nutrients composed of monosaccharides.
  carbo = carbon
  hydrate = water

simple sugars: the monosaccharides (glucose, fructose, and galactose) and the disaccharides (sucrose, lactose, and maltose).

monosaccharides (mon-oh-SACK-uh-rides): single sugar units.
  mono = one
  saccharide = sugar

---

* Monosaccharides and disaccharides (sugars) are sometimes called *simple carbohydrates,* and the polysaccharides (starch and fiber) are sometimes called *complex carbohydrates.*

weak; if it rises substantially above normal, the person may become fatigued. ■ Left untreated, fluctuations to the extremes—either high or low—can be fatal. Blood glucose **homeostasis** is regulated primarily by two hormones: **insulin**, which moves glucose from the blood into the cells, and **glucagon**, which brings glucose out of storage when blood glucose falls (as occurs between meals).

Fructose Fructose is the sweetest of the sugars. Fructose occurs naturally in fruits, honey, and saps. Other sources include soft drinks, ready-to-eat cereals, and other products sweetened with **high-fructose corn syrup** (defined in the Glossary on p. 45). Glucose and fructose are the most common monosaccharides in nature.

Galactose The third single sugar, galactose, occurs mostly as part of lactose, a disaccharide also known as *milk sugar*. During digestion galactose is freed as a single sugar.

Disaccharides

In disaccharides, pairs of single sugars are linked together. Three disaccharides are important in nutrition: maltose, sucrose, and lactose. All three have glucose as one of their single sugars. As Table 2–1 shows, the other monosaccharide is either another glucose (in maltose), or fructose (in sucrose), or galactose (in lactose). The shapes of the sugars in Table 2–1 reflect their chemical structures as drawn on paper.

Sucrose Sucrose (table, or white, sugar) is the most familiar of the three disaccharides and is what people mean when they speak of "sugar." This sugar is usually obtained by refining the juice from sugar beets or sugarcane to provide the brown, white, and powdered sugars available in the supermarket, but it occurs naturally in many fruits and vegetables.

When a person eats a food containing sucrose, enzymes in the digestive tract split the sucrose into its glucose and fructose components. Because the body can convert fructose to glucose, one molecule of sucrose can ultimately yield two molecules of glucose.

Lactose Lactose is the principal carbohydrate of milk. Most human infants are born with the digestive enzymes necessary to split lactose into its two monosaccharide parts, glucose and galactose, so as to absorb it. Breast milk thus provides a simple, easily digested carbohydrate that meets an infant's energy needs; many formulas do, too, because they are made from milk.

Many people lose the ability to digest lactose after infancy. This condition, known as **lactose intolerance**, varies widely among ethnic groups, indicating that the trait is genetically determined. The prevalence of lactose intolerance is lowest (<10%) among Scandinavians and other northern Europeans and highest (≥80%) among native North Americans and Southeast Asians. Lactose intolerance is not the same as milk allergy, which is caused by an immune reaction to the protein in milk. Lactose intolerance is discussed further in Chapter 19, and milk allergy in Chapter 12.

TABLE 2–1			
The Major Sugars			
Monosaccharides		**Disaccharides**	
Glucose	(blue hexagon)	Sucrose (glucose + fructose)	(blue hexagon + purple pentagon)
Fructose	(purple pentagon)	Lactose (glucose + galactose)	(blue hexagon + green hexagon)
Galactose	(green hexagon)	Maltose (glucose + glucose)	(blue hexagon + blue hexagon)
(found only as part of lactose)			

■ Diabetes, a disorder characterized by elevated blood glucose, is the topic of Chapter 21.

disaccharides (dye-SACK-uh-rides): a pair of sugar units bonded together.
di = two

polysaccharides: long chains of monosaccharide units arranged as starch, glycogen, or fiber.
poly = many

glucose: a monosaccharide, the sugar common to all disaccharides and polysaccharides; also called *blood sugar* or *dextrose*.

fructose: a monosaccharide; sometimes known as *fruit sugar*. It is abundant in fruits, honey, and saps.
fruct = fruit

galactose: a monosaccharide; part of the disaccharide lactose.

homeostasis (HOME-ee-oh-STAY-sis): the maintenance of constant internal conditions (such as chemistry, temperature, and blood pressure) by the body's control system.
homeo = the same
stasis = staying

insulin: a hormone secreted by the pancreas in response to high blood glucose. It promotes cellular glucose uptake for use or storage.

glucagon (GLOO-ka-gon): a hormone that is secreted by special cells in the pancreas in response to low blood glucose concentration and elicits release of glucose from storage.

sucrose: a disaccharide composed of glucose and fructose; commonly known as *table sugar*, *beet sugar*, or *cane sugar*.
sucro = sugar

lactose: a disaccharide composed of glucose and galactose; commonly known as *milk sugar*.
lact = milk

lactose intolerance: a condition that results from inability to digest the milk sugar lactose; characterized by bloating, gas, abdominal discomfort, and diarrhea. Lactose intolerance differs from milk allergy, which is caused by an immune reaction to the protein in milk.

Maltose The third disaccharide, **maltose**, is a plant sugar that consists of two glucose units. Maltose is produced whenever starch breaks down—as happens in plants when they break down their stored starch for energy and start to sprout and in human beings during carbohydrate digestion.

Polysaccharides

Unlike the sugars, which contain the three monosaccharides—glucose, fructose, and galactose—in different combinations, the polysaccharides are composed almost entirely of glucose (and, in some cases, other monosaccharides). Three types of polysaccharides are important in nutrition: starch, glycogen, and fibers.

Glycogen is a storage form of energy for human beings and animals; starch plays that role in plants; and fibers provide structure in stems, trunks, roots, leaves, and skins of plants. Both glycogen and starch are built of entirely glucose units; fibers are composed of a variety of monosaccharides and other carbohydrate derivatives.

Starch **Starch** is a long, straight or branched chain of hundreds of glucose units linked together. These giant molecules are packed side by side in a rice grain or potato root—as many as a million per cubic inch of food. When a person eats the plant, the body splits the starch into glucose units ■ and uses the glucose for energy.

All starchy foods are plant foods. Grains are the richest food source of starch. In most human societies, people depend on a staple grain for much of their food energy: rice in Asia; wheat in Canada, the United States, and Europe; corn in much of Central and South America; and millet, rye, barley, and oats elsewhere. A second important source of starch is the legume (bean and pea) family. Legumes include peanuts and "dry" beans such as butter beans, kidney beans, "baked" beans, black-eyed peas (cowpeas), chickpeas (garbanzo beans), and soybeans. Root vegetables (tubers) such as potatoes and yams are a third major source of starch, and in many non-Western societies, they are the primary starch sources. Grains, legumes, and tubers not only are rich in starch, but may also contain abundant dietary fiber, protein, and other nutrients.

Glycogen **Glycogen** molecules, which are also made of chains of glucose, are more highly branched than starch molecules. Glycogen is found in meats only to a limited extent and not at all in plants.* For this reason, glycogen is not a significant food source of carbohydrate, but it does play an important role in the body. The human body stores much of its glucose as glycogen in the liver and muscles.

After a meal, as blood glucose rises, the pancreas is the first organ to respond. It releases the hormone insulin, which signals the body's tissues to take up surplus glucose. Muscle and liver cells use some of this excess glucose to build glycogen. The muscles hoard two-thirds of the body's total glycogen and use it just for themselves during physical activity. The brain stores a tiny fraction of the total, thought to provide an emergency glucose reserve sufficient to fuel the brain for an hour or two in severe glucose deprivation.[2] The liver stores the remainder and is generous with its glycogen, making it available as blood glucose for the brain or other tissues when the supply runs low.

Fibers **Dietary fibers** are the structural parts of plants and thus are found in all plant-derived foods—vegetables, fruits, whole grains, and legumes. Most dietary fibers are polysaccharides—chains of sugars—just as starch is, but in fibers the sugar units are held together by bonds that human digestive enzymes cannot break. Consequently, most dietary fibers pass through the body providing little or no energy for its use. Figure 2–2 shows the difference between starch and the plant fiber cellulose. In addition to cellulose, fibers include the polysaccharides hemicellulose, pectins, gums, and mucilages, as well as the nonpolysaccharide lignins.

■ The short chains of glucose units that result from the breakdown of starch are known as *dextrins*. The word sometimes appears on food labels because dextrins can be used as thickening agents in foods.

---

* Glycogen in animal muscles rapidly breaks down after slaughter.

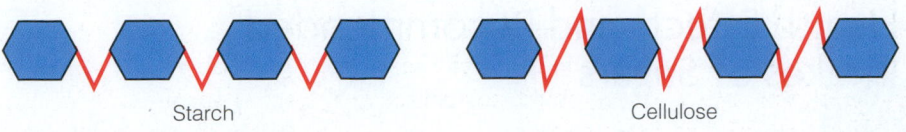

Starch   Cellulose

FIGURE 2-2

Starch and Cellulose Molecules Compared (Small Segments)
The bonds that link the glucose units together in cellulose are different from the bonds in starch (and glycogen). Human enzymes cannot digest cellulose.

■ Soluble, viscous, fermentable fibers are often gummy or add thickness to foods.

■ Insoluble, nonviscous, less fermentable fibers are often tough, stringy, or gritty in foods.

Cellulose is the main constituent of plant cell walls, so it is found in all vegetables, fruits, and legumes. Hemicellulose is the main constituent of cereal fibers. Pectins are abundant in vegetables and fruits, especially citrus fruits and apples. The food industry uses pectins to thicken jelly and keep salad dressing from separating. Gums and mucilages have similar structures and are used as additives or stabilizers by the food industry. Lignins are the tough, woody parts of plants; few foods people eat contain much lignin.

A few starches are classified as fibers. Known as **resistant starches**, these starches escape digestion and absorption in the small intestine. Starch may resist digestion for several reasons, including the individual's efficiency in digesting starches and the food's physical properties. Resistant starch is common in whole legumes, raw potatoes, and unripe bananas.

Although human enzymes do not break down cellulose and other dietary fibers, bacteria in the human digestive tract can digest some fibers. Bacterial digestion of fibers can generate some absorbable products that can yield energy when metabolized. Food fibers, therefore, can contribute some energy (1.5 to 2.5 kcalories per gram), depending on the extent to which they break down in the body.

Fibers are divided into two general groups by their chemical and physical properties.* In the first group are fibers that dissolve in water (**soluble fibers**). These form gels (are **viscous**) and are easily digested by bacteria in the human large intestine (are easily fermented). ■ Commonly found in barley, legumes, fruits, oats, and vegetables, these fibers are often associated with lower risks of chronic diseases (as discussed in a later section). In foods, soluble fibers add pleasing consistency, such as the pectin that puts the gel in jelly and the gums that are added to salad dressings and other foods to thicken them.

Other fibers do not dissolve in water (**insoluble fibers**), do not form gels (are not viscous), and are less readily fermented. ■ Insoluble fibers, such as cellulose and many hemicelluloses, are found in the outer layers of whole grains (bran), the strings of celery, the hulls of seeds, and the skins of corn kernels. These fibers retain their structure and rough texture even after hours of cooking. In the body, they aid the digestive system by easing elimination.[3]

IN SUMMARY

- Carbohydrate is the body's preferred energy source. Six simple sugars are important in nutrition: the three monosaccharides (glucose, fructose, and galactose) and the three disaccharides (sucrose, lactose, and maltose).

- The three disaccharides are pairs of monosaccharides; each contains glucose paired with one of the three monosaccharides. The polysaccharides (chains of monosaccharides) are glycogen, starches, and fibers.

- Both glycogen and starch are storage forms of glucose—glycogen in the body and starch in plants—and both yield energy for human use.

- The dietary fibers also contain glucose (and other monosaccharides), but human digestive enzymes cannot break their bonds, so they yield little, if any, energy.

*The DRI committee has proposed these fiber definitions: the term *dietary fibers* refers to naturally occurring fibers in intact foods, and *functional fibers* refers to added fibers that have health benefits. *Total fiber* refers to the sum of fibers from both sources.

maltose: a disaccharide composed of two glucose units; sometimes known as *malt sugar*.

starch: a plant polysaccharide composed of glucose and digestible by human beings.

glycogen (GLY-co-gen): a polysaccharide composed of glucose, made and stored by liver and muscle tissues of human beings and animals as a storage form of glucose. Glycogen is not a significant food source of carbohydrate and is not counted as one of the polysaccharides in foods.

dietary fibers: a general term denoting in plant foods the polysaccharides cellulose, hemicellulose, pectins, gums, and mucilages, as well as the non-polysaccharide lignins, that are not digested by human digestive enzymes, although some are digested by GI tract bacteria.

resistant starches: starches that escape digestion and absorption in the small intestine of healthy people.

soluble fibers: indigestible food components that readily dissolve in water and often impart gummy or gel-like characteristics to foods. An example is pectin from fruit, which is used to thicken jellies.

viscous: a gel-like consistency.

insoluble fibers: the tough, fibrous structures of fruits, vegetables, and grains; indigestible food components that do not dissolve in water.

Health Effects and Recommended Intakes of Sugars

Fiber-rich carbohydrate foods such as vegetables, whole grains, legumes, and fruits should predominate in people's diets; concentrated sweets such as candy, cola beverages and other soft drinks, cookies, pies, cakes, and other foods with **added sugars** add kcalories, but few, if any, other nutrients. These foods therefore contribute discretionary kcalories to the diet. ■

The **naturally occurring sugars** in vegetables, fruits, grains, and milk are acceptable because they are accompanied by many nutrients.

Health Effects of Sugars

Estimates are that each man, woman, and child in the United States consumes over 100 pounds of sugar per year, or a little less than 2 pounds per week. The steady upward trend in consumption of added sugars shown in Figure 2–3 is largely the result of a dramatic increase in consumption of commercially prepared foods and beverages that contain added sugars. Food manufacturers are adding the sugars to foods during processing. By one account, consumption of regular soft drinks and sugar-sweetened fruit drinks is responsible for 80 percent of this increase.[4] Desserts and jams and jellies make up most of the remainder. In contrast, people are adding less sugar in the kitchen. The committee that drafted the *Dietary Guidelines for Americans 2005* offers clear advice on added sugars: treat them as discretionary kcalories.[5] In other words, fill the day's nutrient needs with foods contributing little or no added sugar. Most people can afford only a little added sugar in their diets if they are to meet nutrient needs within kcalorie limits.

The U.S. population is not alone in increasing sugar consumption. The trend is a worldwide phenomenon. In response, the World Health Organization has also taken a stand on sugar intake: consume no more than 10 percent of total kcalories from added sugars.[6]

The increase in sugar consumption has raised many questions about sugar's effects on health. Many accusations have been made against sugar, but the Food and Drug

■ The USDA Food Guide distinguishes between naturally occurring and added sugars, designating the kcalories from added sugars as discretionary kcalories.

© Brian Leatart/Foodpix/Jupiter Images

Starch- and fiber-rich foods are the foods to emphasize.

FIGURE 2–3

Added Sugars: Average Supply per Person in the United States, 1890– Present

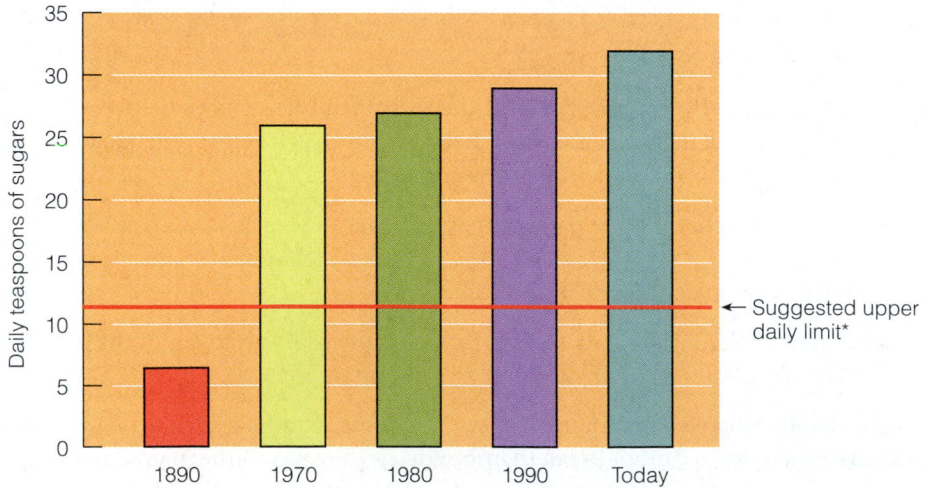

Daily teaspoons of sugars: 1890, 1970, 1980, 1990, Today

← Suggested upper daily limit*

* The World Health Organization recommends an upper limit of about 12 teaspoons in an adequate 2,200-kcalorie diet.

Source: Data from J. Putnam and S. Haley, Estimating consumption of caloric sweeteners, *Amber Waves*, April 2003, available at www.ers. usda.gov/Amberwaves/April03/Indicators/Behind.Data.htm.

CHAPTER TWO

Administration (FDA) and the National Academy of Sciences have concluded that in moderate amounts, sugars pose no major health risk. In excess, however, they can be detrimental in two ways. One, sugars can contribute to nutrient deficiencies by supplying energy (kcalories) without providing nutrients. Two, sugars contribute to tooth decay.

Sugars and Nutrient Deficiencies Empty-kcalorie foods that contain lots of added sugar such as cakes, candies, and sodas deliver glucose and energy with few, if any, other nutrients. By comparison, foods such as whole grains, vegetables, legumes, and fruits that contain some natural sugars and lots of starches and fibers deliver protein, vitamins, and minerals along with their glucose and energy.

A person spending 200 kcalories of a day's energy allowance on a 16-ounce soda gets little of value for those kcaloric "dollars." In contrast, a person using 200 kcalories on three slices of whole wheat bread gets 9 grams of protein, 6 grams of fiber, plus several of the B vitamins with those kcalories. For the person who wants something sweet, perhaps a reasonable compromise would be to have two slices of bread with a teaspoon of jam on each. The amount of sugar a person can afford to eat depends on how many kcalories are available beyond those needed to deliver indispensable vitamins and minerals.

With careful food selections, an adult can obtain all the needed nutrients within an allowance of about 1,500 kcalories. Some people have more generous energy allowances with which to "purchase" nutrients. For example, an active teenage boy may need as many as 3,000 kcalories per day. If he eats mostly nutritious foods, then the "empty kcalories" of cola beverages may be an acceptable addition to his diet. In contrast, an inactive older woman who is limited to fewer than 1,500 kcalories per day can afford only the most nutrient-dense foods.

Sugar can contribute to nutrient deficiencies only by displacing nutrients. For nutrition's sake, the appropriate attitude to take is not that sugar is "bad" and must be avoided, but that nutritious foods must come first. If the nutritious foods end up crowding sugar out of the diet, that is fine—but not the other way around. As always, the goals to seek are balance, variety, and moderation.

Sugars and Dental Caries Both sugars and starches begin breaking down to sugars in the mouth and so can contribute to the development of **dental caries**. Any carbohydrate-containing food, including bread, bananas, or milk, as well as sugar, can support bacterial growth in the mouth.[7] These bacteria produce the acid that eats away tooth enamel. Of major importance is the length of time the food stays in the mouth. This, in turn, depends on the composition of the food, how sticky the food is, how often a person eats the food, and especially whether the teeth are brushed afterward. ■ Total sugar intake still plays a major role in caries incidence; populations whose diets provide no more than 10 percent of kcalories from sugar have a low prevalence of dental caries.[8]

Controversies Surrounding Sugars

Sugars have been blamed for a variety of other health problems.[9] Most commonly, reports accuse sugar of causing obesity and hyperactivity and aggressive behavior in children. ■

Obesity On the first accusation—that sugar causes obesity—the evidence shows a supportive role, but not a direct cause-and-effect relationship. The incidence of obesity often rises as a population's sugar consumption increases, but this evidence is insufficient to name sugar as the cause. Often, a rise in a nation's sugar intake mirrors a rise in income and with it greater consumption of kcalories, processed foods, meats, and fats.[10] Another confounding factor is that as people begin to gain weight, their

Sugary soft drinks are the leading source of added sugars in the United States; cakes, cookies, pies, and other baked goods come next; and sweetened fruit drinks and punches follow closely behind.

© Polara Studios, Inc.

■ Nutrition in Practice 2 discusses nutrition and dental health.

■ Professionally, hyperactivity is called attention-deficit/hyperactivity disorder (see Chapter 12).

added sugars: sugars and syrups added to a food for any purpose, such as to add sweetness or bulk or to aid in browning (baked foods). Also called *carbohydrate sweeteners,* they include glucose, fructose, corn syrup, concentrated fruit juice, and other sweet carbohydrates.

naturally occurring sugars: sugars that are not added to a food but are present as its original constituents, such as the sugars of fruit or milk.

dental caries: the gradual decay and disintegration of a tooth.

■ Reduce the incidence of dental caries by practicing good oral hygiene and consuming sugar- and starch-containing foods and beverages less frequently.

physical activity often declines, making it impossible to tell whether the sugar, excess kcalories from other sources, or too little physical activity is at fault.

Concentrated sweets and soft drinks do make it easy to exceed energy needs quickly, however, and an excessive intake of food energy can cause obesity. In the United States, obesity has reached epidemic proportions as intakes of both food energy and sugar-sweetened beverages, especially soft drinks, have skyrocketed.[11] Adolescents who drink upward of 26 ounces (about two cans) of sugar-sweetened soft drinks daily consume 400 more kcalories per day than teens who abstain from such drinks. Soft drinks have been the focus of attention because the increase in intake of added sugars is accounted for largely by a dramatic rise in consumption of high-fructose corn syrup, mostly in soft drinks. The trend toward greater use of this sweetener parallels unprecedented gains of body fatness in the population.[12] Between 1977 and 2001, as people grew fatter, their intake of kcalories from soft drinks nearly tripled and that from fruit drinks and punches doubled.[13] Thus, to the extent that sugar contributes to an excessive energy intake, it can play a role in the development of obesity.

Behavior The accusation that sugar causes hyperactive or aggressive behavior in children has not been proved. Scientific research has failed to demonstrate any consistent effect of sugar on behavior in either normal or hyperactive children.[14] ■ If sugar is related to behavior problems in children, it may be because the sugary foods replace nutrient-dense foods in children's diets, making nutrient deficiencies likely. Many different nutrient deficiencies adversely affect behavior. A lack of nutrients in children's diets, not sugar itself, can in some cases contribute to undesirable behavior.

■ Chapter 12 offers further discussion of children's nutrition.

Recommended Sugar Intakes

Moderate sugar intakes—enough for pleasure, but not enough to displace more nutritious foods—are not harmful. Sugar is a delicious, concentrated source of food energy, but it contains no protein, vitamins, or minerals. Eaten in the place of nutrient- and fiber-rich foods, it makes malnutrition likely.

The *Dietary Guidelines for Americans 2005* urge people to "choose and prepare foods and beverages with little added sugars." The USDA Food Guide suggests that, within kcalorie limits, small amounts of added sugars can be enjoyed as part of the discretionary kcalorie allowance in a nutrient-dense diet that provides no more than 30 percent of kcalories from fat. ■ The USDA Food Guide recommendations represent about 5 to 10 percent of the day's total energy intake. As noted earlier, the World Health Organization agrees that people should restrict their consumption of added sugars to 10 percent or less of total energy.

The DRI committee did not set a Tolerable Upper Intake Level for added sugars, but as mentioned, excessive intakes can interfere with sound nutrition and dental health. Few people can eat lots of sugary treats and still meet all of their nutrient needs without exceeding their kcalorie allowance. Instead, the DRI committee suggests, as a rather high maximum, that added sugars should account for no more than 25 percent of the day's total energy intake.[15] For a person consuming 2,000 kcalories per day, 25 percent represents 500 kcalories from added sugars—quite a lot of sugar. ■ Perhaps an athlete in training whose energy needs are high can afford the added sugars

■ The USDA Food Guide suggests:*

- ■ 3 tsp for 1,600 kcal
- ■ 5 tsp for 1,800 kcal
- ■ 8 tsp for 2,000 kcal
- ■ 9 tsp for 2,200 kcal
- ■ 12 tsp for 2,400 kcal

■ For perspective, each of these sources of concentrated sugars provides about 500 kcal:

- ■ 40 oz cola
- ■ $\frac{1}{2}$ cup honey
- ■ 125 jelly beans
- ■ 23 marshmallows
- ■ 30 tsp sugar

* The amounts of added sugars suggested here are *not* specific recommendations for amounts of added sugars to consume, but rather represent the amounts that can be included in the diet at each kcalorie level.

from sports drinks without compromising nutrient intake, but most people would do better following recommendations to limit added sugar consumption to less than 10 percent of the day's total energy intake.

People often fail to recognize sugar in all its forms and so do not realize how much they consume. To help your clients estimate how much sugar they consume, tell them to treat all of the following concentrated sweets as equivalent to 1 teaspoon of white sugar (4 grams of carbohydrate):

■ 1 teaspoon brown sugar, candy, jam, jelly, any corn sweetener, syrup, honey, molasses, or maple sugar

■ 1 tablespoon catsup

■ 1½ ounces carbonated soft drink

These portions of sugar all provide about the same number of kcalories. Some are closer to 10 kcalories (for example, 14 kcalories for molasses), while some are over 20 (22 kcalories for honey), so an average figure of 16 kcalories is an acceptable approximation. The accompanying glossary presents the multitude of names that denote sugar on food labels.

People often ask: What is the difference between honey and white sugar? Is honey, by virtue of being natural, more nutritious? Honey, like white sugar, contains glucose

GLOSSARY OF SUGARS

brown sugar: white sugar with molasses added, 95% pure sucrose.

concentrated fruit juice sweetener: a concentrated sugar syrup made from dehydrated, deflavored fruit juice, commonly grape juice; used to sweeten products that can then claim to be "all fruit."

confectioner's sugar: finely powdered sucrose, 99.9% pure.

corn sweetners: corn syrup and sugar solutions derived from corn.

corn syrup: a syrup, mostly glucose, partly maltose, produced by the action of enzymes on cornstarch.

dextrose: an older name for glucose.

evaporated cane juice: raw sugar from which impurities have been removed.

fructose, galactose, glucose: the monosaccharides.

granulated sugar: common table sugar, crystalline sucrose, 99.9% pure.

high-fructose corn syrup (HFCS): a form of corn syrup used by the food industry to sweeten many foods and beverages, including soft drinks. The sweetener is similar in composition and sweetness to table sugar.

honey: a concentrated solution primarily composed of glucose and fructose, produced by enzymatic digestion of the sucrose in nectar by bees.

invert sugar: a mixture of glucose and fructose formed by the splitting of sucrose in an industrial process. Sold only in liquid form and sweeter than sucrose, invert sugar forms during certain cooking procedures and works to prevent crystallization of sucrose in soft candies and sweets.

lactose, maltose, sucrose: the disaccharides.

levulose: an older name for fructose.

maple sugar: a concentrated solution of sucrose derived from the sap of the sugar maple tree, mostly sucrose. This sugar was once common but is now usually replaced by sucrose and artificial maple flavoring.

molasses: a syrup left over from the refining of sucrose from sugarcane; a thick, brown syrup. The major nutrient in molasses is iron, a contaminant from the machinery used in processing it.

raw sugar: the first crop of crystals harvested during sugar processing. Raw sugar cannot be sold in the United States because it contains too much filth (dirt, insect fragments, and the like). Sugar sold as "raw sugar" is actually evaporated cane juice.

turbinado (ter-bih-NOD-oh) **sugar:** raw sugar from which the filth has been washed; legal to sell in the United States.

white sugar: pure sucrose, produced by dissolving, concentrating, and recrystallizing raw sugar.

TABLE 2–2

Sample Nutrients in Sugars and Other Foods

The indicated portion of any of these foods provides approximately 100 kcalories. Notice that for a similar number of kcalories and grams of carbohydrate, milk, legumes, fruits, grains, and vegetables offer more of the other nutrients than do the sugars.

	Size of 100 kcal Portion	Carbohydrate (g)	Protein (g)	Calcium (mg)	Iron (mg)	Vitamin A (µg)	Vitamin C (mg)
Foods							
Milk, 1% low-fat	1 c	12	8	300	0.1	144	2
Kidney beans	½ c	20	7	30	1.6	0	2
Apricots	6	24	2	30	1.1	554	22
Bread, whole wheat	1½ slices	20	4	30	1.9	0	0
Broccoli, cooked	2 c	20	12	188	2.2	696	148
Sugars							
Sugar, white	2 tbs	24	0	trace	trace	0	0
Molasses, blackstrap	2½ tbs	28	0	343	12.6	0	0.1
Cola beverage	1 c	26	0	6	trace	0	0
Honey	1½ tbs	26	trace	2	0.2	0	trace

You receive about the same amount and kinds of sugars from an orange as from a tablespoon of honey, but the packaging makes a big nutrition difference.

■ A 12-ounce can of cola contains the equivalent of 8 teaspoons of sugar.

and fructose. The difference is that in white sugar, the glucose and fructose are bonded together in pairs, whereas in honey some of them are paired and some are free single sugars. When you eat either white sugar or honey, though, your body breaks all of the sugars apart into single sugars. It ultimately makes no difference whether you eat single sugars linked together, as in white sugar, or the same sugars unlinked, as in honey; they will end up as single sugars in your body. True, honey contains trace amounts of a few vitamins and minerals, but to say that honey is nutritious is misleading.

Some sugar sources are more nutritious than others, though. Consider a fruit such as an orange. The orange provides the same sugars and about the same energy as a tablespoon of sugar or honey, but the packaging makes a big difference in nutrient density. The sugars of the orange are diluted in a large volume of fluid that contains valuable vitamins and minerals, and the flesh and skin of the orange are supported by fibers that also offer health benefits. A tablespoon of honey offers no such bonuses.

Of course, a cola beverage, containing many teaspoons of sugar, offers no advantages either. Table 2–2 shows sample nutrients supplied by some sugar sources and other foods; note the "0s" and "traces" by honey, sugar, and the cola beverage and the substantial numbers by the others. ■

IN SUMMARY

- In moderate amounts, sugars pose no major health threat. Excessive sugar intake may increase the risk of nutrient deficiencies and dental caries, however.

- A person deciding to limit daily sugar intake should recognize that not all sugars need to be restricted, just concentrated sweets with added sugars, which are high in kcalories and relatively lacking in other nutrients. Sugars that occur naturally in fruits, vegetables, and milk are acceptable.

Health Effects of Alternative Sweeteners

Many consumers turn to alternative sweeteners to help them control kcalories and limit their use of sugar. In doing so, they encounter two sets of alternative sweeteners: sugar alcohols and artificial sweeteners.

TABLE 2–3

Sugar Alcohols

Sugar Alcohols	Relative Sweetness[a]	Energy (kcal/g)	Approved Uses
Isomalt	0.5	2.0	Candies, chewing gum, ice cream, jams and jellies, frostings, beverages, baked goods
Lactitol	0.4	2.0	Candies, chewing gum, frozen dairy desserts, jams and jellies, frostings, baked goods
Maltitol	0.9	2.1	Particularly good for candy coating
Mannitol	0.7	1.6	Bulking agent, chewing gum
Sorbitol	0.5	2.6	Special dietary foods, candies, gums
Xylitol	1.0	2.4	Chewing gum, candies, pharmaceutical and oral health products

[a] The relative sweetness depends on the temperature, acidity, and other flavors of the foods in which the substance occurs. The sweetness of pure sucrose is the standard with which the approximate sweetness of sugar substitutes is compared.

Alternative Sweeteners: Sugar Alcohols

The **sugar alcohols** are carbohydrates, but they yield slightly less energy (2 to 3 kcalories per gram) than sucrose (4 kcalories per gram) because they are not absorbed completely.[16] The sugar alcohols are sometimes called **nutritive sweeteners** because they do yield some energy.

The sugar alcohols occur naturally in fruits; they are also used by manufacturers to provide sweetness and bulk to cookies, sugarless gum, hard candies, and jams and jellies. Their sweetness relative to sugar is shown in Table 2–3. Unlike sucrose, sugar alcohols are fermented in the large intestine by intestinal bacteria. Consequently, side effects such as gas, abdominal discomfort, and diarrhea make the sugar alcohols less attractive than the artificial sweeteners.

The advantage of using sugar alcohols is that they do not contribute to dental caries. Bacteria in the mouth metabolize sugar alcohols much more slowly than sucrose, thereby inhibiting the production of acids that promote caries formation. They are therefore valuable in chewing gums, breath mints, and other products that people keep in their mouths awhile. The FDA allows food labels to carry a health claim (see p. 29 in Chapter 1) about the relationship between sugar alcohols and the nonpromotion of dental caries as long as the FDA criteria for sugar-free status and other criteria are met.

Alternative Sweeteners: Artificial Sweeteners

The **artificial sweeteners** are not carbohydrates. They yield virtually no energy in the amounts typically used and are sometimes called nonnutritive sweeteners. Like the sugar alcohols, artificial sweeteners make foods taste sweet without promoting tooth decay. The Glossary and Table 2–4 (p. 48) offer details on artificial sweeteners of interest.

Saccharin Saccharin has been used for more than 100 years in the United States and is currently used by millions of people, mainly in soft drinks and as the tabletop sweetener Sweet'n Low or Sugar Twin. Questions about the safety of saccharin arose in 1977 when experiments suggested

sugar alcohols: sugarlike compounds. Like sugars, they are sweet to taste but yield 2 to 3 kcal per gram, slightly less than sucrose. Examples are maltitol, mannitol, sorbitol, isomalt, lactitol, and xylitol.

nutritive sweeteners: sweeteners that yield energy, including both the sugars and the sugar alcohols.

artificial sweeteners: noncarbohydrate, nonkcaloric synthetic sweetening agents; sometimes called *nonnutritive sweeteners*.

These chewing gums contain sugar alcohols, which are better than sugar for the teeth but are not kcalorie-free.

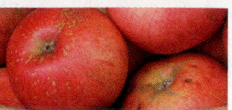

GLOSSARY OF ARTIFICIAL SWEETENERS

acesulfame (AY-sul-fame) potassium, also called **acesulfame-K:** a zero-kcalorie sweetener approved by the FDA and Health Canada.

alitame: a nonkcaloric sweetener formed from the amino acids L-aspartic acid and L-alanine. In the United States, the FDA is considering its approval.

aspartame: a compound of phenylalanine and aspartic acid that tastes like the sugar sucrose but is much sweeter. It is used in both the United States and Canada.

cyclamate: a zero-kcalorie sweetener under consideration for use in the United States and used with restrictions in Canada.

neotame (NEE-oh-tame): an artificial sweetener composed of two amino acids (phenylalanine and aspartic acid) linked in such a way as to make them indigestible by human enzymes.

saccharin: a zero-kcalorie sweetener used freely in the United States but restricted in Canada.

sucralose: a nonkcaloric sweetener derived from a chlorinated form of sugar that travels through the digestive tract unabsorbed. Approved for use in the United States and Canada.

tagatose: an incompletely absorbed monosaccharide sweetener derived from lactose with a kcaloric value of 1.5 kcalories per gram. About 80 percentage of the ingested tagatose travels to the large intestine where bacterial colonies ferment it. Tagatose is not readily used by mouth bacteria and so does not promote dental caries.

that it caused bladder tumors in second-generation rats fed high doses. The FDA proposed banning saccharin as a result. Public outcry in favor of saccharin was so loud that Congress declared a moratorium on the ban, a moratorium that has been repeatedly extended. In 1991, the FDA withdrew its proposal to ban saccharin, but labels still had to carry a consumer warning about saccharin as a cancer hazard. Saccharin remained on the government's roster of "anticipated carcinogens" until the year 2000. At that time, government officials reviewed the research and reversed their opinion, removing saccharin from the carcinogen list and freeing it from the labeling requirement.[17] Opponents to these changes, however, maintain that saccharin causes cancer in mice and rats and so should be avoided.

TABLE 2–4

U.S.-Approved Sweeteners

Artificial Sweeteners	Energy (kcal/g)	Acceptable Daily Intake (ADI)	Average Amount to Replace 1 tsp Sucrose[a]	Approved Uses
Saccharin (SugarTwin, Sweet'n Low, others)	0	5 mg/kg body weight (341 mg for a 150 lb person)	12 mg	Tabletop sweeteners, wide range of foods, beverages, cosmetics, and pharmaceutical products
Aspartame (NutraSweet, Equal, others)	4	50 mg/kg body weight (3,409 mg for a 150 lb person)	18 mg	General-purpose sweetener in all foods and beverages Warning to population with PKU
Acesulfame-potassium (Sunette, Sweet One)	0	15 mg/kg body weight (1,023 mg for a 150 lb person)	25 mg	Alcoholic beverages, baked goods, candies, chewing gum, desserts, gelatins, puddings, tabletop sweeteners
Sucralose (Splenda)	0	5 mg/kg body weight (341 mg for a 150 lb person)	6 mg	Baked goods, carbonated beverages, chewing gum, coffee and tea, dairy products, frozen desserts, fruit spreads, salad dressing, syrups, tabletop sweeteners
Neotame	0	18 mg/day	0.5 µg	Baked goods, beverages (nonalcoholic), candies, chewing gum, frostings, frozen desserts, gelatins, puddings, jams and jellies, syrups
Tagatose	1.5	7.5 g/day	1 tsp	Bakery products, beverages, cereals, chewing gum, confections, dairy products, dietary supplements, health bars, tabletop sweetener

[a]Rounded Values

Does saccharin cause cancer? The largest population study to date, involving 9,000 men and women, showed overall that saccharin use did not raise the risk of cancer. Among certain small groups of the population, however, such as those who both smoked heavily and used saccharin, the risk of bladder cancer was slightly greater. Other studies involving more than 5,000 people with bladder cancer showed no association between bladder cancer and saccharin use. Common sense dictates that consuming large amounts of saccharin is probably not safe, but at moderate intake levels, saccharin is currently assumed to be safe for most people. The FDA has set **Acceptable Daily Intake (ADI)** levels for the artificial sweeteners used in the United States.

Aspartame **Aspartame** is the active ingredient in NutraSweet, which is used in many commercially prepared foods, and in Equal, a tabletop sweetener. Aspartame is 200 times sweeter than sucrose and was approved by the FDA in 1981. Aspartame is one of the most studied of all food additives: extensive animal and human studies document its safety. Long-term consumption of aspartame is safe and is not associated with any adverse health effects.[18] Aspartame is approved for use in more than 100 countries.

The FDA's approval of aspartame is based on the assumption that no one will consume more than the ADI of 50 milligrams per kilogram of body weight in a day. This daily intake is indeed a lot: for a 132-pound person, it adds up to 80 packets of Equal or 15 soft drinks sweetened only with aspartame. Most users of aspartame consume between 2 and 10 milligrams per kilogram of body weight per day. Still, a child who drinks a quart of Kool-Aid sweetened with aspartame on a hot day and also has pudding, chewing gum, cereal, and other products sweetened with aspartame takes in more than the ADI. Although this presents no proven hazard, it seems wise to offer children other foods so as not to exceed the limit. Infants or toddlers under two years old should not be fed artificially sweetened foods and drinks.

Aspartame and PKU Although aspartame is considered safe for most people, individuals with the metabolic disorder phenylketonuria (PKU) are an exception. The labels of products that contain aspartame must include information for individuals with PKU. Aspartame contains the amino acid phenylalanine, and people with PKU cannot dispose of it efficiently (see Nutrition in Practice 16).

Acesulfame Potassium The FDA approved **acesulfame potassium (acesulfame-K)** in 1988 after reviewing more than 90 safety studies, conducted over 15 years. Marketed under the trade names Sunette and Sweet One, acesulfame-K is about as sweet as aspartame. It is used in chewing gum, beverages, instant coffee and tea, gelatins, and puddings, as well as for table use. Unlike aspartame, acesulfame-K holds up well during cooking.

Sucralose **Sucralose** received FDA approval in 1998 for use as a sweetener in the United States. Sucralose is the only artificial sweetener made from sucrose. Many years of testing have deemed sucralose safe to use and, specifically, not a cause of cancer. Sucralose is not recognized by the body as sugar and therefore passes through unchanged. Sucralose is heat stable and so is useful for cooking and baking; it is used in commercially prepared products and as the tabletop sweetener Splenda. Its sugarlike taste and versatility are earning sucralose some popularity among consumers.

Neotame In 2002, the FDA approved **neotame** on the strength of findings from over 110 safety studies conducted on both animals and human beings. Neotame is so intensely sweet—about 7,000 to 13,000 times sweeter than sugar—that very little is needed to sweeten foods and beverages. Currently, neotame is available to food manufacturers for sweetening processed foods but not to consumers for home use.

Acceptable Daily Intake (ADI): the amount of an artificial sweetener that individuals can safely consume each day over the course of a lifetime without adverse effect. It includes a 100-fold safety factor.

A chemical cousin of aspartame, neotame also contains the amino acids phenylalanine and aspartic acid. Unlike aspartame, however, neotame is indigestible. Whereas digestive enzymes can separate the amino acids in aspartame, neotame's chemistry is sufficiently different to block the enzymes' action. This slight chemical difference makes neotame a better choice for people with phenylketonuria because it provides no phenylalanine to the body.

Tagatose The FDA has granted **tagatose**, a relative of fructose, the status of "generally recognized as safe," making it available as a lower-kcalorie sweetener for foods, beverages, confections, dietary supplements, and other uses. Tagatose is derived from lactose, but unlike fructose or lactose, 80 percent of tagatose remains unabsorbed until it reaches the large intestine. There the normal bacterial colonies of the intestine ferment tagatose, releasing gases and small products that are absorbed. At high doses, tagatose causes flatulence, rumbling, and loose stools. Otherwise, no adverse side effects have been noted by the maker.

Other Artificial Sweeteners FDA approval for two other sweeteners—**cyclamate** and **alitame**—is still pending. To date, no safety issues have been raised for alitame. Cyclamate, on the other hand, has been battling safety issues for 50 years. Approved by the FDA in 1949, cyclamate was banned in 1970 because of evidence indicating that it caused bladder cancer in rats. In 1985, the National Academy of Sciences concluded that evidence to date indicated that cyclamate did not cause cancer in human beings but that further studies were warranted. In Canada, cyclamate is restricted to use as a tabletop sweetener on the advice of a physician. In the United States, the FDA is currently reviewing a petition to reapprove the use of cyclamate.

Artificial Sweeteners and Weight Control Many people eat and drink products sweetened with artificial sweeteners to help them control weight. Does this work? Ironically, a few studies have reported that after consuming such products, people experience heightened feelings of hunger. Despite these reports, most studies find that artificial sweeteners do not heighten feelings of hunger, enhance food intake, or cause weight gain in people. When people reduce their energy intakes by replacing sugar in their diets with artificial sweeteners and then compensate for the reduced energy at later meals, energy intake may stay the same or increase. Using artificial sweeteners will not automatically lower energy intake; to successfully control energy intake, a person needs to make informed diet and activity decisions throughout the day.

IN SUMMARY

- Two types of alternative sweeteners are sugar alcohols and artificial sweeteners.

- Sugar alcohols are carbohydrates, but they yield slightly less energy than sucrose.

- Sugar alcohols do not contribute to dental caries.

- The artificial sweeteners are not carbohydrates and yield no energy.

- Like the sugar alcohols, artificial sweeteners do not promote tooth decay.

Health Effects of Starch and Dietary Fibers

Despite dietary recommendations that people should eat generous servings of starch and fiber-rich carbohydrate foods for their health, many people still believe that carbohydrate is the "fattening" component of foods. Gram for gram, carbohydrates contribute fewer kcalories to the body than do dietary fats, so a moderate diet based on starch- and fiber-rich carbohydrate foods is likely to be lower in kcalories than a diet based on high-fat foods.

For health's sake, most people should increase their intakes of carbohydrate-rich foods such as whole grains, vegetables, legumes, and fruits—foods noted for their starch, fiber, and naturally occurring sugars. In addition, most people should also limit their intakes of foods with added sugars and the types of fats associated with heart disease (see Chapter 3). A diet that emphasizes whole grains, vegetables, legumes, and fruits is almost invariably moderate in food energy, low in fats that can harm health, and high in dietary fiber, vitamins, and minerals. All these factors working together can help reduce the risks of obesity, cancer, cardiovascular disease, diabetes, dental caries, gastrointestinal disorders, and malnutrition.

It is difficult to determine which carbohydrates contribute to which health benefits. Starch and fibers almost always occur together in foods (except refined foods), so it is hard to distinguish their effects. Some health effects appear to be especially closely associated with fibers, however.

Carbohydrates: Disease Prevention and Recommendations

Fiber-rich carbohydrate foods benefit health in many ways. Foods such as whole grains, legumes, vegetables, and fruits supply valuable vitamins, minerals, and phytochemicals, along with abundant dietary fiber and little or no fat. The following sections describe some of the health benefits of diets that emphasize a variety of these foods each day.

Heart Disease Diets rich in whole grains, legumes, and vegetables, especially those rich in whole grains, may protect against heart disease, although sorting out the exact reasons why has proved difficult.[19] Such diets are low in animal fat and cholesterol and high in dietary fibers, vegetable proteins, and phytochemicals—all factors associated with a lower risk of heart disease. ■ It is relatively certain that soluble fibers (such as oat bran, barley, and legumes) lower blood cholesterol by binding cholesterol compounds and carrying them out of the body with the feces.[20] High-fiber foods may also lower blood cholesterol indirectly by displacing fatty, cholesterol-raising foods from the diet. Even when dietary fat intake is low, research shows that high intakes of soluble fiber exert separate and significant blood cholesterol–lowering effects.

■ The role of animal fat and cholesterol in heart disease is discussed in Chapter 3. The role of vegetable proteins in heart disease is presented in Chapter 4. The benefits of phytochemicals in disease prevention are presented in Nutrition in Practice 8.

Diabetes Some fibers delay the passage of nutrients from the stomach to the small intestine. This delay slows glucose absorption, thus eliciting a moderate insulin response and a moderate rise in blood glucose. This slow, sustained rise in blood glucose is desirable. The term **glycemic effect** refers to how quickly glucose is absorbed after a person eats, how high blood glucose rises, and how quickly it returns to normal. A high glycemic effect reflects fast glucose absorption and a surge in blood glucose, which is undesirable. A study of more than 90,000 women showed that women whose diets had the highest glycemic effect and the lowest cereal fiber content were most vulnerable to **type 2 diabetes** independent of other dietary constituents or known risk factors.[21] ■

■ Diabetes is the topic of Chapter 21.

GI Health Fibers that both enlarge and soften stools such as cellulose (in cereal brans, fruits, and vegetables) ease elimination for the rectal muscles and thereby alleviate or prevent constipation and hemorrhoids. Other fibers help to solidify watery stools.

Some fibers (again, such as cereal bran) help keep the contents of the intestinal tract moving easily. This action helps prevent compaction of the intestinal contents, which could obstruct the appendix and permit bacteria to invade and infect it.

In addition, fibers stimulate the muscles of the gastrointestinal (GI) tract so that they retain their health and tone. This prevents the muscles from becoming weak and the lining of the digestive tract from bulging out in places, as occurs in diverticulosis. Insoluble fiber, particularly cellulose, seems to be most beneficial in lowering the risk of diverticulosis.[22] Chapter 18 describes diverticulosis.

glycemic (gligh-SEEM-ic) **effect:** the extent to which a food raises the blood glucose concentration and elicits an insulin response.

type 2 diabetes: the type of diabetes that accounts for 90 to 95 percent of diabetes cases and usually results from insulin resistance coupled with insufficient insulin secretion.

■ Choose fiber-rich fruits, vegetables, and whole grains often.

Cancer Many, but not all, research studies suggest that increasing dietary fiber protects against colon cancer.[23] On completing a study of almost 520,000 people, researchers concluded that doubling the naturally occurring fiber in diets of populations with low fiber intakes could reduce the risks of colon cancer by 40 percent.[24] Importantly, the study focused on dietary fiber, not fiber supplements or additives. Fiber supplements and additives lack valuable nutrients and phytochemicals that also help protect against cancer.

How fiber may help prevent colon cancer is under investigation. One focus of research is fiber's ability to dilute and speed removal of potential cancer-causing agents from the large intestine (colon). In addition, some fibers stimulate bacterial fermentation of fiber in the colon, a process that produces small fatlike molecules that lower the **pH**.[25] These small fatlike molecules and the lower pH inhibit cancer growth in the colon.[26]

Other processes may also be at work. As research progresses, the *Dietary Guidelines for Americans 2005* and the USDA Food Guide recommend a high-fiber diet that includes $4\frac{1}{2}$ cups of fruits and vegetables and at least 3 ounces of whole grains each day for those whose energy intakes are about 2,000 kcalories per day. ■

■ Recommended amounts of foods from each group for varying kcalorie levels are shown in Table 1–6 on page 20.

Weight Management Fiber-rich foods tend to be low in fats and added sugars and therefore promote weight loss by delivering less energy per bite. High-fiber foods also promote a feeling of fullness as they absorb water. In addition, soluble fibers in a meal slow the movement of food through the upper digestive tract, so a person feels fuller longer. In a study of more than 27,000 men, those who ate the most whole-grain foods, such as certain cooked and cold breakfast cereals, whole wheat bread, brown rice, and popcorn, gained the least amount of weight over eight years.[27] The researchers observed a dose-response relationship: for every 40-gram increase in whole grains from all foods per day, weight gain was reduced by about one pound. This inverse relationship between consumption of whole-grain foods and weight gain also seems to hold true for women. Over a span of 12 years, women who ate more whole-grain foods gained significantly less weight than those who ate less whole-grain foods.[28] In contrast, intake of refined grain foods was positively associated with weight gain.

Many weight-loss products on the market today contain bulk-inducing fibers such as methylcellulose, but buying pure fiber compounds like this is neither necessary nor advisable. Besides adding bulk to the diet, high-fiber foods supply other nutrients as well. Most experts agree that the health benefits attributed to fiber may come from other constituents of fiber-containing foods and not from the fiber alone.[29] Therefore, to use fiber in a weight-loss plan, select fresh fruits, vegetables, legumes, and whole-grain foods. As a bonus, fiber-rich foods are often economical as well as nutritious. Table 2–5 summarizes fibers and their health benefits.

Harmful Effects of Excessive Fiber Intake Despite fiber's benefits to health, when too much fiber is consumed, some minerals may bind to it and be excreted with it without becoming available for the body to use. When mineral intake is adequate, however, a reasonable intake of high-fiber foods does not seem to compromise mineral balance.

People with marginal intakes who eat mostly high-fiber foods may not be able to take in enough food to meet energy or nutrient needs. The malnourished, the elderly, and young children adhering to all-plant (vegan) diets are especially vulnerable to

TABLE 2-5

Characteristics, Food Sources, and Health Effects of Fibers

Fiber Characteristics	Major Food Sources	Actions in the Body	Health Benefits
Viscous, soluble, more fermentable			
■ Gums ■ Pectins ■ Psyllium ■ Some hemicelluloses	Barley, oats, oat bran, rye, fruits (apples, citrus), legumes (especially young green peas and black-eyed peas), seaweeds, seeds and husks, vegetables; fibers used as food additives	■ Lower blood cholesterol by binding bile ■ Slow glucose absorption ■ Slow transit of food through upper GI tract, lending satiety ■ Hold moisture in stools, softening them ■ Yield small fatlike molecules after fermentation that the large intestine can use for energy	■ Lower risk of heart disease ■ Lower risk of diabetes
Nonviscous, insoluble, less fermentable			
■ Cellulose ■ Lignin ■ Resistant starch ■ Many hemicelluloses	Brown rice, fruits, legumes, seeds, vegetables (cabbage, carrots, brussels sprouts), wheat bran, whole grains; extracted fibers used as food additives	■ Increase fecal weight and speed fecal passage through large intestine ■ Provide bulk and feelings of fullness	■ Alleviate constipation ■ Lower risks of diverticulosis, hemorrhoids, and appendicitis ■ May help with weight management

this problem. Fibers also carry water out of the body and can cause dehydration. Advise clients to add an extra glass or two of water to go along with the fiber added to their diets. Athletes may want to avoid bulky, fiber-rich foods just prior to competition. Adding purified fibers, such as oat bran or wheat bran, to foods can be taken to extremes. Too much fiber and too little fluid can obstruct the GI tract. Also, purified fiber may not affect the body the same way as the fiber in its original food product.

Recommendations The DRI committee advises that carbohydrates should contribute about half (45 to 65 percent) of the energy requirement. A person consuming 2,000 kcalories a day should therefore have 900 to 1,300 kcalories of carbohydrate, or between 225 and 325 grams. ■ This amount is more than adequate to meet the RDA for carbohydrate, which is set at 130 grams per day based on the average minimum amount of glucose used by the brain. ■ [30]

When it established the Daily Values that appear on food labels, the FDA used a 60 percent of kcalories guideline in setting the Daily Value for carbohydrate at 300 grams per day. ■ For most people, this means increasing total carbohydrate intake. To this end, as mentioned earlier, the *Dietary Guidelines for Americans 2005* encourages people to choose fiber-rich fruits, vegetables, and whole grains often.

Recommendations for fiber encourage the same foods just mentioned: whole grains, vegetables, fruits, and legumes, which also provide vitamins, minerals, and phytochemicals. The FDA set the Daily Value for fiber at 25 grams or 11.5 grams per 1,000-kclaorie intake. ■ The DRI recommendation is slightly higher—14 grams per 1,000-kcalorie intake. ■ Similarly, the American Dietetic Association suggests 20 to 35 grams of fiber daily, which is about twice the average intake in the United States. [31]

As health care professionals, you can advise your clients that an effective way to add dietary fiber while lowering fat is to substitute plant sources of proteins (legumes) for some of the animal sources of protein (meats and cheeses) in the diet. Another way to add fiber is to encourage clients to consume the recommended amounts of fruits and vegetables each day. People choosing high-fiber foods are wise to seek out a variety of fiber sources and to drink extra fluids to help the fiber do its job. Many foods provide fiber in varying amounts as Figure 2-4 (p. 54) shows.

As mentioned earlier, too much fiber is no better than too little. The World Health Organization recommends an upper limit of 40 grams of dietary fiber per day.

■ 45% of 2,000 kcal:
2,000 × .45 = 900 kcal
900 kcal ÷ 4 kcal/g = 225 g

65% of 2,000 kcal:
2,000 × .65 = 1,300 kcal
1,300 kcal ÷ 4 kcal/g = 325 g

■ RDA for carbohydrate:
130 g/day
45 to 65% of energy intake

■ Daily Values:
25 g fiber (based on 11.5g/1,000 kcal)

■ Fiber AI:
14 g/1,000 kcal/day

Men:
19–50 yr: 38 g/day
51+ yr: 30 g/day

Women:
19–50 yr: 25 g/day
51+ yr: 21 g/day

■ Reminder: An AI (Adequate Intake) is used as a guide for nutrient intake when a RDA cannot be established (see Chapter 1).

pH: the unit of measure expressing a substance's acidity or alkalinity (Chapter 4 provides a more detailed definition.)

Grains

Whole-grain products provide 1 to 2 g of fiber or more per serving:

- 1 slice whole wheat or rye bread (1 g)
- 1 slice pumpernickel bread (2 g)
- $\frac{1}{2}$ c ready-to-eat 100% bran cereal (10 g)
- $\frac{1}{2}$ c cooked barley, bulgur, grits, oatmeal (2 to 3 g)

© Polara Studios, Inc.

Vegetables

Most vegetables contain 2 to 3 g of fiber per serving:

- 1 c raw bean sprouts
- $\frac{1}{2}$ c cooked broccoli, brussels sprouts, cabbage, carrots, cauliflower, collards, corn, eggplant, green beans, green peas, kale, mushrooms, okra, parsnips, potatoes, pumpkin, spinach, sweet potatoes, swiss chard, winter squash
- $\frac{1}{2}$ c chopped raw carrots, peppers

© Polara Studios, Inc.

Fruits

Fresh, frozen, and dried fruits have about 2 g of fiber per serving:

- 1 medium apple, banana, kiwi, nectarine, orange, pear
- $\frac{1}{2}$ c applesauce, blackberries, blueberries, raspberries, strawberries
- Fruit juices contain very little fiber

© Polara Studios, Inc.

Legumes and Nuts

Many legumes provide about 8 g of fiber per serving:

- $\frac{1}{2}$ c cooked baked beans, black beans, black-eyed peas, kidney beans, navy beans, pinto beans

Some legumes provide about 5 g of fiber per serving:

- $\frac{1}{2}$ c cooked garbanzo beans, great northern beans, lentils, lima beans, split peas

Most nuts and seeds provide 1 to 3 g of fiber per serving:

- 1 oz almonds, cashews, hazelnuts, peanuts, pecans, pumpkin seeds, sunflower seeds

© Polara Studios, Inc.

FIGURE 2–4

Fiber in Selected Foods

- A diet rich in starches and dietary fibers helps prevent heart disease, diabetes, GI disorders, and possibly some types of cancer. It also supports efforts to manage body weight.

- For these reasons, recommendations urge people to eat plenty of whole grains, vegetables, legumes, and fruits—enough to provide 45 to 65 percent of the daily energy from carbohydrate.

Carbohydrates: Food Sources

A day's meals based on the USDA Food Guide not only meet carbohydrate recommendations, but also provide abundant fiber, too. Grains, vegetables, fruits, and legumes deliver fiber to the diet, and like milk as well, are noted for their valuable energy-yielding starches and dilute sugars. Each class of foods makes its own typical carbohydrate contribution. The USDA Food Guide in Chapter 1 can help you and your clients obtain carbohydrate-rich foods.

Grains Most foods in this group—a slice of whole wheat bread, half an English muffin or bagel, a 6-inch tortilla, or $1/2$ cup of rice, pasta, or cooked cereal—provide about 15 grams of carbohydrate, mostly as starch.* Most grain choices should be low in fat and sugar. When extra kcalories are needed to meet energy needs, some selections higher in fat, preferably unsaturated fat (see Chapter 3), and sugar can supply discretionary kcalories. These choices include biscuits, muffins, and snack crackers.

Vegetables Some vegetables are major contributors of starch in the diet. Just a small white or sweet potato or $1/2$ cup of cooked dry beans, corn, peas, plantain, or winter squash provides 15 grams of carbohydrate, as much as in a slice of bread, though as a mixture of sugars and starch. A $1/2$-cup portion of carrots, okra, onions, tomatoes, cooked greens, or most other nonstarchy vegetables or a cup of salad greens provides about 5 grams as a mixture of starch and sugars. Each of these foods also contributes a little protein, some fiber, and no fat.

Fruits The size of a typical serving of fruit varies depending on the form of the fruit: $1/2$ cup of juice; a small banana, apple, or orange; $1/2$ cup of most canned or fresh fruit; or $1/4$ cup of dried fruit. A typical fruit serving contains an average of about 15 grams of carbohydrate, mostly as sugars, including the fruit sugar fructose. Fruits vary greatly in their water and fiber contents, and therefore their sugar concentrations vary also. No more than a third of the day's fruit should come from juice. With the exception of avocado, which is high in fat, the fruits contain insignificant amounts of fat and protein.

Milk, Cheese, and Yogurt One cup of milk or yogurt or the equivalent (1 cup of buttermilk, $1/2$ cup of dry milk powder, or $1/2$ cup of evaporated milk) provides a generous 12 grams of carbohydrate. Among cheeses, cottage cheese provides about 6 grams of carbohydrate per cup, while most other types contain little, if any, carbohydrate. These foods also contribute high-quality protein as well as several important vitamins and minerals. Calcium-fortified soy beverages are options for providing calcium and about the same amount of carbohydrate as milk. All milk products vary in fat content, an important consideration in choosing among them; Chapter 3 provides the details.

---

* Gram values in this section are adapted from the U.S. Food Exchange System.

Cream and butter, although dairy products, are not equivalent to milk because they contain little or no carbohydrate and insignificant amounts of the other nutrients important in milk. They are appropriately placed with the solid fats and added sugars.

Meat, Poultry, Fish, Legumes, Eggs, and Nuts With two exceptions, foods of this group provide almost no carbohydrate to the diet. The exceptions are nuts, which provide a little starch and fiber along with their abundant fat, and dry beans, which are excellent sources of both starch and fiber. Just a ¹/₂-cup serving of beans provides 15 grams of carbohydrate, an amount equal to the richest carbohydrate sources. Among sources of fiber, beans and other legumes are outstanding, providing as much as 8 grams in ¹/₂ cup. The carbohydrate content of a diet can be determined by using a nutrient composition table such as that found in Appendix A, the exchange list system described in Chapter 21, or a computer diet analysis program.

Carbohydrates: Food Labels and Health Claims

Food labels list the amount, in grams, of total carbohydrate—including starch, fibers, and sugars—per serving. Fiber grams are also listed separately, as are the grams of sugars. (With this information, consumers can calculate starch grams by subtracting the grams of fibers and sugars from the total carbohydrate.) Sugars on the Nutrition Facts panel of a food label reflect both added sugars and those that occur naturally in foods. Total carbohydrate and dietary fiber are also expressed in the "% Daily Values" column for a person consuming 2,000 kcalories; there is no Daily Value for sugars.

The FDA authorizes four health claims on food labels concerning fiber-rich carbohydrate foods. One is for "fiber-containing grain products, fruits, and vegetables and reduced risk of cancer." Another is for "fruits, vegetables, and grain products that contain fiber and reduced risk of coronary heart disease," a third is for "soluble fiber from whole oats and from psyllium seed husk and reduced risk of coronary heart disease," and a fourth is for "whole grains and reduced risk of heart disease and certain cancers." Chapter 1 describes the criteria foods must meet to bear these health claims.

IN SUMMARY

- Grains, vegetables, fruits, and legumes contribute dietary fiber to people's diets and, like milk, also contribute energy-yielding starches and dilute sugars.

- Food labels list grams of total carbohydrate and also provide separate listings of grams of fiber and sugar.

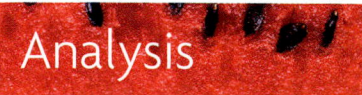

Analysis

YOUR DIET

Carbohydrates

Most of the energy people receive from foods comes from carbohydrates. Healthy choices provide carbohydrates rich in fiber, starches, vitamins, minerals, and naturally occurring sugars. A diet that is consistently low in dietary fiber and high in added sugar can lead to health problems. The table below displays related foods in categories that roughly indicate their fiber and sugar contents.

Look back to the table you completed at the end of Chapter 1 that compares the foods you ate with the USDA Food Guide. In the table below, find the appropriate column for *five* of your carbohydrate choices. If your food is not listed in the first column (High in Fiber/Low in Added Sugar), suggest an alternative selection to place in this column. As an example, if you consumed a Rice Krispies Treat® (low in fiber, high in added sugar), you could list the food in the third column and write "brown rice" in the first column. If the food has no obvious alternatives, like candy or sweetened soda, circle the food and list an alternative choice below the table (for example, you can replace regular soda with diet soda or low-fat milk).

High in Fiber/Low in Added Sugar	Intermediate Fiber and/or Sugar	Low in Fiber/High in Added Sugar
Apple with peel	Applesauce, sweetened	Fruit drink, 10% apple juice
Oatmeal	Granola	Granola breakfast bar
Whole wheat bread	Bagel, plain	Danish pastry
Corn on the cob	Creamed corn	Corn flakes, added sugar
Baked sweet potato	Candied sweet potato casserole	Sweet potato pie

CLINICAL APPLICATIONS

1. Considering the health benefits of carbohydrate-rich foods, especially those that provide starch and fiber, what suggestions would you offer to a client who reports the following:

   ■ Eats only three servings of refined, sugary breads or cereals each day.

   ■ Eats one serving of vegetables (usually french fries) each day.

   ■ Drinks fruit juice once a day, but never eats fruit.

   ■ Eats cheese at least twice a day, but does not drink milk.

   ■ Eats large servings of meat at least twice a day.

   ■ Eats hard candy two or three times a day.

Self Check

1. Carbohydrates are found in virtually all foods **except:**
   a. milks.
   b. meats.
   c. breads.
   d. vegetables.

2. Polysaccharides include:
   a. galactose, starch, and glycogen.
   b. starch, glycogen, and fiber.
   c. lactose, maltose, and glycogen.
   d. sucrose, fructose, and glucose.

3. The chief energy source of the body is:
   a. sucrose.
   b. starch.
   c. glucose.
   d. fructose.

4. The primary form of stored glucose in animals is:
   a. glycogen.
   b. cellulose.
   c. starch.
   d. lactose.

5. The polysaccharide that helps form the cell walls of plants is:
   a. cellulose.
   b. starch.
   c. glycogen.
   d. lactose.

6. Which of the following items may denote sugar on food labels?
   a. corn syrup
   b. aspartame
   c. xylitol
   d. cellulose

7. The two types of alternative sweeteners are:
   a. saccharin and cyclamate.
   b. sugar alcohols and artificial sweeteners.
   c. sorbitol and xylitol.
   d. sucrose and fructose.

8. A diet high in carbohydrate-rich foods such as whole grains, vegetables, fruits, and legumes is:
   a. most likely low in fat.
   b. most likely low in fiber.
   c. most likely poor in vitamins and minerals.
   d. most likely disease promoting.

9. A fiber-rich diet may help to prevent or control:
   a. diabetes.
   b. heart disease.
   c. constipation.
   d. all of the above.

10. The DRI fiber recommendation is:
    a. 10 grams per 1,000 kcalories.
    b. 15 to 25 grams per day.
    c. 14 grams per 1,000 kcalories.
    d. 40 to 55 grams per day.

Answers are in Appendix H.

Notes

1. U.S. Food and Drug Administration, *Counting Calories: Report of the Working Group on Obesity,* March 12, 2004, available at www.cfsan.fda.gov/~dms/owg-rpt.html; Standing Committee on the Scientific Evaluation of Dietary Reference Intakes, Food and Nutrition Board, Institute of Medicine, *Dietary Reference Intakes for Energy, Carbohydrate, Fiber, Fat, Fatty Acids, Cholesterol, Protein, and Amino Acids* (Washington, D.C.: National Academies Press, 2005), pp. 265–338; A. Trichopoulou and coauthors, Lipid, protein, and carbohydrate intake in relation to body mass index, *European Journal of Clinical Nutrition* 56 (2002): 37–43.

2. R. Gruetter, Glycogen: The forgotten cerebral energy store, *Journal of Neuroscience Research* 74 (2003): 179–183; I. Y. Choi, E. R. Seaquist, and R. Gruetter, Effect of hypoglycemia on brain glycogen metabolism in vivo, *Journal of Neuroscience Research* 72 (2003): 25–32.

3. B. V. McCleary, Dietary fibre analysis, *Proceedings of the Nutrition Society* 62 (2003): 3–9.

4. B. M. Popkin and S. J. Nielsen, The sweetening of the world's diet, *Obesity Research* 11 (2003): 1325–1332.

5. U.S. Department of Agriculture and U.S. Department of Health and Human Services, *2005 Dietary Guidelines Committee Report,* 2005, available online at www.healthierus.gov/dietaryguidelines.

6. M. Nestle, as quoted by O. Dyer, U.S. government rejects WHO's attempts to improve diet, *British Medical Journal* 328 (2004): 185.

7. Position of the American Dietetic Association: Oral health and nutrition, *Journal of the American Dietetic Association* 103 (2003): 615–625.

8. Joint WHO/FAO Expert Consultation, *Diet, Nutrition, and the Prevention of Chronic Diseases* (Geneva, Switzerland: World Health Organization, 2003), p. 119.

9. J. M. Jones and K. Elam, Sugars and health: Is there an issue? *Journal of the American Dietetic Association* 103 (2003): 1058–1060.

10. Popkin and Nielsen, 2003.

11. M. B. Schulze and coauthors, Sugar-sweetened beverages, weight gain, and incidence of type 2 diabetes in young and middle-aged women, *Journal of the American Medical Association* 292 (2004): 927–934; A. M. Coulston and R. K. Johnson, Sugar and sugars: Myths and realities, *Journal of the American Dietetic Association* 102 (2002): 351–353.

12. V. S. Malik, M. B. Schulze, and F. B. Hu, Intake of sugar-sweetened beverages and weight gain: A systematic review, *American Journal of Clinical Nutrition* 84 (2006): 274–288.

13. G. A. Bray, S. J. Nielsen, and B. M. Popkin, Consumption of high-fructose corn syrup in beverages may play a role in the epidemic of obesity, *American Journal of Clinical Nutrition* 79 (2004): 537–543; S. J. Nielsen and B. M. Popkin, Changes in beverage intake between 1977 and 2001, *American Journal of Preventive Medicine* 27 (2004): 205–210.

14. Standing Committee on the Scientific Evaluation of Dietary Reference Intakes, 2005, pp. 294–338.

15. Standing Committee on the Scientific Evaluation of Dietary Reference Intakes, 2005, p. 770.

16. Position of the American Dietetic Association: Use of nutritive and nonnutritive sweeteners, *Journal of the American Dietetic Association* 104 (2004): 255–275.

17. Fact Sheet: The Report on Carcinogens, 9th ed., National Institutes of Health News Release, available at www.nih.gov/news/pr/may2000/niehs-15.htm.

18. Position of The American Dietetic Association, 2004.

19. M. K. Jensen and coauthors, Intakes of whole grains, bran, and germ and the risk of coronary heart disease in men, *American Journal of Clinical Nutrition* 80 (2004): 1492–1499; F. B. Hu and W. C. Willett, Optimal diets for prevention of coronary heart disease, *Journal of the American Medical Association* 288 (2002): 2569–2578; N. M. McKeown and coauthors, Whole-grain intake is favorably associated with metabolic risk factors for type 2 diabetes and cardiovascular disease in the Framingham Offspring Study, *American Journal of Clinical Nutrition* 76 (2002): 390–398.

20. K. M. Behall, D. J. Scholfield, and J. Hallfrisch, Diets containing barley significantly reduce lipids in mildly hypercholesterolemic men and women, *American Journal of Clinical Nutrition* 80 (2004): 1185–1193; Position of the American Dietetic Association, Health implications of dietary fiber, *Journal of the American Dietetic Association* 102 (2002): 992–1000; L. Van Horn and N. Ernst, A summary of the science supporting the new National Cholesterol Education program dietary recommendations: What dietitians should know, *Journal of the American Dietetic Association* 101 (2001): 1148–1154.

21. M. B. Schulze and coauthors, Glycemic index, glycemic load, and dietary fiber intake and incidence of type 2 diabetes in younger and middle-aged women, *American Journal of Clinical Nutrition* 80 (2004): 348–356.

22. E. Cunningham and W. Marcason, What role does fiber play in diverticular disease? *Journal of the American Dietetic Association* 102 (2002): 225.

23. J. G. Muir and coauthors, Combining wheat bran with resistant starch has more beneficial effects on fecal indexes than does wheat bran alone, *American Journal of Clinical Nutrition* 79 (2004): 1020–1028; Editorial board, The facts on fiber and colon cancer, *Nutrition & the M.D.* March 2004, pp. 6–7; U. Peters and

coauthors, Dietary fibre and colorectal adenoma in a colorectal cancer early detection programme, *Lancet* 361 (2003): 1491–1495; S. A. Bingham and coauthors, Dietary fibre in food and protection against colorectal cancer in the European Prospective Investigation into Cancer and Nutrition (EPIC): An observational study, *Lancet* 361 (2003): 1496–1501; T. Asano and R. S. McLeod, Dietary fibre for the prevention of colorectal adenomas and carcinomas, *Cochrane Database of Systematic Reviews* 2 (2002): CD003430.

24. Bingham and coauthors, 2003.

25. Muir and coauthors, 2004.

26. Muir and coauthors, 2004; A. Andoh, T. Tsujikasw, and Y. Fujiyama, Role of dietary fiber and short-chain fatty acids in the colon, *Current Pharmaceutical Design* 9 (2003): 347–358; I. McMillan and coauthors, Opposing effects of butyrate and bile acids on apoptosis of human colon adenoma cells: Differential activation of PKC and MAP kinases, *British Journal of Cancer* 88 (2003): 748–753.

27. P. Koh-Banerjee and coauthors, Changes in whole-grain, bran, and cereal fiber consumption in relation to 8-y weight gain among men, *American Journal of Clinical Nutrition* 80 (2004): 1237–1245.

28. S. Liu and coauthors, Relation between changes in intakes of dietary fiber and grain products and changes in weight and development of obesity among middle-aged women, *American Journal of Clinical Nutrition* 78 (2003): 920–927.

29. Standing Committee on the Scientific Evaluation of Dietary Reference Intakes, 2005, pp. 391–399.

30. Standing Committee on the Scientific Evolution of Dietary Reference Intakes, 2005, p. 265.

31. Position of the American Dietetic Association, 2002.

NUTRITION AND DENTAL HEALTH

Chapter 2 emphasized the health benefits of eating carbohydrate-rich foods, especially those containing starch and fiber. Carbohydrates may support overall health, but they do not necessarily promote dental health. The carbohydrates people eat and the times they eat them play a major role in the development of dental caries—a pervasive health problem throughout the world.

What is dental caries?

Dental caries is an infectious oral disease that develops in the tooth **enamel** (see the accompanying Glossary and Figure NP2–1). Caries develops when bacteria that reside in the **dental plaque** consume and metabolize carbohydrates, producing acids that attack the tooth enamel. Thus at least two main ingredients are required to make dental caries: bacteria and carbohydrates. In addition, factors such as heredity, nutrition status during early tooth development, dental hygiene practices, and fluoride intake influence a person's susceptibility to caries. Poor nutrition during pregnancy, infancy, or early childhood can impair the development of healthy teeth, making caries likely.[1] Table NP2–1 shows the effects of specific nutrient deficiencies on tooth development.

DENTAL CARIES GLOSSARY

cariogenic (KARE-ee-oh-JEN-ik): conducive to dental decay.

dental caries (KARE-eez): the gradual decay and disintegration of a tooth.

dental plaque (PLACK): a gummy mass of bacteria that grows on teeth and can lead to dental caries and gum disease.

enamel: the hard, white, dense substance that forms a covering for the crown of a tooth.

How do carbohydrate-rich foods promote caries development?

The bacteria that promote dental caries thrive on food particles that contain carbohydrate. Both sugar and starch can support bacterial growth. Equally important is the length of time the food stays in the mouth, and this depends on how soon the teeth are brushed after eating and how sticky the food is. The damage a food does relates to both its carbohydrate content and its stickiness. For example, raisins and granola, which adhere to the teeth, cause more caries than a food that is easily rinsed off such as a sugary beverage.

Sugar can be eaten without inviting tooth decay if it is removed from tooth surfaces promptly. Bacterial action is maximal in the first 20 minutes after the first contact. If immediate brushing is not possible, water or other beverages

FIGURE NP2–1

A Tooth
The inner layer of dentin is bonelike material. The outer layer is enamel, which is harder than bone. Caries begin when acid dissolves the enamel that covers the tooth. If it is not repaired, the decay may penetrate the dentin and spread into the pulp of the tooth, causing inflammation and an abscess.

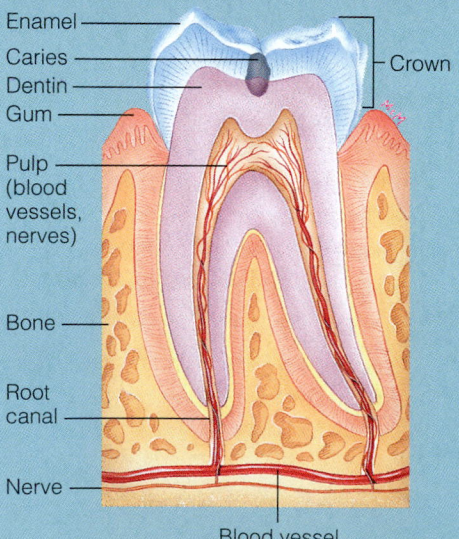

Enamel
Caries
Dentin — Crown
Gum
Pulp (blood vessels, nerves)
Bone
Root canal
Nerve
Blood vessel

TABLE NP2–1

Nutrient Deficiencies Affecting Tooth Development

Nutrient Deficiency	Effect on Tooth Development
Protein	Small, irregularly shaped teeth; delayed eruption; high caries susceptibility
Vitamin C	Disturbance of dentin formation
Vitamin A	Disturbance of enamel formation, delayed eruption
Vitamin D	Poor mineralization, pitting, striations
Calcium	Poor mineralization
Phosphorus	Poor mineralization
Magnesium	Enamel underdeveloped
Iron	High caries susceptibility
Zinc	High caries susceptibility
Fluoride	High caries susceptibility

swished in the mouth after a meal can effectively rinse the teeth. Once-a-day flossing may also effectively control formation of caries, regardless of the carbohydrate content of the diet. Some people may never get caries because they have inherited resistance to them.

Does saliva rinse the mouth and protect the teeth?

Yes, and some foods stimulate more saliva flow than others. Saliva protects against caries formation in several ways. In addition to rinsing the mouth, it also dilutes the caries-causing acid produced by bacteria, exerts antibacterial action, and provides protective minerals such as calcium and phosphorus that promote the repair (remineralization) of tooth enamel.[1] Foods that elicit saliva flow may therefore defend against caries formation, but not all of them are protective. Foods that also contain sugar may promote acid formation. Apples are an example: they stimulate saliva flow, but they also release sugar, so they have both caries-preventing and caries-promoting effects. Clearly, many different factors influence caries development, making it difficult to predict exactly which foods are **cariogenic**.

Do any foods prevent caries?

Yes. Some foods stimulate saliva flow and do not contribute to acid formation in the mouth: cheese is an example. Such foods are good choices to eat at the end of a meal. Cheese is a powerful saliva stimulant and does not promote acid formation, so it reduces the cariogenicity of a meal. Furthermore, the high calcium and phosphorus content of cheese supports dental health.

High-fiber foods are, in general, anticariogenic, especially if their sugar content is low. For example, raw vegetables do not stick to the teeth, and because they require vigorous chewing, they stimulate saliva flow. Cocoa products (including chocolate), coffee, tea, and beer all contain tannin, an acid that prevents caries formation. Table NP2–2 lists some dietary recommendations for controlling dental caries.

Besides foods, what other factors protect against dental caries development?

Research shows that when fluoride is added to the water supply, the children in the community have fewer dental caries than children who drink nonfluoridated water. Water fluoridation is the most effective, least expensive way to provide dental care to everyone.[3] Fluoride increases the mineralization of teeth, helps prevent tooth decay, and promotes tooth enamel remineralization throughout life.[4] The following recommendations will maximize protection against dental caries:

- Use sugars sparingly; watch for hidden sugars in foods; and use low-sugar or sugar-free products whenever possible.

- Restrict sweets to mealtimes.

- After eating a meal or a between-meal snack, brush and floss or, at least, rinse with water.

- Limit the time that teeth are exposed to sticky foods.

- In any case, brush at least twice daily, and floss at least once daily.

- Visit a dentist regularly; repair damaged teeth.

- Drink fluoridated water; if advised by a pediatrician, provide infants and children with fluoride supplements when such water is not available.

- Eat a balanced diet composed of a variety of foods that will maintain adequate nutrition status.

- Eat foods that are rich in calcium and phosphorus.

- Eat a variety of firm, fibrous foods that will stimulate saliva flow.

In summary, learning and practicing sound dental hygiene habits, as well as developing eating habits that are consistent with both dental health and nutritional health, will serve a person throughout life.

TABLE NP2–2

Dietary Recommendations for Controlling Dental Caries

Food Group	Low Cariogenicity (Use When Teeth Cannot Be Brushed Immediately)	High Cariogenicity (Do Not Use unless Followed by Prompt and Thorough Dental Hygiene)
Milk/yogurt/cheese	Milk, cheese, plain yogurt, cottage cheese	Ice cream, ice milk, milk shakes, fruited yogurts, eggnog
Meats/poultry/fish/legumes/eggs/nuts	Meat, fish, poultry, eggs, legumes, nuts	Peanut butter with added sugar, luncheon meats with added sugar, meats with sugared glazes
Fruits[a]	Fresh, packed in water	Dried (raisins, figs, dates), packed in syrup or juice, jams, jellies, preserves, fruit juices and drinks
Vegetables	Most vegetables	Candied sweet potatoes, glazed carrots
Grains[b]	Popcorn, toast, hard rolls, pretzels, pizza, bagels granola bars	Cookies, sweet rolls, pies, cakes, dry sugared cereals as between-meal snacks, doughnuts, potato chips,
Other	Sugarless gum, coffee or tea without sugar	Sugared soft drinks, candy, fudge, caramels, honey, sugars, syrups

[a] Tiny particles of bananas can get lodged between teeth and decompose, increasing risk of caries.
[b] Tiny particles of breads, crackers, and chips can also become lodged in teeth, promoting caries formation.

Notes

1. Position of the American Dietetic Association: Oral health and nutrition, *Journal of the American Dietetic Association* 103 (2003): 615–625.

2. D. P. DePaola and coauthors, Nutrition and dental medicine, in M. E. Shils and coeditors, *Modern Nutrition in Health and Disease*, 10th ed. (Philadelphia: Lippincott Williams & Wilkins, 2006), pp. 1152–1178; A. Van Nieuw Amerongen, J. G. Bolscher, and E. C. Veerman, Salivary proteins: Protective and diagnostic value in cariology? *Caries Research* 38 (2004): 247–253.

3. Position of the American Dietetic Association: The impact of fluoride on health, *Journal of the American Dietetic Association* 105 (2005): 1620–1628.

4. Position of the American Dietetic Association, 2005.

© Richard Hutchings/PhotoEdit

3 Lipids

M ost people know that too much fat in the diet, especially certain kinds of fat, imposes health risks, but they may be surprised to learn that too little does, too. People in the United States, however, are more likely to eat too much fat than too little.

Fat is a member of the class of compounds called **lipids**. The lipids in foods and in the human body include triglycerides (**fats** and **oils**), phospholipids, and sterols.

Roles of Body Fat

L ipids perform many tasks in the body, but most importantly, they provide energy. A constant flow of energy is so vital to life that any other function is sacrificed to maintain it. Chapter 2 described one safeguard against such an emergency—the stores of glycogen in the liver that provide glucose to the blood whenever the supply runs short. The body's stores of glycogen are limited, however. In contrast, the body's capacity to store fat for energy is virtually unlimited due to the fat-storing cells of the **adipose tissue**. The fat cells of the adipose tissue readily take up and store fat, growing in size as they do so. When extra energy storage is needed, new fat cells are readily produced. Fat cells are more than just storage depots, however; fat cells produce enzymes and secrete hormones that help regulate the appetite and influence other body functions.[1] Figure 3–1 shows a fat cell.

The fat stored in fat cells supplies 60 percent of the body's ongoing energy needs during rest. During some types of physical activity or prolonged periods of food deprivation, ■ fat stores may make an even greater energy contribution. The brain and nerves, however, need their energy as glucose, and the body cannot convert fat to glucose. After a long period of glucose deprivation (during fasting or starvation), brain and nerve cells develop the ability to derive about half of their energy from a special form of fat known as **ketone bodies**, but they still require glucose as well. This means that people wanting to lose weight need to eat a certain minimum amount of carbohydrate to meet their energy needs, even when they are limiting their food intakes.

■ Chapter 6 discusses fat use during fasting.

FIGURE 3–1

A Fat Cell
Within the fat, or adipose, cell, lipid is stored in a droplet. This droplet can greatly enlarge, and the fat cell membrane will expand to accommodate its swollen contents.

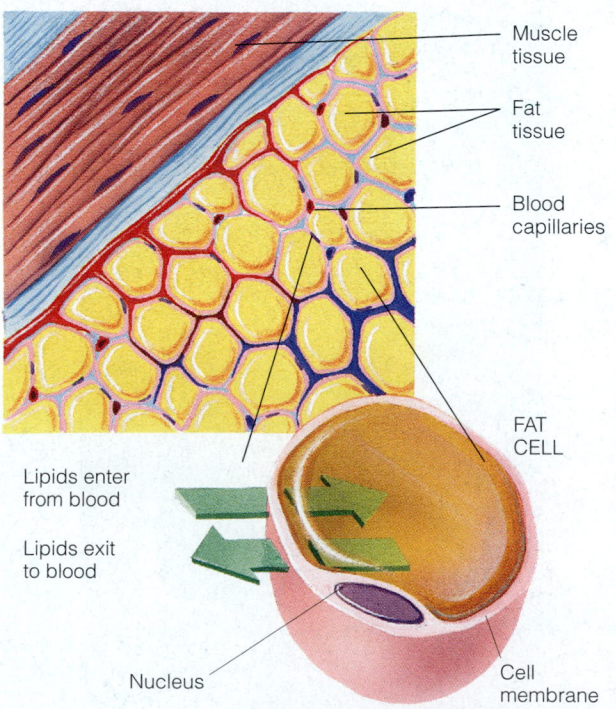

Muscle tissue

Fat tissue

Blood capillaries

FAT CELL

Lipids enter from blood

Lipids exit to blood

Nucleus

Cell membrane

In addition to supplying energy, fat serves other roles in the body. Natural oils in the skin provide a radiant complexion; in the scalp, they help nourish the hair and make it glossy. The layer of fat beneath the skin insulates the body from extremes of temperature. A pad of hard fat beneath each kidney protects it from being jarred and damaged, even during a motorcycle ride on a bumpy road. The soft fat in a woman's breasts protects her mammary glands from heat and cold and cushions them against shock. The fat embedded in muscle tissue shares with muscle glycogen the task of providing energy when the muscles are active. The phospholipids and the sterol cholesterol are cell membrane constituents that help maintain the structure and health of all cells. Table 3–1 summarizes the major functions of fats in the body.

Body fat supplies much of the fuel that muscles need to do their work.

IN SUMMARY

- Lipids in the body not only serve as energy reserves, but also protect the body from temperature extremes, cushion the vital organs, and provide the major material of cell membranes.

The Chemist's View of Lipids

The diverse and vital functions that lipids play in the body reveal why eating too little fat can be harmful. As mentioned earlier, too much fat in the diet seems to be the greater problem for most people. To understand both the beneficial and harmful effects that fats exert on the body, a closer look at the structure and function of members of the lipid family is in order.

Triglycerides

When people talk about fat—for example, "I'm too fat" or "That meat is fatty"—they are usually referring to triglycerides. Among lipids, **triglycerides** predominate—both in the diet and in the body. The name *triglyceride* almost explains itself: three (*tri*) **fatty acids** attached to a **glycerol** "backbone." Figure 3–2 shows how three fatty acids combine with glycerol to make a triglyceride.

TABLE 3–1

The Functions of Fats in the Body

Fats in the body:
- Provide energy
- Insulate the body against temperature extremes
- Protect the body's vital organs from shock
- Form the major material of cell membranes

lipids: a family of compounds that includes triglycerides (fats and oils), phospholipids, and sterols.

fats: lipids that are solid at room temperature (70°F or 25°C).

oils: lipids that are liquid at room temperature (70°F or 25°C).

adipose tissue: the body's fat, which consists of masses of fat-storing cells called adipose cells.

ketone (KEY-tones) **bodies:** acidic, fat-related compounds formed from the incomplete breakdown of fat when carbohydrate is not available.

triglycerides (try-GLISS-er-rides): one of the main classes of lipids; the chief form of fat in foods and the major storage form of fat in the body; composed of glycerol with three fatty acids attached.
  tri = three
  glyceride = a compound of glycerol

fatty acids: organic compounds composed of a chain of carbon atoms with hydrogens attached and an acid group at one end.

glycerol (GLISS-er-ol): an organic compound, three carbons long, that can form the backbone of triglycerides and phospholipids.

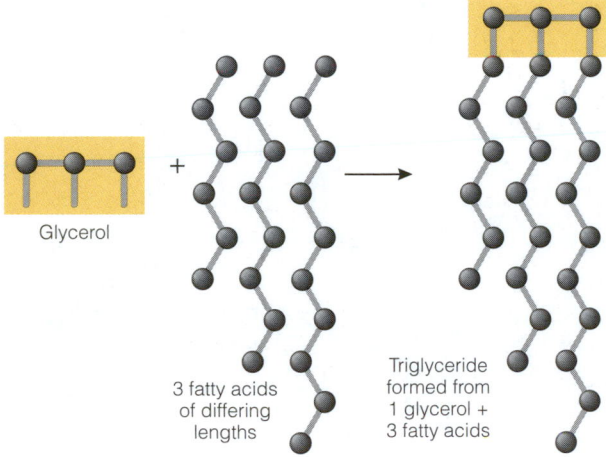

Glycerol

3 fatty acids of differing lengths

Triglyceride formed from 1 glycerol + 3 fatty acids

FIGURE 3–2

Triglyceride Formation
Glycerol, a small, water-soluble compound, plus three fatty acids, equals a triglyceride.

Fatty Acids

When energy from any energy-yielding nutrient is to be stored as fat, the nutrient is first broken into small fragments. Then the fragments are linked together into chains known as fatty acids. The fatty acids are then packaged, three at a time, with glycerol to make triglycerides.

Chain Length and Saturation Fatty acids may differ from one another in two ways—in chain length and in degree of saturation. The chain length refers to the number of carbons in a fatty acid. Saturation also refers to its chemical structure—specifically, to the number of hydrogens the carbons in the fatty acid are holding. If every available carbon is filled to capacity with hydrogen atoms, the chain is called a **saturated fatty acid**. A saturated fatty acid is fully loaded with hydrogens and has only single bonds between the carbons. The first zigzag structure in Figure 3–3 represents a saturated fatty acid.

Unsaturated Fatty Acids In some fatty acids, including most of those in plants and fish, hydrogens are missing in the fatty acid chains. The places where the hydrogens are missing are called points of unsaturation, and a chain containing such points is called an **unsaturated fatty acid**. An unsaturated fatty acid has at least one double bond between its carbons. If there is one point of unsaturation, the chain is a **monounsaturated fatty acid**. The second structure in Figure 3–3 is an example. If there are two or more points of unsaturation, then the fatty acid is a **polyunsaturated fatty acid** (see the third structure in Figure 3–3).

Hard and Soft Fat A triglyceride can contain any combination of fatty acids—long chain or short chain and saturated, monounsaturated, or polyunsaturated. The degree of saturation of the fatty acids in a fat influences the health of the body (discussed in a later section) and the characteristics of foods. Fats that contain the shorter-chain or the more unsaturated fatty acids are softer at room temperature and melt more readily. A comparison of three fats—lard (which comes from pork),

Three Types of Fatty Acids
The more carbon atoms in a fatty acid, the longer it is. The more hydrogen atoms attached to those carbons, the more saturated the fatty acid is.

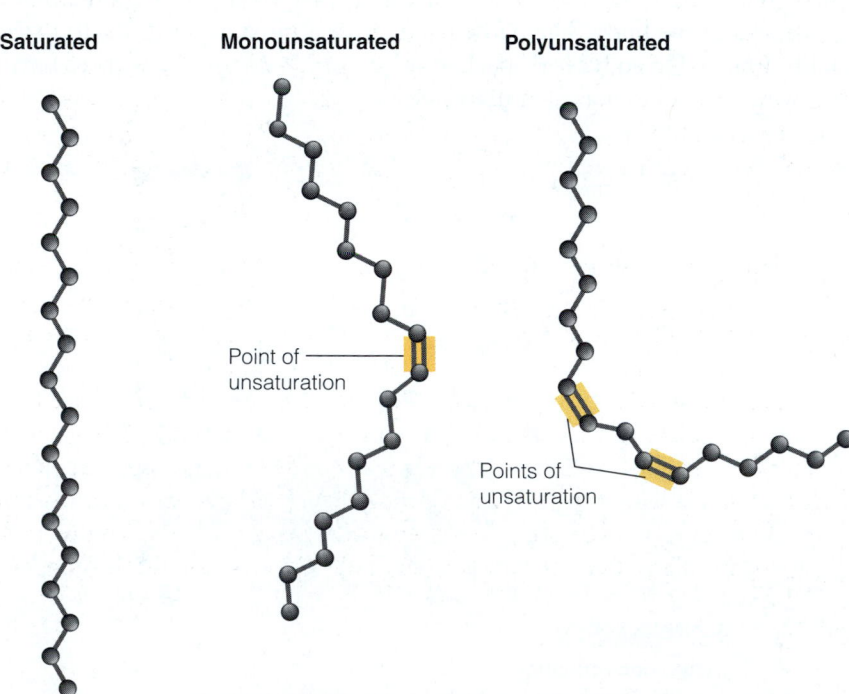

Saturated Monounsaturated Polyunsaturated

Point of unsaturation

Points of unsaturation

chicken fat, and safflower oil—illustrates these differences: lard is the most saturated and the hardest; chicken fat is less saturated and somewhat soft; and safflower oil, which is the most unsaturated, is a liquid at room temperature.

Stability Saturation also influences stability. Fats can become **rancid** when exposed to oxygen. Polyunsaturated fatty acids spoil most readily because their double bonds are unstable. The **oxidation** of unsaturated fats produces a variety of compounds that smell and taste rancid; saturated fats are more resistant to oxidation and thus less likely to become rancid. Other types of spoilage can occur due to microbial growth, however.

Manufacturers can protect fat-containing products against rancidity in three ways—none of them perfect. First, products may be sealed airtight and refrigerated—an expensive and inconvenient storage system. Second, manufacturers may add **antioxidants** to compete for the oxygen and thus protect the oil (examples are the additives **BHA** and **BHT** and vitamins C and E).* Third, manufacturers may saturate some or all of the points of unsaturation by adding hydrogen atoms—a process known as hydrogenation.

Hydrogenation Hydrogenation offers two advantages: it protects against oxidation (thereby prolonging shelf life) and also alters the texture of foods by increasing the solidity of fats. When partially hydrogenated, vegetable oils become spreadable margarine. Hydrogenated fats make piecrusts flaky and puddings creamy. A disadvantage is that hydrogenation makes polyunsaturated fats more saturated. Consequently, any health advantages of using polyunsaturated fats instead of saturated fats are lost in hydrogenation.

***Trans*-Fatty Acids** Another disadvantage of hydrogenation is that some of the molecules that remain unsaturated after processing change shape from *cis* to *trans*. In nature, most unsaturated fatty acids are *cis*-fatty acids—meaning that the hydrogens next to the double bonds are on the same side of the carbon chain. Only a few fatty acids in nature (notably in milk and butter) are ***trans*-fatty acids**—meaning that the hydrogens next to the double bonds are on opposite sides of the carbon chain (see Figure 3–4 on p. 68). These arrangements result in different configurations for the fatty acids, and this difference affects function: in the body, *trans*-fatty acids that derive from hydrogenation behave more like saturated fats than like unsaturated fats. The relationship between *trans*-fatty acids and heart disease has been the subject of much recent research, as a later section describes. In contrast, naturally occurring fatty acids such as **conjugated linoleic acid** that have a *trans* configuration may have health benefits.[2]

Essential Fatty Acids The human body can synthesize all the fatty acids it needs from carbohydrate, fat, or protein except for two—**linoleic acid** and **linolenic acid**. Both linoleic acid and linolenic acid are polyunsaturated fatty acids. Because they cannot be made from other substances in the body, they must be obtained from food and are therefore called **essential fatty acids**. Linoleic acid and linolenic acid are found in small amounts in plant oils, and the body readily stores them, making deficiencies unlikely. From both of these essential fatty acids, the body makes important substances that help regulate a wide range of body functions: blood pressure, clot formation, blood lipid concentration, the immune response, the inflammatory response to injury, and many others.[3] These two essential nutrients also serve as structural components of cell membranes.

* BHA is butylated hydroxyanisole; BHT is butylated hydroxytolvene.

---

saturated fatty acid: a fatty acid carrying the maximum possible number of hydrogen atoms (having no points of unsaturation).

unsaturated fatty acid: a fatty acid with one or more points of unsaturation where hydrogens are missing (includes monounsaturated and polyunsaturated fatty acids).

monounsaturated fatty acid (MUFA): a fatty acid that has one point of unsaturation; for example, the oleic acid found in olive oil.

polyunsaturated fatty acids (PUFA): fatty acids with two or more points of unsaturation. For example, linoleic acid has two such points, and linolenic acid has three. Thus polyunsaturated *fat* is composed of triglycerides containing a high percentage of PUFA.

rancid: the term used to describe fats when they have deteriorated, usually by oxidation. Rancid fats often have an "off" odor.

oxidation (OKS-ee-day-shun): the process of a substance combining with oxygen.

antioxidants: compounds that protect others from oxidation by being oxidized themselves.

BHA, BHT: preservatives commonly used to slow the development of "off" flavors, odors, and color changes caused by oxidation.

hydrogenation (high-dro-gen-AY-shun): a chemical process by which hydrogens are added to monounsaturated or polyunsaturated fats to reduce the number of double bonds, making the fats more saturated (solid) and more resistant to oxidation (protecting against rancidity). Hydrogenation produces *trans*-fatty acids.

***trans*-fatty acids:** fatty acids in which the hydrogens next to the double bond are on opposite sides of the carbon chain.

conjugated linoleic acid: a collective term for several fatty acids that have the same chemical formulas as linoleic acid but with different configurations.

linoleic acid, linolenic acid: polyunsaturated fatty acids that are essential for human beings.

essential fatty acids: fatty acids that the body requires but cannot make in amounts sufficient to meet its physiological needs.

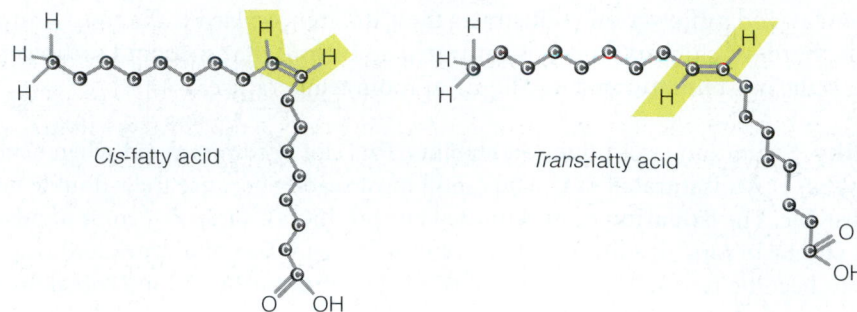

FIGURE 3–4

Cis- and *Trans-*Fatty Acids
Compared

Cis-fatty acid

Trans-fatty acid

■ Chemists use the term *omega*, the last letter of the Greek alphabet, to refer to the position of the last double bond in a fatty acid.

Linoleic Acid: An Omega-6 Fatty Acid Linoleic acid is an **omega-6 fatty acid** ■ found in the seeds of plants and in the oils produced from the seeds. Any diet that contains vegetable oils, seeds, nuts, and whole-grain foods provides enough linoleic acid to meet the body's needs. Researchers have long known and appreciated the importance of the omega-6 fatty acid family.

Linolenic Acid and Other Omega-3 Fatty Acids Linolenic acid belongs to a family of polyunsaturated fatty acids known as **omega-3 fatty acids**, a family that also includes **EPA** and **DHA**. EPA and DHA are found primarily in fish oils. As mentioned, the human body cannot make linolenic acid, but given dietary linolenic acid, it can make EPA and DHA, although the process is slow.

The importance of omega-3 fatty acids was first recognized during the 1980s when research began to unveil impressive roles for EPA and DHA in metabolism and disease prevention. The brain has a high content of DHA, and both EPA and DHA are needed for normal brain development.[4] DHA is also especially active in the rods and cones of the retina of the eye.[5] Today, researchers know that these omega-3 fatty acids are essential for normal growth and development and that they may play an important role in the prevention and treatment of heart disease, diabetes, hypertension, arthritis, and cancer.[6]

Phospholipids

Up to now, this discussion has focused on one class of lipids, the triglycerides (fats and oils), and their component parts, the fatty acids (see Table 3–2). Two other classes of lipids, the **phospholipids** and sterols, make up only 5 percent of the lipids in the diet, but they are nevertheless worthy of attention. Among the phospholipids, the lecithins are of particular interest.

Structure of Phospholipids Like the triglycerides, the **lecithins** and other phospholipids have a backbone of glycerol; they differ from the triglycerides in having only two fatty acids attached to the glycerol. In place of the third fatty acid, they have a phosphate group (a phosphorus-containing acid) and a molecule of **choline** or a similar compound. The fatty acids make the phospholipids soluble in fat; the phosphate group enables them to dissolve in water. Such versatility benefits the food industry, which uses phospholipids as **emulsifiers** to mix fats with water in such products as mayonnaise and candy bars.

Roles of Phospholipids Lecithins and other phospholipids are important constituents of cell membranes. They also act as emulsifiers in the body, helping to keep other fats in solution in the watery blood and body fluids.

Phospholipids: Not Essential Lecithins periodically receive attention in the popular press. People may hear that lecithins are a major constituent of cell membranes (true),

TABLE 3–2
The Lipid Family
Triglycerides (fats and oils)
■ Glycerol (1 per triglyceride)

■ Fatty acids (3 per triglyceride)
  Saturated
  Monounsaturated
  Polyunsaturated
    Omega-6
    Omega-3

Phospholipids (such as the lecithins)
Sterols (such as cholesterol)

that the functioning of all cells depends on the integrity of the cell membranes (true), and that consumers must therefore take lecithin supplements (false). The body digests lecithins before it absorbs them, so the lecithins people eat do not reach the body tissues intact. Instead, the lecithins used for building cell membranes are made from scratch by the liver. In other words, the lecithins are not essential nutrients.

Sterols

Sterols are large, complex molecules consisting of interconnected rings of carbon. Cholesterol is the most familiar sterol, but others, such as vitamin D and the sex hormones (for example, testosterone), are important, too.

Sterols in Foods Both plant and animal-derived foods contain sterols, but only animal-derived foods contain significant amounts of cholesterol: meats, eggs, fish, poultry, and dairy products. Organ meats, such as liver and kidneys, and eggs are richest in cholesterol; cheeses and meats have less. Shellfish contain many sterols, but much less cholesterol than was previously thought.

Cholesterol Synthesis Like the lecithins, cholesterol can be made by the body, so it is not an essential nutrient. Your liver is manufacturing it now, as you read. The raw materials that the liver uses to make cholesterol can all be taken from glucose or fatty acids. In other words, cholesterol can be made from either carbohydrate or fat. Most of the body's cholesterol ends up in the membranes of cells, where it performs vital structural and metabolic functions.

Cholesterol's Two Routes in the Body After being made, cholesterol leaves the liver by two routes:

1. It may be made into **bile** (a compound that prepares fat for digestion), stored in the gallbladder, and delivered to the intestine.
2. It may travel, via the bloodstream, to all the body's cells.

The bile that is made from cholesterol in the liver is released into the intestine to aid in the digestion and absorption of fat. After bile does its job, most of it is reabsorbed into the body and recycled; the rest is excreted in the feces.

Cholesterol Excreted While bile is in the intestine, some of it may be trapped by soluble fibers or by some medications, which carry it out of the body in feces. The excretion of bile reduces the total amount of cholesterol remaining in the body.

Cholesterol Transport Some cholesterol, packaged with other lipids and protein, leaves the liver via the arteries and is transported to the body tissues by the blood. ■ These packages of lipids and proteins are called **lipoproteins**. As the lipoproteins travel through the body, tissues can extract lipids from them. Cholesterol's harmful effects in the body occur when it forms deposits in the artery walls. These deposits contribute to **atherosclerosis**, a disease that can cause heart attacks and strokes.

IN SUMMARY

- Table 3–2 summarizes the members of the lipid family.

- The predominant lipids both in foods and in the body are triglycerides, which have glycerol backbones with three fatty acids attached.

- Fatty acids vary in the length of their carbon chains and their degree of saturation. Those that are fully loaded with hydrogens are saturated; those that are missing hydrogens and therefore have double bonds are unsaturated (monounsaturated or polyunsaturated).

■ Both the intestine and the liver make lipoproteins. Chapter 5 tells the story of lipid transport.

omega-6 fatty acid: a polyunsaturated fatty acid with its endmost double bond six carbons back from the end of its carbon chain; long recognized as important in nutrition. Linoleic acid is an example.

omega-3 fatty acids: polyunsaturated fatty acids in which the endmost double bond is three carbons back from the end of the carbon chain; relatively newly recognized as important in nutrition. Linolenic acid is an example.

EPA, DHA: omega-3 fatty acids made from linolenic acid. The full name for EPA is *eicosapentaenoic* (EYE-cosa-PENTA-ee-NO-ick) *acid*. The full name for DHA is *docosahexaenoic* (DOE-cosa-HEXA-ee-NO-ick) *acid*.

phospholipids: one of the main classes of lipids; compounds similar to triglycerides but have choline (or another compound) and a phosphorus-containing acid in place of one of the fatty acids.

lecithins: one type of phospholipid.

choline (KOH-leen): a nitrogen-containing compound found in foods and made in the body from an amino acid.

emulsifiers: substances that mix with both fat and water and that disperse the fat in the water, forming an emulsion.

sterols: one of the main classes of lipids; includes cholesterol, vitamin D, and the sex hormones (such as testosterone).

bile: a compound that prepares fats and oils for digestion; made by the liver, stored in the gallbladder, and released into the small intestine.

lipoproteins: clusters of lipids associated with proteins that serve as transport vehicles for lipids in the lymph and blood.

atherosclerosis (ath-er-oh-scler-OH-sis): a type of artery disease characterized by accumulations of lipid-containing material on the inner walls of the arteries (see Chapter 22).

- Most triglycerides contain more than one type of fatty acid.

- Fatty acid saturation affects the physical characteristics and storage properties of fats.

- Hydrogenation, which makes polyunsaturated fats more saturated, gives rise to *trans*-fatty acids, altered fatty acids that may have health effects similar to those of saturated fatty acids.

- Linoleic acid and linolenic acid are essential nutrients. In addition to serving as structural parts of cell membranes, they make powerful substances that help regulate blood pressure, blood clot formation, and the immune response.

- Phospholipids, including the lecithins, have a unique chemical structure that allows them to be soluble in both water and fat.

- In the body, phospholipids are major constituents of cell membranes; the food industry uses phospholipids as emulsifiers.

- Sterols include cholesterol, bile, vitamin D, and the sex hormones.

- Only animal-derived foods contain significant amounts of cholesterol.

Health Effects and Recommended Intakes of Fats

Of all the dietary factors related to chronic diseases prevalent in developed countries, high intakes of certain fats are by far the most significant. The person who chooses a diet too high in saturated fats or *trans* fats invites the risk of **cardiovascular disease (CVD)**, and heart disease is the number one killer of adults in the United States and Canada. As for cancer, evidence is less compelling than for heart disease, but it does suggest that a diet high in certain kinds of fat is associated with a greater-than-average risk of developing some types of cancer.[7] ■ Conversely, some research suggests that omega-3 fatty acids from fish may protect against some cancers.[8] Obesity carries serious risks to health, and the high energy density of fatty foods makes it easy for people to exceed their energy needs and so gain unneeded weight. ■ The links between diet and disease are the focus of much research. Some points about fats and heart health are presented here because they underlie dietary recommendations concerning fats. ■

Fats and Heart Health

As noted earlier and described in detail in Chapter 5, cholesterol travels in the blood within lipoproteins. Two of the lipoproteins, **low-density lipoproteins (LDL)** and **high-density lipoproteins (HDL)**, play major roles with regard to heart health and are the focus of most recommendations made for reducing the risk of heart disease. A high blood LDL cholesterol concentration is a predictor of the likelihood of suffering a fatal heart attack or stroke, and the higher the LDL, the earlier the episode is expected to occur. Conversely, high HDL cholesterol signifies a *lower* disease risk.

Most people realize that elevated blood cholesterol is an important risk factor for heart disease. Most people may not realize, though, that cholesterol in *food* is not the main influential factor in raising *blood* cholesterol.

Saturated Fats and Blood Cholesterol The main dietary factors associated with elevated blood LDL cholesterol are high saturated fat and high *trans* fat intakes.* LDL

***

* It should be noted that not all saturated fatty acids have the same cholesterol-raising effect. Stearic acid, an 18-carbon fatty acid, does not seem to raise blood cholesterol.

■ Nutrition and cancer is a topic of Chapter 25.

■ The health risks of obesity are described in Chapter 6, and Chapter 7 focuses on weight management.

■ Nutrition and heart disease is the topic of Chapter 22.

cholesterol indicates a risk of heart disease because high LDL concentrations promote the uptake of cholesterol in the blood vessel walls.

Fats from animal sources are the main contributors of saturated fats in most people's diets. ■ Some vegetable fats (coconut oil, palm kernel oil, and palm oil) and hydrogenated fats such as shortening or stick margarine provide smaller amounts of saturated fats. To minimize intake of saturated fat, most people need to eat less meat. When eating meat, choose the leanest cuts and trim away the visible fat.[9] Selecting fat-free milk and using nonhydrogenated margarine and unsaturated cooking oil such as olive oil, safflower oil, or canola oil are other simple changes that can dramatically lower saturated fat intake and heart disease risk.[10] ■

Trans-Fatty Acids and Blood Cholesterol Consuming fats with *trans*-fatty acids poses a risk to the health of the heart and arteries by raising LDL and lowering HDL cholesterol and by producing inflammation.[11] When news of *trans*-fatty acids' effects on heart health was first emerging, some people hastily switched from using margarine back to butter, believing oversimplified reports that margarine provided no heart health advantage over butter. It is true that most margarines and virtually all shortenings are made from mostly hydrogenated fats and therefore contain substantial *trans*-fatty acids—up to 40 percent by weight. ■ Some margarines, however, especially the soft or liquid varieties, are made from unhydrogenated oils. These have long proved to be less likely to elevate blood cholesterol than the saturated fats of butter. When oils (but not hydrogenated oils) are the first ingredient listed on a margarine label, that margarine is probably low in both *trans*-fatty acids and saturated fat. In general, margarine contributes less *trans* fat to the average diet than do other contributors such as processed foods.

In addition to soft and liquid margarine choices, some margarines are now specially formulated to contain few or no *trans*-fatty acids. Other types contain **sterol esters**, a functional food ingredient that reduces blood cholesterol when consumed in addition to a low-saturated fat diet.* Sterol esters are not recognized by the intestine and therefore are not absorbed, and they also block the absorption of cholesterol. Simply adding the margarine to a high-fat diet is unlikely to bring benefits, however. Sterol esters work only when people cut their fat intakes as well. Drawbacks include the price (three or four times higher than regular margarine), a high fat content (the full-fat kind equals the fat in regular margarine), and an unproven record of safety for use by certain populations, such as growing children.

Many foods other than margarine also contribute *trans*-fatty acids. Fast foods, chips, baked goods, and other commercially prepared foods are high in fats, containing up to 50 percent *trans*-fatty acids. Fast-food chains fry foods in hydrogenated vegetable oil that contains abundant *trans*-fatty acids, although healthier commercial fats are now being developed. Overall, consumers are now eating more fats containing *trans*-fatty acids than ever before, amounting to about 3 percent of daily kcalories, and they are eating these *trans*-fatty acids in the form of processed foods (see the margin list).[12] ■

All food labels must list grams of *trans* fat in foods to help consumers make informed choices.[13] Reducing total fat and replacing both saturated and *trans* fats with monounsaturated and polyunsaturated fats may be the wisest strategies for preventing heart disease. To this end, many manufacturers are reformulating foods to reduce their contents of harmful *trans* fats. At the same time, as a later section describes, food scientists are perfecting fat replacers intended to eliminate added fats altogether.

Dietary Cholesterol and Blood Cholesterol Dietary cholesterol has also been implicated in raising blood cholesterol and increasing the risk of heart disease, although its effect is not as strong as that of saturated fat or *trans* fat. Still, health experts advise limiting cholesterol intake.

---

* Two brand names of margarines with sterol esters currently on the market are *Benecol* and *Take Control*.

■ Major sources of saturated fats:
  - Whole milk (and even reduced-fat milk), cheese, butter, cream, cream cheese, sour cream, and ice cream
  - Fatty cuts of beef and pork and processed meats such as bacon, sausage, and hot dogs
  - Tropical oils (coconut, palm, and palm kernel) and products that contain them such as cakes, cookies, doughnuts, and pastries
  - Shortening and lard

■ Nutrition in Practice 3 examines various types of fats and their roles in supporting or harming heart health.

■ The words *hydrogenated vegetable oil* or *shortening* in an ingredients list indicate *trans*-fatty acids in the product.

■ Major food sources of *trans* fats:
  - Cakes, cookies, pies, doughnuts, and crackers
  - Meat and dairy products
  - Hard margarine
  - Fried potatoes and other commercial fried foods such as chicken
  - Potato chips and corn chips
  - Shortening

cardiovascular disease (CVD): a general term for all diseases of the heart and blood vessels (see Chapter 22).

low-density lipoproteins (LDL): the type of lipoproteins that carry cholesterol and triglycerides from the liver to the cells of the body and are composed primarily of cholesterol.

high-density lipoproteins (HDL): the type of lipoproteins that transport cholesterol back to the liver from peripheral cells; composed primarily of protein.

sterol esters: compounds, derived from plants, that belong to the sterol family of lipids and have been shown experimentally to reduce blood cholesterol when consumed in place of other fats in a low-saturated fat diet.

- Major sources of cholesterol:
  - Eggs
  - Meat, poultry, and shellfish
  - Cheese
  - Milk

Recall that cholesterol is found primarily in foods derived from animals. Consequently, eating less fat from meat, eggs, and milk products helps lower dietary cholesterol intake (as well as total and saturated fat intakes). ■

Monounsaturated Fatty Acids and Blood Cholesterol Replacing saturated and *trans* fats with monounsaturated fat such as olive oil may be an effective dietary strategy to prevent heart disease. The lower rates of heart disease among people in the Mediterranean region of the world are often attributed to their liberal use of olive oil, a rich source of monounsaturated fatty acids.[14] Olive oil also delivers valuable phytochemicals that help to protect against heart disease. Nutrition in Practice 3 examines the role of olive oil and other fats in supporting or harming heart health.

Polyunsaturated Fatty Acids, Blood Cholesterol, and Heart Disease Risk Polyunsaturated fatty acids (PUFA) of the omega-6 and omega-3 families are potent protectors against heart disease. The primary omega-6 fatty acid, linoleic acid, which is found in vegetable oils such as corn and safflower oils, exerts most of its beneficial effect by lowering both total blood cholesterol and LDL cholesterol.[15] In fact, linoleic acid is the only dietary fatty acid that effectively lowers LDL cholesterol.[16] Possibly, linoleic acid works by slowing production of LDL and by enhancing LDL clearance.

The omega-3 fatty acids, EPA and DHA, which are found mainly in fatty fish, exert their beneficial effects by influencing the function of both the heart and the blood vessels. Specifically, EPA and DHA protect heart health by:[17]

- Lowering blood triglycerides

- Preventing blood clots

- Protecting against irregular heartbeats

- Lowering blood pressure

- Defending against inflammation

The primary member of the omega-3 family, linolenic acid, may benefit heart health as well, but evidence for this effect is much less certain than for EPA and DHA.[18]

Balance Omega-6 and Omega-3 Intakes The U.S. diet is high in omega-6 fatty acids (due to the increased production and use of vegetable cooking oils such as corn and cottonseed oils) and low in omega-3 fatty acids. Experts recommend a more balanced intake. The best way for people to increase their intakes of omega-3 fatty acids is to follow the advice of the American Heart Association: eat at least two servings of fatty fish each week (see Table 3–3).[19] Even one fish meal per week has been associated with a reduced risk of heart disease.[20]

Greater heart health benefits can be expected when fish is grilled, baked, or broiled, partly because the varieties prepared this way often contain more EPA and DHA than species used for fried fish in fast-food restaurants and frozen products. Additionally, benefits are attained by avoiding commercial frying fats, which are often laden with *trans* fat and saturated fat.[21] Further benefits arise when fish replaces high-fat meat or other foods rich in saturated fats in several meals each week. Although not every study supports a benefit from eating fish and fish oil, results from many population studies and controlled clinical trials support a recommendation to eat fish.[22]

Some species of fish and shellfish, however, may contain significant levels of mercury or other environmental contaminants.[23] Most healthy people can safely consume most species of ocean fish several times per week, but some people face greater risks. Women who may become pregnant, pregnant and lactating women, and children are more sensitive to contaminants than others, but even they can benefit from safer fish varieties within recommended limits (see Chapter 11 for details).

Grilling or broiling fish, instead of frying them, preserves their beneficial omega-3 fatty acids while adding little or no saturated fat.

TABLE 3–3

Food Sources of Omega-6 and Omega-3 Fatty Acids

Omega-6	
Linoleic acid	Seeds, nuts, vegetable oils (corn, cottonseed, safflower, sesame, soybean, sunflower), poultry fat
Omega-3	
Linolenic acid[a]	Oils (canola, flaxseed, soybean, walnut, wheat germ; liquid or soft margarine made from canola or soybean oil)
	Nuts and seeds (butternuts, flaxseeds, walnuts, soybean kernels) Vegetables (soybeans)
EPA and DHA	Human milk
	Fatty coldwater fish[b] (mackerel, salmon, bluefish, mullet, sablefish, menhaden, anchovy, herring, lake trout, sardines, tuna)

[a] Alpha-linolenic acid. Also found in the seed oil of the herb *evening primrose*.
[b] All of these fish except tuna provide at least 1 gram of omega-3 fatty acids in 100 grams of fish (3.5 ounces); the fish oil content of each species varies with the season and site of harvest. Tuna provides fewer omega-3 fatty acids, but because it is commonly consumed, its contribution can be significant.

For everyone, consuming a variety of different types of fish is a good idea to minimize exposure to any single toxin that may accumulate in a favored species. The margin lists the species most heavily contaminated with mercury and those that are low in mercury. ■ As for freshwater fish in some areas, consumers should check local advisories about the safety of fish caught by family and friends.

Omega-3 Supplements Omega-3 fatty acid supplements are aggressively marketed as a cure-all for many different diseases without regard for consumer safety. The Food and Drug Administration (FDA) does not permit labels to claim that fish oil supplements can prevent or cure diseases, but it does allow the claim that research is suggestive but inconclusive regarding heart disease. Experts agree that adding *fish* to the diet two or three times per week may help to prevent disease, but the idea that fish oil *supplements* are beneficial and safe in any amount is erroneous.

Most importantly, high intakes of omega-3 fatty acids may increase bleeding time, raise LDL cholesterol, and suppress immune function.[24] Fish oil supplements themselves may carry hazards. They may contain high levels of the two most potentially toxic vitamins, A and D. Fish oil supplements are made from fish skins and livers that may have accumulated toxic concentrations of pesticides, heavy metals such as mercury, and other industrial contaminants. Unless the oils are refined to eliminate them, such contaminants can become further concentrated in the pills.[25] Furthermore, omega-3 and omega-6 fatty acids compete for the same slots in the body. Consequently, taking supplements of one can interfere with normal body functions that depend on a proper balance between the two. Finally, supplements may not contain the quantities of omega-3 fatty acids that the labels claim.

Recommendations

Some fat in the diet is essential for good health, but too much fat, especially saturated fat and *trans* fat, increases the risks of chronic diseases. Defining the exact level

■ ■ Fish most heavily contaminated with mercury: shark, swordfish, king mackerel, and tilefish (also called golden bass or golden snapper).

■ Fish or shellfish low in mercury: shrimp, canned light tuna, salmon, pollock, and catfish. Canned albacore ("white") tuna contains more mercury than light tuna.

Dietary Guidelines for Americans 2005

■ Keep total fat intake between 20 and 35 percent of kcalories, with most fats coming from sources of polyunsaturated and monounsaturated fatty acids, such as fish, nuts, and vegetable oils.

- DRI for fat:
  - 20 to 35% of energy intake.

- Daily Value:
  - 65 g fat (based on 30% of 2,000 kcal diet)

- Linoleic acid AI:
  - 5 to 10% of energy intake

  Men:
  - 19–50 yr: 17 g/day
  - 51+ yr: 14 g/day

  Women:
  - 19–50 yr: 12 g/day
  - 51+ yr: 11 g/day

  Linolenic acid AI:
  - 0.6 to 1.2% of energy intake

  Men: 1.6 g/day

  Women: 1.1 g/day

- Maximum saturated fat intakes set by the *Dietary Guidelines for Americans 2005*:

  1,600 kcal diet: 18 g

  2,000 kcal diet: 20 g

  2,200 kcal diet: 24 g

  2,500 kcal diet: 25 g

  2,800 kcal diet: 31 g

- Daily Values:
  - 20 g saturated fat (based on 10% of 2,000 kcal diet)
  - 300 mg cholesterol

of total fat intake at which risk of inadequacy or prevention of disease occurs is not possible; for this reason, no RDA or upper limit has been set.[26] Instead, the recommendation for total fat is set at 20 to 35 percent of the daily energy intake. ■ The top end of this range is slightly higher than previous recommendations. Diets with up to 35 percent of kcalories from fat can be compatible with good health if energy intake is reasonable and saturated fat intake is low. When total fat intake exceeds 35 percent of kcalories, saturated fat intakes increase to unhealthy levels. Fat and oil intakes below 20 percent of kcalories increase the risk of inadequate essential fatty acid intakes. The FDA established Daily Values on food labels using 30 percent of energy intake as the guideline for fat. ■

Part of the allowance for total fat should provide for the essential fatty acids— linoleic acid and linolenic acid. Recommendations suggest that linoleic acid provide 5 to 10 percent of the daily energy intake and linolenic acid, 0.6 to 1.2 percent. ■

Saturated fats, *trans* fats, and cholesterol increase total blood cholesterol and LDL cholesterol and therefore the risk of heart disease. Even low intakes of each may elevate heart disease risk.[27] Thus recommendations urge people to eat diets that are low in saturated fat, *trans* fat, and cholesterol. Specifically, consume less than 10 percent of kcalories from saturated fat, keep *trans* fat intakes as low as possible, and consume less than 300 milligrams of cholesterol each day.[28] ■ To help consumers meet these goals, the FDA established Daily Values on food labels using 10 percent of energy intake for saturated fat; the Daily Value for cholesterol is 300 milligrams, regardless of energy intake. ■ There is no Daily Value for *trans* fat.

Nutrition in Practice 3 suggests substituting monounsaturated or polyunsaturated fats for harmful saturated fats. No beneficial change in blood lipids occurs when monounsaturated or polyunsaturated fat is *added* to a diet rich in saturated or *trans* fat, however. The best diet for heart health is also rich in fruits, vegetables, nuts, and whole grains that offer many health advantages by supplying abundant nutrients, fiber, and phytochemicals.

IN SUMMARY

- High intakes of saturated or *trans* fats contribute to heart disease, obesity, and other health problems.

- High blood cholesterol, specifically, poses a risk of heart disease, and high intakes of saturated fat contribute most to high blood cholesterol. High intakes of *trans*-fatty acids also appear to raise blood cholesterol. Cholesterol in foods presents less of a risk.

- Polyunsaturated fatty acids of the omega-6 and omega-3 families protect against heart disease.

- When monounsaturated fat such as olive oil replaces saturated and *trans* fats in the diet, the risk of heart disease may be lessened.

- Though some fat in the diet is necessary, health authorities recommend a diet moderate in total fat and low in saturated fat, *trans* fat, and cholesterol.

Fats in Foods

TABLE 3–4
The Functions of Fats in Foods

Fats in foods:
- Contribute flavor and aroma
- Influence the texture, adding creaminess, smoothness, moistness, or crispness
- Help make foods tender
- Carry fat-soluble vitamins

Fats are important in foods as well as in the body. Many of the compounds that give foods their flavors and aromas are found in fats and oils. The delicious aromas associated with bacon, ham, and other meats, as well as with onions being sautéed, come from fats. Fats also influence the texture of many foods, enhancing smoothness, creaminess, moistness, or crispness. In addition, four vitamins—A, D, E, and K—are soluble in fat. When the fat is removed from a food, many fat-soluble compounds, including these vitamins, are also removed. Table 3–4 summarizes the roles of fats in foods.

Fats are also an important part of most people's ethnic or national cuisines. Each culture has its own favorite food sources of fats and oils. In Canada, canola oil (also known as rapeseed oil) is widely used. In the Mediterranean area, Greeks, Italians, and Spaniards rely heavily on olive oil. Both canola oil and olive oil are rich sources of monounsaturated fatty acids. Asians use the polyunsaturated oil of soybeans. Jewish people traditionally employ chicken fat. Everywhere in North America, butter and margarine are widely used.

Finding the Fats in Foods

The remainder of this chapter and the Nutrition in Practice show you how to choose fats wisely with the goals of providing optimal health and pleasure in eating. To achieve such goals, you need to know which foods offer unsaturated fats that provide the essential fatty acids and which offer harmful saturated and *trans* fats. Perhaps most important for many people is learning to control portion sizes, particularly portions of fatty foods that can pack hundreds of kcalories into just a few bites.

The *Dietary Guidelines for Americans 2005* urge that, beyond a healthy minimum, people should limit their fat intakes in order to limit intakes of saturated fat, *trans* fat, and kcalories. To do this requires consistently making nutrient-dense choices, such as fat-free milk, fat-free cheese, and the leanest meats, and refraining from adding solid fats, such as butter, hard margarine, or shortening, to foods during preparation or at the table. Higher-fat foods may be included in the diet, but the fat kcalories they provide must fit within a person's discretionary kcalorie allowance. Exceptions are the fats of fatty fish, nuts, and vegetable oils: they provide EPA and DHA, linoleic acid and linolenic acid, and vitamin E needed for a healthful diet. ■ Within kcalorie limits, the kcalories these fats provide are necessary, not discretionary. Many people, especially sedentary people, have few or no discretionary kcalories to spend on high-fat foods. ■

■ Table 1–6 of Chapter 1 specifies how much oil is required to meet nutrient needs at several kcalorie levels.

■ The concept of discretionary kcalories was discussed in Chapter 1.

Added Fats in Foods A dollop of dessert topping, a spread of butter on bread, oil or shortening in a recipe, dressing on a salad—all of these are examples of *added* fats. Indeed, all sorts of fats can be added to foods during commercial or home preparation or at the table. The following amounts of these fats contain about 5 grams of pure fat, providing 45 kcalories and negligible protein and carbohydrate:

- 1 teaspoon of oil or shortening

- 1½ teaspoons of mayonnaise, butter, or margarine

- 1 tablespoon of regular salad dressing, cream cheese, or heavy cream

- 1½ tablespoons of sour cream

The majority of added fats in the diet are invisible. They are the hidden fats of fried foods and baked goods, sauces and mixed dishes, and dips and spreads. Other invisible fats include the fats in the marbling of meat and the fat ground into lunch meats and hamburger.

- 1 c whole milk:
  - 8 g fat
  - 5 g saturated fat
  - 24 mg cholesterol
- 1 c reduced-fat milk:
  - 5 g fat
  - 2 g saturated fat
  - 20 mg cholesterol
- 1 c low-fat milk:
  - 2 g fat
  - 1.5 g saturated fat
  - 10 mg cholesterol
- 1 c fat-free milk:
  - 0 g fat
  - 0 g saturated fat
  - 5 mg cholesterol

Milk, Yogurt, and Cheese The fat in whole milk is about 63 percent saturated fat; the cholesterol content is 24 milligrams per cup for whole milk or 5 milligrams for fat-free milk. ■ Thus choosing fat-free in place of whole milk reduces your intake of cholesterol as well as of saturated fat.

Note that cream and butter do not appear in the milk group. Milk and yogurt are rich in calcium and protein, but cream and butter are not. Cream and butter are fats, as are whipped cream, sour cream, and cream cheese, so they are grouped together with the solid fats. That is why the food group that includes milk is carefully labeled the "milk, yogurt, and cheese group," not the "dairy group." Cheeses are major contributors of saturated fat in people's diets.

Meat, Poultry, Fish, Legumes, Eggs, and Nuts Meats conceal a good deal of the fat—and much of the saturated fat—that people consume. To help "see" the fat in meats, it is useful to think of them in four categories according to their fat contents: very lean, lean, medium-fat, and high-fat meats, as the exchange lists do in Appendix C. Meats in all four categories contain about equal amounts of protein, but their fat contents differ and their saturated fat and kcalorie amounts vary significantly. Table 1–10 on pages 27 and 28 in Chapter 1 provides some definitions concerning the fat contents of meats.

The 2005 USDA Food Guide suggests that most adults limit a day's intake of meats or equivalents to about 5 to 7 ounces. For comparison, the smallest fast-food hamburger weighs about 3 ounces. A steak served in a restaurant often runs 8, 12, or 16 ounces, more than a whole day's meat allowance. You may have to weigh a serving or two of meat to see how much you are eating.

People think of meat as protein food, but calculation of its nutrient content reveals a surprising fact. A big (3-ounce), fast-food hamburger sandwich contains 23 grams of protein and 20 grams of fat. Because protein offers 4 kcalories per gram and fat offers 9, the sandwich provides 92 kcalories from protein and about twice that amount from fat. The kcalorie total, counting carbohydrates from the bun and condiments, is over 400 kcalories, with more than 50 percent of them from fat. Hot dogs, fried chicken sandwiches, and fried fish sandwiches are also high-fat choices. Because so much of the energy in a meat eater's diet is hidden from view, people can easily overeat on high-fat food, making weight control difficult.

When choosing beef or pork, look for lean cuts named *loin* or *round* from which the fat can be trimmed. Eat small portions, too. As for chicken and turkey, these meats are naturally lean, but commercial processing and frying add fats, especially in "patties," "nuggets," "fingers," or "wings." Chicken wings are mostly skin, and a chicken stores most of its fat just under its skin. The tastiest wing snacks have also been fried in cooking fat (often a hydrogenated, saturated type with *trans*-fatty acids), smothered with a buttery, spicy sauce, and then dipped in blue cheese dressing, making wings an extraordinarily high-fat snack. People who snack on wings may want to plan on eating low-fat foods at several other meals to balance them out.

Vegetables, Fruits, and Grains Choosing vegetables, fruits, whole grains, and legumes also helps lower the saturated fat, cholesterol, and total fat content of the diet. Most vegetables and fruits naturally contain little or no fat. Avocados and olives are exceptions, but most of their fat is unsaturated, which is not harmful to heart health. Most grains contain only small amounts of fat. Some grain *products* such as fried taco shells, croissants, and biscuits are high in saturated fat, so consumers need to read food labels. Similarly, many people add butter, margarine, or cheese sauce to grains and vegetables, which raises their saturated and *trans* fat contents. Because fruits are often eaten without added fat, a diet that includes several servings of fruit daily can help a person meet the dietary recommendations for fat.

A diet rich in vegetables, fruits, whole grains, and legumes offers abundant vitamin C, folate, vitamin A, vitamin E, and dietary fiber—all important in supporting health. Consequently, such a diet protects against disease by reducing saturated fat, cholesterol, and total fat and increasing nutrients. It also provides valuable phytochemicals that help defend against heart disease.

Cutting Fat Intake and Choosing Unsaturated Fats

Knowing which foods contain the most fat is the first step toward meeting the recommendation to limit dietary fat in general and saturated fat in particular. As a general rule, a person who eats meat and wishes to reduce both saturated fat and cholesterol intake can eat fewer high-fat meats and dairy foods, fewer eggs, and more poultry (without the skin), fish, and fat-free dairy products. A vegetarian who eats dairy products and eggs can shift to fat-free milk and fat-free cheeses and limit butter and egg intake. Vegetarians who omit animal-derived foods generally eat less saturated fat and little or no cholesterol because plant foods do not contain significant amounts of cholesterol. The "How To" (p. 78) offers strategies for making heart-healthy choices, food group by food group.

Fats and kcalories Removing fat from food also removes energy as Figure 3–5 shows. A pork chop with the fat trimmed to within a half-inch of the lean provides 340 kcalories; with the fat trimmed off completely, it supplies 230 kcalories. A baked potato with butter and sour cream (1 tablespoon each) has 350 kcalories; a plain baked potato has 220 kcalories. The single most effective step you can take to reduce the energy value of a food is to eat it with less fat.

Choosing Unsaturated Fats When a person does eat fats, those to choose are the unsaturated ones. Remember, the softer a fat is, the more unsaturated it is. Generally

FIGURE 3–5

Cutting Fat Cuts kCalories—and Saturated Fat

Pork chop with fat (340 kcal, 19 g fat, 7 g saturated fat)

Potato with 1 tbs butter and 1 tbs sour cream (350 kcal, 14 g fat, 10 g saturated fat)

Whole milk, 1 c (150 kcal, 8 g fat, 5 g saturated fat)

Pork chop with fat trimmed off (230 kcal, 9 g fat, 3 g saturated fat)

Plain potato (220 kcal, <1 g fat, 0 g saturated fat)

Fat-free milk, 1 c (90 kcal, <1 g fat, <1 g saturated fat)

MAKE HEART-HEALTHY CHOICES—BY FOOD GROUP

Breads and Cereals

- Select breads, cereals, and crackers that are low in saturated and *trans* fat (for example, bagels instead of croissants).

- Prepare pasta with a tomato sauce instead of a cheese or cream sauce.

- Use fruit butters or jellies instead of butter or margarine on bread.

Vegetables and Fruits

- Enjoy the natural flavor of steamed vegetables (without butter) for dinner and fruits for dessert.

- Eat at least two vegetables (in addition to a salad) with dinner.

- Snack on raw vegetables or fruits instead of high-fat items like potato chips.

- Buy frozen vegetables without sauce.

Milk and Milk Products

- Switch from whole milk to reduced-fat, from reduced-fat to low-fat, and from low-fat to fat-free (nonfat) milk.

- Use fat-free and low-fat cheeses (such as part-skim ricotta and low-fat mozzarella) instead of regular cheeses.

- Use fat-free or low-fat yogurt or sour cream instead of regular sour cream.

- Use evaporated fat-free milk instead of cream.

- Enjoy fat-free frozen yogurt, sherbet, or ice milk instead of ice cream.

Meat and Legumes

- Fat adds up quickly, even with lean meat; limit intake to about 6 ounces (cooked weight) daily.

- Eat at least two servings of fish per week (particularly fish such as mackerel, lake trout, herring, sardines, and salmon).

- Choose fish, poultry, or lean cuts of pork or beef; look for unmarbled cuts named *round* or *loin* (eye of round, top round, bottom round, round tip, tenderloin, sirloin, center loin, and top loin).

- When choosing processed meats such as lunch meats and hot dogs, choose those that are low in saturated fat and cholesterol.

- Trim the fat from pork and beef; remove the skin from poultry.

- Grill, roast, broil, bake, stir-fry, stew, or braise meats; do not fry. When possible, place food on a rack so that fat can drain.

- Use lean ground turkey or lean ground beef in recipes; brown ground meats without added fat and then drain off fat.

- Select tuna, sardines, and other canned meats packed in water; rinse oil-packed items with hot water to remove much of the fat.

- Fill kabob skewers with lots of vegetables and slivers of meat; create main dishes and casseroles by combining a little meat, fish, or poultry with a lot of pasta, rice, or vegetables.

- Use legumes often.

- Eat a meatless meal or two daily.

- Use egg substitutes in recipes instead of whole eggs or use two egg whites in place of each whole egg.

Fats and Oils

- Use butter or stick margarine sparingly; select soft margarines instead of hard margarines.

- Use fruit butters, reduced-kcalorie margarines, or butter replacers instead of butter.

- Use low-fat or fat-free mayonnaise and salad dressing instead of regular.

- Limit use of lard and meat fat.

- Limit use of products made with coconut oil, palm kernel oil, and palm oil (read labels on bakery goods, processed foods, popcorn oils, and nondairy creamers).

- Reduce use of hydrogenated shortenings and stick margarines and products that contain them (read labels on crackers, cookies, and other commercially prepared baked goods); use vegetable oils instead.

Miscellaneous

- Use a nonstick pan or coat the pan lightly with vegetable oil.

- Refrigerate soups and stews; when the fat solidifies, remove it.

- Use wine; lemon, orange, or tomato juice; herbs; spices; fruits; or broth instead of butter or margarine when cooking.

- Stir-fry in a small amount of oil; add moisture and flavor with broth, tomato juice, or wine.

- Use variety to enhance enjoyment of the meal: vary colors, textures, and temperatures—hot cooked versus cool raw foods—and use garnishes to complement food.

- Omit high-fat meat gravies and cheese sauces.

Source: Adapted from Expert Panel on Detection, Evaluation, and Treatment of High Blood Cholesterol in Adults (Adult Treatment Panel III), *Third Report of the National Cholesterol Education Program (NCEP)*, NIH publication no. 02-5215 (Bethesda, Md.: National Heart, Lung, and Blood Institute, 2002), pp. V-25–V-27.

speaking, vegetable and fish oils are rich in polyunsaturates, olive oil and canola oil are rich in monounsaturates, and the harder fats—animal fats—are more saturated (see Figure 3–6).

Do Not Overdo Fat Restriction Some people actually manage to eat too *little* fat—to their detriment. Among them are young women and men with eating disorders, described in Nutrition in Practice 6. As a practical guideline, it is wise to include the equivalent of at least a teaspoon of fat in every meal.

Cautions People who wish to make choices consistent with current recommendations need to learn how to read food labels, limit fat in general, and seek out the polyunsaturated and monounsaturated fats in preference to the saturated ones. Remember, however, vegetable fat or vegetable oil does not always mean unsaturated fat. Both coconut oil and palm oil, for example, which are often used in nondairy creamers, are saturated fats, and both raise blood cholesterol.

Fat Replacers Today, consumers can choose from thousands of fat-reduced products. Many bakery goods, lunch meats, cheeses, spreads, frozen desserts, and other products made with **fat replacers** offer less than half a gram of fat, saturated fat, and *trans* fat in a serving. Some of these products contain **artificial fats**, and others use conventional ingredients in unconventional ways to reduce fats and kcalories. Among the latter, manufacturers can:

- Add water or whip air into foods.

- Add fat-free milk to creamy foods.

- Use lean meats and soy protein to replace high-fat meats.

- Bake foods instead of frying them.

At room temperature, unsaturated fats (such as those found in oil) are usually liquid, whereas saturated fats (such as those found in butter) are solid.

© Polara Studios, Inc.

Key:

- ■ Saturated fatty acids
- ■ Monounsaturated fatty acids
- ■ Polyunsaturated, omega-6 fatty acids
- ■ Polyunsaturated, omega-3 fatty acids

Animal fats and the tropical oils of coconut and palm contain mostly saturated fatty acids.

| Coconut oil |
| Butter |
| Beef tallow (beef fat) |
| Palm oil |
| Lard (pork fat) |

Some vegetable oils, such as olive and canola, are rich in monounsaturated fatty acids.

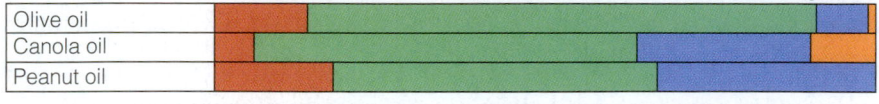

| Olive oil |
| Canola oil |
| Peanut oil |

Many vegetable oils are rich in polyunsaturated fatty acids.

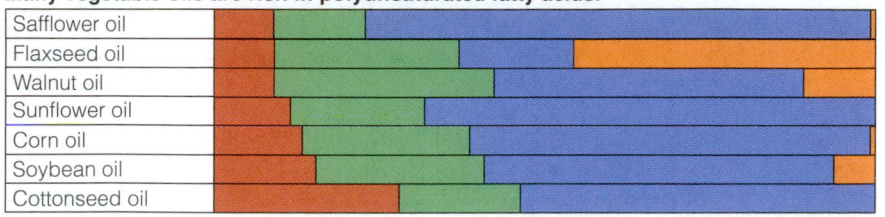

| Safflower oil |
| Flaxseed oil |
| Walnut oil |
| Sunflower oil |
| Corn oil |
| Soybean oil |
| Cottonseed oil |

FIGURE 3–6

Comparison of Dietary Fats
Most fats are a mixture of saturated, monounsaturated, and polyunsaturated fatty acids.

fat replacers: ingredients that replace some or all of the functions of fat in foods and may or may not provide energy.

artificial fats: zero-energy fat replacers that are chemically synthesized to mimic the sensory and cooking qualities of naturally occurring fats but are totally or partially resistant to digestion.

Common food ingredients such as fibers, sugars, or proteins can also take the place of fats in some foods. In particular, fat replacers made from oats or barley not only cut down on fats in foods, but also introduce beneficial viscous fibers, while imparting desirable tastes and textures associated with real fats.[29] Products made from sugars or proteins still provide kcalories, but far fewer kcalories from fats. Manufactured fat replacers consist of chemical derivatives of carbohydrate, protein, or fat, or modified versions of foods rich in those constituents.

A familiar example of an artificial fat that has been approved for use in snack foods such as potato chips, crackers, and tortilla chips is **olestra**. Olestra's chemical structure is similar to that of a regular fat (a triglyceride) but with important differences. A triglyceride is composed of a glycerol molecule with three fatty acids attached, whereas olestra is made of a sucrose molecule with six to eight fatty acids attached. Enzymes in the digestive tract cannot break the bonds of olestra, so unlike sucrose or fatty acids, olestra passes through the system unabsorbed.

The FDA's evaluation of olestra's safety addressed two questions. First, is olestra toxic? Research on both animals and humans supports the safety of olestra as a partial replacement for dietary fats and oils, with no reports of cancer or birth defects. Second, does olestra affect either nutrient absorption or the health of the digestive tract? When olestra passes through the digestive tract unabsorbed, it binds with the fat-soluble vitamins A, D, E, and K and carries them out of the body, robbing the person of these valuable nutrients. To compensate for these losses, the FDA requires the manufacturer to fortify olestra with vitamins A, D, E, and K. Saturating olestra with these vitamins does not make the product a good source of vitamins, but it does block olestra's ability to bind with the vitamins from other foods. An asterisk in the ingredients list informs consumers that these added vitamins are "dietarily insignificant."

Some consumers experience digestive distress with olestra consumption: cramps, gas, bloating, and diarrhea. The FDA initially required a label warning stating that "olestra may cause abdominal cramping and loose stools" and that it "inhibits the absorption of some vitamins and other nutrients," but recently concluded that such a statement is no longer warranted.

Consumers need to keep in mind that low-fat and fat-free foods still deliver kcalories. Decades ago, consumers hailed the arrival of artificial sweeteners as a weight-loss wonder, but in reality, kcalories saved by using artificial sweeteners were readily replaced by kcalories from other foods. Alternatives to fat can help to lower energy intake and support weight loss only when they actually *replace* fat and energy in the diet.

Read Food Labels Labels list total fat, saturated fat, *trans* fat, and cholesterol contents of foods in addition to fat kcalories per serving. Because each package provides information for a single serving and serving sizes are standardized, consumers can easily compare similar products.

IN SUMMARY

- Fats in foods contribute to sensory appeal—enhancing the flavor, aroma, and texture of foods.

- Fats in foods deliver fat-soluble vitamins, energy, and essential fatty acids.

- While some fat in the diet is necessary, the *Dietary Guidelines for Americans 2005* recommend limiting fat intakes in order to limit intakes of saturated fat and *trans* fat.

- Fats added to foods during preparation or at the table are a major source of fat in the diet.

- The choice between whole and fat-free milk products can make a big difference to the fat, saturated fat, and cholesterol content of a diet.

- Meats account for a large proportion of the hidden fat and saturated fat in many people's diets.

- Most people consume more meat than is recommended.

- Most vegetables and fruits naturally contain little or no fat.

- Grain products such as croissants and biscuits can be high in saturated fat, so consumers need to read food labels to learn which foods in this group contain fats.

- Consumers today can choose from an array of fat-reduced products, and many bakery goods and other foods made with fat replacers offer less than a half a gram of fat, saturated fat, and *trans* fat in a serving.

- Some products use artificial fats such as olestra, while others use conventional ingredients such as water or fat-free milk to reduce fat and kcalories.

- Food labels list total fat, saturated fat, cholesterol, and *trans* fat, as well as fat kcalories per serving.

Chapters 2 and 3 look briefly at the two major energy fuels in the body—carbohydrate and fat. When used for energy, each has desirable characteristics. The glucose derived from carbohydrate is needed by the brain and nerve tissues and is easily used for energy in other cells. Fat is a particularly useful fuel because the body stores it efficiently and in generous amounts. Chapter 4 looks at protein, a nutrient that can be used as fuel, but whose primary role is to provide machinery for getting things done.

olestra: a synthetic fat made from sucrose and fatty acids that provides zero kcalories per gram; also known as *sucrose polyester*.

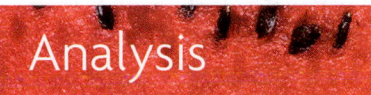

Analysis

YOUR DIET

Lipids

To maintain good health, eat enough, but not too much, fat and select the right kinds. For example, moderate intakes of foods rich in polyunsaturated and monounsaturated fatty acids such as olive oil, nuts, and fish, lower the risk of heart disease. Conversely, diets high in saturated fats and *trans* fats increase the risk of heart disease. Table NP3–1 in the Nutrition in Practice that follows this chapter lists food sources of healthful and harmful fatty acids. As Chapter 3 describes, the major sources of saturated fats in most people's diets are whole milk and whole milk products such as cheese and ice cream, fatty cuts of beef and pork, processed meats such as bacon and sausage, and products that contain tropical oils (coconut oil, palm kernel oil, and palm oil) such as cookies, pastries, and donuts. Major sources of *trans* fats include fast foods, chips, baked goods, and other commercially prepared foods.

- Make a list of all the fast foods and packaged foods such as chips and cookies that you ate today, or that you might typically eat in a day.

- Use Appendix A to look up the saturated fat content of the fast foods you ate (*trans* fat is not listed in Appendix A).

- Check the labels of the packaged foods you ate and list the saturated and *trans* fat amounts listed.

Fast food or packaged food consumed	Saturated fat (grams)	*Trans* fat (grams)

- The recommended saturated fat intake is less than 10 percent of total kcalories per day. The recommendation for *trans* fat is to keep it as low as possible. Thus, for a 2,000-kcalorie diet, the recommended saturated fat intake is 22 grams or less. (Recall that 1 gram of fat is equal to 9 kcalories).

  Example:

  2000 kcal × 0.10 saturated fat = 200 kcal.
  200 kcal ÷ 9 kcal/gram fat = 22 gram saturated fat.

- Based on your own estimated kcalorie needs (see Table 1–7), is your saturated fat intake 10 percent or less of total kcalories?

  Estimated kcalorie need: _____ kcalories
  Saturated fat recommendation: _____ grams
  Your saturated fat intake: _____ grams

- If not, list several foods you consumed that supplied high amounts of saturated fat and suggest alternative foods in the same food group that contain lower amounts.

If your food choices included packaged or fast foods that contain significant amounts of *trans* fats, suggest alternative foods in the same food group that contain minimal amounts.

CLINICAL APPLICATIONS

1. The connection between the overconsumption of fats, especially saturated and *trans* fat, and chronic diseases (obesity, diabetes, cancer, and cardiovascular disease) underscores the importance of being alert to a client's fat intake. What advice would you offer a client who reports the following?

- Eats two or more 6-ounce servings of meat each day.
- Drinks whole milk and eats regular cheddar cheese each day.
- Eats four to five servings of breads and cereals each day, including a bagel with cream cheese for breakfast, a bologna sandwich on white bread for lunch, and biscuits or cornbread with butter to accompany dinner.
- Eats one serving of fruit each day and eats vegetables only on occasion.

2. Make a list of foods, beverages, and seasonings that your client can substitute for foods high in saturated fat.

Self Check

1. Three classes of lipids in the body are:
   a. triglycerides, fatty acids, and cholesterol.
   b. triglycerides, phospholipids, and sterols.
   c. fatty acids, phospholipids, and cholesterol.
   d. glycerol, fatty acids, and triglycerides.

2. A triglyceride consists of:
   a. three glycerols attached to a lipid.
   b. three fatty acids attached to a glucose.
   c. three fatty acids attached to a glycerol.
   d. three phospholipids attached to a cholesterol.

3. A fatty acid that has the maximum possible number of hydrogen atoms is known as a(n):
   a. saturated fatty acid.
   b. monounsaturated fatty acid.
   c. PUFA.
   d. essential fatty acid.

4. The difference between *cis*- and *trans*-fatty acids is:
   a. the number of double bonds.
   b. the length of their carbon chains.
   c. the location of the first double bond.
   d. the configuration around the double bond.

5. Essential fatty acids:
   a. are used to make substances that regulate blood pressure, among other functions.
   b. can be made from carbohydrates.
   c. include lecithin and cholesterol.
   d. cannot be found in commonly eaten foods.

6. Lecithins and other phospholipids in the body function as:
   a. emulsifiers.
   b. enzymes.
   c. temperature regulators.
   d. shock absorbers.

7. To minimize saturated fat intake and lower the risk of heart disease, most people need to:
   a. eat less meat.
   b. select fat-free milk.
   c. use nonhydrogenated margarines and cooking oils such as olive oil or canola oil.
   d. all of the above.

8. To include omega-3 fatty acids in the diet, the American Heart Association recommends eating:
   a. cholesterol-free margarine.
   b. fish oil supplements.
   c. hydrogenated margarine.
   d. at least two fish meals per week.

9. Some examples of foods with hidden fats are:
   a. cheese, lettuce, and fruit juices.
   b. fried foods, sauces, dips, and lunch meats.
   c. fish, rice, and potatoes.
   d. baked potatoes, vegetables, and fruits.

10. Generally speaking, vegetable and fish oils are rich in:
    a. polyunsaturated fat.
    b. saturated fat.
    c. cholesterol.
    d. *trans*-fatty acids.

Answers to these questions are found in Appendix H.

Notes

1. P. E. Scherer, Adipose tissue: From lipid storage compartment to endocrine organ, *Diabetes* 55 (2006): 1537–1545; G. Fruhbeck, The adipose tissue as a source of vasoactive factors, *Current Medicinal Chemistry Cardiovascular and Hematological Agents* 3 (2004): 197–208.

2. M. A. Zulet and coauthors, Inflammation and conjugated linoleic acid: Mechanisms of action and implications for human health, *Journal of Physiology and Biochemistry* 61 (2005): 483–494; S. Tricon and coauthors, The effects of conjugated linoliec acid on human health-related outcomes, *Proceedings of the Nutrition Society* 64 (2005): 171–182.

3. J. L. Breslow, n-3 fatty acids and cardiovascular disease, *American Journal of Clinical Nutrition* 83 (2006): 1477S–1482S; K. C. Hayes, Relative cardiovascular benefits of n-6 and n-3 fatty acids, *Nutrition and the M.D.*, January 2005, pp. 1–4; H. Tapiero and coauthors, Polyunsaturated fatty acids (PUFA) and eicosanoids in human health and pathologies, *Biomedicine and Pharmacotherapy* 56 (2002): 215–222.

4. C. L. Cheatham, J. Colombo, and S. E. Carlson, n-3 fatty acids and cognitive and visual acuity development: Methodologic and conceptual considerations, *American Journal of Clinical Nutrition* 83 (2006): 1458S–1466S; J. M. Alessandri and coauthors, Polyunsaturated fatty acids in the central nervous system: Evolution of concepts and nutritional implications throughout life, *Reproduction, Nutrition, Development* 44 (2004): 509–538.

5. Cheatham, Colombo, and Carlson, 2006; D. R. Hoffman and coauthors, Maturation of visual acuity is accelerated in breast-fed term infants fed baby food containing DHA-enriched egg yolk, *Journal of Nutrition* 134 (2004): 2307–2313; Alessandri and coauthors, 2004.

6. Breslow, 2006; Hayes, 2005; J. A. Nettleton and R. Katz, n-3 long-chain polyunsaturated fatty acids in type 2 diabetes: A review, *Journal of the American Dietetic Association* 105 (2005): 428–440; V. Wijendran and K. C. Hayes, Dietary n-6 and n-3 fatty acid balance and cardiovascular health, *Annual Review of Nutrition* 24 (2004): 597–615; M. F. Leitzmann and coauthors, Dietary intake of n-3 and n-6 fatty acids and the risk of prostate cancer, *American Journal of Clinical Nutrition* 80 (2004): 203–216; Tapiero and coauthors, 2002; P. M. Kris-Etherton and coauthors, Fish consumption, fish oil, omega-3 fatty acids, and cardiovascular disease, *Circulation* 106 (2002): 2747–2757.

7. R. Stoeckli and U. Keller, Nutritional fats and the risk of type 2 diabetes and cancer, *Physiology and Behavior* 83 (2004): 611–615; Leitzmann, and coauthors, 2004; S. A. Bingham and coauthors, Are imprecise methods obscuring a relation between fat and breast cancer? *Lancet* 362 (2003): 212–214; C. E. Spiegelman and coauthors, Premenopausal fat intake and risk of breast cancer, *Journal of the National Cancer Institute* 95 (2003): 1079–1085.

8. W. E. Hardman, (n-3) fatty acids and cancer therapy, *Journal of Nutrition* 134 (2004): 3427S–3430S; National Cancer Policy Board, Institute of Medicine, S. J. Curry, T. Byers, and M. Hewitt, eds., *Fulfilling the Potential of Cancer Prevention and Early Detection* (Washington, D.C.: National Academies Press, 2003), p. 77; P. D. Terry, T. E. Rohan, and A. Wolk, Intakes of fish and marine fatty acids and the risks of cancers of the breast and prostate and of other hormone-related cancers: A review of the epidemologic evidence, *American Journal of Clinical Nutrition* 77 (2003): 532–543.

9. Standing Committee on the Scientific Evaluation of Dietary Reference Intakes, Food and Nutrition Board, Institute of Medicine, *Dietary Reference Intakes for Energy, Carbohydrate, Fiber, Fat, Fatty Acids, Cholesterol, Protein, and Amino Acids* (Washington, D.C.: National Academies Press, 2005), pp. 835–836.

10. Expert Panel on Detection, Evaluation, and Treatment of High Blood Cholesterol in Adults (Adult Treatment Panel III), *Third Report of the National Cholesterol Education Program (NCEP)*, NIH publication no. 02-5215 (Bethesda, Md.: National Heart, Lung, and Blood Institute, 2005): pp. V-25–V-27.

11. D. J. Baer and coauthors, Dietary fatty acids affect plasma markers of inflammation in healthy men fed controlled diets: A randomized crossover study, *American Journal of Clinical Nutrition* 79 (2004): 969–973; D. Mozaffarian and coauthors, Dietary intake of *trans* fatty acids and systemic inflammation in women, *American Journal of Clinical Nutrition* 79 (2004): 606–612; Standing Committee on the Scientific Evaluation of Dietary Reference Intakes, 2005, pp. 494–505.

12. S. L. Elias and S. M. Innis, Bakery foods are the major dietary source of *trans*-fatty acids among pregnant women with diets providing 30 percent energy from fats, *Journal of the American Dietetic Association* 102 (2002): 46–51.

13. Revealing *trans* fats, *FDA Consumer*, September/October 2003, pp. 20–26.

14. D. B. Panagiotakos and coauthors, Can a Mediterranean diet moderate the development and clinical progression of coronary heart disease? *Medical Science Monitor* 10 (2004): RA193–198; A. H. Stark and Z. Madar, Olive oil as a functional food: Epidemiology and nutritional approaches, *Nutrition Reviews* 60 (2002): 170–176.

15. V. Wijendran and K. C. Hayes, Dietary n-6 and n-3 fatty acid balance and cardiovascular health, *Annual Review of Nutrition* 24 (2004): 597–615.

16. Wijendran and Hayes, 2004.

17. Wijendran and Hayes, 2004; A. T. Erkkila and coauthors, Fish intake is associated with a reduced progression of coronary artery atherosclerosis in postmenopausal women with coronary artery

disease, *American Journal of Clinical Nutrition* 80 (2004): 626–632; P. M. Kris-Etherton and coauthors, Fish consumption, fish oil, omega-3 fatty acids, and cardiovascular disease, *Circulation* 106 (2002): 2747–2757; F. B. Hu and coauthors, Fish and omega-3 fatty acid intake and risk of coronary heart disease in women, *Journal of the American Medical Association* 287 (2002): 1815–1821; C. M. Albert and coauthors, Blood levels of long-chain n-3 fatty acids and the risk of sudden death, *New England Journal of Medicine* 346 (2002): 1113–1118.

18. Breslow, 2006.

19. American Heart Association Scientific Statement, Diet and lifestyle recommendations revision 2006, *Circulation* 114 (2006): 82–96.

20. Erkkila and coauthors, 2004.

21. Kris-Etherton and coauthors, 2002.

22. D. Mozaffarian, R. N. Lemaitre, and L. H. Kuller, Fish consumption and stroke risk in elderly individuals, The Cardiovascular Health Study, *Archives of Internal Medicine* 165 (2005): 200–206; Hayes, 2005; V. Wijendran and Hayes, 2004; Erkkila and coauthors, 2004.

23. Backgrounder for the 2004 FDA/EPA Consumer Advisory: What you need to know about mercury in fish and shellfish, 2004, available at www.fda.gov/oc/opacom/hottopics/mercury/back-grounder.html.

24. H. E. Theobald and coauthors, LDL cholesterol-raising effect of low-dose docosahexaenoic acid in middle-aged men and women, *American Journal of Clinical Nutrition* 79 (2004): 558–563; S. Kew and coauthors, Effects of oils rich in eicosapentaenoic and docosahexaenoic acids on immune cell composition and function in healthy humans, *American Journal of Clinical Nutrition* 79 (2004): 674–681; S. Bechoua and coauthors, Influence of very low dietary intake of marine oil on some functional aspects of immune cells in healthy elderly people, *British Journal of Nutrition* 89 (2003): 523–532; Standing Committee on the Scientific Evaluation of Dietary Reference Intakes, 2005, pp. 486–494.

25. Backgrounder for the 2004 FDA/EPA Consumer Advisory: What you need to know about mercury in fish and shellfish, 2004.

26. Standing Committee on the Scientific Evaluation of Dietary Reference Intakes, 2005, pp. 769–770.

27. Standing Committee on the Scientific Evaluation of Dietary Reference Intakes, 2005, pp. 835–836.

28. U.S. Department of Agriculture and U.S. Department of Health and Human Services, *Dietary Guidelines for Americans 2005*, 6th ed., available online at www.healthierus.gov.

29. Position of the American Dietetic Association: Fat replacers, *Journal of the American Dietetic Association* 105 (2005): 266–275.

FIGURING OUT FATS

To consumers, advice about dietary fat appears to change almost daily. "Eat less fat." "Eat more fatty fish." "Give up butter—use margarine instead." "Give up margarine—replace it with olive oil." "Steer clear of saturated." "Seek out omega-3." "Stay away from *trans*." "Stick with mono- and polyunsaturated." No wonder people feel confused about dietary fat. This Nutrition in Practice begins with a look at the latest dietary fat guidelines. It continues by identifying which foods provide which fats. It closes with strategies to help consumers choose the right amounts of the right kinds of fats for a healthy diet.

Why do today's fat messages seem to change constantly and become more confusing?

The confusion stems in part from the complexities of fat and in part from the nature of recommendations. As Chapter 3 explains, "dietary fat" refers to several kinds of fats; some fats support health whereas others damage it. Foods typically provide a mixture of fats in varying proportions. It has taken researchers decades to sort through the relationships among the various kinds of fat and their roles in supporting or harming health. Translating these research findings into dietary recommendations is a challenging process. Too little information can mislead consumers, but too much detail can overwhelm them. As scientific understanding has grown, recommendations have evolved to become less general and more specific. Recommendations may seem to "change constantly and become more confusing," but in fact they are becoming more meaningful.

How exactly have dietary recommendations for fat changed to become more meaningful for consumers?

Dietary recommendations for fat have changed by shifting the emphasis from lowering total fat, in general, to limiting saturated and *trans* fat, specifically. For decades, health experts urged consumers to limit total fat intake to 30 percent or less of energy intake. This advice was straightforward—cut the fat, improve your health. Health experts recognized that saturated fats and *trans* fats were the ones that raise blood cholesterol, but they reasoned that by limiting total fat, saturated and *trans* fat intake would decline as well. People were simply advised to cut back on all fat so that they would cut back on saturated and *trans* fat. Such advice may have oversimplified the message and unnecessarily restricted total fat.

But low-fat diets have been recommended for years to help people manage weight and reduce the risk of heart disease. Are you saying that low-fat diets are no longer recommended?

Low-fat diets have a place in treatment plans for people with elevated blood lipids or heart disease, but researchers question the wisdom of such diets for healthy people as a means of controlling weight and preventing diseases. Several problems accompany low-fat diets. For one, many people find low-fat diets difficult to maintain over time. For another, low-fat diets are not necessarily low-kcalorie diets; if energy intake exceeds energy needs, weight gain follows, and obesity brings a host of health problems, including heart disease. For still another, diets extremely low in fat may exclude fatty fish, nuts, seeds, and vegetable oils—all valuable sources of many essential fatty acids, phytochemicals, vitamins, and minerals. Importantly, the fats from these sources protect against heart disease, as later sections explain.

How have today's recommendations for fat been revised?

Today, health experts have revised dietary recommendations to acknowledge that not all fats have damaging health consequences. In fact, higher intakes of some kinds of fats (for example, the omega-3 fatty acids) support good health. Instead of urging people to cut back on all fats, current recommendations suggest carefully replacing the "bad" saturated fats with the "good" unsaturated fats and enjoying them in moderation.[1] The goal is to create a diet moderate in kcalories that provides enough of the fats that support good health, but not too much of those that harm health. (Turn to pp. 70–74 for a review of the health consequences of each type of fat.)

With these findings and goals in mind, the DRI committee concluded that a diet containing 20 to 35 percent of energy intake from fat, but reduced in saturated fat and *trans* fat and moderate in energy, is compatible with low rates of heart disease, diabetes, obesity, and cancer.[2] Heart-healthy recommendations suggest that within this range, consumers should try to minimize their intakes of saturated fats, *trans* fat, and cholesterol and use monounsaturated and polyunsaturated fats instead.[3]

How can people distinguish between the fats in foods that support health and those that might harm it?

Asking consumers to limit their total fat intake was less than perfect advice, but it was straightforward—find the fat and cut back. Asking consumers to keep their intakes of saturated fats, *trans* fats, and cholesterol low and to use monounsaturated and polyunsaturated fats instead may be more on target with heart health, but it also makes diet planning more complicated. To make appropriate selections, consumers must first learn which foods contain which fats. For example, avocados, bacon, walnuts, potato chips, and mackerel are all high-fat foods, yet some of these foods have detrimental effects on heart health when consumed in excess, whereas others seem neutral or even beneficial.

Is there evidence to clarify why some high-fat foods are compatible with a heart-healthy diet and others are not?

Yes. The traditional diets of Greece and other countries in the Mediterranean region are exemplary in their use of "good" fats, especially olives and their oil. A classic study of the world's people, the Seven Countries Study, found that death rates from heart disease were strongly associated with diets high in saturated fats, but only weakly linked with total fat.[4] In fact, the two countries with the highest fat intakes, Finland and the Greek island of Crete, had the highest (Finland) and lowest (Crete) rates of heart disease deaths. In both countries, the people consumed 40 percent or more of their kcalories from fat. Clearly, a high-fat diet was not the primary problem, so researchers refocused their attention on the type of fat. They began to notice the benefits of olive oil.

When dark green (virgin) olive oil replaces saturated fats, such as those of butter, coconut oil or palm oil, hydrogenated stick margarine, lard, or shortening, it may offer numerous health benefits.[5] Olive oil helps to protect against heart disease by:

Olives and their oil may benefit heart health.

© Matthew Farruggio

- Lowering total and LDL cholesterol and not lowering HDL cholesterol or raising triglycerides[6]

- Reducing LDL cholesterol's susceptibility to oxidation[7]

- Reducing blood-clotting factors

- Providing phytochemicals that act as antioxidants (see Nutrition in Practice 8)[8]

- Lowering blood pressure[9]

The labels of olive oil and foods containing it are now allowed to state the following qualified health claim: "Limited and not conclusive scientific evidence suggests that eating 2 tablespoons of olive oil daily may reduce the risk of coronary heart disease due to the monounsaturated fat in olive oil. To achieve this possible benefit, olive oil is to replace a similar amount of saturated fat and not increase the total number of kcalories you eat in a day. One serving of this product contains X grams of olive oil."

People who hope that olive oil will be a magic potion against heart disease are bound to be disappointed. Drizzling olive oil on a high-saturated fat food, such as a cheese and sausage pizza, does not make the food healthier. Like other fats, olive oil delivers 9 kcalories per gram, which can contribute to weight gain in people who fail to balance their energy intake with their energy output. Its role in a healthy diet is to *replace* the saturated fats. Other vegetable oils, such as canola or safflower oil, in their liquid unhydrogenated states, are also generally low in saturated fats and high in unsaturated fats, but they lack the phytochemicals that olive oil provides. When choosing olive oils, go for the darker "extra virgin" kind because it contains the highest levels of potentially beneficial phytochemicals.

Good olive oil may help protect against heart disease. Are there other food fats that may also be protective?

Possibly so. Tree nuts and peanuts are traditionally excluded from low-fat diets, and for good reason. Nuts provide up to 80 percent of their kcalories from fat, and a quarter cup (about an ounce) of mixed nuts provides over 200 kcalories. In a review of the literature, however, researchers found that people who ate a 1-ounce serving of nuts on five or more days per week had a reduced risk of heart disease compared with people consuming no nuts.[10] A smaller positive association was noted for any amount greater than one serving of nuts per week. The nuts were those commonly eaten in the United States: almonds, Brazil nuts, cashews, hazelnuts, macadamia nuts, pecans, pistachios, walnuts, and even peanuts. On average, these nuts contain mostly monounsaturated fat (59 percent), some polyunsaturated fat (27 percent), and little saturated fat (14 percent).

Research has shown a benefit from walnuts and almonds in particular. In study after study, walnuts, when substituted for other fats in the diet, produce favorable effects on blood lipids—even in people with elevated total and LDL cholesterol. Results are similar for almonds.

Studies on peanuts, macadamia nuts, pecans, and pistachios follow suit, indicating that including nuts may be a wise strategy against heart disease. Nuts may protect against heart disease because they provide:

- Monounsaturated and polyunsaturated fats in abundance, but few saturated fats

- Fiber, vegetable protein, and other valuable nutrients, including the antioxidant vitamin E

- Phytochemicals, concentrated in their brown papery skins, that act as antioxidants (see Nutrition in Practice 8)[11]

In addition to their heart benefits, nuts may also benefit other body organs—people who frequently consume nuts and other healthy fats suffer fewer gallbladder problems.[12]

Stay mindful of kcalories when snacking on nuts.

Before advising your clients to include nuts in their diets, a caution is in order. As mentioned, most of the energy nuts provide comes from fats. Consequently, they deliver many kcalories per bite. In studies examining the effects of nuts on heart disease, researchers carefully adjust diets to make room for the nuts without increasing the total kcalories—that is, they use nuts *instead of, not in addition to,* other foods (such as meats, potato chips, oils, margarine, and butter). People who do not make similar replacements could end up gaining weight if they simply add nuts on top of their regular diets. Weight gain, in turn, elevates blood lipids and raises the risks of heart disease.

What about fish? I have a friend whose doctor told her that eating fish is good for the heart. Is this true?

Yes. The preceding chapter made clear that fish oils hold the potential to improve health, and particularly the health of the heart. Research studies have provided strong evidence that increasing omega-3 fatty acids in the diet supports heart health and lowers the risk of death from heart disease.[13] For this reason, the American Heart Association and other authorities recommend including two fatty fish servings per week in a heart-healthy diet.[14] People who eat some fish each week can lower their risks of heart attack and stroke.[15] Table 3–3 on p. 73 lists fish that provide at least 1 gram of omega-3 fatty acids per serving.

Fish is the best source of EPA and DHA in the diet, but it is also a major source of mercury and other environmental contaminants. Most fish contain at least trace amounts of mercury, but tilefish, swordfish, king mackerel, and shark have especially high levels. Freshwater fish may contain PCBs and other pollutants, so local advisories warn sport fishers of species that can pose problems. The chapter lists safer species of fish. To minimize risks while obtaining fish benefits, vary your choices among fatty fish species often.

If olive oil, nuts, and fatty fish are protective against heart disease, which fats are harmful?

The number one dietary determinant of LDL cholesterol is saturated fat. Figure NP3–1 (p. 88) shows that each 1 percent

Fish is a good source of the omega-3 fatty acids.

increase in energy from saturated fatty acids in the diet may produce a 2 percent jump in heart disease risk by elevating blood LDL cholesterol. Conversely, reducing saturated fat intake by 1 percent can be expected to produce a 2 percent drop in heart disease risk by the same mechanism. Even a 2 percent drop in LDL represents a significant improvement for the health of the heart.[16] Like saturated fats, *trans* fats also raise heart disease risk by elevating LDL cholesterol. A heart-healthy diet limits foods rich in these two types of fat.

Which foods are highest in saturated and *trans* fats?

The major sources of saturated fats in the U.S. diet are fatty meats, whole-milk products, tropical oils, and products made from any of these foods. To limit saturated fat intake, consumers must choose carefully among these high-fat foods. Over a third of the fat in most meats is saturated. Similarly, over half of the fat is saturated in whole milk and other high-fat dairy products, such as cheese, butter, cream, half-and-half, cream cheese, sour cream, and ice cream. Consumers rarely use the tropical oils of palm, palm kernel, and coconut in the kitchen, but these oils are used heavily by food manufacturers and so are commonly found in many commercially prepared foods.

When choosing meats, milk products, and commercially prepared foods, look for those lowest in saturated fat. Labels

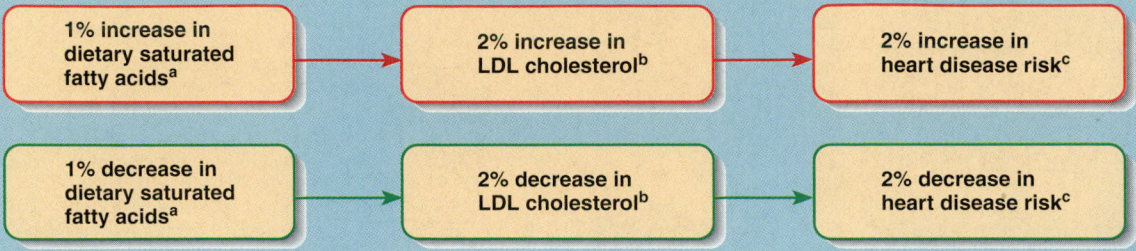

FIGURE NP3–1

Potential Relationships among Dietary Saturated Fatty Acids, LDL Cholesterol, and Heart Disease Risk

Source: Third Report of the National Cholesterol Education Program (NCEP) Expert Panel on Detection, Evaluation, and Treatment of High Blood Cholesterol in Adults (Adult Treatment Panel III), NIH publication no. 02-5215 (Bethesda, Md. National Heart, Lung, and Blood Institute, 2002), p. V-8 and II-4.

provide a useful guide for comparing products in this regard, and Appendix A lists the saturated fat in several thousand foods.

Even with careful selections, a nutritionally adequate diet will provide some saturated fat. Zero saturated fat is not possible even when experts design menus with the mission to keep saturated fat as low as possible.[17] Diets based on fruits, vegetables, legumes, nuts, soy products, and whole grains can, and often do, deliver less saturated fat than diets that depend heavily on animal-derived foods, however.

As for *trans* fats, Chapter 3 explains that solid shortening and margarine are made from vegetable oil that has been hardened through hydrogenation. This process both saturates some of the unsaturated fatty acids and introduces *trans*-fatty acids. Many convenience foods contain *trans* fats. Table NP3–1 summarizes which foods provide which fats. Substituting unsaturated fats for saturated fats at each meal and snack can help protect against heart disease. Table NP3–2 provides several examples and shows how such substitutions can lower saturated fat and raise unsaturated fat—even when total fat and kcalories remain unchanged.

TABLE NP3–1

Food Sources of Fatty Acids

Healthful Fatty Acids		
Monounsaturated	**Omega-6 Polyunsaturated**	**Omega-3 Polyunsaturated**
Avocado	Margarine (nonhydrogenated)	Fatty fish (herring, mackerel, salmon, tuna)
Nuts (almonds, cashews, filberts, hazelnuts, macadamia nuts, peanuts, pecans, pistachios)	Mayonnaise	Flaxseed
	Nuts (walnuts)	Nuts
Oils (canola, olive, peanut, sesame)	Oils (corn, cottonseed, safflower, soybean)	
Olives	Salad dressing	
Peanut butter (old-fashioned)	Seeds (pumpkin, sunflower)	
Seeds (sesame)		

Harmful Fatty Acids	
Saturated	**Trans**
Bacon	Commercial baked goods, including cookies, cakes, pies, or other products made with margarine or vegetable shortening
Butter	
Cheese	Fried foods, particularly restaurant and fast foods such as french fries and chicken
Chocolate	Many fried or processed snack foods, including microwave popcorn, chips, and crackers
Coconut	
Cream, half-and-half	Margarine (hydrogenated or partially hydrogenated)
Cream cheese	Nondairy creamers
Lard	Shortening
Meat	
Milk fat (whole-milk products)	
Oils (coconut, palm, palm kernel)	
Shortening	
Sour cream	

Note: Keep in mind that foods contain a mixture of fatty acids; see Figure 3–6, p. 79.

Choosing Unsaturated Fat instead of Saturated Fat

Examples of ways to replace saturated fats with unsaturated fats include sautéing foods in olive oil instead of butter, garnishing salads with sunflower seeds instead of bacon, snacking on mixed nuts instead of potato chips, using avocado instead of cheese on a sandwich, and eating salmon instead of steak. Portion sizes have been adjusted so that each of these foods provides approximately 100 kcalories. Notice that for a similar number of kcalories and grams of fat, the first choices offer less saturated fat and more unsaturated fat.

Foods (100 kcal portions)	Saturated Fat (g)	Unsaturated Fat (g)	Total Fat (g)
Olive oil 1 tbs vs. butter 1 tbs	2 vs. 7	9 vs. 4	11 vs. 11
Sunflower seeds 2 tbs vs. bacon 2 slices	1 vs. 3	7 vs. 6	8 vs. 9
Mixed nuts 2 tbs vs. potato chips 10 chips	1 vs. 2	8 vs. 5	9 vs. 7
Avocado 6 slices vs. cheese 1 slice	2 vs. 4	8 vs. 4	10 vs. 8
Salmon 2 oz vs. steak 1 oz	1 vs. 2	3 vs. 3	4 vs. 5
Totals	**7 vs. 18**	**35 vs. 22**	**42 vs. 40**

So it seems that some fats are "good" and others are "bad" from the body's point of view. Is that right?

The saturated and *trans* fats do indeed seem mostly bad for the health of the heart. Aside from providing energy, which unsaturated fats can do equally well, saturated and *trans* fats bring no indispensable benefits to the body. Furthermore, no harm can come from consuming diets low in them.

In contrast, the unsaturated fats are mostly good for the health of the heart when consumed in moderation. To date, their one proven fault seems to be that they, like all fats, provide abundant energy to the body and so may promote obesity if they drive kcalorie intakes higher than energy needs.[18] Obesity, in turn, often begets many body ills, as Chapter 6 makes clear.

When judging foods by their fatty acids, keep in mind that the fat in foods is a mixture of "good" and "bad," providing both saturated and unsaturated fatty acids. Even predominantly monounsaturated olive oil delivers some saturated fat. Consequently, even when a person chooses foods with mostly unsaturated fats, saturated fat can still add up if total fat is high. For this reason, fat must be kept below 35 percent of total kcalories if the diet is to be moderate in saturated fat.

Additionally, food manufacturers may come to the assistance of consumers wishing to avoid the health threats from saturated and *trans* fats. A margarine maker has announced that it will no longer offer products containing *trans* fats; a major snack manufacturer will soon reduce the saturated and *trans* fats in some of its products and offer snack foods in single-serving packages. A new oil with no *trans* fats, less saturated fat, and more unsaturated fat has been introduced for commercial production of french fries. Other companies are likely to follow if consumers respond favorably.

Basing one's diet on vegetables, fruits, and legumes as part of a balanced daily diet is a good idea, as is *replacing* saturated fats such as butter, shortening, and meat fat with unsaturated fats such as olive oil and the oils from nuts and fish. These foods provide vitamins, minerals, and phytochemicals—all valuable in protecting the body's health. To further protect health, you may want to reduce fats from convenience foods and fast foods; choose small portions of meat, fish, and poultry; and include fresh foods from all the groups each day. Take care to select portion sizes that will best meet your energy needs. Also, be physically active each day.

Notes

1. Standing Committee on the Scientific Evaluation of Dietary Reference Intakes, Food and Nutrition Board, Institute of Medicine, *Dietary Reference Intakes for Energy, Carbohydrate, Fiber, Fat, Fatty Acids, Cholesterol, Protein, and Amino Acids* (Washington, D.C.: National Academies Press, 2005); Expert Panel on Detection, Evaluation, and Treatment of High Blood Cholesterol in Adults (Adult Treatment Panel III), *Third Report of the National Cholesterol Education Program (NCEP)*, NIH publication no. 02-5215 (Bethesda, Md.: National Heart, Lung, and Blood Institute, 2002).

2. Standing Committee on the Scientific Evaluation of Dietary Reference Intakes, 2005, pp. 769–770.

3. American Heart Association Scientific Statement, Diet and lifestyle recommendations revision 2006, *Circulation* 114 (2006): 82–96; U.S. Department of Agriculture and U.S. Department of Health and Human Services, *Dietary Guidelines for Americans 2005*, 6th ed., available online at www.heathierus.gov.

4. A. Keys, Seven Countries: A Multivariate Analysis of Death and Coronary Heart Disease (Cambridge, Mass.: Harvard University Press, 1980).

5. F. Visioli and coauthors, Virgin Olive Oil Study (VOLOS): Vasoprotective potential of extra virgin olive oil in mildly dislipidemic patients, *European Journal of Nutrition* 44 (2005): 121–127; A. H. Stark and Z. Madar, Olive oil as a functional food: Epidemiology and nutritional approaches, *Nutrition Reviews* 60 (2002): 170–176.

6. M. I. Covas and coauthors, The effect of polyphenols in olive oil on heart disease risk factors, *Annals of Internal Medicine* 145 (2006): 333–341.

7. Visioli and coauthors, 2005.

8. Visioli and coauthors, 2005; F. Visioli and C. Galli, Biological properties of olive oil phytochemicals, *Critical Reviews in Food Science and Nutrition* 42 (2002): 209–221; M. N. Vissers and coauthors, Olive oil phenols are absorbed in humans, *Journal of Nutrition* 132 (2002): 409–417.

9. T. Psaltopoulou and coauthors, Olive oil, the Mediterranean diet, and arterial blood pressure: The Greek European Prospective Investigation into Cancer and Nutrition (EPIC) study, *American Journal of Clinical Nutrition* 80 (2004): 1012–1018.

10. J. H. Kelly and J. Sabate, Nuts and coronary heart disease: An epidemiological perspective, *British Journal of Nutrition* 96 (2006): S61–S67.

11. S. S. Wijeratne, M. M. Abou-Zaid, and F. Shahidi, Antioxidant polyphenols in almond and its coproducts, *Journal of Agricultural and Food Chemistry* 54 (2006): 312–318; P. E. Milbury and coauthors, Determination of flavonoids and phenolics and their distribution in almonds, *Journal of Agricultural and Food Chemistry* 54 (2006): 5027–5033.

12. C.-J. Tsai and coauthors, Frequent nut consumption and decreased risk of cholecystectomy in women, *American Journal of Clinical Nutrition* 80 (2004): 76–81; C.-J. Tsai and coauthors, A prospective cohort study of nut consumption and the risk of gallstone disease in men, *American Journal of Epidemiology* 160 (2004): 961–968.

13. J. L. Breslow, n-3 fatty acids and cardiovascular disease, *American Journal of Clinical Nutrition* 83 (2006): 1477S–1482S; K. C. Hayes, Relative cardiovascular benefits of n-6 and n-3 fatty acids, *Nutrition and the M.D.*, January 2005, pp. 1–4; V. Wijendran and K. C. Hayes, Dietary n-6 and n-3 fatty acid balance and cardiovascular health, *Annual Review of Nutrition* 24 (2004): 597–615; A. T. Erkkila and coauthors, Fish intake is associated with a reduced progression of coronary artery atherosclerosis in postmenopausal women with coronary artery disease, *American Journal of Clinical Nutrition* 80 (2004): 626–632; P. M. Kris-Etherton and coauthors, Fish consumption, fish oil, omega-3 fatty acids, and cardiovascular disease, *Circulation* 106 (2002): 2747–2757; F. B. Hu and coauthors, Fish and omega-3 fatty acid intake and risk of coronary heart disease in women, *Journal of the American Medical Association* 287 (2002): 1815–1821; C. M. Albert and coauthors, Blood levels of long-chain n-3 fatty acids and the risk of sudden death, *New England Journal of Medicine* 346 (2002): 1113–1118.

14. American Heart Association Scientific Statement, Diet and lifestyle recommendations revision 2006, *Circulation* 114 (2006): 82–96.

15. D. Mozaffarian, R. N. Lemaitre, and L. H. Kuller, Fish consumption and stroke risk in elderly individuals, *Archives of Internal Medicine* 165 (2005): 200–206; Wijendran and Hayes, 2004; H. Iso and coauthors, Intake of fish and omega-3 acids and risk of stroke in women, *Journal of the American Medical Association* 285 (2001): 304–312.

16. Expert Panel on Detection, Evaluation, and Treatment of High Blood Cholesterol in Adults (Adult Treatment Panel III), 2002, p.V-8.

17. Standing Committee on the Scientific Evaluation of Dietary Reference Intakes, 2005, p. 835.

18. Standing Committee on the Scientific Evaluation of Dietary Reference Intakes, 2005, pp. 796–797.

4 Proteins and Amino Acids

P eople think of proteins as bodybuilding nutrients, the material of strong muscles, and rightly so. No new living tissue can be built without them. Some proteins form structures such as muscle, bone, skin, and other tissues. Other proteins do the cells' work. The energy to fuel that work comes primarily from carbohydrates and fats.

The Chemist's View of Proteins

P roteins are chemical compounds that contain the same atoms as carbohydrates and lipids—carbon (C), hydrogen (H), and oxygen (O)—but proteins are different in that they also contain nitrogen (N) atoms. These nitrogen atoms give the name *amino* (nitrogen containing) to the amino acids that form the links in the chains we call proteins.

The Structure of Proteins

About 20 different **amino acids** may appear in proteins.* All amino acids share a common chemical "backbone," and it is these backbones that are linked together to form proteins. Each amino acid also carries a side group, which varies from one amino acid to another (see Figure 4–1). The side group makes the amino acids differ in size, shape, and electrical charge. The side groups on amino acids are what make proteins so varied in comparison with either carbohydrates or lipids.

Protein Chains The 20 amino acids can be linked end-to-end in a virtually infinite variety of sequences to form proteins. When two amino acids bond together, the resulting structure is known as a **dipeptide**. Three amino acids bonded together form a **tripeptide**. As additional amino acids join the chain, the structure becomes a **polypeptide**. Most proteins are a few dozen to several hundred amino acids long.

Protein Shapes Polypeptide chains twist into complex shapes. Each amino acid has special characteristics that attract it to, or repel it from, the surrounding fluids and other amino acids. Because of these interactions, polypeptide chains fold and intertwine into intricate coils (see Figure 4–2) and other shapes. The amino acid sequence of a protein determines the specific way the chain will fold.

Protein Functions The dramatically different shapes of proteins enable them to perform different tasks in the body. Some, such as hemoglobin in the blood (see Figure 4–3), are globular in shape; some are hollow balls that can carry and store

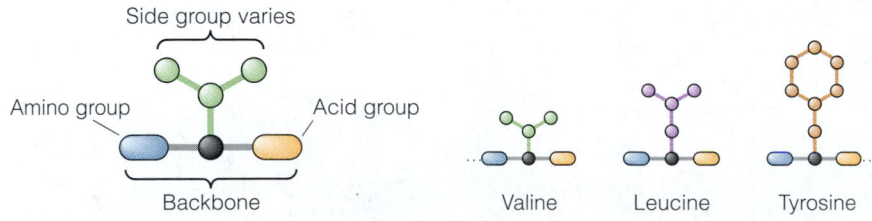

Side group varies

Amino group   Acid group

Backbone

Valine   Leucine   Tyrosine

All amino acids have a "backbone" made of an amino acid group (which contains nitrogen) and an acid group. The side group varies from one amino acid to the next.

Note that the side group is a unique structure that differentiates one amino acid from another.

* Besides the 20 common amino acids, which can all be components of proteins, others occur individually (for example, ornithine).

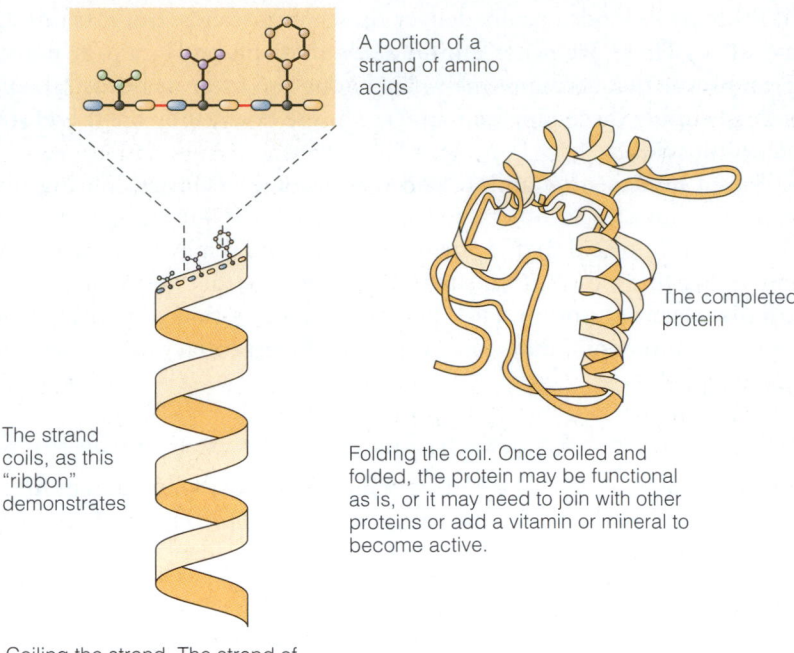

A portion of a strand of amino acids

FIGURE 4–2

The Coiling and Folding of a Protein Molecule

The completed protein

The strand coils, as this "ribbon" demonstrates

Folding the coil. Once coiled and folded, the protein may be functional as is, or it may need to join with other proteins or add a vitamin or mineral to become active.

Coiling the strand. The strand of amino acids takes on a spring-like shape as their side groups variously attract and repel each other.

materials within them; and some, such as those that form tendons, are more than ten times as long as they are wide, forming stiff, sturdy, rodlike structures.

Essential Amino Acids

Proteins in foods do not provide body proteins directly, but rather supply the amino acids from which the body makes its own proteins. More than half of the amino acids are **nonessential amino acids**, meaning that the body can make them

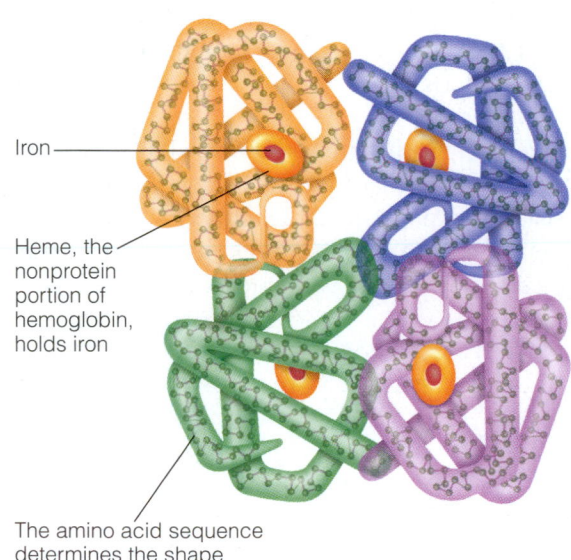

Iron

Heme, the nonprotein portion of hemoglobin, holds iron

The amino acid sequence determines the shape of the polypeptide chain

FIGURE 4–3

The Structure of Hemoglobin
Four highly folded polypeptide chains form the globular hemoglobin protein.

proteins: compounds composed of carbon, hydrogen, oxygen, and nitrogen atoms made from strands of amino acids. Some amino acids also contain sulfur atoms.

amino (a-MEEN-oh) **acids:** building blocks of protein. Each has a hydrogen atom, an amino group, and an acid group attached to a central carbon, which also carries a distinctive side chain.
amino = containing nitrogen

dipeptide: two amino acids bonded together.
di = two
peptide = amino acid

tripeptide: three amino acids bonded together.
tri = three

polypeptide: ten or more amino acids bonded together. An intermediate strand of between four and ten amino acids is an *oligopeptide*.
poly = many
oligo = few

nonessential amino acids: amino acids that the body can synthesize; also called **dispensable** amino acids.

for itself. Proteins in foods usually deliver these amino acids, but it is not essential that they do so. There are other amino acids that the body cannot make at all, however, and some that it cannot make fast enough to meet its needs. The proteins in foods must supply these nine amino acids to the body; they are therefore called **essential amino acids**. ■

Sometimes a nonessential amino acid can become essential. During illness or conditions of trauma or in other special circumstances such as premature birth, the need for an amino acid that is normally nonessential may become greater than the body's ability to produce it. In such circumstances, that amino acid becomes a **conditionally essential amino acid**. Research suggests that glutamine, normally a nonessential amino acid, may be a conditionally essential amino acid for some critically ill people.[1]

IN SUMMARY

- Chemically speaking, proteins are more complex than carbohydrates or lipids; proteins are made of some 20 different amino acids, 9 of which the body cannot make (they are essential).

- Each amino acid contains a central carbon atom with an amino group, an acid group, a hydrogen atom, and a unique side group attached to it.

- The distinctive sequence of amino acids in each protein determines its shape and function.

Proteins in the Body

What distinguishes you chemically from any other human being are minute differences in your particular body proteins (enzymes, antibodies, and others). These differences are determined by your proteins' amino acid sequences, which are written into the genes you inherited from your parents and ancestors. The genes direct the making of all the body's proteins.

The human body contains an estimated 30,000 or more different kinds of proteins. The roles of more than 3,000 of these proteins are now known. Only a few of the many roles proteins play are described here, but these should serve to illustrate proteins' versatility, uniqueness, and importance.

Enzymes **Enzymes** are catalysts that are essential to all life processes. Enzymes in the cells of plants or animals put together the pairs of sugars that make disaccharides and the long strands of sugars that make starch, cellulose, and glycogen. Enzymes also dismantle these compounds to free their constituent parts and release energy. Enzymes also assemble and disassemble lipids, assemble all other compounds that the body

FIGURE 4–4

Enzyme Action
Each enzyme facilitates a specific chemical reaction. In this diagram, an enzyme enables two compounds to make a more complex structure, but the enzyme itself remains unchanged.

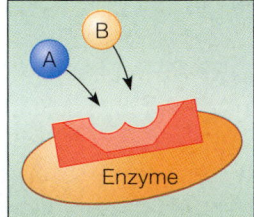

The separate compounds, A and B, are attracted to the enzyme's active site, making a reaction likely.

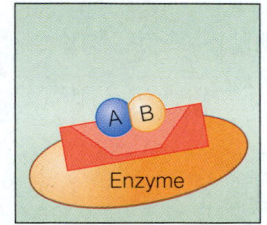

The enzyme forms a complex with A and B.

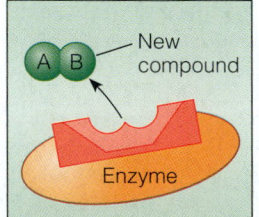

The enzyme is unchanged, but A and B have formed a new compound, AB.

makes, and disassemble all compounds that the body can use for building tissue and other metabolic work. As Figure 4–4 shows, enzymes themselves are not altered by the reactions they facilitate. All enzymes are proteins, and when amino acids have to be put together to make proteins, it is enzymes that put them together, too. In other words, these proteins can even make other proteins.

The protein story moves in a circle. To follow the circle in nutrition, start with a person eating food proteins. The food proteins are broken down by digestive enzymes, proteins themselves, into amino acids. The amino acids enter the cells of the body, where other proteins (enzymes) put the amino acids together in long chains whose sequences are specified by the genes. The chains fold and twist back on themselves to form proteins, and some of these proteins become enzymes themselves. Some of these enzymes break apart compounds; others put compounds together. Day by day, in billions of reactions, these processes repeat themselves, and life goes on. Only living systems can achieve such self-renewal. A toaster cannot produce another toaster; a car cannot fix a broken-down car. Only living creatures and the parts they are composed of—the cells—can duplicate and repair themselves.

Fluid and Electrolyte Balance Proteins help maintain the body's **fluid and electrolyte balance**. As Figure 4–5 shows, the body's fluids are contained in three major body compartments: (1) the spaces inside the blood vessels; (2) the spaces within the cells; and (3) the spaces between the cells (the interstitial spaces outside the blood vessels). Fluids flow back and forth between these compartments, and proteins in the fluids, together with minerals, ■ help to maintain the needed distribution of these fluids.

Proteins are able to help determine the distribution of fluids in living systems for two reasons: first, proteins cannot pass freely across the membranes that separate the body compartments, and second, they are attracted to water. A cell that "wants" a certain amount of water in its interior space cannot move the water around directly, but it can manufacture proteins, and these proteins will hold water. Thus the cell can use proteins to help regulate the distribution of water indirectly. Similarly, the body makes proteins for the blood and the interstitial (intercellular) spaces. These proteins help maintain the fluid volume in those spaces. Excess fluid accumulation in the interstitial spaces is called **edema**.

Not only is the quantity of the body fluids vital to life, but so is their composition. Special transport proteins in the membranes of cells continuously transfer substances into and out of cells to maintain balance. For example, sodium is concentrated outside the cells, and potassium is concentrated inside. The balance of these two minerals is critical to nerve transmission and muscle contraction. Any disturbance in this balance triggers a major medical emergency. Such imbalances can cause irregular heartbeats, kidney failure, muscular weakness, and even death.

Acid-Base Balance Proteins also help maintain the balance between **acids** and **bases** within the body's fluids. Normal body processes continually produce acids and bases, which must be carried by the blood to the kidneys and lungs for excretion. The blood must do this without upsetting its own **acid-base balance**. Blood **pH** is one of the most tightly controlled conditions in the body. If the blood becomes too acidic, vital proteins may undergo **denaturation**, losing their shape and ability to function. A similar situation arises when the balance tips too far toward base. These imbalances are known as **acidosis** and **alkalosis**, respectively, and both can be fatal. Figure 4–6 (p. 96) shows the normal and abnormal pH ranges of body fluids, as well as the pHs of some common substances.

Proteins such as albumin in blood help to prevent acid-base imbalances. In a sense, the proteins protect one another by gathering up extra acid (hydrogen) ions when there are too many in the surrounding medium and by releasing them when there are too few. By accepting and releasing hydrogen ions, proteins act as **buffers**, maintaining the acid-base balance of the blood and body fluids.

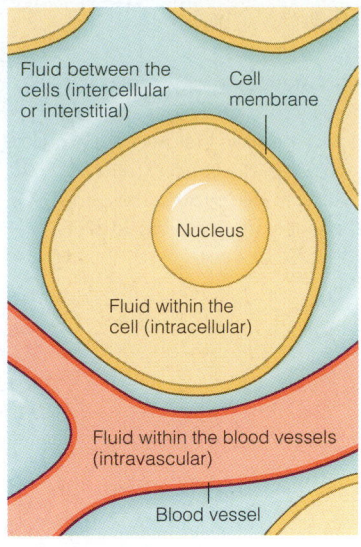

Fluid between the cells (intercellular or interstitial)

Cell membrane

Nucleus

Fluid within the cell (intracellular)

Fluid within the blood vessels (intravascular)

Blood vessel

FIGURE 4–5

One Cell and Its Associated Fluids

■ Minerals are helper nutrients. The attraction of protein and mineral particles to water is due to osmotic pressure (see Chapter 9).

essential amino acids: amino acids that the body cannot synthesize in amounts sufficient to meet physiological need; also called **indispensable** amino acids. Nine amino acids are known to be essential for human adults:

- *histidine* (HISS-tuh-deen)

- *isoleucine* (eye-so-LOO-seen)

- *leucine* (LOO-seen)

- *lysine* (LYE-seen)

- *methionine* (meh-THIGH-oh-neen)

- *phenylalanine* (fen-il-AL-uh-neen)

- *threonine* (THREE-oh-neen)

- *tryptophan* (TRIP-toe-fane, TRIP-toe-fan)

- *valine* (VAY-leen)

conditionally essential amino acid: an amino acid that is normally nonessential but must be supplied by the diet in special circumstances when the need for it becomes greater than the body's ability to produce it.

enzymes: protein catalysts. A catalyst is a compound that facilitates chemical reactions without itself being changed in the process.

fluid and electrolyte balance: maintenance of the necessary amounts and types of fluid and minerals in each compartment of the body fluids.

FIGURE 4-6

The pH Scale
A substance's acidity or alkalinity is measured in pH units. Each step down the scale indicates a tenfold increase in the concentration of hydrogen ions. Notice how small the range of normal blood pH is.

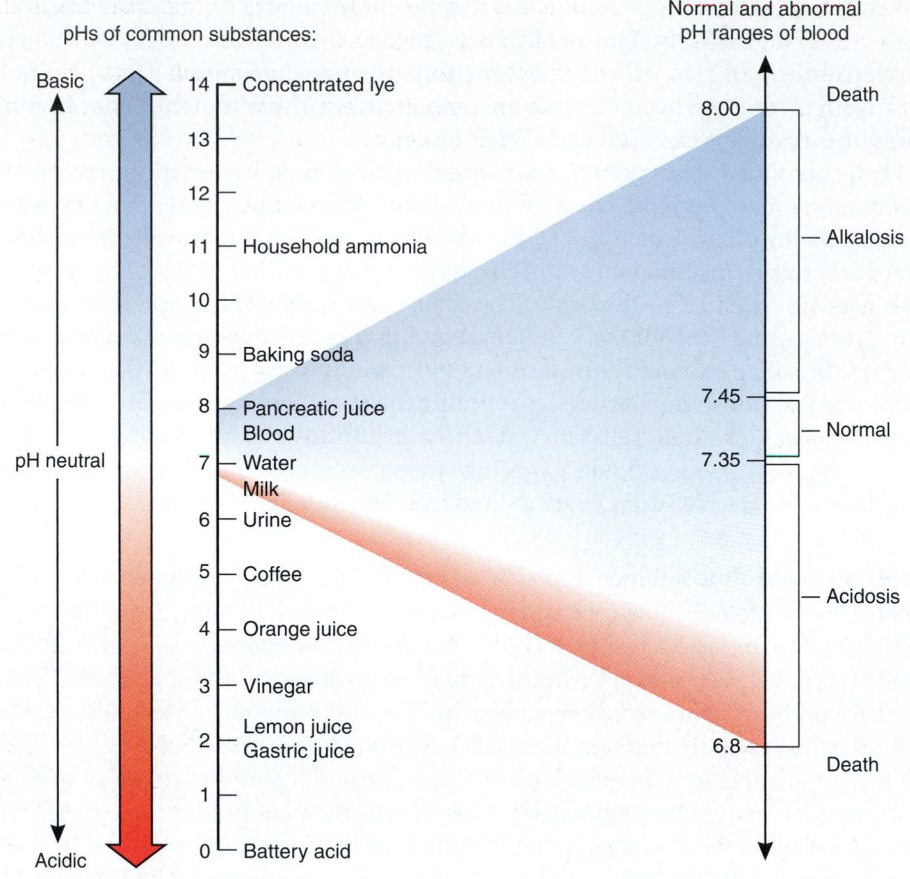

pHs of common substances:

Normal and abnormal pH ranges of blood

Basic

14 — Concentrated lye

13

12

11 — Household ammonia

10

9 — Baking soda

8 — Pancreatic juice
Blood

pH neutral 7 — Water
Milk

6 — Urine

5 — Coffee

4 — Orange juice

3 — Vinegar

2 — Lemon juice
Gastric juice

1

Acidic 0 — Battery acid

8.00 — Death

— Alkalosis

7.45 —
— Normal
7.35 —

— Acidosis

6.8 — Death

Antibodies Other proteins in the blood—the **antibodies**—defend against viruses, bacteria, and other disease agents. The antibodies work so efficiently that if a million bacterial cells are injected into the skin of a healthy person, fewer than ten are likely to survive for five hours, which explains why most diseases never have a chance to get started. Without sufficient protein, the body cannot maintain its resistance to disease.

Hormones The blood also carries messenger molecules known as **hormones**, and *some* hormones are proteins. (Recall that some hormones are sterols, members of the lipid family.) Among the proteins that act as hormones are glucagon and insulin. Hormones have many profound effects, which will become evident in subsequent chapters.

Transport Proteins Some proteins move about in the body fluids, transporting nutrients and other molecules from one organ to another. The protein hemoglobin, which carries oxygen from the lungs to the body's cells, is a prime example. The lipoproteins transport lipids around the body, as Chapter 5 describes. In addition, special proteins also carry vitamins and minerals.

Growth, Maintenance, and Repair The body uses amino acids to build the proteins of all its new tissues. The new tissues may be in an embryo, in a growing child, or in new hair and nails. Proteins also help replace worn-out cells in everyone's body all the time. For example, the millions of cells that line the intestinal tract live for three to five days; they are constantly being shed and must be replaced. The cells of the skin die and rub off, and new ones grow from underneath. The body uses proteins to repair damaged tissues, too. The protein collagen serves as the mending material of torn tissue, forming scars to hold the separated parts together.

Both inside and outside the body, cells constantly make and break down their proteins. When proteins break down, their component amino acids are liberated within the cells or released into the general circulation. Some of these amino acids may be promptly recycled into other proteins; others may be stripped of their nitrogen and used for energy. By reusing amino acids to build proteins, however, the body conserves and recycles a valuable commodity. The entire process of breakdown, recovery, and synthesis is called **protein turnover**.

People need to eat protein-rich foods every day to replace the protein they continuously lose. If the body is growing, it needs more protein than is necessary just for maintenance. Children end each day with more blood cells, more muscle cells, and more skin cells than they had at the beginning of the day. So protein is needed both for routine maintenance (replacement) and for growth (addition).

Providing Energy and Glucose Even though amino acids are needed to do the work that only they can perform—build vital proteins—they will be sacrificed to provide energy and glucose if need be. For most people eating a normal, mixed diet, protein provides about 10 to 15 percent of the daily need for energy.[2] Under conditions of inadequate energy or carbohydrate, protein use speeds up.[3] Keeping energy and glucose available is one of the body's highest priorities: without energy, cells die; without glucose, the brain and nervous system falter. When glucose or fatty acids are limited, cells are forced to use amino acids for energy and glucose. The body does not make a specialized storage form of protein as it does for carbohydrate and fat. Glucose is stored as glycogen in the liver and muscles, fat as triglycerides in the adipose tissue, but body protein is available only as the working and structural components of the tissues. When the need arises, the body dismantles its tissue proteins and uses them for energy. Thus, over time, energy deprivation (starvation) always incurs wasting of lean body tissue as well as fat loss.

© Ariel Skelley/Corbis

Growing children end each day with more bone, blood, muscle, and skin cells than they had at the beginning of the day.

edema (eh-DEEM-uh): the swelling of body tissue caused by leakage of fluid from the blood vessels and accumulation of the fluid in the interstitial spaces.

acids: compounds that release hydrogen ions in a solution.

bases: compounds that accept hydrogen ions in a solution.

acid-base balance: the balance maintained between acid and base concentrations in the blood and body fluids.

pH: the concentration of hydrogen ions. The lower the pH, the stronger the acid. Thus pH 2 is a strong acid; pH 6 is a weak acid; pH 7 is neutral; and a pH above 7 is alkaline.

denaturation (dee-nay-cher-AY-shun): the change in a protein's shape brought about by heat, acid, or other agents. Past a certain point, denaturation is irreversible.

acidosis: acid accumulation in the blood and body fluids; depresses the central nervous system and can lead to disorientation and, eventually, coma.

alkalosis: excessive base in the blood and body fluids.

buffers: compounds that can reversibly combine with hydrogen ions to help keep a solution's acidity or alkalinity constant.

antibodies: large proteins of the blood and body fluids, produced in response to invasion of the body by unfamiliar molecules (mostly proteins) called *antigens.* Antibodies inactivate the invaders and so protect the body.
 anti = against

hormones: chemical messengers. Hormones are secreted by a variety of glands in the body in response to altered conditions. Each travels to one or more target tissues or organs and elicits specific responses to restore normal conditions.

protein turnover: the continuous breakdown and synthesis of body proteins involving the recycling of amino acids.

<table>
<tr><td colspan="1">TABLE 4–1</td></tr>
<tr><td>Summary of Functions of Proteins</td></tr>
<tr><td>Enzymes. Proteins facilitate chemical reactions.</td></tr>
<tr><td>Fluid and electrolyte balance. Proteins help to maintain the distribution and composition of various body fluids.</td></tr>
<tr><td>Acid-base balance. Proteins help maintain the acid-base balance of body fluids by acting as buffers.</td></tr>
<tr><td>Antibodies. Proteins act against disease agents to fight diseases.</td></tr>
<tr><td>Hormones. Proteins regulate body processes. Some, but not all, hormones are made of protein.</td></tr>
<tr><td>Transportation. Proteins transport substances such as lipids, vitamins, minerals, and oxygen around the body.</td></tr>
<tr><td>Growth and maintenance. Proteins form integral parts of most body structures such as skin, tendons, ligaments, membranes, muscles, organs, and bones. As such, they support the growth and repair of body tissues.</td></tr>
<tr><td>Energy and glucose. Proteins provide some fuel, and glucose if needed, for the body's energy needs.</td></tr>
</table>

IN SUMMARY

- The list of protein functions discussed here and summarized in Table 4–1 is by no means exhaustive. Nevertheless, it does give some sense of the immense variety of proteins and their importance in the body.

Protein and Health

During the time that scientists have been studying nutrition, no nutrient has been more intensely scrutinized than protein. As you know by now, it is indispensable to life. It should come as no surprise that protein deficiency can have devastating effects on people's health. But, as with the other nutrients, protein in excess can be harmful, too, so this section also discusses the consequences of protein excess.

Protein-Energy Malnutrition

When people are deprived of food and suffer an energy deficit, they degrade their own body protein for energy and indirectly suffer a protein deficiency, as well as an energy deficiency. Because protein and energy deprivation go hand in hand, public health officials have adopted an abbreviation for the overlapping pair: **protein-energy malnutrition (PEM)**. PEM often strikes early in childhood, but it endangers many adults as well. ■ PEM is the most widespread form of malnutrition in the world today. Most of the 33,000 children who die each day are malnourished.[4] PEM is prevalent in Africa, Central America, South America, the Middle East, and South and East Asia, but developed countries including the United States are not immune to it. PEM is common among some population groups in the United States: impoverished people living on U.S. Indian reservations, in inner cities, and in rural areas; many elderly people; homeless children; and those suffering from the eating disorder anorexia nervosa.

Of all population groups, children are most seriously affected by malnutrition. Children who are thin for their heights may have recently developed PEM, whereas children who are short for their ages may have experienced PEM for extended periods of time. Stunted growth due to PEM is easy to overlook because a small child may look quite normal, but it may be the most common sign of malnutrition in the developing countries.

■ PEM can be a consequence of many different conditions. PEM has been recognized in people with many chronic diseases such as cancer and AIDS and in those who have severe stresses such as burns or extensive infections (see Chapter 24). The consequences of PEM as a world malnutrition problem are considered here; the problems associated with PEM and illness are described throughout later chapters.

Marasmus and Kwashiorkor PEM takes two different forms, with some cases exhibiting a combination of the two. In one form, the person is shriveled and emaciated—this disease is called **marasmus**. In the second, a swollen belly and skin rash are present, and the disease is named **kwashiorkor**. In the combination, some features of each type are present. Marasmus reflects a chronic inadequate food intake and therefore inadequate energy, vitamins, and minerals, as well as too little protein. Kwashiorkor may result from severe acute malnutrition, with too little protein to support body functions.

Marasmus Marasmus commonly occurs in children from 6 to 18 months of age in all the overpopulated and impoverished areas of the world. Children in impoverished nations subsist on a weak cereal drink that supplies scant energy and protein of low quality: such food can barely sustain life, much less support growth. Consequently, marasmic children look like little old people—just skin and bones.

Without adequate nutrition, muscles, including the heart muscle, waste and weaken. Because the brain normally grows to almost its full adult size within the first two years of life, marasmus impairs brain development and learning ability. Reduced synthesis of key hormones leads to a metabolism so slow that body temperature drops below normal. There is little or no fat under the skin to insulate against cold. Some hospital workers find that the primary need of marasmic children is to be clothed, covered, and kept warm.

The starving child faces this threat to life by engaging in as little activity as possible—not even crying for food. The body gathers all its forces to meet the crisis, so it cuts down on any expenditure of energy not needed for the heart, lungs, and brain to function. Growth ceases; the child is no larger at age four than at age two. Digestive enzymes are in short supply, the digestive tract lining deteriorates, and absorption fails. The child cannot assimilate what little food is eaten.

Blood proteins, including hemoglobin, are no longer synthesized. Antibodies to fight off invading bacteria are degraded to provide amino acids for other uses, rendering the child vulnerable to infection. Then **dysentery**, an infection of the digestive tract, causes diarrhea, further depleting the body of nutrients. In the marasmic child, once infection has set in, kwashiorkor often follows and the immune response weakens further.[5] The infection that occurs with malnutrition ■ is responsible for two-thirds of the deaths of young children in developing countries.

If caught in time, a child's starvation may be reversed by careful nutrition therapy. In the severe cases, fluid balances are most critical. Diarrhea will have depleted the body's potassium and disturbed other electrolyte balances. Careful correction of fluid and electrolyte imbalances usually raises the blood pressure and strengthens the heart. After the first 24 to 48 hours, protein and energy may be given in small quantities, with intakes gradually increased as tolerated. Years after PEM is corrected, however, a child may still experience deficits in thinking and achievement in school compared with well-nourished peers.

Kwashiorkor Kwashiorkor was originally a Ghanaian word meaning a "sickness that infects the first child when the second child is born." If you consider how kwashiorkor often develops, you can easily see how the Ghanaians arrived at this name for the disease. When a mother who has been nursing her first child bears a second child, she weans the first child and puts the second one on the breast. The first child, suddenly switched from nutrient-dense, protein-rich breast milk to a starchy, protein-poor gruel, soon begins to sicken and die. Kwashiorkor typically sets in at about the age of two. Though rare in the United States, kwashiorkor is not entirely unknown, usually occurring when well-meaning but misinformed parents offer their young children a protein-poor "health-food" rice drink instead of cow's milk.[6]

■ When two variables interact so that each increases the other, synergism (SIN-er-jiz-um) is said to be occurring. Malnutrition and infection are a deadly combination because they work in this way.

> syn = with, together
> ergism = work

protein-energy malnutrition (PEM): a deficiency of protein and food energy; the world's most widespread malnutrition problem, including both marasmus and kwashiorkor.

> mal = bad, poor

marasmus (ma-RAZZ-mus): the most common form of severe PEM before one year of age. Marasmus is characterized by generalized muscle wasting associated with extreme deprivation, or impaired absorption, of energy, protein, vitamins, and minerals.

kwashiorkor (kwash-ee-OR-core or kwash-ee-or-CORE): a severe form of PEM that occurs more frequently after 18 months of age. Kwashiorkor is characterized by failure to grow and develop, changes in the pigmentation of the hair and skin, edema, and fatty liver. Kwashiorkor is associated with inadequate protein intake and infections.

dysentery (DIS-en-terry): an infection of the gastrointestinal tract caused by an amoeba or bacterium that gives rise to severe diarrhea.

> dys = bad
> entery = intestine

Features of Marasmus and Kwashiorkor in Children

Separating PEM into two classifications oversimplifies the condition, but at the extremes, marasmus and kwashiorkor exhibit marked differences. Marasmus-kwashiorkor mix presents symptoms common to both marasmus and kwashiorkor. In all cases, children are likely to develop diarrhea, infections, and multiple nutrient deficiencies.

Marasmus	Kwashiorkor
Infancy (less than 2 yr)	Older infants and young children (1 to 3 yr)
Severe deprivation or impaired absorption of protein, energy, vitamins, and minerals	Inadequate protein intake or, more commonly, infections
Develops slowly; chronic PEM	Rapid onset; acute PEM
Severe weight loss	Some weight loss
Severe muscle wasting with fat loss	Some muscle wasting, with retention of some body fat
Growth: <60% weight-for-age	Growth: 60 to 80% weight-for-age
No detectable edema	Edema
No fatty liver	Enlarged, fatty liver
Anxiety, apathy	Apathy, misery, irritability, sadness
Appetite may be normal or impaired	Loss of appetite
Hair is sparse, thin, and dry; easily pulled out	Hair is dry and brittle; easily pulled out; changes color; becomes straight
Skin is dry, thin, and wrinkled	Skin develops lesions

Some symptoms of kwashiorkor resemble those of marasmus (see Table 4–2). Proteins and hormones that previously maintained fluid balance diminish, and fluid leaks into the interstitial spaces. The child's limbs and belly become swollen with edema, a distinguishing feature of kwashiorkor; a **fatty liver** develops due to a lack of the protein carriers that transport fat out of the liver. The child's hair loses

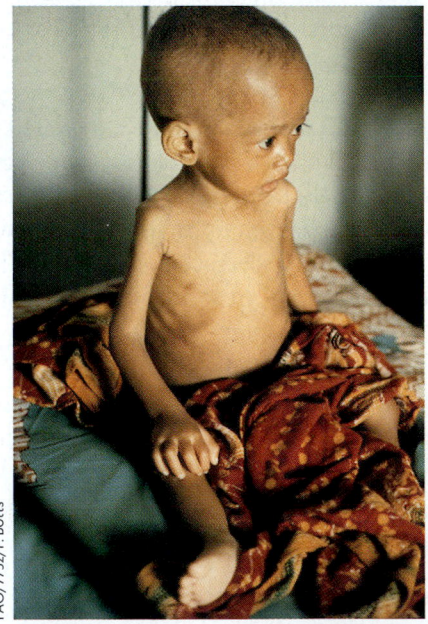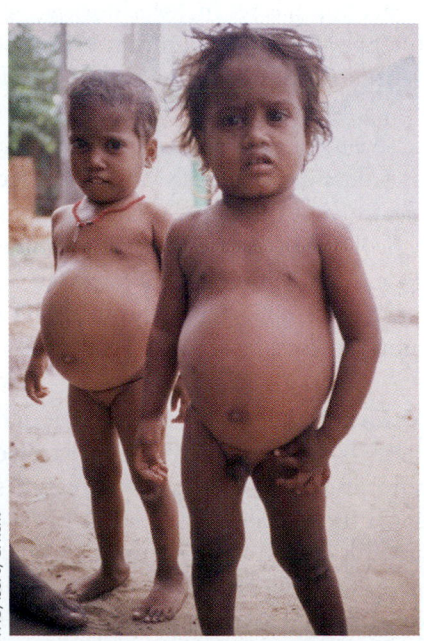

In the photo on the left, the extreme loss of muscle and fat characteristic of marasmus is apparent in the child's matchstick arms and legs. In contrast, the edema is characteristic of kwashiorkor is apparent in the swollen bellies of the children in the photo on the right.

its color; the skin becomes patchy and scaly, sometimes with ulcers and sores that fail to heal.

Clearly, preventing marasmus and kwashiorkor from developing in the first place is more desirable than attempting to treat the conditions after they have set in. Experts assure us that we possess the knowledge, technology, and resources to end the hunger that leads to PEM. Programs that have involved the local people in the process of identifying the problem and devising its solution have met with some success. But until those who have the food, technology, and resources make fighting hunger a priority, the war on hunger will not be won.

Protein Excess

While many of the world's people struggle to obtain enough food and enough protein to survive, in the developed nations protein is so abundant that problems of protein excess are seen. Overconsumption of protein offers no benefits and may pose health risks for the heart, weakened kidneys, and the bones.[7] Selecting too many protein-rich foods, such as meat and milk, adds saturated fat and crowds out fruits, vegetables, and whole grains, a problem from the standpoint of chronic disease risks.

Heart Disease Foods rich in animal protein tend to be high in saturated fat, a known contributor to atherosclerosis and heart disease. In a study of more than 40,000 women, the higher the intake of protein from red meat and dairy products, the greater the risk of heart disease.[8] As this chapter's Nutrition in Practice explains, people who substitute plant protein for animal protein lower their risks of dying from heart disease.[9]

Kidney Disease Excretion of the end products of protein metabolism depends, in part, on an adequate fluid intake and healthy kidneys. A high protein intake increases the work of the kidneys but does not appear to cause kidney disease. A high-protein diet does worsen existing kidney disease and may accelerate a decline in only mildly impaired kidneys.[10] One of the most effective ways to slow the progression of kidney disease is to restrict dietary protein.

Adult Bone Loss Research clearly shows that high protein intakes increase urinary calcium excretion.[11] It is less clear, however, whether high protein intakes deplete the bones of calcium and thereby compromise bone health.[12] It has been speculated that the effects of protein intake on bone health may depend on dietary calcium intake.[13] Increasing calcium intake may compensate for the effects of protein on urinary calcium excretion. In a three-year study of healthy elderly men and women, protein intake was positively associated with a gain in bone density in those who received calcium supplements (500 milligrams per day).[14] In contrast, the unsupplemented group lost bone over the three-year period. The researchers concluded that in many elderly men and women, bone health may be improved by increasing protein intake as long as calcium (and vitamin D) intakes meet recommendations. In establishing protein recommendations, the DRI committee considered protein's effect on bone health but did not find sufficient evidence to set a Tolerable Upper Intake Level.[15]

IN SUMMARY

- Protein deficiencies arise from both energy-poor and protein-poor diets and lead to the devastating diseases of marasmus and kwashiorkor.

- Together, these diseases are known as PEM (protein-energy malnutrition), a major form of malnutrition causing death in children worldwide.

- Excesses of protein offer no advantage; in fact, overconsumption of protein-rich foods may incur health problems as well.

fatty liver: an accumulation of fat in the liver. In PEM, fat accumulates in the liver because no protein is available to form the lipoproteins that normally escort fat molecules in the blood (see Chapter 20).

Protein and Amino Acid Supplements

In view of the high protein intakes of people in the United States and other developed nations, it is surprising that many people feel compelled to take protein. Why do people take protein supplements? Athletes take them hoping to build muscle. Dieters take them to spare their bodies' protein while losing weight. Some women take them to strengthen their fingernails. People take individual amino acid supplements too—to make themselves sleep better and to relieve depression. Do protein and amino acid supplements really do these things? Probably not. Are they safe? Not always.

Protein Supplements Protein supplements, including **whey protein** are popular with athletes, but well-fed athletes do not need them. A by-product of cheese manufacturing, whey protein from milk is a common low-cost ingredient in many protein powders. When combined with strength training, whey supplements may increase protein synthesis slightly, but they do not seem to improve athletic performance.[16] If supplements create a surplus of protein or certain amino acids, the excess must be metabolized, placing a burden on the kidneys to excrete excess nitrogen. To build stronger muscles, athletes need to eat food with adequate energy and protein to support the weight-training work that does increase muscle mass.

Amino Acid Supplements Popular reports that the amino acid tryptophan could relieve pain and cure depression and insomnia led to widespread public use about 20 years ago. More than 1,500 people who elected to take tryptophan developed an illness called EMS (short for *eosinophilia-myalgia syndrome*). EMS is characterized by severe muscle and joint pain, limb swelling, an elevated white blood cell count, extremely high fever, and, in some cases, death. Contaminants in the supplement were thought to be the cause of the disease, and the Food and Drug Administration (FDA) issued a recall of all products containing tryptophan except for specific medical formulas.

The DRI committee reviewed the available research on amino acids, but with next to no safety research in existence, the committee was unable to set Tolerable Upper Intake Levels for supplemental doses. Until research becomes available, no level of amino acid supplementation can be assumed to be safe for all people. Growth or altered metabolism makes the following groups of people especially likely to suffer harm from amino acid supplements:

- All women of childbearing age
- Pregnant or lactating women
- Infants, children, and adolescents
- Elderly people
- People with inborn errors of metabolism that affect their bodies' handling of amino acids
- Smokers
- People on low-protein diets for therapeutic reasons
- People with chronic or acute mental or physical illnesses who take amino acids without medical supervision

Protein Recommendations and Nitrogen Balance

The committee that established the RDA states that a generous daily protein allowance for a healthy adult is 0.8 gram per kilogram (2.2 pounds) of healthy body weight. The protein RDA is adjusted to cover additional needs for building new tissue and so is slightly higher for infants, children, and pregnant and lactating women. ■

In setting the RDA, the committee assumes that the protein eaten will be of good quality, that it will be consumed together with adequate energy from carbohydrate and fat, and that other nutrients in the diet will be adequate. The committee also assumes that the RDA will be applied only to healthy individuals with no unusual alteration of protein metabolism. Most people in the United States receive much more protein than they need.

Nitrogen Balance Underlying the protein recommendation are **nitrogen balance** studies, which compare nitrogen lost by excretion with nitrogen eaten in food. If the body maintains the same amount of protein in its tissues from day to day, it is in nitrogen equilibrium (zero nitrogen balance). If the body adds protein, it is in positive nitrogen balance; if it loses protein, it is in negative nitrogen balance. ■

Normally, healthy adults are in nitrogen equilibrium; that is, their nitrogen intakes equal their nitrogen outputs. In other words, protein intake from food balances with nitrogen excretion in the urine, feces, and sweat. Growing children, adolescents, and pregnant women are in positive nitrogen balance because they are adding new blood, bone, and muscle cells to their bodies. People who are fasting or starving and people who suffer from traumas, such as burns (see Chapter 24), are in negative nitrogen balance because their bodies are forced to use protein for energy.

Protein on Food Labels All food labels must state the *quantity* of protein in grams. The "percent Daily Value" for protein is not mandatory on all labels, but it is required whenever a food makes a protein claim or is intended for consumption by children under four years old. ■ *

IN SUMMARY

- Normal, healthy people do not need amino acid or protein supplements.

- Optimally, an adult's diet will be adequate in energy from carbohydrate and fat and will deliver at least 0.8 grams of protein per kilogram of healthy body weight each day.

Protein in Foods

In the United States and Canada, where nutritious foods are abundant, most people eat protein in such large quantities that they receive all the amino acids they need. In countries where food is scarce and the people eat only marginal amounts of protein-rich foods, however, the *quality* of the protein becomes crucial.

Protein Quality

The protein quality of the diet determines, in large part, how well children grow and how well adults maintain their health. Put simply, **high-quality proteins** provide enough of all the essential amino acids needed to support the body's work, and low-quality proteins do not. Two factors influence protein quality—the protein's digestibility and its amino acid composition.

---

* For labeling purposes, the Daily Values for protein are as follows: for infants, 14 grams; for children under age four, 16 grams; for older children and adults, 50 grams; for pregnant women, 60 grams; and for lactating women, 65 grams.

Sidebar

■ RDA for protein (adult) = 0.8 g/kg
   ■ 10 to 35% of energy intake
To calculate your protein need:
1. Find your body weight in pounds (or kilograms).
2. Convert pounds to kilograms, if necessary (pounds divided by 2.2).
3. Multiply kilograms by 0.8 to find total grams of protein recommended.
For example:
1. Weight = 150 lb
2. 150 lb ÷ 2.2 lb/kg = 68 kg
3. 68 kg × 0.8 g/kg = 54 g

■ Nitrogen equilibrium (zero nitrogen balance): N in = N out
Positive nitrogen balance:
   N in > N out
Negative nitrogen balance:
   N in < N out

■ Daily Values:
   ■ 50 g protein (based on 10% of 2,000 kcal diet)

whey protein: a by-product of cheese production; falsely promoted as increasing muscle mass. Whey is the watery part of milk that separates from the curds.

nitrogen balance: the amount of nitrogen consumed (N in) as compared with the amount of nitrogen excreted (N out) in a given period of time. The laboratory scientist can estimate the protein in a sample of food, body tissue, or excreta by measuring the nitrogen in it.

high-quality proteins: dietary proteins containing all the essential amino acids in relatively the same amounts that human beings require. They may also contain nonessential amino acids.

Digestibility   As explained earlier, proteins must be digested before they can provide amino acids. **Protein digestibility** depends on such factors as the protein's source and the other foods eaten with it. The digestibility of most animal proteins is high (90 to 99 percent); plant proteins are less digestible (70 to 90 percent for most, but over 90 percent for soy).

Amino Acid Composition   To make proteins, cells must have all the needed amino acids available simultaneously. The liver can produce any nonessential amino acid that may be in short supply so that the cells can continue linking amino acids into protein strands. If an essential amino acid is missing, however, a cell must dismantle its own proteins to obtain it. Therefore, to prevent protein breakdown, dietary protein must supply at least the nine essential amino acids plus enough nitrogen-containing amino groups and energy for the synthesis of the others. If the diet supplies too little of any essential amino acid, protein synthesis will be limited. The body makes whole proteins only; if one amino acid is missing, the others cannot form a "partial" protein. An essential amino acid that is available in the shortest supply relative to the amount needed to support protein synthesis is called a **limiting amino acid**.

High-Quality Proteins   A high-quality protein contains all the essential amino acids in amounts adequate for human use; it may or may not contain all the others. Generally, proteins derived from animal foods (meats, fish, poultry, cheese, eggs, yogurt, and milk) are high quality, although gelatin is an exception. Proteins derived from plant foods (legumes, grains, and vegetables) tend to be limiting in one or more essential amino acids. Some plant proteins are notoriously low quality—for example, corn protein. Others are high quality—for example, soy protein. As discussed in Nutrition in Practice 4, the educated vegetarian can design a diet that is adequate in protein by choosing a variety of legumes, whole grains, nuts, and vegetables. Table 4–3 lists

TABLE 4–3
Protein-Containing Foods

Milk, Yogurt, and Cheese

Each of the following provides about 8 grams of protein:

1 c milk, buttermilk, or yogurt (choose low-fat or fat-free)

1 oz regular cheese (for example, cheddar or Swiss; choose low-fat)

¼ c cottage cheese (choose low-fat or fat-free)

Meat, Poultry, Fish, Legumes, Eggs, and Nuts

Each of the following provides about 7 grams of protein:

1 oz meat, poultry, or fish (choose lean meats to limit saturated fat intake)

½ c legumes (navy beans, pinto beans, black beans, lentils, soybeans, and other dried beans and peas)

1 egg

½ c tofu (soybean curd)

2 tbs peanut butter

1 to 2 oz nuts or seeds

Grains

Each of the following provides about 3 grams of protein:

1 slice of bread

½ c cooked rice, pasta, cereals, or other grain foods

Vegetables

Each of the following provides about 2 grams of protein:

½ c cooked vegetables

1 c raw vegetables

Vegetarians obtain their protein from whole grains, legumes, nuts, vegetables, and, in some cases, eggs and milk products.

the protein contents of foods based on the food groups of the USDA Food Guide in Chapter 1. Fruits are not included in Table 4–3 because they contribute only small amounts of protein.

Complementary Proteins If the body does not receive all the essential amino acids it needs, the supply of essential amino acids will dwindle until body organs are compromised. Obtaining enough essential amino acids presents no problem to people who regularly eat high-quality proteins, such as those of meat, fish, poultry, cheese, eggs, milk, and many soybean products. The proteins of these foods contain ample amounts of all the essential amino acids. An equally sound choice is to eat two different protein foods from plants so that each supplies the amino acids missing in the other. In this strategy, the two protein-rich foods are combined to yield **complementary proteins** (see Figure 4–7)— proteins containing all the essential amino acids in amounts sufficient to support health. This concept is illustrated in Figure NP4–1 in this chapter's Nutrition in Practice. The two proteins need not even be eaten together, as long as the day's meals supply them both and the diet provides enough energy and total protein from a variety of sources.

Protein Sparing

Dietary protein—no matter how high the quality—will not be used efficiently and will not support growth when energy from carbohydrate and fat is lacking. The body assigns top priority to meeting its energy need and, if necessary, will break down protein to meet this need. After stripping off and excreting the nitrogen from the amino acids, the body will use the remaining carbon skeletons in much the same way it uses those from glucose or fat. A major reason why people must have ample carbohydrate and fat in the diet is to prevent this wasting of protein. ■

IN SUMMARY

- A diet inadequate in any of the essential amino acids limits protein synthesis.

- The best guarantee of amino acid adequacy is to eat foods containing high-quality proteins or combinations of foods containing complementary proteins so that each can supply the amino acids missing in the other.

- Vegetarians who consume no foods of animal origin can meet their protein needs by eating a variety of whole grains, legumes, seeds, nuts, and vegetables.

	Ile	Lys	Met	Trp
Legumes	■	■		
Grains			■	■
Together	■	■	■	■

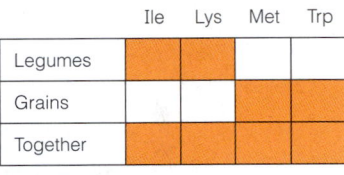

FIGURE 4–7

Complementary Proteins

In general, legumes provide plenty of isoleucine (Ile) and lysine (Lys) but fall short in methionine (Met) and tryptophan (Trp). Grains have the opposite strengths and weaknesses, making them a perfect match for legumes.

■ Reminder: Carbohydrate and fat allow amino acids to be used to build body proteins. This is known as the *protein-sparing effect* of carbohydrate and fat.

protein digestibility: a measure of the amount of amino acids absorbed from a given protein intake.

limiting amino acid: an essential amino acid that is present in dietary protein in the shortest supply relative to the amount needed for protein synthesis in the body.

complementary proteins: two or more proteins whose amino acid assortments complement each other in such a way that the essential amino acids missing from one are supplied by the other.

Analysis

YOUR DIET

Proteins and Amino Acids

Most people in the United States and Canada consume more protein than they need. This is not surprising considering the abundance of food eaten and the central role meats hold in the North American diet. Foods that derive from animals—meats, fish, poultry, eggs, and milk products—provide plenty of protein but are often accompanied by fat. Those that derive from plants—whole grains, vegetables, and legumes—may provide less protein but also less fat.

- Calculate your daily protein needs: Look up the healthy weight range for a person of your height on the Body Mass Index (BMI) table on the inside back cover. If your present weight falls within that range, use it for the following calculations. If your present weight falls outside the range, use the midpoint of the healthy weight range as your reference weight.

- Convert pounds to kilograms, if necessary (pounds divided by 2.2 equals kilograms).

- Multiply kilograms by 0.8 to get your RDA in grams per day. (Older teens 14 to 18 years old, multiply by 0.85.) Example:

$$\text{Weight} = 150 \text{ pounds}$$
$$150 \text{ pounds} \div 2.2 \text{ pounds/kilogram} = 68 \text{ kilograms (rounded off)}$$
$$68 \text{ kilograms} \times 0.8 \text{ grams/kilogram} = 54 \text{ grams protein}$$
$$\text{(rounded off)}$$

- Protein RDA: _____ grams

- Using the table you made in Chapter 1 which compares the foods you ate in a day with the USDA Food Guide recommendations, estimate your protein intake for the day in the accompanying table. Multiply the amount of food you consumed by the estimated protein in the food to guesstimate your total protein intake.

  Total protein intake:_____ grams

- Do you eat enough protein to meet your protein RDA?

- Next, determine whether your protein intake came mostly from plant-based foods or animal-based foods. To estimate your intake from plant-based foods, add up the grams of protein in the first two rows (grains, vegetables), and the grams of protein derived from plant-based foods in the last row (legumes, tofu, nuts, and others). Your total protein intake from plant-based foods: _____ grams.

- What percentage of your total protein intake derived from plant-based foods? For example, if your total protein intake for the day was 60 grams and your protein intake from plant-based foods was 20 grams:

$$20 \div 60 = .33$$
$$.33 \times 100 = 33\% \text{ protein intake from plant-based foods}$$

  Does your diet consist of mostly plant-based or animal-based protein foods? _____

- If the majority of protein in your diet derives from animal-based foods, suggest some strategies to include more plant-based protein foods in your diet. Legumes, whole grains, vegetables, nuts, and seeds are rich in fiber and lower in saturated fat than some animal-based protein foods such as red meat, whole milk, and cheese.

Food Groups[a]	Amount Consumed	Estimated Protein	Totals (grams)
Grains	_____ ounce(s)	3 grams/ounce (1 ounce equals 1 slice bread, ½ cup cooked rice, cereal, or pasta, or 1 cup ready-to-eat cereal)	_____
Vegetables	_____ cup(s)	2 grams/½ cup cooked vegetables or 1 cup raw vegetables	_____
Milk, Yogurt, and Cheese	_____ cup(s)	8 grams/cup of milk or yogurt, or 1 ounce of cheese, or ¼ cup cottage cheese	_____
Meat, Poultry, Fish, Legumes, Eggs, and Nuts	_____ ounce(s)	7 grams/ounce of meat, poultry, or fish; ½ cup of legumes; 1 egg; ½ cup tofu; 2 tablespoons peanut butter; 1 to 2 ounces of nuts or seeds	_____ meat, poultry, fish, eggs _____ legumes, tofu, nuts, or other plant-based protein foods

[a] Fruits and oils are not included here because they contribute little, if any, protein.

1. Considering the health effects of too little dietary protein, what suggestions would you have for a teenage girl who reports the following information about her food intake:

   ■ She never eats any meat or other animal-derived foods because she is a vegan. On a typical day, she eats toast and juice for breakfast; chips, a soft drink, and a piece of fruit for lunch; and a small serving of plain pasta with tomato sauce or steamed vegetables for dinner, along with a glass of water or tea.

   ■ She takes amino acid supplements because a friend told her that the only way to get amino acids if she does not eat meat is to take them as supplements.

2. Considering the health effects of excess dietary protein, what advice would you have for a college athlete who tells you he wants to bulk up his muscles and reports the following information about his food intake:

   ■ He eats large servings of meat (usually red meat) at least twice a day. He drinks whole milk two or three times a day and eats eggs and bacon for breakfast almost every day.

   ■ He avoids breads, cereals, and pasta in order to save room for protein-rich foods such as meat, milk, and eggs.

   ■ He eats a piece of fruit once in awhile but seldom eats vegetables because they are too time-consuming to prepare.

Self Check

1. Proteins are chemically different from carbohydrates and fats because they also contain:
   a. iron.
   b. sodium.
   c. nitrogen.
   d. phosphorus.

2. The basic building blocks for protein are:
   a. side groups.
   b. amino acids.
   c. glucose units.
   d. saturated bonds.

3. Enzymes are proteins that, among other things:
   a. defend the body against disease.
   b. regulate fluid and electrolyte balance.
   c. facilitate chemical reactions by changing themselves.
   d. help assemble disaccharides into starch, cellulose, or glycogen.

4. Functions of proteins in the body include:
   a. supplying omega-3 fatty acids for growth, lowering serum cholesterol, and helping with weight control.
   b. supplying fiber to aid digestion, digesting cellulose, and providing the main fuel source for muscles.
   c. protecting organs against shock, helping the body use carbohydrate efficiently, and providing triglycerides.
   d. supporting growth and maintenance, supplying hormones to regulate body processes, and maintaining fluid and electrolyte balance.

5. The swelling of body tissue caused by the leakage of fluid from the blood vessels into the interstitial spaces is called:
   a. edema.
   b. anemia.
   c. acidosis.
   d. sickle-cell anemia.

6. Major proteins in the blood that protect against bacteria and other disease agents are called:
   a. acids.
   b. buffers.
   c. antigens.
   d. antibodies.

7. Marasmus can be distinguished from kwashiorkor because in marasmus:
   a. only adults are victims.
   b. the cause is usually an infection.
   c. severe wasting of body fat and muscle occurs.
   d. the limbs and face swell with edema, and the belly bulges with a fatty liver.

8. The RDA for protein for a healthy adult is ___ gram(s) per kilogram of appropriate or average body weight for height.
   a. 0.5
   b. 0.8
   c. 1.1
   d. 1.4

9. Generally speaking, from which of the following foods are complete proteins derived?
   a. milk, gelatin, and soy
   b. rice, potatoes, and eggs
   c. meats, fish, and poultry
   d. vegetables, grains, and fruits

10. An incomplete protein lacks one or more:
    a. hydrogen bonds.
    b. essential fatty acids.
    c. saturated fatty acids.
    d. essential amino acids.

Answers to these questions can be found in Appendix H.

Notes

1. T. R. Ziegler and coauthors, Trophic and cytoprotective nutrition for intestinal adaptation, mucosal repair, and barrier function, *Annual Review of Nutrition* 23 (2003): 229–261.

2. Standing Committee on the Scientific Evaluation of Dietary Reference Intakes, Food and Nutrition Board, Institute of Medicine, *Dietary Reference Intakes for Energy, Carbohydrate, Fiber, Fat, Fatty Acids, Cholesterol, Protein, and Amino Acids* (Washington, D.C.: National Academies Press, 2005), p. 605.

3. Standing Committee on the Scientific Evaluation of Dietary Reference Intakes, 2005, p. 605.

4. Data from www.unicef.org, posted April 2005 and May 2006.

5. M. Reid and coauthors, The acute-phase protein response to infection in edematous and nonedematous protein-energy malnutrition, *American Journal of Clinical Nutrition* 76 (2002): 1409–1415.

6. K. A. Katz and coauthors, Rice nightmare; Kwashiorkor in 2 Philadelphia-area infants fed Rice Dream beverage, *Journal of the American Academy of Dermatology* 52 (2005): S69–S72.

7. Standing Committee on the Scientific Evaluation of Dietary Reference Intakes, 2005, p. 694.

8. L. E. Kelemen and coauthors, Associations of dietary protein with disease and mortality in a prospective study of postmenopausal women, *American Journal of Epidemiology* 161 (2005): 239–249.

9. B. L. McVeigh and coauthors, Effect of soy protein varying in isoflavone content on serum lipids in healthy young men, *American Journal of Clinical Nutrition* 83 (2006): 244–251; Kelemen and coauthors, 2005; S. Tonstad, K. Smerud, and L. Hoie, A comparison of the effects of 2 doses of soy protein or casein on serum lipids, serum lipoproteins, and plasma total homocysteine in hypercholesterolemic subjects, *American Journal of Clinical Nutrition* 76 (2002): 78–84.

10. E. L. Knight and coauthors, The impact of protein intake on renal function decline in women with normal renal function or mild renal insufficiency, *Annals of Internal Medicine* 138 (2003): 460–467.

11. B. Dawson-Hughes, Calcium and protein in bone health, *The Proceedings of the Nutrition Society* 62 (2003): 505–509; F. Ginty, Dietary protein and bone health, *The Proceedings of the Nutrition Society* 62 (2003): 867–876; J. E. Kerstetter, K. O. O'Brien, and K. L. Insogna, Low protein intake: The impact on calcium and bone homeostasis in humans, *Journal of Nutrition* 133 (2003): 855S–861S.

12. Dawson-Hughes, 2003; Ginty, 2003; Kerstetter, O'Brien, and Insogna, 2003.

13. B. Dawson-Hughes, Interaction of dietary calcium and protein in bone health in humans, *Journal of Nutrition* 133 (2003): 852S–854S.

14. B. Dawson-Hughes and S. S. Harris, Calcium intake influences the association of protein intake with rates of bone loss in elderly men and women, *American Journal of Clinical Nutrition* 75 (2002): 773–779.

15. Standing Committee on the Scientific Evaluation of Dietary Reference Intakes, 2005, p. 694.

16. K. D. Tipton, Ingestion of casein and whey proteins result in muscle anabolism after resistance exercise, *Medicine and Science in Sports and Exercise* 36 (2004): 33–38.

VEGETARIAN DIETS

Eating patterns all along the continuum of dietary choices—from one end, where people eat no foods of animal origin, to the other end, where they eat generous quantities of meat every day—can support or compromise nutritional health. The quality of the diet depends not on whether it consists of all plant foods or centers on meat, but on whether the eater's food choices are based on sound nutrition principles: adequacy of nutrient intakes, balance and variety of foods chosen, appropriate energy intake, and moderation in intakes of substances such as saturated fat, sodium, alcohol, and caffeine that are harmful when consumed in excess. As mentioned in Chapter 4, however, because vegetarian diets exclude at least some animal-derived foods, they are usually lower in saturated fat and cholesterol than many meat-based diets.

People choose to exclude meat and other animal-derived foods from their diets for various reasons—concern for animal welfare, beliefs about health, religious convictions, or convenience. Some believe that vegetarianism is better for the environment; some, that it is healthier; and some, that it is less costly than the meat-eating alternative. Some just like it better. Whatever the reasons, vegetarians and health professionals who work with them should be aware of the nutrition and health implications of vegetarian diets.

Because vegetarian diets vary in both the types and the amounts of animal-derived foods they include, these differences must be considered when evaluating the health status of vegetarians. The accompanying glossary defines the various kinds of vegetarian diets.

Are vegetarian diets nutritionally sound?

The American Dietetic Association takes the position that well-planned vegetarian diets offer nutrition and health benefits to adults in general.[1] Research suggests that meat-eating adults who switch to vegetarian diets reduce their risks of heart disease, hypertension, diabetes, some types of cancer, and obesity.[2]

What should be my main concerns when planning a nutritionally sound vegetarian diet?

A vegetarian diet planner faces the same task as other diet planners—obtaining a variety of foods that provide all the needed nutrients within an energy allowance that maintains a healthy body weight. The challenge is to do so using at least one less food group. Since all vegetarians omit meat and some omit other animal-derived foods, protein, the nutrient that meat is famous for, merits some discussion here.

Is protein a problem in vegetarian diets?

No, protein is not the problem it was once thought to be in vegetarian diets. People who include animal-derived foods such as milk and eggs in their diets need not worry at all about

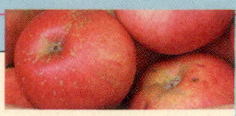

GLOSSARY OF VEGETARIAN TERMS

lacto-ovo vegetarians: people who include milk or milk products and eggs, but omit meat, fish, shellfish, and poultry from their diets.

lacto-vegetarians: people who include milk or milk products, but exclude meat, poultry, fish, shellfish, and eggs from their diets.

semivegetarians: people who include some, but not all, groups of animal-derived foods in their diets; they usually exclude meat and may occasionally include poultry, fish, and shellfish; also called *partial vegetarians*.

vegans: people who exclude all animal-derived foods (including meat, poultry, fish, shellfish, eggs, cheese, and milk) from their diets; also called *strict vegetarians* or *total vegetarians*.

protein deficiency. Even for those who eat only plant-derived foods, protein intakes are usually satisfactory as long as energy intakes are adequate and protein sources are varied.[3] A mixture of proteins from whole grains, legumes, seeds, nuts, and vegetables can provide adequate amounts of high-quality protein.

The idea persists that **vegans** must carefully combine their plant-protein foods in order to obtain the protein they need, but this is not necessary. Plant foods can provide more than enough high-quality protein and can sustain people in good health, as long as the diet supplies sufficient energy and does not include too many empty-kcalorie foods. As Chapter 4 explains, however, a meal delivers higher-quality protein when it combines two or more different individual plant-protein sources. Figure NP4–1 (p. 110) shows combinations of plant proteins that provide higher-quality protein than the individual foods alone could supply.

What sorts of food energy intakes do vegetarian diets provide?

Researchers find that vegetarians as a group are closer to a healthy body weight than nonvegetarians.[4] Because obesity impairs health in a number of ways, vegetarians therefore have a health advantage. Vegetarian diets tend to be high in starch- and fiber-rich carbohydrates and low in saturated fat, characteristics that are consistent with current dietary recommendations aimed at reducing the incidence of obesity and other degenerative diseases in this country.

Not all vegetarians fit the average pattern, though. Obesity does threaten vegetarians who include milk, eggs, and cheese in their diets. They can easily consume excess saturated fat and food energy and so must be careful to select fat-free and low-fat dairy foods and to avoid relying too heavily on these foods in general.

In contrast, people who exclude all animal-derived foods (vegans) may have trouble obtaining *enough* food energy. This is

Vegetarians who eat no foods from animal sources select foods from two or more of these columns to create high-quality protein combinations:

Grains	Legumes	Seeds and Nuts	Vegetables
Barley	Dried beans (pinto, kidney, navy, etc.)	Almonds	Broccoli
Bulgur	Dried lentils	Cashews	Cabbage
Oats	Dried peas	Nut butters	Peppers
Pasta	Peanuts	Sesame seeds	Spinach
Rice	Soy products	Sunflower seeds	Squash
Whole-grain breads		Walnuts	

Black beans and rice, a favorite Hispanic combination.

Tofu and stir-fried vegetables with rice, an Asian dish.

FIGURE NP4–1

Nonmeat Mixtures That Provide High-Quality Protein

especially true for children. Vegan diets can fail to provide food energy sufficient to support the growth of a child within a bulk of food small enough for the child to eat. Frequent meals of fortified breads, cereals, or pastas with legumes, nuts, nut butters, and sources of unsaturated fats can help to meet protein and energy needs in a smaller volume at each sitting.[5] The MyPyramid resources, introduced in Chapter 1, include tips for planning vegetarian diets using the USDA Food Guide. In addition, several food guides have been developed specifically for vegetarian diets.[6] They all address the particular nutrition

FIGURE NP4–2

An Example of a Vegetarian Food Pyramid
Review Figure 1–4 and Table 1–6 to find recommended daily amounts from each food group, serving size equivalents, examples of common foods within each group, and the most notable nutrients for each group. Tips for planning a vegetarian diet can be found at MyPyramid.gov.

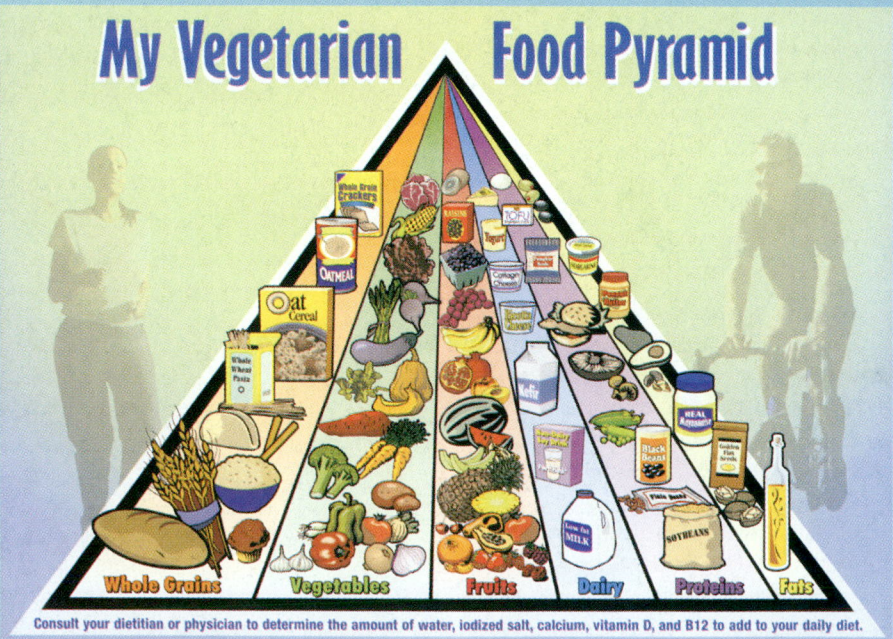

Source: © GC Nutrition Council, 2006, adapted from USDA 2005 Dietary Guidelines and www.mypyramid.gov. Copies can be ordered from 301-680-6717.

concerns of vegetarians, but differ slightly. Figure NP4–2 presents one version.

Tell me about vitamins and minerals. Does a person eating a vegetarian diet need to take vitamin supplements?

That depends on the kind of vegetarian diet. The diet of **lacto-ovo vegetarians** can be complete in all vitamins, but for vegans, several vitamins may be a problem. One such vitamin is B_{12}. Because vitamin B_{12} occurs only in animal-derived foods, supplements are necessary to prevent deficiency. Women who have adhered to all-plant diets for many years are especially likely to have low vitamin B_{12} stores. Pregnant vegan women, whose needs for vitamin B_{12} are especially high, find it virtually impossible to maintain adequate vitamin B_{12} status without taking supplements or including a reliable food source of the nutrient.

A vitamin B_{12} deficiency can take a long time to develop in adults because up to four years' worth of the vitamin can be stored in the body. But when the deficiency sets in, it does severe damage to the nervous system (see Chapter 8). In infants, deficiencies set in more rapidly and so threaten their nervous systems earlier. All vegan mothers must be sure to take the appropriate supplements or to use vitamin B_{12}-fortified products such as soy milk or breakfast cereals.

What other vitamins do vegans need?

Another vitamin of concern is vitamin D. The milk drinker is protected, provided the milk is fortified with vitamin D, but there is no practical source of vitamin D in plant foods. Regular exposure to the sun will prevent a deficiency, but vegans who are homebound or live in a northern climate or smoggy city probably should take vitamin D supplements to defend against bone loss. This is particularly important for infants, children, and older adults.

Riboflavin, another vitamin often obtained from milk, is not a problem for the vegan who eats dark greens frequently in ample servings. The vegan who does not consume a lot of greens, however, may not meet riboflavin needs. Nutritional yeast is a rich source of riboflavin for the vegetarian.

On a vegan diet, vitamin B_{12}, vitamin D, and riboflavin can be problems if a person is not careful. What about minerals?

For *all* vegetarians, not just the vegan, two minerals may be of concern—iron and zinc. Legumes are an important source of iron in the vegetarian diet. The iron in legumes, however, is not as absorbable as that in meat. In fact, people absorb three times as much iron from a meal that includes meat as from one that does not. For this reason, the iron recommendation for adult vegetarian men, premenopausal women, and adolescent girls is almost double the recommendation for meat eaters of the same gender and age.[7] Because vitamin C in fruits and vegetables can triple iron absorption from other foods eaten at the same meal, vegetarian meals should be rich in foods offering vitamin C.

Zinc may also be a problem nutrient for vegetarians. It is widespread in plant foods, but its availability may be hindered by the fibers and other binders found in fruits and vegetables. The zinc needs of vegetarians and the effects of mineral binders are subjects of intensive study at the present time. While research continues, vegetarians are advised to eat varied diets that include legumes such as navy beans and kidney beans, zinc-enriched cereals, and whole-grain breads well leavened with yeast, which improves the availability of their minerals. For those who include seafood in their diets, oysters, crabmeat, and shrimp are rich in zinc.

What about calcium for the vegan?

Yes, calcium is of concern. The milk-drinking vegetarian is protected from deficiency, but the vegan must find other sources of calcium. Some good calcium sources are regular and ample servings of dark green leafy vegetables such as kale and collard; legumes; calcium-fortified foods such as breakfast cereals, soy milk, and orange juice; some nuts such as almonds; and certain seeds such as sesame seeds. The choices should be varied because binders in some of these foods may hinder calcium absorption. The vegan is urged to use calcium-fortified soy milk in ample quantities regularly. This is especially important for children. Infant formula based on soy is fortified with calcium and can easily be used in cooking foods, even for adults.

Do vegetarian diets provide adequate amounts of the essential fatty acids?

Vegetarian diets typically provide enough omega-6 fatty acids but lack omega-3 fatty acids. This imbalance slows production of EPA and DHA in the body, and without fish or eggs in the diet, intake of EPA and DHA falls short as well. To compensate for this inadequacy, vegetarians need to include good sources of linolenic acid, such as flaxseed, walnuts, soybeans, and their oils, in their diets daily.

Are there any other health advantages to the vegetarian diet?

Yes. Vegetarian protein foods are often higher in fiber, richer in certain vitamins and minerals, and lower in fat, especially saturated fat, than meats. Vegetarians can enjoy a nutritious diet low in saturated fat provided they limit foods such as butter, cream cheese, and sour cream. If vegetarians follow the guidelines presented here and plan carefully, they can support their health as well as, or perhaps better than, nonvegetarians.

Abundant evidence supports the idea that vegetarians may actually be healthier than meat eaters. Informed vegetarians are not only more likely to be at the desired weights for their heights, but also to have lower blood cholesterol levels, lower rates of certain kinds of cancer, better digestive function, and more. Even among people who are health conscious, generally vegetarians experience fewer deaths from cardiovascular disease than meat eaters do. Because many vegetarians also abstain from smoking and the consumption of alcohol, dietary practices alone probably do not account for all the aspects of improved health. Clearly, however, they contribute significantly to it.

Notes

1. Position of the American Dietetic Association and Dietitians of Canada: Vegetarian diets, *Journal of the American Dietetic Association* 103 (2003): 748–765.

2. S. E. Berkow and N. D. Barnard, Blood pressure regulation and vegetarian diets, *Nutrition Reviews* 63 (2005): 1–8; J. Sabate, The contribution of vegetarian diets to health and disease: A paradigm shift? *American Journal of Clinical Nutrition* 78 (2003): 502S–507S; P. N. Singh, J. Sabate, and G. E. Fraser, Does low meat consumption increase life expectancy in humans? *American Journal of Clinical Nutrition* 78 (2003): 526S–532S; F. B. Hu, Plant-based foods and prevention of cardiovascular disease: An overview, *American Journal of Clinical Nutrition* 78 (2003): 544S–551S.

3. Position of the American Dietetic Association and Dietitians of Canada, 2003.

4. S. E. Berkow and N. Barnard, Vegetarian diets and weight status, *Nutrition Reviews* 64 (2006): 175–188; P. K. Newby, K. L. Tucker, and A. Wolk, Risk of overweight and obesity among semivegetarian, lactovegetarian, and vegan women, *American Journal of Clinical Nutrition* 81 (2005): 1267–1274; E. H. Haddad and J. S. Tanzman, What do vegetarians in the United States eat? *American Journal of Clinical Nutrition* 78 (2003): 626S–632S; N. Brathwaite and coauthors, Obesity, diabetes, hypertension, and vegetarian status among Seventh-Day Adventists in Barbados: Preliminary results, *Ethnicity and Disease* 13 (2003): 34–39.

5. Position of the American Dietetic Association and Dietitians of Canada, 2003.

6. M. Virginia, V. Melina, and A. R. Mangels, A new food guide for North American vegetarians, *Journal of the American Dietetic Association* 103 (2003): 771–775; C. A. Venti and C. S. Johnston, Modified food guide pyramid for lactovegetarians and vegans, *Journal of Nutrition* 132 (2002): 1050–1054.

7. Standing Committee on the Scientific Evaluation of Dietary Reference Intakes, Food and Nutrition Board, Institute of Medicine, *Dietary Reference Intakes for Vitamin A, Vitamin K, Arsenic, Boron, Chromium, Copper, Iodine, Iron, Manganese, Molybdenum, Nickel, Silicon, Vanadium, and Zinc* (Washington, D.C.: National Academy Press, 2001), pp. 9–45.

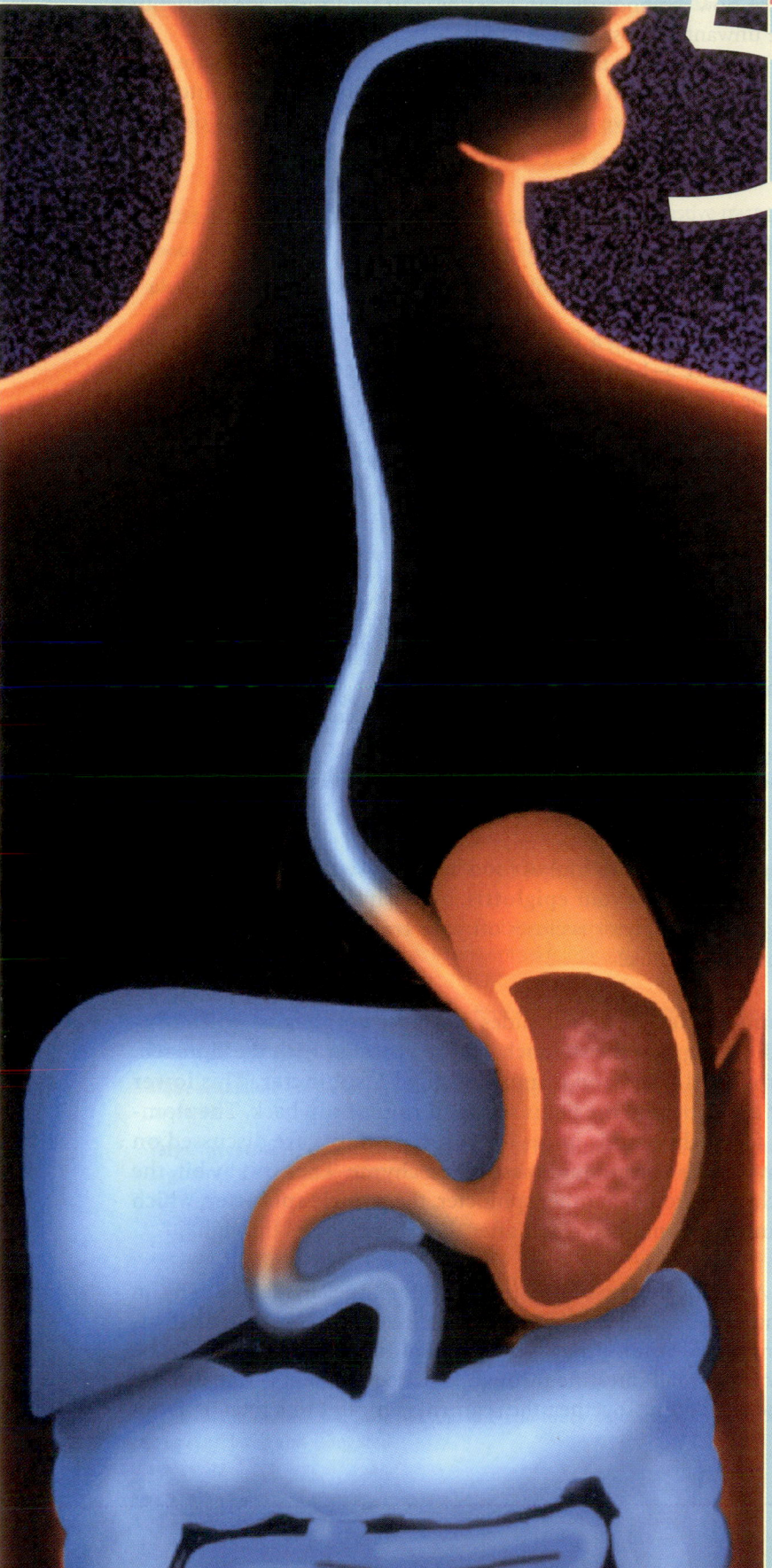

5 Digestion and Absorption

© SuperStock, Inc./SuperStock

FOOD SAFETY

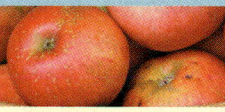

The Food and Drug Administration (FDA) lists **foodborne illness** as the leading food safety concern in the United States because episodes of food poisoning far outnumber episodes of any other kind of food contamination. The **CDC** estimates that 76 million people experience foodborne illness each year in the United States.[1] For some 5,000 people each year, the symptoms (Table NP5–1) can be so severe as to cause death. Most vulnerable are pregnant women; very young, very old, sick, or malnourished people; and those with a weakened immune system (as in AIDS). By taking the proper precautions, people can minimize their chances of contracting foodborne illnesses. The accompanying glossary defines related terms.

What is foodborne illness?

Foodborne illness can be caused by either an infection or an intoxication. Table NP5–2 summarizes the most common or severe foodborne illnesses, their food sources, general symptoms, and prevention methods.

What is the difference between foodborne infections and food intoxications?

Foodborne infections are caused by eating foods contaminated by infectious microbes. Two of the most common foodborne **pathogens** are *Campylobacter jejuni* and *Salmonella,* which enter the GI tract in contaminated foods such as undercooked poultry and unpasteurized milk. Symptoms generally include abdominal cramps, fever, vomiting, and diarrhea.

Food intoxications are caused by eating foods containing natural toxins or, more likely, microbes that produce toxins. The most common food toxin is produced by *Staphylococcus aureus;* it affects more than one million people each year. Less common, but more infamous, is *Clostridium botulinum,* an organism that produces a deadly toxin in anaerobic conditions such as improperly canned (especially home-canned) foods and homemade garlic or herb-flavored oils stored at room temperature. Botulism paralyzes muscles; consequently, seeing,

GLOSSARY

CDC (Centers for Disease Control): a branch of the Department of Health and Human Services that is responsible for, among other things, monitoring foodborne diseases.

cross-contamination: the contamination of food by bacteria that occurs when the food comes into contact with surfaces previously touched by raw meat, poultry, or seafood.

foodborne illness: illness transmitted to human beings through food and water, caused by either an infectious agent (foodborne infection) or a poisonous substance (food intoxication); commonly known as **food poisoning**.

Hazard Analysis Critical Control Points (HACCP): a systematic plan to identify and correct potential microbial hazards in the manufacturing, distribution, and commercial use of food products; commonly referred to as "HASS-ip."

pasteurization: heat processing of food that inactivates some, but not all, microorganisms in the food; not a sterilization process. Bacteria that cause spoilage are still present.

pathogens (PATH-oh-jens): microorganisms capable of producing disease.

sushi: vinegar-flavored rice and seafood, typically wrapped in seaweed and stuffed with colorful vegetables. Some sushi is stuffed with raw fish; other varieties contain cooked seafood.

traveler's diarrhea: nausea, vomiting, and diarrhea caused by consuming food or water contaminated by any of several organisms, most commonly, *E. coli, Shigella, Campylobacter jejuni,* and *Salmonella.*

speaking, swallowing, and breathing become difficult.[2] Because death can occur within 24 hours of onset, botulism demands immediate medical attention. Even then, survivors may suffer the effects for months or years.

How do people get foodborne illness?

Transmission of foodborne illness has changed as our food supply and lifestyles have changed.[3] In the past, foodborne illness was caused by one person's error in a small setting, such as improperly refrigerated egg salad at a family picnic, and affected only a few victims. Today, we are eating more foods prepared and packaged by others. Consequently, when a food manufacturer or restaurant chef makes an error, foodborne illness can become epidemic. An estimated 80 percent of reported foodborne illnesses are caused by errors in a commercial setting, such as the improper **pasteurization** of milk at a large dairy.

TABLE NP5–1

Symptoms of Foodborne Illness

Get medical help when these symptoms occur

- Bloody diarrhea
- Diarrhea lasting more than 3 days
- Difficulty breathing
- Difficulty swallowing
- Double vision
- Fever lasting more than 24 hours
- Headache accompanied by muscle stiffness and fever
- Numbness, muscle weakness, and tingling sensations in the skin
- Rapid heart rate, fainting, and dizziness

Foodborne Illnesses

Disease and Organism That Causes it	Most Frequent Food Sources	Onset and General Symptoms	Prevention Methods[a]
Foodborne Infections			
Campylobacteriosis (KAM-pee-loh-BAK-ter-ee-OH-sis) *Compylobacter jejuni* bacterium	Raw and undercooked poultry, unpasteurized milk, contaminated water	Onset: 2 to 5 days. Diarrhea, vomiting, abdominal cramps, fever; sometimes bloody stools; lasts 2 to 10 days	Cook foods thoroughly; use pasteurized milk; use sanitary food-handling methods.
Cryptosporidiosis (KRIP-toe-spo-rid-ee-OH-sis) *Cryptosporidium parvum* parasite	Commonly contaminated swimming or drinking water, even from treated sources; highly chlorine-resistant; contaminated raw produce and unpasteurized juices and ciders	Onset: 2 to 10 days. Diarrhea, stomach cramps, upset stomach, slight fever. Symptoms may come and go for weeks or months.	Wash all raw vegetables and fruits before peeling. Use pasteurized milk and juice. Do not swallow drops of water while using pools, hot tubs, ponds, lakes, rivers, or streams for recreation.
Cyclosporiasis (sigh-clo-spore-EYE-uh-sis) *Cyclospora cayetanensis* parasite	Contaminated water; contaminated fresh produce	Onset: 1 to 14 days. Watery diarrhea, loss of appetite, weight loss. stomach cramps, nausea, vomiting, fatigue. Symptoms may come and go for weeks or months.	Use treated, boiled, or bottled water; cook foods thoroughly; peel fruit.
E. coli infection *Escherichia coli*[b] bacterium	Undercooked ground beef, unpasteurized milk and juices, raw fruits and vegetables, contaminated water, and person-to-person contact	Onset: 1 to 8 days. Severe bloody diarrhea, abdominal cramps, vomiting; lasts 5 to 10 days	Cook ground beef thoroughly; use pasteurized milk; use sanitary food-handling methods; use treated, boiled, or bottled water.
Gastroenteritis[c] Norwalk virus	Person-to-person contact; raw foods, salads, sandwiches	Onset: 1 to 2 days. Vomiting; lasts 1 to 2 days.	Use sanitary food-handling methods.
Glardiasis (JYE-are-DYE-ah-sis) *Giardio lamblia* parasite	Contaminated water; uncooked foods	Onset: 7 to 14 days. Diarrhea (but occasionally constipation), abdominal pain, gas	Use sanitary food-handling methods; avoid raw fruits and vegetables where protozoa are endemic; dispose of sewage properly.
Hepatitis (HEP-ah-TIE-tis) Hepatitis A virus	Undercooked or raw shellfish	Onset: 15 to 50 days (28 days average). Diarrhea, dark urine, fever, headache, nausea, abdominal pain, jaundice (yellowed skin and eyes from buildup of wastes); lasts to 2 to 12 weeks	Cook foods thoroughly.
Listeriosis (lis-TER-ee-OH-sis) *Listeria monocytogenes* bacterium	Unpasteurized milk; fresh soft cheeses; luncheon meats, hot dogs	Onset: 1 to 21 days. Fever, muscle aches, nausea, vomiting, blood poisoning, complications in pregnancy, and meningitis (stiff neck, severe headache, and fever)	Use sanitary food-handling methods; cook foods thoroughly; use pasteurized milk.
Perfringens (per-FRINCE-enz) **food poisoning** *Clostridium perfringens* bacterium	Meats and meat products stored at between 120° and 130°F	Onset: 8 to 16 hours. Abdominal pain, diarrhea, nausea; lasts 1 to 2 days	Use sanitary food-handling methods; cook foods thoroughly; refrigerate foods promptly and properly.

continued

Disease and Organism That Causes it	Most Frequent Food Sources	Onset and General Symptoms	Prevention Methods[a]
Salmonellosis (sal-moh-neh-LOH-sis) *Salmonella* bacteria (>2,300 types)	Raw or undercooked eggs, meats, poultry, raw milk and other dairy products, shrimp, frog legs, yeast, coconut, pasta, and chocolate	Onset: 1 to 3 days. Fever, vomiting, abdominal cramps, diarrhea; lasts 4 to 7 days; can be fatal	Use sanitary food-handling methods; use pasteurized milk; cook foods thoroughly; refrigerate foods promptly and properly.
Shigellosis (Shi-gel-LOH-sis) *Shigella* bacteria (>30 types)	Person-to-person contact; raw foods, salads, sandwiches, and contaminated water	Onset: 1 to 2 days. Bloody diarrhea, cramps, fever; lasts 4 to 7 days	Use sanitary food-handling methods; cook foods thoroughly; use proper refrigeration.
Vibrio (VIB-ree-oh) **bacteria** *Vibrio vulnificus*[d] bacterium	Raw or undercooked seafood and contaminated water	Onset: 1 to 7 days. Diarrhea, abdominal cramps, nausea, vomiting; lasts 2 to 5 days; can be fatal	Use sanitary food-handling methods; cook foods thoroughly.
Yersiniosis (yer-SIN-ee-OH-sis) *Yersinia enterocolitica* bacteruim	Diarrhea, vomiting fever, abdominal pain	Onset: 1 to 2 days. Fever, stomach pain, diarrhea; lasts 1 to 3 weeks	Cook foods thoroughly; use pasteurized milk; use treated, boiled or bottled water.
Food Intoxications			
Botulism (BOT-chew-lizm) Botulinum toxin [produced by *Clostridium botulinum* bacterium, which grows without oxygen, in low-acid foods, and at temperatures between 40° and 120°F; the **botulinum** (BOT-chew-line-um) **toxin** responsible for botulism is called **botulin** (BOT-chew-lin)]	Anaerobic environment of low acidity (canned corn, peppers, green beans, soups, beets, asparagus, mushrooms, ripe olives, spinach, tuna, chicken, chicken liver, liver pâté, luncheon meats, ham, sausage, stuffed eggplant, lobster, and smoked and salted fish)	Onset: 4 to 36 hours. Nervous system symptoms, including double vision, inability to swallow, speech difficulty, and progressive paralysis of the respiratory system; often fatal; leaves prolonged symptoms in survivors	Use proper canning methods for low-acid foods; refrigerate homemade garlic and herb oils; avoid commercially prepared foods with leaky seals or with bent, bulging, or broken cans.
Staphylococcal (STAF-il-oh-KOK-al) food poisoning Staphylococcal toxin (produced by *Staphylococcus aureus* bacterium)	Toxin produced in improperly refrigerated meats; egg, tuna, potato, and macaroni salads; cream-filled pastries	Onset: 1 to 6 hours. Diarrhea, nausea, vomiting, abdominal cramps, fever; lasts 1 to 2 days	Use sanitary food-handling methods; cook food thoroughly; refrigerate foods promptly and properly; use proper home canning methods.

Note: Travelers' diarrhea is most commonly caused by *E.coli, Compylobacter jejuni, Shigella*, and *Salmonella*.

[a] Table NP5–3 provides more details on the proper handling, cooking and refrigeration of foods.

[b] The most serious strain is *E.coli* STEC O157.

[c] Gastroenteritis refers to an inflammation of the stomach and intestines but is the most common name used for illnesses caused by Norwalk viruses.

[d] Most cases of *Vibrio vulnificus* occur in persons with underlying illness, particularly those with fever disorders, diabetes, cancer, and AIDS, and those who require long-term steroid use. The fatality rate is 50 percent for this population.

In the mid-1990s, when a fast-food restaurant served undercooked burgers tainted with an infectious strain of *Escherichia coli*, hundreds of patrons became ill, and at least three people died. In the early 2000s, a national food company had to recall over 4 million pounds of poultry products after *Listeria* poisoning killed 7 people and made over 50 others sick. In the 2006 *E. coli* outbreak from fresh spinach, nearly 200 people became sick, and two elderly women and a two-year old boy died before consumers got the FDA message to not eat fresh spinach. These incidents and others since have focused the national spotlight on two important safety issues: disease-causing organisms are commonly found in raw foods, and thorough cooking kills most of these foodborne pathogens. This heightened awareness sparked a much needed overhaul of national food safety programs.

What kinds of programs are in place to help keep foods safe?

To make our food supply safe for consumers, the United States Department of Agriculture (USDA), the FDA, and the food-processing industries have developed and implemented programs to control foodborne illness.* The **Hazard Analysis Critical Control Points (HACCP)** system requires food manufacturers to identify points of contamination and implement controls to prevent foodborne disease. For example, after tracing two large outbreaks of salmonellosis to imported cantaloupe, producers began using chlorinated water to wash the melons and to make ice for packing and shipping. Safety procedures such as this prevent hundreds of thousands of foodborne illnesses each year and are responsible for the decline in infections over the past decade.[4]

This example raises another issue regarding the safety of imported foods. FDA inspectors cannot keep pace with the increasing numbers of imported foods; they inspect fewer than 2 percent of the almost 3 million shipments of fruits, vegetables, and seafood coming into more than 300 ports in the United States each year. The FDA is working with other countries to adopt the safe food-handling practices used in the United States.

Are foods bought in grocery stores and foods eaten in restaurants safe?

Canned and packaged foods sold in grocery stores are easily controlled, but rare accidents do happen. Batch numbering makes it possible to recall contaminated foods through public announcements via newspapers, television, and radio. In the grocery store, consumers can buy items before the "sell by" date and inspect the safety seals and wrappers of packages. A broken seal, bulging can lid, or mangled package fails to protect the consumer against microbes, insects, spoilage, or even vandalism.

State and local health regulations provide guidelines on the cleanliness of facilities and the safe preparation of foods for restaurants, cafeterias, and fast-food establishments. Even so, consumers can also take these actions to help prevent foodborne illnesses when dining out:

- Wash hands with hot, soapy water before meals.
- Expect clean tabletops, dinnerware, utensils, and food preparation areas.
- Expect cooked foods to be served piping hot and salads to be fresh and cold.
- Refrigerate doggy bags within two hours.

Improper handling of foods can occur anywhere along the line from commercial manufacturers to large supermarkets to small restaurants to private homes. Maintaining a safe food supply requires everyone's efforts.

What can people do to protect themselves from foodborne illness?

Whether microbes multiply and cause illness depends, in part, on a few key food-handling behaviors in the kitchen—whether

* In addition to HACCP, these programs include the Emerging Infections Program (EIP), the Foodborne Diseases Active Surveillance Network (FoodNet), and the Food Safety Inspection Service (FSIS).

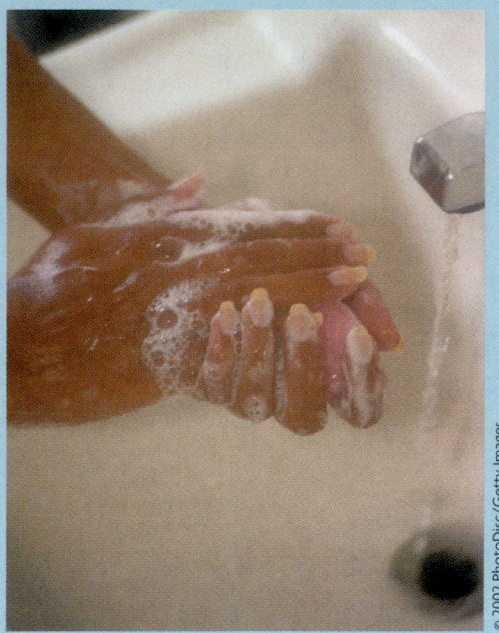

© 2002 PhotoDisc/Getty Images

Wash your hands with warm water and soap for at least 20 seconds before preparing or eating food to reduce the chance of microbial contamination.

the kitchen is in your home, a school cafeteria, a gourmet restaurant, or a canning manufacturer.[5] For the most part, foodborne illness can be prevented by doing four simple things:

- **Keep a clean, safe kitchen.** Wash countertops, cutting boards, hands, sponges, and utensils in hot, soapy water before and after each step of food preparation.

- **Avoid cross-contamination.** Keep raw eggs, meat, poultry, and seafood separate from other foods. Wash all utensils and surfaces (such as cutting boards or platters) that have been in contact with these foods with hot, soapy water be-

Dietary Guidelines

- To avoid microbial foodborne illness, clean hands, food contact surfaces, and fruits and vegetables.

fore using them again. Bacteria inevitably left on the surfaces from the raw meat can recontaminate the cooked meat or other
foods—a problem known as **cross-contamination.** Washing raw eggs, meat, and poultry is not recommended as the extra handling increases the risk of cross-contamination.

Dietary Guidelines

- To avoid microbial foodborne illness, separate raw, cooked, and ready-to-eat foods while shopping, preparing, or storing foods.

- **Keep hot foods hot.** Cook foods long enough to reach internal temperatures that will kill microbes, and maintain adequate temperatures to prevent bacterial growth until the foods are served.

Dietary Guidelines

■ To avoid microbial foodborne illness, cook foods to a safe temperature to kill microorganisms.

- **Keep cold foods cold.** Go directly home upon leaving the grocery store and immediately unpack foods into the refrigerator or freezer upon arrival. After a meal, refrigerate any leftovers immediately.

Dietary Guidelines

■ To avoid microbial foodborne illness, chill (refrigerate) perishable food promptly and defrost foods properly.

Unfortunately, consumers commonly fail to follow these simple food-handling recommendations.[6] See Table NP5–3 for additional food safety tips.

What precautions need to be taken when preparing meat and poultry?

Figure NP5–1 presents label instructions for the safe handling of meat and poultry and two types of USDA seals. Meats and

---

TABLE NP5–3

Strategies to Prevent Foodborne Illnesses

Most foodborne illnesses can be prevented by following four simple rules: keep a clean kitchen, avoid cross-contamination, keep hot foods hot, and keep cold foods cold.

Keep a Clean Kitchen

■ Wash fruits and vegetables in a clean sink with a scrub brush and warm water; store washed and unwashed produce separately.
■ Use hot, soapy water to wash hands, utensils, dishes, nonporous cutting boards, and countertops before handling food and between tasks when working with different foods. Use a bleach solution on cutting boards (one capful per gallon of water).
■ Cover cuts with clean bandages before food preparation; dirty bandages carry harmful microorganisms.
■ Mix foods with utensils, not hands; keep hands and utensils away from mouth, nose, and hair.
■ Anyone may be a carrier of bacteria and should avoid coughing or sneezing over food. A person with a skin infection or infectious disease should not prepare food.
■ Wash or replace sponges and towels regularly.
■ Clean up food spills and crumb-filled crevices.

Avoid Cross-Contamination

■ Wash all surfaces that have been in contact with raw meats, poultry, eggs, fish, and shellfish before reusing.
■ Serve cooked foods on a clean plate. Separate raw foods from those that have been cooked.
■ Do not use marinade that was in contact with raw meat for basting or sauces.

Keep Hot Foods Hot

■ When cooking meats or poultry, use a thermometer to test the internal temperature. Insert the thermometer between the thigh and the body of a turkey or into the thickest part of other meats, making sure the tip of the thermometer is not in contact with bone or the pan. Cook to the temperature indicated for that particular meat (see Figure NP5–2); cook hamburgers to at least medium well-done. If you have safety question, call the USDA Meat and Poultry Hotline: (800) 535-4555.
■ Cook stuffing separately, or stuff poultry just prior to cooking.
■ Do not cook large cuts of meat or turkey in a microwave oven; it leaves some parts undercooked while overcooking others.
■ Cook eggs before eating them (soft-boiled for at least $3\frac{1}{2}$ minutes; scrambled until set, not runny; fried for at least 3 minutes on one side and 1 minute on the other).
■ Cook seafood thoroughly. If you have safety questions about seafood call the FDA hotline: (800) FDA-4010.
■ When serving foods, maintain temperatures at 140°F or higher.
■ Heat leftovers thoroughly to at least 165°F.

Keep Cold Foods Cold

■ When running errands, stop at the grocery store last. When you get home, refrigerate the perishable groceries (such as meats and dairy products) immediately. Do not leave perishables in the car any longer than it takes for ice cream to melt.
■ Put packages of raw meat, fish, or poultry on a plate before refrigerating to prevent juices from dripping on food stored below.
■ Buy only foods that are solidly frozen in store freezers.
■ Keep cold foods at 40°F or less; keep frozen foods at 0°F or less (keep a thermometer in the refrigerator).
■ Marinate meats in the refrigerator, not on the counter.
■ Refrigerate leftovers promptly; use shallow containers to cool foods faster; use leftovers within 3 to 4 days.
■ Thaw meats or poultry in the refrigerator, not at room temperature. If you must hasten thawing, use cool water (changed every 30 minutes) or a microwave oven.
■ Freeze meat, fish, or poultry immediately if not planning to use within a few days.

continued

In General

- Do not reuse disposable containers; use nondisposable containers or recycle instead.
- Do not taste food that is suspect. "If in doubt, throw it out."
- Throw out foods with danger-signaling odors. Be aware, though, that most food-poisoning bacteria are odorless, colorless, and tasteless.
- Do not buy or use items that have broken seals or mangled packaging; such containers cannot protect against microbes, insects, spoilage, or even vandalism. Check safety seals, buttons, and expiration dates.
- Follow label instructions for storing and preparing packaged and frozen foods; throw out foods that have been thawed or refrozen.
- Discard foods that are discolored, moldy, or decayed or that have been contaminated by insects or rodents.

For Specific Food Items

- *Canned goods.* Carefully discard food from cans that leak or bulge so that other people and animals will not accidentally ingest it; before canning, seek professional advice from the USDA Extension Service (check your telephone book under U.S. government listings, or ask directory assistance).
- *Milk and cheeses.* Use only pasteurized milk and milk products. Aged cheeses, such as cheddar and Swiss, do well for an hour or two without refrigeration, but they should be refrigerated or stored in an ice chest for longer periods.
- *Eggs.* Use clean eggs with intact shells. Do not eat eggs, even pasteurized eggs, raw; raw eggs are commonly found in Caesar salad dressing, eggnog, cookie dough, hollandaise sauce, and key lime pie. Cook eggs until whites are firmly set and yolks begin to thicken.
- *Honey.* Honey may contain dormant bacterial spores, which can awaken in the human body to produce botulism. In adults, this poses little hazard, but infants under one year of age should never be fed honey. Honey can accumulate enough toxin to kill an infant; it has been implicated in several cases of sudden infant death. (Honey can also be contaminated with environmental pollutants picked up by the bees.)
- *Mayonnaise.* Commercial mayonnaise may actually help a food to resist spoilage because of the acid content. Still, keep it cold after opening.
- *Mixed salads.* Mixed salads of chopped ingredients spoil easily because they have extensive surface area for bacteria to invade, and they have been in contact with cutting boards, hands, and kitchen utensils that easily transmit bacteria to food (regardless of their mayonnaise content). Chill them well before, during, and after serving.
- *Picnic foods.* Choose foods that last without refrigeration such as fresh fruits and vegetables, breads and crackers, and canned spreads and cheeses that can be opened and used immediately. Pack foods cold, layer ice between foods, and keep foods out of water.
- *Seafood.* But only fresh seafood that has been properly refrigerated or iced. Cooked seafood should be stored separately from raw seafood to avoid cross-contamination.

FIGURE NP5–1

Meat and Poultry Safety, Grading, and Inspection Seals

Inspection is mandatory; grading is voluntary. Neither guarantees that the product will not cause foodborne illnesses, but consumers can help to prevent foodborne illnesses by following the safe handling instructions.

The voluntary "Graded by USDA" seal indicates that the product has been graded for tenderness, juiciness, and flavor. Beef is graded Prime (abundant marbling of the meat muscle), Choice (less marbling), and Select (lean). Similarly, poultry is graded A, B, and C.

The mandatory "Inspected and Passed by the USDA" seal ensures that meat and poultry products are safe, wholesome, and correctly labeled. Inspection does not guarantee that the meat is free of potentially harmful bacteria.

Safe Handling Instructions

THIS PRODUCT WAS PREPARED FROM INSPECTED AND PASSED MEAT AND/OR POULTRY. SOME FOOD PRODUCTS MAY CONTAIN BACTERIA THAT CAN CAUSE ILLNESS IF THE PRODUCT IS MISHANDLED OR COOKED IMPROPERLY. FOR YOUR PROTECTION, FOLLOW THESE SAFE HANDLING INSTRUCTIONS.

KEEP REFRIGERATED OR FROZEN. THAW IN REFRIGERATOR OR MICROWAVE.

KEEP RAW MEAT AND POULTRY SEPARATE FROM OTHER FOODS. WASH WORKING SURFACES (INCLUDING CUTTING BOARDS), UTENSILS, AND HANDS AFTER TOUCHING RAW MEAT OR POULTRY.

COOK THOROUGHLY.

KEEP HOT FOODS HOT. REFRIGERATE LEFTOVERS IMMEDIATELY OR DISCARD.

The USDA requires that safe handling instructions appear on all packages of meat and poultry.

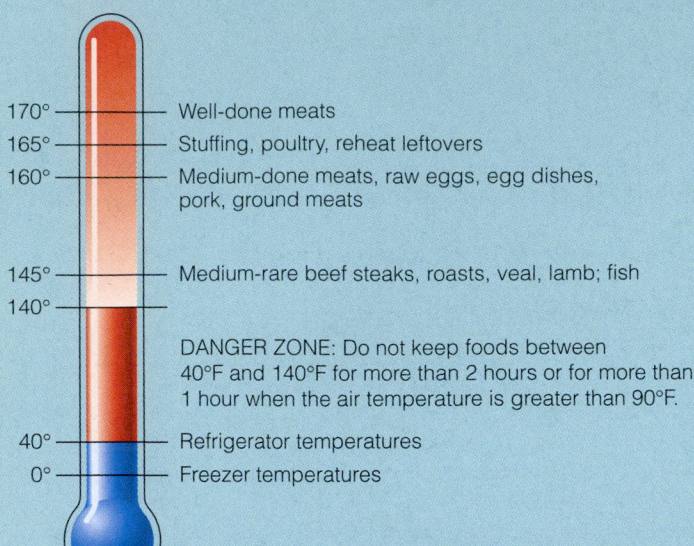

170° — Well-done meats
165° — Stuffing, poultry, reheat leftovers
160° — Medium-done meats, raw eggs, egg dishes, pork, ground meats

145° — Medium-rare beef steaks, roasts, veal, lamb; fish
140°

DANGER ZONE: Do not keep foods between 40°F and 140°F for more than 2 hours or for more than 1 hour when the air temperature is greater than 90°F.

40° — Refrigerator temperatures
0° — Freezer temperatures

FIGURE NP5–2

Recommended Safe Temperatures (Fahrenheit)
Bacteria multiply rapidly at temperatures between 40°F and 140°F. Cook foods to the temperatures shown on this thermometer and hold them at 140°F or higher.

poultry contain bacteria and provide a moist, nutrient-rich environment that favors microbial growth. Ground meat is especially susceptible because it receives more handling than other kinds of meat and has more surface exposed to bacterial contamination. Consumers cannot detect the harmful bacteria in or on meat. For safety's sake, cook meat thoroughly, using a thermometer to test the internal temperature (see Figure NP5–2).

Cook hamburgers to 160°F; color alone cannot determine doneness. Some burgers will turn brown before reaching 160°F, whereas others may retain some pink color, even when cooked to 175°F.

David Chasey/Photodisc Red/Getty Images

Eating raw seafood is a risky proposition.

How can a person enjoy seafood safely?

Most seafood available in the United States and Canada is safe, but eating it undercooked or raw can cause severe illnesses—hepatitis, worms, parasites, viral intestinal disorders, and other diseases.* Rumor has it that freezing fish will make it safe to eat raw, but this is only partly true. Commercial freezing will kill mature parasitic worms, but only cooking can kill all worm eggs and other microorganisms that can cause illness. For safety's sake, all seafood should be cooked until it is opaque. Even **sushi** can be safe to eat when chefs combine cooked seafood and other ingredients into delicacies.

Eating raw oysters can be dangerous for anyone, but people with liver disease and weakened immune systems are most vulnerable. At least ten species of bacteria found in raw oysters can cause serious illness and even death.[†] Raw oysters may also carry the hepatitis A virus, which can cause liver disease. Some hot sauces can kill many of these bacteria, but not the virus; alcohol may also protect some people against some oyster-borne illnesses, but not enough to guarantee protection (or to recommend drinking alcohol). Pasteurization of raw oysters—holding them at a specified temperature for a specified time—holds promise for killing bacteria without cooking the oyster or altering its texture or flavor.

As population density increases along the shores of seafood-harvesting waters, pollution inevitably invades the sea life there. Preventing seafood-borne illness is in large part a task of controlling water pollution. To help ensure a safe seafood market, the FDA requires processors to adopt food safety practices based on the HACCP system mentioned earlier.

* Diseases caused by toxins from the sea include ciguatera poisoning, scombroid poisoning, and paralytic and neurotoxic shellfish poisoning.
† Raw oysters can carry the bacterium *Vibrio vulnificus;* see Table NP5–2 for details.

Chemical pollution and microbial contamination lurk not only in the water, but also in the boats and warehouses where seafood is cleaned, prepared, and refrigerated. Seafood is one of the most perishable foods: time and temperature are critical to its freshness and flavor. To keep seafood as fresh as possible, people in the industry "keep it cold, keep it clean, and keep it moving." Wise consumers eat it cooked.

Do foods that are unsafe to eat smell bad?

Fresh food generally smells fresh. Not all types of food poisoning are detectable by odor, but some bacterial wastes produce "off" odors. If an abnormal odor exists, the food is spoiled. Throw it out or, if it was recently purchased, return it to the grocery store. Do not taste it. Table NP5–4 lists safe refrigerator storage times for selected foods.

Local health departments and the USDA Extension Service can provide additional information about food safety. Should precautions fail and mild foodborne illness develop, drink clear liquids to replace fluids lost through vomiting and diarrhea. If serious foodborne illness is suspected, first call a physician. Then wrap the remainder of the suspected food and label the container so that the food cannot be mistakenly eaten, place it

Dietary Guidelines

- To avoid microbial foodborne illness, avoid raw (unpasteurized) milk or any products made from unpasteurized milk, raw or partially cooked eggs or foods containing raw eggs, raw or undercooked meat and poultry, unpasteurized juices, and raw sprouts.

TABLE NP5–4

Safe Refrigerator Storage Times ($\leq 40\,^\circ$F)

1 to 2 Days
Raw ground meats, breakfast or other raw sausages, raw fish or poultry; gravies

3 to 5 Days
Raw steaks, roasts, or chops; cooked meats, poultry, vegetables, and mixed dishes; lunch meats (packages opened); mayonnaise salads (chicken, egg, pasta, tuna)

1 Week
Hard-cooked eggs, bacon or hot dogs (opened packages); smoked sausages or seafood

2 to 4 Weeks
Raw eggs (in shells); lunch meats, bacon, or hot dogs (packages unopened); dry sausages (pepperoni, hard salami); most aged and processed cheeses (Swiss, brick)

2 Months
Mayonnaise (opened jar); most dry cheese (Parmesan, Romano)

TABLE NP5–5

Strategies to Achieve Food Safety while Traveling

Foodborne illnesses contracted while traveling are colloquially known as travelers' diarrhea. A bout of this ailment can ruin the most enthusiastic tourist's trip. To avoid foodborne illness, follow the food safety tips outlined in Table NP5–3. In addition, while traveling:

- Wash your hands often with soap and hot water, especially before handling food or eating. Use antiseptic gel or hand wipes.
- Eat only well-cooked and hot or canned foods. Eat raw fruits or vegetables only if you have washed them in purified water and peeled them yourself. Skip salads and raw fish and shellfish.
- Be aware that water and ice made from it may be unsafe. Use safe, bottled water for drinking, making ice cubes, and brushing teeth. Alternatively, take along disinfecting tablets or a device to boil water. Do not use ice unless it was made from purified or bottled water.
- Drink no beverages made with tap water. Drink only treated, boiled, canned, or bottled beverages, and drink them without ice, even if they are not chilled to your liking.
- Refuse dairy products unless they have been properly pasteurized and refrigerated.
- Do not buy food and drinks from street vendors.
- Before you leave on the trip, ask you physician to recommend an antimotility agent and an antibiotic to take with you in case your efforts to avoid illness fail.

To sum up these recommendations, "Boil it, cook it, peel it, or forget it." Chances are excellent that if you follow these rules, you will remain well.

in the refrigerator, and hold it for possible inspection by health authorities.

How can a person defend against foodborne illness when traveling to foreign countries?

People who travel to other countries have a 50–50 chance of contracting a foodborne illness, commonly described as **travelers' diarrhea**.[7] Like many other foodborne illnesses, travelers' diarrhea is a sometimes serious, always annoying bacterial infection of the digestive tract. The risk is high because some countries' cleanliness standards for food and water may be lower than those in the United States and Canada. Also, every region's microbes are different, and while people are immune to those in their own neighborhoods, they have had no chance to develop immunity to the pathogens in places they are visiting for the first time. Table NP5–5 offers strategies for food safety while traveling.

Notes

1. Centers for Disease Control and Prevention, *FoodNet Surveillance Report for 2004*, June 2006.

2. E. A. Coleman and M. E. Yergler, Botulism, *American Journal of Nursing* 102 (2002): 44–47.

3. Position of the American Dietetic Association: Food and water safety, *Journal of the American Dietetic Association* 103 (2003): 1203–1218.

4. Centers for Disease Control and Prevention, *FoodNet Surveillance Report for 2004*, June 2006.

5. B. J. McCabe-Sellers and S. E. Beattie, Food safety: Emerging trends in foodborne illness surveillance and prevention, *Journal of the American Dietetic Association* 104 (2004): 1709–1717.

6. J. B. Anderson and coauthors, A camera's view of consumer food-handling behaviors, *Journal of the American Dietetic Association* 104 (2004): 186–191.

7. E. T. Ryan, M. E. Wilson, and K. C. Kain, Illness after international travel, *New England Journal of Medicine* 347 (2002): 505–516.

6

Metabolism, Energy Balance, and Body Composition

© Blend Images/SuperStock

very organ, every tissue, and every cell of the body engages in **metabolism**, the chemical reactions involved in releasing energy, breaking down compounds, and making new compounds. Much like a factory, the body works efficiently to manufacture needed products and dispose of wastes. All these processes are regulated by hormonal signals that coordinate supply and demand.

In disease, metabolic processes can become disturbed, and some diseases are caused by metabolic disturbances. This chapter introduces the principal organs and their metabolic roles, the metabolism of the energy nutrients, energy balance, body composition, and the health risks associated with too much body fat. The next chapter offers strategies toward solving the problems of too much or too little body fat.

The Organs and Their Metabolic Roles

he metabolic reactions of every organ contribute to the body's ability to function normally and maintain health. Metabolic reactions also use or release energy and therefore affect body weight, with consequences for health.

The Principal Organs

Of particular concern to metabolism are the digestive organs, liver, pancreas, circulatory system, and kidneys. Together, they perform much of the work of breaking down compounds, making new ones, transporting nutrients and oxygen throughout the body, and removing the wastes generated by metabolic processes.

The Digestive Organs As Chapter 5 describes, the digestive system transports foods through the GI tract; produces digestive juices and enzymes; absorbs nutrients; provides transport proteins to carry lipids and vitamins to other sites in the body; and reabsorbs salts and fluids. The digestive system also possesses the body's most rapidly multiplying cells: when healthy, they replace themselves every few days. Disorders affecting the GI tract interfere with the ingestion, digestion, absorption, and metabolism of nutrients, as described in Chapters 17 and 18.

The Liver Nutrients absorbed into the bloodstream are taken first to the liver, as described in Chapter 5. The liver is one of the body's most active metabolic factories. It receives nutrients and metabolizes, packages, stores, or ships them out for use by other organs. It metabolizes and stores most vitamins and many minerals. It manufactures bile, which the body uses to emulsify fat for digestion and absorption. It metabolizes and detoxifies drugs, prepares waste products for excretion, and participates in iron recycling and blood cell manufacture. It also makes many proteins necessary for health, including immune factors, transport proteins, and clotting factors. When liver disorders disrupt metabolism, they profoundly affect both nutrition and health status, as described in Chapter 20.

The Pancreas The pancreas contributes digestive juices to the GI tract, but also has another metabolic function: it produces the hormones insulin and glucagon that regulate the body's use of glucose. After a meal, as blood glucose rises, the pancreas secretes insulin. Insulin prompts cells to take up glucose and use it as fuel; insulin also prompts liver cells to store glucose as glycogen. When blood glucose falls (as occurs between meals), the pancreas responds by secreting glucagon into the blood. Glucagon raises blood glucose by signaling the liver to dismantle its glycogen stores and release glucose into the blood for use by all the other body cells. Glucose is an indispensable fuel for brain cells, nerve cells, and red blood cells. Its availability is therefore crucial to normal nervous system activity and blood chemistry. Abnormalities associated with the digestive functions of the pancreas are described in Chapter 20, and those associated with its hormonal functions are described in Chapter 21.

The Heart and Blood Vessels The heart and blood vessels conduct blood with its cargo of nutrients and oxygen to the all the other body cells and carry wastes away from them. Diseases of the heart and arteries therefore affect the health of the whole body. Metabolic reactions that affect the heart and blood vessels include, most importantly, the making and transport of lipoproteins, the carriers of cholesterol and other lipids from the liver to the tissues and back again. High blood levels of low-density lipoproteins (LDL) and very-low-density lipoproteins (VLDL) promote atherosclerosis which increases the risk of disability or death from heart attacks and strokes. Chapter 22 is devoted to these conditions.

The Kidneys The kidneys are also active metabolic organs. Unceasingly, for 24 hours of every day, they filter waste products from the blood to be excreted in the urine and reabsorb needed nutrients, thereby maintaining the blood's delicate chemical balances. The kidneys' cells also produce compounds that help to regulate blood pressure and convert a precursor compound to active vitamin D, thereby helping to maintain the bones. Disorders of the kidneys nearly always involve the heart and the skeleton; kidney disorders are the subject of Chapter 23.

The Body's Metabolic Work

The metabolic work that the body's cells do, like all work, requires energy, and foods supply that energy. Foods, in turn, get their energy from the sun, either directly (in the case of photosynthesizing plants that make carbohydrate) or indirectly (in the case of animals that eat plants). When chemical reactions in cells release stored energy from energy-yielding nutrients, that energy becomes available to do the cells' work.

Heat Energy and Body Temperature The cells of each organ conduct metabolic activities specific to that organ. In addition, all cells must maintain themselves, and many must reproduce. To do this, they must have all the essential nutrients available to them: energy nutrients, vitamins, and minerals, as well as water. As cells do their metabolic work, the chemical reactions involved release heat, and this heat keeps the body warm. By regulating the rates at which these metabolic reactions release heat energy, the body maintains its constant normal temperature of about 98.6°F.

Accelerated Metabolism During severe stress to the body, metabolism speeds up. Fever sometimes develops. An accelerated metabolism signifies that fuels are being used at a rate more rapid than normal; this may lead to wasting of body organs and loss of weight, including loss of vital lean tissue. Chapter 24 describes the metabolic consequences of severe stress, and Chapter 25 describes the metabolic consequences of the wasting syndrome.

Building Up and Breaking Down Compounds When not needed by the cells for energy, the basic units of the energy-yielding nutrients are used to build body compounds. The building up of body compounds is known as **anabolism**; this book represents anabolic reactions, wherever possible, by "up" arrows in chemical diagrams (such as those shown in Figure 6–1 on p. 142). Glucose units can be strung together to make glycogen chains. Glycerol and fatty acids can be assembled into triglycerides. Amino acids can be linked together to make proteins. These anabolic reactions, in which simple compounds are put together to form larger, more complex structures, involve doing work and so require the energy provided by ATP.

metabolism: the sum total of all the chemical reactions that go on in living cells.

anabolism (an-ABB-o-lism): reactions in which small molecules are put together to build larger ones. Anabolic reactions require energy.
  ana = up

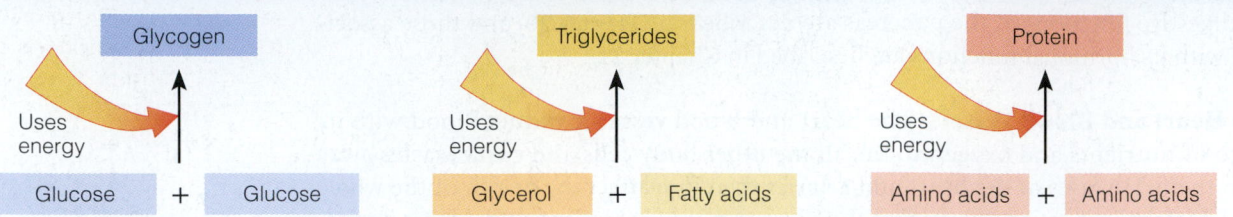

ANABOLIC REACTIONS

Anabolic reactions include the making of glycogen, triglycerides, and protein; these reactions require differing amounts of energy.

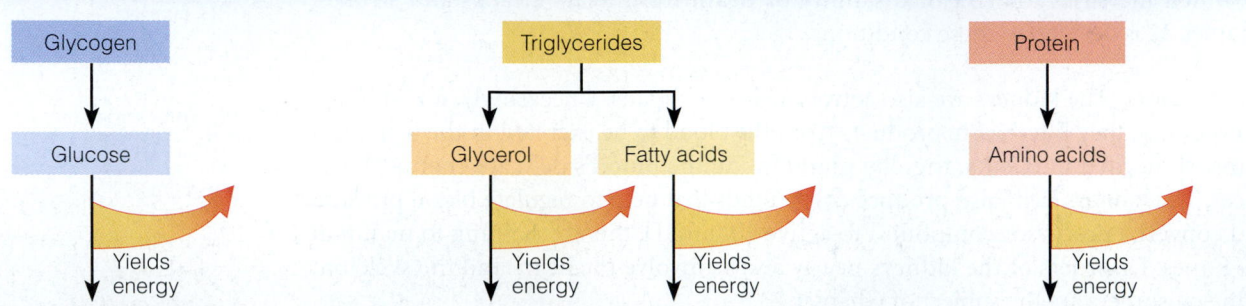

CATABOLIC REACTIONS

Catabolic reactions include the breakdown of glycogen, triglycerides, and protein; the further catabolism of glucose, glycerol, fatty acids, and amino acids releases differing amounts of energy. Much of the energy released is captured in the bonds of adenosine triphosphate (ATP).

FIGURE 6–1

Anabolic and Catabolic Reactions Compared

The breaking down of body compounds is known as **catabolism**. Catabolic reactions usually release energy and are represented, wherever possible, by "down" arrows in chemical diagrams (review Figure 6–1). Glycogen can be broken down to glucose, triglycerides to fatty acids and glycerol, and protein to amino acids.

IN SUMMARY

- Metabolism occurs throughout the body, all the time, and supports normal health. Foods supply the energy required for the metabolic work of the body.

- When not needed for energy, the body's cells use the basic units of the energy-yielding nutrients to build body compounds (anabolism).

- When the body needs energy, the cells break the basic units of the energy-yielding nutrients down further to release energy (catabolism).

The Body's Use of Fuels

How does your body get the energy needed to maintain all of its cellular activities from the foods you eat? The answer to this question lies in an understanding of metabolism, the chemical reactions that occur in all living cells.

Energy Metabolism

Although every aspect of our lives depends on energy, the concept of energy can be difficult to grasp because it cannot be seen or touched, and it manifests in various

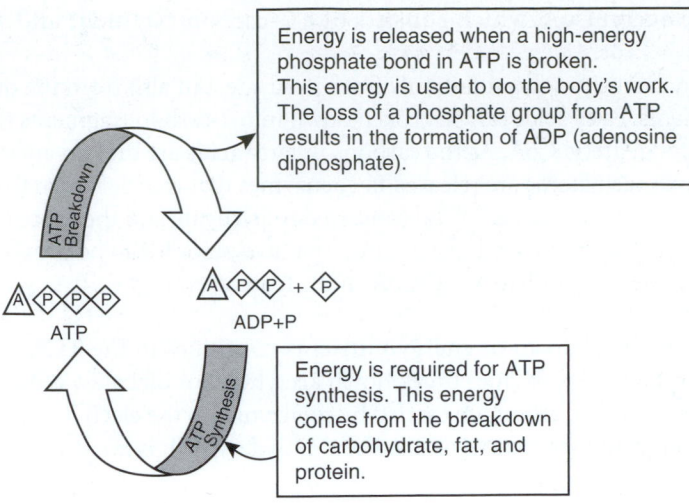

Energy is released when a high-energy phosphate bond in ATP is broken. This energy is used to do the body's work. The loss of a phosphate group from ATP results in the formation of ADP (adenosine diphosphate).

Energy is required for ATP synthesis. This energy comes from the breakdown of carbohydrate, fat, and protein.

FIGURE 6–2

ATP

ATP (adenosine triphosphate) is a high-energy compound that can capture and transfer energy to catalyze chemical reactions.

forms, including heat, mechanical, electrical, and chemical energy. In the body, heat energy maintains a constant body temperature, and electrical energy sends nerve impulses, for example. Energy is stored in foods and in the body as chemical energy. **Energy metabolism** is the sum total of all the chemical reactions that the body uses to obtain or expend energy from foods. Earlier chapters described how the energy-yielding nutrients—carbohydrate, fat, and protein—are broken down into basic units that are absorbed into the blood:

- From carbohydrates: glucose

- From fat (triglycerides): glycerol and fatty acids

- From proteins: amino acids

This section picks up from there, describing what becomes of these nutrients.

The high-energy compound **ATP (adenosine triphosphate)** is able to transfer small amounts of usable energy to move our muscles and supply our enzymes with the energy they need to catalyze chemical reactions. Figure 6–2 illustrates how ATP does this. When ATP breaks down and releases one of its phosphate groups, a small amount of energy is released and used in the body to build compounds. With the loss of the phosphate group, the ATP becomes ADP (adenosine diphosphate). During energy metabolism, ATP is re-created by attaching a phosphate group to ADP. ATP is produced continuously throughout the day using the energy from the breakdown of the energy-yielding nutrients.

Breaking Down Nutrients for Energy To produce ATP, the body breaks down any or all of the four basic units—glucose, fatty acids, glycerol, and amino acids—into even smaller units. Each of these nutrients travels down a different pathway, but all can eventually become acetyl CoA, enter the TCA cycle, and provide hydrogens for the electron transport chain. Details follow.

The breakdown of glucose (a 6-carbon compound) into two **pyruvate** (a 3-carbon compound) is called **glycolysis**, and it produces just two usable ATP. As the carbons in glucose are broken apart to produce pyruvate, the hydrogen atoms attached to the carbons are transferred by **coenzymes** to the electron transport chain. ■ Thus, the reactions of glycolysis produce a small amount of ATP, pyruvate, and hydrogen-rich coenzymes that are used later in energy metabolism. After glycolysis, pyruvate is

■ The coenzymes used in energy metabolism contain B vitamins; hence, B vitamins play critical roles in ATP production.

catabolism (ca-TAB-o-lism): reactions in which large molecules are broken down to smaller ones. Catabolic reactions release energy.
kata = down

energy metabolism: all the reactions by which the body obtains and expends the energy from food.

ATP or adenosine (ah-DEN-oh-seen) **triphosphate** (tri-FOS-fate): a common high-energy compound that contains three phosphate groups. The bonds between the phosphate groups are often described as "high-energy" because of their readiness to release their energy.

pyruvate (PIE-roo-vate): a 3-carbon compound that plays a key role in energy metabolism.

glycolysis (gligh-COLL-ih-sis): the metabolic breakdown of glucose to pyruvate.
glyco = glucose
lysis = breakdown

coenzymes (co-EN-zime): small molecules that work with enzymes to facilitate the enzymes' activity. Most coenzymes have B vitamins as part of the structures.

converted to **acetyl CoA**, which consists of a 2-carbon fragment and a coenzyme called **CoA**.

Acetyl CoA can be produced not only from pyruvate, but also from the other energy-yielding nutrients. Fatty acids can be broken down into 2-carbon fragments that combine with CoA to form acetyl CoA. As the carbons in fatty acids are broken apart to produce acetyl CoA, hydrogen atoms are released to coenzymes that transfer them to the electron transport chain. Glycerol can easily be converted to pyruvate and therefore can also produce acetyl CoA. The amino acids have various pathways; some can be converted to pyruvate, others can be converted to acetyl CoA, and a few can enter the TCA cycle directly.

TCA Cycle   The breakdown of energy nutrients continues in the **TCA cycle (tricarboxylic acid cycle)**, ■ as enzymes break down acetyl CoA molecules. With each turn of the TCA cycle, hydrogen atoms are carried by coenzymes to the electron transport chain. The waste product of these reactions is carbon dioxide, which is eventually exhaled.

Electron Transport Chain   The final step in energy metabolism occurs at the **electron transport chain**. In this process, enzymes attach a phosphate group to ADP, creating ATP (review Figure 6–2). The hydrogen atoms that were collected by coenzymes during glycolysis, fat breakdown, and the TCA cycles provide the chemical energy that drives ATP production. Finally, the same hydrogen atoms are linked with oxygen to produce water. Figure 6–3 summarizes the metabolic pathways just described.

Aerobic and Anaerobic Metabolism   The production of ATP via the electron transport chain requires oxygen in the final step and is called *aerobic* metabolism. Glycolysis produces ATP without the need for oxygen and is therefore called *anaerobic* metabolism. When exercise intensity requires more ATP than can be provided by the electron transport chain (due to limited oxygen or other factors), ATP production from glycolysis is stepped up, making glucose a critical fuel for the exercising muscles. Chapter 10 provides information about anaerobic and aerobic activities and their effects on fuel use.

Glucose Production

When glucose levels drop, glucose can be produced from several other compounds in a process called **gluconeogenesis**. In Figure 6–3, you can see two-way arrows between glucose and pyruvate, but only a one-way arrow between pyruvate and acetyl CoA. The arrows show that pyruvate can be reconverted to glucose, but that acetyl CoA cannot. Any compound that can be converted to pyruvate can be used to make glucose. Any compound that is converted to acetyl CoA cannot be used to make glucose. ■

Triglycerides and Glucose Production   Recall that triglycerides (the primary form of fat in the body) consist of three fatty acids and a glycerol. Because fatty acids break down to acetyl CoA, they cannot be used to make glucose. The glycerol portion of a triglyceride, however, can be converted to pyruvate and thus can yield glucose. Because glycerol represents only about 5 percent of the weight of a triglyceride molecule, fat is an inefficient source of glucose.* About 95 percent of a triglyceride cannot be converted to glucose at all. The task of producing glucose is left to amino acids, which are obtained by breaking down the body's proteins.

Amino Acids and Glucose Production   As Chapter 4 explains, the primary role of amino acids is to maintain supplies of needed body proteins. If amino acids are needed for energy or if they are consumed in excess, they first undergo **deamination**,

----

* Figure 3–2 in Chapter 3 showed that glycerol (3 carbons) plus 3 fatty acids (most often 16 to 18 carbons) equals a triglyceride. Thus the small glycerol molecule represents only 3 of the 50 or so carbons in the triglyceride.

■ TCA is the abbreviation for *tricarboxylic acid,* named after the chemical structures of the compounds in this pathway.

■ Glucose ⟷ pyruvate → acetyl CoA → energy

acetyl CoA (ASS-eh-teel, or ah-SEET-il, coh-AY): a 2-carbon compound (acetate, or acetic acid) with a molecule of CoA attached to it.

CoA (Coh-AY) coenzyme A: the coenzyme derived from the B vitamin pantothenic acid and central to energy metabolism.

TCA cycle or tricarboxylic (try-car-box-ILL-ick) **acid cycle:** a series of metabolic reactions that break down molecules of acetyl CoA to carbon dioxide and hydrogen atoms; also called the **Kreb's cycle** after the biochemist who elucidated its reactions.

electron transport chain: the final pathway in energy metabolism that transports electrons from hydrogen to oxygen and captures the energy released in the bonds of a high-energy compound, ATP.

gluconeogenesis (gloo-co-nee-oh-GEN-ih-sis): the making of glucose from a noncarbohydrate source.

deamination: removal of the amino group (NH_2) from a compound such as an amino acid.

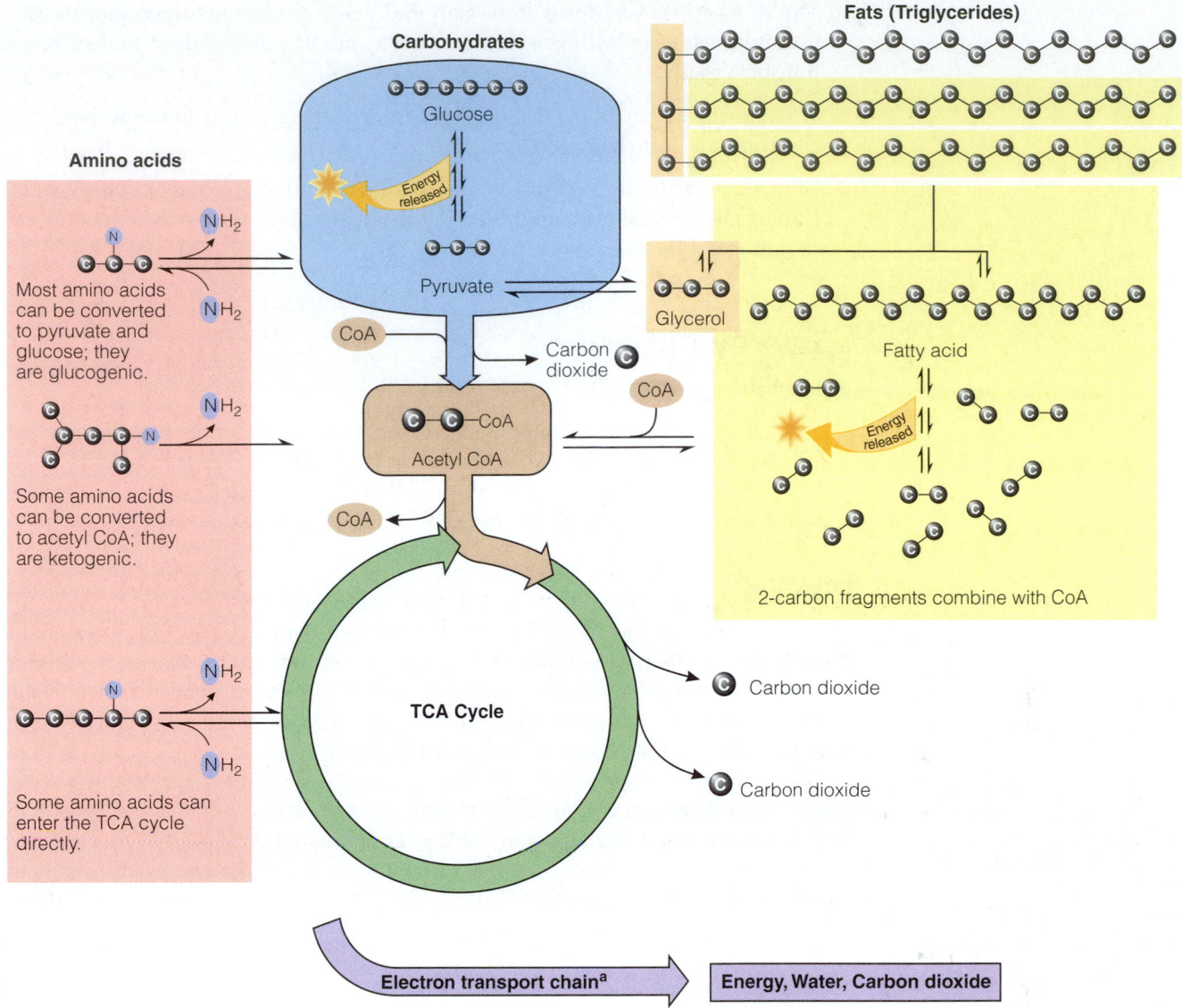

Carbohydrates

Glucose

Energy released

Pyruvate

Amino acids

NH_2

Most amino acids can be converted to pyruvate and glucose; they are glucogenic.

NH_2

NH_2

Some amino acids can be converted to acetyl CoA; they are ketogenic.

NH_2

NH_2

Some amino acids can enter the TCA cycle directly.

Fats (Triglycerides)

CoA

Carbon dioxide

CoA

CoA—CoA

Acetyl CoA

Glycerol

Fatty acid

Energy released

2-carbon fragments combine with CoA

TCA Cycle

Carbon dioxide

Carbon dioxide

Electron transport chain[a] → **Energy, Water, Carbon dioxide**

[a] Coenzymes carry hydrogens from the breakdown of the energy nutrients to the electron transport chain.

FIGURE 6–3

The Central Pathways of Energy Metabolism

a reaction in which they are stripped of their nitrogen. The nitrogen can be used to make other compounds, including the nonessential amino acids, or it can be excreted. ■ With nitrogen removed, most of the amino acids can be converted to pyruvate and can therefore provide glucose. Several of the amino acids can only be converted to acetyl CoA and therefore cannot supply glucose. Thus protein, unlike fat, is a fairly efficient source of glucose when carbohydrate is not available.

■ The principal nitrogen-excretion product of metabolism is **urea** (you-REE-uh).

IN SUMMARY

- ATP provides chemical energy for our bodies' cells and is produced by breaking down the energy-yielding nutrients.

- The chemical pathways that produce ATP are glycolysis, the TCA cycle, and the electron transport chain.

- Glycolysis breaks down glucose to pyruvate, which can be converted to acetyl CoA. Fatty acids can be broken down to acetyl CoA, as can many amino acids.

urea (yoo-REE-uh): the principal nitrogen-excretion product of protein metabolism.

- The TCA cycle breaks down acetyl CoA molecules, producing carbon dioxide as waste and releasing hydrogen atoms to coenzymes that deliver them to the electron transport chain.

- When the hydrogen atoms that originate from energy-yielding nutrients enter the electron transport chain, ATP is made.

- Glucose can be produced from compounds that can be converted to pyruvate, including glycerol and most amino acids. Fatty acids can be used only for energy and cannot make glucose.

Energy Imbalance

When a person takes in and expends roughly the same number of kcalories, body weight remains stable. In other words, the body's energy is in balance. Many people, however, overeat and gain weight; others eat too little and lose weight. This section examines the two extremes of energy imbalance—feasting and fasting.

Feasting

When people consume more energy than they expend, much of the excess is stored as body fat. Fat can be made from an excess of any energy-yielding nutrient. In addition, excess energy from alcohol is also stored as fat.[1] Alcohol has also been shown to slow down the body's use of fat for fuel, causing more fat to be stored.[2] The fat cells of the adipose tissue enlarge as they fill with fat, as Figure 6–4 shows.

Excess Carbohydrate Surplus carbohydrate (glucose) is first stored as glycogen in the liver and muscles, but the glycogen-storing cells have limited capacity. Once glycogen stores are filled, most of the additional carbohydrate is used for energy, displacing the body's use of fat for energy and allowing body fat to accumulate. Thus excess carbohydrate can contribute to obesity.

Excess Fat Surplus dietary fat contributes easily to the body's fat stores. After a meal, fat is sent to the body's adipose tissue where it is stored until needed for energy. Thus, excess fat from food easily adds to body fat.

Excess Protein Surplus protein may also contribute to body fat. If not needed to build body protein or to meet energy needs, amino acids will lose their nitrogens and be converted, through a series of reactions, to triglycerides. These, too, swell the fat cells and add to body weight. Figure 6–5 shows the metabolic events of feasting.

FIGURE 6–4

Fat Cell Enlargement

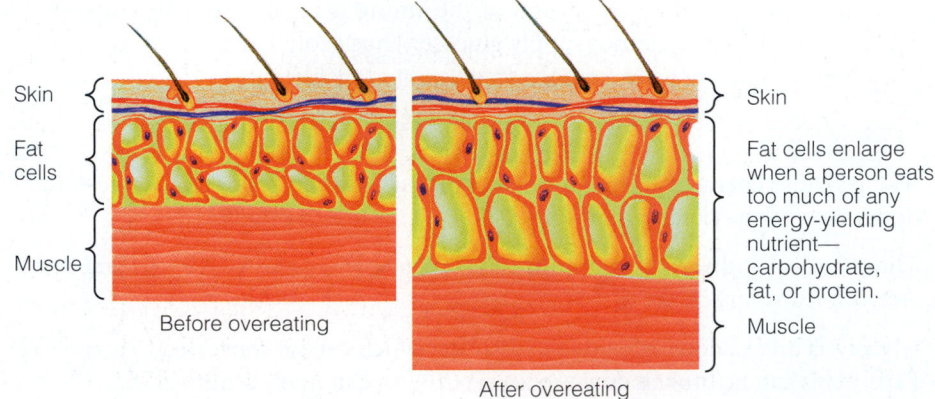

Skin

Fat cells

Muscle

Before overeating

Skin

Fat cells enlarge when a person eats too much of any energy-yielding nutrient— carbohydrate, fat, or protein.

Muscle

After overeating

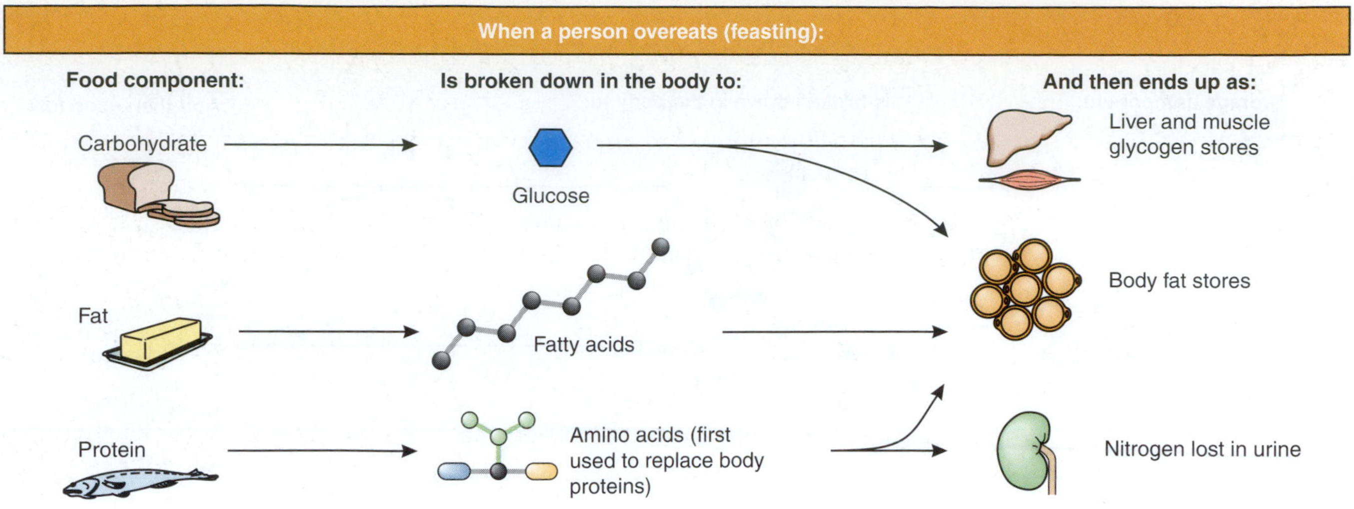

When a person overeats (feasting):

Food component:	Is broken down in the body to:	And then ends up as:
Carbohydrate	Glucose	Liver and muscle glycogen stores
Fat	Fatty acids	Body fat stores
Protein	Amino acids (first used to replace body proteins)	Nitrogen lost in urine

FIGURE 6–5

Feasting
When people overeat, they store energy.

IN SUMMARY

- Excess energy from carbohydrate, fat, protein, and alcohol leads to storage as body fat in adipose tissue.

Fasting

The body expends energy all the time. Even when a person is asleep and totally relaxed, the cells of many organs are hard at work.

Energy Deficit The body's top priority is to meet the energy needs for its ongoing cellular activity. Its normal way of meeting its energy needs is by periodic refueling, that is, by eating several times per day. When food is not available, the body uses fuel reserves from its own tissues. If people voluntarily choose not to eat, we say they are fasting; if they have no choice (as in a famine), we say they are starving. The body, however, makes no distinction between the two—metabolically, fasting and starvation are identical. In either case, the body switches to a wasting metabolism, drawing on its stores of carbohydrate and fat and, within a day or so, on its vital protein tissues as well.

Glycogen Used First As fasting begins, glucose from the liver's glycogen stores and fatty acids from the body's adipose tissue flow into the cells to fuel their work. Several hours later, liver glycogen is exhausted, and most of the glucose is used up. Low blood glucose concentrations serve as a signal to promote further fat breakdown.

Glucose Needed for the Brain At this point, a few hours into a fast, most of the cells depend on fatty acids to continue providing fuel. But as mentioned earlier, the nervous system (brain and nerves) and red blood cells cannot use fatty acids; they need glucose. Glucose has to be present to permit the brain's energy-metabolizing machinery to work. Normally, the nervous system consumes a little more than half of the total glucose used each day—about 400 to 600 kcalories' worth.

Protein Breakdown and Ketosis Because fat stores cannot provide the glucose needed by the brain, nerves, and red blood cells, body protein tissues (such as liver and muscle) always break down to some extent during fasting. In the first few days of a fast, body protein provides about 90 percent of the needed glucose, and glycerol provides about 10 percent. ■ If body protein losses were to continue at this rate,

■ Reminder: *Glycerol* represents only about 5 percent of the weight of a triglyceride molecule. About 95 percent of a triglyceride (the fatty acids attached to the glycerol) cannot be converted to glucose.

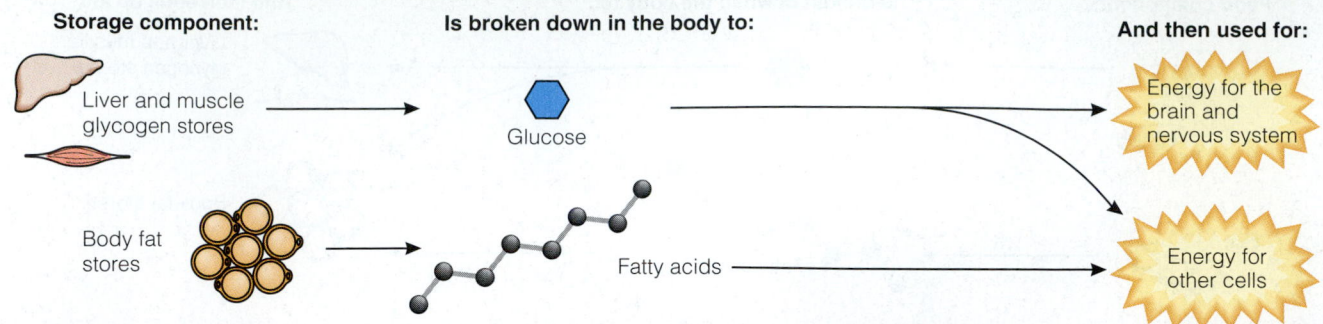

When a person draws on stores (fasting):

Storage component: Is broken down in the body to: And then used for:

Liver and muscle glycogen stores → Glucose → Energy for the brain and nervous system

Body fat stores → Fatty acids → Energy for other cells

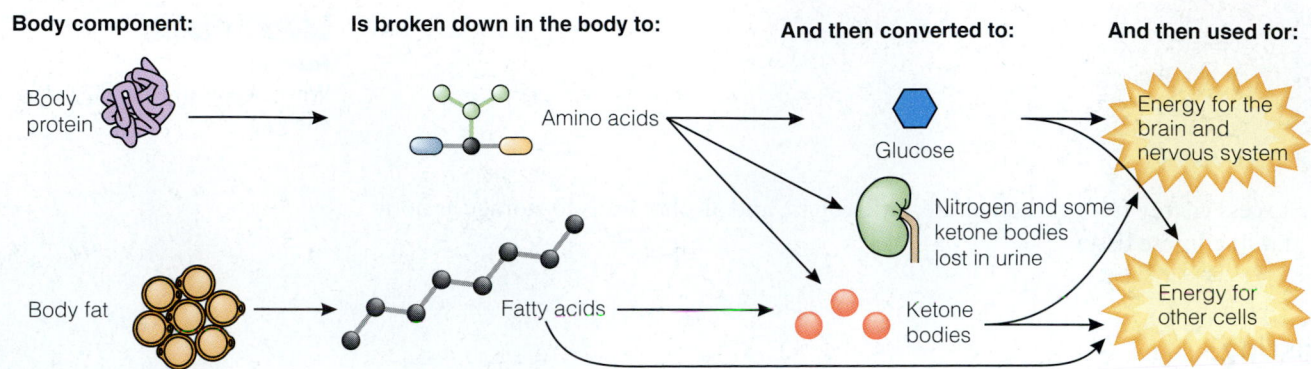

If the fast continues beyond glycogen depletion:

Body component: Is broken down in the body to: And then converted to: And then used for:

Body protein → Amino acids → Glucose → Energy for the brain and nervous system

Nitrogen and some ketone bodies lost in urine

Body fat → Fatty acids → Ketone bodies → Energy for other cells

FIGURE 6–6

Fasting

When people are fasting, they draw on stored energy.

■ Reminder: *Ketone bodies* are acidic, fat-related compounds formed from the incomplete breakdown of fat when carbohydrate is not available. Small amounts of ketone bodies are normally produced during energy metabolism, but when their blood concentration rises, they spill into the urine. The combination of a high blood concentration of ketone bodies (*ketonemia*) and ketone bodies in the urine (*ketonuria*) is called *ketosis*.

■ Fasting = living on the body's fat and protein

■ In fasting, muscle and lean tissues give up protein to supply amino acids for conversion to glucose. This glucose, with ketone bodies produced from fat, fuels the brain's activities.

death would ensue within about three weeks. As the fast continues, however, the body finds a way to use its fat to fuel the brain. It adapts by condensing together fragments derived from fatty acids to produce ketone bodies, ■ which can serve as fuel for some brain cells. Ketone body production rises until, after several weeks of fasting, it is meeting much of the nervous system's energy needs. Still, many areas of the brain rely exclusively on glucose, and body protein continues to be sacrificed to produce it. ■ Figure 6–6 shows the metabolic events that occur during fasting. ■

Slowed Metabolism As fasting continues and the nervous system shifts to partial dependence on ketone bodies for energy, the body simultaneously reduces its energy output (metabolic rate) and conserves both fat and lean tissue. Because of the slowed metabolism, energy use falls to a bare minimum.

Hazards of Fasting The body's adaptations to fasting are sufficient to maintain life for a long period. Mental alertness need not be diminished. Even physical energy can remain sufficient for a surprisingly long time. Still, fasting is not without its hazards. Among the many changes that take place in the body are:

■ Wasting of lean tissues

■ Impairment of disease resistance

■ Lowering of body temperature

■ Disturbances of the body's fluid and electrolyte balances

For the person who wants to lose weight, fasting is not the best way to go. The body's lean tissue continues to be degraded, sometimes amounting to as much as 50 percent of the weight lost over the first week. Over the long term, a diet only moderately restricted in energy promotes primarily *fat* loss and the retention of more lean tissue than a severely restricted fast.

IN SUMMARY

- When fasting, the body makes a number of adaptations: increasing the breakdown of fat to provide energy for most of the cells, using glycerol and amino acids to make glucose for the red blood cells and central nervous system, producing ketones to fuel the brain, and slowing metabolism.

- All of these measures conserve energy and minimize losses.

- Over the long term, a diet moderately restricted in energy promotes primarily fat loss and the retention of lean tissue.

Energy Balance

f a person maintains a healthy weight over time, the person is in energy balance. Food energy intake equals energy expenditure: deposits of fat made at one time are compensated for by withdrawals made at another. In other words, the body uses fat as a savings account for energy. But, unlike money, having more fat is not better; there is an optimum.

A day's energy balance can be stated like this: Change in energy stores equals the food energy taken in (kcalories) minus the energy expended on metabolism and physical activities (kcalories). More simply:

$$\text{Change in energy stores} = \text{energy in (kcalories)} - \text{energy out (kcalories)}$$

Energy In

The energy provided by foods and beverages is the only contributor to the "energy in" side of the energy balance equation. One way to become familiar with the amounts of energy provided by foods and beverages is to look up the kcalories in the Table of Food Composition (Appendix A). Alternatively, diet analysis computer programs can readily provide this information for those with computer access.

Food composition data would reveal that an apple provides about 70 kcalories from carbohydrate and a candy bar supplies about 250 kcalories mostly from fat and carbohydrate. You may already know that for each 3,500 kcalories you eat in excess of expenditures, you store approximately 1 pound of body fat. ■ Similarly, a pound of fat is lost for each 3,500 kcalories expended beyond those consumed. The fat stores of even a healthy-weight adult represent an ample reserve of energy—50,000 to 200,000 kcalories.

Energy Out

The body expends energy in two major ways: to fuel its **basal metabolism** and to fuel its **voluntary activities.** People can change their voluntary activities to expend more or less energy in a day, and over time they can also change their basal metabolism by building up the body's metabolically active lean tissue, as explained in Chapter 7.

Energy for Basal Metabolism Basal metabolism supports the body's work that goes on all the time without the person's conscious awareness. The beating of the heart, inhaling and exhaling of air, maintenance of body temperature, and transmission of nerve and hormonal messages to direct these activities are the basal processes that maintain life. As

■ 1 lb body fat = 3,500 kcal
Body fat, or adipose tissue, is composed of a mixture of mostly fat, some protein, and water. A pound of body fat (454 g) is approximately 87% fat, or (454 × 0.87) 395 g, and 395 g × 9 kcal/g = 3,555 kcal.

basal metabolism: the energy needed to maintain life when a person is at complete digestive, physical, and emotional rest. Basal metabolism is normally the largest part of a person's daily energy expenditure.

voluntary activities: the component of a person's daily energy expenditure that involves conscious and deliberate muscular work—walking, lifting, climbing, and other physical activities. Voluntary activities normally require less energy per day than basal metabolism does.

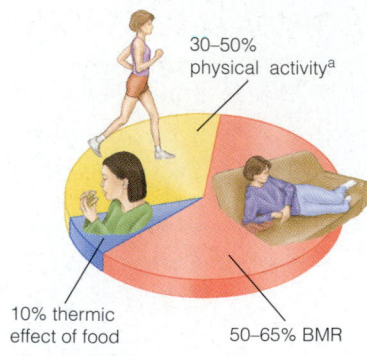

30–50% physical activity[a]

10% thermic effect of food

50–65% BMR

FIGURE 6–7

Components of Energy Expenditure
In most people, basal metabolism represents the person's largest expenditure of energy, followed by physical activity and the thermic effect of food.

[a] For a sedentary person, physical activities may account for less than half as much energy as basal metabolism, whereas a very active person's activities may equal the energy cost of basal metabolism.

Source: Data from the Standing Committee on the Scientific Evaluation of Dietary Reference Intakes, Food and Nutrition Board, Institute of Medicine. *Dietary Reference Intakes for Energy, Carbohydrate, Fiber, Fat, Fatty Acids, Cholesterol, Protein, and Amino Acids* (Washington, DC: National Academic Press, 2005). p. 115.

Figure 6–7 shows, basal metabolism represents about two-thirds of the total energy a sedentary person spends in a day. In practical terms, a person whose total energy needs are 2,000 kcalories per day may expend 1,000 to 1,300 of them to support basal metabolism.

The **basal metabolic rate (BMR)** is the rate at which the body expends energy for these activities. This rate varies from person to person and may vary for an individual with a change in circumstance or physical condition. The rate is slowest when a person is sleeping undisturbed, but it is usually measured when the person is awake, but lying still, in a room with a comfortable temperature after a restful sleep and an overnight (12- to 14-hour) fast. A similar measure of energy output—called the **resting metabolic rate (RMR)**—is slightly higher than the BMR because its criteria for recent food intake and physical activity are not as strict.

Table 6–1 summarizes the factors that raise and lower the BMR. For the most part, the BMR is highest in people who are growing (children, adolescents, and pregnant women) and in those with considerable lean body mass (physically fit people and males). One way to increase the BMR is to maximize lean body tissue by participating regularly in endurance and strength-building activities. The BMR is also fast in people who are tall and so have a large surface area for their weight, in people with fever or under stress, in people taking certain medications, and in people with highly active thyroid glands. The BMR is

TABLE 6–1

Factors That Affect the BMR

Factor	Effect on BMR
Age	Lean body mass diminishes with age, slowing the BMR.[a]
Height	In tall, thin people, the BMR is higher.[b]
Growth	In children and pregnant women, the BMR is higher.
Body composition (gender)	The more lean tissue, the higher the BMR (which is why males usually have a higher BMR than females). The more fat tissue, the lower the BMR.
Fever	Fever raises the BMR.[c]
Stresses	Stresses (including many diseases and certain drugs) raise the BMR.
Environmental temperature	Both heat and cold raise the BMR.
Fasting/starvation	Fasting/starvation lowers the BMR.[d]
Malnutrition	Malnutrition lowers the BMR.
Hormones (gender)	The thyroid hormone thyroxin, for example, can speed up or slow down the BMR.[e] Premenstrual hormones slightly raise the BMR.
Smoking	Nicotine increases energy expenditure.
Caffeine	Caffeine increases energy expenditure.
Sleep	BMR is lowest when sleeping.

[a] The BMR begins to decrease in early adulthood (after growth and development cease) at a rate of about 2 percent/decade. A reduction in voluntary activity as well brings the total decline in energy expenditure to 5 percent/decade.

[b] If two people weigh the same, the taller, thinner person will have the faster metabolic rate, reflecting the greater skin surface, through which heat is lost by radiation, in proportion to the body's volume.

[c] Fever raises the BMR by 7 percent for each degree Fahrenheit.

[d] Prolonged starvation reduces the total amount of metabolically active lean tissue in the body, although the decline occurs sooner and to a greater extent than body losses alone can explain. More likely, the neural and hormonal changes that accompany fasting are responsible for changes in the BMR.

[e] The thyroid gland releases hormones that travel to the cells and influence cellular metabolism. Thyroid hormone activity can speed up or slow down the rate of metabolism by as much as 50 percent.

slowed by loss of lean tissue and by depression of thyroid hormone activity due to disease, inactivity, fasting, or malnutrition.

Energy for Physical Activities The number of kcalories expended on voluntary physical activities depends on three factors: muscle mass, body weight, and activity. The larger the muscle mass required for the activity and the heavier the body part being moved, the more kcalories are expended. The activity's duration, frequency, and intensity also influence energy costs: the longer, the more frequent, and the more intense the activity, the more kcalories expended. Table 6–2 (p. 152) gives average energy expenditures for people of different body weights engaged in various activities and shows that a heavy person uses more energy per minute than a light person does in the same activity.

Energy for Thermic Effect of Food When food is taken into the body, many cells that have been dormant become active. The muscles that move the food through the intestinal tract speed up their rhythmic contractions, and the cells that manufacture and secrete digestive juices begin their tasks. All these and other cells need extra energy as they participate in the digestion, absorption, and metabolism of food. This cellular activity produces heat and is known as the **thermic effect of food**. The thermic effect of food is generally thought to represent about 10 percent of the total food energy taken in. For purposes of rough estimates, the thermic effect of food is not always included.

Physical activity expends energy and benefits health in many ways.

Estimating Energy Requirements

In estimating energy requirements, the DRI committee developed equations that consider how the following factors influence energy expenditure: ■

- **Gender.** In general, women have a lower BMR than men, in large part because men typically have more lean body mass. In addition, menstrual hormones influence the BMR in women, raising it just prior to menstruation. Two sets of energy equations—one for men and one for women—were developed to accommodate the influence of gender on energy expenditure.

- **Growth.** The BMR is high in people who are growing. For this reason, pregnant women, children, and adolescents have their own sets of energy equations.

- **Age.** The BMR declines during adulthood as lean body mass diminishes. Physical activities tend to decline as well, bringing the average reduction in energy expenditure to about 5 percent per decade. The decline in the BMR that occurs when a person becomes less active reflects the loss of lean body mass and may be prevented with ongoing physical activity. Because age influences energy expenditure, it is also factored into the energy equations.

- **Physical activity.** Using individual values for various physical activities (as in Table 6–2) is time-consuming and impractical for estimating the energy needs of a population. Instead, various activities are clustered according to the typical intensity of a day's efforts. Energy equations include a physical activity factor for various levels of intensity for each gender (see the "How to" on p. 153).

- **Body composition and body size.** The BMR is high in people who are tall and so have a large surface area. Similarly, the more a person weighs, the more energy is expended on basal metabolism. For these reasons, the energy equations include a factor for both height and weight.

■ Note that Table 6–1 (p. 150) listed these factors among those that influence BMR and consequently energy expenditure.

basal metabolic rate (BMR): the rate of energy use for metabolism under specified conditions: after a 12-hour fast and restful sleep, without any physical activity or emotional excitement, and in a comfortable setting. It is usually expressed as kcalories per kilogram of body weight per hour.

resting metabolic rate (RMR): a measure of the energy use of a person at rest in a comfortable setting; similar to the BMR but with less stringent criteria for recent food intake and physical activity. Consequently, the RMR is slightly higher than the BMR.

thermic effect of food: an estimation of the energy required to process food (digest, absorb, transport, metabolize, and store ingested nutrients).

TABLE 6–2

Energy Expended on Various Activities

The values listed in this table reflect both the energy spent in physical activity and the amount used for BMR.

Activity	kcal/lb/min[a]	kcalories per Minute at Different Body Weights				
		110 lb	125 lb	150 lb	175 lb	200 lb
Aerobic dance (vigorous)	.062	6.8	7.8	9.3	10.9	12.4
Basketball (vigorous, full court)	.097	10.7	12.1	14.6	17.0	19.4
Bicycling						
13 mph	.045	5.0	5.6	6.8	7.9	9.0
15 mph	.049	5.4	6.1	7.4	8.6	9.8
17 mph	.057	6.3	7.1	8.6	10.0	11.4
19 mph	.076	8.4	9.5	11.4	13.3	15.2
21 mph	.090	9.9	11.3	13.5	15.8	18.0
23 mph	.109	12.0	13.6	16.4	19.0	21.8
25 mph	.139	15.3	17.4	20.9	24.3	27.8
Canoeing, flat water, moderate pace	.045	5.0	5.6	6.8	7.9	9.0
Cross-country skiing						
8 mph	.104	11.4	13.0	15.6	18.2	20.8
Golf (carrying clubs)	.045	5.0	5.6	6.8	7.9	9.0
Handball	.078	8.6	9.8	11.7	13.7	15.6
Horseback riding (trot)	.052	5.7	6.5	7.8	9.1	10.4
Rowing (vigorous)	.097	10.7	12.1	14.6	17.0	19.4
Running						
5 mph	.061	6.7	7.6	9.2	10.7	12.2
6 mph	.074	8.1	9.2	11.1	13.0	14.8
7.5 mph	.094	10.3	11.8	14.1	16.4	18.8
9 mph	.103	11.3	12.9	15.5	18.0	20.6
10 mph	.114	12.5	14.3	17.1	20.0	22.9
11 mph	.131	14.4	16.4	19.7	22.9	26.2
Soccer (vigorous)	.097	10.7	12.1	14.6	17.0	19.4
Studying	.011	1.2	1.4	1.7	1.9	2.2
Swimming						
20 yd/min	.032	3.5	4.0	4.8	5.6	6.4
45 yd/min	.058	6.4	7.3	8.7	10.2	11.6
50 yd/min	.070	7.7	8.8	10.5	12.3	14.0
Table tennis (skilled)	0.45	5.0	5.6	6.8	7.9	9.0
Tennis (beginner)	.032	3.5	4.0	4.8	5.6	6.4
Walking (brisk pace)						
3.5 mph	.035	3.9	4.4	5.2	6.1	7.0
4.5 mph	.048	5.3	6.0	7.2	8.4	9.6
Weight lifting						
light-to-moderate effort	.024	2.6	3.0	3.6	4.2	4.8
vigorous effort	.048	5.2	6.0	7.2	8.4	9.6
Wheelchair basketball	.084	9.2	10.5	12.6	14.7	16.8
Wheeling self in wheelchair	.030	3.3	3.8	4.5	5.3	6.0

[a] To calculate kcalories spent per minute of activity for your own body weight, multiply kcal/lb/min by your exact weight and then multiply that number by the number of minutes spent in the activity. For example, if you weigh 142 pound and you want to know how many kcalories you spent doing 30 minutes of vigorous aerobic dance: 0.062 × 142 = 8.8 kcalories per minute; 8.8 × 30 (minutes) = 264 total kcalories spent.

ESTIMATE ENERGY REQUIREMENTS

To determine your estimated energy requirements (EER), use the appropriate equation, inserting your age in years, weight (wt) in kilograms, height (ht) in meters, and physical activity (PA) factor from the accompanying table. (To convert pounds to kilograms, divide by 2.2; to convert inches to meters, divide by 39.37.)

- For men 19 years and older:

$$EER = [662 - (9.53 \times age)] + PA \times [(15.91 \times wt) + (539.6 \times ht)]$$

- For women 19 years and older:

$$EER = [354 - (6.91 \times age)] + PA \times [(9.36 \times wt) + (726 \times ht)]$$

For example, consider an active 30-year-old male who is 5 feet 11 inches tall and weighs 178 pounds. First, he converts his weight from pounds to kilograms and his height from inches to meters, if necessary:

$$178 \text{ lb} \div 2.2 = 80.9 \text{ kg}$$
$$71 \text{ in} \div 39.37 = 1.8 \text{ m}$$

Next, he considers his level of daily physical activity and selects the appropriate PA factor from the accompanying table (in this example, 1.25 for an active male). Then, he inserts his age, PA factor, weight, and height into the appropriate equation:

$$EER = [662 - (9.53 \times 30)] + 1.25 \times [(15.91 \times 80.9) + (539.6 \times 1.8)]$$

(A reminder: do calculations within the parentheses first, and multiplication before addition and subtraction.) He calculates:

$$EER = [662 - (9.53 \times 30)] + 1.25 \times (1287 + 971)$$
$$EER = [662 - (9.53 \times 30)] + (1.25 \times 2258)$$
$$EER = 662 - 286 + 2823$$
$$EER = 3199$$

The estimated energy requirement for an active 30-year-old male who is 5 feet 11 inches tall and weighs 178 pounds is about 3,200 kcalories/day. His actual requirement probably falls within a range of 200 kcalories above and below this estimate.

Physical Activity (PA) Factors for EER Equations			
	Men	**Women**	**Physical Activity**
Sedentary	1.0	1.0	Typical daily living activities
Low active	1.11	1.12	Plus 30 to 60 minutes moderate activity
Active	1.25	1.27	Plus ≥ 60 minutes moderate activity
Very active	1.48	1.45	Plus ≥ 60 minutes moderate activity and 60 minutes vigorous activity or 120 minutes moderate activity

Note: Moderate activity is equivalent to walking at 3½ to 4½ miles per hour.

As just explained, energy needs vary between individuals depending on such factors as gender, growth, age, physical activity, and body composition and body size. Even when two people are similarly matched, however, their energy needs will still differ because of genetic differences. Perhaps one day genetic research will reveal how to estimate requirements for each individual. For now, the "How to" provides instructions on calculating estimated energy requirements using the DRI equations and physical activity factors. ■

IN SUMMARY

- A person takes in energy from food and, on average, expends most of it on basal metabolic activities, some of it on physical activities, and about 10 percent on the thermic effect of food.

- Because energy requirements vary from person to person, such factors as gender, growth, age, and body composition and body size must be considered when calculating energy expended on basal metabolism, and the intensity and duration of the activity must be taken into account when calculating expenditures on physical activities.

- Appendix D presents tables that provide a shortcut to estimating total energy expenditure and instructions to help you determine the appropriate physical activity factor to use in the equations.

Body Weight, Body Composition, and Health

The body's weight reflects its composition—the proportions of its bone, muscle, fat, fluid, and other tissue. All of these body components can vary in quantity and quality: the bones can be dense or porous, the muscles can be well developed or underdeveloped, fat can be abundant or scarce, and so on. By far the most variable tissue is body fat. For health's sake, weight management efforts should focus on eating and activity habits to improve body composition.

Defining Healthy Body Weight

How much should a person weigh? How can a person know if her weight is appropriate for her height and age? How can a person know if his weight is jeopardizing his health? Questions such as these seem so simple, yet the answers can be complex—and different depending on whom you ask.

The Criterion of Fashion In asking what an ideal body weight is, people often mistakenly turn to fashion for the answer. Without a doubt, our society sets unrealistic ideals for body weight, especially for women. Magazines, movies, and television all convey the message that to be thin is to be beautiful. As a result, the media have a great influence on the weight concerns and dieting patterns of people of all ages, but most tragically on young, impressionable children and adolescents.[3] Even five-year-olds are concerned about their body weight.[4] One-half of preteen girls and one-third of preteen boys are dissatisfied with their body weight and shape.[5]

Importantly, perceived body image has little to do with actual body weight or size. People of all shapes, sizes, and ages—including extremely thin fashion models with anorexia nervosa and fitness instructors with ideal body composition—have learned to be unhappy with their "overweight" bodies. Such dissatisfaction can lead to damaging behaviors, such as starvation diets, diet pill abuse, and failure to seek health care.[6] The first step toward making healthy changes may be self-acceptance. Keep in mind that fashion is fickle; the body shapes valued by our society change with time. Furthermore, body shapes valued by our society differ from those of other societies. The standards defining "ideal" are subjective and frequently have little in common with health. Table 6–3 offers some tips for adopting health as an ideal, rather than society's misconceived image of beauty.

The Criterion of Health Even if our society were to accept fat as beautiful, obesity would still be a major risk factor for several life-threatening diseases as discussed later in this chapter. For this reason, the most important criterion for determining how much a person should weigh and how much body fat a person needs is not appearance but good health and longevity. A range of healthy body weights has been identified using a common measure of weight and height—the body mass index.

Body Mass Index The **body mass index (BMI)** describes relative weight for height and often correlates with degree of body fatness and disease risks. The equation that is used to derive BMI values is provided in the margin. ■ A person who takes measurements in pounds and inches can convert them to metric units ■ or can use this modified equation:

$$ \text{BMI} = \frac{\text{weight (lb)}}{\text{height (in)}^2} \times 703 $$

■ $\text{BMI} = \dfrac{\text{weight (kg)}}{\text{height (m)}^2}$

- ■ ■ To convert pounds to kilograms, divide by 2.2.
- ■ To convert inches to meters, divide by 39.37.

body mass index (BMI): an index of a person's weight in relation to height; determined by dividing the weight (in kilograms) by the square of the height (in meters).

Tips for Accepting a Healthy Body Weight

Value yourself and others for human attributes other than body weight. Realize that prejudging people by weight is as harmful as prejudging them by race, religion, or gender.

Use positive, nonjudgmental descriptions of your body.

Accept positive comments from others.

Focus on your whole self including your intelligence, social grace, and professional and scholastic achievements.

Accept that no magic diet exists.

Stop dieting to lose weight. Adopt a lifestyle of healthy eating and physical activity permanently.

Follow the USDA Food Guide. Never restrict food intake below the minimum levels that meet nutrient needs.

Become physically active, not because it will help you get thin but because it will make you feel good and enhance your health.

Seek support from loved ones. Tell them of your plan for a healthy life in the body you have been given.

Seek professional counseling, *not* from a weight-loss counselor, but from someone who can help you make gains in self-esteem without weight as a factor.

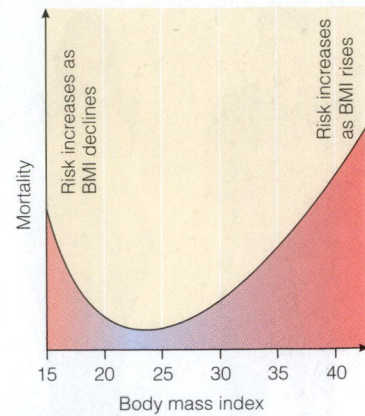

FIGURE 6–8

Body Mass Index and Mortality
This J-shaped curve describes the relationship between body mass index (BMI) and mortality and shows that both underweight and overweight present risks of a premature death.

The table on the inside back cover shows weights for various heights using the BMI to define underweight, healthy weight, overweight, and obesity. As the table shows, healthy weight falls between a BMI of 18.5 and 24.9. Most people with a BMI within this range have few of the health risks typically associated with too-low or too-high body weight. Risks increase as BMI falls below 18.5 or rises above 24.9 (see Figure 6–8), reflecting the reality that both underweight and overweight impair health status.

The BMI values are most accurate in assessing degrees of obesity and are less useful for evaluating nonobese people's body fatness. BMI values fail to reveal two valuable pieces of information in assessing disease risk. They do not reveal how much of the weight is fat, and they do not indicate where the fat is located. For this knowledge, measures of body composition are needed.

Body Composition

For many people, being overweight compared with the standard means that they are over*fat*. This is not the case for athletes with dense bones and well-developed muscles; they may be over*weight* but carry little body fat. Conversely, inactive people may seem to have acceptable weights but still carry too much body fat. In addition, the distribution of fat on the body may be even more critical than overfatness alone.

Central Obesity Even more than total body fat, fat that collects deep within the central abdominal area of the body may be especially likely to lead to diabetes, stroke, hypertension, and coronary artery disease (see Figure 6–9).[7] The risk of death from all causes may be

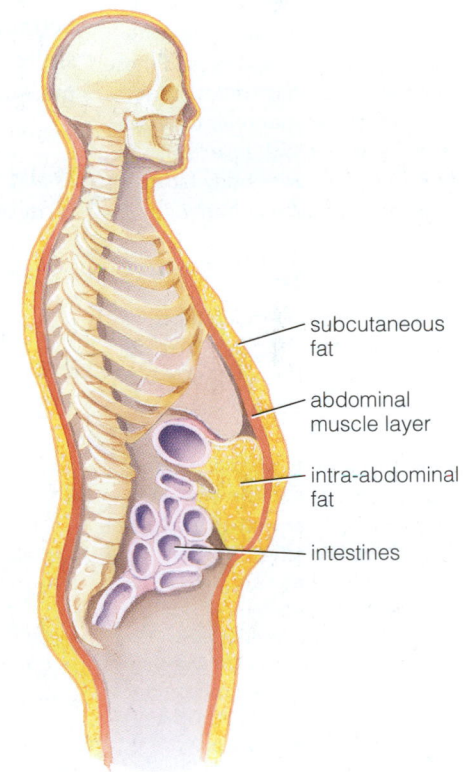

FIGURE 6–9

Intra-abdominal Fat and Subcutaneous Fat
The fat deep within the body's abdominal cavity may pose an especially high risk to health.

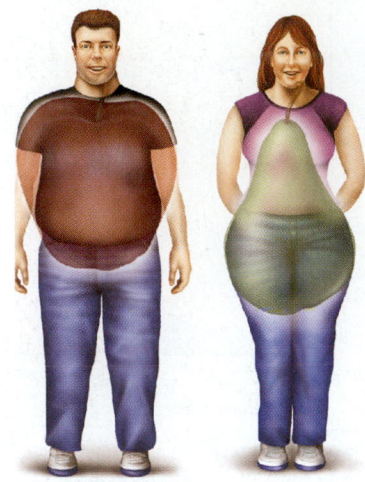

FIGURE 6–10

"Apple" and "Pear" Body Shapes Compared
Upper-body fat is more common in men than in women and is closely associated with heart disease, stroke, diabetes, hypertension, and some types of cancer. In contrast, lower-body fat is more common in women than in men and is not usually associated with chronic diseases. Popular articles sometimes call bodies with upper-body fat as "apples" and those with lower-body fat, "pears." Researchers sometimes refer to upper-body fat as "android" (manlike) obesity and to lower-body fat as "gynoid" (womanlike) obesity.

FIGURE 6–11

Measuring Waist Circumference
Using a nonstretching tape measure, measure the body round the point just above the iliac crest. Take the measure at the end of a normal expiration. A healthy waist circumference for men is no larger than 102 cm (40 inches); for women, no larger than 88 cm (35 inches).
Source: National Heart, Lung, and Blood Institute, National Institutes of Health. *The Practical Guide: Identification, Evaluation, and Treatment of Overweight and Obesity in Adults,* NIH publication no. 00-4084 (Washington, DC: Government Printing Office, 2000). p. 9.

■ The higher percentage of body fat in women compared to men is normal and necessary for reproduction.

higher for those with **central obesity** than for those whose fat accumulates elsewhere in the body. Unlike the subcutaneous fat layers lying just beneath the skin of the abdomen and elsewhere, **intra-abdominal fat**, when mobilized, goes directly to the liver where it is made into triglyceride-carrying VLDL (very-low-density lipoproteins).[8] Fat from elsewhere may arrive in the liver eventually, but it takes a circuitous route that first allows other tissues the chance to pull it from the circulation and metabolize it.

Abdominal fat creates the "apple" profile of central obesity. Fat around the hips and thighs creates more of a "pear" profile (see Figure 6–10). Abdominal fat is common in women past menopause and even more common in men. Even when total body fat is similar, men have more abdominal fat than either premenopausal or postmenopausal women. For those women with abdominal fat, the risks of cardiovascular disease and mortality are increased, just as they are for men. Smokers, too, may carry more of their body fat centrally. A smoker may weigh less than the average nonsmoker, but the smoker's waist circumference may be greater, leading researchers to think that smoking may directly affect body fat distribution.[9] Two other factors that may affect body fat distribution are intakes of alcohol and physical activity. Moderate-to-high alcohol consumption may favor central obesity.[10] In contrast, regular physical activity seems to prevent abdominal fat accumulation.[11]

Skinfold Measures Skinfold measurements provide an accurate estimate of total body fat and a fair assessment of the fat's location. About half of the fat in the body lies directly beneath the skin, so the thickness of this subcutaneous fat is assumed to reflect total body fat. Measures taken from central-body sites (around the abdomen) better reflect changes in fatness than those taken from upper sites (arm and back). A skilled assessor can obtain an accurate **skinfold measure** and then compare the measurement with standards (see Appendix E).

Waist Circumference **Waist circumference** serves as an indicator of abdominal fatness (see Figure 6–11). Previously, a ratio of waist-to-hip measurements served this purpose, but waist circumference alone has been deemed a valid indicator for both men and women. In general, women with a waist circumference greater than 35 inches and men with a waist circumference greater than 40 inches have a high risk of central obesity-related health problems.[12]

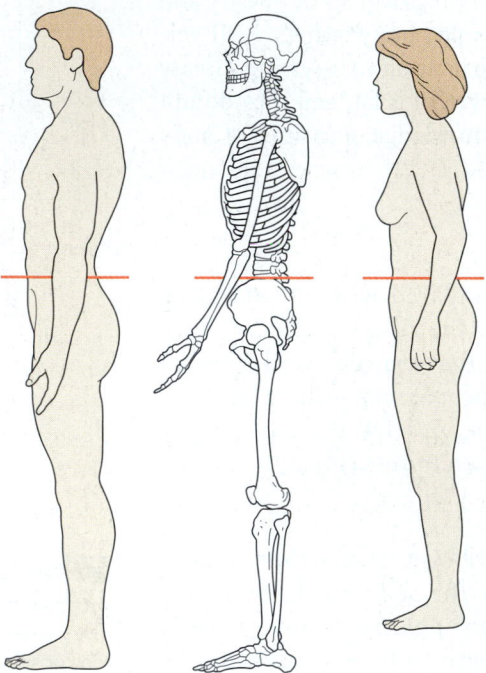

How Much Body Fat Is Too Much?
People often ask exactly how much fat is too fat for health. Ideally, a person has enough fat to meet basic needs but not so much as to incur health risks. The ideal amount of body fat depends partly on the person. A man with a BMI within the recommended range may have between 13 and 21 percent body fat; a woman, 23 to 31 percent. ■

Many athletes have a lower percentage of body fat—just enough fat to provide fuel, insulate and protect the body, assist in nerve impulse transmissions, and support normal hormone activity, but not so much as to burden the muscles. For athletes, body fat might be 5 to 10 percent for men and 15 to 20 percent for women.

For an Alaskan fisherman, a higher-than-average percentage of body fat is probably beneficial because fat helps prevent heat loss in cold weather. A woman starting a pregnancy needs sufficient body fat to support conception and fetal growth. Below a certain threshold for body fat, individuals may become infertile, develop depression, experience abnormal hunger regulation, or be unable to keep warm. These thresholds differ for each function and for each individual; much remains to be learned about them.

IN SUMMARY

- Clearly, the most important criterion of appropriate fatness is health.

- Current standards for body weight are based on the BMI (body mass index), which describes a person's weight in relation to height.

- Health risks increase with a BMI below 18.5 or above 24.9.

- Central obesity, in which excess fat is distributed around the trunk of the body, presents greater health risks than excess fat distributed on the lower body.

- Researchers use a number of techniques to assess body composition including waist circumference and skinfold measures.

Risks of Overweight and Obesity

Despite our nation's preoccupation with body image and weight loss, the prevalence of overweight and obesity continues to rise dramatically (see Figure 6–12 p. 158).[13] During the previous decade, obesity increased in every state, in both genders, and across all ages, races, and education levels. Almost 65 percent of U.S. adults are **overweight** (BMI of 25 or greater) or obese (BMI of 30 or greater).[14] The prevalence of overweight is especially high among women, the poor, and some ethnic groups. In short, obesity is a major public health problem that is becoming more prevalent.

Health Risks of Obesity The growing prevalence of overweight and obesity is a matter of concern because both conditions present risks to health. Indeed the health risks of **obesity** are so many that it has been declared a disease. Excess weight contributes to up to half of all cases of hypertension, thereby increasing the risk of heart attack and stroke.[15] Often weight loss alone can normalize the blood pressure of an overfat person.

Excess body weight also increases the risk of type 2 diabetes. Most adults with type 2 diabetes are overweight or obese.[16] Diabetes (type 2) is three times more likely to develop in an obese person than in a nonobese person. Furthermore, the person with type 2 diabetes often has central obesity. Central-body fat cells appear to be larger and more insulin-resistant than lower-body fat cells.[17] The association between **insulin resistance** and obesity is strong, and both are major risk factors for the development of type 2 diabetes.

Diabetes appears to be influenced by weight gains as well as by body weight. A weight gain of more than 10 pounds since age 18 doubles the risk of developing diabetes, even in women of average weight. In contrast, weight loss effectively improves glucose tolerance and insulin resistance.[18]

© Rich Shaff

At 6 feet 4 inches tall and 250 pounds, this runner would be considered overweight by most standards. Yet he is clearly not overfat.

central obesity: excess fat around the trunk of the body; also called **abdominal fat** or **upper-body fat.**

intra-abdominal fat: fat stored within the abdominal cavity in association with the internal abdominal organs, as opposed to fat stored directly under the skin (subcutaneous fat); also called visceral fat.

skinfold measure: a clinical estimate of total body fatness in which the thickness of a fold of skin on the back of the arm (over the triceps muscle), below the shoulder blade (subscapular), or in other places is measured with a caliper.

waist circumference: a measurement used to assess a person's abdominal fat.

overweight: overfatness of a moderate degree; defined as a body mass index (BMI) of 25.0 through 29.9.

obesity: overfatness with adverse health effects, as determined by reliable measures and interpreted with good medical judgment. Obesity is officially defined as a body mass index (BMI) of 30 or higher.

insulin resistance: the condition in which a normal amount of insulin produces a subnormal effect in muscle, adipose, and liver cells, resulting in an elevated fasting glucose; a metabolic consequence of obesity that precedes type 2 diabetes.

FIGURE 6–12

**The Increasing Prevalence of
Obesity among U.S. Adults by State**
Source: http://www.cdc.gov/nccdphp/
dnpa/obesity/trend/maps/index.htm

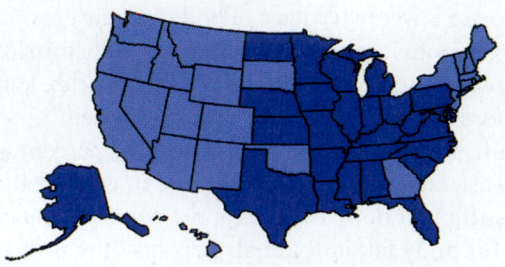

1995: Over half the states had prevalence rates ≥15 percent, but no state had prevalence rates ≥20 percent.

Key:
- 10%–14%
- 15%–19%
- 20%–24%
- 25%–29%
- ≥30%

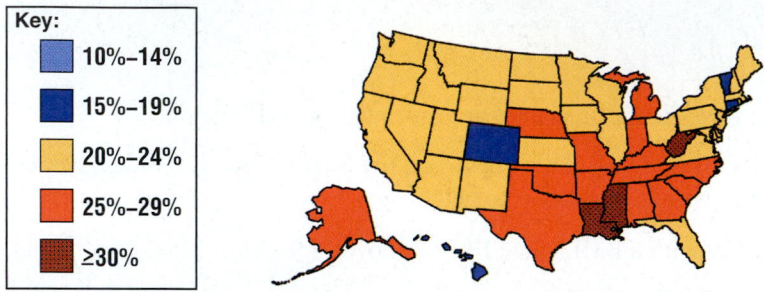

2005: Only four states had prevalence rates <20 percent, about one-third of the states had prevalence rates ≥25 percent, with three states having prevalence rates ≥30 percent.

In addition to diabetes and hypertension, other risks threaten obese adults. Among them are high blood lipids, cardiovascular disease, sleep apnea (abnormal ceasing of breathing during sleep), osteoarthritis, abdominal hernias, some cancers, varicose veins, gout, gallbladder disease, kidney stones, respiratory problems (including Pickwickian syndrome, a breathing blockage linked with sudden death), liver malfunction, complications in pregnancy and surgery, flat feet, and even a high accident rate.[19] Each year these obesity-related illnesses cost our nation billions of dollars. The cost in terms of lives is also great. People with lifelong obesity are twice as likely to die prematurely as others. In the United States, obesity is second only to tobacco use as the most significant cause of preventable death.

Some obese people, however, seem to remain healthy and live long despite their body fatness. Physical fitness, despite body fatness, may protect against premature death from disease.[20] It may be that genetics and other risk factors such as smoking help determine who among the overweight will be susceptible to diseases and who will stay healthy. Still, the majority of obese people do develop associated health problems. To help in identifying those most at risk, obesity experts have developed guidelines, described next.

National Guidelines for Identifying Those at Risk from Obesity The U.S. guidelines for identifying and evaluating the risks to health from overweight and obesity rely on three indicators. The first indicator is a person's BMI (see Table 6–4). As a general guideline, overweight for adults is defined as BMI of 25.0 through 29.9, and obesity is defined as BMI equal to or greater than 30.

The second indicator is waist circumference, which, as discussed earlier, reflects the degree of intra-abdominal fatness in proportion to body fatness. As Table 6–4 shows, women with a waist circumference greater than 35 inches and men with a waist circumference greater than 40 inches are at greater risk of type 2 diabetes, hypertension,

TABLE 6-4

Disease Risks Based on BMI and Waist Circumference[a]

The degree of risk is heightened by the presence of specific disease or other risk factors, such as elevated blood LDL cholesterol or smoking (see the margin).

BMI	Waist ≤ 40 Inches (Men) or ≤ 35 Inches (Women)	Waist > 40 Inches (Men) or > 35 Inches (Women)
18.5 or less (Underweight)	Low	—
18.5–24.9 (Normal)	Low	—
25.0–29.9 (Overweight)	Increased	High
30.0–34.9 (Obese, class I)	High	Very high
35.0–39.9 (Obese, class II)	Very high	Very high
40 or greater (Extremely obese, class III)	Extremely high	Extremely high

[a] Risk for type 2 diabetes, hypertension, and cardiovascular disease.

Source: National Heart, Lung, and Blood Institute, National Institutes of Health, *The Practical Guide: Identification, Evaluation, and Treatment of Overweight and Obesity in Adults,* NIH publication no. 00-4084 (Washington, DC: Government Printing Office, 2000).

and cardiovascular disease than women or men with waist circumferences equal to, or below, these measures. In other words, waist circumference is an independent predictor of disease risk.

The third indicator is the person's disease risk profile. People who have one or more of the diseases listed in the margin, ■ or three or more of the cardiovascular disease risk factors listed, have a very high risk for disease complications and mortality that requires aggressive treatment to manage the disease or modify the risk factors.

Other Risks of Obesity Although some obese people seem to escape health problems, few in our society can avoid the social and economic handicaps. Our society places enormous value on thinness. Obese people are often bypassed for romance, employment, or college admission. They pay more for insurance premiums and for clothing. This is especially true for women. These social and economic disadvantages have psychological consequences, too—fat people often feel rejected and embarrassed, and this hurts self-esteem.

IN SUMMARY

- The health risks of obesity are many and serious.

- Guidelines for identifying the health risks of overweight and obesity are based on a person's BMI, waist circumference, and disease risk profile.

- Obesity also incurs social, economic, and psychological risks.

This chapter described energy metabolism, energy balance, and the health risks of obesity. The next chapter discusses the health risks of underweight, the causes of obesity, and strategies to lose, maintain, or gain weight.

■ The National Heart, Lung, and Blood Institute states that aggressive treatment is urgently needed for a clinically obese person (BMI ≥ 30) who also has any one of the following:

- Established cardiovascular disease (CVD)
- Established type 2 diabetes, or impaired glucose tolerance (prediabetes)
- Sleep apnea, including temporary stopping of breathing

The same urgency for treatment exists for an obese person with any *three* of the following CVD risk factors:

- Hypertension
- Smoking
- High LDL cholesterol
- Low HDL cholesterol
- Sedentary lifestyle
- Age older than 45 years (men) or 55 years (women)
- Heart disease of an immediate family member before age 55 (male) or 65 (female)

SOURCE: National Heart, Lung, and Blood Institute, National Institutes of Health, *The Practical Guide: Identification, Evaluation, and Treatment of Overweight and Obesity in Adults,* NIH publication no. 00-4084 (Washington, DC: Government Printing Office, 2000).

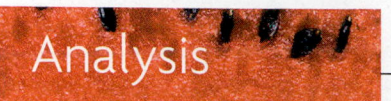

Analysis

CLINICAL APPLICATIONS

1. Compare the energy a person might spend on various physical activities. Refer to Table 6–2 on p. 152, and compute how much energy a person who weighs 142 pounds would spend doing each of the following activities. An example using aerobic dance has been provided for you. You may want to compare various activities based on your own weight.

   30 minutes of vigorous aerobic dance:

   0.062 kcal/lb/min × 142 lb = 8.8 kcal/min
       8.8 kcal/min × 30 min = 264 kcal

   a. 2 hours of golf, carrying clubs
   b. 20 minutes running at 9 mph

   c. 45 minutes of swimming at 20 yd/min
   d. 1 hour of walking at 3.5 mph

2. Using the "How to" on p. 153 as a guide, determine your estimated energy requirements (EER).

Answers

1. (a.) 0.045 kcal/lb/min × 142 lb = 6.4 kcal/min, 6.4 kcal/min × 120 min = 768 kcal; (b) 0.103 kcal/lb/min × 142 lb = 14.6 kcal/min, 14.6 kcal/min × 20 min = 292 kcal; (c) 0.032 kcal/lb/min × 142 lb = 4.5 kcal/min, 4.5 kcal/min × 45 = 202.5 kcal; (d) 0.035 kcal/lb/min × 142 lb = 5 kcal/min, 5 kcal/min × 60 min = 300 kcal.

Self Check

1. Before entering the TCA cycle, each of the energy-yielding nutrients is broken down to:
   a. ammonia
   b. pyruvate
   c. electrons
   d. acetyl CoA

2. As carbohydrate and fat stores are depleted during fasting or starvation, the body then uses ___ as its fuel source.
   a. alcohol
   b. protein
   c. glucose
   d. triglycerides

3. When carbohydrate is not available to provide energy for the brain, as in starvation, the body produces ketone bodies from:
   a. glucose.
   b. glycerol.
   c. fatty acid fragments.
   d. amino acids.

4. Three hazards of fasting are:
   a. water weight loss, decrease in mental alertness, and wasting of lean tissue.
   b. water weight gain, impairment of disease resistance, and lowering of body temperature.
   c. water weight gain, decrease in mental alertness, and impairment of disease resistance.
   d. wasting of lean tissue, impairment of disease resistance, and disturbances of the body's salt and water balance.

5. Two activities that contribute to the basal metabolic rate are:
   a. walking and running.
   b. maintenance of heartbeat and running.

   c. maintenance of body temperature and walking.
   d. maintenance of heartbeat and body temperature.

6. Three factors that affect the body's basal metabolic rate are:
   a. height, weight, and energy intake.
   b. age, body composition, and height.
   c. fever, body composition, and altitude.
   d. weight, fever, and environmental temperature.

7. The largest component of energy expenditure is:
   a. basal metabolism.
   b. physical activity.
   c. indirect calorimetry.
   d. thermic effect of food.

8. Which of the following reflects height and weight?
   a. body mass index
   b. central obesity
   c. waist circumference
   d. body composition

9. The BMI range that correlates with the fewest health risks is:
   a. 16.5 to 20.9.
   b. 18.5 to 24.9.
   c. 25.5 to 30.9.
   d. 30.5 to 34.9.

10. Which of the following health risks is *not* associated with being overweight?
    a. hypertension
    b. heart disease
    c. type 1 diabetes
    d. gallbladder disease

Answers to these questions can be found in Appendix H.

Notes

1. S. G. Wannamethee, A. G. Shaper, and P. H. Whincup, Alcohol and adiposity: Effects of quantity and type of drink and time relation with meals, *International Journal of Obesity and Related Metabolic Disorders*, online print, August 2, 2005; R. A. Breslow and B. A. Smothers, Drinking patterns and body mass index in never smokers, *American Journal of Epidemiology* 161 (2005): 368–376; A. Raben and coauthors, Meals with similar energy densities but rich in protein, fat, carbohydrate, or alcohol have different effects on energy expenditure and substrate metabolism but not on appetite and energy intake, *American Journal of Clinical Nutrition* 77 (2003): 91–100.

2. Raben and coauthors, 2003.

3. L. C. Andrist, Media images, body dissatisfaction, and disordered eating in adolescent women, *American Journal of Maternal Child Nursing* 28 (2003): 119–123; K. K. Davison, C. N. Markey, and L. L. Birch, A longitudinal examination of patterns in girls' weight concerns and body dissatisfaction from ages 5–9 years, *International Journal of Eating Disorders* 33 (2003): 320–332; J. Wardle, J. Waller, and E. Fox, Age of onset and body dissatisfaction in obesity, *Addictive Behaviors*, 27 (2002): 561–573.

4. J. A. Shunk and L. L. Birch, Girls at risk for overweight at age 5 are at risk for dietary restraint, disinhibited overeating, weight concerns, and greater weight gain from 5 to 9 years, *Journal of the American Dietetic Association* 104 (2004): 1120–1126; Davison, Markey, and Birch, 2003.

5. H. Truby and S. J. Paxton, Development of the Children's Body Image Scale, *British Journal of Clinical Psychology* 41 (2002): 185–203; H. A. Hausenblas and coauthors, Body image in middle school children, *Eating and Weight Disorders* 7 (2002): 244–248.

6. C. A. Drury and M. Louis, Exploring the association between body weight, stigma of obesity, and health care avoidance, *Journal of the American Academy of Nurse Practitioners* 14 (2002): 554–561.

7. Y. Wang and coauthors, Comparison of abdominal adiposity and overall obesity in predicting risk of type 2 diabetes among men, *American Journal of Clinical Nutrition* 81 (2005): 555–563; G. De Simone, and coauthors, Body composition and fat distribution influence systemic hemodynamics in the absence of obesity: The HyperGEN Study, *American Journal of Clinical Nutrition* 81 (2005): 757–761; I. Janssen, P. T. Katzmarzyk, and R. Ross, Waist circumference and not body mass index explains obesity-related health risk, *American Journal of Clinical Nutrition* 79 (2004): 379–384; S. Zhu and coauthors, Waist circumference and obesity-associated risk factors among whites in the Third National Health and Nutrition Survey: Clinical action thresholds, *American Journal of Clinical Nutrition* 76 (2002): 743–749.

8. F. X. Pi-Sunyer, The epidemiology of central fat distribution in relation to disease, *Nutrition Reviews* 62 (2004): S120–S126.

9. D. Canoy and coauthors, Cigarette smoking and fat distribution in 21,828 British men and women: A population-based study, *Obesity Research* 13 (2005): 1466–1475.

10. Wannamethee, Shaper, and Whincup, 2005; J. M. Dorn and coauthors, Alcohol drinking patterns differentially affect central adiposity as measured by abdominal height in women and men, *Journal of Nutrition* 133 (2003): 2655–2662.

11. C. A. Holcomb, D. L. Heim, and T. M. Loughin, Physical activity minimizes the association of body fatness with abdominal obesity in white, premenopausal women: Results from the Third National Health and Nutrition Examination Survey, *Journal of the American Dietetic Association* 104 (2004): 1859–1862.

12. Janssen, Katzmarzyk, and Ross, 2004; Zhu and coauthors, 2002.

13. G. L. Blackburn and W. A. Walker, Science-based solutions to obesity: What are the roles of academia, government, industry, and health care? *American Journal of Clinical Nutrition* 82 (2005): 207S–210S.

14. Food and Drug Administration, *Counting Calories: Report of the Working Group on Obesity,* March 12, 2004, available at www.cfsan.fda.gov/~dms/owg-rpt.html.; A. A. Hedley and coauthors, Prevalence of overweight and obesity among US children, adolescents, and adults, 1999–2002, *Journal of the American Medical Association* 291 (2004): 2847–2850.

15. U.S. Preventive Services Task Force, *Screening for Obesity in Adults,* update of *Guide to Clinical Preventive Service,* 2nd ed. (2003), available at www.ahrq.gov/clinic/upstf/uspsobes.htm#summary.

16. Prevalence of overweight and obesity among adults with diagnosed diabetes—United States, 1988–1994 and 1999–2002, *Morbidity and Mortality Weekly Report* 53 (2004): 1066–1068.

17. E. H. Livingston, Lower body subcutaneous fat accumulation and diabetes mellitus risk, *Surgery for Obesity and Related Diseases* 2 (2006): 362–368.

18. G. M. Reaven, The insulin resistance syndrome: Definition and dietary approaches to treatment, *Annual Review of Nutrition* 25 (2005): 391–406; S. Klein and coauthors, Weight management through lifestyle modification for the prevention and management of type 2 diabetes: Rationale and strategies, A statement of the American Diabetes Association, the North American Association for the Study of Obesity, and the American Society for Clinical Nutrition, *American Journal of Clinical Nutrition* 80 (2004): 257–263.

19. G. R. Dagenais and coauthors, Prognostic impact of body weight and abdominal obesity in women and men with cardiovascular disease, *American Heart Journal* 149 (2005): 54–60; E. N. Taylor, M. J. Stampfer, and G. C. Curhan, Obesity, weight gain, and the risk of kidney stones, *Journal of the American Medical Association* 293 (2005): 455–462; C. Tsai and coauthors,

Prospective study of abdominal adiposity and gallstone disease in US men, *American Journal of Clinical Nutrition* 80 (2004): 38–44; E. E. Calle and coauthors, Overweight, obesity, and mortality from cancer in a prospectively studied cohort of U.S. adults, *New England Journal of Medicine* 348 (2003): 1625–1638.

20. J. M. Jakicic and A. D. Otto, Physical activity considerations for the treatment and prevention of obesity, *American Journal of Clinical Nutrition* 82 (2005): 226S–229S; P. T. Katzmarzyk and coauthors, Fitness, fatness, and estimated coronary heart disease risk: The HERITAGE Family Study, *Medicine and Science in Sports and Exercise* 33 (2001): 585–590.

Nutrition in Practice

EATING DISORDERS

The exact number of people in the United States afflicted with some form of an **eating disorder** (see the accompanying Glossary for the relevant terms) is unknown because many cases never get reported.[1] About 3 percent of young women between 18 and 30 years of age and a smaller percent of men have **anorexia nervosa, bulimia nervosa,** or **binge eating disorder**. Many more suffer from other related conditions that do not meet the strict criteria for anorexia nervosa, bulimia nervosa, or binge eating disorder but still imperil a person's well-being. Characteristics of disordered eating such as restrained eating, binge eating, purging, fear of fatness, and distortion of body image are common, especially among young middle-class girls. In most other societies, these behaviors and attitudes are much less prevalent.

Why do so many young people in our society suffer from eating disorders?

Most experts agree that the causes are multifactorial: sociocultural, psychological, and perhaps neurochemical.[2] Excessive pressure to be thin is at least partly to blame. When low body weight becomes an important goal, people begin to view normal, healthy body weight as too fat. Healthy people then take unhealthy actions to lose weight. Severe restriction of food intake may create intense hunger that leads to binges. Research confirms this theory, showing that unhealthy or dangerous diets often precede binge eating in adolescent girls.[3] Energy restriction followed by bingeing can set in motion a pattern of **weight cycling**, which may make weight loss and maintenance more difficult over time.

People who attempt extreme weight loss are dissatisfied with their bodies to begin with; they may also be depressed or suffer social anxiety.[4] As weight loss becomes more and more difficult, psychological problems worsen, and the likelihood of developing full-blown eating disorders intensifies.

People with anorexia nervosa suffer from an extreme preoccupation with weight loss that seriously endangers their health and even their lives. People with bulimia engage in episodes of binge eating alternating with periods of severe dieting or self-starvation. Some bulimics also follow binge eating with self-induced vomiting, laxative abuse, or diuretic abuse to "undo the damage."

Are there other groups, besides girls and young women, who are vulnerable to anorexia nervosa and bulimia nervosa?

Yes. Athletes who participate in sports that emphasize leanness are at special risk for developing eating disorders.[5] Athletes must often meet stringent weight requirements to compete in their sport. Many athletes report that they engage in behaviors that are typical of people with eating disorders. Female

GLOSSARY

amenorrhea (ay-MEN-oh-REE-ah): the absence of or cessation of menstruation. **Primary amenorrhea** is menarche delayed beyond 16 years of age. **Secondary amenorrhea** is the absence of three to six consecutive menstrual cycles.

anorexia nervosa: an eating disorder characterized by a refusal to maintain a minimally normal body weight, self-starvation to the extreme, and a disturbed perception of body weight and shape; seen (usually) in adolescent girls and young women.
> *anorexia* = without appetite
> *nervosa* = of nervous origin

binge eating disorder: an eating disorder whose criteria are similar to those of bulimia nervosa, excluding purging or other compensatory behaviors.

bulimia (byoo-LEEM-ee-uh) **nervosa:** recurring episodes of binge eating combined with a morbid fear of becoming fat, usually followed by self-induced vomiting or purging.

cathartic: a strong laxative.

cognitive therapy: psychological therapy aimed at changing undesirable behaviors by changing underlying thought processes contributing to these behaviors. In anorexia nervosa, a goal is to replace false beliefs about body weight, eating, and self-worth with health-promoting beliefs.

eating disorder: a disturbance in eating behavior that jeopardizes a person's physical and psychological health.

emetic (em-ETT-ic): an agent that causes vomiting.

female athlete triad: a potentially fatal triad of medical problems: disordered eating, amenorrhea, and osteoporosis.

weight cycling: repeated rounds of weight loss and subsequent regain, with reduced ability to lose weight with each attempt; also called *yo-yo dieting*.

competitors often report being terrified of becoming fat, being obsessed with food, and using laxatives in attempting to control weight. Dancers, jockeys, wrestlers, distance runners, bodybuilders, divers, figure skaters, gymnasts, and others whose body weight and appearance are frequently judged in comparison with an "ideal" are especially prone to develop problems.[6] These athletes may engage in extreme weight-loss practices such as overtraining, prolonged fasting, vomiting, taking diet pills, and using steam baths and saunas to induce sweating.

Men account for about 1 in 20 cases in the general population, but among male athletes and dancers, eating disorders are much more common. Male teenagers normally average about 15 percent of body weight as fat, but some high school athletes strive to carry only 5 percent or so of their body weight as fat.

For example, wrestlers are required to "make weight" to compete in the lowest weight class to face the smallest possible opponents. To that end, wrestlers starve themselves, don rubber suits, sweat in steam rooms, and take diuretics to shed water weight before weighing in for competition. These practices were responsible for the deaths of three college athletes in recent years and have caused untold misery and harm to many others. Athletes engaging in these practices actually compromise their athletic abilities. The diminished anaerobic strength, reduced endurance, decreased oxygen capacity, and general weakness caused by food deprivation and dehydration can impair performance, an effect lasting days after food and water are replenished.

Even among athletes, however, women are most vulnerable to developing eating disorders. Many female athletes appear healthy but in fact may develop the three interrelated components of the **female athlete triad:** disordered eating, **amenorrhea** (the absence of three or more consecutive menstrual cycles), and osteoporosis.

How does the female athlete triad develop?

Many athletic women engage in self-destructive eating behaviors (disordered eating) because they and their coaches have adopted unsuitable weight standards. An athlete's body must be heavier for height than a nonathlete's body because the athlete's bones and muscles are denser. Weight standards that may be appropriate for others are inappropriate for athletes. Indicators such as skinfold measures yield more useful information about body composition.

Many young female athletes severely restrict energy intakes to improve performance, enhance the aesthetic appeal of their performance, or meet the weight guidelines of their specific sports. They fail to realize that the loss of lean tissue that accompanies energy restriction actually impairs their physical performance. Risk factors for the female athlete triad include the following:

- Young age (adolescence)
- Pressure to excel at a chosen physical activity
- Focus on achieving or maintaining an "ideal" body weight or body fat percentage
- Participation in sports or competitions that judge performance on aesthetic appeal such as gymnastics, figure skating, or dance
- Dieting at an early age
- Unsupervised dieting

As for amenorrhea, its prevalence among premenopausal women in the United States is about 2 to 5 percent overall, but among female athletes, it may be as high as 66 percent. Contrary to previous notions, amenorrhea is not a normal adaptation to strenuous physical training: it is a symptom of

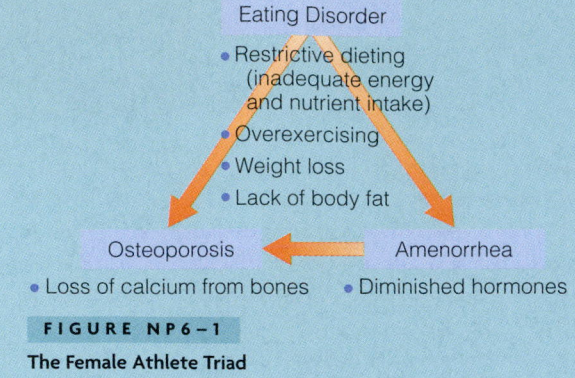

FIGURE NP6–1

The Female Athlete Triad

something going wrong. Amenorrhea is characterized by low blood estrogen, infertility, and often bone mineral losses.

In general, weight-bearing physical activity, dietary calcium, and the hormone estrogen protect against the bone loss of osteoporosis, but in women with disordered eating and amenorrhea, strenuous activity may impair bone health.[7] Vigorous training combined with low food energy intakes and other life stresses may promote bone loss even without obvious menstrual irregularities.[8] Such bone losses may increase the risks of stress fractures today and osteoporosis in later life (see Figure NP6–1). Many underweight young athletes have decreased bone density, similar to that of 50- to 60-year-old women, when they should have dense, strong bones. Young athletes should be encouraged to consume at least 1,300 milligrams of calcium each day, to eat nutrient-dense foods, and to obtain enough food energy to cover the energy expended in physical activity.

What can be done to prevent eating disorders in athletes and dancers?

To prevent eating disorders in athletes and dancers, both the performers and their coaches must be educated about links between inappropriate body weight ideals, improper weight-loss techniques, eating disorder development, adequate nutrition, and safe weight-control methods. Coaches and dance instructors should never encourage unhealthy weight loss to qualify for competition or to conform to distorted artistic ideals. Frequent weighings can push young people who are striving to lose weight into a cycle of starving to confront the scale, then bingeing uncontrollably afterwards. The erosion of self-esteem that accompanies these events can interfere with the normal psychological development of the teen years and set the stage for serious problems later on.

The accompanying "How to" provides some suggestions to help athletes and dancers protect themselves against developing eating disorders. The next sections describe eating disorders that anyone, athlete or nonathlete, may experience.

What are the characteristics of anorexia nervosa?

Most anorexia nervosa victims are females who come from middle- or upper-class families. The person with anorexia nervosa is often a perfectionist who works hard to please her parents. She may identify so strongly with her parents' ideals and

COMBAT EATING DISORDERS

The following guidelines may be useful in combating eating disorders:

- Never restrict food intakes to below the amounts suggested for adequacy by the USDA Food Guide.

- Eat frequently. People often do not eat frequent meals because of time constraints, but eating can be incorporated into other activities, such as snacking while studying or commuting. The person who eats frequently never gets so hungry as to allow hunger to dictate food choices.

- If not at a healthy weight, establish a reasonable weight goal based on a healthy body composition.

- Allow a reasonable time to achieve the goal. A reasonable loss of excess fat can be achieved at the rate of about 1 percent of body weight per week.

- Establish a weight-maintenance support group with people who share interests.

Specific guidelines for athletes and dancers include:

- Replace weight-based goals with performance-based goals.

- Remember that eating disorders impair physical performance. Seek confidential help in obtaining treatment if needed.

- Restrict weight-loss activities to the off-season.

- Focus on proper nutrition as an important facet of your training—as important as proper technique.

---

goals for her that she sometimes feels she has no identity of her own. She is respectful of authority but sometimes feels like a robot, and she may act that way, too: polite but controlled, rigid, and unspontaneous. She earnestly desires to control her own destiny, but she feels controlled by others. When she does not eat, she gains control.

How does a person know when dieting is going too far?

When a person loses weight to well below the average for her height and is no longer slim, but too slim, and still doesn't stop, she has gone too far. Regardless of how thin she is, she looks in the mirror and sees herself as fat. Central to the diagnosis of anorexia nervosa is a distorted body image that overestimates body fatness. Table NP6–1 shows the criteria that professionals use to diagnose anorexia nervosa. Anorexia nervosa resembles an addiction. The characteristic behavior is obsessive and compulsive. Before drawing conclusions about someone who is extremely thin or who eats very little, remember that diagnosis of anorexia nervosa requires professional assessment.

What is the harm in being very thin?

Anorexia nervosa damages the body much as starvation does. In young people, growth ceases and normal development falters. They lose so much lean tissue that basal metabolic rate slows. Additionally, the heart pumps inefficiently and irregularly, the heart muscle becomes weak and thin, the heart chambers diminish in size, and the blood pressure falls.[9] Minerals that help to regulate heartbeat become unbalanced. Many deaths occur due to multiple organ system failure: the heart, kidneys, and liver cease to function.

Starvation brings other physical consequences as well: loss of brain tissue, impaired immune response, anemia, and a loss of

---

TABLE NP6–1

Criteria for Diagnosis of Anorexia Nervosa

A person with anorexia nervosa demonstrates the following:

A. Refusal to maintain body weight at or above a minimal normal weight for age and height, e.g., weight loss leading to maintenance of body weight less than 85 percent of that expected; or failure to make expected weight gain during period of growth, leading to body weight less than 85 percent of that expected.

B. Intense fear of gaining weight or becoming fat, even though underweight.

C. Disturbance in the way in which one's body weight or shape is experienced; undue influence of body weight or shape on self-evaluation, or denial of the seriousness of the current low body weight.

D. In females past puberty, amenorrhea, i.e., the absence of at least three consecutive menstrual cycles. (A woman is considered to have amenorrhea if her periods occur only following hormone, e.g., estrogen, administration.)

Two types:

- **Restricting type:** During the episode of anorexia nervosa, the person does not regularly engage in binge eating or purging behavior (i.e., self-induced vomiting or the misuse of laxatives, diuretics, or enemas).

- **Binge eating/purging type:** During the episode of anorexia nervosa, the person regularly engages in binge eating or purging behavior (i.e., self-induced vomiting or the misuse of laxatives, diuretics, or enemas).

Source: Reprinted with permission from the *Diagnostic and Statistical Manual of Mental Disorders*, 4th ed., Text Revision, American Psychiatric Association, 2000.

© Tony Freeman/Photoedit, Inc.

Women with anorexia nervosa see themselves as fat, even when they are dangerously underweight.

are treated can maintain their body weight within 15 percent of healthy weight; at that weight, many of them begin menstruating again. The other half have poor or fair treatment outcomes, and two-thirds of those treated fight an ongoing mental battle with recurring morbid thoughts about food and body weight. Many relapse into abnormal eating behaviors or die. Anorexia nervosa has one of the highest mortality rates among psychiatric disorders.[12] An estimated 1,000 women die each year of anorexia nervosa—most commonly from cardiac complications due to malnutrition or from suicide.[13]

How does bulimia nervosa differ from anorexia nervosa?

Bulimia nervosa is distinct from anorexia nervosa and is more prevalent. More men suffer from bulimia nervosa than from anorexia, but bulimia is still more common in women. The secretive nature of bulimic behaviors makes recognition of the problem difficult, but once it is recognized, diagnosis is based on the criteria listed in Table NP6–2.

TABLE NP6–2
Criteria for Diagnosis of Bulimia Nervosa

A person with bulimia nervosa demonstrates the following:

A. Recurrent episodes of binge eating. An episode of binge eating is characterized by both of the following:
   1. Eating, in a discrete period of time (e.g., within any two-hour period), an amount of food that is definitely larger than most people would eat during a similar period of time and under similar circumstances
   2. A sense of lack of control over eating during the episode (e.g., a feeling that one cannot stop eating or control what or how much one is eating).
B. Recurrent inappropriate compensatory behavior in order to prevent weight gain, such as self-induced vomiting; misuse of laxatives, diuretics, enemas, or other medications; fasting; or excessive exercise.
C. Binge eating and inappropriate compensatory behaviors that both occur, on average, at least twice a week for three months.
D. Self-evaluation unduly influenced by body shape and weight.
E. The disturbance does not occur exclusively during episodes of anorexia nervosa.

Two types:

■ **Purging type:** The person regularly engages in self-induced vomiting or the misuse of laxatives, diuretics, or enemas.
■ **Nonpurging type:** The person uses other inappropriate compensatory behaviors, such as fasting or excessive exercise, but does not regularly engage in self-induced vomiting or the misuse of laxatives, diuretics, or enemas.

Source: Reprinted with permission from the *Diagnostic and Statistical Manual of Mental Disorders*, 4th ed., Text Revision, American Psychiatric Association, 2000.

digestive function that worsens malnutrition.[10] Digestive functioning becomes sluggish, the stomach empties slowly, and the lining of the intestinal tract shrinks. The ailing digestive tract fails to sufficiently digest any food the victim may eat. The pancreas slows its production of digestive enzymes. The person may suffer from diarrhea, further worsening malnutrition.

What kind of treatment helps people with anorexia nervosa?

Treatment of anorexia nervosa requires a multidisciplinary approach that addresses two sets of issues and behaviors: those relating to food and weight and those involving relationships with oneself and others.[11] Teams of physicians, nurses, psychiatrists, family therapists, and dietitians work together to treat people with anorexia nervosa. Appropriate diet is crucial for normalizing body weight and must be tailored individually to each client's needs. Seldom are clients willing to eat for themselves, but if they are, chances are they can recover without other interventions.

Professionals classify clients based on the risks posed by the degree of malnutrition present. Clients with low risks may benefit from family counseling, **cognitive therapy**, behavior modification, and nutrition guidance; those with greater risks may also need other forms of psychotherapy and supplemental formulas to provide extra nutrients and energy.

High-risk clients may require hospitalization and may need to be force-fed by tube at first to forestall death. This step causes psychological trauma. Medications are commonly prescribed, but to date, they play a limited role in treatment.

Denial runs high among those with anorexia nervosa. Few seek treatment on their own. Almost half of the women who

The typical person with bulimia is well educated, in her early twenties, and close to ideal body weight. She is a high achiever, with a strong feeling of dependence on her parents. She experiences considerable social anxiety and has difficulty establishing personal relationships. She is sometimes depressed and often exhibits impulsive behavior.

Like the person with anorexia nervosa, the person with bulimia spends much time thinking about her body weight and food. Her preoccupation with food manifests itself in secretive binge eating episodes followed by self-induced vomiting, fasting, or the use of laxatives or diuretics. Such behaviors typically begin in late adolescence after a long series of various unsuccessful weight-reduction diets. People with bulimia commonly follow a pattern of restrictive dieting interspersed with bulimic behaviors and experience weight fluctuations of more than 10 pounds up and down over short periods of time.

Unlike the person with anorexia nervosa, the person with bulimia is aware of the consequences of her behavior, feels that it is abnormal, and is deeply ashamed of it. She feels inadequate and unable to control her eating, so she tends to be passive and to look to men for confirmation of her sense of self-worth. When she is rejected, either in reality or in her imagination, her bulimia becomes worse. If her depression deepens, she may seek solace in drug or alcohol abuse or other addictive behaviors. Many studies show a link between bulimia nervosa and drug and alcohol dependency.[14]

What exactly is binge eating?

Binge eating is unlike normal eating, and the food is not consumed for its nutritional value. The binge eater has a compulsion to eat. A typical binge occurs periodically, is done in secret, usually at night, and lasts an hour or more. A binge frequently

For many people with bulimia, guilt, depression, and self-condemnation follow a binge eating episode.

© Michael Newman/Photoedit

follows a period of rigid dieting, so the binge eating is accelerated by hunger. During a binge, the person with bulimia may consume from one thousand to several thousand kcalories of food. The food typically contains little fiber or water, has a smooth texture, and is high in sugar and fat, so it is easy to consume vast amounts rapidly with little chewing.

What are the consequences of binge eating?

After a binge, to purge the food from the body, the person may use a **cathartic**—a strong laxative that can injure the lower intestinal tract, or the person may induce vomiting, using an **emetic**—a drug intended as first aid for poisoning. On first glance, purging seems to offer a quick and easy solution to the problems of unwanted kcalories and body weight. Many people perceive such behavior as neutral or even positive, when, in fact, bingeing and purging have serious physical consequences. Fluid and electrolyte imbalances caused by vomiting or diarrhea can lead to metabolic alkalosis, a condition characterized by apathy, confusion, and muscle spasms. Vomiting causes irritation and infection of the pharynx, esophagus, and salivary glands; erosion of the teeth; and dental caries. The esophagus may rupture or tear, as may the stomach. Overuse of emetics depletes potassium concentrations and can lead to death by heart failure.

What is the treatment for bulimia nervosa?

As for people with anorexia nervosa, a team approach provides the most effective treatment for people with bulimia nervosa. Bulimia nervosa is easier to treat than anorexia nervosa in many respects because it seems to be more of a chosen behavior. People with bulimia know that their behavior is abnormal, and many are willing to try to cooperate.

The goal of the dietary plan to treat bulimia is to help the client gain control, establish regular eating patterns, and restore nutritional health. Energy intake should not be severely restricted. The person needs to learn to eat a quantity of nutritious food sufficient to nourish her body and to satisfy hunger (at least 1,600 kcalories per day). The person may also benefit from a regular exercise program.[15] The "How to" (p. 168) offers some ways to begin correcting bulimia nervosa.

Anorexia nervosa and bulimia nervosa are distinct eating disorders, yet they sometimes overlap. Anorexia victims may purge, and victims of both conditions share an overconcern with body weight and the tendency to drastically undereat. The two disorders can also appear in the same person, or one can lead to the other. Other people have eating disorders that fall short of anorexia nervosa or bulimia nervosa but share some of their features, such as fear of body fatness. One such condition is binge eating disorder (defined earlier in the Glossary).

How does binge eating disorder differ from bulimia nervosa?

Up to half of all people who restrict eating to lose weight periodically binge without purging, including about one-third of obese

How To

The following advice has proved useful for people fighting bulimia nervosa.

Planning principles:

- Plan meals and snacks; record plans in a food diary prior to eating.
- Plan meals and snacks that require eating at the table and using utensils.
- Refrain from finger foods.
- Refrain from "dieting" or skipping meals.

Nutrition principles:

- Eat a well-balanced diet and regularly timed meals consisting of a variety of foods.
- Include raw vegetables, salad, or raw fruit at meals to prolong eating times.
- Choose whole-grain, high-fiber breads, pasta, rice, and cereals to increase bulk.
- Consume adequate fluid, particularly water.

Other tips:

- Choose meals that provide protein and fat for satiety and bulky, fiber-rich carbohydrates for immediate feelings of fullness.
- Consume the amounts of food specified in the USDA Food Guide (pp. 18–19).
- For convenience (and to reduce temptation) select foods that naturally divide into portions. Select one potato, rather than rice or pasta that can be overloaded onto the plate; purchase yogurt and cottage cheese in individual containers; look for small packages of precut steak or chicken; choose frozen dinners with metered portions.
- Include 30 minutes of physical activity every day—exercise may be an important tool in controlling bulimia.

---

people who regularly engage in binge eating. Obesity itself, however, does not constitute an eating disorder. Table NP6–3 lists the official diagnostic criteria for binge eating disorder.

Clinicians note differences between people with bulimia nervosa and those with binge eating disorder.[16] People with binge eating disorder consume less during a binge, rarely purge, and exert less restraint during times of dieting. Similarities also exist, including feeling out of control, disgusted, depressed, embarrassed, guilty, or distressed because of their self-perceived gluttony.[17] Binge eating behavior responds more readily to treatment than other eating disorders, and resolving such behaviors can be a first step to authentic weight control. Successful treatment also improves physical health, mental health, and the chances of breaking the cycle of rapid weight losses and gains.

At so young an age as 12 years, beautifully growing, normal-weight females are already worried that they are too fat. Many are "on diets." Magazines, newspapers, and television all present the message that to be thin is to be beautiful and happy. Anorexia nervosa and bulimia nervosa are not a form of rebellion against these unreasonable expectations, but rather the exaggerated acceptance of them. Body dissatisfaction is a primary factor in the development of eating disorders.[18] Perhaps a person's best defense against these disorders is to learn to appreciate his or her own uniqueness.

Notes

1. Position of the American Dietetic Association: Nutrition intervention in the treatment of anorexia nervosa, bulimia nervosa, and other eating disorders, *Journal of the American Dietetic Association* 106 (2006): 2073-2082.

TABLE NP6–3

Criteria for Diagnosis of Binge Eating Disorder

A person with a binge eating disorder demonstrates the following:

A. Recurrent episodes of binge eating. An episode of binge eating is characterized by both of the following:
   1. Eating, in a discrete period of time (e.g., within any two-hour period), an amount of food that is definitely larger than most people would eat in a similar period of time under similar circumstances.
   2. A sense of lack of control over eating during the episode (e.g., a feeling that one cannot stop eating or control what or how much one is eating).
B. Binge eating episodes are associated with at least three of the following:
   1. Eating much more rapidly than normal.
   2. Eating until feeling uncomfortably full.
   3. Eating large amounts of food when not feeling physically hungry.
   4. Eating alone because of being embarrassed by how much one is eating.
   5. Feeling disgusted with oneself, depressed, or very guilty after overeating.
C. The binge eating causes marked distress.
D. The binge eating occurs, on average, at least twice a week for six months.
E. The binge eating is not associated with the regular use of inappropriate compensatory behaviors (e.g., purging, fasting, excessive exercise) and does not occur exclusively during the course of anorexia nervosa or bulimia nervosa.

Source: Reprinted with permission from the *Diagnostic and Statistical Manual of Mental Disorders*, 4th ed., Text Revision, American Psychiatric Association, 2000.

2. H. Steiner and coauthors, Risk and protective factors for juvenile eating disorders, *European Child and Adolescent Psychiatry* 12 (2003): Suppl. 1: 38–46.

3. A. E. Field and coauthors, Relation between dieting and weight change among preadolescents and adolescents, *Pediatrics* 112 (2003): 900–906.

4. W. H. Kaye and coauthors, Comorbidity of anxiety disorders with anorexia and bulimia nervosa, *American Journal of Psychiatry* 161 (2004): 2215–2221.

5. M. K. Torstveit and J. Sundgot-Borgen, The female athlete triad: Are elite athletes at increased risk? *Medicine and Science in Sports and Exercise* 37 (2005): 184–193.

6. Torstveit and Sundgot-Borgen, 2005.

7. C. L. Zanker and C. B. Cooke, Energy balance, bone turnover, and skeletal health in physically active individuals, *Medicine and Science in Sports and Exercise* 36 (2004): 1372–1381.

8. K. L. Cobb and coauthors, Disordered eating, menstrual irregularity, and bone mineral density in female runners, *Medicine and Science in Sports and Exercise* 35 (2003): 711–719.

9. C. Romano and coauthors, Reduced hemodynamic load and cardiac hypotrophy in patients with anorexia nervosa, *American Journal of Clinical Nutrition* 77 (2003): 308–312.

10. J. Yager and A. E. Andersen, Anorexia nervosa, *New England Journal of Medicine* 353 (2005): 1481–1488.

11. Committee on Adolescence, Identifying and treating eating disorders, *Pediatrics* 111 (2003): 204–211.

12. P. K. Keel and coauthors, Predictors of mortality in eating disorders, *Archives of General Psychiatry* 60 (2003): 179–183.

13. M. B. Tamburrino and R. A. McGinnis, Anorexia nervosa: A review, *Panminerva Medica* 44 (2002): 313–311.

14. C. M Bulik and coauthors, Alcohol use disorder comorbidity in eating disorders: A multicenter study, *Journal of Clinical Psychiatry* 65 (2004): 1000–1006; B. Vastag, What's the connection? No easy answers for people with eating disorders and drug abuse, *Journal of the American Medical Association* 285 (2001): 1006–1007.

15. J. Sundgot-Borgen and coauthors, The effect of exercise, cognitive therapy, and nutritional counseling in treating bulimia nervosa, *Medicine and Science in Sports and Exercise* 34 (2002): 190–195.

16. A. E. Dingemans, M. J. Bruna, and E. F. Van Furth, Binge eating disorder: A review, *International Journal of Obesity and Related Metabolic Disorders* 26 (2002): 299–307.

17. D. M. Ackard and coauthors, Overeating among adolescents: Prevalence and associations with weight-related characteristics and psychological health, *Pediatrics* 111 (2003): 66–74.

18. J. Polivy and C. P. Herman, Causes of eating disorders, *Annual Review of Psychology* 53 (2002): 186–213.

7 Weight Management: Overweight and Underweight

© BananaStock/Alamy

Are you pleased with your body weight? If you answered yes, you are a rare individual. Nearly all people in our society think they should weigh more or less (mostly less) than they do. Usually, their primary reason is appearance, but they often perceive, correctly, that their weight is also related to physical health. Chapter 6 addressed the health risks of being overweight. As discussed in a later section of this chapter, health risks also accompany being underweight.

Overweight and underweight both result from energy imbalance. The simple picture is as follows. Overweight people have consumed more food energy than they have expended and have banked the surplus in their body fat. To reduce body fat, overweight people need to expend more energy than they take in from food. In contrast, underweight people have consumed too little food energy to support their activities and so have depleted their bodies' fat stores and possibly some of their lean tissues as well. To gain weight, they need to take in more food energy than they expend.

This chapter's missions are to present strategies toward solving the problems of excessive and deficient body fatness and to point out how appropriate body composition, once achieved, can be maintained. The chapter emphasizes overweight because it has been more intensively studied and is a more widespread health problem in the developed countries.

Causes of Obesity

Henceforth, this chapter will use the term *obesity* to refer to excess body fat. Excess body fat accumulates when people take in more food energy than they expend. Why do they do this? Is it genetic? Metabolic? Psychological? Behavioral? All of these? Most likely, obesity has many interrelated causes.

Genetics and Weight A person's genetic makeup influences the body's tendency to consume or store too much energy or to expend too little.[1] When both parents are obese, the chances that their children will be obese are quite high (up to 80 percent), whereas when neither parent is obese, the chances are relatively small (less than 10 percent). Adoption studies find a similarity in obesity between biological parents and their natural children, but not between adoptive parents and their adopted children.

To determine the relative contributions of genetic and environmental factors to body weight, one group of researchers studied identical and fraternal twins, some of whom were reared together and some apart. Like previous studies, this study found that identical twins were twice as likely to have similar weights as fraternal twins were—even when reared apart. These findings suggest an important role for genetics in determining a person's *susceptibility* to obesity.[2]

Genetics also influences the way energy is *stored*. When identical twins were given an extra 1,000 kcalories per day for 100 days, some pairs gained less than 10 pounds whereas others gained up to 30 pounds. Within each pair, the amount of weight gained, percentage of body fat, and distribution of fat were similar.

Genetic factors also influence how much energy the body *expends*. For example, the differences in basal metabolic rate (BMR) between individuals are greater than can be explained by age, gender, and body composition alone. Similarities within families suggest a genetic influence on BMR. A low metabolic rate is a major risk factor for weight gain.

Lipoprotein Lipase Some of the research investigating genetic influence on obesity focuses on the enzyme **lipoprotein lipase (LPL)**, which promotes fat storage in fat cells and muscle cells. People with high LPL activity are especially efficient at storing fat. Obese people generally have much more LPL activity in their fat cells than lean people do (their muscle cell LPL activity is similar, though). This high LPL activity makes fat storage especially efficient. Consequently, even modest excesses in energy intake have a more dramatic impact on obese people than on lean people.

Leptin Researchers have discovered a gene in humans called the obesity (*ob*) gene. The obesity gene codes for the protein **leptin**. ■ Leptin is a hormone primarily produced and secreted by the fat cells in proportion to the amount of fat stored.[3] A gain in body fatness stimulates the production of leptin, which, by way of the hypothalamus, suppresses the appetite, increases energy expenditure, and produces fat loss. Fat loss produces the opposite effect—suppression of leptin production, increased appetite, and decreased energy expenditure. As the accompanying photo shows, mice with a defective obesity gene do not produce leptin and can weigh up to three times as much as normal mice. When injected with leptin, the mice lose weight. (Because leptin is a protein, it would be destroyed during digestion if given orally; consequently, it must be given by injection.)

Although it is extremely rare, researchers have identified a genetic deficiency of leptin in human beings as well. An error in the gene that codes for leptin was discovered in two extremely obese children whose blood levels of leptin were barely detectable. Without leptin, the children had little appetite control; they were constantly hungry and ate considerably more than their siblings or peers. Given daily injections of leptin, these children lost a substantial amount of weight, confirming leptin's role in regulating appetite and body weight.

Most obese people do not have leptin deficiency, however. In fact, in obese people, the more body fat, the more leptin.[4] Researchers speculate that leptin rises in an effort to suppress appetite and inhibit fat storage when fat cells are ample. Obese people with elevated leptin concentrations may be resistant to its satiating effect.[5] The absence of or resistance to leptin in obesity parallels the scenario of insulin in diabetes: some people have an insulin deficiency (type 1), whereas many others have elevated insulin but are resistant to its glucose-storing effect (type 2).

Research on leptin is ongoing, and scientists are exploring the possibility that leptin may one day help treat human obesity. So far, efforts to treat obesity with leptin have been disappointing. Even if leptin never proves useful as an antiobesity drug, its discovery has contributed much to our understanding of the complexities of the human body. For example, scientists no longer view adipose tissue as a metabolically sluggish storage depot for fat, but rather as a hormonally active regulatory tissue with widespread effects on the body.[6] In addition to its appetite function, leptin may have roles in immunity, reproduction, bone formation, and sexual maturation.[7]

Ghrelin Researchers recently discovered another protein that also acts as a hormone primarily in the hypothalamus, but it works in the opposite direction of leptin. Known as **ghrelin**, this protein is synthesized and secreted primarily by the stomach cells and promotes a positive energy balance by stimulating appetite and promoting efficient energy storage.[8] The role ghrelin plays in regulating food intake and body weight is the subject of much intense research.[9] Pharmaceutical companies are eager to develop products that mimic ghrelin to treat wasting conditions, as well as products that oppose ghrelin's actions to treat obesity.

Ghrelin powerfully triggers the desire to eat. Blood levels of ghrelin typically rise before a meal and fall rapidly after it—reflecting the hunger and satiety that precede and follow eating.[10] Furthermore, the rise and fall of ghrelin concentrations are

■ Genes instruct cells to make proteins, and each protein performs a unique function.

Courtesy Amgen, Inc. Photo © John Sholtis/The Rockefeller University

The mouse on the left is genetically obese—it lacks the gene for producing leptin. The mouse on the right is *also* genetically obese, but because it receives leptin, it eats less, expends more energy, and is less obese than it would be had it not received the leptin.

lipoprotein lipase (LPL): an enzyme that hydrolyzes triglycerides in the blood into fatty acids and glycerol for absorption into the cells. There they are metabolized or reassembled for storage.

leptin: a hormone produced by fat cells under the direction of the (*ob*) gene. It decreases appetite and increases energy expenditure.
leptos = thin

ghrelin (GRELL-in): a hormone produced primarily by the stomach cells. It signals the hypothalamus of the brain to stimulate appetite and food intake.

dose-dependent on the number of kcalories ingested.[11] In other words, large meals suppress ghrelin (and hunger) to a greater extent than small meals do.[12]

In general, fasting blood levels of ghrelin correlate inversely with body weight: lean people have high ghrelin levels and obese people have low levels.[13] Interestingly, while ghrelin levels are high in underweight people, they are exceptionally high in anorexia nervosa ■ and return to normal with nutrition intervention—indicating that both body weight and nutrition status influence ghrelin levels.[14] Also noteworthy, ghrelin levels are markedly high in Prader-Willi syndrome, a rare inherited disorder often characterized by severe obesity, and they remain elevated even after a meal, which helps to explain the excessive appetite and food consumption commonly seen in this disorder.[15] Similarly, ghrelin levels do not seem to decline as much after a meal in obese people, as they do for lean people.[16]

Ghrelin fights to maintain a stable body weight.[17] On average, ghrelin levels are high whenever the body is in negative energy balance, as occurs during low-kcalorie diets. This response may help explain why weight loss is so difficult to maintain. Ghrelin levels decline again whenever the body is in positive energy balance, as occurs with weight gains.

Like leptin, ghrelin plays roles in the body beyond energy regulation. In fact, it was first recognized for its participation in growth hormone activity.[18] The focus of intense, ongoing ghrelin research, however, is on its actions in stimulating appetite and short-term food intake and in regulating long-term body weight.

Fat Cell Development   Another cause of obesity may be the development of excess fat cells during childhood. The amount of fat on a person's body reflects both fat cell *number* and fat cell *size*. The number of fat cells increases most rapidly during the growing years of late childhood and early puberty. Fat cell numbers increase more rapidly in obese children than in lean children, and obese children entering their teen years may already have as many fat cells as do adults of normal weight.

Fat cells can also expand in size. After they reach their maximum size, more cells can develop to store more fat. Thus obesity develops when a person's fat cells increase in number, in size, or quite often both. Figure 7–1 illustrates fat cell development.

With fat loss, the size of the fat cells shrinks, but not their number. For this reason, people with extra fat cells may tend to regain lost weight rapidly. Prevention of obesity is most critical during the growing years when fat cell number is increasing.

Set-Point Theory   One popular theory of why a person may store too much fat is the **set-point theory**. The set-point theory proposes that body weight, like body temperature, is physiologically regulated. Researchers have noted that many people who lose weight on reducing diets quickly regain all their lost weight. This suggests that somehow the body chooses a weight that it wants to be and defends that weight by regulating eating behaviors and hormonal actions. Research confirms that the body

■ Anorexia nervosa is an eating disorder discussed in Nutrition in Practice 6.

FIGURE 7–1

Fat Cell Development
Fat cells are capable of increasing their size by 20-fold and their number by several thousandfold.

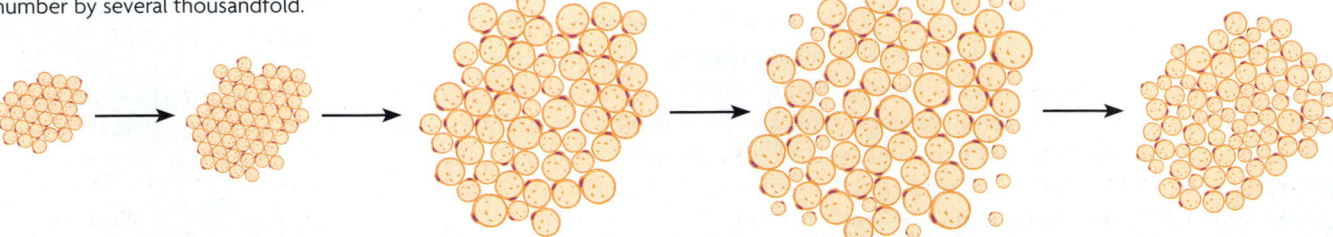

During growth, fat cells increase in number.

When energy intake exceeds expenditure, fat cells increase in size.

When fat cells have enlarged and energy intake continues to exceed energy expenditure, fat cells increase in number again.

With fat loss, the size of the fat cells shrinks, but not the number.

adjusts its metabolism whenever it gains or loses weight—in the direction that returns to the initial body weight: energy expenditure increases with weight gain and decreases with weight loss. These changes in energy expenditure are greater than those predicted based on body composition and help to explain why it is so difficult for an obese person to maintain weight losses. An individual's set point for body weight may be adjustable, shifting over the life span in response to physiological changes and to genetic, dietary, and other factors.

Environmental Stimuli As discussed earlier, genetic factors play a partial role in determining a person's susceptibility to obesity, but they do not fully explain obesity. Obesity rates have risen dramatically during recent decades wherever people enjoy conditions of prosperity and abundance, but the human gene pool has remained unchanged. The environment must therefore play a role as well. In other words, although a person's genetic inheritance may make obesity likely, lifestyle and environmental conditions will determine the extent to which the disease is expressed.[19]

People may overeat in response to stimuli in their surroundings—primarily, the availability of many delectable foods. Most people in the United States find high-kcalorie foods readily available, relatively inexpensive, heavily advertised, and reasonably tasty. Food is available everywhere, all the time—thanks largely to fast food. Fast-food restaurants line the highways and crowd out mom-and-pop restaurants and businesses in small towns and big cities. Convenience stores, service stations, malls, airports, and even schools offer fast food as well. Most alarming are the extraordinarily large serving sizes and ready-to-go meals offered in supersize combinations.[20] People buy the large sizes and combinations, perceiving them to be a good value, but then they eat more than they need. Research shows that people eat more if they are served more.[21] Portion sizes of virtually all foods and beverages have increased markedly in the past several decades, most notably at fast-food restaurants.[22] The increase in portion sizes parallels the growing prevalence of overweight and obesity in the United States, beginning in the 1970s, rising sharply in the 1980s and 1990s, and continuing today.[23]

Fast food is often high in fat, and fat is perceived as especially palatable. People can easily overeat fat-rich foods because their delicious tastes stimulate eating and each bite of food is kcalorie-dense. Not only does fat deliver more than twice as many kcalories, gram for gram, as protein and carbohydrate, but it also seems to be stored preferentially by the body and with great efficiency in many people. A steady diet of high-kcalorie, high-fat fast food encourages obesity.[24]

The combination of large portions and energy-dense foods strikes a double blow. Reducing portion sizes is helpful, but the real kcalorie savings come from lowering the energy density.[25] Satisfying portions of foods with low energy density such as fruits and vegetables can help with weight loss.[26]

Learned Behavior Psychological stimuli also trigger inappropriate eating behaviors in some people. Appropriate eating behavior is a response to **hunger**. Hunger is a drive programmed into people by their heredity. **Appetite**, in contrast, is learned and can lead people to ignore hunger or to overrespond to it. Hunger is physiological, whereas appetite is psychological, and the two do not always coincide.

Food behavior is also intimately connected to deep emotional needs such as the primitive fear of starvation. Yearnings, cravings, and addictions with profound psychological significance can express themselves in people's eating behavior. An emotionally insecure person might eat rather than call a friend and risk rejection. Another person might eat to relieve boredom or to ward off depression.

Physical Inactivity The possible causes of obesity mentioned so far all relate to the input side of the energy equation. What about output? People may be obese not because they eat too much but because they expend too little energy. More than

set-point theory: the theory that proposes that the body tends to maintain a certain weight by means of its own internal controls.

hunger: the physiological need to eat, experienced as a drive to obtain food; an unpleasant sensation that demands relief.

appetite: the psychological desire to eat; a learned motivation that is experienced as a pleasant sensation that accompanies the sight, smell, or thought of appealing foods.

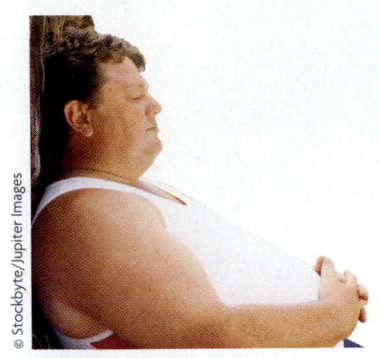
Lack of physical activity fosters obesity.

one-third of the overweight population reports no physical activity during their leisure time. Obese people observed closely are often seen to eat less than lean people, but they are sometimes so extraordinarily inactive that they still manage to accumulate an energy surplus. Reducing their food intake further would jeopardize health and incur nutrient deficiencies. Physical activity is a necessary component of nutritional health. People must be physically active if they are to eat enough food to deliver all the nutrients needed without unhealthy weight gain.

Our environment, however, fosters inactivity.[27] One hundred years ago, 30 percent of the energy used in farm and factory work came from muscle power; today, only 1 percent does. Modern technology has replaced physical activity at home, at work, and in transportation. Inactivity contributes to obesity and poor health.[28] In turn, television watching may contribute most to physical inactivity.

Watching television contributes to obesity in several ways.[29] First, television viewing requires little energy beyond the resting metabolic rate. Second, it replaces time spent in more vigorous activities. Third, television influences family food purchases; viewers are more likely to engage in between-meal snacking and eat the high-kcalorie foods most heavily advertised on programs.

IN SUMMARY

- Genetics, fat cell development, set-point theory, and overeating all offer possible, but still incomplete, explanations of obesity.

- Most likely, obesity has not one cause but different causes and combinations of causes in different people.

- Some causes may be within a person's control, and some may be beyond it.

- Like all the other "causes" of obesity, inactivity alone fails to explain it fully.

Obesity Treatment: Who Should Lose?

An estimated 35 to 45 percent of all U.S. women (and 20 to 30 percent of all U.S. men) are trying to lose weight at any given time, spending up to $40 billion each year to do so. Some of these people do not even need to lose weight. Others need to lose weight but are not successful. Still others are successful. Despite the general notion that practically no one maintains weight loss over the long term, research shows otherwise. Approximately 20 percent of overweight people do achieve successful long-term weight loss, meaning that they "intentionally lost at least 10 percent of their body weight and kept it off at least one year."[30]

Many people assume that every overweight person can achieve slenderness and should pursue that goal. Consider, however, that most overweight people cannot become slender. People vary in their weight tendencies just as they vary in their potentials for height and degrees of health. The question of whether a person should lose weight depends on many factors: the extent of overweight, age, health, and genetics, to name a few. Weight-loss advice does not apply equally to all overweight people. Some people may risk more in the process of losing weight than in remaining overweight. Others may reap significant health benefits with just modest weight loss.

IN SUMMARY

- Weight-loss advice does not apply equally to all overweight people.

- Some people may risk more through misguided efforts to lose weight than by remaining overweight, whereas others may benefit from just a modest weight loss.

Inappropriate Obesity Treatments

The risks people incur in attempting to lose weight often depend on how they go about it. Weight-loss plans and obesity treatments abound—some are adequate, but many are ineffective and possibly dangerous. Fad diets are the topic of Nutrition in Practice 7. This section addresses other inappropriate obesity interventions. Aggressive approaches to obesity for those obese people who face high risks of medical problems and must lose weight rapidly are discussed in the next section. Reasonable approaches to overweight for those seeking safe, gradual weight losses are saved for the last part of the obesity treatment discussion.

Over-the-Counter Weight-Loss Supplements and Drugs

Millions of people in the United States use nonprescription weight-loss products, spending close to $2 billion per year for such products.[31] Most of the people who use over-the-counter weight-loss supplements are women, especially young obese women, but almost 10 percent are of normal weight. Promoters and marketers of weight-loss products make all kinds of claims for their products with only one intention—profit. Such claims as "eat all you want and lose weight," "take 3 pills before bedtime and watch the fat disappear," "blocks carbs," "blocks fat," and many more lure people into believing that maybe this time a product will really work.

In 2000, the FDA recommended that manufacturers voluntarily discontinue marketing over-the-counter products containing phenylpropanolamine, ■ an ingredient commonly used in products to suppress appetite.* Reported side effects include dry mouth, rapid pulse, nervousness, sleeplessness, hypertension, irregular heartbeats, kidney failure, seizures, and strokes.

The FDA recently approved the first nonprescription weight-loss drug for use in the United States. The drug is a low-dose version of the prescription drug orlistat, discussed in the next section.†

■ Read labels of over-the-counter products to determine if they contain *phenylpropanolamine* (fen-ill-pro-pa-NOLE-a-mean).

Herbal Products and Dietary Supplements

In their search for weight-loss magic, some consumers turn to "natural" herbal products and dietary supplements, even though few have proved to be effective.[32] People falsely believe that "natural" herbs are not harmful to the body, but many herbs contain toxins. Belladonna and hemlock are infamous examples, but many lesser-known herbs, such as sassafras, contain toxins as well. Furthermore, because herbs are marketed as "dietary supplements," manufacturers need not present scientific evidence of their safety or effectiveness to the FDA before marketing them. Evidence about their safety is gathered only through reports of consumers who sicken or die after using the remedies.

A now familiar example is ephedra (also called ma huang), a herb that showed promise as a weight-loss drug in preliminary studies. Immediately, ephedra-containing products for dieters and athletes flooded the market. Many consumers of these products reported ill effects including cardiac arrest, abnormal heartbeats, hypertension, strokes, and seizures; the supplements have been linked to some deaths as well.[33] For this reason, the FDA has banned the sale of dietary supplements containing ephedra and its active constituent, ephedrine.‡ ■ Table 7–1 (p. 178) presents the claims and the dangers behind ephedrine and several other weight-loss supplements.

■ Ephedrine is an amphetamine-like substance extracted from the Chinese ephedra herb ma huang.

---

* Phenylpropanolamine is not commercially available in Canada.

† The low-dose, over-the-counter version of orlistat is marketed under the trade name Alli (AL-eye).

‡ Ma huang (ephedrine) is illegal in Canada.

TABLE 7-1

Selected Herbal and Other Dietary Supplements Marketed for Weight Loss

Product	Manufacturers' Claims	Research Findings	Adverse Effects
Bitter orange[a] (Citrus aurantium, a natural flavoring that contains synephrine, a compound structurally similar epinephrine)	Stimulates weight loss; provides an alternative to ephedra	Little evidence available	May increase blood pressure; may interact with drugs
Chitosan[b] (pronounced KITE-oh-san; derived from chitin, the substance that forms the hard shells of lobsters, crabs, and other crustaceans)	Binds to dietary fat, preventing digestion and absorption	Ineffective	Impaired absorption of fat-soluble vitamins
Chromium (trace mineral)	Eliminates body fat	Ineffective; weight gain reported when not accompanied by exercise	Headaches, sleep disturbances, and mood swings; hexavalent form is toxic and carcinogenic.
Conjugated linoleic acid (CLA; a group of fatty acids related to linoleic acid, but with different *cis*- and *trans*-configurations)	Reduces body fat and suppresses appetite	Some evidence in animal studies, but ineffective in human studies	None known
Ephedrine[c] (amphetamine-like substance derived from the Chinese ephedra herb ma huang)	Speeds body's metabolism	Weight loss and dangerous side effects	Insomnia, tremors, heart attacks, strokes, and death; the FDA has banned the sale of these products.
Hydroxycitric acid[d] (active ingredient derived from the rind of the tropical fruit *garcinia cambogia*)	Inhibits the enzyme that converts citric acid to fat; suppresses appetite	Ineffective	Toxicity symptoms reported in animal studies; headache, respiratory and gastrointestinal symptoms in humans
Pyruvate[e] (3-carbon compound produced during glycolysis)	Speeds body's metabolism	Modest weight loss with high doses	GI distress
Yohimbine (derived from the bark of a West African tree)	Promotes weight loss	Ineffective	Nervousness, insomnia, anxiety, dizziness, tremors, headaches, nausea, vomiting, hypertension

Note: The FDA has not approved the use of any of these products; most products are used in conjunction with a 1,000- to 1,800-kcalorie diet.

[a] Marketed under the trade names Xenadsine EFX, metabolite ultra, NOW Diet Support
[b] Marketed under the trade names Chitorich, Exofat, Fat Breaker, Fat Blocker, Fat Magnet, Fat Trapper, and Fatsorb
[c] Marketed under the trade names Diet Fuel, Metabolite, and Nature's Nutrition Formula One
[d] Marketed under the trade names Ultra Burn, Citralean, CitraMax, Citrin, Slim Life, Brindleslim, Medislim, and Beer Belly Busters
[e] Marketed under the trade names Exercise in a Bottle, Pyruvate Punch, Pyruvate-c, and Provate

Herbal laxatives containing senna, aloe, rhubarb root, cascara, castor oil, and buckthorn (or various combinations) are commonly sold as "dieter's tea." Such concoctions cause nausea, vomiting, diarrhea, cramping, and fainting and may have contributed to the deaths of four women who had drastically reduced their food intakes. Consumers mistakenly believe that laxatives will diminish nutrient absorption and reduce kcalorie intake. But remember that absorption occurs primarily in the upper small intestine, and these laxatives act on the lower large intestine. Nutrition in Practice 15 explores the possible benefits and potential dangers of herbal products and other alternative therapies. Anyone using dietary supplements for weight loss should first consult a physician.

Other Gimmicks

Other gimmicks do not help with weight loss either. Hot baths do not speed up metabolism so that pounds can be lost in hours. Steam and sauna baths do not melt the fat off the body, although they may dehydrate people so that they lose water weight. Brushes, sponges, wraps, creams, and massages intended to move, burn, or break up **"cellulite"** do nothing of the kind because there is no such thing as cellulite.

Aggressive Treatments of Obesity

For some obese people, the medical problems caused by their obesity demand treatment approaches that may themselves incur some risks. The health benefits to be gained by weight loss, however, may make these risks worth taking.

Obesity Drugs

Several prescription medications for weight loss have been tried over the years. When used as part of a long-term comprehensive weight-loss program, medications can help obese people to lose approximately 10 percent of their weight and maintain that loss for at least one year. Because weight regain commonly occurs with the discontinuation of drug therapy, treatment is long term. The long-term use of medications poses risks. Medical experts do not yet know whether a person would benefit more from maintaining a 20-pound excess or from taking a medication for a decade to keep the 20 pounds off.

The challenge is to develop an effective medication that can be used over time without adverse side effects or the potential for abuse. No such medication currently exists.[34] Two prescription medications, however, are currently in use; others, including leptin, are being studied.

Sibutramine Sibutramine is an appetite suppressant that works on the brain's neurotransmitters.* The drug is most effective when used in combination with a reduced-kcalorie diet and increased physical activity. Sibutramine enhances satiety and elevates energy expenditure.[35] Side effects include dry mouth, rapid heart rate, insomnia, headache, and high blood pressure.[36] The FDA cautions those with hypertension against using it. Anyone taking sibutramine should monitor blood pressure carefully. As more information becomes known about the molecular chemistry of appetite control, safer appetite suppressants may be developed.

Orlistat Orlistat takes a different approach to weight loss.† Not an appetite suppressant, orlistat inhibits the action of fat-digesting enzymes in the GI tract and so reduces fat digestion and absorption by about 30 percent.[37] As a result of absorbing less fat, people often lose weight. The problem with undigested fat is that it leaves the digestive tract intact, carrying with it fat-soluble vitamins and phytochemicals that would otherwise have been absorbed by the body. Possible side effects of orlistat resemble those of the artificial fat olestra—diarrhea and digestive distress.[38]

Surgery

Surgery as an approach to weight loss is justified in some specific cases of **clinically severe obesity**. Two procedures, **gastric bypass** and **gastric banding**, have gained wide acceptance. Both procedures limit food intake by effectively reducing the capacity of the stomach and suppressing hunger by reducing the production of ghrelin.[39] The results are dramatic: after gastric bypass surgery, most people achieve

---

* Sibutramine's trade name is Meridia.
† Orlistat's trade name is Xenical.

cellulite (SELL-you-light or SELL-you-leet): supposedly, a lumpy form of fat; actually, a fraud. Fatty areas of the body may appear lumpy when the strands of connective tissue that attach the skin to underlying muscles pull tight where the fat is thick. The fat itself is the same as fat anywhere else in the body. If the fat in these areas is lost, the lumpy appearance disappears.

clinically severe obesity: a BMI of 40 or greater or a BMI of 35 or greater and one or more serious conditions such as hypertension. Another term used to describe the same condition is *morbid obesity*.

gastric bypass: surgery that restricts stomach size and reroutes food from the stomach to the lower part of the small intestine; creates a chronic, lifelong state of malabsorption by preventing normal digestion and absorption of nutrients.

gastric banding: a surgical means of producing weight loss by restricting stomach size with a constricting band; used in people whose severe obesity brings extreme health risks.

a lasting weight loss of more than 50 percent of their excess body weight.[40] Research shows that weight-loss surgery also helps to improve to blood lipids, diabetes, sleep apnea, and hypertension.[41]

Laparoscopic weight-loss surgery techniques are now used to perform gastric bypass and other weight loss surgeries. Laparoscopic weight-loss surgery produces significant, long-term weight loss, shortens recovery, and is less invasive than open surgery.[42]

The long-term safety and effectiveness of gastric surgery depend, in large part, on compliance with dietary instructions. Common immediate postsurgical complications include infections, nausea, vomiting, and dehydration; in the long term, vitamin and mineral deficiencies and psychological problems are common. Lifelong medical supervision is necessary for those who choose the surgical route, but in suitable candidates, the health benefits of weight loss may prove worth the risks.[43]

IN SUMMARY

- Obese people with high risks of medical problems may need aggressive treatment, including drugs or surgery.

Reasonable Strategies for Weight Loss

Successful weight-loss strategies embrace small changes, moderate losses, and reasonable goals.[44] A person who loses 10 to 20 pounds in one year by consistently choosing nutrient-dense foods and engaging in regular physical activity is much more likely to maintain those losses and reap health benefits than if more weight were lost in less time by adopting the latest fad diet. Modest weight loss, even when a person is still overweight, can improve control of diabetes and reduce the risks of heart disease by lowering blood pressure and blood cholesterol, especially for those with abdominal fat.

Of course, the same eating and activity habits that improve health often lead to a healthier body weight and composition as well. Successful weight loss is defined not by pounds lost, but by health gained. People less concerned with disease risks may prefer to set goals for personal fitness, such as being able to play with children or climb stairs without becoming short of breath.

Whether the goal is health or fitness, weight-loss expectations need to be reasonable. Unreachable targets ensure frustration and failure. Setting reasonable goals helps to achieve the desired result in managing weight. For example, obese people who must reduce their weight to lower their disease risks might set three broad goals:

1. Reduce body weight by about 10 percent over half a year's time.
2. Maintain a lower body weight over the long term.
3. At a minimum, prevent further weight gain.

Such goals may be achieved or even exceeded, providing a sense of accomplishment instead of disappointment.

A healthy body contains enough lean tissue to support health and the right amount of fat to meet body needs.

© Lori Adamski Peek/Stone/Getty Images

Dietary Guidelines for Americans 2005

- For those who need to lose weight: Aim for a slow, steady weight loss by decreasing kcalorie intake while maintaining an adequate nutrient intake and increasing physical activity.

A Healthful Eating Plan

Contrary to the claims of many fad diets, no particular eating plan is magical, and no specific food must be either included or avoided for weight management. You are the one who will have to live with the plan, so you better be involved in its planning. The diet is successful only if you can maintain a healthy weight. Think of it as an eating plan that you will adopt for life. It must consist of foods that you like, that are available to you, and that are within your means.

A Realistic Energy Intake The main characteristic of a weight-loss diet is that it provides less energy than the person needs to maintain present body weight. If food energy is restricted too severely, dieters may not receive sufficient nutrients and may lose lean tissue. Rapid weight loss usually means excessive loss of lean tissue, a lower BMR, and a rapid weight gain to follow. Restrictive eating may also set in motion the unhealthy behaviors of eating disorders (described in Nutrition in Practice 6).

Table 7–2 outlines the recommendations of a weight-loss diet. Energy intake should provide nutritional adequacy without excess—that is, somewhere between deprivation and complete freedom to eat whatever, whenever. A reasonable suggestion is that an adult needs to increase activity and reduce food intake to create a deficit of 500 to 1,000 kcalories per day. ■ Such a deficit produces a weight loss of 1 to 2 pounds per week.[45] This amounts to an intake of 1,000 to 1,200 kcalories per day for most women and 1,200 to 1,600 kcalories per day for most men.[46] Table 7–3 (p. 182) suggests daily food amounts from which to build balanced 1,000- to 1,600-kcalorie diets. Diets providing energy intakes lower than 800 kcalories, called very low-calorie diets (VLCD), are notoriously unsuccessful at achieving lasting weight loss and can be dangerous and so are not recommended.

■ 1 lb body fat = 3,500 kcal
To lose 1 pound per week, cut 500 kcal/day.

TABLE 7–2

Recommendations for a Weight-Loss Diet

Nutrient	Recommended Intake
kcalories	
For people with BMI ≥ 35	Approximately 500 to 1,000 kcalories per day reduction from usual intake
For people with BMI between 27 and 35	Approximately 300 to 500 kcalories per day reduction from usual intake
Total fat	30% or less of total kcalories
Saturated fatty acids[a]	8 to 10% of total kcalories
Monounsaturated fatty acids	Up to 15% of total kcalories
Polyunsaturated fatty acids	Up to 10% of total kcalories
Cholesterol[a]	< 300 mg per day
Protein[b]	Approximately 15% of total kcalories
Carbohydrate[c]	55% or more of total kcalories
Sodium chloride[d]	No more than 2,300 mg of sodium or approximately 6 g of sodium chloride (salt) per day
Calcium	1,000 to 1,500 mg per day
Fiber[c]	20 to 30 g per day

[a] People with high blood cholesterol should aim for less than 7 percent kcalories from saturated fat and 200 milligrams cholesterol per day.
[b] Protein should be derived from plant sources and lean sources of animal protein.
[c] Carbohydrates and fiber should be derived from vegetables, fruits, and whole grains.
[d] Value from the *Dietary Guidelines for Americans 2005*; the Tolerable Upper Intake Level for sodium is 2,300 milligrams.

Source: National Heart, Lung, and Blood Institute, National Institutes of Health, *The Practical Guide: Identification, Evaluation, and Treatment of Overweight and Obesity in Adults*, NIH publication no. 00-4084 (Washington, DC: Government Printing Office, 2000), p. 27.

laparoscopic weight-loss surgery: a procedure in which surgeons gain access to the abdomen via several small incisions. A tiny video camera is inserted through one of the incisions and surgical instruments through the others. The surgeons watch their work on a large-screen monitor.

TABLE 7–3

Daily Amounts from Each Food Group for 1,000- to 1,600-kcalorie Diets

FOOD GROUP	1,000 kcal	1,200 kcal	1,400 kcal	1,600 kcal
Fruit	1 c	1 c	1½ c	1½ c
Vegetables	1 c	1½ c	1½ c	2 c
Grains	3 oz	4 oz	5 oz	5 oz
Meat and legumes	2 oz	3 oz	4 oz	5 oz
Milk	3 c	3 c	3 c	3 c
Oils	3 tsp	3 tsp	3 tsp	4 tsp

Note: The USDA Food Guide patterns for 1,000-, 1,200-, and 1,400-kcalorie diets were designed for children and provide 2 cups of milk. They are modified here to include an additional cup of milk, since 3 cups per day is recommended for all adults. The discretionary kcalorie allowance for these patterns is about 100 kcalories.

Nutritional Adequacy Nutritional adequacy is difficult to achieve on fewer than 1,200 kcalories per day, and most healthy adults need never consume any less than that. A plan that provides an adequate intake supports a healthier and more successful weight loss than a restrictive plan that creates feelings of starvation and deprivation, which can lead to an irresistible urge to binge.

Take a look at the 1,200-kcalorie diet in Table 7–3. Such an intake would allow most people to lose weight and still meet their nutrient needs with careful, nutrient-dense food selections. (Women might need calcium or iron supplements.) Keep in mind that well-balanced diets that emphasize fruits, vegetables, whole grains, lean meats or meat alternates, and low-fat or fat-free milk products offer many health rewards even when they do not result in weight loss.

Small Portions Overweight people usually need to learn to eat less food at each meal—one piece of chicken for dinner instead of two, a teaspoon of butter on the vegetables instead of a tablespoon, and one cookie for dessert instead of six. Chew foods slowly and thoroughly. The goal is to eat enough food for energy, nutrients, and pleasure, but not more. This amount should leave a person feeling satisfied—not necessarily full. Keep in mind that even low-fat foods can deliver a lot of kcalories when a person eats large quantities.

Balancing Carbohydrates, Fats, and Protein Healthy diets based on abundant fresh fruits and vegetables, low-fat milk products, legumes, lean meats, fish, poultry, or meat alternates, and whole grains are high in carbohydrates, adequate in fiber and protein, and low in the kinds of fats associated with diseases. They are also best for managing weight. Earlier chapters described the importa nce of each of the energy-yielding nutrients to health. Therefore, diets for weight management should provide all three within the ranges recommended by the DRI committee. ■

Wholesome, high-fiber, unprocessed or lightly processed foods offer bulk and satiety for fewer kcalories than smooth, quickly consumed refined foods. Thus choosing whole grains and fiber-rich vegetables in place of most refined grains and added fats and sugars benefits both weight and nutrition. Choose fats sensibly by avoiding sources of saturated and *trans* fats and including enough of the health-supporting fats (details in Nutrition in Practice 3) to provide satiety but not so much as to oversupply kcalories. Lean meats or other low-fat protein sources also provide satiety. Limit these foods but do not eliminate them.

Lower Energy Density As discussed earlier, to lower energy intake, people can choose smaller portion sizes, or they can reduce the energy density of the foods they eat. Research shows that eating satisfying portions of foods that are low in energy density

■ The DRI committee recommends:
- 45 to 65% kcalories from carbohydrate
- 20 to 35% kcalories from fat
- 10 to 35% kcalories from protein

Selecting grapes with their high water content instead of raisins increases the volume and cuts the energy intake in half.

Even at the same weight and similar serving sizes, the fiber-rich broccoli delivers twice the fiber of the potatoes for about one-fourth the energy.

By selecting the water-packed tuna (on the right) instead of the oil-packed tuna, a person can enjoy the same amount for fewer kcalories.

FIGURE 7–2

Energy Density
Decreasing the energy density (kcal/g) of foods allows a person to eat satisfying portions while still reducing energy intake. To lower energy density, select foods high in water or fiber and low in fat.

(such as fruits, vegetables, and broth-based soups) maintains satiety while reducing energy intake.[47] Foods containing substantial water or fiber and those low in fat help to lower a meal's energy density (see Figure 7–2). The Clinical Applications feature at the end of this chapter describes how to calculate the energy density of foods.

Sugar and Alcohol A person trying to achieve or maintain a healthy weight needs to pay attention to sugar and alcohol, as well as fat. Using them for pleasure on occasion is compatible with health as long as most daily choices are of nutrient-dense foods.

Meal Spacing Three meals per day is standard in our society, but no law says you cannot have four or five—be sure they are smaller, of course. People who eat small, frequent meals are reported to be successful at weight loss and maintenance. Make sure that mild hunger, not appetite, is prompting you to eat. Eat regularly, and eat before you become extremely hungry.

Adequate Water Learn to satisfy thirst with water. Water fills the stomach between meals and meets the fluid needs that were formerly met by eating extra food (remember that food provides water). An estimated 75 to 150 kcalories per day come from drinking sweetened beverages.[48] Simply replacing nutrient-poor, energy-dense beverages with water could save a person up to 15 pounds per year. Water also helps the gastrointestinal tract adapt to a high-fiber diet.

■ DRI for physical activity: 60 min/day (moderate intensity)

Physical Activity

Either dieting or physical activity alone can produce some weight loss. Clearly, however, the combination is most effective.[49] People who combine diet and physical activity are more likely to lose more fat, retain more muscle, and regain less weight than those who only diet.[50] To prevent weight gain and support weight loss, the *Dietary Guidelines for Americans 2005,* the DRI committee, and other fitness experts advise 60 minutes of moderately intense physical activity (walking/jogging at $3^{1}/_{2}$ to $4^{1}/_{2}$ miles per hour) per day in addition to the activities of daily life.[51] ■ Physical activity may also help counteract the negative effects of excess body weight on health.[52] For example, physical activity reduces abdominal obesity, and this change improves blood pressure, insulin resistance, and fitness of the heart and lungs, even without weight loss.[53]

Being active—even if overweight—is healthier than being sedentary. With a BMI of 36, aerobics instructor Jennifer Portnick is considered obese, but her daily workout routine helps to keep her in good health.

Dietary Guidelines for Americans 2005

■ To help manage body weight and prevent gradual unhealthy body weight gain in adulthood: Engage in approximately 60 minutes of activity of moderate-to-vigorous intensity on most days of the week while not exceeding caloric intake requirements.

Energy Expenditure Physical activity makes many contributions to weight loss and maintenance. For one thing, it directly increases energy output by the muscles and cardiovascular system. A 150-pound person walking a brisk 4 miles per hour for 30 minutes spends an extra 185 kcalories on that activity. A football player may spend several thousand extra kcalories on a day of heavy training.

BMR Activity also contributes to energy output in an indirect way—by speeding up basal metabolism.[54] It does this both immediately and over the long term. On any given day, after intense and prolonged exercise, basal metabolism remains elevated for several hours. Over the long term, daily vigorous activity for many weeks gradually shifts body composition toward more lean tissue, which is more active metabolically than fat tissue. The ongoing metabolic rate rises accordingly, and this makes a contribution toward continued weight loss or maintenance.

The raised metabolic rate continues for as long as the person is physically active on a regular basis. The more energy expended in metabolic activities, the greater the energy requirement. This means that a person can eat more without gaining weight.

Appetite Control Physical activity also helps to control appetite. People think that exercising will make them want to eat, but this is not entirely true. Yes, active people do have healthy appetites, but immediately after a good workout, most people do not feel like eating. They want to shower and may be thirsty, but they do not want to eat. The reason is that the body has responded to the stress of activity by mobilizing fuels from storage: glucose and fatty acids are abundant in the blood. At the same time, the body has suppressed its digestive functions. Hard physical work and eating are not compatible.

Psychological Benefits Physical activity helps especially to curb the inappropriate appetite that prompts a person to eat when bored, anxious, or depressed. Weight management programs encourage people to go out and be active when they are tempted to eat but are not really hungry.

Physical activity also helps to reduce stress. Since stress itself is a cue to inappropriate eating behavior for many people, activity can help here, too.

Activity offers still more psychological advantages. The fit person looks and feels healthy, and high self-esteem accompanies these benefits. High self-esteem tends to support a person's resolve to persist in a weight-control effort, rounding out a beneficial cycle. ■

■ Benefits of physical activity in a weight management program:
- Improved body composition
- Favorable effects on disease risks
- Short-term increase in energy expenditure (from exercise and from a slight rise in BMR)
- Long-term increase (slight) in BMR
- Appetite control
- Stress reduction and control of stress eating
- Physical, and therefore psychological, well-being
- High self-esteem

Choosing Activities What kind of physical activity is best? People seeking to lose weight should choose activities that they enjoy and are willing to do regularly. Health care professionals frequently advise people who want to manage their body weight and lose fat to engage in activities of low-to-moderate intensity for a long duration, such as an hour-long fast-paced walk. The reasoning behind such advice is that people exercising at low-to-moderate intensity are likely to stick with their activity for longer times and are less likely to injure themselves. People who regularly engage in more *vigorous* physical activities (for example, fast bicycling or endurance running), however, have less body fat than those who engage in moderately intense activities. The conditioned body that is adapted to strenuous and prolonged aerobic activity uses more fat all day long, not just during activity. The bottom line on physical activity and weight and/or fat loss seems to be that total energy expenditure is the main factor, regardless of how a person does it.

In addition to activities such as walking or cycling, there are hundreds of ways to incorporate energy-spending activities into daily routines: take the stairs instead of the elevator, walk to the neighbor's apartment instead of making a telephone call, and

FIGURE 7–3

Time	Place	Activity or food eaten	People present	Mood
10:30– 10:40	School vending machine	6 peanut butter crackers and 12 oz. cola	by myself	Starved
12:15– 12:30	Restaurant	Sub sandwich and 12 oz. cola	friends	relaxed & friendly
3:00– 3:45	Gym	Weight training	workout partner	tired
4:00– 4:10	Snack bar	Small frozen yogurt	by myself	OK

Food and Activity Diary
The entries in a food and activity diary should include the times and places of meals and snacks, the types and amounts of foods eaten, and a description of the individual's feelings when eating. The diary should also record physical activities: the kind, the intensity level, the duration, and the person's feelings about them.

rake the leaves instead of using a blower. These activities burn only a few kcalories each, but over one year's time they become significant.

Spot Reducing People sometimes ask about "spot reducing." Unfortunately, no one part of the body gives up fat in preference to another. Fat cells all over the body release fat in response to demand, and the fat is then used by whatever muscles are active. No exercise can remove the fat from any one particular area—and, incidentally, neither can a massage machine that claims to break up fat on trouble spots.

Physical activity can help with trouble spots in another way. Strengthening muscles in a trouble area can help to improve their tone; stretching to gain flexibility can help with posture problems. Thus cardiorespiratory endurance, strength, and flexibility workouts all have a place in fitness programs.

Behavior and Attitude

Behavior modification therapy provides ways to overcome barriers to making dietary changes and increasing physical activity. Behavior modification therapy does more than help people decide which behaviors to change; it also teaches them how to change.[55] Behavior and attitude are important supporting factors in achieving and maintaining appropriate body weight and composition. Changing the behaviors of overeating and underexercising that lead to and perpetuate obesity requires time and effort. A person must commit to take action.

Becoming Aware of Behaviors A person who is aware of all the behaviors that create a problem has a head start on developing a solution. First, the person needs to establish a baseline (a record of present eating and physical activity behaviors) against which to measure future progress. It is best to keep a diary (see Figure 7–3) that includes the time and place of meals and snacks, the type and amount of foods eaten, the persons present when food is eaten, and a description of the individual's feelings when eating. The diary should also record physical activities: kind, intensity level, duration, and the person's feelings about them. These entries will help the individual identify possible behaviors to change.

Making Small Changes The "How To" (p. 186) describes behavioral strategies to support weight management. A particularly attractive feature of these strategies is that they do not involve blaming oneself or putting oneself down—an important element in fostering self-esteem.

behavior modification: the changing of behavior by the manipulation of *antecedents* (cues or environmental factors that trigger behavior), the behavior itself, and *consequences* (the penalties or rewards attached to behavior).

How To

1. Eliminate inappropriate cues:

   - Do not buy problem foods.

   - Eat only in one room at the designated time.

   - Shop when not hungry.

   - Avoid vending machines, fast-food restaurants, and convenience stores.

   - Turn off television, video games, and computers.

2. Suppress the cues you cannot eliminate:

   - Serve individual plates; do not serve "family style."

   - Make small portions look large by spreading them over the plate.

   - Create obstacles to consuming problem foods—wrap them and freeze them, making them less quickly accessible.

   - Control deprivation; plan and eat regular meals.

   - Plan the time spent in sedentary activities, such as watching television or using a computer—do not use these activities just to fill time.

3. Strengthen cues to appropriate behaviors:

   - Share appropriate foods with others.

   - Store appropriate foods in convenient spots in the refrigerator.

   - Learn appropriate portion sizes.

   - Plan appropriate snacks.

   - Keep sports and play equipment by the door.

4. Repeat desired behaviors:

   - Slow down eating—put down utensils between bites.

   - Always use utensils.

   - Leave some food on your plate.

   - Move more—shake a leg, pace, stretch often.

   - Join groups of active people and participate.

5. Arrange negative consequences for negative behavior:

   - Ask that others respond neutrally to your deviations (make no comments—even negative attention is a reward).

   - If you slip, do not punish yourself.

6. Reward yourself personally and immediately for positive behaviors:

   - Buy tickets to sports events, movies, concerts, or other nonfood amusement.

   - Indulge in a new small purchase.

   - Get a massage; buy some flowers.

   - Take a hot bath; read a good book.

   - Treat yourself to a lesson in a new active pursuit such as horseback riding, handball, or tennis.

   - Praise yourself; visit friends.

   - Nap; relax.

Maintaining Weight Finally, be aware that it can be hard to maintain weight loss. On arriving at the goal weight after months of self-discipline and new habit formation, the victorious weight loser must not "celebrate" by resuming old eating habits. Membership in an ongoing weight control organization and regular, continued physical activity can provide indispensable support for the formerly overweight person who wants to remain trim.

Personal Attitude For many people, overeating and being overweight may have become an integral part of their identity. Changing diet and activity behaviors without attention to a person's self-concept invites failure.

Many people overeat to cope with the stresses of life. To break out of that pattern, they must first identify the particular stressors that trigger their urges to overeat. Then, when faced with these situations, they must learn to practice problem-solving skills. When the problems that trigger the urge to overeat are dealt with in alternative

Dietary Guidelines for Americans 2005

- To sustain weight loss in adulthood, participate in at least 60 to 90 minutes of moderately intense physical activity daily while not exceeding caloric intake requirements. Some people may need to consult with a health care provider before participating in this level of activity.

ways, people may find that they eat less. The message is that sound emotional health supports the ability to take care of health in all ways—including nutrition, weight management, and fitness.

IN SUMMARY

- A person who adopts a lifelong "eating plan for good health" rather than a "diet for weight loss" will be more likely to keep the lost weight off. Table 7–4 offers several tips for successful weight management.

- Physical activity should be an integral part of a weight management program.

- Physical activity can increase energy expenditure, improve body composition, help control appetite, reduce stress and stress eating, and enhance physical and psychological well-being.

- Behavior modification provides ways to overcome barriers to successful weight management.

TABLE 7–4

Weight Management Strategies

In General

Focus on healthy eating and activity habits, not on weight losses or gains.

Adopt reasonable expectations about health and fitness goals and about how long it will take to achieve them.

Make nutritional adequacy a high priority.

Learn, practice, and follow a healthful eating plan for the rest of your life.

Participate in some form of physical activity regularly.

Adopt permanent lifestyle changes to achieve and maintain a healthy weight.

For Weight Loss

Energy out should exceed energy in by about 500 kcalories/day. Increase your physical activity enough to spend more energy than you consume from foods.

Emphasize foods with a low energy density and a high nutrient density.

Eat small portions. Share a restaurant meal with a friend or take home half for lunch tomorrow.

Eat slowly.

Limit high-fat foods. Make legumes, whole grains, vegetables, and fruits central to your diet plan.

Limit low-fat treats to the serving size on the label.

Limit concentrated sweets and alcoholic beverages.

Drink a glass of water before you begin to eat and another while you eat. Drink plenty of water throughout the day (8 glasses or more per day).

Keep a record of diet and exercise habits; it reveals problem areas, the first step toward improving behaviors.

Learn alternative ways to deal with emotions and stresses.

Attend support groups regularly or develop supportive relationships with others.

For Weight Gain

Energy in should exceed energy out by at least 500 kcalories/day. Increase your food intake enough to store more energy than you spend in exercise. Exercise and eat to build muscles.

Expect weight gain to take time (1 pound per month would be reasonable).

Emphasize energy-dense foods.

Eat at least three meals per day.

Eat large portions of foods and expect to feel full.

Eat snacks between meals.

Drink plenty of juice and milk.

Underweight

Underweight is far less prevalent than overweight, affecting no more than 5 percent of U.S. adults. The health risks associated with underweight are fewer than those that accompany overweight. Both underweight women and those who have lost a significant amount of weight, however, are more susceptible to osteoporosis. Underweight women may become infertile or may give birth to unhealthy infants. An underweight woman can improve her chances of bearing a healthy infant by gaining weight prior to conception, during pregnancy, or both.

Underweight becomes more hazardous when accompanied by undernutrition. An inadequate supply of nutrients and energy leaves the body underprepared to handle its many metabolic and physical tasks. A person without reserves has a particularly difficult battle when faced with medical stresses such as surgery or the wasting diseases of cancer and AIDS. Thus underweight people are urged to gain lean tissue and body fat (as an energy reserve) and to acquire protective amounts of all the nutrients that can be stored.

An extreme underweight condition known as anorexia nervosa is sometimes seen in young people who exercise unreasonable self-denial in order to control their weight. They go to such extremes that they become severely undernourished and underweight. The distinguishing feature of a person with anorexia nervosa, as opposed to other thin people, is that the starvation is intentional. Anorexia nervosa is a major eating disorder seen in our society today. Another is bulimia nervosa—compulsive overeating, usually with purging. Eating disorders are the subject of Nutrition in Practice 6.

IN SUMMARY

- Both the incidence of underweight and the health problems associated with it are less prevalent than overweight and its associated problems.

Strategies for Weight Gain

Weight gain, like weight loss, is an individual matter. People who are healthy at their present weights may stay there; those who are unhealthy might try to gain weight.

Some people are unalterably thin by reasons of genetics or early physical influences. Those who wish to gain weight for appearance's sake or to improve athletic performance should be aware that a healthful weight can be achieved only through physical activity, particularly strength training, combined with a high energy intake. Eating many high-kcalorie foods can bring about weight gain, but it will be mostly fat. This can be as detrimental to health as being slightly underweight. In an athlete, such a weight gain can impair performance. Therefore, in weight gain, as in weight loss, physical activity is an essential component of a sound plan.

Physical Activity to Build Muscles The person who wants to gain weight should use **weight training** primarily. As activity is increased, energy intake must be increased to support that activity. Eating extra food will then support a gain of both muscle and fat.

Energy-Dense Foods Energy-dense foods (the very ones eliminated from a successful weight-loss diet) hold the key to weight gain. Pick the highest-kcalorie items from each food group—that is, milk shakes instead of fat-free milk, peanut butter instead of lean meat, avocados instead of cucumbers, and whole wheat muffins instead of whole wheat bread. Because fat contains more than twice as many kcalories per teaspoon as sugar does, fat adds kcalories without adding much bulk.

Be aware that health experts routinely recommend a moderate-fat diet for the general U.S. population because the biggest health problems are overweight and heart disease. Eating high-kcalorie, high-fat foods is not healthy for most people but may be essential for an underweight individual who needs to gain weight. An underweight person who is physically active and eating a nutritionally adequate diet can afford a few extra kcalories from fat. For health's sake, it would be wise to select foods with monounsaturated and polyunsaturated fats instead of those with saturated or *trans* fats: for example, sautéing vegetables in olive oil instead of butter or hydrogenated margarine.

Three Meals Daily People wanting to gain weight should eat at least three hearty meals per day. Many people who are underweight have simply been too busy (sometimes for months) to eat enough to gain or maintain weight. Therefore, they need to make meals a priority and plan them in advance. Taking time to prepare and eat each meal can help, as can learning to eat more food within the first 20 minutes of a meal before you begin to feel full. Another suggestion is to eat meaty appetizers or the main course first and leave the soup or salad until later.

Large Portions It is also important for the underweight person to learn to eat more food at each meal: have two sandwiches for lunch instead of one, drink milk from a larger glass, and eat cereal from a larger bowl. Expect to feel full. Most underweight individuals are accustomed to small quantities of food. When they begin eating significantly more, they feel uncomfortable. This is normal and passes over time.

Extra Snacks Because a substantially higher energy intake is needed each day, in addition to eating more food at each meal, it is necessary to eat more frequently. Between-meal snacking offers a solution. For example, a student might make three sandwiches in the morning and eat them between classes in addition to the day's three regular meals.

Juice and Milk Beverages provide an easy way to increase energy intake. Consider that 6 cups of cranberry juice add almost 1,000 kcalories to the day's intake. kcalories can be added to milk by mixing in powdered milk or packets of instant breakfast.

For people who are underweight due to illness, concentrated liquid formulas are often recommended because a weak person can swallow them easily. A registered dietitian can recommend high-protein, high-kcalorie formulas to help the underweight person maintain or gain weight. Used in addition to regular meals, these formulas can help considerably.

IN SUMMARY

- To gain weight, a person must train physically and increase energy intake by selecting energy-dense foods, eating regular meals, taking larger portions, and consuming extra snacks and beverages. Review the tips for weight gain in Table 7–4 (p. 187).

weight training: the use of free weights or weight machines to provide resistance for developing muscle strength and endurance; also called *resistance training*. A person's own body weight may also be used to provide resistance as when a person does push-ups, pull-ups, or abdominal crunches.

Analysis

YOUR DIET

To enjoy good health and maintain a reasonable body weight, you need to combine sensible eating habits and regular physical activity. This exercise allows you to evaluate your current body weight and consider some lifestyle factors important for weight management.

First, check to see whether your current body weight falls within the "healthy weight" range in the BMI table at the back of this text. Find your height in the left-hand column of the table and look across the row to find your weight. Is your BMI in the healthy range, which is highlighted in green?

Alternatively, you can calculate your BMI using the equation shown on p. 153. A healthy weight usually falls within a BMI of 18.5 and 24.9.

Your BMI:_____

If your BMI falls within the underweight, overweight, or obese range, you may want to gain or lose weight to improve your health. Even if your BMI is within the healthy range, you may wish to improve your eating habits or fitness level. Review Figure 7–3 (p. 185) and complete your own food and activity record for at least a 24-hour period to reveal information about daily habits that may support or undermine achieving a healthy weight.

Time	Place	Activity or Food Eaten	People Present	Mood

- Make a list of the habits that support maintaining or achieving a healthy weight. For example, do you drink water rather than soda throughout the day and limit the time you spend watching television or playing computer games?

- Make a list of the habits that do not support maintaining or achieving a healthy weight. For example, do you skip breakfast, and then eat fast food or vending machine food?

- What changes would you like to make in your daily habits to improve your health and nutrition?

CLINICAL APPLICATIONS

1. Chapter 1 discussed the nutrient density of foods—their nutrient contribution per kcalorie. Another way to evaluate foods is to consider their energy density—their energy contribution per gram:

   - A carrot weighing 72 grams delivers 31 kcalories.

   - To calculate the energy density, divide kcalories by grams: 31 kcalories divided by 72 grams = 0.43 kcalories per gram

2. Do the same for french fries weighing 50 grams and contributing 167 kcalories: 167 kcalories divided by 50 grams = _____ kcalories per gram

   - The more kcalories per gram, the greater the energy density.

   - Which food is more energy dense? The conclusion is no surprise, but understanding the mathematics may offer valuable insight into the concept of energy density.

   - French fries are more energy dense, providing 3.34 kcalories per gram. They provide more energy per gram—and per bite.

3. Considering a food's energy density is especially useful in planning diets for weight management. Foods with a high energy density can help with weight gain, whereas those with a low energy density can help with weight loss. Give some examples of foods that you might suggest for a client who wants to gain weight and some that might be appropriate for a client who is trying to lose weight.

© Matthew Farruggio

1. Two causes of obesity in humans are:
   a. set-point theory and BMI.
   b. genetics and physical inactivity.
   c. genetics and low-carbohydrate diets.
   d. mineral imbalances and fat cell
      imbalance.

2. The protein produced by the fat cells under the direction of the *ob* gene is called:
   a. leptin.
   b. orlistat.
   c. sibutramine.
   d. lipoprotein lipase.

3. All of the following describe the behavior of fat cells *except*:
   a. the number decreases when fat is lost from the body.
   b. the storage capacity for fat depends on both cell number and
      cell size.
   c. the size is larger in obese people than in normal-weight
      people.
   d. the number increases severalfold during the growth years and
      tapers off when adult status is reached.

4. The obesity theory that suggests the body chooses to be at a specific weight is the:
   a. fat cell theory.
   b. enzyme theory.
   c. set-point theory.
   d. external cue theory.

5. The biggest problem associated with the use of prescription drugs in the treatment of obesity is:
   a. cost
   b. the necessity for long-term dosage.
   c. ineffectiveness.
   d. adverse side effects.

6. Television watching contributes to obesity for all of the following reasons, *except*:
   a. it promotes inactivity.
   b. it promotes between-meal snacking.
   c. it replaces time that could be spent eating.
   d. it gives high exposure to energy-dense foods featured in the
      advertisements.

7. What is the best approach to weight loss?
   a. Avoid foods containing carbohydrates.
   b. Eliminate all fats from the diet and decrease water intake.
   c. Greatly increase protein intake to prevent body protein loss.
   d. Reduce daily energy intake and increase energy expenditure.

8. To help prevent body fat gain, the DRI committee suggests daily, moderately intense, physical activities totaling:
   a. 20 minutes.
   b. 60 minutes.
   c. $1\frac{1}{2}$ hours.
   d. 3 hours.

9. Suggestions to change behaviors for successful weight control include:
   a. shop only when hungry.
   b. eat in front of the television for distraction.
   c. learn appropriate portion sizes.
   d. eat quickly.

10. Which strategy would *not* help an underweight person to gain weight?
    a. Exercise.
    b. Drink plenty of water.
    c. Eat snacks between meals.
    d. Eat large portions of foods.

Answers to these questions can be found in Appendix H.

Notes

1. H. N. Lyon and J. N. Hirschhorn, Genetics of common forms of obesity: A brief overview, *American Journal of Clinical Nutrition* 82 (2005): 215S–217S; R. J. F. Loos and T. Rankinen, Gene-diet interactions on body weight changes, *Journal of the American Dietetic Association* 105 (2005): 29S–34S; S. Tholin and coauthors, Genetic and environmental influences on eating behavior: The Swedish Young Male Twins Study, *American Journal of Clinical Nutrition* 81 (2005): 564–569; R. J. F. Loos and C. Bouchard, Obesity—Is it a genetic disorder? *Journal of Internal Medicine* 254 (2003): 401–425.

2. Lyon and Hirschhorn, 2005; Loos and Rankinen, 2005; Tholin and coauthors, 2005; Loos and Bouchard, 2003.

3. D. E. Cummings and M. W. Schwartz, Genetics and pathophysiology of human obesity, *Annual Review of Medicine* 54 (2003): 453–471.

4. V. Popovic and L. H. Duntas, Brain somatic cross-talk: Ghrelin, leptin, and ultimate challenges of obesity, *Nutritional Neuroscience* 8 (2005): 1–5.

5. Popovic and Duntas, 2005.

6. G. Fruhbeck, The adipose tissue as a source of vasoactive factors, *Current Medicinal Chemistry Cardiovascular and Hematological Agents* 3 (2004): 197–208; R. L. Bradley, K. A. Cleveland, and B. Cheatham, The adipocyte as a secretory

organ: Mechanism of vesicle transport and secretory pathways, *Recent Progress in Hormone Research* 56 (2001): 329–358.

7. P. Fietta, Focus on leptin, a pleiotropic hormone, *Minerva Medica* 96 (2005): 65–75; J. Harvey and M. L. Ashford, Leptin in the CNS: Much more than a satiety signal, *Neuropharmacology* 44 (2003): 845–854; S. Takeda, F. Elefteriou, and G. Karsenty, Common endocrine control of body weight, reproduction, and bone mass, *Annual Review of Nutrition* 23 (2003): 403–411.

8. D. E. Cummings, K. E. Foster-Schubert, and J. Overduin, Ghrelin and energy balance: Focus on current controversies, *Current Drug Targets* 6 (2005): 153–169.

9. Cummings, Foster-Schubert, and Overduin, 2005; Popovic and Duntas, 2005; J. Eisenstein and A. Greenberg, Ghrelin: Update 2003, *Nutrition Reviews* 61 (2003): 101–104.

10. Cummings, Foster-Schubert, and Overduin, 2005; J. Orr and B. Davy, Dietary influences on peripheral hormones regulating energy intake: Potential applications for weight management, *Journal of the American Dietetic Association* 105 (2005): 1115–1124.

11. H. S. Callahan and coauthors, Postprandial suppression of plasma ghrelin level is proportional to ingested caloric load but does not predict intermeal interval in humans, *Journal of Clinical Endocrinology and Metabolism* 89 (2004): 1319–1324.

12. Cummings, Foster-Schubert, and Overduin, 2005.

13. Popovic and Duntas, 2005; M. Tanaka and coauthors, Habitual binge/purge behavior influences circulating ghrelin levels in eating disorders, *Journal of Psychiatric Research* 37 (2003): 17–22.

14. M. Tanaka and coauthors, Effect of nutritional rehabilitation on circulating ghrelin and growth hormone levels in patients with anorexia nervosa, *Regulatory Peptides* 122 (2004): 163–168; V. Tolle and coauthors, Balance in ghrelin and leptin plasma levels in anorexia nervosa patients and constitutionally thin women, *Journal of Clinical Endocrinology and Metabolism* 88 (2003): 109–116.

15. A. M. Haqq and coauthors, Serum ghrelin levels are inversely correlated with body mass index, age, and insulin concentrations in normal children and are markedly increased in Prader-Willi syndrome, *Journal of Clinical Endocrinology and Metabolism* 88 (2003): 174–178; A. DelParigi and coauthors, High circulating ghrelin: A potential cause for hyperphagia and obesity in Prader-Willi syndrome, *Journal of Clinical Endocrinology and Metabolism* 87 (2002): 5461–5464.

16. N. Tentolouris and coauthors, Differential effects of high-fat and high-carbohydrate content isoenergetic meals on plasma active ghrelin concentrations in lean and obese women, *Hormone and Metabolic Research* 36 (2004): 559–563; P. J. English and coauthors, Food fails to suppress ghrelin levels in obese humans, *Journal of Clinical Endocrinology and Metabolism* 87 (2002): 2984.

17. D. E. Cummings and coauthors, Plasma ghrelin levels after diet-induced weight loss or gastric bypass surgery, *New England Journal of Medicine* 346 (2002): 1623–1630.

18. Cummings, Foster-Schubert, and Overduin, 2005.

19. Loos and Rankinen, 2005.

20. L. R. Young and M. Nestle, The contribution of expanding portion sizes to the US obesity epidemic, *American Journal of Public Health* 92 (2002): 246–249.

21. B. J. Rolls and coauthors, Increasing the portion size of a packaged snack increases energy intake in men and women, *Appetite* 42 (2004): 63–69.

22. S. J. Nielsen and B. M. Popkin, Patterns and trends in food portion sizes, 1977–1998, *Journal of the American Medical Association* 289 (2003): 450–453.

23. Young and Nestle, 2002.

24. S. A. Bowman and B. T. Vinyard, Fast food consumption of U.S. adults: Impact on energy and nutrient intakes and overweight status, *Journal of the American College of Nutrition* 23 (2004): 163–168.

25. B. J. Rolls, A. Drewnowski, and J. H. Ledikwe, Changing the energy density of diet as a strategy for weight management, *Journal of the American Dietetic Association* 105 (2005): 98S–103S.

26. J. A. Ello-Martin, J. H. Ledikwe, and B. J. Rolls, The Influence of food portion size and energy density on energy intake: Implications for weight management, *American Journal of Clinical Nutrition* 82 (2005): 236S–241S.

27. K. M. Booth, M. M. Pinkston, and W. S. C. Poston, Obesity and the built environment, *Journal of the American Dietetic Association* 105 (2005): S110–S117.

28. J. M. Jakicic and A. D. Otto, Physical activity considerations for the treatment and prevention of obesity, *American Journal of Clinical Nutrition* 82 (2005): 226S–229S; L. L. Frank and coauthors, Effects of exercise on metabolic risk variables in overweight postmenopausal women, *Obesity Research* 13 (2005): 615–625; D. L. Thompson, J. Rakow, and S. M. Perdue, Relationship between accumulated walking and body composition in middle-aged women, *Medicine and Science in Sports and Exercise* 36 (2004): 911–914; R. Jurca and coauthors, Associations of muscle strength and aerobic fitness with metabolic syndrome in men, *Medicine and Science in Sports and Exercise* 36 (2004): 1301–1307; R. G. Ketelhut, I. W. Franz, and J. Scholze, Regular exercise as an effective approach in antihypertensive therapy, *Medicine and Science in Sports and Exercise* 36 (2004): 4–8; F. B. Hu and coauthors, Adiposity as compared with physical activity in predicting mortality among women, *New England Journal of Medicine* 351 (2004): 2694–2703; S. L. Wong and coauthors, Cardiorespiratory fitness is associated with lower abdominal fat independent of body mass index, *Medicine and Science in Sports and Exercise* 36 (2004): 286–291.

29. S. Gable, Y. Chang, and J. L. Krull, Television watching and frequency of family meals are predictive of overweight onset and persistence in a national sample of school-aged children, *Journal of the American Dietetic Association* 107 (2007): 53–61; F. B. Hu and coauthors, Television watching and other sedentary behaviors in relation to risk of obesity and type 2 diabetes mellitus in women, *Journal of the American Medical Association* 289 (2003): 1785–1791.

30. R. R. Wing and S. Phelan, Long-term weight loss maintenance, *American Journal of Clinical Nutrition* 82 (2005): 222S–225S; R. R. Wing and J. O. Hill, Successful weight loss maintenance, *Annual Review of Nutrition* 21 (2001): 323–341.

31. NBJ's annual overview of the nutrition industry VII, *Nutrition Business Journal* 7 (2002): 1–10, as cited in P. R. Thomas, Dietary supplements for weight loss? *Nutrition Today* 40 (2005): 6–12.

32. P. R. Thomas, Dietary supplements for weight loss? *Nutrition Today* 40 (2005): 6–12; M. H. Pittler and E. Ernst, Dietary supplements for body-weight reduction: A systematic review, *American Journal of Clinical Nutrition* 79 (2004): 529–536.

33. P. G. Shekelle and coauthors, Efficacy and safety of ephedra and ephedrine for weight loss and athletic performance: A meta-analysis, *Journal of the American Medical Association* 289 (2003): 1537–1545.

33. F. Greenway, Another type of intervention: Treating obesity with medication, *Journal of the American Dietetic Association* 105 (2005): 895–898; Position of the American Dietetic Association: Weight management, *Journal of the American Dietetic Association* 102 (2002): 1145–1155.

35. S. B. Moyers, Medications as adjunct therapy for weight loss: Approved and off-label agents in use, *Journal of the American Dietetic Association* 105 (2005): 948–959.

36. Greenway, 2005.

37. Greenway, 2005; Position of the American Dietetic Association, 2002.

38. Moyers, 2005.

39. Cummings, Foster-Schubert and Overduin, 2005.

40. G. L. Blackburn, Solutions in weight control: Lessons from gastric surgery, *American Journal of Clinical Nutrition* 82 (2005): 248S–252S; H. Buchwald and coauthors, Bariatric surgery: a systematic review and meta-analysis, *Journal of the American Medical Association* 292 (2004): 1724–1737.

41. Buchwald and coauthors, 2004.

42. Blackburn, 2005.

43. R. E. Brolin, Bariatric surgery and long-term control of morbid obesity, *Journal of the American Medical Association* 288 (2002): 2793–2796.

44. C. A. Nonas and G. D. Foster, Setting achievable goals for weight loss, *Journal of the American Dietetic Association* 105 (2005): S118–S123.

45. National Heart, Lung, and Blood Institute, National Institutes of Health, *The Practical Guide: Identification, Evaluation, and Treatment of Overweight and Obesity in Adults,* NIH publication no. 00-4084 (Washington, D.C.: Government Printing Office, 2000), p. 23.

46. National Heart, Lung, and Blood Institute, National Institutes of Health, 2000.

47. Ello-Martin, Ledikwe, and Rolls, 2005.

48. B. M. Popkin and coauthors, A new proposed guidance system for beverage consumption in the United States, *American Journal of Clinical Nutrition* 83 (2006): 529–542.

49. Jakicic and Otto, 2005; Wing and Phelan, 2005; Position of the American Dietetic Association, 2002.

50. Wing and Phelan, 2005; American College of Sports Medicine, Position Stand, Appropriate intervention strategies for weight loss and prevention of weight regain for adults, *Medicine and Science in Sports and Exercise* 33 (2001): 2145–2156.

51. Standing Committee on the Scientific Evaluation of Dietary Reference Intakes, Food and Nutrition Board, Institute of Medicine, *Dietary Reference Intakes for Energy, Carbohydrate, Fiber, Fat, Fatty Acids, Cholesterol, Protein, and Amino Acids* (Washington, D.C.: National Academies Press, 2005), p. 880; U.S. Department of Agriculture and U.S. Department of Health and Human Services, *2005 Dietary Guidelines Committee Report,* 2005, available online at www.healthierus.gov/dietaryguidelines/; S. N. Blair, M. J. LaMonte, and M. Z. Nichaman, The evolution of physical activity recommendations: How much is enough? *American Journal of Clinical Nutrition* 79 (2004): 913S–920S; American College of Sports Medicine, Position Stand, 2001.

52. Jakicic and Otto, 2005.

53. Wong and coauthors, 2004; Ketelhut, Franz, and Scholze, 2004.

54. Standing Committee on the Scientific Evaluation of Dietary Reference Intakes, 2005, p. 115.

55. A. N. Fabricatore, Behavior therapy and cognitive-behavioral therapy of obesity: Is there a difference? *Journal of the American Dietetic Association* 107 (2007): 92–99; G. D. Foster, A. P. Makris, and B. A. Bailer, Behavioral treatment of obesity, *American Journal of Clinical Nutrition* 82 (2005): 230S–235S.

FAD DIETS

To paraphrase William Shakespeare, "a fad diet by any other name would still be a fad diet." The names are legion: Atkins New Diet Revolution, South Beach Diet, Eat Right 4 Your Type Diet, Ultimate Weight Solution Diet, and Zone Diet.* Year after year, "new and improved" diets appear on bookstore shelves and circulate among friends. Sometimes fad diets seem to work for awhile, but more often than not, their success is short-lived. Then another fad diet takes the spotlight. Here is how Dr. K. Brownell, an obesity researcher at Yale University, describes this phenomenon: "When I get calls about the latest diet fad, I imagine a trick birthday cake candle that keeps lighting up and we have to keep blowing it out."

Why do health professionals not speak out against the relentless promotion of fad diets?

Realizing that many fad diets do not offer a safe and effective plan for weight loss, health professionals speak out, but they never get the candle blown out permanently. New fad diets can keep making outrageous claims because no one requires their advocates to prove what they say. Fad diet gurus do not have to conduct credible research on the benefits or dangers of their diets. They can simply make recommendations and then later, if questioned, search for bits and pieces of research that support the conclusions they have already reached. That is backwards. Diet and health recommendations should *follow* years of sound research that has been reviewed by panels of scientists *before* being offered to the public.

How can the promoters of fad diets get away with exaggerated claims?

Because anyone can publish anything—in books or on the Internet—peddlers of fad diets can make unsubstantiated statements that fall far short of the truth, but sound impressive to the uninformed. They often offer distorted bits of legitimate research. They may start with one or more actual facts, but then leap from one erroneous conclusion to the next. Anyone who wants to believe these claims has to wonder how the thousands of scientists working on obesity research over the past century could possibly have missed such obvious connections. Table NP7–1 presents some of the claims and truths of fad diets.

© Geri Engberg

*The following sources offer comparisons and evaluations of various fad diets for your review: Battle of the diet books, II, *Nutrition Action Healthletter*, July/August 2006, pp. 10–11; B. Liebman, Weighing the diet books, *Nutrition Action Healthletter*, January/February 2004, pp. 1–8; S. T. St. Jeor and coauthors, Dietary protein and weight reduction: A statement for healthcare professionals from the nutrition committee of the Council on Nutrition, Physical Activity, and Metabolism of the American Heart Association, *Circulation* 104 (2001): 1869–1874.

TABLE NP7–1

The Claims and Truths of Fad Diets

The Claim:	You can lose weight with "exceptionally easy rules."
The Truth:	Most fad diet plans have complicated rules that require you to calculate protein requirements, count carbohydrate grams, combine certain foods, time meal intervals, purchase special products, plan daily menus, and measure serving sizes.
The Claim:	You can lose weight by eating a specific ratio of carbohydrates, protein, and fat.
The Truth:	Weight loss depends on expending more energy than you take in, not on the proportion of energy nutrients.
The Claim:	This "revolutionary diet" can "reset your genetic code."
The Truth:	You inherited your genes and cannot alter your genetic code.
The Claim:	High-protein diets are popular, selling more than 20 million books, because they work.
The Truth:	Weight-loss books are popular because people grasp for quick fixes and simple solutions to their weight problems. If book sales were an indication of weight-loss success, we would be a lean nation—but they are not, and neither are we.
The Claim:	People gain weight on low-fat diets.
The Truth:	People can gain weight on low-fat diets if they over-indulge in carbohydrates and proteins while cutting fat; low-fat diets are not necessarily low-kcalorie diets. But people can also lose weight on low-fat diets if they cut kcalories as well as fat.
The Claim:	High-protein diets energize the brain.

continued

TABLE NP7–1 (continued)

The Truth:	The brain depends on glucose for its energy; the primary dietary source of glucose is carbohydrate, not protein.
The Claim:	Thousands of people have been successful with this plan.
The Truth:	Authors of fad diets have not published their research findings in scientific journals. Success stories are anecdotal, and failures are not reported.
The Claim:	Carbohydrates raise blood glucose levels, triggering insulin production and fat storage.
The Truth:	Insulin promotes fat storage when energy intake exceeds energy needs. Furthermore, insulin is only one hormone involved in the complex processes of maintaining the body's energy balance and health.
The Claim:	You will lose weight fast without counting kcalories or exercising because the diet alters metabolism.
The Truth:	No known trick of metabolism produces weight loss without diet or exercise.

Most of the popular diets I hear about seem to target carbohydrates—eat less carbohydrate, more protein, or eat "good carbs" instead of "bad carbs." Is this true?

Yes, to a point, although once you start sifting through the stacks of diet books available today, you begin to realize that fad diets come in almost as many shapes and sizes as the people who search them out. Some restrict fats or carbohydrates, some limit portion sizes, some focus on which foods can or cannot be eaten with other foods, and some claim that your genetic type or blood type determines which foods you should or should not eat to manage your weight and prevent disease. Table NP7–2 compares some of the more popular diets and shows that many of them focus on restricting carbohydrate intake. Exceptions include diets such as the Ornish Diet and the Pritikin Program, which advocate a *high*-carbohydrate, very-low-fat diet. These diets are not as popular today as they once were because information about eating diets high in carbohydrate-rich foods that support health and restricting certain kinds of fat-rich foods is available at no cost by way of national

TABLE NP7–2

Popular Diets Compared

Diet	Major Premise Promoted	Strong Point(s)	Weak Point(s)
High-Carbohydrate, Low-Fat			
Ornish Diet	■ By strictly limiting fat (both animal and vegetable), you eat fewer kcalories without eating less food.	■ High-fiber, low-fat foods in this plan can lower cholesterol and blood pressure.	■ So little fat that essential fatty acids may be lacking. ■ Excludes fish, nuts, and olive oil, which may protect against heart disease.
Pritikin Program	■ By eating low-fat, mainly plant-based foods, you can eat more food and still feel satisfied.	■ No food group is completely eliminated in this high-fiber, low-fat diet program. ■ Some use of foods rich in omega-3 fatty acids is encouraged.	■ For some people, very low-fat diets may be unsatisfying and therefore difficult to adhere to.
Low-Carbohydrate, High-Protein			
Atkins Diet	■ People are overweight or obese because they have metabolic imbalances caused by eating too many carbohydrates. ■ By restricting carbohydrates, these metabolic imbalances can be corrected. ■ You can lose weight without lowering kcalorie intake.	■ Quick, short-term weight loss is achieved.	■ Restricts carbohydrates to a level that induces ketosis. ■ Ketosis can cause nausea, light-headedness, and fatigue. ■ Ketosis can worsen existing medical problems such as kidney disease. ■ A diet high in fat such as Atkins can increase the risk of heart disease and some cancers.
Low-Carbohydrate			
Zone Diet	■ Eating the correct proportions of carbohydrate, fat, and protein leads to hormonal balance, weight loss, disease prevention, and increased vitality.	■ Promotes weight loss because it is a low-kcalorie diet.	■ The diet is rigid, restrictive, and complicated, making it difficult for most people to follow accurately. ■ The overblown health claims of the diet's proponents are based on misinterpreted science and remain unsubstantiated.

TABLE NP7–2 (continued)

Diet	Major Premise Promoted	Strong Point(s)	Weak Point(s)
Carbohydrate-Modified South Beach Diet	■ Eating "good carbohydrates" such as vegetables, whole-wheat pastas, and brown rice will maintain satiety and resist cravings for "bad carbohydrates" such as white rice and potatoes.	■ The diet encourages consumption of vegetables, lean meats, and fish and the use of mono-unsaturated oils when cooking. ■ Restricts fatty meats and cheeses as well as sweets. ■ The diet emphasizes mostly healthy foods.	■ Starchy carbohydrates and all fruits are completely excluded during the first two weeks. ■ The benefits of exercise for health are not emphasized.
The Ultimate Weight Solution Diet	■ Foods that require a great deal of effort to prepare and to eat are nutrient dense. ■ Eating these kinds of foods (raw vegetables, vegetable soup, whole grains, beans, meat, poultry, and fish) will lead to weight loss. ■ Foods that take little effort to prepare and to eat provide excess kcalories relative to nutrients. ■ Eating these kinds of foods (fast foods, puddings, high-kcalorie convenience foods, easy-to-prepare processed foods) leads to uncontrolled eating and weight gain.	■ The diet encourages consumption of lean meats and fish, whole grains, vegetables, fruit, and low-fat milk, yogurt, and cheese. ■ Restricts fatty meats and cheeses as well as sweets. ■ The diet emphasizes mostly healthy foods. Exercise is encouraged.	■ Confusing as to exactly what to eat or how much.
Metabolic Type Eat Right 4 Your Type	■ Your blood type determines which foods you should eat or not eat.	■ None	■ Food groups or individual foods are excluded, depending on blood type. ■ No scientific data on the relationship between blood type and food choices.

recommendations and guidelines such as the USDA Food Guide and the *Dietary Guidelines for Americans 2005*.

Thus, many of the most popular diets today, regardless of what their names are, espouse a low-carbohydrate diet. Most low-carbohydrate diets by design are relatively high in protein.

Low-carbohydrate meals overemphasize meat, fish, poultry, eggs, and cheeses and shun breads, pastas, fruits, and vegetables.

Some of the diets severely restrict carbohydrate while emphasizing protein (low-carbohydrate, high-protein), whereas others focus more on replacing certain "bad" carbohydrates with "good" carbohydrates (carbohydrate-modified), while still restricting overall carbohydrate intake to some degree.

What do you mean by "bad" carbohydrates and "good" carbohydrates?

This is a trend in fad diet books, and in short, here are the claims:

■ "Bad" carbohydrates, such as sugar, white flour, and potatoes, cause a rapid rise in blood sugar which raises blood insulin concentrations.

■ High blood insulin concentrations make people gain weight because insulin promotes fat storage.

■ People gain weight because high blood insulin concentrations lower blood sugar so much that people become hungry and so eat more food.

■ "Good" carbohydrates such as whole grains, vegetables, and beans cause a slow rise in blood sugar and a moderate insulin response and do not promote weight gain or hunger.

Now, here is what research reveals so far:

- The "good" carb versus "bad" carb diet premise is based on the glycemic effect of foods discussed in Chapters 2 and 21. The glycemic effect of a food depends on how the food is ripened, processed, and cooked; the time of day the food is eaten; the other foods eaten with it; and the presence or absence of certain diseases such as type 2 diabetes in the person eating the food.[1] Thus the glycemic effect of a particular food varies—fad diet books mislead people by claiming that each food has a set glycemic effect.

- All whole-grain foods do not have a low glycemic effect, and all refined grains do not have a high glycemic effect. For example, pasta has a low glycemic effect whether it is whole wheat or white, but thin pasta has a higher glycemic effect than thick pasta. Thus fad diet books mislead consumers in this respect as well.

- There is no clear evidence that higher blood insulin concentrations enhance food intake or weight gain in healthy people.[2]

- Evidence that foods with a low glycemic effect promote weight loss is lacking. A review of the evidence thus far concludes that the ideal long-term study has not yet been conducted.[3]

Are the diets that promote "good" carbs versus "bad" carbs dangerous to health?

In general, no. In fact, most of the foods that such diets promote are healthy foods—lean meats, fat-free or low-fat milk and yogurt, vegetables, whole grains, beans, and fruit. The *Dietary Guidelines for Americans 2005* encourage consumers to eat the same foods. The *Dietary Guidelines* also advise consumers to eat less saturated fat, however, and some low-carbohydrate diets can be high in saturated fat. Like some of the carbohydrate-modified diets, the *Dietary Guidelines*, also encourage people to eat a diet high in fiber-rich carbohydrate foods, rather than one that is restricted in such foods. The diet books claim that people lose weight because they switch from eating some types of carbohydrate foods to eating others, when, in truth, people lose weight on the diets because they are eating fewer kcalories.

Some of the diets exclude or restrict healthful foods such as carrots and bananas because the importance of the glycemic effect of a single food is inflated. Worse, because some foods are restricted or excluded, fad diet books often encourage readers to purchase special supplements, "energy bars," or other products to go along with the diets. Nevertheless, carbohydrate-modified diets such as the South Beach Diet or Dr. Phil's Ultimate Weight Solution Diet do include mostly healthy foods.

What is the appeal of very low-carbohydrate, high-protein diets?

Probably the greatest appeal of a very low-carbohydrate, high-protein diet such as the Atkins Diet is that it turns current diet recommendations upside down. Foods such as meats and milk products that need to be selected carefully to limit saturated fat can now be eaten with abandon. Grains, legumes, vegetables, and fruits that people are told to eat in abundance can now be ignored. For some people, this is a dream come true: steaks without the potatoes, ribs without the coleslaw, and meatballs without the pasta. Who can resist the promise of weight loss while eating freely from a list of favorite foods?

I have heard several people say that they successfully lose weight on these diets. Is this true?

If low-carbohydrate, high-protein diets were as successful as some people claim, then consumers who tried them would lose lots of weight, and our obesity problems would be solved. Obviously, this is not the case. Similarly, if high-protein diets were as worthless as others claim, then consumers would eventually stop pursuing them. Clearly, this is not happening either. These diets have enough going for them that they work for some people at least for a short time, but they fail to produce long-lasting results for most people. Studies report that people following low-carbohydrate, high-protein diets do lose weight.[4] In fact, they lose more than people following conventional high-carbohydrate, low-fat diets—but only for the first six months. Their later gains make up the difference, so total weight loss is no different after one year.[5]

People who have followed very low-carbohydrate, high-protein diet plans for several months have lost weight—at least temporarily. But can these diets also be harmful?

Diets that overemphasize protein and fall short on carbohydrate may not harm healthy people if used for only a little while, but they cannot support optimal health for long. Some fad diets focus so intently on promoting protein and curbing carbohydrate that they fail to account for the fat that accompanies many high-protein foods. A breakfast of bacon and eggs, lunch of ham and cheese, and dinner of barbecued short ribs would provide 100 grams of protein—and 121 grams of fat! Yet this day's meals, even with a snack of peanuts, provide only 1,600 kcalories. Without careful selection, protein-rich diets can be extraordinarily high in saturated fat and cholesterol—dietary factors that raise LDL cholesterol and the risks for heart disease.

Overall, studies report that people following low-carbohydrate, high-protein diets have little or no change in blood pressure or blood lipids—risk factors for heart disease.[6] Some researchers speculate that the weight loss that occurs on these diets offsets the adverse effects of a diet high in saturated fat and low in fruits and vegetables.[7] Others point out that different sources of protein have different effects on risk factors for heart disease.[8] For example, the effects of white meat from chicken or fish differ from those of red meat. Diets containing large amounts of red meat appear to increase the risk of heart disease. In contrast, replacing animal sources of protein with plant sources of protein may benefit health.

What are the other drawbacks to low-carbohydrate, high-protein diets?

The quality of the diet suffers when carbohydrates are restricted.[9] Without fruits, vegetables, and whole grains, high-protein diets lack not only carbohydrate but fiber, vitamins, minerals, and phytochemicals as well—all dietary factors protective against disease. To help shore up some of these inadequacies, fad diets often recommend a daily supplement. Conveniently, many of the companies selling fad diets also peddle these supplements. But, as Nutrition in Practice 9 explains, foods offer many more health benefits than any supplement can provide. Quite simply, if the diet is inadequate, it needs to be improved, not supplemented.

Is it true that people do not get hungry on low-carbohydrate, high-protein diets, and if so, is there any scientific basis for this?

Of the three energy-yielding nutrients, protein is the most satiating. Consequently, high-protein meals may suppress hunger and delay the start of the next meal. Furthermore, studies have reported that people tend to eat less after a high-protein meal than after a low-protein meal.[10] In one study, when protein intake increased from 15 percent of total energy to 30 percent but carbohydrate was held constant at 50 percent of total energy, people decreased their energy intakes and lost body weight and body fat.[11] The researchers suggest that less emphasis should be placed on carbohydrate restriction, without regard for the accompanying increase in fat intake that often occurs with low-carbohydrate, high-protein diets. Other research shows that when *fat* is held to 30 percent of total energy and protein is either 12 percent or 25 percent of total energy, greater weight loss and greater loss of intra-abdominal fat occur on the higher-protein diet.[12] In real-life situations, there is a strong association between a person's protein intake and BMI—the higher the intake, the higher the BMI.[13] This association remains apparent even after adjusting for energy intake and physical activity.

Thus, even though guidelines from the DRI committee include higher protein intakes (10 to 35 percent of total energy) than recommended previously, long-term studies of high-protein intakes are needed to ascertain the health consequences of such diets. One such study is under way: The DiOGenes (Diet, Obesity, and Genes) project is examining the interactions among a high dietary protein intake, the glycemic effect of foods, and genetic and behavioral factors in preventing weight gain and regain.[14] The study focuses on about 700 overweight or obese adults and their children in eight different countries across Europe and may involve the United States as well.

Is a person's metabolism altered while following a low-carbohydrate, high-protein diet?

When a person consumes a low-carbohydrate diet, a metabolism similar to that of fasting prevails (see Chapter 6 for a review of fasting). With little dietary carbohydrate coming in, the body uses its glycogen stores to provide glucose for the cells of the brain, nerves, and blood. Once the body depletes its glycogen reserves,

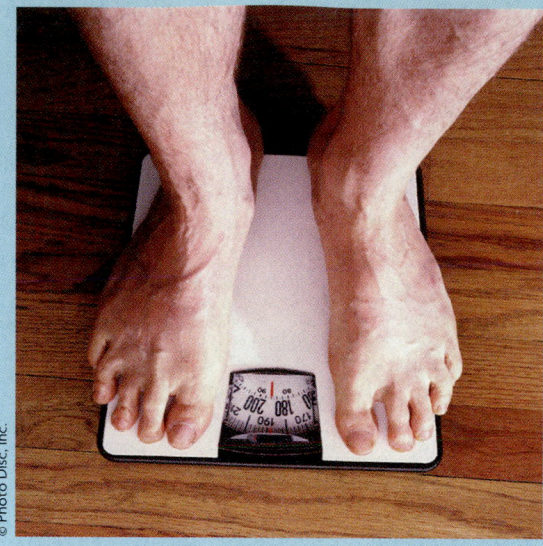

© Photo Disc, Inc.

The cautious consumer distinguishes between loss of fat and loss of weight.

it begins making glucose from the amino acids of protein. A low-carbohydrate diet may provide abundant protein from food, but the body still uses some protein from body tissues.

Low-carbohydrate diets also induce ketosis, and ketones can be detected in the urine. Ketones form whenever glucose is lacking and fat breakdown is incomplete. Many fad diets regard ketosis as the key to losing weight, but a study comparing weight-loss diets found no relation between ketosis and weight loss.[15] People in ketosis may experience a loss of appetite and a dramatic weight loss within the first few days. They would be disillusioned if they were aware that much of this weight loss reflects the loss of glycogen and protein together with large quantities of body fluids and important minerals.[16] They need to learn to appreciate the difference between loss of *fat* and loss of *weight*. Fat losses on ketogenic diets are no greater than on other diets providing the same number of kcalories. Once the dieter returns to well-balanced meals that provide adequate energy, carbohydrate, fat, protein, vitamins, and minerals, the body avidly retains these needed nutrients. The weight will return, quite often to a level higher than the starting point. Table NP7–3 lists other consequences of a ketogenic diet.

TABLE NP7–3
Adverse Side Effects of Low-Carbohydrate, Ketogenic Diets
Nausea
Fatigue (especially if physically active)
Constipation
Low blood pressure
Elevated uric acid (which may exacerbate kidney disease and cause inflammation of the joints in those predisposed to gout)
Stale, foul taste in the mouth (bad breath)
In pregnant women, fetal harm and stillbirth

Why are consumers so obsessed with diet books and products?

With over half of our nation's adults overweight and many more concerned about their weight, the market for a weight-loss book, product, or program is huge. Americans spend an estimated $33 billion per year on weight-loss books and products. Even a plan that offers only minimal weight-loss success easily attracts a following. Carbohydrate-modified and low-carbohydrate, high-protein diet plans offer a little success to some people for a short time. Here is why.

Who wants to count kcalories? Even experienced dieters find counting kcalories burdensome, not to mention timeworn. They want a new, easy way to lose weight, and carbohydrate-modified and low-carbohydrate, high-protein diet plans seem to offer this boon. But, though these diets often claim to disregard kcalories, their design typically ensures a low energy intake. Most of the sample menu plans provided by these diets, especially in the early stages, are designed to deliver an average of 1,200 kcalories per day.

Weight loss occurs because of the low energy intake—not the proportion of energy nutrients.[17] Success depends on the restricted intake, not on protein's magical powers or carbohydrate's evil forces. This is an important point. Any diet can produce weight loss, at least temporarily, if intake is restricted. The real value of a diet is determined by its ability to maintain weight loss and support good health over the long term. The goal is not simply weight loss, but health gains—and whether carbohydrate-modified or low-carbohydrate, high-protein diets can support optimal health over time remains unknown. Table NP7–4 offers guidelines for identifying fad diets and other weight-loss scams; it includes the hallmarks of a reasonable weight-loss program as well.

Many people seem confused about what kinds of dietary changes they need to make to lose weight. Perhaps diet books that tell them what to do are simply what they are looking for. Do you agree?

Yes, most people need specific instructions and examples to make dietary changes. Popular diets offer dieters a plan. The user does not have to decide what foods to eat, how to prepare them, or how much to eat. Unfortunately, these instructions serve short-term weight-loss needs only. They do not provide for long-term changes in lifestyle that will support weight maintenance or health goals.

Chapter 7 includes reasonable approaches to weight management and concludes that the ideal diet is one you can live with for the rest of your life. Keep that criterion in mind when you evaluate the next "latest and greatest weight-loss diet" that comes along.

TABLE NP7–4

Guidelines for Identifying Fad Diets and Other Weight-Loss Scams

1. They promise dramatic, rapid weight loss. Weight loss should be gradual and not exceed 2 pounds per week.

2. They promote diets that are nutritionally unbalanced or extremely low in kcalories. Diets should provide:
   - A reasonable number of kcalories (not fewer than 1,000 kcalories per day for women and 1,200 kcalories per day for men)
   - Enough, but not too much, protein (between the RDA and twice the RDA)
   - Enough, but not too much, fat (between 20 and 35 percent of daily energy intake from fat)
   - Enough carbohydrate to spare protein and prevent ketosis (at least 100 grams per day) and 20 to 30 grams of fiber from food sources
   - A balanced assortment of vitamins and minerals from a variety of foods from each of the food groups
   - At least 1 liter (about 1 quart) of water daily or 1 milliliter per kcalorie daily—whichever is more

3. They use liquid formulas rather than foods. Foods should accommodate a person's ethnic background, taste preferences, and financial means.

4. They attempt to make clients dependent upon special foods or devices. Programs should teach clients how to make good choices from the conventional food supply.

5. They fail to encourage permanent, realistic lifestyle changes. Programs should provide physical activity plans that involve spending at least 300 kcalories per day and behavior modification strategies that help to correct poor eating habits.

6. They misrepresent salespeople as "counselors" supposedly qualified to give guidance in nutrition and/or general health. Even if adequately trained, such "counselors" would still be objectionable because of the obvious conflict of interest that exists when providers profit directly from products they recommend and sell.

7. They collect large sums of money at the start or require that clients sign contracts for expensive, long-term programs. Programs should be reasonably priced and run on a pay-as-you-go basis.

8. They fail to inform clients of the risks associated with weight loss in general or the specific program being promoted. They should provide information about dropout rates, the long-term success of their clients, and possible side effects.

9. They promote unproven or spurious weight-loss aids such as human chorionic gonadotropin hormone (HCG), starch blockers, diuretics, sauna belts, body wraps, passive exercise, ear stapling, acupuncture, electric muscle-stimulating (EMS) devices, spirulina, amino acid supplements (e.g., arginine, ornithine), glucomannan, methylcellulose (a "bulking agent"), "unique" ingredients, and so forth.

10. They fail to provide for weight maintenance after the program ends.

Notes

1. R. Clemens and P. Pressman, Clinical value of glycemic index unclear, *Food Technology* 58 (2004): 18; F. X. Pi-Sunyer, Glycemic index and disease, *American Journal of Clinical Nutrition* 76 (2002): 290S–298S.

2. Pi-Sunyer, 2002.

3. R. Clemens and P. Pressman, Clinical value of glycemic index unclear, *Food Technology* 58 (2004): 18; M. A. Pereira and coauthors, Effects of a low-glycemic load diet on resting energy expenditure and heart disease risk factors during weight loss, *Journal of the American Medical Association* 292 (2004): 2482–2490; A. Raben, Should obese patients be counselled to follow a low-glycaemic index diet? No. *Obesity Reviews* 3 (2002): 245–256; D. B. Pawlak, C. B. Ebbeling, and D. S. Ludwig, Should obese patients be counselled to follow a low-glycaemic index diet? Yes, *Obesity Reviews* 3 (2002): 235–243.

4. A. Astrup, Atkins and other low-carbohydrate diets: Hoax or an effective tool for weight loss? *Lancet* 364 (2004): 897–899; E. C. Westman and coauthors, Effect of 6-month adherence to a very low carbohydrate diet program, *American Journal of Medicine* 113 (2002): 30–36.

5. L. Stern and coauthors, The effects of low-carbohydrate versus conventional weight loss diets in severely obese adults: One-year follow-up of a randomized trial, *Annals of Internal Medicine* 140 (2004): 778–785; F. F. Samaha and coauthors, A low-carbohydrate diet as compared with a low-fat diet in severe obesity, *New England Journal of Medicine* 348 (2003): 2074–2081; B. J. Brehm and coauthors, A randomized trial comparing a very low-carbohydrate diet and a calorie-restricted low fat diet on body weight and cardiovascular risk factors in healthy women, *Clinics in Endocrinology and Metabolism* 88 (2003): 1617–1623; G. D. Foster and coauthors, A randomized trial of a low-carbohydrate diet for obesity, *New England Journal of Medicine* 348 (2003): 2082–2090.

6. D. M. Bravata and coauthors, Efficacy and safety of low-carbohydrate diets, *Journal of the American Medical Association* 289 (2003): 1837–1850.

7. Foster and coauthors, 2003.

8. F. B. Hu, Protein, body weight, and cardiovascular health, *American Journal of Clinical Nutrition* 82 (2005): 242S–247S.

9. Atkins and other low-carbohydrate diets: Hoax or an effective tool for weight loss? *Lancet* 364 (2004): 16886–16889; E. T. Kennedy and coauthors, Popular diets: Correlation to health, nutrition, and obesity, *Journal of the American Dietetic Association* 101 (2001): 411–420.

10. A. Astrup, The satiating power of protein—A key to obesity prevention? *American Journal of Clinical Nutrition* 82 (2005): 1–2; D. S. Weigle and coauthors, A high-protein diet induces sustained reductions in appetite, ad libitum caloric intake, and body weight despite compensatory changes in diurnal plasma leptin and ghrelin concentrations, *American Journal of Clinical Nutrition* 82 (2005): 41–48.

11. Weigle and coauthors, 2005.

12. A. Due and coauthors, Effect of normal-fat diets, either medium or high in protein, on body weight in overweight subjects: A randomised 1-year trial, *International Journal of Obesity and Related Metabolic Disorders* 28 (2004): 1283–1290.

13. A. Trichopoulou and coauthors, Lipid, protein and carbohydrate intake in relation to body mass index, *European Journal of Clinical Nutrition* 56 (2002): 37–43.

14. W. H. M. Saris and A. Harper, DiOGenes: A multidisciplinary offensive focused on the obesity epidemic, *Obesity Reviews* 6 (2005): 175–176.

15. Foster and coauthors, 2003.

16. S. T. St. Jeor and coauthors, Dietary protein and weight reduction: A statement for healthcare professionals from the nutrition committee of the Council on Nutrition, Physical Activity, and Metabolism of the American Heart Association, *Circulation* 104 (2001): 1869–1874.

17. D. K. Layman and coauthors, A reduced ratio of dietary carbohydrate to protein improves body composition and blood lipid profiles during weight loss in adult women, *Journal of Nutrition* 133 (2003): 411–417; Bravata and coauthors, 2003; M. R. Freedman, J. King, and E. Kennedy, Popular diets: A scientific review, *Obesity Research* 9 (2001): 1S–5S.

8

The Vitamins

altrendo images/Getty Images

arlier chapters focused primarily on the energy-yielding nutrients—carbohydrate, fat, and protein. This chapter and Chapter 9 discuss the nutrients everyone thinks of when nutrition is mentioned—vitamins and minerals.

Vitamins—An Overview

Vitamins occur in foods in much smaller quantities than do the energy-yielding nutrients, ■ and they themselves contribute no energy to the body. Instead, they serve mostly as facilitators of body processes. They are a powerful group of substances, as their absence attests. Vitamin A deficiency can cause blindness, a lack of niacin can cause dementia, and a lack of vitamin D can retard bone growth. The consequences of deficiencies are so dire and the effects of restoring the needed nutrients so dramatic that people spend billions of dollars each year on vitamin supplements to cure many different ailments. Vitamins certainly contribute to sound nutritional health, but supplements do not cure all ills. Actually, a vitamin can cure only the disease caused by a deficiency of that vitamin. Vitamins' roles in supporting optimal health extend far beyond preventing deficiency diseases, however. Emerging evidence points to relationships between low intakes of vitamins and chronic diseases such as cancer and heart disease.

A child once defined vitamins as "what, if you don't eat, you get sick." The description is both insightful and accurate. A more prosaic definition is that vitamins are potent, essential, noncaloric, organic nutrients needed from foods in trace amounts to perform specific functions that promote growth, reproduction, and the maintenance of health and life. Two characteristics distinguish vitamins from energy nutrients:

1. Vitamins do not yield energy when broken down but assist the enzymes that release energy from carbohydrate, fat, and protein.
2. Vitamins are needed in much smaller amounts than the energy nutrients.

As the individual vitamins were discovered, they were named or given letters, numbers, or both. This led to the confusion that still exists today. This chapter uses the names shown in Table 8–1; alternative names are given in Tables 8–2 and 8–3, which appear later in the chapter.

Bioavailability The availability of vitamins from foods depends on two factors: the quantity provided by a food and the amount absorbed and used by the body (the vitamin's **bioavailability**). Researchers analyze foods to determine their vitamin contents and publish the results in tables of food composition such as Appendix A. Determining the bioavailability of a vitamin is more difficult because it depends on many factors, including:

- The efficiency of digestion
- A person's previous nutrient intake and nutrition status
- Other foods eaten at the same time
- The method of food preparation (raw or cooked, for example)
- The source of the nutrient (naturally occurring, synthetic, or fortified)

Experts consider these factors when estimating recommended intakes.

Precursors Some of the vitamins are available from foods in inactive forms known as **precursors**, or provitamins. Once inside the body, the precursor is converted to the

TABLE 8–1
Vitamin Names
Fat-Soluble Vitamins
Vitamin A
Vitamin D
Vitamin E
Vitamin K
Water-Soluble Vitamins
B vitamins
Thiamin
Riboflavin
Niacin
Pantothenic acid
Biotin
Vitamin B_6
Folate
Vitamin B_{12}
Vitamin C

active form of the vitamin. Thus, in measuring a person's vitamin intake, it is important to count both the amount of the actual vitamin and the potential amount available from its precursors.

Solubility Vitamins fall naturally into two classes—fat soluble and water soluble. The solubility of a vitamin confers on it many characteristics and determines how it is absorbed, transported, stored, and excreted. This discussion of vitamins begins with the fat-soluble vitamins.

IN SUMMARY

- Vitamins are essential, nonkcaloric nutrients that are needed in trace amounts in the diet to help facilitate body processes.

Fat-Soluble Vitamins

The fat-soluble vitamins—A, D, E, and K—usually occur together in the fats and oils of foods, and the body absorbs them in the same way it absorbs lipids. Therefore, any condition that interferes with fat absorption can precipitate a deficiency of the fat-soluble vitamins. Once absorbed, fat-soluble vitamins are stored in the liver and fatty tissues until the body needs them. They are not readily excreted, and unlike most of the water-soluble vitamins, they can build up to toxic concentrations. Excesses of vitamins A and D from supplements can reach toxic levels easily.

The capacity to store fat-soluble vitamins affords a person some flexibility in dietary intake. When blood concentrations begin to decline, the body can retrieve the vitamins from storage. Thus a person need not eat a day's allowance of each fat-soluble vitamin every day but need only make sure that over time, average daily intakes approximate recommended intakes. In contrast, most water-soluble vitamins must be consumed more regularly because the body does not store them to any great extent.

Vitamin A and Beta-Carotene

Vitamin A has the distinction of being the first fat-soluble vitamin to be recognized. Today, after almost a century of revelations, vitamin A and its plant-derived precursor, **beta-carotene**, continue to intrigue researchers with their diverse roles and profound effects on health.

Vitamin A is a versatile vitamin, with roles in gene expression, vision, cell differentiation (thereby maintaining the health of body linings and skin), immunity, and reproduction and growth.[1] Three different forms of vitamin A are active in the body: retinol, retinal, and retinoic acid. Each form of vitamin A performs specific tasks. Retinol supports reproduction and is the major transport and storage form of the vitamin; the cells convert retinol to retinal or retinoic acid as needed. Retinal is active in vision, and retinoic acid acts as a hormone, regulating cell differentiation, growth, and embryonic development. A special transport protein, **retinol-binding protein (RBP)**, picks up retinol from the liver where it is stored and carries it in the blood. ■

Vitamin A's Role in Gene Expression Vitamin A exerts considerable influence on body functions through its regulation of the activities of the genes.[2] Genes direct the synthesis of proteins, including enzymes, and enzymes perform the metabolic work of the tissues (see Chapter 4). Hence, factors that influence gene expression also affect the metabolic activities of the tissues and the health of the body. Hundreds of genes have been suggested as regulatory targets of the retinoic acid form of vitamin A.[3]

Researchers have long known that simply possessing the genetic equipment needed to make a particular protein does not guarantee that the protein will be produced, any

■ Measurement of the blood concentration of RBP is a sensitive test of vitamin A status.

vitamins: essential, nonkcaloric, organic nutrients needed in tiny amounts in the diet.

bioavailability: the rate and extent to which a nutrient is absorbed and used.

precursors: compounds that can be converted into other compounds; with regard to vitamins, compounds that can be converted into active vitamins; also known as **provitamins**.

vitamin A: a fat-soluble vitamin. Its three chemical forms are *retinol* (the alcohol form), *retinal* (the aldehyde form), and *retinoic acid* (the acid form).

beta-carotene: a vitamin A precursor made by plants and stored in human fat tissue; an orange pigment.

retinol-binding protein (RBP): the specific protein responsible for transporting retinol.

FIGURE 8–1

Vitamin A's Role in Vision

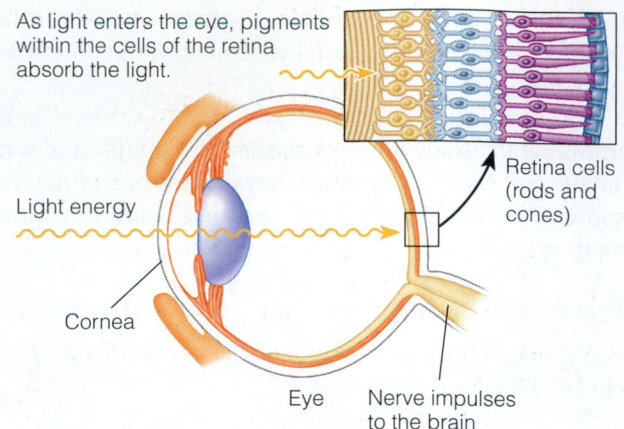

As light enters the eye, pigments within the cells of the retina absorb the light.

Retina cells (rods and cones)

Light energy

Cornea

Eye   Nerve impulses to the brain

more than owning a car guarantees you a trip across town. To get the car rolling, you must also use the right key to trigger the events that start up its engine or turn it off at the appropriate time. Some dietary components, including the retinoic acid form of vitamin A, are now known to be such keys—they help to activate or deactivate certain genes and thus affect the production of specific proteins.[4]

Vitamin A's Role in Vision Vitamin A plays two indispensable roles in the eye. It helps maintain a healthy, crystal-clear outer window, the **cornea**; and it participates in the events of light detection at the **retina**. Figure 8–1 shows vitamin A's site of action inside the eye.

When vitamin A is lacking, the eye has difficulty adapting to changing light levels. At night, after the eye has adapted to darkness, a lag occurs before the eye can see again after a flash of bright light. This lag in the recovery of night vision is known as **night blindness**. Either vitamin A-rich foods or supplements effectively treat night blindness.[5] Because night blindness is easy to test, it aids in the diagnosis of vitamin A deficiency. Night blindness is only a symptom, however, and may indicate a condition other than vitamin A deficiency.

Vitamin A's Role in Protein Synthesis and Cell Differentiation The role that vitamin A plays in vision is undeniably important, but only one-thousandth of the body's vitamin A is in the retina. Much more is in the skin and the linings of organs, where it works behind the scenes at the genetic level to promote protein synthesis and cell **differentiation**. The process of cell differentiation allows each type of cell to mature so that it is capable of performing a specific function.

All body surfaces, both inside and out, are covered by layers of cells known as **epithelial cells**. The **epithelial tissue** on the outside of the body is the skin. The epithelial tissues inside the body include the linings of the mouth, stomach, and intestines; the linings of the lungs and the passages leading to them; the lining of the bladder; the linings of the uterus and vagina; and the linings of the eyelids and sinus passageways. The epithelial tissues on the inside of the body must be kept smooth. To ensure that they are, the epithelial cells on their surfaces secrete a smooth, slippery substance (mucus) that coats the tissues and protects them from invasive microorganisms and other harmful particles. The **mucous membrane** that lines the stomach also shields its cells from digestion by gastric juices. Vitamin A, by way of its role in cell differentiation, helps to maintain the integrity of the epithelial cells.

Vitamin A's Role in Immunity Vitamin A's role in protecting the body's immune response was recognized more than 70 years ago when researchers noticed that vitamin A deficiency reduces resistance to infection.[6] Immune function depends on the

cornea (KOR-nee-uh): the hard, transparent membrane covering the outside of the eye.

retina (RET-in-uh): the layer of light-sensitive nerve cells lining the back of the inside of the eye; consists of rods and cones.

night blindness: the slow recovery of vision after exposure to flashes of bright light at night; an early symptom of vitamin A deficiency.

differentiation: the development of specific functions different from those of the original.

epithelial (ep-i-THEE-lee-ul) **cells**: cells on the surface of the skin and mucous membranes.

epithelial tissue: tissue composing the layers of the body that serve as selective barriers between the body's interior and the environment (examples are the cornea, skin, respiratory lining, and lining of the digestive tract).

mucous membrane: membrane composed of mucus-secreting cells that lines the surfaces of body tissues. (Reminder: *Mucus* is the smooth, slippery substance secreted by these cells.)

growth, differentiation, and activation of the cells that defend the body against infectious agents.[7]

Vitamin A's Role in Reproduction, Growth, and Development Vitamin A is crucial to normal reproduction and growth. In men, vitamin A participates in sperm development, and in women, vitamin A promotes normal fetal growth and development.[8] During pregnancy, vitamin A is transferred to the fetus and is essential to the development of the nervous system, lungs, heart, kidneys, skeleton, eyes, and ears.[9] ■

Beta-Carotene's Role as an Antioxidant For many years scientists believed beta-carotene to be of interest solely as a vitamin A precursor. Eventually, researchers began to recognize that beta-carotene is an extremely effective **antioxidant** in the body. Antioxidants are compounds that protect other compounds (such as lipids in cell membranes) from attack by oxygen. Oxygen triggers the formation of compounds known as **free radicals** that can start chain reactions in cell membranes. If left uncontrolled, these chain reactions can damage cell structures and impair cell functions. Oxidative and free-radical damage to cells is suspected of instigating some early stages of cancer and heart disease. Research has identified links between oxidative damage and the development of many other diseases, including age-related blindness, Alzheimer's disease, arthritis, cataracts, diabetes, and kidney disease.[10]

Studies of populations suggest that people whose diets are low in foods rich in beta-carotene have higher incidences of certain types of cancer than those whose diets contain generous amounts of such foods. Based on findings that beta-carotene in foods may protect against cancer, researchers designed a study to determine the effects of beta-carotene *supplements* on the incidence of lung cancer among smokers. The researchers expected to see a beneficial effect, but instead they found that smokers taking the beta-carotene supplements suffered a *greater* incidence of lung cancer than those taking placebos. ■ In a different study, the Physicians' Health Study, researchers tested the effects of beta-carotene supplements versus a placebo on cancer and heart disease risk in more than 20,000 male physicians for 12 years. Beta-carotene neither decreased nor increased cancer or heart disease risk. Beta-carotene is just one of many nutrients and compounds present in foods. As discussed in Nutrition in Practice 8, many other **phytochemicals** are also present in foods and may be responsible for some of the protective effects attributed to beta-carotene. Until more is known, eating beta-carotene–rich foods, not supplements, is in the best interests of health.[11] Based on research so far, the DRI committee has not established a recommended intake value for beta-carotene.

Vitamin A Deficiency Up to a year's supply of vitamin A can be stored in the body, 90 percent of it in the liver. If a healthy adult were to stop eating vitamin A-rich foods, deficiency symptoms would not begin to appear until after stores were depleted, which would take one to two years. Then, the consequences would be profound and severe. Table 8–2, later in this chapter, lists some of them.

In vitamin A deficiency, cell differentiation and maturation are impaired. The epithelial cells flatten and begin to produce **keratin**—the hard, inflexible protein of hair and nails. In the eye, this process leads to drying and hardening of the cornea, which may progress to permanent blindness. ■ ■ Vitamin A deficiency is the major cause of childhood blindness in the world, causing as many as one-half million children to lose their sight every year.[12] More than 200 million children worldwide endure less severe forms of vitamin A deficiency, making them vulnerable to infectious diseases.

All body surfaces, both inside and out, maintain their integrity with the help of vitamin A. When vitamin A is lacking, cells of the skin harden and flatten, making it dry, rough, scaly, and hard. An accumulation of keratin makes a lump around each hair **follicle** (keratinization). ■

■ Nutrition during pregnancy is discussed in Chapter 11.

■ A placebo is an inactive substance used in research studies.

■ The progressive blindness caused by vitamin A deficiency is called *xerophthalmia* (zer-off-THAL-mee-uh).
  xero = dry
  ophthalm = eye

■ An early sign of xerophthalmia is *xerosis* (drying of the cornea); the last and most severe stage is *keratomalacia* (kerr-uh-to-mal-AY-shuh), or total blindness.
  malacia = softening, weakening

■ The accumulation of the hard material keratin around each hair follicle is *follicular hyperkeratosis*.

antioxidant (anti-OX-ih-dant): a compound that protects other compounds from oxygen by itself reacting with oxygen. *Oxidation* is a potentially damaging effect of normal cell chemistry involving oxygen.
  anti = against
  oxy = oxygen

free radicals: highly reactive chemical forms that can cause destructive changes in nearby compounds, sometimes setting up a chain reaction.

phytochemicals: nonnutrient compounds in plant-derived foods that have biological activity in the body.

keratin (KERR-uh-tin): a water-insoluble protein; the normal protein of hair and nails. Keratin-producing cells may replace mucus-producing cells in vitamin A deficiency.

follicle (FOLL-i-cul): a group of cells in the skin from which a hair grows.

In the mouth, a vitamin A deficiency results in drying and hardening of the salivary glands, making them susceptible to infection. Secretions of mucus in the stomach and intestines are reduced, hindering normal digestion and absorption of nutrients. Infections of other mucous membranes also become likely.

Vitamin A's role in maintaining the body's defensive barriers may partially explain the relationship between vitamin A deficiency and susceptibility to infection.[13] In several studies, when children with measles complicated by diarrhea, infections such as pneumonia, or both were given vitamin A supplements, their overall survival rates were significantly higher than those of similar children who did not receive vitamin A.

The evidence that vitamin A reduces the severity of measles and measles-related infections and diarrhea has prompted the World Health Organization (WHO) and UNICEF (United Nations International Children's Emergency Fund) to make control of vitamin A deficiency a major goal in their quest to improve child survival throughout the developing world. The American Academy of Pediatrics recommends vitamin A supplementation for certain groups of measles-infected infants and children in the United States.

Vitamin A Toxicity Vitamin A toxicity is a real possibility when people consume concentrated amounts of **preformed vitamin A** in foods derived from animals, fortified foods, or supplements.[14] Plant foods contain the vitamin only as beta-carotene, its inactive precursor form. The precursor does not convert to active vitamin A rapidly enough to cause toxicity.

Children are most vulnerable to vitamin A toxicity because, being smaller, they need less than adults, and it is easy to give them too much in pill form. The availability of breakfast cereals, instant meals, fortified milk, and chewable candylike vitamins, each containing 100 percent or more of the recommended daily intake of vitamin A, makes it possible for a well-meaning parent to provide several times the daily allowance of the vitamin to a child within a few hours.[15] Serious toxicity is seen in infants and young children when they are given more than ten times the recommended amount every day for weeks at a time.

Excessive vitamin A also poses a **teratogenic** risk.[16] In pregnant women, chronic use of vitamin A supplements providing three to four times the amount recommended for pregnancy can cause malformations of the fetus. The Tolerable Upper Intake Level of 3,000 micrograms for women of childbearing age is based on the teratogenic effect of vitamin A.[17]

Excessive amounts of vitamin A over the years may weaken the bones and contribute to osteoporosis (see the later discussion of vitamin D for more on osteoporosis).[18] In some studies, people consuming large amounts of vitamin A either from supplements or from foods containing retinol have a significantly greater risk of hip fractures.[19] More research is needed to clarify the relationship between vitamin A intake and bone health, but such findings suggest that most people should not take vitamin A supplements. Even multivitamin supplements provide more vitamin A than most people need. ■

Certain vitamin A relatives are available by prescription as acne treatments. When applied directly to the skin surface, these preparations help relieve the symptoms of acne. Taking massive doses of vitamin A internally will *not* cure acne, however, and may cause the symptoms itemized in Table 8–2. In most cases, foods are a better choice than supplements for needed nutrients. The best way to ensure a safe vitamin A intake is to eat generous servings of vitamin A-rich foods.

Beta-Carotene Conversion and Toxicity Nutrition scientists do not use micrograms to specify the quantity of beta-carotene in foods. Instead, they use a value known as **retinol activity equivalents (RAE)**, ■ which express the amount of retinol the body actually derives from a plant food after conversion. The body can make one unit of retinol from about 12 units of beta-carotene.

■ Multivitamin supplements typically provide:
- 750 μg (2,500 IU)
- 1,500 μg (5,000 IU)

For perspective, the RDA for vitamin A is 700 μg for women and 900 μg for men. The IU (*international unit*) was used earlier to express the amount of vitamins in foods and is still used on supplement labels.

■ 1 μg RAE = 1 μg retinol
= 12 μg beta-carotene from food

preformed vitamin A: vitamin A in its active form.

teratogenic (ter-AT-oh-jen-ik): causing abnormal fetal development and birth defects.
terato = monster
genic = to produce

retinol activity equivalents (RAE): a measure of vitamin A activity; the amount of retinol that the body will derive from a food containing preformed retinol or its precursor beta-carotene.

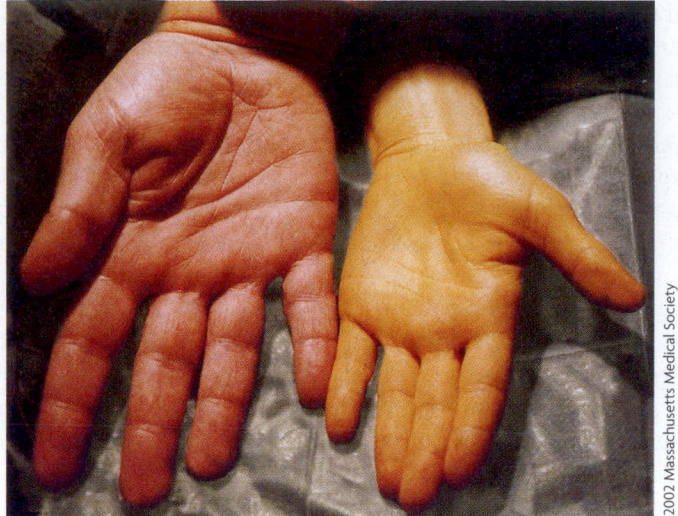

FIGURE 8–2

Symptom of Beta-Carotene Excess—Discoloration of the Skin
The hand on the right shows the skin discoloration that occurs from excess beta-carotene. (The hand on the left belongs to someone else and is shown for comparison.)

© 2002 Massachusetts Medical Society

As mentioned earlier, beta-carotene from plant foods is not converted to the active form of vitamin A rapidly enough to be hazardous. It has, however, been known to turn people bright orange-yellow if they eat too much. ▪ Beta-carotene builds up in the fat just beneath the skin and imparts a yellowish cast (see Figure 8–2).

Vitamin A in Foods Preformed vitamin A is found only in foods of animal origin. The richest sources of vitamin A are liver and fish oil, but milk, cheese, and fortified cereals are also good sources. Healthy people can eat vitamin A-rich foods in large amounts without risking toxicity with the possible exception of liver. Eating liver once every week or so is enough. Butter and eggs also provide some vitamin A to the diet.

Because vitamin A is fat soluble, it is lost when milk is skimmed. Fat-free milk is thus often fortified with vitamin A to compensate. Margarine is also usually fortified so as to provide the same amount of vitamin A as butter. Snapshot 8–1 shows a sampling of the richest food sources of both preformed vitamin A and beta-carotene.

▪ Yellowing of the skin caused by excess carotene in the blood is known as *carotenemia* (KAR-oh-teh-NEE-me-ah). Carotenemia can be distinguished from jaundice because the mucous membranes lining the eyelids do not turn yellow as they do in jaundice.

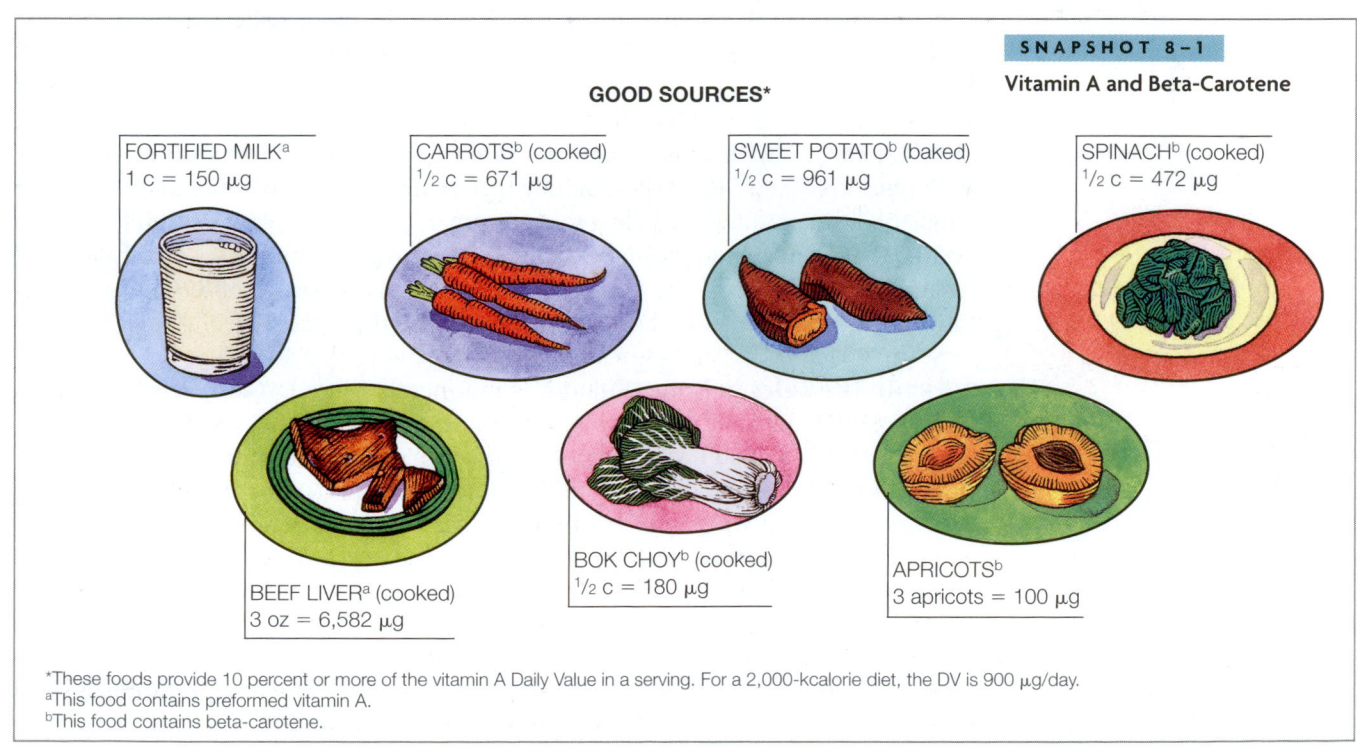

Vitamin A and Beta-Carotene

GOOD SOURCES*

FORTIFIED MILK[a]
1 c = 150 µg

CARROTS[b] (cooked)
$^1/_2$ c = 671 µg

SWEET POTATO[b] (baked)
$^1/_2$ c = 961 µg

SPINACH[b] (cooked)
$^1/_2$ c = 472 µg

BEEF LIVER[a] (cooked)
3 oz = 6,582 µg

BOK CHOY[b] (cooked)
$^1/_2$ c = 180 µg

APRICOTS[b]
3 apricots = 100 µg

*These foods provide 10 percent or more of the vitamin A Daily Value in a serving. For a 2,000-kcalorie diet, the DV is 900 µg/day.
[a]This food contains preformed vitamin A.
[b]This food contains beta-carotene.

Sunlight promotes vitamin D synthesis in the skin. Exposure to the sun should be moderate, however; excessive exposure may cause skin cancer.

■ The precursor of vitamin D made in the liver is 7-dehydroc-holesterol, which is made from cholesterol. This is one of the body's many "good" uses for cholesterol.

■ The final, active vitamin is 1-25 dihydroxycholecalciferol or, more simply, dihydroxy vitamin D.

■ Chapter 9 offers a detailed discussion on minerals.

Fast-food meals often lack vitamin A. When fast-food restaurants offer salads with cheese, carrots, and other vitamin A-rich foods, the nutritional quality of their meals greatly improves.

Beta-Carotene in Foods Many foods from plants contain beta-carotene, the orange pigment responsible for the bold colors of many fruits and vegetables. Carrots, sweet potatoes, pumpkins, cantaloupe, and apricots are all rich sources, and their bright orange color enhances the eye appeal of the plate. Another colorful group, *dark* green vegetables, such as spinach, other greens, and broccoli, owe their color to both chlorophyll and beta-carotene. The orange and green pigments together impart a deep, murky green color to the vegetables. Other colorful vegetables, such as iceberg lettuce, beets, and sweet corn, can fool you into thinking they contain beta-carotene, but these foods derive their color from other pigments and are poor sources of beta-carotene. As for "white" plant foods such as rice and potatoes, they have little or none. Recommendations to eat *dark* green or *deep* orange vegetables and fruits at least every other day help people to meet their vitamin A needs.

Vitamin D

Vitamin D is different from all the other nutrients in that the body can synthesize it in significant quantities with the help of sunlight. Therefore, in a sense, vitamin D is not an essential nutrient. Given enough sun, people need no vitamin D from foods.

Vitamin D's Metabolic Conversions The liver manufactures a vitamin D precursor, which migrates to the skin where it is converted to a second precursor with the help of the sun's ultraviolet rays. Next, the liver and then the kidneys alter the second precursor to produce the active vitamin. ■ Vitamin D precursors from plants require the same two conversions by the liver and kidneys to become active. The biological activity of the active vitamin is 500- to 1,000-fold greater than that of its precursor. Diseases that affect either the liver or the kidneys may impair the transformations of precursor vitamin D to active vitamin D and therefore produce symptoms of vitamin D deficiency.

Vitamin D's Actions Although known as a vitamin, vitamin D is actually a hormone—a compound manufactured by one organ of the body that has effects on another. The best-known vitamin D target organs are the small intestine, kidneys, and bones, but scientists have discovered many other vitamin D target tissues, including the brain, pancreas, skin, some cancer cells, and the reproductive organs.[20] These discoveries suggest that numerous additional functions for vitamin D may surface, including regulation of the immune system.[21] Research is hinting that to incur a deficit of vitamin D is to invite problems of many kinds, including some common forms of cancers, infectious diseases, heart disease, high blood pressure, rheumatoid arthritis, inflammatory conditions, and even multiple sclerosis.[22] The well-established problems, however, concern impaired calcium balance and the bones, both during growth and throughout life.

Vitamin D's Roles in Bone Vitamin D is a member of a large, cooperative bone-making and maintenance team composed of nutrients and other compounds, including vitamins A, C, and K; the hormones parathormone (parathyroid hormone) and calcitonin; the protein collagen; and the minerals calcium, phosphorus, magnesium, and fluoride. Many of these interactions take place at the genetic level in ways that are under investigation.[23] Vitamin D's special role in bone growth is to make calcium and phosphorus available in the blood that bathes the bones. ■ The bones grow denser and stronger as the minerals are deposited from the blood. Vitamin D acts in three ways to maintain blood concentrations of calcium and phosphorus: it stimulates their absorption from the gastrointestinal (GI) tract; it mobilizes calcium and phosphorus from bones into the blood; and it stimulates their retention by the kidneys.

Vitamin D Deficiency The symptoms of vitamin D deficiency are those of calcium deficiency. The bones fail to calcify normally and may grow so weak that they become bent when they have to support the body's weight. A child with the vitamin D-deficiency disease **rickets** who is old enough to walk characteristically develops bowed legs, often the most obvious sign of the disease (see Figure 8–3). Rickets was a major pediatric health problem in the United States before vitamin D fortification of commercially prepared milk was introduced decades ago. Unfortunately, rickets seems to be making a comeback among African-American breastfed infants who are exposed to little sunlight and receive no vitamin D supplements.[24] Worldwide, rickets afflicts a large number of children who receive inadequate food and little exposure to sunlight. The adult form of rickets is called **osteomalacia**.

Adolescents, who often abandon vitamin D-fortified milk in favor of soft drinks, may also prefer indoor pastimes such as computer games to outdoor activities during daylight hours. Such teens often lack vitamin D and so fail to develop the bone density needed to prevent bone loss in later life.[25]

Inadequate vitamin D is recognized as a risk factor in **osteoporosis** (reduced bone density). Without sufficient vitamin D, absorption of calcium is limited, and bone remodeling is impaired. This combination leads to a loss of bone mass.

Vitamin D Toxicity Whereas vitamin D deficiency depresses calcium absorption, blood calcium, and bone mineralization, an excess of vitamin D does the opposite, as shown in Table 8–2. It enhances calcium absorption, produces high blood calcium, and promotes return of bone calcium into the blood. The excess calcium then tends to precipitate in the soft tissues, forming stones, including kidney stones. Calcification may also harden the blood vessels and is especially dangerous in the major arteries of the heart and lungs, where it can cause death.

Vitamin D in excess is the most toxic of all the vitamins. The amounts of vitamin D in foods available in the United States and Canada are well within safe limits, but supplements containing the vitamin in concentrated form are not. Adults should use caution when taking vitamin D supplements and keep them out of the reach of children. The DRI committee has set a Tolerable Upper Intake Level for vitamin D at 50 micrograms per day (2,000 IU on supplement labels). ■

Vitamin D from the Sun Most of the world's population relies on natural exposure to sunlight to maintain adequate vitamin D nutrition. The sun imposes no risk of vitamin D toxicity. Prolonged exposure to sunlight degrades the vitamin D precursor in the skin, preventing its conversion to the active vitamin. Even lifeguards on southern beaches are safe from vitamin D toxicity from the sun.

Prolonged exposure to sunlight has other undesirable consequences, however, such as premature wrinkling of the skin and the risk of skin cancer. These risks may be reduced by using sunscreens. Unfortunately, sunscreens with sun protection factors (SPF) of 8 and above also retard vitamin D synthesis.[26] A strategy to avoid this dilemma is to apply sunscreen after enough time has elapsed to provide sufficient vitamin D. For most people, exposing hands, face, and arms on a clear summer day for 10 minutes, a few times per week, should be sufficient to maintain vitamin D nutrition.[27] Dark-skinned people require longer exposure than light-skinned people, but by three hours, vitamin D synthesis in heavily pigmented skin arrives at the same plateau as in fair skin after 30 minutes.

The ultraviolet rays from tanning lamps and tanning booths may also stimulate vitamin D synthesis, but the hazards outweigh any possible benefits. The Food and Drug Administration (FDA) warns that if the lamps are not properly filtered, people using tanning booths risk burns, damage to the eyes and blood vessels, and skin cancer.

The ultraviolet rays of the sun that promote vitamin D synthesis cannot penetrate clouds, smoke, smog, heavy clothing, window glass, or even window screens. In the

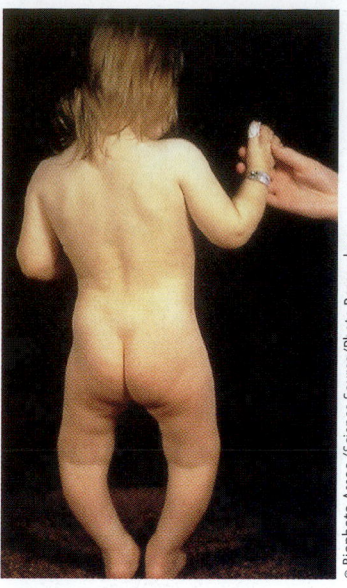

© Biophoto Assoc./Science Source/Photo Researchers

FIGURE 8–3

Rickets
This child has the bowed legs of the vitamin D-deficiency disease rickets.

■ Vitamin D activity was previously expressed in international units (IU), but as of 1980, it is expressed in micrograms of cholecalciferol, the active form of vitamin D. To convert, use the following factor:

- 100 IU = 2.5 µg
- 400 IU = 10 µg

rickets: the vitamin D-deficiency disease in children.

osteomalacia (os-tee-oh-mal-AY-shuh): a bone disease characterized by softening of the bones. Symptoms include bending of the spine and bowing of the legs. The disease occurs most often in adults with renal failure or malabsorption disorders.
  osteo = bone
  mal = bad (soft)

osteoporosis (os-tee-oh-pore-OH-sis): literally, porous bones; reduced density of the bones, also known as *adult bone loss*.

United States and Canada, dark-skinned people who live in smoggy northern cities or who lack exposure to sunlight are at greatest risk of vitamin D deficiency.[28] A surprisingly high number of otherwise healthy northern adults have been found to have low blood levels of vitamin D, most often at the end of the winter season, even though they were drinking milk fortified with vitamin D.[29] It may be that milk does not deliver enough vitamin D to prevent a drop in blood levels through the winter months. In addition, people who are housebound or institutionalized and those who work at night may incur (over years) a vitamin D deficiency, as do many elderly people, who have limited exposure to sunlight and become less efficient at activating vitamin D as they age. For these people, dietary vitamin D is essential.[30] Because of increased risk with age, the DRI committee set Adequate Intakes (AI) for vitamin D that increase over the years (see AIs in the margin). ■

- Vitamin D AI:
  - Adults (19–50 yr): 5 µg/day
  - Adults (51–70 yr): 10 µg/day
  - Adults (> 70 yr): 15 µg/day

Vitamin D in Foods   Only a few animal foods, notably eggs, liver, butter, some fatty fish, and fortified milk, supply significant amounts of vitamin D. For those who use margarine in place of butter, fortified margarine is a significant source. Infant formulas are fortified with vitamin D in amounts adequate for daily intake as long as infants consume at least 500 milliliters (15 ounces) of formula. Breast milk is low in vitamin D, so vitamin D supplements (5 micrograms daily) are recommended for infants who are breastfed exclusively and for those who do not receive at least 500 milliliters of vitamin D-fortified formula per day.[31] These sources, plus any exposure to the sun, provide infants with more than enough of this vitamin.

The fortification of milk with vitamin D is the best guarantee that children will meet their vitamin D needs and underscores the importance of milk in children's diets. Vitamin D supplements (5 micrograms daily) are recommended for children and adolescents who do not drink at least 2 cups of milk per day, are not regularly exposed to sunlight, or do not take a multivitamin supplement containing at least 5 micrograms of vitamin D.[32] Unlike milk, cheese and yogurt are not fortified with vitamin D. Vegans, and especially their children, may have low vitamin D intakes because few fortified plant sources exist. Exceptions include margarine and some soy milks. In the United States, breakfast cereals may be fortified with vitamin D, as their labels indicate.

Vitamin E

More than 80 years ago, researchers discovered a compound in vegetable oils necessary for reproduction in rats. The compound was named **tocopherol**, which means "offspring." Eventually, the compound was named vitamin E. When chemists isolated four tocopherol compounds, they designated them by the first four letters of the Greek alphabet: alpha, beta, gamma, and delta. Of these, alpha-tocopherol is the gold standard for vitamin E activity; recommended intakes are based on it. Table 8–2 later in the chapter summarizes important information about vitamin E.

Vitamin E as an Antioxidant   Like beta-carotene, vitamin E is a fat-soluble antioxidant. It protects other substances from oxidation ■ by being oxidized itself. If there is plenty of vitamin E in the membranes of cells exposed to an oxidant, chances are this vitamin will take the brunt of any oxidative attack, protecting the lipids and other vulnerable components of the membranes. Vitamin E is especially effective in preventing the oxidation of polyunsaturated fatty acids (PUFA), but it protects all other lipids (for example, vitamin A) as well.

■ Reminder: *Oxidation* is a type of chemical reaction, so named because oxygen is one of the agents that often brings it about.

Vitamin E exerts an especially important antioxidant effect in the lungs, where the cells are exposed to high concentrations of oxygen. Vitamin E also protects the lungs from air pollutants that are strong oxidants.

Some evidence suggests that vitamin E may also offer protection against heart disease by protecting LDL (low-density lipoproteins) from oxidation.[33] The oxidation of LDL encourages the development of atherosclerosis. Vitamin E may also benefit heart health by inhibiting several other processes involved in the development of heart disease.[34] Despite this theoretical understanding of how vitamin E may exert protection against heart disease, the results of controlled clinical studies in which human beings were given vitamin E supplements have been disappointing. Two extensive long-term studies, one with healthy people and one with people with vascular disease or diabetes, showed no benefit of vitamin E supplementation to heart health.[35] One review of vitamin E supplementation and heart disease concluded that in people with chronic diseases, vitamin E supplements may be harmful.[36] Several leading vitamin E researchers point out that clinical trials of vitamin E supplementation differ in many important ways, such as the selection of subjects, source of the vitamin, dose of the vitamin, and outcomes studied. Such differences may partly explain the inconsistent findings.

Future research that addresses such issues may clarify the relationship between vitamin E and heart disease. In the meantime, the American Heart Association and a U.S. Preventive Services Task Force of scientists concluded that insufficient support exists to recommend taking vitamin E or other antioxidant supplements to prevent cardiovascular disease.[37] The American Heart Association does support the consumption of antioxidant-rich fruits and vegetables, as well as whole grains and nuts, to reduce the risk of heart disease.[38]

Vitamin E Myth Although research continues to reveal possible roles for vitamin E, it has also clearly discredited claims that vitamin E improves athletic skill, enhances sexual performance, or cures sexual dysfunction in males. Vitamin E also does not prevent or cure hereditary **muscular dystrophy**, nor does it slow or prevent processes of aging, such as graying of the hair, wrinkling of the skin, or reduced activity of body organs.

Vitamin E Deficiency When blood concentrations of vitamin E fall below a certain critical level, the red blood cells tend to break open and spill their contents, probably because the PUFA in their membranes oxidize. This classic vitamin E-deficiency symptom, known as **erythrocyte hemolysis**, is seen in premature infants born before the transfer of vitamin E from the mother to the fetus that takes place in the last weeks of pregnancy. Vitamin E treatment corrects erythrocyte hemolysis.

The few symptoms of vitamin E deficiency that have been observed in adults include loss of muscle coordination and reflexes with impaired movement, vision, and speech. All of these symptoms may be caused by oxidative damage; vitamin E treatment corrects them.

In adults, vitamin E deficiency is usually associated with diseases, notably those that cause malabsorption of fat. These include diseases of the liver, gallbladder, and pancreas, as well as various hereditary diseases involving digestion and use of nutrients. ■

On rare occasions, vitamin E deficiencies develop in people without diseases. Most likely, such deficiencies occur after years of eating diets extremely low in fat; using fat substitutes, such as diet margarines and salad dressings, as the only sources of fat; or consuming diets composed of highly processed or "convenience" foods. Extensive heating in the processing of foods destroys vitamin E.

Vitamin E Toxicity Vitamin E supplement use has increased in recent years as its antioxidant action against disease has been recognized. As a result, signs of toxicity are now known or suspected, although vitamin E toxicity is not nearly as common, and its effects are not as serious as vitamin A or vitamin D toxicity. Extremely high doses of vitamin E interfere with the blood-clotting action of vitamin K and enhance the action of anticoagulant medications, leading to hemorrhage.

Pooled results from studies involving almost 136,000 people revealed that those taking vitamin E in doses greater than 400 IU (268 milligrams) per day were at an increased risk of death from all causes compared with people taking smaller doses.[39] In contrast, three other reports of pooled results from vitamin E supplement trials found no evidence that vitamin E supplementation up to 800 IU (536 milligrams) increased

■ Nutrition and upper GI disorders are discussed in Chapter 17. Nutrition and lower GI disorders are discussed in Chapter 18. Nutrition and liver disorders are discussed in Chapter 20.

tocopherol (tuh-KOFF-er-ol): a general term for several chemically related compounds, one of which has vitamin E activity.

muscular dystrophy (DIS-tro-fee): a hereditary disease in which the muscles gradually weaken, with the most debilitating effects occurring in the lungs. This disease should not be confused with *nutritional* muscular dystrophy, a vitamin E-deficiency disease of animals characterized by gradual paralysis of the muscles.

erythrocyte (er-REETH-ro-cite) **hemolysis** (he-MOLL-uh-sis): rupture of the red blood cells, caused by vitamin E deficiency.
erythro = red
cyte = cell
hemo = blood
lysis = breaking

Vegetable oils, some nuts and seeds such as almonds and sunflower seeds, and wheat germ are rich in vitamin E.

or decreased mortality.[40] Other researchers point out that vitamin E supplements in amounts greater than the RDA are widely used in the United States and other industrialized countries and that reports of adverse effects are rare, thereby giving credence to the safety of supplemental vitamin E in amounts below the Tolerable Upper Intake Level (UL).[41] To err on the safe side, until more is known about the safety of vitamin E supplements, intakes should probably be kept low, and they certainly should not exceed the UL of 1,000 milligrams of alpha-tocopherol per day. The UL for vitamin E is more than 65 times greater than the recommended intake for adults (15 milligrams).

Vitamin E in Foods Vitamin E is widespread in foods. About 20 percent of the vitamin E in the diet comes from vegetable oils and the products made from them, such as margarine, salad dressings, and shortenings. (Soybean oils and wheat germ oils have especially high concentrations of vitamin E.) Another 20 percent comes from fruits and vegetables although none of these is a good source by itself. Fortified cereals and other grain products contribute about 15 percent of vitamin E in the diet, and meats, poultry, fish, eggs, milk products, nuts, and seeds contribute smaller percentages. Because vitamin E is readily destroyed by heat processing and oxidation, fresh or lightly processed foods are the best sources of this vitamin.

Prior to 2000, values of vitamin E in food reflected all of the different tocopherols and were expressed in "milligrams of tocopherol equivalents." These measures overestimated the amount of alpha-tocopherol. To estimate the alpha-tocopherol content of foods stated in tocopherol equivalents, multiply by 0.8.

Vitamin K

■ K stands for the Danish word *koagulation* (coagulation or "clotting").

Vitamin K has long been known for its role in blood clotting, ■ where its presence can make the difference between life and death. The vitamin also participates in the synthesis of several bone proteins.[42] Without vitamin K, the bones produce an abnormal protein that cannot bind to the minerals that normally form bones, so bone density is low.[43] Short-term vitamin K depletion increases the rate of bone turnover, and the rate of bone turnover is a major determinant of bone mineral density.[44] In young girls, better vitamin K status is associated with decreased bone turnover.[45] Thus vitamin K may influence the risk of fracture: people who consume abundant vitamin K, often in the form of green leafy vegetables, suffer fewer hip fractures than those with lower intakes.[46]

Blood Clotting At least 13 different proteins and the mineral calcium are involved in making blood clot. Vitamin K is essential for the activation of seven of these proteins, among them prothrombin, the precursor of the enzyme thrombin (see Figure 8–4). When any of the blood-clotting factors is lacking, **hemorrhagic disease** results. If an artery or vein is cut or broken, bleeding goes unchecked. Of course, this is not to say that hemorrhaging is always caused by a vitamin K deficiency.

FIGURE 8–4

Blood-Clotting Process

When blood is exposed to air, foreign substances, or secretions from injured tissues, platelets (small, cell-like structures in the blood) release a phospholipid known as thromboplastin. Thromboplastin catalyzes the conversion of the inactive protein prothrombin to the active enzyme thrombin. Thrombin then catalyzes the conversion of the precursor protein fibrinogen to the active protein fibrin that forms the clot.

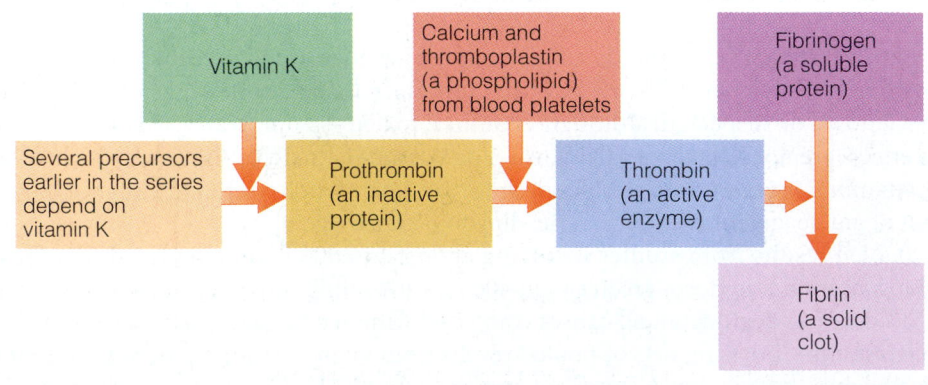

Intestinal Synthesis Like vitamin D, vitamin K can be obtained from a nonfood source. Bacteria in the intestinal tract ■ synthesize vitamin K that the body can absorb, but people cannot depend on this source alone for their vitamin K.

Vitamin K Deficiency Vitamin K deficiency is rare, but it may occur in two circumstances. First, it may arise in conditions of fat malabsorption. Second, some medications interfere with vitamin K's synthesis and action in the body: antibiotics kill the vitamin K-producing bacteria in the intestine, and anticoagulant medications interfere with vitamin K metabolism and activity. When vitamin K deficiency does occur, it can be fatal.

Vitamin K for Newborns Newborn infants present a unique case of vitamin K nutrition. An infant is born with a **sterile** digestive tract, and some weeks pass before the vitamin K-producing bacteria become fully established in the infant's intestines. At the same time, plasma prothrombin concentrations are low (this helps prevent blood clotting during the stress of birth, which might otherwise be fatal). A single dose of vitamin K, usually in a water-soluble form, is given at birth to prevent hemorrhagic disease in the newborn.

Vitamin K Toxicity Vitamin K toxicity is rare, and no adverse effects have been reported with high intakes. Therefore, a Tolerable Upper Intake Level has not been established. High doses of vitamin K can reduce the effectiveness of anticoagulant medications used to prevent blood clotting.[47] People taking these medications should eat vitamin K-rich foods in moderation and keep their intakes consistent from day to day.

Vitamin K in Foods Many foods contain ample amounts of vitamin K, notably green leafy vegetables, members of the cabbage family, and some vegetable oils. ■ Other vegetables such as iceberg lettuce and green beans provide smaller amounts.

Notable food sources of vitamin K include green vegetables such as collards, spinach, bib lettuce, brussels sprouts, and cabbage and vegetable oils such as soybean oil and canola oil.

■ Reminder: The bacterial inhabitants of the digestive tract are known as the *intestinal flora*. *flora* = plant inhabitants

■ Vitamin K AI:
   ■ Men: 120 µg/day
   ■ Women: 90 µg/day

IN SUMMARY

- The fat-soluble vitamins are vitamins A, D, E, and K.

- Vitamin A is essential to gene expression, vision, cell differentiation and integrity of epithelial tissues, immunity, and reproduction and growth.

- Vitamin A deficiency can cause blindness, sickness, and death and is a major problem worldwide.

- Overdoses of vitamin A are possible and dangerous.

- Vitamin D raises calcium and phosphorus levels in the blood. A deficiency can cause rickets in children or osteomalacia in adults.

- Vitamin D is the most toxic of all the vitamins.

- People exposed to the sun make vitamin D in their skin; fortified milk is an important food source.

- Vitamin E acts as an antioxidant in cell membranes and is especially important in the lungs where cells are exposed to high concentrations of oxygen.

- Vitamin E deficiency is rare in healthy human beings. Vitamin E is widely distributed in plant foods.

- Vitamin K is necessary for blood to clot and for bone health.

- The bacterial inhabitants of the digestive tract produce vitamin K, but people need vitamin K from foods as well.

- Dark green, leafy vegetables are good sources of vitamin K.

hemorrhagic (hem-oh-RAJ-ik) **disease:** the vitamin K-deficiency disease in which blood fails to clot.

sterile: free of microorganisms, such as bacteria.

TABLE 8–2

The Fat-Soluble Vitamins—A Summary

Vitamin Name	Chief Functions	Deficiency Symptoms	Toxicity Symptoms	Significant Sources
Vitamin A (Retinol, retinal, retinoic acid; main precursor is beta-carotene)	Vision, maintenance of cornea, epithelial cells, mucous membranes, skin; bone and tooth growth; reproduction; regulation of gene expression; immunity	Infectious diseases, night blindness, blindness (xerophthalmia), keratinization	Reduced bone mineral density, liver abnormalities, birth defects	Retinol: milk and milk products; eggs; liver Beta-carotene: spinach and other dark, leafy greens; broccoli; deep orange fruits (apricots, cantaloupe) and vegetables (carrots, winter squashes, sweet potatoes, pumpkin)
Vitamin D (Calciferol, cholecalciferol, dihydroxy vitamin D; precursor is cholesterol)	Mineralization of bones (raises blood calcium and phosphorus by increasing absorption from digestive tract, withdrawing calcium from bones, stimulating retention by kidneys)	Rickets, osteomalacia	Calcium imbalance (calcification of soft tissues and formation of stones)	Synthesized in the body with the help of sunshine; fortified milk, margarine, butter, and cereals; eggs; liver; fatty fish (salmon, sardines)
Vitamin E (Alpha-tocopherol, tocopherol)	Antioxidant (stabilization of cell membranes, regulation of oxidation reactions, protection of polyunsaturated fatty acids [PUFA] and vitamin A)	Erythrocyte hemolysis, nerve damage	Hemorrhagic effects	Polyunsaturated plant oils (margarine, salad dressings, shortenings), green and leafy vegetables, wheat germ, whole-grain products, nuts, seeds
Vitamin K (Phylloquinone, menaquinone, naphthoquinone)	Synthesis of blood-clotting proteins and bone proteins	Hemorrhage	None known	Synthesized in the body by GI bacteria; green leafy vegetables; cabbage-type vegetables; vegetable oils

Table 8–2 offers a complete summary of the fat-soluble vitamins.

Water-Soluble Vitamins

The B vitamins and vitamin C are the water-soluble vitamins. These vitamins, found in the watery compartments of foods, are distributed into water-filled compartments of the body. They are easily absorbed into the bloodstream and are just as easily excreted if their blood concentrations rise too high. Thus water-soluble vitamins are less likely to reach toxic concentrations in the body than are fat-soluble vitamins. Foods never deliver excessive amounts of water-soluble vitamins, but the large doses concentrated in vitamin supplements can reach toxic levels.

B Vitamins

Despite advertisements that claim otherwise, B vitamins do not give people energy. Carbohydrate, fat, and protein—the *energy-yielding* nutrients—supply the fuel for energy. B vitamins help to burn that fuel but do not serve as fuel themselves.

Coenzymes The eight B vitamins were listed in Table 8–1. Each is part of an enzyme helper known as a **coenzyme**. Some B vitamins have other important functions in the body as well, but the roles these vitamins play as parts of coenzymes are the best understood. A coenzyme is a small molecule that combines with an enzyme to make it active. With the coenzyme in place, a substance is attracted to the enzyme, and the reaction proceeds instantaneously. Figure 8–5 illustrates coenzyme action.

Active forms of five of the B vitamins—thiamin, riboflavin, niacin, pantothenic acid, and biotin—participate in the release of energy from carbohydrate, fat, and protein. A coenzyme containing vitamin B_6 assists enzymes that metabolize amino acids. The making of new cells depends on a folate coenzyme, and the making of this coenzyme depends on vitamin B_{12}.

The eight B vitamins play many specific roles in helping the enzymes to perform thousands of different molecular conversions in the body. They must be present in every cell continuously for the cells to function as they should. As for vitamin C, its primary role, discussed later, is as an antioxidant.

B Vitamin Deficiencies In academic and clinical discussions of vitamins, different sets of deficiency symptoms are ascribed to each individual vitamin. Such clear-cut symptoms, however, are found only in laboratory animals that have been fed contrived diets that lack just one nutrient. In reality, a deficiency of any single B vitamin seldom shows up in isolation because people do not eat nutrients one by one; they eat foods containing mixtures of many nutrients. If a major class of foods is missing from the diet, all of the nutrients delivered by those foods will be lacking to various extents.

In only two cases have dietary deficiencies associated with single B vitamins been observed on a large scale in human populations. Diseases have been named for these deficiency states. One of them, **beriberi**, was first observed in Southeast Asia when the custom of polishing rice became widespread. Rice contributed 80 percent of the energy intake of the people in these areas, and rice bran was their principal source of thiamin. When the bran was removed to make the rice whiter, beriberi spread like wildfire.

The niacin-deficiency disease, **pellagra,** became widespread in the southern United States in the early part of the twentieth century among people who subsisted on a low-protein diet with a staple grain of corn. This diet was unusual in that it supplied neither enough niacin nor enough tryptophan, its amino acid precursor, to make the niacin intake adequate.

Even in the cases of beriberi and pellagra, the deficiencies were probably not pure. When foods were provided containing the one vitamin known to be needed, other vitamins that may have been in short supply came as part of the package.

Major deficiency diseases such as pellagra and beriberi no longer occur in the United States and Canada, but more subtle deficiencies of nutrients, including the B vitamins, are sometimes observed. When they do occur, it is usually in people whose food choices are poor because of poverty, ignorance, illness, or poor health habits such as alcohol abuse.

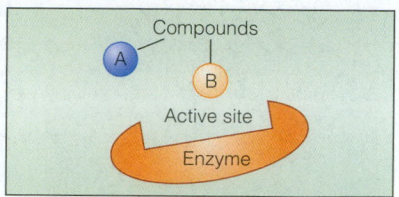

Without the coenzyme, compounds A and B do not respond to the enzyme.

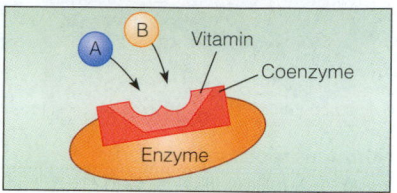

With the coenzyme in place, compounds A and B are attracted to the active site on the enzyme, and they react.

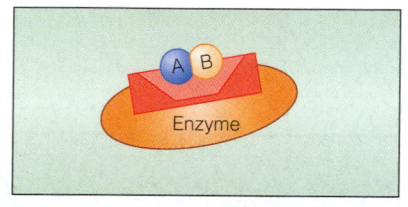

The reaction is completed with the formation of a new product. In this case the product is AB.

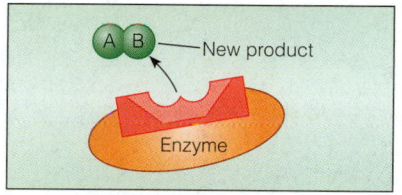

The product AB is released.

FIGURE 8–5

Coenzyme Action

---

coenzyme (co-EN-zime): a small molecule that works with an enzyme to promote the enzyme's activity. Many coenzymes have B vitamins as part of their structure.
  co = with

beriberi: the thiamin-deficiency disease; characterized by loss of sensation in the hands and feet, muscular weakness, advancing paralysis, and abnormal heart action.

pellagra (pell-AY-gra): the niacin-deficiency disease. Symptoms include the "4 Ds": diarrhea, dermatitis, dementia, and, ultimately, death.
  pellis = skin
  agra = seizure

Nutritious foods such as pork, legumes, sunflower seeds, and enriched and whole-grain breads are valuable sources of thiamin.

■ Note: The terms *fortified* and *enriched* may be used interchangeably.

■ Thiamin RDA:
  ■ Men: 1.2 mg/day
  ■ Women: 1.1 mg/day

refined grain: a product from which the bran, germ, and husk have been removed, leaving only the endosperm.

fortification: the addition to a food of nutrients that were either not originally present or present in insignificant amounts. Fortification can be used to correct or prevent a widespread nutrient deficiency, to balance the total nutrient profile of a food, or to restore nutrients lost in processing.

enrichment: the addition to a food of nutrients to meet a specified standard. In the case of refined bread or cereal, five nutrients have been added: thiamin, riboflavin, niacin, and folate in amounts approximately equivalent to, or higher than, those originally present and iron in amounts to alleviate the prevalence of iron-deficiency anemia.

niacin equivalents (NE): the amount of niacin present in food, including the niacin that can theoretically be made from tryptophan, its precursor, present in the food.

Interdependent Systems Table 8–3, at the end of this chapter, sums up a few of the better-established facts about B vitamin deficiencies. A look at the table will make another generalization possible. Different body systems depend to different extents on these vitamins. Processes in nerves and in their responding tissues, the muscles, depend heavily on glucose metabolism and hence on thiamin, so paralysis sets in when this vitamin is lacking; but thiamin is important in all cells, not just in nerves and muscles. Similarly, because the red blood cells and GI tract cells divide the most rapidly, two of the first symptoms of a deficiency of folate are a type of anemia and GI tract deterioration—but again, all systems depend on folate, not just these. The list of symptoms in Table 8–3 is far from complete.

B Vitamin Enrichment of Foods If the staple food of a region is made from **refined grain**, vitamin B deficiencies are especially likely. One way to protect people from deficiencies is to add nutrients to their staple food, a process known as **fortification** or **enrichment.** ■ The enrichment of refined breads and cereals has drastically reduced the incidence of iron and B vitamin deficiencies.

The preceding discussion has shown both the great importance of B vitamins in promoting normal, healthy functioning of all body systems and the severe consequences of deficiency. Now you may want to know how to be sure you and your clients are getting enough of these vital nutrients. The next sections present information on each B vitamin. Keep in mind that *foods* can provide all the needed nutrients and that supplements are a poor second choice. Some supplements are absurdly costly, but even if they are inexpensive, most people do not need them. Nutrition in Practice 9 discusses uses and choices of supplements in more detail.

Thiamin

All cells use thiamin, which plays a critical role in their energy metabolism. Thiamin also occupies a special site on nerve cell membranes. Consequently, as mentioned earlier, thiamin is critical to the normal functioning of the nerves and muscles.

Thiamin Need As long as people consume enough food to meet their energy needs—and obtain that energy from nutritious foods—thiamin needs will be met. People who derive a large proportion of their energy from empty-kcalorie items, such as sugar or alcohol, risk thiamin deficiency, a condition that seems to be reappearing as the population of malnourished and homeless people rises. A person who is fasting or who has adopted a very low-kcalorie diet needs as much thiamin as when eating enough to meet energy needs.

In developed countries today, abuse of alcohol may lead to a severe form of thiamin deficiency, Wernicke-Korsakoff syndrome.[48] Alcohol contributes energy but carries almost no nutrients with it and often displaces food. In addition, alcohol impairs absorption of thiamin from the digestive tract and hastens its excretion in urine, tripling the risk of deficiency. Wernicke-Korsakoff syndrome is characterized by symptoms that are almost indistinguishable from alcohol abuse itself: mental confusion, disorientation, loss of memory, jerky eye movements, and staggering gait. Unlike alcohol toxicity, the syndrome responds quickly to an injection of thiamin, and some experts recommend a precautionary dose for any patients suspected of having the syndrome.[49]

Thiamin in Foods Thiamin occurs in small quantities in virtually all nutritious foods, but it is concentrated in only a few foods, of which pork is the most commonly eaten. If you keep empty-kcalorie foods to a minimum and emphasize nutritious foods each day, you will easily meet your thiamin needs. Foods chosen from the bread and cereal group should be either whole grain or enriched. Thiamin is not stored in the body to any great extent, so daily intake is important. ■

Riboflavin

Like thiamin, riboflavin facilitates energy production in the body. The needs of infants, children, and pregnant women rise rapidly during periods of active growth.

Riboflavin in Foods Unlike thiamin, riboflavin is not evenly distributed among the food groups. ■ The major contributors of riboflavin to people's diets are milk, milk products, meats, and green vegetables (broccoli, turnip greens, asparagus, and spinach). The riboflavin richness of milk and milk products is a good reason to include these foods in every day's meals. No other commonly eaten food can make such a substantial contribution. People who omit milk and milk products from their diets can substitute generous servings of dark green, leafy vegetables. Among the meats, liver and heart are the richest sources, but all lean meats, as well as eggs, offer some riboflavin.

Effects of Light Riboflavin is light sensitive; the ultraviolet rays of the sun or of fluorescent lamps can destroy it. For this reason, milk is sold in cardboard or opaque plastic containers to protect the riboflavin in the milk from light. In contrast, riboflavin is heat stable, so ordinary cooking does not destroy it.

Niacin

Like thiamin and riboflavin, niacin participates in the energy metabolism of every body cell. Niacin is unique among the B vitamins in that the body can make it from protein. The amino acid tryptophan can be converted to niacin in the body: 60 milligrams of tryptophan yield 1 milligram of niacin. Recommended intakes are therefore stated in **niacin equivalents (NE),** ■ reflecting the body's ability to convert tryptophan to niacin. ■

Certain forms of niacin supplements in amounts ten times or more than the dietary recommendation cause "niacin flush," a dilation of the capillaries of the skin with perceptible tingling that, if intense, can be painful. The Tolerable Upper Intake Level (35 milligrams NE) is based on flushing as the critical adverse effect.

Niacin Used as a Medication Physicians sometimes use diet and large doses of a form of niacin (nicotinic acid) to lower blood cholesterol in the treatment of atherosclerosis. When used this way, niacin leaves the realm of nutrition to become a pharmacological agent, a drug. ■ As with any medication, self-dosing with niacin is ill advised; large doses may injure the liver and produce some symptoms of diabetes.[50]

Niacin in Foods Meat, poultry, and fish contribute about half the niacin equivalents most people consume; enriched breads and cereals contribute about one-fourth. Among the vegetables, mushrooms, asparagus, and green leafy vegetables are the richest niacin sources. Niacin is less vulnerable to losses during food preparation and storage than other water-soluble vitamins. Being fairly heat-resistant, niacin can withstand reasonable cooking times, but like other water-soluble vitamins, it will leach into cooking water.

Pantothenic Acid and Biotin

Two other B vitamins—pantothenic acid and biotin—are also important in energy metabolism. Pantothenic acid was first recognized as a substance that stimulates growth. It is a component of a key enzyme that makes possible the release of energy from the energy nutrients. Pantothenic acid is involved in more than 100 different steps in the synthesis of lipids, neurotransmitters, steroid hormones, and hemoglobin. Biotin plays an important role in metabolism as a coenzyme that carries carbon dioxide. Emerging evidence indicates that biotin participates in other processes such as

Milk and milk products supply much (about 50 percent) of the riboflavin in people's diets, but meats, eggs, green vegetables, and enriched and whole-grain breads and cereals are good sources, too.

■ Riboflavin RDA:
  ■ Men: 1.3 mg/day
  ■ Women: 1.1 mg/day

■ A food containing 1 mg of niacin and 60 mg of tryptophan contains the niacin equivalent of 2 mg, or 2 mg NE.

■ Niacin RDA:
  ■ Men: 16 mg NE/day
  ■ Women: 14 mg NE/day

■ When a normal dose of a nutrient clears up a deficiency condition, the effect is a *physiological* one. When a large dose of a nutrient overwhelms a body system and acts like a drug, the effect is a *pharmacological* one.

Niacin-rich foods include meat, fish, poultry, and peanut butter, as well as enriched breads and cereals and a few vegetables.

gene expression and cell signaling and in the structure of DNA-binding proteins in the cell nucleus.[51]

■ Pantothenic acid AI:
  ■ Adults: 5 mg/day

■ Biotin AI:
  ■ Adults: 30 µg/day

■ The protein *avidin* in egg whites binds biotin.

Pantothenic Acid and Biotin in Foods Both pantothenic acid ■ and biotin ■ are more widespread in foods than the other vitamins discussed so far. There seems to be no danger that people who consume a variety of foods will suffer deficiencies. Claims that pantothenic acid and biotin are needed in pill form to prevent or cure disease conditions are at best unfounded and at worst intentionally misleading.

Biotin Deficiency Biotin deficiencies are rare but have been reported in adults fed artificially by vein without biotin supplementation. Researchers can induce biotin deficiency in animals or human beings by feeding them raw egg whites, which contain a protein that binds biotin and prevents its absorption. ■ Long-term use of anticonvulsant medication may also lead to biotin deficiency, as may alcohol abuse.[52]

Vitamin B$_6$

Vitamin B$_6$ has been called the "sleeping giant" of vitamins. A surge of research interest in the last two decades has not only revealed new knowledge, but has also raised new questions. For example, unlike the other water-soluble vitamins, vitamin B$_6$ is stored extensively in muscle tissue. Most recently, research interest has centered on a possible role for vitamin B$_6$ in the treatment of disease.

Metabolic Roles of Vitamin B$_6$ Vitamin B$_6$ has long been known to play roles in protein and amino acid metabolism. In the cells, vitamin B$_6$ helps to convert one kind of amino acid, which the cells have in abundance, to another, which they need in larger amounts. It also aids in the conversion of the amino acid tryptophan to niacin and plays important roles in the synthesis of hemoglobin and neurotransmitters, the communication molecules of the brain. Vitamin B$_6$ also assists in releasing stored glucose from glycogen and thus contributes to the regulation of blood glucose. Research suggests roles for vitamin B$_6$ in immune function, cognitive performance, and hormone response.[53] The association between vitamin B$_6$ and immune function is related to the critical role the vitamin plays in protein metabolism. Vitamin B$_6$ deficiency can significantly impair the immune response, perhaps by way of impaired antibody production.

Vitamin B$_6$ status may be related to cardiovascular disease risk. Elevated blood levels of the amino acid homocysteine correlate with a high incidence of heart disease.[54] Evidence suggests that low blood concentrations of vitamin B$_6$, vitamin B$_{12}$, and folate are associated with elevated homocysteine concentrations.[55] In a study of more than 7,000 people, the use of B vitamin supplements over a six-year period lowered homocysteine concentrations.[56] Lowering homocysteine, however, may not help in preventing heart attacks.[57] Supplements of the B vitamins do not always benefit those with heart disease and, in fact, may actually increase the risks.[58] Thus far, a decline in heart disease has not emerged from controlled studies of supplementation in healthy people.[59]

Vitamin B$_6$ Deficiency Besides a weakening immune response, vitamin B$_6$ deficiency is expressed in general symptoms, such as weakness, irritability, and insomnia. Other symptoms include a greasy, flaky dermatitis; anemia; and, in advanced cases, convulsions.

Vitamin B$_6$ Toxicity For years it was believed that vitamin B$_6$, like other water-soluble vitamins, could not reach toxic concentrations in the body. Toxic effects of vitamin B$_6$ became known when a physician reported them in women who had been taking more than 2 grams of vitamin B$_6$ daily (20 times the current Tolerable Upper Intake Level) for two months or more. Most of these women had been attempting to relieve premenstrual syndrome (PMS), the cluster of physical, emotional, and psychological symptoms that some women experience prior to menstruation. The first symptom of toxicity was numb feet; then the women lost sensation in their hands; and then they became unable to walk. The women recovered after they discontinued the supplements.

Vitamin B$_6$ Recommendations Because vitamin B$_6$ coenzymes play many roles in amino acid metabolism, previous RDA were expressed in terms of protein intakes; the current RDA ■ for vitamin B$_6$, however, are not. Research does not support claims that large doses of vitamin B$_6$ enhance muscle strength or physical endurance.

Vitamin B$_6$ in Foods The richest food sources of vitamin B$_6$ are protein-rich meat, fish, and poultry. Potatoes, a few other vegetables, and some fruits are good sources, too. Foods lose vitamin B$_6$ when heated.

Folate

The B vitamin folate is active in cell division. During periods of rapid growth and cell division, such as pregnancy and adolescence, folate needs increase, and deficiency is especially likely. When a deficiency occurs, the replacement of the rapidly dividing cells of the blood and the GI tract falters. Not surprisingly, two of the first symptoms of a folate deficiency are a type of anemia and GI tract deterioration (see Table 8–3 on p. 225).

Folate, Alcohol, and Drugs Of all the vitamins, folate appears to be the most vulnerable to interactions with alcohol and other drugs.[60] As Nutrition in Practice 20 describes, alcohol-addicted people risk folate deficiency because alcohol impairs folate's absorption and increases its excretion. Furthermore, as people's alcohol intakes rise, their folate intakes decline. Many medications, including aspirin, oral contraceptives, and anticonvulsants, also impair folate status. Smoking exerts a negative effect on folate status as well.

Folate and Neural Tube Defects Research studies confirm the importance of folate in preventing **neural tube defects (NTD)**. ■ The brain and spinal cord develop from the neural tube, and defects in its orderly formation during the early weeks of pregnancy may result in various central nervous system disorders and death. Folate supplements taken before conception and continued throughout the first trimester of pregnancy can prevent NTD. For this reason, the American Academy of Pediatrics and the Public Health Services recommend that all women of childbearing age who are capable of becoming pregnant take 0.4 milligram (400 micrograms) of folate daily.

Folate status improves more with supplementation or fortification than with a dietary intake that meets recommendations. ■ Neural tube defects arise early in pregnancy before most women realize they are pregnant, and most women eat too few fruits and vegetables to supply even half the folate needed to prevent NTD. For these reasons, in the late 1990s, the FDA mandated that enriched grain products (flour, cornmeal, pasta, and rice) be fortified with an especially absorbable synthetic form of folate, folic acid. ■ Since this fortification began, typical folate intakes from fortified foods have increased dramatically—by more than double the predicted levels—and observers report an almost 25 percent drop in the national incidence of NTD, even among women receiving late or no prenatal care.[61] Researchers expect to see declines in some other birth defects and miscarriages as well.[62] Folate fortification also raises safety concerns, however.[63] High doses of folate can complicate the diagnosis of vitamin B$_{12}$ deficiency, as discussed later. The DRI committee set a Tolerable Upper Intake Level of 1,000 micrograms per day from fortified foods or supplements.

Folate status, like vitamin B$_6$ status, may be related to cardiovascular disease risk.[64] As discussed earlier, elevated levels of homocysteine are associated with a greater risk

Most protein-rich foods such as meat, fish, and poultry provide ample vitamin B$_6$; some vegetables and fruits are good sources too.

■ Vitamin B$_6$ RDA:
- Adults (19–50): 1.3 mg/day
- Women (51– >70): 1.5 mg/day
- Men (51– >70): 1.7 mg/day

■ The two main types of neural tube defects are *spina bifida* (literally, "split spine") and *anencephaly* ("no brain").

■ Bread products, flour, corn grits, and pasta must be fortified with 140 µg per 100 g of food (about $\frac{1}{2}$ c cooked food or 1 slice of bread).

■ Folate RDA:
- Adults: 400 µg/day

neural tube defects (NTD): malformations of the brain, spinal cord, or both during embryonic development.

of cardiovascular disease. One of folate's key roles in the body is to metabolize homocysteine. Without folate, homocysteine accumulates. Fortified foods and folate supplements raise blood folate and reduce homocysteine levels.[65]

Folate in Foods As Snapshot 8–2 shows, the best food sources of folate are liver, legumes, beets, and leafy green vegetables (the vitamin's name suggests the word *foliage*). Among the fruits, oranges, orange juice, and cantaloupe are the best sources. With fortification, grain products are good sources of folate, too. Heat and oxidation during cooking and storage can destroy up to half of the folate in foods.

The difference in absorption between naturally occurring food folate and synthetic folate that enriches foods and is added to supplements necessitates compensation when measuring folate. **Dietary folate equivalents,** or **DFE,** convert all forms of folate into units that are equivalent to the folate in foods. Most food labels and tables of food composition express folate values in micrograms, however; the accompanying "How to" describes how to estimate dietary folate equivalents.

Vitamin B$_{12}$

Vitamin B$_{12}$ and folate share a special relationship: vitamin B$_{12}$ assists folate in cell division. Their roles intertwine, but each performs a specific task that the other cannot accomplish.

Vitamin B$_{12}$, Folate, and Cell Division Vitamin B$_{12}$ (in coenzyme form) stands by to accept carbon groups from folate as folate removes them from other compounds. The passing of these carbon groups from folate to vitamin B$_{12}$ regenerates the active form of folate so that it can continue its dismantling tasks. In the absence of vitamin B$_{12}$, folate is trapped in its inactive, metabolically useless form, unable to do its job. When folate is either trapped due to a vitamin B$_{12}$ deficiency or unavailable due to a deficiency of folate itself, cells that are growing most rapidly, notably the blood cells, are the first to be affected. Thus a deficiency of either nutrient—vitamin B$_{12}$ or folate—impairs maturation of the blood cells and produces anemia. The anemia is identifiable by microscopic examination of the blood, which reveals many large, immature red blood cells. ■ Either vitamin B$_{12}$ or folate will clear up the anemia.

Vitamin B$_{12}$ and the Nervous System Although either vitamin will clear up the anemia caused by vitamin B$_{12}$ deficiency, if folate is given when vitamin B$_{12}$ is needed, the result

■ Large-cell anemia is known as *macrocytic* or *megaloblastic* anemia.
    macro = large
    cyte = cell
    mega = large
    blast = immature cell

How To

ESTIMATE DIETARY FOLATE EQUIVALENTS

Folate is expressed in terms of DFE (dietary folate equivalents) because synthetic folate from supplements and fortified foods is absorbed at almost twice (1.7 times) the rate of naturally occurring folate from other foods. Use the following equation to calculate:

DFE = µg food folate + (1.7 × µg synthetic folate)

Consider, for example, a pregnant woman who takes a supplement and eats a bowl of fortified corn flakes, two slices of fortified bread, and one cup of fortified pasta. From the supplement and fortified foods, she obtains synthetic folate:

Supplement	100 µg folate
Fortified corn flakes	100 µg folate
Fortified bread	40 µg folate
Fortified pasta	60 µg folate
	300 µg folate

To calculate the DFE, multiply the amount of synthetic folate by 1.7:

300 µg × 1.7 = 510 mg DFE

Now add the naturally occurring folate from the other foods in her diet—in this example, another 90 micrograms of folate.

510 µg DFE + 90 µg = 600 µg DFE

Notice that if we had not converted synthetic folate from supplements and fortified foods to DFE, this woman's intake would appear to fall short of the 600 micrograms recommended for pregnancy (300 µg + 90 µg = 390 µg). But as this example shows, her intake does meet the recommendation. At this time, supplement and fortified food labels list folate in micrograms only, not micrograms DFE, making such calculations necessary.

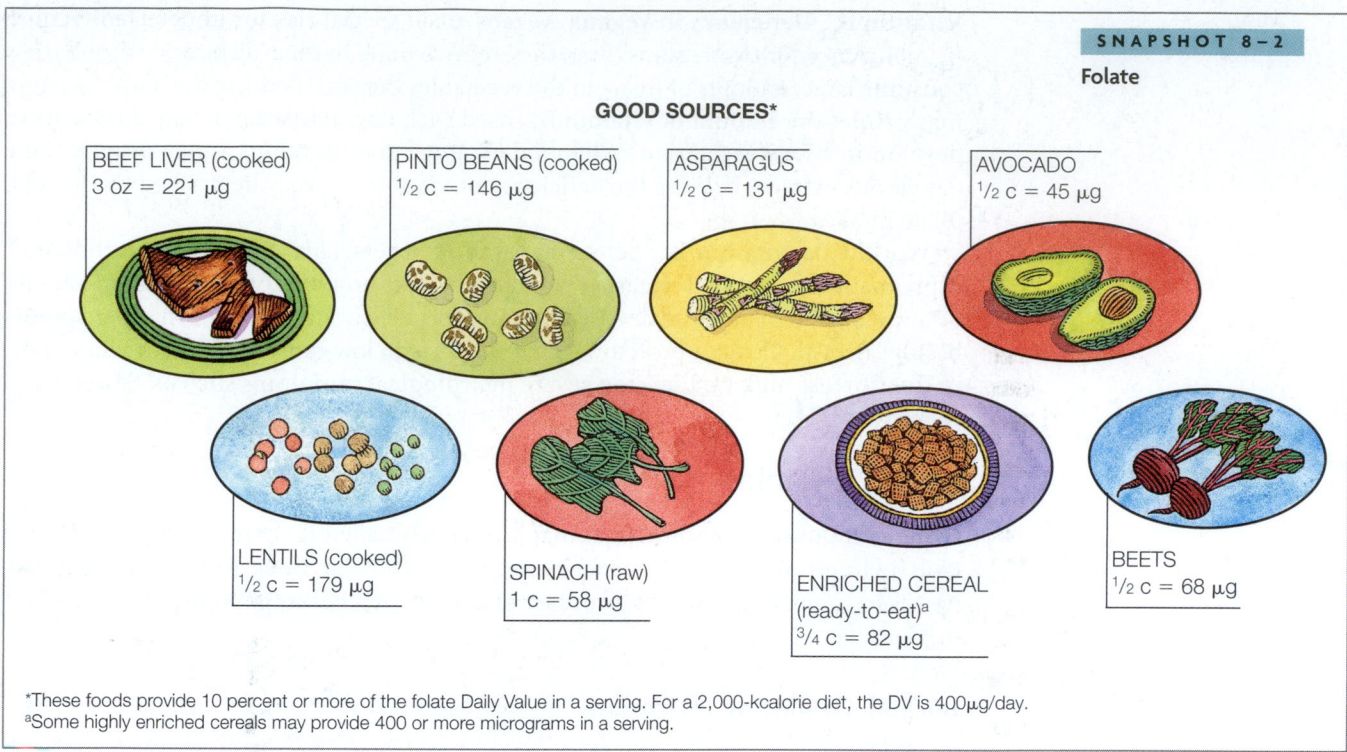

GOOD SOURCES*

BEEF LIVER (cooked)
3 oz = 221 µg

PINTO BEANS (cooked)
1/2 c = 146 µg

ASPARAGUS
1/2 c = 131 µg

AVOCADO
1/2 c = 45 µg

LENTILS (cooked)
1/2 c = 179 µg

SPINACH (raw)
1 c = 58 µg

ENRICHED CEREAL
(ready-to-eat)[a]
3/4 c = 82 µg

BEETS
1/2 c = 68 µg

*These foods provide 10 percent or more of the folate Daily Value in a serving. For a 2,000-kcalorie diet, the DV is 400 µg/day.
[a]Some highly enriched cereals may provide 400 or more micrograms in a serving.

is disastrous, not to the blood but to the nervous system. The reason: vitamin B_{12} also helps maintain nerve fibers. A vitamin B_{12} deficiency can ultimately result in devastating neurological symptoms, undetectable by a blood test. A deceptive folate "cure" of the anemia in vitamin B_{12} deficiency allows the nerve deterioration to progress, leading to paralysis and permanent nerve damage. This interaction between folate and vitamin B_{12} raises safety concerns about the use of folate supplements and fortification of foods.

The way folate masks vitamin B_{12} deficiency underlines a point already made several times: it takes a skilled diagnostician to make a correct diagnosis. A person who self-diagnoses on the basis of a single observed symptom takes a serious risk.

Vitamin B_{12} Absorption Vitamin B_{12} requires an **"intrinsic factor"**—a compound made inside the body—for absorption from the intestinal tract into the bloodstream. This intrinsic factor is made in the stomach, where it attaches to the vitamin; the complex then passes to the small intestine and is gradually absorbed.

Loss of Intrinsic Factor In some cases, intrinsic factor production becomes inadequate or ceases altogether—for example, after surgical removal of the stomach. Some people inherit a defective gene for intrinsic factor. Because vitamin B_{12} deficiency in the body may be caused either by a lack of the vitamin in the diet or by the body's inability to absorb the vitamin, a change in diet alone may not correct the deficiency. When absorption failure is the problem, vitamin B_{12} must be supplied by injection.

Vitamin B_{12} in Foods A unique characteristic of vitamin B_{12} is that it is found almost exclusively in foods derived from animals. People who eat meat are guaranteed an adequate intake, ■ and lacto-ovo vegetarians (who use milk, cheese, and eggs) are also protected from deficiency. It is a myth, however, that fermented soy products, such as miso (a soybean paste), or sea algae, such as spirulina, provide vitamin B_{12} in its active form. Extensive research shows that the amounts of vitamin B_{12} listed on the labels of these plant products are inaccurate and misleading because the vitamin B_{12} in these products occurs in an inactive, unavailable form. Vegans must take vitamin B_{12} supplements or find other sources of active vitamin B_{12}. Some loss of vitamin B_{12} occurs when foods are heated in microwave ovens.

■ Vitamin B_{12} RDA:
  ■ Adults: 2.4 µg/day

dietary folate equivalents (DFE): the amount of folate available to the body from naturally occurring sources, fortified foods, and supplements, accounting for differences in bioavailability from each source.

intrinsic factor: inside the system. Anemia that reflects a vitamin B_{12} deficiency caused by lack of intrinsic factor is known as *pernicious anemia*.

Vitamin B$_{12}$ Deficiency in Vegans Vegans are at special risk for undetected vitamin B$_{12}$ deficiency for two reasons: first, they receive none in their diets, and second, they consume large amounts of folate in the vegetables they eat. Because the body can store many times the amount of vitamin B$_{12}$ used each day, a deficiency may take years to develop in a new vegetarian. When a deficiency does develop, it may progress to a dangerous extreme because the deficiency of vitamin B$_{12}$ may be masked by the high folate intake.

Worldwide, vitamin B$_{12}$ deficiency among vegetarians is a growing problem.[66] A pregnant or lactating vegetarian woman who eats no foods of animal origin should be aware that her infant can develop a vitamin B$_{12}$ deficiency, even if the mother appears healthy. Breastfed infants born to vegan mothers with low concentrations of vitamin B$_{12}$ in their breast milk can develop severe neurological symptoms such as seizures and cognitive problems.[67]

Non-B Vitamins

Other compounds are sometimes inappropriately called B vitamins because, like the true B vitamins, they serve as coenzymes in metabolism. Even if they were essential, however, supplements would be unnecessary because these compounds are abundant in foods.

Inositol, Choline, and Carnitine Among the non-B vitamins are a trio of substances known as inositol, choline, and carnitine. Researchers are exploring the possibility that these substances may be essential. Thus far, only choline has been assigned an Adequate Intake value. ■

Other Non-B Vitamins Other substances have also been mistaken for essential nutrients. They include para-aminobenzoic acid (PABA), bioflavonoids (vitamin P or hesperidin), and ubiquinone. Other names you may hear are "vitamin B$_{15}$" (a hoax) and "vitamin B$_{17}$" (laetrile, a fake cancer-curing drug and not a vitamin). There is, however, one other water-soluble vitamin of great interest and importance—vitamin C.

Vitamin C

Three hundred years ago, any man who joined the crew of a seagoing ship knew he had only half a chance of returning alive—not because he might be slain by pirates or die in a storm, but because he might contract the dread disease **scurvy**. Then a physician with the British navy found that citrus fruits could cure the disease, and thereafter, all ships were required to carry lime juice for every sailor. (This is why British sailors are still called "limeys" today.) In the 1930s, the antiscurvy factor in citrus fruits was isolated from lemon juice and named **ascorbic acid**. Today, hundreds of millions of vitamin C pills are produced in pharmaceutical laboratories.

Metabolic Roles of Vitamin C Vitamin C's action defies a simple, tidy description. It plays many important roles in the body, and its modes of action differ in different situations.

Vitamin C's Role in Collagen Formation The best-understood action of vitamin C is its role in helping to form **collagen**, the single most important protein of connective tissue. Collagen serves as the matrix on which bone is formed, the material of scars, and an important part of the "glue" that attaches one cell to another. This latter function is especially important in the artery walls, which must expand and contract with each beat of the heart, and in the walls of the capillaries, which are thin and fragile. Vitamin C also plays a role in the production of carnitine, important for transporting fatty acids within cells.

■ Choline AI:
■ Men: 550 mg/day
■ Women: 425 mg/day

Vitamin C as an Antioxidant Vitamin C is also an important antioxidant. Recall that the antioxidants beta-carotene and vitamin E protect fat-soluble substances from oxidizing agents; vitamin C protects water-soluble substances the same way. By being oxidized itself, vitamin C regenerates already-oxidized substances such as iron and copper to their original, active form. In the intestines, it protects iron from oxidation and so enhances iron absorption. In the cells and body fluids, it helps to protect other molecules, including the fat-soluble compounds vitamin A, vitamin E, and the polyunsaturated fatty acids.

Vitamin C in Amino Acid Metabolism Vitamin C is also involved in the metabolism of several amino acids. Some of these amino acids end up being used to make hormones of great importance in body functioning, among them norepinephrine and thyroxine.

Role of Stress During stress, the adrenal glands release large quantities of vitamin C together with the stress hormones epinephrine and norepinephrine. What the vitamin has to do with the stress reaction is unclear, but it is known that stress increases vitamin C needs somewhat.

Vitamin C as a Possible Antihistamine Newspaper headlines touting vitamin C as a cure for colds and cancer have appeared frequently over the years. Some research suggests that vitamin C (2 grams per day for two weeks) may reduce the severity and duration of cold and allergy symptoms by reducing blood histamine concentrations. In other words, vitamin C acts as an antihistamine. If further research confirms vitamin C's antihistamine effect, its use may permit people to rely less heavily on antihistamine drugs when suffering from cold and allergy symptoms.

Vitamin C's Role in Cancer Prevention and Treatment The role of vitamin C in the prevention and treatment of cancer is still being studied. In a dozen or so different well-controlled studies, researchers identified individuals with and without cancer and assessed their dietary intakes of vitamin C. They found that people with high vitamin C intakes had lower risks of cancers than did people with low intakes. The correlation may reflect not just an association with vitamin C, but also the broader benefits of a diet rich in fruits and vegetables and low in fat. It does not support the taking of vitamin C supplements to prevent or treat cancer.

Vitamin C Deficiency When intake of vitamin C is inadequate, the body's vitamin C pool dwindles, and **latent** scurvy appears. The blood vessels show the first deficiency signs. The gums around the teeth begin to bleed easily, and capillaries under the skin break spontaneously, producing pinpoint hemorrhages. Then the symptoms of **overt** scurvy appear. Muscles, including the heart muscle, may degenerate. The skin becomes rough, brown, scaly, and dry. Wounds fail to heal because scar tissue will not form without collagen. Bone rebuilding falters; the ends of the long bones become softened, malformed, and painful; and fractures occur. The teeth may become loose in the jawbone and fall out. Anemia and infections are common. Sudden death is likely, perhaps because of massive bleeding into the joints and body cavities.

It takes only 10 or so milligrams of vitamin C per day to prevent overt scurvy, and not much more than that to cure it. Once diagnosed, scurvy is readily reversible with moderate doses, in the neighborhood of 100 milligrams per day. Such an intake is easily achieved by including vitamin C-rich foods in the diet.

Vitamin C Toxicity The easy availability of vitamin C in pill form and the publication of books recommending vitamin C to prevent everything from the common cold

scurvy: the vitamin C-deficiency disease.

ascorbic acid: one of the two active forms of vitamin C. Many people refer to vitamin C by this name.
   a = without
   scorbic = having scurvy

collagen: the characteristic protein of connective tissue.
   kolla = glue
   gennan = produce

latent: the period in the course of a disease when the condition is present but the symptoms have not begun to appear.
   latens = lying hidden

overt: out in the open, full-blown.
   ouvrire = to open

■ Doses of 10 to 30 or more times the recommended intake of a nutrient are termed *megadoses*. In the case of vitamin C, current recommendations are 75 mg/day for women and 90 mg/day for men. The Tolerable Upper Intake Level for vitamin C is 2,000 mg/day.

■ The anticoagulants with which vitamin C interferes are warfarin and dicumarol.

to life-threatening cancer have led thousands of people to take large doses of vitamin C. ■ Not surprisingly, instances of vitamin C causing harm have surfaced.

Some of the suspected toxic effects of vitamin C megadoses have not been confirmed, but others have been seen often enough to warrant concern. Nausea, abdominal cramps, and diarrhea are often reported. Several instances of interference with medical regimens are known. Large amounts of vitamin C excreted in the urine obscure the results of tests used to detect diabetes. People taking anticoagulants ■ may unwittingly counteract the effect of these medications if they also take massive doses of vitamin C. Vitamin C megadoses can also enhance iron absorption too much, resulting in iron overload (see Chapter 9).

People with sickle-cell anemia may be especially vulnerable to megadoses of vitamin C. Those who have a tendency toward **gout**, as well as those who have a genetic abnormality that alters the way they metabolize vitamin C, are more prone to forming kidney stones if they take megadoses of vitamin C.

Recommended Intakes of Vitamin C The vitamin C recommendation is 90 milligrams for men and 75 milligrams for women. These amounts are far higher than the 10 milligrams per day needed to prevent the symptoms of scurvy. In fact, they are close to the amount at which the body's pool of vitamin C is full to overflowing: about 100 milligrams per day.

Special Needs for Vitamin C As is true of all nutrients, unusual circumstances may raise vitamin C needs. Among the stresses known to do so are infections; burns; surgery; extremely high or low temperatures; toxic doses of heavy metals, such as lead, mercury, and cadmium; and the chronic use of certain medications, including aspirin, barbiturates, and oral contraceptives. Smoking, too, has adverse effects on vitamin C status. Cigarette smoke contains oxidants, which deplete this potent antioxidant. Accordingly, the vitamin C recommendation for smokers is set high, at 125 milligrams for men and 105 milligrams for women.

Safe Limits Few instances warrant the taking of more than 100 to 300 milligrams of vitamin C per day. The risks may not be great for adults who dose themselves with 1 to 2 grams per day, but those taking more than 2 grams per day and especially those taking above 3 grams per day should be aware of the distinct possibility of harm.[68]

Vitamin C in Foods The inclusion of intelligently selected fruits and vegetables in the daily diet guarantees a generous intake of vitamin C. Even those who wish to ingest amounts well above the RDA can easily meet their goals by eating certain foods (see Snapshot 8–3). Citrus fruits are rightly famous for their vitamin C contents. Certain other fruits and vegetables are also rich sources: cantaloupe, strawberries, broccoli, and brussels sprouts. No animal foods other than organ meats, such as chicken liver and kidneys, contain vitamin C. The humble potato is an important source of vitamin C in Western countries, where potatoes are eaten so frequently that they make substantial vitamin C contributions overall. They provide about 20 percent of all the vitamin C in the average diet. Vitamin C in foods is easily oxidized, so store cut produce and juices in airtight containers.

■ Iron is discussed in Chapter 9.

gout (GOWT): a metabolic disease in which crystals of uric acid precipitate in the joints.

Vitamin C and Iron Absorption Eating foods containing vitamin C at the same meal with foods containing iron can double or triple the absorption of iron from those foods. ■ This strategy is highly recommended for women and children, whose energy intakes are not large enough to guarantee that they will get enough iron from the foods they eat.

GOOD SOURCES*

SWEET RED PEPPER (raw)
½ c = 142 mg

BRUSSELS SPROUTS (cooked)
½ c = 48 mg

GRAPEFRUIT
½ grapefruit = 36 mg

SWEET POTATO
½ c = 20 mg

ORANGE JUICE
½ c = 62 mg

GREEN PEPPERS (raw)
½ c = 60 mg

BROCCOLI (cooked)
½ c = 51 mg

STRAWBERRIES
½ c = 43 mg

BOK CHOY
½ c = 22 mg

*These foods provide 10 percent or more of the Vitamin C Daily Value in a serving. For a 2,000-kcalorie diet, the DV is 60 mg/day.

IN SUMMARY

- The B vitamins and vitamin C are the water-soluble vitamins.

- Each B vitamin is part of an enzyme helper known as a coenzyme.

- As parts of coenzymes, the B vitamins assist in the release of energy from glucose, amino acids, and fats and help in many other body processes.

- Folate and vitamin B_{12} are important in cell division.

- Vitamin C's primary role is as an antioxidant.

- Historically, famous B vitamin-deficiency diseases are beriberi (thiamin) and pellagra (niacin). The vitamin C-deficiency disease is known as scurvy.

Table 8–3 summarizes functions, deficiency and toxicity symptoms, and food sources of the water-soluble vitamins.

TABLE 8–3

The Water-Soluble Vitamins—A Summary

Vitamin Name	Chief Functions	Deficiency Symptoms	Toxicity Symptoms	Significant Sources
Thiamin (Vitamin B_1)	Part of a coenzyme used in energy metabolism	Beriberi (edema or muscle wasting), anorexia and weight loss, neurological disturbances, muscular weakness, heart enlargement and failure	None reported	Enriched, fortified, or whole-grain products; pork
Riboflavin (Vitamin B_2)	Part of coenzymes used in energy metabolism	Inflammation of the mouth, skin, and eyelids; sensitivity to light; sore throat	None reported	Milk products; enriched, fortified, or whole-grain products; liver

continued

TABLE 8-3 (continued)

Vitamin Name	Chief Functions	Deficiency Symptoms	Toxicity Symptoms	Significant Sources
Niacin (Nicotinic acid, nicotinamide, niacinamide, vitamin B_3; precursor is dietary tryptophan, an amino acid)	Part of coenzymes used in energy metabolism	Pellagra (diarrhea, dermatitis, and dementia)	Niacin flush, liver damage, impaired glucose tolerance	Milk, eggs, meat, poultry, fish, whole-grain and enriched breads and cereals, nuts, and all protein-containing foods
Biotin	Part of a coenzyme used in energy metabolism	Skin rash, hair loss, neurological disturbances	None reported	Widespread in foods; GI bacteria synthesis
Pantothenic acid	Part of a coenzyme used in energy metabolism	Digestive and neurological disturbances	None reported	Widespread in foods
Vitamin B_6 (Pyridoxine, pyridoxal, pyridoxamine)	Part of coenzymes used in amino acid and fatty acid metabolism	Scaly dermatitis, depression, confusion, convulsions, anemia	Nerve degeneration, skin lesions	Meats, fish, poultry, potatoes, legumes, noncitrus fruits, fortified cereals, liver, soy products
Folate (Folic acid, folacin, pteroylglutamic acid)	Activates vitamin B_{12}; helps synthesize DNA for new cell growth	Anemia; smooth, red tongue; mental confusion; elevated homocysteine	Masks vitamin B_{12} deficiency	Fortified grains, leafy green vegetables, legumes, seeds, liver
Vitamin B_{12} (Cobalamin)	Activates folate; helps synthesize DNA for new cell growth; protects nerve cells	Anemia; nerve damage and paralysis	None reported	Foods derived from animals (meat, fish, poultry, shellfish, milk, cheese, eggs), fortified cereals
Vitamin C (Ascorbic acid)	Synthesis of collagen, carnitine, hormones, neurotransmitters; antioxidant	Scurvy (bleeding gums, pinpoint hemorrhages, abnormal bone growth, and joint pain)	Diarrhea, GI distress	Citrus fruits, cabbage-type vegetables, dark green vegetables (such as bell peppers and broccoli), cantaloupe, strawberries, lettuce, tomatoes, potatoes, papayas, mangoes

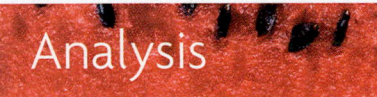

Analysis

YOUR DIET

Vitamins

A diet that supplies nutrient-dense foods from each food group can provide ample vitamins. To learn whether your food choices include good sources of vitamins, insert the recommended amounts of foods for the food groups listed in the table below using information from Tables 1-6 and 1-8 (pp. 20 and 21). Next, estimate your intake for each food group listed (you can use the food lists generated for previous diet projects to help you assess your intakes). In the third column, calculate the percentages of recommended intakes you consumed (Your Intake ÷ Recommended Intake × 100). Circle the percentages that are less than 100%.

■ Do you eat an adequate amount of fruits? Citrus fruits offer abundant vitamin C and some folate. Some deep orange fruits such as cantaloupe (not oranges) supply vitamin A.

Food Choices	Suggested Daily Intake	Your Daily Intake	% of Recommended Intake
Fruits	_____ c	_____ c	—
Grains	_____ oz	_____ oz	—
Milk and milk products	_____ c	_____ c	—

Vegetable Subgroups	Suggested Weekly Intake	Your Weekly Intake	
Dark green vegetables	_____ c	_____ c	—
Orange and deep yellow	_____ c	_____ c	—
Legumes	_____ c	_____ c	—

- Do you choose whole or enriched grains often? These choices supply substantial amounts of thiamin, riboflavin, niacin, and folate.

- Do you consume vitamin A- and D-fortified milk products regularly? Milk products supply riboflavin and vitamins A, D, and B_{12}.

- Do you make frequent choices from the vegetable subgroups? Most dark green vegetables provide significant folate and vitamins A and E. Many deep orange vegetables supply vitamin A. Legumes supply substantial folate.

- If you take supplements, list the vitamin content of each supplement. Then compare your intake with the upper levels (UL) for vitamins listed on the first page of this book.

CLINICAL APPLICATIONS

1. How might a vitamin deficiency weaken a client's resistance to disease?

2. Pull together information from Chapter 1 about the different food groups and the significant sources of vitamins shown in the photos and Snapshots throughout this chapter. Consider which vitamins might be lacking in the diet of a client who reports the following:

   - Dislikes leafy green vegetables

   - Never uses milk, milk products, or cheese

   - Follows a very low-fat diet

   - Eats a fruit or vegetable once per day

What additional information would help you pinpoint problems with vitamin intake?

Self Check

1. Which of the following vitamins are fat soluble?
   a. vitamins B, C, and E
   b. vitamins B, C, D, and E
   c. vitamins A, C, E, and K
   d. vitamins A, D, E, and K

2. Which of the following describes fat-soluble vitamins?
   a. They include thiamin, vitamin A, and vitamin K.
   b. They cannot be stored to any great extent and so must be consumed daily.
   c. Toxic levels can be reached by consuming citrus fruits and vegetables.
   d. They can be stored in the liver and fatty tissues and can build up toxic concentrations.

3. Night blindness and susceptibility to infection are the result of a deficiency of which vitamin?
   a. niacin
   b. vitamin C
   c. vitamin A
   d. vitamin K

4. Good sources of vitamin D include:
   a. eggs, fortified milk, and sunlight.
   b. citrus fruits, sweet potatoes, and spinach.
   c. leafy green vegetables, cabbage, and liver.
   d. breast milk, polyunsaturated plant oils, and citrus fruits.

5. Which of the following describes water-soluble vitamins?
   a. They include vitamins D and E.
   b. They are frequently toxic.
   c. They are stored extensively in tissues.
   d. They are easily absorbed and excreted.

6. A coenzyme is:
   a. a fat-soluble vitamin.
   b. an energy-yielding nutrient.
   c. a source of vitamin K.
   d. a molecule that combines with an enzyme to make it active.

7. Good food sources of folate include:
   a. citrus fruits, dairy products, and eggs.
   b. liver, legumes, and leafy green vegetables.
   c. dark green vegetables, corn, and cabbage.
   d. potatoes, broccoli, and whole-wheat bread.

8. Which vitamin is present only in foods of animal origin?
   a. riboflavin
   b. pantothenic acid
   c. vitamin B_{12}
   d. the inactive form of vitamin A

9. Which of the following nutrients is an antioxidant that protects water-soluble substances from oxidizing agents?
   a. beta-carotene
   b. thiamin
   c. vitamin C
   d. vitamin D

10. Eating foods containing vitamin C at the same meal can increase the absorption of which mineral?
    a. iron
    b. calcium
    c. magnesium
    d. folate

Answers to these questions appear in Appendix H.

Notes

1. A. C. Ross, Vitamin A and carotenoids, in M. E. Shils and coeditors, *Modern Nutrition in Health and Disease*, 10th ed. (Philadelphia: Lippincott, Williams, & Wilkins, 2006), pp. 351–375.

2. Ross, 2006; J. Bastien and C. Rochette-Egly, Nuclear retinoid receptors and the transcription of retinoid-target genes, *Gene* 328 (2004): 1–16; J. E. Balmer and R. Blomhoff, Gene expression regulation by retinoic acid, *Journal of Lipid Research* 43 (2002): 1773–1808.

3. Balmer and Blomhoff, 2002.

4. A. Li and coauthors, All-trans retinoic acid negatively regulates cytotoxic activities of nature killer cell line 92, *Biochemical and Biophysical Research Communications* 352 (2007): 42–47; A. Malaspina and F. Turkheimer, A review of the functional role and the expression profile of retinoid signaling and of nuclear receptors in human spinal cord, *Brain Research Bulletin* 71 (2007): 437–446; O. Sorg and coauthors, Proposed mechanisms of action for retinoid derivatives in the treatement of skin aging, *Journal of Cosmetic Dermatology* 4 (2005): 237–244; G. L. Johanning and C. J. Piyathilake, Retinoids and epigenetic silencing in cancer, *Nutrition Reviews* 61 (2003): 284–289.

5. M. J. Haskell and coauthors, Recovery from impaired dark adaptation in nightblind pregnant Nepali women who receive small daily doses of vitamin A as amaranth leaves, carrots, goat liver, vitamin A-fortified rice, or retinyl palmitate, *American Journal of Clinical Nutrition* 81 (2005): 461–471.

6. Ross, 2006.

7. Standing Committee on the Scientific Evaluation of Dietary Reference Intakes, Food and Nutrition Board, Institute of Medicine, *Dietary Reference Intakes for Vitamin A, Vitamin K, Arsenic, Boron, Chromium, Copper, Iodine, Iron, Manganese, Molybdenum, Nickel, Silicon, Vanadium, and Zinc* (Washington, DC: National Academy Press, 2001), pp. 82– 161.

8. C. Debier and Y. Larondelle, Vitamins A and E: Metabolism, roles, and transfer to offspring, *British Journal of Nutrition* 93 (2005): 153–174.

9. Debier and Larondelle, 2005; M. Clagett-Dame and H. F. DeLuca, The role of vitamin A in mammalian reproduction and embryonic development, *Annual Review of Nutrition* 22 (2002): 347–381; Standing Committee on the Scientific Evaluation of Dietary Reference Intakes, 2001, pp. 84–86.

10. P. I. Moreira and coauthors, Oxidative stress: The old enemy in Alzheimer's disease pathophysiology, *Current Alzheimer Research* 4 (2005): 403–408; F. Grodstein, J. Chen, and W. C. Willett, High-dose antioxidant supplements and cognitive function in community-dwelling elderly women, *American Journal of Clinical Nutrition* 77 (2003): 975–984; S. F. Clark, The biochemistry of antioxidants revisited, *Nutrition in Clinical Practice* 1

(2002): 5–17; M. J. Engelhart and coauthors, Dietary intake of antioxidants and risk of Alzheimer disease, *Journal of the American Medical Association* 287 (2002): 3223–3229.

11. National Institutes of Health State-of-the-Science Conference Statement: Multivitamin/mineral supplements and chronic disease prevention, *Annals of Internal Medicine* 145 (2006): 364–371.

12. Standing Committee on the Scientific Evaluation of Dietary Reference Intakes, 2001.

13. Ross, 2006; Debier and Larondelle, 2005.

14. K. L. Penniston and S. A. Tanumihardjo, The acute and chronic toxic effects of vitamin A, *American Journal of Clinical Nutrition* 83 (2006): 191–201.

15. H. S. Lam and coauthors, Risk of vitamin A toxicity from candy-like chewable vitamin supplements for children, *Pediatrics* 118 (2006): 820–824.

16. Standing Committee on the Scientific Evaluation of Dietary Reference Intakes, 2001, pp. 128–129.

17. Standing Committee on the Scientific Evaluation of Dietary Reference Intakes, 2001, pp. 126–133.

18. K. L. Penniston and S. A. Tanumihardjo, The acute and chronic toxic effects of vitamin A, *American Journal of Clinical Nutrition* 83 (2006): 191–201; H. A. Jackson and A. H. Sheehan, Effect of vitamin A on fracture risk, *Annals of Pharmacotherapy*, October 25, 2005, e-pub ahead of print; P. Genaro and L. A. Martini, Vitamin A supplementation and risk of skeletal fracture, *Nutrition Reviews* 62 (2004): 65–72.

19. K. Michaelsson and coauthors, Serum retinol levels and the risk of fractures, *New England Journal of Medicine* 348 (2003): 287–294; D. Feskanich and coauthors, Vitamin A intake and hip fractures among postmenopausal women, *Journal of the American Medical Association* 287 (2002): 47–54.

20. A. W. Norman and H. H. Henry, Vitamin D, in B. A. Bowman and R. M. Russell, eds., *Present Knowledge in Nutrition,* 9th ed. (Washington, DC: International Life Sciences Institute Press, 2006), pp. 198–210.

21. Norman and Henry, 2006; M. F. Holick, Vitamin D, in M. E. Shils and coeditors, *Modern Nutrition in Health and Disease,* 10th ed. (Philadelphia: Lippincott, Williams, & Wilkins, 2006), pp. 376–395.

22. P. T. Liu and coauthors, Toll-like receptor triggering of a vitamin D-mediated human antimicrobial response, *Science* 311 (2006): 1770–1773; Holick, 2006; T. Dietrich and coauthors, Association between serum concentrations of 25-hydroxyvitamin D and gingival inflammation, *American Journal of Clinical Nutrition*

82 (2005): 575–580; M. T. Cantorna and coauthors, Vitamin D status, 1, 25-dihydroxyvitamin D$_3$, and the immune system, *American Journal of Clinical Nutrition* 80 (2004): 1717S–1720S; M. F. Holick, Vitamin D: Importance in the prevention of cancers, type 1 diabetes, heart disease, and osteoporosis, *American Journal of Clinical Nutrition* 79 (2004): 362–371; J. B. Zella and H. F. DeLuca, Vitamin D and autoimmune diabetes, *Journal of Cellular Biochemistry* 88 (2003): 216–222; A. Zitterman, Vitamin D in preventive medicine: Are we ignoring the evidence? *British Journal of Nutrition* 89 (2003): 552–572; I. A. van der Mei and coauthors, Past exposure to sun, skin phenotype, and risk of multiple sclerosis: Case-control study, *British Medical Journal* 327 (2003): 316–332.

23. Holick, 2006.

24. J. M. Lee and coauthors, Viamin D deficiency in a healthy group of mothers and newborn infants, *Clinical Pediatrics* 46 (2007): 42–44; E. E. Ziegler and coauthors, Vitamin D deficiency in breast-fed infants in Iowa, *Pediatrics* 118 (2006): 603–610; P. Weisberg and coauthors, Nutritional rickets among children in the United States: Review of cases reported between 1986 and 2003, *American Journal of Clinical Nutrition* 80 (2004): 1697S–1705S.

25. S. S. Sullivan and coauthors, Adolescent girls in Maine are at risk for vitamin D insufficiency, *Journal of the American Dietetic Association* 105 (2005): 971–974; M. K. M. Lehtonen-Veromaa and coauthors, Vitamin D and attainment of peak bone mass among peripubertal Finnish girls: a 3-y prospective study, *American Journal of Clinical Nutrition* 76 (2002): 1446–1453.

26. Holick, 2006.

27. M. A. Johnson and M. G. Kimlin, Vitamin D, aging, and the 2005 Dietary Guidelines for Americans, *Nutrition Reviews* 64 (2006): 410–421.

28. Johnson and Kimlin, 2006; M. S. Calvo and S. J. Whiting, Prevalence of vitamin D insufficiency in Canada and the United States: Importance to health status and efficacy of current food fortification and dietary supplement use, *Nutrition Reviews* 61 (2003): 107–113.

29. Calvo and Whiting, 2003; V. Tangpricha and coauthors, Vitamin D insufficiency among free-living adults, *American Journal of Medicine* 112 (2002): 659–662.

30. M. F. Holick, Sunlight and vitamin D for bone health and prevention of autoimmune diseases, cancers, and cardiovascular disease, *American Journal of Clinical Nutrition* 80 (2004): 1678–1688; R. P. Heaney and coauthors, Human serum 25-hydroxycholecalciferol response to extended oral dosing with cholecalciferol, *American Journal of Clinical Nutrition* 77 (2003): 204–210.

31. L. M. Gartner, F. R. Greer, and the Section on Breastfeeding and Committee on Nutrition, Prevention of rickets and vitamin D deficiency: New guidelines for vitamin D intake, *Pediatrics* 111 (2003): 908–910.

32. Gartner, Greer, and the Section on Breastfeeding and Committee on Nutrition, 2003.

33. U. Singh, S. Devaraj, and I. Jialal, Vitamin E, oxidative stress, and inflammation, *Annual Review of Nutrition* 25 (2005): 151–174; Standing Committee on the Scientific Evaluation of Dietary Reference Intakes, Food and Nutrition Board, Institute of Medicine, *Dietary Reference Intakes for Vitamin C, Vitamin E, Selenium, and Carotenoids* (Washington, DC: National Academy Press, 2000), pp. 211–216.

34. Singh, Devaraj, and Jialal, 2005.

35. I. Lee and coauthors, Vitamin E in the primary prevention of cardiovascular disease and cancer, The Women's Health Study: A randomized controlled trial, *Journal of the American Medical Association* 294 (2005): 56–65; E. Lonn and coauthors, Effects of long-term vitamin E supplementation on cardiovascular events and cancer, *Journal of the American Medical Association* 293 (2005): 1338–1347.

36. E. R. Miller and coauthors, Meta-analysis: High-dosage vitamin supplementation may increase all-cause mortality, *Annals of Internal Medicine* 142 (2005): 37–46.

37. National Institutes of Health State-of-the-Science Conference Statement, 2006; P. M. Kris-Etherton and coauthors, Antioxidant vitamin supplements and cardiovascular disease, *Circulation* 110 (2004): 637–641; U.S. Preventive Task Force, Routine vitamin supplementation to prevent cancer and cardiovascular disease: Recommendations and rationale, *Annals of Internal Medicine* 139 (2003): 51–55.

38. AHA Scientific Statement, Diet and lifestyle recommendations revision 2006, A Scientific statement from the American Heart Association Nutrition Committee, *Circulation* 114 (2006): 82–96; Kris-Etherton and coauthors, 2004.

39. Miller and coauthors, 2005.

40. R. S. Eidelman and coauthors, Randomized trials of vitamin E in the treatment and prevention of cardiovascular disease, *Archives of Internal Medicine* 164 (2004): 1552–1556; P. G. Shekelle and coauthors, Effect of supplemental vitamin E for the prevention and treatment of cardiovascular disease, *Journal of General Internal Medicine* 19 (2004): 380–389; Vivekananthan and coauthors, Use of antioxidant vitamins for the prevention of cardiovascular disease: Meta-analysis of randomised trials, *Lancet* 361 (2003): 2017-2023.

41. J. N. Hathcock and coauthors, Vitamins E and C are safe across a broad range of intakes, *American Journal of Clinical Nutrition* 81 (2005): 736–745.

42. J. W. Suttie, Vitamin K, in M. E. Shils and coeditors, *Modern Nutrition in Health and Disease*, 10th ed. (Philadelphia: Lippincott, Williams, & Wilkins, 2006), pp. 412–425.

43. S. L. Booth and coauthors, Vitamin K intake and bone mineral density in women and men, *American Journal of Clinical Nutrition* 77 (2003): 512–516.

44. K. D. Cashman, Vitamin K status may be an important determinant of childhood bone health, *Nutrition Reviews* 63 (2005): 284–293.

45. H. J. Kalkwarf and coauthors, Vitamin K, bone turnover, and bone mass in girls, *American Journal of Clinical Nutrition* 80 (2004): 1075–1080.

46. N. C. Binkley and coauthors, A high phylloquinone intake is required to achieve maximal osteocalcin γ-carboxylation, *American Journal of Clinical Nutrition* 76 (2002): 1055–1060.

47. M. A. Johnson, Influence of vitamin K on anticoagulant therapy depends on vitamin K status and the source and chemical forms of vitamin K, *Nutrition Reviews* 63 (2005): 91–100.

48. R. F. Butterworth, Thiamin, in M. E. Shils and coeditors, *Modern Nutrition in Health and Disease*, 10th ed. (Philadelphia: Lippincott, Williams, & Wilkins, 2006), pp. 426–433.

49. A. D. Thomson and E. J. Marshall, The natural history and pathophysiology of Wernicke's Encephalopathy and Korsakoff's Psychosis, *Alcohol and Alcoholism* 41 (2006): 151–158.

50. C. Bourgeois, D. Cercantes-Laurean, and J. Moss, Niacin, in M. E. Shils and coeditors, *Modern Nutrition in Health and Disease*, 10th ed. (Philadelphia: Lippincott, Williams, & Wilkins, 2006), pp. 426–433.

51. J. Zempleni, Uptake, localization, and noncarboxylase roles of biotin, *Annual Review of Nutrition* 25 (2005): 175–196.

52. D. M. Mock, Biotin, in M. E. Shils and coeditors, *Modern Nutrition in Health and Disease*, 10th ed. (Philadelphia: Lippincott, Williams, & Wilkins, 2006), pp. 498–506.

53. A. D. Mackey, S. R. Davis, and J. F. Gregory, Vitamin B_6, in M. E. Shils and coeditors, *Modern Nutrition in Health and Disease*, 10th ed. (Philadelphia: Lippincott, Williams, & Wilkins, 2006), pp. 452–461; K. L. Tucker and coauthors, High homocysteine and low B vitamins predict cognitive decline in aging men: The Veterans Affairs Normative Aging Study, *American Journal of Clinical Nutrition* 82 (2005): 627–635; J. Bryan, E. Calvaresi, and D. Hughes, Short-term vitamin B-12 or vitamin B-6 supplementation slightly affects memory performance but not mood in women of various ages, *Journal of Nutrition* 132 (2002): 1345–1356.

54. M. Haim and coauthors, Serum homocysteine and long-term risk of myocardial infarction and sudden death in patients with coronary heart disease, *Cardiology* 107 (2006): 52–56; M. B. Kazemi and coauthors, Homocysteine level and coronary artery disease, *Angiology* 57 (2006): 9–14; D. S. Wald, M. Law, and J. K. Morris, Homocysteine and cardiovascular disease: Evidence on causality from a meta-analysis, *British Medical Journal* 325 (2002): 1202–1217; The Homocysteine Studies Collaboration, Homocysteine and risk of ischemic heart disease and stroke, *Journal of the American Medical Association* 288 (2002): 2015–2022.

55. D. Genser and coauthors, Homocysteine, folate, and vitamin B_{12} in patients with coronary heart disease, *Annals of Nutrition and Metabolism* 50 (2006): 413–419; Ø. Bleie and coauthors, Changes in basal and post-methionine load concentrations of total homocysteine and cystathionine after B vitamin intervention, *American Journal of Clinical Nutrition* 80 (2004): 641–648; E. Nurk and coauthors, Changes in lifestyle and plasma total homocysteine: The Hordaland Homocysteine Study, *American Journal of Clinical Nutrition* 79 (2004): 812–819; K. L. Tucker and coauthors, Breakfast cereal fortified with folic acid, vitamin B-6, and vitamin B-12 increases vitamin concentrations and reduces homocysteine concentrations: A randomized trial, *American Journal of Clinical Nutrition* 79 (2004): 805–811; G. Schnyder and coauthors, Effect of homocysteine-lowering therapy with folic acid, vitamin B12, and vitamin B6 on clinical outcome after percutaneous coronary intervention: The Swiss Heart Study: A randomized controlled trial, *Journal of the American Medical Association* 288 (2002): 973–979; K. M. Fairfield and R. H. Fletcher, Vitamins for chronic disease prevention in adults, *Journal of the American Medical Association* 287 (2002): 3116–3126.

56. E. Nurk and coauthors, 2004.

57. B-Vitamin Treatment Trialists' Collaboration, Homocysteine-lowering trials for prevention of cardiovascular events: A review of the design and power of the large randomized trials, *American Heart Journal* 151 (2006): 282–287.

58. E. Lonn and coauthors, Homocysteine lowering with folic acid and B vitamins in vascular disease, *New England Journal of Medicine* 354 (2006): 1567–1577; K. H. Bonaa and coauthors, Homocysteine lowering and cardiovascular events after acute myocardial infarction, *New England Journal of Medicine* 354 (2006): 1578–1588; G. Schnyder and coauthors, 2002.

59. National Institutes of Health State-of-the-Science Conference Statement, 2006.

60. R. Carmel, Folic acid, in M. E. Shils and coeditors, *Modern Nutrition in Health and Disease*, 10th ed. (Philadelphia: Lippincott, Williams, & Wilkins, 2006), pp. 470–481.

61. Centers for Disease Control and Prevention, Spina bifida and anencephaly before and after folic acid mandate—United States, 1995–1996 and 1999–2000, *Morbidity and Mortality Weekly Report* 53 (2004): 362–365; E. A. Yetley and J. I. Rader, Modeling the level of fortification and post-fortification assessments: U.S. experience, *Nutrition Reviews* 62 (2004): S50–S59.

62. L. B. Bailey and R. J. Berry, Folic acid supplementation and the occurrence of congenital heart defects, orofacial clefts, multiple births, and miscarriage, *American Journal of Clinical Nutrition* 81 (2005): 1213S–1217S.

63. M. S. Morris and coauthors, Folate and vitamin B-12 status in relation to anemia, macrocytosis, and cognitive impairment in older Americans in the age of folic acid fortification, *American Journal of Clinical Nutrition* 85 (2007): 193–200; J. L. Rader and B. O. Schneeman, Prevalence of neural tube defects, folate status, and folate fortification of enriched cereal-grain products in the United States, *Pediatrics* 117 (2006): 1394–1399.

64. S. Voutilainen and coauthors, Serum folate and homocysteine and the incidence of acute coronary events: The Kuopio Ischaemic Heart Disease Risk Factor Study, *American Journal of Clinical Nutrition* 80 (2004): 317–323; G. Schnyder and coauthors, 2002; D. S. Wald, M. Law, and J. K. Morris, Homocysteine and cardiovascular disease: Evidence on causality from a meta-analysis, *British Medical Journal* 325 (2002): 2002–2008.

65. Homocysteine Lowering Trialists' Collaboration, Dose-dependent effects of folic acid on blood concentrations of homocysteine: A meta-analysis of the randomized trials, *American Journal of Clinical Nutrition* 82 (2005): 806–812; C. M. Pfeiffer and coauthors, Biochemical indicators of B vitamin status in the US population after folic acid fortification: Results from the National Health and Nutrition Examination Survey 1999–2000, *American Journal of Clinical Nutrition* 82 (2005): 442–450; K. L. Tucker and coauthors, 2004; F. V. van Oort, and coauthors, Folic acid and reduction of plasma homocysteine concentrations in older adults: A dose-response study, *American Journal of Clinical Nutrition* 77 (2003): 1318–1323.

66. S. P. Stabler and R. H. Allen, Vitamin B_{12} deficiency as a worldwide problem, *Annual Review of Nutrition* 24 (2004): 299–326.

67. Carmel, 2006.

68. Standing Committee on the Scientific Evaluation of Dietary Reference Intakes, 2000, p. 155.

PHYTOCHEMICALS AND FUNCTIONAL FOODS

The wisdom of the familiar advice, "eat your vegetables, they're good for you," stands on firmer scientific ground today than ever before as population studies around the world suggest that diets rich in vegetables and fruits protect against heart disease, cancer, and other chronic diseases.[1] We now know that the "goodness" of vegetables, fruits, and other whole foods such as legumes and grains comes not only from the nutrients they contain, but also from the **nonnutrients** known as **phytochemicals** that they offer.[2]

Vegetables, fruits, and other whole foods are the simplest examples of foods now known as **functional foods**. Functional foods provide health benefits beyond basic nutrition by altering one or more physiological processes. Modified foods, such as those that have been fortified with nutrients, phytochemicals, herbs, or other food components, also are functional foods.[3] Functional foods that fit this description include orange juice fortified with calcium, folate-enriched cereal, beverages with herbal additives, and margarine enhanced with sterol esters. This Nutrition in Practice begins with a look at the evidence concerning the effectiveness and safety of a few selected phytochemicals in the simplest of functional foods—vegetables, fruits, and other whole foods. Then the discussion turns to examine the most controversial of functional foods—novel foods to which phytochemicals have been added to promote health. How these foods fit into a healthy diet is still unclear.[4] Table NP8–1 begins by defining some terms.

What are phytochemicals, and what do they do?

Phytochemicals are nonnutrient compounds found in plants. In foods, phytochemicals impart tastes, aromas, colors, and other characteristics. They give hot peppers their burning sensation, garlic and onions their pungent flavor, chocolate its bitter tang, and tomatoes their dark red color. In the body, phytochemicals can have profound physiological effects, acting as antioxidants, mimicking hormones, and suppressing the development of diseases.[5] Notably, cancer and heart disease are linked to processes involving oxygen compounds in the body, and antioxidants are thought to oppose these actions. Table NP8–2 introduces the names, possible physiological effects, and food sources of phytochemicals.

Why are phytochemicals receiving so much attention these days, and what are some examples of those in the spotlight?

Diets rich in whole grains, legumes, vegetables, and fruits seem to be protective against heart disease and cancer, but identify-

TABLE NP8–1
Phytochemical and Functional Food Terms

carotenoids (kah-ROT-eh-noyds): pigments commonly found in plants and animals, some of which have vitamin A activity. The carotenoid with the greatest vitamin A activity is beta-carotene.

flavonoids (FLAY-von-oyds): yellow pigments in foods; phytochemicals that may exert physiological effects on the body.

flaxseed: the small brown seed of the flax plant; valued as a source of linseed oil, fiber, and omega-3 fatty acids.

functional foods: foods that contain physiologically active compounds that provide health benefits beyond basic nutrition.

lignans: phytochemicals present in flaxseed, but not in flax oil, that are converted to phytosterols by intestinal bacteria and are under study for potential health benefits.

lutein (LOO-teen): a plant pigment of yellow hue; a phytochemical believed to play roles in eye functioning and health.

lycopene (LYE-koh-peen): a pigment responsible for the red color of tomatoes and other red-hued vegetables; a phytochemical that may act as an antioxidant in the body.

nonnutrients: compounds in foods that do not fit within the six classes of nutrients.

phytochemicals (FIGH-toe-CHEM-ih-cals): biologically active compounds of plants believed to confer resistance to diseases.

phytoestrogens: plant-derived compounds that have structural and functional similarities to human estrogen. Phytoestrogens include the isoflavones genistein, daidzein, and glycitein.

phytosterols: plant-derived compounds that have structural similarities to cholesterol and lower blood cholesterol by competing with cholesterol for absorption. Phytosterols include sterol esters and stanol esters.

tofu: a white curd made of soybeans, popular in Asian cuisines, and considered to be a functional food.

ing the specific foods or components of foods that are responsible is difficult.[6] Whenever bits of research news surface, however, new supplements appear—and terms like *antioxidants* and *phytochemicals* become buzzwords again. Meanwhile, scientists are conducting extensive research studies to discover phytochemical connections to disease prevention, but so far, solid evidence is generally lacking. Some of the likeliest candidates include **flavonoids** and **carotenoids** (including **lycopene**).

What are flavonoids, and in which foods are they found?

Flavonoids, a large group of phytochemicals known for their health-promoting qualities, are found in whole grains, vegetables,

A Sampling of Phytochemicals—Possible Effects and Food Sources

Name	Possible Effects	Food Sources
Capsaicin	May modulate blood clotting, may reduce the risk of fatal clots in heart and artery disease.	Hot peppers
Carotenoids (including beta-carotene, lutein, lycopene, and hundreds of related compounds)[a]	Act as antioxidants; possibly reduce risks of heart disease, age-related eye disease,[b] cancer, and other diseases.	Deeply pigmented fruits and vegetables (apricots, broccoli, cantaloupe, carrots, pumpkin, spinach, sweet potatoes, tomatoes)
Curcumin	May inhibit enzymes that activate carninogens.	Turmeric, a yellow-colored spice
Flavonoids (including flavones, flavonols, and isoflavones, catechins, others)[c,d]	Act as antioxidants; may scavenge carcinogens; bind to nitrates in the stomach, preventing conversion to nitrosamines; inhibit cell proliferation; flavonoids of blueberries may improve memory.	Berries, black tea, celery, chocolate, citrus fruits, green tea, olives, onions, oregano, purple grapes, purple grape juice, soybeans and soy products, vegetables, whole wheat, wine
Indoles	May trigger production of enzymes that block DNA damage from carcinogens; may inhibit estrogen action.	Broccoli and other cruciferous vegetables (brussels sprouts, cabbage, cauliflower), horseradish, mustard greens
Isothiocyanates (including sulforaphane)	May inhibit enzymes that activate carcinogens; trigger production of enzymes that detoxify carcinogens.	Broccoli and other cruciferous vegetables (brussels sprouts, cabbage, cauliflower), horseradish, mustard greens
Monoterpenes (including limonene)	May trigger enzyme production to detoxify carcinogens; may inhibit cancer promotion and cell proliferation.	Citrus fruit peels and oils
Organosulfur compounds (including allicin)	May speed production of carcinogen-destroying enzymes or slow production of carcinogen-activating enzymes.	Chives, garlic, leeks, onions
Phenolic acids[d] (including ellagic acid)	May trigger enzyme production to make carcinogens water soluble, facilitating excretion.	Coffee beans, fruits (apples, blueberries, cherries, grapes, oranges, pears, prunes, strawberries), oats, potatoes, soybeans
Phytic acid	Binds to minerals, preventing free-radical formation, possibly reducing cancer risk.	Whole grains
Phytoestrogens (members of the flavonoid family, genistein and diadzein)	May inhibit estrogen and produce these actions: inhibit cell replication in GI tract; reduce risk of breast, colon, ovarian, prostate, and other estrogen-sensitive cancers; reduce cancer cell survival; may reduce risk of osteoporosis. May also alter blood lipids favorably and reduce heart disease risk when consumed in soy foods.	Soybeans, soy flour, soy milk, tofu, textured vegetable protein, other legume products
Phytoestrogens (lignans)	Block estrogen activity in cells, possibly reducing the risk of cancer of the breast, colon, ovaries, and prostate.	Flaxseed, whole grains
Protease inhibitors	May suppress enzyme production in cancer cells, slowing tumor growth; inhibit hormone binding; inhibit malignant changes in cells.	Broccoli sprouts, potatoes, soybeans and other legumes, soy products
Resveratrol[e]	May offset artery-damaging effects of high-fat diets.	Red wine, peanuts
Saponins	May interfere with DNA replication, preventing cancer cells from multiplying; stimulate immune response.	Alfalfa sprouts, other sprouts, green vegetables, potatoes, tomatoes
Tannins[d]	May inhibit carcinogen activation and cancer promotion; act as antioxidants.	Black-eyed peas, grapes, lentils, red and white wine, tea

[a] Other carotenoids include alpha-carotene, beta-cryptoxanthin, and zeaxanthin.
[b] The age-related eye disease is macular degeneration.
[c] Other flavonoids of interest include ellagic acid and ferulic acid.
[d] A subset of the larger group *polyphenolic phytochemicals*.
[e] A member of the chemical group stilbene, which is a subset of the larger group *polyphenolic phytochemicals*.

fruits, herbs, spices, teas, and red wine. A large body of population evidence spanning many countries reveals that deaths from cancer, heart disease, and heart attacks are less common where these foods are plentiful in the diet, where tea is a beverage, or where red wine is consumed in moderation.[7] Flavonoids are powerful antioxidants that may help to protect LDL against oxidation and reduce blood platelet stickiness, making blood clots less likely.[8] Nevertheless, more evidence is needed before any claims can be made for flavonoids themselves as the protective factors in foods, particularly when they are extracted from foods or herbs and sold as supplements.[9] Furthermore, studies evaluating potentially adverse health effects of flavonoids must be conducted as well.[10]

Flavonoids impart a bitter taste to foods, so manufacturers often refine away the natural flavonoids to please consumers who usually prefer milder flavors.[11] For example, the hearty taste of whole wheat foods vanishes when whole wheat is refined into white flour by removing the tough brown parts that contain flavonoids. For white grape juice or white wine, manufacturers remove the red, flavonoid-rich grape skins to lighten the flavor and color of the product, while greatly reducing its beneficial flavonoid content. For example, one flavonoid of grapes and wine may have anticancer activity.*[12]

What about carotenoids?

In addition to flavonoids, fruits and vegetables are rich in carotenoids—the red and yellow pigments of plants. Some carotenoids, such as beta-carotene, are vitamin A precursors. Some research suggests that a diet rich in carotenoids is associated with a lower risk of heart disease.[13] Among the carotenoids that may defend against heart disease is lycopene. Researchers are investigating a tentative link between low levels of lycopene in the blood and elevated incidence of heart disease, heart attack, and stroke.[14] Lycopene may also protect against certain types of cancer.

What is lycopene, and what foods contain it?

Lycopene is a red pigment with powerful antioxidant activity found in guava, papaya, pink grapefruit, tomatoes (especially cooked tomatoes and tomato products), and watermelon. More than 80 percent of the lycopene consumed in the United States comes from tomato products such as tomato sauce, tomato juice, and catsup. Around the world, people who eat five or more tomato-containing meals per week are less likely to suffer from cancers of the esophagus, prostate, or stomach than those who avoid tomatoes.[15] Lycopene is a leading candidate for this protective effect.

Lycopene may inhibit cancer cell reproduction. In some studies, women who consumed a diet rich in fruits and vegetables had high blood lycopene concentrations and a reduced cervical cancer risk.[16]

Are lycopene supplements a good idea?

Stick with the tried and true. Eat more fruits and vegetables. Cancer research favors eating foods high in lycopene, but

*The flavonoid is resveratrol.

research does not support the consumption of purified supplements of lycopene. Recall from Chapter 8 that diets high in the carotenoid beta-carotene often correlate with low rates of lung cancer. When given to smokers in studies, however, purified beta-carotene in supplement form *increases* lung cancer rates. The only safe option is to eat lycopene-rich foods such as tomatoes and tomato products and avoid concentrated supplements of lycopene until safety studies are completed.

Do foods contain other phytochemicals that may help to protect people from cancer or other diseases?

Foods contain thousands of different phytochemicals, and so far only a few have been researched at all. There are still many questions about the phytochemicals that have been studied and only tentative answers about their roles in human health. For example, compared with people in the West, Asians living in Asia suffer less frequently from osteoporosis (adult bone loss); cancers, especially of the breast, colon, and prostate; and heart disease. Among many differences between the diets of the two regions, Asians consume far more soybeans and soy products such as **tofu** than do Westerners.

Soybeans contain phytochemicals known as **phytoestrogens**. Researchers suspect that the phytoestrogens of soy foods, their protein content, or a combination of these factors may be responsible for health effects in soy-eating peoples. Nevertheless, research, though ongoing, is limited. So far, we know with certainty that phytoestrogens are plant-derived chemical relatives of the human hormone estrogen, they weakly mimic or modulate the hormone's effects on some body tissues, and they also act as antioxidants. We also know that breast cancer, colon cancer, and prostate cancer are estrogen-sensitive—that is, they grow when exposed to estrogen. Whether any of these or other actions of phytoestrogens may alter the course of estrogen-sensitive cancers remains unknown, but results from breast cancer studies do not support the idea.[17] Until more is known, the safest way to obtain soy phytoestrogens is to include moderate amounts of soy-based foods in the diet as generations of Asian people have done through the ages.[18]

Other foods under study for potential health benefits include **flaxseed** and its oil. Flaxseed is found as a whole seed or ground meal or as flaxseed oil. Flaxseed is of interest for its possible benefits to heart health because it is the richest known source of both the omega-3 fatty acid linolenic acid and the phytoestrogen **lignans**, as well as being a good source of soluble fiber.[19] Flaxseed oil, though rich in linolenic acid, does not contain fiber or lignans. Studies of the direct effect of giving flaxseed to people are lacking, and some risks are possible with its use. Large quantities of flaxseed can cause digestive distress, and severe allergic reactions to flaxseed have been reported.

What about other phytochemical supplements?

Even when people do not eat in the best interest of health, taking supplements of purified phytochemicals is not the way to go. Phytochemicals can alter body functions, sometimes powerfully. Researchers are just beginning to understand how a handful of phytochemicals work, and what is current today may change tomorrow. The body is equipped to handle phytochemicals in diluted form, mixed with all of the other constituents of foods, but it is not adapted to phytochemicals in concentrated form. Fruits, vegetables, legumes, whole grains, nuts, and seeds contain a wide array of beneficial nutrients and thousands of phytochemicals. The best way to reap the benefits of phytochemicals is by eating foods, not supplements (see Figure NP8–1).

How do whole foods compare with processed foods that have been enriched with phytochemicals?

Good question. The U.S. food supply is being transformed by a proliferation of functional foods—foods claimed to provide health benefits beyond those of the traditional nutrients. In truth, all the foods mentioned so far in this Nutrition in Practice, from tomatoes to tofu, are functional—they stand out as having a potentially greater impact on the health of the body (see Table NP8–3 on p. 236). Cranberries may protect against urinary tract infections because cranberries contain a phytochemical that dislodges bacteria from the tract.[20] Cooked tomatoes provide lycopene, along with **lutein** (an antioxidant associated with healthy eye function), vitamin C (an antioxidant vitamin), and many other healthful attributes. This has not stopped food manufacturers from trying to create functional foods as well. As consumer demand for healthful foods continues to grow, so will the development of func-

---

FIGURE NP8–1

An Array of Phytochemicals in a Variety of Fruits and Vegetables

Broccoli and broccoli sprouts contain an abundance of sulforaphane.

An apple a day—rich in flavonoids.

The phytoestrogens genistein and diadzein are found in soybeans and soy products.

Garlic, leeks, and onions, with their abundant organosulfur compounds, may benefit health.

The phytochemicals of grapes, red wine, and peanuts include resveratrol.

Strawberries are a source of flavonoids.

Citrus fruits provide limonene.

The flavonoids in black tea differ from those in green tea, with different potential effects on health.

Cooked tomatoes are the best source of lycopene, a carotenoid that may reduce cancer risk.

The flavonoids in cocoa, chocolate, fruits and vegetables, legumes, and tea may benefit health.

Spinach and other green, orange, red, and yellow fruits and vegetables contain carotenoids.

Flaxseed is the richest source of lignans.

Blueberries are among the richest sources of flavonoids.

Garlic, Citrus fruits, © EyeWire, Inc.; Flaxseed, Courtesy of Flax Council of Canada; Broccoli, Courtesy of Brassica Protection Products; Apples, Strawberries, Tomatoes, Blueberries, Black Tea, Soybeans, Grapes, Digital Imagery © 2001 PhotoDisc, Inc.; Orange, Spinach, Cocoa by Matthew Farruggio

TABLE NP8-3

Categories of Functional Foods

These categories of functional foods have been identified by the American Dietetic Association:[a]

1. Foods with naturally occurring beneficial compounds, which qualify to bear on their labels FDA "A" level health claim. These claims are listed in Table 1-11 on p. 29.
2. Foods with naturally occurring beneficial compounds, such as fish that provide fish oils, which have significant evidence supporting their benefits but lack the degree of certainty necessary for an "A" health claim.
3. Foods with biological plausibility for an effect, such as chocolate, garlic, and other foods discussed earlier in this Nutrition in Practice.
4. Foods fortified with nutrients or other constituents associated with the prevention or treatment of a disease, such as calcium-fortified orange juice marketed to reduce the risk of osteoporosis. Other examples include folate-enriched cereal to reduce birth defects and margarine enhanced with sterol esters to lower blood cholesterol.
5. Fortified foods marketed as "dietary supplements." Examples include beverages, candies, and bars enhanced with antioxidants, minerals, or herbs.

[a]Position of the American Dietetic Association: Functional foods, *Journal of the American Dietetic Association* 104 (2004): 814–826.

tional foods.[21]

What are some examples of manufactured functional foods?

Many processed foods become functional foods when they are fortified with nutrients or enhanced with phytochemicals or herbs (calcium-fortified orange juice, for example). Less frequently, an entirely new food is created, as in the case of a meat substitute made of mycoprotein—a protein derived from a fungus.*[22] This functional food not only provides dietary fiber, polyunsaturated fats, and high-quality protein, but it also lowers LDL cholesterol, raises HDL cholesterol, improves glucose response, and prolongs satiety after a meal. Such a novel functional food raises the question—is it a food or a drug?

Is the distinction between a food and a drug pretty clear?

Not too long ago, most of us could agree on what was a food and what was a drug. Today, functional foods blur the distinctions. They have characteristics similar to both foods and drugs but do not fit neatly into either category. For example, food companies have already developed products with added phytochemicals. Consider margarine, for example.

Eating nonhydrogenated margarine sparingly instead of butter generously may lower blood cholesterol slightly over several months and clearly falls into the food category. Taking the drug Lipitor, on the other hand, lowers blood cholesterol significantly within weeks and clearly falls into the drug category. But margarine enhanced with **phytosterols** that lower blood cholesterol is in a gray area between the two. The margarine looks and tastes like a food, but it acts like a drug.

The use of functional foods as drugs creates a whole new set of diet-planning problems. Not only must foods provide an adequate intake of all the nutrients to support good health, but they must also deliver druglike ingredients to protect against disease. Like drugs used to treat chronic diseases, functional foods may need to be eaten several times per day for several months or years to have a beneficial effect. Sporadic users may be disappointed in the results. When used four times per day for four weeks, margarine enriched with sterol esters reduces cholesterol by 8 percent, much more than regular margarine does, but not nearly as much as the 32 percent reduction seen with cholesterol-lowering drugs.[23] For this reason, functional foods may be more useful for prevention and mild cases of disease than for intervention and more severe cases.

How do the costs of functional foods compare with those of conventional foods?

Foods and drugs differ dramatically in cost. Functional foods such as fruits and vegetables incur no added costs, but foods that have been manufactured with added phytochemicals can be expensive, costing up to six times as much as their conventional counterparts. The price of functional foods typically falls between that of traditional foods and medicines.

What about using sterol-enhanced margarines to lower cholesterol or other manufactured functional foods to help prevent cancer?

To achieve a desired health effect, which is the better choice: to eat a food designed to affect some body function or simply to adjust the diet? Does it make more sense to use a margarine enhanced with a sterol ester that lowers blood cholesterol or simply to limit the amount of butter eaten?* Is it smarter to eat eggs enriched with omega-3 fatty acids or to restrict egg consumption? Might functional foods offer a sensible solution for improving our nation's health—if done correctly? Perhaps so—but there is a problem with functional foods: the food industry is moving too fast for either scientists or the Food and Drug Administration (FDA) to keep up. Consumers were able to buy soup with St. John's wort that claimed to enhance mood and fruit juice with echinacea that was supposed to fight colds while scientists were still conducting their studies on these ingredients. Research to determine the safety and effectiveness of these substances is still in progress. Until this work is complete,

* This mycoprotein product is marketed under the trade name Quorn (pronounced KWORN).

* Margarine products that lower blood cholesterol contain either sterol esters from vegetable oils, soybeans, and corn or stanol esters from wood pulp.

consumers are on their own in finding the answers to the following questions:

- **Does it work?** Research is generally lacking and findings are often inconclusive.

- **How much does it contain?** Food labels are not required to list the quantities of added phytochemicals. Even if they were, consumers have no standard for comparison and cannot deduce whether the amounts listed are a little or a lot. Most importantly, until research is complete, food manufacturers do not know what amounts (if any) are most effective—or most toxic.

- **Is it safe?** Functional foods can act like drugs. They contain ingredients that can alter body functions and cause allergies, drug interactions, drowsiness, and other side effects. Yet, unlike drug labels, food labels do not provide instructions for the dosage, frequency, or duration of treatment.

- **Has the FDA issued warnings about any of the ingredients?** Check the FDA's MedWatch Web site (www.fda.gov/medwatch) or call the FDA (1-888-INFO-FDA) to find out.

- **Is it healthy?** Adding phytochemicals to a food does not magically make it a healthy choice. A candy bar may be fortified with phytochemicals, but it is still made mostly of sugar and fat.

Critics suggest that the designation "functional foods" may be nothing more than a marketing tool. After all, even the most experienced researchers cannot yet identify the perfect combination of nutrients and phytochemicals to support optimal health. Yet manufacturers are freely experimenting with various concoctions as if they possessed that knowledge. Is it okay for them to sprinkle phytochemicals on fried snack foods and label them "functional," thus implying health benefits? Do we want our children receiving their nourishment from fortified caramel candies and chocolate cakes?

What is the final word regarding phytochemicals and functional foods?

Functional foods currently on the market promise to "enhance mood," "promote relaxation and good karma," "increase alertness," and "improve memory," among other claims.

Nature has elegantly designed foods to provide us with a complex array of dozens of nutrients and thousands of additional compounds that may benefit health—most of which we have yet to identify or understand. Over the years, we have taken those foods and first deconstructed them and then reconstructed them in an effort to "improve" them. With new scientific understandings of how nutrients—and the myriad of other compounds in foods—interact with genes, we may someday be able to design foods to meet the *exact* health needs of *each* individual.[24] If the present trend continues, then someday physicians may be able to prescribe the perfect foods to enhance a person's health, and farmers will be able to grow them. In the meantime, however, it seems clear that a moderate approach to phytochemicals and functional foods is warranted. People who eat the recommended amounts of a variety of fruits and vegetables may cut their risk of many diseases by as much as half. Replacing some meat with soy foods or other legumes may also lower heart disease and cancer risks. Beneficial constituents are widespread among foods. Take a no-nonsense approach where your health is concerned: choose a wide variety of whole grains, legumes, fruits, and vegetables in the context of an adequate, balanced, and varied diet and receive all of the health benefits that these foods offer.

Notes

1. M. Pavia and coauthors, Association between fruit and vegetable consumption and oral cancer: a meta-analysis of observational studies, *American Journal of Clinical Nutrition* 83 (2006): 1126–1134; L. E. Kelemen and coauthors, Vegetables, fruit, and antioxidant-related nutrients and risk of non-Hodgkin lymphoma: A National Cancer Institute Surveillance, Epidemiology, and End Results population-based case-control study, *American Journal of Clinical Nutrition* 83 (2006): 1401–1410; D. P. Hayes, The protective role of fruits and vegetables against radiation-induced cancer, *Nutrition Reviews* 63 (2005): 303–311; J. M. Genkinger and coauthors, Fruit, vegetables, and antioxidant intake and all-cause cancer, and cardiovascular disease mortality in a community-dwelling population in Washington County, Maryland, *American Journal of Epidemiology* 160 (2004): 1223–1333; L. M. Steffen and coauthors, Associations of whole-grain, refined-grain, and fruit and vegetable consumption with risks of all-cause mortality and incident coronary artery disease and ischemic stroke: The Aterosclerosis Risk in communities (ARIC) Study, *American Journal of Clinical Nutrition* 78 (2003): 383–390; L. A. Bazzano and coauthors, Fruit and vegetable intake and risk of cardiovascular disease in US adults: The First National Health and Nutrition Examination Survey Epidemiologic Follow-up Study, *American Journal of Clinical Nutrition* 76 (2002): 93–99.

2. D. F. Birt, Phytochemicals and cancer prevention: From epidemiology to mechanism of action, *Journal of the American Dietetic Association* 106 (2006): 20–21.

3. Position of the American Dietetic Association: Functional foods, *Journal of the American Dietetic Association* 104 (2004): 814–826.

4. C. H. Halsted, Dietary supplements and functional foods: 2 sides of a coin? *American Journal of Clinical Nutrition* 77 (2003): 1001S–1007S.

5. A. Smeltzer and Y. S. Kim, The effects of bioactive food components on *p53* pathway in cancer prevention, *Nutrition Today* 40 (2005): 50–53; P. M. Kris-Etherton and coauthors, Bioactive compounds in foods: Their role in the prevention of cardiovascular disease and cancer, *American Journal of Medicine* 113 (2002): 71S–88S.

6. Kris-Etherton and coauthors, 2002.

7. J. A. Vita, Polyphenols and cardiovascular disease: Effects on endothelial and platelet function, *American Journal of Clinical Nutrition* 81 (2005): 292S–297S; A. H. Wu and coauthors, Green tea and risk of breast cancer in Asian Americans, *International Journal of Cancer* 106 (2003): 574–579; J. M. Geleijnse and coauthors, Inverse association of tea and flavonoid intakes with incident myocardial infarction: The Rotterdam Study, *American Journal of Clinical Nutrition* 75 (2002): 880–886; P. Knekt and coauthors, Flavonoid intake and risk of chronic disease, *American Journal of Clinical Nutrition* 76 (2002): 560–568.

8. M. J. Davies and coauthors, Black tea consumption reduces total and LDL cholesterol in mildly hypercholesterolemic adults, *Journal of Nutrition* 133 (2003): 3298S–3302S.

9. A. Scalbert, I. T. Johnson, and M. Saltmarsh, Polyphenols: Antioxidants and beyond, *American Journal of Clinical Nutrition* 81 (2005): 215S–217S; J. A. Ross and C. M. Kasum, Dietary flavonoids: Bioavailability, metabolic effects, and safety, *Annual Review of Nutrition* 22 (2002): 19–34.

10. L. I. Mennen and coauthors, Risks and safety of polyphenol consumption, *American Journal of Clinical Nutrition* 81 (2005): 326S–329S.

11. I. Lesschaeve and A. C. Noble, Polyphenols: factors influencing their sensory properties and their effects on food and beverage preferences, *American Journal of Clinical Nutrition* 81 (2005): 330S–335S.

12. P. Signorelli and R. Ghidoni, Resveratrol as an anticancer nutrient: Molecular basis, open questions and promises, *Journal of Nutritional Biochemistry* 16 (2005): 449–466.

13. S. K. Osganian and coauthors, Dietary carotenoids and risk of coronary artery disease in women, *American Journal of Clinical Nutrition* 77 (2003): 1390–1399.

14. T. H. Rissanen and coauthors, Serum lycopene concentrations and carotid atherosclerosis: The Kuopio Ischaemic Heart Disease Risk Factor Study, *American Journal of Clinical Nutrition* 77 (2003): 133–138.

15. E. Giovannucci and coauthors, A prospective study of tomato products, lycopene, and prostate cancer risk, *Journal of the National Cancer Institute* 94 (2002): 391–398; D. Herber and Q. Y. Lu, Overview of mechanisms of action of lycopene, *Experimental Biology and Medicine* 227 (2002): 920–923; T. M. Vogt and coauthors, Serum lycopene, other serum carotenoids, and risk of prostate cancer in US blacks and whites, *American Journal of Epidemiology* 155 (2002): 1023–1032; Q. Y. Lu and coauthors, Inverse associations between plasma lycopene and other carotenoids and prostate cancer, *Cancer Epidemiology, Biomarkers, and Prevention* 10 (2001): 749–756.

16. R. Garcia-Closa and coauthors, The role of diet and nutrition in cervical carcinogenesis: A review of recent evidence, *International Journal of Cancer* 117 (2005): 629–637; R. L. Sedjo and coauthors, Vitamin A, carotenoids, and risk of persistent oncogenic human papillomavirus infection, *Cancer Epidemiology, Biomarkers and Prevention* 11 (2002): 876–884.

17. L. Keinan-Boker and coauthors, Dietary phytoestrogens and breast cancer risk, *American Journal of Clinical Nutrition* 79 (2004): 282–288; A. Cassidy, Potential risks and benefits of phytoestrogen-rich diets, *International Journal of Vitamin Nutrition Research* 73 (2003): 120–126.

18. C. Munro and coauthors, Soy isoflavones: A safety review, *Nutrition Reviews* 61 (2003): 1–33.

19. L. T. Bloedon and P. O. Szapary, Flaxseed and cardiovascular risk, *Nutrition Reviews* 62 (2004): 18–27.

20. A. B. Howell and B. Foxman, Cranberry juice and adhesion of antibiotic resistant uropathogens, *Journal of the American Medical Association* 287 (2002): 3082–3083.

21. Position of the American Dietetic Association, 2004.

22. T. Peregrin, Mycoprotein: Is America ready for a meat substitute derived from a fungus? *Journal of the American Dietetic Association* 102 (2002): 628.

23. L. A. Simons, Additive effect of plant sterol-ester margarine and cerivastatin in lowering low-density lipoprotein cholesterol in primary hypercholesterolemia, *American Journal of Cardiology* 90 (2002): 737–740.

24. J. A. Milner, Functional foods and health: A US perspective, *British Journal of Nutrition* 88 (2002): S151–S158.

9

Water
and the Minerals

Water and Body Fluids

© BananaStock/SuperStock

239

Water is the most indispensable nutrient of all.

The body's water cannot be considered separately from the minerals dissolved in it. A person can drink pure water, but in the body, water mingles with minerals to become fluids in which all life processes take place. This chapter begins by discussing the body's fluids and their chief minerals. The focus then shifts to other functions of the minerals.

Water and Body Fluids

Water constitutes about 60 percent of an adult's body weight and a higher percentage of a child's body weight. Every cell in the body is bathed in a fluid of the exact composition that is best for that cell. The body fluids bring to each cell the ingredients it requires and carry away the end products of the life-sustaining reactions that take place within the cell's boundaries. The water in the body fluids:

- Carries nutrients and waste products throughout the body

- Participates in metabolic reactions

- Serves as the solvent for minerals, vitamins, amino acids, glucose, and many other small molecules

- Aids in maintaining the body's blood pressure and temperature

- Maintains blood volume

- Acts as a lubricant and cushion around joints and inside the eyes, spinal cord, and amniotic sac surrounding a fetus in the womb

To support these and other vital functions, the body actively regulates its **water balance**.

Water Balance

The cells themselves regulate the composition and amounts of fluids within and surrounding them. The entire system of cells and fluids remains in a delicate but firmly maintained state of dynamic equilibrium. Imbalances such as **dehydration** (see Table 9–1) and **water intoxication** can occur, but the body quickly restores the balance to normal if it can. The body controls both water intake and water excretion.

Water Intake Regulation The body can survive for only a few days without water. In healthy people, thirst governs water intake. Thirst is finely adjusted to ensure a water intake that meets the body's needs. When the blood becomes too concentrated (having lost water but not salt and other dissolved substances), the mouth becomes dry, and the brain center known as the **hypothalamus** initiates drinking behavior.

Thirst lags behind the lack of water. A water deficiency that develops slowly can switch on drinking behavior in time to prevent serious dehydration, but a deficiency that develops quickly may not. Also, thirst itself does not remedy a water deficiency; a person must pay attention to the thirst signal and take the time to get a drink. With aging, thirst sensations may diminish. Dehydration can threaten elderly people who do not develop the habit of drinking water regularly.

Water Excretion Regulation Water excretion is regulated by the brain and the kidneys. The cells of the brain's hypothalamus, which monitor blood salts, stimulate the **pituitary gland** to release **antidiuretic hormone (ADH)** whenever the salts are too

TABLE 9–1

Signs of Mild and Severe Dehydration

Mild Dehydration (Loss of <5% Body Weight)	Severe Dehydration (Loss of >5% Body Weight)
Thirst	Pale skin
Sudden weight loss	Bluish lips and fingertips
Rough dry skin	Confusion; disorientation
Dry mouth, throat, body linings	Rapid, shallow breathing
Rapid pulse	Weak, rapid, irregular pulse
Low blood pressure	Thickening of blood
Lack of energy; weakness	Shock; seizures
Impaired kidney function	Coma; death
Reduced quantity of urine; concentrated urine	
Decreased mental functioning	
Decreased muscular work and athletic performance	
Fever or increased internal temperature	
Fainting	

Source: Standing Committee on the Scientific Evaluation of Dietary Reference Intakes, Food and Nutrition Board, Institute of Medicine, *Dietary Reference Intakes: Water, Potassium, Sodium, Chloride, and Sulfate* (Washington, DC: National Academies Press, 2005), pp. 90–122.

■ The enzyme *renin* (REN-in), released by the kidneys in response to low blood pressure, aids the kidneys in retaining water through the *renin-angiotensin mechanism.*

■ 500 mL = about ½ qt

concentrated or the blood volume or blood pressure is too low. ADH stimulates the kidneys to reabsorb water rather than excrete it. Thus, the more water you need, the less you excrete.

If too much water is lost from the body, blood volume and blood pressure fall. Cells in the kidneys respond to the low blood pressure by releasing an enzyme. ■ Through a complex series of events, involving the hormone **aldosterone**, this enzyme also causes the kidneys to retain more water. Again, the effect is that when more water is needed, less is excreted.

Minimum Water Needed These mechanisms can maintain water balance only if a person drinks enough water. The body must excrete a minimum of about 500 milliliters each day as urine—enough to carry away the waste products generated by a day's metabolic activities. Above this amount, excretion adjusts to balance intake, so the more a person drinks, the more dilute the urine becomes. In addition to urine, some water is lost from the lungs as vapor, some is excreted in feces, and some evaporates from the skin. A person's water losses from all of these routes total about 2½ liters (about 2½ quarts) per day on the average. ■ Table 9–2 shows how fluid intake and output naturally balance out.

water balance: the balance between water intake and water excretion that keeps the body's water content constant.

dehydration: the loss of water from the body that occurs when water output exceeds water input. The symptoms progress rapidly from thirst to weakness, to exhaustion and delirium, and end in death if not corrected.

water intoxication: the rare condition in which body water contents are too high. The symptoms may include confusion, convulsion, coma, and even death in extreme cases.

hypothalamus (high-poh-THALL-uh-mus): a part of the brain that helps regulate many body balances, including fluid balance.

pituitary (pit-TOO-ih-tary) **gland:** in the brain, the "king gland" that regulates the operation of many other glands.

antidiuretic hormone (ADH): a hormone released by the pituitary gland in response to high salt concentrations in the blood. The kidneys respond by reabsorbing water.

aldosterone (al-DOS-ter-own): a hormone secreted by the adrenal glands that stimulates the reabsorption of sodium by the kidneys; also regulates chloride and potassium concentrations.

TABLE 9–2

Water Balance

Water Sources	Amount (mL)	Water Losses	Amount (mL)
Liquids	550 to 1,500	Kidneys (urine)	500 to 1,400
Foods	700 to 1,000	Skin (sweat)	450 to 900
Metabolic water	200 to 300	Lungs (breath)	350
		GI tract (feces)	150
Total	1,450 to 2,800	Total	1,450 to 2,800

Water Recommendations and Sources

Water Recommendations and Sources Water needs vary greatly depending on the foods a person eats, the environmental temperature and humidity, the person's activity level, and other factors. Accordingly, a general water requirement is difficult to establish. In the past, recommendations for adults were expressed in proportion to the amount of energy expended under normal environmental conditions. For the person who expends about 2,000 kcalories per day, this works out to 2 to 3 liters, or about 8 to 12 cups. This recommendation is in line with the Adequate Intake (AI) for *total* water ■ set by the DRI committee. Total water includes not only drinking water, but water in other beverages and in foods as well.[1]

Because a wide range of water intakes will prevent dehydration and its harmful consequences, the AI is based on average intakes. Strenuous physical activity and heat stress can increase water needs considerably, however.[2] In general, you can tell from the color of the urine whether a person needs more water. Pale yellow urine reflects appropriate dilution.[3]

The obvious dietary sources of water are water itself and other beverages, but nearly all foods also contain water. Most fruits and vegetables contain up to 90 percent water; many meats and cheeses contain at least 50 percent. The energy nutrients in foods also give up water during metabolism.

People often ask whether caffeine-containing beverages such as coffee, tea, or soda can help to meet water needs. When people who normally abstain from caffeine drink a caffeine-containing beverage, their urine output increases somewhat more than it would for a similar amount of water. This occurs because caffeine acts a mild diuretic. Research is mixed on whether any but the highest caffeine intakes (four or five cups of coffee) cause a net water deficit in the body; most people make up for small water losses by drinking additional fluid later. The DRI committee considered such findings in making its recommendations for water intake and concluded that "caffeinated beverages contribute to the daily total water intake similar to that contributed by non-caffeinated beverages."[4] In other words, it does not seem to matter whether people rely on caffeine-containing beverages or other beverages to meet their fluid needs.

In contrast, alcohol should probably not be used to meet fluid needs. As Nutrition in Practice 20 explains, alcohol has many adverse effects on health and nutrition status.

Fluid and Electrolyte Balance

When mineral **salts** dissolve in water, they separate (dissociate) into charged particles known as ions, which can conduct electricity. For this reason, a salt that dissociates in water is known as an **electrolyte.** ■ The body fluids, which contain water and partly dissociated salts, are **electrolyte solutions**.

The body's electrolytes are vital to the life of the cells and therefore must be closely regulated to help maintain the appropriate distribution of body fluids. The major minerals form salts that dissolve in the body fluids; the cells direct where these salts go; and the movement of the salts determines where the fluids flow because water follows salt. ■ Cells use this force to move fluids back and forth across their membranes. Thanks to the electrolytes, water can be held in compartments where it is needed.

Proteins in the cell membranes move ions in or out of the cells. These protein pumps tend to concentrate sodium and chloride outside cells and potassium and other ions inside. By maintaining specific amounts of sodium outside and potassium inside, cells can regulate the exact amounts of water inside and outside their boundaries.

Margin notes

■ AI for *total* water:
- Men: 3.7 L/day
- Women: 2.7 L/day
Conversion factors:
- 1 L = 1 qt = 32 oz = 4 c

■ Exceptions: A compound in which the positive ions are hydrogen ions (H^+) is an acid (example: hydrochloric acid, or H^+Cl^-); a compound in which the negative ions are hydroxyl ions (OH^-) is a base (example: potassium hydroxide, or K^+OH^-).

■ The simple statement that water follows salt describes the force that chemists call *osmosis*.

Water follows salt. Notice the beads of "sweat," formed on the right-hand slices of eggplant, which were sprinkled with salt. Cellular water moves across each cell's membrane (water-permeable divider) toward the higher concentration of salt (dissolved particles) on the surface.

© Craig M. Moore

Healthy kidneys regulate the body's sodium, as well as its water, with remarkable precision. The intestinal tract absorbs sodium readily, and it travels freely in the blood. But the kidneys excrete unneeded amounts. The kidneys actually filter all of the sodium out of the blood; then, they return to the bloodstream the exact amount the body needs to retain. Thus the body's total electrolytes remain constant, while the urinary electrolytes fluctuate according to what is eaten.

In some cases, the body's mechanisms for maintaining fluid and electrolyte balances cannot compensate for a sudden loss of large amounts of fluid and electrolytes. Vomiting, diarrhea, heavy sweating, fever, burns, wounds, and the like may incur great fluid and electrolyte losses, precipitating an emergency that demands medical intervention. ■

■ The body's responses to severe stress and trauma are discussed in Chapter 24.

Acid-Base Balance

The body uses ions not only to help maintain water balance, but also to regulate the acidity (pH) of its fluids. Like proteins, electrolyte mixtures in the body fluids protect the body against changes in acidity by acting as **buffers**—substances that can accommodate excess acids or bases.

The body's buffer systems serve as a first line of defense against changes in the fluids' acid-base balance. The lungs, skin, gastrointestinal (GI) tract, and kidneys provide other defenses. Of these organ systems, the kidneys play the primary role in maintaining acid-base balance. Thus, disorders of the kidneys impair the body's ability to regulate its acid-base balance, as well as its fluid and electrolyte balances.

IN SUMMARY

- Water makes up about 60 percent of the body's weight.

- Water helps transport nutrients and waste products throughout the body, participates in metabolic reactions, acts as a solvent, assists in maintaining blood pressure and body temperature, acts as a lubricant and cushion around joints, and serves as a shock absorber.

- To maintain water balance, intake from liquids, foods, and metabolism must equal losses from kidneys, skin, lungs, and feces.

- Electrolytes help maintain the appropriate distribution of body fluids and help to maintain acid-base balance as well.

Major Minerals

Table 9–3 lists the major minerals and trace minerals in the body, and Figure 9–1 (p. 244) shows the amounts found in the body. The most prevalent minerals are calcium and phosphorus, the chief minerals of bone. The distinction between the major minerals and the trace minerals does not mean that one group is more important than the other. A deficiency of the few micrograms of iodine needed daily is just as serious as a deficiency of the several hundred milligrams of calcium. The major minerals are so named because they are present, and needed, in larger amounts in the body than the trace minerals.

Although all the major minerals influence the body's fluid balance, sodium, chloride, and potassium are most noted for that role. For this reason, these three minerals are discussed first. Each major mineral also plays other specific roles in the body. Sodium, potassium, calcium, and magnesium are critical to nerve transmission and muscle contractions. Phosphorus and magnesium are involved in energy metabolism.

TABLE 9-3
Major Minerals and Trace Minerals

Major Minerals	Trace Minerals
Calcium	Arsenic
Chloride	Boron
Magnesium	Chromium
Phosphorus	Cobalt
Potassium	Copper
Sodium	Fluoride
Sulfur	Iodine
	Iron
	Manganese
	Molybdenum
	Nickel
	Selenium
	Silicon
	Zinc

salts: compounds composed of charged particles (ions). An example of a salt is potassium chloride (K^+Cl^-).

electrolyte: a salt that dissolves in water and dissociates into charged particles called *ions*.

electrolyte solutions: solutions that can conduct electricity.

buffers: compounds that can reversibly combine with hydrogen ions to help keep a solution's acidity or alkalinity constant.

FIGURE 9–1

Amounts of Minerals in a 60-kilogram (132-pound) Human Body

Not only are the major minerals present in the body in larger amounts than the trace minerals, but they are also needed by the body in large amounts. Recommended intakes for the major minerals are stated in *hundreds of milligrams* or *grams*, whereas those for the trace minerals are listed in *tens of milligrams* or even *micrograms*.

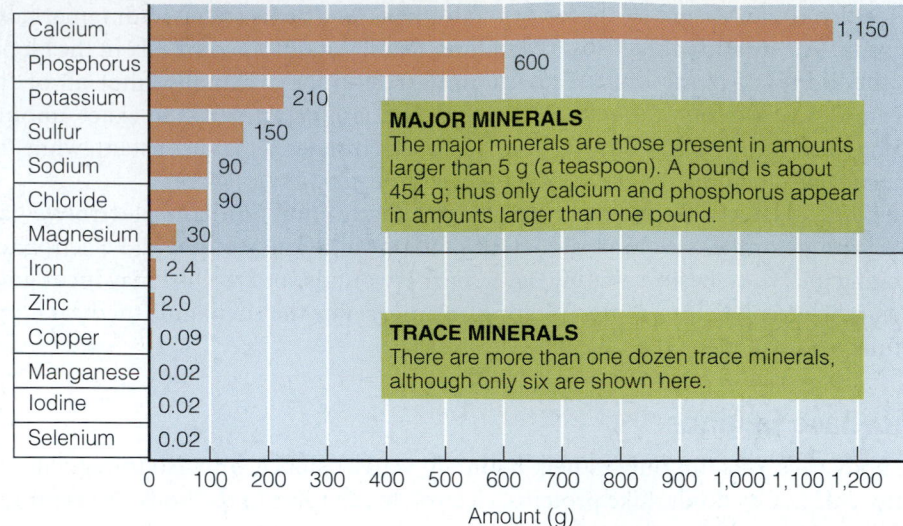

Mineral	Amount (g)
Calcium	1,150
Phosphorus	600
Potassium	210
Sulfur	150
Sodium	90
Chloride	90
Magnesium	30
Iron	2.4
Zinc	2.0
Copper	0.09
Manganese	0.02
Iodine	0.02
Selenium	0.02

MAJOR MINERALS
The major minerals are those present in amounts larger than 5 g (a teaspoon). A pound is about 454 g; thus only calcium and phosphorus appear in amounts larger than one pound.

TRACE MINERALS
There are more than one dozen trace minerals, although only six are shown here.

Calcium, phosphorus, and magnesium contribute to the structure of the bones. Sulfur helps determine the shape of proteins. Table 9–6 later in the chapter provides a summary of information about the major minerals.

Sodium

Sodium is the principal electrolyte in the **extracellular fluid** (the fluid outside the cells) and the primary regulator of the extracellular fluid volume. When the blood concentration of sodium rises, as when a person eats salted foods, thirst prompts the person to drink water until the appropriate sodium-to-water ratio is restored. Sodium also helps maintain acid-base balance and is essential to muscle contraction and nerve transmission. Too much sodium, however, can contribute to high blood pressure.

Sodium Recommendations and Food Sources Diets rarely lack sodium, and even when intakes are low, the body adapts by reducing sodium losses in urine and sweat, thus making deficiencies unlikely. Sodium recommendations ■ are set low enough to protect against high blood pressure, but high enough to allow an adequate intake of other nutrients. Because high sodium intakes correlate with high blood pressure, the Tolerable Upper Intake Level (UL) for adults is set at 2,300 milligrams per day, slightly lower than the Daily Value used on food labels (2,400 milligrams). The UL corresponds to slightly more than 1 teaspoon of salt (sodium chloride). ■

Cultures vary in their use of salt. In the United States, men consume an average of 3,300 milligrams of sodium (equivalent to about 8 grams of salt) per day. Asian people, whose staple sauces and flavorings are based on soy sauce and monosodium glutamate (MSG), consume the equivalent of about 20 to 30 grams of salt per day. In China, Japan, and Korea, high blood pressure is as prevalent as in the United States, or more so.

Sodium intakes also vary widely. People who eat mostly processed foods have the highest sodium intakes, whereas those who eat mostly whole, unprocessed foods, such as fresh fruits and vegetables, have the lowest intakes. In fact, about three-fourths of the sodium in people's diets comes from salt added to foods by manufacturers. Figure 9–2 shows that processed foods contain not only more sodium, but also less potassium than their less processed counterparts.

Sodium and Blood Pressure More than 60 million people in the United States have high blood pressure, or **hypertension**, or are taking antihypertensive medications.[5] Hypertension is a major risk factor for cardiovascular disease. As blood pressure rises, the risk of death from cardiovascular disease climbs steadily.[6]

■ Sodium AI:
■ 1,500 mg/day (19–50 yr)
■ 1,300 mg/day (51–70 yr)
■ 1,200 mg/day (>70 yr)

■ Salt (sodium chloride) is about 40 percent sodium.
1 g salt contributes 400 mg sodium.
5 g salt = 1 tsp.
1 tsp salt contributes 2,000 mg sodium.

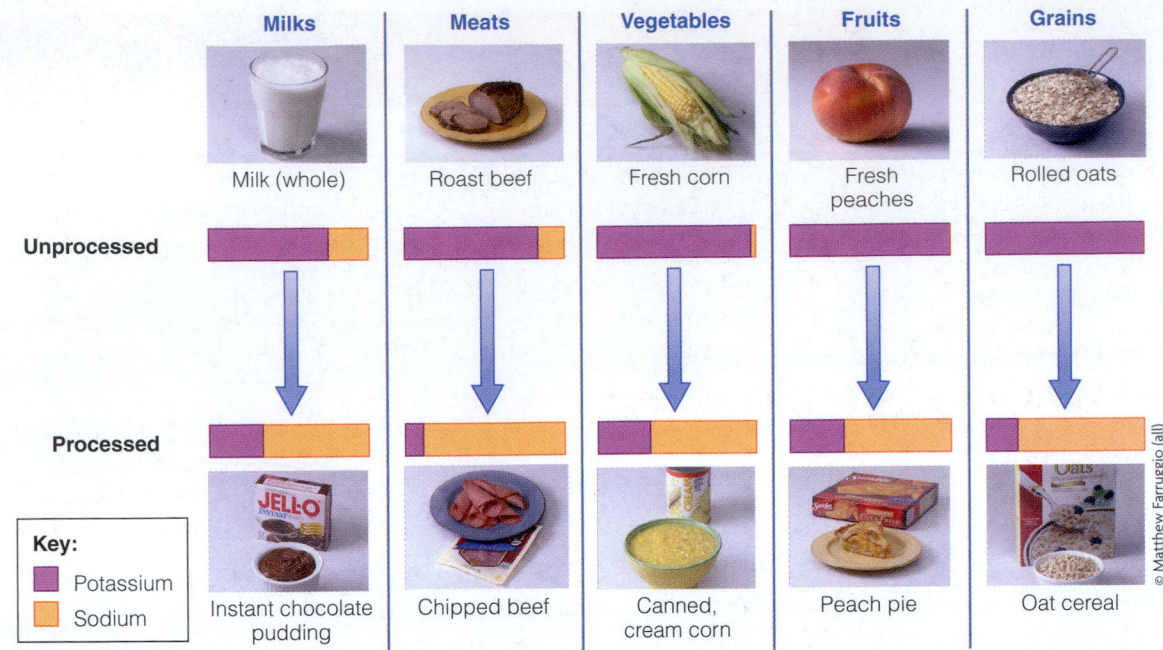

Milks   **Meats**   **Vegetables**   **Fruits**   **Grains**

Milk (whole)   Roast beef   Fresh corn   Fresh peaches   Rolled oats

Unprocessed

Processed

Key:
- Potassium
- Sodium

Instant chocolate pudding   Chipped beef   Canned, cream corn   Peach pie   Oat cereal

FIGURE 9–2

What Processing Does to the Sodium and Potassium Contents of Foods
People who eat foods high in salt often happen to be eating fewer potassium-containing foods at the same time. Notice how potassium is lost and sodium is gained as foods become more processed, causing the potassium-to-sodium ratio to fall dramatically. Even when potassium is not lost, the addition of sodium still lowers the potassium-to-sodium ratio. Limiting sodium intake may help in two ways—by lowering blood pressure in salt-sensitive individuals and by indirectly raising potassium intakes in all individuals.

For years, a high *sodium* intake was considered the primary factor responsible for high blood pressure. Then research pointed to *salt* (sodium chloride) as the dietary culprit. Salt has a greater effect on blood pressure than either sodium or chloride alone or in combination with other ions.[7] The relationship between salt intake and blood pressure is direct—the more salt a person eats, the higher the blood pressure goes.[8] This effect occurs more strongly among people who are **salt-sensitive**. People who tend to be salt-sensitive include those with hypertension, chronic kidney disease, or diabetes; African-Americans; and people over 50 years of age because the blood pressure responds to salt more dramatically in older age.[9] Overweight people also appear to be particularly sensitive to the effect of salt on blood pressure. For them, a high salt intake correlates strongly with heart disease, and salt restriction helps to lower blood pressure.

In fact, a salt-restricted diet lowers blood pressure in people without hypertension as well. Because reducing salt intake causes no harm and diminishes the risk of hypertension and heart disease, the *Dietary Guidelines for Americans 2005* advise limiting daily *sodium* intakes to less than 2,300 milligrams (approximately 1 teaspoon of *salt*). Higher intakes seem to be well tolerated in most healthy people, however. The "How to" (p. 246) offers strategies for cutting salt (and, therefore, sodium) intake.

One diet plan, known as the DASH (Dietary Approaches to Stop Hypertension) diet, also lowers blood pressure.[10] The DASH approach emphasizes fruits, vegetables, and low-fat dairy products; includes whole grains, nuts, poultry, and fish; and calls for reduced intakes of red meat, butter, and other high-fat foods. The DASH diet in combination with a reduced sodium intake is even more effective at lowering blood pressure than either strategy alone.[11] Chapter 22 offers a complete discussion of hypertension and the dietary recommendations for its prevention and treatment.

extracellular fluid: fluid residing outside the cells; includes the fluid between the cells (*interstitial fluid*), plasma, and the water of structures such as the skin and bones. Extracellular fluid accounts for about one-third of the body's water.

hypertension: high blood pressure.

salt-sensitive: an individual who responds to a high salt intake with an increase in blood pressure or to a low salt intake with a decrease in blood pressure.

CUT SALT (AND SODIUM) INTAKE

Most people eat more salt (and therefore sodium) than they need. Some people can lower their blood pressure by avoiding highly salted foods and removing the saltshaker from the table. Foods eaten without salt may seem less tasty at first, but with repetition, people can learn to enjoy the natural flavors of many unsalted foods. Strategies to cut salt intake include:

- Select fresh, unprocessed foods.

- Cook with little or no added salt.

- Prepare foods with sodium-free spices such as basil, bay leaves, curry, garlic, ginger, mint, oregano, pepper, rosemary, and thyme; lemon juice; vinegar; or wine.

- Add little or no salt at the table; taste foods before adding salt.

- Read labels with an eye open for sodium. (See the Glossary on p. 27–28 for terms used to describe the sodium contents of foods on labels.)

- Select low-salt or salt-free products when available.

Use these foods sparingly:

- Foods prepared in brine, such as pickles, olives, and sauerkraut.

- Salty or smoked meats, such as bologna, corned or chipped beef, bacon, frankfurters, ham, lunch meats, salt pork, sausage, and smoked tongue.

- Salty or smoked fish, such as anchovies, caviar, salted and dried cod, herring, sardines, and smoked salmon.

- Snack items such as potato chips, pretzels, salted popcorn, salted nuts, and crackers.

- Condiments such as bouillon cubes; seasoned salts; MSG; soy, teriyaki, Worcestershire, and barbeque sauces; prepared horseradish, catsup, and mustard.

- Cheeses, especially processed types.

- Canned and instant soups.

Chloride

The chloride ion is the major negative ion of the extracellular fluids, where it occurs primarily in association with sodium. Like sodium, chloride is critical to maintaining fluid, electrolyte, and acid-base balances in the body. In the stomach, the chloride ion is part of hydrochloric acid, which maintains the strong acidity of the gastric fluids.

Salt is a major food source of chloride, and as with sodium, processed foods are a major contributor of this mineral to people's diets. Because salt contains a higher proportion of chloride (by weight) than sodium, chloride recommendations ■ are slightly higher than, but still equivalent to, those of sodium. In other words, ¾ teaspoon of salt will deliver some sodium, more chloride, and still meet the AI for both.

- Chloride AI:
  - 2,300 mg/day (19–50 yr)
  - 2,000 mg/day (51–70 yr)
  - 1,800 mg/day (>70 yr)

Potassium

Potassium is the principal positively charged ion inside the body cells. It plays a major role in maintaining fluid and electrolyte balance and cell integrity. Potassium is also critical to keeping the heartbeat steady. The sudden deaths that occur in severe diarrhea and in children with kwashiorkor or people with eating disorders are likely due to heart failure caused by potassium loss.

Potassium Deficiency and Toxicity Potassium deficiency results more often from excessive losses than from deficient intakes. Deficiency arises in abnormal conditions such as diabetic acidosis, dehydration, or prolonged vomiting or diarrhea; potassium deficiency can also result from the regular use of certain medications, including **diuretics**, **steroids**, and **cathartics**. Potassium deficiency is characterized by an increase in blood pressure, salt sensitivity, kidney stones, and bone turnover. As deficiency progresses, symptoms include irregular heartbeats, muscle weakness, and glucose intolerance.[12] Inadequate potassium intakes are possible with diets low in fresh fruits and vegetables, but out-and-out deficiencies of potassium are unlikely in healthy people.

Potassium toxicity does not result from overeating foods high in potassium; therefore, a Tolerable Upper Intake Level was not set. Toxicity can result from overconsumption of

diuretics (dye-yoo-RET-ics): medications that promote the excretion of water through the kidneys. Not all diuretics increase the urinary loss of potassium. Some, called potassium-sparing diuretics, are less likely to result in a potassium deficiency (see Chapter 23).

steroids (STARE-oids): medications used to reduce tissue inflammation, to suppress the immune response, or to replace certain steroid hormones in people who cannot synthesize them.

cathartics (ca-THART-ics): strong laxatives.

cofactor: a mineral element that, like a coenzyme, works with an enzyme to facilitate a chemical reaction.

potassium salts or supplements and from certain diseases or medications.[13] Given more potassium than the body needs, the kidneys accelerate their excretion. If the GI tract is bypassed, however, and potassium is injected directly into a vein, it can stop the heart.

Potassium Recommendations and Food Sources In healthy people, almost any reasonable diet provides enough potassium to prevent the dangerously low blood potassium that indicates a severe deficiency. Potassium is abundant inside all living cells, both plant and animal, and because cells remain intact until foods are processed, the richest sources of potassium are *fresh* foods of all kinds—especially fruits and vegetables. A typical U.S. diet, however, with its low intakes of fruits and vegetables, provides only about half the recommended intake.[14] ■ Although blood potassium may remain normal on such a diet, chronic diseases are more likely to occur.

Calcium

Calcium occupies more space in this chapter than any other major mineral. Other minerals are revisited later in this book, where their key roles in heart disease and kidney disease are discussed. Calcium deserves emphasis here in the normal nutrition part of the book because an adequate intake of calcium early in life helps grow a healthy skeleton and prevent bone disease in later life.

Calcium Roles in the Body Calcium owns the distinction of being the most abundant mineral in the body. Ninety-nine percent of the body's calcium is stored in the bones, where it plays two important roles. First, it is an integral part of bone structure. Second, it serves as a calcium bank available to the body fluids should a drop in blood calcium occur.

Calcium in Bone As bones begin to form, calcium salts form crystals on a matrix of the protein collagen. As the crystals become denser, they give strength and rigidity to the maturing bones. As a result, the long leg bones of children can support their weight by the time they have learned to walk. Figure 9–3 shows the lacy network of calcium-containing crystals in the bone.

Many people have the idea that bones are inert, like rocks. Not so. Bones continuously gain and lose minerals in an ongoing process of remodeling. Growing children gain more bone than they lose, and healthy adults maintain a reasonable balance. When withdrawals substantially exceed deposits, however, problems such as osteoporosis develop.

From birth to approximately age 20, the bones are actively growing by modifying their length, width, and shape (see Figure 9–4 on p. 248). This rapid growth phase overlaps with the next period of peak bone mass development, which occurs between the ages of 12 and 30. During this period, skeletal mass increases. Bones grow thicker and denser by remodeling, a maintenance and repair process involving the loss of existing bone and the deposition of new bone. In the final phase, which begins between 30 and 40 years of age and continues throughout the remainder of life, bone loss exceeds new bone formation.

Calcium in Body Fluids The 1 percent of the body's calcium that circulates in the fluids as ionized calcium is vital to life. It helps regulate muscle contractions, transmit nerve impulses, clot blood, and secrete hormones, digestive enzymes, and neurotransmitters. In most of these processes, the calcium ion operates in two ways: it conveys signals received at the cell surface to the inside of the cell and it activates the proteins involved in the function.[15] Calcium is a **cofactor** for several enzymes as well.

Calcium Balance Blood calcium concentration is tightly controlled. Whenever blood calcium rises too high, a system of hormones and vitamin D promotes its deposit into

Fresh fruits and vegetables provide potassium in abundance.

■ Potassium AI:
  ■ Adults: 4,700 mg

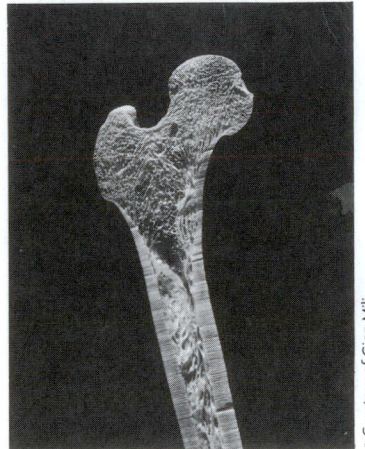

FIGURE 9–3

Cross Section of Bone
The lacy structural elements are *trabeculae* (tra-BECK-you-lee), which can be drawn on to replenish blood calcium.

FIGURE 9–4

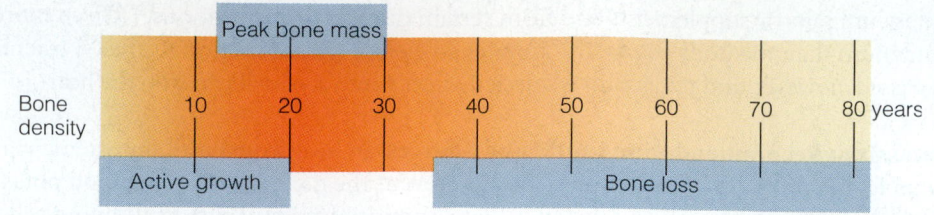

■ The regulators are hormones from the thyroid and parathyroid glands, as well as vitamin D. One hormone, *parathormone*, raises blood calcium. Another hormone, *calcitonin*, lowers blood calcium by inhibiting release of calcium from bone. The hormonelike *vitamin D* raises blood calcium by acting at the three sites listed.

■ Reminder: *Osteoporosis* is a condition characterized by reduced density of the bones. The bones become porous and fragile and fracture easily.

bone. Whenever blood calcium falls too low, the regulatory system ■ acts in three locations to raise it:

1. The small intestine absorbs more calcium.
2. The bones release more calcium.
3. The kidneys excrete less calcium.

Thus blood calcium rises to normal.

The calcium stored in bone provides a nearly inexhaustible source of calcium for the blood. Even in a calcium deficiency, blood calcium remains normal. Blood calcium changes only in response to abnormal regulatory control, not to diet. Blood calcium above normal causes **calcium rigor**: the muscles contract and cannot relax. Blood calcium below normal causes **calcium tetany**—also characterized by uncontrolled muscle contraction. These conditions are caused by a lack of vitamin D or by abnormal concentrations of the hormones that regulate calcium homeostasis.

Although a chronic *dietary* deficiency of calcium or a chronic deficiency due to poor absorption does not change blood calcium, it does deplete the savings account in the bones. This is an important concept: it is the bones, not the blood, that are robbed by calcium deficiency.

Calcium and Osteoporosis As mentioned earlier, bone mass peaks at the time of skeletal maturity (about age 30), and a high peak bone mass is the best protection against later age-related bone loss and fracture. Adequate calcium nutrition during the growing years is essential to achieving optimal peak bone mass.[16] Following menopause, women lose about 15 percent of their bone mass, as do middle-aged and older men. When bone loss has reached such an extreme that bones fracture under even common, everyday stresses, the condition is known as osteoporosis. ■ Osteoporosis afflicts more than 40 million people, mostly women 50 years of age or older.[17] Men, however, are not immune to osteoporosis. Each year, one million and a half people—30 percent of them men—suffer broken hips, pelvises, legs, arms, hands, and ankles attributable to osteoporosis.

Both genetic and environmental factors contribute to osteoporosis. Table 9–4 summarizes risk and protective factors for osteoporosis. Osteoporosis is more prevalent in women than men for several reasons. First, women consume less dietary calcium than men do. Second, at all ages, women's bone mass is lower than men's because women generally have smaller bodies. Finally, bone loss accelerates after menopause and exceeds men's losses for up to ten years.

In addition to calcium, many other minerals and vitamins, including phosphorus, magnesium, fluoride, and vitamin D, help to form and stabilize the structure of bones. Any or all of these elements are needed to prevent bone loss. The first, most obvious lines of defense are to maintain a lifelong adequate intake of calcium and to "exercise it into place." Active bones are denser than sedentary bones.[18] Weight-bearing physical activity, such as walking, running, dancing, and weight training, prompts the bones to deposit minerals. It has long been known that when people are confined to bed, both their muscles and their bones lose strength. Muscle strength and bone strength go together: when muscles work, they pull on the bones, and both are stimulated to grow stronger.

TABLE 9-4

Risk Factors and Protective Factors for Osteoporosis

Risk Factors	Protective Factors
Older age	Younger age
Low body mass index (BMI)	High BMI
Caucasian, Asian, or Hispanic heritage	African-American heritage
Cigarette smoking	No smoking
Alcohol consumption in excess	Alcohol consumption in moderation
Sedentary lifestyle	Regular weight-bearing exercise
Use of glucocorticoids or anticonvulsants	Use of diuretics
Female gender	Male gender
Maternal history of osteoporosis fracture or personal history of fracture	Bone density assessment and treatment (if necessary)
Estrogen deficiency in women (amenorrhea or menopause, especially early or surgically induced); testosterone deficiency in men	Use of estrogen therapy
Lifetime diet inadequate in calcium and vitamin D	Lifetime diet rich in calcium and vitamin D

Calcium and Hypertension Some evidence suggests that calcium may protect against hypertension.[19] For this reason, restricting salt intake to treat hypertension is narrow advice, especially considering the success of the DASH diet in lowering blood pressure. The DASH diet is not particularly low in sodium, but it is rich in calcium, as well as in potassium and magnesium. As mentioned earlier, the DASH diet, together with a reduced sodium intake, is more effective at lowering blood pressure than either strategy alone.

Calcium and Obesity Calcium may also play a role in maintaining a healthy body weight.[20] Analyses of national survey data as well as small clinical studies show an inverse relationship between calcium intake and body fatness: the higher the calcium intake, the lower the body fatness.[21] In particular, calcium from dairy foods, but *not* from supplements, seems to influence body weight.[22] Not all research suggests that calcium or dairy foods are associated with body weight.[23] Large, well-designed clinical studies are needed to clarify the effects of dietary calcium intake on body weight.

Calcium Recommendations As mentioned earlier, blood calcium concentration does not reflect calcium status. Calcium recommendations ■ are therefore based on balance studies, which measure daily intake and excretion. An optimal calcium intake reflects the amount needed to retain the most calcium. The more calcium retained, the greater the bone density (within genetic limits) and, potentially, the lower the risk of osteoporosis. Calcium recommendations during adolescence are set high (1,300 milligrams per day) to help ensure that the skeleton will be strong and dense. Between the ages of 19 and 50, recommendations are lowered to 1,000 milligrams per day. For those over 50, recommendations are raised again to 1,200 milligrams, to minimize bone loss. Some authorities advocate calcium recommendations as high as 1,500 milligrams per day for women over 50. Many women have intakes well below recommendations, however.

Calcium in Foods Calcium is found most abundantly in a single food group—milk and milk products. For this reason, dietary recommendations advise daily consumption of low-fat or fat-free milk products. One cup of milk offers about 300 milligrams of calcium, so an adult who drinks 3 cups of milk per day (or eats the

■ Calcium AI:
- Adults (19—50 yr): 1,000 mg/day
- Adults (51 and older): 1,200 mg/day

calcium rigor: hardness or stiffness of the muscles caused by high blood calcium.

calcium tetany: intermittent spasms of the extremities due to nervous and muscular excitability caused by low blood calcium.

equivalent in yogurt) is well on the way to meeting daily calcium needs (see Table 9–5). The other dairy food that contains comparable amounts of calcium is cheese. One slice of cheese (1 ounce) contains about two-thirds as much calcium as one cup of milk. Cottage cheese contains much less. Snapshot 9–1 shows foods that are rich in calcium, and the accompanying "How to" suggests ways of adding calcium to meals.

Some foods offer large amounts of calcium because of fortification. Calcium-fortified juice, high-calcium milk (milk with extra calcium added), and calcium-fortified cereals are examples. Some mineral waters provide as much as 500 milligrams of calcium per liter, offering a convenient way to meet both calcium and water needs.[24]

Among the vegetables, mustard greens, kale, parsley, watercress, and broccoli are good sources of available calcium. Some dark green, leafy vegetables—notably spinach and Swiss chard—appear to be calcium-rich but actually provide very little, if any, calcium to the body. These foods contain **binders** that prevent calcium absorption.

Aided by vitamin D, the body is able to regulate its absorption of calcium by altering its production of the calcium-binding protein. More of this protein is made if more calcium is needed. Infants and children absorb up to 60 percent of the calcium they ingest, and pregnant women absorb about 50 percent. Other adults, who are not growing, absorb about 25 percent.

People may think that taking a calcium supplement is preferable to getting calcium from food, but foods offer important fringe benefits. For example, drinking 3 cups of milk fortified with vitamins A and D will supply substantial amounts of other nutrients. Furthermore, the vitamin D and possibly other nutrients in the milk enhance calcium absorption. Some people absorb calcium better from milk and milk products than from even the most absorbable supplements. The National Institutes of Health concludes that foods are the best sources of calcium and recommends supplements only when intake from food is insufficient.

SNAPSHOT 9–1

Calcium

GOOD SOURCES*

SARDINES (with bones)
3 oz = 324 mg

MILK
1 c = 300 mg

TOFU (calcium set)
1/2 c = 275 mg

BLACK-EYED PEAS (cooked)
1/2 c = 105 mg

CHEDDAR CHEESE
$1^1/_2$ oz = 306 mg

TURNIP GREENS (cooked)
1 c = 197 mg

WAFFLE (whole grain)
1 waffle = 196 mg

BROCCOLI[a] (cooked)
$1^1/_2$ c = 93 mg

* These foods provide 10 percent or more of the calcium Daily Value in a serving. For a 2,000-kcalorie diet, the DV is 1,000 mg/day.
[a] Although broccoli, kale, and some other cooked green leafy vegetables fall short of supplying 10 percent of calcium in a serving, these foods are important sources of bioavailable calcium. Almonds also supply calcium. Other greens, such as spinach and chard, contain calcium in an unabsorbable form. Note that the amounts for green vegetables exceed the ½ c servings of the USDA Food Guide. Some calcium-rich mineral waters may also be good sources.

ADD CALCIUM TO DAILY MEALS

For those who tolerate milk, many cooks slip extra calcium into meals by sprinkling a tablespoon or two of fat-free dry milk into almost everything. The added kcalorie value is small, and changes to the taste and texture of the dish are practically nil. Yet each 2 tablespoons adds about 100 extra milligrams of calcium and moves people closer to meeting the recommendation to obtain 3 cups of milk each day.

Here are some more tips for including calcium-rich foods in your meals.

At Breakfast
- Choose calcium-fortified orange or vegetable juice.
- Serve tea or coffee, hot or iced, with milk.
- Choose cereals, hot or cold, with milk.
- Cook hot cereals with milk instead of water; then mix in 2 tablespoons of fat-free dry milk.
- Make muffins or quick breads with milk and extra fat-free powdered milk.
- Add milk to scrambled eggs.
- Moisten cereals with flavored yogurt.

At Lunch
- Add low-fat cheeses to sandwiches, burgers, or salads.
- Use a variety of green vegetables, such as watercress or kale, in salads and on sandwiches.
- Drink fat-free milk or calcium-fortified soy milk as a beverage or in a smoothie.
- Drink calcium-rich mineral water as a beverage (studies suggest significant calcium absorption).
- Marinate cabbage shreds or broccoli spears in low-fat Italian dressing for an interesting salad that provides calcium.
- Choose coleslaw over potato and macaroni salads.
- Mix the mashed bones of canned salmon into salmon salad or patties.
- Eat sardines with their bones.
- Stuff potatoes with broccoli and low-fat cheese.
- Try pasta such as ravioli stuffed with low-fat ricotta cheese instead of meat.
- Sprinkle Parmesan cheese on pasta salads.

At Supper
- Toss a handful of thinly sliced green vegetables, such as kale or young turnip greens, with hot pasta; the greens wilt pleasingly in the steam of the freshly cooked pasta.
- Serve a green vegetable every night and try new ones—how about kohlrabi? It tastes delicious when cooked like broccoli.
- Learn to stir-fry Chinese cabbage and other Asian foods.
- Try tofu (the calcium-set kind); this versatile food has inspired whole cookbooks devoted to creative uses.
- Add fat-free powdered milk to almost anything—meat loaf, sauces, gravies, soups, stuffings, casseroles, blended beverages, puddings, quick breads, cookies, brownies. Be creative.
- Choose frozen yogurt, ice milk, or custards for dessert.

Phosphorus

Phosphorus is the second most abundant mineral in the body. About 85 percent of it is found combined with calcium in the crystals of the bones and teeth. As part of one of the body's buffer systems (phosphoric acid), phosphorus is also found in all body tissues. Phosphorus is a part of DNA and RNA, the genetic material present in every cell. Thus phosphorus is necessary for all growth. Phosphorus also plays many key roles in the transfer of energy that occurs during cellular metabolism. Phosphorus-containing lipids (phospholipids) help transport other lipids in the blood. Phospholipids are also principal components of cell membranes.

Animal protein is the best source of phosphorus because the mineral is so abundant in the cells of animals. Diets that provide adequate energy and protein also supply adequate phosphorus. ■ Dietary deficiencies are rare. A summary of facts about phosphorus appears in Table 9–6 later in the chapter.

- Phosphorus RDA:
  - Adults: 700 mg/day

binders: chemical compounds in foods that combine with nutrients (especially minerals) to form complexes the body cannot absorb. Examples include *phytates* and *oxalates*.

Magnesium

Magnesium barely qualifies as a major mineral. Only about 1 ounce of magnesium is present in the body of a 130-pound person, over half of it in the bones.[25] Most of the rest is in the muscles, heart, liver, and other soft tissues, with only 1 percent in the body

coauthors, Dairy products do not lead to alterations in body weight or fat mass in young women in a 1-y intervention, *American Journal of Clinical Nutrition* 81 (2005): 751–756; S. I. Barr, Increased dairy product or calcium intake: Is body weight or composition affected in humans? *Journal of Nutrition* 133 (2003): 245S–248S.

24. P. Galan and coauthors, Contribution of mineral waters to dietary calcium and magnesium intake in a French adult population, *Journal of the American Dietetic Association* 102 (2002): 1658–1662.

25. R. K. Rude and M. E. Shils, Magnesium, in M. E. Shils and coeditors, *Modern Nutrition in Health and Disease*, 10th ed. (Philadelphia: Lippincott, Williams, & Wilkins, 2006), pp. 223–247.

26. Rude and Shils, 2006; S. H. Jee and coauthors, The effect of magnesium supplementation on blood pressure: A meta-analysis of randomized clinical trials, *American Journal of Hypertension* 15 (2002): 691–696.

27. S. C. Larsson, L. Berfkvist, and A. Wolk, Magnesium intake in relation to risk of colorectal cancer in women, *Journal of the American Medical Association* 293 (2005): 86–89.

28. M. W. Hentze, M. U. Muckenthaler, and N. C. Andrews, Balancing acts: Molecular control of mammalian iron metabolism, *Cell* 117 (2004): 285–297.

29. E. Nemeth and T. Ganz, Regulation of iron metabolism by hepcidin, *Annual Review of Nutrition* 26 (2006): 323–342; R. J. Wood and A. G. Ronnenberg, Iron, in M. E. Shils and coeditors, *Modern Nutrition in Health and Disease*, 10th ed. (Philadelphia: Lippincott, Williams, & Wilkins, 2006), pp. 248–270; R. E. Fleming and B. R. Bacon, Orchestration of iron homeostasis, *New England Journal of Medicine* 352 (2005): 1741–1744.

30. Wood and Ronnenberg, 2006; Fleming and Bacon, 2005.

31. J. L. Beard and J. R. Connor, Iron status and neural functioning, *Annual Review of Nutrition* 23 (2003): 41–58.

32. K. C. White, Anemia is a poor predictor of iron deficiency among toddlers in the United States: For heme the bell tolls, *Pediatrics* 115 (2005): 315–320.

33. T. Brownlie and coauthors, Marginal iron deficiency without anemia impairs aerobic adaptation among previously untrained women, *American Journal of Clinical Nutrition* 75 (2002): 734–742.

34. Standing Committee on the Scientific Evaluation of Dietary Reference Intakes, Food and Nutrition Board, National Institute of Health, *Dietary Reference Intakes for Vitamin A, Vitamin K, Arsenic, Boron, Chromium, Copper, Iodine, Iron, Manganese, Molybdenum, Nickel, Silicon, Vanadium, and Zinc* (Washington, DC: National Academy Press, 2001), pp. 311–316.

35. L. E. Caulfield, Maternal zinc deficiency and maternal and child health in Peru, *Nutrition Today* 39 (2004): 78–87.

36. J. C. King and R. J. Cousins, Zinc, in M. E. Shils and coeditors, *Modern Nutrition in Health and Disease*, 10th ed. (Philadelphia: Lippincott, Williams, & Wilkins, 2006), pp. 271–285.

37. Standing Committee on the Scientific Evaluation of Dietary Reference Intakes, 2001, pp. 479–480.

38. J. R. Hunt, Bioavailability of iron, zinc, and other trace minerals from vegetarian diets, *American Journal of Clinical Nutrition* 78 (2003): 633S–639S.

39. Standing Committee on the Scientific Evaluation of Dietary Reference Intakes, 2001, p. 442.

40. A. J. Duffield-Lillico, I. Shureiqi, and S. M. Lippman, Can selenium prevent colorectal cancer? A signpost from epidemiology, *Journal of the National Cancer Institute* 96 (2004): 1645–1647.

41. R. F. Burk and O. A. Levander, Selenium, in M. E. Shils and coeditors, *Modern Nutrition in Health and Disease*, 10th ed. (Philadelphia: Lippincott, Williams, & Wilkins, 2006), pp. 312–325.

42. International Council for the Control of Iodine Deficiency, Iodine deficiency disorder, available online at www.iccidd.org. Site updated August 18, 2005; visited on November 28, 2005.

43. Standing Committee on the Scientific Evaluation of Dietary Reference Intakes, 2001, pp. 278–282.

44. Food and Drug Administration, Center for Drug Evaluation and Research, Guidance: Potassium iodide as a thyroid blocking agent in radiation emergencies, November 2001, available at www.fda.gov/cder/guidance/index.htm.

45. J. R. Turnlund, Copper, in M. E. Shils and coeditors, *Modern Nutrition in Health and Disease*, 10th ed. (Philadelphia: Lippincott, Williams, & Wilkins, 2006), pp. 286–299.

46. Standing Committee on the Scientific Evaluation of Dietary Reference Intakes, 2001, pp. 224–257.

47. Position of the American Dietetic Association: The impact of fluoride on health, *Journal of the American Dietetic Association* 105 (2005): 1620–1628.

48. D. P. DePaola and coauthors, Nutrition and dental medicine, in M. E. Shils and coeditors, *Modern Nutrition in Health and Disease*, 10th ed. (Philadelphia: Lippincott, Williams, & Wilkins, 2006), pp. 1152–1178; Position of the American Dietetic Association, 2005.

49. Position of the American Dietetic Association, 2005.

50. Standing Committee on the Scientific Evaluation of Dietary Reference Intakes, 2001, pp. 197–223.

51. M. D. Althuis and coauthors, Glucose and insulin responses to dietary chromium supplements: A meta-analysis, *American Journal of Clinical Nutrition* 76 (2002): 148–155; D. Ghosh and coauthors, Role of chromium supplementation in Indians with type 2 diabetes mellitus, *Journal of Nutritional Biochemistry* 13 (2002): 690–697; S. M. Bahijri and A. M. Mufti, Beneficial effects of chromium in people with type 2 diabetes, and urinary chromium response to glucose load as a possible indicator of status, *Biological Trace Element Research* 85 (2002): 97–109.

Nutrition in Practice

VITAMIN AND MINERAL SUPPLEMENTS

At least half of U.S. adults collectively spend billions of dollars per year on vitamin and mineral supplements.[1] This trend is accelerating as scientists discover more and more links between nutrition and disease prevention. Most people take a single pill containing a multitude of nutrients; others take huge doses of single nutrients in an attempt to ward off diseases.[2] In many cases, taking supplements is a costly but harmless practice; sometimes, it is both costly and harmful to health. The main message of this Nutrition in Practice is that most healthy people can get the nutrients they need from foods. Supplements cannot substitute for a healthy diet. For some people, however, certain nutrient supplements may be desirable. In some cases, they can correct deficiencies; in others, they can reduce the risks of disease.

Do foods really contain enough vitamins and minerals to supply all that most people need?

Emphatically, yes, for both healthy adults and children who choose a variety of foods. The USDA Food Guide described in Chapter 1 is the guide to follow to achieve adequate intakes. People who meet their nutrient needs from foods, rather than supplements, have little risk of deficiency or toxicity.

Do some people need supplements?

Yes, some people may suffer marginal nutrient deficiencies due to illness, alcohol or drug addiction, or other conditions that limit food intake.[3] People who may benefit from nutrient supplements in amounts consistent with the RDA include:

© Tom Carter/PhotoEdit

- People with nutrient deficiencies.

- People with low food energy intakes (less than 1,200 kcalories per day), such as habitual dieters.

- People with illnesses that take away the appetite.

- People with illnesses that impair nutrient absorption, such as diseases of the gallbladder, pancreas, and digestive system.

- People taking medications that interfere with nutrient metabolism.

- People who are lactose intolerant, have milk allergies, or otherwise do not consume enough dairy products to forestall extensive bone loss need calcium.

- People with limited milk intake and sun exposure need vitamin D.

- Elderly people who may have difficulty chewing or swallowing and so do not eat enough food to meet nutrient needs.

- Women who bleed excessively during menstruation need iron supplements.

- People in certain stages of the life cycle who have increased nutrient needs (for example, infants need iron, and some may need fluoride and vitamin D; women of childbearing age need folate; pregnant women need iron; and the elderly need vitamin D).

- People who eat all-plant diets (vegans) and those with atrophic gastritis need vitamin B_{12}.

- Newborn infants need a single dose of vitamin K at birth under the direction of a physician.

- People who have infections or injuries or who have undergone major surgery. (The increased metabolic needs associated with these severe stresses are discussed in Chapter 24.)

Most adults can get all the nutrients they need by eating a varied diet of nutrient-dense foods. Nutrients are potentially toxic when taken in large doses, and individual tolerances vary depending on health and age. Whenever a health care professional finds a person's diet inadequate, the right corrective step is to improve the person's food choices and eating patterns, not to begin supplementation.

Why do so many people take supplements?

People frequently take supplements for mistaken reasons, such as "they give me energy" or "they make me strong." Other invalid reasons why people may take supplements include:

- Their feeling of insecurity about the nutrient content of the food supply

- Their belief that extra vitamins and minerals will help them cope with stress

- Their belief that supplements can enhance athletic performance or build lean body tissue without physical work

- Their desire to prevent, treat, or cure symptoms or diseases ranging from the common cold to cancer[4]

Ironically, many supplement users eat more nutrient-dense diets than nonusers and therefore need supplements less. In addition, little relationship exists between the nutrients people need and the ones they take in supplements. In fact, an argument against supplements is that they may lull people into a false sense of security. A person might eat irresponsibly, thinking "my supplement will cover my needs."

Do antioxidant supplements prevent cancer and heart disease?

Again, it is better advice to eat a very nutritious diet. Evidence from population studies shows a correlation between low intakes of antioxidant nutrients such as vitamin C, vitamin E, beta-carotene and other carotenoids, and the mineral selenium and a high incidence of disease. Other studies show that low intakes of vegetables and fruits, the richest sources of antioxidant nutrients, are linked with an increased incidence of certain types of cancer.[5] More than 200 population studies have examined the effects of fruits and vegetables on cancer risk, and many show that people who eat more of these foods are less likely to develop certain cancers. However, findings from other types of studies, such as intervention studies and clinical trials, show weaker associations between fruit and vegetable intake and reduced risk of cancer.[6] Some researchers speculate that fruit and vegetable intake may play a smaller role in total cancer protection than previously thought, but fruits and vegetables do contain protective factors for specific cancers. For example, research suggests lycopene-rich tomatoes protect against prostate cancer.[7] When research combines many different fruits and vegetables, specific protective factors may not stand out. Many experts agree that the antioxidant vitamins in these foods are probably important protective factors, but they also note that other constituents of fruits and vegetables (see Table NP8–2 on p. 233) certainly have not been ruled out as contributing factors.

The way to apply this information is to eat nutritious foods. Before supplementation is recommended as a strategy to prevent cancer or other diseases, researchers must determine the optimal doses to reduce risk and the potential adverse effects of long-term supplementation.

When a person needs a vitamin-mineral supplement, what kind should be used?

Take your health care professional's advice, if it is offered. If you are selecting a supplement yourself, a single, balanced vitamin-mineral supplement should suffice. Choose the kind that provides all the nutrients in amounts less than, equal to, or very close to the RDA (remember, you get some nutrients from foods). For those who require a higher dose, such as young

women who need supplemental folate in the childbearing years, choose a supplement with just the needed nutrient or in combination with a reasonable dose of others. Avoid any preparations that, in a daily dose, provide more than the recommended intake of vitamin A, vitamin D, or any mineral or more than the Tolerable Upper Intake Level for any nutrient. In addition, avoid these:

- High doses of iron (more than 10 milligrams per day) except for menstruating women. People who menstruate need more iron than people who do not.

- "Organic" or "natural" preparations with added substances. They are no better than standard types, but they cost more.

- "High-potency" or "therapeutic dose" supplements. More is not better.

- Items not needed in human nutrition such as inositol and carnitine. These particular items will not harm you, but they reveal a marketing strategy that makes the whole mix suspect.

As for price, be aware that local or store brands may be just as good as or better than nationally advertised brands.[8] If they are less expensive, it may be because the price does not have to cover the cost of national advertising. Finally, be aware that if you see a USP symbol on the label, it means that the manufacturer has voluntarily paid an independent laboratory to test the product and affirm that it contains the ingredients listed and that it will dissolve or disintegrate in the digestive tract to make the ingredients available for absorption. The symbol does not imply that the supplement has been tested for safety or effectiveness with regard to health, however.

Can supplement labels help consumers make informed choices?

Yes, to some extent. To enable consumers to make more informed choices about nutrient supplements, the Food and Drug

This symbol means that a supplement contains the nutrients stated and that it will dissolve in the digestive system—the symbol does not guarantee safety or health advantages.

Administration (FDA), with the encouragement of the American Dietetic Association (ADA), published labeling regulations for supplements. The Dietary Supplement Health and Education Act subjects supplements to the same general labeling requirements that apply to foods. Specifically:

- Nutrition labeling for dietary supplements is required. The nutrition panel on supplements is called "Supplement Facts" (see Figure NP9–1). The Supplement Facts panel lists the quantity and the percentage of the Daily Value for each nutrient in the supplement. Ingredients that have no Daily Value—for example, sugars and gelatin—appear in a list below the Supplement Facts panel.

- Labels may make nutrient claims (as "high" or "low") according to specific criteria (for example, "an excellent source of vitamin C").

- The FDA authorizes health claims on supplement labels about the relationship between folate and the risk of neural tube defects, calcium and osteoporosis, soluble fiber from whole oats and psyllium husks and heart disease, and omega-3 fatty acids and heart disease.

- Supplement labels are not allowed to include health claims on a number of other nutrient-disease relationships that have been approved for foods.

- Products may not bear claims to diagnose, treat, cure, or relieve a specific disease.

- Labels may make structure-function claims (see Chapter 1) about the role a nutrient plays in the body, explain how the nutrient performs its function, and indicate that consuming the nutrient is associated with general well-being.

- Labels may claim a substance benefits common complaints such as memory loss or menstrual cramps without proof of effectiveness.

However, note that despite these requirements, in effect, the Dietary Supplement Health and Education Act resulted in the deregulation of the supplement industry. Unlike food additives or drugs, supplements do not need the FDA's approval before being marketed. Manufacturers alone decide whether their products are safe and effective. Should a problem arise, the burden falls to the FDA to prove that the supplement poses an unreasonable risk and should be removed from the market. To do this, the FDA needs scientific evidence of supplement safety from manufacturers and the records of adverse health effects reported by consumers, but just a few ethical companies provide this information voluntarily. Consumers can report adverse reactions from supplements directly to the FDA via its hotline or Web site, but most people are unaware of these options.*

* Consumers should report suspected harm from dietary supplements to their health care providers or to the FDA's MedWatch program at (800) FDA-1088 or on the Internet at www.fda.gov/medwatch/.

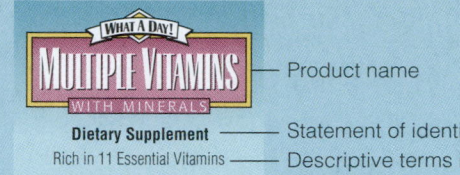

MULTIPLE VITAMINS WITH MINERALS	Product name
Dietary Supplement	Statement of identity
Rich in 11 Essential Vitamins	Descriptive terms if product meets criteria
100 TABLETS	Contents or weight

FOR YOUR PROTECTION, DO NOT USE IF PRINTED FOIL SEAL UNDER CAP IS BROKEN OR MISSING.
DIRECTIONS FOR USE: One tablet daily for adults. WARNING: CLOSE TIGHTLY AND KEEP OUT OF REACH OF CHILDREN. CONTAINS IRON, WHICH CAN BE HARMFUL OR FATAL TO CHILDREN IN LARGE DOSES. IN CASE OF ACCIDENTAL OVERDOSE, SEEK PROFESSIONAL ASSISTANCE OR CONTACT A POISON CONTROL CENTER IMMEDIATELY.
Store in a dry place at room temperature (59°–86°F).

Supplement Facts
Serving Size 1 Tablet

Amount Per Tablet	% Daily Value
Vitamin A 5000 IU (40% Beta Carotene)	100%
Vitamin C 60 mg	100%
Vitamin D 400 IU	100%
Vitamin E 30 IU	100%
Thiamin 1.5 mg	100%
Riboflavin 1.7 mg	100%
Niacin 20 mg	100%
Vitamin B6 2 mg	100%
Folate 400 mcg	100%
Vitamin B12 6 mcg	100%
Biotin 30 mcg	10%
Pantothenic Acid 10 mg	100%
Calcium 130 mg	13%
Iron 18 mg	100%
Phosphorus 100 mg	10%
Iodine 150 mcg	100%
Magnesium 100 mg	25%
Zinc 15 mg	100%
Selenium 10 mcg	14%
Copper 2 mg	100%
Manganese 2.5 mg	71%
Chromium 10 mcg	8%
Molybdenum 10 mcg	6%
Chloride 34 mg	1%
Potassium 37.5 mg	1%

Supplement Facts panel
The suggested dose

The name, quantity per tablet, and "% Daily Value" for all nutrients listed; nutrients without a Daily Value may be listed below.

INGREDIENTS: Dicalcium Phosphate, Magnesium Hydroxide, Microcrystalline Cellulose, Potassium Chloride, Ascorbic Acid, Ferrous Fumarate, Modified Cellulose Gum, Zinc Sulfate, Gelatin, Stearic Acid, Vitamin E Acetate, Hydroxypropyl Methylcellulose, Niacinamide, Calcium Silicate, Citric Acid, Magnesium, Stearate, Calcium Pantothenate, Artificial Colors (FD&C Red No. 40, Titanium Dioxide, FD&C Yellow No. 6 and FD&C Blue No. 2), Selenium Yeast, Manganese Sulfate, Polyethylene Glycol, Cupric Sulfate, Molybdenum Yeast, Chromium Yeast, Vitamin A Acetate, Pyridoxine Hydrochloride, Riboflavin, Sodium Lauryl Sulfate, Thiamin Mononitrate, Beta Carotene, Folic Acid, Polysorbate 80, Vitamin D, Potassium Iodide, Gluten, Biotin, Cyanocobalamin.

 **Supplements, Inc.**
1234 Fifth Avenue
Anywhere, USA

All ingredients must be listed on the label, but not necessarily in the ingredient list nor in descending order of predominance; ingredients named in the nutrition panel need not be repeated here.

Name and address of manufacturer

FIGURE NP9–1

An Example of a Supplement Label

Notes

1. National Institutes of Health State-of-the-Science Conference Statement: Multivitamin/mineral supplements and chronic disease prevention, *Annals of Internal Medicine* 145 (2006): 364–371.

2. National Institutes of Health State-of-the-Science Conference Statement, 2006; B. Babgaleh and coauthors, Dietary supplements in a National Survey: Prevalence of use and reports of adverse events, *Journal of the American Dietetic Association* 106 (2006): 1966–1974.

3. Position of the American Dietetic Association: Fortification and nutritional supplements, *Journal of the American Dietetic Association* 105 (2005): 1300–1311.

4. Practice Paper of the American Dietetic Association: Dietary supplements, *Journal of the American Dietetic Association* 105 (2005): 460–470.

5. W. C. Willett and E. Giovannucci, Epidemiology of diet and cancer, in M.E. Shils and coeditors, *Modern Nutrition in Health and Disease*, 10th ed. (Philadelphia: Lippincott, Williams, & Wilkins, 2006), pp. 1267–1279; M. R. Forman and coauthors, Nutrition and cancer prevention: A multidisciplinary perspective on human trials, *Annual Review of Nutrition* 24 (2004): 223–254; H. Chen and coauthors, Dietary patterns and adenocarcinoma of the esophagus and distal stomach, *American Journal of Clinical Nutrition* 75 (2002): 137–144.

6. Willett and Giovannucci, 2006; P. Terry, J. B. Terry, and A. Wolk, Fruit and vegetable consumption in the prevention of cancer: An update, *Journal of Internal Medicine* 250 (2001): 280–290.

7. E. Giovannucci and coauthors, A prospective study of tomato products, lycopene, and prostate cancer risk, *Journal of the National Cancer Institute* 94 (2002): 391–398: D. Herber and Q. Y. Lu, Overview of mechanisms of action of lycopene, *Experimental Biology and Medicine* 227 (2002): 920–923; T. M. Vogt and coauthors, Serum lycopene, other serum carotenoids, and risk of prostate cancer in US blacks and whites, *American Journal of Epidemiology* 155 (2002): 1023–1032; Q. Y. Lu and coauthors, Inverse associations between plasma lycopene and other carotenoids and prostate cancer, *Cancer Epidemiology, Biomarkers, and Prevention* 10 (2001): 749–756.

8. B. Liebman, Spin the bottle: How to pick a multivitamin, *Nutrition Action Health Letter*, January/February 2003, pp. 3–9.

10 Fitness and Nutrition

© Silverstock/Digital Vision/Getty Images

273

n the body, nutrition and physical activity are interactive—each influences the other. The working body demands all three energy-yielding nutrients—carbohydrate, lipids, and protein—to fuel activity. The body also needs protein and a host of supporting nutrients to build lean tissue. Physical activity benefits the body's nutrition by helping to regulate the use of fuels, by pushing the body composition toward the lean, and by increasing the daily kcalorie allowance. With more kcalories come more nutrients and other beneficial constituents of foods.

For those just beginning to increase **fitness**, be assured that improvement is not only possible, but also an inevitable result of becoming more active. As you improve your physical fitness, you not only *feel* better and stronger, but you also *look* better. Physically fit people walk with confidence and purpose because posture and self-image improve along with physical fitness.

If you are already physically fit, the following description applies. You move with ease and balance. You have endurance, and your energy lasts for hours. You are strong and meet daily physical challenges without strain. What is more, you are prepared to meet mental and emotional challenges because physical fitness also supports mental and emotional energy and resilience.

This chapter is written for athletes and for active people who train like athletes. Casual athletes (those who compete only with their own goals) and competitive athletes (those who compete with others) are the same with regard to their food and fluid needs. The chapter refers to "you" to make the connection between academic thinking and personal choices. To understand the interactions between physical activity and nutrition, you must first know a few things about fitness, its benefits, and **training** to develop fitness.

■ Each comparison influences the risks associated with chronic disease and death similarly:

Vigorous exercise versus minimal exercise

Healthy weight versus 20 percent overweight

Nonsmoking versus smoking (one pack per day)

© Zoran Milich/Masterfile

Physical activity, or its lack, exerts a significant and pervasive influence on everyone's nutrition and overall health.

Fitness

Fitness depends on a certain minimum amount of **physical activity** or **exercise**. Physical activity and exercise both involve bodily movement, muscle contraction, and enhanced energy expenditure, but by definition, a distinction is made between the two terms. Exercise is often considered to be a vigorous, structured, and planned type of physical activity. Because this chapter focuses on the active body's use of energy nutrients—whether that body is pedaling a bike across campus or pedaling a stationary bike in a gym—for our purposes, the terms *physical activity* and *exercise* will be used interchangeably.

Benefits of Fitness

People who regularly engage in just moderate physical activity live longer on average than those who are physically inactive.[1] Yet, despite an increasing awareness of the health benefits that physical activity confers, 25 percent of adults in the United States are completely inactive.[2] A **sedentary** lifestyle ranks with smoking and obesity as a powerful risk factor for developing the major killer diseases of our time—cardiovascular disease, some forms of cancer, stroke, diabetes, and hypertension.[3] ■

As a person becomes physically active, the health of the entire body improves. Compared with unfit people, physically fit people enjoy:

- **More restful sleep.** Rest and sleep occur naturally after periods of physical activity. During rest, the body repairs injuries, disposes of wastes generated during activity, and builds new physical structures.

- **Better nutritional health.** Physical activity expends energy and thus allows people to eat more food. If they choose wisely, active people will consume more nutrients and be less likely to develop nutrient deficiencies.

- **Improved body composition.** A balanced program of physical activity limits body fat and increases or maintains lean tissue. Thus, physically active people have relatively less body fat than sedentary people at the same body weight.[4]

- **Improved bone density.** Weight-bearing physical activity builds bone strength and protects against osteoporosis.[5]

- **Enhanced resistance to colds and other infectious diseases.** Fitness enhances immunity.[*6]

- **Lower risks of some types of cancers.** Lifelong physical activity may help to protect against colon cancer, breast cancer, and some other cancers.[7]

- **Stronger circulation and lung function.** Physical activity that challenges the heart and lungs strengthens both the circulatory and the respiratory system.

- **Lower risks of cardiovascular disease.** Physical activity lowers blood pressure, slows resting pulse rate, lowers total blood cholesterol, and raises HDL cholesterol, thus reducing the risks of heart attacks and strokes.[8] Some research suggests that physical activity may reduce the risk of cardiovascular disease in another way as well—by reducing intra-abdominal fat stores.[9]

- **Lower risks of type 2 diabetes.** Physical activity normalizes glucose tolerance.[10] Regular physical activity reduces the risk of developing type 2 diabetes and benefits those who already have the condition.

- **Reduced risk of gallbladder disease (women).** Regular physical activity reduces women's risk of gallbladder disease—perhaps by facilitating weight control and lowering blood lipid levels.[11]

- **Lower incidence and severity of anxiety and depression.** Physical activity may improve mood and enhance the quality of life by reducing depression and anxiety.[12]

- **Stronger self-image.** The sense of achievement that comes from meeting physical challenges promotes self-confidence.

- **Longer life and higher quality of life in the later years.** Active people live longer, healthier lives than sedentary people.[13] Even a two-mile walk daily can add years to a person's life. In addition to extending longevity, physical activity supports independence and mobility in later life by reducing the risk of falls and minimizing the risk of injury should a fall occur.[14]

You do not have to run marathons to reap the health rewards of physical activity.[15] For health's sake, the *Dietary Guidelines for Americans 2005* specify that people need to spend an accumulated minimum of 30 minutes in some sort of physical activity on most days of each week (see Figure 10–1 on p. 276).[16] However, both the *Dietary Guidelines for Americans 2005* and the DRI committee advise that 30 minutes of physical activity each day is not enough for adults to maintain a healthy body weight (BMI of 18.5 to 24.9) and recommend at least 60 minutes of moderately intense activity such as walking or jogging each day.[17] The hour or more of activity can be split into shorter sessions throughout the day—two 30-minute sessions or four 15-minute sessions, for example.[18]

---

* Moderate physical activity can stimulate immune function. Intense, vigorous, prolonged activity such as marathon running, however, may compromise immune function.

fitness: the characteristics that enable the body to perform physical activity; more broadly, the ability to meet routine physical demands with enough reserve energy to rise to a physical challenge; the body's ability to withstand stress of all kinds.

training: regular practice of an activity, which leads to physical adaptations of the body with improvement in flexibility, strength, or endurance.

physical activity: bodily movement produced by muscle contractions that substantially increase energy expenditure.

exercise: planned, structured, and repetitive bodily movement that promotes or maintains physical fitness.

sedentary: physically inactive (literally "sitting down a lot").

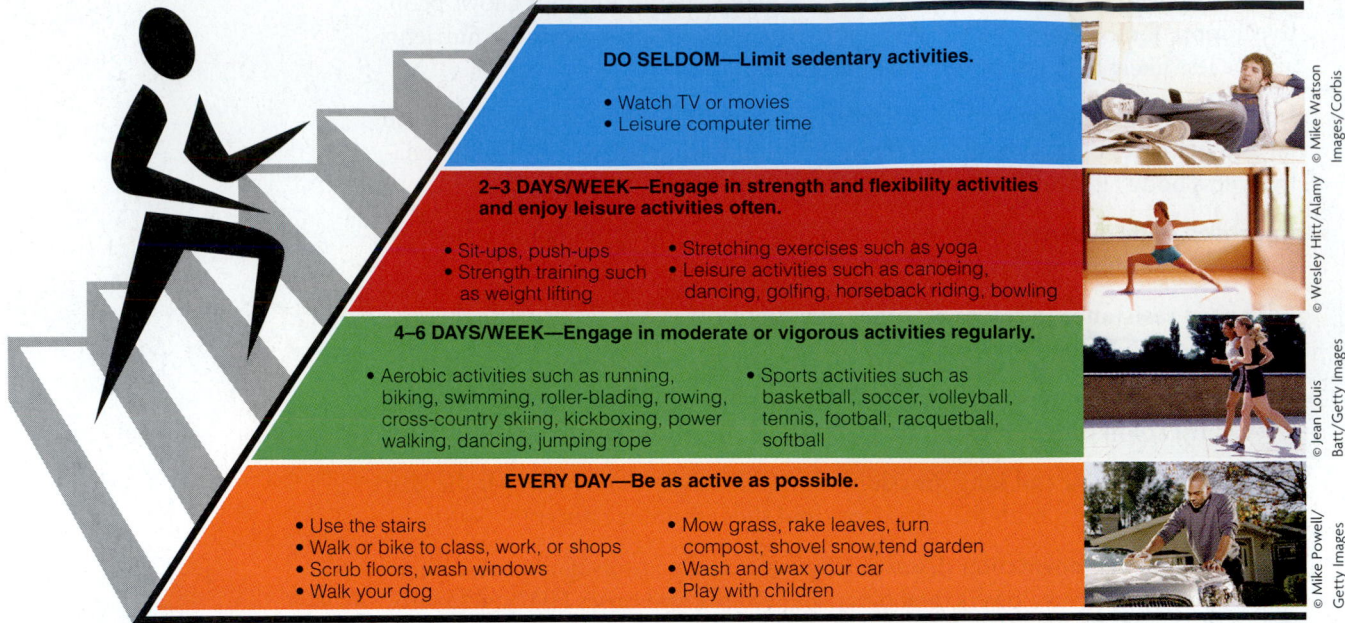

Note: Tips for increasing physical activity every day can be found at MyPyramid.gov.

FIGURE 10-1

Physical Activity Pyramid

For many people, the health benefits of regular, moderate physical activity are reward enough. Others seek the kinds and amount of physical activity that will not only benefit health, but also improve their physical fitness or their performance in sports. To develop and maintain *fitness,* the American College of Sports Medicine (ACSM) recommends the types and amounts of physical activity presented in Table 10–1. The kinds and amounts of physical activity that improve physical fitness also provide still greater health benefits (further reduction of cardiovascular disease risk and improved body composition, for example).[19]

IN SUMMARY

- Physical activity and fitness benefit people's physical and psychological well-being and improve their resistance to disease.

- Physical activity to improve physical fitness offers additional personal benefits.

flexibility: the capacity of the joints to move through a full range of motion; the ability to bend and recover without injury.

muscle strength: the ability of muscles to work against resistance.

muscle endurance: the ability of a muscle to contract repeatedly within a given time without becoming exhausted.

cardiorespiratory endurance: the ability to perform large-muscle dynamic exercise of moderate-to-high intensity for prolonged periods.

Dietary Guidelines for Americans 2005

- Engage in regular physical activity and reduce sedentary activities to promote health, psychological well-being, and a healthy body weight.

- Engage in at least 30 minutes of moderate-intensity physical activity, above usual activity, at work or home on most days of the week.

- For most people, greater health benefits can be obtained by engaging in physical activity of more vigorous intensity or longer duration.

- Achieve physical fitness by including cardiovascular conditioning, stretching exercises for flexibility, and resistance exercises for muscle strength and endurance.

TABLE 10–1
Guidelines for Physical Fitness

	Cardiorespiratory	Strength	Flexibility
	© PhotoDisc, Inc.	© David Hanover Photography	© David Hanover Photography
Type of Activity	Aerobic activity that uses large-muscles groups and can be maintained continuously	Resistance activity that is performed at a controlled speed and through a full range of motion	Stretching activity that uses the major muscle groups
Frequency	3 to 5 days per week	2 to 3 days per week	2 to 7 days per week
Intensity	55 to 90% of maximum heart rate	Enough to enhance muscle strength and improve body composition	Enough to develop and maintain a full range of motion
Duration	20 to 60 minutes	8 to 12 repetitions of 8 to 10 different exercises (minimum)	2 to 4 repetitions of 15 to 30 seconds per muscle group
Examples	Running, cycling, swimming, inline skating, rowing, power walking, cross-country skiing, kickboxing, jumping rope; sports activities such as basketball, soccer, raquetball, tennis, volleyball	Pull-ups, push-ups, weight lifting, pilates	Yoga

Source: Adapted from American College of Sports Medicine, General principles of exercise prescription, in *ACSM's Guidelines for Exercise Testing and Prescription*, 7th ed. (Philadelphia, PA: Lippincott Williams & Wilkins, 2006), pp. 133–173.

The Essentials of Fitness

To be physically fit, you need to develop enough **flexibility, muscle strength, muscle endurance,** and **cardiorespiratory endurance** to allow you to meet the everyday demands of life with some to spare, and you need to achieve a reasonable body composition. A person who practices a physical activity *adapts* by becoming better able to perform it after each session—with more flexibility, more strength, and more endurance.

Developing Fitness People shape their bodies by what they choose to do and not do. Muscle cells and tissues respond to an **overload** of physical activity by gaining strength and size, a response called **hypertrophy**. The opposite is also true: if not called on to perform, muscles diminish in size and weaken, a response called **atrophy**. Thus, cyclists often have well-developed legs but less arm or chest strength; a tennis player may have one superbly strong arm, while the other is just average. A variety of physical activities produces the best overall fitness, and to this end, people need to work different muscle groups from day to day. For balanced fitness (see Table 10–2 on p. 278), stretching enhances flexibility, weight training develops muscle strength and endurance, and **aerobic** activity improves cardiorespiratory endurance. It makes sense to give muscles a rest, too, because it takes a day or two to replenish muscle fuel supplies and to repair wear and tear incurred through physical activity.

overload: an extra physical demand placed on the body; an increase in the frequency, duration, or intensity of an activity. A principle of training is that for a body system to improve, it must be worked at frequencies, durations, or intensities that increase by increments.

hypertrophy (high-PURR-tro-fee): an increase in size (for example, of a muscle) in response to use.

atrophy (AT-tro-fee): a decrease in size (for example, of a muscle) because of disuse.

aerobic (air-ROH-bic): requiring oxygen. Aerobic activity strengthens the heart and lungs by requiring them to work harder than normal to deliver oxygen to the tissues.

People's bodies are shaped by the activities they perform.

Periodic rest also gives muscles time to adapt to an activity. During rest, muscles build more of the cellular structures required to perform the activity that preceded the rest. For example, the muscle cells of a superbly trained weight lifter store extra granules of glycogen, build up strong connective tissues, and add bulk to the special proteins that contract the muscles, thereby increasing the muscles' ability to perform.* In the same way, the muscle cells of a distance swimmer adapt to burn fat and to sustain prolonged exertion. Therefore, if you wish to become a better jogger, swimmer, or biker, you should train mostly by jogging, swimming, or biking. Your performance will improve as your muscles adapt to do the activity.

IN SUMMARY

- The components of fitness are flexibility, muscle strength, muscle endurance, and cardiorespiratory endurance.

- To build fitness, a person must engage in physical activity. Muscles adapt to activities they are called upon to perform repeatedly.

Weight Training **Weight training** has long been recognized as a method to build lean body mass and develop and maintain muscle strength and endurance. Additional benefits of weight training have emerged: progressive weight training also helps prevent and manage several chronic diseases, including cardiovascular disease, and enhances psychological well-being.[20]

By promoting strong muscles in the back and abdomen, weight training can improve posture and reduce the risk of back injury. Weight training can also help prevent the decline in physical mobility that often accompanies aging.[21] Older adults, even those in their eighties, who participate in weight training programs not only gain muscle strength, but also improve their muscle endurance, which enables them

weight training: the use of free weights or weight machines to provide resistance for developing muscle strength and endurance. A person's own body weight may also be used to provide resistance, as when a person does push-ups, pull-ups, or sit-ups; also called *resistance training*.

---

* All muscles contain a variety of muscle fibers, but there are two main types—slow-twitch (also called *red fibers*) and fast-twitch (also called *white fibers*). Slow-twitch fibers contain extra metabolic equipment to perform fat-burning aerobic work; the fast-twitch type store extra glycogen for anaerobic work. Muscle fibers of one type take on some of the characteristics of the other as an adaptation to exercise.

to walk significantly longer before exhaustion. Leg strength and walking endurance are powerful indicators of an older adult's physical abilities.

Yet another benefit is that weight training can help to maximize and maintain bone mass.[22] Research shows that even in women past menopause (when most women are losing bone), a one-year program of weight training improves bone density; the more weight lifted, the greater the improvement.[23] ■

Weight training can emphasize either muscle strength or muscle endurance. To emphasize muscle strength, combine high resistance (heavy weight) with a low number of repetitions. To emphasize muscle endurance, combine less resistance (lighter weight) with more repetitions. Weight training enhances performance in other sports, too. When they train with weights, swimmers can develop a more efficient stroke, and tennis players can develop a more powerful serve.[24]

The key to regular physical activity is finding an activity that you enjoy.

IN SUMMARY

- Weight training offers health and fitness benefits to adults.
- Weight training reduces the risk of cardiovascular disease, improves older adults' physical mobility, and helps maximize and maintain bone mass.

Cardiorespiratory Endurance   Although weight training provides some cardiovascular benefits, the kind of physical activity most beneficial to the health of the heart is cardiorespiratory endurance training. Cardiorespiratory endurance determines how long you can remain active with an elevated heart rate—it is the ability of the heart and lungs to sustain a given physical demand. Cardiorespiratory training enhances the capacity of the heart, lungs, and blood to deliver oxygen to, and remove wastes from, the body's cells. Thus, cardiorespiratory endurance training is aerobic.

The body's adaptation to the demands of regular aerobic activity involves a complex sequence of heart-healthy events. As cardiorespiratory endurance improves, the body delivers oxygen more efficiently. In fact, the accepted measure of a person's cardiorespiratory fitness is maximal oxygen uptake (**VO$_2$max**).[25] With cardiorespiratory endurance, the total blood volume and the number of red blood cells increase, so the blood can carry more oxygen. The heart muscle becomes stronger, and its **cardiac output** increases.[26] Each beat empties the heart's chambers more completely, so the heart pumps more blood per beat—its **stroke volume** increases. This makes fewer beats necessary, so the pulse rate falls. The muscles that inflate and deflate the lungs gain strength and endurance, so breathing becomes more efficient. Blood moves easily through the blood vessels because the muscles of the heart contract powerfully, and contraction of the skeletal muscles pushes the blood through the veins. Such improvements keep resting blood pressure normal.[27] Figure 10–2 (p. 280) shows the major relationships among the heart, lungs, and muscles. The improvements that come with cardiorespiratory endurance also raise blood HDL, the lipoprotein associated with lower heart disease risk. ■

Which activities produce these beneficial changes? Effective activities elevate the heart rate, are sustained for longer than 20 minutes, and use most of the large-muscle groups of the body (legs, buttocks, and abdomen). Examples are swimming, cross-country skiing, rowing, fast walking, jogging, fast bicycling, soccer, hockey, basketball, in-line skating, lacrosse, and rugby. The ACSM guidelines for developing and maintaining cardiorespiratory fitness were given in Table 10–1 on page 277.

An informal pulse check can give you some indication of how conditioned your heart is. The average resting pulse rate for adults is around 70 beats per minute. Active

■ Extremes in physical activity, together with severely restricted energy intakes, may be detrimental to bone health in some young women. Such women risk developing the "female athlete triad," discussed in Nutrition in Practice 6.

■ Cardiorespiratory endurance is characterized by:
  - Increased cardiac output and oxygen delivery
  - Increased heart strength and stroke volume
  - Slowed resting pulse
  - Increased breathing efficiency
  - Improved circulation
  - Reduced blood pressure

VO$_2$ max: the maximum rate of oxygen consumption by an individual (measured at sea level).

cardiac output: the volume of blood discharged by the heart each minute.

stroke volume: the amount of oxygenated blood ejected from the heart toward body tissues at each beat.

FIGURE 10–2

Delivery of Oxygen by the Heart and Lungs to the Muscles

The cardiorespiratory system responds to the muscles' demand for oxygen by building up its capacity to deliver oxygen. Researchers can measure cardiorespiratory fitness by measuring the maximum amount of oxygen a person consumes per minute while working out, a measure called VO_2max.

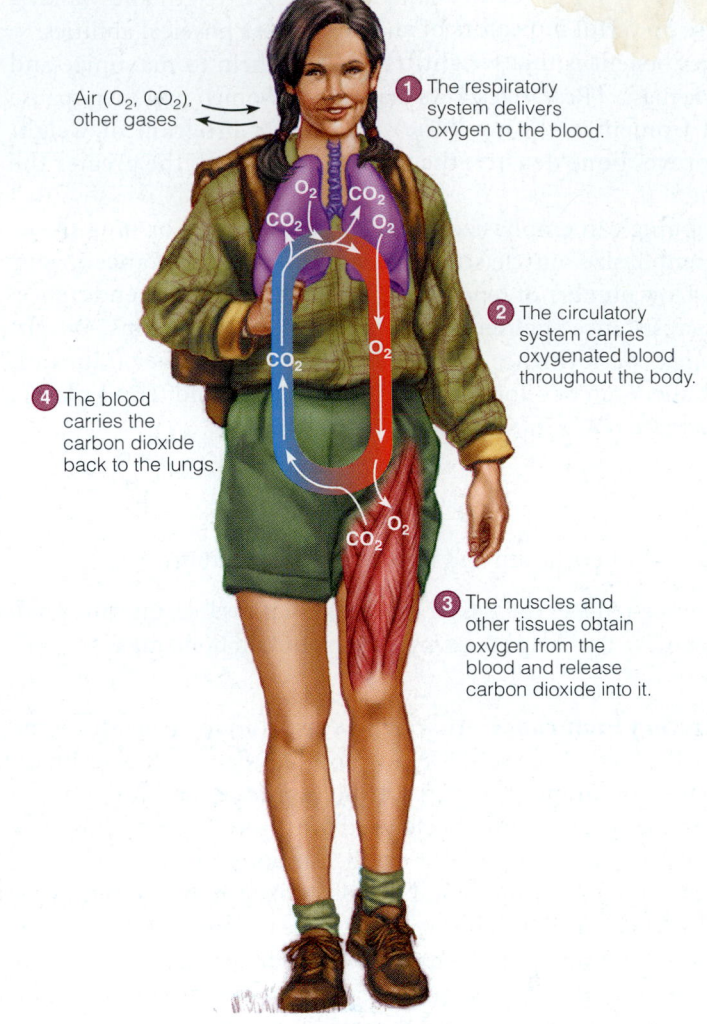

Air (O_2, CO_2), other gases

1 The respiratory system delivers oxygen to the blood.

2 The circulatory system carries oxygenated blood throughout the body.

4 The blood carries the carbon dioxide back to the lungs.

3 The muscles and other tissues obtain oxygen from the blood and release carbon dioxide into it.

■ To take your resting pulse: Sit down and relax for five minutes before you begin. Using a watch or clock with a second hand, place your hand over your heart or your finger firmly over an artery at the underside of the wrist or side of the throat under the jawbone. Start counting your pulse at a convenient second, and continue counting for ten seconds. If a heartbeat occurs exactly on the tenth second, count it as one-half beat. Multiply by 6 to obtain the beats per minute.

To ensure a true count: Use only fingers, not your thumb, on the pulse point (the thumb has a pulse of its own). Press just firmly enough to feel the pulse. Too much pressure can interfere with the pulse

people can have resting pulse rates of 50 or even lower. To take your pulse, follow the directions in the margin. ■

IN SUMMARY

■ Cardiorespiratory endurance training enhances the ability of the heart and lungs to deliver oxygen to the muscles.

■ With cardiorespiratory endurance training, the heart becomes stronger, breathing becomes more efficient, and the health of the entire body improves.

The rest of this chapter describes the interactions between nutrients and physical activity. Nutrition alone cannot endow you with fitness or athletic ability, but along with the right mental attitude, it complements your effort to obtain them. Conversely, unwise food selections can stand in your way.

The Active Body's Use of Fuels

The fuels that support physical activity are glucose (from carbohydrate), fatty acids (from fat), and, to a small extent, amino acids (from protein). The body uses different mixtures of fuels depending on the intensity and duration of its activities and depending on its own prior training.

FIGURE 10–3

The Effect of Diet on Physical Endurance
A high-carbohydrate diet can in-crease an athlete's endurance. In this study, the high-fat diet pro-vided 94 percent of kcalories from fat and 6 percent from protein; the normal mixed diet provided 55 per-cent of kcalories from carbohydrate; and the high-carbohydrate diet pro-vided 83 percent of kcalories from carbohydrate.
Source: © BonnieKamin/PhotoEdit

During rest, the body derives a little more than half of its energy from fatty acids, most of the rest from glucose, and a little from amino acids. During physical activity, the body adjusts its fuel mix to use the stored glucose of muscle glycogen. In the early minutes of an activity, muscle glycogen provides the majority of energy the muscles use to go into action. As activity continues, messenger molecules, including the hor-mone **epinephrine**, flow into the bloodstream to signal the liver and fat cells to liber-ate their stored energy nutrients, primarily glucose and fatty acids. Thus, hormones set the table for the muscles' energy feast, and the muscles help themselves to the fuels passing by in the blood.

Glucose Use and Storage

Both the liver and muscles store glucose as glycogen; the liver can also make glucose from fragments of other nutrients. Muscles conserve their glycogen stores—they do not release their glucose into the bloodstream to share with other body tissues, as the liver does. This is fortunate because a muscle that shared its glycogen reserves with other tissues might lack glucose at a critical time such as when running from danger. A muscle that conserves its glycogen is prepared to act in emergencies because muscle glucose fuels quick action. As activity continues, glucose from the liver's stored glyco-gen and dietary glucose absorbed from the digestive tract also become important sources of fuel for muscle activity.

Diet Affects Glycogen Storage and Use The body constantly uses and replenishes its glycogen. The more carbohydrate a person eats, the more glycogen muscles store (up to a limit) and the longer the stores will last to support physical activity.

A classic report compared fuel use during physical activity by three groups of run-ners, each on a different diet. For several days before testing, one of the groups ate a normal mixed diet (55 percent of kcalories from carbohydrate); a second group ate a high-carbohydrate diet (83 percent of kcalories from carbohydrate); and the third group ate a high-fat diet (94 percent of kcalories from fat). As Figure 10–3 shows, the high-carbohydrate diet enabled the athletes to work longer before exhaustion. This study and many others established that a high-carbohydrate diet enhances an athlete's endurance by ensuring ample glycogen stores.

Intensity of Activity Affects Glycogen Use The body's glycogen stores are much more limited than its fat stores. Glycogen supplies can easily support everyday activi-ties but are limited to less than 2,000 kcalories of energy. However, fat stores can usu-ally provide more than 70,000 kcalories and fuel hours of activity without running

epinephrine (EP-ih-NEFF-rin): the major hormone that elicits the stress response.

out. How long a person's glycogen will last during physical activity depends not only on diet, but also on the intensity of the activity.

Moderate activities such as easy jogging, during which breathing is steady and easy, use glycogen slowly. The lungs and circulatory system have no trouble keeping up with the muscles' need for oxygen. The individual breathes easily, and the heart beats steadily—the activity is aerobic. The muscles derive their energy from both glucose and fatty acids. By depending partly on fatty acids, moderate aerobic activity conserves glycogen stores. Joggers still use glycogen, however, and eventually they can run out of it.

Intense activities—the kind that make it difficult "to catch your breath," such as a quarter-mile race—use glycogen quickly. Muscles must begin to rely more heavily on glucose, which can be partially broken down by **anaerobic** metabolism. Thus, the muscles begin drawing more heavily on their limited glycogen supply. Anaerobic breakdown of glycogen yields energy to muscle tissue when energy demands outstrip the body's ability to provide energy aerobically, but it does so by spending the muscles' glycogen reserves.

Lactate During intense activity, anaerobic breakdown of glucose produces **lactate**. Muscles release lactate formed during exercise into the blood and it travels to the liver. There, liver enzymes convert the lactate back into glucose. Glucose can then return to the muscles to fuel additional activity. At low intensities, lactate is readily cleared from the blood by the liver, but at higher intensities, lactate accumulates. When the rate of lactate production exceeds the rate of clearance, intense activity can be maintained for only one to three minutes (as in a 400- or 800-meter race or a boxing match). Accumulation of lactate was long blamed for a type of muscle fatigue, but recent thought disputes this idea. It is true that muscles produce lactate during a type of fatigue, but the lactate does not cause the fatigue.[28]

Activity Duration Affects Glycogen Use Glycogen use during physical activity depends on the *duration* of the activity as well as its *intensity*. In the first 10 minutes or so of an activity, the active muscles rely almost entirely on their own stores of glycogen. Within the first 20 minutes or so of moderate activity, a person uses up about one-fifth of the available glycogen. As the muscles devour their own glycogen, they become ravenous for more glucose, and the liver responds by emptying out its glycogen stores.

A person who exercises moderately for longer than 20 minutes begins to use less glucose and more fat for fuel. Still, glucose use continues, and if the activity lasts long enough and is intense enough, blood glucose declines and muscle and liver glycogen stores are depleted. Physical activity can continue for a short time thereafter only because the liver produces, from lactate and certain amino acids, the minimum amount of glucose needed to briefly forestall total depletion. Glycogen depletion generally occurs after about two hours of vigorous exercise.* When glycogen depletion occurs, it brings nervous system function to a near halt, making continued exertion at the same intensity almost impossible. Marathon runners call this "hitting the wall."

Maintaining Blood Glucose for Activity To avoid such debilitation, endurance athletes try to maintain their blood glucose concentrations for as long as they can. The following guidelines will help endurance athletes maximize glucose supply:

- Eat a high-carbohydrate diet (approximately 8 grams of carbohydrate per kilogram of body weight or about 70 percent of energy intake) regularly.†

---

* Here "vigorous exercise" means exercise at 75 percent of VO₂ max.

† Percentage of energy intake is meaningful only when total energy intake is known. Consider that at high energy intakes (5,000 kcalories/day), even a moderate carbohydrate diet (40 percent of energy intake) supplies 500 grams of carbohydrate—enough for a 137-pound (62 kilogram) athlete in heavy training. By comparison, at a moderate energy intake (2,000 kcalories/day), a high-carbohydrate intake (70 percent of energy intake) supplies 350 grams—plenty of carbohydrate for most people, but not enough for athletes in heavy training.

anaerobic (AN-air-ROH-bic): not requiring oxygen. Anaerobic activity may require strength but does not work the heart and lungs very hard for a sustained period.

lactate: a compound produced during the breakdown of glucose in anaerobic metabolism.

carbohydrate loading: a regimen of moderate exercise, followed by eating a high-carbohydrate diet, which enables muscles to temporarily store glycogen beyond their normal capacity; also called *glycogen loading* or *glycogen supercompensation*.

How to

MAXIMIZE GLYCOGEN STORES: CARBOHYDRATE LOADING

Some athletes use a technique called carbohydrate loading to maximize their muscle glycogen before a competition. Carbohydrate loading can nearly double muscle glycogen concentrations. In general, the athlete tapers training during the week before the competition and then eats a high-carbohydrate diet during the three days just prior to the event.[a] Specifically, the athlete follows the plan in the accompanying table. In this carbohydrate loading plan, glycogen storage occurs slowly, and athletes must alter their training for several days before the event.

In contrast, a group of researchers have designed a quick method of carbohydrate loading that has produced promising preliminary results. The researchers found

Before the Event	Training Intensity	Training Duration	Dietary Carbohydrate
6 days	Moderate (70% VO₂ max)	90 min	Normal (5 g/kg body weight)
4–5 days	Moderate (70% VO₂ max)	40 min	Normal (5 g/kg body weight)
2–3 days	Moderate (70% VO₂ max)	20 min	High-carbohydrate (10 g/kg body weight)
1 day	Rest	—	High-carbohydrate (10 g/kg body weight)

that athletes could attain above-normal concentrations of muscle glycogen by eating a high-carbohydrate diet (10 g/kg body weight) after a short (3 minutes) but very intense bout of exercise.[b] More studies are needed to confirm these findings and to determine whether an exercise session of less intensity and shorter

duration would accomplish the same results.

Extra glycogen gained through carbohydrate loading can benefit an athlete who must keep going for 90 minutes or longer. Those who exercise for shorter times simply need a regular high-carbohydrate diet. In a hot climate, extra glycogen confers an additional advantage: as glycogen breaks down, it releases water, which helps to meet the athlete's fluid needs.

[a] E. Coleman, Carbohydrate and exercise, in *Sports Nutrition: A Practice Manual for Professionals*, 4th ed., ed. M. Dunford (Chicago: The American Dietetic Association, 2006). pp. 14–32.

[b] T.J. Fairchild and coauthors. Rapid carbohydrate loading after a short bout of near maximal-intensity exercise. *Medicine and Science in Sports and Exercise* 34 (2002): 980–986.

■ Take glucose (usually in sports drinks) periodically during activities that last for 45 minutes or more.

■ Eat carbohydrate-rich foods (approximately 60 grams of carbohydrate) immediately following activity. ■

■ Train the muscles to store as much glycogen as possible.

The last section of this chapter, "Diets for Physically Active People," discusses how to design a high-carbohydrate diet for performance, and the accompanying "How to" describes **carbohydrate loading**—a technique used to maximize glycogen stores for long endurance competitions.

Glucose during Activity Glucose ingested before and during exhausting endurance activities (lasting more than 45 minutes) makes its way from the digestive tract to the working muscles, augmenting dwindling internal glucose supplies from the muscle and liver glycogen stores.[29] Especially during games such as soccer or hockey, which last for hours and demand repeated bursts of intense activity, athletes benefit from carbohydrate-containing drinks taken during the activity.[30]

Before concluding that glucose might be good for your own performance, consider first whether you engage in *endurance* activity. Do you run, swim, bike, or ski nonstop at a rapid pace for more than 45 minutes at a time, or do you compete in games lasting for hours? If not, the glucose picture changes. For an everyday jog or swim lasting less than 45 minutes, glucose probably will not help (or harm) performance. Even in athletes, extra carbohydrate does not benefit those who engage in sports in which fatigue is unrelated to blood glucose, such as 100-meter sprinting, baseball, casual basketball, and weight lifting.

■ For perspective, snack ideas providing 60 grams of carbohydrate include:

16 oz sports drink and a small bagel

16 oz milk and four oatmeal-raisin cookies

8 oz pineapple juice and a granola bar

Those who compete in endurance activities require fluid and carbohydrate fuel.

© Cleo Photography/PhotoEdit, Inc.

■ Foods with a high glycemic index:
 Cornflakes
 Mashed potatoes
 Short-grain rice
 Waffles
 Watermelon
 White bread
■ In popular magazine articles and on the Internet, foods with a high glycemic index are sometimes called "high impact carbs," and those with a low glycemic index are sometimes called "low impact carbs."

■ Factors that affect glucose use during physical activity:
 ■ Carbohydrate intake
 ■ Intensity and duration of the activity
 ■ Degree of training

Glucose after Activity Eating high-carbohydrate foods *after* physical activity also enlarges glycogen stores. Train normally; then, within two hours after physical activity, consume a high-carbohydrate meal, such as a glass of orange juice and some graham crackers, toast, or cereal. This method accelerates the rate of glycogen storage by 300 percent for awhile. This is especially important to athletes who train hard more than once per day. Timing is important—eating the meal after two hours have passed reduces the glycogen synthesis rate by almost half. For athletes who do not feel like eating right after exercise, **high-carbohydrate energy drinks** are available. These fruit-flavored drinks are higher in kcalories and carbohydrate than the regular sports drinks discussed later in this chapter (p. 292).

Chapter 2 introduced the glycemic effect and discussed some possible health benefits of eating a diet ranking low on the glycemic index. For athletes wishing to maximize muscle glycogen synthesis after strenuous training, however, eating foods with a high glycemic index (see the margin) ■ may be more beneficial.[31] Foods with a high glycemic index elicit greater rates of glycogen resynthesis compared to foods with a low glycemic index.[32] ■

Degree of Training Affects Glycogen Use Training affects glycogen use during activity in at least two ways. First, muscles that deplete their glycogen stores through work adapt to store greater amounts of glycogen to support that work. Second, trained muscles burn more fat and at higher intensities than untrained muscles, so they require less glucose to perform the same amount of work.[33] A person attempting an activity for the first time uses up much more glucose per minute than an athlete trained to perform it. A trained person can work at high intensities for longer periods than an untrained person while using the same amount of glycogen. ■

IN SUMMARY

■ Glucose is supplied by dietary carbohydrate or made by the liver.

■ Glucose is stored in both liver and muscle tissue as glycogen.

■ Total glycogen stores affect an athlete's endurance.

■ The more intense an activity, the more glucose it demands.

■ During anaerobic metabolism, the body spends glucose rapidly and accumulates lactate.

■ Physical activity of long duration places demands on the body's glycogen stores.

■ Carbohydrate ingested before and during long-duration activity may help to forestall fatigue.

■ Carbohydrate loading is a regimen of physical activity and diet that enables an athlete's muscles to store larger-than-normal amounts of glycogen to extend endurance.

■ After strenuous training, eating foods with a high glycemic index may help restore glycogen most rapidly.

■ Highly trained muscles use less glucose and more fat than do untrained muscles to perform the same work, so their glycogen lasts longer.

Fat Use During Activity

As Figure 10–3 showed, researchers have long recognized the importance of a high-carbohydrate diet for endurance performance. When endurance athletes "fat load" by consuming high-fat, low-carbohydrate diets for one to three days, their performance is impaired because their small glycogen stores are depleted quickly.[34] Endurance athletes who adhere to a high-fat, low-carbohydrate diet for more than one week, however, adapt by relying more on fat to fuel activity. Even with fat adaptation, however, performance benefits are not consistently evident.[35] In some cases, athletes on high-fat diets experience greater fatigue and perceive the activity to be more strenuous than athletes on high-carbohydrate diets.[36]

Diets high in saturated fat carry risks of heart disease, too. Physical activity offers some protection against cardiovascular disease, but athletes, like everyone else, can suffer heart attacks and strokes. Most nutrition experts agree that the potential for adverse health effects from prolonged high-fat diets makes them an unwise choice for athletes.

A diet that overly restricts fat is not recommended either. Athletes who restrict fat below 20 percent of total energy intake may fail to consume adequate energy and nutrients. Sports nutrition experts recommend that endurance athletes consume 20 to 30 percent of their energy from fat.[37] One expert says the message is "not that high-fat diets improve performance, but rather that very low-fat diets inhibit performance."

In contrast to *dietary* fat, *body* fat stores are extremely important during physical activity, as long as the activity is not too intense. Unlike the body's limited glycogen stores, fat stores can fuel hours of activity without running out.

Early in activity, muscles begin to draw on fatty acids from two sources—fats stored within the working muscles and fats from fat deposits such as the fat under the skin. Areas with the most fat to spare donate the greatest amounts of fatty acids to the blood (although they may not be the areas from which one might choose to lose fat). This is why "spot reducing" does not work: muscles do not own the fat that surrounds them. Fat cells release fatty acids into the blood for all the muscles to share. Proof is found in a tennis player's arms: the fatfolds measure the same in both arms, even though one arm has better-developed muscles than the other. A balanced fitness program that includes strength training, however, will tighten muscles underneath the fat, improving the overall appearance. Keep in mind that some body fat is essential to good health.

Intensity and Duration Affect Fat Use The *intensity* of physical activity affects fat use. As the intensity of activity increases, fat makes less and less of a contribution to the fuel mixture. Remember that fat can be broken down for energy only by aerobic metabolism. For fat to fuel activity, oxygen must be abundantly available. If a person is breathing easily during activity, the muscles are getting all the oxygen they need and are able to use more fat in fuel mixture.

The *duration* of activity also influences fat use. Early in an activity, as the muscles draw on fatty acids, blood levels fall. If the activity continues for more than a few minutes, the hormone epinephrine signals the fat cells to begin breaking down their stored triglycerides and liberating fatty acids into the blood. After about 20 minutes of activity, the blood fatty acid concentration surpasses the normal resting concentration. Thereafter, sustained, moderate activity uses body fat stores as its major fuel. ■

Degree of Training Affects Fat Use Training—repeated aerobic activity—produces the adaptations that permit the body to draw more heavily on fat for fuel. Aerobically trained muscles burn fat more readily than untrained muscles. With aerobic training, the heart and lungs also become stronger and better able to deliver oxygen to the muscles during high-intensity activities. In turn, this improved oxygen supply enables the muscles to burn more fat. These adaptations reward not only trained athletes, but also all active people; a person who trains by way of aerobic activities such as distance running or cycling becomes well suited to the activity.

© Jim Cummin/CORBIS

Low- to moderate-intensity aerobic exercises that can be sustained for a long time (more than 20 minutes) use some glucose but more fat for fuel.

■ Factors that affect fat use during physical activity:
 Fat intake
 Intensity and duration of the activity
 Degree of training

high-carbohydrate energy drinks: fruit-flavored commercial beverages used to restore muscle glycogen after exercise or as a pregame beverage.

IN SUMMARY

- Athletes who eat high-fat diets may burn more fat during endurance activity, but the risks to health outweigh any possible performance benefits.

- The intensity and duration of activity, as well as the degree of training, affect fat use.

Protein Use During Activity

Athletes use protein to build and maintain muscle and other lean tissue structures and, to a small extent, to fuel activity. The body handles protein differently during activity than during rest.

Protein for Building Muscle Tissue In the hours of rest that follow physical activity, muscles speed up their rate of protein synthesis—they build more of the proteins they need to perform the activity. Research shows that eating high-quality protein, either by itself or together with carbohydrate, immediately following physical activity stimulates muscle protein synthesis.[38] Whenever the body rebuilds a part of itself, it must tear down the old structures to make way for the new ones. Physical activity, with just a slight overload, calls into action both the protein-dismantling and the protein-synthesizing equipment of individual muscle cells that work together to remodel muscles.

Dietary protein provides the needed amino acids for synthesis of new muscle proteins. As Chapter 4 pointed out, the true director of synthesis of muscle protein is physical activity itself. Repeated activity signals the muscle cells' genetic material to begin producing more of the proteins needed to perform the work at hand.

The genetic protein-making equipment inside the nuclei of muscle cells seems to "know" when proteins are needed. Furthermore, it knows *which* proteins are needed to support each type of physical activity. Apparently, the intensity and pattern of muscle contractions initiate signals that direct the muscles' genetic material to make particular proteins. For example, a weight lifter's workout sends the information that muscle fibers need added bulk for strength and more enzymes for making and using glycogen. A jogger's workout stimulates production of proteins needed for aerobic oxidation of fat and glucose. Muscle cells are exquisitely responsive to the need for proteins, and they build them conservatively only as needed.

Finally, after muscle cells have made all the decisions about which proteins to build and when, protein nutrition comes into play. During active muscle-building phases of training, a weight lifter might add between $1/4$ ounce and 1 ounce (between 7 and 28 grams) of protein to existing muscle mass each day. This extra protein comes from ordinary food.

Protein for Fuel Not only do athletes retain more protein, but they also use a little more protein as fuel. Muscles speed up their use of amino acids for energy during physical activity, just as they speed up their use of glucose and fatty acids. Still, protein contributes at most about 10 percent of the total fuel used, both during activity and during rest. ■

Diet Affects Protein Use during Activity The factors that regulate how much protein is used during activity seem to be the same ones that regulate the use of glucose and fat—one factor is diet. People who consume diets adequate in energy and rich in *carbohydrate* use less protein than those who eat protein- and fat-rich diets. Recall that carbohydrates spare proteins from being broken down to make glucose when needed. Because physical activity requires glucose, a diet lacking in carbohydrate necessitates the conversion of amino acids to glucose.

For perfect functioning, every nutrient is needed.

■ Factors that affect protein use during physical activity:

Carbohydrate intake

Intensity and duration of the activity

Degree of training

Intensity and Duration Affect Protein Use The intensity and duration of the activity also affect protein use. Endurance athletes who train for over one hour per day, engaging in aerobic activity of moderate intensity and long duration, may deplete their glycogen stores by the end of their training and become more dependent on body protein for energy. In contrast, anaerobic strength training does not use more protein for energy but does demand more protein to build muscle. Thus, the protein needs of both endurance and strength athletes are higher than those of sedentary people, but not as high as the protein intakes many athletes consume.

Degree of Training Affects Protein Use Finally, the extent of training also affects the use of protein. Particularly in strength athletes such as bodybuilders, the higher the degree of training, the less protein a person uses during activity at a given intensity.

Protein Recommendations for Active People Although most athletes need somewhat more protein than do sedentary people, average protein intakes in the United States are high enough to cover those needs. Therefore, athletes in training should attend to protein needs but should back up the protein with ample carbohydrate. Otherwise, they will burn off as fuel the very protein they wish to retain in muscle.

The DRI does not recommend greater-than-normal protein intakes for athletes, but other authorities do.[39] These recommendations specify different protein intakes for athletes pursuing different activities (see Table 10–3).[40] A later section translates protein recommendations into a diet plan and shows that no one needs protein supplements or even large servings of meat to obtain the highest recommended protein intakes.

IN SUMMARY

- Physical activity stimulates muscle cells to break down and synthesize protein, resulting in muscle adaptation to activity.

- Athletes use protein both for building muscle tissue and for energy. Diet, intensity and duration of activity, and training affect protein use during activity.

- Although athletes need more protein than sedentary people, a balanced, high-carbohydrate diet provides sufficient protein to cover an athlete's needs.

TABLE 10–3

Recommended Protein Intakes for Athletes

	Recommendations (g/kg/day)	Protein Intakes (g/day)	
		Males	Females
RDA for adults	0.8	56	44
Recommended intake for power (strength or speed) athletes	1.6–1.7	112–119	88–94
Recommended intake for endurance athletes	1.2–1.6	84–112	66–88
U.S. average intake	—	95	65

Note: Daily protein intakes are based on a 70-kilogram (154-pound) man and 55-kilogram (121-pound) woman.

Sources: Committee on Dietary Reference Intakes, *Dietary Reference Intakes for Energy, Carbohydrate, Fiber, Fat, Fatty Acids, Cholesterol, Protein, and Amino Acids* (Washington DC: National Academies Press, 2005), pp. 660–661; Position of the American Dietetic Association, Dietitians of Canada, and the American College of Sports Medicine: Nutrition and athletic performance, *Journal of the American Dietetic Association* 100 (2000): 1543–1556.

Vitamins and Minerals to Support Activity

Many vitamins and minerals assist in releasing energy from fuels and transporting oxygen. This knowledge has led many people to believe, mistakenly, that vitamin and mineral *supplements* offer physically active people both health benefits and athletic advantages. (Review Nutrition in Practice 9 for a discussion of vitamin and mineral supplements and see Nutrition in Practice 10, which explores supplements and other products people use in the hope of enhancing athletic performance.)

Supplements

Many athletes take supplements in the hope of improving their performance. A meta-analysis of more than 10,000 athletes involved in 15 sports at all levels found that about half of the athletes use vitamin-mineral supplements.[41] Elite athletes use supplements to a greater extent than college athletes, who use supplements more than high school athletes. Supplement use by women exceeds that of men, and use by athletes exceeds that of the general population. One of the most common reasons athletes give for supplement use is "to improve performance."

Nutrient supplements do not enhance the performance of well-nourished athletes or active people. However, deficiencies of vitamins and minerals do impede performance. Regular, strenuous physical activity increases the demand for energy, but athletes and active people who eat enough nutrient-dense foods to meet energy needs also meet their vitamin and mineral needs. Active people eat more food; it stands to reason that with the right choices, they will get more nutrients.

■ Stringent weight requirements pose a risk of developing eating disorders. See Nutrition in Practice 6.

Athletes who lose weight to meet low body-weight requirements, however, may consume so little food that they fail to obtain all the nutrients they need.[42] ■ The practice of "making weight" is opposed by many health and fitness organizations, but for athletes who choose this course of action, a single daily multivitamin-mineral tablet that provides no more than the DRI recommendations for nutrients can be beneficial. In addition, some athletes do not eat enough food to maintain body weight during times of intense training or competition. For these athletes, too, a daily multivitamin-mineral supplement can be helpful.

Some athletes believe that taking vitamin or mineral supplements just before competition will enhance performance. These beliefs are contrary to scientific reality. Most vitamins and minerals function as small parts of larger working units. After entering the blood, they have to wait for the cells to combine them with their appropriate other parts so that they can do their work. This takes time—hours or days. Vitamins or minerals taken right before an event do not improve performance, even if the person is actually suffering deficiencies of those nutrients.

Nutrients of Special Concern

In general, active people who eat well-balanced meals do not need vitamin or mineral supplements. Two nutrients, vitamin E and iron, do merit special attention, however, for different reasons. Vitamin E is addressed because so many athletes take supplements of it. Iron is discussed because some female athletes may be unaware that they need supplements.

Vitamin E   During prolonged, high-intensity physical activity, the muscles' consumption of oxygen increases tenfold or more, enhancing the production of damaging free radicals in the body.[43] Vitamin E is a potent fat-soluble antioxidant that vigorously defends cell membranes against oxidative damage. Some athletes take megadoses of vitamin E in hope of preventing such oxidative damage to muscles. In some studies, supplementation with vitamin E does seem to protect against oxidative stress; others show no effect on oxidative stress; and a few show enhanced stress after vitamin E supplementation.[44] There is little evidence that vitamin E supplements can improve performance.[45] Clearly, more research is needed, but in the meantime, physically active people can benefit by using vegetable oils and eating generous servings of antioxidant-rich fruits and vegetables regularly. ■

Iron and Performance Physically active young women, especially those who engage in endurance activities such as distance running, are prone to iron deficiency.[46] Habitually low intakes of iron-rich foods, high iron losses through menstruation, and the high demands of muscles for the iron-containing molecules of aerobic metabolism and the muscle protein myoglobin can contribute to iron deficiency in young female athletes.

Adolescent female athletes who eat vegetarian diets may be particularly vulnerable to iron deficiency.[47] The bioavailability of iron is often poor in plant-based diets because such diets are high in fiber and phytic acid and because the nonheme iron in plant foods is not absorbed as well as the heme iron in animal-derived foods. Vegetarian diets are usually rich in vitamin C, however, which enhances iron absorption. To protect against iron deficiency, vegetarian athletes need to pay close attention to their intake of good dietary sources of iron (fortified cereals, legumes, nuts, and seeds) and include vitamin C-rich foods with each meal.[48] As long as vegetarian athletes, like all athletes, consume enough nutrient-dense foods, they can perform as well as anyone.

Iron deficiency impairs performance because iron helps deliver the muscles' oxygen. Insufficient oxygen delivery reduces aerobic work capacity, so the person tires easily. Whether marginal deficiency without clinical signs of anemia hinders physical performance is less clear.[49]

Early in training, athletes may develop low blood hemoglobin. This condition, sometimes called "sports anemia," is not a true iron-deficiency condition. Strenuous training promotes destruction of the more fragile, older red blood cells, and the resulting cleanup work reduces the blood's iron content temporarily. Strenuous activity also promotes increases in the fluid of the blood; with more fluid, the red blood cell count in a unit of blood drops. Most researchers view sports anemia as an *adaptive,* temporary response to endurance training. True iron-deficiency anemia requires treatment with prescribed iron supplements, but sports anemia goes away by itself, even with continued training.

The best strategy concerning iron is to determine individual needs. Many menstruating women border on iron deficiency even without the additional iron demand and losses incurred by physical activity. Active teens of both genders have high iron needs because they are growing. For women and teens, prescribed supplements may be needed to correct a deficiency of iron that is confirmed by tests. (Medical testing is needed to eliminate nondietary causes of anemia, such as internal bleeding or cancer.)

Female athletes may be at special risk of iron deficiency.

■ The Tolerable Upper Intake Level (UL) for vitamin E is 1,000 mg per day.

IN SUMMARY

- With the possible exception of iron, well-nourished active people and athletes do not need nutrient supplements.

- Female athletes need to pay special attention to their iron needs.

- Iron-deficiency anemia impairs physical performance because iron is the blood's oxygen handler.

- Sports anemia is a harmless temporary adaptation to physical activity.

Foods like these are packed with the nutrients that active people need.

Fluids and Electrolytes in Physical Activity

The body's need for water far surpasses its need for any other nutrient. If the body loses too much water, its life-supporting chemistry is compromised.

The exercising body loses water primarily via sweat; second to that, breathing uses water, exhaled as vapor. During physical activity, both routes can be significant, and dehydration is a real threat. The first symptom of dehydration is fatigue. A water loss of greater than 2 percent of body weight can reduce a person's capacity to do muscular work.[50] A person with a water loss of about 7 percent is likely to collapse. The athlete who arrives at an event even slightly dehydrated starts out at a competitive disadvantage.

Temperature Regulation

As Chapter 9 pointed out, sweat cools the body. The conversion of water to vapor uses up a great deal of heat, so as sweat evaporates, it cools the skin's surface and the blood flowing beneath it.

Hyperthermia In hot, humid weather, sweat may fail to evaporate because the surrounding air is already laden with water. In **hyperthermia**, body heat builds up and triggers maximum sweating, but without sweat evaporation, little cooling takes place. In such conditions, active people must take precautions to avoid **heat stroke**. Heat stroke is an especially dangerous accumulation of body heat with accompanying loss of body fluid. To reduce the risk of heat stroke, drink enough fluid before and during the activity, rest in the shade when tired, and wear lightweight clothing that allows sweat to evaporate.[51] The rubber or heavy suits sold with promises of weight loss during physical activity are dangerous because they promote profuse sweating, prevent sweat evaporation, and invite heat stroke. If you experience any of the symptoms of heat stroke listed in the margin, stop your activity, sip cold fluids, seek shade, and ask for help. ■ The condition demands medical attention—it can kill.

■ Symptoms of heat stroke:[52]
Clumsiness

Confusion, other mental changes, loss of consciousness

Dizziness

Headache

Internal (rectal) temperature above 104° Fahrenheit

Nausea

Stumbling

Hot, flushed, dry skin, or pale, sweat-soaked skin

Hypothermia In cold weather, **hypothermia**, or loss of body heat, can pose as serious a threat as heat stroke does in hot weather. Inexperienced runners participating in long races on cold or wet, chilly days are especially vulnerable to hypothermia. Slow runners can produce too little heat to keep warm, especially if their clothing is inadequate. Early symptoms of hypothermia include feeling cold, shivering, apathy, and social withdrawal.[53] As body temperature continues to fall, shivering stops, and disorientation, slurred speech, and change in behavior or appearance set in. People with these symptoms soon become helpless to protect themselves from further body heat losses. Even in cold weather, the body still sweats and needs fluids, but the fluids should be warm or at room temperature to help prevent hypothermia.

Fluid Needs during Physical Activity

Endurance athletes can lose 1.5 quarts or more of fluid during *each hour* of activity. To prepare for fluid losses, the athlete must hydrate before activity. To replace fluid losses, the person must rehydrate during and after activity. (Table 10–4 presents one schedule of hydration for physical activity.) Even then, in hot weather, the digestive tract may not be able to absorb enough water fast enough to keep up with an athlete's sweat losses, and some degree of dehydration may be inevitable. Athletes who know their body's **hourly sweat rate** can strive to replace the total amount of fluid lost during activity to prevent dehydration.[54]

Athletes who are preparing for competition are often advised to drink extra fluids in the last few days of training before the event. The extra fluid is not stored in the

TABLE 10-4

Hydration Schedule for Physical Activity

When to Drink	Amount of Fluid
2 hr before activity	2 to 3 c
15 min before activity	1 to 2 c
Every 15 min during activity	½ to 2 c (Drink enough to minimize loss of body weight, but don't overdrink.)
After activity	2 c for each pound of body weight loss[a]

[a] Drinking 2 cups of fluid every 20 to 30 minutes after exercise until the total amount required is consumed is more effective for rehydration than drinking the needed amount all at once. Rapid fluid replacement after exercise stimulates urine production and results in less body water retention.

Source: R. Murray, Fluid, electrolytes, and exercise in *Sports Nutrition: A Practice Manual for Professionals*, 4th ed., ed. M. Dunford (Chicago: The American Dietetic Association, 2005). pp. 94–115: D. J. Casa, P.M. Clarkson, and W.O. Roberts, American College of Sports Medicine Roundtable on Hydration and Physical Activity: Consensus Statements, *Current Sports Medicine Reports* 4 (2005): 115–127.

body, but drinking extra ensures maximum tissue hydration at the start of the event. Full hydration is imperative for every athlete both in training and in competition. Any coach or athlete who withholds fluids during practice for any reason takes a great risk and is subject to sanctions by the American College of Sports Medicine.

Athletes who rely on thirst to govern fluid intake can easily become dehydrated. During activity thirst becomes detectable only *after* fluid stores are depleted. Do not wait to feel thirsty before drinking.

Water What is the best fluid to support physical activity? The best drink for most active bodies is just plain cool water, for two reasons: (1) water rapidly leaves the digestive tract to enter the tissues, and (2) it cools the body from the inside out. Endurance athletes are an exception: they need more from their fluids than water alone. The first priority for endurance athletes should always be replacement of fluids to prevent life-threatening heat stroke. But endurance athletes also need carbohydrate to supplement their limited glycogen stores, so glucose is important, too. The "How to" (p. 292) compares water and sports drinks as fluid sources for endurance athletes.

Electrolyte Losses and Replacement During physical activity, the body loses electrolytes—the minerals sodium, potassium, and chloride—in sweat. Beginners lose these electrolytes to a much greater extent than do trained athletes. The body's adaptation to physical activity includes better conservation of these electrolytes. To replenish lost electrolytes, a person ordinarily needs only to eat a regular diet that meets energy and nutrient needs. In events lasting more than 45 minutes, sports drinks may be needed to replace fluids and electrolytes. Salt tablets can worsen dehydration and impair performance; they increase potassium losses, irritate the stomach, and cause vomiting. Athletes should avoid them.

Sodium Depletion When athletes compete in endurance sports lasting longer than three hours, replenishing electrolytes becomes crucial. If athletes sweat profusely over a long period of time and do not replace lost sodium, a dangerous condition of sodium depletion, known as **hyponatremia**, may result. The symptoms of hyponatremia are similar to, but not the same as, those of dehydration (see the margin). ■ Recent research shows that some athletes who sweat profusely may also lose more sodium in their sweat than others—and are prone to debilitating heat cramps.[55] These athletes lose twice as much sodium in sweat as athletes who do not cramp. Depending on individual variation, exercise intensity, and changes in ambient temperature and humidity, sweat rates for these athletes can exceed 2 quarts per hour.[56]

© Thinkstock/Getty Images

Active people need extra fluid, even in cold weather.

■ Symptoms of hyponatremia:
  Severe headache
  Vomiting
  Bloating, puffiness from water retention (shoes tight, rings tight)
  Confusion
  Seizure

hyperthermia: an above-normal body temperature.

heat stroke: an acute and life-threatening reaction to heat buildup in the body.

hypothermia: a below-normal body temperature.

hourly sweat rate: the amount of weight lost plus fluid consumed during exercise per hour.

hyponatremia (HIGH-poh-na-TREE-mee-ah): a decreased concentration of sodium in the blood (*hypo* means "below"; *natrium* means "sodium"; *emia* means "blood").

How to

EVALUATE SPORT DRINKS

Hydration is critical to optimal performance. Water best meets the fluid needs of most people, yet manufacturers market many good-tasting sports drinks for active people. More than 20 "power beverages" compete for their share of the more than $1 billion market. What do sports drinks have to offer?

- *Fluid.* Sports drinks offer fluids to help offset the loss of fluids during physical activity, but plain water can do this, too. Alternatively, diluted fruit juices or flavored water can be used if preferred to plain water.

- *Glucose.* Sports drinks offer simple sugars or glucose polymers that help maintain hydration and blood glucose and enhance performance as effectively as, or maybe even better than, water. Such measures are especially beneficial for strenuous endurance activities lasting longer than 45 minutes, during intense activities, or during prolonged competitive games that demand repeated intermittent activity.[a] Sports drinks are also suitable for events lasting less than 45 minutes although plain water is appropriate as well.[b]

Fluid transport to the tissues from beverages containing up to 8 percent glucose is rapid. Most sports drinks contain about 7 percent carbohydrate (about half the sugar of ordinary soft drinks, or about 5 teaspoons in each 12 ounces). Less than 6 percent may not enhance performance, and more than 8 percent may cause abdominal cramps, nausea, and diarrhea.

Although glucose does enhance endurance performance in strenuous competitive events, for the moderate exerciser, it can be counterproductive if weight loss is the goal. Glucose is sugar, and like candy, it provides only empty kcalories—no vitamins or minerals. Most sports drinks provide between 50 and 100 kcalories per cup.

- *Sodium and other electrolytes.* Sports drinks offer sodium and other electrolytes to help replace those lost during physical activity. Sodium in sports drinks also helps to increase the rate of fluid absorption from the GI tract and maintain plasma volume during activity and recovery.

Most physically active people do not need to replace the minerals lost in sweat immediately; a meal eaten within hours of competition replaces these minerals soon enough. Most sports drinks are relatively low in sodium, however, so those who choose to use these beverages run little risk of excessive intake.

- *Good taste.* Manufacturers reason that if a drink tastes good, people will drink more, thereby ensuring adequate hydration. For athletes who prefer the flavors of sports drinks over water, it may be worth paying for good taste to replace lost fluids.

- *Psychological edge.* Sports drinks provide a psychological edge for some people who associate the drinks with athletes and sports. The need to belong is valid. If the drinks boost morale and are used with care, they may do no harm.

For athletes who exercise for 45 minutes or more, sports drinks provide an advantage over water. For most physically active people, though, water is the best fluid to replenish lost fluids. The most important thing to do is drink—even if you don't feel thirsty.

[a] D. J. Casa, P.M. Clarkson, and W.O. Roberts, American College of Sports Medicine Roundtable on Hydration and Physical Activity: Consensus Statements, *Current Sports Medicine Reports* 4 (2005): 115-127; E. Coleman, Fluid replacement for athletes, *Sports Medicine Digest* 25 (2003): 76–77; Inter-association Task Force on External Heat Illness Consensus Statements, *NATA News*, June 2003.
[b] Position of the American Dietetic Association, Dietitians of Canada, and the American College of Sports Medicine: Nutrition and athletic performance, *Journal of the American Dietetic Association* 100 (2000): 1543–1556.

Hyponatremia may also occur when endurance athletes drink such large amounts of water over the course of a long event that they overhydrate, diluting the body fluids to such an extent that the sodium concentration becomes extremely low.[57] During long competitions, when athletes lose sodium through heavy sweating *and* consume excessive amounts of liquids, especially water, hyponatremia becomes likely.

Some athletes may be vulnerable to hyponatremia even when they drink sports drinks during an event.[58] Sports drinks do contain sodium, but as the accompanying "How to" points out, their sodium content is low. In some cases, it is too low to replace sweat losses. Still, sports drinks do offer more sodium than plain water.

To prevent hyponatremia, endurance athletes need to replace sodium during prolonged events. They should favor sports drinks over water and eat pretzels in the last half of a long race.[59] Some may need beverages with higher sodium concentrations than commercial sports drinks. In the days before the event, especially an event in the heat, athletes should not restrict salt in their diets.

Other Beverages

Carbonated beverages are not a good choice for meeting an athlete's fluid needs. Although they are composed largely of water, the air bubbles from the carbonation make a person feel full quickly and so may limit fluid intake. Caffeine's effects on athletic performance are discussed in this chapter's Nutrition in Practice, and the amounts of caffeine in foods and beverages are listed at the beginning of Appendix A.

Athletes sometimes drink beverages that contain alcohol, but these beverages are inappropriate as fluid replacements. Alcohol is a diuretic. It promotes the excretion of water; of vitamins such as thiamin, riboflavin, and folate; and of minerals such as calcium, magnesium, and potassium—exactly the wrong effects for fluid balance and nutrition. Alcohol has many detrimental effects on physical activity. It impairs temperature regulation, making hypothermia or heat stroke much more likely. It alters perceptions and slows reaction time. It depletes strength and endurance and deprives people of their judgment, thereby compromising their safety in sports. Many sports-related fatalities and injuries each year involve alcohol. ■

IN SUMMARY

- Hyperthermia and heat stroke can be a threat to physically active people in hot, humid weather.

- Hypothermia threatens those who exercise in the cold.

- Physically active people lose fluids and must replace them to avoid dehydration.

- Water is the best drink for most physically active people, but endurance athletes need beverages that supply glucose as well as fluids.

- During events lasting longer than three hours, athletes need to pay special attention to replacing sodium losses to prevent hyponatremia.

- Caffeine-containing drinks within limits may not impair performance, but water and fruit juice are preferred.

- Alcohol use can impair performance in many ways and is not recommended.

Diets for Physically Active People

No one diet best supports physical performance. Active people who choose foods within the framework of the diet-planning principles presented in Chapter 1 can design many excellent diets.

Nutrient Density A physically active person needs a diet composed mostly of nutrient-dense foods, the kind that supply a maximum of vitamins and minerals for the energy they provide. When athletes eat mostly refined, processed foods that have suffered nutrient losses and contain too much added sugar and solid fat, their nutrition status suffers. Even if foods are fortified or enriched, manufacturers cannot replace the whole range of nutrients and nonnutrients lost in refining. For example, manufacturers mill out much of a food's original magnesium and chromium but do not replace them. This does not mean that athletes can never choose a white bread, bologna, and mayonnaise sandwich, but only that later they should eat a large salad or big portions of vegetables and whole grains and drink a glass of milk to compensate. The nutrient-dense foods will provide the magnesium and chromium; the bologna sandwich provides extra energy, mostly from fats. ■

■ Beer facts:
Beer is not carbohydrate-rich. Beer is kcalorie-rich, but only one-third of its kcalories are from carbohydrates. The other two-thirds are from alcohol.

Beer is mineral-poor. Beer contains a few minerals, but to replace those lost in sweat, athletes need good sources such as fruit juices.

Beer is vitamin-poor. Beer contains tiny traces of some B vitamins, but it cannot compete with rich food sources.

Beer causes fluid losses. Beer is a fluid, but alcohol is a diuretic and causes the body to lose more fluid in urine than is provided by the beer.

■ Small daily choices, when made consistently, enhance an athlete's nutritional health.

Carbohydrate Athletes must eat for energy, and their energy needs can be immense. Athletes need full glycogen stores, and they need to strive to prevent heart disease and cancer by limiting fat, especially saturated fat. To serve these special needs, a diet that is high in carbohydrate (60 to 70 percent of total kcalories), moderate in unsaturated fats (20 to 30 percent), and adequate in protein (10 to 20 percent) works best. Even if you do not compete in glycogen-depleting events, such a diet provides adequate fiber while supplying abundant nutrients and energy.

■ Compare and decide which best meets your needs:
1 sandwich of 2 slices bologna, 2 slices white bread, 2 tbs mayonnaise (525 kcal, 9% protein, 23% carbohydrate, 68% fat)

or

2 sandwiches of 2 slices lean ham, 4 slices whole-wheat bread, 2 tsp mayonnaise (503 kcal, 20% protein, 51% carbohydrate, 29% fat)

With these principles in mind, compare the two 500-calorie sandwich meals in the margin. ■ The trick to getting enough carbohydrate energy is easy, at least in theory: just reduce the amount of fat and meat in a meal, and let carbohydrate-rich foods fill in for them.

Adding carbohydrate-rich foods is a sound and reasonable option for increasing energy intake, up to a point. It becomes unreasonable when the person cannot eat enough food to meet energy needs. At that point, the person can add more food energy into the diet by adding refined sugars, oils, or liquid meals. Still, these energy-rich additions must be superimposed on nutrient-rich choices; energy alone is not enough.

Some athletes use commercial high-carbohydrate liquid supplements to obtain the carbohydrate and energy needed for heavy training and top performance. Most of these products contain **glucose polymers** and about 18 to 24 percent carbohydrate. These supplements do not *replace* regular food; they are meant to be used in *addition* to it. Unlike the sports beverages discussed in the "How to" on p. 292, these high-carbohydrate supplements are too concentrated in carbohydrate to be used for fluid replacement.

Protein In addition to carbohydrate and some fat (and the energy they provide) athletes need protein. Meats and milk products head the list of protein-rich foods, but suggesting that athletes and active people eat more than the recommended servings of meat would be shortsighted advice. As mentioned repeatedly, active people need diets rich in carbohydrate, and meats have none to offer. Legumes, whole grains, and vegetables provide some protein with abundant carbohydrate. Table 10–3 (p. 287) shows recommended protein intakes.

A Performance Diet Example A person weighing 70 kilograms who engages in vigorous physical activity on a daily basis could easily require 3,000 to 5,000 kcalories per day. To meet this need, the person can choose a variety of nutrient-dense foods. Figure 10–4 shows examples of meals that provide about 3,300 kcalories. These meals supply about 125 grams of protein, equivalent to the highest recommended intake for an athlete weighing 160 pounds. Obviously, the higher a person's energy intake, the more protein that person will receive, assuming the foods chosen are nutrient dense. This relationship breaks down only when people meet their energy needs with high-fat, high-sugar confections.

The meals in Figure 10–4 provide 63 percent of their kcalories from carbohydrate. Athletes who train exhaustively for endurance events may want to aim for somewhat higher carbohydrate levels—from 65 to 75 percent. Notice that breakfast, though low in saturated fat, is filling and hearty. Current thinking supports the idea that athletes benefit from such a morning start. If you train early in the morning, try splitting breakfast into two parts. An hour or so before training, eat some toast, juice, and fruit. Later, after your workout, come back for the cereal and milk.

A strategy used by professional sports nutritionists to maximize athletes' intakes of energy and carbohydrates is to make sure that vegetable and fruit choices are as dense as possible in both nutrients and energy. One cup of iceberg lettuce supplies few kcalories or nutrients, but a half-cup portion of cooked sweet potatoes is a powerhouse of vitamins, minerals, and carbohydrate energy. Similarly, it takes one cup of cubed melon to equal the kcalories and carbohydrate in a half-cup of canned fruit. Small

Breakfast
1 c shredded wheat with
  low-fat milk and banana
2 slices whole-wheat toast
  with jelly
1½ c orange juice

Lunch
2 turkey sandwiches
1½ c low-fat milk
Large bunch of grapes

Snack
3 c plain popcorn
A smoothie made from:
  1½ c apple juice
  1½ frozen bananas

Dinner
Salad: 1 c spinach, carrots,
  and mushrooms with
  ½ c garbanzo beans,
  1 tbs sunflower seeds, and
  1 tbs ranch salad dressing
1 c spaghetti with meat sauce
1 c green beans
1 corn on the cob
2 slices Italian bread
4 tsp butter
1 piece angel food cake with
  fresh strawberries and
  whipping cream
1 c low-fat milk

> **Total kcal: about 3,300**
> 63% kcal from carbohydrate
> 22% kcal from fat
> 15% kcal from protein
> All vitamin and mineral intakes exceed
> the RDA for both men and women.

FIGURE 10–4

High-Carbohydrate Meals for Athletes
This sample menu provides about 3,300 kcalories, with almost 520 grams of carbohydrate (63 percent of total kcalories) and about 125 grams of protein (15 percent of total kcalories). In addition to meeting the carbohydrate and protein needs of an athlete who has a 3,300-kcalorie intake, these meals also meet or exceed recommendations for all vitamins and minerals.

choices like these, made consistently, can contribute significantly to nutrient, energy, and carbohydrate intakes.

Meals before Competition Before competition, athletes may eat particular foods or practice rituals that convey psychological advantages. One eats steak the night before; another spoons up honey at the start of the event. As long as these foods or rituals remain harmless, they should be respected. Still, scientists have made recommendations for the **pregame meal.** The foods should be carbohydrate-rich and the meal light (300 to 800 kcalories). It should be easy to digest and should contain fluids. Breads, potatoes, pasta, and fruit juices—carbohydrate-rich foods low in fat, protein, and fiber—form the basis of the pregame meal (see Figure 10–5 on p. 296 for some examples). Bulky, fiber-rich foods such as raw vegetables and high-fiber cereals, although usually desirable, are best avoided just before competition. Such foods can cause stomach discomfort during performance. The competitor should finish eating three to four hours before competition to allow time for the stomach to empty before exertion.

What about drinks or candylike sport bars claiming to provide "complete" nutrition? These mixtures of carbohydrate, protein (usually amino acids), fat, some fiber, and certain vitamins and minerals may taste good and provide additional food energy for a game or for weight gain. They fall short of providing "complete" nutrition, however, because they lack many of real food's nutrients and the nonnutrients that benefit health. These products may provide one advantage for active people—they are easy to eat in the hours before competition. They are also expensive.

Liquid meals are easy to digest, ■ and many such meals are commercially available. Alternatively, athletes can mix fat-free milk or juice, frozen fruits, and flavorings in a blender. Do not drop a raw egg in the blender because raw eggs often carry bacteria that cause food poisoning.

■ High-carbohydrate, liquid pregame meal ideas:
Apple juice, frozen banana, and cinnamon

Papaya juice, frozen strawberries, and mint

Fat-free milk, frozen banana, and vanilla

glucose polymers: compounds that supply glucose, not as single molecules, but linked in chains somewhat like starch. The objective is to attract less water from the body into the digestive tract.

pregame meal: a meal eaten three to four hours before athletic competition.

300-kcalorie meal
1 large apple
4 saltine crackers
1½ tbs reduced-fat
   peanut butter

500-kcalorie meal
1 large whole-wheat bagel
2 tbs jelly
1½ c low-fat milk

750-kcalorie meal
1 large baked potato
2 tsp soft margarine
1 c steamed broccoli
1 c mixed carrots and green
   peas
5 vanilla wafers
1½ c apple or pineapple juice

FIGURE 10–5

Examples of High-Carbohydrate Pregame Meals
Pregame meals should be eaten three to four hours before the event and provide 300 to 800 kcalories, primarily from carbohydrate-rich foods. Each of these sample meals provides at least 65 percent of total kcalories from carbohydrate.

If you want to excel physically, apply the most accurate nutrition knowledge along with dedication to rigorous training. A diet that provides ample fluid and consists of a variety of nutrient-dense foods in quantities to meet energy needs will enhance not only athletic performance, but also overall health as well. Training and genetics being equal, who would win a competition—the person who habitually consumes less than the amounts of nutrients needed or the one who arrives at the event with a long history of full nutrient stores and well-met metabolic needs?

Analysis

YOUR DIET

Fitness and Nutrition

Fitness depends on a certain minimum amount of physical activity. Ideally, the quantity and quality of the physical activity you select will improve your cardiorespiratory endurance, body composition, strength, and flexibility. Examine your activity choices by keeping an activity diary for one week. For each physical activity, be sure to record the type of activity, the level of intensity, and the duration. In addition, record the times and places of beverage consumption and the types and amounts of beverages consumed. Now compare the choices you made in your one-week activity diary to the guidelines for physical fitness (see Table 10–1).

- How often were you engaged in aerobic activity to improve cardiorespiratory endurance? Was the intensity of aerobic activity between 55 and 90 percent of your maximum heart rate? Did each session last at least 20 minutes?

- How often did you participate in resistance activities to develop strength? Was the intensity enough to enhance muscle strength

and improve body composition? Did you perform 8 to 10 different exercises, repeating each one 8 to 12 times?

- How often did you stretch to improve your flexibility? Was the intensity enough to develop and maintain a full range of motion? Did you hold each stretch 15 to 30 seconds and repeat each stretch two to four times?

- Do you drink plenty of fluids daily, especially water, before, during, and after physical activity?

- What changes could you make to improve your fitness?

CLINICAL APPLICATIONS

1. During her freshman year in college, Kim spent much of her free time bonding with new friends over pizzas, burgers, and fried chicken. Now in her sophomore year, she has decided that it is time to shed the "freshman 15" she gained and get fit quickly. She is replacing her pizzas and burgers with protein shakes and diet sodas; she avoids carbohydrates; and she takes vitamin supplements because she's heard they'll give her energy.

Kim's demanding sophomore academic load leaves little time during the week to do anything but go to class, study, and work on assignments. She has therefore decided to spend much of her weekend time running, working out with weights, swimming, and playing pick-up soccer on Sunday afternoons. After a few weeks of her new eating and fitness plan, however, Kim is feeling tired and run down much of the time, her muscles ache, and she falls asleep when she's trying to study.

■ What dietary advice would you suggest to help Kim feel healthier and more energetic?

■ What fitness strategies would you offer Kim to prevent her fatigue and sore muscles?

2. Zak, a junior in college, has been weight training four days a week for the past year to gain strength and bulk up his muscles. Along with his training, he has changed his diet from a normal mixed diet to one that emphasizes large servings of meat, eggs, and milk. Zak also takes amino acid supplements in hopes of building more muscle, faster. What advice would you offer Zak about his dietary habits? What would you tell him about the supplements he is taking?

Self Check

1. Regular physical activity helps protect against:
   a. backaches, cancer, and emphysema.
   b. cancer, diabetes, and heart disease.
   c. obesity, kidney disease, and anemia.
   d. high blood pressure, cancer, and osteopenia.

2. Fitness benefits health by:
   a. increasing lean body tissue and enhancing resistance to colds and other infectious diseases.
   b. lowering the risk of heart disease, decreasing muscle mass, and improving nutritional health.
   c. building bone strength, lowering the risk of some cancers, and increasing anxiety.
   d. reducing diabetes risk, compromising lung function, and promoting a strong self-image.

3. Which of the following characteristics is not a component of fitness?
   a. muscle endurance
   b. conditioning
   c. flexibility
   d. muscle strength

4. Which of the following provides most of the energy the muscles use in the early minutes of activity?
   a. fat
   b. protein
   c. glycogen
   d. vitamins

5. "Hitting the wall" is a term runners sometimes use to describe:
   a. dehydration.
   b. competition.
   c. indigestion.
   d. glucose depletion

6. Conditioned muscles rely less on _____ and more on _____ for energy.
   a. protein; fat
   b. fat; protein

   c. glycogen; fat
   d. fat; glycogen

7. Vitamin or mineral supplements taken right before an event are useless for improving performance because the:
   a. athlete sweats the nutrients out during the event.
   b. stomach can't digest supplements during physical activity.
   c. nutrients are diluted by all the fluids the athlete drinks.
   d. body needs hours or days for the nutrients to do their work.

8. Physically active young women, especially those who are endurance athletes, are prone to:
   a. energy excess.
   b. iron deficiency.
   c. protein overload.
   d. iodine deficiency.

9. Plain, cool water is the best fluid for everyday active people because it:
   a. rapidly leaves the digestive tract to enter the tissues and cool the body.
   b. tastes good.
   c. provides carbohydrate.
   d. leaves the digestive tract slowly.

10. A recommended pregame meal includes plenty of fluids and provides between:
    a. 300 and 800 kcalories, mostly from fat-rich foods.
    b. 50 and 100 kcalories, mostly from fiber-rich foods.
    c. 1000 and 2000 kcalories, mostly from protein-rich foods.
    d. 300 and 800 kcalories, mostly from carbohydrate-rich foods.

Answers to these questions appear in Appendix H.

Notes

1. T. M. Manini and coauthors, Daily activity energy expenditure and mortality among older adults, *Journal of the American Medical Association* 296 (2006): 171–179; I. Janssen and C. J. Jo-liffe, Influence of physical activity on mortality in elderly with coronary artery disease, *Medicine and Science in Sports and Exercise*, 38 (2006): 418–423.

2. Centers for Disease Control and Prevention, Trends in leisure-time physical inactivity by age, sex, and race/ethnicity—United States, 1994–2004, *Morbidity and Mortality Weekly Report* 54 (2005): 991–994.

3. D. E. Warburton, C. W. Nicol, and S. S. Bredin, Health benefits of physical activity: The evidence, *Canadian Medical Association Journal* 174 (2006): 801–809; B. Tehard and coauthors, Effect of physical activity on women at increased risk of breast cancer: Results from the E3N cohort study, *Cancer Epidemiology, Bio-markers, and Prevention* 15 (2006): 57–64; L Bernstein and coau-thors, Lifetime recreational exercise activity and breast cancer risk among black women and white women, *Journal of the Na-tional Cancer Institute* 97 (2005): 1671–1679; M. L. Slattery, Physi-cal activity and colorectal cancer, *Sports Medicine* 34 (2004): 239–252; F. B. Hu and coauthors, Adiposity as compared with physical activity in predicting mortality among women, *New England Journal of Medicine* 351 (2004): 2694–2703; K. R. Even-son and coauthors, The effect of cardiorespiratory fitness and obesity on cancer mortality in women and men, *Medicine and Science in Sports and Exercise* 35 (2003): 270–277; J. Dorn and coauthors, Lifetime physical activity and breast cancer risk in pre- and postmenopausal women, *Medicine and Science in Sports and Exercise* 35 (2003): 278–285; C. D. Lee and S. N. Blair, Cardiorespiratory fitness and stroke mortality in men, *Medicine and Science in Sports and Exercise* 34 (2002): 592–595.

4. J. M. Jakicic and A. D. Otto, Physical activity considerations for the treatment and prevention of obesity, *American Journal of Clinical Nutrition* 82 (2005): 226S–229S; C. A. Slentz and coauthors, Effects of the amount of exercise on body weight, body composition, and measures of central obesity: STRRIDE—a randomized controlled study, *Archives of Internal Medicine* 164 (2004): 31–39.

5. American College of Sports Medicine, Position stand: Physical activity and bone health, *Medicine and Science in Sports and Exercise* 36 (2004): 1985–1996; T. Lloyd and coauthors, Lifestyle factors and the development of bone mass and bone strength in young women, *Journal of Pediatrics* 144 (2004): 776–782.

6. R. Jankord and B. Jemiolo, Influence of physical activity on serum IL-6 and IL-10 levels in healthy older men, *Medicine and Science in Sports and Exercise* 36 (2004): 960–964; D. C. Nieman, Current perspectives on exercise immunology, *Current Sports Medicine Reports* 2 (2003): 239–242; C. E. Matthews and coauthors, Mod-erate to vigorous physical activity and risk of upper-respiratory tract infection, *Medicine and Science in Sports and Exercise* 34 (2002): 1242–1248.

7. B. Tehard and coauthors, 2006; L. Bernstein and coauthors, 2005; M. L. Slattery, 2004; K. R. Evenson and coauthors, 2003.

8. M. B. Conroy and coauthors, Past physical activity, current physi-cal activity, and risk of coronary heart disease, *Medicine and Science in Sports and Exercise* 37 (2005): 1251–1256; C. Richardson and coauthors, Physical activity and mortality across cardiovas-cular disease risk groups, *Medicine and Science in Sports and Exercise* 36 (2004): 1923–1929; American College of Sports Medi-cine, Position stand, Exercise and hypertension, *Medicine and Science in Sports and Exercise* 36 (2004): 533–553; M. R. Car-nethon and coauthors, Cardiorespiratory fitness in young adult-hood and the development of cardiovascular disease risk factors, *Journal of the American Medical Association* 290 (2003): 3092–3100; J. E. Manson and coauthors, Walking com-pared with vigorous exercise for the prevention of cardiovascu-lar events in women, *New England Journal of Medicine* 347 (2002): 716–725; Lee and Blair, 2002.

9. S. L. Wong and coauthors, Cardiorespiratory fitness is associated with lower abdominal fat independent of body mass index, *Medicine and Science in Sports and Exercise* 36 (2004): 286–291; P. T. Katzmarzyk and coauthors, Targeting the Metabolic Syn-drome with exercise: Evidence from the HERITAGE Family Study, *Medicine and Science in Sports and Exercise* 35 (2003): 1703–1709.

10. A. L. Macdonald and coauthors, Monitoring exercise-induced changes in glycemic control in type 2 diabetes, *Medicine and Science in Sports and Exercise* 38 (2006): 201–207; American Dia-betes Association, Position statement; Physical activity/exercise and diabetes mellitus, *Diabetes Care* 26 (2003): S73–S77; R. M. van Dam and coauthors, Physical activity and glucose tol-erance in elderly men: The Zutphen Elderly Study, *Medicine and Science in Sports and Exercise* 34 (2002): 1132–1136; K. J. Stewart, Exercise training and the cardiovascular consequences of type 2 diabetes and hypertension: Plausible mechanisms for improving cardiovascular health, *Journal of the American Medical Associa-tion* 288 (2002): 1622–1631.

11. K. L. Storti and coauthors, Physical activity and decreased risk of clinical gallstone disease among post-menopausal women, *Pre-ventive Medicine* 41 (2005): 772–777.

12. D. I. Galper and coauthors, Inverse association between physi-cal inactivity and mental health in men and women, *Medicine and Science in Sports and Exercise* 38 (2006): 173–178; J. B. Bartholomew, D. Morrison, and J. T. Ciccolo, Effects of acute exercise on mood and well-being in patients with major depressive disorder, *Medicine and Science in Sports and Exercise* 37 (2005): 2032–2037; W. J. Strawbridge and coauthors, Physical activity reduces the risk of subsequent depression for older adults, *American Journal of Epidemiol-ogy* 156 (2002): 328–334.

13. T. M. Manini and coauthors, 2006; G. Huang and coauthors, Resting heart rate changes after endurance training in older adults: A meta-analysis, *Medicine and Science in Sports and Exercise* 37 (2005): 1381–1386; American College of Sports Medicine, Expert panel, Physical activity programs and behavior counseling in older populations, *Medicine and Science in Sports and Exercise* 36 (2004): 1997–2003; J. Myers and coauthors, Exercise capacity and mortality among men referred for exercise testing, *New England Journal of Medicine* 346 (2002): 793–801.

14. W. W. N. Tsang and C. W. Y. Hui-Chan, Comparison of muscle torque, balance, and confidence in older Tai Chi and healthy adults, *Medicine and Science in Sports and Exercise* 37 (2005): 280–289; F. Li, and coauthors, Tai Chi: Improving functional balance and predicting subsequent falls in older persons, *Medicine and Science in Sports and Exercise* 36 (2004): 2046–2052.

15. T. M. Manini and coauthors, 2006; J. Myers and coauthors, 2002.

16. U.S. Department of Agriculture and U.S. Department of Health and Human Services, *Nutrition and Your Health: Dietary Guidelines for Americans 2005,* 6th ed., Home and Garden Bulletin no. 232 (Washington D C: 2005), available at www.healthierus.gov or call (888) 878-3256.

17. U.S. Department of Agriculture and U. S. Department of Health and Human Services, 2005; Standing Committee on the Scientific Evaluation of Dietary Reference Intakes, Food and Nutrition Board, Institute of Medicine, *Dietary Reference Intakes for Energy, Carbohydrate, Fiber, Fat, Fatty Acids, Cholesterol, Protein, and Amino Acids* (Washington, DC: National Academies Press, 2005), pp. 880–935.

18. T. S. Altena and coauthors, Lipoprotein subfraction changes after continuous or intermittent exercise training , *Medicine and Science in Sports and Exercise* 38 (2006): 367–372; M. Murphy and coauthors, Accumulating brisk walking for fitness, cardiovascular risk, and psychological health, *Medicine and Science in Sports and Exercise* 34 (2002): 1468–1474.

19. D. E. Warburton, C. W. Nicol, and S. S. Bredin, 2006; C. Rosenbloom and M. Bahns, What can we learn about diet and physical activity from master athletes? *Nutrition Today* 40 (2005): 267–272; Manson and coauthors, 2002; Lee and Blair, 2002.

20. American College of Sports Medicine, Position stand: Progression models in resistance training for healthy adults, *Medicine and Science in Sports and Exercise,* 34 (2002): 364–380.

21. J. A. Katula and coauthors, Strength training in older adults: An empowering intervention, *Medicine and Science in Sports and Exercise* 38 (2006): 106–111; American College of Sports Medicine, Expert panel, Physical activity programs and behavior counseling in older populations, *Medicine and Science in Sports and Exercise* 36 (2004): 1997–2003; R. Seguin and M. E. Nelson, The benefits of strength training for older adults, *American Journal of Preventive Medicine* 25 (2003): 141–149.

22. American College of Sports Medicine, Position Stand: Physical activity and bone health, *Medicine and Science in Sports and Exercise* 36 (2004): 1985–1996.

23. E. C. Cussler and coauthors, Weight lifted in strength training predicts bone change in postmenopausal women, *Medicine and Science in Sports and Exercise* 35 (2003): 10–17.

24. W. J. Kraemer and coauthors, Physiological changes with periodized resistance training in women tennis players, *Medicine and Science in Sports and Exercise* 35 (2003): 157–168.

25. P. G. Snell and coauthors, Maximal oxygen uptake as a parametric measure of cardiorespiratory capacity, *Medicine and Science in Sports and Exercise* 39 (2007): 103–107.

26. S. Sharma, Athlete's heart—Effect of age, sex, ethnicity and sporting discipline, *Experimental Physiology* 88 (2003): 665–669.

27. American College of Sports Medicine, Position stand: Exercise and Hypertension, *Medicine and Science in Sports and Exercise* 36 (2004): 533–553; R. G. Ketelhut, I. W. Franz, and J. Scholze, Regular exercise as an effective approach in antihypertensive therapy, *Medicine and Science in Sports and Exercise* 36 (2004): 4–8.

28. S. P. Cairns, Lactic acid and exercise performance: Culprit or friend? *Sports Medicine* 36 (2006): 279–291; T. H. Pedersen and coauthors, Intracellular acidosis enhances the excitability of working muscle, *Science* 305 (2004): 1144–1147; D. Allen and H. Westerblad, Enhanced: Lactic acid—the latest performance-enhancing drug, *Science* 305 (2004): 1112–1113; R. A. Roberts, F. Ghiasvand, and D. Parker, Biochemistry of exercise-induced metabolic acidosis, *American Journal of Physiology-Regulatory, Integrative and Comparative Psychology* 287 (2004): R502–R516.

29. G. A. Wallis and coauthors, Dose-response effects of ingested carbohydrate on exercise metabolism in women, *Medicine and Science in Sports and Exercise* 39 (2007): 131–138; A. C. Utter and coauthors, Carbohydrate supplementation and perceived exertion during prolonged running, *Medicine and Science in Sports and Exercise* 36 (2004): 1036–1041.

30. J. J. Winnick and coauthors, Carbohydrate feedings during team sport exercise preserve physical and CNS function, *Medicine and Science in Sports and Exercise* 37 (2005): 306–315; R. S. Welsh and coauthors, Carbohydrates and physical/mental performance during intermittent exercise to fatigue, *Medicine and Science in Sports and Exercise* 34 (2002): 723–731.

31. E. J. Coleman, Carbohydrate and exercise, in *Sports Nutrition: a Practice Manual for Professionals,* 4th ed., ed., M. Dunford (Chicago, Il.: American Dietetic Association, 2006), pp. 14–32; P. M. Siu and S. H. S. Wong, Use of the glycemic index: Effects on feeding patterns and exercise performance, *Journal of Physiological Anthropology and Applied Human Science* 23 (2004): 1–6.

32. S. L. Wee and coauthors, Ingestion of a high-glycemic index meal increases muscle glycogen storage at rest but augments its utilization during subsequent exercise, *Journal of Applied Physiology* 99 (2005): 707–714; Siu and Wong, 2004.

33. J. Manetta and coauthors, Fuel oxidation during exercise in middle-aged men: Role of training and glucose disposal, *Medicine and Science in Sports and Exercise* 34 (2002): 423–429.

34. L. M. Burke and J. A. Hawley, Effects of short-term fat adaptation on metabolism and performance of prolonged exercise, *Medicine and Science in Sports and Exercise* 34 (2002): 1492–1498.

35. L. Havemann and coauthors, Fat adaptation followed by carbohydrate loading compromises high-intensity sprint performance, *Journal of Applied Physiology* 100 (2006): 194–202; Burke and Hawley, 2002; L. M. Burke and coauthors, Adaptations to short-term high-fat diet persist during exercise despite high carbohydrate availability, *Medicine and Science in Sports and Exercise* 34 (2002): 83–91.

36. J. W. Helge, Long-term fat diet adaptation, effects on performance, training capacity, and fat utilization, *Medicine and Science in Sports and Exercise* 34 (2002): 1499–1504; N. D. Stepto and coauthors, Effect of short-term fat adaptation on high-intensity training, *Medicine and Science in Sports and Exercise* 34 (2002): 449–455.

37. Position of the American Dietetic Association, Dietitians of Canada, and the American College of Sports Medicine: Nutrition and athletic performance, *Journal of the American Dietetic Association* 100 (2000): 1543–1556.

38. N. R. Rodriquez, L. M. Vislocky, and P. C. Gaine, Dietary protein, endurance exercise, and human skeletal-muscle protein turnover, *Current Opinion in Clinical Nutrition and Metabolic Care* 10 (2007): 40–45; R. R. Wolfe, Skeletal muscle protein metabolism and resistance exercise, *Journal of Nutrition* 136 (2006): 525S–528S; J. S. Volek, Influence of nutrition on response to resistance training, *Medicine and Science in Sports and Exercise* 36 (2004): 689–696; M. Suzuki, Glycemic carbohydrates consumed with amino acids or protein right after exercise enhance muscle formation, *Nutrition Reviews* 61 (2003): S88–S94; D. K. Levenhagen and coauthors, Postexercise protein intake enhances body and leg accretion in humans, *Medicine and Science in Sports and Exercise* 34 (2002): 828–837.

39. Standing Committee on the Scientific Evaluation of Dietary Reference Intakes, 2005, pp. 660–661.

40. Position of the American Dietetic Association, Dietitians of Canada, and the American College of Sports Medicine, 2000.

41. T. L. Schwenk and C. D. Costley, When food becomes a drug: Nonanabolic nutritional supplement use in athletes, *American Journal of Sports Medicine* 30 (2002): 907–916.

42. Position of the American Dietetic Association, Dietitians of Canada, and the American College of Sports Medicine, 2000.

43. J. Finaud, G. Lac, and E. Filaire, Oxidative stress: Relationship with exercise and training, *Sports Medicine* 36 (2006): 327–358; T. A. Watson and coauthors, Antioxidant restriction and oxidative stress in short-duration exhaustive exercise, *Medicine and Science in Sports and Exercise* 37 (2005): 63–71.

44. S. L. Williams and coauthors, Antioxidant requirements of endurance athletes: Implications for health, *Nutrition Reviews* 64 (2006): 93–108.

45. M. L. Urso and P. M. Clarkson, Oxidative stress, exercise, and antioxidant supplementation, *Toxicology* 189 (2003): 41–54.

46. S. L. Akabas and K. R. Dolins, Micronutrient requirements of physically active women: What can we learn from iron? *American Journal of Clinical Nutrition* 81 (2005): 1246S–1251S.

47. S. I. Barr and C. A. Rideout, Nutritional considerations for vegetarian athletes, *Nutrition* 20 (2004): 696–703; Position of the American Dietetic Association, Dietitians of Canada, and the American College of Sports Medicine, 2000.

48. Position of the American Dietetic Association and Dietitians of Canada: Vegetarian diets, *Journal of the American Dietetic Association* 103 (2003): 748–765.

49. P. S. Hinton and L. M. Sinclair, Iron supplementation maintains ventilatory threshold and improves energetic efficiency in iron-deficient nonanemic athletes, *European Journal of Clinical Nutrition* 61 (2007): 30–39; T. Brownlie and coauthors, Marginal iron deficiency without anemia impairs aerobic adaptation among previously untrained women, *American Journal of Clinical Nutrition* 75 (2002): 734–742.

50. American College of Sports Medicine, Position stand: Exercise and fluid replacement, *Medicine & Science in Sports & Exercise* 39 (2007): 377–390; D. J. Casa, P. M. Clarkson, and W. O. Roberts, American College of Sports Medicine Roundtable on Hydration and Physical Activity: Consensus Statements, *Current Sports Medicine Reports* 4 (2005): 115–127; Standing Committee on the Scientific Evaluation of Dietary Reference Intakes, Food and Nutrition Board, Institute of Medicine, *Dietary Reference Intakes for Water, Potassium, Sodium, Chloride, and Sulfate* (Washington, DC: National Academies Press, 2005), pp. 108–110.

51. C. K. Seto, D. Way, and N. O'Connor, Environmental illness in athletes, *Clinics in Sports Medicine* 24 (2005): 695–718.

52. American College of Sports Medicine, Position stand: Exertional heat illness during training and competition, *Medicine & Science in Sports & Exercise* 39 (2007): 556-572.

53. American College of Sports Medicine, Position stand: Prevention of cold injuries during exercise, *Medicine and Science in Sports and Exercise* 38 (2006): 2012–2029.

54. Casa, Clarkson, and Roberts, 2005.

55. Seto, Way, and O'Connor, 2005; J. R. Stofan and coauthors, Sweat and sodium losses in NCAA Division 1 football players with a history of whole-body muscle cramping, paper presented at the annual meeting of the American College of Sports Medicine, 2003.

56. Standing Committee on the Scientific Evaluation of Dietary Reference Intakes, Food and Nutrition Board, Institute of Medicine, *Dietary Reference Intakes for Water, Potassium, Sodium, Chloride, and Sulfate* (Washington, DC: National Academies Press, 2005), pp. 127–132.

57. J. W. Gardner, Death by water intoxication, *Military Medicine* 168 (2003): 432–434.

58. M. Hsieh and coauthors, Hyponatremia in runners requiring on-site medical treatment at a single marathon, *Medicine and Science in Sports and Exercise* 34 (2002): 185–189.

59. E. R. Eichner, Exertional hyponatremia: Why so many women? *Sports Medicine Digest* 24 (2002): 54, 56.

SUPPLEMENTS AND ERGOGENIC AIDS ATHLETES USE

In a world where body conditioning and skill are hard won, athletes gravitate to promises that they can easily improve their performance by taking pills, powders, or potions. Athletes often hear well-intended, but unsubstantiated, advice from their coaches and peers, who advise them to use nutrient supplements, take drugs, or follow procedures that claim to deliver results with little effort.[1] The wish to win is strong, but no amount of wishing can change the fact that a large majority of supplements sold for athletes are frauds. If the products that are tried have no effect and are harmless, they are only a waste of money. But some products sold as performance-enhancing aids are harmful or actually impair performance, and these are a waste of athletic potential as well. This Nutrition in Practice looks at scientific evidence for and against some of the most common dietary supplements and hormonal preparations available to athletes, and the accompanying Glossary defines them. Chapter 4 included a discussion on protein powders and amino acid supplements (see p. 102).

What does ergogenic mean?

Ergogenic means work enhancing or work producing. In connection with athletic performance, ergogenic aids are substances or treatments that purportedly improve athletic performance above and beyond what is possible through training alone. Research findings do not, for the most part, support the claims made for **ergogenic aids**.[2] When you hear a claim that a product is ergogenic, remember to consider the source of the claim and ask who may gain from the sale. Nutrition in Practice 1 described ways to recognize misinformation and quackery.

I know that some athletes, especially endurance athletes, are taking carnitine supplements. What is carnitine?

Carnitine is a nonessential nutrient. Endurance athletes believe carnitine will help them burn more fat, thereby sparing glycogen during endurance events. Carnitine is also promoted to body-builders as a "fat burner."

In the body, carnitine facilitates the transfer of fatty acids across the mitochrondrial membrane. Supplement manufacturers suggest that with more carnitine available, fat oxidation will be enhanced, but this does not seem to be the case. Carnitine supplementation neither raises muscle carnitine concentrations nor enhances athletic performance.[3] It does, however, produce diarrhea in about half of the people who use it. Milk and meat products are good sources of carnitine, and supplements are not needed.

GLOSSARY

The Glossary on p. 303 includes additional supplements commonly used to enhance performance.

anabolic steroids: drugs related to the male sex hormone, testosterone, that stimulate the development of lean body mass.
  anabolic = promoting growth
  sterols = compounds chemically related to cholesterol

caffeine: a natural stimulant found in many common foods and beverages, including coffee, tea, and chocolate. High doses cause headaches, trembling, rapid heart rate, and other undesirable side effects.

chromium picolinate (CROW-mee-um pick-oh-LYN-ate): a trace mineral supplement; falsely promoted as building muscle, enhancing energy, and burning fat. Picolinate is a derivative of the amino acid tryptophan that seems to enhance chromium absorption.

creatine (KREE-ah-tin): a nitrogen-containing compound that combines with phosphate to form the high-energy compound creation phosphate (or phosphocreatine) in muscles. Claims that creatine enhances energy use and muscle strength need further confirmation.

DHEA (dehydroepiandrosterone) and **androstenedione:** hormones made in the adrenal glands that serve as precursors to the male hormone testosterone; falsely promoted as burning fat, building muscle, and slowing aging. Side effects include acne, aggressiveness, and liver enlargement.

ergogenic (ER-go-JEN-ick) **aids:** substances or techniques used in an attempt to enhance physical performance.
  ergo = work
  genic = gives rise to

carnitine: a nonessential nonprotein amino acid made in the body from lysine that helps transport fatty acids across the mitochondrial membrane. Carnitine supposedly "burns" fat and spares glycogen during endurance events, but in reality it does neither.

conjugated linoleic acid: a naturally occuring *trans* fatty acid with 18 carbons and 2 double bonds; sometimes taken as a supplement to improve body composition.

My friend who is a bodybuilder takes a supplement called chromium picolinate. Advertisements in magazines and health food stores make all kinds of impressive claims about it. Are any of the claims true?

Chapter 9 introduced chromium as an essential trace mineral involved in carbohydrate and lipid metabolism. Advertisements in bodybuilding magazines claim that **chromium picolinate**, which is more easily absorbed than chromium alone, builds muscle, enhances energy, and burns fat. Such claims derive from one or two initial studies on this mineral. Most subsequent studies, however, show no effects of chromium picolinate supplementation on strength, lean body mass, or body fat.[4]

A lot of my friends are taking creatine. Why is it so popular?

Interest in and use of creatine monohydrate supplements to enhance energy production during intense activity has grown dramatically in the last few years. Power athletes such as weight lifters use **creatine** supplements to enhance stores of the high-energy compound creatine phosphate (or phosphocreatine) in muscles. Theoretically, the more creatine phosphate in muscles, the higher the intensity at which an athlete can train. High-intensity training stimulates the muscles to adapt, which, in turn, improves performance.

The results of some studies suggest that creatine supplementation enhances performance of short-term, high-intensity strength activity such as weight lifting or repeated sprinting.[5] In contrast, creatine supplementation has not been shown to benefit endurance activity.[6] The question of whether short-term use of creatine supplements is safe continues to be studied, but so far, the supplements are viewed to be safe for healthy adults.[7]

Some medical and fitness experts voice concern that, like many performance enhancement supplements before it, creatine is being taken in huge doses (5 to 30 grams per day) before evidence of its value or safety has been ascertained. Even people who eat red meat, which is a creatine-rich food, do not consume near the amount some athletes are taking. Despite the uncertainties, creatine supplements are not illegal in international competition. The American Academy of Pediatrics strongly discourages the use of creatine supplements, as well as the use of any performance-enhancing substance in adolescents less than 18 years of age.[8]

What is conjugated linoleic acid, and why do some athletes take it?

Conjugated linoleic acid (CLA) derives from the essential fatty acid, linoleic acid. CLA is part of a group of naturally occurring polyunsaturated fatty acids found in beef, lamb, and dairy products. In animal studies, CLA has been shown to reduce body fat and increase lean body mass—findings that have sparked interest in CLA as a performance-enhancing aid.[9] Only a few studies of the effects of supplemental CLA on body composition in human beings have been conducted, and the results seem less promising.[10] When researchers studied the combined effects of supplemental CLA and resistance training on body composition in men and women, they found small increases in lean body mass and reductions in body fat, but no improvements in strength.[11] The researchers noted that although the effects were statistically significant, they were nevertheless small and should be weighed against the relatively high cost of supplemental CLA.

What about caffeine? I have heard that it can improve endurance performance.

Some research supports the use of **caffeine** to enhance endurance and, to some extent, to enhance short-term, high-intensity exercise performance.[12] Caffeine is a stimulant that elicits a number of physiological and psychological effects in the body. Caffeine enhances alertness and reduces fatigue.[13] The possible benefits of caffeine use must be weighed against its adverse effects—stomach upset, nervousness, irritability, headaches, and diarrhea. Caffeine-containing beverages should be used in moderation, if at all, and *in addition* to other fluids, not as a substitute for them. College, national, and international athletic competitions prohibit the use of caffeine in amounts greater than the equivalent of 5 or 6 cups of coffee consumed in a two-hour period prior to competition. Urine tests that detect more caffeine than this disqualify athletes from competition. (The table at the start of Appendix A provides a list of common caffeine-containing items and the doses they deliver.)

I have heard that anabolic steroids are dangerous, but one of my friends takes them. His mother is a doctor, and she constantly monitors his blood pressure when he is taking steroids. Are they safe in his case?

Anabolic steroids are not safe in your friend's case or in any case; they have dangerous side effects and are illegal. Technically called androgenic-anabolic steroid drugs, they are derivatives of the male sex hormone testosterone. Testosterone promotes the development of male characteristics (androgenic) and lean body mass (anabolic). Athletes take steroids to stimulate muscle bulking.

To athletes struggling to excel, the promise of bigger, stronger muscles than training alone can produce is tempting. Especially in professional circles such as Major League Baseball and the National Football League, where monetary rewards for excellence are sky-high, steroid use is common despite its illegality and side effects.

The American College of Sports Medicine and the American Academy of Pediatrics condemn athletes' use of anabolic steroids, and the International Olympic Committee bans their use. These authorities cite the known toxic side effects and maintain that steroid use is a form of cheating. Competitors who use the drugs put other athletes in the difficult position of either conceding an unfair advantage to abusing competitors or taking steroids and accepting the risk of harmful side effects (see Table NP10–1). Young athletes should never be forced to make such a choice.

The dangers of steroid use cannot be overemphasized. Health care professionals are obligated to warn athletes of these dangers. Speak simply and emphatically: the price for the potential competitive edge that steroids confer is high—sometimes life itself. Steroids are not simple pills that build bigger muscles. They are complex chemicals to which the body reacts in many ways, particularly when bodybuilders and other athletes take large amounts.[14] Among the side effects and

Anabolic Steroids: Side Effects and Adverse Reactions

Mind

- Extreme aggression with hostility ("steroid rage"); mood swings; anxiety; dizziness; drowsiness; unpredictability; insomnia; psychotic depression; personality changes, suicidal thoughts

Face and Hair

- Swollen appearance; greasy skin; severe, scarring acne; mouth and tongue soreness; yellowing of whites of eyes (jaundice)
- In females, male-pattern hair loss and increased growth of face and body hair

Voice

- In females, irreversible deepening of voice

Chest

- In males, breathing difficulty, breast development
- In females, breast atrophy

Heart

- Heart disease; elevated or reduced heart rate; heart attack; stroke; hypertension; increased LDL; reduced HDL

Abdominal Organs

- Nausea; vomiting; bloody diarrhea; pain; edema; liver tumors (possibly cancerous); liver damage, disease, or rupture leading to fatal liver, liver failure; kidney stones and damage; gallstones; frequent urination; possible rupture of aneurysm or hemorrhage

Blood

- Blood clots: high risk of blood poisoning; those who share needles risk contracting HIV (the AIDS virus) or other disease-causing organisms; septic shock (from injections)

Reproductive System

- In males, permanent shrinkage of testes; prostate enlargement with increased risk of cancer; sexual dysfunction; loss of fertility, excessive and painful erections
- In females, loss of menstruation and fertility; permanent enlargement of external genitalia; fetal damage, if pregnant

Muscles, Bones, and Connective Tissues

- Increased susceptibility to injury with delayed recovery times; cramps; tremors; seizure like movements; injury at injection site
- In adolescents, failure to grow to normal height

Other

- Fatigue; increased risk of cancer

---

adverse reactions that steroids produce are cancerous liver tumors that impair liver function, causing it to rupture and hemorrhage; testicular shrinkage in men and masculinization of women; cardiovascular problems; and sterility. Your friend is sure to develop side effects no matter how closely a trainer or doctor monitors him. The safest effective way to build muscle has always been through hard training, and always will be.

What are DHEA and androstenedione, and why do some athletes use them?

Some athletes use **DHEA** (dehydroepiandrosterone) and **androstenedione** as alternatives to anabolic steroids. Androstenedione, or "andro," made headlines in the late 1990s when the media reported its use by baseball great Mark McGwire.

DHEA and androstenedione are hormones made in the adrenal glands that serve as precursors to the male hormone testosterone. Advertisements claim the hormones "burn fat," "build muscle," and "slow aging," but evidence to support such claims is lacking.

Short-term side effects of DHEA and androstenedione include oily skin, acne, body hair growth, liver enlargement, and aggressive behavior. Long-term effects of DHEA and

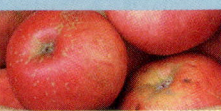

GLOSSARY OF SUBSTANCES PROMOTED AS ERGOGENIC AIDES

The glossary on p. 301 includes supplements mentioned in the text. Chapter 4 includes a discussion on protein and amino acid supplements.

arginine: a nonessential amino acid falsely promoted as enhancing the secretion of human growth hormone, the breakdown of fat, and the development of muscle.

boron: a nonessential mineral that is promoted as a "natural" steroid replacement.

coenzyme Q10: a lipid found in cells (mitochondria) shown to improve exercise performance in heart disease patients, but not effective in improving the performance of healthy athletes.

gamma-oryzanol: a plant sterol that supposedly provides the same physical responses as anabolic steroids without the adverse side effects; also known as ferulic acid, lerulate, or FRAC.

ginseng: a plant whose extract supposedly boosts energy. Side effects of chronic use include nervousness, confusion, and depression.

HMB (beta-hydroxy-beta-methylbutyrate): a metabolite of the branched-chain amine acid leucine. Claims that HMB increases muscle mass and strength are based on the results of two studies from the lab that developed HMB as a supplement.

pyruvate: a 3-carbon compound that plays a key role in energy metabolism. Supplements claim to burn fat and enhance performance.

ribose: a 5-carbon sugar falsely promoted as improving the regeneration of ATP and thereby the speed of recovery after high-power exercise.

royal jelly: the substance produced by worker bees and fed to the queen bee; falsely promoted as increasing strength and enhancing performance.

sodium bicarbonate: baking soda; an alkaline salt believed to neutralize blood lactic acid and thereby to reduce pain and enhance possible workload. "Soda loading" may cause intestinal bloating and diarrhea.

spirulina: a kind of alga ("blue green manna") that supposedly contains large amounts of protein and vitamin B12, suppresses appetite, and improves athletic performance. It does none of these things and is potentially toxic.

androstenedione use remain to be seen and may take years to become evident. The potential for harm from these supplements is great, and athletes, as well as others, should avoid them.

Recently, the Food and Drug Administration (FDA) sent letters to producers of dietary supplements warning that products containing androstenedione are considered to be adulterated and therefore are illegal to sell and that criminal penalties could result from continued sales. The National Collegiate Athletic Association, the National Football League, and the International Olympic Committee have banned the use of androstenedione and DHEA in competition. The American Academy of Pediatrics and many other medical professional groups have spoken out against the use of these and other "hormone replacement" substances.

Carnitine does not enhance athletic performance, results of studies on chromium picolinate are inconsistent, and experts are concerned about the long-term effects of large doses of creatine. Caffeine may or may not be effective, but can have adverse side effects, and is illegal. Steroids pose serious health risks and are illegal. Do any of the substances athletes use to boost performance work?

For the most part, no. Many of these substances have been studied and found to be worthless. The Glossary on p. 303 lists and describes many more substances promoted as ergogenic aids.

Health professionals can positively influence athletes and others interested in boosting athletic performance by stressing the measures that do help to enhance performance. They are regular training and sound nutrition.

Notes

1. K. A. Erdman, T. S. Fung, and R. A. Reimer, Influence of performance level on dietary supplementation in elite Canadian athletes, *Medicine and Science in Sports and Exercise* 38 (2006): 349–356.

2. R. Calfee and P. Fadale, Popular ergogenic drugs and supplements in young athletes, *Pediatrics* 117 (2006): e577-589; F. Brouns and coauthors, Functional foods and food supplements for athletes: From myths to benefit claims substantiation through the study of selected biomarkers, *British Journal of Nutrition* 88 (2002): S177–S186.

3. E. M. Broad, R. J. Maughan, S. D. Galloway, Effects of four weeks L-carnitine L-tartrate ingestion on substrate utilization during prolonged exercise, *International Journal of Sports Nutrition and Exercise Metabolism* 15 (2005): 665–679; E. P. Brass, Carnitine and sports medicine: Use or abuse? *Annals of the New York Academy of Sciences* 1033 (2004): 67–78.

4. J. B. Vincent, The potential value and toxicity of chromium picolinate as a nutritional supplement, weight loss agent and muscle development agent, *Sports Medicine* 33 (2003): 213–230.

5. P. J. Cribb and coauthors, Effects of whey isolate, creatine, and resistance training on muscle hypertrophy, *Medicine & Science in Sports & Exercise* 39 (2007): 298–307; M. C. Peyrebrune and coauthors, Effect of creatine supplementation on training for competition in elite swimmers, *Medicine and Science in Sports and Exercise* 37 (2005): 2140–2147; J. D. Branch, Effect of creatine supplementation on body composition and performance: A meta-analysis, *International Journal of Sports Nutrition and Exercise Metabolism* 13 (2003): 198–226; R. L. Dempsey, M. F. Mazzone, and L. N. Meurer, Does oral creatine supplementation improve strength? A meta-analysis, *Journal of Family Practice* 51 (2002): 945–951.

6. T. L. Schwenk and C. D. Costley, When food becomes a drug: Nonanabolic nutritional supplement use in athletes, *The American Journal of Sports Medicine* 30 (2003): 907–916; Branch, 2003.

7. A. Shao and J. N. Hathcock, Risk assessment for creatine monohydrate, *Regulatory Toxicology and Pharmacology* 45 (2006): 242–251; M. Dunford and M. Smith, Dietary supplements and ergogenic aids, in *Sports Nutrition: A Practice Manual for Professionals*, 4th ed., ed. M. Dunford (Chicago, IL: American Dietetic Association, 2006), pp. 116–141; E. Bizzarini and L. De Angelis, Is the use of oral cratine supplementation safe? *Journal of Sports Medicine and Physical Fitness* 44 (2004): 411–416.

8. American Academy of Pediatrics, Policy Statement, Committee on Sports Medicine and Fitness, Use of performance-enhancing substances, *Pediatrics* 115 (2005): 1103–1106.

9. A. M. Bhattacharya and coauthors, The combination of dietary conjugated linoleic acid and treadmill exercise lowers gain in body fat mass and enhances lean body mass in high fat-fed male Balb/C mice, *Journal of Nutrition* 135 (2005): 1124–1130; M. A. Belury, Dietary conjugated linoleic acid in health: Physiological effects and mechanisms of action, *Annual Review of Nutrition* 22 (2002): 505–531.

10. A. H. M. Terpstra, Effect of conjugated linoleic acid on body composition and plasma lipids in humans: An overview of the literature, *American Journal of Clinical Nutrition* 79 (2004): 352–361.

11. C. Pinkoski and coauthors, The effects of conjugated linoleic acid supplementation during resistance training, *Medicine and Science in Sports and Exercise* 38 (2006): 339–348.

12. K. T. Schneiker and coauthors, Effects of caffeine on prolonged intermittent-sprint ability in team-sport athletes, *Medicine and Science in Sports and Exercise* 38 (2006): 578–585; G. R. Stuart and coauthors, Multiple effects of caffeine on simulated high-intensity team-sport performance, *Medicine and Science in Sports and Exercise* 37 (2005): 1998–2005; S. A. Paluska, Caffeine and exercise, *Current Sports Medicine Reports* 2 (2003): 213–219.

13. Stuart and coauthors, 2005; Paluska, 2003.

14. A. B. Parkinson and N. A. Evans, Anabolic androgenic steroids: A survey of 500 users, *Medicine and Science in Sports and Exercise* 38 (2006): 644–651.

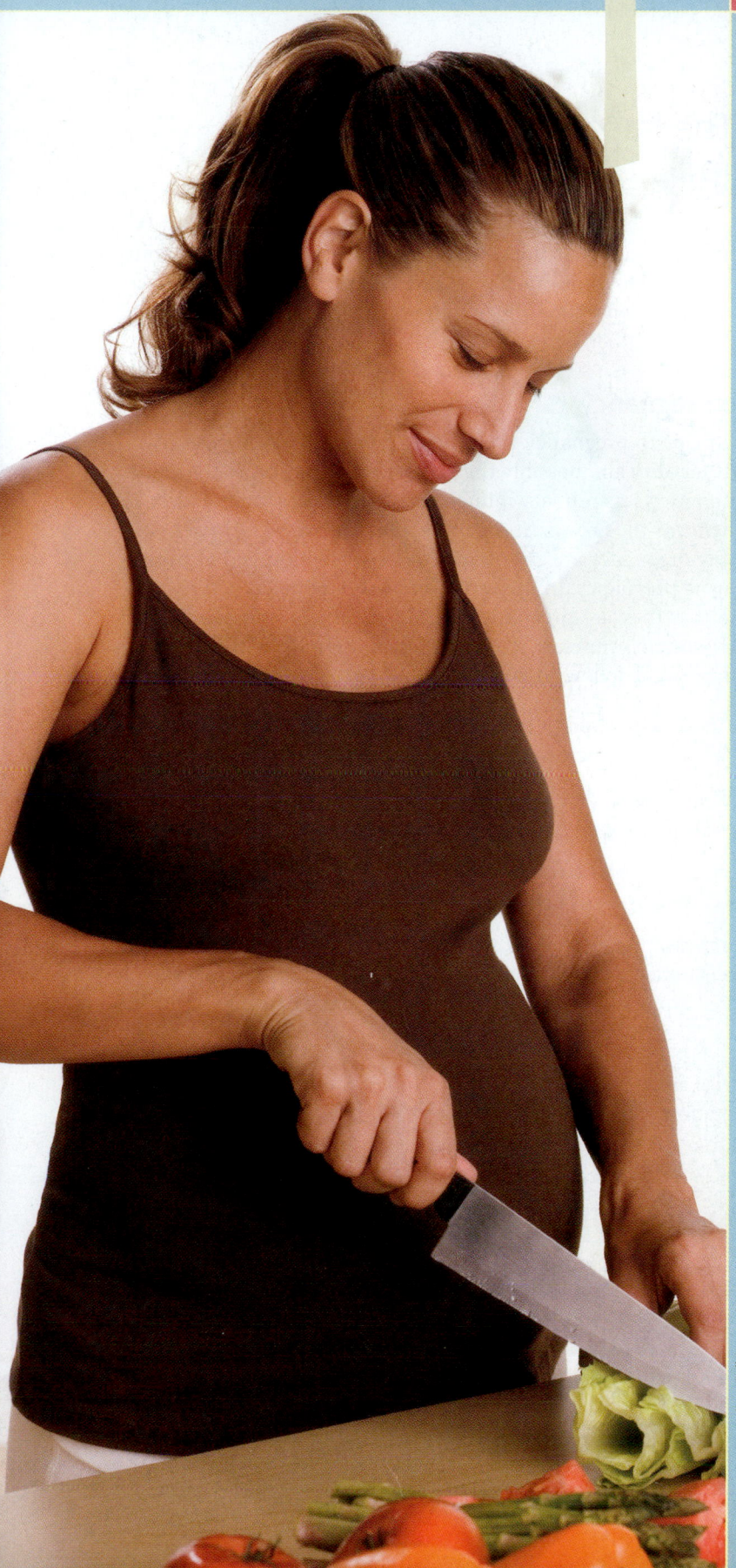

11

Nutrition through the Life Span: Pregnancy and Lactation

© Tom Grill/Iconica/Getty Images

305

All people need the same nutrients, but the amounts they need vary depending on their stage of life. This chapter focuses on nutrition in preparation for and support of pregnancy and lactation. Chapters 12 and 13 address the needs of infants, children, adolescents, and older adults.

Pregnancy: The Impact of Nutrition on the Future

The woman who enters pregnancy with full nutrient stores, sound eating habits, and a healthy body weight has done much to ensure an optimal pregnancy. Then, if she eats a variety of nutrient-dense foods during the pregnancy itself, her own and her infant's health will benefit further.

Nutrition Prior to Pregnancy

A section on nutrition prior to pregnancy must focus mainly on women. A man's nutrition may affect his **fertility** and possibly the genetic contributions he makes to his children, but nutrition exerts its primary influence through the woman. Her body provides the environment for the growth and development of a new human being. Full nutrient stores *before* pregnancy are essential both to conception and to healthy infant development during pregnancy. In the early weeks of pregnancy, before many women are even aware that they are pregnant, significant developmental changes occur that depend on a woman's nutrient stores. In preparation for a healthy pregnancy, a woman can establish the following habits:[1]

- **Achieve and maintain a healthy body weight.** Both underweight and overweight women and their newborns face increased risks of complications.

- **Choose an adequate and balanced diet.** Malnutrition reduces fertility and impairs the early development of an infant should a woman become pregnant.

- **Be physically active.** A woman who wants to be physically active *when* she is pregnant needs to become physically active *beforehand*.

- **Avoid harmful influences.** Both maternal and paternal ingestion of harmful substances (such as cigarettes, alcohol, drugs, or environmental contaminants) can alter genes or their expression, interfering with fertility and causing abnormalities.

Young adults who nourish and protect their bodies do so not only for their own sakes, but also for future generations.

© Masterfile

Both parents can prepare in advance for a healthy pregnancy.

fertility: the capacity of a woman to produce a normal ovum periodically and of a man to produce normal sperm; the ability to reproduce.

Prepregnancy Weight

Appropriate weight prior to pregnancy benefits pregnancy outcome. Being either underweight or overweight presents medical risks during pregnancy and childbirth. Underweight women are therefore advised to gain weight before becoming pregnant, and overweight women are advised to lose excess weight.

Underweight Infant birth weight correlates with prepregnancy weight and weight gain during pregnancy and is the most potent single predictor of the infant's future health and survival. An underweight woman has a high risk of having a **low birth weight (LBW)** infant, especially if she is unable to gain sufficient weight during pregnancy.[2] Compared with normal-weight infants, low-birth weight infants are more likely to contract diseases and nearly forty times more likely to die in the first month of life. Impaired growth and development during pregnancy may have long-term health effects as well. Research suggests that when nutrient supplies fail to meet demands, permanent adaptations take place that may make obesity or chronic diseases such as hypertension more likely in later life (see Nutrition in Practice 12).[3] Other hazards of low birth weight may include lower adult IQ and other brain impairments, and short stature.[4] Underweight women are therefore advised to gain weight before becoming pregnant and to strive to gain adequately during pregnancy.

Nutritional deficiency, coupled with low birth weight, is the underlying cause of more than half of all the deaths worldwide of children under five years of age. In the United States, the infant mortality rate in 2004 was 6.8 deaths per 1,000 live births.[5] This rate, though higher than that of some other developed countries, has seen a significant steady decline over the past two decades and stands as a tribute to public health efforts aimed at reducing infant deaths (see Figure 11–1).

Not all cases of low birth weight reflect poor nutrition. Heredity, disease conditions, smoking, and drug use (including alcohol) during pregnancy all contribute.[6] Even with optimal nutrition and health during pregnancy, some women give birth to small infants for unknown reasons. But poor nutrition is the major factor in low birth weight—and an avoidable one, as later sections make clear.[7]

Overweight and Obesity Obese women are also urged to strive for healthy weights before pregnancy. The infant of an obese mother may be larger than normal and may be large even if born prematurely. The large early infant may not be recognized as premature and thus may not receive the special medical care required. The infant of an obese mother may be twice as likely to be born with a neural tube defect as others.[8] Folate's role has been examined, but a more likely explanation seems to be poor glycemic control.[9] Obese women are more likely to require drugs to induce labor or require surgical intervention for the birth, and they suffer gestational diabetes, hypertension, and infections after the birth more often than do women of healthy weight.[10] ■ In addition, both overweight and obese women have a greater risk of giving birth to infants with heart defects and other abnormalities.[11] An appropriate goal for the obese woman who wishes to become pregnant is to strive to attain a healthy prepregnancy body weight so as to minimize her medical risks and those of her future child.

Healthy Support Tissues

A major reason that the mother's prepregnancy nutrition is so crucial is that it determines whether her **uterus** will be able to support the growth of a healthy **placenta** during the first month of **gestation**. The placenta is both a supply depot and a waste-removal system for the fetus. If the placenta works perfectly, the fetus wants for nothing. If it does not, no alternative source of sustenance is available, and the fetus will fail to thrive. Figure 11–2 (p. 308) shows the placenta, a mass of tissue in which maternal and

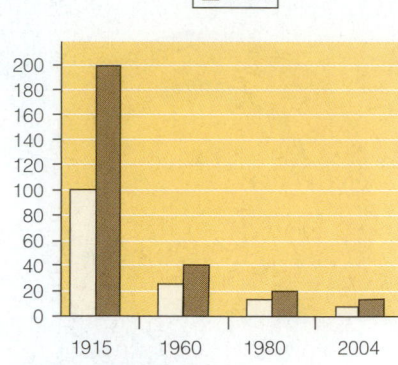

□ White
■ Black

FIGURE 11–1

Infant Mortality Decline over Time
The graph shows infant deaths per 1,000 live births.
Source: B. E. Hamilton and coauthors, Annual summary of vital statistics: 2005, *Pediatrics* 119 (2007): 345–360.

■ Neural tube defects and gestational diabetes are discussed in later sections.

low birth weight (LBW): a birth weight less than 5½ lb (2,500 g); indicates probable poor health in the newborn and poor nutrition status of the mother during pregnancy. Normal birth weight for a full-term infant is 6½ to 9 lb (about 3,000 to 4,000 g). Low-birth weight infants are of two different types. Some are **premature**; they are born early and are of a weight **appropriate for gestational age (AGA)**. Others have suffered growth failure in the uterus; they may or may not be born early, but they are **small for gestational age (SGA)**.

uterus (YOO-ter-us): the womb, the muscular organ within which the infant develops before birth.

placenta (pla-SEN-tuh): an organ that develops inside the uterus early in pregnancy, in which maternal and fetal blood circulate in close proximity and exchange materials. The fetus receives nutrients and oxygen across the placenta; the mother's blood picks up carbon dioxide and other waste materials to be excreted via her lungs and kidneys.

gestation: the period of about 40 weeks (three trimesters) from conception to birth; the term of a pregnancy.

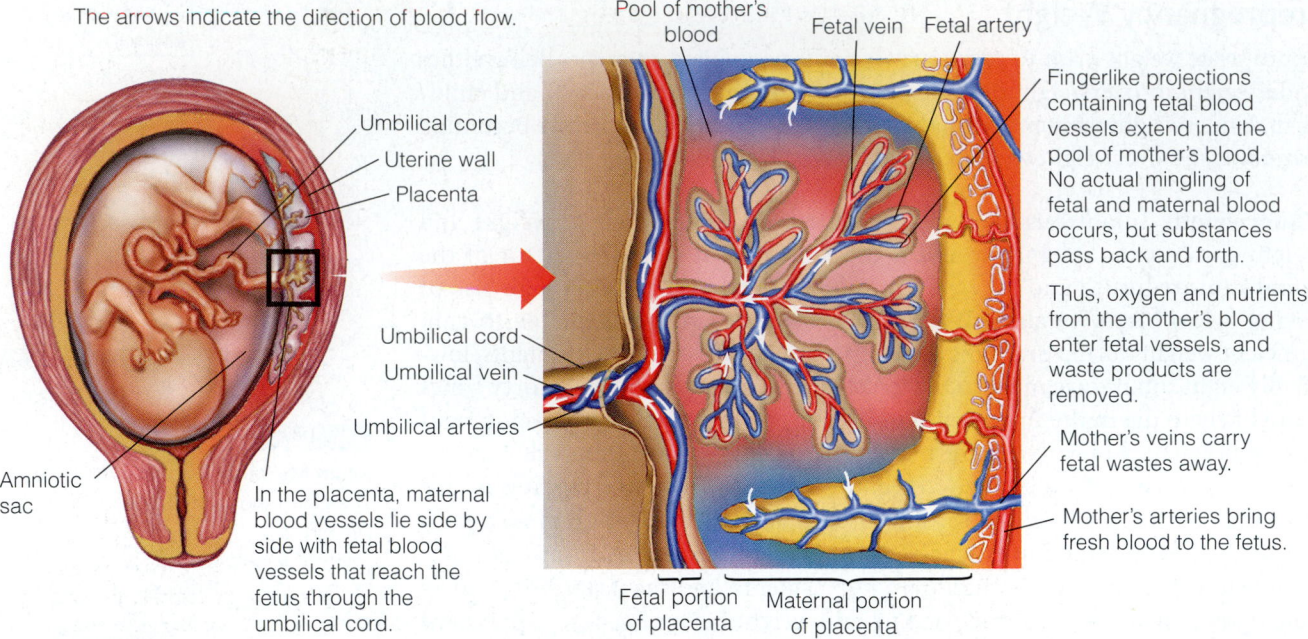

The arrows indicate the direction of blood flow.

Umbilical cord
Uterine wall
Placenta

Amniotic sac

In the placenta, maternal blood vessels lie side by side with fetal blood vessels that reach the fetus through the umbilical cord.

Umbilical cord
Umbilical vein
Umbilical arteries

Pool of mother's blood
Fetal vein   Fetal artery

Fingerlike projections containing fetal blood vessels extend into the pool of mother's blood. No actual mingling of fetal and maternal blood occurs, but substances pass back and forth.

Thus, oxygen and nutrients from the mother's blood enter fetal vessels, and waste products are removed.

Mother's veins carry fetal wastes away.

Mother's arteries bring fresh blood to the fetus.

Fetal portion of placenta   Maternal portion of placenta

FIGURE 11–2

The Placenta

umbilical (um-BIL-ih-cul) cord: the ropelike structure through which the fetus's veins and arteries reach the placenta; the route of nourishment and oxygen into the fetus and the route of waste disposal from the fetus.

amniotic (am-nee-OTT-ic) sac: the "bag of waters" in the uterus in which the fetus floats.

lactation: production and secretion of breast milk for the purpose of nourishing an infant.

ovum (OH-vum): the female reproductive cell, capable of developing into a new organism upon fertilization; commonly referred to as an egg.

zygote (ZY-goat): the product of the union of ovum and sperm; so-called for the first two weeks after fertilization.

implantation: the stage of development in which the zygote embeds itself in the wall of the uterus and begins to develop; occurs during the first two weeks after conception.

fetal blood vessels intertwine and exchange materials. The two bloods never mix, but the barrier between them is notably thin. To grasp how thin, picture your hands as fetal blood vessels, skintight surgical gloves as the tissue-thin placenta, and your gloved hands immersed in water as the pool of maternal blood. Across this thin barrier, nutrients and oxygen move from the mother's blood into the fetus's blood, and wastes move out of the fetal blood to be excreted by the mother. The **umbilical cord** is the pipeline from the placenta to the fetus. The **amniotic sac** surrounds and cradles the fetus, cushioning it with fluids.

The placenta is an active metabolic organ with many responsibilities of its own. It actively gathers up hormones, nutrients, and protein molecules such as antibodies and transfers them into the fetal bloodstream. The placenta also produces a broad range of hormones that act in many ways to maintain pregnancy and prepare the mother's breasts for **lactation**.[12] A healthy placenta is essential for the developing fetus to attain its full potential.

IN SUMMARY

- Adequate nutrition before pregnancy establishes physical readiness and nutrient stores to support fetal growth.

- Both underweight and overweight women should strive for appropriate body weights before pregnancy.

- Newborns who weigh less than 5½ pounds face greater health risks than normal-weight infants.

- The healthy development of the placenta depends on adequate nutrition before pregnancy.

The Events of Pregnancy

The newly fertilized **ovum**, called a **zygote**, begins as a single cell and divides into many cells during the days after fertilization. Within two weeks, the zygote embeds itself in the uterine wall in a process known as **implantation**, and the placenta begins to grow inside the uterus. Minimal growth in size takes place at this time, but it is a

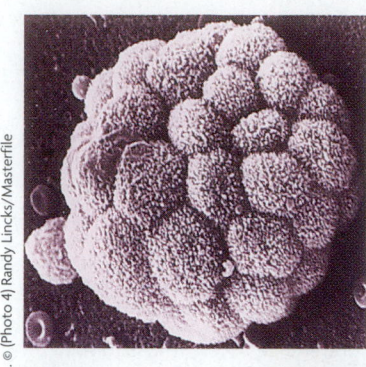

(1) A newly fertilized ovum is about the size of the period at the end of this sentence. This zygote at less than one week after fertilization is not much bigger and is ready for implantation.

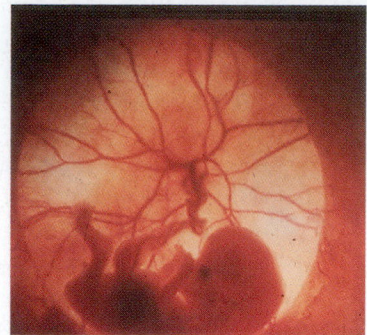

(3) A fetus after 11 weeks of development is just over one inch long. Notice the umbilical cord and blood vessels connecting the fetus with the placenta.

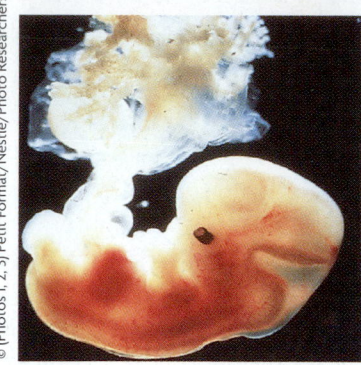

(2) After implantation, the placenta develops and begins to provide nourishment to the developing embryo. An embryo five weeks after fertilization is about 1/2 inch long.

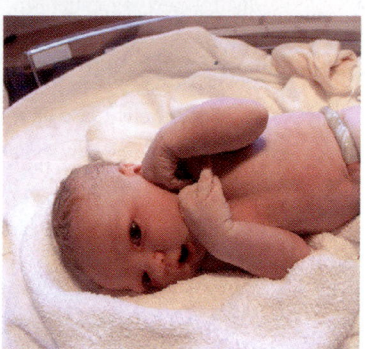

(4) A newborn infant after nine months of development measures close to 20 inches in length. The average birth weight is about 7 1/2 pounds. From eight weeks to term, this infant grew 20 times longer and 50 times heavier.

FIGURE 11–3

Stages of Embryonic and Fetal Development

crucial period in development. Adverse influences such as smoking, drug abuse, and malnutrition at this time lead to failure to implant or to abnormalities such as neural tube defects that can cause the loss of the zygote, possibly before the woman knows she is pregnant.

The Embryo and Fetus During the next six weeks of development, the **embryo** registers astonishing physical changes (see Figure 11–3). At eight weeks, the **fetus** has a complete central nervous system, a beating heart, a fully formed digestive system, well-defined fingers and toes, and the beginnings of facial features.

In the last seven months of pregnancy, the fetal period, the fetus grows fifty times heavier and twenty times longer. Critical periods of cell division and development occur in organ after organ. Most successful pregnancies last 39 to 41 weeks and produce a healthy infant weighing between 6½ and 9 pounds. The 40 or so weeks of pregnancy are divided into thirds, each of which is called a **trimester**.

A Note about Critical Periods Each organ and tissue type grows with its own characteristic pattern and timing. The development of each takes place only at a certain time—the **critical period**. Whatever nutrients and other environmental conditions are necessary during this period must be supplied on time if the organ is to reach its full potential. If the development of an organ is limited during a critical period, recovery is impossible. For example, the fetus's heart and brain are well developed at 14 weeks; the lungs, 10 weeks later. Therefore, early malnutrition impairs the heart and brain; later malnutrition impairs the lungs.

The effects of malnutrition during critical periods of pregnancy are seen in defects of the nervous system of the embryo (explained later), in the child's poor dental health, and in the adolescent's and adult's vulnerability to infections and possibly higher risks of diabetes, hypertension, stroke, or heart disease.[13] The effects of malnutrition during critical periods are irreversible; abundant and nourishing food, consumed after the critical time, cannot remedy harm already done.

embryo (EM-bree-oh): the developing infant from two to eight weeks after conception.

fetus (FEET-us): the developing infant from eight weeks after conception until its birth.

trimester: a period representing one-third of the term of gestation. A trimester is about 13 to 14 weeks.

critical period: a finite period during development in which certain events occur that will have irreversible effects on later developmental stages; usually a period of rapid cell division.

TABLE 11–1
Factors Placing Pregnant Women at Nutritional Risk
Women likely to develop nutrient deficiencies include those who:
Are young (adolescents)
Have had many previous pregnancies (3 or more to mothers under age 20; 4 or more to mothers age 20 or older)
Have short intervals between pregnancies (<18 months)
Lack nutrition knowledge, have too little money to purchase adequate food, or have too little family support
Ordinarily consume an inadequate diet due to food faddism, preferences, weight-loss "dieting," uninformed vegetarianism, or eating disorders
Smoke cigarettes or use alcohol or drugs
Are lactose intolerant or suffer chronic health conditions requiring special diets
Are underweight or overweight at conception
Are carrying twins or triplets
Gain insufficient or excessive weight during pregnancy
Have a low level of education

Table 11–1 provides a list of factors that make nutrient deficiencies and complications likely during pregnancy. Notice that young age heads the list; a later section explains why pregnant adolescents are especially prone to malnutrition.

IN SUMMARY

- Placental development, implantation, and early critical periods depend on maternal nutrition before and during pregnancy.

Nutrient Needs during Pregnancy

■ Nutrient and energy intake recommendations for pregnant women are listed on the inside front cover of this book.

Nutrient needs during pregnancy increase more for certain nutrients than for others. ■ Figure 11–4 shows the percentage increase in nutrient intakes recommended for pregnant women compared to nonpregnant women. To meet the high nutrient demands of pregnancy, a woman must make careful food choices, but her body will also do its part by maximizing nutrient absorption and minimizing losses.[14]

Energy, Carbohydrate, Protein, and Fat Energy needs vary with the progression of pregnancy. In the first trimester, the pregnant woman needs no additional energy, but as pregnancy progresses, her energy needs rise. She requires an additional 340 kcalories daily during the second trimester and an extra 450 kcalories each day during the third trimester.[15] Well-nourished pregnant women meet these demands for more energy in several ways: some eat more food, some reduce their activity, and some store less of their food energy as fat. A woman can easily meet the need for extra kcalories by selecting more nutrient-dense foods from the five food groups. Table 1–6 (on p. 20) provides suggested eating patterns for several kcalorie levels, and Table 11–2 offers a sample menu for pregnant and lactating women.

If a woman chooses less nutritious options such as sugary soft drinks or fatty snack foods to meet her energy needs, she will undoubtedly come up short on nutrients. The increase in the need for nutrients is even greater than that for energy, so the mother-to-be should choose nutrient-dense foods such as whole-grain breads and cereals, legumes, dark green vegetables, citrus fruits, low-fat milk and milk products, lean meats, fish, poultry, and eggs.

Ample carbohydrate (ideally, 175 grams or more per day and certainly no less than 135 grams) is necessary to fuel the fetal brain and spare the protein needed for fetal

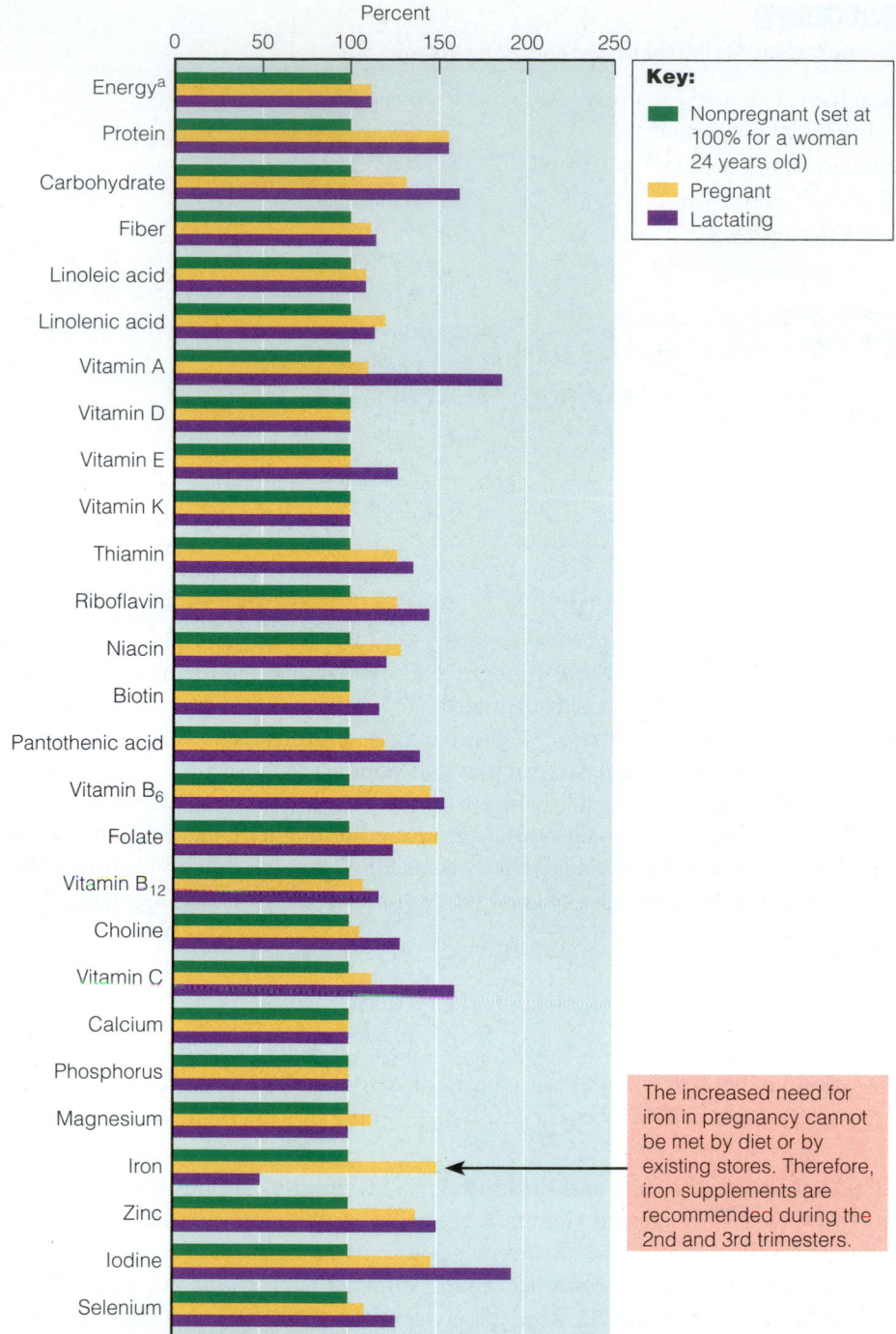

Percent

FIGURE 11–4

Comparison of Nutrient Recommendations for Nonpregnant, Pregnant, and Lactating Women

[a] Energy allowance during pregnancy is for 2nd trimester; energy allowance during the 3rd trimester is slightly higher; no additional allowance is provided during the 1st trimester. Energy allowance during lactation is for the first 6 months; energy allowance during the second 6 months is slightly higher.

Key:
- Nonpregnant (set at 100% for a woman 24 years old)
- Pregnant
- Lactating

The increased need for iron in pregnancy cannot be met by diet or by existing stores. Therefore, iron supplements are recommended during the 2nd and 3rd trimesters.

growth. Fiber in carbohydrate-rich foods such as whole grains, vegetables, and fruit can help alleviate the constipation that many pregnant women experience.

The protein RDA for pregnancy is 25 grams per day higher than for nonpregnant women. However, most women in the United States do not need to add protein-rich foods to their diets because they already exceed the recommended protein intake for pregnancy. Excess protein may also have adverse effects, as Chapter 4 explained.

Some vegetarian women limit or omit protein-rich meats, eggs, and dairy products from their diets. For them, meeting the recommendation for food energy each day and including several generous servings of plant-protein foods such as legumes, tofu,

TABLE 11–2

Sample Menu for Pregnant and Lactating Women

Breakfast	Dinner
1 whole wheat English muffin	Chicken cacciatore
2 tbs peanut butter	3 oz chicken
1 c low-fat vanilla yogurt	$\frac{1}{2}$ c stewed tomatoes
$\frac{1}{2}$ c fresh strawberries	1 c rice
1 c orange juice	$\frac{1}{2}$ c summer squash
Midmorning snack	$1\frac{1}{2}$ c salad (spinach, mushrooms, carrots)
$\frac{1}{2}$ c cranberry juice	1 tbs salad dressing
1 oz pretzels	1 slice Italian bread
Lunch	2 tsp soft margarine
Sandwich (tuna salad on whole wheat bread)	1 c low-fat milk
$\frac{1}{2}$ carrot (sticks)	—
1 c low-fat milk	—
	—

Note: This sample meal plan provides about 2,500 kcalories (55 percent from carbohydrate, 20 percent from protein, and 25 percent from fat) and meets most of the vitamin and mineral needs of pregnant and lactating women.

whole grains, nuts, and seeds are imperative. Protein supplements during pregnancy can be harmful, and their use is discouraged.

The high nutrient requirements of pregnancy leave little room in the diet for excess energy from added purified fats such as oil, margarine, and butter. However, the essential fatty acids are particularly important to the growth and development of the fetus.[16] The brain contains a substantial amount of lipid material and depends heavily on long-chain omega-3 and omega-6 fatty acids for its growth, function, and structure. (See Table 3–3 on p. 73 for a list of good food sources of the essential fatty acids.)

- Folate RDA during pregnancy: 600 μg/day

- Reminder: Neural tube defects (NTD) are malformations of the brain, spinal cord, or both during embryonic development.

- A pregnancy affected by a neural tube defect can occur in any woman, but these factors make it more likely:
  - Inadequate folate intake
  - A previous pregnancy affected by a neural tube defect
  - Maternal diabetes (type 1)
  - Maternal use of antiseizure medications
  - Maternal obesity
  - Exposure to high temperatures early in pregnancy (prolonged fever or hot tub use)
  - Race/ethnicity (neural tube defects are more common among whites and Hispanics than among others)
  - Low socioeconomic status

IN SUMMARY

- Pregnancy brings physiological adjustments that demand increased intakes of energy and nutrients.

- A balanced diet that includes more nutrient-dense foods from each of the five food groups can help to meet these needs.

Of Special Interest: Folate and Vitamin B₁₂ The vitamins famous for their roles in cell reproduction—folate and vitamin B_{12}—are needed in large amounts during pregnancy. New cells are laid down at a tremendous pace as the fetus grows and develops. At the same time, because the mother's blood volume increases, the number of her red blood cells must rise, requiring more cell division and therefore more vitamins. To accommodate these needs, the recommendation for folate during pregnancy increases from 400 to 600 micrograms per day.

As described in Chapter 8, folate plays an important role in preventing neural tube defects. To review, the early weeks of pregnancy are a critical period for the formation and closure of the **neural tube** that will later develop to form the brain and spinal cord. By the time a woman suspects she is pregnant, usually around the sixth week of pregnancy, the embryo's neural tube normally has closed. A neural tube defect (NTD) occurs when the tube fails to close properly. In the United States, about 30 of every 100,000 newborns are born with a neural tube defect.[17] When the neural tube fails to close properly and brain development fails, a rare but lethal defect known as **anencephaly** occurs. All infants with anencephaly die shortly after birth.

In a more common NTD, the spinal cord and backbone do not develop normally, and the result is **spina bifida** (see Figure 11–5). The membranes covering the spinal

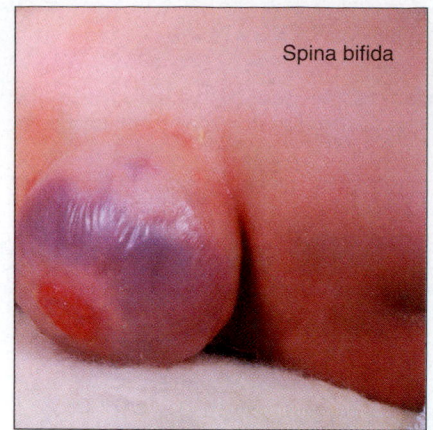

Spina bifida

© Biophoto Associates/ Photo Researchers, Inc.

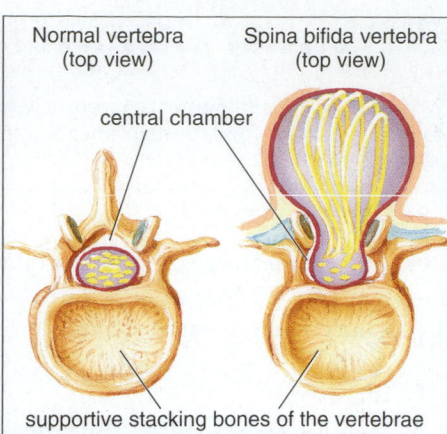

Normal vertebra (top view)   Spina bifida vertebra (top view)

central chamber

supportive stacking bones of the vertebrae

Normally, the bony central chamber closes fully to encase the spinal cord and its surrounding membranes and fluid. In spina bifida, the two halves of the slender bones that should complete the casement of the cord fail to join.

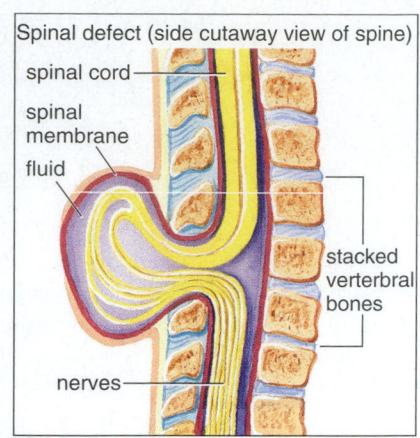

Spinal defect (side cutaway view of spine)

spinal cord
spinal membrane
fluid
stacked verterbral bones
nerves

In the serious form shown here, membranes and fluid have bulged through the gap and nerves are exposed, invariably leading to some degree of paralysis and often to mental retardation.

FIGURE 11–5

Spina Bifida—A Neural Tube Defect

cord often protrude from the spine as a sac, and sometimes a portion of the spinal cord is contained in the sac. Spina bifida is often accompanied by varying degrees of paralysis, depending on the extent of spinal cord damage. Mild cases may not be noticed. Moderate cases may involve curvature of the spine, muscle weakness, mental handicaps, and other ills, while severe cases can lead to death.

To reduce the risk of neural tube defects, women who are capable of becoming pregnant should obtain 400 micrograms of folic acid daily from supplements, fortified foods, or both, *in addition* to eating folate-rich foods (see Table 11–3). The DRI committee recommends intake of synthetic folate, called folic acid, in supplements and fortified food because it is absorbed better than the folate naturally present in foods. Foods that naturally contain folate are still important, however, because they contribute to folate intakes while providing other needed vitamins, minerals, fiber, and phytochemicals.

All enriched grain products (cereal, grits, pasta, rice, bread, and the like) sold commercially in the United States are fortified with folic acid. This measure has improved folate status in women of childbearing age and lowered the number of neural tube defects that occur each year.[18] Researchers expect to see declines in some other birth defects (cleft lip and/or cleft palate) and miscarriages as well.[19] Folate fortification does raise one safety concern, however. The pregnant woman needs a greater amount of vitamin B_{12} to assist folate in the manufacture of new cells. Because high doses of folate complicate the diagnosis of a vitamin B_{12} deficiency, ■ quantities of 1 milligram or more require a prescription.

■ Chapter 8 describes how excessive folate intakes can mask the symptoms of a vitamin B_{12} deficiency.

TABLE 11–3	
Rich Folate Sourcesᵃ	
Natural Folate Sources	**Fortified Folate Sources**
Liver (3 oz) 221 µg	Multi-Grain Cheerios Plus cereal (1 c) 400 µgᵇ
Lentils (½ c) 179 µg	Product 19 cereal (1 c) 400 µgᵇ
Chickpeas or pinto beans (½ c) 145 µg	Total cereal (1 c) 400 µgᵇ
Asparagus (½ c) 131 µg	Pasta, cooked (1 c) 110 µg
Spinach (1 c raw) 131 µg	Rice, cooked (1 c) 134 µg
Avocado (½ c) 45 µg	Bagel (1 small whole) 75 µg
Orange juice (1 c) 74 µg	Waffles, frozen (2) 36 µg
Beets (½ c) 68 µg	Bread, white (1 slice) 28 µg

ᵃ Folate amounts for these and 2,000 other foods are listed in the Table of Food Composition in Appendix A.
ᵇ Folate in cereals varies; read the Nutrition Facts panel of the label.

neural tube: the embryonic tissue that later forms the brain and spinal cord.

anencephaly (AN-en-SEF-a-lee): an uncommon and always fatal type of neural tube defect; characterized by the absence of a brain.
an = not (without)
encephalus = brain

spina (SPY-nah) **bifida** (BIFF-ih-dah): one of the most common types of neural tube defects; characterized by the incomplete closure of the spinal cord and its bony encasement.
spina = spine
bifida = split

■ Vitamin B$_{12}$ RDA during pregnancy:
 2.6 μg/day

■ Calcium AI during pregnancy:
  ■ 1,300 mg/day (14–18 yr)
  ■ 1,000 mg/day (19–50 yr)

Phosphorus RDA during pregnancy:
  ■ 1,250 mg/day (14–18 yr)
  ■ 700 mg/day (19–50 yr)

Magnesium RDA during pregnancy:
  ■ 400 mg/day (14–18 yr)
  ■ 350 mg/day (19–30 yr)
  ■ 360 mg/day (31–50 yr)

Three cups of milk per day will supply 900 mg of calcium. For other food sources of calcium, see Chapter 9.

■ Fluoride AI during pregnancy:
  ■ 3.0 mg/day

People who eat meat, eggs, or dairy products receive all the vitamin B$_{12}$ they need, even for pregnancy. ■ Those who exclude all animal products from the diet need vitamin B$_{12}$-fortified foods or supplements.

IN SUMMARY

■ Due to their key roles in cell reproduction, folate and vitamin B$_{12}$ are needed in increased amounts during pregnancy.

■ Folate plays an important role in preventing neural tube defects.

Vitamin D and Calcium for Bones Vitamin D and the minerals involved in building the skeleton—calcium, phosphorus, and magnesium—are in great demand during pregnancy. ■ Insufficient intakes may produce abnormal fetal bone development.

Intestinal absorption of calcium doubles early in pregnancy, when the mother's bones store the mineral. Later, as the fetal bones begin to calcify, there is a dramatic shift of calcium across the placenta.[20] Whether the calcium added to the mother's bones early in pregnancy is withdrawn to build the fetus's bones later is unclear. In the final weeks of pregnancy, more than 300 milligrams per day are transferred to the fetus. Recommendations to ensure an adequate calcium intake during pregnancy are aimed at conserving the mother's bone mass while supplying fetal needs.

For women whose prepregnancy calcium intakes are below recommendations, as most are, increased calcium intakes may be especially important. Milk products offer many advantages over supplements, as emphasized in earlier chapters. Because bones are still actively depositing minerals until about age 25, adequate calcium is especially important for young women. Pregnant women under age 25 who consume less than 600 milligrams of calcium per day need to increase their intakes of milk, cheese, yogurt, and other calcium-rich foods. Alternatively, and less preferably, they may need a daily supplement of 600 milligrams of calcium.

Women who exclude milk products need calcium-fortified foods such as soy milk. It is worth noting that not all soy milk is fortified with calcium; products fortified with calcium and vitamin D are recommended. Calcium-fortified orange juice offers folate and vitamin C as well as calcium.

Fluoride Mineralization of the fetus's teeth begins in the fifth month after conception. For this and for bone development, fluoride may be needed. ■ Fluoride crosses the placenta, and whether the placenta can defend against excess intakes is questionable. Therefore, fluoride supplements are not recommended for pregnant women who drink fluoridated water. For women who live in communities without fluoridated water, a fluoride supplement may protect fetal teeth.

IN SUMMARY

■ All pregnant women, but especially those who are less than 25 years of age, need to pay special attention to calcium to ensure adequate intakes.

■ Fluoride supplements are not recommended for pregnant women who drink fluoridated water, but for those who live in communities where the water is not fluoridated, a fluoride supplement may protect fetal teeth.

Iron The body conserves iron especially well during pregnancy: menstruation ceases, and absorption of iron increases up to threefold due to a rise in the blood's iron-absorbing and iron-carrying protein transferrin. Still, iron needs are so high that stores dwindle during pregnancy. ■

The developing fetus draws heavily on the mother's iron stores to create stores of its own to last through the first four to six months of life. Even women with inadequate iron stores transfer significant amounts of iron to the fetus, suggesting that the iron needs of the fetus have priority over those of the mother.[21] Iron losses also occur with the bleeding that is inevitable at birth.

Few women enter pregnancy with adequate iron stores. Women who enter pregnancy with iron-deficiency anemia have a greater-than-normal risk of delivering low-birth weight or preterm infants.[22] For all women not taking supplements containing iron, a daily iron supplement containing 30 milligrams is recommended during the second and third trimesters of pregnancy. When a low hemoglobin or hematocrit is confirmed by a repeat test, more than 30 milligrams of iron may be prescribed. To enhance iron absorption, the supplement should be taken between meals and with liquids other than milk, coffee, or tea, which inhibit iron absorption.

Zinc Zinc is required for DNA and RNA synthesis and thus for protein synthesis. ■ Typical zinc intakes for pregnant women are lower than recommendations, but fortunately, zinc absorption increases when zinc intakes are low.[23] Routine supplementation during pregnancy is not advised.[24] However, women taking iron supplements (more than 30 milligrams per day) may need zinc supplementation because large doses of iron can interfere with the body's absorption and use of zinc. Most supplements for pregnancy provide about 30 to 60 milligrams of iron a day. Zinc is most abundant in foods of high-protein content, such as shellfish, meat, and nuts.

IN SUMMARY

- A daily iron supplement is recommended for all pregnant women during the second and third trimesters.

- Iron interferes with zinc absorption, so women taking iron supplements (more than 30 milligrams per day) may need zinc supplements as well.

Nutrient Supplements Physicians usually recommend daily multivitamin-mineral supplements for pregnant women. These prenatal supplements typically provide more folate, iron, and calcium than regular supplements. Prenatal supplements are especially beneficial for women who do not eat adequately and for those in high-risk groups: women carrying multiple fetuses, cigarette smokers, and alcohol and drug abusers.[25] For these women, prenatal supplements may be of some help in reducing the risks of preterm delivery, low birth weights, and birth defects. Be aware that supplements cannot prevent the vast majority of destruction from tobacco, alcohol, and drugs, which continues unopposed, as later sections explain. Figure 11–6 presents a label from a standard prenatal supplement.

IN SUMMARY

- Women most likely to benefit from multivitamin-mineral supplements during pregnancy include those who do not eat adequately, those carrying twins or triplets, and those who smoke cigarettes or are alcohol or drug abusers.

Food Assistance Programs

Women of limited financial means may eat diets too low in calcium, iron, vitamins A and C, and protein. Often, they and their children need help in obtaining food and

- Iron RDA during pregnancy: 27 mg/day

In pregnancy, hemoglobin values of 12 g are not unusual, and 11 g is where the line defining "too low" is often drawn. Appendix E discusses more sensitive measures of iron status.

Food sources of iron:

- Liver, oysters
- Red meat, fish, other meat
- Dried fruits (raisins, prunes)
- Legumes (dried beans, peas, lima beans)
- Dark green vegetables

Reminder: Vitamin C-rich foods enhance iron absorption from foods.

- Zinc RDA during pregnancy:
  - 12 mg/day (≤18 yr)
  - 11 mg/day (19–50 yr)

Prenatal Vitamins

Supplement Facts
Serving Size 1 Tablet

Amount Per Tablet	% Daily Value for Pregnant/ Lacting Women
Vitamin A 4000 IU	50%
Vitamin C 100 mg	167%
Vitamin D 400 IU	100%
Vitamin E 11 IU	37%
Thiamin 1.84 mg	108%
Riboflavin 1.7 mg	85%
Niacin 18 mg	90%
Vitamin B_6 2.6 mg	104%
Folate 800 mcg	100%
Vitamin B_{12} 4 mcg	50%
Calcium 200 mg	15%
Iron 27 mg	150%
Zinc 25 mg	167%

INGREDIENTS: calcium carbonate, microcrystalline cellulose, dicalcium phosphate, ascorbic acid, ferrous fumarate, zinc oxide, acacia, sucrose ester, niacinamide, modified cellulose gum, di-alpha tocopheryl acetate, hydroxypropyl methylcellulose, hydroxypropyl cellulose, artificial colors (FD&C blue no. 1 lake, FD&C red no. 40 lake, FD&C yellow no. 6 lake, titanium dioxide), polyethylene glycol, starch, pyridoxine hydrochloride, vitamin A acetate, riboflavin, thiamin mononitrate, folic acid, beta carotene, cholecalciferol, maltodextrin, gluten, cyanocobalamin, sodium bisulfite.

FIGURE 11–6

Example of a Prenatal Supplement Label

Notice that vitamin A is reduced to guard against birth defects, while extra amounts of folate, iron, and other nutrients are provided to meet the specific needs of pregnant women.

benefit from nutrition counseling. At the federal level, the **Special Supplemental Food Program for Women, Infants, and Children (WIC)** provides nutritious foods, along with nutrition education to low-income pregnant and lactating women and their children.[26] The foods offered are milk and cheese, iron-fortified cereals, fruit or vegetable juices, carrots, eggs, dried beans, tuna fish, and peanut butter. These foods provide nutrients often lacking in diets of low-income women and children. For infants given infant formula, WIC also provides iron-fortified formula. However, WIC encourages mothers to breastfeed their infants and offers incentives such as longer eligibility to participate in WIC and breast pumps and other materials to support breastfeeding to those who do.[27]

More than 7 million people—most of them infants and young children—receive WIC benefits each month. Participation in the WIC program benefits both the nutrient status and the growth and development of infants and children.[28] WIC participation during pregnancy can effectively reduce infant mortality, low birth weight, and maternal and newborn medical costs.

The Food Stamp Program provides a debit card that can also help to stretch the low-income pregnant woman's grocery dollars. Many communities provide educational services and materials, including nutrition, food budgeting, and shopping information through the local agricultural extension service. Organizations such as the American Dietetic Association, the American Diabetes Association, and local hospitals also provide nutrition information.

IN SUMMARY

- Food assistance programs such as WIC can provide nutritious food for pregnant women of limited financial means.

Weight Gain

Women must gain weight during pregnancy—fetal and maternal well-being depends on it. Ideally, a woman will have begun her pregnancy at a healthy weight, but even more importantly, she will gain within the recommended weight range based on her prepregnancy body mass index (BMI). Table 11–4 presents recommended weight gains for pregnancy. For the normal-weight woman, the ideal pattern is about 3½ pounds total during the first trimester and one pound per week thereafter. ■ Pregnancy weight gains within the recommended ranges are associated with fewer surgical births, a greater number of healthy birth weights, and other positive outcomes for both mothers and infants, but many women do not gain within these ranges.

Dieting during pregnancy is not recommended. Even an obese woman should gain at least 15 pounds for the best chance of delivering a healthy infant.[29] Weight gain for a pregnant adolescent must be adequate enough to accommodate her own growth and that of her fetus. Women who are carrying twins must gain more still. A sudden, large weight gain is a danger signal, however, because it may indicate the onset of preeclampsia (discussed later in the chapter).

The weight the pregnant woman gains is nearly all lean tissue: placenta, uterus, blood, milk-producing glands, and the fetus itself (see Figure 11–7). The fat she gains is needed later for lactation. Physical activity can help a pregnant woman cope with the extra weight, as the next section explains. Some weight is lost at delivery, but many women retain a few pounds with each pregnancy.

■ A prenatal weight-gain grid (Appendix E) plots the rate of weight gain during pregnancy.

Dietary Guidelines for Americans 2005

- Pregnant women should ensure appropriate weight gain as specified by a health care provider.

TABLE 11-4

Recommended Weight Gains Based on Prepregnancy Weight

Prepregnancy Weight	Recommended Weight Gain
Underweight (BMI <18.5)	28 to 40 lb (12.5 to 18.0 kg)
Healthy weight (BMI 18.5 to 24.9)	25 to 35 lb (11.5 to 16.0 kg)
Overweight (BMI 25.0 to 29.9)	15 to 25 lb (7.0 to 11.5 kg)
Obese (BMI ≥30)	15 lb minimum (6.8 kg minimum)

Note: These classifications for BMI are slightly different from those developed in 1990 by the Committee on Nutritional Status during Pregnancy and Lactation for the publication *Nutrition during Pregnancy* (Washington, DC: National Academy Press 1990). That committee acknowledged that because such classifications had not been validated by research on pregnancy outcome, "any cut off points will be arbitrary for women of reproductive age." For these reasons, it seems appropriate to use the values developed for adults in 1998 by the National Institutes of Health (see Chapter 6).

IN SUMMARY

- Weight gain is essential for a healthy pregnancy.

- A woman's prepregnancy BMI, her own nutrient needs, and the number of fetuses she is carrying help to determine appropriate weight gain.

Physical Activity

Physical activity is important to the pregnant woman, not only to help her carry the extra weight of pregnancy without strain, but also to help ease her upcoming childbirth. Staying active during the course of a normal, healthy pregnancy improves the fitness of the mother-to-be, facilitates labor, helps to prevent or manage gestational diabetes, and reduces psychological stress.[30] Women who remain active during pregnancy report fewer discomforts throughout their pregnancies and retain habits that help in losing excess weight and regaining fitness after the birth.

Pregnant women should take care in choosing their physical activities, however. They should participate in "low-impact" activities and avoid sports in which they might fall or be hit by other people or objects. As is true for everyone, the frequency, duration, and intensity of the activity affect the likelihood of the benefits or risks. A pregnant woman should consult her health care provider before taking up additional activity. A few guidelines are offered in Figure 11–8 (p. 318). Several of the guidelines are aimed

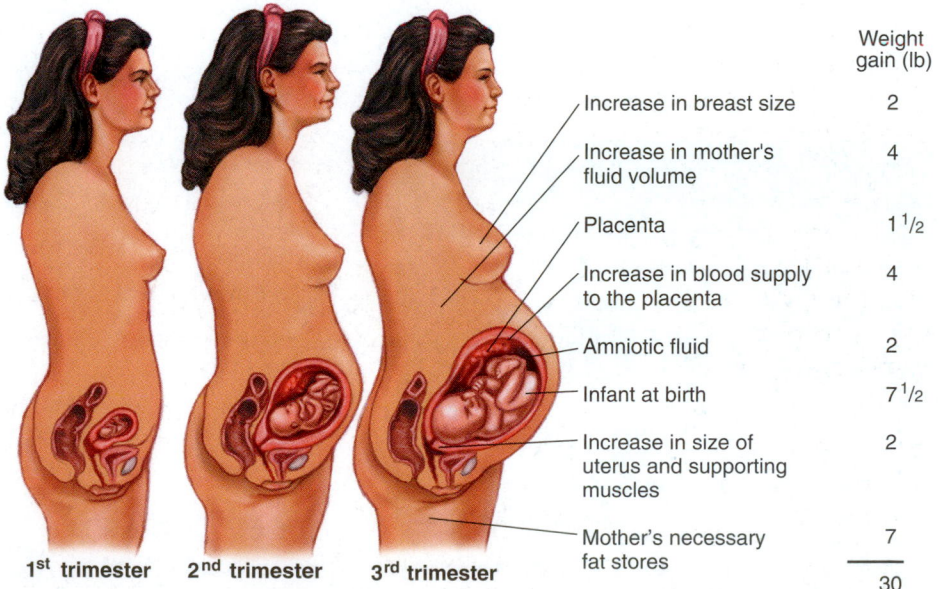

	Weight gain (lb)
Increase in breast size	2
Increase in mother's fluid volume	4
Placenta	1 1/2
Increase in blood supply to the placenta	4
Amniotic fluid	2
Infant at birth	7 1/2
Increase in size of uterus and supporting muscles	2
Mother's necessary fat stores	7
	30

1st trimester   2nd trimester   3rd trimester

FIGURE 11-7

Components of Weight Gain during Pregnancy

Special Supplemental Food Program for Women, Infants, and Children (WIC): a high quality, cost-effective health care and nutrition services program administered by the U.S. Department of Agriculture for low-income women, infants, and children who are nutritionally at risk. WIC provides supplemental foods, nutrition education, and referrals to health care and other social services.

at preventing excessively high internal body temperature and dehydration, both of which can harm fetal development. To this end, a pregnant woman should also stay out of saunas, steam rooms, and hot whirlpools.

IN SUMMARY

■ By remaining active throughout pregnancy, a woman can develop the strength she needs to carry the extra weight and maintain habits that will help her lose it after the birth.

Common Nutrition-Related Concerns of Pregnancy

Food sensitivities, nausea, heartburn, and constipation are common during pregnancy. A few simple strategies can help alleviate maternal discomforts (see Table 11–5).

Food Cravings and Aversions Some women develop cravings for or aversions to certain foods and beverages during pregnancy. Individual **food cravings** during pregnancy do not seem to reflect real physiological needs. In other words, a woman who craves pickles does not necessarily need salt. Similarly, cravings for ice cream are common during pregnancy but do not signify a calcium deficiency. **Food aversions** and cravings that arise during pregnancy are probably due to hormone-induced changes in taste and sensitivities to smells.

Nonfood Cravings Some pregnant women develop cravings for nonfood items such as laundry starch, clay, soil, or ice—a practice known as pica.[31] Pica may be practiced for cultural reasons that reflect a society's folklore; it is especially common among African-American women. Pica is often associated with iron deficiency, but whether iron deficiency leads to pica or pica leads to iron deficiency is unclear. Eating clay or soil may interfere with iron absorption and displaces iron-rich foods from the diet. Furthermore, if the soil or clay contains environmental contaminants such as lead or parasites, health and nutrition suffer.

FIGURE 11–8

Guidelines for Physical Activity during Pregnancy

DO		DON'T
Do exercise regularly (most, if not all, days of the week).		Don't exercise vigorously after long periods of inactivity.
Do warm up with 5 to 10 minutes of light activity.		Don't exercise in hot, humid weather
Do 30 minutes or more of moderate physical activity; 20 to 60 minutes of more intense activity on 3 to 5 days per week will provide greater fitness benefits.		Don't exercise when sick with fever.
		Don't exercise while lying on your back after the first trimester of pregnancy or stand motionless for prolonged periods.
Do cool down with 5 to 10 minutes of slow activity and gentle stretching.		Don't exercise if you experience any pain or discomfort.
Do drink water before, during, and after exercise.		Don't participate in activities that may harm the abdomen or involve jerky, bouncy movements.
Do eat enough to support the additional needs of pregnancy plus exercise.		Don't scuba dive.

© 2002 tracy Frankel/Image Bank/Getty Images

Pregnant women can enjoy the benefits of physical activity.

TABLE 11–5

Strategies to Alleviate Maternal Discomforts

To alleviate the nausea of pregnancy:

On waking, get up slowly.

Eat dry toast or crackers.

Chew gum or suck hard candies.

Eat small, frequent meals whenever hunger strikes.

Avoid foods with offensive odors.

When nauseated, do not drink citrus juice, water, milk, coffee, or tea.

To prevent or alleviate constipation:

Eat foods high in fiber.

Exercise daily.

Drink at least 8 glasses of liquids per day.

Respond promptly to the urge to defecate.

Use laxatives only as prescribed by a physician; avoid mineral oil—it carries needed fat-soluble vitamins out of the body.

To prevent or relieve heartburn:

Relax and eat slowly.

Eat small, frequent meals.

Drink liquids between meals.

Avoid spicy or greasy foods.

Sit up while eating.

Wait 1 hour after eating before lying down.

Wait 2 hours after eating before exercising.

Morning Sickness The nausea of "morning" (actually, anytime) sickness is usually benign, although it is distressing to some women. It arises from the hormonal changes taking place early in pregnancy, ranges from mild queasiness to debilitating nausea, and afflicts more than half of all pregnant women. Many women complain that smells, especially cooking smells, make them sick. Thus minimizing odors is a key to alleviating morning sickness. Traditional strategies for quelling nausea are listed in Table 11–5, but many women benefit most from simply eating the foods they want when they feel like eating.

Heartburn Heartburn, a burning sensation in the lower esophagus near the heart, is common during pregnancy and is also benign. As the growing fetus puts increasing pressure on the woman's stomach, acid may back up and create a burning sensation in her throat. Tips to relieve heartburn are also listed in Table 11–5.

Constipation As the hormones of pregnancy alter muscle tone and the thriving infant crowds intestinal organs, an expectant mother may complain of constipation, another harmless but annoying condition. A high-fiber diet, physical activity, and plentiful fluids will help relieve this condition. Also, responding promptly to the urge to defecate can help. Laxatives should be used only as prescribed by the physician. Mineral oil should not be used because it interferes with the absorption of fat-soluble vitamins.

IN SUMMARY

- Food cravings typically do not reflect physiological needs.
- The nausea, heartburn, and constipation that sometimes accompany pregnancy can usually be alleviated with a few simple strategies.

food cravings: deep longings for particular foods.

food aversions: strong desires to avoid particular foods.

Problems in Pregnancy

Just as adequate nutrition and normal weight gain support the health of the mother and growth of the fetus, maternal diseases can have an adverse effect. If discovered early, many diseases can be controlled—another reason why early prenatal care is recommended. Some nutrition measures can help alleviate the most common problems encountered during pregnancy.

Gestational Diabetes Some women are prone to developing a pregnancy-related form of diabetes, **gestational diabetes**. Gestational diabetes usually resolves after the infant is born, but some women go on to develop diabetes (type 2) later in life, especially if they are overweight.[32] Gestational diabetes can lead to fetal or infant sickness or death. When it is identified early and managed properly, the most serious risks fall dramatically.[33] More commonly, gestational diabetes leads to surgical birth and high infant birth weight.[34] The American Diabetes Association recommends that all women be assessed for risk of gestational diabetes at their first prenatal examination. ■ Those with elevated risks should undergo further testing and treatment.[35] Chapter 21 provides information about medical nutrition therapy for gestational diabetes.

Hypertension Hypertension complicates pregnancy and affects its outcome in different ways, depending on when the hypertension first develops and on how severe it becomes. Hypertension can be a preexisting chronic condition that develops before a woman becomes pregnant or a transient condition that develops during the pregnancy and subsides after childbirth. In some cases, hypertension that develops during pregnancy warns of the ominous disorder preeclampsia.

Preexisting Chronic Hypertension In addition to the health risks normally imposed by hypertension (heart attack and stroke), high blood pressure increases the risks of having a low-birth weight infant or of having the placenta separate from the wall of the uterus before the birth, resulting in stillbirth. Ideally, before a woman with hypertension becomes pregnant, her blood pressure will be under control.

Transient Hypertension of Pregnancy Some women first develop hypertension during the second half of pregnancy. Most often, the rise in blood pressure is mild and does not affect the pregnancy adversely. Blood pressure usually returns to normal during the first few weeks after childbirth. This transient hypertension of pregnancy differs from the life-threatening hypertension diseases of pregnancy—preeclampsia and eclampsia.

Preeclampsia Hypertension may signal the onset of **preeclampsia**, a condition characterized not only by high blood pressure, but also by protein in the urine and fluid retention (edema). ■ Preeclampsia, which affects less than 10 percent of pregnant women, usually occurs with first pregnancies and almost always appears after 20 weeks of gestation.[36] Symptoms typically regress within 48 hours of delivery. The edema of preeclampsia is a whole-body edema, distinct from the localized fluid retention women normally experience late in pregnancy. ■

Preeclampsia affects almost all of the woman's organs—the circulatory system, liver, kidneys, and brain. If it progresses, she may experience convulsions; when this occurs, the condition is called **eclampsia**. Maternal mortality during pregnancy is rare in developed countries, but eclampsia is one of the most common causes. Preeclampsia demands prompt medical attention. Treatment focuses on regulating blood pressure and preventing convulsions.

■ Risk factors for gestational diabetes:
- Obesity
- Personal history of gestational diabetes
- Strong family history of diabetes
- Glucose in the urine

These racial and ethnic groups are more prone to gestational diabetes:
- Hispanic American
- Native American
- Asian-American
- African-American
- Pacific Islander

■ The normal edema of pregnancy responds to gravity: blood pools in the ankles. The edema of preeclampsia is a generalized edema. The distinction helps with diagnosis.

■ Warning signs of preeclampsia:
- Hypertension
- Protein in the urine
- Upper abdominal pain
- Severe and constant headaches
- Swelling, especially of the face
- Dizziness
- Blurred vision
- Sudden weight gain (1 lb/day)

- Conditions such as gestational diabetes, hypertension, and preeclampsia can threaten the health and life of both mother and infant.

- Such conditions require medical and nutrition treatment.

Practices to Avoid

A general guideline for the pregnant woman is to eat a normal, healthy diet and practice moderation. A woman's daily choices during pregnancy take on enormous importance. Forewarned, pregnant women can choose to abstain from or avoid potentially harmful practices.

Cigarette Smoking One practice to be avoided during pregnancy is cigarette smoking. A surgeon general's warning states that parental smoke can kill an otherwise healthy fetus or newborn. Constituents of cigarette smoke, such as nicotine and cyanide, are toxic to a fetus. Research shows that smoking during pregnancy can cause damage to fetal chromosomes, which could lead to developmental defects or genetic disorders, including cancer.[37] Smoking also restricts the blood supply to the growing fetus and so limits the delivery of oxygen and nutrients and the removal of wastes. It slows growth, thus retarding physical development of the fetus, and it may cause behavioral or intellectual problems later.

A mother who smokes is more likely to have a complicated birth, and her infant is more likely to be of low birth weight.[38] The more a mother smokes, the smaller her infant will be. Of all preventable causes of low birth weight in the United States, smoking has the greatest impact. Sudden infant death syndrome (SIDS), the unexplained death that sometimes occurs in an otherwise healthy infant, has been linked to the mother's cigarette smoking during pregnancy.[39] Research suggests that even in women who do not smoke, exposure to **environmental tobacco smoke (ETS**, or secondhand smoke) during pregnancy increases the risk of low birth weight and the likelihood of SIDS.[40]

Unfortunately, an estimated one out of nine pregnant women smokes, and rates are even higher for unmarried women and those who have not graduated from high school.[41] In addition to the harms already described, cigarette (and cigar) smoking adversely affects the pregnant woman's nutrition status and thus impairs fetal nutrition and development. Smokers tend to have lower intakes of dietary fiber, vitamin A, beta-carotene, folate, and vitamin C.

Medicinal Drugs and Herbal Supplements Medicinal drugs taken during pregnancy can cause serious birth defects. Pregnant women should not take over-the-counter drugs or any medications not prescribed by a physician. Drug labels warn: "As with any drug, if you are pregnant or nursing a baby, seek the advice of a health professional before using this product." For aspirin and ibuprofen, there is an additional warning: "It is especially important not to use aspirin (or ibuprofen) during the last three months of pregnancy unless specifically directed to do so by a doctor because it may cause problems in the unborn child or excessive bleeding during delivery." Such warnings should be taken seriously.

Some pregnant women mistakenly consider herbal supplements to be safe alternatives to medicinal drugs and take them to relieve nausea, promote water loss, alleviate depression, help them sleep, or for other reasons. Some herbal products may be safe, but almost none have been tested for safety or effectiveness during pregnancy. Pregnant women should stay away from herbal supplements, teas, or other products unless their safety during pregnancy has been ascertained.[42] The American Dietetic Association's Web site lists more than 100 herbal supplements that may not be safe to use

gestational diabetes: the presence of abnormal glucose tolerance during pregnancy.

preeclampsia: a condition characterized by hypertension, fluid retention, and protein in the urine.

eclampsia: a severe complication during pregnancy in which convulsions occur.

environmental tobacco smoke (ETS): the combination of exhaled smoke (mainstream smoke) and smoke from lighted cigarettes, pipes, or cigars (sidestream smoke) that enters the air and may be inhaled by other people.

■ Fetal effects of abused drugs:

- Amphetamines: Suspected nervous system damage; behavioral abnormalities
- Barbiturates: Drug withdrawal symptoms in the newborn, lasting up to six months
- Cocaine: Uncontrolled jerking motions; paralysis; permanent mental and physical damage
- Marijuana: Short-term irritability at birth
- Opiates (including heroin): Drug withdrawal symptoms in the newborn; permanent learning disability (attention deficit/hyperactivity disorder)

■ To protect their fetuses and newborns from listeriosis, pregnant women should:

- Avoid the following Mexican soft cheeses: queso blanco, queso fresco, queso de hoja, queso de crema, and asadero. Also avoid feta cheese, Brie, Camembert, and blue-veined cheeses like Roquefort.
- Use only pasteurized dairy products.
- Eat only thoroughly cooked meat, poultry, and seafood.
- Thoroughly reheat until steaming hot all hot dogs, luncheon meats, and deli meats, including cured meats like salami, before eating them.
- Wash all fruits and vegetables.
- Do not eat refrigerated smoked seafood such as salmon or trout or any fish labeled "nova-style," "lox," or "kippered," unless it is an ingredient in a cooked dish.
- Do not eat refrigerated pâté or meat spreads. Canned or shelf-stable pâté and meat spreads are safer.

during pregnancy.* Chapter 15 offers more information about herbal supplements and other alternative therapies.

Drugs of Abuse Women who abuse drugs such as marijuana and cocaine during pregnancy inflict serious health consequences, including nervous system disorders, on their fetuses. Drugs of abuse such as cocaine easily cross the placenta and impair fetal growth and development.[43] Infants born to mothers who abuse crack and other forms of cocaine face low birth weight, heartbeat abnormalities, the pain of withdrawal, or even death as they first experience life outside the womb. Some other effects of drugs of abuse on the fetus are listed in the margin. ■

Environmental Contaminants Infants and young children of pregnant women exposed to environmental contaminants such as lead show signs of delayed mental and psychomotor development. During pregnancy, lead readily moves across the placenta, inflicting severe damage on the developing fetal nervous system.[44]

Mercury is a contaminant of concern as well. As discussed in Chapter 3, fatty fish are a good source of omega-3 fatty acids, but some fish contain large amounts of the pollutant mercury, which can harm the developing brain and nervous system.[45] Because the benefits of moderate fish consumption outweigh the risks, women who may become pregnant, pregnant women, lactating women, and children up to the age of 12 are advised to do the following:[46]

- Avoid shark, swordfish, king mackerel, and tilefish (also called golden snapper or golden bass).
- Limit average weekly consumption to 12 ounces (cooked or canned) of seafood *or* to 6 ounces (cooked or canned) of white (albacore) tuna.

Foodborne Illness The vomiting and diarrhea caused by many foodborne illnesses can leave a pregnant woman exhausted and dangerously dehydrated. Particularly threatening is **listeriosis**, which can cause miscarriage, stillbirth, or severe brain or other infections to fetuses and newborns. According to the Centers for Disease Control and Prevention, pregnant women are "about 20 times more likely than other healthy adults to get listeriosis."[47] A woman with listeriosis may develop symptoms such as fever, vomiting, and diarrhea in about 12 hours after eating a contaminated food, and serious symptoms may develop one week to six weeks later. A blood test can reliably detect listeriosis, and antibiotics given promptly to the pregnant sufferer can often prevent infection of the fetus or newborn. Preventive measures pregnant women can take to avoid contracting listeriosis are listed in the margin. ■

Vitamin-Mineral Megadoses Many vitamins and minerals are toxic when taken in excess. Among vitamins, a single massive dose of preformed vitamin A (100 times the recommended intake) has caused birth defects. Chronic use of lower doses of vitamin A supplements (three to four times the recommended intake) may also cause birth defects.

Dietary Guidelines for Americans 2005

- Pregnant women should not eat or drink unpasteurized milk, milk products, or juices; raw or undercooked eggs, meat, or poultry; or raw sprouts.
- Pregnant women should only eat certain deli meats and frankfurters that have been reheated to steaming hot.

---

* Go to www.eatright.org and click on "Position Papers," then on "Life Span," and then on "pregnancy/breastfeeding."

Intakes before the seventh week of pregnancy appear to be the most damaging. For this reason, additional vitamin A is not recommended during pregnancy, and the vitamin is prescribed in the first trimester of pregnancy only upon evidence of deficiency, which is rare.

Dieting Weight-loss dieting, even for short periods, is hazardous during pregnancy. Low-carbohydrate diets or fasts that cause ketosis deprive the growing fetal brain of needed glucose and may impair its development. Such diets are also likely to be deficient in other nutrients vital to fetal growth. Energy restriction during pregnancy is dangerous, regardless of the woman's prepregnancy weight or the amount of weight gained in the previous month.

Sugar Substitutes Artificial sweeteners have been studied extensively and found to be acceptable during pregnancy if used within the FDA's guidelines (see Chapter 2).[48] Women with phenylketonuria should not use aspartame, as Chapter 2 explains.

Caffeine Caffeine crosses the placenta, and the developing fetus has only a limited ability to metabolize it. Hundreds of researchers over several decades have studied the effects of caffeine during pregnancy, often with conflicting results. As some researchers note, women who drink more coffee than other women may differ from other women in other ways as well.[49] For example, heavy coffee drinkers are also likely to be smokers, so separating the effects of caffeine from those of smoking is often difficult. Despite such difficulties, at least two conclusions about caffeine intake during pregnancy have emerged.[50] First, research studies have not indicated that caffeine (even in high doses) causes birth defects in human infants (as it does in animals).[51] Second, moderate caffeine intake (3 cups of coffee per day) during pregnancy has no effect on infant birth weight or length of gestation.[52]

Some recent evidence does suggest that drinking more than 3 cups of coffee per day increases the risk of fetal death.[53] This is a dose-response relationship: as the number of cups of coffee increases, so does the risk of fetal death (the researchers found no association with tea or cola consumption). More research is needed to confirm this finding, but in light of this evidence, the most sensible course is to limit caffeine consumption to the equivalent of one cup of coffee or two 12-ounce cola beverages per day. Caffeine amounts in food and beverages are listed in Appendix A.

Alcohol Drinking alcohol during pregnancy threatens the fetus with irreversible brain damage, growth retardation, mental retardation, facial abnormalities, vision abnormalities, and many more health problems—a spectrum of symptoms known as **fetal alcohol spectrum disorders (FASD)**. Children at the most severe end of the spectrum (those with all of the symptoms) are defined as having **fetal alcohol syndrome (FAS)**.[54] The fetal brain is extremely vulnerable to a glucose or oxygen deficit, and alcohol causes both by disrupting placental functioning. The lifelong mental retardation and other tragedies of FAS can be prevented by abstaining from drinking alcohol during pregnancy. However, once the damage is done, the child remains impaired.

The photo on p. 324 shows the facial abnormalities of FAS, which are easy to depict. A visual picture of the internal harm is impossible, but that damage seals the fate of the child. An estimated 2 to 15 of every 10,000 children are victims of FAS, making it one of the leading known preventable causes of mental retardation in the world.[55]

One out of 10 pregnant women drinks alcohol sometime during her pregnancy; 1 out of 40 pregnant women reports "frequent" drinking (seven or more drinks per week) or binge drinking (five or more drinks on one occasion).[56] Almost half of all pregnancies are unintended, and many are conceived during a binge drinking episode.[57]

listeriosis: a serious foodborne infection that can cause severe brain infection or death in a fetus or newborn; caused by the bacterium *Listeria monocytogenes*, which is found in soil and water.

fetal alcohol spectrum disorders (FASD): a spectrum of physical, behavioral, and cognitive disabilities caused by prenatal alcohol exposure.

fetal alcohol syndrome (FAS): the cluster of symptoms seen in an infant or child whose mother consumed excessive alcohol during her pregnancy. FAS includes, but is not limited to, brain damage, growth retardation, mental retardation, and facial abnormalities.

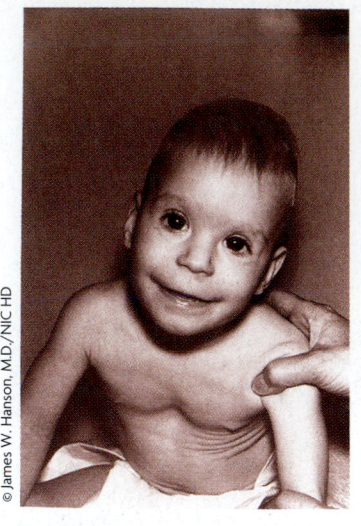

These facial traits are typical of fetal alcohol syndrome, caused by drinking alcohol during pregnancy—low nasal bridge, short eyelid opening, underdeveloped groove in the center of the upper lip, small midface, short nose, and small head circumference.

For women who know they are pregnant and choose to drink alcohol, the question is how much alcohol is too much. Compared to women who drink less than one drink *per week*, a sizable and significant increase in stillbirths occurs in women who drink five or more drinks per week.[58] Low birth weight is reported among infants born to women who drink 1 ounce (two drinks) of alcohol per day during pregnancy, and FAS is also known to occur with as few as two drinks per day. Birth defects have been reliably observed among the children of women who drink 2 ounces (four drinks) of alcohol daily during pregnancy. The most severe impact is likely to occur in the first two months, before the woman may even be aware that she is pregnant.

Research using animals shows that one-fifth of the amount of alcohol needed to produce major visible defects will produce learning impairment or other defects in the offspring.[59] The cluster of mental problems associated with prenatal alcohol exposure and FASD is known as **alcohol-related neurodevelopmental disorder (ARND)**, and the physical malformations are referred to as **alcohol-related birth defects (ARBD)**.

For every child diagnosed with full-blown FAS, many more with FASD go undiagnosed until problems develop in the preschool years. Upon reaching adulthood, such children are ill equipped for employment, relationships, and the other facets of life most adults take for granted. Anyone exposed to alcohol before birth may respond differently to it, and also to certain drugs, than if no exposure had occurred, making addictions likely.

The American Academy of Pediatrics takes the position that women should stop drinking as soon as they *plan* to become pregnant.[60] Researchers have looked for a "safe" alcohol intake limit during pregnancy and have found none. Their conclusion: abstinence from alcohol is the best policy for pregnant women.

For pregnant women who have already drunk alcohol, the advice is "stop now." A woman who has drunk heavily during the first two-thirds of her pregnancy can still prevent some organ damage by stopping heavy drinking during the third trimester.

IN SUMMARY

■ Abstaining from smoking and other drugs, including alcohol, limiting intake of foods known to contain unsafe levels of contaminants such as mercury, taking precautions against foodborne illness, avoiding large doses of nutrients, refraining from dieting, using artificial sweeteners in moderation, and limiting caffeine use are recommended during pregnancy.

Adolescent Pregnancy

Each year in the United States, an estimated 900,000 adolescent girls become pregnant.[61] Of these, about half choose to continue their pregnancies. A pregnant adolescent presents a special case of intense nutrient needs. Young teenage girls have a hard enough time meeting nutrients needs for their own rapid growth and development, let alone those of pregnancy. Many teens enter pregnancy deficient in vitamins A and D, folate, iron, calcium, and zinc, which places both mother and fetus at risk. Smoking also presents risks, and teens are more likely to smoke while pregnant than older women. Pregnant adolescents have more miscarriages, premature births, stillbirths, and low-birth weight infants than do pregnant adult women.[62] Their greatest risk is death of the infant: mothers under age 16 bear more infants who die within the first year than do women in any other age group. These factors combine to make adolescent pregnancy a major public health problem.

Adequate nutrition is an indispensable component of prenatal care for adolescents and can substantially improve the outlook for both mother and infant. To support the needs of both mother and fetus, a pregnant teenager with a BMI in the normal range is encouraged to gain about 35 pounds to reduce the likelihood of a low-birth weight infant.[63] Pregnant and lactating adolescents would do well to follow the eating pattern presented in Table 1–6 on p. 30, making sure to choose a kcalorie level high enough to support weight gain.

IN SUMMARY

- Proper nutrition and adequate weight gain are especially important in reducing the risk of poor pregnancy outcome in adolescents.

Breastfeeding

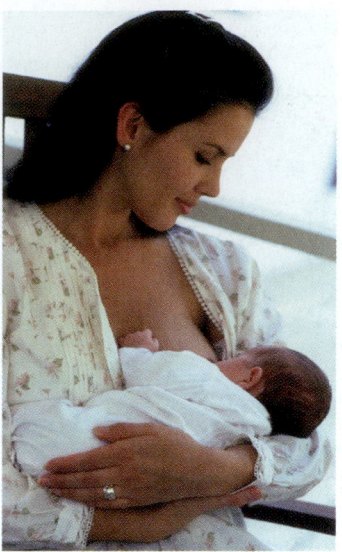

A woman who decides to breast-feed offers her infant a full array of nutrients and protective factors to support optimal health and development.

The American Academy of Pediatrics (AAP) recommends that infants receive breast milk for at least the first 12 months of life and beyond for as long as mutually desired by mother and child.[64] The American Dietetic Association (ADA) advocates breastfeeding for the nutritional health it confers on the infant as well as for the physiological, social, economic, and other benefits it offers the mother.[65] The AAP and the ADA recognize exclusive breastfeeding for 6 months and breastfeeding with complementary foods for at least 12 months as an optimal feeding pattern for infants.[66] Breast milk's unique nutrient composition and protective factors promote optimal infant health and development. The only acceptable alternative to breast milk is iron-fortified formula. Adequate nutrition of the mother supports successful lactation, and without it, lactation is likely to falter or fail.

Nutrition during Lactation

By continuing to eat nutrient-dense foods, not restricting weight gain unduly, and enjoying ample food and fluid at frequent intervals throughout lactation, the mother who chooses to breastfeed her infant will be nutritionally prepared to do so. An inadequate diet does not support the stamina, patience, and self-confidence that nursing an infant demands. Figure 11–4 (on p. 311) shows how a lactating woman's nutrient needs differ from those of a nonpregnant woman, and Table 11–2 (on p. 312) presents a sample menu that meets those needs.

Energy A nursing woman produces about 25 ounces of milk per day, with considerable variation from woman to woman and in the same woman from time to time, depending primarily on the infant's demand for milk. Producing this milk costs a woman almost 500 kcalories per day above her regular need during the first six months of lactation. To meet this energy need, the woman is advised to eat an extra 330 kcalories of food each day. The other 170 kcalories can be drawn from the fat stores she accumulated during pregnancy. The food energy consumed by the nursing mother should carry with it abundant nutrients. Severe energy restriction hinders milk production and can compromise the mother's health.

Weight Loss A question often raised is whether breastfeeding promotes a more rapid loss of the extra body fat accumulated during pregnancy. Results of studies about the relationship between feeding method and loss of body fat and body weight are inconsistent. When breastfeeding continues for three months or longer, lactation does seem to accelerate a woman's weight loss, but factors such as percentage of body fat and weight gain during pregnancy also play a role.[67] This does not mean that a breastfeeding woman can eat unlimited food and still effortlessly return to her prepregnancy weight. In general, most women lose 1 to two pounds per month during the first four to six months of lactation. Some women may lose more, and others may maintain or even gain weight. Neither the quality nor the quantity of breast milk is adversely

alcohol-related neurodevelopmental disorder (ARND): a condition caused by prenatal alcohol exposure. ARND is diagnosed when there is a confirmed history of substantial, regular maternal alcohol intake or heavy episodic drinking and behavioral, cognitive, or central nervous system abnormalities known to be associated with alcohol exposure.

alcohol-related birth defects (ARBD): a condition caused by prenatal alcohol exposure. ARBD is diagnosed when there is a history of substantial, regular maternal alcohol intake or heavy episodic drinking and birth defects known to be associated with alcohol exposure.

■ Moderate weight reduction is safe and does not compromise weight gain of the nursing infant.

affected by moderate weight loss, and infants grow normally. However, too large an energy deficit, especially soon after birth, will inhibit lactation.

Breastfeeding costs energy, but carefully chosen programs of diet and physical activity are still the cornerstones of weight control. Physical activity in particular helps to reduce body fatness and improve fitness while having little effect on a woman's milk production or her infant's weight gain.

Vitamins and Minerals Another question often raised is whether a mother's milk may lack a nutrient if she fails to get enough in her diet. The answer differs from one nutrient to the next, but in general, nutritional deprivation of the mother reduces the *quantity,* not the *quality,* of her milk. Women can produce milk with adequate protein, carbohydrate, fat, folate, and most minerals, even when their own supplies are limited. For these nutrients, milk quality is maintained at the expense of maternal stores. This is most evident in the case of calcium. Dietary calcium has no effect on the calcium concentration of breast milk, but maternal bones lose some of their density during lactation when calcium intakes are inadequate.[68] Such losses are generally made up quickly when lactation ends, and breastfeeding has no long-term harmful effects on women's bones.[69]

Nutrients in breast milk most likely to decline in response to prolonged inadequate intakes are the vitamins—especially vitamins B_6, B_{12}, A, and D. Vitamin supplementation of undernourished women appears to help normalize the vitamin concentrations in their milk and may be beneficial.

Water Despite misconceptions, a mother who drinks more fluid does not produce more breast milk. The nursing mother is nevertheless advised to drink plenty of liquids each day (about 13 cups) to protect herself from dehydration. ■ To help themselves remember to drink enough liquid, many women make a habit of drinking a glass of milk, juice, or water each time the infant nurses as well as at mealtimes.

■ The DRI recommendation for *total* water intake during lactation is 3.8 L/day. This includes 3.1 L or about 13 c as total beverages, including drinking water.

Particular Foods Some infants may be sensitive to foods such as cow's milk, onions, or garlic in the mother's diet and become uncomfortable when she eats them. Nursing mothers should not automatically avoid such foods, however. A mother who is breastfeeding her infant is advised to eat whatever nutritious foods she chooses. Then, if a particular food seems to cause the infant discomfort, she can try eliminating that food from her diet for a few days and see if the problem goes away.

Contraindications to Breastfeeding

Some substances may impair maternal milk production or enter the breast milk and interfere with normal infant development. In addition, some medical conditions prohibit breastfeeding.

Alcohol Alcohol easily enters breast milk and can adversely affect the production, volume, composition, and ejection of breast milk as well as overwhelm an infant's

■ Neither acute nor regular exercise adversely affects the mother's ability to successfully breastfeed.

immature alcohol-degrading system.[70] Alcohol concentration in breast milk peaks within one hour after ingestion of even moderate amounts (equivalent to a can of beer). It may alter the taste of the milk to the disapproval of the nursing infant, who may, in protest, drink less milk than normal.

Caffeine Caffeine can make an infant jittery and wakeful. As during pregnancy, caffeine consumption should be moderate.

Cigarette Smoke Health care professionals should actively discourage smoking by lactating women. Research shows that lactating women who smoke produce less milk and milk with a lower fat content than mothers who do not smoke. Consequently, their infants gain less weight than infants of nonsmokers.

A lactating woman who smokes not only exposes her infant to nicotine and other chemicals via her breast milk, but may also expose the infant to sidestream smoke. Infants who are "smoked over" experience a wide array of health problems—poor growth, hearing impairment, vomiting, breathing difficulties, and even unexplained death.

Medications and Illicit Drugs If a nursing mother must take medication that is secreted in breast milk and is known to affect the infant, then breastfeeding must be put off for the duration of treatment. Meanwhile, the flow of milk can be sustained by pumping the breasts and discarding the milk. Many prescription medications do not reach nursing infants in sufficient quantities to affect them adversely and so have no impact on breastfeeding. Other drugs are not at all compatible with breastfeeding, either because they are secreted into the milk and can harm the infant or because they suppress lactation.[71] A nursing mother should consult with the prescribing physician before taking medicines. Breastfeeding is also contraindicated if the mother uses illicit drugs. Drug addicts, including alcohol abusers, are capable of taking such high doses that their infants can become addicts by way of breast milk.

Many women wonder about using oral contraceptives during lactation. One type that combines the hormones estrogen and progestin seems to suppress milk output, lower the nitrogen content of the milk, and shorten the duration of breastfeeding. In contrast, progestin-only pills have no effect on breast milk or breastfeeding and are considered appropriate for lactating women.

Maternal Illness If a woman has an ordinary cold, she can go on nursing without worry. If susceptible, the infant will catch it from her anyway, and thanks to immunological protection, a breastfed infant may be less susceptible than a formula-fed infant would be. If a woman has active untreated tuberculosis or is receiving therapeutic radioactive isotopes, breastfeeding is contraindicated.[72]

The human immunodeficiency virus (HIV), responsible for causing AIDS, can be passed from an infected mother to her infant during pregnancy, at birth, or through breast milk, especially during the early months of breastfeeding. Thus women in developed countries who have tested positive for HIV should not breastfeed if the infant is not infected. They should choose a safe alternative feeding method, such as breast milk from a milk bank. Milk banks in the United States pasteurize donated human milk and make it available to infants who lack access to milk from their own mothers. Pasteurization destroys harmful organisms, such as HIV, but leaves intact most of the beneficial constituents of the milk.

Throughout the world, breastfeeding prevents millions of infant deaths each year. In developing countries, where the feeding of inappropriate or contaminated formulas causes 1.5 million infant deaths each year, breastfeeding can be critical to infant survival. Thus the question of whether HIV-infected women in developing countries should breastfeed comes down to a delicate balance between risks and benefits.

For HIV-positive women in developing countries who are literate, have access to safe water, and have an uninterrupted supply of infant formula, replacement feeding may reduce the risk of infant illness and death from AIDS. For those mothers without safe water and with minimal education, the risk of replacement feeding may be substantial in terms of infant mortality. The World Health Organization and UNICEF (United Nations Children's Fund), in acknowledging the transmission of HIV by way of breast milk, recommend formula feeding for infants of HIV-positive mothers in developing countries if they can be ensured uninterrupted access to safely prepared, nutritionally adequate breast milk substitutes.

IN SUMMARY

- The lactating woman needs enough energy and nutrients to produce about 25 ounces of milk per day. She also needs extra fluid.

- Alcohol, caffeine, smoking, and drugs may reduce milk production or enter breast milk and impair infant development.

- Some maternal illnesses are incompatible with breastfeeding.

Nutrition Assessment Checklist

FOR PREGNANT AND LACTATING WOMEN

Medical History

Check the medical record for:

- ☐ Alcohol or illicit drug abuse
- ☐ Chronic diseases
- ☐ Gestational diabetes
- ☐ History of previous pregnancies (number, intervals, outcomes, multiple births, and gestational age birth weights)
- ☐ Hypertension
- ☐ Neural tube defect in an infant born previously
- ☐ Preeclampsia

Note risk factors for complications during pregnancy, including:

- ☐ Cigarette smoking
- ☐ Food faddism
- ☐ Lactose intolerance
- ☐ Low socioeconomic status
- ☐ Significant or prolonged vomiting
- ☐ Very young or old age
- ☐ Weight-loss dieting

Note any complaints of:

- ☐ Constipation
- ☐ Heartburn
- ☐ Morning sickness

Medications

For pregnant women who are using drug therapy for medical conditions, note:

- ☐ Potential for contraindication to breastfeeding
- ☐ GI tract side effects that might reduce food intake or change nutrient needs

Dietary Intake

For all pregnant and lactating women, especially those considered at risk nutritionally, assess the diet for:

- ☐ Total energy
- ☐ Protein
- ☐ Calcium, phosphorus, magnesium, iron, and zinc
- ☐ Folate and vitamin B$_{12}$
- ☐ Vitamin D

Anthropometric Data

Measure baseline height and weight:

- ☐ Prepregnancy weight

Reassess weight at each medical checkup and determine whether gains are appropriate. Note:

- ☐ Weight gain during pregnancy
- ☐ Gestational age

Laboratory Tests

Monitor the following laboratory tests for pregnant women:

- ☐ Hemoglobin, hematocrit, or other tests of iron status
- ☐ Blood glucose

Physical Signs

Blood pressure measurement is a routine measurement in physical exams but is especially important for pregnant women. Look for physical signs of:

- ☐ Iron deficiency
- ☐ Edema
- ☐ Protein-energy malnutrition
- ☐ Folate deficiency

This chapter has focused on nutrition during pregnancy and lactation. The Nutrition Assessment Checklist (p. 328) helps to identify nutrition-related factors that may help prevent or correct potential problems in pregnant or lactating women. Chapter 12 explores the dietary needs of infants, children, and adolescents.

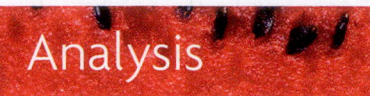

Analysis

CASE STUDY

Women in Her First Pregnancy

Ellen Cassidy is a 24-year-old woman who is four months pregnant. This is her first pregnancy, and she is eager to learn how to feed herself during pregnancy as well as her infant after birth. She is 5 feet 3 inches tall and currently weighs 150 pounds. Her prepregnancy weight was 148 pounds. Ellen is very concerned about her 2-pound weight gain.

1. Consult the BMI table (inside back cover), and using the "Healthy Weight" section, find a healthy weight in the middle of the range appropriate for a woman of Ellen's height.

2. Do you think that Ellen's weight at the start of her pregnancy was appropriate for her height? Why or why not? Should Ellen be concerned about her 2-pound weight gain? Why or why not?

3. What advice should you give Ellen about her weight gain during pregnancy? What other dietary advice would you give her?

4. Discuss methods of infant feeding with Ellen and describe some of the advantages breastfeeding would offer her. What advice will you give Ellen if she decides to breastfeed?

CLINICAL APPLICATIONS

1. Consider the different factors in a pregnant woman's history that can affect her nutrition status and the outcome of her pregnancy. Describe what steps you would take to remedy potential problems for the following clients:
   a. A 15-year-old adolescent of low socioeconomic status is in her first trimester of pregnancy. She began the pregnancy at a normal, healthy weight, but her weight gain during pregnancy so far has been less than expected. Her favorite beverages are soft drinks; her favorite foods are French fries and boxed macaroni and cheese.
   b. A lactose-intolerant, 22-year-old pregnant woman has been eating a vegan diet for the past year or so. She began the pregnancy slightly underweight (BMI = 18.4), but her weight gain has been adequate and consistent during the four months of her pregnancy. She complains of feeling tired all the time.

2. What information would you give to a pregnant woman who is considering breastfeeding her infant, but isn't quite sure why she should?

Self Check

1. The most important single predictor of an infant's future health and survival is:
   a. the infant's birth weight.
   b. the infant's iron status at birth.
   c. the mother's weight at delivery.
   d. the mother's prepregnancy weight.

2. A mother's prepregnancy nutrition is important to a healthy pregnancy because it determines the development of:
   a. the largest baby possible.
   b. adequate maternal iron stores.
   c. an adequate fat supply for the mother.
   d. healthy support tissues—the placenta, amniotic sac, umbilical cord, and uterus.

3. A pregnant woman needs an extra 450 calories above the allowance for nonpregnant women during which trimester(s)?
   a. first
   b. second
   c. third
   d. first, second, and third

4. Two nutrients needed in large amount during pregnancy for rapid cell proliferation are:
   a. vitamin B_{12} and vitamin C.
   b. calcium and vitamin B_6.
   c. folate and vitamin B_{12}.
   d. copper and zinc.

5. For a woman who is at the appropriate weight for height and is carrying a single fetus, the recommended weight gain during pregnancy is:
   a. 40 to 60 pounds.
   b. 25 to 35 pounds.
   c. 10 to 20 pounds.
   d. 20 to 40 pounds.

6. Rewards of physical activity during pregnancy may include:
   a. weight loss.
   b. decreased incidence of pica.
   c. relief from morning sickness.
   d. reduced stress and easier labor.

7. During pregnancy, the combination of high blood pressure, protein in the urine, and edema signals:
   a. jaundice.
   b. preeclampsia.
   c. gestational diabetes.
   d. gestational hypertension.

8. Which of the following preventative measures should pregnant women take to avoid contracting listeriosis?
   a. avoid feta cheese
   b. avoid pasteurized milk
   c. thoroughly heat hot dogs
   d. (a) and (c)

9. To facilitate lactation, a mother needs:
   a. about 5000 kcalories a day.
   b. adequate nutrition and fluid intake.
   c. vitamin and mineral supplements.
   d. a glass of wine or beer before each feeding.

10. A woman who breastfeeds her infant should drink plenty of water to:
    a. produce more milk.
    b. suppress lactation.
    c. prevent dehydration.
    d. dilute nutrient concentrations.

Answers to these questions can be found in Appendix H.

Notes

1. Recommendations to improve preconception health and health care—United States, *Morbidity and Mortality Weekly Report* 55 (2006): 1–23.

2. R. L. Goldenberg and J. F. Culhane, Low birth weight in the United States, *American Journal of Clinical Nutrition* 85 (2007): 584S-590S; M. S. Kramer, The epidemiology of adverse pregnancy outcomes: An overview, *Journal of Nutrition* 133 (2003): 1592–1596.

3. L. Adair and D. Dahly, Developmental determinants of blood pressure in adults, *Annual Review of Nutrition* 25 (2005): 407–434; M. Hanson and coauthors, Report on the 2nd World Congress on Fetal Origins of Adult Disease, Brighton, U.K., June 7–10, 2003, *Pediatric Research* 55 (2004): 894–897; G. Wu and coauthors, Maternal nutrition and fetal development, *Journal of Nutrition* 134 (2004): 2169–2172; C. M. Law and coauthors, Fetal, infant, and childhood growth and adult blood pressure: A longitudinal study from birth to 22 years of age, *Circulation* 105 (2002): 1088–1092.

4. M. Hack and coauthors, Outcomes in young adulthood for very-low-birth weight infants, *New England Journal of Medicine* 346 (2002): 149–157.

5. B. E. Hamilton and coauthors, Annual summary of vital statistics: 2005, *Pediatrics* 119 (2007): 345–360.

6. Kramer, 2003.

7. U. Ramakrishnan, Nutrition and low birth weight: From research to practice, *American Journal of Clinical Nutrition* 79 (2004): 17–21.

8. J. C. King, Maternal obesity, metabolism, and pregnancy outcomes, *Annual Review of Nutrition* 26 (2006): 271–291; M. L. Watkins and coauthors, Maternal obesity and risk for birth defects, *Pediatrics* 111 (2003): 1152–1158.

9. King, 2006.

10. King, 2006; Position of the American Dietetic Association: Nutrition and lifestyle for a healthy pregnancy outcome, *Journal of the American Dietetic Association* 102 (2002): 1479–1490.

11. T. Henriksen, Nutrition and pregnancy outcomes, *Nutrition Reviews* 64 (2006): S19–S23; Watkins and coauthors, 2003.

12. J. C. Cross and L. Mickelson, Nutritional influences on implantation and placental development, *Nutrition Reviews* 64 (2006): S12–S18; M. C. Lacroix and coauthors, Placental growth hormones, *Endocrine* 1 (2002): 73–79.

13. C. Yajnik, Nutritional control of fetal growth, *Nutrition Reviews* 64 (2006): S50–S51; L. Adair and D. Dahly, Developmental determinants of blood pressure in adults, *Annual Review of Nutrition* 25 (2005): 407–434; C. A. Singhal and coauthors, Programming of lean body mass: A link between birth weight, obesity, and cardiovascular disease? *American Journal of Clinical Nutrition* 77 (2003): 726–730; B. E. Birgisdottir and coauthors, Size at birth and glucose intolerance in a relatively genetically homogeneous, high-birth weight population, *American Journal of Clinical Nutrition* 76 (2002): 399–403.

14. M. F. Picciano, Pregnancy and lactation: Physiological adjustments, nutritional requirements and the role of dietary supplements, *Journal of Nutrition* 133 (2003): 1997S–2002S.

15. Standing Committee on the Scientific Evaluation of Dietary Reference Intakes, Food and Nutrition Board, Institute of Medicine, *Dietary Reference Intakes for Energy, Carbohydrate, Fiber, Fat, Fatty Acids, Cholesterol, Protein, and Amino Acids* (Washington, DC: National Academies Press, 2005), pp. 185–194.

16. R. Uauy and A. D. Dangour, Nutrition in brain development and aging: Role of essential fatty acids, *Nutrition Reviews* 64 (2006): S24–S33; W. C. Heird and A. Lapillonne, The role of essential fatty acids in development, *Annual Review of Nutrition* 25 (2005): 549–571.

17. T. J. Mathews, Trends in spina bifida and anencephalus in the United States, 1991–2001, *National Center for Health Statistics Health and Stats*, www.cdc.gov/nchs/products/pubs/pubd/hestats/spine-anen.htm, site visited February 28, 2006.

18. Centers for Disease Control and Prevention, Spina bifida and anencephaly before and after folic acid mandate—United States, 1995–1996 and 1999–2000, *Morbidity and Mortality Weekly Report* 53 (2004): 362–365; E. A. Yetley and J. I. Rader, Modeling the level of fortification and post-fortification assessments: U.S. experience, *Nutrition Reviews* 62 (2004): S50–S59.

19. A. J. Wilcox and coauthors, Folic acid supplements and risk of facial clefts: National population based case-control study, *British Medical Journal* 334 (2007): 464; L. B. Bailey and R. J. Berry, Folic acid supplementation and the occurrence of congenital heart defects, orofacial clefts, multiple births, and miscarriage, *American Journal of Clinical Nutrition* 81 (2005): 1213S–1217S.

20. K. O'Brien, Maternal and fetal mineral partitioning: Whose needs predominate? *Nutrition Today* 40 (2005): 130–137.

21. K. O'Brien and coauthors, Maternal iron status influences iron transfer to the fetus during the third trimester of pregnancy, *American Journal of Clinical Nutrition* 77 (2003): 924–930.

22. T. O. Scholl, Iron status during pregnancy: Setting the stage for mother and infant, *American Journal of Clinical Nutrition* 81 (2005): 1218S–1222S.

23. C. M. Donangelo and coauthors, Zinc absorption and kinetics during pregnancy and lactation in Brazilian women, *American Journal of Clinical Nutrition* 82 (2005): 118–124.

24. D. Shah and H. P. S. Sachdev, Zinc deficiency in pregnancy and fetal outcome, *Nutrition Reviews* 64 (2006): 15–30.

25. Position of the American Dietetic Association, 2002.

26. M. M. Black and coauthors, Special Supplemental Nutrition Program for Women, Infants, and Children participation and infants' growth and health: A multisite surveillance study, *Pediatrics* 114 (2004): 169–176.

27. A. S. Ryan and W. Zhou, Lower breastfeeding rates persist among the Special Supplemental Nutrition Program for Women, Infants, and Children participants, 1978-2003, *Pediatrics* 117 (2006): 1136-1146; V. Oliveira and M. Prell, Sharing the economic burden: Who pays for WIC's infant formula? U.S. Department of Agriculture Economic Research Service, *Amber Waves*, available at www.ers.usda.gov/AmberWaves/Septmeber04/Features/infantfomula.htm.

28. Black and coauthors, 2004.

29. Position of the American Dietetic Association, 2002.

30. R. Artal and M. O'Toole, Guidelines of the American College of Obstetricians and Gynecologists for exercise during pregnancy and the postpartum period, *British Journal of Sports Medicine* 37 (2003): 6–12; American College of Obstetricians and Gynecologists, Exercise during pregnancy and postpartum period, ACOG Committee Opinion no. 267, *Obstetrics and Gynecology* 99 (2002): 171–173.

31. R. W. Corbett, C. Ryan, and S. P. Weinrich, Pica in pregnancy: Does it affect pregnancy outcomes? *American Journal of Maternal and Child Nursing* 28 (2003): 183–189.

32. R. J. Kaaja and I. A. Greer, Manifestations of chronic disease during pregnancy, *Journal of the American Medical Association* 294 (2005): 2751–2757; Position statement from the American Diabetes Association: Gestational diabetes mellitus, *Diabetes Care* 26 (2003): S103–S105.

33. C. A. Crowther and coauthors, Effect of treatment of gestational diabetes mellitus on pregnancy outcomes, *New England Journal of Medicine* 352 (2005): 2477–2486.

34. Position statement from the American Diabetes Association, 2003.

35. American Diabetes Association, Position statement, Diagnosis and classification of diabetes mellitus, *Diabetes Care* 29 (2006): S43–S48.

36. C. G. Solomon and E. W. Seely, Preeclampsia—searching for the cause, *New England Journal of Medicine* 350 (2004): 641–642.

37. R. A. de la Chica and coauthors, Chromosomal instability in amniocytes from fetuses of mothers who smoke, *Journal of the American Medical Association* 293 (2005): 1212–1222.

38. G. P. Bogler and L. T. Kozlowski, Differential influence of maternal smoking on infant birth weight: Gene-environment interaction and targeted intervention, *Journal of the American Medical Association* 287 (2002): 241–242.

39. American Academy of Pediatrics, Task Force on Sudden Infant Death Syndrome, The changing concept of sudden infant death syndrome: Diagnostic coding shifts, controversies regarding the sleeping environment, and new variables to consider in reducing risk, *Pediatrics* 116 (2005): 1245–1255; Centers for Disease Control and Prevention, Smoking during pregnancy—United States, 1990–2002, *Morbidity and Mortality Weekly Report* 53 (2004): 911–915.

40. J. R. DiFranza, C. A. Aligne, and M. Weitzman, Prenatal and postnatal environmental tobacco smoke exposure and children's health, *Pediatrics* 113 (2004): 1007–1015; K. I. McMartin and coauthors, Lung tissue concentrations of nicotine in sudden infant death syndrome (SIDS), *Journal of Pediatrics* 140 (2002): 205–209.

41. Centers for Disease Control and Prevention, Smoking during pregnancy—United States, 2004; S. J. Ventura and coauthors, Trends and variations in smoking during pregnancy and low birth weight: Evidence from the birth certificate, 1990–2000, *Pediatrics* 111 (2003): 1176–1180.

42. Position of the American Dietetic Association, 2002.

43. L. T. Singer and coauthors, Cognitive outcomes of preschool children with prenatal cocaine exposure, *Journal of the American Medical Association* 291 (2004): 2448–2456; L. T. Singer and coauthors, Cognitive and motor outcomes of cocaine-exposed infants, *Journal of the American Medical Association* 287 (2002): 1952–1960; E. S. Bandstra and coauthors, Intrauterine growth of full-term infants: Impact of prenatal cocaine exposure, *Pediatrics* 108 (2001): 1309–1319.

44. A. Gomaa and coauthors, Maternal bone lead as an independent risk factor for fetal neurotoxicity: A prospective study, *Pediatrics* 110 (2002): 110–118.

45. S. E. Schober and coauthors, Blood mercury levels in US children and women of childbearing age, 1999-2000, *Journal of the American Medical Association* 289 (2003): 1667–1674.

46. J. R. Hibbeln and coauthors, Maternal seafood consumption in pregnancy and neurodevelopmental outcomes in childhood (ALSPAC study): an observational cohort study, *Lancet* 369 (2007): 578-585; D. Mozaffarian and E. B. Rimm, Fish intake, contaminants, and human health: Evaluating the risks and the benefits, *Journal of the American Medical Association* 296 (2006): 1885–1899; Institute of Medicine report brief, *Seafood Choices: Balancing Benefits and Risks*, October, 2006.

47. Center for the Evaluation of Risks to Human Reproduction, http://cerhr.niehs.nih.gov/genpub/topics/listeria-ccae.html.

48. Position of the American Dietetic Association: Use of nutritive and nonnutritive sweeteners, *Journal of the American Dietetic Association* 104 (2004): 255–275.

49. E. Hey, Coffee and pregnancy (editorial), *British Medical Journal* 334 (2007): 377; A. Leviton and L. Cowan, A review of the literature relating caffeine consumption by women to their risk of reproductive hazards, *Food and Chemical Toxicology* 40 (2002): 1271–1310.

50. E. Hey, Coffee and pregnancy (editorial), *British Medical Journal* 334 (2007): 377; A. Leviton and L. Cowan, A review of the literature relating caffeine consumption by women to their risk of reproductive hazards, *Food and Chemical Toxicology* 40 (2002): 1271–1310.

51. M. L. Brownie, Maternal exposure to caffeine and risk of congenital anomalies: A systematic review, *Epidemiology* 17 (2006): 323–333.

52. B. H. Bech and coauthors, Effect of reducing caffeine intake on birth weight and length of gestation: Randomised controlled trial, *British Medical Journal* 334 (2007): 409.

53. A. Matijasevich and coauthors, Maternal caffeine consumption and fetal death: A case-control study in Uruguay, *Paediatric and Perinatal Epidemiology* 20 (2006): 100–109; B. H. Bech and coauthors, Coffee and fetal death: a cohort study with prospective data, *American Journal of Epidemiology* 162 (2005): 983–990.

54. H. E. Hoyme and coauthors, A practical approach to diagnosis of fetal alcohol spectrum disorders: Clarification of the 1996 Institute of Medicine criteria, *Pediatrics* 115 (2005): 39–47.

55. Centers for Disease Control, Fetal alcohol syndrome, www.cdc.gov/ncbddd/fas/fasask.htm, site visited February 28, 2006.

56. J. Tsai and R. L. Floyd, Alcohol consumption among women who are pregnant or who might become pregnant—United States, 2002, *Morbidity and Mortality Weekly Report* 53 (2004): 1178–1179.

57. T. S. Naimi and coauthors, Binge drinking in the preconception period and the risk of unintended pregnancy: Implications for women and their children, *Pediatrics* 111 (2003): 1136–1141.

58. U. Kesmodel and coauthors, Moderate alcohol intake during pregnancy and the risk of stillbirth and death in the first year of life, *American Journal of Epidemiology* 155 (2002): 305–312.

59. Committee on Substance Abuse and Committee on Children with Disabilities, American Academy of Pediatrics, Fetal alcohol syndrome and alcohol-related neurodevelopmental disorders, *Pediatrics* 106 (2000): 358–361.

60. Committee on Substance Abuse and Committee on Children with Disabilities, 2000.

61. J. D. Klein and the Committee on Adolescence, American Academy of Pediatrics, Adolescent pregnancy: Current trends and issues, *Pediatrics* 116 (2005): 281–286.

62. D. S. Elfenbein and M. E. Felice, Adolescent pregnancy, *Pediatric Clinics of North America* 50 (2003): 781–800, viii.

63. D. J. Hunt and coauthors, Effects of nutrition education programs on anthropometric measurements and pregnancy outcomes of adolescents, *Journal of the American Dietetic Association* 102 (2002): S100–S102.

64. American Academy of Pediatrics, Policy statement: Breastfeeding and the use of human milk, *Pediatrics* 115 (2005): 496–506.

65. Position of the American Dietetic Association: Promoting and supporting breastfeeding, *Journal of the American Dietetic Association* 105 (2005): 811–818.

66. American Academy of Pediatrics, 2005; Position of the American Dietetic Association, 2005.

67. G. Kac and coauthors, Breastfeeding and postpartum weight retention in a cohort of Brazilian women, *American Journal of Clinical Nutrition* 79 (2004): 487–493.

68. K. O'Brien and coauthors, Bone calcium turnover during pregnancy and lactation in women with low calcium diets is associated with calcium intake and circulating insulin-like growth factor 1 concentrations, *American Journal of Clinical Nutrition* 83 (2006): 317–323.

69. F. F. Bezerra and coauthors, Bone mass is recovered from lactation to postweaning in adolescent mothers with low calcium intakes, *American Journal of Clinical Nutrition* 80 (2004): 1322–1326; L. M. Paton and coauthors, Pregnancy and lactation have no long-term deleterious effect on measures of bone mineral in healthy women: A twin study, *American Journal of Clinical Nutrition* 79 (2003): 707–714.

70. American Academy of Pediatrics, 2005.

71. S. Ito and A. Lee, Drug excretion into breast milk—overview, *Advanced Drug Delivery Reviews* 55 (2003): 617–627; Committee on Drugs, American Academy of Pediatrics, The transfer of drugs and other chemicals into human milk, *Pediatrics* 108 (2001): 776–789.

72. American Academy of Pediatrics, 2005.

ENCOURAGING SUCCESSFUL BREASTFEEDING

As discussed in Chapter 11, breastfeeding offers benefits to both mother and infant. The American Academy of Pediatrics (AAP), the American Dietetic Association, and the U.S. Department of Health and Human Services all advocate breastfeeding as the preferred means of infant feeding.[1] Promotion of breastfeeding is an integral part of the WIC program's nutrition education component. During the late 1980s, breastfeeding was on the decline after reaching a high of about 60 percent of mothers in 1984. National efforts to promote breastfeeding seem to be working, at least to some extent: an encouraging trend of breastfeeding is emerging, with more than 70 percent of mothers initiating breastfeeding in 2003.[2] Nevertheless, only about one in three infants is still being breastfed at six months of age. The AAP and many other health organizations recommend exclusive breastfeeding for the first six months of life.[3] Exclusive breastfeeding is defined as an infant's consumption of human milk with no supplementation of any kind (no water, no other type of milk, no juice, and no other foods), except for vitamins, minerals, and medications. The AAP recommends that breastfeeding continue for at least one year and thereafter for as long as mutually desired.[4] In the United States, only about one in five infants is still breastfeeding at one year of age. Increasing the rates of breastfeeding initiation and duration is one of the goals of *Healthy People 2010*:

> Increase the proportion of mothers who breastfeed immediately after birth, for the first six months, and preferably, through the infant's first year of life. Increase the proportion of mothers who breastfeed exclusively.[5]

Despite the trend toward increasing breastfeeding, the percentage of mothers choosing to breastfeed their infants and continuing to do so still falls short of goals.

Why don't more women choose to breastfeed their infants?

Many experts cite two major deterrents: public advertising encouraging the use of infant formula and the medical community's failure to encourage breastfeeding. As an example of the medical lack of encouragement, some hospitals routinely separate mother and infant soon after birth. The child's first feeding then comes from the bottle rather than the breast. Furthermore, many hospitals send new mothers home with free samples of infant formula. The World Health Organization opposes this practice because it sends a misleading message that medical authorities favor infant formula over breast milk for infants. Even in hospitals where women are encouraged to breastfeed and are supported in doing so, little, if any, assistance is available after hospital discharge when many breastfeeding women still need assistance. Up to half of mothers who initially breastfeed their infants stop within one month—seemingly due to lack of knowledge.

Women who receive early and repeated breastfeeding information and support breastfeed their infants longer than other women do.[6] Information and instruction are especially important during the *prenatal* period when most women decide whether to breastfeed or to feed formula. Nurses and other health care professionals can play a crucial role in encouraging successful breastfeeding by offering women adequate, accurate information about breastfeeding that permits them to make informed choices. Table NP11–1 lists ten steps maternity facilities and health care professionals can take to promote successful breastfeeding among new mothers.

If breastfeeding is a natural process, what do mothers need to learn?

Although lactation is an automatic physiological process, breastfeeding requires some learning. This learning is most successful in a supportive environment. It begins with preparatory steps taken before the infant is born.

What are these preparatory steps?

Toward the end of pregnancy and throughout lactation, a woman who intends to breastfeed should stop using soap and lotions on her breasts. The natural secretions of the breasts themselves lubricate the nipple area best. A woman who plans to breastfeed should also acquire at least two nursing bras before her infant is born. The bras should provide good support and have drop-flaps so that either breast can be freed for nursing.

TABLE NP11–1

Ten Steps to Successful Breastfeeding

To promote breastfeeding, every maternity facility should:

- Develop a written breastfeeding policy that is routinely communicated to all health care staff.
- Train all health care staff in the skills necessary to implement the breastfeeding policy.
- Inform all pregnant women about the benefits and management of breastfeeding.
- Help mothers initiate breastfeeding within half an hour of birth.
- Show mothers how to breastfeed and how to maintain lactation, even if they need to be separated from their infants.
- Give newborn infants no food or drink other than breast milk, unless medically indicated.
- Practice rooming-in, allowing mothers and infants to remain together 24 hours per day.
- Encourage breastfeeding on demand.
- Give no artificial nipples or pacifiers to breastfeeding infants.
- Foster the establishment of breastfeeding support groups and refer mothers to them at discharge from the facility.

Sources: U.S. Department of Health and Human Services, Office on Women's Health, *Breastfeeding: HHS Blueprint for Action on Breastfeeding* (Washington, DC: U.S. Department of Health and Human Services, 2000).

How soon after birth should breastfeeding start?

As soon as possible. Immediately after the delivery, for a short period, the infant is intensely alert and intent on suckling. This is the ideal time for the first breastfeeding and facilitates successful lactation.

What does the new mother need to know in order to continue breastfeeding her infant successfully?

She needs to learn how to relax and position herself so that she and the infant will be comfortable and the infant can breathe freely while nursing. She also needs to understand that infants have a **rooting reflex** that makes them turn toward any touch on the face. (The accompanying glossary defines this and other relevant terms.) Consequently, she should touch the infant's cheek to her nipple so that the infant will turn the right way to nurse. The mother can then place four fingers under the breast and her thumb on top to support the breast and present the nipple to the infant. The mother's fingers and thumb should be behind the areola, the colored ring around the nipple so as not to interfere with the infant latching onto the breast (see Figure NP11–1). With the breast supported, the mother tickles the infant's lips with the breast until the infant's mouth opens wide. The mother can then gently bring the infant forward onto the breast. The nipple must rest well back on the infant's tongue so that the infant's gums will squeeze on the glands that release the milk and swallowing will be effortless. To break the suction, if necessary, the mother can slip a finger between the infant's mouth and her breast.

Does it hurt to have the infant sucking so hard on the breast?

Breastfeeding should not be painful if the infant is positioned correctly. The mother has a **letdown reflex** that forces milk to the front of her breast when the infant begins to nurse, virtually propelling the milk into the infant's mouth. Letdown is necessary for the infant to obtain milk easily, and the mother needs to relax for letdown to occur. The mother who assumes a comfortable position in an environment without interruptions will find it easiest to relax.

How long should the infant be allowed to nurse at each feeding?

Although the infant sucks half the milk from the breast within the first 2 minutes and 80 to 90 percent of it within 4 minutes,

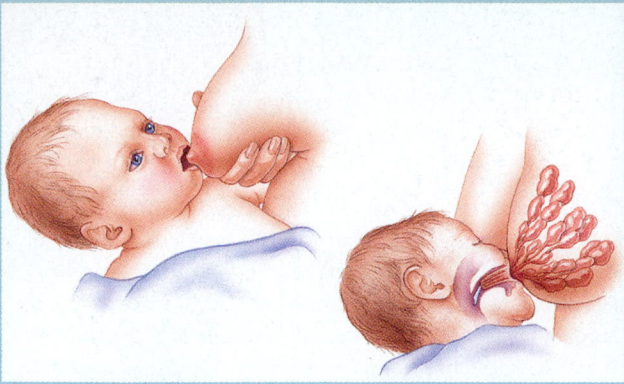

FIGURE NP11–1

Infant's Grasp on Mother's Breast
The mother supports the breast with her fingers and thumb behind the areola to present the nipple to the infant. Once the infant latches onto the breast, the infant's lips and gums pump the areola, releasing milk from the mammary glands into the milk ducts that lie beneath the areola.

sucking on each breast for 10 to 15 minutes is encouraged. The sucking itself, as well as the complete removal of milk from the breast, stimulates the mammary glands to produce milk for the next nursing session. Successive sessions should start on alternate breasts to ensure that each breast is emptied regularly. This pattern maintains the same supply and demand for each breast and thus prevents either breast from overfilling.

Infants should be fed "on demand" and not be held to a rigid schedule. The breastfed infant may average 8 to 12 feedings per 24-hour period during the first month or so. Once the mother's milk supply is well established and the infant's capacity has increased, the intervals between feedings will become longer.

What if a mother wants to skip one or two feedings daily—for example, because she works outside the home?

The mother can express breast milk into a bottle ahead of time, freeze the breast milk, and, when needed, substitute the expressed breast milk for a nursing session. Breast milk can be kept refrigerated for 48 hours or frozen. Frozen milk can be kept for one month in a freezer attached to a refrigerator or for three to six months in a zero-degree deep freezer.[7]

The mother can hand express her breast milk or use one of several different breast pumps available. The bicycle-horn type of manual breast pump is difficult to keep clean and is not recommended. Cylinder-type manual pumps or electric breast pumps are safer and are also more efficient. Alternatively, a mother can substitute formula for those missed feedings and continue to breastfeed at other times.

What about problems associated with breastfeeding such as sore nipples or infection of the breast?

Most problems associated with breastfeeding can be resolved. Many mothers experience sore nipples during the initial days of breastfeeding. Sore nipples need to be treated kindly, but nursing can continue. Improper feeding position is a frequent cause

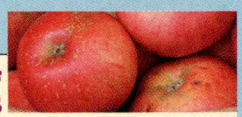

GLOSSARY OF BREASTFEEDING TERMS

engorgement: overfilling of the breasts with milk.

letdown reflex: the reflex that forces milk to the front of the breast when the infant begins to nurse.

mastitis: infection of a breast.

rooting reflex: a reflex that causes an infant to turn toward whichever cheek is touched in search of a nipple.

of sore nipples: the mother should make sure the infant is taking the entire nipple and part of the areola onto the tongue. She should nurse on the less sore breast first to get letdown going while the infant is sucking hardest; then she can switch to the sore breast. Between times, she should expose her nipples to light and air to heal them.

Before lactation is well established, when the schedule changes, or when a feeding is missed, the breasts may become full and hard—an uncomfortable condition known as **engorgement**. The infant cannot grasp an engorged nipple and so cannot provide relief by nursing. A gentle massage or warming the breasts with a cloth soaked in warm water or in a shower helps to initiate letdown and to release some of the accumulated milk; then the mother can pump out some of her milk and allow the infant to nurse.

Infection of the breast, known as **mastitis**, is best managed by continuing to breastfeed. By drawing off the milk, the infant helps to relieve pressure in the infected area. The infant is safe because the infection is between the milk-producing glands, not inside them.

Even if everything is going smoothly, the nursing mother should ideally have enough help and support so that she can rest in bed a few hours each day for the first week or so. Successful breastfeeding requires the support of all those who care.[8] This, plus adequate nutrition, ample fluids, fresh air, and physical activity, will do much to enhance the well-being of mother and infant.

Notes

1. American Academy of Pediatrics, Policy statement: Breastfeeding and the use of human milk, *Pediatrics* 115 (2005): 496–506; Position of the American Dietetic Association: Promoting and supporting breastfeeding, *Journal of the American Dietetic Association* 105 (2005): 811–818; U.S. Department of Health and Human Services, Office on Women's Health, *Breastfeeding: HHS Blueprint for Action on Breastfeeding* (Washington, DC: U.S. Department of Health and Human Services, 2000).

2. Centers for Disease Control and Prevention, 2003 National Immunization Survey, available at www.cdc.gov/breastfeeding/NIS_data, site visited January 16, 2006.

3. American Academy of Pediatrics, 2005.

4. American Academy of Pediatrics, 2005.

5. *Healthy People 2010: National Health Promotion and Disease Prevention Objectives* (Washington, DC: U.S. Department of Health and Human Services, 2000).

6. A. Ekström and E. Nissen, A mother's feelings for her infant are strengthened by excellent breastfeeding counseling and continuity of care, *Pediatrics* 118 (2006): e309–e314; I. B. Ahluwalia, B. Morrow, and J. Hsia, Why do women stop breastfeeding? Findings from the Pregnancy Risk Assessment and Monitoring System, *Pediatrics* 116 (2005): 1408–1412; J. Laberere and coauthors, Efficacy of breastfeeding support provided by trained clinicians during an early, routine, preventive visit: A prospective, randomized, open trial of 226 mother-infant pairs, *Pediatrics* 115 (2005): e139–e146.

7. Breastfeeding beyond infancy, in *American Academy of Pediatrics, New Mother's Guide to Breastfeeding*, eds. J. Y. Meek and S. Tippins (New York: Bantam Books, 2002), pp. 158–188.

8. A. Pisacane and coauthors, A controlled trial of the father's role in breastfeeding promotion, *Pediatrics* 116 (2005): e494–e498.

12 Nutrition Through the Life Span: Infancy, Childhood, and Adolescence

© Rubberball/Jupiter Images

337

Nutrient needs change throughout life, depending on rates of growth, activity, and many other factors. Nutrient needs also vary from individual to individual, but generalizations are possible and useful. The first year of life is a time of phenomenal growth and development. After the first year, a child continues to grow and change, but more slowly. Sound nutrition throughout infancy and childhood promotes normal growth and development; facilitates academic and physical performance; and helps prevent obesity, diabetes, heart disease, cancer, and other degenerative diseases in adulthood. As children enter the teen years, a foundation built by years of eating nutritious foods best prepares them to meet the upcoming demands of rapid growth. This chapter examines the special nutrient needs of infants, children, and adolescents.

Nutrition of the Infant

Early nutrition affects later development, and early feeding sets the stage for eating habits that will influence nutrition status for a lifetime. Trends change, and experts argue about the fine points, but properly nourishing an infant is relatively simple, overall. Common sense in the selection of infant foods and a nurturing, relaxed environment go far to promote an infant's health and well-being.

In developed countries with well-nourished populations, such as the United States and Canada, the dietary practices that have the most influence on an infant's nutrition status are the type of milk the infant receives and the age at which solid foods are introduced. The first part of this chapter is devoted to feeding the infant and identifying the nutrients most often deficient in infant diets.

Nutrient Needs during Infancy

An infant grows faster during the first year than ever again, as Figure 12–1 shows. The growth of infants and children directly reflects their nutritional well-being and is an important parameter in assessing their nutrition status. Health care professionals use growth charts to evaluate the growth and development of children from birth to 20 years of age (see Appendix E).

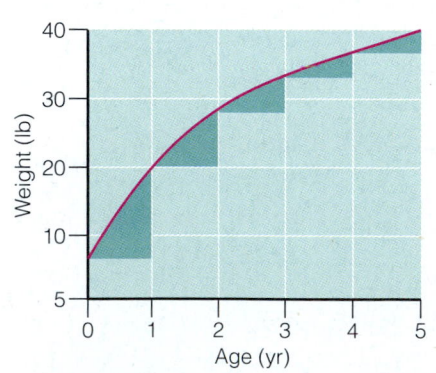

FIGURE 12–1

Weight Gain of Human Infants in Their First Five Years of Life

In the first year, an infant's birth weight may triple, but over the following several years, the rate of weight gain gradually diminishes.

Nutrients to Support Growth An infant's birth weight doubles by about four to six months of age and triples by the age of one year. (Consider that if an adult, starting at 120 pounds, were to do this, the person's weight would increase to 360 pounds in a single year.) The infant's length changes more slowly than weight, increasing about 10 inches from birth to one year. By the end of the first year, the growth rate slows considerably. An infant typically gains less than 10 pounds during the second year and grows about 5 inches in height.

Not only do infants grow rapidly, but in proportion to body weight, their basal metabolic rate is remarkably high—about twice that of an adult. The rapid growth and metabolism of the infant demand an ample supply of all the nutrients. Of special importance during infancy are the energy nutrients and the vitamins and minerals critical to the growth process, such as vitamin A, vitamin D, and calcium.

Because they are small, infants need smaller *total* amounts of these nutrients than adults do, but as a percentage of body weight, infants need more than twice as much of most nutrients. Infants require about 100 kcalories per kilogram of body weight per day; most adults require fewer than 40 (see Table 12–1). Figure 12–2 compares a five-month-old infant's needs (per unit

TABLE 12–1

Infant and Adult Heart Rate, Respiration Rate, and Energy Needs Compared

	Infants	Adults
Heart rate (beats/minute)	120 to 140	70 to 80
Respiration rate (breaths/minute)	20 to 40	15 to 20
Energy needs (kcal/body weight)	45/lb (100/kg)	<18/lb (<40/kg)

of body weight) with those of an adult man. Some of the differences are extraordinary. Around six months of age, energy needs begin to increase less rapidly as the growth rate begins to slow, but some of the energy saved by slower growth is spent on increased activity. When their growth slows, infants spontaneously reduce their energy intakes. Parents should expect their infants to adjust their food intakes downward when appropriate and should not force or coax them to eat more.

Vitamin K nutrition for newborns presents a unique case. A newborn's digestive tract is sterile, and vitamin K-producing bacteria take weeks to establish themselves in the infant's intestines. To prevent uncontrolled bleeding in the newborn, a single dose of vitamin K is given at birth.[1]

Water One of the most important nutrients for infants, as for everyone, is water. The younger a child is, the more of its body weight is water. Breast milk or infant formula

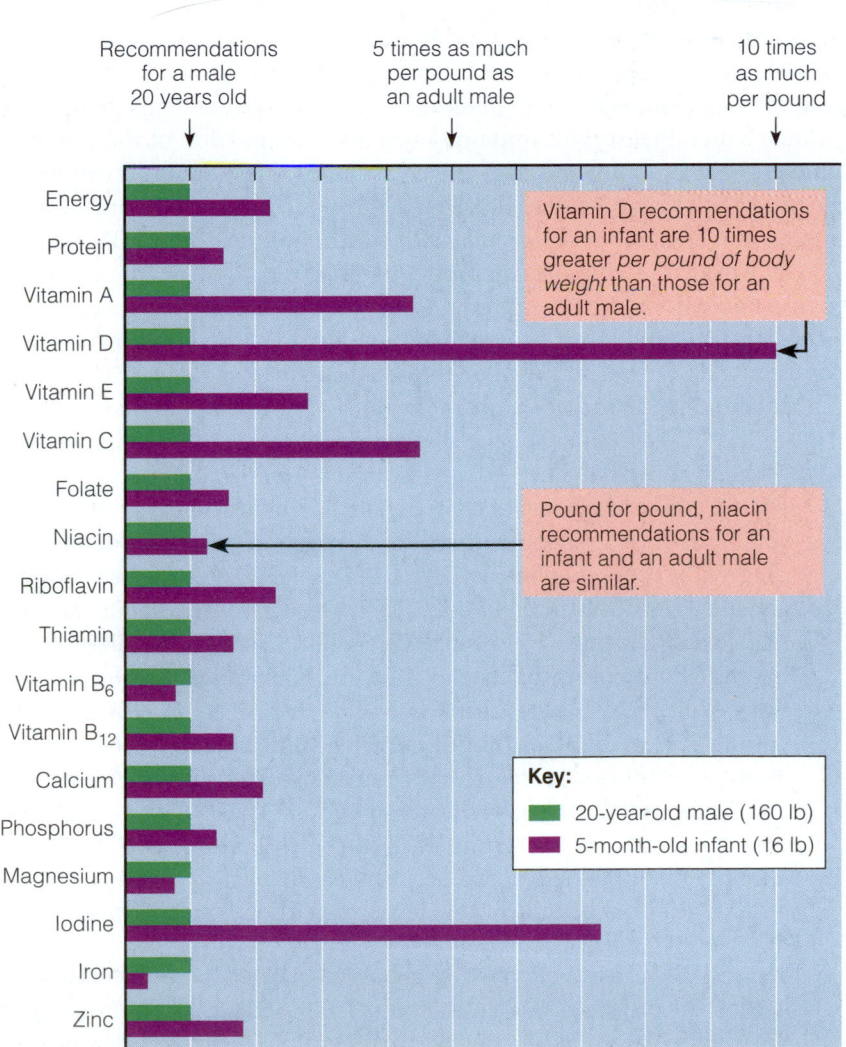

FIGURE 12–2

Nutrient Recommendations for a Five-Month-Old Infant and an Adult Male Compared on the Basis of Body Weight

Infants may be relatively small and inactive, but they use large amounts of energy and nutrients in proportion to their body size to keep all their metabolic processes going.

Recommendations for a male 20 years old

5 times as much per pound as an adult male

10 times as much per pound

Vitamin D recommendations for an infant are 10 times greater *per pound of body weight* than those for an adult male.

Pound for pound, niacin recommendations for an infant and an adult male are similar.

Energy, Protein, Vitamin A, Vitamin D, Vitamin E, Vitamin C, Folate, Niacin, Riboflavin, Thiamin, Vitamin B₆, Vitamin B₁₂, Calcium, Phosphorus, Magnesium, Iodine, Iron, Zinc

Key:
- 20-year-old male (160 lb)
- 5-month-old infant (16 lb)

After six months of age, the energy saved by slower growth is spent on increased activity.

normally provides enough water to replace fluid losses in a healthy infant. Even in hot, dry climates, neither breastfed nor formula-fed infants need supplemental water.[2] However, because proportionately more of an infant's body water than an adult's is *outside* the cells—between the cells and in the vascular space—rapid fluid losses and the resulting dehydration can be life-threatening. Conditions that cause rapid fluid loss, such as vomiting or diarrhea, require an electrolyte solution designed for infants.

IN SUMMARY

■ Infants' rapid growth and development depend on adequate nutrient supplies, including water from breast milk and formula.

Breast Milk

Breast milk excels as a source of nutrients for the young infant. With the possible exception of vitamin D (discussed later), breast milk provides all the nutrients a healthy infant needs for the first six months of life.[3] It provides many other health benefits as well.

Energy Nutrients The balance of energy nutrients in breast milk differs dramatically from the balance recommended for adults (see Figure 12–3). Yet, for infants, breast milk is the most nearly perfect food, proving that people at different stages of life have different nutrient needs.

The carbohydrate in breast milk (and standard infant formula) is lactose. In addition to being easily digested, lactose enhances calcium absorption.

The lipids in breast milk—and infant formula—provide the main source of energy in the infant's diet. Breast milk contains a generous proportion of the essential fatty acids linoleic acid and linolenic acid, as well as their longer-chain derivatives arachidonic acid and docosahexaenoic acid (DHA). Until recently, infant formula provided only linoleic and linolenic acid. Formula with arachidonic acid and DHA added is now commercially available.[4] Infants can produce some arachidonic acid and DHA from linoleic and linolenic acid, but some infants may need more than they can make.

Arachidonic acid and DHA are found abundantly in the developing brain and the retina of the eye. Research has focused on the visual and mental development of breastfed infants and infants fed standard formula without DHA and arachidonic acid added.[5] Breastfed infants generally score higher on tests of mental development than formula-fed infants do, and researchers are investigating whether this difference can be attributed to DHA and arachidonic acid in breast milk.[6] So far, results are mixed. In one study, researchers found no developmental or visual differences between infants fed standard formula and those fed formula with added DHA and arachidonic acid.[7] In two other studies, infants were given either standard formula or formula with added DHA and arachidonic acid. The infants fed formula fortified with DHA and arachidonic acid had better visual function at one year of age than those who were fed standard formula.[8]

The protein in breast milk is largely **alpha-lactalbumin**, a protein the human infant can easily digest. Another breast milk protein, **lactoferrin**, indirectly benefits the baby's iron nutrition and also acts as an antibacterial agent. Lactoferrin is an iron-gathering compound that helps absorb iron into the infant's bloodstream, keeps intestinal bacteria from getting enough iron to grow out of control, and also works directly to kill some bacteria.[9]

Vitamins and Minerals With the exception of vitamin D, the vitamin content of the breast milk of a well-nourished mother is ample. Even vitamin C, for which cow's milk is a poor source, is supplied generously. However, the

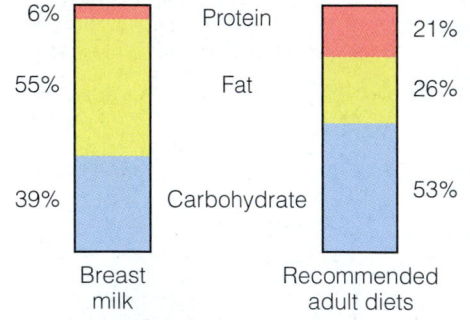

6%	Protein	21%
55%	Fat	26%
39%	Carbohydrate	53%
Breast milk		Recommended adult diets

FIGURE 12–3

Percentages of Energy-Yielding Nutrients in Breast Milk and in Recommended Adult Diets
The proportions of energy-yielding nutrients in human breast milk differ from those recommended for adults.[a]

[a] The values listed for adults represent approximate midpoints of the acceptable ranges for protein (10 to 35 percent), fat (20 to 35 percent), and carbohydrate (45 to 65 percent).

concentration of vitamin D in breast milk is low, and vitamin D deficiency impairs bone mineralization. Vitamin D deficiency is most likely in infants who are not exposed to sunlight regularly, have darkly pigmented skin, and receive breast milk without vitamin D supplementation.[10] Reports of infants in the United States developing the vitamin D–deficiency disease rickets and recommendations by the American Academy of Pediatrics to keep infants under six months of age out of direct sunlight have prompted updated vitamin D guidelines. The AAP now recommends a vitamin D supplement for all infants who are breastfed exclusively and for any infants who do not receive at least 500 milliliters (15 ounces) per day of vitamin D-fortified formula.[11]

As for minerals, the calcium content of breast milk is ideal for infant bone growth, and the calcium is well absorbed. Breast milk is also low in sodium. The limited amount of iron in breast milk is highly absorbable, and its zinc, too, is absorbed better than from cow's milk, thanks to the presence of a zinc-binding protein.

Supplements for Infants Pediatricians may prescribe supplements containing vitamin D, iron, and fluoride (after six months of age). Table 12–2 offers a schedule of supplements during infancy.

Immunological Protection Breast milk offers the infant unsurpassed protection against infection.[12] Protective factors include antiviral agents, antibacterial agents, and other infection inhibitors.

During the first two or three days of lactation, the breasts produce **colostrum**, a premilk substance containing antibodies and white cells from the mother's blood. Colostrum is relatively sterile as it leaves the breast, and the infant cannot contract a bacterial infection from it even if the mother has one. Colostrum contains maternal immune factors that inactivate harmful bacteria within the digestive tract. Later, breast milk also delivers immune factors, although not as many as colostrum. Among them are **bifidus factors** and lactoferrin.

Breast milk also contains several enzymes, several hormones (including thyroid hormone and prostaglandins), and lipids, all of which protect the infant against infection. Breastfed babies are less prone to develop stomach and intestinal disorders during the first few months of life and so experience less vomiting and diarrhea than formula-fed infants do. In fact, research shows that breast milk contains not only antibodies against the most common cause of diarrhea in infants and young children, but also the protein **lactadherin** which binds to and inhibits replication of the

alpha-lactalbumin (lackt-AL-byoo-min): the chief protein in human breast milk, as **casein** (CAY-seen) is the chief protein in cow's milk.

lactoferrin (lack-toe-FERR-in): a protein in breast milk that binds iron and keeps it from supporting the growth of the infant's intestinal bacteria.

colostrum (co-LAHS-trum): a milklike secretion from the breasts that is rich in protective factors. Colostrum is present during the first day or so after delivery, before milk appears.

bifidus (BIFF-id-us, by-FEED-us) **factors:** factors in colostrum and breast milk that favor the growth of the "friendly" bacterium *Lactobacillus* (lack-toe-ba-SILL-us) *bifidus* in the infant's intestinal tract. These bacteria prevent other, less desirable intestinal inhabitants from flourishing.

lactadherin (lack-tad-HAIR-in): a protein in breast milk that attacks diarrhea-causing viruses.

TABLE 12–2

Supplements for Full-Term Infants

	Vitamin D[a]	Iron[b]	Fluoride[c]
Breastfed infants:			
Birth to six months of age	✓	—	—
Six months to one year	✓	✓	✓
Formula-fed infants:			
Birth to six months of age	—	—	—
Six months to one year	—	✓	✓

[a] Vitamin D supplements are recommended for all infants who are exclusively breastfed and for any infants who do not receive at least 500 milliliters (15 ounces) of vitamin D–fortified formula.
[b] Infants four to six months of age need additional iron, preferably in the form of iron-fortified cereal for both breastfed and formula-fed infants and iron-fortified infant formula for formula-fed infants.
[c] The pediatrician's decision to recommend a fluoride supplement should be based on the total amount of fluoride from all sources available to a child daily.

Source: Adapted from Committee on Nutrition, American Academy of Pediatrics, *Pediatric Nutrition Handbook*, 5th ed., ed. R. E. Kleinman (Elk Grove Village, IL: American Academy of Pediatrics, 2004).

The infant thrives on infant formula offered with affection.

infective agent.*[13] Breastfeeding reduces the severity and duration of symptoms associated with this infection. Breastfeeding also protects against other common illnesses of infancy such as middle ear infection and respiratory illness.[14] Nutrition in Practice 11 offers suggestions for successful breastfeeding.

Other Potential Benefits Researchers are investigating whether breastfeeding may also help protect against obesity in childhood and later years, but so far results are inconsistent. For example, a well-controlled study of more than 15,000 adolescents and their mothers indicated that those who were mostly breastfed for the first six months of life were less likely to become overweight than those who were fed formula.[15] However, a study of much younger children (three to five years of age) found no clear evidence that breastfeeding influences body weight.[16] A review of more than sixty published studies investigating the relationship between infant feeding and obesity suggests that initial breastfeeding protects against obesity in later life.[17] The authors cautioned that further review of studies that control for confounding factors such as maternal obesity, smoking, age of assessment, and socioeconomic background is needed.

Breast milk may also offer protection against the development of cardiovascular disease. Compared with formula-fed infants, breastfed infants have lower blood cholesterol as adults.[18]

Breastfeeding may also have a positive effect on intelligence later in life.[19] In one study, young adults who had been breastfed as long as nine months scored higher on two different intelligence tests than those who had been breastfed less than one month. Many other studies suggest a beneficial effect of breastfeeding on intelligence, but when subjected to strict standards of methodology (for example, large sample size and appropriate intelligence testing), the evidence is less convincing.[20]

Infant Formula

Breastfeeding offers many benefits to both mother and infant, and it should be encouraged whenever possible. However, formula feeding offers an acceptable alternative to breastfeeding. The mother who feeds her infant formula can offer the same closeness, warmth, and stimulation during feedings as the breastfeeding mother can.

Many mothers choose to breastfeed at first but wean their children within the first 1 to 12 months. Before infants reach one year of age, mothers must wean them onto *infant formula,* ■ not onto plain cow's milk of any kind—whole, reduced fat, low fat, or fat-free.

Infant Formula Composition Manufacturers can prepare formulas from cow's milk in such a way that they do not differ significantly from human milk in nutrient content. Figure 12–4 illustrates the energy nutrient balance of both. Formulas contain no protective antibodies for human infants, but preventive medical care (vaccinations) and reliable public health measures (clean water) help minimize this disadvantage. The educated mother whose water supply is reliable can prepare safe, sanitary formulas. However, lead-contaminated water is a major source of lead poisoning in infants.

Infant Formula Standards National and international standards have been set for the nutrient contents of infant formulas. U.S. standards are based on AAP recommendations, and the FDA mandates quality control procedures to ensure that these standards are met. All standard formulas are therefore nutritionally similar. Small differences in nutrient content are sometimes confusing but usually unimportant.

Special Formulas Standard formulas may be inappropriate for some infants. For example, premature babies require special formulas. Infants allergic to milk

* The most common cause of diarrhea in the United States is rotavirus. More children are hospitalized for rotavirus infection than for any other single cause.

■ Formula preparation:
   ■ Liquid concentrate (moderately expensive, relatively easy)—mix with equal part water.
   ■ Powdered formula (least expensive, lightest for travel)—read label directions.
   ■ Ready-to-feed (easiest, most expensive)—pour directly into clean bottles.

	Breast milk		Infant formula
Protein	6%		9%
Fat	55%		49%
Carbohydrate	39%		42%

FIGURE 12–4

Percentages of Energy-Yielding Nutrients in Breast Milk and in Infant Formula
The average proportions of energy-yielding nutrients in human breast milk and formula differ slightly. In contrast, cow's milk provides too much protein and too little carbohydrate.

protein can drink special **hypoallergenic formulas** or formulas based on soy protein.[21] Soy formulas are lactose-free and so can be used for infants with lactose intolerance as well. They are also useful as an alternative to milk-based formulas for vegetarian families. For infants with other special needs, many other variations are available.

Risks of Formula Feeding In developing countries and in poor areas of the United States, formula may be unavailable, overdiluted in an attempt to save money, or prepared with contaminated water. Overdilution of formula can cause malnutrition and growth failure. Contaminated formula often causes infections leading to diarrhea, dehydration, and failure to absorb nutrients. Wherever sanitation is poor, breastfeeding should take priority over feeding formula. Breast milk is sterile, and its antibodies enhance an infant's resistance to disease.

Iron in Formula The AAP recommends iron-fortified formulas for all formula-fed infants.[22] Low-iron formulas have no role in infant feeding. Use of iron-fortified formulas has risen in recent decades and is credited with the decline of iron-deficiency anemia in U.S. infants.

Nursing Bottle Tooth Decay Dentists advise against putting an infant to bed with a bottle. Salivary flow, which normally cleanses the mouth, diminishes as the infant falls asleep. Sucking for long times pushes the jawline out of shape and causes a bucktoothed profile (protruding upper and receding lower teeth). Furthermore, prolonged sucking on a bottle of formula, milk, or juice bathes the upper teeth in a carbohydrate-rich fluid that nourishes decay-producing bacteria. (The tongue covers and protects most of the lower teeth, but they, too, may be affected.) The result can be extensive and rapid tooth decay. To prevent **nursing bottle tooth decay**, no child should be put to bed with a bottle as a pacifier.

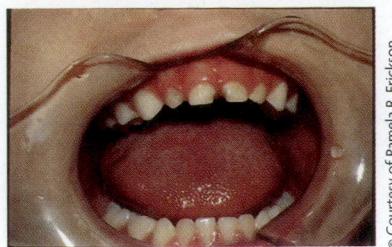

Nursing bottle syndrome in an early stage.

© Courtesy of Pamela R. Erickson

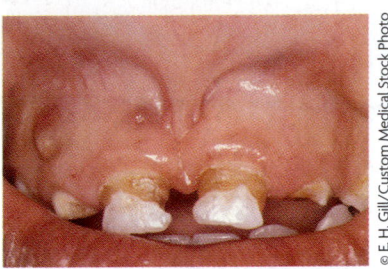

Nursing bottle syndrome, an extreme example. The upper teeth have decayed all the way to the gum line.

© E. H. Gill/Custom Medical Stock Photo

The Transition to Cow's Milk

The age at which whole cow's milk can be introduced to the infant's diet has long been a source of controversy. The AAP advises that whole cow's milk is not appropriate during the first year.[23] Children one to two years of age should not be given reduced-fat, low-fat, or fat-free milk routinely; they need the fat of whole milk. Between the ages of two and five years, a gradual transition from whole milk to the lower-fat milks can take place but care should be taken to avoid excessive restriction of dietary fat.

In some infants, particularly those younger than six months of age, whole cow's milk causes intestinal bleeding, which can lead to iron deficiency. Cow's milk is also a poor source of iron. Consequently, it both causes iron loss and fails to replace iron. Furthermore, the bioavailability of iron from infant cereal and other foods is reduced when cow's milk replaces breast milk or iron-fortified formula during the first year. Compared to breast milk or iron-fortified formula, cow's milk is higher in calcium and lower in vitamin C, characteristics that reduce iron absorption. Furthermore, the higher protein concentration of cow's milk can stress the infant's kidneys. In short, cow's milk is a poor choice during the first year of life; infants need breast milk or iron-fortified infant formula.

Introducing First Foods

Changes in the body organs during the first year affect the infant's readiness to accept solid foods. Until the child is several months old, the immature stomach and intestines

hypoallergenic formulas: clinically tested infant formulas that do not provoke reactions in 90 percent of infants or children with confirmed cow's milk allergy. Like all infant formulas, hypoallergenic formulas must demonstrate nutritional suitability to support infant growth and development. Extensively hydrolyzed and free amino acid–based formulas are examples.

nursing bottle tooth decay: extensive tooth decay due to prolonged tooth contact with formula, milk, fruit juice, or other carbohydrate-rich liquid offered to an infant in a bottle.

> **Dietary Guidelines for Americans 2005**
>
> ■ Children 2 to 8 years should consume 2 cups of fat-free or low-fat milk or equivalent milk products.

can digest milk sugar (lactose), but not starch. This is one of the many reasons why breast milk and formula are such good foods for an infant; they provide simple, easily digested carbohydrate that supplies energy for the infant's growth and activity.

When to Introduce Solid Food The AAP supports exclusive breastfeeding for approximately six months but recognizes that infants are often developmentally ready to accept complementary foods between four and six months of age.[24] Thus foods may be started gradually beginning sometime between four and six months, depending on the infant's readiness. Indications of readiness for solid foods include:

- The infant can sit with support and control head movements.

- The infant is six months old.

Table 12–3 presents a suggested sequence for introducing new foods.

The addition of foods to an infant's diet should be governed by three considerations: the infant's nutrient needs, the infant's physical readiness to handle different forms of foods, and the need to detect and control allergic reactions. With respect to nutrient needs, the nutrient needed earliest is iron, then vitamin C.

TABLE 12–3

Infant Development and Recommended Foods

Age (mo)	Feeding Skill	Foods Introduced into the Diet
0–4	Turns head toward any object that brushes cheek.	Feed breast milk or infant formula.
	Initially swallows using back of tongue; gradually begins to swallow using front of tongue as well.	
	Strong reflex (extrusion) to push food out during first 2 to 3 months.	
4–6	Extrusion reflex diminishes, and the ability to swallow nonliquid foods develops.	Begin iron-fortified cereal mixed with breast milk, formula, or water.
	Indicates desire for food by opening mouth and leaning forward.	Begin pureed vegetables and fruits.
	Indicates satiety or disinterest by turning away and leaning back.	
	Sits erect with support at 6 months.	
	Begins chewing action.	
	Brings hand to mouth.	
	Grasps objects with palm of hand.	
6–8	Able to feed self with fingers.	Begin mashed vegetables and fruits.
	Develops pincher (finger to thumb) grasp.	Begin plain baby food meats.
	Begins to drink from cup.	Begin plain, unsweetened fruit juices from cup.
8–10	Begins to hold own bottle.	Begin breads and cereals from table.
	Reaches for and grabs food and spoon.	Begin yogurt.
	Sits unsupported.	Begin pieces of soft, cooked vegetables and fruit from table.
		Gradually begin finely cut meats, fish, casseroles, cheese, eggs, and mashed legumes.
10–12	Begins to master spoon but still spills some.	Add variety. Gradually increase portion sizes.[a]

[a] Portions of foods for infants and young children are smaller than those for an adult. For example, a grain serving might be $1/2$ slice of bread instead of 1 slice, or $1/4$ cup rice instead of $1/2$ cup.

Note: Because each stage of development builds on the previous stage, the foods from an earlier stage continue to be included in all later stages.

Source: Adapted in part from Committee on Nutrition, American Academy of Pediatrics, *Pediatric Nutrition Handbook*, 5th ed., ed. R. E. Kleinman (Elk Grove Village, IL: American Academy of Pediatrics, 2004), pp. 103–115.

Foods to Provide Iron and Vitamin C Iron deficiency is common in young children throughout the world, especially between the ages of six months and three years when they are growing fast and milk, which is a poor source of iron, has a large place in their diets. The iron an infant has stored from before birth typically runs out after the birth weight doubles, long before the end of the first year. Infants can derive adequate iron first from breast milk or formula with iron, then from iron-fortified cereals, and from meat or meat alternates such as legumes. Once infants are consuming iron-fortified cereals, parents or caregivers should begin selecting vitamin C-rich foods to go with meals to enhance iron absorption. The best sources of vitamin C are fruits and vegetables (see p. 225).

Fruit juice is a source of vitamin C, but excessive juice intake can lead to diarrhea in infants and young children.[25] AAP recommendations limit juice consumption for infants and young children (one to six years of age) to between 4 and 6 ounces per day.[26] Beyond these limits, fruit juices contribute excessive kcalories and displace other nutrient-rich foods. Fruit juices can be diluted with water and should be served in a cup, not a bottle, once the infant is six months of age or older.

Physical Readiness for Solid Foods The ability to swallow solid food develops at four to six months for most infants, and food offered by spoon helps to develop swallowing ability. At eight months to one year, an infant can sit up, handle finger foods, and begins to teethe. At that time, hard crackers and other hard finger foods may be introduced to promote the development of manual dexterity and control of the jaw muscles. These feedings must occur under the watchful eye of an adult because the infant can also choke on such foods.

Some parents want to feed solids at an earlier age, on the theory that "stuffing the baby" at bedtime promotes sleeping through the night. There is no proof for this theory. On average, infants start to sleep through the night at about the same age (three to four months) regardless of when solid foods are introduced.

Allergy-Causing Foods New foods should be introduced singly and at intervals spaced to permit detection of allergies. For example, when cereals are introduced, rice cereal is offered first for several days because it is unlikely to cause an allergic response. Wheat cereal is offered last because it is the most common offender. If a cereal causes an allergic reaction (irritability due to skin rash, digestive upset, or respiratory discomfort), its use should be discontinued before introducing the next food.

Choice of Infant Foods Baby foods commercially prepared in the United States and Canada are safe, and except for mixed dinners and heavily sweetened desserts, they generally have high nutrient density. An alternative for the parent who wants the infant to have family foods is to "blenderize" a small portion of the table food (cooked without salt or sugar) at each meal.

Foods to Omit Sweets of any kind (including baby food "desserts") have no place in an infant's diet. The added food energy conveys few, if any, nutrients to support growth and can promote obesity. Canned vegetables are also inappropriate for infants; they often contain too much sodium. Honey and corn syrup should never be fed to infants because of the risk of botulism. Infants and even young children cannot safely chew and swallow any of the foods listed in the margin; ▪ they can easily choke on

Foods such as iron-fortified cereals and formulas, mashed legumes, and strained meats provide iron.

▪ To prevent choking, do not give infants or young children:
  - Gum
  - Popcorn
  - Whole grapes
  - Cherries
  - Raw celery
  - Carrots
  - Whole beans
  - Hot dog slices
  - Hard or gel-type candies
  - Marshmallows
  - Nuts
  - Peanut butter

Keep these nonfood items out of their reach:
  - Coins
  - Balloons
  - Small balls
  - Pen tops
  - Other items of similar size

Dietary Guidelines for Americans 2005

▪ Infants and young children should not eat or drink unpasteurized milk, milk products, or juices; raw or undercooked eggs, meat, poultry, shellfish, or fish; or raw sprouts.

Ideally, a one-year-old eats many of the same healthy foods as the rest of the family.

■ Reminder: *Milk anemia* develops when an excessive milk intake displaces iron-rich foods from the diet.

TABLE 12–4

Sample Menu for a One-Year-Old

Breakfast
$\frac{1}{2}$ c whole milk
$\frac{1}{2}$ c iron-fortified unsweetened cereal
$\frac{1}{2}$ c orange juice

Morning snack
$\frac{1}{2}$ c yogurt
$\frac{1}{2}$ c fruit[a]

Lunch
$\frac{1}{2}$ c whole milk
$\frac{1}{2}$ c vegetables[b]
1 egg or $\frac{1}{4}$ c tofu
$\frac{1}{2}$ c noodles

Afternoon snack
$\frac{1}{2}$ c whole milk
$\frac{1}{2}$ slice whole wheat toast
1 tbs apple butter

Dinner
$\frac{1}{2}$ c whole milk
1 oz chopped meat or $\frac{1}{4}$ c well-cooked mashed legumes
$\frac{1}{4}$ c potato, rice, or pasta
$\frac{1}{2}$ c vegetables[b]

[a] Include citrus fruits, melons, and berries.
[b] Include dark green, leafy, and deep yellow vegetables.

these foods. Nonfood items of small size should always be kept out of the infant's reach to prevent choking. Also, an infant's caregiver must be on guard against food poisoning and take precautions against it as described in Nutrition in Practice 5. The *Dietary Guidelines for Americans 2005* address this risk.

Foods at One Year   At one year of age, whole cow's milk can become a primary source of most of the nutrients the infant needs; 2 to 3 cups per day meet those needs sufficiently. More milk than this displaces iron-rich foods and can lead to the iron-deficiency anemia known as milk anemia. ■ Other foods—meat and meat alternates, iron-fortified cereal, enriched or whole-grain bread, fruits, and vegetables—should be supplied in variety and in amounts sufficient to round out total energy needs. Ideally, a one-year-old will sit at the table, eat many of the same foods everyone else eats, and drink liquids from a cup—not a bottle. Table 12–4 shows a sample menu that meets a one-year-old's requirements.

Looking Ahead

Probably the most important single measure to undertake during the first year is to encourage eating habits that will support continued normal weight as the child grows. This means introducing a variety of nutritious foods in an inviting way, not forcing the infant to finish the bottle or the baby food jar, avoiding concentrated sweets and empty-kcalorie foods, and encouraging physical activity. Parents should avoid teaching infants to seek food as a reward, to expect food as comfort for unhappiness, or to associate food deprivation with punishment. Infants seem to have no internal "kcalorie counter," and they stop eating when their stomachs feel full. Nutrient-dense, low-kcalorie foods will satisfy as long as they provide bulk.

Normal dental development is also promoted by supplying nutritious foods, avoiding sweets, and discouraging the association of food with reward or comfort. Dental health is the subject of Nutrition in Practice 2.

The AAP recommends against a fat-modified diet during infancy. The available evidence does not warrant dietary manipulation to lower blood cholesterol in infants.

Mealtimes

The wise parent of a one-year-old offers nutrition and love together. Both promote growth. Children "fed with love" grow more in both weight and height than children fed the same food in an emotionally negative climate.

The person feeding a one-year-old should be aware that exploring and experimenting are normal and desirable behaviors at this time in a child's life. The child is developing a sense of autonomy that, if allowed to develop, will lay the foundation for later confidence and effectiveness as an individual. The child's impulses, if consistently denied, can turn to shame and self-doubt. In light of the developmental and nutrient needs of one-year-olds and in the face of their often contrary and willful behavior, a few feeding guidelines may be helpful:

■ **Discourage unacceptable behavior (such as standing at the table or throwing food) by removing the child from the table to wait until later to eat.** Be consistent and firm, not punitive. The child will soon learn to sit and eat.

■ **Let the child explore and enjoy food.** This may mean the child eats with fingers for awhile. Use of the spoon will come in time.

■ **Do not force food on children.** Provide children with nutritious foods, and let them choose which ones and how much they will eat. Gradually, they will acquire a taste for different foods. If children refuse milk, provide cheese, cream soups, and yogurt.

■ **Limit sweets strictly.** Infants have little room in their 1,000-kcalorie daily energy allowance for empty-kcalorie sweets, except occasionally.

These recommendations reflect a spirit of tolerance that serves the best interests of the child emotionally as well as physically.

IN SUMMARY

■ The primary food for infants during the first 12 months is either breast milk or iron-fortified formula.

■ In addition to nutrients, breast milk also offers immunological protection.

■ At about four to six months, infants should gradually begin eating solid foods. By one year, they are drinking from a cup and eating many of the same foods as the rest of the family.

Early and Middle Childhood

After the age of one, growth rate slows, but the body continues to change dramatically (see Figure 12–5). At one, infants have just learned to stand and toddle; by two, they walk confidently and are learning to run, jump, and climb. Nutrition and physical activity have helped them prepare for these new accomplishments by adding to the mass and density of their bone and muscle tissue. Thereafter, their bones continue to grow longer, and their muscles to gain size and strength, though unevenly and more slowly, until adolescence.

Energy and Nutrient Needs

An infant's appetite declines markedly around the first birthday, consistent with the slowed growth rate. Thereafter, the appetite fluctuates. At times children seem to be insatiable, and at other times they seem to live on air and water. Parents and other caregivers need not worry about this—a child will need and demand much more food during periods of rapid growth than during slow periods. The perfect regulation of appetite in children of normal weight guarantees that their food energy intakes will be right for each stage of growth.

Children's Appetites Many people mistakenly believe that they must "make" their children eat the right amounts of food, and children's erratic appetites often reinforce this belief. Although children's food energy intakes vary widely from meal to meal, total daily energy intake remains remarkably constant.[27] If children eat less at one meal, they eat more at the next, and vice versa.

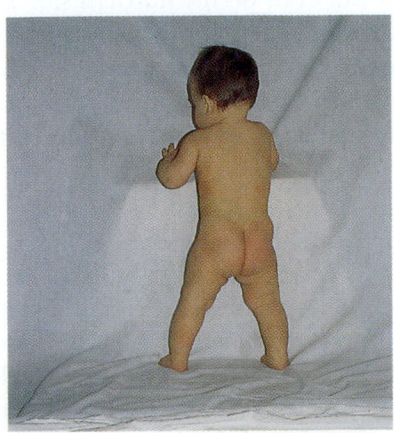 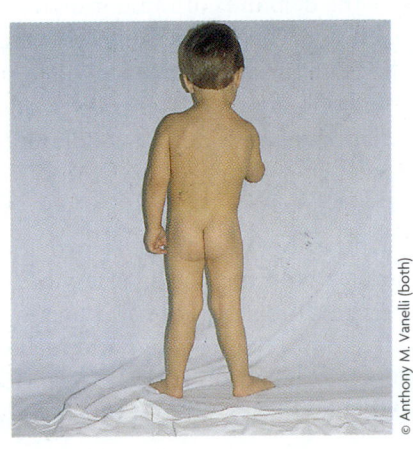

© Anthony M. Vanelli (both)

FIGURE 12–5

Body Shape of a One-Year-Old and a Two-Year-Old Compared
The body shape of a one-year-old (left) changes dramatically by age two (right). The two-year-old has lost much of the baby fat; the muscles (especially in the back, buttocks, and legs) have firmed and strengthened; and the leg bones have lengthened.

Dietary Guidelines for Americans 2005

■ Children should engage in at least 60 minutes of physical activity on most, preferably all, days of the week.

However, parents do need to help children choose the right foods, and with overweight children, they may need to help more, as described later. Overweight children may not adjust their energy intakes appropriately and may disregard appetite-regulation signals and eat in response to external cues, such as television commercials.

Energy Individual children's energy needs vary widely, depending on their growth and physical activity. A one-year-old child needs approximately 800 kcalories per day; an active six-year-old needs twice as many kcalories per day. By age ten, an active child needs about 2,000 kcalories per day. Total energy needs increase gradually with age, but energy needs per kilogram of body weight actually decline. Physically active children of any age need more energy because they expend more, and inactive children can become obese even when they eat less food than the average. Unfortunately, the prevalence of overweight in preschool children 2 to 5 years of age and adolescents 12 to 19 years of age has more than doubled over the past 30 years, and it has more than tripled for children 6 to 11 years of age.[28] An estimated 9 million U.S. children over the age of six are obese. Obesity poses hazards to the health of children both now and in the future (Nutrition in Practice 12 provides details). Strategies to prevent obesity in children must focus on balancing energy intake and energy expenditure. In other words, factors that influence both eating and physical activity must be addressed. Schools would serve our children well by offering activities to promote physical fitness.[29] Children who learn to enjoy physical play and exercise, both at home and at school, are best prepared to maintain active lifestyles as adults.

Some children, notably those adhering to a vegan diet, may have difficulty meeting their energy needs. Grains, vegetables, and fruits provide plenty of fiber, adding bulk, but may provide too few kcalories to support growth. Soy products, other legumes, and nut or seed butters offer more concentrated sources of energy to support optimal growth and development.

Nutrients Steady growth during childhood necessitates a gradual increase in intakes of most nutrients. Nutrient recommendations cluster children into age groupings that reflect similarities in growth rate, biological changes, and hormone status (see inside front cover of this text).

Ideally, children accumulate stores of nutrients before adolescence. Then, when they take off on the adolescent growth spurt and their nutrient intakes cannot keep pace with the demands of rapid growth, they can draw on the nutrient stores accumulated earlier. This is especially true of calcium; the denser the bones are in childhood, the better prepared they will be to support teen growth and still withstand the inevitable bone losses of later life.[30] Consequently, the way children eat influences their nutritional health during childhood, during their teen years, and for the rest of their lives.

Food Patterns for Children To provide all the needed nutrients, a child's meals and snacks should include a variety of foods from each food group—in amounts suited to the child's appetite and needs. Figure 12–6 shows MyPyramid designed for children 6 to 11 years of age and includes the recommended amounts of food for an 1,800-kcalorie intake. Table 12–5 (p. 350) lists amounts of food for several kcalorie levels below 1,800 kcalories, which are appropriate for most younger children and

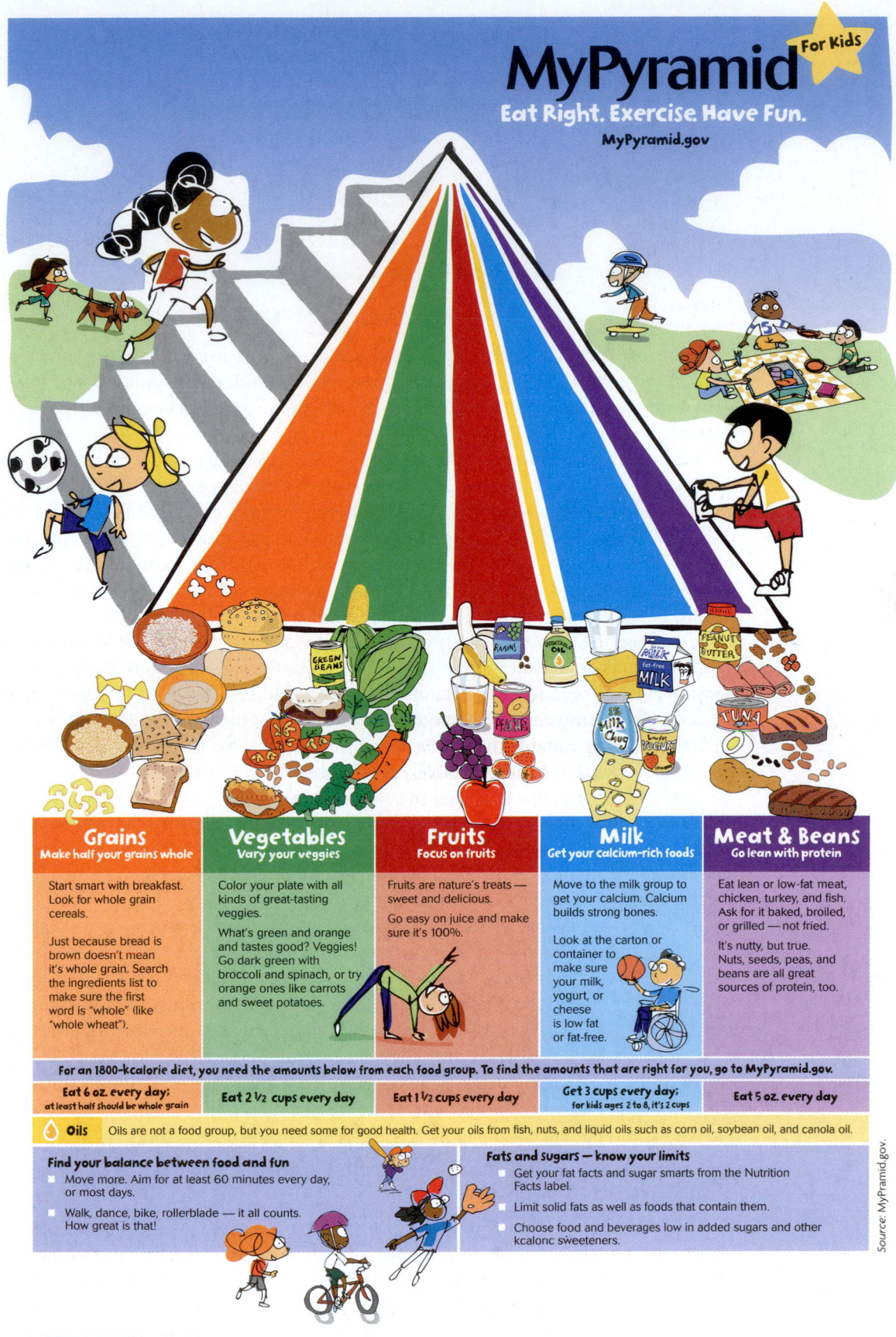

FIGURE 12–6

MyPyramid for Children

TABLE 12–5

Recommended Daily Amounts for Each Food Group (1,000 to 1,600 kCalories)

Food Group	1,000 kcal	1,200 kcal	1,400 kcal	1,600 kcal
Fruit	1 c	1 c	1½ c	1½ c
Vegetables	1 c	1½ c	1½ c	2 c
Grains	3 oz	4 oz	5 oz	5 oz
Meat and legumes	2 oz	3 oz	4 oz	5 oz
Milk	2 c	2 c	2 c	3 c
Oils	3 tsp	3 tsp	3 tsp	4 tsp

Note: The discretionary kcalorie allowance for these patterns is about 100 kcalories.

sedentary older children. Estimated daily kcalorie needs for children are shown in Table 12–6. Table 1–6 in Chapter 1 provides recommended daily amounts of foods from each group for higher kcalorie levels, which are appropriate for active older children.

Based on an assessment tool developed by the U.S. Department of Agriculture (USDA), the great majority of children two to nine years of age consume a diet ranked "poor" or "needs improvement."[31] A comprehensive survey, called the Feeding Infants and Toddlers Study (FITS), assessed the food and nutrient intakes of more than 3,000 infants and toddlers.[32] The survey found fruit and vegetable intakes of infants and toddlers are limited, and in fact, about 25 percent of infants and toddlers older than 9 months of age did not eat a single serving of fruits or vegetables in a day.[33] By 15 to 18 months of age, the most commonly consumed vegetable was french fries, and the most commonly consumed fruit was bananas—neither particularly rich sources of vitamins or minerals. Parents and caregivers of infants and toddlers thus need to offer a much greater variety of nutrient-dense vegetables and fruits at meals and snacks to help insure nutrient adequacy. Among other nutrition concerns for U.S. children are inadequate intakes of calcium and fiber and excessive intakes of saturated fat.[34]

Children's Food Choices Parents and other caregivers can do much to foster the development of healthy eating habits in a child. The challenge is to deliver nutrients in the form of meals and snacks that are both nutritious and appealing so that children will learn to enjoy a variety of nutritious foods.

Candy, cola, and other concentrated sweets must be limited in children's diets. If such foods are permitted in large quantities, the only possible outcomes are nutrient deficiencies, obesity, or both. Children cannot be trusted to choose nutritious foods on the basis of taste alone. The preference for sweets is innate, and children naturally gravitate to them. Overweight children, especially, need help in selecting nutrient-dense foods that will meet their nutrient needs within their energy allowances. Underweight children or active, normal-weight children can enjoy higher-kcalorie foods, but these should still be nutritious. Examples are ice cream and pudding in the milk group and whole-grain or enriched pancakes and crackers in the bread group.

Malnutrition in Children

Most children in the United States and Canada have access to regular meals, but hunger and malnutrition do appear in certain circumstances. For example, children in very low-income families are more likely to be hungry and malnourished. An estimated 12 million U.S. children are hungry at least some of the time and are living in poverty.[35]

Effects of Hunger Both short-term and long-term hunger exerts negative effects on behavior and health. Short-term hunger, such as when a child misses a meal, impairs

TABLE 12–6

Estimated Daily kCalorie Needs for Children

Children	Sedentary[a]	Active[b]
2 to 3 yr	1,000	1,400
Females		
4 to 8 yr	1,200	1,800
9 to 13 yr	1,600	2,200
Males		
4 to 8 yr	1,400	2,000
9 to 13 yr	1,800	2,600

[a] *Sedentary* describes a lifestyle that includes only the activities typical of day-to-day life.
[b] *Active* describes a lifestyle that includes at least 60 minutes per day of moderate physical activity (equivalent to walking more than 3 miles per day at 3 to 4 miles per hour) in addition to the activities of day-to-day life.

the child's ability to pay attention and to be productive. Hungry children are irritable, apathetic, and uninterested in their environment. Long-term hunger impairs growth and immune defenses. Food assistance programs such as the WIC program (discussed in Chapter 11) and the School Breakfast and National School Lunch Programs (discussed later in this chapter) are designed to protect against hunger and improve the health of children.

Healthy, well-nourished children are alert in the classroom and energetic at play.

Hunger and School Performance Children who eat nutritious breakfasts function better than their peers who do not. A nutritious breakfast is a central feature of a diet that meets the needs of children and supports their healthy growth and development.[36] When a child consistently skips breakfast or is allowed to choose sugary foods (candy or marshmallows) in place of nourishing ones (whole-grain cereals), the child will fail to get enough of several nutrients. Nutrients missed from a skipped breakfast will not be "made up" at lunch and dinner but will be left out completely that day.

Children who eat no breakfast are more likely to be overweight, perform poorly in tasks requiring concentration, have shorter attention spans, score lower on tests, and be tardy or absent more often than their well-fed peers.[37] Common sense dictates that it is unreasonable to expect anyone to learn and perform work when no fuel has been provided. By the late morning, discomfort from hunger may become distracting even if a child has eaten breakfast. Chronically underfed children suffer all the more. Teachers aware of the late-morning slump in their classrooms wisely request that a midmorning snack be provided; it improves classroom performance all the way to lunchtime.

Iron Deficiency and Behavior In U.S. children and adolescents, as in infants after six months, iron deficiency remains common despite iron fortification of foods and other programs to combat this deficiency. Iron deficiency has well-known and widespread effects on children's behavior and intellectual performance.[38] Most people are familiar with the role of iron in carrying oxygen in the blood. Iron also works as part of large molecules to release energy within cells and plays key roles in many molecules of the brain and nervous system. A lack of iron not only causes an energy crisis, but also directly affects behavior, mood, attention span, and learning ability.

Iron deficiency is usually diagnosed by a deficit of iron in the *blood*, after the deficiency has progressed all the way to anemia. However, a child's *brain* is sensitive to slightly lowered iron concentrations long before the blood deficits appear. Iron's effects are hard to distinguish from the effects of other factors in children's lives, but it is likely that iron deficiency manifests itself in a lowering of "motivation to persist in intellectually challenging tasks," a shortening of the attention span, and a reduction of overall intellectual performance. Iron-deficient children perform less well on tests and have more conduct disturbances than their nonanemic classmates. When combined with other nutrient deficiencies, iron-deficiency has synergistic effects that are especially detrimental to learning. Furthermore, children who had iron-deficiency anemia *as infants* continue to perform poorly as they grow older, even if their iron status improves.[39] The long-term damaging effects on mental development make prevention of iron deficiency during infancy and early childhood a high priority.

Preventing Iron Deficiency To avert iron deficiency, children's foods must deliver 7 to 10 milligrams of iron per day. To achieve this goal, milk intakes must be limited after infancy because milk is a poor source of iron. Children should receive enough milk products to ensure adequate calcium and riboflavin intakes, but no more. That means 2 cups of milk per day up to age eight, increasing to 3 cups per day from age nine on. After age two, if reduced-fat milk is used instead of whole milk, the saved kcalories can

TABLE 12–7

Iron-Rich Foods Children Like[a]

Breads, cereals, and grains

Canned macaroni ($\frac{1}{2}$ c)

Canned spaghetti ($\frac{1}{2}$ c)

Cream of wheat ($\frac{1}{4}$ c)

Fortified dry cereals (1 oz)[b]

Noodles, rice, or barley ($\frac{1}{2}$ c)

Tortillas (1 flour or whole wheat, 2 corn)

Whole wheat, enriched, or fortified bread (1 slice)

Vegetables

Baked flavored potato skins ($\frac{1}{2}$ skin)

Cooked mung bean sprouts or snow peas ($\frac{1}{2}$ c)[c]

Cooked mushrooms ($\frac{1}{2}$ c)

Green peas ($\frac{1}{2}$ c)

Mixed vegetable juice (1 c)

Fruits

Canned plums (3 plums)

Cooked dried apricots ($\frac{1}{4}$ c)

Dried peaches (4 halves)

Raisins (1 tbs)

Meats and legumes

Bean dip ($\frac{1}{4}$ c)

Canned pork and beans ($\frac{1}{3}$ c)

Lean chopped roast beef or cooked ground beef (1 oz)

Liverwurst on crackers ($\frac{1}{2}$ oz)

Meat casseroles ($\frac{1}{2}$ c)

Mild chili or other bean/meat dishes ($\frac{1}{4}$ c)

Peanut butter and jelly sandwich ($\frac{1}{2}$ sandwich)

Sloppy joes ($\frac{1}{2}$ sandwich)

[a] Each serving provides at least 1 milligram of iron. Vitamin C-rich foods included with these snacks increase iron absorption.
[b] Some fortified breakfast cereals contain more than 10 milligrams of iron per half-cup serving (read the labels).
[c] Raw sprouts may pose a bacterial hazard to young children.

be invested in iron-rich foods such as lean meats, fish, poultry, eggs, and legumes. Whole-grain or enriched breads and cereals also contribute iron. Table 12–7 lists iron-rich foods children like.

Other Nutrient Deficiencies Iron is only one of several dozen nutrients that can be displaced by a diet of nutrient-poor foods. Any of the other nutrients may be lacking as well, and the deficiencies of those nutrients may also cause both behavioral and physical symptoms (see Table 12–8). A child with behavioral symptoms of nutrient deficiencies may be irritable, aggressive, disagreeable, or sad and withdrawn. Such a child may be labeled "hyperactive," "depressed," or "unlikable," but in fact these traits may arise from simple, albeit marginal, malnutrition. Parents and medical practitioners often overlook the possibility that malnutrition may account for abnormalities of appearance and behavior. Any departure from normal healthy appearance and behavior is a sign of possible poor nutrition. In any such case, inspection of the child's diet

TABLE 12–8

Physical Signs of Malnutrition in Children

	Well-Nourished	Malnourished	Possible Nutrient Deficiencies
Hair	Shiny, firm in the scalp	Dull, brittle, dry, loose; falls out	Protein-energy malnutrition (PEM)
Eyes	Bright, clear pink membranes; adjust easily to light	Pale membranes; spots; redness; adjust slowly to darkness	Vitamin A, B vitamins, zinc, and iron
Teeth and gums	No pain or caries, gums firm, teeth bright	Missing, discolored, decayed teeth. Gums bleed easily and are swollen and spongy.	Minerals and vitamin C
Face	Clear complexion without dryness or scaliness	Off-color, scaly, flaky, cracked skin	PEM, vitamin A, and iron
Glands	No lumps	Swollen at front of neck, cheeks	PEM and iodine
Tongue	Red, bumpy, rough	Sore, smooth, purplish, swollen	B vitamins
Skin	Smooth, firm, good color	Dry, rough, spotty; "sandpaper" feel or sores; lack of fat under skin	PEM, essential fatty acids, vitamin A, B vitamins, and vitamin C
Nails	Firm, pink	Spoon-shaped, brittle, ridged	Iron
Internal systems	Regular heart rhythm, heart rate, and blood pressure; no impairment of digestive function, reflexes, or mental status	Abnormal heart rate, heart rhythm, or blood pressure; enlarged liver, spleen; abnormal digestion; burning, tingling of hands, feet; loss of balance, coordination; mental confusion, irritability, fatigue	PEM and minerals
Muscles and bones	Muscle tone; posture, long bone development appropriate for age	"Wasted" appearance of muscles; swollen bumps on skull or ends of bones; small bumps on ribs; bowed legs or knock-knees	PEM, minerals, and vitamin D

by a registered dietitian or other qualified health care professional is in order. Any suspicion of dietary inadequacies, *no matter what other causes may be implicated*, should prompt steps to correct those inadequacies immediately.

Lead Poisoning in Children

The damage caused by malnutrition may be compounded by environmental factors such as lead poisoning. A two-way interaction is typical: lead poisoning can cause an iron deficiency, and an iron deficiency can impair the body's defenses against lead absorption. Adequate calcium may slow lead's absorption or interfere with its toxic effects in the body. Like iron deficiency, mild lead toxicity has nonspecific effects, including diarrhea, irritability, reduced ability of the blood to carry oxygen, and fatigue. The symptoms may be reversible if exposure stops soon enough. With higher levels of lead, the signs become more pronounced, yet pinpointing a cause may still be difficult. Children lose their general cognitive, verbal, and perceptual abilities and develop learning disabilities and behavior problems. Still more severe lead toxicity can cause irreversible nerve damage, paralysis, mental retardation, and death.

Lead toxicity is most prevalent among children under age six—more than 300,000 may have blood concentrations high enough to cause mental, behavioral, and other health problems.[40] Lead aggressively attacks fetuses, infants, and children because the body absorbs lead most efficiently during times of rapid growth. Blood concentrations of lead generally reach a peak in two-year-old children; age two is the typical time of exploring surroundings "hand to mouth" and ingesting lead-tainted dirt and paint chips. Children's behaviors and activities—putting their hands in their mouths, playing in dirt, and eating nonfood items—favor their chances of exposure to lead.[41]

Paint is the primary source of lead in children's lives.

A ban on leaded gasoline and reductions in other uses of lead have dramatically decreased the amount of lead in the environment.[42] The result has been a gratifying decline in children's blood lead concentrations since the 1970s. The decline follows the reduction in the nation's use of leaded gasoline, leaded house paint, and lead-soldered food cans. A nationwide lead-monitoring system is now in place, and aggressive community programs are testing and treating children for lead poisoning. Despite such safeguards, exposure to lead still remains a threat to about 25 percent of U.S. children, especially those in low-income families who live in older houses that still contain deteriorating lead-based paint.[43] The accompanying "How to" suggests strategies to protect against lead poisoning.

How To

PROTECT AGAINST LEAD TOXICITY

Researchers simultaneously made three major discoveries about lead toxicity: lead poisoning has *subtle* effects, the effects are *permanent*, and they occur at *low levels of exposure*. The amount of lead recognized to cause harm is only 10 micrograms per 100 milliliters of blood. Some research shows that blood lead concentration *below* this amount may adversely affect children's scores on intelligence tests.[a] Consequently, consumers should take ultraconservative measures to protect themselves and especially their infants and young children from lead poisoning. The American Academy of Pediatrics and the Centers for Disease Control recommend screening in communities with a substantial number of houses built before 1950 and in those with a substantial number of children with elevated lead levels. In addition to screening children most likely to be exposed, pediatricians should alert all parents to the possible dangers of lead exposure and explain prevention strategies.

Preventive strategies include:

- In contaminated environments, keep small children from putting dirty or old painted objects in their mouths, and make sure children wash their hands before eating. Similarly, keep small children from eating any nonfood items. Lead poisoning has been reported in young children who have eaten crayons or pool cue chalk.

- Wet-mop floors and damp-sponge walls regularly. Children's blood lead levels decline when the homes they live in are cleaned regularly.

- Be aware that other countries do not have the same regulations protecting consumers against lead. Children have been poisoned by eating crayons made in China and drinking fruit juice canned in Mexico.

- Do not use lead-contaminated water to make infant formula.

- Once you have opened canned food, store it in a lead-free container to prevent lead migration into the food.

- Do not store acidic foods or beverages (such as vinegar or orange juice) in ceramic dishware or alcoholic beverages in pewter or crystal decanters.

- Many manufacturers are now making lead-safe products.[b] Old, handmade, or imported ceramic cups and bowls may contain lead and should not be used to heat coffee or tea or acidic foods such as tomato soup.

- Feed children nutritious meals regularly.

- Before using your newspaper to wrap food, mulch garden plants, or add to your compost, confirm with the publisher that the paper uses no lead in its ink.

The Environmental Protection Agency (EPA) also publishes a booklet, *Lead and Your Drinking Water*, in which the following cautions appear:

- Have the water in your home tested by a competent laboratory.

- Use only cold water for drinking, cooking, and making formula (cold water absorbs less lead).

- When water has been standing in pipes for more than two hours, flush the cold-water pipes by running water through them for 30 seconds before using it for drinking, cooking, or mixing formulas.

- If lead contamination of your water supply seems probable, obtain additional information and advice from the EPA and your local public health agency.

By taking these steps, parents can protect themselves and their children from this preventable danger.[c]

[a] R. L. Canfield and coauthors, Intellectual impairment in children with blood lead concentrations below 10 µg per deciliter, *New England Journal of Medicine* 348 (2003): 1517–1526.

[b] *A Shopper's Guide to Low-Lead China* is available from the Environmental Defense Fund, 257 Park Avenue South, New York, NY 10010; telephone (800) 284-3322.

[c] The National Lead Information Center provides two hotlines: call (800) LEAD-FYI (532-3394) for general information or (800) 424-LEAD (424-5323) with specific questions.

Food Allergies

Food allergies are frequently blamed for physical and behavioral abnormalities in children, but just 6 percent of children are diagnosed with true food allergies.[44] Food allergies diminish with age, until in adulthood they affect about 1 or 2 percent of the population.[45] A true food allergy occurs when a whole food protein or other large molecule enters the body and elicits an immunologic response. (Recall that large molecules of food are normally dismantled in the digestive tract to smaller ones before absorption.) The body's immune system reacts to a food protein or other large molecule as it does to an antigen—by producing antibodies or other defensive agents. ■ ■ A problem that results from exposure to food substances but does not involve the immune system is known as **food intolerance.**

Asymptomatic and Symptomatic Allergies Allergies may have one or two components. They always involve antibodies; they may or may not involve symptoms. A person may produce antibodies without having any symptoms (known as *asymptomatic allergy*) or may produce antibodies and have symptoms (known as *symptomatic allergy*). However, a person who experiences symptoms without producing antibodies does not have an allergy. This means that allergies have to be diagnosed by testing for antibodies. Once a food allergy has been diagnosed, therapy requires strict elimination of the offending food. Children with allergies, like all children, need all their nutrients, so it is important to include other foods that offer the same nutrients as the omitted foods.[46]

Allergy Symptoms A symptomatic allergy will exhibit different symptoms depending on the location of the reaction. In the digestive tract, the allergy may cause nausea or vomiting; in the skin, it may cause rashes; and in the nasal passages and lungs, it may cause inflammation or asthma. A dangerous, generalized, all-systems shock reaction known as **anaphylactic shock** can also occur.

Immediate and Delayed Reactions Allergic reactions to food can occur with different timings, simply classified as immediate and delayed. In both, the antigen interacts immediately with the immune system, but symptoms may appear within minutes or after several (up to 24) hours. Identifying the food that causes an immediate allergic reaction is easy because symptoms correlate closely with the time of eating the food. Identifying the food that caused a delayed reaction is more difficult because the symptoms may not appear until a day after the offending food was eaten; by this time, many other foods will have been eaten, too, complicating the picture. Allergic reactions to single foods are common. Reactions to multiple foods are the exception, not the rule.

Anaphylactic Shock The life-threatening food allergy reaction of anaphylactic shock is most often caused by peanuts, tree nuts, milk, eggs, wheat, soybeans, fish, or shellfish. Among these foods, eggs, milk, soy, and peanuts most often cause problems in children. Children are more likely to outgrow allergies to eggs, milk, and soy than allergies to peanuts. Peanuts cause more life-threatening reactions than do all other food allergies combined. Research is currently under way to help those with peanut allergies tolerate small doses, thus saving lives and minimizing reactions.[47] One possible solution depends on finding a natural, hypoallergenic peanut among the 14,000 varieties of peanuts that grow. Families of children with a life-threatening food allergy and school personnel who supervise them must guard them against any exposure to the allergen. The child must learn to identify which foods pose a problem and then learn and use refusal skills for all foods that may contain the allergen.

Parents of children with allergies can pack safe foods for lunches and snacks and ask school officials to strictly enforce a "no swapping" policy in the lunchroom. The

■ **Reminder:** *Antigens* are substances foreign to the body that elicit the formation of antibodies or an inflammation reaction from immune system cells. Food antigens are usually glycoproteins (large proteins with glucose molecules attached).

■ **Reminder:** *Antibodies* are large proteins that are produced in response to antigens and then inactivate the antigens.

© Polara Studios, Inc.

These eight normally wholesome foods—milk, shellfish, fish, peanuts, tree nuts, eggs, wheat, and soybeans (and soy products)—may cause life-threatening symptoms in people with allergies.

food allergies: adverse reactions to foods that involve an immune response; also called **food-hypersensitivity reactions**.

food intolerance: an adverse response to a food or food additive that does not involve the immune system.

anaphylactic (AN-ah-feh-LAC-tic) **shock:** a life-threatening whole-body allergic reaction to an offending substance.

■ Symptoms of impending ana-
phylactic shock:
  ■ Tingling sensation in mouth
  ■ Swelling of the tongue and
    throat
  ■ Irritated, reddened eyes
  ■ Difficulty breathing, asthma
  ■ Hives, swelling, rashes
  ■ Vomiting, abdominal
    cramps, diarrhea
  ■ Drop in blood pressure
  ■ Loss of consciousness
  ■ Death

child must be able to recognize the symptoms of impending anaphylactic shock, ■ such as a tingling of the tongue, throat, or skin or difficulty breathing. Any person with food allergies severe enough to cause anaphylactic shock should wear a medical alert bracelet or necklace. Finally, the responsible child and the school staff should be prepared to administer injections of **epinephrine**, which prevents anaphylaxis after exposure to the allergen. Many preventable deaths occur each year when people with food allergies accidentally ingest the allergen but have no epinephrine.

Food Labeling Food labels must announce the presence of common allergens in plain language, using the names of the eight most common allergy-causing foods.[48] For example, a food containing "textured vegetable protein" must say "soy" on its label. Similarly, "casein" must be identified as "milk" and so forth. Food producers must also prevent cross-contamination during production and clearly label the foods in which it is likely to occur.[49] For example, equipment used for making peanut butter must be scrupulously clean before being used to pulverize cashew nuts for cashew butter to protect unsuspecting cashew butter consumers from peanut allergens.

Technology may soon offer new solutions. Drugs under development may interfere with the immune response that causes allergic reactions.[50] Through genetic engineering, scientists may one day eliminate allergens from peanuts, soybeans, and other foods to make them safer.

Other Adverse Reactions to Foods Adverse reactions to foods that are not true food allergies include:

■ A reaction specific to the flavor enhancer monosodium glutamate, or MSG

■ Reactions to chemicals in foods, such as the natural laxative in prunes

■ Symptoms of digestive diseases, such as hernias and ulcers, that are aggravated by eating any food

■ Enzyme deficiencies, such as lactose intolerance, that cause symptoms superficially indistinguishable from those of food allergy

■ Psychological reactions based on the belief that certain foods cause certain symptoms

The simple dislike of a food may be a clue to allergy or to any of these reactions.

Food Dislikes Parents are advised to watch for signs of food dislikes and take them seriously. Children's food aversions may be the result of nature's efforts to protect them from allergic or other adverse reactions. Test for allergies, and then apply nutrition knowledge conscientiously in deciding how to alter the diet.

Hyperactivity

Hyperactivity affects behavior and learning in about 5 to 10 percent of young school-aged children. Left untreated, it can interfere with a child's social development and ability to learn. Treatment focuses on relieving the symptoms and controlling the associated problems; there is no cure. Physicians often manage hyperactivity through behavior modification, special educational techniques, psychological counseling, and, in some cases, drug therapy.[51]

Parents of hyperactive children sometimes seek help from alternative therapies, including special diets. They mistakenly believe a solution may lie in manipulating the diet—most commonly, by excluding sugar or food additives. Adding carrots or eliminating candy is such a simple solution that many parents eagerly give such diet advice a try. These dietary changes will not solve the problem of true hyperactivity. Studies have consistently found no convincing evidence that sugar causes hyperactivity or worsens behavior. Recommendations to restrict sugar in children's diets to prevent or treat behavior problems are groundless. The Case Study at the end of the chapter offers an opportunity to think about these issues in relation to a specific child.

Children can become excitable, rambunctious, and unruly as a result of a desire for attention, lack of sleep, overstimulation, watching too much television or playing too many video games, too much caffeine from colas or chocolate, or a lack of physical activity. Such behaviors may suggest that more consistent care is needed. It helps to insist on regular hours of sleep, regular mealtimes, and regular outdoor activity.

IN SUMMARY

- Children's appetites and nutrient needs reflect their stage of growth.

- Long-term hunger and malnutrition impair growth and health.

- Short-term hunger exerts more subtle effects on children's health and behavior—such as poor academic performance.

- Iron deficiency is widespread and has many physical and behavioral consequences.

- Lead toxicity is prevalent among young children and can have irreversible effects on health and behavior.

- True food allergies are somewhat rare in children, and children can outgrow some food allergies.

- Some allergies, however, can cause dangerous, life-threatening reactions in both children and adults.

- "Hyper" behavior is not caused by poor nutrition; misbehavior may reflect inconsistent care.

Food Choices and Eating Habits of Children

The childhood years are the parents' last chance to influence their children's food choices. Parents who want to promote nutritious choices and healthful habits provide access to nutrient-dense, delicious foods and opportunities for active play at home. Food choices and regular physical activity can not only promote healthy growth, but, as mentioned earlier, can also help prevent the degenerative diseases of later life. Many experts agree that early childhood is the time to put into effect practices that, until recently, were recommended only for adults. Childhood Obesity and the Early Development of Chronic Diseases is the title of the Nutrition in Practice that follows this chapter.

Mealtimes at Home

Feeding children requires not only providing a variety of nutritious foods, but also nurturing the children's self-esteem and well-being. Parents face a number of challenges in preparing meals that both appeal to their children's tastes and provide needed nutrients. Because the interactions between parents and children can set the stage for lifelong attitudes and habits, a child's preferences should be treated with respect, even when nutrient needs must take precedence.

Honoring Children's Preferences Researchers attempting to explain children's food preferences encounter many contradictions. Children say they like colorful foods, yet most often reject green and yellow vegetables while favoring brown peanut butter and white potatoes, apple wedges, and bread. They do like raw vegetables better than cooked ones, so it is wise to offer vegetables that are raw or slightly undercooked and crunchy

epinephrine: one of the stress hormones secreted whenever emergency action is needed; prescribed therapeutically to relax the bronchioles during allergy or asthma attacks.

hyperactivity: inattentive and impulsive behavior that is more frequent and severe than is typical of others a similar age; professionally called **attention deficit/hyperactivity disorder (ADHD)**.

Eating is more fun when friends are there.

and bright in color. They should be warm, not hot, because a child's mouth is much more sensitive than an adult's. The flavor should be mild (a child has more taste buds), and smooth foods such as mashed potatoes or pea soup should have no lumps (a child wonders, with some disgust, what the lumps might be).

Young children like to eat at little tables and to be served little portions of food. They also love to eat with other children and have been observed to stay at the table longer and eat more food when in the company of their peers. Parents who serve food in a relaxed and casual manner, without anxiety, provide an environment in which a child's negative emotions will be minimized.

Avoiding Power Struggles   It is not surprising that problems over food often arise during the second or third year, when children begin asserting their independence. Many of these problems stem from the conflict between children's developmental stages and capabilities and parents who, in attempting to do what they think is best for their children, try to control every aspect of eating. Such conflicts can disrupt children's abilities to regulate their own food intakes or to determine their own likes and dislikes. For example, many people share the misconception that children must be persuaded or coerced to try new foods. In fact, the opposite is true. When children are forced to try new foods, especially when offered rewards for eating a particular food (if you eat your vegetables, you can play video games), they are less likely to try those foods again than are children who are left to decide for themselves. Similarly, when children are restricted from eating their favorite foods, they are more likely to want those foods. As dietitian and family therapist Ellyn Satter notes, the parent is responsible for *what* the child is offered to eat, but the child is responsible for *how much* and even *whether* to eat.

When introducing new foods at the table, parents are advised to offer them one at a time and only in small amounts at first. The more often a food is presented to a young child, the more likely the child will like that food. Between five and ten exposures to a new food are necessary before a toddler shows an enhanced preference for the food. Whenever possible, the new food should be presented at the beginning of the meal, when the child is hungry, but the child should make the decision to accept or reject it. Parents have their own inclinations and dislikes; so do children. It is best never to make an issue of food acceptance. A power struggle almost invariably sets a firm pattern of resistance and permanently closes the child's mind.

Television's Influence   Watching television adversely affects children's nutritional health.[52] As Chapter 7 pointed out, watching television contributes to obesity.[53] Children who have television sets in their bedrooms spend more time watching television and are more likely to be overweight than children who do not have televisions in their rooms.[54] Children who watch television for more than four hours per day or during meals are least likely to eat fruits and vegetables and most likely to be obese.[55] Not only are they inactive, but they often also snack on the fattening foods that are advertised.[56]

The average child sees about 30,000 commercials per year, and almost all of them urge viewers to purchase sugarcoated breakfast cereals, candy bars, chips, fast foods, and carbonated beverages. Those foods add sugar, fat, and salt to the diet and displace foods that provide needed nutrients. Many parents and pediatricians believe that food ads aimed at children should be banned because they support corporate profits rather than children's health. Alternatively, parents can teach their children how to evaluate food ads and make healthful choices.[57]

Preventing Choking   When feeding children, parents must always be alert to the dangers of choking. A choking child is a silent child—an adult should be present whenever a child is eating. Make sure the child sits when eating; choking is more likely when a child is running or falling (see p. 345 for a list of foods and nonfood items most likely to cause choking).

Play First Ideally, each meal is preceded, not followed, by the activity the child looks forward to the most. A number of schools have discovered that children eat a much better lunch if it is served after, rather than before, recess. Otherwise children "hurry up and eat" so that they can go play.

Child Participation Allowing children to help plan and prepare the family's meals provides enjoyable learning experiences and encourages children to eat the foods they have prepared. Vegetables are pretty, especially when fresh, and provide opportunities for children to learn about color, growing things and their seeds, and shapes and textures—all of which are fascinating to young children. Measuring, stirring, decorating, and arranging foods are skills that even a very young child can practice with enjoyment and pride (see Table 12–9).

Snacks Parents may find that their children often snack so much that they are not hungry at mealtimes. Instead of teaching children *not* to snack, teach them *how* to snack. Provide snacks that are as nutritious as the foods served at mealtime. Snacks can even be mealtime foods that are served individually over time, instead of all at once on one plate. When providing snacks to children, think of the food groups and offer such snacks as pieces of cheese, sliced strawberries, cooked baby carrots, and egg salad on whole wheat crackers (see Table 12–10). Snacks that are easy to prepare should be readily available to children, especially if the children arrive home after school before their parents.

TABLE 12–9
Food Skills of Preschoolers[a]
Age 1–2 years, when large muscles develop, the child:
Uses short-shanked spoon
Helps feed self
Lifts and drinks from cup
Helps scrub, tear, break, or dip foods
Age 3 years, when medium hand muscles develop, the child:
Spears food with fork
Feeds self independently
Helps wrap, pour, mix, shake, or spread foods
Helps crack nuts with supervision
Age 4 years, when small finger muscles develop, the child:
Uses all utensils and napkin
Helps roll, juice, mash, or peel foods
Cracks egg shells
Age 5 years, when fine coordination of fingers and hands develops, the child:
Helps measure, grind, grate, and cut (soft foods with a dull knife)
Uses hand-cranked egg beater with supervision

[a] These ages are approximate. Healthy, normal children develop at their own pace.

TABLE 12–10
Healthful Snack Ideas—Think Food Groups, Alone and in Combination
Selecting two or more foods from different food groups adds variety and nutrient balance to snacks. The combinations are endless, so be creative.
Grain Products
Grain products are filling snacks, especially when combined with other foods:
Cereal with fruit and milk
Crackers and cheese
Whole-grain toast with peanut butter
Popcorn with grated cheese
Oatmeal raisin cookies with milk
Vegetables
Cut-up fresh, raw vegetables make great snacks alone or in combination with foods from other food groups:
Celery with peanut butter
Broccoli, cauliflower, and carrot sticks with a flavored cottage cheese dip
Fruits
Fruits are delicious snacks and can be eaten alone—fresh, dried, or juiced—or combined with other foods:
Apples and cheese
Bananas and peanut butter
Peaches with yogurt
Raisins mixed with sunflower seeds or nuts
Meats and Meat Alternates
Meat and meat alternates add protein to snacks:
Refried beans and cheese on tortillas
Tuna on crackers
Luncheon meat on whole-grain bread
Milk and Milk Products
Milk can be used as a beverage with any snack, and many other milk products, such as yogurt and cheese, can be eaten alone or with other foods as listed previously.

Preventing Dental Caries Children frequently snack on sticky, sugary foods that stay on the teeth and provide an ideal environment for the growth of bacteria that cause dental caries. Teach children to brush and floss after meals, to brush or rinse after eating snacks, to avoid sticky foods, and to select crisp or fibrous foods frequently.

Serving as Role Models In an effort to practice these many tips, parents may overlook perhaps the single most important influence on their children's food habits—themselves.[58] Parents who do not eat oranges should not be surprised when their children refuse to eat oranges. Likewise, parents who dislike the smell of brussels sprouts may not be able to persuade children to try them. Children learn much through imitation. Parents, older siblings, and other caregivers set an irresistible example by sitting with younger children, eating the same foods, and having pleasant conversations during mealtime.

While serving and enjoying food, caregivers can promote both physical and emotional health at every stage of a child's life. They can help their children to develop both a positive self-concept and a positive attitude toward food. If the beginnings are right, children will grow without the conflicts and confusions over food that lead to nutrition and health problems.

Nutrition at School

While parents are doing what they can to establish good eating habits in their children at home, others are preparing and serving foods to their children at day care centers and schools. In addition, children begin learning about food and nutrition in the classroom. Meeting the nutrition and education needs of children is critical to supporting their healthy growth and development.[59] The American Dietetic Association states that "schools and communities have a shared responsibility to provide all students with access to high-quality foods and school-based nutrition services as an integral part of the total education program."

The U.S. government funds programs to provide nutritious, high-quality meals for children at school. Both the School Breakfast Program and the National School Lunch Program provide meals at a reasonable cost to children from families with the financial means to pay. Meals are available free or at reduced cost to children from low-income families.

■ The school breakfast must contain at a minimum:
  ■ One serving of fluid milk
  ■ One serving of fruit or vegetable or full-strength juice
  ■ Two servings of bread or bread alternates; or two servings of meat or meat alternates; or one of each

School Breakfast The School Breakfast Program ■ is available in more than 80 percent of the nation's schools that offer school lunch, and close to 9 million children participate in it.[60] Nevertheless, for many children who need it, the School Breakfast Program is either unavailable, or the children do not participate in it.[61] The majority of children who eat school breakfasts are from low-income families. As research results continue to emphasize the positive impact breakfast has on school performance and health, vigorous campaigns to expand school breakfast programs are under way.

School Lunch More than 28 million children receive lunches through the National School Lunch Program—more than half of them free or at a reduced price.[62] School lunches are designed to provide at least one-third of the recommendation for energy, protein, vitamin A, vitamin C, iron, and calcium. They must also include specified

TABLE 12–11

School Lunch Patterns for Different Ages[a]

Food Group	Preschool (Age)		Grade School through High School (Grade)		
	1 to 2	3 to 4	K to 3	4 to 6	7 to 12
Meat or meat alternate					
1 serving:					
Lean meat, poultry, or fish	1 oz	$1^1/_2$ oz	$1^1/_2$ oz	2 oz	3 oz
Cheese	1 oz	$1^1/_2$ oz	$1^1/_2$ oz	2 oz	3 oz
Large egg(s)	$^1/_2$	$^3/_4$	$^3/_4$	1	$1^1/_2$
Cooked dry beans or peas	$^1/_4$ c	$^3/_8$ c	$^3/_8$ c	$^1/_2$ c	$^3/_4$ c
Peanut butter	2 tbs	3 tbs	3 tbs	4 tbs	6 tbs
Yogurt	$^1/_2$ c	$^3/_4$ c	$^3/_4$ c	1 c	$1^1/_2$ c
Peanuts, soynuts, tree nuts, or seeds[b]	$^1/_2$ oz	$^3/_4$ oz	$^3/_4$ oz	1 oz	$1^1/_2$ oz
Vegetable and/or fruit					
2 or more servings, both to total	$^1/_2$ c	$^1/_2$ c	$^1/_2$ c	$^3/_4$ c	$^3/_4$ c
Bread or bread alternate[c]					
Servings	5 per week	8 per week	8 per week	8 per week	10 per week
Milk					
1 serving of fluid milk	3/4 c	3/4 c	1 c	1 c	1 c

[a] The quantities listed represent per lunch minimums for each age and grade except those for the oldest group, which are recommendations. Schools unable to serve the recommended quantities for grades 7 to 12 must provide at least the amount shown for grades 4 to 6.

[b] These meat alternates may be used to meet no more than half of the meat or meat alternate requirement; therefore, they must be used in a meal with another meat or meat alternate.

[c] Schools must serve daily at least $^1/_2$ serving of bread or bread alternate to the youngest age group and at least 1 serving to older children.

Source: U.S. Department of Agriculture, National School Lunch Program Regulations, revised January 1, 1998.

■ The American Dietetic Association has set nutrition standards for child care programs. Among them, meal plans should:

- Be nutritionally adequate and consistent with the *Dietary Guidelines for Americans 2005*.
- Involve parents in planning.
- Follow recommended meal patterns that balance energy and nutrients with children's ages, appetites, activity levels, and special needs while respecting cultural and ethnic differences.
- Minimize added fat, sugar, and sodium.
- Emphasize fresh fruit, fresh and frozen vegetables, and whole grains.
- Provide furniture and eating utensils that are age appropriate and developmentally suitable to encourage children to accept and enjoy mealtime.

SOURCE: Position of the American Dietetic Association: Benchmarks for nutrition programs in child care settings, *Journal of the American Dietetic Association* 105 (2005): 979–986.

numbers of servings from each food group. In an effort to help reduce cardiovascular disease risk, all government-funded meals served at schools must follow the *Dietary Guidelines for Americans 2005*. ■ Table 12–11 shows school lunch patterns for children of different ages.

Parents often rely on school lunches to meet a significant part of their children's nutrient needs on school days. Indeed, students who regularly eat school lunches have higher intakes of many nutrients and fiber than students who do not.[63] However, children do not always like what they are served, and school lunch programs must strike a balance between what children want to eat and what will nourish them and guard their health.

Competing Influences at School Serving nutritious lunches is only half the battle; students need to eat them, too. Short lunch periods and long waiting lines prevent some students from eating a school lunch and leave others with too little time to complete their meals.[64] Nutrition efforts at schools are also undermined when students can buy what the USDA labels "competitive foods"—meals from fast-food restaurants or a la carte foods such as pizza or snack foods and carbonated beverages from snack bars, school stores, and vending machines.[65] These foods and beverages compete with nutritious school lunches. Children receive a mixed message when they are left on their own to choose between the health-supporting school lunch and the high-fat, high-salt foods with low nutrient density that their taste buds may prefer. In one study, students who selected competitive foods in addition to, or instead of, school meals consumed more energy and fat and less calcium and vitamin A than those who selected only the school lunch.[66] Increasingly, school-based nutrition issues are being addressed by legislation. Some states restrict the sale of competitive foods and have higher rates of participation in school meal programs than the national average. Federal legislation mandates that all school districts that participate in the USDA's National School Lunch Program develop and put in place a local wellness policy.[67] Nutrition professionals advocate further legislative measures that would prohibit sales of food and beverages from vending machines or school stores in middle and high schools until 30 minutes after the end of the last meal unless they are part of the school food service and meet *Dietary Guidelines* standards.[68] Reducing the prices of nutritious foods also greatly increases the likelihood that students will purchase them.[69]

IN SUMMARY

- Adults at home and at school need to provide children with nutrient-dense foods and teach them how to make healthful choices.

- Adults also need to provide ample opportunity for children to be physically active.

The Teen Years

As children pass through **adolescence** on their way to becoming adults, they change in many ways. Their physical changes make their nutrient needs high, and their emotional, intellectual, and social changes make meeting those needs a challenge.

Teenagers make many more choices for themselves than they did as children. They are not fed; they eat. Food choices made during the teen years profoundly affect health, both now and in the future. At the same time, social pressures thrust choices at them: whether to drink alcoholic beverages and whether to develop their bodies to meet extreme ideals of slimness or athletic prowess. Their interest in nutrition—both valid information and misinformation—derives from personal, immediate experiences. They are concerned with how diet can improve their lives now—they try the latest fad diet in order to fit into a new bathing suit, avoid greasy foods in an effort to clear acne, or eat a plate of pasta to prepare for a big sporting event. In presenting information on the nutrition and health of adolescents, this chapter includes topics of interest to teens.

Growth and Development during Adolescence

With the onset of adolescence, the steady growth of childhood speeds up abruptly and dramatically, and the growth patterns of females and males become distinct. Hormones direct the intensity and duration of the adolescent growth spurt, profoundly affecting every organ of the body, including the brain. After two to three years of intense growth and a few more at a slower pace, physically mature adults emerge.

In general, a female's adolescent growth spurt begins at age 10 or 11, and a male's at 12 or 13. The spurt's duration is about two and a half years. Before **puberty**, the differences between male and female body composition are minimal. During the adolescent spurt, gender differences become apparent in the skeletal system, lean body mass, and fat stores. In males, the lean body mass—muscle and bone—becomes much greater, and in females, fat becomes a larger percentage of the total body weight. On average, males grow 8 inches taller, and females grow 6 inches taller. Males gain approximately 45 pounds, and females about 35 pounds.

Energy and Nutrient Needs

The energy needs of adolescents vary greatly, depending on the current rate of growth, gender, body composition, and physical activity.[70] Boys' energy needs may be especially high; they experience a more intense growth spurt and, as mentioned, develop more lean body mass than girls do. An active teenage boy of fifteen may need 3,500 kcalories or more per day just to maintain his weight. In general, because girls enter their growth spurts earlier and grow less than boys, their energy needs peak sooner and decline earlier than those of their male peers. An inactive girl of fifteen whose growth is nearly at a standstill may need fewer than 1,800 kcalories per day if she is to avoid excessive weight gain. Thus teenage girls need to pay special attention to being physically active and selecting foods of high nutrient density so as to meet their nutrient needs without exceeding their energy needs.

Obesity The insidious problem of obesity becomes ever more apparent in adolescence and often continues into adulthood. Energy balance is often difficult to regulate in this society—an estimated 15 percent of U.S. children and adolescents 6 to 19 years of age are overweight.[71] The problem is most evident in African-American females and in Hispanic children of both genders. Without intervention, overweight teens will face numerous physical and socioeconomic consequences for years to come. The consequences of obesity are so dramatic and our society's attitude toward obese people is so negative that even teens of normal weight perceive a need to control their weight. Healthy, normal-weight teenagers are often "on diets" and make all kinds of unhealthy weight-loss attempts—even taking up smoking.[72] Some adolescents may benefit from lower-kcalorie diets that increase fruits, vegetables, fat-free milk, and other nutritious foods while limiting cookies, cakes, soft drinks, fried snacks, and other less healthy choices. However, most weight-loss dieting undertaken by adolescents, particularly females, is self-prescribed and generally unhealthful and can easily lead to nutrient deficiencies.[73] When taken to extremes, restrictive diets bring dramatic physical consequences of their own, as Nutrition in Practice 6 explains.

Vitamins Recommendations for most vitamins increase during the teen years (see the tables on the inside front cover of this text). Several of the vitamin recommendations for adolescents are similar to those for adults, including the recommendation for vitamin D. During puberty, both the activation of vitamin D and the absorption of calcium are enhanced, thus supporting the intense skeletal growth of the adolescent years without additional vitamin D.

Iron The need for iron increases during adolescence for both females and males, but for different reasons. Iron needs increase for females as they start to menstruate, and for males as their lean body mass develops. Hence, the RDA increases at age fourteen for both males and females. Because menstruation continues throughout a woman's childbearing years, the RDA for iron remains high into late adulthood. For males, the RDA returns to preadolescent values in early adulthood.

In addition, iron needs increase when the adolescent growth spurt begins, whether that occurs before or after age fourteen.[74] This shifting requirement makes pinpointing an adolescent's need somewhat complicated.

Furthermore, iron recommendations for girls before age fourteen do not reflect the iron losses of menstruation. The average age of menarche (first menstruation) in the United States is 12.5 years, however.[75] Therefore, for girls under the age of fourteen

adolescence: the period of growth from the beginning of puberty until full maturity. Timing of adolescence varies from person to person.

puberty: the period in life in which a person becomes physically capable of reproduction.

who have started to menstruate, an additional 2.5 milligrams of iron per day is recommended.[76] Thus the RDA for iron depends not only on age and gender, but also on whether the individual is in a growth spurt or has begun to menstruate, as listed in the margin.

Iron intakes often fail to keep pace with increasing needs, especially for females, who typically consume fewer iron-rich foods such as meat and fewer total kcalories than males. Not surprisingly, iron deficiency is most prevalent among adolescent girls. Iron-deficient children and teens score lower on standardized tests than those who are not iron deficient.

Calcium Adolescence is a crucial time for bone development, and the requirement for calcium reaches its peak during these years.[77] Unfortunately, low calcium intakes among adolescents have reached crisis proportions: 90 percent of females and 70 percent of males ages 12 to 19 years have calcium intakes below recommendations.[78] Paired with physical inactivity, low calcium intakes can compromise the development of peak bone mass, greatly increasing the risk of osteoporosis and other bone disease later.[79] Increasing milk products in the diet to meet calcium recommendations greatly increases bone density.[80] Once again, however, teenage girls are most vulnerable, for their milk—and therefore their calcium—intakes begin to decline at the time when their calcium needs are greatest.[81] Furthermore, women have much greater bone losses than men in later life. In addition to dietary calcium, bones grow stronger with physical activity.

Food Choices and Health Habits

Teenagers like the freedom to come and go as they choose and eat what they want when they have time. In the face of many demands on their time, including after school jobs, social activities, and home responsibilities, they almost inevitably fall into irregular eating habits, relying on quick snacks or fast foods for meals.[82] Only about one-third of adolescents eat evening meals at home with their families, but almost 80 percent say that family meals are important to them.[83] The adolescent who does eat at home with family members consumes more nutritious fruits, vegetables, whole grains, and calcium-rich foods and fewer soft drinks than others.[84]

Many adolescents also begin to skip breakfast on a regular basis, missing out on important nutrients that are not made up at later meals during the day. Compared with those who skip breakfast, teenagers who do eat breakfast have higher intakes of vitamins A and C and riboflavin, as well as calcium, iron, and zinc.[85] Teenagers who eat breakfast are therefore more likely to meet their nutrient intake recommendations.

Ideally, in light of adolescents' busy schedules and desire for freedom, the adult becomes a **gatekeeper**, controlling the type and availability of food in the teenager's environment. Teenage sons and daughters and their friends should find plenty of nutritious, easy-to-grab food in the refrigerator (meats for sandwiches; low-fat cheeses; fresh, raw vegetables and fruits; fruit juices; and milk) and more in the kitchen cabinets (whole-grain breads, peanut butter, nuts, popcorn, and cereal). In many households today, all the adults work outside the home, and teenagers perform some of the gatekeeper's roles, such as shopping for groceries or choosing fast or prepared foods.

Snacks On average, about one-fourth of an adolescent's total daily energy intake comes from snacks, which, if chosen carefully, can contribute some of the needed nutrients (see Table 12–10 on p. 359). Often, however, adolescents choose foods that are

© Steve Dunwell/Index Stock Imagery

Nutritious snacks contribute valuable nutrients to an active teen's diet.

too high in saturated fat and sodium and too low in fiber to support the future health of their arteries.[86] Their calcium intakes often fall short unless they snack on dairy products, and they may fail to obtain enough iron and vitamin A. For iron and other nutrients, a teenager might snack on iron-containing meat sandwiches or tuna fish, low-fat bran muffins, or tortillas with spicy bean spread along with a glass of orange juice to help maximize the iron's absorption.

Beverages Increasingly, as Figure 12–7 shows, adolescents are drinking soft drinks with lunch, supper, and snacks. About the only time they select fruit juices is at breakfast. When they drink milk, they are more likely to consume it with a meal (especially breakfast) than as a snack. Soft drinks, when chosen as the primary beverage, may affect the density of the bones because they displace milk from the diet.[87] Because of their greater food intakes, boys are more likely to drink enough milk to meet their calcium needs, whereas girls typically fall short of calcium recommendations. Regular soft drink consumption is also linked with overweight in adolescents.[88]

Soft drinks present another problem, when caffeine intake becomes excessive. However, caffeine seems to be relatively harmless when used in moderate doses (the equivalent of fewer than three 12-ounce cola beverages per day). ■ In greater amounts, it can cause the symptoms associated with anxiety—sweating, tenseness, and inability to concentrate.

Eating Away from Home Adolescents eat about one-third of their meals away from home, and their nutritional welfare is enhanced or hindered by the choices they make. A lunch of a hamburger, a chocolate shake, and french fries supplies substantial quantities of many nutrients at a kcalorie cost of about 800, an energy intake many adolescents can afford. When they eat this sort of lunch, teens can adjust their breakfast and dinner choices to include fruits and vegetables for vitamin A, vitamin C, folate, and fiber and lean meats and legumes for iron and zinc. Fortunately, many fast-food restaurants are offering more nutritious choices than the standard hamburger meal.

Peer Influence Teenagers are intensely engaged in day-to-day life with their peers and preparing for their future lives as adults. Adults need to remember that adolescents have the right to make their own decisions—even if those decisions are not in line with the adults' own views. Gatekeepers can set up the environment so that nutritious foods are available and can stand by with reliable nutrition information and advice, but the rest is up to the teenagers. Ultimately, they make the choices.

IN SUMMARY

- Nutrient needs rise dramatically as children enter the rapid growth phase of the teen years.

- The busy lifestyles of teenagers add to the challenge of meeting their nutrient needs, especially for iron and calcium.

Assessment of nutrition status in healthy infants, children, and adolescents can confirm that development is normal or can catch potential problems early. ■ The Nutrition Assessment Checklist (p. 366) highlights problems to look for when working with infants, children, and adolescents.

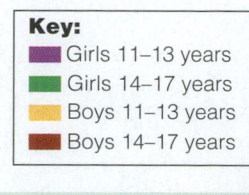

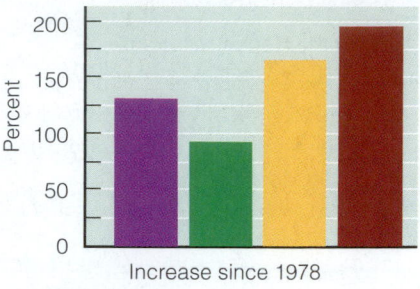

Key:
- Girls 11–13 years
- Girls 14–17 years
- Boys 11–13 years
- Boys 14–17 years

FIGURE 12–7

Increase in Soft Drink Consumption of U.S. Adolescents

Average soft drink consumption by adolescents more than doubled between 1978 and the present.

Source: S. A. French, B. H. Lin, and J. F. Guthrie. National trends in soft drink consumption among children and adolescents age 6–17 years: Prevalence, amounts, and sources, 1977/1978 to 1994/1998, *Journal of the American Dietetic Association* 103 (2003): 1326–1331.

■ Appendix A provides a table of the caffeine contents of beverages, foods, and medications.

■ Chapter 14 offers details about nutrition assessment.

gatekeeper: with respect to nutrition, a key person who controls other people's access to foods and thereby exerts a profound impact on their nutrition. Examples are the spouse who buys and cooks the food, the parent who feeds the children, and the caregiver in a day care center.

Nutrition Assessment Checklist

FOR INFANTS, CHILDREN, AND ADOLESCENTS

Medical History

Check the medical record for:

- Alcohol, tobacco, or illicit drug abuse
- Attention deficit/hyperactivity disorder (ADHD)
- Diabetes or other chronic disorders
- Eating disorders
- Food allergies
- Lactose intolerance
- Obesity
- Pregnancy

Medications

For children or adolescents being treated with drug therapy for medical conditions, note:

- Side effects that might reduce food intake or change nutrient needs
- Proper administration of medication with respect to food intake

Dietary Intake

For infants, note:

- Method of feeding (breastfeeding, formula, or both)
- Frequency and duration of breastfeeding
- Amount of infant formula
- Practice of putting infant to bed with bottle
- Solid foods the infant is fed, if any
- Amount of food the infant is fed

For all children and adolescents, especially those considered at risk nutritionally, assess the diet for:

- Total energy
- Protein
- Calcium and iron
- Vitamin A, vitamin C, and folate
- Fiber

Note the following:

- Number of days each week a nutritious breakfast is eaten
- Number of hours the child or teen sleeps each day
- Number of soft drinks the child or teen drinks each day
- Number of fast-food meals eaten each day
- Number and type of snacks eaten each day
- Type and amount of physical activity
- Amount of caffeine consumed

Anthropometric Data

Measure baseline height and weight.

- Note infant birth weight.
- Reassess height, weight, and growth patterns at each medical checkup.
- Note weight, length, and head circumference of infants.
- Note significant obesity or underweight and intervention strategies employed.

Laboratory Tests

Monitor the following laboratory tests for infants:

- Blood glucose of infants born to mothers with gestational diabetes
- Results of tests for inborn errors

Monitor the following laboratory tests for children and adolescents:

- Hemoglobin, hematocrit, or other tests of iron status
- Blood glucose for children or adolescents with diabetes
- Blood lead concentrations

Physical Signs

Look for physical signs of:

- Protein-energy malnutrition
- Iron deficiency
- Vitamin A deficiency
- Vitamin C deficiency
- Folate deficiency

Analysis

CASE STUDY

Boy with Disruptive Behavior

Freddie Willis is a six-year-old boy who seldom sits still, often misbehaves, and is frequently sick. Freddie's eating habits are erratic and poor, as is his appetite. He often misses breakfast because he is too tired to get up in time to eat before school. By midmorning, Freddie is irritable and disruptive in the classroom. At lunchtime he trades the peanut butter and banana sandwich his mother packed in his lunchbox for a piece of cake. After school he hurries home to watch television while he eats his favorite snack—cola and potato chips. At dinnertime Freddie picks at his food because he isn't very hungry. Later on, when it's time for bed, Freddie complains that he's hungry. His parents let him stay up to have a bowl of cereal (the kind with marshmallows) before he finally falls asleep.

1. What factors in Freddie's daily routine might be contributing to his restless behavior?

2. Discuss some changes in diet that might improve Freddie's health and disposition.

1. At two and a half years old, Travis is healthy, though slightly underweight, and headstrong. Travis's mother hovers over him at every meal and insists that he take several bites of every food on his plate, even if he dislikes the food or is not familiar with it. Even though Travis is hungry when he sits down to a meal, his mother's constant urging to get him to eat quickly quells any interest he had in eating. Travis simply folds his arms across his chest, closes his mouth tightly, and refuses to eat any more food. After more begging, pleading, and nagging, Travis's mother becomes angry and sends him away from the table. Travis is not allowed to snack between meals because his mother is concerned that snacks will ruin his appetite.

   ■ What factors might be contributing to Travis's refusal to eat?

   ■ Travis's mother is concerned about her son's underweight. What strategies would you suggest to help Travis gain weight?

   ■ What advice would you offer Travis's mother to help her improve mealtimes with her son?

2. Loni is a physically inactive, slightly overweight 16-year-old who usually skips breakfast because she is too tired to get up early enough to eat it and get to school on time. At lunch, she would rather socialize with friends than stand in the cafeteria line, so she grabs a cola and crackers or chips from the vending machine. She stays after school to work on the yearbook with friends and then joins them at a nearby fast-food restaurant for a quick meal of chicken nuggets, fries, and another cola drink. When Loni gets home, her family is often sitting down to dinner, but because she is no longer hungry, she picks at her food until she is excused to do her homework. Later, she snacks on a cola drink and popcorn before going to bed. Loni has noticed that she is gaining weight, having a difficult time falling asleep, becoming easily fatigued, and getting sick frequently.

   ■ What would you tell Loni to do to help her sleep better?

   ■ What dietary advice would you suggest to help Loni stop gaining weight and start feeling healthier?

   ■ What else might you suggest to help Loni feel stronger and healthier?

Self Check

1. Breast milk is recommended for the first 12 months of life because it offers complete nutrition and ———— to the infant.
   a. fluoride
   b. fructose
   c. immunological protection
   d. pica

2. Which of the following is a characteristic of iron deficiency in children?
   a. it rarely develops in those with high intakes of milk.
   b. it affects brain function before anemia sets in.
   c. it is a primary factor in hyperactivity.
   d. mild deficiency enhances mental performance by lowering physical activity level thereby leading to increased attention span.

3. Three symptoms of lead toxicity are:
   a. diarrhea, irritability, and fatigue.
   b. low blood sugar, hair loss, and skin rash.
   c. increased heart rate, hyperactivity, and dry skin.
   d. bleeding gums, brittle fingernails, and swollen glands.

4. Allergic reactions to foods are most often caused by:
   a. corn, rice, or meats.
   b. eggs, peanuts, or milk.
   c. red meats, milk, or MSG.
   d. seafood, dark greens, or lactose.

5. When introducing new foods to children:
   a. reward children as they try new foods.
   b. offer many choices to encourage variety.
   c. offer one new food at the end of the meal.
   d. offer one new food at the beginning of the meal.

6. Which of the following is not true? Children who watch a lot of television are likely to:
   a. become obese.
   b. spend less time being physically active.
   c. learn healthy eating tips from programs.
   d. eat the foods most often advertised on television.

7. Which of the following strategies is not effective?
   a. Play first, eat later.
   b. Provide small portions.
   c. Encourage children to help prepare meals.
   d. Use dessert as a reward for eating vegetables.

8. During the growth spurt of adolescence:
   a. females gain more weight than males.
   b. males gain more fat, proportionately, than females.
   c. differences in body composition between males and females become apparent.
   d. similarities in body composition between males and females become apparent.

9. Two nutrients that are usually lacking in adolescents' diets are:
   a. zinc and fat.
   b. iron and calcium.
   c. protein and thiamin.
   d. vitamin A and riboflavin.

10. To help teenagers consume a balanced diet, parents can:
    a. monitor the teens' food intake.
    b. give up—parents can't influence teenagers.
    c. keep the cupboards and refrigerator well stocked.
    d. forbid snacking and insist on regular, well-balanced meals.

Answers to these questions are in Appendix H.

Notes

1. American Academy of Pediatrics, Policy statement, Controversies concerning vitamin K and the newborn, *Pediatrics* 112 (2003): 191–192.

2. Committee on Nutrition, American Academy of Pediatrics, *Pediatric Nutrition Handbook*, 5th ed., ed. R. E. Kleinman (Elk Grove Village, IL: American Academy of Pediatrics, 2004), pp. 110–111.

3. American Academy of Pediatrics, Policy statement: Breastfeeding and the use of human milk, *Pediatrics* 115 (2005): 496–506.

4. J. D. Carver, Advances in nutritional modifications of infant formulas, *American Journal of Clinical Nutrition* 77 (2003): 1550S–1554S.

5. W. C. Heird and A. Lapillonne, The role of essential fatty acids in development, A*nnual Review of Nutrition* 25 (2005): 549–571; J. C. McCann and B. N. Ames, Is docosahexaenoic acid, an n-3 long-chain polyunsaturate fatty acid, required for development of normal brain function? An overview of evidence from cognitive and behavioral tests in humans and animals, *American Journal of Clinical Nutrition* 82 (2005): 281–295; N. Auestad and coauthors, Visual, cognitive, and language assessments at 39 months: A follow-up study of children fed formulas containing long-chain polyunsaturated fatty acids to 1 year of age, *Pediatrics* 112 (2003): e177–183.

6. C. L. Cheatham, J. Columbo, and S. E. Carlson, n-3 fatty acids and cognitive and visual acuity development: Methodologic and conceptual considerations, *American Journal of Clinical Nutrition* 83 (2006): 1458S–1466S; W. W. Koo, Efficacy and safety of docosahexaenoic acid and arachidonic acid addition to infant formulas: Can one buy better vision and intelligence? *Journal of the American College of Nutrition* 22 (2003): 101–107; E. E. Birch and coauthors, A randomized controlled trial of long-chain polyunsaturated fatty acid supplementation of formula in term infants after weaning at 6 wk of age, *American Journal of Clinical Nutrition* 75 (2002): 570–580.

7. Auestad and coauthors, 2003.

8. E. E. Birch and coauthors, Visual maturation of term infants fed long-chain polyunsaturated fatty acid–supplemented or control formula for 12 mo, *American Journal of Clinical Nutrition* 81 (2005): 871–879; Birch and coauthors, 2002.

9. B. Lonnerdal, Nutritional and physiologic significance of human milk proteins, *American Journal of Clinical Nutrition* 77 (2003): 1537S–1543S.

10. L. M. Gartner, F. R. Greer, and the Section on Breastfeeding and Committee on Nutrition, Prevention of rickets and vitamin D deficiency: New guidelines for vitamin D intake, *Pediatrics* 111 (2003): 908–910.

11. Gartner, Greer, and the Section on Breastfeeding and Committee on Nutrition, 2003.

12. American Academy of Pediatrics, 2005; Position of the American Dietetic Association: Promoting and supporting breastfeeding, *Journal of the American Dietetic Association* 105 (2005): 811–818.

13. D. S. Newburg, G. M. Ruiz-Palacios, and A. L. Morrow, Human milk glycans protect infants against enteric pathogens, *Annual Review of Nutrition* 25 (2005): 37–58.

14. American Academy of Pediatrics, 2005; Position of the American Dietetic Association, 2005.

15. M. W. Gillman and coauthors, Risk of overweight among adolescents who were breastfed as infants, *Journal of the American Medical Association* 285 (2001): 2461–2467.

16. M. L. Hediger and coauthors, Association between infant breastfeeding and overweight in young children, *Journal of the American Medical Association* 285 (2001): 2453–2460.

17. C. G. Owen and coauthors, Effect of infant feeding on the risk of obesity across the life course: A quantitative review of published evidence, *Pediatrics* 115 (2005): 1367–1377.

18. R. M. Martin and coauthors, Breastfeeding and atherosclerosis: Intima-media thickness and plaques at 65-year follow-up of the Boyd Orr cohort, *Arteriosclerosis, Thrombosis, and Vascular Biology* 25 (2005): 1482–1488; C. G. Owen and coauthors, Infant feeding and blood cholesterol: A study in adolescents and a systematic review, *Pediatrics* 110 (2002): 597–608.

19. M. C. Daniels and L. S. Adair, Breastfeeding influences cognitive development in Filipino children, *Journal of Nutrition* 135 (2005): 2589–2595; E. L. Mortensen and coauthors, The association between duration of breastfeeding and adult intelligence, *Journal of the American Medical Association* 287 (2002): 2365–2371.

20. G. Der, G. D. Batty, and I. J. Deary, Effect of breast feeding on intelligence in children: prospective study, sibling pairs analysis, and meta-analysis, *British Medical Journal* doi:10.1136/bmj.38978.699583.55 (published October 4, 2006); A. Jain, J. Concato, and J. M. Leventhal, How good is the evidence linking breastfeeding and intelligence? *Pediatrics* 109 (2002): 1044–1053.

21. L. Seppo and coauthors, A follow-up study of nutrient intake, nutritional status, and growth in infants with cow milk allergy fed either a soy formula or an extensively hydrolyzed whey formula, *American Journal of Clinical Nutrition* 82 (2005): 140–145; Committee on Nutrition, American Academy of Pediatrics, Formula feeding of term infants, in R. E. Kleinman ed., *Pediatric Nutrition Handbook*, 5th ed., (Elk Grove Village, IL: American Academy of Pediatrics, 2004), pp. 87–97.

22. Committee on Nutrition, American Academy of Pediatrics, *Pediatric Nutrition Handbook*, 2004, pp. 87–97.

23. Committee on Nutrition, American Academy of Pediatrics, *Pediatric Nutrition Handbook*, 2004, p. 111.

24. Committee on Nutrition, American Academy of Pediatrics, *Pediatric Nutrition Handbook*, 2004, p. 105.

25. Committee on Nutrition, American Academy of Pediatrics, Complementary feeding, in R. E. Kleinman, ed. *Pediatric Nutrition Handbook*, 5th ed., (Elk Grove Village, IL: American Academy of Pediatrics, 2004), pp. 103–115.

26. Committee on Nutrition, American Academy of Pediatrics, Complementary feeding, 2004; J. D. Skinner and B. R. Carruth, A longitudinal study of children's juice intake and growth: The juice controversy revisited, *Journal of the American Dietetic Association* 101 (2001): 432–437; Committee on Nutrition, American Academy of Pediatrics, The use and misuse of fruit juice in pediatrics, *Pediatrics* 107 (2001): 1210–1213.

27. M. K. Fox and coauthors, Relationship between portion size and energy intake among infants and toddlers: Evidence of self-regulation, *Journal of the American Dietetic Association* 106 (2006): S77–S83.

28. Institute of Medicine, Overview of IOM's childhood obesity prevention study, *Fact Sheet*, September, 2004 available at www.iom.edu.

29. American Academy of Pediatrics, Council on Sports Medicine and Fitness and Council on School Health, Active healthy living: Prevention of childhood obesity through increased physical activity, *Pediatrics* 117 (2006): 1834–1842.

30. F. R. Greer, N. F. Krebs, and the Committee on Nutrition, Optimizing bone health and calcium intakes of infants, children, and adolescents, *Pediatrics* 117 (2006): 578–585; H. Vatanparast and coauthors, Positive effects of vegetable and fruit consumption and calcium intake on bone mineral accrual in boys during growth from childhood to adolescence: The University of Saskatchewan Pediatric Bone Mineral Accrual Study, *American Journal of Clinical Nutrition* 82 (2005): 700–706; J. O. Fisher and coauthors, Meeting calcium recommendations during middle childhood reflects mother-daughter beverage choices and predicts bone mineral status, *American Journal of Clinical Nutrition* 79 (2004): 698–706.

31. A. Carlson and coauthors, U.S. Department of Agriculture Center for Nutrition Policy and Promotion, Report card on the diet quality of children ages 2 to 9, *Nutrition Insights* 25 (2001), available at www.usda.gov/cnpp/Insights/Insight25.pdf.

32. P. Ziegler and coauthors, Feeding infants and toddlers study (FITS): Development of the FITS Survey in comparison to other dietary survey methods, *Journal of the American Dietetic Association* 106 (2006): S12–S27.

33. J. Stang, Improving the eating patterns of infants and toddlers, *Journal of the American Dietetic Association* 106 (2006): S7–S9; M. K. Fox and coauthors, Feeding infants and toddlers study: What foods are infants and toddlers eating? *Journal of the American Dietetic Association* 104 (2004): S22–S30.

34. Position of the American Dietetic Association: Dietary guidance for healthy children ages 2 to 11 years, *Journal of the American Dietetic Association* 104 (2004): 660–677.

35. Forum on Child and Family Statistics, *America's Children: Key National Indicators of Well-Being, 2005*, available at www.childstats.gov.

36. G. C. Rampersaud and coauthors, Breakfast habits, nutritional status, body weight, and academic performance in children and adolescents, *Journal of the American Dietetic Association* 105 (2005): 743–760; S. G. Affenito and coauthors, Breakfast consumption by African-American and white adolescent girls correlates positively with calcium and fiber intake and negatively with body mass index, *Journal of the American Dietetic Association* 105 (2005): 938–945; Position of the American Dietetic Association, 2004.

37. Rampersaud and coauthors, 2005; Affenito and coauthors, 2005; Position of the American Dietetic Association, 2004; C. S. Berkey and coauthors, Longitudinal study of skipping breakfast and weight change in adolescents, *International Journal of Obesity and Related Metabolic Disorders* 27 (2003): 1258–1266.

38. J. L. Beard and J. R. Connor, Iron status and neural functioning, *Annual Review of Nutrition* 23 (2003): 41–58.

39. B. Lozhoff and coauthors, Long-lasting neural and behavioral effects of iron deficiency in infancy, *Nutrition Reviews* 64 (2006): S34–S43.

40. Centers for Disease Control and Prevention, Blood lead levels—United States, 1999–2002, *Morbidity and Mortality Weekly Report* 54 (2005): 513–616.

41. Committee on Environmental Health, American Academy of Pediatrics, Policy statement: Lead exposure in children: Prevention, detection, and management, *Pediatrics* 116 (2005): 1036–1046.

42. Centers for Disease Control and Prevention, 2005.

43. Committee on Environmental Health, American Academy of Pediatrics, 2005.

44. U.S. Department of Health and Human Services, National Institutes of Health, National Institute of Allergy and Infectious Diseases, *Food Allergy: An Overview*, NIH publication no. 04-5518 (July 2004), available at www.niaid.nih.gov; R. Formanek, Food allergies: When food becomes the enemy, *FDA Consumer*, July/August 2001, pp. 12–16.

45. H. Skolnick and coauthors, The natural history of peanut allergy, *Journal of Allergy and Clinical Immunology* 107 (2001): 367–374; Formanek, 2001.

46. L. Christie and coauthors, Food allergies in children affect nutrient intake and growth, *Journal of the American Dietetic Association* 102 (2002): 1648–1651.

47. H. Metzger, Two approaches to peanut allergy, *New England Journal of Medicine* 348 (2003): 1046–1048.

48. Food Allergen Labeling and Consumer Protection Act of 2004, available at http://thomas.loc.gov/cgi-bin/query/F?c108:6:./temp/~c108Dz8zuL:e48634.

49. Formanek, 2001.

50. B. Merz, Studying peanut anaphylaxis, *New England Journal of Medicine* 348 (2003): 975–976; Metzger, 2003; X. M. Li and coauthors, Persistent protective effect of heat-killed *Escherichia coli* producing "engineered," recombinant peanut proteins in a murine model of peanut allergy, *Journal of Allergy and Clinical Immunology* 112 (2003): 159–167.

51. M. L. Wolraich and coauthors, Attention-deficit/hyperactivity disorder among adolescents: A review of the diagnosis, treatment, and clinical implications, *Pediatrics* 115 (2005): 1734–1746.

52. S. J. Salmon, K. J. Campbell, and D. A. Crawford, Television viewing habits associated with obesity risk factors: A survey of Melbourne schoolchildren, *Medical Journal of Australia* 184 (2006): 64–67; D. M. Matheson and coauthors, Children's food consumption during television viewing, *American Journal of Clinical Nutrition* 79 (2004): 1088–1094.

53. R. M. Viner and T. J. Cole, Television viewing in early childhood predicts adult body mass index, *Journal of Pediatrics* 147 (2005): 429–435.

54. B. A. Dennison, T. A. Erb, and P. L. Jenkins, Television viewing and television in bedroom associated with overweight risk among low-income preschool children, *Pediatrics* 109 (2002): 1028–1035.

55. K. A. Coon and coauthors, Relationships between use of television during meals and children's food consumption patterns, *Pediatrics* 107 (2001): 167 [www.pediatrics.org/cgi/content/full/107/1/e7]; C. J. Crespo and coauthors, Television watching, energy intake, and obesity in U.S. children: Results from the Third National Health and Nutrition Examination Survey, 1988–1994, *Archives of Pediatrics and Adolescent Medicine* 155 (2001): 360–365.

56. J. L. Wiecha and coauthors, The hidden and potent effects of television advertising, *Archives of Pediatrics and Adolescent Medicine* 160 (2006): 436–442.

57. T. J. Hindin, I. R. Contento, and J. D. Gussow, A media literacy nutrition education curriculum for head start parents about the effects of television advertising on their children's food requests, *Journal of the American Dietetic Association* 104 (2004): 192–198.

58. J. Wardle, S. Carnell, and L. Cooke, Parental control over feeding and children's fruit and vegetable intake: How are they related? *Journal of the American Dietetic Association* 105 (2005): 227–232; A. T. Galloway and coauthors, Parental pressure, dietary patterns, and weight status among girls who are "picky eaters," *Journal of the American Dietetic Association* 105 (2005): 541–548; L. J. Cooke and coauthors, Demographic, familial and trait predictors of fruit and vegetable consumption by preschool children, *Public Health Nutrition* 2 (2004): 251–252.

59. Position of the American Dietetic Association: Benchmarks for nutrition programs in child care settings, *Journal of the American Dietetic Association* 105 (2005): 979–986.

60. Position of the American Dietetic Association: Local support for nutrition integrity in schools, *Journal of the American Dietetic Association* 106 (2006): 122–133.

61. Position of the American Dietetic Association, 2006.

62. Position of the American Dietetic Association, 2006.

63. Position of the American Dietetic Association, 2004; K. W. Cullen and I. Zakeri, Fruits, vegetables, milk, and sweetened beverages consumption and access to a la carte/snack bar meals at school, *American Journal of Public Health* 94 (2004): 463–467; P. M. Gleason and C. W. Suitor, Eating at school: How the National School Lunch Program affects children's diets, *American Journal of Agricultural Economics* 85 (2003): 1047–1051.

64. Position of the American Dietetic Association, 2006.

65. Position of the American Dietetic Association, 2006; Committee on School Health, American Academy of Pediatrics, Soft drinks in schools, *Pediatrics* 113 (2004): 152–154; J. L. Kramer-Atwood and coauthors, Fostering healthy food consumption in schools: Focusing on the challenges of competitive foods, *Journal of the American Dietetic Association* 102 (2002): 1228–1233.

66. S. B. Templeton and coauthors, Competitive foods increase the intake of energy and decrease the intake of certain nutrients by adolescents consuming school lunch, *Journal of the American Dietetic Association* 105 (2005): 215–220.

67. Position of the American Dietetic Association, 2006.

68. Position of the American Dietetic Association, Society for Nutrition Education, and American School Food Service Association, Nutrition Services: An essential component of comprehensive school health programs, *Journal of the American Dietetic Association* 103 (2003): 505–514.

69. S. A. French, Public health strategies for dietary change: Schools and workplaces, *Journal of Nutrition* 135 (2005): 910-912; S. A. French, Pricing effects on food choices, *Journal of Nutrition* 133 (2003): 841S–843S.

70. Standing Committee on the Scientific Evaluation of Dietary Reference Intakes, Food and Nutrition Board, Institute of Medicine, *Dietary Reference Intakes for Energy, Carbohydrate, Fiber, Fat, Fatty Acids, Cholesterol, Protein, and Amino Acids* (Washington, DC: National Academies Press, 2005), pp. 177–182.

71. Food and Nutrition Board, Institute of Medicine, *Preventing Childhood Obesity: Health in the Balance* (Washington, DC: National Academies Press, 2005), p. 56.

72. S. E. Saarni and coauthors, Intentional weight loss and smoking in young adults, *International Journal of Obesity and Related Metabolic Disorders* 28 (2004): 796–802; J. Cawley, S. Markowitz, and J. Tauras, Lighting up and slimming down:

The effects of body weight and cigarette prices on adolescent smoking initiation, *Journal of Health Economics* 23 (2004): 293–311; D. Neumark-Sztainer and coauthors, Weight-control behaviors among adolescent girls and boys: Implications for dietary intake, *Journal of the American Dietetic Association* 104 (2004): 913–920.

73. Neumark-Sztainer and coauthors, 2004.

74. Standing Committee on the Scientific Evaluation of Dietary Reference Intakes, Food and Nutrition Board, Institute of Medicine, *Dietary Reference Intakes for Vitamin A, Vitamin K, Arsenic, Boron, Chromium, Copper, Iodine, Iron, Manganese, Molybdenum, Nickel, Silicon, Vanadium, and Zinc* (Washington, DC: National Academy Press, 2001), pp. 290–393.

75. W. C. Chumlea and coauthors, Age at menarche and racial comparisons in US girls, *Pediatrics* 111 (2003): 112–113.

76. Committee on the Scientific Evaluation of Dietary Reference Intakes, 2001, pp. 290–393.

77. F. R. Greer, N. F. Krebs, and the Committee on Nutrition, Optimizing bone health and calcium intakes of infants, children, and adolescents, *Pediatrics* 117 (2006): 578–585.

78. Greer, Krebs, and the Committee on Nutrition, 2006.

79. Federal Update, Milk matters, *Journal of the American Dietetic Association* 102 (2002): 469.

80. H. J. Kalkwarf, J. C. Khoury, and B. P. Lanphear, Milk intake during childhood and adolescence, adult bone density, and osteoporotic fractures in US women, *American Journal of Clinical Nutrition* 77 (2003): 257–265.

81. S. A. Bowman, Beverage choices of young females: Changes and impact on nutrient intakes, *Journal of the American Dietetic Association* 102 (2002): 1234–1239.

82. M. Story, D. Neumark-Sztainer, and S. French, Individual and environmental influences on adolescent eating behaviors, *Journal of the American Dietetic Association* 102 (2002): S40–S51.

83. Story, Neumark-Sztainer, and French, 2002.

84. D. Neumark-Sztainer and coauthors, Family meal patterns: Associations with sociodemographic characteristics and improved dietary intake among adolescents, *Journal of the American Dietetic Association* 103 (2003): 317–322.

85. Rampersaud and coauthors, 2005.

86. American Heart Association, S. S. Gidding and coauthors, Dietary recommendations for children and adolescents: A guide for practitioners, *Pediatrics* 117 (2006): 544–559.

87. Greer, Krebs, and the Committee on Nutrition, 2006; Vatanparast and coauthors, 2005; G. Mrdjenovic and D. A. Levitsky, Nutritional and energetic consequences of sweetened drink consumption in 6- to 13-year-old children, *Journal of Pediatrics* 142 (2003): 604–610.

88. J. James and coauthors, Preventing childhood obesity by reducing consumption of carbonated drinks: Cluster randomised controlled trial, *British Medical Journal* 328 (2004): 1237.

CHILDHOOD OBESITY AND THE EARLY DEVELOPMENT OF CHRONIC DISEASES

When most people think of health problems in children and adolescents, they typically think of ear infections, colds, and acne, not type 2 diabetes and hypertension. Today, unprecedented numbers of U.S. children are being diagnosed with obesity and the serious "adult diseases" such as type 2 diabetes that accompany overweight.[1] For children born in the United States in the year 2000, the risk of developing type 2 diabetes sometime in their lives is estimated to be 30 percent for boys and 40 percent for girls.[2] U.S. children are not alone—rapidly rising rates of obesity are threatening the health of an alarming number of children around the globe.[3] Without immediate intervention, some 60 million children are destined to suffer type 2 diabetes and hypertension in childhood followed by **cardiovascular disease (CVD)** in early adulthood.

Over the past three decades, researchers have been observing how changes in body weight, blood lipids, blood pressure, and individual behaviors correlate with the development of CVD over time—from infancy to childhood through adolescence and into young adulthood. Some major findings have emerged from this research:

- Changes inside the arteries—changes predictive of CVD—are evident in childhood.

- Obesity in children affects these changes.

- Behaviors that influence the development of obesity and of CVD are learned and begin early in life. These behaviors include overeating, physical inactivity, and cigarette smoking.

This Nutrition in Practice focuses on efforts to prevent childhood obesity, type 2 diabetes, and CVD (see the accompanying glossary for definitions of the relevant terms). The years of childhood are emphasized here, for the earlier in life health-promoting habits become established, the better they will stick.

What about genetics? Do some people inherit the tendency to become obese or develop diabetes or CVD regardless of the lifestyle habits they adopt?

For obesity, as well as for CVD, hypertension, and type 2 diabetes, genetics does not appear to play a *determining* role; that is, a person is not simply destined at birth to develop them. Instead, genetics appears to play a *permissive* role—the potential is inherited and will then develop, if given a push by factors in the environment such as poor diet, sedentary lifestyle, and cigarette smoking. Researchers note that the relationship between genes and the environment is a synergistic one—their combined effects are greater than the sum of their individual effects.[4] The Pima Indians exemplify this gene-environment

GLOSSARY

atherosclerosis (ATH-er-oh-scler-OH-sis): a type of artery disease characterized by plaques (accumulations of lipid-containing material) on the inner walls of the arteries (see Chapter 22).
   athero = porridge or soft
   scleros = hard
   osis = condition

cardiovascular disease (CVD): a general term for all diseases of the heart and blood vessels. Atherosclerosis is the main cause of CVD. When the arteries that carry blood to the heart muscle become blocked, the heart suffers damage known as **coronary heart disease (CHD)**.
   cardio = heart
   vascular = blood vessels

fatty streaks: accumulations of cholesterol and other lipids along the walls of the arteries.

plaque (PLACK): an accumulation of fatty deposits, smooth muscle cells, and fibrous connective tissue that develops in the artery walls in atherosclerosis. Plaque associated with atherosclerosis is known as **atheromatous** (ATH-er-OH-ma-tus) **plaque**.

interaction. Their population seems to be genetically predisposed to obesity and type 2 diabetes—they have high rates of both diseases.[5] However, those who live in the restrictive, remote environment of the Mexican mountains have a lower prevalence of obesity and type 2 diabetes than those who live in the obesity-promoting environment of Arizona.[6]

What about events that take place during fetal development—malnutrition for example? Can they affect a person's tendency to develop diseases later in life?

A theory called *fetal programming* or *fetal origins of disease* states that maternal malnutrition or other harmful conditions at a critical period of fetal development may have lifelong effects on an individual's pattern of genetic expression and therefore on the tendency to develop obesity and certain diseases.[7] Blood pressure is the most studied outcome in researching this theory because it is easy to measure and is a known risk factor for CVD.[8] Most studies use infant birth weight as the indicator of fetal nutrition status. For the most part, research shows that lower birth weight increases the risk of adult hypertension.[9] Researchers' efforts to explain why fetal nutrient insufficiency and lower birth weight enhance the risk of adult hypertension center on factors that affect the function of the kidneys, vascular system, nervous system, and other organs or systems involved in the complex process of blood pressure regulation. For example, fetal growth restriction may impair kidney growth and development as well as the function of blood vessels.

Research also suggests that *postnatal* growth influences adult blood pressure.[10] People with relatively low birth weights who grow to be relatively large adults seem to have the highest risk of hypertension as adults.[11] This type of growth pattern—low birth weight followed by rapid "catch-up" growth—may be a risk factor for later hypertension, although more research is needed to confirm or refute these findings.

Why has type 2 diabetes become so prevalent?

Type 2 diabetes, a chronic disease closely linked with obesity, has been on the rise among children and adolescents as the prevalence of obesity in U.S. youth has increased in recent years. Obesity is the most important risk factor for type 2 diabetes—most of the children diagnosed with type 2 diabetes are obese.[12] Most are diagnosed during puberty, but as children become more obese and less active, the disease is appearing in younger and younger children. Type 2 diabetes is most likely to occur in those who are obese and sedentary and have a family history of diabetes.

How does type 2 diabetes develop?

In type 2 diabetes, the body's cells become insulin-resistant—that is, the cells become less sensitive to insulin, reducing the amount of glucose entering the cells from the blood. The combination of obesity and insulin resistance produces a cluster of symptoms, including high blood pressure and high blood lipids, which in turn promotes the development of atherosclerosis and the early development of CVD.[13] Other common problems evident by early adulthood include kidney disease, blindness, and miscarriages. The complications of diabetes, especially when encountered at a young age, can shorten life expectancy. Chapter 21 offers a detailed discussion of diabetes.

Prevention and treatment of type 2 diabetes depend on weight management, which can be particularly difficult in a young person's world of food advertising, video games, and pocket money for candy bars. The activity and dietary suggestions to help defend against heart disease later in this discussion apply to type 2 diabetes as well.

How does CVD develop, and when does its development begin?

Most CVD involves **atherosclerosis**—the accumulation of cholesterol and other blood lipids along the walls of the arteries. Frequently, atherosclerosis and its complications interfere with the flow of blood to the heart and can lead to coronary heart disease (CHD), which, in turn, raises the likelihood of a heart attack. When atherosclerosis interferes with blood flow to the brain, a stroke can result. Infants are born with healthy, smooth, clear arteries, but within the first decade of life, **fatty streaks** may begin to appear. During adolescence, these fatty streaks

may begin to turn to **plaques** (Figure 22–2 in Chapter 22 shows the formation of plaques in atherosclerosis). By early adulthood, the fibrous plaques may begin to calcify and become raised lesions, especially in boys and young men. As the lesions grow more numerous and thicken, the heart disease rate begins to rise, and the rise becomes dramatic at about age 45 in men and 55 in women. From this point on, arterial damage and blockage progress rapidly, and heart attacks and strokes threaten life. In short, the consequences of atherosclerosis, which become apparent only in adulthood, have their beginnings in the first decades of life.[14]

Children with the highest risks of developing heart disease are sedentary and obese, with diabetes, high blood pressure, and high blood cholesterol.[15] In contrast, children with the lowest risks of heart disease are physically active and of normal weight, with low blood pressure and favorable lipid profiles.

Parents do not need to worry about their children's blood cholesterol, do they?

Atherosclerotic lesions reflect blood cholesterol: as blood cholesterol increases, lesion coverage increases. Cholesterol values at birth are similar in all populations; differences emerge in early childhood. Standard values for cholesterol screening in children and adolescents are listed in Table NP12–1.

In general, blood cholesterol tends to rise as saturated fat intakes increase. Blood cholesterol also correlates with childhood obesity, especially central obesity.[16] In obese children, the LDL (low-density lipoprotein) cholesterol value is often too high, and the HDL (high-density lipoprotein) value is too low for health. These relationships are apparent throughout childhood, and their magnitude increases with age.

Children who are both overweight and have high blood cholesterol are likely to have parents who developed CVD early.[17] For this reason, screening is recommended for children and adolescents whose parents or grandparents have CVD, those whose parents have elevated blood cholesterol, and those whose family history is unavailable, especially if other risk factors are evident.[18] Because blood cholesterol in children is a good predictor of adult values, some experts recommend universal screening for all children, particularly for those who are overweight, smoke, are sedentary, or consume diets high in saturated fat. Early—but not advanced—atherosclerotic lesions are reversible, making screening and education a high priority.

Cholesterol Values for Children and Adolescents

Disease Risk	Total Cholesterol (mg/dL)	LDL Cholesterol (mg/dL)
Acceptable	<170	<100
Borderline	170–199	100–129
High	≥200	≥130

Note: Adult values appear in Table 22–2 on p. 606.

Is hypertension a concern for children and adolescents?

Pediatricians routinely monitor blood pressure in children and adolescents. High blood pressure may signal an underlying disease or the early onset of hypertension. Hypertension accelerates the development of atherosclerosis.[19] Hypertension may develop in the first decades of life, especially among obese children, and worsen with time. Children with hypertension can often make dramatic improvements by participating in regular aerobic activity and by losing weight or maintaining their weight as they grow taller. Restricting dietary sodium also causes an immediate drop in most children's and adolescents' blood pressure.[20]

Why is there an epidemic of childhood obesity in the United States?

Over the past three decades, as the prevalence of childhood obesity throughout the United States has more than doubled for young children and adolescents and tripled for children 6 to 11 years of age, the society our children live in has changed considerably.[21] In many families today, both parents work outside the home and they work longer hours; more emphasis is placed on convenience foods and foods eaten away from home; meal choices at school are more diverse and often less nutritious; sedentary activities such as watching television and playing video or computer games occupy much of children's free time; and opportunities for physical activity and outdoor play both during school and outside of school have declined. All of these factors and many others influence children's eating and activity patterns.

Children learn food behaviors from their families, and entire families may be eating too much, dieting inappropriately, and exercising too little. Research shows that one in four toddlers (19 to 24 months of age) exceeds estimated energy requirements as a result of eating such foods as candy, pizza, chicken nuggets, sodas, sweet tea, and salty snacks like cheese puffs and chips.[22] Thus, when researchers ask, "Are today's children eating more kcalories than those of 30 years ago?" the answer comes back, "Yes." Some researchers report an increase of 100 to 200 kcalories per day for all age groups, enough to account for significant weight gain.[23]

Research links added sugars, especially high-fructose corn syrup—the easily consumed, energy-dense liquid sugar added to soft drinks—with excess body fatness in children.[24] Each 12-ounce can of soft drink provides the equivalent of about 10 teaspoons of sugar and 150 kcalories. More than half of children consume at least one soft drink each day at school; adolescent males consume the most—four or more cans daily.[25] According to one estimate, the risk of obesity increases by 60 percent with each sugared soft drink consumed daily.[26]

Although the tremendous increase in soft drink consumption may play some role in the obesity epidemic, much of the epidemic can be explained by lack of physical activity. Children have become more sedentary, and sedentary children are more often overweight. A child who spends more than an hour or two each day in front of a television, computer monitor, or other media can become obese and develop unhealthy blood lipids even while eating fewer kcalories than a more active child.[27] Physically active children have higher HDL, lower LDL, and lower blood pressure than sedentary children, and these positive findings often persist into adulthood.

Just as blood cholesterol and obesity track over the years, so does a person's level of physical activity. Compared with inactive adolescents, those who are physically active weigh less, smoke less, eat diets lower in saturated fats, and have better blood lipid profiles. The message is clear: physical activity offers numerous health benefits, and children and adolescents who are active today are most likely to be active for years to come.

What can concerned adults do to help prevent childhood obesity?

In light of all these findings, parents and teachers are encouraged to make major efforts to prevent childhood obesity. Suggestions include the following: encourage children to eat slowly, to pause and enjoy the company of their table companions, and to stop eating when they are full. Teach them how to select nutrient-dense snacks and to serve themselves appropriate portions. Never force children to clean their plates.

Above all, be sensitive in teaching children nutrition principles that can help to prevent obesity. Children can easily get the idea that their worth is tied to their body weight. Some parents fail to realize that society's ideal of slimness can be perilously close to starvation and that a child encouraged to "diet" cannot obtain the energy and nutrients required for normal growth and development. Even healthy children without eating disorders have been observed to limit their growth through "dieting." Pediatricians warn parents to avoid extremes; they caution that while intentions may be good, excessive food restriction may create nutrient deficiencies and impair growth. Furthermore, parental control over eating may instigate battles and foster attitudes about foods that can lead to inappropriate eating behaviors. Weight gain in truly overweight children can be controlled safely without compromising growth, but the process should be overseen by a health care professional.

Are adult dietary recommendations appropriate for children?

Regardless of family history, all children over age two should eat a variety of foods and maintain desirable weight (see Table NP 12–2). Children (4 to 18 years of age) should receive at least 25 percent and no more than 35 percent of total energy from fat, less than 10 percent from saturated fat, and less than 300 milligrams of cholesterol per day.[28]

Dietary Guidelines for Americans 2005

- Keep total fat intake between 30 to 35 percent of kcalories for children 2 to 3 years of age and between 25 to 35 percent of kcalories for children and adolescents 4 to 18 years of age, with most fats coming from sources of polyunsaturated and monounsaturated fatty acids, such as fish, nuts, and vegetable oils.

TABLE NP12–2

American Heart Association Dietary Guidelines and Strategies for Children[a]

- Balance dietary kcalories with physical activity to maintain normal growth.
- Every day, engage in 60 minutes of moderate to vigorous play or physical activity.
- Eat vegetables and fruits daily—use fresh, frozen, and canned vegetables and fruits and serve at every meal; limit those with added fats, salt, and sugar.
- Limit juice intake (4 to 6 ounces per day for children 1 to 6 years of age, 8 to 12 ounces per day for children 7 to 18 years of age).
- Use vegetable oils (canola, soybean, olive, safflower, or other unsaturated oils) and soft margarines low in saturated fat and *trans* fatty acids instead of butter or most other animal fats in the diet.
- Choose whole-grain breads and cereals rather than refined products; read labels and make sure that "whole grain" is the first ingredient.
- Reduce the intake of sugar-sweetened beverages and foods.
- Consume low-fat and nonfat milk and milk products daily.
- Include two servings of fish per week, especially fatty fish such as broiled or baked salmon.
- Choose legumes and tofu in place of meat for some meals.
- Choose only lean cuts of meat and reduced-fat meat products; remove the skin from poultry.
- Eat less salt, including salt from processed foods; breads, breakfast cereals, and soups may be high in salt and/or sugar, so read food labels and choose high-fiber, low-salt, low-sugar alternatives.
- Limit the intake of high-kcalorie add-ons such as gravy, Alfredo sauce, cream sauce, cheese sauce, and hollandaise sauce.
- Serve age-appropriate portion sizes on appropriately sized plates and bowls.

[a] These guidelines are for children 3 years of age and older.

Source: Adapted from American Heart Association, Samuel S. Gidding, and coauthors, Dietary recommendations for children and adolescents: A guide for practitioners, *Pediatrics* 117 (2006): 544–559.

Recommendations limiting fat and cholesterol are not intended for infants or children under two years old. Infants and toddlers need a higher percentage of fat to support their rapid growth.

Healthy children over age two can begin the transition to eating according to recommendations by eating fewer foods high in saturated fat and selecting more fruits and vegetables. Healthy meals can occasionally include moderate amounts of a child's favorite foods, even if they are high in saturated fat such as french fries and ice cream. However, a steady diet of offerings from some "children's menus" in restaurants, such as chicken nuggets, hot dogs, and french fries, easily exceeds a prudent intake of saturated fat, *trans* fat, and kcalories and invites both nutrient shortages and gains of body fat.[29] Most restaurant chains are changing children's menus to include steamed vegetables, fruit, and broiled or grilled poultry—additions welcomed by busy parents who often dine out or purchase take-out foods.

Other fatty foods, such as nuts, vegetable oils, and safer varieties of fish, such as light canned tuna or salmon, are important for their essential fatty acids. Low-fat milk and milk products deserve special attention in a child's diet for the needed calcium and other nutrients they supply.

Can parents or caregivers do anything else to help children reduce their risks of CVD?

Even though the focus of this text is nutrition, another risk factor for CVD that starts in childhood and carries over into adulthood must also be addressed—cigarette smoking. The prevalence of cigarette smoking among U.S. adolescents is on the decline, but every day, 3,000 young people start smoking. Approximately 80 percent of all adult smokers began smoking before the age of 18.

Of those adolescents who continue smoking, half will eventually die of smoking-related causes. Efforts to teach children about the dangers of smoking need to be aggressive. Children are not likely to consider the long-term health consequences of tobacco use. They are more likely to be struck by the immediate health consequences, such as shortness of breath when playing sports, or social consequences, such as having bad breath. Whatever the context, the message to all children and teens should be clear: do not start smoking. If you have already started, quit now.

In conclusion, treatment of established obesity is notoriously unsuccessful, making prevention of childhood obesity and the diseases it engenders a high national priority.[30] Education is clearly needed. Classroom lessons that are reinforced by lunchroom offerings and other school policies often help children change their eating habits for the better.[31] However, when school nutrition policy conflicts with classroom teachings and adults set poor examples, children are likely to follow their taste buds, not nutrition teachings.

The best advice to help turn the tide of obesity and related diseases is the easiest to give and perhaps the most difficult to follow: do not smoke, choose a diet in accord with the *Dietary Guidelines for Americans 2005,* follow the USDA Food Guide (Chapter 1), and make it a habit to be physically active each day. Last, parents and other significant adults can help mold children's behaviors by the examples they set.

Notes

1. C. L. Ogden and coauthors, Prevalence of overweight and obesity in the United States, 1999–2004, *Journal of the American Medical Association* 295 (2006): 1549–1555; J. P. Kaplan, C. T. Liverman, and V. I. Kraak, eds., *Preventing Childhood Obesity: Health in the Balance* (Washington,

DC: National Academies Press, 2005), pp. 1–20; M. L. Cruz
and coauthors, Pediatric obesity and insulin resistance:
Chronic disease risk and implications for treatment and
prevention beyond body weight modification, *Annual
Review of Nutrition* 25 (2005): 435–468; T. Lobstein,
L. Baur, and R. Uauy, Obesity in children and young people:
A crisis in public health, *Obesity Reviews* 5 (2004): 4–85.

2. Kaplan, Liverman, and Kraak, 2005.

3. A. M. Prentice, The emerging epidemic of obesity in
developing countries, *International Journal of
Epidemiology* 35 (2006): 93–99; M. Kohn and M. Booth,
The worldwide epidemic of obesity in adolescents,
Adolescent Medicine 14 (2003): 1–9; L. S. Lieberman,
Dietary, evolutionary, and modernizing influences on
the prevalence of type 2 diabetes, *Annual Review of
Nutrition* 23 (2003): 345–377; Committee on Nutrition,
American Academy of Pediatrics, Prevention of
pediatric overweight and obesity, *Pediatrics* 112 (2003):
424–430.

4. R. J. F. Loos and T. Rankinen, Gene-diet interactions on
body weight changes, *Journal of the American Dietetic
Association* 105 (2005): S29–S34.

5. L. F. Baier and R. L. Hanson, Genetic studies of the
etiology of type 2 diabetes in Pima Indians, *Diabetes* 53
(2004): 1181–1186.

6. Loos and Rankinen, 2005; A. M. Kriska and coauthors,
Physical activity, obesity, and the incidence of type 2
diabetes in a high-risk population, *American Journal of
Epidemiology* 158 (2003): 669–675; J. Esparza and
coauthors, Daily energy expenditure in Mexican and USA
Pima Indians: Low physical activity as a possible cause of
obesity, *International Journal of Obesity and Related
Metabolic Disorders* 24 (2000): 55–59.

7. L. Adair and D. Dahly, Developmental determinants of
blood pressure in adults, *Annual Review of Nutrition* 25
(2005): 407–434; G. Wu and coauthors, Maternal nutrition
and fetal development, *Journal of Nutrition* 134 (2004):
2169–2172.

8. Adair and Dahly, 2005.

9. Adair and Dahly, 2005.

10. Adair and Dahly, 2005.

11. D. J. Barker and coauthors, Growth and living conditions in
childhood and hypertension in adult life: A longitudinal
study, *Journal of Hypertension* 20 (2002): 1951–1956;
S. M. Robinson and D. J. Barker, Coronary heart disease:
A growth disorder, *Proceedings of the Nutrition Society* 61
(2002): 537–542.

12. T. S. Hannon, F. Rao, and S. A. Arslanian, Childhood obesity
and type 2 diabetes mellitus, *Pediatrics* 116 (2005):
473–480; Cruz and coauthors, 2005.

13. G. S. Boyd and coauthors, Effect of obesity and high
blood pressure on plasma lipid levels in children and
adolescents, *Pediatrics* 116 (2005): 442–446; R. Kohen-
Avramoglu, A. Theriault, and K. Adeli, Emergence of the
metabolic syndrome in childhood: An epidemiological
overview and mechanistic link to dyslipidemia, *Clinical
Biochemistry* 36 (2003): 413–420.

14. S. Li and coauthors, Childhood cardiovascular risk factors
and carotid vascular changes in adulthood: The Bogalusa
Heart Study, *Journal of the American Medical
Association* 290 (2003): 2271–2276; K. B. Keller and L.
Lemberg, Obesity and the metabolic syndrome, *American
Journal of Clinical Care* 12 (2003): 167–170.

15. V. N. Muratova and coauthors, The relation of obesity to
cardiovascular risk factors among children: The CARDIAC
project, *West Virginia Medical Journal* 98 (2002): 263–267.

16. I. Janssen and coauthors, Combined influence of body
mass index and waist circumference on coronary artery
disease risk factors among children and adolescents,
Pediatrics 115 (2005): 1623–1630; O. Fiedland and coauthors,
Obesity and lipid profiles in children and adolescents,
Journal of Pediatric Endocrinology and Metabolism 15
(2002): 1011–1016; T. Dwyer and coauthors, Syndrome X in
8-y-old Australian children: Stronger associations with
current body fatness than with infant size or growth,
*International Journal of Obesity and Related Metabolic
Disorders* 26 (2002): 1301–1309.

17. B. Glowinska, M. Urban, and A. Koput, Cardiovascular risk
factors in children with obesity, hypertension and
diabetes: Lipoprotein (a) levels and body mass index
correlate with family history of cardiovascular disease,
European Journal of Pediatrics 161 (2002): 511–518.

18. A Wiegman and coauthors, Family history and
cardiovascular risk in familial hypercholesterolemia: Data in
more than 1000 children, *Circulation* 107 (2003): 1473–1478.

19. National High Blood Pressure Education Program Working
Group on High Blood Pressure in Children and
Adolescents, The fourth report on the diagnosis,
evaluation, and treatment of high blood pressure in
children and adolescents, *Pediatrics* 114 (2004): 555S–576S.

20. F. J. He and G. A. MacGregor, Importance of salt in
determining blood pressure in children: meta-analysis of
controlled trials, *Hypertension* 48 (2006): 861-869.

21. Kaplan, Liverman, and Kraak, 2005.

22. S. A. Lederman and coauthors, Summary of the
presentations at the Conference on Preventing Childhood
Obesity, December 8, 2003, *Pediatrics* 114 (2004): 1146–1173.

23. S. Kranz, A. M. Siega-Riz, and A. H. Herring, Changes in diet
quality of American preschoolers between 1977 and 1998,
American Journal of Public Health 94 (2004): 1525–1530;
S. J. Nielsen, A. M. Siega-Riz, and B. M. Popkin, Trends in

energy intake in U.S. between 1977 and 1996: Similar shifts seen across age groups, *Obesity Research* 10 (2002): 370–378.

24. J. A. Welsh and coauthors, Overweight among low-income preschool children associated with the consumption of sweet drinks: Missouri, 1999–2002, *Pediatrics* 115 (2005): 223–229.

25. American Academy of Pediatrics, Committee on School Health, Soft drinks in schools, *Pediatrics* 113 (2004): 152–154.

26. D. S. Ludwig, K. E. Peterson, and L. S. Gortmaker, Relation between consumption of sugar-sweetened drinks and childhood obesity: A prospective, observational analysis, *Lancet* 357 (2001): 505–508.

27. M. H. Proctor and coauthors, Television viewing and change in body fat from preschool to early adolescence: The Framingham Children's Study, *International Journal of Obesity and Related Metabolic Disorders* 27 (2003): 827–833.

28. Standing Committee on the Scientific Evaluation of Dietary Reference Intakes, Food and Nutrition Board, Institute of Medicine, *Dietary Reference Intakes for Energy, Carbohydrate, Fiber, Fat, Fatty Acids, Cholesterol, Protein, and Amino Acids* (Washington, DC: National Academies Press, 2005), pp. 769–879.

29. J. Hurley and B. Liebman, Kids' cuisine: "What would you like with your fries?" *Nutrition and Action Healthletter* 31 (2004): 12–15.

30. American Academy of Pediatrics, Policy Statement: Active healthy living: Prevention of childhood obesity through increased physical activity, *Pediatrics* 117 (2006): 1834–1842; S. Caprio and M. Genel, Confronting the epidemic of childhood obesity, *Pediatrics* 115 (2005): 494–495.

31. S. M. Gross and B. Cinelli, Coordinated school health program and dietetics professionals: Partners in promoting healthful eating, *Journal of the American Dietetic Association* 104 (2004): 793–798.

13

Nutrition Through The Life Span: Later Adulthood

© BananaStock/Alamy

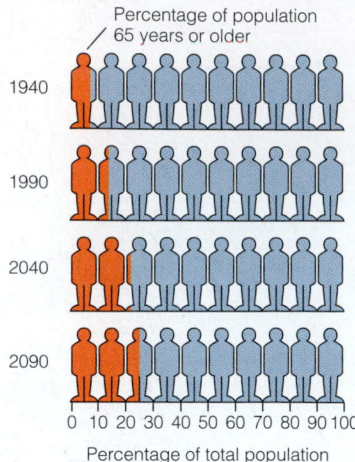

Percentage of population
65 years or older

1940

1990

2040

2090

0 10 20 30 40 50 60 70 80 90 100
Percentage of total population

FIGURE 13–1

The Aging of the U.S. Population
In 1940, 6.8 percent of the population was 65 or older. In 1990, 12.7 percent of us had reached age 65; by 2040, 21.7 percent will have reached age 65; and by 2090, nearly one out of four Americans will be 65 or older.

Chapters 11 and 12 were devoted to stages of the life cycle that require special nutrition attention: pregnancy, lactation, infancy, childhood, and adolescence. Much of the text before that focused on nutrition to support wellness during adulthood. This chapter describes the special nutrition needs of the later adult years.

The most urgent nutrition need of older people is to have made good food choices in the past! All of life's nutrition choices incur health consequences for better or for worse. A single day's intakes of nutrients may exert only a minute effect on body organs and their functions, but over years and decades, the repeated effects accumulate to have major impacts. This being the case, it is of great importance for everyone, of every age, to pay close attention today to nutrition.

The U.S. population is growing older. The majority of citizens are now middle-aged, and the ratio of old people to young is increasing, as Figure 13–1 shows. Our society uses the arbitrary age of 65 to define the transition point between middle age and old age, but growing "old" happens day by day, with changes occurring gradually over time. Since 1950, the population of people over 65 has almost tripled, and people over 85 years old are the fastest-growing age group.[1] The U.S. Bureau of the Census projects that by 2040 more than one million Americans will be 100 years old or older.

Life expectancy in the United States is 78 years, up from about 47 years in 1900.[2] Women today live about five years longer than men. Advances in medical science—antibiotics and other treatments—are largely responsible for almost doubling the life expectancy since 1900. Improved nutrition and an abundant food supply have also contributed to lengthening life expectancy. The human **life span**, currently estimated at 130 years, is the upper limit of human **longevity**, even given optimal nutrition. With recent advances in medical technology and genetic knowledge, researchers may one day be able to extend the life span even further by slowing, or perhaps preventing, aging and its accompanying diseases.[3]

The study of the aging process is among the youngest of the scientific disciplines. Not until the twentieth century did human beings achieve a life expectancy worthy of a science devoted to studying it. The idea that nutrition can influence the way the human body ages is particularly appealing because diet is a factor that people can control and change.

Nutrition and Longevity

What has been learned so far about the effects of nutrition and environment on longevity provides incentive for researchers to keep asking questions about how and why human beings age. Among their questions are:

- To what extent is aging inevitable, and can it be slowed through changes in lifestyle and environment?

- What roles does nutrition play in aging, and what roles can it play in slowing aging?

With respect to the first question, it seems that aging is an inevitable, natural process, programmed into the genes at conception.[4] However, people can slow the process within the natural limits set by heredity. They can adopt healthy lifestyle habits such as eating nutritious food and engaging in physical activity. In fact, an estimated 70 to 80 percent of the average person's life expectancy may depend on individual health-related behaviors; genes determine the remaining 20 to 30 percent.[5]

With respect to the second question, good nutrition helps to maintain a healthy body and can therefore ease the aging process in many significant ways.[6] Clearly, nutrition can improve the **quality of life** in the later years.

Slowing the Aging Process

One approach researchers use to search out the secret of long life has been to study older people. Some people are young for their ages, whereas others are old for their ages. What makes the difference?

Healthy Habits Six lifestyle habits seem to have a profound influence on people's health and therefore on their **physiological age**:

- Sleeping regularly and adequately
- Eating well-balanced meals, including breakfast, regularly
- Maintaining a healthy body weight
- Engaging in regular physical activity
- Not smoking
- Not using alcohol, or using it in moderation

Over the years, the effects of these lifestyle choices accumulate—that is, those who follow all of these practices live longer and have fewer disabilities as they age.[7] They are in better health, even if older in **chronological age**, than people who do not adopt these behaviors. Even though people cannot alter their birth dates, they may be able to add years to and enhance the quality of their lives. Physical activity seems to be most influential in preventing or slowing the many changes that many people seem to accept as an inevitable consequence of old age. In other words, physical activity and long life seem to go together.[8]

Physical Activity The many and remarkable benefits of regular physical activity outlined in Chapter 10 are not limited to the young. Compared to those who are inactive, older adults who are active weigh less; have greater flexibility, more endurance, better balance, and better health; and live longer.[9] They reap additional benefits from various activities as well: aerobic activities improve cardiorespiratory endurance, blood pressure, and blood lipid concentrations; moderate endurance activities improve the quality of sleep; and strength training significantly improves posture and mobility. In fact, regular physical activity is a powerful predictor of a person's mobility in the later years. Physical activity also increases blood flow to the brain, thereby preserving mental ability, alleviating depression, supporting independence, and improving the quality of life.[10]

Muscle mass and muscle strength tend to decline with aging, making older people vulnerable to falls and immobility. Falls are a major cause of fear, injury, disability, dependence, and even death among older adults.[11] Regular physical activity tones, firms, and strengthens muscles, helping to improve confidence, reduce the risk of falling, and minimize the risk of injury should a fall occur. Strength training, even in frail, elderly people over 85 years of age, has been shown not only to improve balance, muscle strength, and mobility, but also to increase energy expenditure and energy intake. This finding highlights another reason to be physically active: a person spending energy on physical activity can afford to eat more food and with it, more nutrients. People who

---

Dietary Guidelines for Americans 2005

- Older adults should participate in regular physical activity to reduce the functional declines associated with aging and to achieve the other benefits of physical activity identified for all adults.

---

life expectancy: the average number of years lived by people in a given society.

life span: the maximum number of years of life attainable by a member of a species.

longevity: long duration of life.

quality of life: a person's perceived physical and mental well-being.

physiological age: a person's age as estimated from his or her body's health and probable life expectancy.

chronological age: a person's age in years from his or her date of birth.

Regular physical activity promotes a healthy, independent lifestyle.

are committed to an ongoing fitness program have higher energy and nutrient intakes than more sedentary people.

Activities of all kinds are recommended to maintain and promote health. Strength training improves muscle strength, which enhances a person's ability to perform many of life's daily tasks such as climbing stairs and carrying packages.[12] Muscle strength during midlife (between the ages of 45 and 65) predicts health and disability 25 years later. Greater muscle strength during midlife may act as a strength *reserve* later on, protecting older adults from disability even when chronic conditions develop. In short, improving overall strength during early and middle adulthood could potentially lower the risk of later physical disability.

Ideally, physical activity should be part of each day's schedule and should be intense enough to prevent muscle atrophy and to speed up the heartbeat and respiration rate. Although aging affects both speed and endurance to some degree, older adults can still train and achieve exceptional performances. Healthy older adults who have not been active can ease into a suitable routine. They can start by walking short distances until they are walking at least 10 minutes continuously and then gradually increase their distance to a 30- to 45-minute walk at least five days per week. With persistence, people can achieve great improvements at any age. Table 13–1 provides exercise guidelines for seniors; people with medical conditions should check with a physician before beginning an exercise routine, as should sedentary men over 40 and women over 50 who want to participate in a vigorous program.

Restriction of kCalories In their efforts to understand longevity, researchers have not only observed people, but have also manipulated influencing factors, such as diet, in animals. This research has produced some interesting and suggestive findings. For example, animals live longer and have fewer age-related diseases when their energy intakes are restricted.

Several mechanisms to explain how energy restriction prolongs life in animals have been proposed but not proved. Research suggests that food restriction may extend the life span by preventing damaging lipid oxidation, thereby delaying the onset of

TABLE 13–1				
Exercise Guidelines for Older Adults				
	Endurance	**Strength**	**Balance**	**Flexibility**
Examples				
Start easy	Be active 5 minutes on most or all days.	Using 0- to 2-pound weights, do one set of 8 repetitions twice a week.	Hold onto table or chair with one hand, then with one finger.	Hold stretch 10 seconds; do each stretch three times.
Progress gradually to goal	Be active 30 minutes (minimum) on most or all days.	Increase weight as able; do two sets of 8–15 repetitions twice a week.	Do not hold onto table or chair; then close eyes.	Hold stretch 30 seconds; do each stretch five times.
Cautions and comments	Stop if you are breathing so hard you can't talk or if you feel dizziness or chest pain.	Breathe out as you contract and in as you relax (do not hold breath); use smooth, steady movements.	Incorporate balance techniques with strength exercises as you progress.	Stretch after strength and endurance exercises for 20 minutes, three times a week; use slow, steady movements; bend joints slightly.

Source: Exercise: A Guide from the National Institute on Aging, www.nia.nih.gov.

age-related diseases, such as atherosclerosis.[13] Gene activity also appears to play a key role. Energy restriction in animals prevents alterations in gene expression that are associated with aging.[14] Experiments with food restriction and longevity in animals have *not* suggested any direct applications to human nutrition.

Many of the physiological responses to energy restriction seen in animals also occur in people whose intakes are *moderately* restricted. When people cut back on their usual energy intake by 10 to 20 percent, ■ body weight, body fat, and blood pressure drop, and blood lipids and insulin response improve—favorable changes for preventing chronic diseases.[15] The reduction in oxidative damage that occurs with energy restriction in animals also occurs in people whose diets include antioxidant nutrients and phytochemicals. Diets, such as the Mediterranean diet, that include an abundance of fruits, vegetables, olive oil, whole grains, and legumes—with their array of antioxidants and phytochemicals—support good health and long life.[16]

■ For perspective, a person with a usual energy intake of 2,000 kcalories might cut back to 1,600 to 1,800 kcalories.

Nutrition and Disease Prevention

Nutrition alone, even if ideal, cannot ensure a long and robust life. Nevertheless, nutrition clearly affects aging and longevity in human beings by way of its role in disease prevention. Among the better-known relationships between nutrition and disease are the following:

- Appropriate energy intake helps prevent *obesity, diabetes*, and related *cardiovascular diseases* such as atherosclerosis and hypertension (Chapters 6, 21, and 22) and may influence the development of some forms of *cancer* (Chapter 25).

- Adequate intakes of essential nutrients prevent *deficiency diseases* such as scurvy, goiter, anemia, and the like (Chapters 8 and 9).

- Variety in food intake, as well as ample intakes of certain fruits and vegetables, may be protective against certain types of *cancer* (Chapter 25).

- Moderation in sugar intake helps prevent *dental caries* (Nutrition in Practice 2).

- Appropriate fiber intakes help prevent disorders of the digestive tract such as *constipation, diverticulosis*, and possibly *colon cancer* (Chapters 2 and 18).

- Moderate sodium intake and adequate intakes of potassium, calcium, and other minerals help prevent *hypertension* (Chapters 9 and 22).

- An adequate calcium intake throughout life helps protect against *osteoporosis* (Chapter 9).

Other, less well-established links between nutrition and disease are being discovered each day. Research that focuses on how life factors affect aging and disease processes is vital to ensuring that more and more people can look forward to long, healthy lives.

IN SUMMARY

- Life expectancy in the United States increased dramatically in the twentieth century.

- Factors that enhance longevity include limited or no alcohol use, regular balanced meals, weight control, adequate sleep, abstinence from smoking, and regular physical activity.

- Nutrition alone, even if ideal, cannot guarantee a long and robust life. At the very least, however, nutrition—especially when combined with regular physical activity—can influence aging and longevity in human beings by supporting good health and preventing disease.

Nutrition-Related Concerns during Late Adulthood

Nutrition through the prime years may play a greater role than has been realized in preventing many changes once thought to be inevitable consequences of growing older. The following discussions of cataracts and macular degeneration, arthritis, and the aging brain show that nutrition may provide at least some protection against some of the conditions commonly associated with aging.

Cataracts and Macular Degeneration

Cataracts are age-related thickenings in the lenses of the eye that impair vision. If not surgically removed, they ultimately lead to blindness. Oxidative stress appears to play a significant role in the development of cataracts, and the antioxidant nutrients and phytochemicals in fruits and vegetables may help minimize the damage.[17] Some studies suggest that a diet providing ample carotenoids, vitamin C, and vitamin E may be especially important for preventing early onset of cataracts.[18] Also, people who follow the *Dietary Guidelines for Americans 2005* are reported to have fewer cataracts, but cataracts can occur even in well-nourished individuals due to exposure to ultraviolet light, oxidative damage, viral infections, toxic substances, genetic disorders, injury, or other trauma.[19] Most cataracts are vaguely called senile cataracts, meaning "caused by aging." In the United States, more than half of all adults age 65 and older have a cataract.

One other diet-related factor may play a role in cataract development: obesity. In a study of more than 100,000 men and women, those with a body mass index (BMI) above 30 had a significantly greater risk of cataracts compared to those with a BMI less than 23.[20] How obesity may influence the development of cataracts is not known, but researchers speculate that conditions that typically accompany obesity such as glucose intolerance and insulin resistance may provide clues.

Another cause of visual loss among older people is **macular degeneration**, a deterioration of the macular region of the retina. Age-related macular degeneration (AMD) is the leading cause of blindness in individuals over 65 years of age in the United States; nearly 2 million people are afflicted with the disease.[21] AMD is more prevalent among white people than among black people, and the prevalence of AMD increases dramatically with age—more than 15 percent of white women over the age of 80 have AMD. As with cataracts, risk factors for AMD include oxidative stress from sunlight; some studies have found that supplements of antioxidant nutrients plus zinc and of the carotenoids lutein and zeaxanthin may help to protect against AMD.[22] Diets high in omega-3 fatty acids from fish may also offer some protection against AMD.[23] The omega-3 fatty acid docosahexaenoic acid (DHA) is a major structural lipid in the membrane of the retina, and DHA also affects signaling pathways in the retina.[24]

Arthritis

Arthritis is the leading cause of disability among older adults—affecting close to 60 percent of all older people.[25] The most common type of **arthritis** that disables older people is **osteoarthritis**, a painful swelling of the joints. During movement, the ends of bones are normally protected from wear by cartilage and by small sacs of fluid that lubricate the joint. With age, the cartilage sometimes disintegrates, and the joints become malformed and painful to move. Osteoarthritis afflicts millions of people around the world, especially the elderly.

One known connection between osteoarthritis and nutrition is overweight. Weight loss may relieve some of the pain for overweight people with osteoarthritis, partly because the joints affected are often weight-bearing joints that are stressed and irritated by having to carry excess poundage. Interestingly, weight loss often relieves the worst pain of osteoarthritis in the hands as well, even though they are not weight-bearing joints. Jogging and other weight-bearing activities do not worsen osteoarthritis. In fact, both aerobic activity and strength training offer improvements in physical performance and pain relief, especially when accompanied by even modest weight loss.[26]

Nutrition quackery to treat arthritis is abundant, but no one universally effective diet for arthritis relief is known. Table 13–2 presents some of the many *noneffective* dietary treatments for osteoarthritis. Traditional medical intervention for arthritis includes medication and surgery. Two popular supplements for treating osteoarthritis—glucosamine and chondroitin—may indeed alleviate pain and improve mobility, and they may even slow the progression of osteoarthritis if substantial additional research supports findings to date.[27]

Another type of arthritis, known as **rheumatoid arthritis**, has a possible link to diet through the immune system. In rheumatoid arthritis, the immune system mistakenly attacks the bone coverings as if they were made of foreign tissue. In some individuals, certain foods, notably vegetables and olive oil, may moderate the inflammatory responses and provide some relief.[28]

Another nutrient linked to rheumatoid arthritis is the omega-3 fatty acid found in fish oil, eicosapentaenoic acid (EPA). Research shows that the same diet recommended for heart health—one low in saturated fat from meats and milk products and high in oils from fish—helps prevent or reduce the inflammation in the joints that makes arthritis so painful.[29] Researchers theorize that EPA probably interferes with the action of prostaglandins, compounds involved in inflammation.

Another possible link between nutrition and rheumatoid arthritis involves the oxidative damage to the membranes within joints that causes inflammation and swelling. The antioxidant vitamins C and E and the carotenoids defend against oxidation, and increased intakes of these nutrients may help prevent or relieve the pain of rheumatoid arthritis.[30]

The Aging Brain

The brain, like all of the body's organs, responds to both genetic and environmental factors that can enhance or diminish its amazing capacities. One of the challenges researchers face when studying the aging of the human brain is to distinguish among changes caused by normal, age-related, physiological processes; changes caused by diseases; and changes caused by cumulative, extrinsic factors such as diet.

TABLE 13–2	
Noneffective Dietary Strategies for Arthritis	
Alfalfa tea	Garlic
Aloe vera liquid	Honey
Amino acid supplements	Inositol
Blackstrap molasses	Kelp
Burdock root	Lecithin
Calcium	Para-amino benzoic acid (PABA)
Celery juice	Raw liver
Cod liver oil	Superoxide dismutase (SOD)
Copper supplements	Vitamin D
Dimethyl sulfoxide (DMSO)	Vitamin megadoses
Fasting	Watercress
Fresh fruit	Yeast

cataracts: thickenings of the eye lenses that impair vision and can lead to blindness.

macular (MACK-you-lar) **degeneration:** deterioration of the macular area of the eye that can lead to loss of central vision and eventual blindness. The **macula** is a small, oval, yellowish region in the center of the retina that provides the sharp, straight-ahead vision so critical to reading and driving.

arthritis: inflammation of a joint, usually accompanied by pain, swelling, and structural changes.

osteoarthritis: a painful, chronic disease of the joints that occurs when the cushioning cartilage in a joint breaks down; joint structure is usually altered, with loss of function; also called **degenerative arthritis**.

rheumatoid arthritis: a disease of the immune system involving painful inflammation of the joints and related structures.

Both foods and mental challenges nourish the brain.

The brain normally changes in some characteristic ways as it ages. For one thing, its blood supply decreases. For another, the number of **neurons**, the brain cells that specialize in transmitting information, diminishes as people age. When the number of nerve cells in one part of the **cerebral cortex** diminishes, hearing and speech are affected. Losses of neurons in other parts of the cortex can impair memory and cognitive function. When the number of neurons in the hindbrain diminishes, balance and posture are affected. Losses of neurons in other parts of the brain affect still other functions.

Nutrient Deficiencies and Brain Function Clinicians now recognize that much of the cognitive loss and forgetfulness generally attributed to aging is due in part to extrinsic, and therefore controllable, factors such as nutrient deficiencies. The ability of neurons to synthesize specific neurotransmitters depends in part on the availability of precursor nutrients that are obtained from the diet.[31] For example, the neurotransmitter serotonin derives from the amino acid tryptophan. To function properly, the enzymes involved in neurotransmitter synthesis require vitamins and minerals. Thus nutrient deficiencies may contribute to the loss of memory and cognition that some older adults experience.[32] Such losses may be preventable or at least diminished or delayed through diet and physical activity.[33] Table 13–3 summarizes some of the better-known connections between brain function and nutrients.

In some instances, the degree of cognitive loss is extensive. Such **senile dementia** may be attributable to a specific disorder such as a brain tumor or Alzheimer's disease.

Alzheimer's Disease In **Alzheimer's disease**, the most prevalent form of senile dementia, brain cell death occurs in the areas of the brain that coordinate memory and cognition. Alzheimer's disease afflicts 4.5 million people in the United States, and that number is expected to almost triple by the year 2050.[34] Diagnosis of Alzheimer's depends on its characteristic symptoms: the victim gradually loses memory and reasoning, the ability to communicate, physical capabilities, and eventually life itself.

Researchers are closing in on the cause of Alzheimer's disease.* Clearly, genetic factors are involved.[35] Free radicals and oxidative stress may also be involved.[36] Nerve cells in the brains of people with Alzheimer's disease show evidence of free-radical attack—damage to DNA, cell membranes, and proteins. They also show evidence of the minerals that trigger free-radical attacks—iron, copper, zinc, and aluminum. Some research suggests that the antioxidant nutrients can limit free-radical damage and delay or prevent Alzheimer's disease.[37]

Most people have heard of an association between aluminum and the development of Alzheimer's, although a causal connection seems unlikely. Brain concentrations of

TABLE 13–3	
Summary of Nutrient-Brain Relationships	
Brain Function	**Depends on an Adequate Intake of:**
Short-term memory	Vitamin B$_{12}$, vitamin C, vitamin E
Performance in problem-solving tests	Riboflavin, folate, vitamin B$_{12}$, vitamin C
Mental health	Thiamin, niacin, zinc, folate
Cognition	Folate, vitamin B$_6$, vitamin B$_{12}$, iron, vitamin E
Vision	Essential fatty acids, vitamin A
Neurotransmitter synthesis	Tyrosine, tryptophan, choline

* A report on the genetic and other aspects of Alzheimer's is available from Alzheimer's Disease Education and Referral Center, P.O. Box 8250, Silver Springs, MD 20907-8250.

aluminum in people with Alzheimer's exceed normal brain concentrations by some 10 to 30 times, but blood and hair aluminum remains normal, indicating that the accumulation is caused by something in the brain itself, not by an overload of aluminum in the body. Thus the high brain aluminum must be at least partly a result, rather than a cause, of Alzheimer's.

Research suggests that cardiovascular disease risk factors such as high blood pressure, diabetes, and elevated levels of the amino acid homocysteine may be related to the development of Alzheimer's.[38] Diets designed to support a healthy heart, including omega-3 fatty acids and light-to-moderate alcohol intake, may benefit a healthy brain as well.[39]

Treatment for Alzheimer's disease involves providing care to clients and support to their families. There is no cure for Alzheimer's disease, but medications are used to treat some of the symptoms. The Food and Drug Administration has approved several prescription medications for people with mild-to-moderate Alzheimer's disease.[40] The medications act to increase the level in the brain of acetylcholine, which is essential to memory. The drugs have effects on the symptoms of the disease, but there is no evidence that they slow its progression.

Maintaining appropriate body weight may be the most important nutrition concern for the person with Alzheimer's. Depression and forgetfulness can lead to poor food intake, and restlessness may increase energy needs. Perhaps the best that a caregiver can do nutritionally for a person with Alzheimer's is to supervise food planning and mealtimes. Providing well-liked and well-balanced meals and snacks in a cheerful atmosphere encourages food consumption. To minimize confusion, offer a few ready-to-eat foods, in bite-size pieces, with seasonings and sauces. To avoid mealtime disruptions, control distractions such as television, children, and the telephone.

IN SUMMARY

- Cataracts, age-related macular degeneration, and arthritis afflict millions of older adults, while others face senile dementia and other losses of brain function.

- Some of these problems may be inevitable, but others are preventable, and good nutrition may play a key preventive role.

Energy and Nutrient Needs during Late Adulthood

Knowledge about the nutrient needs and nutrition status of older adults has grown considerably in the last decade or so. The Dietary Reference Intakes (DRI) cluster people over the age of 50 into two age categories—51 to 70 years old and 71 years and older. Increasingly, research is showing that the nutrition needs of people 50 to 70 years old may be very different from those of people over 70—a pattern that makes sense considering the wide age span involved.

Setting standards for older people is difficult because individual differences become more pronounced as people grow older.[41] One person may tend to omit vegetables from his diet, and by the time he is old, he will have an associated set of nutrition problems. Another may have omitted milk and milk products all her life—her nutrition problems will be different. Also, as people age, they may suffer different chronic diseases and take different medications—both have impacts on nutrient needs. Table 13–4 (p. 388) lists some changes of aging that can affect nutrition. Even before all this, people start out with different genetic predispositions and ways of handling nutrients, and the effects of these differences become magnified over the years. Researchers have difficulty even defining "healthy aging," a prerequisite to developing recommendations that

neurons: nerve cells; the structural and functional units of the nervous system. Neurons initiate and conduct nerve transmissions.

cerebral cortex: the outer surface of the cerebrum, which is the largest part of the brain.

senile dementia: the loss of brain function beyond the normal loss of physical adeptness and memory that occurs with aging.

Alzheimer's disease: a progressive, degenerative disease that attacks the brain and impairs thinking, behavior, and memory.

TABLE 13-4

Examples of Physical Changes of Aging That Affect Nutrition

Mouth	Tooth loss, gum disease, and reduced salivary output impede chewing and swallowing. Swallowing disorders and choking may become likely. Discomfort and pain associated with eating may reduce food intake.
Digestive tract	Intestines lose muscle strength resulting in sluggish motility that leads to constipation (see Chapter 18). Stomach inflammation, abnormal bacterial growth, and greatly reduced acid output impair digestion and absorption. Pain may cause food avoidance or reduced intake.
Hormones	For example, the pancreas secretes less insulin, and cells become less responsive, causing abnormal glucose metabolism.
Sensory organs	Diminished senses of smell and taste can reduce appetite; diminished sight can make food shopping and preparation difficult.
Body composition	Weight loss and decline in lean body mass lead to lowered energy requirements. May be preventable or reversible through physical activity.
Urinary tract	Increased frequency of urination may limit fluid intake.

are designed to meet the "needs of practically all healthy persons." The next sections give special attention to a few nutrients of concern.

Energy and Energy Nutrients

Energy needs decline with advancing age. As a general rule, adult energy needs decline an estimated 5 percent per decade. For one thing, as people age, they usually reduce their physical activity, although they need not do so. For another, basal metabolic rate declines 1 to 2 percent per decade in part because lean body mass and thyroid hormones diminish.[42]

Loss of muscle mass, known as **sarcopenia**, can be significant in the later years (its prevalence is more than 50 percent among those older than 80), and its consequences can be dramatic (see Figure 13–2).[43] As skeletal muscle mass diminishes, people lose their ability to move and to maintain balance, making falls likely. The limitations that accompany the loss of muscle mass and strength play a key role in the diminishing health that often accompanies aging.[44] To some extent, however, declines in lean body mass and energy needs may not be entirely inevitable. Physical activity may be a key, as may a nutrient-dense diet.[45] Optimal nutrition and regular physical activity can help maintain muscle mass and strength and minimize the changes in body composition associated with aging.[46] Physical activity not only increases energy expenditure, but, along with sound nutrition, also enhances bone density and supports many body functions as well.

FIGURE 13-2

Sarcopenia

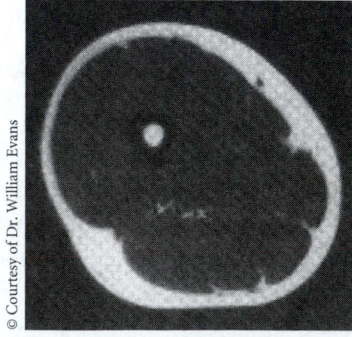
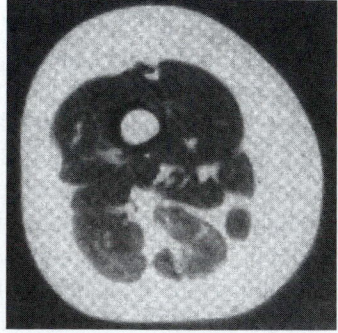

© Courtesy of Dr. William Evans

These cross sections of two women's thighs may appear to be about the same size from the outside, but the 20-year-old woman's thigh (left) is dense with muscle tissue. The 64-year-old woman's thigh (right) has lost muscle and gained fat, changes that may be largely preventable with strength-building physical activities.

The lower energy expenditures of many older adults require that they eat less food energy to maintain their weights. Accordingly, the energy RDA for adults decreases slightly, beginning at age 51. Energy intakes typically decline in parallel with needs. Still, many older adults are overweight, indicating that their food intakes do not decline enough to compensate for their reduced energy expenditures. Overweight and obesity in older adults increase the risk for many diseases and for disabilities as well.[47]

On limited energy allowances, people must select mostly nutrient-dense foods. There is little leeway for added sugars, solid fats, or alcohol. Older adults can follow the USDA Food Guide (see p. 18–19), making sure to choose the recommended amounts of food from each food group daily that are appropriate to their energy needs (see Table 1–7 on p. 20).

Protein The protein needs of older adults appear to be about the same as those of younger people. Protein is especially important for older adults to support a healthy immune system and to prevent muscle wasting. However, since energy needs decrease, the protein has to be obtained from low-kcalorie sources of high-quality protein, such as lean meats, poultry, fish, and eggs; fat-free and low-fat milk products; legumes; and grains.

Underweight or malnourished older adults need protein- and energy-dense snacks such as hard-boiled eggs, tuna fish and crackers, peanut butter on graham crackers, and hearty soups. Drinking liquid nutritional formulas between meals can also boost energy and nutrient intakes. Importantly, the diet should provide enjoyment as well as nutrients.[48]

Carbohydrate and Fiber As always, abundant carbohydrate is needed to protect protein from being used as an energy source. The recommendation to obtain ample amounts of mostly whole-grain breads, cereals, rice, and pasta holds true for older people. As for younger people, a steady supply of carbohydrate is essential for optimal brain functioning.[49] With age, fiber takes on extra importance for its role in alleviating constipation, a common complaint among older adults and among nursing home residents in particular. Fruits and vegetables supply viscous (soluble) fibers thought to help ward off diseases of aging, but factors such as transportation problems, limited cooking facilities, and chewing problems limit some elderly people's intakes of fresh fruits and vegetables.[50] Most older adults do not obtain the recommended daily 25 or more grams of fiber (14 grams per 1,000 kcalories).[51] When low fiber intakes are combined with low fluid intakes, inadequate physical activity, and constipating medications, constipation becomes inevitable.

Fat As is true for people of all ages, fat intake needs to be moderate in the diets of most older adults—enough to enhance flavors and provide valuable nutrients, but not so much as to raise risks of cancer, atherosclerosis, and other degenerative diseases. The recommendation should not be taken too far; limiting fat too severely may lead to nutrient deficiencies and weight loss—two problems that carry greater health risks in the elderly than overweight.

Water

Dehydration is a risk for older adults, who may not notice or pay attention to their thirst or who find it difficult and bothersome to get a drink or to get to a bathroom.[52] Older adults who have lost bladder control may be afraid to drink too much water. Despite real fluid needs, older people do not seem to feel as thirsty or notice mouth dryness as readily as younger people. Many nursing home employees say it is hard to persuade their elderly clients to drink enough water and fruit juices.

Total body water decreases as people age, so even mild stresses such as fever or hot weather can precipitate rapid dehydration in older adults. Dehydrated older adults

sarcopenia (SAR-koh-PEE-nee-ah): loss of skeletal muscle mass, strength, and quality.

seem to be more susceptible to urinary tract infections, pneumonia, **pressure ulcers**, confusion, and disorientation.[53] An intake of 9 cups per day of total beverages, including water, is recommended for women; for men, the recommendation is 13 cups per day of total beverages, including water.[54]

Vitamins and Minerals

As research reveals more about how specific vitamins and minerals influence disease prevention and how age-related physiological changes affect nutrient metabolism, optimal intakes of vitamins and minerals for different groups of older adults are being defined. This section highlights the vitamins and minerals of greatest concern to older adults.

Vitamin D Older adults face a greater risk of vitamin D deficiency than younger people do. Only vitamin D-fortified milk provides significant vitamin D, and many older adults drink little or no milk. Consequently, the vitamin D intakes of many older adults are less than half of recommendations. Further compromising the vitamin D status of many older people, especially those in nursing homes, is their limited exposure to sunlight. Finally, aging reduces the skin's capacity to make vitamin D and the kidneys' ability to convert it to its active form. Vitamin D recommendations to prevent bone loss and to maintain vitamin D status in older people, especially those who engage in minimal outdoor activity, are listed in the margin. ■

Vitamin B$_{12}$ The DRI committee recommends that adults aged 51 years and older obtain 2.4 micrograms of vitamin B$_{12}$ daily *and* that vitamin B$_{12}$-fortified foods (such as fortified cereals) or supplements be used to meet much of the DRI recommended intake.[55] ■ The committee's recommendation reflects the finding that between 10 and 30 percent of people older than 50 years lose the ability to produce enough stomach acid to make the protein-bound form of vitamin B$_{12}$ available for absorption. Synthetic vitamin B$_{12}$ is reliably absorbed, however. Given the poor cognition, anemia, and devastating neurological effects associated with a vitamin B$_{12}$ deficiency, an adequate intake is imperative.

One cause of the malabsorption of protein-bound vitamin B$_{12}$ is a condition known as **atrophic gastritis**. The prevalence of atrophic gastritis among those 60 years of age and older is high.

Folate As is true of vitamin B$_{12}$, folate intakes of older adults typically fall short of recommendations. ■ The elderly are also more likely to have medical conditions or to take medications that can compromise folate status.

Iron Among the minerals, iron deserves first mention. ■ Iron-deficiency anemia is less common in older adults than in younger people, but it still occurs in some, especially in those with low food energy intakes. Aside from diet, other factors in many older people's lives make iron deficiency likely: chronic blood loss from disease conditions and medicines and poor iron absorption due to reduced secretion of stomach acid and antacid use. Iron deficiency impairs immunity and leaves older adults vulnerable to infectious diseases.[56] Anyone concerned with older people's nutrition should keep these possibilities in mind.

- Vitamin D AI during late adulthood:
  - 10 µg/day (51–70 yr)
  - 15 µg/day (>70 yr)

- Vitamin B$_{12}$ RDA during late adulthood:
  - 2.4 µg/day

- Folate RDA during late adulthood:
  - 400 µg/day

- Iron RDA during late adulthood:
  - 8 mg/day

pressure ulcers: damage to the skin and underlying tissues as a result of compression and poor circulation; commonly seen in people who are bedridden or chairbound.

atrophic gastritis (a-TRO-fik gas-TRI-tis): a condition characterized by chronic inflammation of the stomach accompanied by a diminished size and functioning of the mucosa and glands.

> ### Dietary Guidelines for Americans 2005
>
> - Older adults should consume extra vitamin D from vitamin D-fortified foods and/or supplements.

Zinc Zinc intake is commonly low in older people. Zinc deficiency can depress the appetite and blunt the sense of taste, thereby leading to low food intakes and worsening of zinc status. Many medications that older adults commonly use can impair zinc absorption or enhance its excretion and thus lead to deficiency. ■

Calcium The importance of abundant dietary calcium throughout life to protect against osteoporosis has been emphasized throughout this book. ■ The calcium intakes of many people, especially women, in the United States are well below the recommended amount. ■

Nutrient Supplements for Older Adults

People judge for themselves how to manage their nutrition, and some turn to supplements. Advertisers target older people with appeals to take supplements and eat "health" foods, claiming that these products prevent disease and promote longevity. About 60 percent of all women over 60 years of age take some type of nutrient supplement, and about 40 percent of older men do.[57] Quite often those who take supplements are not deficient in the nutrients being supplemented.

Elderly people often benefit from a balanced low-dose vitamin and mineral supplement, however.[58] Such supplements supply many of the needed minerals along with the vitamins often lacking in older people's diets without providing too much of any one nutrient.

However, food is still the best source of nutrients for everybody. Supplements are just that—supplements to foods, not substitutes for them. For anyone who is motivated to obtain the best possible health, it is never too late to learn to eat well, become physically active, and adopt other lifestyle changes such as quitting smoking, moderating alcohol use, and the like. Table 13–5 summarizes the nutrient concerns of aging. Table 13–6 (p. 392) offers strategies for growing old healthfully.

■ Zinc RDA during late adulthood:
- 11 mg/day (men)
- 8 mg/day (women)

■ If milk causes stomach discomfort, as many older adults report, then lactose-modified milk or other calcium-rich foods should take its place.

■ Calcium AI during late adulthood:
- 1,200 mg/day

TABLE 13–5

Summary of Nutrient Concerns of Aging

Nutrient	Effect of Aging	Comments
Water	Lack of thirst and decreased total body water make dehydration likely.	Mild dehydration is a common cause of confusion. Difficulty obtaining water or getting to the bathroom may compound the problem.
Energy	Need decreases as muscle mass decreases (sarcopenia).	Physical activity moderates the decline.
Fiber	Likelihood of constipation increases with low intakes and changes in the GI tract.	Inadequate water intakes and lack of physical activity, along with some medications, compound the problem.
Protein	Needs may stay the same or increase slightly.	Low-fat, high-fiber legumes and grains meet both protein and other nutrient needs.
Vitamin B_{12}	Atrophic gastritis is common.	Deficiency causes neurological damage; supplements may be needed.
Vitamin D	Increased likelihood of inadequate intake; skin synthesis declines.	Daily sunlight exposure in moderation or supplements may be beneficial.
Calcium	Intakes may be low; osteoporosis is common.	Stomach discomfort commonly limits milk intake; calcium substitutes or supplements may be needed.
Iron	In women, status improves after menopause; deficiencies are linked to chronic blood losses and low stomach acid output.	Adequate stomach acid is required for absorption; antacid or other medicine use may aggravate iron deficiency; vitamin C and meat increase absorption.

TABLE 13-6

Strategies for Growing Old Healthfully

- Choose nutrient-dense foods.
- Be physically active. Walk, run, dance, swim, bike, or row for aerobic activity. Lift weights, do calisthenics, or pursue some other activity to tone, firm, and strengthen muscles. Practice balancing on one foot or doing simple movements with your eyes closed. Modify activities to suit changing abilities and tastes.
- Maintain appropriate body weight.
- Reduce stress (cultivate self-esteem, maintain a positive attitude, manage time wisely, know your limits, practice assertiveness, release tension, and take action).
- For women, discuss with a physician the risks and benefits of estrogen replacement therapy.
- For people who smoke, discuss with a physician strategies and programs to help you quit.
- Expect to enjoy sex, and learn new ways of enhancing it.
- Use alcohol only moderately, if at all; use drugs only as prescribed.
- Take care to prevent accidents.
- Expect good vision and hearing throughout life; obtain glasses and hearing aids if necessary.
- Take care of your teeth; obtain dentures if necessary.
- Be alert to confusion as a disease symptom, and seek diagnosis.
- Take medications as prescribed; see a physician before self-prescribing medicines or herbal remedies and a registered dietitian before self-prescribing supplements.
- Control depression through activities and friendships; seek professional help if necessary.
- Drink plenty of water every day.
- Practice mental skills. Keep on solving math problems and crossword puzzles, playing cards or other games, reading, writing, imagining, and creating.
- Make financial plans early to ensure security.
- Accept change. Work at recovering from losses; make new friends.
- Cultivate spiritual health. Cherish personal values. Make life meaningful.
- Go outside for sunshine and fresh air as often as possible.
- Be socially active—play bridge, join an exercise or dance group, take a class, teach a class, eat with friends, volunteer time to help others.
- Stay interested in life—pursue a hobby, spend time with grandchildren, take a trip, read, grow a garden, or go to the movies.
- Enjoy life.

The Effects of Drugs on Nutrients

As people grow older, the use of medicines—from over-the-counter types such as aspirin and laxatives to prescription medications of all kinds—becomes commonplace. Most drugs interact with one or more nutrients in several ways, usually resulting in greater-than-normal needs for these nutrients. Chapter 15 offers a discussion of diet-drug interactions and describes the many reasons why elderly people are vulnerable to such interactions.

The most common drug that can affect nutrition in older people is alcohol. The effects of alcohol on people of all ages are explained in Nutrition in Practice 20.

IN SUMMARY

- Table 13–5 summarizes the nutrient concerns of aging.

- The ever-growing number of older people creates an urgent need to know more about how their nutrient requirements differ from those of others and how such knowledge can enhance their health.

Food Choices and Eating Habits of Older Adults

To provide any benefit, strategies and interventions to improve a person's nutrition status must be based on knowledge of food preferences and eating patterns. Menus and feeding programs for older adults must take into consideration not only the food likes and dislikes, but also the living conditions, economic status, and medical conditions of this diverse group of people. If nutrition intervention is to be successful, it is essential to know what foods people will eat, in what settings they like to eat these foods, and whether they can buy and prepare meals.

For the most part, older people are independent, socially sophisticated, mentally lucid, fully participating members of society who report themselves to be happy and healthy. In fact, chronic disabilities among the elderly have declined dramatically in recent years.

Older people spend more money per person on foods to eat at home and less money on foods away from home than other age groups. Manufacturers would be wise to cater to the preferences of older adults by providing good-tasting, nutritious foods in easy-to-open, single-serving packages with labels that are easy to read. Such products enable older adults to maintain their independence; most of them want to take care of themselves and need to feel a sense of control and involvement in their own lives. As discussed earlier, another way older adults can take care of themselves is by remaining or becoming physically active. Physical activity helps preserve one's ability to perform daily tasks and so promotes independence.

Shared meals can brighten the day and enhance the appetite.

Individual Preferences Familiarity, taste, and health beliefs are most influential on older people's food choices. Eating foods that are familiar, especially ethnic foods that recall family meals and pleasant times, can be comforting. Older adults are choosing poultry and fish, low-fat milk and milk products, and high-fiber breads and grains, indicating that they recognize the importance of diet in supporting good health. However, few older adults consume the recommended amounts of milk products.

Meal Setting The food choices and eating habits of older adults are also affected by the changes in lifestyle that often accompany aging in this society. Whether people live alone, with others, or in institutions affects the way they eat. For example, men living alone are most likely to be poorly nourished. Older adults who live alone do not make poorer food choices than those who live with companions; rather, they consume too little food. Loneliness is directly related to inadequacies, especially of energy intakes.

Depression Another factor affecting food intake and appetite in older people is depression. Loss of appetite and motivation to cook or even to eat frequently accompanies depression. An overwhelming feeling of grief and sadness at the death of a spouse, friend, or family member may leave many people, particularly elderly people, with a feeling of powerlessness to overcome the depression. The support and companionship of family and friends, especially at mealtimes, can help overcome depression and enhance appetite. The Case Study at the end of this chapter presents a man who has several of these problems. Use the suggestions here and in the last section of this chapter to help develop solutions. The Nutrition Assessment Checklist for Older Adults at the end of the chapter helps to pinpoint nutrition-related factors to look for when working with older adults. To *determine* the risk of malnutrition in older clients, health care professionals can keep in mind the characteristics and questions listed in Table 13–7.

TABLE 13–7

Risk Factors for Malnutrition in Older Adults

	These questions help *determine* the risk of malnutrition in older adults:
Disease	■ Do you have an illness or condition that changes the types or amounts of foods you eat?
Eating poorly	■ Do you eat fewer than two meals per day? Do you eat fruits, vegetables, and milk products daily?
Tooth loss or mouth pain	■ Is it difficult or painful to eat?
Economic hardship	■ Do you have enough money to buy the food you need?
Reduced social contact	■ Do you eat alone most of the time?
Multiple medications	■ Do you take three or more different prescribed or over-the-counter medications daily?
Involuntary weight loss or gain	■ Have you lost or gained 10 pounds or more in the last six months?
Needs assistance	■ Are you physically able to shop, cook, and feed yourself?
Elderly person	■ Are you older than 80?

Food Assistance Programs

Federally funded programs can provide food and nutrition services for older adults.[59] The Older Americans Act (OAA) provides many different services and support to help older adults remain independent. An integral component of the OAA is the OAA Nutrition Program, formerly known as the Elderly Nutrition Program, which offers services that promote health to older adults. Table 13–8 summarizes food assistance programs available to the elderly.

Meals for Singles

Many older adults live alone, and singles of all ages face challenges in purchasing, storing, and preparing food. Large packages of meat and vegetables are often intended for families of four or more, and even a head of lettuce can spoil before one person can use it all. Many singles live in small dwellings and have little storage space for foods. A limited income presents additional obstacles. This section offers suggestions that can help to solve some of the problems singles face, beginning with a special note about the dangers of foodborne illness.

Foodborne Illness The risk of older adults getting a foodborne illness is greater than for other adults. The consequences of an upset stomach, diarrhea, fever, vomiting, abdominal cramps, and dehydration are often more severe, sometimes leading to

---

TABLE 13–8

Food Assistance Programs for Older Adults

OAA Nutrition Program

Services: Provides congregate and home-delivered meals to improve older people's nutrition status. Includes transportation to congregate meal sites; shopping assistance; information and referral; and, to some extent, nutrition counseling and education.

Impact: Improves the nutrient content of high-risk older adults' diets and offers socialization and recreation. Many of the nutrition programs around the country go above and beyond federal requirements of congregate and home meals by offering lunch clubs, ethnic meals, and meals for older homeless people.

Food Stamp Program

Services: Supplements income for low-income households by means of a card similar to a debit card that can be used to purchase food.

Impact: Serves more as an income supplement for some elderly participants than as a device to improve nutrition status. For other elderly food stamp participants, nutrient intakes are higher than those of nonparticipants with similar incomes.

Meals on Wheels

Services: Delivers meals directly to the homebound elderly; integrated into the meal delivery services provided by the OAA Program.

Impact: Focuses on filling the need for weekend and holiday meals for homebound elderly people, a service that is limited in the OAA Nutrition Program.

Senior Farmers Market Nutrition Program

Services: Provides low-income older adults with coupons that can be exchanged for fresh fruits, vegetables, and herbs at community-supported farmers' markets and roadside stands; administered by the USDA. State agencies may limit sales to specific foods that are locally grown to encourage recipients to support farmers in their own states.

Impact: Increases fresh fruit and vegetable consumption, provides nutrition information, and even reaches the homebound elderly, a group of people who normally do not have access to farmers' markets.

Dietary Guidelines for Americans 2005

- Older adults should not eat or drink unpasteurized milk, milk products, or juices; raw or undercooked eggs, meat, poultry, fish, or shellfish; or raw sprouts.

- Older adults should only eat certain deli meats and frankfurters that have been reheated to steaming hot.

paralysis, meningitis, or even death. For these reasons, older adults need to carefully follow the food safety suggestions presented in Nutrition in Practice 5.

Spend Wisely People who have the means to shop and cook for themselves can cut their food bills just by being wise shoppers. Large supermarkets are usually less expensive than convenience stores. A grocery list helps reduce impulse buying, and specials and coupons can save money when the items featured are those that the shopper needs and uses.

Buying the right amount so as not to waste any food is a challenge for people eating alone. They can buy fresh milk in the size best suited for personal needs. Pint-size and even cup-size boxes ■ of milk are available and can be stored unopened on a shelf for up to three months without refrigeration.

Many foods that offer a variety of nutrients for practically pennies have a long shelf life; staples such as rice, pasta, dry powdered milk, and dried legumes can be purchased in bulk and stored for months at room temperature. Other foods that are usually a good buy include whole pieces of cheese rather than sliced or shredded cheese, fresh produce in season, variety meats such as chicken livers, and cereals that require cooking instead of ready-to-serve cereals.

A person who has ample freezing space can buy large packages of meat, such as pork chops, ground beef, or chicken, when they are on sale. Then the meat can be immediately wrapped in individual servings for the freezer. All the individual servings can be put in a bag marked appropriately with the contents and the date.

Frozen vegetables are more economical in large bags than in small boxes. The amount needed can be taken out, and the bag can be closed tightly with a twist tie or rubber band. If the package is returned quickly to the freezer each time, the vegetables will stay fresh for a long time.

Finally, breads and cereals usually must be purchased in larger quantities. Again the amount needed for a few days can be taken out, and the rest can be stored in the freezer.

People who do not have freezers can ask the grocer to break open a package of wrapped meat and rewrap the portion needed. Similarly, eggs can be purchased by the half-dozen. Eggs do keep for long periods, if stored properly in the refrigerator.

Fresh fruits and vegetables can be purchased individually. A person can buy fresh fruit at various stages of ripeness: a ripe one to eat right away, a semiripe one to eat soon after, and a green one to ripen on the windowsill. If vegetables are packaged in large quantities, the grocer can break open the package so that a smaller amount can be purchased. Small cans of fruits and vegetables, even though they are more expensive per unit, are a reasonable alternative, considering that it is expensive to buy a regular-size can and let the unused portion spoil.

Be Creative Creative chefs think of various ways to use foods when only large amounts are available. For example, a head of cauliflower can be divided into thirds. Then one-third is cooked and eaten hot. Another third is put into a vinegar and oil marinade for use in a salad. The last third can be used in a casserole or stew.

■ Boxes of milk that can be stored at room temperature have been exposed to temperatures above those of pasteurization just long enough to sterilize the milk—a process called *ultrahigh temperature (UHT)*.

A variety of vegetables and meats can be enjoyed stir-fried; inexpensive vegetables such as cabbage, celery, and onions are delicious when crisp cooked in a little oil with herbs or lemon added. Interesting frozen vegetable mixtures are available in larger grocery stores. Cooked, leftover vegetables can be dropped in at the last minute. A bonus of a stir-fried meal is that there is only one pan to wash. Similarly, a microwave oven allows a home chef to use fewer pots and pans. Meals and leftovers can also be frozen or refrigerated in microwavable containers to reheat as needed.

Many frozen dinners or grocery store take-out foods offer nutritious options. Adding a fresh salad, a whole wheat roll, and a glass of milk can make a nutritionally balanced meal.

Also, single people should not hesitate to invite someone to share meals with them whenever there is enough food. It is likely that the person will return the invitation, and both parties will get to enjoy companionship and a meal prepared by others.

Nutrition Assessment Checklist

FOR OLDER ADULTS

Medical History

Check the medical record for:

- Alcohol abuse
- Alzheimer's disease or other dementia or confusion
- Arthritis
- Cataracts
- Chronic diseases (cancer, heart disease, hypertension diabetes)
- Cigarette, cigar, or pipe smoking; use of other tobacco products
- Constipation
- Dehydration
- Dental disease or tooth loss
- Depression
- Inflammation of the stomach (gastritis)
- Swallowing disorders

Medications

For older adults being treated with drug therapy for medical conditions, note:

- Use of multiple medications—prescription and/or over-the-counter medications such as laxatives and pain relievers
- Side effects that might reduce food intake or change nutrient needs
- Proper administration of medication with respect to food intake
- Malnutrition—is the person's nutrition status questionable even before considering side effects of medications that worsen nutrition status?
- Diminished mental capacity that might interfere with taking correct medications and doses
- Dehydration (can alter effects of medications)

Dietary Intake

For all older adults, especially those at risk nutritionally, assess the diet for:

- Total energy
- Protein

- Calcium, iron, and zinc
- Vitamin B_6, vitamin B_{12}, folate, and vitamin D

Note the following:

- Number of meals eaten each day
- Number and ages of people in household
- Amount of milk consumed each day
- Type and frequency of outdoor activity
- Type and frequency of physical activity
- Financial resources
- Transportation resources
- Physical disabilities
- Mental alertness

Anthropometric Data

Measure baseline height and weight.

- Reassess height and weight at each medical checkup.
- Note significant overweight or underweight, which warrants intervention.
- Use skinfold measures to reveal altered body composition that may indicate malnutrition and loss of lean tissue.

Laboratory Tests

- Hemoglobin, hematocrit, or other tests of iron status
- Serum albumin or other measures of protein status
- Serum folate
- Serum B_{12}

Physical Signs

Look for physical signs of:

- Protein-energy malnutrition
- Iron and zinc deficiency
- Folate deficiency

■ Food choices of older adults are affected by health status and changed life circumstances.

■ Older people can benefit from both the nutrients provided and the social interaction available at congregate meals.

■ Other government programs deliver meals to those who are homebound.

■ With creativity and careful shopping, those living alone can prepare nutritious, inexpensive meals.

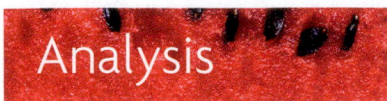

Analysis

CASE STUDY

Elderly Man with a Poor Diet

Mr. Brezenoff is 75 years old and lives alone. He has slowly been losing weight since his wife died one year ago. At 5 feet 8 inches tall, he currently weighs 124 pounds. His previous weight was 150 pounds. In talking with Mr. Brezenoff, you realize that he does not even like to talk about food, let alone eat it. "My wife always did the cooking before, and I ate well. Now I just don't feel like eating." You manage to find out that he skips breakfast, has soup and bread for lunch, and sometimes eats a cold-cut sandwich or a frozen dinner for supper. He seldom sees friends or relatives. Mr. Brezenoff has also lost several teeth and does not eat any raw fruits or vegetables because he finds them hard to chew. He lives on a meager but adequate income.

1. Consult the BMI table (inside the back cover of this text), and judge whether Mr. Brezenoff is at a healthy weight. What other assessments might you use to back up your judgment? Is his weight loss significant?

2. What factors are contributing to Mr. Brezenoff's poor food intake? What nutrients are probably deficient in his diet?

3. Look at Mr. Brezenoff as an individual and suggest ways he can improve his diet and his lifestyle.

4. What other aspects of Mr. Brezenoff's physical and mental health should you consider in helping him to improve his food intake?

CLINICAL APPLICATIONS

1. Ms. Hamilton is an 80-year-old woman in excellent health who lives alone, eats a well-balanced diet, enjoys an active social life, and walks every day. Consider the way Ms. Hamilton's health and nutrition status might be affected by the following situations:

■ Many of Ms. Hamilton's friends pass away or move into extended care facilities.

■ Ms. Hamilton falls and breaks her hip.

■ Ms. Hamilton begins to feel isolated and depressed.

Describe interventions the health care professional can take to help Ms. Hamilton deal with each situation to prevent her from falling into a downward spiral.

Self Check

1. The fastest-growing age group in the United States is:
   a. 21 years of age.
   b. 30 to 45 years of age.
   c. 50 to 70 years of age.
   d. over 85 years of age.

2. Which of the following lifestyle habits can enhance the length and quality of people's lives?
   a. moderate smoking
   b. 6 hours of sleep daily
   c. regular physical activity
   d. skipping breakfast

3. Among the better-known relationships between nutrition and disease prevention are:
   a. appropriate fiber intake helps prevent goiter.
   b. moderate sodium intake helps prevent obesity.
   c. moderate sugar intake helps prevent hypertension.
   d. appropriate energy intake helps prevent diabetes and cardiovascular disease.

4. A disease of the immune system that involves painful inflammation of the joints is:
   a. sarcopenia.
   b. osteoarthritis.
   c. senile dementia.
   d. rheumatoid arthritis.

5. Examples of low-kcalorie, high-quality protein foods include:
   a. cottage cheese, sour cream, and eggs.
   b. green and yellow vegetables and citrus fruits.
   c. potatoes, rice, pasta, and whole-grain breads.
   d. lean meats, poultry, fish, legumes, fat-free milk, and eggs.

6. For malnourished and underweight people, protein- and energy-dense snacks include:
   a. fresh fruits and vegetables.
   b. yogurt and cottage cheese.
   c. whole grains and high-fiber legumes.
   d. hard-boiled eggs and peanut butter and graham crackers.

7. Which of the following does not explain why dehydration is a risk for older adults?
   a. They do not seem to feel thirsty.
   b. Total body water increases with age.
   c. They may find it difficult to get a drink.
   d. They may have difficulty swallowing liquids.

8. Inadequate milk intake and limited exposure to sunlight contribute to older adults' risk of:
   a. vitamin A deficiency.
   b. vitamin D deficiency.
   c. riboflavin deficiency.
   d. vitamin B_6 deficiency.

9. Two risk factors for malnutrition in older adults are:
   a. loneliness and multiple medication use.
   b. increased energy needs and lack of fiber.
   c. decreased mineral absorption and antioxidant intake.
   d. high carbohydrate intake and lack of physical activity.

10. Two strategies to improve nutrition status when growing old include:
    a. increase vitamin A intake and exercise 30 minutes daily.
    b. choose nutrient-dense foods and maintain appropriate weight.
    c. avoid high-fiber foods and take a daily vitamin-mineral supplement.
    d. eat at least one big meal per day and drink at least 10 glasses of water daily.

Answers to these questions appear in Appendix H.

Notes

1. Federal Interagency Forum on Aging-Related Statistics, Older Americans 2004: Key Indicators of Well-Being, November 2006, available at www.agingstats.gov; Trends in aging—United States and worldwide, *Morbidity and Mortality Weekly Report* 52 (2003): 101–106.

2. B. E. Hamilton and coauthors, Annual summary of vital statistics: 2005, *Pediatrics* 119 (2007): 345-360.

3. Living well to 100: Nutrition, genetics, inflammation, supplement to *American Journal of Clinical Nutrition* 83 (2006): 401S–490S.

4. W. S. Browner and coauthors, The genetics of human longevity, *American Journal of Medicine* 117 (2004): 851–860.

5. T. Perls, Genetic and environmental influences on exceptional longevity and the AGE nomogram, *Annals of the New York Academy of Sciences* 959 (2002): 1–13.

6. Position of the American Dietetic Association: Nutrition across the spectrum of aging, *Journal of the American Dietetic Association* 105 (2005): 616–633.

7. D. K. Houston and coauthors, Dairy, fruit, and vegetable intakes and functional limitations and disability in a biracial cohort: The Atherosclerosis Risk in Communities Study, *American Journal of Clinical Nutrition* 81 (2005): 515–522; K. T. B. Knoops and coauthors, Mediterranean diet, lifestyle factors, and 10-year mortality in elderly European men and women, *Journal of the American Medical Association* 292 (2004): 1433–1439; G. E. Fraser and D. J. Shavlik, Ten years of life: Is it a matter of choice? *Archives of Internal Medicine* 161 (2001): 1645–1652.

8. C. R. Richardson and coauthors, Physical activity and mortality across cardiovascular disease risk groups, *Medicine and Science in Sports and Exercise* 36 (2004): 1923–1929; F. B. Hu and coauthors, Adiposity as compared with physical activity in predicting mortality among women, *New England Journal of Medicine* 351 (2004): 2694–2703; M. E. Cress and coauthors, Physical activity programs and behavior counseling in older adult populations, *Medicine and Science in Sports and Exercise* 36 (2004): 1997–2003.

9. Position of the American Dietetic Association, 2005; Cress and coauthors, 2004; E. W. Gregg and coauthors, Relationship of changes in physical activity and mortality among older women, *Journal of the American Medical Association* 289 (2003): 2379–2386.

10. C. H. Hillman and coauthors, Physical activity and executive control: Implications for increased cognitive health during older adulthood, *Research Quarterly for Exercise and Sport* 75 (2004): 176–185; W. J. Strawbridge and coauthors, Physical activity reduces the risk of subsequent depression for older adults, *American Journal of Epidemiology* 156 (2002): 328–334.

11. Centers for Disease Control and Prevention, Fatalities and injuries from falls among older adults—United States, 1993–2003

and 2001–2005, *Morbidity and Mortality Weekly Report* 55 (2006): 1221–1224.

12. K. A. Martin Ginis and coauthors, Weight training to activities of daily living: Helping older adults make a connection, *Medicine and Science in Sports and Exercise* 38 (2006): 116–121; P. S. Ades and coauthors, Resistance training on physical performance in disabled older female cardiac patients, *Medicine and Science in Sports and Exercise* 35 (2003): 1265–1270.

13. L. K. Heilbronn and E. Ravussin, Calorie restriction and aging: Review of the literature and implications for studies in humans, *American Journal of Clinical Nutrition* 78 (2003): 361–369.

14. Position of the American Dietetic Association, 2005.

15. G. Wolf, Calorie restriction increases life span: A molecular mechanism, *Nutrition Reviews* 64 (2006): 89–92; L. Fontana and coauthors, Long-term calorie restriction is highly effective in reducing the risk for atherosclerosis in humans, *Proceedings of the National Academy of Sciences* 101 (2004): 6659–6663.

16. Knoops and coauthors, 2004.

17. W. G. Christen and coauthors, Fruit and vegetable intake and the risk of cataract in women, *American Journal of Clinical Nutrition* 81 (2005): 1417–1422.

18. C. Chitchumroonchokchai and coauthors, Xanthophylls and α-tocopherol decrease UVB-induced lipid peroxidation and stress signaling in human lens epithelial cells, *Journal of Nutrition* 134 (2004): 3225–3232; A. Taylor, and coauthors, Long-term intake of vitamins and carotenoids and odds of early age-related cortical and posterior subcapsular lens opacities, *American Journal of Clinical Nutrition* 75 (2002): 540–549.

19. S. M. Moeller and coauthors, Overall adherence to the *Dietary Guidelines for Americans* is associated with reduced prevalence of early age-related nuclear lens opacities in women, *Journal of Nutrition* 134 (2004): 1813–1819.

20. J. M. Weintraub and coauthors, A prospective study of the relationship between body mass index and cataract extraction among US women and men, *International Journal of Obesity and Related Metabolic Disorders* 26 (2002): 1588–1595.

21. Eye Diseases Prevalence Research Group, Prevalence of age-related macular degeneration in the United States, *Archives of Ophthalmology* 122 (2004): 564–572.

22. P. R. Trumbo and K. C. Ellwood, Lutein and zeaxanthin intakes and risk of age-related macular degeneration and cataracts: An evaluation using the Food and Drug Administration's evidence-based review system for health claims, *American Journal of Clinical Nutrition* 84 (2006): 971–974; R. Van Leeuwen and coauthors, Dietary intake of antioxidants and risk of age-related macular degeneration, *Journal of the American Medical Association* 294 (2005): 3101–3107; D. Hartman and coauthors, Plasma kinetics of zeaxanthin and 3'-dehydro-lutein after multiple oral doses of synthetic zeaxanthin, *American Journal of Clinical Nutrition* 79 (2004): 410–417; N. I. Krinsky, J. T. Landrum, and R. A. Bone, Biologic mechanisms of the protective role of lutein and zeaxanthin in the eye, *Annual Review of Nutrition* 23 (2003): 171–201.

23. B. M. Kuehn, Studies probe diet's role in eye disease, *Journal of the American Medical Association* 294 (2005): 32–33.

24. J. P. SanGiovanni and E. Y. Chew, The role of omega-3 long-chain polyunsaturated fatty acids in health and disease of the retina, *Progress in Retinal and Eye Research* 24 (2005): 87–138.

25. Position of the American Dietetic Association, 2005.

26. L. Devos-Comby, T. Cronan, and S. C. Roesch, Do exercise and self-management interventions benefit patients with osteoarthritis of the knee? A metaanalytic review, *Journal of Rheumatology* 33 (2006): 744–756; S. P. Messier and coauthors, Exercise and dietary weight loss in overweight and obese older adults with knee osteoarthritis: The Arthritis, Diet, and Activity Promotion Trial, *Arthritis and Rheumatism* 50 (2004): 1501–1510.

27. R. F. Bruyere and coauthors, Structural and symptomatic efficacy of glucosamine and chondroitin in knee osteoarthritis: A comprehensive meta-analysis, *Archives of Internal Medicine* 163 (2003): 1514–1522.

28. L. Skoldstam, L. Haffors, and G. Johansson, An experimental study of a Mediterranean diet intervention for patients with rheumatoid arthritis, *Annals of the Rheumatic Diseases* 62 (2003): 208–214.

29. O. Adam, Dietary fatty acids and immune reactions in synovial tissue, *European Journal of Medical Research* 8 (2003): 381–387; L. Cleland, M. James, and S. Proudman, The role of fish oils in the treatment of rheumatoid arthritis, *Drugs* 63 (2003): 845–853.

30. D. J. Pattison and coauthors, Dietary β-cryptoxanthin and inflammatory polyarthritis: Results from a population-based prospective study, *American Journal of Clinical Nutrition* 82 (2005): 451–455; J. R. Cerhan and coauthors, Antioxidant micronutrients and risk of rheumatoid arthritis in a cohort of older women, *American Journal of Epidemiology* 157 (2003): 345–354.

31. R. J. Wurtman and coauthors, Effects of normal meals rich in carbohydrates or proteins on plasma tryptophan and tyrosine ratios, *American Journal of Clinical Nutrition* 77 (2003): 128–132.

32. K. L. Tucker and coauthors, High homocysteine and low B vitamins predict cognitive decline in aging men: The Veterans Affairs Normative Aging Study, *American Journal of Clinical Nutrition* 82 (2005): 627–635; G. Ravaglia and coauthors, Homocysteine and folate as risk factors for dementia and Alzheimer disease, *American Journal of Clinical Nutrition* 82 (2005): 636–643; P. Quadri and coauthors, Homocysteine, folate, and vitamin B-12 in mild cognitive impairment, Alzheimer disease, and vascular dementia, *American Journal of Clinical Nutrition* 80 (2004): 114–122.

33. R. D. Abbott and coauthors, Walking and dementia in physically capable elderly men, *Journal of the American Medical Association* 292 (2004): 1447–1453; J. Weuve and coauthors, Physical activity, including walking, and cognitive function in older women, *Journal of the American Medical Association* 292 (2004): 1454–1461.

34. L. E. Hebert and coauthors, Alzheimer disease in the US population: Prevalence estimates using the 2000 census, *Archives of Neurology* 60 (2003): 1119–1122.

35. T. D. Bird, Genetic factors in Alzheimer's disease, *New England Journal of Medicine* 352 (2005): 862–864.

36. P. I. Moreira and coauthors, Oxidative stress: The old enemy in Alzheimer's disease pathophysiology, *Current Alzheimer Research* 2 (2005): 403–408.

37. P. P. Zandi and coauthors, Reduced risk of Alzheimer disease in users of antioxidant vitamin supplements: The Cache County Study, *Archives of Neurology* 61 (2004): 82–88; M. J. Engelhart and coauthors, Dietary intake of antioxidants and risk of Alzheimer disease, *Journal of the American Medical Association* 287 (2002): 3223–3229; M. C. Morris, Dietary intake of antioxidant nutrients and the risk of incident Alzheimer disease in a biracial community study, *Journal of the American Medical Association* 287 (2002): 3230–3237.

38. Ravaglia and coauthors, 2005; Tucker and coauthors, 2005; Quadri and coauthors, 2004.

39. R. Uauy and A. D. Dangour, Nutrition in brain development and aging: Role of essential fatty acids, *Nutrition Reviews* 64 (2006): S24–S33; T. Den Heijer and coauthors, Alcohol intake in relation to brain magnetic resonance imaging findings in older persons without dementia, *American Journal of Clinical Nutrition* 80 (2004): 992–997; F. Calon and coauthors, Docosahexaenoic acid protects from dendritic pathology in an Alzheimer's disease mouse model, *Neuron* 43 (2004): 633–645.

40. L. Bren, Alzheimer's: Searching for a cure, *FDA Consumer*, July/August 2003, pp. 19–25.

41. R. Chernoff, Micronutrient requirements in older women, *American Journal of Clinical Nutrition* 81 (2005): 1240S–1245S.

42. N. Meunier and coauthors, Basal metabolic rate and thyroid hormones of late-middle-aged and older human subjects: The ZENITH study, *European Journal of Clinical Nutrition* 59 (2005): S53–S57.

43. H. K. Kamel, Sarcopenia and aging, *Nutrition Reviews* 61 (2003): 157–167; I. Janssen, S. B. Heymsfield, and R. Ross, Low relative skeletal muscle mass (sarcopenia) in older persons is associated with functional impairment and physical disability, *Journal of the American Geriatric Society* 50 (2002): 889–896; C. W. Bales and C. S. Ritchie, Sarcopenia, weight loss, and nutritional frailty in the elderly, *Annual Review of Nutrition* 22 (2002): 309–323.

44. M. Cesari and coauthors, Frailty syndrome and skeletal muscle: Results from the Invecchiare in Chianti study, *American Journal of Clinical Nutrition* 83 (2006): 1142–1148.

45. K. S. Nair, Aging muscle, *American Journal of Clinical Nutrition* 81 (2005): 953–963; D. K. Houston and coauthors, Dairy, fruit, and vegetable intakes and functional limitations and disability in a biracial cohort: The Atherosclerosis Risk in Communities Study, *American Journal of Clinical Nutrition* 81 (2005): 515–522.

46. Nair, 2005; Houston and coauthors, 2005; P. Szulc and coauthors, Hormonal and lifestyle determinants of appendicular skeletal muscle mass in men: The MINOS Study, *American Journal of Clinical Nutrition* 80 (2004): 496–503.

47. Position of the American Dietetic Association, 2005.

48. Position of the American Dietetic Association: Liberalization of the diet prescription improves quality of life for older adults in long-term care, *Journal of the American Dietetic Association* 105 (2005): 1955–1965.

49. D. Benton and S. Nabb, Carbohydrate, memory, and mood, *Nutrition Reviews* 61 (2003): S61–S67.

50. Position of the American Dietetic Association: Nutrition across the spectrum of aging, *Journal of the American Dietetic Association* 105 (2005): 616–633; N. R. Sahyoun and E. Krall, Lower dietary quality among older adults with self-perceived ill-fitting dentures, *Journal of the American Dietetic Association* 103 (2003): 1494–1499.

51. Standing Committee on the Scientific Evaluation of Dietary Reference Intakes, Food and Nutrition Board, Institute of Medicine, *Dietary Reference Intakes for Energy, Carbohydrate, Fiber, Fat, Fatty Acids, Cholesterol, Protein, and Amino Acids* (Washington DC: National Academies Press, 2005), pp. 387–389.

52. Position of the American Dietetic Association: Nutrition across the spectrum of aging, *Journal of the American Dietetic Association* 105 (2005): 616–633; D. R. Thomas, Dehydration in older adults, *Nutrition and the M.D.*, November 2004, pp. 1–3.

53. Standing Committee on the Scientific Evaluation of Dietary Reference Intakes, Food and Nutrition Board, Institute of Medicine, *Dietary Reference Intakes for Water, Potassium, Sodium, Chloride, and Sulfate* (Washington, DC: National Academies Press, 2005), pp. 118–127; Position of the American Dietetic Association: Liberalization of the diet prescription improves quality of life for older adults in long-term care, *Journal of the American Dietetic Association* 105 (2005): 1955–1965.

54. Standing Committee on the Scientific Evaluation of Dietary Reference Intakes, *Dietary Reference Intakes for Water, Potassium, Sodium, Chloride, and Sulfate*, 2005, pp. 149–150.

55. R. Chernoff, 2005.

56. N. Ahluwalia and coauthors, Immune function is impaired in iron-deficient, homebound, older women, *American Journal of Clinical Nutrition* 79 (2004): 516–521.

57. Executive Summary: Conference on dietary supplement use in the elderly—proceedings of the conference held January 14–15, 2003, Natcher Auditorium, National Institutes of Health, Bethesda, MD, *Nutrition Reviews* 62 (2004): 160–175.

58. Position of the American Dietetic Association: Nutrition across the spectrum of aging, *Journal of the American Dietetic Association* 105 (2005): 616–633.

59. Position of the American Dietetic Association: Nutrition across the spectrum of aging, *Journal of the American Dietetic Association* 105 (2005): 616–633.

Nutrition in Practice

HUNGER AND COMMUNITY NUTRITION

One person in every eight worldwide experiences persistent hunger—not the healthy appetite triggered by anticipation of a hearty meal, but the painful sensation caused by a lack of food. Hunger deprives a person of the physical and mental energy needed to enjoy a full life and often leads to severe malnutrition and death. Tens of thousands die of starvation each day—one child every five seconds.[1]

In the United States, where most people enjoy a life of relative abundance, about one in every nine households has one or more members who experience pain from hunger caused by lack of food. In these households, 12 million children do not know where their next meal is coming from, or when it will come.[2] Given the agricultural bounty and enormous wealth in this country, do these numbers surprise you? The limited or uncertain availability of nutritionally adequate and safe foods is known as **food insecurity** and is a major social problem in our nation today.[3] The "How to" box describes how national surveys identify food insecurity in the United States, Figure NP13–1 (p. 402) presents the most recent findings, and the Glossary (p. 402) defines related terms. Surveys such as these provide crude, but necessary, data to estimate the degree of hunger in this country.

How To

IDENTIFY FOOD INSECURITY IN A U.S. HOUSEHOLD

To determine the extent of food insecurity in a household, surveys ask questions about behaviors and conditions known to characterize households having difficulty meeting basic food needs during the past 12 months. Most often, adults tend to protect their children from hunger. In the most severe cases, children also suffer from hunger and eat less.

1. Did you worry whether your food would run out before you got money to buy more?

2. Did you find that the food you bought just didn't last, and you didn't have money to buy more?

3. Were you unable to afford to eat balanced meals?

4. Did you or other adults in your household ever cut the size of your meals or skip meals because there wasn't enough money for food?

5. Did this happen in three or more months during the previous year?

6. Did you ever eat less than you felt you should because there wasn't enough money for food?

7. Were you ever hungry but didn't eat because you couldn't afford enough food?

8. Did you lose weight because you didn't have enough money for food?

9. Did you or other adults in your household ever not eat for a whole day because there wasn't enough money for food?

10. Did this happen in three or more months during the previous year?

11. Did you rely on only a few kinds of low-cost food to feed your children because you were running out of money to buy food?

12. Were you unable to feed your children a balanced meal because you couldn't afford it?

13. Were your children not eating enough because you just couldn't afford enough food?

14. Did you ever cut the size of your children's meals because there wasn't enough money for food?

15. Were your children ever hungry but you just couldn't afford more food?

16. Did your children ever skip a meal because there wasn't enough money for food?

17. Did this happen in three or more months during the previous year?

18. Did your children ever not eat for a whole day because there wasn't enough money for food?

The more positive responses, the greater the food insecurity. Households with children answer all of the questions and are categorized as follows:

$\leq$2 positive responses = food secure

3–7 responses = low food security

$\geq$8 positive responses = very low food security

Households without children answer the first 10 questions and are categorized as follows:

$\leq$2 positive responses = food secure

3–5 responses = low food security

$\geq$6 positive responses = very low food security

Source: United States Department of Agriculture, *Household Food Security in the United States, 2005,* available at www.ers.usda.gov/publications/err29.

GLOSSARY

emergency shelters: facilities that are used to provide temporary housing.

food bank: a facility that collects and distributes food donations to authorized organizations feeding the hungry.

food insecurity: limited or uncertain access to foods of sufficient quality or quantity to sustain a healthy and active life. Food insecurity categories:
Low food security: reduced quality of diet with little or no indication of reduced food intake.
Very low food security: multiple indications of disrupted eating patterns and reduced food intake.

food pantries: programs that provide groceries to be prepared and eaten at home.

food poverty: hunger resulting from inadequate access to available food for various reasons, including inadequate resources, political obstacles, social disruptions, poor weather conditions, and lack of transportation.

food recovery: collecting wholesome food for distribution to low-income people who are hungry. Four common methods of food recovery are:
Field gleaning: collecting crops from fields that either have already been harvested or are not profitable to harvest.
Perishable food rescue or salvage: collecting perishable produce from wholesalers and markets.
Prepared food rescue: collecting prepared foods from commercial kitchens.
Nonperishable food collection: collecting processed foods from wholesalers and markets.

food security: access to enough food to sustain a healthy and active life.
High food security: no indications of food-access problems or limitations.
Marginal food security: one or two indications of food-access problems but with little or no change in food intake.

soup kitchens: programs that provide prepared meals to be eaten on-site.

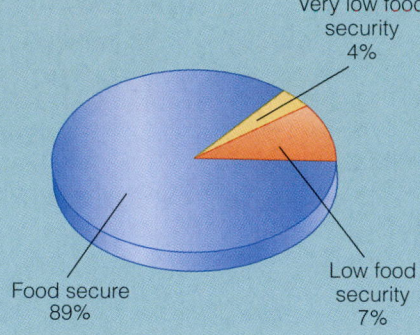

FIGURE NP13–1

Food Security of U.S. Households, 2005
Source: USDA Economic Research Service. Food Security in the United States: Conditions and Trends, 2006, available at www.ers.usda.gov/Briefing/Food Security/trends.htm.

Why is hunger a problem in developed countries such as the United States where food is abundant?

Hunger has many causes, but in developed countries, the primary cause is **food poverty.** People are hungry not because there is no food nearby to purchase, but because they lack sufficient money with which to buy nutritious food and pay for other necessities, such as housing, clothing, medicines, and utilities. More than 12 percent of the population of the United States lives in poverty. Even those above the poverty line may not have **food security**. Physical and mental illnesses and disabilities, sudden job losses, and high living expenses threaten their financial stability. Further contributing to food poverty are other problems such as abuse of alcohol and other drugs; lack

of awareness of available food assistance programs; and the reluctance of people, particularly the elderly, to accept what they perceive as "welfare" or "charity." Lack of resources remains the major cause of food poverty, and solving this problem would do a lot to relieve hunger.

In the United States, food poverty and hunger reach into many segments of society, affecting not only the chronic poor (migrant workers, the unskilled and unemployed, the homeless, and some elderly), but also the so-called working poor. Some are displaced farm families. Some are former blue-collar and white-collar workers forced out of their trades and professions into minimum-wage jobs. These people outnumber the chronic poor, and they are not on welfare. They have jobs, but the pay is too low to meet their needs. Families with incomes below a certain level are simply unable to buy sufficient amounts of nourishing foods, even if they are skilled in food shopping. For many of the children in these families, school lunch is their only meal of the day. Otherwise they go hungry, waiting for an adult to find money for food. Not surprisingly, these children are more likely to have health problems than those who eat regularly.[4] They also tend to perform poorly in school and in social situations.[5]

What U.S. food programs are directed at relieving hunger in the United States?

The American Dietetic Association (ADA) calls for aggressive action to bring an end to domestic hunger and to achieve food and nutrition security for all residents of the United States.[6] Many federal and local programs aim to prevent or relieve malnutrition and hunger in the United States.

An extensive network of federal assistance programs provides life-giving food daily to millions of U.S. citizens. One out of every six Americans receives food assistance of some kind, at a total cost of almost $40 billion per year. Even so, the programs are not fully successful in preventing hunger, but they do seem to improve the nutrient intakes of those who participate. Such programs include the WIC program for low-

These people and many others like them in the United States face food insecurity daily.

income pregnant women, breastfeeding mothers, and their young children (Chapter 11); the school lunch and breakfast programs for children (Chapter 12); and the food assistance programs for older adults such as congregate meals and Meals on Wheels (Chapter 13).

The centerpiece of food programs for low-income people in the United States is the Food Stamp Program, administered by the U.S. Department of Agriculture (USDA). The USDA issues debit cards through state agencies to households—people who buy and prepare food together. The amount a household receives depends on its size, resources, and income. Recipients may use the cards to purchase food and food-bearing plants and seeds, but not to buy tobacco, cleaning items, alcohol, or other nonfood items. The accompanying "How to" offers shopping tips for those on a limited budget.

The Food Stamp Program is the largest of the federal food assistance programs, both in amount of money spent and in number of people participating. Almost 24 million people receive food stamps at a cost of over $25 billion per year; over half of the recipients are children.[7]

Although food assistance programs improve nutrient intakes significantly, hunger continues to plague the United States. Of the estimated 2 million homeless people in the United States who are eligible for food assistance, only 15 percent of single adults and 50 percent of families receive food stamps.

Why do health care professionals need to know about food assistance programs?

Health care professionals who work in public health are generally well acquainted with food assistance programs, and often many of their clients receive such assistance. Regardless of the setting in which health care professionals see clients, it is important to encourage those who may be having financial

How To

PLAN HEALTHY, THRIFTY MEALS

Chapter 1 introduced the USDA Food Guide and principles for planning a healthy diet. Meeting that goal on a limited budget adds to the challenge. To save money and spend wisely, plan and shop for healthy meals with the following tips in mind:

Planning

- Make a grocery list before going to the store to avoid expensive "impulse" items.

- Do not shop when hungry.

- Use leftovers.

- Center meals on rice, noodles, and other grains.

- Use small quantities of meat, poultry, fish, or eggs.

- Use legumes instead of meat, poultry, fish, or eggs several times per week.

- Use cooked cereals such as oatmeal instead of ready-to-eat breakfast cereals.

- Cook large quantities when time and money allow.

- Check for sales and clip coupons for products you need; plan meals to take advantage of sale items.

Shopping

- Buy day-old bread and other products from the bakery outlet.

- Select whole foods instead of convenience foods (potatoes instead of instant mashed potatoes, for example).

- Try store brands.

- Buy fresh produce that is in season; buy canned or frozen items at other times.

- Buy only the amount of fresh foods that you will eat before it spoils. Buy large bags of frozen items or dry goods; when cooking, take out the amount needed and store the remainder.

- Buy fat-free dry milk; mix and refrigerate quantities needed for a day or two. Buy fresh milk by the gallon or half-gallon.

- Buy less expensive cuts of meat. Chuck and bottom round roast are usually inexpensive; cover during cooking and cook long enough to make meat tender. Buy whole chickens instead of pieces.

- Compare the unit price (cost per ounce, for example) of similar foods so that you can select the least expensive brand or size.

- Buy nonfood items such as toilet paper and laundry detergent at discount stores instead of grocery stores.

For daily menus and recipes for healthy, thrifty meals, visit the USDA Center for Nutrition Policy and Promotion: www.usda.gov/cnpp

problems to talk with a social worker who can assess their eligibility for food assistance programs. The subject of food assistance must be approached in a nonjudgmental and tactful manner—the client may feel uncomfortable about seeking assistance.

Are there other programs aimed at reducing hunger in the United States?

Efforts to resolve the problem of hunger in the United States do not depend solely on federal assistance programs. National **food recovery** programs have made a dramatic difference; the largest program, America's Second Harvest, coordinates the efforts of more than 40,000 **food pantries, emergency shelters,** and **soup kitchens** that feed over 25 million people per year. Table NP13–1 lists addresses, telephone numbers, and Web sites for America's Second Harvest and other hunger relief organizations.

Each year, an estimated one-fifth of our food supply is wasted in fields, commercial kitchens, grocery stores, and restaurants—that is enough food to feed 49 million people. Food recovery programs collect and distribute good food that would otherwise go to waste. For example, volunteers might pick corn left in an already harvested field, a grocer might deliver ripe bananas to a local **food bank,** and a caterer might take leftover chicken salad to a community shelter. All of these efforts help to feed the hungry in the United States.

Feeding the hungry in the United States.

© Skjold/The Image Works

What about local efforts and community nutrition programs?

Food recovery programs depend on volunteers. Concerned citizens work through local agencies and churches to feed the hungry. Community-based food pantries provide groceries, and soup kitchens serve prepared meals. Meals often deliver adequate nourishment. But most homeless people receive fewer than one and a half meals per day, so many are still inadequately nourished. Health care professionals can serve as valuable members of community groups seeking to provide food assistance.

TABLE NP13–1

Hunger Relief Organizations

Action without Borders
350 Fifth Ave., Suite 6614
New York, NY 10118
(212) 843-3973
www.idealist.org

America's Second Harvest
35 E. Wacker Dr. #2000
Chicago, IL 60601
(800) 771-2303
www.secondharvest.org

Bread for the World
50 F St. NW, Suite 500
Washington, DC 20010
(800) 82-BREAD or
(800) 822-7323
(202) 639-9400;
fax (202) 639-9401
www.bread.org

Children's Hunger
Relief Fund
182 Farmer's Lane
Suite 200
Santa Rosa, CA 95405
(888) 781-1585
www.childrenshungerrelief.org

Congressional Hunger Center
229 ¹⁄₂ Pennsylvania Ave.
Washington, DC 20003
(202) 547-7022
www.hungercenter.org

Food Research and Action Center
1875 Connecticut Ave.
Suite 540
Washington, DC 20009
www.frac.org

OXFAM America
26 West St.
Boston, MA 02111-1206
(800) 77-OXFAM or
(800) 776-9326
www.oxfam.america.org

Pan American Health Organization
525 23rd St. NW
Washington, DC 20037
(202) 974-3000
www.paho.org

Society of St. Andrew
3383 Sweet Hollow Rd.
Big Island, VA 24526
(800) 333-4597
www.endhunger.org

United Nations Food and Agriculture
Organization (FAO)
1001 22nd St. NW, Suite 300
Washington, DC 20437
(202) 653-2400
www.fao.org

United Nations International Children's
Emergency Fund (UNICEF)
3 United Nations Plaza
New York, NY 10017-4414
(212) 326-7035
www.unicef.org

United Nations World Food Program
Via Cesare Giulio
Viola, 68
Parco dé Medici
Rome, Italy 00148
www.wfp.org

World Health Organization (WHO)
525 23rd St. NW
Washington, DC 20037
(202) 861-3200
www.who.org

World Hunger Year
505 Eighth Ave., 21st Floor
New York, NY 10018-6582
(800) Gleanlt
www.worldhungeryear.org

Community-based efforts to feed citizens include food pantries that provide groceries.

© AP/Wide World Photos

Notes

1. Food and Agriculture Organization of the United Nations, *State of Food Insecurity in the World 2005*.

2. U.S. Department of Agriculture, *Household Food Security in the United States, 2005*, available at www.ers.usda.gov/publications/err29.

3. Position of the American Dietetic Association: Food insecurity and hunger in the United States, *Journal of the American Dietetic Association* 106 (2006): 446–458.

4. J. T. Cook and coauthors, Food insecurity is associated with adverse health outcomes among human infants and toddlers, *Journal of Nutrition* 134 (2004): 1432–1438.

5. D. F. Jyoti, E. A. Frongillo, and S. J. Jones, Food insecurity affects school children's academic performance, weight gain, and social skills, *Journal of Nutrition* 135 (2005): 2831–2839.

6. Position of the American Dietetic Association, 2006.

7. USDA Food and Nutrition Service, www.fns.usda.gov/fsp, site visited November 27, 2006.

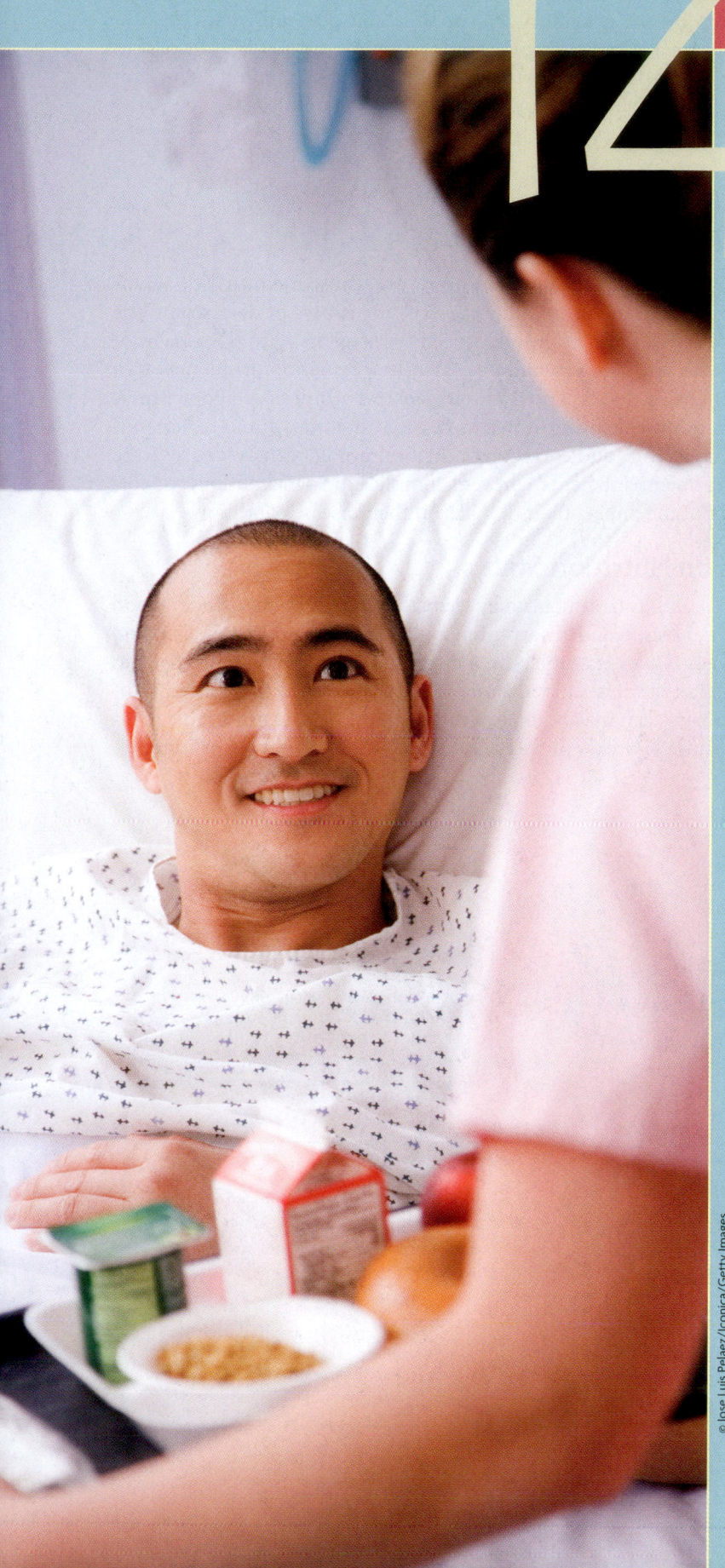

14

Illness and Nutrition Care

© Jose Luis Pelaez/Iconica/Getty Images

Previous chapters of this book introduced the nutrients and described how the appropriate dietary choices can support good health. Turning now to clinical nutrition, this chapter describes the nutrition care process and the implementation of nutrition care in clinical practice. Ensuring that nutrient needs are met is a key part of this process, so the chapter also describes typical methods for estimating energy requirements and provides examples of common dietary modifications.

Nutrition in Health Care

Many medical problems can alter nutrient needs and cause malnutrition. Conversely, poor nutrition status can influence both the course of disease and the body's response to treatment. Malnutrition has been reported in 40 to 60 percent of patients hospitalized with acute illness, and those with no nutrition problems often exhibit a decline in nutrition status within three weeks of admission.[1] Poor nutrition status can eventually weaken immune function and compromise a person's healing ability. Thus, preventing and correcting nutrition problems may improve the outcome of medical treatments and help to prevent complications. In addition, patients are often concerned about the impact their diet has on their disease condition.

Effects of Illness on Nutrition Status

An illness, its symptoms, and its treatments may lead to malnutrition by reducing food intake, interfering with digestion and absorption, or altering nutrient metabolism and excretion (see Figure 14–1). For example, the nausea associated with some illnesses and disease treatments can diminish appetite and reduce food intake; similarly, an inflamed mouth or esophagus can make the physical act of eating uncomfortable. Certain medications can cause anorexia or gastrointestinal (GI) discomfort or interfere

FIGURE 14–1

Ways in Which Illness Can Affect Nutrition Status

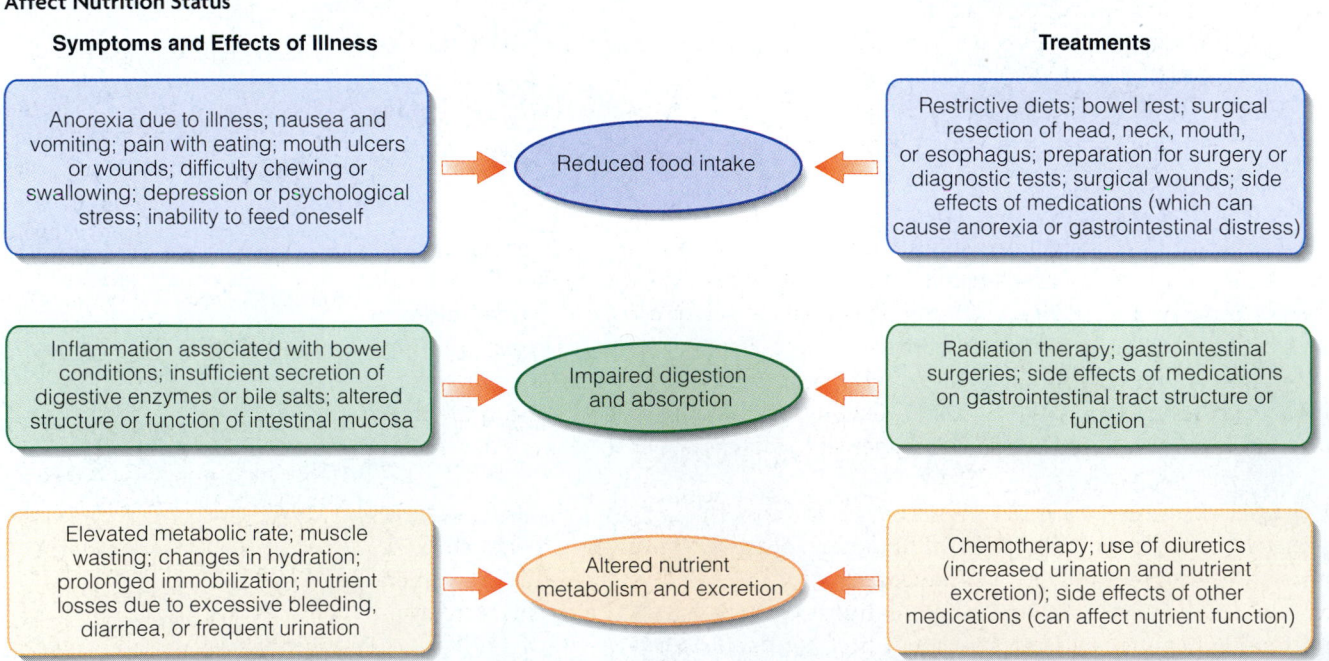

Symptoms and Effects of Illness

Treatments

Anorexia due to illness; nausea and vomiting; pain with eating; mouth ulcers or wounds; difficulty chewing or swallowing; depression or psychological stress; inability to feed oneself → **Reduced food intake** ← Restrictive diets; bowel rest; surgical resection of head, neck, mouth, or esophagus; preparation for surgery or diagnostic tests; surgical wounds; side effects of medications (which can cause anorexia or gastrointestinal distress)

Inflammation associated with bowel conditions; insufficient secretion of digestive enzymes or bile salts; altered structure or function of intestinal mucosa → **Impaired digestion and absorption** ← Radiation therapy; gastrointestinal surgeries; side effects of medications on gastrointestinal tract structure or function

Elevated metabolic rate; muscle wasting; changes in hydration; prolonged immobilization; nutrient losses due to excessive bleeding, diarrhea, or frequent urination → **Altered nutrient metabolism and excretion** ← Chemotherapy; use of diuretics (increased urination and nutrient excretion); side effects of other medications (can affect nutrient function)

with nutrient function and metabolism. Prolonged bed rest often results in **pressure sores**, which increase metabolic stress and raise protein and energy needs. ■

The dietary changes required during an acute illness are usually temporary and can be tailored to accommodate an individual's preferences and lifestyle. However, chronic illnesses may require long-term dietary modifications. For example, diabetes treatment requires lifelong changes in diet and lifestyle that some people may find difficult to maintain. The challenge for health professionals is to help their patients appreciate the potential benefits of treatment and accept dietary changes that can improve their health.

Responsibility for Nutrition Care

The members of a health care team work together to ensure that the nutritional needs of patients are met during illness. Their roles in nutrition care may overlap, and job descriptions in different institutions may vary somewhat. In some cases, nutrition care is incorporated into the medical care plan developed by the entire health care team. Such plans, called **critical pathways**, outline coordinated plans of care for specific medical diagnoses, treatments, or procedures.

Physicians Physicians are responsible for meeting all of a patient's medical needs, including nutrition. They prescribe **diet orders** and other orders related to nutrition care, including those for nutrition assessment and dietary counseling. Physicians rely on nurses, registered dietitians, and other health professionals to alert them to nutrition problems, suggest strategies for handling nutrition problems, and provide nutrition services.

Registered Dietitians Registered dietitians ■ are food and nutrition experts who are qualified to provide **medical nutrition therapy**. They conduct nutrition and dietary assessments; diagnose nutritional problems; develop, implement, and evaluate **nutrition care plans** (described in a later section); plan and approve menus; and provide nutrition education. Registered dietitians may also work as managers of food and cafeteria services in health care institutions.

Registered Dietetic Technicians Registered dietetic technicians often work in partnership with registered dietitians and assist in the implementation and monitoring of nutrition services. Depending on their background and experience, they may screen patients for nutrition problems, provide patient education and counseling, develop menus and recipes, ensure appropriate meal delivery, and monitor patients' food choices and intakes. Dietetic technicians sometimes supervise food service operations and may have roles in purchasing, inventory, quality control, sanitation, or safety.

Nurses Nurses interact closely with patients and thus are in an ideal position to identify people who would benefit from nutrition services. Nurses often screen patients for nutrition problems and may participate in nutrition assessments. Nurses also provide direct nutrition care, such as encouraging patients to eat, finding practical solutions to food-related problems, recording a patient's food intake, and answering questions about special diets. As members of **nutrition support teams,** nurses are responsible for administering tube and intravenous feedings. In facilities that do not employ registered dietitians, nurses often assume responsibility for much of the nutrition care.

Other Health Care Professionals Other heath care professionals may also assist with nutrition care. Pharmacists, physical therapists, occupational therapists, speech therapists, social workers, nursing assistants, and home health care aides can be instrumental in alerting dietitians or nurses to nutrition problems or may share relevant information about a patient's health status or personal needs.

■ Chapter 24 discusses the nutrition needs of patients undergoing acute metabolic stress.

■ Reminder: A *registered dietitian* has met the academic and professional requirements to qualify for the RD credential conferred by the American Dietetic Association (or Dietitians of Canada), including a bachelor's degree in nutrition or dietetics, a supervised internship, and the successful completion of a national examination.

pressure sores: regions of damaged skin and tissue due to prolonged pressure on the affected area by an external object, such as a bed, wheelchair, or cast; vulnerable areas of the body include buttocks, hips, and heels. Also called **decubitus** (deh-KYU-bih-tus) **ulcers.**

critical pathways: coordinated programs of treatment that merge the care plans of different health practitioners; also called **clinical pathways.**

diet orders: specific instructions concerning dietary management; also called **diet prescriptions.**

medical nutrition therapy: nutrition care provided by a registered dietitian; includes diagnosing nutrition problems, prescribing diet plans, and providing dietary counseling.

nutrition care plans: strategies for meeting an individual's nutritional needs.

nutrition support teams: health care professionals responsible for the provision of nutrients by tube feeding or intravenous infusion.

TABLE 14-1
Criteria for Identifying Malnutrition Risk
Age, medical diagnosis, severity of illness
Height and weight, body mass index (BMI), recent unintentional weight changes
Results of laboratory tests that indicate poor health status
Recent changes in appetite or food intake
Problems or symptoms that make eating difficult (such as chewing or swallowing difficulty or nausea and vomiting)
Food allergies or intolerances or extensive dietary restrictions
Presence of anemia, tissue wasting, or pressure sores
History of diabetes, renal disease, or other chronic illness
Use of medications that can impair nutrition status
Depression or social isolation

Nutrition Screening

A **nutrition screening** is an assessment tool that helps to identify patients who are malnourished or at risk for malnutrition. The information used in screening includes the admitting diagnosis, information from the medical record, physical measurements and laboratory test results collected during the admission process, and responses given by the patient or caregiver to an interview or questionnaire. Table 14–1 lists examples of information collected during screening. The **Joint Commission on Accreditation of Healthcare Organizations (JCAHO)**, a nonprofit organization that confers accreditation to health care institutions, recommends that a nutrition screening be conducted within 24 hours of a patient's admission to a hospital or other extended-care facility. Nutrition screening is also frequently included in outpatient services and community health programs. A nutrition screening often leads to a referral for nutrition care.

The Nutrition Care Process

Registered dietitians use a systematic approach to medical nutrition therapy called the **nutrition care process.**[2] Figure 14–2 presents the four distinct, yet interrelated, steps of the nutrition care process:

1. Nutrition assessment

2. Nutrition diagnosis

3. Nutrition intervention

4. Nutrition monitoring and evaluation

Although the nutrition care process is easiest to visualize as a series of steps, the steps are frequently revisited in order to reassess and revise diagnoses and intervention strategies.

Nutrition Assessment Nutrition assessment involves the collection of information needed to evaluate a patient's nutrition status and dietary needs. This information can be obtained from medical, social, and diet histories; anthropometric data; biochemical analyses; and physical examination. The assessment data helps the dietitian to identify specific nutrition problems and their underlying causes and determine a plan of action to prevent or correct nutrient imbalances. Assessment data also allows the dietitian to determine whether a care plan is working. The next section of this chapter describes the components of nutrition assessment in more detail.

FIGURE 14–2

The Nutrition Care Process

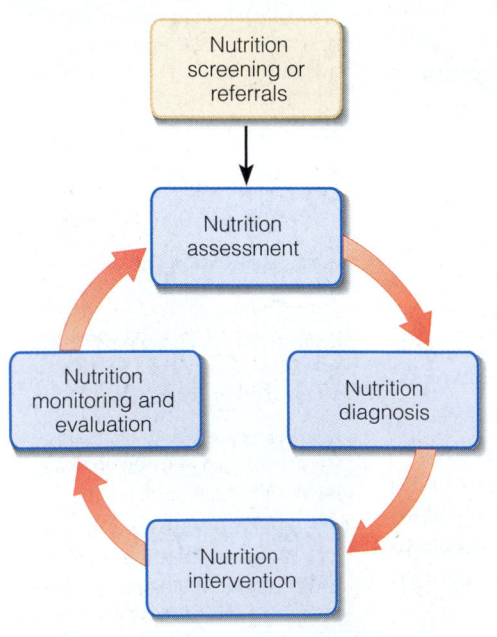

Nutrition screening or referrals → Nutrition assessment → Nutrition diagnosis → Nutrition intervention → Nutrition monitoring and evaluation

Nutrition Diagnosis Each nutrition problem receives a separate diagnosis, which includes the specific nutrition problem, etiology or cause, and signs and symptoms that provide evidence of the problem.[3] For example, a potential nutrition diagnosis might state: "unintentional weight loss *(problem)* related to insufficient kcaloric intake *(etiology or cause)* as evidenced by a 10-pound weight loss (8 percent of body weight) in the past few months *(sign or symptom)*." A nutrition diagnosis can change during the course of an illness.

Nutrition Intervention After nutrition problems are identified, the appropriate treatments can be determined. An intervention may include dietary changes, nutrition education, or a change in medication. It considers an individual's food habits, lifestyle, and other personal factors. Goals are stated in terms of measurable outcomes; for example, goals for an overweight person with diabetes might include changes in blood glucose levels and body weight. Desirable outcomes may also include changes in dietary behaviors and lifestyle; for example, an interview with a diabetes patient may reveal that he or she has learned to control carbohydrate intake and has started to exercise regularly.

Nutrition Monitoring and Evaluation The effectiveness of the nutrition care plan must be evaluated periodically: the original goals and outcome measures are reviewed and compared with earlier assessment data and diagnoses. Sometimes a change in a person's situation alters nutritional needs; for example, a new medication may alter a person's tolerance to certain foods. A nutrition care plan must be flexible enough to adapt to the new situation.

If a patient is unable or unwilling to make the suggested changes, the care plan should be redesigned and take into account the reasons that the earlier plan was not successful. This new plan may need to include motivational techniques or additional patient education. If the patient remains unwilling to modify behaviors despite the expected benefits, the health care provider can try again at a later time when the patient may be more receptive.

IN SUMMARY

- Illnesses and their treatments can affect food intake and nutrient needs, leading to malnutrition. In turn, poor nutrition can reduce the effectiveness of medical treatment.

- The combined efforts of each member of the health care team ensure that patients receive optimal nutrition care.

- Nutrition screening identifies individuals who can benefit from nutrition assessment and follow-up nutrition care.

- The nutrition care process includes four interrelated steps: nutrition assessment, nutrition diagnosis, nutrition intervention, and nutrition monitoring and evaluation.

Nutrition Assessment

A nutrition assessment provides the information needed for identifying nutrition problems and designing a nutrition care plan; follow-up assessments help to determine if the care plan has been effective. Ideally, the assessment should be sensitive enough to detect subtle nutrition problems and specific enough to identify problem nutrients. This section describes the types of information and measures that are most commonly obtained in a nutrition assessment.

nutrition screening: an assessment tool that quickly identifies patients at risk of developing malnutrition.

Joint Commission on Accreditation of Healthcare Organizations (JCAHO): a nonprofit organization that sets standards for health care performance and safety and confers accreditation to health care organizations that meet those standards.

nutrition care process: a problem-solving method that dietetics professionals use to evaluate and treat nutrition-related problems.

TABLE 14-2

Historical Information Used in Nutrition Assessment

Medical History	Social History	Diet History
Current complaint(s)	Socioeconomic status	Dietary pattern
Past medical condition(s)	Cultural/ethnic identity	Dietary restrictions
Family history of illness	Educational level	Usual and present weight
Surgical history	Living situation	Food allergies or intolerances
Medication history	Use of tobacco/street drugs	Use of alcohol
Allergies	Shopping arrangements	Chewing/swallowing ability
Use of dietary/herbal supplements	Cooking facilities	Need for feeding assistance

Historical Information

Historical information provides clues about the patient's nutrition status and nutrient needs and uncovers personal preferences that need to be considered when developing a nutrition care plan. Table 14–2 summarizes the elements of medical, social, and diet histories that contribute to a nutrition assessment. This information can be obtained from the medical record or by interviewing the patient or caregiver.

Medical History The medical history helps the health practitioner identify health problems that may affect food intake or require dietary changes. Table 14–3 lists medical problems that are frequently associated with nutrition problems. Many other conditions can lead to malnutrition, especially if nutrition problems are not recognized and corrected before they become serious.

A medical history includes a survey of prescription drugs, over-the-counter medications, and dietary supplements used by an individual. Certain medications can have detrimental effects on nutrition status, and some dietary components can alter the absorption and metabolism of drugs. Chapter 15 provides examples of the diet-drug interactions that need consideration when planning nutrition care.

Social History Social factors influence food choices as well as a person's ability to deal with health and nutrition problems. For example, cultural background can affect food preferences. Financial concerns may restrict access to health care and nutritious foods. Some individuals depend on others to prepare or procure food. A person who lives alone or is depressed may eat poorly or be unable to follow complex dietary instructions.

Diet History A **diet history** is a detailed account of a person's dietary practices. It includes information about food intake, meal patterns, lifestyle habits, and physical problems that influence dietary choices (see Table 14–2). Although various methods exist for obtaining data, the procedure often includes an interview about recent food intake (for example, a *24-hour recall*) and a survey of usual food choices (such as a *food frequency questionnaire*). The following section describes the most common methods of gathering food intake information.

Food Intake Data

Obtaining accurate food intake data is challenging, since results may vary depending on both the individual's memory and honesty and the assessor's skill and training. In addition, each method has its own strengths and weaknesses, so best results

TABLE 14-3

Medical Problems Often Associated with Malnutrition

Acquired immune deficiency syndrome (AIDS)

Alcoholic liver disease

Anorexia nervosa

Aspiration pneumonia

Bulimia

Burns, extensive or severe

Cancer and cancer treatments

Celiac disease

Dehydration

Diabetes, newly diagnosed or uncontrolled

Gestational diabetes

Head trauma

Hypoglycemia

Inflammatory bowel disease

Jaw fracture

Multiple trauma

Liver disease: hepatic encephalopathy

Pregnancy-induced hypertension

Renal disease, end-stage

Skin ulcer

Swallowing difficulty

Vomiting, excessive

TABLE 14–4

Methods for Obtaining Food Intake Data

Method	Description	Advantages	Disadvantages
24-hour recall	Guided interview in which the foods and beverages consumed in a 24-hour period are described in detail.	■ Results are not dependent on literacy or educational level of respondent. ■ Interview occurs after food is consumed, so it does not interfere with food choices. ■ Relatively easy and quick assessment method.	■ Reliant on memory. ■ Food items that cause embarrassment (alcohol, desserts) may be omitted. ■ Under- and overestimation of food intakes are common. ■ Skill of interviewer affects outcome. ■ Data from a single day cannot represent the respondent's usual intake accurately. ■ Seasonal variations may not be addressed.
Food frequency questionnaire	Written survey of food consumption during a specific period of time, often a one-year period.	■ Examines long-term food intake, so day-to-day and seasonal variability should not affect results. ■ Completed after food is consumed, so does not interfere with food choices. ■ Low-cost method.	■ Reliant on memory. ■ Not good for monitoring short-term changes in food intake. ■ Serving sizes are often difficult for respondents to evaluate without assistance. ■ Calculated nutrient intakes may not be accurate. ■ Food lists include common foods only. ■ Food lists for the general population are of limited value in special populations.
Food record	Written account of food consumed during a specified period, usually several consecutive days. Accuracy is improved by including weights or measures of foods.	■ Process does not rely on memory. ■ Recording foods as they are consumed improves likelihood of obtaining accurate food intake data. ■ Useful for controlling intake because keeping records can increase awareness of food choices.	■ Recording process itself influences food intake. ■ Time-consuming and burdensome for respondent; requires high degree of motivation. ■ Underreporting is common. ■ Requires literacy and the physical ability to write. ■ Seasonal changes in diet are not taken into account.
Direct observation	Observation of meal trays or shelf inventories before and after eating; possible only in residential facilities.	■ Process does not rely on memory. ■ Does not interfere with person's food intake. ■ Can be used to evaluate acceptability of prescribed diet.	■ Possible only in residential situations. ■ Labor-intensive.

are obtained from using a combination of methods. Table 14–4 summarizes the methods commonly used and each method's advantages and disadvantages.

The 24-Hour Recall The **24-hour recall** is a guided interview in which an individual recounts all of the foods and beverages consumed in the past 24 hours or during the previous day. The interviewer includes questions about the times when meals or snacks were eaten, amounts consumed, and ways in which foods were prepared. The assessor may begin by asking: "What is the first thing you ate or drank yesterday morning?" After the first food items are described, the follow-up questions might be: "What time

diet history: a comprehensive record of a person's food intake and dietary practices.

24-hour recall: a record of foods consumed in the previous 24 hours; sometimes modified to include foods consumed in a typical day.

Food models and measuring utensils can help an individual visualize portion sizes.

was that?" and "How much did you eat?" Questioning continues until the intake record for the day is complete. Food models or measuring cups and spoons can be used to help the individual visualize and describe the amounts consumed. After the day's intake is recounted, the interviewer asks whether the intake that day is fairly typical and, if not, how it varies from the person's usual intake. A recall interview may be conducted on several nonconsecutive days to get a better representation of a person's usual diet.

A recall interview can yield useful data for developing an acceptable nutrition care plan and identifying food items that may need to be restricted due to illness. However, it is a poor technique for determining the adequacy of a diet because it does not take into account fluctuations in food intake or seasonal variations. Moreover, food intakes are often underestimated because the process relies on an individual's memory and reporting accuracy. People often forget to mention beverages, condiments, and snack foods unless specifically prompted to do so, and some find it embarrassing to report consumption of foods such as chocolate, butter, and red meat.[4]

Food Frequency Questionnaire A **food frequency questionnaire** surveys the foods and beverages regularly consumed during a specific time period. Some questionnaires are qualitative only: food lists contain common foods, organized by food group, with boxes to check to indicate frequency of consumption. Other types of questionnaires provide semiquantitative information by including portion sizes as well. Figure 14–3 shows a sample section of a semiquantitative questionnaire that surveys fruit intake over the previous year. Because the respondent is often asked to estimate food intakes over a one-year period, the results should not be affected by seasonal changes in diet. Conversely, a disadvantage of this method is its inability to determine recent changes in food intake.

Simple versions of food frequency questionnaires focus on food categories relevant to a person's medical condition. For example, a questionnaire designed to evaluate calcium

FIGURE 14–3

Sample Section of a Food Frequency Questionnaire

FRUIT	HOW OFTEN								HOW MUCH			
	Never or less than once per month	1 per mon.	2–3 per mon.	1 per week	2 per week	3–4 per week	5–6 per week	Every day	MEDIUM SERVING	YOUR SERVING SIZE		
										S	M	L
EXAMPLE: Bananas	○	○	○	●	○	○	○	○	1 medium	○ 1/2	● 1	○ 2
Bananas	○	○	○	○	○	○	○	○	1 medium	○ 1/2	○ 1	○ 2
Apples, applesauce	○	○	○	○	○	○	○	○	1 medium or 1/2 cup	○ 1/2	○ 1	○ 2
Oranges (not including juice)	○	○	○	○	○	○	○	○	1 medium	○ 1/2	○ 1	○ 2
Grapefruit (not including juice)	○	○	○	○	○	○	○	○	1/2 medium	○ 1/4	○ 1/2	○ 1
Cantaloupe	○	○	○	○	○	○	○	○	1/4 medium	○ 1/8	○ 1/4	○ 1/2
Peaches, apricots (fresh, in season)	○	○	○	○	○	○	○	○	1 medium	○ 1/2	○ 1	○ 2
Peaches, apricots (canned or dried)	○	○	○	○	○	○	○	○	1 medium or 1/2 cup	○ 1/2	○ 1	○ 2
Prunes, or prune juice	○	○	○	○	○	○	○	○	1/2 cup	○ 1/4	○ 1/2	○ 1
Watermelon (in season)	○	○	○	○	○	○	○	○	1 slice	○ 1/2	○ 1	○ 2
Strawberries, other berries (in season)	○	○	○	○	○	○	○	○	1/2 cup	○ 1/4	○ 1/2	○ 1
Any other fruit, including kiwi, fruit cocktail, grapes, raisins, mangoes	○	○	○	○	○	○	○	○	1/2 cup	○ 1/4	○ 1/2	○ 1

intake may include only milk products, fortified foods, certain fruits and vegetables, and dietary supplements that contain calcium. A computer analysis can then quickly estimate the individual's calcium intake and compare it to recommendations.

Food Record A **food record** is a written account of foods and beverages consumed during a specified time period, usually several consecutive days. Foods are recorded as they are consumed in order to obtain the most complete and accurate record possible; thus the process does not rely on memory. A detailed food record includes the types and amounts of foods and beverages consumed, times of consumption, and methods of preparation. It provides valuable information about food intake as well as a person's response to and compliance with medical nutrition therapy. Unfortunately, food records require a great deal of time to complete, and people need to be highly motivated to keep accurate records. Another drawback is that the recording process itself may influence food intake. Furthermore, day-to-day and seasonal variations in food intake make it difficult to obtain accurate estimates of nutrient values in just a few days or even a week.

Direct Observation In facilities that serve meals, food intakes can be directly observed and analyzed. This method can also reveal a person's food preferences, changes in appetite, and any problems with a prescribed diet. Health practitioners use direct observation to conduct patients' **kcalorie counts** to determine the food energy (and often, protein) consumed by patients during a single day or several consecutive days. To perform a kcalorie count, the clinician estimates food intake by recording the dietary items that a patient is given at meals and subtracting the amounts remaining after meals are completed; this procedure allows an estimate of the kcaloric content of foods and beverages actually consumed. Although a useful means of discerning patients' intakes, direct observation requires regular and careful documentation and can be labor-intensive and costly.

Anthropometric Data

Measures of body size, known as **anthropometric** measurements, can reveal problems related to both overnutrition and protein-energy malnutrition (PEM). ■ Height (or length) and weight are the most common anthropometric measurements and help to evaluate growth in children and nutrition status in adults. Other helpful data include an individual's percentage of body fat and circumferences of the head, waist, and limbs. ■

Height (or Length) Poor growth in children can signify malnutrition. In adults, height measurements alone do not reflect current nutrition status but can be used for estimating a person's energy needs or appropriate body weight. Length is measured in infants and children younger than 24 months of age, and height is usually measured in older children and adults. ■ Length can also be measured in adults and children who cannot stand unassisted due to physical or medical reasons. The "How to" on p. 416 describes some standard techniques for measuring length and height.

In adults, height can be estimated from equations that include either the knee height or the full arm span, both of which correlate well with height. Knee height, which extends from the top of the knee to the heel, is measured in a sitting position. The measure of knee height is frequently used in bedridden patients; specific formulas are available for different age, gender, and ethnic groups. The full arm span is the distance from the tip of one middle finger to the other when the arms are extended horizontally. In children with disabilities that affect stature, alternative measures of linear growth include the full arm span, lower-leg lengths (knee to heel, similar to the knee height measure), and upper-arm lengths (shoulder to elbow), which can be compared with reference percentiles.

Body Weight During clinical care, health care providers monitor body weights carefully. Weight changes may reflect changes in hydration status, and an involuntary loss

■ Reminder: *Protein-energy malnutrition* is a deficiency of protein and food energy and is characterized by weight loss and loss of muscle mass.

■ Appendix E provides information about methods for determining the percentage of body fat.

■ *Length* is measured while a person is recumbent (lying down), whereas *height* is measured while a person is standing upright.

food frequency questionnaire: a survey of foods routinely consumed. Some questionnaires ask about the types of food eaten and yield only qualitative information, whereas others include questions about portions consumed and yield semiquantitative data as well.

food record: a detailed log of food eaten during a specified time period, usually several days. A food record may also include information regarding disease symptoms, physical activity, and medication use; also called a **food diary.**

kcalorie counts: the determination of food energy (and often, protein) consumed by patients for one or more days.

anthropometric (AN-throw-poe-MEH-trik): related to physical measurements of the human body, such as height, weight, body circumferences, and percentage of body fat.

How To

MEASURE LENGTH AND HEIGHT

To improve the accuracy of length and height measurements, keep the following in mind:

- Always measure—never ask! Self-reported heights are less accurate than measured heights. If height is not measured, document that the height is self-reported.

- Measure the length of infants and young children by using a measuring board with a fixed headboard and a movable footboard. It generally takes two people to measure length. One person gently holds the infant's head against the headboard; the other straightens the infant's legs and moves the footboard to the bottom of the infant's feet.

- Measure height next to a wall on which a nonstretchable measuring tape or board has been fixed. Ask the person to stand erect without shoes and with heels together. The person's eyes and head should be facing forward, with heels, buttocks, and shoulder blades touching the wall. Place a ruler or other flat, stiff object on the top of the head at a right angle to the wall and carefully note the height measurement.

- Immediately record length and height measurements to the nearest 1/8 inch or 0.1 centimeter.

- For evaluating growth rate in young children, use the appropriate growth chart (Appendix E) when plotting results. If length is measured, use the growth chart for children between 0 and 36 months; if height is measured, use the chart for individuals between 2 and 20 years.

- Higher values are obtained from supine measurements than from vertical height measurements due to gravity.

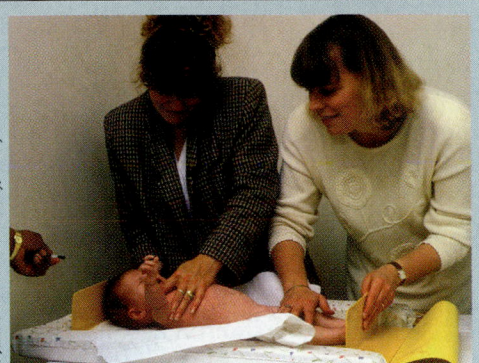

It takes two people to measure the length of an infant.

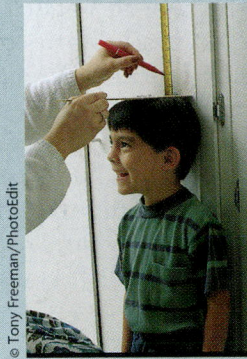

Standing erect allows for an accurate height measurement.

---

of body weight can signify PEM. Body weights can be compared with healthy ranges on height-weight tables and growth charts or used to calculate the Body Mass Index (BMI). ■ The "How to" includes suggestions for improving the accuracy of weight measurements. Table 14–5 describes a quick method for estimating desirable weight.

■ Reminder: BMI = $\dfrac{\text{weight (kg)}}{\text{height (m)}^2}$

A healthy BMI typically falls between 18.5 and 25.

TABLE 14–5

Quick Estimate of Desirable Body Weight[a] (Hamwi Equation)

Men

For first 5 feet, consider 106 pounds a reasonable weight. For each inch over 5 feet, add 6 pounds. For each inch under 5 feet, subtract 6 pounds.

Add 10% for a large-framed individual; subtract 10% for a small-framed individual.

Example: For a man 5 feet 8 inches tall (medium frame), a desirable weight would be 154 pounds. (106 lb + 48 = 154 lb)

Women

For first 5 feet, consider 100 pounds a reasonable weight. For each inch over 5 feet, add 5 pounds. For each inch under 5 feet, subtract 5 pounds.

Add 10% for a large-framed individual; subtract 10% for a small-framed individual.

Example: For a woman 5 feet 6 inches tall (medium frame), a desirable weight would be 130 pounds. (100 lb + 30 = 130 lb)

[a] This method does not account for differences in age or race.

Source: G. J. Hamwi, Changing dietary concepts, in T. S. Donowski, ed., *Diabetes Mellitus: Diagnosis and Treatment* (New York: American Diabetes Association, 1964), pp. 73–78.

CHAPTER FOURTEEN

MEASURE WEIGHT

Tips for measuring weight include:

- Always measure—never ask! Self-reported weights are often inaccurate. If weight is not measured, document that the weight is self-reported.

- Valid weight measurements require scales that have been carefully maintained, calibrated, and checked for accuracy at regular intervals. Beam balance and electronic scales are the most accurate. Bathroom scales are inaccurate and inappropriate in the clinical setting.

- Measure an infant's weight with a scale that allows the infant to sit or lie down. The tray should be large enough to support an infant or young child up to 40 pounds, and the scale should weigh in ½-ounce or 10-gram increments. For accurate results, weigh infants without clothes or diapers. Excessive movement by the infant can reduce accuracy.

- Children who can stand are weighed in the same way as adults, using beam balance or electronic scales with platforms large enough for standing comfortably. If repeated weight measurements are needed, each weighing should take place at the same time of day (preferably before breakfast), in the same amount of clothing, after the person has voided, and on the same scale. Record weights to the nearest ¼ pound or 0.1 kilogram.

- Special scales and hospital beds with built-in scales are available for weighing people who are bedridden.

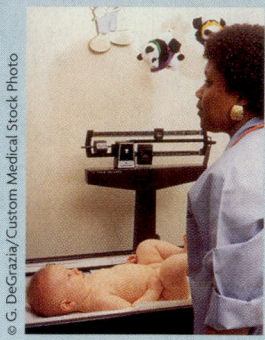

© G. DeGrazia/Custom Medical Stock Photo

Infants are weighed on scales that allow them to sit or lie down.

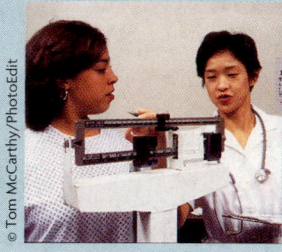

© Tom McCarthy/PhotoEdit

Beam balance scales allow accurate weight measurements for older children and adults.

Weight data is often expressed as percent of "ideal body weight" (%IBW) or percent of "usual body weight" (%UBW) in order to assess the degree of nutritional risk associated with illness. The %UBW is more effective for interpreting weight changes that occur in overweight and obese individuals, since %IBW may fail to identify significant weight loss. Conversely, in patients who have been underweight throughout life, %IBW can overstate the degree of weight loss due to illness. The "How to" below describes how to estimate %IBW and %UBW, and Table 14–6 on p. 418 shows how to interpret these values.

How To

ESTIMATE AND EVALUATE %IBW AND %UBW

To estimate %IBW, compare an individual's current weight with a reasonable (ideal) weight obtained from height-weight tables or calculated using the quick method in Table 14–5:

$$\%IBW = \frac{current\ weight}{ideal\ weight} \times 100$$

For example, suppose you wish to calculate %IBW for a man who is 5 feet 8 inches tall and weighs 123 pounds. Using Table 14–5, you estimate that a reasonable weight for this man would be 154 pounds.

$$\%IBW = \frac{123\ lb}{154\ lb} \times 100 = 80\%$$

The man in this example weighs 80 percent of his ideal body weight. A look at Table 14–6 indicates that at 80 percent of IBW he is mildly underweight.

To estimate %UBW, compare a person's current weight with the weight that the person generally maintains:

$$\%UBW = \frac{current\ weight}{usual\ weight} \times 100$$

For example, if a woman loses 32 pounds during illness and her usual weight is 145 pounds, her current weight would be 113 pounds. These values can be incorporated into the preceding equation:

$$\%UBW = \frac{113\ lb}{145\ lb} \times 100 = 78\%$$

The woman in this example weighs 78 percent of her usual weight. A look at Table 14–6 shows that a person at 78 percent of UBW is moderately underweight.

TABLE 14–6

Use of Body Weight for Assessing Nutritional Risk

%IBW	%UBW	Nutritional Risk
>120	—	Obesity
110–120	—	Overweight
90–109	—	Adequate weight—not at risk
80–89	85–95	Risk of mild malnutrition
70–79	75–84	Risk of moderate malnutrition
<70	<75	Risk of severe malnutrition

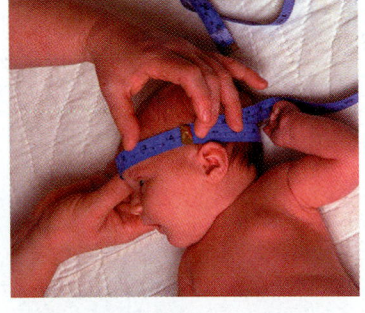

Head circumference measurements can help to assess brain growth.

■ Blood test results are reported in terms of either *plasma* or *serum* levels. *Plasma* is the yellow fluid that remains after cells are removed and still contains clotting factors. *Serum* is the fluid remaining after both cells and clotting factors are removed.

Head Circumference A measurement of head circumference helps to assess brain growth and malnutrition in children up to three years of age, although this measure is not necessarily reduced in a malnourished child. Head circumference values can also track brain development in premature and small-for-gestational-age infants. To measure head circumference, the assessor encircles the largest circumference measure of a child's head with a nonstretchable measuring tape: the tape is placed just above the eyebrows and ears and around the occipital prominence at the back of the head (see the photo). The measurement is read to the nearest 1/8 inch or 0.1 centimeter.

Circumferences of Waist and Limbs Circumferences of the waist and limbs help the assessment of body fat and muscle mass, respectively. Waist circumference correlates with visceral fat and can help in assessing overnutrition (see Figure 6–11, p. 156). Circumferences of the mid-upper arm, mid-thigh, and mid-calf regions can help in evaluating the effects of illness, aging, and PEM on skeletal muscle content. For improved accuracy, circumference measurements are often used together with skinfold measurements to correct for the subcutaneous fat in limbs.

Anthropometric Assessment in Infants and Children To evaluate growth patterns, periodic measurements of height (or length), weight, and head circumference are plotted on growth charts, such as those provided in Appendix E. The most commonly used growth charts compare height (or length) to age, weight to age, head circumference to age, weight to length, and BMI to age. Although individual growth patterns vary, a child's growth will generally stay at about the same percentile throughout childhood; a sharp drop in a previously steady growth pattern suggests malnutrition. Growth patterns that fall below the 5th percentile may also be cause for concern, although genetic influences must be considered when interpreting low values. Growth charts with BMI-for-age percentiles can be used to assess risk of underweight and overweight in children over two years of age: the 10th and 85th percentiles are used as cutoffs to identify children who may be malnourished or overweight, respectively.[5]

Anthropometric Assessment in Adults Health practitioners routinely record and monitor weight and height during illness. Weight changes must be evaluated carefully: although unintentional weight *loss* can indicate malnutrition, weight *gain* may result from fluid retention rather than overnutrition. Fluid retention often accompanies worsening disease in patients with heart failure, liver cirrhosis, and kidney failure and can mask the weight loss associated with PEM. In assessing the significance of weight loss, the rate should be considered as well as the amount: a 10 percent involuntary weight loss within a six-month period suggests risk of PEM. Some medications can also lead to weight loss or gain.

Many of the illnesses discussed in later chapters are associated with lean tissue losses that resist nutrition intervention. Losses in both lean tissue and height are common with aging even though body weights may remain stable. Including anthropometric measures such as skinfold measurements and limb circumferences can help the health practitioner identify changes in body composition that need to be addressed in the treatment plan.

Biochemical Analyses

Biochemical analyses provide information about protein-energy nutrition, vitamin and mineral status, fluid and electrolyte balance, and organ function. Most tests are based on analyses of blood and urine samples, which contain proteins, nutrients, and metabolites that reflect nutrition and health status. Table 14–7 lists and describes common blood tests ■ with nutritional implications. Laboratory tests relevant to specific diseases will be discussed in the chapters that follow.

TABLE 14-7

Routine Laboratory Tests with Nutritional Implications

This table presents a partial listing of some uses of commonly performed lab tests that have implications for nutritional problems.

Laboratory Test	Acceptable Range	Description
Hematology		
Red blood cell (RBC) count	Male: 4.3–5.7 million/μL Female: 3.8–5.1 million/μL	Number of RBC; aids anemia diagnosis.
Hemoglobin (Hb)	Male: 13.5–17.5 g/dL Female: 12.0–16.0 g/dL	Hemoglobin content of RBC; aids anemia diagnosis.
Hematocrit (Hct)	Male: 39–49% Female: 35–45%	Percentage RBC in total blood volume; aids anemia diagnosis.
Mean corpuscular volume (MCV)	80–100 fL	RBC size, helps to distinguish between microcytic and macrocytic anemias.
Mean corpuscular hemoglobin concentration (MCHC)	31–37% Hb/cell	Hb concentration within RBCs, helps to distinguish iron-deficiency anemia.
White blood cell (WBC) count	4,500–11,000 cells/μL	Number of WBC; general assessment of immunity.
Blood Chemistry		
Serum Proteins		
Total protein	6.4–8.3 g/dL	Protein levels are not specific to disease or highly sensitive; they can reflect body protein, illness or infections, changes in hydration or metabolism, pregnancy, or medications.
Albumin	3.4–4.8 g/dL	May reflect illness or PEM; slow to respond to improvement or worsening of disease.
Transferrin	200–400 mg/dL >60 yr: 180–380 mg/dL	May reflect illness, PEM, or iron deficiency; slightly more sensitive to changes than albumin.
Prealbumin (transthyretin)	10–40 mg/dL	May reflect illness or PEM; more responsive to health status changes than albumin or transferrin.
C-reactive protein	68–8,200 ng/mL	Indicator of inflammation or disease.
Serum Enzymes		
Creatine kinase (CK)	Male: 38–174 U/L Female: 26–140 U/L	Different forms of CK are found in muscle, brain, and heart. High levels in blood may indicate heart attack, brain tissue damage, or skeletal muscle injury.
Lactate dehydrogenase (LDH)	208–378 U/L	LDH is found in many tissues. Specific types may be elevated after heart attack, lung damage, or liver disease.
Alkaline phosphatase	25–100 U/L	Found in many tissues; often measured to evaluate liver function.
Aspartate aminotransferase (AST, formerly SGOT)	10–30 U/L	Usually monitored to assess liver damage; elevated in most liver diseases. Levels are somewhat increased after muscle injury.
Alanine aminotransferase (ALT, formerly SGPT)	Male: 10–40 U/L Female: 7–35 U/L	Usually monitored to assess liver damage; elevated in most liver diseases. Levels are somewhat increased after muscle injury.

continued

Interpreting laboratory values can be challenging when a number of different factors influence test results. For example, serum protein values can be affected by fluid imbalances, ■ pregnancy, medications, and exercise. Similarly, serum levels of vitamins and minerals are often poor indicators of nutrient deficiency because the values are affected by multiple physiological factors; therefore, a variety of tests are generally needed to diagnose a nutrition problem. Taken together with other assessment data, laboratory test results help to present a clearer picture than is possible otherwise.

■ Fluid retention can produce lab results that are deceptively low. Dehydration may cause lab results to be deceptively high.

TABLE 14–7 (continued)

Laboratory Test	Acceptable Range	Description
Serum Electrolytes		
Sodium	136–146 mEq/L	Helps to evaluate hydration status or neuromuscular, kidney, and adrenal functions.
Potassium	3.5–5.1 mEq/L	Helps to evaluate acid-base balance and kidney function; can detect potassium imbalances.
Chloride	98–106 mEq/L	Helps to evaluate hydration status and detect acid-base and electrolyte imbalances.
Other		
Glucose (fasting)[a]	74–106 mg/dL >60 yrs: 80–115 mg/dL	Detects risk of glucose intolerance, diabetes mellitus, and hypoglycemia; helps to monitor diabetes treatment.
Glycosylated hemoglobin (Hb A_{1c})	5.0–7.5% of Hb	Used to monitor long-term blood glucose control (approximately 1 to 3 months prior).
Blood urea nitrogen (BUN)	6–20 mg/dL	Primarily used to monitor kidney function; value is altered by liver failure, dehydration, or shock.
Uric acid	Male: 3.5–7.2 mg/dL Female: 2.6–6.0 mg/dL	Used for detecting gout or changes in kidney function; levels affected by age and diet; varies among different ethnic groups.
Creatinine (serum or plasma)	Male: 0.7–1.3 mg/dL Female: 0.6–1.1 mg/dL	Used to monitor renal function.

[a] Fasting glucose levels that repeatedly exceed 100 mg/dL suggest prediabetes.

Note: μL = microliter; dL = deciliter; fL = femtoliter; ng = nanogram; U/L = units per liter; mEq = milliequivalents.

Source: L. Goldman and D. Ausiello, eds., *Cecil Textbook of Medicine* (Philadelphia: Saunders, 2004).

■ Plasma proteins are synthesized in the liver, so plasma levels can reflect liver function.

Plasma Proteins Plasma protein levels can help in the assessment of protein status, although levels may fluctuate for different reasons. For example, both PEM and liver disease can reduce plasma proteins. ■ Metabolic stress causes the release of hormones that alter plasma protein levels. Values are also influenced by pregnancy, kidney function, zinc status, and some medications. Because plasma proteins are affected by so many factors, their values must be considered along with other data to evaluate nutrition status. The following paragraphs describe several of the plasma proteins commonly measured during illness.

■ The term *half-life* defines the length of time that a substance remains in plasma. The albumin in plasma has a 3-week half-life, meaning that half of the amount circulating in plasma is degraded in a 3-week period.

Albumin Albumin is the most abundant plasma protein, and its levels are routinely measured during illness. Although many medical conditions influence albumin, it is slow to reflect changes in nutrition status because of its large body pool and slow rate of degradation. ■ In people with chronic PEM, albumin levels remain normal for long periods of time despite depletion of body proteins; levels fall only after prolonged malnutrition. Likewise, when malnutrition is treated, albumin concentrations increase slowly, so albumin is not a sensitive indicator of effective treatment.

■ Transferrin's half-life in plasma is approximately 8 to 10 days.

Transferrin Transferrin transports iron, so its concentrations respond to both PEM and iron status. Transferrin breaks down in the body more rapidly than albumin, ■ but it responds relatively slowly to nutrition therapy. In addition, evaluating protein-energy status using transferrin is difficult if an iron deficiency is also present. Transferrin levels rise as iron deficiency worsens and fall as iron status improves.

■ Half-lives of prealbumin and retinol-binding protein are 2 days and 12 hours, respectively.

Prealbumin and Retinol-Binding Protein Levels of prealbumin (also called transthyretin) and retinol-binding protein decrease rapidly during PEM and respond quickly to improved protein intakes. ■ Thus these proteins are more sensitive than albumin to changes in protein status. Like other plasma proteins, their usefulness in nutrition assessment is limited because they are affected by a number of different factors, including metabolic stress, zinc deficiency, and various medical conditions. Prealbumin and retinol-binding protein are more expensive to measure than albumin, so they are not routinely included during nutrition assessment.

Physical Examinations

As with the other assessment methods, interpreting physical signs of malnutrition requires skill and clinical judgment. Most physical signs are nonspecific; they can reflect any of several nutrient deficiencies as well as conditions unrelated to nutrition. For example, cracked lips may be caused by several B vitamin deficiencies but may also be caused by sunburn, windburn, or dehydration. Dietary and laboratory data are usually needed as additional evidence to confirm suspected nutrient deficiencies.

Physical signs of malnutrition are often evident in parts of the body where the cells are replaced at a rapid rate.

Signs of malnutrition appear most often in parts of the body where cell replacement occurs at a rapid rate, such as the hair, skin, and digestive tract (including the mouth and tongue). Table 14–8 lists some clinical signs of nutrient deficiencies. Many of the symptoms listed occur only in advanced stages of deficiency. Chapters 8 and 9 provide additional examples of the clinical signs of nutrient imbalances.

Fluid imbalances Fluid imbalances accompany many illnesses and can also result from the use of certain medications. Attention to the physical signs of fluid retention or dehydration can help health care providers correctly interpret the results of blood tests and the body weight measurement.

Fluid retention (also called *edema*) can accompany malnutrition, infection, or injury. It can be caused by impaired blood circulation and is often associated with diseases of the heart, kidney, liver, and lungs. Physical signs include weight gain, facial puffiness, swelling of limbs, abdominal distention, and tight-fitting shoes.

TABLE 14–8

Clinical Signs of Nutrient Deficiencies

Body System	Acceptable	Signs of Malnutrition	Other Possible Causes
Hair	Shiny, firm in scalp	Dull, brittle, dry, loose; falls out (PEM); corkscrew hair (copper)	Excessive hair bleaching; hair loss from aging, chemotherapy, or radiation therapy
Eyes	Bright, clear pink membranes; adjust easily to light	Pale membranes (iron); spots, dryness, nightblindness (vitamin A); redness at corners of eyes (B vitamins)	Anemia, unrelated to nutrition; eye disorders; allergies
Lips	Smooth	Dry, cracked, or with sores in the corner of the lips (B vitamins)	Sunburn, windburn, excessive salivation from ill-fitting dentures or other disorders
Mouth and gums	Red tongue without swelling, normal sense of taste; teeth without caries; gums without bleeding, swelling, or pain	Smooth or magenta tongue (B vitamins), decreased taste sensations (zinc); swollen, bleeding gums (vitamin C)	Medications, periodontal disease (poor oral hygiene)
Skin	Smooth, firm, good color	Poor wound healing (PEM, vitamin C, zinc); dry, rough, lack of fat under skin (essential fatty acids, PEM, B vitamins); bruising, bleeding under skin (vitamins C and K)	Poor skin care, diabetes mellitus, aging, medications
Nails	Smooth, firm, pink	Ridged (PEM); spoon shaped, pale (iron)	
Other	—	Dementia, peripheral neuropathy (B vitamins); swollen glands at front of neck (PEM, iodine); bowed legs (vitamin D)	Disorders of aging (dementia), diabetes mellitus (peripheral neuropathy)

Dehydration can result from fever, sweating, vomiting, diarrhea, excessive urination, and skin injury or burns (due to fluid loss through skin lesions). Symptoms include thirst, dry skin or mouth, and reduced skin tension. The urine may be dark yellow or amber colored, and urine volume may be unusually low. Dehydration risk is greatest in the elderly, who have reduced thirst responses to water deprivation.

Functional Assessment Nutrient deficiencies can impair normal physiological functions; for example, zinc deficiency can depress immunity and slow wound healing. Protein-energy malnutrition can lead to **wasting**, the breakdown and loss of body tissues. Functional tests help health practitioners evaluate the changes in physiological functions and losses in body strength that accompany malnutrition or disease. For example, assessment of immunity may include testing the skin's response to antigens that cause redness and swelling when immune function is adequate. Exercise tolerance might be assessed by measuring the distance a patient can walk in several minutes. The chapters that follow include additional examples of functional assessment. The accompanying Case Study can help you review the different components of a nutrition assessment.

Integrating Assessment Data

Combining the results from several methods can improve the accuracy of nutrition assessment. One technique for doing so, known as Subjective Global Assessment (SGA), combines historical information with the results of a physical examination to predict the nutrition status of acute care patients (see Table 14–9). Elements of SGA may include weight and dietary changes, gastrointestinal symptoms, work capacity, level of metabolic stress, degree of muscle wasting and fat loss, and presence of edema or ascites.■ SGA is widely used to evaluate nutrition status and has been found to be applicable to a variety of patient populations.

■ *Ascites*, the abnormal accumulation of fluid in the abdominal cavity, is discussed in Chapter 20.

Case Study

NUTRITION SCREENING AND ASSESSMENT

Elise Walden is an 85-year-old retired businesswoman who has been a widow for 10 years. She uses a walker and has poorly fitting dentures. She was recently admitted to the hospital with pneumonia and also has congestive heart failure and diabetes. She routinely takes several medications to control blood glucose, hypertension, and heart function, and, in addition to these, the physician ordered antibiotics to treat the pneumonia. During an initial nutrition screening, Mrs. Walden stated that she had been eating very poorly over the past two weeks. She said that she usually weighs about 125 pounds; a fact that was documented in her medical chart from a previous visit. Although she felt she was losing weight, she didn't know how much weight she may have lost or when she started losing weight. Upon admission to the hospital, Mrs. Walden weighed 115 pounds and was 5 feet, 2 inches tall. Her serum albumin level was 3.0 g/dL. A physical exam revealed edema, and several other laboratory tests confirmed that she was retaining fluid. As a result of the nutrition screening, Mrs. Walden was referred to a registered dietitian for a complete nutrition assessment.

1. From the brief description provided, which items in Mrs. Walden's medical, social, and diet histories might alert the dietitian that she is at risk of malnutrition?

2. Identify a healthy body weight for Mrs. Walden and calculate her %IBW and %UBW. What do the results reveal? What effect does fluid retention have on Mrs. Walden's weight?

3. How can fluid retention alter Mrs. Walden's serum protein levels? What physical symptoms may have suggested that she was retaining excess fluid?

4. What tools can be used to estimate Mrs. Walden's usual food intake? What medical, physical, and social factors are likely to affect her dietary intake?

5. Describe other types of assessment information the dietitian may need before developing a nutrition care plan.

TABLE 14-9

Elements of Subjective Global Assessment

Medical and diet histories	■ Body weight changes in past six months and past two weeks ■ Change in dietary intake and duration of change ■ Current diet: whether suboptimal, low-kcalorie, liquid, or starvation diet ■ Gastrointestinal symptoms: nausea, diarrhea, vomiting, or anorexia ■ Functional ability: full capacity or suboptimal, walking or bedridden ■ Current medical diagnosis ■ Degree of metabolic stress: low, moderate, or high
Physical examination	■ Loss of subcutaneous fat: in triceps or chest ■ Muscle wasting: in quadriceps or deltoids ■ Ankle edema ■ Sacral (lower spine) edema ■ Ascites
SGA rating	■ Well-nourished if recent weight gain, mild fat/muscle loss, improvement in histories ■ Moderate malnutrition suspected if >5% weight loss (not from hydration change), decreased food intake, mild fat/muscle wasting ■ Severe malnutrition if >10% weight loss (not from hydration change), severe fat/muscle wasting, some edema

Source: A. S. Detsky and coauthors, What is subjective global assessment of nutritional status? *Journal of Parenteral and Enteral Nutrition* 11 (1987): 8–13.

IN SUMMARY

■ Nutrition assessments include historical information, anthropometric data, biochemical analyses, and physical examinations. Health care providers assess food intake using 24-hour recall interviews, food frequency questionnaires, food records, and direct observation.

■ Anthropometric measurements help to evaluate growth patterns, overnutrition and undernutrition, and body composition.

■ Plasma proteins such as albumin, transferrin, and prealbumin can help in the assessment of protein status but are influenced by various medical conditions.

■ Physical exams can detect signs of nutrient deficiency, fluid imbalances, and functional impairments related to nutritional problems.

■ By combining the data from different assessment methods, health practitioners can better identify patients who are most likely to develop nutritional problems.

Implementing Nutrition Care

Once the health care professional has collected assessment information, the next steps of the nutrition care process can be carried out. A nutrition care plan often includes both dietary modifications and nutrition education. Note that some aspects of nutrition care fall within the scope of dietetics practice whereas others require the assistance of other health professionals.

wasting: the gradual atrophy (loss) of body tissues; associated with protein-energy malnutrition or chronic illness.

Dietary Modifications

During many illnesses, a person can meet energy and nutrient needs by following a **standard diet**. In other cases, a **modified diet** is prescribed. Modified diets are usually altered by changing the consistency of foods, altering nutrient content, or including or eliminating certain foods.

Table 14–10 lists examples of modified diets that are often prescribed during illness.[6] Mechanically altered diets and blenderized liquid diets change the texture and consistency of foods and are prescribed for individuals with chewing or swallowing impairments. Other dietary modifications may relieve the symptoms of disease. For example, restricting dietary fat or fiber may be helpful for people with certain kinds of intestinal disorders, and restricting dietary sodium can help to control fluid accumulation. A high-kcalorie, high protein diet may help to prevent or reverse malnutrition. Note that a person may have several medical problems and a number of modifications may be needed.

Modified diets should be adjusted to satisfy individual preferences and tolerances. They may also need to be altered as a patient's condition changes. Later chapters include more specific information about modified diets and additional dietary strategies for treating nutritional problems.

TABLE 14–10

Examples of Modified Diets

Type of Diet	Description of Diet	Appropriate Uses
Modified Texture and Consistency		
Mechanically altered diets	Contain foods that are modified in texture. Pureed diets include only pureed foods; mechanical soft diets may include solid foods that are mashed, minced, ground, or soft.	Pureed diets are used for people with swallowing difficulty, poor lip and tongue control, or oral hypersensitivity. Mechanical soft diets are appropriate for people with limited chewing ability or certain swallowing impairments.
Blenderized liquid diet	Contains fluids and foods that are blenderized to liquid form.	For people who cannot chew, swallow easily, or tolerate solid foods.
Clear liquid diet	Contains clear fluids or foods that are liquid at room temperature and leave minimal residue in the colon.	For preparation for bowel surgery or colonoscopy, for acute GI disturbances (such as after GI surgeries), or as a transition diet after intravenous feeding. For short-term use only.
Therapeutic Diets		
Fat-restricted diet	Restricts fat to low (<50 g/day) or very low (<25 g/day) levels in the diet.	For people who have certain malabsorptive disorders or symptoms of diarrhea, flatulence, or steatorrhea (fecal fat) resulting from dietary fat intolerance.
Fiber-restricted diet	Restricts fiber to low levels in the diet (<10 g/day).	For acute phases of intestinal disorders or to reduce fecal output before surgery. Not recommended for long-term use.
Sodium-restricted diet	Restricts sodium; degree of restriction depends on symptoms and disease severity.	To prevent fluid retention or induce fluid loss; used in hypertension, congestive heart failure, renal disease, and liver disease.
High-kcalorie, high-protein diet	Contains foods that are kcalorie and protein dense.	Used for increased kcalorie and protein requirements (in cancer, AIDS, burns, trauma, and other illnesses); also used to reverse malnutrition, improve nutritional status, or promote weight gain.

Sources: American Dietetic Association, *Nutrition Care Manual* (Chicago: American Dietetic Association, 2005); American Dietetic Association, *Manual of Clinical Dietetics* (Chicago: American Dietetic Association, 2000).

Diet Manuals For each modified diet, the exact foods to include or exclude are detailed in a **diet manual**. The dietetics staff at a health care institution may compile the manual using resources from the American Dietetic Assocation or other dietetics organizations. Small facilities may adopt the diet manual prepared by another hospital or a dietetics organization.

Alternative Feeding Routes In most cases, patients meet their nutrient needs by consuming regular foods. If their nutrient needs are high or their appetites poor, liquid formulas can be added to their diets to supplement their intakes. Sometimes, however, a person's medical condition makes it difficult to meet nutrient needs orally. Two options remain: **tube feedings** and **intravenous feedings**, described more fully in Chapter 16.

- **Tube feedings.** Nutritionally complete formulas can be delivered through a tube placed directly into the stomach or intestine. Tube feedings are preferred to intravenous feedings if the GI tract is functioning. For example, a person in a coma is unable to eat but may be able to digest foods and absorb nutrients normally. In this situation, a tube feeding would be the appropriate option.

- **Intravenous feedings.** A person's medical condition sometimes prohibits the use of the GI tract to deliver nutrients. If the person is malnourished and the GI tract cannot be used for a long period of time, intravenous feedings can meet nutritional needs.

The Diet Order As discussed earlier, the physician has the primary responsibility for prescribing an appropriate diet for a patient in a medical facility. Diet orders must be precise to avoid confusion; for example, a "low-sodium diet" should specify the amount of sodium permitted, since "low-sodium" could be interpreted to mean any amount between 500 and 3,000 milligrams. The physician often relies on the dietitian or nurse to recommend changes in the diet order when warranted, such as when a diet order seems inappropriate or outdated.

Nothing by Mouth (NPO) An order to not give a patient anything at all—food, beverages, or medications—is indicated by NPO, an abbreviation for *non per os,* meaning "nothing by mouth." For example, an order may read "NPO for 24 hours" or "NPO until after X-ray." The NPO order is commonly used during certain acute illnesses or diagnostic tests involving the GI tract.

Determining Energy Requirements

Energy needs are affected by a patient's health problem, treatments, current nutrition status, and activity level. In hospital patients, the "gold standard" for calculating energy requirements—**indirect calorimetry**—determines resting metabolic rate (RMR) by measuring the person's oxygen consumption and carbon dioxide elimination during a period of rest. The procedure is labor intensive, so clinicians more often use predictive equations that yield similar results. Once RMR is estimated, the value is multiplied by a factor that accounts for the altered energy demands imposed by the person's medical problem, activity level, and stress level. Several RMR equations in common use are listed in Table 14–11 p. 426, and the "How to" presents an example of this method. In critical care patients, energy needs may be raised further due to fever, mechanical ventilation, restlessness, or the presence of open wounds. ■ Note that patients who are critically ill are usually bedridden and inactive, so the energy needed for physical activity is minimal.

Another popular method of estimating energy needs is to multiply a person's body weight by a factor appropriate for the medical condition. For example, energy needs for critical care patients often fall within the range of 25 to 30 kcalories per kilogram of body weight daily;[7] a patient weighing 160 pounds (72.7 kilograms) may therefore

■ Energy requirements for patients with severe metabolic stress are discussed in Chapter 24.

standard diet: a diet that includes all foods and meets the nutrient needs of healthy people; sometimes called a **regular diet**.

modified diet: a diet that is adjusted in consistency, energy or nutrient content, or by the inclusion or elimination of certain foods; sometimes called a **therapeutic diet**.

diet manual: a resource that specifies the foods allowed and restricted in modified diets and provides sample menus.

tube feedings: liquid formulas delivered through a tube placed in the stomach or intestine.

intravenous feedings: the provision of nutrients through a vein, bypassing the intestine. The intravenous provision of nutrients is called **parenteral nutrition**.

indirect calorimetry (kal-uh-RIM-eh-tree): a determination of resting energy expenditure by measuring a person's oxygen consumption and carbon dioxide elimination during a period of rest.

TABLE 14–11

Selected Equations for Estimating Resting Metabolic Rate (RMR)

Harris-Benedict[a]

Women: RMR = $655.1 + [9.563 \times$ weight (kg)$] + [1.85 \times$ height (cm)$] - [4.676 \times$ age (years)$]$

Men: RMR = $66.5 + [13.75 \times$ weight (kg)$] + [5.003 \times$ height (cm)$] - [6.755 \times$ age (years)$]$

Mifflin-St. Jeor

Women: RMR = $[9.99 \times$ weight (kg)$] + [6.25 \times$ height (cm)$] - [4.92 \times$ age (years)$] - 161$

Men: RMR = $[9.99 \times$ weight (kg)$] + [6.25 \times$ height (cm)$] - [4.92 \times$ age (years)$] + 5$

WHO/FAO/UNU[a]

Girls and women (age range, years):

10–18: RMR = $[7.4 \times$ weight (kg)$] + [482 \times$ height (m)$] + 217$

18–30: RMR = $[13.3 \times$ weight (kg)$] + [334 \times$ height (m)$] + 35$

30–60: RMR = $[8.7 \times$ weight (kg)$] - [25 \times$ height (m)$] + 865$

>60: RMR = $[9.2 \times$ weight (kg)$] + [637 \times$ height (m)$] - 302$

Men and boys (age range, years):

10–18: RMR = $[16.6 \times$ weight (kg)$] + [77 \times$ height (m)$] + 572$

18–30: RMR = $[15.4 \times$ weight (kg)$] - [27 \times$ height (m)$] + 717$

30–60: RMR = $[11.3 \times$ weight (kg)$] + [16 \times$ height (m)$] + 901$

>60: RMR = $[8.8 \times$ weight (kg)$] + [1,128 \times$ height (m)$] - 1,071$

[a] Although these equations are sometimes used for estimating basal metabolic rate (BMR), they were derived from data measured during resting conditions in most cases.

Sources: M. D. Mifflin and coauthors, A new predictive equation for resting energy expenditure in healthy individuals, *American Journal of Clinical Nutrition* 51 (1990): 241–247; World Health Organization, *Energy and Protein Requirements: Report of a Joint FAOAVHO/UNU Expert Consultation* (Geneva: World Health Organization, 1985); J. A. Harris and F. G. Benedict, A biometric study of human basal metabolism, *Proceedings of the National Academy of Sciences USA* 4 (1918): 370–373.

How To

ESTIMATE THE ENERGY REQUIREMENTS OF A HOSPITAL PATIENT

To determine the energy requirements of a hospital patient, the dietitian or nurse first calculates a patient's resting metabolic rate (RMR) and then applies a "stress factor" to accommodate the additional energy needs imposed by illness. The stress factor 1.25 has been shown to be reasonably accurate for many hospitalized patients; other examples are listed in Chapter 24, Table 24–2.

This example uses the WHO/FAO/UNU equation (shown in Table 14–11) and the stress factor 1.25 to determine the energy needs of a 54-year-old patient who is 5 feet 3 inches tall and weighs 115 pounds.

Step 1: The patient's weight and height are converted to the units used in the equation:

Weight in kilograms = 115 lb ÷ 2.2 lb/kg
= 52.3 kg
Height in meters = 63 in × 0.0254 m/in
= 1.6 m

Step 2: Using the WHO/FAO/UNU equation for estimating RMR in women between 30 and 60 years old:

RMR = [8.7 × weight (kg)]
− [25 × height (m)] + 865
= (8.7 × 52.3) − (25 × 1.6) + 865
= 455 − 40 + 865 = 1,280 kcal

Step 3: The RMR value is multiplied by the appropriate stress factor:

RMR × stress factor
= 1,280 × 1.25 = 1,600 kcal

Thus, an appropriate kcaloric intake for this patient would be approximately 1,600 kcalories. Her weight can be monitored to determine if her actual needs are higher or lower. In addition, her energy needs are likely to change as her condition improves.

require between 1,818 and 2,181 kcalories per day. The energy intake can be started within this range and then adjusted as the patient's body weight and other determinants of nutrition status change.

Approaches to Nutrition Care

As described in this chapter, a nutrition care plan often involves dietary modification and nutrition education. The care plan should be compatible with the desires and abilities of the person it is designed to help. The challenge is greater if dietary changes are required for extended periods.

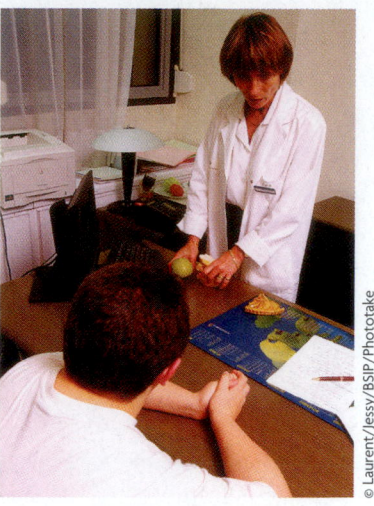

Dietary counseling requires sensitivity to cultural orientation, educational background, and motivation for change.

Long-Term Dietary Intervention   When long-term changes are necessary, a care plan must take into account a person's current food habits, lifestyle, and degree of motivation. Behavior change is a process that occurs in stages; therefore, more than one consultation is usually necessary. The following approaches may be helpful in implementing long-term dietary changes:[8]

- **Determine the individual's readiness for change.** Some people have little desire to change their dietary behaviors, and even those who are willing may not be fully prepared to take the necessary steps. The health practitioner needs to consider a patient's readiness to adopt new dietary behaviors before attempting to implement an ambitious care plan.

- **Emphasize what to eat, rather than what not to eat.** Emphasizing foods to include in the diet, rather than those to restrict, can make dietary changes more appealing. For example, encouraging additional fruits and vegetables is a more attractive message than telling the patient to restrict butter, cream sauces, and ice cream.

- **Suggest only one or two changes at a time.** People are more likely to adopt a dietary plan that does not deviate too much from their usual diet. If they succeed in adopting one or two changes, they are more likely to stick to the plan and be open to additional suggestions. Stricter plans may yield quicker results but are useful only for highly motivated people.

Nutrition Education   Nutrition education allows patients to learn about the dietary factors that affect their particular medical condition. Ideally, this knowledge motivates them to change their diet and lifestyle to improve their health status.

A nutrition education program should be tailored to a person's age, level of literacy, and cultural background. Learning style should also be considered: some people learn best by discussion supplemented with written materials, whereas others prefer visual examples, such as food models and measuring devices.[9] Information can be provided in one-on-one sessions or group discussions. The meeting should also assess the person's understanding of the material and commitment to making changes. Follow-up sessions can reveal whether the person has successfully adopted a dietary plan. For example, a dietitian who counsels a woman who is lactose intolerant and hesitant to use milk products can proceed as follows:

- The dietitian can provide sample menus of a nutritionally adequate diet that that limits milk and milk products. Together the dietitian and the woman can design menus that consider her food preferences.

- Using diet analysis software, the dietitian can demonstrate how altering food choices changes the calcium content of a meal.

- The dietitian can explain how to use the Daily Value on food labels to estimate the calcium content of packaged foods.

- The dietitian can provide information about the advantages and disadvantages of different calcium supplements.

■ The dietitian can assess the woman's understanding by having her identify non-milk products that are high in calcium.

Ideally, the dietitian would be able to monitor the woman's progress in a subsequent counseling session.

Documentating Nutrition Care

Nutrition care is described in the medical record in various ways, but one of the most popular is the SOAP note. The letters represent the types of information included: *Subjective*, *Objective*, *Assessment*, and the *Plan* for care. *Subjective* information is obtained in an interview with the patient or patient's family and includes the main symptoms and complaints related to a particular medical problem. *Objective* information is available from the nutrition screening or assessment data; it includes results of biochemical analyses, anthropometric tests, and physical examinations. *Assessment* is a brief evaluation of the subjective and objective data and provides a concise diagnosis of the nutrition problem. The *Plan* describes recommendations, including dietary prescriptions, special equipment, nutrition education, and referrals that can help solve the problem. Figure 14–4 shows an example of a SOAP note, although there are many possible variations.

FIGURE 14–4

Example of a SOAP Note

SOAP NOTE

Patient Name: *Arthur Jones*　　　　　　　　　Date: *Aug. 10, 2001*

Age: *58*　　　Gender: *Male*　　　Medical diagnosis: *Hypercholesterolemia*

Subjective:

Patient recently learned of his hypercholesterolemia, has no obvious symptoms. Wants trial of dietary/lifestyle changes to reduce need for medication. Willing to attempt weight loss.

Objective:

Total cholesterol: 288 mg/dL　　　　　　*Glucose (fasting): 101 mg/dL*
*　　　　HDL-C: 48 mg/dL*　　　　　　　*Hb$_{A1c}$: 5.7%*
*　　　　LDL-C: 214 mg/dL*　　　　　　*Weight: 268 lb*
Triglycerides: 132 mg/dL　　　　　　*Height: 6'1"*
*　　　　　　　　　　　　　　　　　Waist circumference: 45"*
*　　　　　　　　　　　　　　　　　BMI: 35.4*

Assessment:

Abdominal obesity; elevated LDL cholesterol. Weight loss and lifestyle changes may improve hypercholesterolemia; Mr. Jones is highly motivated to try these before resorting to meds.

Plan:

Goal: 15 lb weight loss over next 6 months; instruction about food portion sizes and lower-kcalorie food choices. Patient to start 30-minute walking program, evenings.

Follow-up visit: One month; patient will identify food portion sizes.

Referral: Heart-healthy workshop on August 17 (one week); patient to attend with wife.

Form completed by: *Carmen Cordova, MPH, R.D.* **Position:** *Dietitian, Nutrition Services*

Other charting styles are in use, but the content is more important than the particular format used. Health care professionals need to learn the charting procedures preferred by a medical facility before making entries in patients' medical records. Generally, entries in the medical record should be as succinct as possible so that they can be easily read and quickly understood by the other members of the health care team. The standardized templates used in electronic data systems generally require concise language.

Improving Food Intake

People in hospitals and other medical facilities often lose their appetites as a result of their medical condition, treatment, or emotional distress. In addition, some medications and other treatments can dramatically alter taste perceptions. Patients usually receive meals at specified times whether they are hungry or not and often must eat in bed without companionship; under these conditions, eating can be more of a chore than a pleasurable experience. Meals may also be unwelcome if the person is in pain or has been sedated.

Nurses and dietetic technicians often have central roles in helping patients to eat. If either the appetite or the sense of taste is affected by illness, the health practitioner should work with the patient to identify foods that are the most enjoyable. When

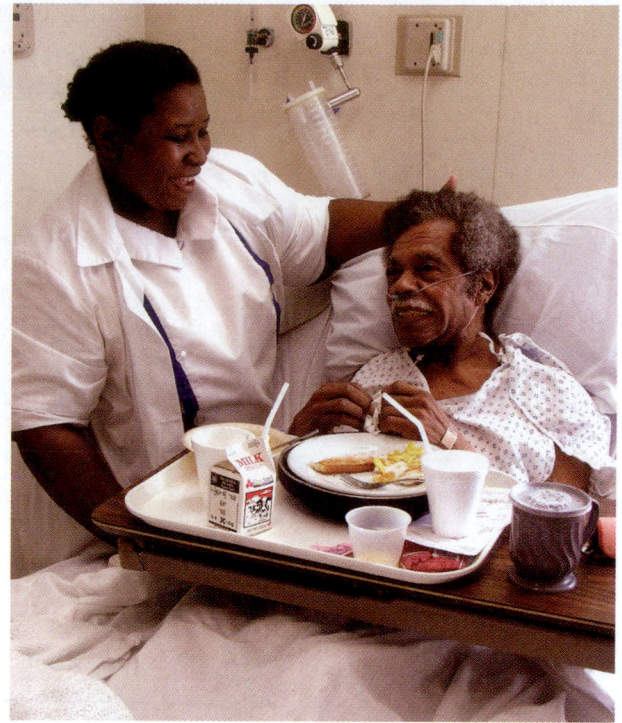
© Sonda Dawes/The Image Works

People enjoy eating when they feel comfortable and cared for.

meals are served, the nurse can check to see that foods and utensils are arranged attractively and help the patient wash up before a meal. The "How to" lists additional suggestions that may help to improve food intake at mealtimes. The accompanying Case Study provides an opportunity for you to review the implementation of nutrition care.

How To

HELP HOSPITAL PATIENTS IMPROVE
THEIR FOOD INTAKES

1. Empathize with the patient. Show that you understand how difficult eating may be. Imagine feeling too sick to move or too tired to sit up.

2. Motivate. Be sure the patient understands how important nutrition is to recovery.

3. Help patients select the foods they like and mark menus appropriately. When appropriate and permissible, let friends or family members bring favorite foods from outside the hospital.

4. For patients who are weak, suggest foods that require little effort to eat. Eating a roast beef sandwich, for example, requires less effort than cutting and eating a steak. Drinking soup from a cup may be easier than eating it with a spoon.

5. Help patients prepare for meals. Help them get comfortable, either in bed or in a chair. Adjust the extension table to a comfortable distance and height, and make sure it is clean. Take these steps before the tray arrives, so the meal can be served promptly and at the right temperature.

6. When the food cart arrives, check the patient's tray. Confirm that the patient is receiving the right diet, that the foods on the tray are those selected from the menu, and that the foods look appealing. Order a new tray if foods are not appropriate.

7. Help with eating, if necessary. Help patients to open containers or cut foods, and assist with feeding if patients cannot feed themselves.

8. Try to solve eating problems. Encourage patients with little appetite to eat the most nutritious foods first and to drink liquids between meals.

9. Take a positive attitude toward the hospital's food. Never say something like "I couldn't eat this either." Instead, say, "The food service department really tries to make foods appetizing. I'm sure we can find a solution."

Case Study

IMPLEMENTING NUTRITION CARE

Max is an eleven-year-old boy who was admitted to the hospital after he passed out while playing with friends. Tests confirm a diagnosis of type 1 diabetes mellitus. Max remains in the hospital for several days until his blood glucose and ketone levels are under control. During this time, he and his family learn about diabetes, the diet Max needs to follow, the use of insulin, how to monitor blood glucose levels, and the required coordination of diet, insulin, and physical activity. The details of diabetes mellitus are reserved for Chapter 21, but for now you can consider the steps that are necessary for implementing nutrition care.

1. Given the chronic nature of Max's illness and his age, what approaches should be used when discussing the required dietary and medical treatments with Max and his family?

2. What factors need consideration when designing a nutrition education program for Max and his parents?

3. Max will need additional care to learn more about diabetes and to make the adjustments that will allow him to cope with his condition. Why is it important to plan follow-up care before Max leaves the hospital?

IN SUMMARY

- Diets prescribed during illness can be modified in consistency or in nutrient content. Diet manuals specify the foods to include in modified diets. Some medical conditions may require the use of tube feedings or intravenous feedings.

- Energy needs can be estimated by measuring or calculating RMR and multiplying the value by a factor that accounts for the altered energy demands imposed by the medical problem.

- A care plan should be compatible with a person's food preferences and willingness to make dietary changes. Nutrition education must be individualized to accommodate a patient's needs and learning style.

- Hospital patients may need assistance at mealtime and encouragement to consume adequate amounts of food.

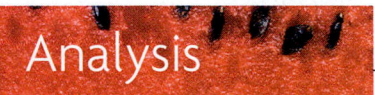

Analysis

CLINICAL APPLICATIONS

1. Describe the possible nutritional implications of these findings from a patient's medical and social histories: age 77, lives alone, recently lost spouse, uses a walker, no natural teeth or dentures, history of hypertension and diabetes, uses medications that cause frequent urination.

2. Calculate the %IBW and %UBW for a man who is 5 feet 11 inches tall with a current weight of 160 pounds and a usual body weight of 180 pounds. Use the method in Table 14–5 (p. 416) to estimate a desirable ("ideal") body weight. What additional information do you need to interpret the implications of this amount of weight loss?

3. James is a 49-year-old male who is 6 feet 2 inches tall and has a usual body weight of 180 pounds. He was admitted to the hospital following an automobile accident and was treated for minor injuries. Using the method described in the "How to" on p. 426, estimate his energy requirement using the appropriate WHO/FAO/UNU equation and the stress factor 1.25.

4. A nutrition screening of an elderly woman admitted to the hospital for minor surgery revealed that she was not at risk for poor nutrition status. She was given a standard diet and was not referred to a dietitian for nutrition assessment. Following surgery, the woman developed several complications, and recovery was slower than expected. You notice that she has eaten only minimal amounts of food for several days and seems disinterested in meals. Describe several steps that can be taken to uncover and address problems that the woman might be having with food.

1. Mr. Hom experiences loss of appetite, difficulty swallowing, and mouth pain as a consequence of illness. Mr. Hom is at risk of malnutrition due to:
   a. altered metabolism.
   b. reduced food intake.
   c. altered excretion of nutrients.
   d. altered digestion and absorption.

2. The nutrition care process is a systematic approach for:
   a. identifying the nutrient content of foods.
   b. ordering special diets.
   c. conducting nutrition screening.
   d. meeting the nutrition needs of patients.

3. All of the following factors place a person at risk for malnutrition *except:*
   a. having a health problem that is frequently associated with PEM.
   b. the use of prescription medications that affect nutrient needs.
   c. a social history that reveals that the individual lives with a spouse in a middle-income neighborhood.
   d. a significant reduction in food intake over the past five or more days.

4. Which dietary assessment method does a health practitioner use to conduct a kcalorie count?
   a. 24-hour recall interview
   b. food frequency questionnaire
   c. food record
   d. direct observation

5. The %IBW of a person who weighs 185 pounds and has a healthy body weight of 150 pounds is:
   a. 123 percent.
   b. 150 percent.
   c. 23 percent.
   d. 81 percent.

6. A malnourished patient has just begun to eat after days without significant amounts of food. Which of the following blood tests would change most quickly as the patient's nutrition status improves?
   a. albumin
   b. transferrin
   c. serum electrolytes
   d. retinol-binding protein

7. Which sign of PEM would be unlikely to show up in a physical examination?
   a. low plasma protein levels
   b. dull, brittle hair
   c. poor wound healing
   d. wasted appearance

8. A nurse notices a food on a patient's tray that she thinks should be excluded from the patient's prescribed diet. What resource can she check to learn whether the food is appropriate?
   a. nutrition care plan
   b. diet order
   c. diet manual
   d. medical record

9. A successful nutrition intervention would include a long list of:
   a. dietary changes that the patient should consider making.
   b. foods that the patient should avoid.
   c. appetizing meals and foods that the patient can include in his or her diet.
   d. reasons why the patient should make dietary changes.

10. The most important factor(s) that affect how nutrition education is presented is (are):
    a. the person's nutrient needs and nutrition status.
    b. the person's abilities and motivation.
    c. the person's medical history.
    d. the entries in the medical record.

Answers to these questions can be found in Appendix H.

Notes

1. D. R. Thomas and coauthors, Malnutrition in subacute care, *American Journal of Clinical Nutrition* 75 (2002): 308–313.

2. K. Lacey and E. Pritchett, Nutrition care process and model: ADA adopts road map to quality care and outcomes management, *Journal of the American Dietetic Association* 103 (2003): 1061–1072.

3. K. Lacey and N. Cross, A problem-based nutrition care model that is diagnostic driven and allows for monitoring and managing outcomes, *Journal of the American Dietetic Association* 102 (2002): 578–589.

4. L. C. Tapsell, V. Brenninger, and J. Barnard, Applying conversation analysis to foster accurate reporting in the diet history interview, *Journal of the American Dietetic Association* 100 (2000): 818–824.

5. K. M. Flegal, R. Wei, and C. Ogden, Weight-for-stature compared with body mass index-for-age growth charts for the United

States from the Centers for Disease Control and Prevention, *American Journal of Clinical Nutrition* 75 (2002): 761–766.

6. American Dietetic Association, *Nutrition Care Manual* (Chicago: American Dietetic Association, 2005).

7. K. A. Kudsk and G. S. Sacks, Nutrition in the care of the patient with surgery, trauma, and sepsis, in M. E. Shils and coeditors, *Modern Nutrition in Health and Disease* (Baltimore: Lippincott Williams & Wilkins, 2006), pp. 1414–1435.

8. K. Glanz, Current theoretical bases for nutrition intervention and their uses, in A. M. Coulston, C. L. Rock, and E. R. Monsen,

eds., *Nutrition in the Prevention and Treatment of Disease* (San Diego, CA: Academic Press, 2001), pp. 83–93; M. C. Rosal and coauthors, Facilitating dietary change: The patient-centered counseling model, *Journal of the American Dietetic Association* 101 (2001): 332–338, 341.

9. J. M. Heins and L. Delahanty, Tools and techniques to facilitate eating behavior change, in A. M. Coulston, C. L. Rock, and E. R. Monsen, eds., *Nutrition in the Prevention and Treatment of Disease* (San Diego, CA: Academic Press, 2001), pp. 105–122.

NUTRITIONAL GENOMICS

Imagine this situation: a physician scrapes a sample of cells from inside your cheek and submits it to a **genomics** lab. In a short time, you receive a report that reveals your disease susceptibilities and recommends dietary and lifestyle changes that can maintain health. You may even be given a prescription for a dietary supplement to prevent the diseases that you are most likely to develop. Unlikely? Perhaps, but these possibilities are being explored by scientists working in the new field of **nutritional genomics**, the study of dietary effects on **gene expression**. Recent research suggests that some dietary factors may be more helpful (or more harmful) in people who have particular genetic variations. The promise of nutritional genomics is a custom-designed dietary prescription that fits each person's specific needs. The Glossary on p. 434 defines genomics and related terms.

What is a genome?

Genetic information is encoded in DNA molecules within the nuclei of almost all of the cells in our bodies. Figure NP14–1 shows how the genetic material is organized within the **genome**, the complete set of genetic information within our cells. The DNA molecules are tightly packed along with associated proteins within the 46 **chromosomes**. Segments of a DNA strand that can eventually be translated into proteins are called **genes**. The sequence of **nucleotides** within each gene encodes the amino acid sequence of a particular protein. Scientists estimate that there are between 20,000 and 25,000 genes in the human genome.[1] However, only a small percentage of the genome codes for proteins: most DNA consists of **noncoding sequences**, whose function, if any, is unclear.

When proteins are made, the information in the DNA sequence is first transcribed (copied) to messenger RNA molecules, which carry the genetic information out of the nucleus. Gene expression can be measured by determining the amounts of messenger RNA in a tissue sample. The expression of thousands of genes can be measured simultaneously using **microarray technology** (see photo).

FIGURE NP14–1

The Human Genome

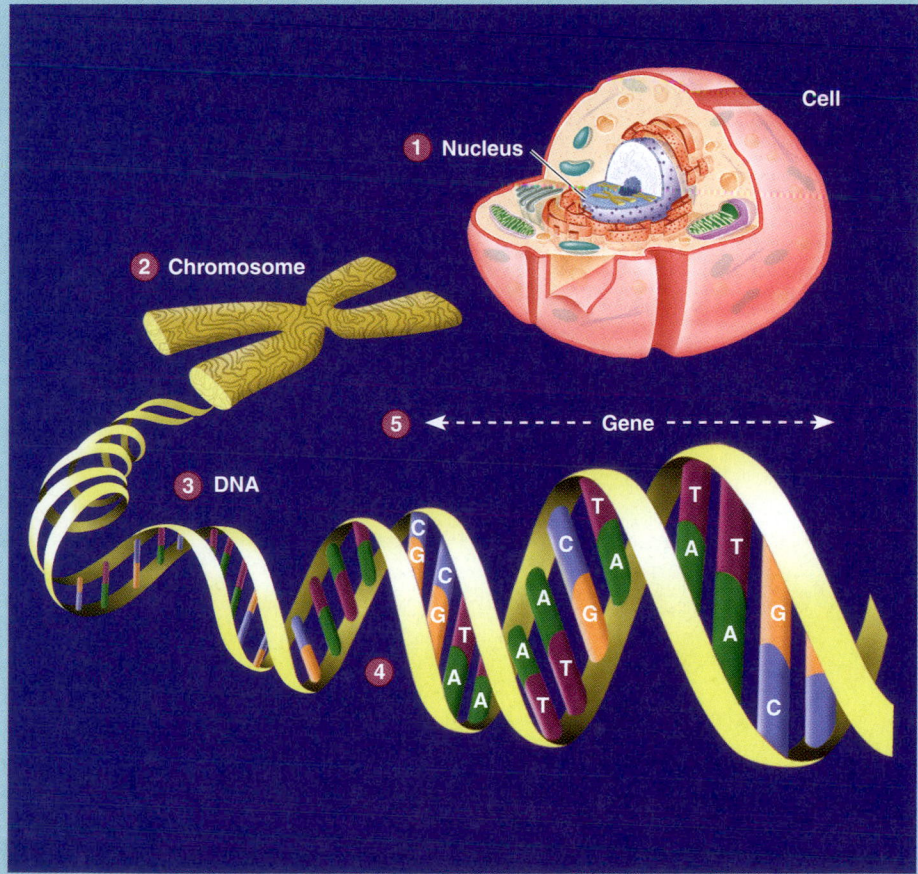

1 The human genome is a complete set of genetic material organized into 46 chromosomes, located within the nucleus of a cell.

2 A chromosome is made of DNA and associated proteins.

3 The double helical structure of a DNA molecule is made up of two long chains of nucleotides. Each nucleotide is composed of a phosphate group, a 5-carbon sugar, and a base.

4 The sequence of nucleotide bases (C, G, A, T) determines the amino acid sequence of proteins. These bases are connected by hydrogen bonding to form base pairs: adenine (A) with thymine (T) and guanine (G) with cytosine (C).

5 A gene is a segment of DNA that includes the information needed to synthesize one or more proteins.

Source: Adapted from "A Primer: From DNA to Life," Human Genome Project, U.S. Department of Energy Office of Science, www.ornl.gov/sci/techresources/Human_Genome/primer_pic.shtml.

GLOSSARY

chromosomes: structures within the nucleus of a cell that contain the cell's DNA and associated proteins.

gene expression: the process by which a cell converts the genetic code into RNA and protein.

genes: segments of DNA that contain the information needed to make proteins.

genome (JEE-nome): the full complement of genetic material in the chromosomes of a cell.

genomics (jee-NO-miks): the study of genomes.

inherited disorders: medical conditions resulting from genetic defects.

microarray technology: research technology that monitors the expression of thousands of genes simultaneously.

multigene or polygenic: involving a number of genes, rather than a single gene.

noncoding sequences: regions of DNA that do not code for proteins. Some noncoding sequences may have regulatory or structural properties, but most have no known function.

nucleotides: the subunits of DNA and RNA molecules. These compounds—cytosine (C), thymine (T), uracil (U), guanine (G), and adenine (A)—are each composed of a phosphate group, a 5-carbon sugar (ribose), and a nitrogen-containing base. A DNA molecule is made up of two long chains of nucleotides held together by hydrogen bonding between nucleotide bases on opposing strands; each hydrogen-bonded nucleotide couple is called a **base pair**.

nutritional genomics: the study of dietary effects on genetic expression; also known as **nutrigenomics**.

polymorphisms: differences in the DNA sequences among individuals. A **single-nucleotide polymorphism** involves a single nucleotide at a particular area in the DNA strand.
> poly = many
> morph = form
> ism = condition

promoter: a region of DNA involved with gene activation.

transcription factors: proteins that bind DNA at specific sequences to regulate gene expression.

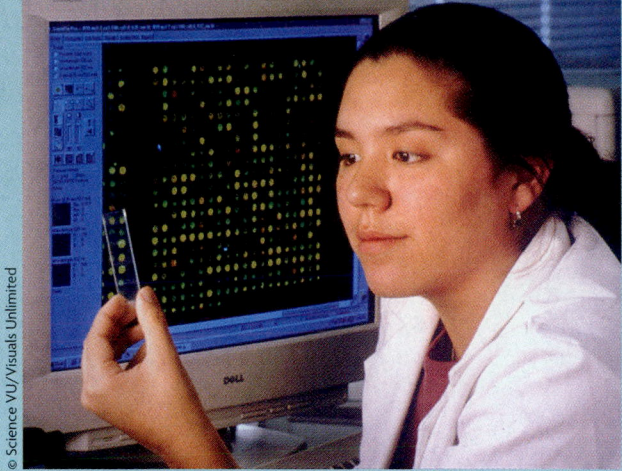

A DNA microarray allows researchers to monitor the expression of thousands of genes simultaneously.

How did research in nutritional genomics begin?

The Human Genome Project, an international effort by industry and government scientists to sequence the human genome, was completed in April 2003. The project led to enormous advances in the research technologies needed to study genes and genetic variation. Knowing the DNA sequences within human chromosomes allows researchers to study how alterations in diet and lifestyle can alter the expression of a multitude of genes. The next steps are to identify the individual genes in the genome and the roles of their protein products, the genes and proteins associated with diseases, and the dietary and lifestyle choices that influence the expression of genes involved in disease.

Genetic differences among individuals have been studied for years, as have the specialized dietary therapies used to treat various **inherited disorders**. For example, an individual may inherit a genetic defect that inhibits the normal metabolism of an essential nutrient and may therefore need to consume a diet that contains either more or less of this nutrient. An example of this type of condition is phenylketonuria (PKU), discussed in Nutrition in Practice 16. Genomic research takes this concept further: instead of focusing on alterations in one or two genes, researchers study the expression of *multiple* genes.

How do nutrients alter gene expression?

Some nutrients can switch gene expression on or off.[2] The **promoter** region of a gene (a DNA region involved with gene activation) acts as the master switch. A large variety of proteins known as **transcription factors** bind to areas on the promoter and either enhance or inhibit gene expression. A combination of dietary factors and hormones influences the types of transcription factors that reach the nucleus and their tendency to bind to DNA. Specific examples of how nutrients can influence gene expression include:

- The transcription factor that enhances gene expression of enzymes required for cholesterol synthesis enters the nucleus only when the cellular cholesterol content is low.

- The transcription factor that inhibits expression of ferritin, an iron-storage protein, changes its affinity for DNA based on the iron content of the cell.

How much genetic variation is there among people?

Except for identical twins, no two individuals are genetically identical. However, the variation in the genomes of any two persons is only about 0.1 percent, a difference of only one base in every 1,000.[3] The most common genetic differences, known as **polymorphisms**, are changes in single nucleotides (**single-nucleotide polymorphisms**). Such variations are significant only if they affect the amino acid sequence of a protein in such a way that protein function is altered.

Genetic variation gives rise to the diversity among human beings—it explains most of the differences in our physical appearances and metabolic characteristics. Along with environmental factors, it also determines our susceptibilities to disease. Diseases affected by a single gene tend to be relatively rare and usually exert their effects early in life. In contrast, common diseases such as heart disease and cancer are influenced by many genes and typically develop over several decades or even longer. In these more complex **multigene, or polygenic**, disorders, many genes can contribute to disease risk, but no single gene may be sufficient to cause the disease on its own.

What are some examples of single-gene disorders?

Examples of single-gene disorders include PKU, sickle-cell anemia, and the iron-overload disease hemochromatosis. Single-gene disorders may seriously disrupt metabolism and often require significant dietary or medical intervention. However, not all single-gene disorders have life-threatening ramifications. For example, lactose intolerance can result from an alteration in the promoter of the lactase gene; it may cause gastrointestinal discomfort but is readily managed by simple dietary changes.

How are multigene disorders different from single-gene disorders?

Multigene diseases are usually sensitive to a number of environmental influences, including diet and lifestyle.[4] These environmental factors can directly influence the expression of the genes involved. Also, multigene diseases tend to develop over many years, so determining genetic susceptibility may allow a person to modify diet and lifestyle appropriately and reduce the risk of developing the disease.

Heart disease is an example of a disease with multiple gene influences. Its many risk factors represent the involvement of an assortment of genes, which affect disparate aspects of physiology and metabolism. Consider that the major risk factors for heart disease include elevated blood cholesterol levels, obesity, diabetes, and hypertension. The underlying cause of any of these risk factors is rarely known; currently, clinicians screen for the presence of risk factors, but not the reasons why they occur. Should genomic research prove successful, a future assessment approach might be to identify specific genetic alterations and changes in metabolism that lead to the development of individual risk factors. For example, tests may determine whether blood cholesterol levels are high due to excessive cholesterol absorption, excessive liver production, or reduced cholesterol degradation.[5] This information could then guide health care providers to the most appropriate intervention, allowing a better match between treatment recommendations and a person's genetic profile.

Can genomic research be used to explore the differences in nutrient needs among people?

Even though most people apparently can meet their nutrient needs by consuming nutrients at recommended levels, it would be useful to learn more about genetic variations within healthy populations. The techniques that have emerged from genomic research may provide a means for fine-tuning nutrient recommendations for different individuals.[6] Moreover, ideal indicators of nutrient status are still lacking for several of the minerals, such as zinc, magnesium, and chromium. Scientists hope to eventually produce genomic maps that will indicate how various nutrient deficiencies and combinations of deficiencies affect gene expression. These maps may eventually provide data that can help diagnose nutrient deficiencies.

Will knowledge about the human genome substantially change the manner in which health care is provided?

The enthusiasm surrounding genomic research must be put into perspective in terms of both the status of clinical medicine at present and people's willingness to make difficult lifestyle choices. Critics have questioned whether genetic markers for disease would be more useful than simple and inexpensive clinical measurements, which reflect both genetic *and* environmental influences. In other words, knowing that a person is genetically predisposed toward high cholesterol levels is not necessarily more useful than knowing the person's actual blood cholesterol level.[7] Furthermore, if a disease has many genetic risk factors, each gene that contributes to susceptibility may have little influence on its own, so the benefits of identifying an individual genetic marker would be small. The long-range possibility is that many genetic markers will eventually be identified; the hope is that the combined information will be a more useful and accurate predictor of disease and its effective treatments.

Obtaining additional knowledge about disease risk may not be useful unless people are motivated to make serious lifestyle changes. For example, despite the present abundance of disease prevention recommendations, many people seem unwilling to make the changes known to improve health. Researchers have estimated that heart disease and type 2 diabetes are 80 percent and 90 percent preventable, respectively, by changing one's lifestyle to include an appropriate diet, a healthy body weight, and regular exercise, among other factors.[8] Given the difficulty that people have with current recommendations, it is unlikely that they will enthusiastically adopt an even more detailed list of dietary and lifestyle modifications.

What ethical concerns are raised by having extensive knowledge about an individual's genome?

The ability to obtain detailed genetic information raises several concerns. A primary consideration is confidentiality: Should information about a person's susceptibility to disease be released to others without that person's consent? Because environmental factors play such an important role in disease risk, genetic predisposition usually cannot predict whether a person will develop a particular disease. Nevertheless, future insurers of medical services may attempt to charge higher rates or base acceptance criteria on applicants' disease susceptibilities as evidenced by genetic testing. Another important concern is whether genetic testing is always in the best interest of children. Although early knowledge of a child's predisposition to illnesses may be useful for parents who want to provide optimal care, the release of this information could threaten the child's privacy and increase the potential for "genetic discrimination" in the future.

Although genomic research has the potential to improve our ability to diagnose and treat disease, it is still unclear how knowledge of the genome will be translated into useful medical treatments. Still, health care professionals will need to keep informed of the ethical, legal, and social implications of genomics in the fields of medicine and nutrition as this remarkable research continues.

Notes

1. International Human Genome Sequencing Consortium, Finishing the euchromatic sequence of the human genome, *Nature* 431 (2004): 931–945.

2. R. J. Cousins, Nutritional regulation of gene expression and nutritional genomics, in M. E. Shils and coeditors, *Modern Nutrition in Health and Disease* (Baltimore: Lippincott Williams & Wilkins, 2006), pp. 615–626.

3. P. J. Stover and C. Garza, Polymorphisms: Effect on nutrient utilization and metabolism, in M. E. Shils and coeditors, *Modern Nutrition in Health and Disease* (Baltimore: Lippincott Williams & Wilkins, 2006), pp. 627–635.

4. W. C. Willett, Balancing life-style and genomics research for disease prevention, *Science* 296 (2002): 695–698.

5. J. B. German, M.-A. Roberts, and S. M. Watkins, Personal metabolomics as a next generation nutritional assessment, *Journal of Nutrition* 133 (2003): 4260–4266.

6. P. J. Stover, Influence of human genetic variation on nutritional requirements, *American Journal of Clinical Nutrition* 83 (2006): 436S–442S.

7. Willett, 2002.

8. Willett, 2002.

© Bill Aron/PhotoEdit

15

Medications, Herbal Supplements, and Diet-Drug Interactions

acupuncture (AK-you-PUNK-cher): a therapy that involves inserting thin needles into the skin at specific anatomical points, allegedly to correct disruptions in the flow of energy within the body.

aromatherapy: inhalation of oil extracts from plants to cure illness or enhance health.

ayurveda: a traditional medical system from India that promotes the use of diet, herbs, meditation, massage, and yoga for preventing and treating illness.

bioelectrical or bioelectromagnetic therapies: therapies that involve the unconventional use of electric or magnetic fields to cure illness.

biofeedback: a technique in which individuals are trained to gain voluntary control of certain physiological processes, such as skin temperature or brain wave activity, to help reduce stress and anxiety.

biofield therapies: healing methods based on the belief that illnesses can be healed by manipulating energy fields that purportedly surround and penetrate the body. Examples include *acupuncture, qi gong,* and *therapeutic touch.*

chiropractic (KYE-roh-PRAK-tic): an alternative medical system based on the unproven theory that spinal manipulation can restore health.

- A *subluxation* is a misaligned vertebra or other spinal alteration that may cause illness.

- *Adjustment* is the manipulative therapy practiced by chiropractors.

faith healing: the use of prayer or belief in divine intervention to promote healing.

homeopathic (HO-mee-oh-PATH-ic) **medicine:** a practice based on the theory that "like cures like." Substances believed to cause certain symptoms are prescribed for curing the same symptoms but are given in extremely diluted amounts.

hypnotherapy: a technique that uses hypnosis and the power of suggestion to improve health behaviors, relieve pain, and promote healing.

imagery: the use of mental images of things or events to aid relaxation or promote self-healing.

integrative medicine: an approach to medical care that combines conventional therapies and CAM therapies that have been shown to be safe and potentially effective.

massage therapy: manual manipulation of muscles to reduce tension, increase blood circulation, improve joint mobility, and promote healing of injuries.

meditation: a self-directed technique of calming the mind and relaxing the body.

naturopathic (NAY-chur-oh-PATH-ic) **medicine:** an approach to medical care using practices alleged to enhance the body's natural healing abilities. Treatments may include a variety of alternative therapies including dietary supplements, herbal remedies, exercise, and homeopathy.

osteopathic (OS-tee-oh-PATH-ic) **manipulation:** a manipulative technique performed by osteopaths that includes deep tissue massage and manipulation of joints, spine, and soft tissues. Doctors of osteopathic medicine (D.O.s) are fully trained and licensed medical physicians.

qi gong (chee GUNG): a Chinese system that combines movement, meditation, and breathing techniques and allegedly cures illness by enhancing the flow of "qi" energy within the body.

reflexology: a technique that applies pressure or massage on areas of the hands or feet to allegedly cure disease or relieve pain in other areas of the body; sometimes called *zone therapy.*

therapeutic touch: a technique of passing hands over a patient to purportedly identify energy imbalances and transfer healing power from therapist to patient; also called *laying on of hands.*

traditional Chinese medicine (TCM): an approach to medical care based on the concept that illness can be cured by enhancing the flow of "qi" energy within a person's body. Treatments may include herbal therapies, physical exercises, meditation, acupuncture, and remedial massage.

natural healing powers of the body and may include special diets or fasting, herbal remedies and other dietary supplements, **acupuncture**, homeopathy, massage, and various other interventions.

- **Homeopathic medicine** theorizes that "like cures like," meaning that a substance that causes a particular set of symptoms can be used to cure a disease that has similar symptoms. Homeopathic remedies are generally diluted to such an extent that the original substance essentially is no longer present. Homeopaths believe that these remedies have powerful healing effects because the water structure is somehow altered during the dilution process.

- **Traditional Chinese medicine (TCM)** includes a large number of folk practices that originated in China. TCM promotes

Biofeedback training is a stress reduction and relaxation technique.

the theory that the body has pathways (called *meridians*) that conduct energy (called *qi*; pronounced *chee*). The interrupted flow of qi is believed to cause illness. TCM practices allegedly improve the flow of qi and include acupuncture, **qi gong**, herbal remedies, dietary practices, and massage.

What is the theory that underlies mind-body interventions?

Mind-body therapies attempt to improve a person's sense of psychological or spiritual well-being despite the presence of illness. The treatments are also used in the hope of reducing stress, dealing with pain, or lowering blood pressure. Some of these therapies have been incorporated into mainstream medicine for stress reduction or relaxation. For example, **biofeedback** training, in which individuals learn to monitor skin temperature, muscle tension, or brain wave activity while practicing relaxation techniques, is frequently taught by behavioral medicine specialists to help patients reduce stress or anxiety. Other techniques to reduce stress and promote relaxation include **meditation**, art and music therapy, and prayer.

The clinical applications of other mind-body therapies are far more questionable. An example is guided **imagery**, in which a person tries to reverse the disease process (for example, shrink a tumor) by using mental pictures. Another example is the use of **faith healing** to cure disease in place of proven conventional treatments.

Which alternative practices involve physical manipulation, and how do they work?

Manipulative interventions include physical touch, forceful movement of different parts of the body, and the application of pressure. Some practitioners maintain that special energy fields are also manipulated during the physical treatment and that proper energy flow induces healing. The most popular practices include the following:

- **Chiropractic** theory proposes that keeping the nervous system free from obstruction allows the body to heal itself because the process of healing is conducted via the spinal cord and nerves. The main treatment is the "adjust-

ment," a manual manipulation that is said to correct a "subluxation" (a pinched nerve or misaligned vertebra) and restore the body's natural healing ability. Although spinal manipulation has mainly been found to be helpful for improving back pain, most chiropractors still assert that chiropractic can cure disease rather than simply relieve symptoms.[5] For example, many promote spinal manipulation to treat infectious diseases, prevent cancer, and regulate menstrual periods, even though the nervous system and spinal alignment do not play roles in the pathology of these conditions.

- **Massage therapy** is the manipulation of muscle and connective tissue to improve muscle function, reduce pain, or promote relaxation. Massage therapists may also apply heat or cold and give advice about exercises that may improve muscle tone and range of motion. Massage is often integrated into conventional physical therapy, although some massage therapists may incorrectly suggest that massage is a valid treatment for a wide range of medical conditions.

What are the alleged effects of "energy" therapies?

Two categories of therapies involve the alleged curative power of "energy":

- **Biofield therapies** are said to influence the energy that surrounds or pervades the human body. Proponents claim that an energy therapy can strengthen or restore a person's "energy flow" and induce healing. Acupuncture, qi gong, and **therapeutic touch** can be included in this category. Note that CAM adherents often use the term *energy* unscientifically and that there is no objective evidence of this sort of energy flow.

- **Bioelectrical or bioelectromagnetic therapies** use electric or magnetic fields to allegedly promote healing; for example, magnets have been marketed with claims that they can improve circulation, reduce inflammation, and speed recovery from injuries.

Why do so many consumers choose to use CAM, given that the therapies have not been proved to be beneficial?

Surveys suggest that consumers perceive their visits to CAM therapists as far more pleasant than visits to conventional health practitioners. CAM therapists spend more time with patients, are more attentive, and use less invasive interventions.[6] Self-help measures are encouraged, so the consumer has more control over the treatment. The therapies appear to be more "natural" and to have fewer side effects. In addition, many consumers seem satisfied that these treatments "work." Possible explanations for "cures" include:

- A person may seem "cured" because of misdiagnosis; that is, the condition diagnosed by the CAM practitioner may not have actually existed.

- The condition may have been self-limiting, or it may have gone into temporary remission after the treatment.

455

- Undue credit may be inappropriately assigned to the CAM therapy when the improvement was actually due to a previous or concurrent conventional treatment.

- The placebo effect may have had an influence on the course of disease.

The central question remains: Do the CAM therapies merely make people *feel* better, or do they really *get* better? This question can be answered only by well-controlled research studies.

Are any potential dangers associated with the use of CAM?

One of the attractions of alternative therapies is the assumption that they are safe. Recall, however, the concerns associated with the use of herbal products discussed in Chapter 15, which include the possible toxicity of herbal ingredients, product contamination or adulteration, and interactions with conventional medications. Between 1990 and 1999, the FDA recalled more than 100 dietary supplements due to hazards associated with their use.[7]

Another concern is that use of CAM therapies may delay the use of reliable treatments that have demonstrable benefits. Various reports have described how people with treatable medical conditions suffered permanent disability or death when they were misdiagnosed or improperly treated by CAM practitioners. For example, a rare but well-known risk of spinal cord injury or stroke is associated with a type of cervical manipulation performed by chiropractors.[8] Unfortunately, because most CAM therapies are not regulated or monitored, there are no accurate estimates of their adverse effects.

What should health practitioners do if they think their patients are using CAM?

Health practitioners should be aware when their patients are using CAM therapies that may have consequences for the course of their disease and its treatment. It is important to routinely inquire about the use of CAM therapies and to educate patients about the hazards of postponing or stopping conventional treatment. Patients should also be told about potential interactions between conventional treatments and CAM therapies. Some patients may want to learn about differences between evidence-based medical practices and untested CAM theories and may be interested in the **integrative medicine** options available.

All alternative therapies have one characteristic in common: their effectiveness is, for the most part, unproven. As mentioned previously, patients often choose alternative therapies because of their positive interactions with CAM practitioners. Empathizing with patients may go a long way toward winning their trust and improving their compliance with conventional therapy. In addition, health practitioners need to regularly update their knowledge about unconventional practices, using reliable, objective resources, so that they can knowledgeably discuss these options with patients.

Notes

1. P. M. Barnes and E. Powell-Griner, Complementary and alternative medicine use among adults: United States, 2002, *Advance Data from Vital and Health Statistics* 343 (2004): 1–19.

2. J. D. Berman and S. E. Straus, Implementing a research agenda for complementary and alternative medicine, *Annual Review of Medicine* 55 (2004): 239–254.

3. E. Ernst, The role of complementary and alternative medicine, *British Medical Journal* 321 (2000): 1133–1135.

4. B. Barzansky, H. S. Jonas, and S. I. Etzel, Educational programs in US medical schools, 1999–2000, *Journal of the American Medical Association* 284 (2000): 1114–1120.

5. American Medical Association, *Alternative Medicine (Report 12 of the Council on Scientific Affairs, A-97)* (American Medical Association, 1997), www.ama-assn.org/ama/pub/category/13638.html, site visited September 23, 2006.

6. B. Barrett and coauthors, What complementary and alternative medicine practitioners say about health and health care, *Annals of Family Medicine* 2 (2004):253–259; American Medical Association, 1997.

7. Berman and Straus, 2004.

8. W.- L. Chen and coauthors, Vertebral artery dissection and cerebellar infarction following chiropractic manipulation, *Emergency Medical Journal* 23 (2006): e1.doi:10.1136/emj.2004.015636; W. S. Smith and coauthors, Spinal manipulative therapy is an independent risk factor for vertebral artery dissection, *Neurology* 13 (2003):1424–1428; R. Dziewas and coauthors, Cervical artery dissection—clinical features, risk factors, therapy and outcome in 126 patients, *Journal of Neurology* 250 (2003): 1179–1184.

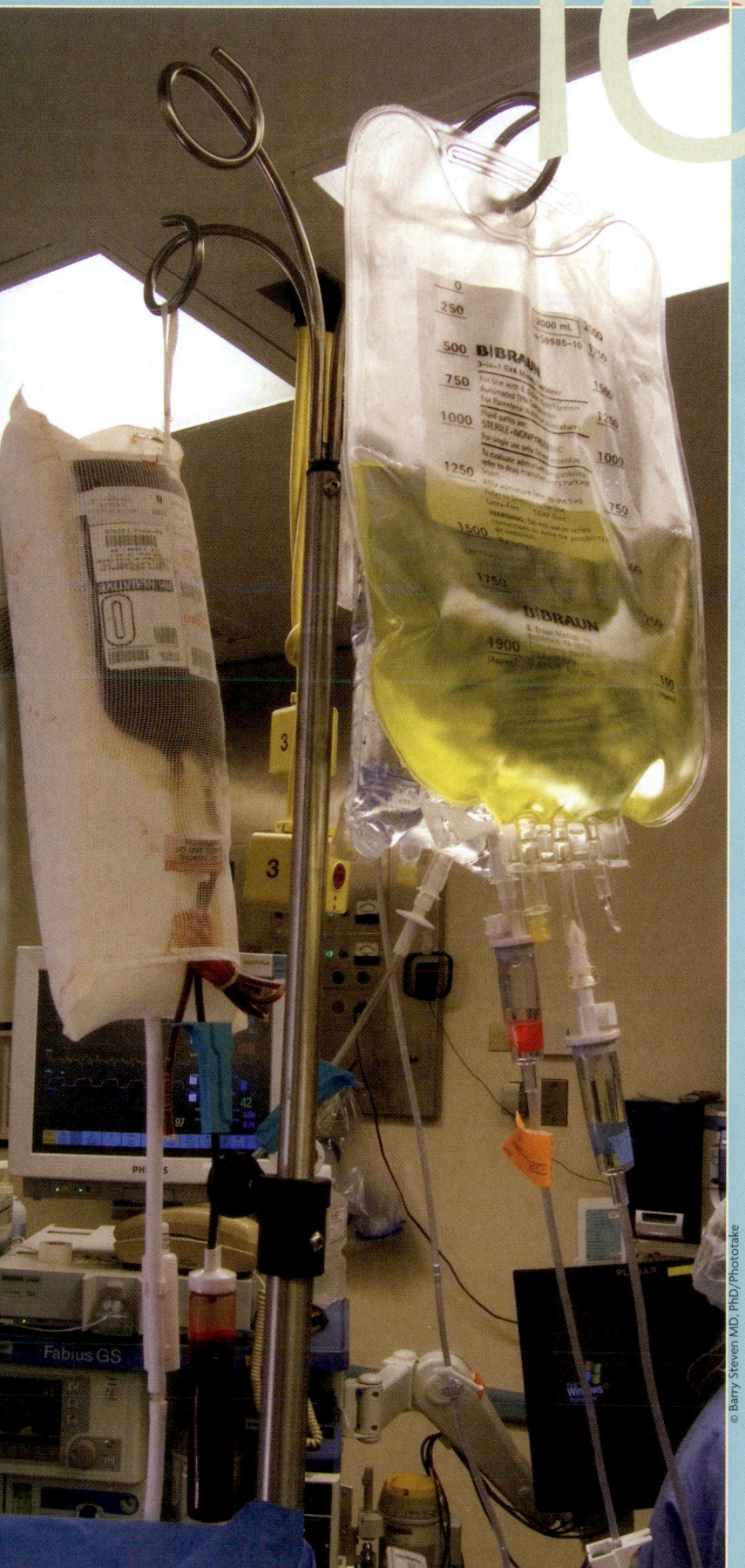

16

Specialized Nutrition Support: Enteral and Parenteral Nutrition

Patients are often too sick to obtain the energy and nutrients they need by consuming foods. Furthermore, some illnesses may interfere with eating, digestion, or absorption to such a degree that conventional foods cannot supply the necessary nutrients. In such cases, **nutrition support**, the delivery of formulated nutrients, can meet a patient's nutritional needs. **Enteral nutrition** provides nutrients using the gastrointestinal (GI) tract. Enteral nutrition includes oral diets or supplements but often refers to the use of tube feedings, which supply nutrients directly to the stomach or intestine via a thin, flexible tube. **Parenteral nutrition** provides nutrients intravenously to patients who do not have adequate gastrointestinal function to handle enteral feedings. If the GI tract remains functional, enteral nutrition support is preferred, partly to avoid the expense and complications associated with intravenous feedings and partly to preserve healthy GI function. Figure 16–1 summarizes the decision-making process for selecting the most appropriate feeding method.

Enteral Nutrition Support

If gastrointestinal function is normal and a poor appetite is the primary nutrition problem, enteral formulas can be provided as an oral supplement to the usual diet. If patients cannot consume enough food or drink enough formula to meet nutrient needs, tube feedings can deliver the required nutrients.

Enteral Formulas

A huge number of enteral formulas that are designed to meet a variety of medical and nutritional needs are on the market; some examples are listed in Appendix G. Formulas can be used alone or given along with other foods. Many formulas can supply all of

FIGURE 16–1

Selecting a Feeding Route

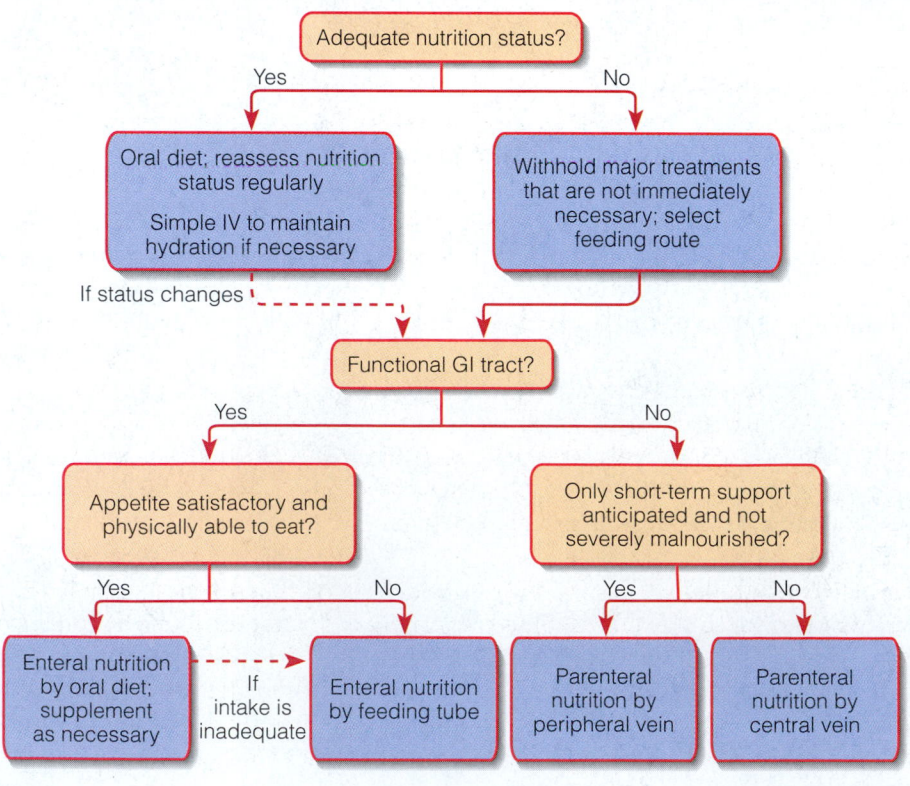

nutrition support: the delivery of formulated nutrients by feeding tube or intravenous infusion.

enteral (EN-ter-al) **nutrition:** the provision of nutrients using the GI tract, including the use of tube feedings and oral diets.

parenteral (par-EN-ter-al) **nutrition:** the intravenous provision of nutrients that bypasses the GI tract.
   par = beside
   entero = intestine

an individual's nutrient requirements when consumed in sufficient volume. Nutritionally complete formulas are essential for a patient who is using a tube feeding or oral liquid diet for more than a few days.

Types of Formulas Formulas are typically classified according to their macronutrient composition. ▪ The main types of formulas and their nutrient composition are as follows:

▪ Reminder: The *macronutrients* are carbohydrates, fats, and proteins.

- **Standard formulas** are used for individuals who can digest and absorb nutrients without difficulty. They contain intact proteins isolated from milk or soybeans or a combination of purified proteins. Carbohydrate sources include modified starches, glucose polymers (such as maltodextrin), and sugars. A few formulas, called **blenderized formulas**, are made from whole foods and derive their protein primarily from pureed meat or poultry.

- **Hydrolyzed formulas** are provided to patients who have compromised digestive or absorptive functions. The formulas contain macronutrients that have been partially or fully broken down to fragments that require little (if any) digestion before absorption. Hydrolyzed formulas are often low in fat and may contain **medium-chain triglycerides (MCT)** to ease digestion and absorption. These formulas are generally lactose-free and result in minimal fecal output.

- **Disease-specific formulas** are designed to meet the specific nutrient needs of patients with particular illnesses. Products have been developed for liver, kidney, and lung diseases; glucose intolerance; and metabolic stress. Disease-specific formulas are generally expensive, and their effectiveness is controversial.

- **Modular formulas** usually contain only one or two macronutrients and are used to enhance other formulas. Several modular formulas may be combined with liquid vitamin and mineral preparations to create individualized formulas for patients with unique nutrient needs.

Nutrient and Energy Densities The percentages of protein, carbohydrate, and fat vary considerably in enteral formulas. Protein content ranges from 8 to 29 percent of total kcalories. Note that protein needs are high in patients with severe metabolic stress, but protein restrictions are necessary in patients with renal failure. Carbohydrate and fat provide most of the energy in enteral formulas; standard formulas often provide 40 to 50 percent of kcalories from carbohydrate and 30 to 45 percent from fat.

Fiber content often influences the selection of an enteral formula. Fiber may be helpful for normalizing intestinal function, treating diarrhea and constipation, and maintaining blood glucose control. Fiber-containing formulas may be avoided, however, during acute intestinal conditions, pancreatitis, or procedures involving the intestines.

The energy density of enteral formulas ranges from 0.5 to 2.0 kcalories per milliliter of fluid. Standard formulas provide 1.0 to 1.2 kcalories per milliliter and are appropriate for patients with average fluid requirements. Formulas that have higher energy densities can meet energy and nutrient needs in a smaller volume of fluid and thus benefit patients with high nutrient needs or fluid restrictions.

Osmolality Osmolality refers to the osmotic property of a solution, that is, a solution's tendency to shift from one fluid compartment to another across a semipermeable membrane. The osmolality depends on a solution's concentrations of molecules and ionic particles. An enteral formula with an osmolality similar to that of blood serum (about 300 milliosmoles per kilogram) is an **isotonic formula**, whereas a **hypertonic formula** has an osmolality greater than that of blood serum.

standard formulas: general-purpose enteral formulas that contain mostly intact proteins and polysaccharides; also called **polymeric formulas**.

blenderized formulas: enteral formulas that are made by blenderizing whole foods.

hydrolyzed formulas: enteral formulas that contain macronutrients that have been partially or fully hydrolyzed; also called **monomeric** or **elemental formulas**.

medium-chain triglycerides (MCT): triglycerides that contain fatty acids that are 8 to 10 carbons in length. MCT do not require digestion and can be absorbed in the absence of lipase or bile.

disease-specific formulas: enteral formulas designed to meet the nutrient needs of patients with specific illnesses.

modular formulas: enteral formulas that contain only one or two macronutrients; used to enhance other formulas, meet specific nutrient needs, or create individualized formulas for people with unique needs.

osmolality (OZ-moe-LAL-ih-tee): the osmotic property of a solution, based on its concentrations of molecules and ionic particles. Osmolality is expressed as milliosmoles (mOsm) per kilogram.

isotonic formula: a formula with an osmolality similar to that of blood serum (300 mOsm/kg).

hypertonic formula: a formula with an osmolality greater than that of blood serum.

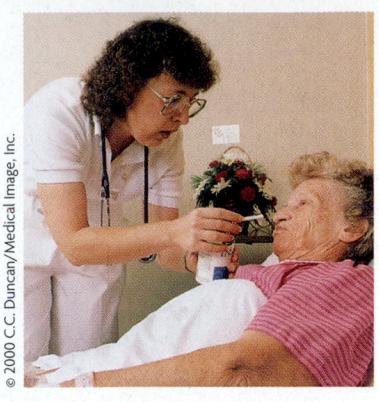

Patients can drink enteral formulas when they are unable to consume enough food from a conventional diet.

Most enteral formulas have osmolalities between 300 and 700 milliosmoles per kilogram; generally, hydrolyzed formulas and nutrient-dense formulas have higher osmolalities than standard formulas. Most people are able to tolerate both isotonic and hypertonic feedings without difficulty.[1] When medications are infused along with enteral feedings, however, the osmotic load increases substantially and may contribute to the diarrhea experienced by many tube-fed patients.

Enteral Nutrition in Medical Care

A person with a functioning GI tract who cannot meet nutrient needs with conventional foods alone may be a candidate for enteral nutrition support. Enteral feedings are preferred over intravenous feedings because they help to stimulate or maintain gut function, cause fewer complications, and are less costly.[2] Similarly, oral feedings are preferred to tube feedings if the person is able to drink enteral formulas, since drinking the formulas eliminates the stress, complications, and expense associated with tube feedings.

Oral Use of Enteral Formulas Enteral formulas can fully meet the nutrient needs of people who can consume only liquids or who require hydrolyzed nutrients. More often, enteral formulas are used to supplement conventional diets when individuals cannot consume enough food to meet their needs. Enteral formulas provide a reliable source of nutrients and can add energy and protein to the diets of malnourished patients. Patients who are weak or debilitated may also find it easier to manage formulas than meals.

When a patient drinks a formula, taste becomes an important consideration. Allowing patients to sample different products and flavors and select the ones they prefer helps to promote acceptance. The "How To" offers additional suggestions for helping patients accept and enjoy oral formulas.

Tube Feedings An individual with a functional GI tract who is unable to consume enough food or formula orally may need to be fed via a feeding tube. A tube feeding

How To

HELP PATIENTS ACCEPT ORAL FORMULAS

People using enteral formulas are often quite ill and have poor appetites. Even when a person enjoys a formula, the taste can become monotonous in time. Hydrolyzed formulas are usually less palatable than standard formulas, and patients may find them difficult to drink. Health professionals can help by trying these suggestions:

- Let the patient sample different formulas that are appropriate for his or her needs, and use only those that the patient enjoys.

- Serve formulas attractively and remind patients to drink them. Formulas offered in a glass on an attractive plate may be more appealing than those served from a can with an unfamiliar name.

- If a patient finds the smell of a formula unappealing, it may help to cover the top of the glass with plastic wrap or a lid, leaving just enough room for a straw.

- Provide easy access. Keep the formula close to the patient's bed where it can be reached with little effort and within sight so that the patient is reminded to drink it. Patients who are very ill may lack the motivation to reach for the formula, let alone drink it.

- Try keeping the formula in an ice bath so that it will be cool and refreshing when the patient drinks it. Check with the patient to make sure the colder temperature is suitable.

- For patients with little appetite, offer the formula in smaller amounts that are easy to tolerate, and serve it more frequently during the day.

- If the patient stops enjoying the formula, recommend different flavors or try other formulas.

delivers a nutritionally complete formula directly to the stomach or intestine. Candidates for tube feedings include:

- People with severe swallowing difficulties

- People who have little or no appetite for extended periods, especially if malnourished

- People with gastrointestinal obstructions, some types of fistulas, or impaired motility in the upper GI tract

- People who have undergone intestinal resections and are beginning enteral feedings

- People who are mentally incapacitated due to confusion, dementia, or neurological difficulties

- People in a coma

- People with extremely high nutrient requirements

- People on mechanical ventilators

Feeding Routes The feeding route chosen depends on the medical condition, expected duration of tube feeding, and potential complications of a particular route. Figure 16–2 illustrates the main feeding routes, and the Glossary on p. 463 describes each route. Table 16–1 (p. 462) summarizes the advantages and disadvantages of each route.

When a patient is expected to be tube-fed for less than four weeks, a **nasogastric** or **nasoenteric** route is generally chosen; ■ for these routes, the feeding tube is passed into the GI tract via the nose. The patient is frequently awake during **transnasal** (through the nose) placement of a feeding tube. While the patient is in a slightly

■ The final location of the feeding tube determines how the feeding route is classified.

FIGURE 16–2

Tube Feeding Routes

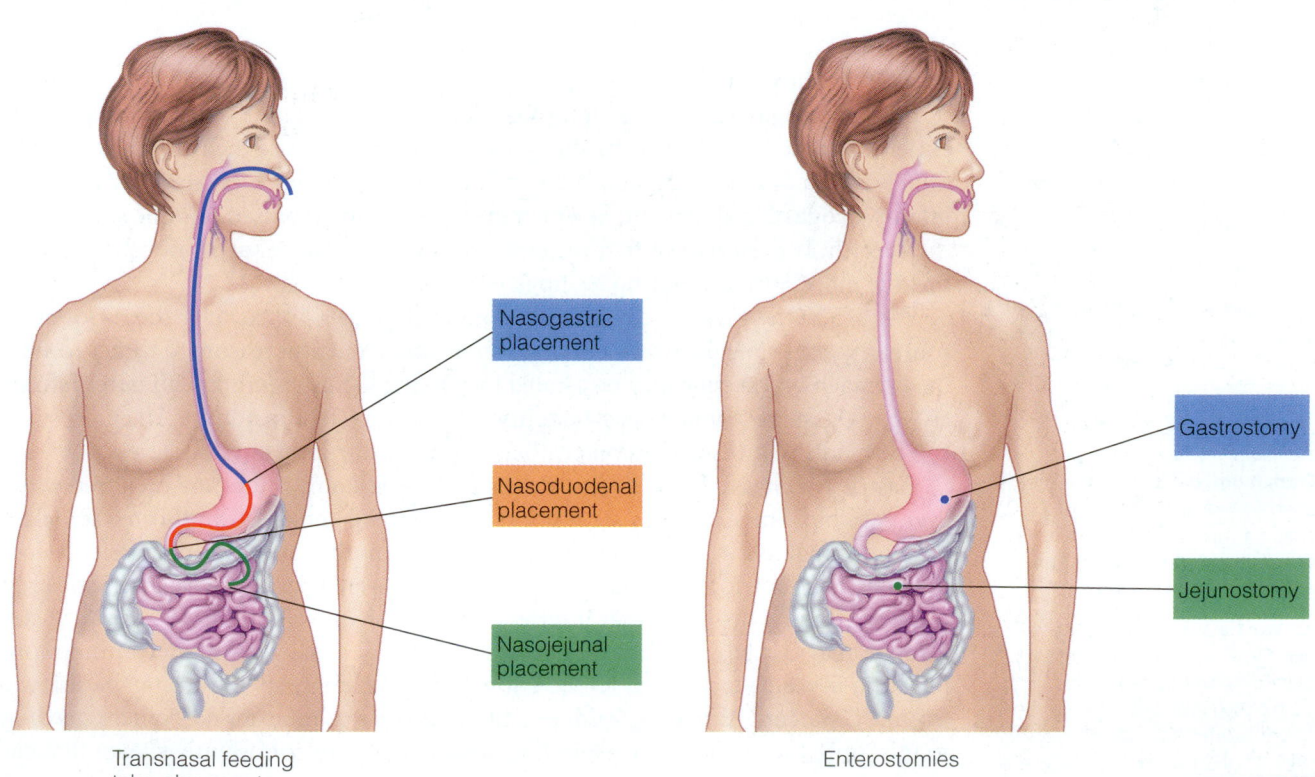

Nasogastric placement

Nasoduodenal placement

Nasojejunal placement

Transnasal feeding tube placements

Gastrostomy

Jejunostomy

Enterostomies

TABLE 16–1

Comparison of Tube Feeding Routes[a]

Insertion Method and Feeding Site	Advantages	Disadvantages
Transnasal	Does not require surgery or incisions for placement.	Easy to remove by disoriented patients, long-term use may irritate the nasal passages, throat, and esophagus.
Nasogastric	Easiest to insert and confirm placement; feedings can often be given intermittently and without an infusion pump.	Highest risk of aspiration in compromised patients.
Nasoduodenal and nasojejunal	Lower risk of aspiration in compromised patients; allow for enteral nutrition earlier than gastric feedings following severe stress; may allow for enteral feeding when obstruction, fistulas, or other medical conditions prevent gastric feeding.	More difficult to insert and confirm placement; feedings require an infusion pump for administration; may take longer to reach nutrition goals.
Tube enterostomies	Allow lower esophageal sphincter to remain closed, reducing the risk of aspiration; more comfortable than transnasal insertion for long-term use; site is not visible under clothing.	May require general anesthesia for insertion; require incisions; greater risk of complications from the insertion procedure; greater risk of infection; may cause skin irritation around the insertion site.
Gastrostomy	Feedings can often be given intermittently and without a pump; easier to insert than a jejunostomy.	Moderate risk of aspiration in high-risk patients.
Jejunostomy	Lowest risk of aspiration; allows for enteral nutrition earlier following severe stress; may allow for enteral feeding when obstructions, fistulas, or medical conditions prevent gastric feeding.	Most difficult to insert; feedings require an infusion pump for administration; may take longer to reach nutrition goals.

[a] Relative to other tube feeding routes. The actual advantages and disadvantages of different insertion procedures depend on the person's medical condition.

■ Aspiration risk is high in patients with esophageal disorders, neurological diseases, and conditions that reduce consciousness or cause dementia.

aspiration: drawing in by suction or breathing; a common complication of enteral feedings in which foreign material enters the lungs, often from reflux of stomach contents.

French units: units of measure used to indicate the size of a feeding tube's outer diameter. One French unit equals 1/3 millimeter.

gastric decompression: the removal of the stomach contents (including swallowed saliva, stomach secretions, and gas) of patients who have motility disorders or obstructions that prevent stomach emptying.

upright position with head tilted, the tube is inserted into a nostril and passed into the stomach (**nasogastric**), duodenum (**nasoduodenal**), or jejunum (**nasojejunal**). If the patient is awake and alert, he or she can swallow water to ease the tube's passage. The final position of the feeding tube tip is verified by abdominal X-ray or other means. In infants, **orogastric** placement, in which the feeding tube is passed into the stomach via the mouth, is preferred over transnasal routes; this placement allows the infant to breathe more normally during feedings.

When a patient will be tube-fed for longer than four weeks or if the nasoenteric route is inaccessible due to an obstruction or other medical reasons, a direct route to the stomach or intestine may be created by passing the tube through an **enterostomy**, an opening in the stomach (**gastrostomy**) or jejunum (**jejunostomy**). An enterostomy can be made by surgical incision or nonsurgically using local anesthesia.

Gastric feedings, such as the nasogastric and gastrostomy routes, are preferred whenever possible. These feedings are more easily tolerated and less complicated to deliver than intestinal feedings because the stomach controls the rate at which nutrients enter the intestine. Gastric feedings are not possible, however, if patients have gastric obstructions or motility disorders that interfere with the stomach's ability to empty. Gastric feedings may also be a problem for patients at high risk of **aspiration**, ■ a common complication in which formula or GI secretions enter the lungs, often from the backflow of stomach contents. Although health practitioners often administer nasoenteric feedings to minimize the possibility of aspiration, studies have not consistently shown that gastric feedings are associated with increased aspiration risk.[3]

For each type of tube placement, the terms are listed in order from the upper to lower organs of the digestive system.

transnasal: through the nose. A **transnasal feeding tube** is one that is inserted through the nose.

nasogastric (NG): tube is placed into the stomach via the nose.

nasoenteric: tube is placed into the GI tract via the nose. (*Nasoenteric feedings* usually refer to *nasoduodenal* and *nasojejunal* feedings.)

nasoduodenal (ND): tube is placed into the duodenum via the nose.

nasojejunal (NJ): tube is placed into the jejunum via the nose.

orogastric: tube is placed into the stomach via the mouth. This method is often used to feed infants because a nasogastric tube can hinder the infant's breathing.

enterostomy (EN-ter-AH-stoe-mee): an opening into the GI tract through which a feeding tube can be passed.

gastrostomy (gah-STRAH-stoe-mee): an opening into the stomach through which a feeding tube can be passed. A nonsurgical technique for creating a gastrostomy under local anesthesia is called *percutaneous endoscopic gastrostomy (PEG).*

jejunostomy (JE-ju-NAH-stoe-mee): an opening in the jejunum through which a feeding tube can be passed. A nonsurgical technique for creating a jejunostomy is called *percutaneous endoscopic jejunostomy (PEJ).* The tube can either be guided into the jejunum via a gastrostomy or passed directly into the jejunum *(direct PEJ).*

Feeding Tubes Feeding tubes are soft and flexible and come in a variety of lengths and diameters. The tube selected largely depends on the patient's age and size, the feeding route, and the formula's viscosity. Once the appropriate length is determined, the tube selected is often the smallest tube through which the formula will flow without clogging.

The outer diameter of a tube is measured in **French units**, in which each unit equals ⅓ millimeter; thus a "12 French" feeding tube has a 4-millimeter diameter. ■ The inner diameter depends on the thickness of the tubing material. Double-lumen tubes are also available; these allow one tube to be used for both intestinal feedings and **gastric decompression,** a procedure in which the stomach contents of patients with motility disorders is removed by suction.

■ 1 French = ⅓ mm
12 French = 12 × ⅓ mm
= 4 mm

Formula Selection The formula is selected after careful assessment of the patient's age, medical problems, nutritional status, and ability to digest and absorb nutrients; some of the considerations are shown in Figure 16–3 (p. 464). Generally, the best formula is one that meets the patient's medical and nutrient needs with the lowest risk of complications and the lowest cost. The vast majority of patients use standard formulas. A person with a functional, but impaired, GI tract may require a hydrolyzed formula. Some nutrition-related factors that influence formula selection include:

- **A patient's energy, protein, and fluid requirements.** Nutrient needs must be met using the volume of formula a patient can tolerate. If fluids need to be restricted, the formula should be able to deliver the necessary nutrients in the volume prescribed.

- **The need for fiber modifications.** The choice of formulas is narrowed if fiber intake needs to be low or high.

- **Individual tolerances (food allergies and sensitivities).** Most formulas are lactose-free because many patients who need enteral formulas have some degree of lactose intolerance. Many formulas are also gluten-free and can accommodate the needs of individuals with celiac disease (gluten sensitivity).

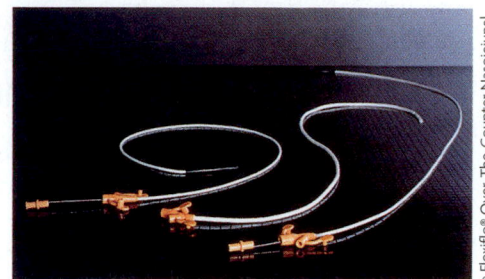

© Flexiflo® Over-The-Counter Nasojejunal Feeding Tube, Courtesy of Ross Products Div., Abbott Laboratories, Columbus OH

Feeding tubes come in various lengths and diameters. The thin wires protruding from the end of the feeding tubes are stylets, which stiffen the tube to ease insertion and are discarded thereafter. The Y-connector (shown here in orange) provides a port for administering water or medications without disrupting the feeding.

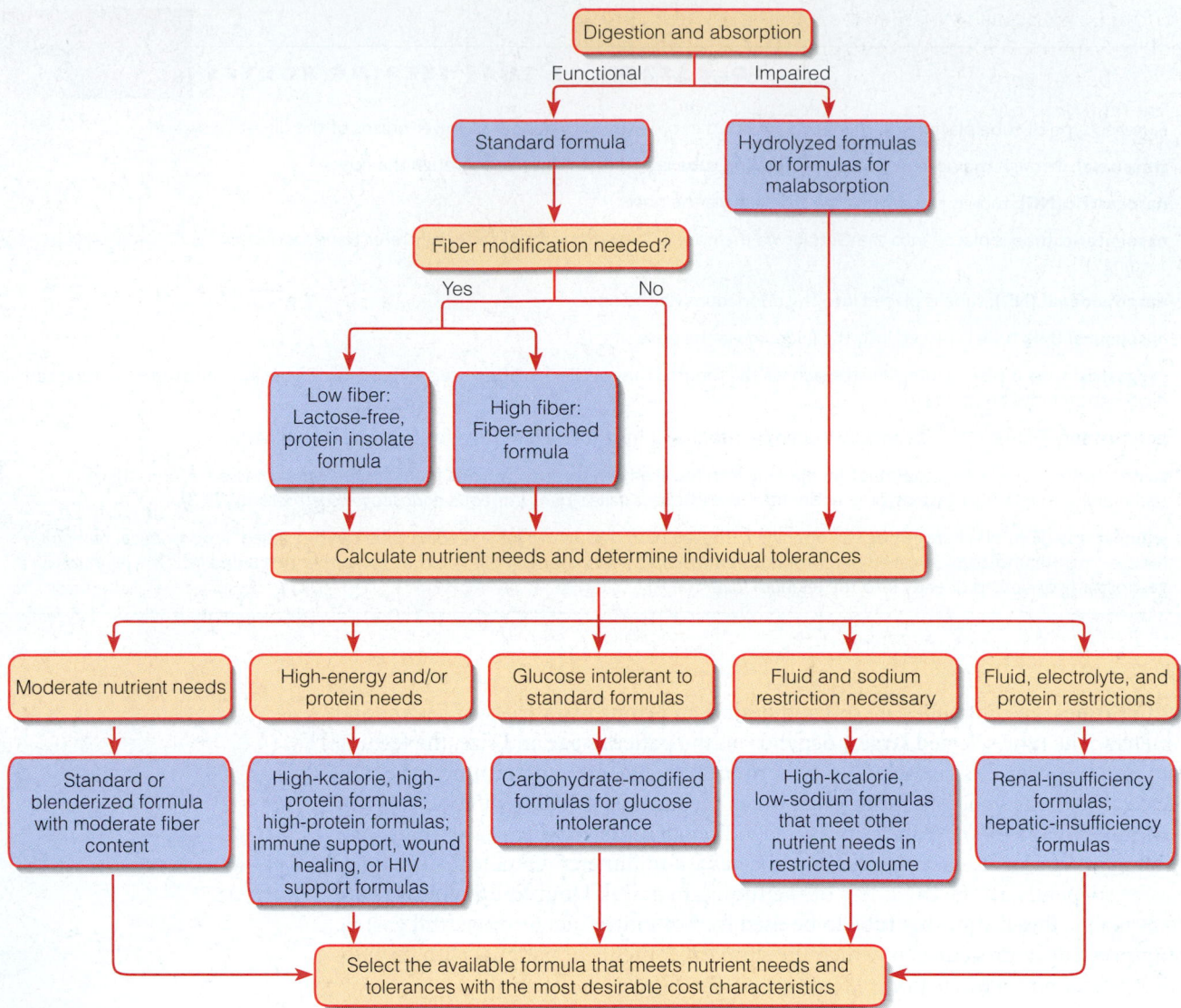

FIGURE 16–3

Selecting a Formula

Health care facilities stock a limited number of formulas, so formula selection is limited by availability. Initially, the dietitian or physician may make an educated guess as to the best formula, based on the criteria previously mentioned. The decision can be reappraised based on the patient's response to the formula.

Administration of Tube Feedings

After the feeding route and formula have been selected, attention is given to delivering the formula. The methods of tube feeding administration vary somewhat from one health care facility to the next. The procedures presented in the following sections are suggested guidelines.

Safe Handling Individuals who are ill or malnourished often have suppressed immune systems, making them vulnerable to infection from foodborne illness. To prevent contamination, the personnel involved in preparing and delivering formulas should work in clean environments, using clean equipment and clean hands. The foodservice department or pharmacy most often assumes responsibility for preparing formulas.

Formulas are available in **open feeding systems** and **closed feeding systems**. With an *open feeding system*, the formula needs to be transferred from its original packaging to a feeding container. Examples include formulas that are packaged in cans or bottles, concentrates that need to be diluted, and powders that require reconstitution. In a *closed feeding system*, the formula is prepackaged in a container that can be connected directly to a feeding tube. Closed systems are less likely to become contaminated, require less nursing time, and can hang for longer periods of time than open systems. Although closed systems cost more initially, they may be less expensive in the long run by preventing bacterial contamination and thus avoiding the costs of treating infections.

Safety Guidelines After the formula reaches the nursing station, the nursing staff assumes responsibility for its safe handling. Hands should be carefully washed before handling formulas and feeding containers. Some facilities require that nonsterile gloves be worn whenever formulas are handled. The following steps can reduce the risk of formula contamination when using open feeding systems:

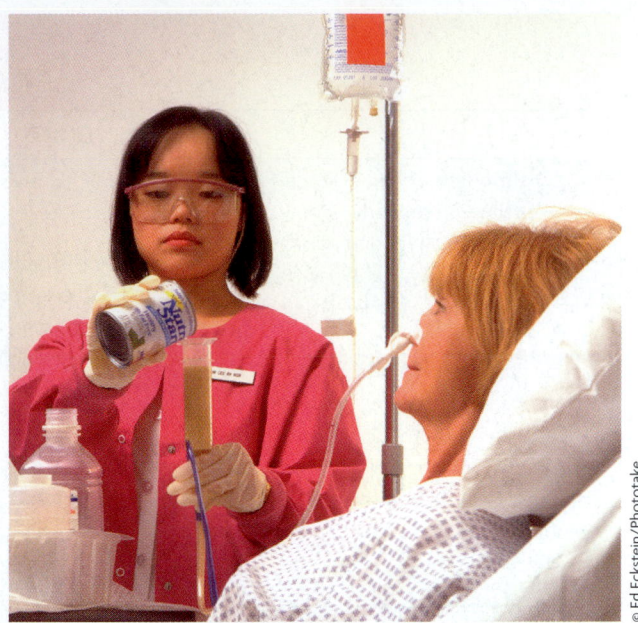

In an open feeding system, the formula is transferred from its original packaging to a feeding container.

- Before opening a can of formula, clean the can opener and the lid. If you do not use the entire can at one feeding, label the can with the time it was opened.

- Store opened cans or mixed formulas in clean, closed containers. Refrigerate the unused portion of formula promptly.

- Discard unlabeled or improperly labeled containers and all opened containers of formula that are not used within 24 hours.

- Hang no more than an 8- to 12-hour supply of formula when using an open feeding system. Discard any formula that remains, rinse out the feeding bag and tubing, and add fresh formula to the feeding bag. Use a new feeding container and tubing (except for the feeding tube itself) every 24 hours.

Closed feeding systems should hang for no longer than 24 to 48 hours; contamination is more likely with longer time periods.

Initiating a Tube Feeding Before starting a tube feeding, health practitioners can ease fears by discussing the procedure with the patient and family members, who may feel anxious about using a feeding tube. The discussion should address the reasons why tube feeding is appropriate and the benefits and risks of the procedure. The "How To" on p. 466 offers suggestions that may help ease the concerns of patients who may benefit from tube feeding.

Serious complications can develop if a transnasal tube is accidentally inserted into the respiratory tract or if formula or GI secretions are aspirated into the lungs. To minimize the risk of incorrect tube placement, clinicians may use X-rays to verify the position of the feeding tube before a feeding is initiated. After the tube's placement has been confirmed, the health practioner secures the tube to the patient's nose and cheek with tape and monitors the position of the tubing throughout the day. Another technique is to test the pH of a sample of bodily fluid drawn into the feeding tube; recall that the pH of stomach fluid is much lower than the pH of fluid obtained from the intestine or respiratory tract. ■

To reduce the risk of aspiration, the patient's upper body is elevated to at least a 30- to 45-degree angle during the feeding and for 30 minutes after the feeding whenever possible. The addition of blue food coloring to formula is sometimes suggested as

■ A gastric sample usually has a pH between 1 and 4. An intestinal sample should have a pH between 6 and 7.

open feeding systems: delivery systems that require formula to be transferred from the original packaging to feeding containers before being administered through feeding tubes.

closed feeding systems: delivery systems in which formula comes prepackaged in containers that are ready to be attached to feeding tubes for administration.

HELP PATIENTS COPE WITH TUBE FEEDINGS

The thought of being "force-fed" is frightening to many people. Some may envision thick feeding tubes or fear that the procedure will be painful. Others may associate tube feedings with disabling injury or irreversible illness. Patients may be less apprehensive once they understand the insertion procedure, the expected duration of the tube feeding, and the strategic role that nutrition plays in recovery from disease. The pointers that follow can help health practitioners prepare patients for transnasal tube feedings:

■ Allow the patient to see and touch the feeding tube. Seeing firsthand that the tube is soft and narrow (only about half the diameter of a pencil) often alleviates anxiety.

■ Show the patient how the feeding apparatus is attached to the feeding tube, and explain how the feeding will work. For young children, use dolls or stuffed toys to demonstrate tube insertion and feeding procedures.

■ Explain that the patient remains fully alert during the procedure and helps pass the tube by swallowing. A numbing solution sprayed on the back of the throat minimizes discomfort and prevents gagging during the procedure.

■ Tell the patient that once the tube has been inserted, most people become accustomed to its presence within a few hours. In most cases, the patient can easily swallow foods and liquids with the tube in place. If permitted, favorite foods or beverages can still be enjoyed.

■ Assure the patient that the tube feeding will be temporary, if such assurance is appropriate.

A tube feeding may be frightening for some patients, but others may be relieved to know that they can receive sound nutrition without any effort. As they feel better and begin to eat again, the volume of the feeding can be reduced and then discontinued when oral intake is adequate.

Tube feedings may cause some patients to feel that they have lost control over an important aspect of their lives. They may also feel self-conscious about how the feeding tube looks or feel awkward when moving around with the equipment. A few measures can help:

■ Involve patients in the decision-making and care process whenever possible. Patients can help to arrange their daily feeding schedules and can perform some of the feeding procedures themselves.

■ Show patients how to manipulate the feeding equipment so that they can get out of bed and move around.

■ Encourage patients to maintain contact with friends and keep busy with the hobbies and activities they enjoy. This measure is especially important for children, teens, and those on long-term feedings.

When caring for infants and children, keep the developmental age of the child in mind and work with parents to ensure that appropriate feeding skills are mastered. Infants can be provided with a pacifier during feedings to help maintain the associations between sucking, swallowing, and fullness. When possible, the formula can be provided by bottle to an infant or by spoon to a child to further develop skills.

The more complex the procedure, the easier it becomes for health care professionals to focus on the procedure and disregard a patient's emotional response. No matter how many technicalities you have to keep in mind, remember to stay focused on the person receiving your care.

---

a means of identifying aspirated formula in lung secretions; however, this practice is now discouraged because several deaths have been attributed to its use.[4]

Formula Delivery A day's nutrient needs can be met by delivering relatively large amounts of formula several times per day (**intermittent feedings**) or smaller amounts continuously throughout the day (**continuous feedings**). A patient may also start on a continuous feeding and gradually transition to an intermittent feeding. Each method has specific uses, advantages, and disadvantages.

Intermittent feedings are best tolerated when they are delivered into the stomach (not the intestine). Generally, about 250 to 400 milliliters is delivered over 20 to 40 minutes using a gravity drip method or an infusion pump. The volume of formula required to meet a patient's nutrient needs is divided into several daily feedings. Because of the relatively high volume of formula delivered at a time, intermittent feedings may be difficult for some patients to tolerate, and the risk of aspiration may be higher than with continuous feedings. An advantage of intermittent feedings is that they are similar to the usual pattern of eating and allow the patient freedom of movement between meals.

PLAN A SCHEDULE FOR ADMINISTERING TUBE FEEDINGS

After selecting a formula that meets the patient's medical and nutrient needs, the clinician determines the volume of formula that meets those needs. Consider a patient who needs 2,000 kcalories daily and is using a standard formula that provides 1.0 kcalorie per milliliter. The total volume of formula required would be 2,000 milliliters per day:

$$x \text{ mL} \times 1.0 \text{ kcal/mL} = 2,000 \text{ kcal}$$

$$x \text{ mL} = \frac{2,000 \text{ kcal}}{1.0 \text{ kcal/mL}} = 2,000 \text{ mL}$$

If the patient is to receive intermittent feedings six times per day, he will need about 330 milliliters of formula at each feeding:

$$2,000 \text{ mL} \div 6 \text{ feedings} = 333 \text{ mL/feeding}$$

Alternatively, if he is to receive intermittent feedings eight times per day, he will need 250 milliliters (or about one can of ready-to-feed formula) at each feeding:

$$2,000 \text{ mL} \div 8 \text{ feedings} = 250 \text{ mL/feeding}$$

He will probably tolerate this volume of formula best if it is delivered over a 30-minute period at each feeding. If the patient is to receive the formula continuously over 24 hours, he will need about 85 milliliters of formula each hour:

$$2,000 \text{ mL} \div 24 \text{ hours} = 83 \text{ mL/hr}$$

Rapid delivery of a large volume of formula into the stomach (250 to 500 milliliters in less than 15 minutes) is called a **bolus feeding**. This type of feeding may be given every 3 to 4 hours using a syringe. Bolus feedings can cause abdominal discomfort, nausea, and cramping in some patients, especially when a feeding is initiated. The risk of aspiration is also greater than with other methods of feeding. For these reasons, bolus feedings are used only in patients who are not critically ill.

Continuous feedings are delivered slowly and at a constant rate over a period of 8 to 24 hours. This type of feeding is recommended for critically ill patients because delivering relatively small volumes at a time may reduce nausea, diarrhea, and the risk of aspiration. Continuous feedings are generally used in patients who receive intestinal feedings. An infusion pump is required to ensure accurate and steady flow rates; consequently, the feedings can limit the patient's freedom of movement and are also more costly.

Formula Volume and Strength Formula administration schedules vary among institutions, so protocols should be reviewed carefully before working with patients. In general, almost all patients can receive undiluted formula (either isotonic or hypertonic) at the start of a feeding.[5] Formulas, especially hypertonic formulas, are sometimes started slowly, and the volume is gradually increased. In rare cases, formulas may be diluted until tolerated by the patient, although this practice is discouraged.

Intermittent feedings may start at about 120 to 150 milliliters at the initial feeding and be increased by 50 to 100 milliliters at each feeding until the goal volume is reached. Continuous feedings may start at about 30 to 40 milliliters per hour and be raised by 30 milliliters per hour every 6 to 8 hours. For both intermittent and continuous feedings, the rate and amount of increase depend on the patient's tolerance to the formula. If the new rate is not tolerated, the rate of delivery progresses more slowly to give the person additional time to adapt. If a patient on an intermittent feeding cannot tolerate the feeding, a continuous feeding may be a better choice. The "How to" describes several ways to plan tube feeding schedules.

Checking Gastric Residuals When a patient receives a gastric feeding, the nurse regularly measures the **gastric residual volume** (the volume of formula remaining in the stomach after feeding) to ensure that the stomach is emptying properly. The gastric residual is measured by gently withdrawing the gastric contents through the feeding tube using a syringe, usually before each intermittent feeding and every 4 to 6 hours

intermittent feedings: delivery of about 250 to 400 mL of formula over 20 to 40 minutes.

continuous feedings: slow delivery of formula at a constant rate over an 8- to 24-hour period.

bolus (BOH-lus) **feeding:** delivery of about 250 to 500 mL of formula in less than 15 minutes.

gastric residual volume: the volume of formula remaining in the stomach from a previous feeding.

during continuous feedings. Although opinions vary, some experts recommend that an evaluation be conducted if the gastric residual exceeds 200 milliliters and that feedings be withheld if it exceeds 500 milliliters.[6] If the tendency to accumulate fluids persists, the physician may recommend intestinal feedings or begin drug therapy to stimulate gastric emptying.

Meeting Water Needs Attention must also be paid to the patient's fluid requirements. Many adults require about 2,000 milliliters (about 2 quarts) of water daily. ■ Fluids may be restricted in persons with kidney, liver, or heart disease. Additional water is required in those with fever, high urine output, diarrhea, excessive sweating, severe vomiting, fistula drainage, high-output ostomies, blood loss, or open wounds.

The water in formulas can meet a substantial portion of water needs. Standard formulas contain about 85 percent water, or about 850 milliliters of water per liter of formula. Nutrient-dense formulas contain about 69 to 72 percent water; exact amounts can be obtained from the product label or manufacturer's information sheet.

In addition to the water from formulas, water needs are partially met by flushes given via feeding tubes. ■ To prevent clogging, feeding tubes are routinely flushed with water before and after each bolus or intermittent feeding and about every 4 hours during continuous feedings. The water used for routine flushes should be included when estimating fluid intakes.

Transition to Table Foods Once the condition requiring a tube feeding resolves, the volume of formula is tapered off, and the patient gradually shifts to an oral diet. Sometimes the patient can begin the transition by drinking the same formula that is delivered by tube. Oral intake should supply about two-thirds of estimated nutrient needs before the tube feeding is discontinued.

■ To estimate fluid requirements in adults and children:
- Adults: allow 30 to 40 mL/kg; 30 mL/kg in older adults.
- Children: allow 50 to 60 mL/kg.
- Infants: allow 150 mL/kg.

■ ■ For intermittent or bolus feedings, feeding tubes should be flushed with 20 to 60 mL of water before and after formula delivery.
- For continuous feedings, feeding tubes should be flushed with 20 to 60 mL of water every 4 hours.

How To

ADMINISTER MEDICATIONS TO PATIENTS RECEIVING TUBE FEEDINGS

The pharmacist is your best resource for learning how and when medications can be administered via feeding tubes, especially when you are dealing with an unfamiliar medication. Check with the pharmacist to learn the following:

- Whether a particular medication is known to be incompatible with formulas.

- The proper timing of medication administration to avoid drug-nutrient interactions.

- For patients using intestinal feedings, whether a medication can be absorbed without exposure to stomach acid.

- Whether a liquid form of a medication is available, and if so, the appropriate dosage of the liquid form.

- If only tablets are available, whether the tablets can be crushed and mixed with water. Enteric-coated and sustained-release medications should not be crushed due to the potential for adverse effects.

In general, it is best to give medications by mouth instead of by tube whenever possible. In some cases, the injectable form of a medication may be the best option. For medications that must be given by feeding tube:

- Do not mix medications with enteral formulas. Do not mix medications together.

- Before administering medications, ensure that the feeding tube is placed correctly, that it is not clogged, and that the gastric residual is not excessive.

- Position the patient in a semiupright position (30 degrees or higher) to prevent aspiration.

- Flush the feeding tube with 15 to 30 milliliters of warm (body temperature) water before and after administering a medication. When more than one medication is administered, flush the tube with 5 milliliters of water after each medication is given.

- Use liquid forms of medications whenever possible. Dilute very viscous or hypertonic liquid medications with 10 to 30 milliliters of water before administering them through the feeding tube.

- If tablets are used, crush tablets to a fine powder and mix with 10 milliliters of warm water before administering.

Delivering Medications Through Feeding Tubes

Patients receiving tube feedings sometimes require one or more medications that need to be delivered through feeding tubes. Because medications can interact with enteral formulas in the same ways that they interact with foods, diet-drug interactions must be considered. Medications can also cause feeding tubes to clog. The "How To" provides some guidelines that may help to prevent complications.

Continuous feedings are ordinarily stopped during the administration of medication so that the components of enteral formulas do not interfere with the medication's absorption. The feeding is typically halted for 15 minutes before and 15 minutes after medication delivery. Some medications may require a longer formula-free interval; for example, feedings need to be stopped for one to two hours before and after administering phenytoin, a medication that controls seizures. In such cases, the delivery rate of formula needs to be increased so that the correct amount of formula can be delivered.

Complications of Tube Feeding

Complications are a frequent occurrence during tube feedings. Common complications include gastrointestinal problems, such as nausea and diarrhea; mechanical problems related to the tube feeding process; and metabolic problems, such as biochemical alterations and nutrient deficiencies. Examples of the complications that may arise and some preventive and corrective measures are summarized in Table 16–2 on p. 470.

Many complications of tube feeding are preventable if the most appropriate feeding route, formula, and delivery method are chosen. Attention to a patient's primary medical condition and medication use is important as well. Health practitioners routinely monitor patients' weights, hydration status, and results of laboratory tests to help them detect problems before complications develop. The Case Study can help you consider the many factors involved in tube feedings.

Case Study

GRAPHICS DESIGNER REQUIRING ENTERAL NUTRITION SUPPORT

Sharyn Eschler is a 24-year-old graphics designer who suffered multiple fractures when she fell from a cliff while hiking. She has been in the hospital for seven days and has no appetite. Sharyn has lost 8 pounds over the course of her hospitalization. Due to the nature of her injuries, she is in traction and is immobile, although the head of her bed can be elevated 45 degrees. From the diet history, it appears that Sharyn's nutrition status was adequate prior to hospitalization. The health care team agrees that a nasoduodenal tube feeding should be instituted before her nutritional status deteriorates further. The standard formula selected for the feeding is lactose-free, and Sharyn's nutrient requirements can be met with 2,200 milliliters of the formula per day.

1. What steps can be taken to prepare Sharyn for tube feeding? Why might nasoduodenal placement of the feeding tube be preferred to nasogastric placement?

2. The physician's orders specify that the feeding should be given continuously over 18 hours. Develop an appropriate tube feeding schedule.

3. After three days of feeding, Sharyn develops diarrhea. Check Table 16–2 to determine the possible causes. What measures can be taken to correct the diarrhea?

4. Describe precautions that should be taken if Sharyn is to receive medications through the feeding tube.

TABLE 16–2

Causes and Prevention or Correction of Tube Feeding Complications

Complications	Possible Causes	Preventive/Corrective Measures
Aspiration of formula	Compromised lower esophageal sphincter, delayed gastric emptying	Use nasoenteric, gastrostomy, or jejunostomy feedings in high-risk patients; check tube placement; elevate head of bed during and for 45 minutes after feeding; check gastric residuals.
Clogged feeding tube	Formula too thick for tube	Select appropriate tube size; flush tubing with water before and after giving formula; use infusion pump to deliver thick formulas. Remedies that may help to unclog feeding tubes include flushes with warm water or solutions that contain pancreatic enzymes or bicarbonate; consult pharmacist for more options.
	Medications delivered through feeding tube	Use oral, liquid, or injectable medications whenever possible; dilute thick or sticky liquid medications with water before administering; crush tablets to a fine powder and mix with water (except enteric-coated or sustained-release medications); flush tubing with water before and after medications are given; give medications individually; do not add medications to the feeding container.
Constipation	Low-fiber formula	Provide additional fluids; use high-fiber formula.
	Lack of exercise	Encourage walking and other activities, if appropriate.
Dehydration and electrolyte imbalance	Excessive diarrhea	See items under *Diarrhea*.
	Inadequate fluid intake	Provide additional fluid.
	Carbohydrate intolerance	Use continuous drip administration of formula; monitor blood glucose; select a formula with a lower amount or different type of carbohydrate; provide a formula with a higher fat content.
	Excessive protein intake	Monitor blood electrolyte levels; reduce protein intake.
Diarrhea, cramps, abdominal distention	Bacterial contamination	Use fresh formula every 24 hours; store opened or mixed formula in a refrigerator; rinse feeding bag and tubing before adding fresh formula; change feeding apparatus every 24 hours; prepare formula with clean hands using clean equipment in a clean environment.
	Lactose intolerance	Use lactose-free formula in patients with current or potential lactose intolerance.
	Hypertonic formula	Use small volume of formula and increase volume gradually.
	Rapid formula administration	Use slow administration rate or use continuous drip feedings.
	Malnutrition/low serum albumin	Use small volume of dilute formula and increase volume and concentration gradually.
Hyperglycemia	Diabetes, hypermetabolism, drug therapy	Check blood glucose; slow administration rate; provide adequate fluids; select a formula with a lower amount or different type of carbohydrate; provide a formula with a higher fat content.
Nausea and vomiting	Obstruction	Discontinue tube feeding.
	Delaying gastric emptying	Check gastric residual; slow administration rate, use continuous drip feedings, or discontinue tube feeding.
	Intolerance to concentration or volume of formula	Use small volume of formula and increase volume and concentration gradually; use continuous drip feedings.
	Psychological reaction to tube feeding	Address patient's concerns.
Skin irritation at enterostomy site	Leakage of GI secretions and friction caused by the tube	Keep site clean; inspect area for redness, tenderness, and drainage; use protective skin cream.

Note: Many of the complications presented here can be caused by the patient's primary disorder or drug therapy rather than the tube feeding itself. In such a case, the corrective measure would include treatment of the disorder or a change in drug therapy. Additionally, other corrective measures that require a physician's order are not shown here.

- Enteral formulas can supplement conventional diets or be used for tube feedings. They differ in their macronutrient composition, energy density, osmolality, and fiber content.

- A nasoenteric feeding route is preferred for short-term tube feedings, whereas enterostomies are used for longer-term feedings. Gastric feedings are preferred, although they are often avoided in patients at risk of aspiration.

- Depending on the feeding route and the patient's medical condition, the formula can be delivered in bolus feedings, intermittently, or continuously.

- Medications should be given separately from formula and accompanied by water flushes to prevent tube clogging.

Parenteral Nutrition Support

The first half of this chapter described how enteral formulas can supplement or replace conventional foods. Because enteral formulas cannot be used if intestinal function is inadequate, the ability to meet nutrient needs by vein is a lifesaving option for critically ill persons. The procedure is costly, however, and is associated with potentially dangerous complications. As previous sections suggested, enteral nutrition support is preferred if the GI tract is functional.

Indications for Parenteral Support

Parenteral nutrition is indicated for patients who do not have functioning GI tracts and who are either malnourished or likely to become so (review Figure 16–1 on p. 458). The following conditions may require use of parenteral nutrition:

- Short-bowel syndrome (part of the small intestine has been removed)

- Severe pancreatitis

- Malabsorption disorders

- Intestinal obstructions or fistulas

- Severe burns or trauma

- Critical illnesses or wasting disorders

- Bone marrow transplants

- Being malnourished and having a high risk of aspiration

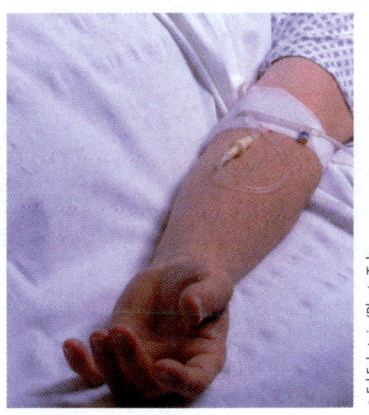

© Ed Eckstein/PhotoTake

The peripheral veins can provide access to the blood for the delivery of parenteral solutions.

Venous Access

Once the decision to use parenteral nutrition has been made, the access site must be selected. The access sites for intravenous feedings fall into two main categories: nutrients may be delivered into the **peripheral veins** located in the arms and legs or into the large-diameter **central veins** located near the heart.

Peripheral Parenteral Nutrition In **peripheral parenteral nutrition (PPN),** nutrient needs are met using only the peripheral veins. Peripheral veins can be damaged by overly concentrated solutions: hypertonic solutions can cause phlebitis (inflammation within the vein), resulting in redness, swelling, and tenderness at the infusion site.

peripheral veins: small-diameter veins that carry blood from the arms and legs.

central veins: large-diameter veins located close to the heart.

peripheral parenteral nutrition (PPN): a type of nutrition support in which intravenous feedings are delivered into peripheral veins.

■ Note that the osmotic property of a solution can be expressed either as *osmolarity* or as *osmolality* (introduced on p. 459).

- ■ *Osmolarity* refers to the number of solutes per liter of solution.
- ■ *Osmolality* refers to the number of solutes per kilogram of solvent.

The term *osmolality* is often preferred when referring to biological solutions such as blood or urine.

Therefore, the *osmolarity* ■ of parenteral solutions is usually kept between 600 and 900 milliosmoles per liter,[7] so PPN can supply only limited amounts of energy and protein. It is used most often in patients who need short-term nutrition support (about 7 to 10 days) and who do not have high nutrient needs or fluid restrictions. PPN is not possible if the peripheral veins are not strong enough to tolerate the procedure. In many cases, it is necessary to rotate venous access sites to prevent inflammation.

Total Parenteral Nutrition (TPN) Most patients meet their nutrient needs using the larger, central veins where blood volume is greater and nutrient concentrations do not need to be limited. Because this method can reliably meet a person's complete nutrient requirements, it is called **total parenteral nutrition (TPN)**. Central veins lie close to the heart where the large volume of blood rapidly dilutes parenteral solutions. Therefore, patients with very high nutrient needs or fluid restrictions are able to receive the nutrient-dense solutions they require. TPN is also preferred for patients who require long-term intravenous feedings.

There are several ways to access central veins. The tip of a central venous **catheter** can be placed directly into a large-diameter central vein or threaded into a central vein through a peripheral vein (see Figure 16–4). Although insertion of peripherally inserted central catheters is less invasive and more easily performed than the direct insertion of catheters into central veins, some studies have found higher complication rates associated with peripheral insertion.[8]

Parenteral Solutions

The pharmacies located within health care institutions are often responsible for preparing parenteral solutions. This arrangement is beneficial because the pharmacist can customize formulations to meet patients' nutrient needs and because the solutions have a limited shelf life. Prescriptions for parenteral solutions are highly individualized and may need to be recalculated daily until the patient's condition is stable. Because the nutrients are provided intravenously, they must be given in forms that are safe to inject directly into the bloodstream.

FIGURE 16–4

Accessing Central Veins for Total Parenteral Nutrition

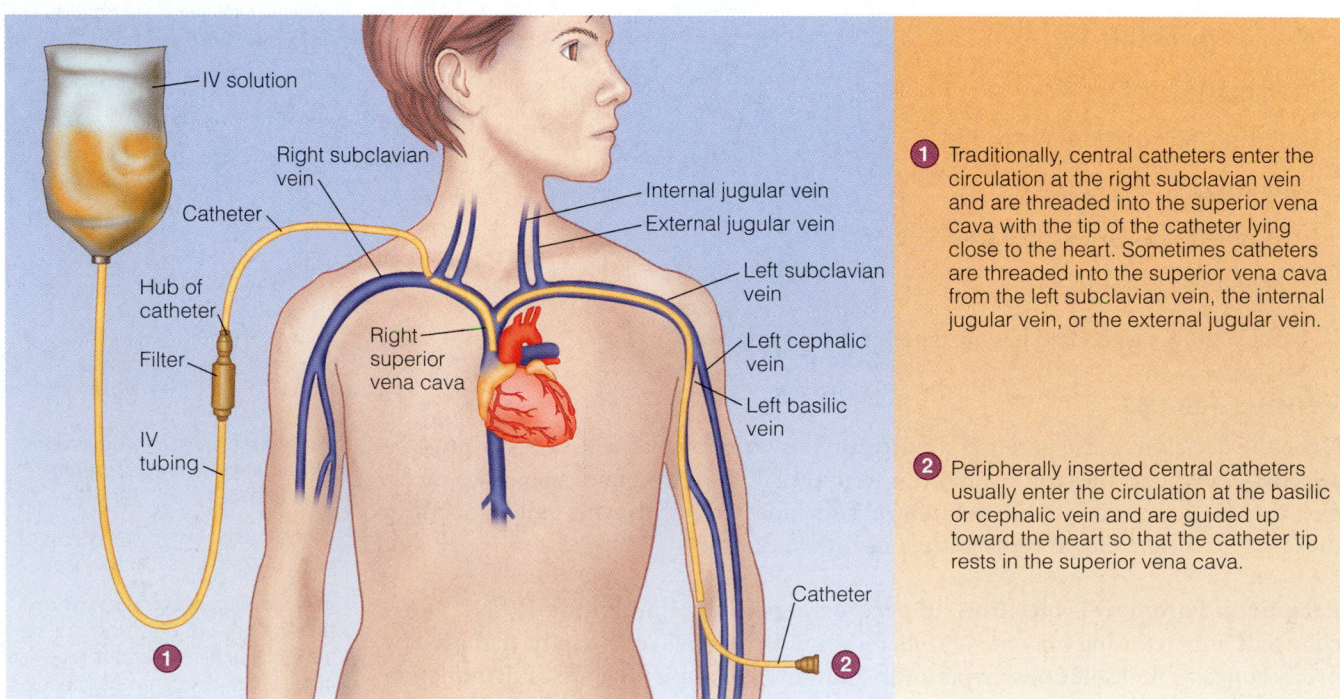

1. Traditionally, central catheters enter the circulation at the right subclavian vein and are threaded into the superior vena cava with the tip of the catheter lying close to the heart. Sometimes catheters are threaded into the superior vena cava from the left subclavian vein, the internal jugular vein, or the external jugular vein.

2. Peripherally inserted central catheters usually enter the circulation at the basilic or cephalic vein and are guided up toward the heart so that the catheter tip rests in the superior vena cava.

Amino Acids Parenteral solutions contain all of the essential amino acids and different combinations of nonessential amino acids. Amino acid concentrations range from 3.5 to 15 percent; ■ the more concentrated solutions are used only for TPN. Just as in regular foods, the amino acids provide 4 kcalories per gram. Disease-specific amino acid solutions are available for patients with liver failure, kidney failure, and metabolic stress.

Carbohydrate Glucose is usually the main source of energy in parenteral feedings. It is provided in the form dextrose monohydrate, in which each glucose molecule is associated with a single water molecule. Dextrose monohydrate provides 3.4 kcalories per gram, slightly less than pure glucose, which provides 4 kcalories per gram. Dextrose solutions are available in concentrations between 2.5 and 70 percent. ■ Concentrations greater than 10 percent are used only in TPN solutions.

In parenteral solutions, the dextrose concentration is indicated by a "D" followed by its concentration in water (W) or normal saline (NS). For example, D5 or D5W indicates that a solution contains 5 percent dextrose in water. Similarly, D5/NS means that a solution contains 5 percent dextrose in normal saline.

Lipids Lipid emulsions supply essential fatty acids and are a significant source of energy. The emulsions usually contain triglycerides from soybean oil and safflower oil, phospholipids to serve as emulsifying agents, and glycerol to make the solutions isotonic. Lipid emulsions are available in 10, 20, and 30 percent solutions, containing 1.1, 2.0, and 3.0 kcalories per milliliter, respectively. Therefore, a 500-milliliter container of 10 percent lipid emulsion would provide 550 kcalories. The same volume of a 20 percent lipid emulsion would provide 1,000 kcalories. ■

Lipid emulsions are often provided daily and may supply 20 to 30 percent of total kcalories. Including lipids as an energy source reduces the need for energy from dextrose and lowers the risk of hyperglycemia in glucose-intolerant patients. Lipid infusions must be restricted in patients with hypertriglyceridemia, however. There is also some concern that lipid emulsions that contain excessive linoleic acid may suppress some aspects of the immune response.

Fluids and Electrolytes Daily fluid needs approximate 30 to 40 milliliters per kilogram of body weight in young adults and 30 milliliters per kilogram of body weight in older adults (averaging between 1,500 and 2,500 milliliters for most people). The amounts are adjusted according to fluid losses and the results of hydration assessment.

The electrolytes added to parenteral solutions include sodium, potassium, chloride, calcium, magnesium, and phosphorus. The amounts in parenteral solutions differ from DRI values because the nutrients are infused directly into the blood and are not influenced by absorption as they are when consumed orally. Blood tests are administered daily to monitor electrolyte levels until patients have stabilized. Electrolyte imbalances can be lethal, so electrolyte management by experienced professionals is necessary whenever intravenous therapies are used.

The electrolyte content of parenteral solutions is expressed in milliequivalents (mEq), which are units indicating the number of ionic charges provided by electrolytes. The body's fluids are neutral solutions that contain equal numbers of positive and negative charges.

Vitamins and Trace Minerals Commercial multivitamin and trace mineral preparations are routinely added to parenteral solutions. All of the water-soluble vitamins and vitamins A, D, and E are supplied; vitamin K is often omitted and must be added separately. The trace minerals added to parenteral solutions include zinc, copper, chromium, selenium, and manganese. Iron is excluded because it alters the stability of other ingredients in parenteral mixtures; special forms of iron need to be injected separately.

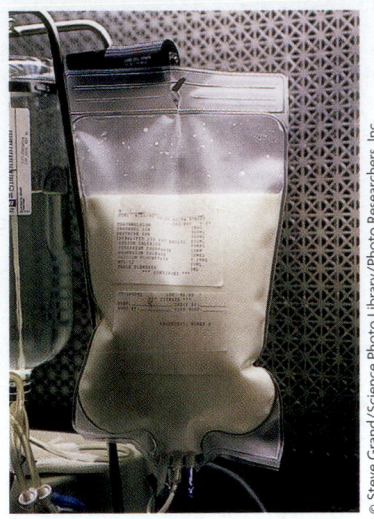

A lipid emulsion gives a parenteral solution a milky white color.

■ A 10 percent amino acid solution supplies 10 g of amino acids per 100 mL of solution.

■ A 10 percent dextrose solution provides 10 g of dextrose monohydrate per 100 mL of solution.

■ For 500 mL of a 10% lipid emulsion: 500 mL × 1.1 kcal/mL = 550 kcal

For 500 mL of a 20% lipid emulsion: 500 mL × 2 kcal/mL = 1,000 kcal

total parenteral nutrition (TPN): a type of nutrition support in which intravenous feedings are delivered into a central vein.

catheter: a thin tube placed within a narrow lumen (such as a blood vessel) or body cavity; can be used to infuse or withdraw fluids or to keep a passage open.

Osmolarity Recall that the **osmolarity** of PPN solutions is limited to 900 milliosmoles per liter, whereas TPN solutions may be as nutrient dense as necessary (see p. 472). The components of a solution that contribute most to its osmolarity are amino acids, dextrose, and electrolytes: as concentrations of these nutrients increase, the osmolarity of a solution increases. Because lipids contribute little to osmolarity, lipid emulsions are used to increase the energy provided in PPN solutions.

Types of Parenteral Solutions When a parenteral solution contains dextrose, amino acids, and lipids, it is called a **total nutrient admixture (TNA)**, a **3-in-1**, or an **all-in-one** solution. A **2-in-1 solution** excludes lipids, and a lipid emulsion is administered separately, often by using a second port in the catheter. The administration of TNA solutions is simpler because only one infusion pump is required. Lipids are usually administered separately when they are not a major energy source and are used only to provide essential fatty acids. The "How To" describes a method for calculating the macronutrient and energy content of a parenteral solution.

Administering Parenteral Nutrition

Providing parenteral nutrition is complex and requires skills from a variety of disciplines. Many hospitals organize nutrition support teams, consisting of physicians, nurses, dietitians, and pharmacists, that specialize in the provision of both intravenous and tube feedings. ■ The nurse, who performs direct patient care, plays a central role in administering and monitoring parenteral feedings.

■ Reminder: A nutrition support team is a multidisciplinary team of health care professionals who are responsible for the provision of nutrients by tube feeding or intravenous infusion.

Insertion and Care of Intravenous Catheters Although a skilled nurse can place catheters into peripheral veins, the insertion of catheters directly into central veins must be performed by a qualified physician. Patients may be awake for the procedure and given local anesthesia. Unnecessary apprehension can be avoided by explaining the procedure to the patient beforehand.

Catheter-related problems frequently cause complications (see Table 16–3). Catheters may be improperly positioned or may dislodge after placement. Air can leak into catheters, obstructing blood flow. Catheters in peripheral veins may cause phlebitis, necessitating reinsertion at an alternate site. A catheter may become clogged from blood clotting or from a buildup of scar tissue around the catheter tip. Catheters are also a leading cause of infection: contamination may be introduced during insertion or may develop at the placement site.

How To

CALCULATE THE MACRONUTRIENT AND ENERGY CONTENT OF PARENTERAL SOLUTIONS

Suppose a patient is receiving 1.25 liters (1,250 milliliters) of a parenteral solution that contains 5 percent amino acids and 30 percent dextrose supplemented by 250 milliliters of a 20 percent lipid emulsion daily. How many grams of protein and carbohydrate is the person receiving and what is the total energy intake for the day?

Amino acids:

$$5\% \text{ amino acids} = \frac{5 \text{ g amino acids}}{100 \text{ mL}}$$

$$\frac{5 \text{ g amino acids}}{100 \text{ mL}} \times 1{,}250 \text{ mL}$$

$$= 62.5 \text{ g of amino acids}$$

62.5 g amino acids × 4.0 kcal/g = 250 kcal

Carbohydrate:

$$30\% \text{ dextrose} = \frac{30 \text{ g dextrose}}{100 \text{ mL}}$$

$$\frac{30 \text{ g dextrose}}{100 \text{ mL}} \times 1{,}250 \text{ mL}$$

$$= 375 \text{ g of dextrose}$$

375 g dextrose × 3.4 kcal/g = 1,275 kcal

Lipid:
Recall that a 20 percent lipid emulsion provides 2.0 kcalories per milliliter. If the patient is given 250 milliliters of the emulsion:

250 mL × 2.0 kcal/mL = 500 kcal

Total energy intake: 250 kcal + 1,275 kcal + 500 kcal = 2,025 kcal

TABLE 16-3

Potential Complications of Parenteral Nutrition

Catheter-Related	Metabolic
Air embolism	Abnormalities in liver function
Blood clotting at catheter tip	Electrolyte imbalances
Clogging of catheter	Gallbladder disease
Dislodgment of catheter	Hyperglycemia, hypoglycemia
Improper placement	Hypertriglyceridemia
Infection, sepsis	Metabolic bone disease
Phlebitis	Nutrient deficiencies
Tissue injury	Refeeding syndrome

To reduce the risk of complications, health practioners use aseptic techniques when inserting catheters, changing tubing, or changing a dressing that covers the catheter site. Unusual bleeding or a wet dressing suggests a problem with catheter placement. A change in infusion rate may indicate a clogged catheter. Infection may be indicated by redness or swelling around the catheter site or by an unexplained fever. Routine inspections of equipment and frequent monitoring of patients' symptoms help to minimize the problems associated with catheter use.

Administration of Parenteral Solutions Infusion protocols vary among institutions. A common approach is to infuse the solution at a slow rate initially and gradually increase the rate over a two- to three-day period. For example, 40 milliliters per hour can be infused during the first 24 hours of administration (supplying 960 milliliters), and the rate can be increased by one liter per day until the goal rate is reached. Another approach is to give the full volume of a nutrient-dilute solution on the first day and advance nutrient concentrations as tolerated. Some protocols suggest starting solutions at full strength unless there is a risk that the patient will become hyperglycemic.

Parenteral solutions can be infused continuously over 24 hours (**continuous parenteral nutrition**) or during 10- to 16-hour periods only (**cyclic parenteral nutrition**). Continuous feedings are given to critically ill and malnourished patients who cannot receive adequate nutrition during shorter time periods. Cyclic feedings are often provided at night so that patients can participate in routine activities during the day. This method is especially suited to patients who require long-term parenteral support or will be infusing parenteral solutions at home. Patients may begin with continuous feedings and transition to cyclic feedings as their condition improves.

Regular monitoring can help to prevent complications. The parenteral solution and tubing are checked daily for signs of contamination. Routine testing of glucose, lipids, and electrolyte levels helps to determine tolerance to solutions. Frequent reassessment of nutritional status may be necessary until a patient has stabilized. Rapid changes in infusion rate are discouraged in some patients due to a risk of developing hyperglycemia or hypoglycemia.[9]

Discontinuing Intravenous Feedings Patients must have adequate GI function before parenteral feedings can be tapered off and enteral feedings begun. During the transition to oral feedings, a combination of feeding methods is often used. By tapering off parenteral feedings at the same time that tube feedings or oral feedings are begun, the two feeding methods together supply the needed nutrients. Clear liquids are usually the first foods offered and include pulp-free fruit juices, soft drinks, and clear broths; small amounts are given initially to determine tolerance. Later feedings include

osmolarity: the concentration of osmotically active particles in a solution, expressed as milliosmoles per liter (mOsm/L). *Osmolality* is an alternative expression of a solution's osmotic properties that is used in clinical practice and uses the unit milliosmoles per kilogram (mOsm/kg).

total nutrient admixture (TNA): a parenteral solution that contains dextrose, amino acids, and lipids; also called a **3-in-1** or an **all-in-one** solution.

2-in-1 solution: a parenteral solution that contains dextrose and amino acids but excludes lipids.

continuous parenteral nutrition: continuous administration of parenteral solutions over a 24-hour period.

cyclic parenteral nutrition: administration of a parenteral solution over a 10- to 16-hour period.

■ Chapter 17 provides more information about the types of foods and meals offered to patients who are beginning oral feedings.

beverages and solid foods that are unlikely to cause discomfort. ■ If gastrointestinal symptoms (such as nausea, vomiting, bloating, or diarrhea) develop, oral feedings are limited in size or frequency until the intestines adapt. Once two-thirds to three-fourths of nutrient needs can be provided enterally, the intravenous feedings may be discontinued.

Transitioning to an oral diet is sometimes difficult because a person's appetite remains suppressed for several weeks after parenteral nutrition is terminated. Patients receiving continuous parenteral feedings may have better appetites during the day if they are switched to nocturnal cyclic feedings before beginning oral intakes.

Managing Metabolic Complications

As discussed in a previous section, the catheters used for parenteral nutrition may cause a number of serious complications. This section describes some metabolic complications that may result from parenteral feedings (review Table 16–3).[10]

Hyperglycemia Hyperglycemia most often occurs in patients who are glucose intolerant or undergoing severe metabolic stress. It can be prevented by providing insulin along with feedings or by restricting the amount of dextrose in a solution. Dextrose infusions are generally limited to less than 5 milligrams per kilogram of body weight per minute in critically ill adult patients so that the carbohydrate intake does not exceed the maximum glucose oxidation rate. ■

■ For most patients receiving parenteral nutrition, blood glucose levels should not exceed 200 mg/dL.

Hypoglycemia Although uncommon, hypoglycemia sometimes occurs when feedings are interrupted or discontinued. In patients at risk, such as young infants, feedings may be tapered off over several hours before discontinuation.

Hypertriglyceridemia Hypertriglyceridemia may develop in critically ill patients who cannot tolerate the amount of lipid emulsion supplied. It may also result from factors that impair lipid clearance, such as excessive carbohydrate feedings or severe infection. If blood triglyceride levels exceed 350 to 400 milligrams per deciliter, lipid infusions should be reduced or stopped.[11]

Refeeding Syndrome Severely malnourished patients who are fed aggressively (parenterally or otherwise) may develop **refeeding syndrome**, characterized by electrolyte and fluid imbalances and hyperglycemia. These effects occur because dextrose infusions raise circulating insulin levels, which promote anabolic processes that quickly remove phosphate, potassium, and magnesium from the blood. The altered electrolyte levels can lead to fluid retention and life-threatening changes in organ systems. To prevent refeeding syndrome, health practitioners usually start parenteral feedings slowly and carefully monitor electrolyte levels when malnourished patients begin receiving nutrition support.

Abnormalities in Liver Function Fatty liver often results from parenteral feedings, but it is usually corrected when the parenteral feedings are discontinued. Long-term parenteral nutrition, however, may result in chronic, irreversible liver disease that may eventually lead to liver failure. The cause of the liver abnormalities is unclear.

Gallbladder Disease When parenteral nutrition continues for more than four weeks, sludge (thickened bile) often builds up in the gallbladder and may eventually lead to gallstone formation. Patients requiring long-term parenteral nutrition may be given cholecystokinin injections (to cause gallbladder contraction and bile release) or have their gallbladders removed surgically.

Metabolic Bone Disease Long-term parenteral nutrition has been associated with lower bone density, which may be related to altered calcium, phosphorus, magnesium,

Case Study

Jerry Huang, a 27-year-old geologist with an inflammatory intestinal disease, underwent a surgical procedure in which a substantial portion of his small intestine was removed. He had received TPN prior to surgery and continued to receive it afterwards. After ten days, tube feeding was begun, which initially delivered very small feedings.

1. List some reasons why the nutrition support team initially chose TPN to provide nutrition support to this patient. How would you explain the need for parenteral feedings to Jerry?

2. Describe the components of a typical TPN solution. Calculate the energy content of 1 liter of a solution that provides 140 grams of dextrose monohydrate, 45 grams of amino acids, and 90 milliliters of 20% lipid emulsion. If Jerry needs 2,100 kcalories per day, how many liters of solution will he need each day?

3. Why is it important that Jerry begin enteral feedings as soon as possible? Assuming that Jerry eventually tolerates a tube feeding, in what ways can the health care team help Jerry make the

transition from parenteral feedings to tube feeding? Consider some of the physiological problems Jerry might face when he begins eating an oral diet.

4. If Jerry is unable to meet nutrient needs orally, he may need to continue tube feeding or TPN at home. As you read through the following sections, consider the factors that would make Jerry a good candidate for a home nutrition support program. Consider both the benefits of a proposed program and the problems he could encounter.

and sodium metabolism. Imbalanced intakes of vitamin D, vitamin K, and phosphorus have also been implicated. The solution to this problem varies among patients.[12]

IN SUMMARY

- Peripheral parenteral nutrition is provided to patients who need short-term nutrition support and who do not have high nutrient needs or fluid restrictions. Total parenteral nutrition can supply nutrient-dense solutions and provide long-term intravenous feedings.

- Parenteral solutions include amino acids, dextrose, electrolytes, vitamins, and minerals. Lipid emulsions may be included in the mixture or may be administered separately.

- Critically ill patients may require continuous feedings, whereas healthier patients and long-term users may prefer cyclic feedings.

- Catheters are frequently the cause of complications, which include improper placement or dislodgment, infection, clotting, embolism, and phlebitis.

- Metabolic complications include hyperglycemia and hypoglycemia, hypertriglyceridemia, fluid and electrolyte imbalances, and diseases affecting the liver, gallbladder, and bone. The Case Study can check your understanding of the concepts introduced in this section.

Nutrition Support at Home

Occasionally, a patient must continue to receive nutrition support (tube feedings or parenteral nutrition) after a medical condition has stabilized. For such a person, home nutrition support might be an option.

refeeding syndrome: a condition that sometimes develops when a severely malnourished person is aggressively fed; characterized by electrolyte and fluid imbalances and hyperglycemia.

Portable pumps and convenient carrying cases allow people who require home nutrition support to move about freely.

The use of home nutrition support is rapidly expanding. Current technology allows for the safe administration of nutrition support in home settings, and insurance coverage often pays a substantial portion of the costs. Home health services and home infusion pharmacies can provide the equipment, enteral formulas or parenteral solutions, and services necessary for home nutrition care. Most importantly, patients using these services can continue to receive specialized nutrition care while leading normal lives.

Candidates for Home Nutrition Support

Individuals referred for home nutrition support usually need long-term nutrition care for chronic medical conditions. Users of home nutrition services must be intellectually capable of learning the necessary procedures, monitoring the treatment, and managing complications as necessary. The home should be clean and have adequate storage for formulas or solutions and equipment. The costs should be clearly explained to families who cannot get insurance reimbursement. Good candidates for home nutrition support include the following:

- People who have functioning GI tracts and illnesses that prevent food from reaching the digestive tract may benefit from home enteral nutrition. Examples include patients with head and neck cancers and patients with neurological impairments that affect swallowing.

- People who have illnesses that severely impair nutrient absorption or cause motility problems in the stomach or intestines may benefit from home parenteral nutrition. Examples include individuals who have had large portions of their small intestine removed and those with intestinal obstructions or malabsorption conditions.

Planning Home Nutrition Care

As with nutrition support provided in health care facilities, planning for home nutrition care involves decisions about access sites, formulas, and nutrient delivery methods. Users of home services should be involved in the decision-making to ensure long-term compliance and satisfaction.

Home Enteral Nutrition   Access to the GI tract is possible using either nasal tubes or enterostomies. People sometimes learn to place nasogastric tubes themselves, which may improve acceptance of the therapy. Active children and adults often prefer low-profile gastrostomy tubes, which allow them to lead a more normal lifestyle. Jejunostomy tubes may be required for some individuals but are less convenient because the frequent feedings required for jejunostomies can interfere with daytime activities.

The choice of formula for home use is influenced by its cost and availability. Insurance reimbursements do not always include the cost of formula, which is considered to be a "food" product. For this reason, some people choose to prepare simple formulas at home. Blenderizing home-cooked foods is possible, but the foods need to be strained to remove particles and clumps that may obstruct the tube. Closed (ready-to-hang) feeding systems are useful for avoiding contamination risk but are not appropriate for intermittent feedings that require smaller amounts of formula.

The advantages and disadvantages associated with different feeding techniques and administration schedules should be fully discussed with patients. For gastric feedings, bolus infusions are simplest and can be quickly delivered. Gravity drip infusions eliminate the need for an infusion pump, but the delivery rates are less reliable. If intermittent feeding schedules are appropriate, they should be tailored to daily routines. Portable pumps can free individuals from the need to infuse formula at home and can also be used when traveling.

Home Parenteral Nutrition Although both peripheral parenteral nutrition and total parenteral nutrition (TPN) can be provided at home, long-term therapy requires access to the larger, central veins that are appropriate for TPN. The catheter can be inserted so that the exit site is in an area that is accessible to the patient.

Parenteral solutions need to be sterile and aseptically prepared, and people who mix their own solutions must be carefully trained. Ready-made parenteral solutions require refrigeration and are stable for limited periods; for example, TNA (total nutrient admixture) solutions may be stable for only one week when refrigerated.

Most people prefer cyclic infusions to continuous infusions and transition to cyclic infusions before discharge from the hospital. Because an infusion pump is required for home TPN, sufficient battery backup should be on hand in case electrical service is interrupted. Portable pumps are helpful for individuals who lead an active lifestyle or prefer to infuse during the day.

Quality of Life Issues

Although home nutrition programs can help to improve health and extend life, consumers of home services and their families may struggle with the lifestyle adjustments required. In addition to the economic impact of nutrition support, home feedings are often time-consuming and inconvenient. Activities and work schedules must be planned around feedings. Extra planning is needed and precautions must be taken

Nutrition Assessment Checklist

FOR PEOPLE RECEIVING TUBE FEEDINGS

Medical History

Check the medical record for medical conditions that:

- [ ] Alter nutrient needs and influence the formula selection
- [ ] Influence the selection of tube placement sites (gastric versus intestinal) and feeding routes
- [ ] Suggest the length of time that tube feeding will be needed

Monitor the medical record for complications that may influence the formula selection or delivery technique, including:

- [ ] Aspiration
- [ ] Constipation
- [ ] Fluid and electrolyte imbalances
- [ ] Diarrhea
- [ ] Hyperglycemia
- [ ] Nausea and vomiting
- [ ] Skin irritation

Medications

Check medications for those that can cause side effects similar to those associated with the tube feeding, such as:

- [ ] Nausea and vomiting
- [ ] Diarrhea
- [ ] Constipation
- [ ] GI discomfort

For medications delivered through the feeding tube, check:

- [ ] Form of medication and possible alternates
- [ ] Viscosity of liquid medications
- [ ] Potential for diet-drug interactions

Dietary Intake

To assess nutritional adequacy, check to see if:

- [ ] Formula is appropriate for patient's needs
- [ ] Supplemental water is provided to meet needs
- [ ] Formula is administered as prescribed

Anthropometric Data

Measure baseline height and weight, and monitor daily weights. If weight is not appropriate:

- [ ] Determine whether energy needs have been correctly assessed.
- [ ] Check to see if formula is being delivered as prescribed.
- [ ] Check for signs of dehydration or overhydration.

Laboratory Tests

Check serum and urine tests for signs of:

- [ ] Fluid and electrolyte imbalances
- [ ] Glucose intolerance
- [ ] Adequacy of protein intake (serum protein levels)
- [ ] Improvement or deterioration of medical condition

Physical Signs

Look for physical signs of:

- [ ] Dehydration or overhydration
- [ ] Delayed gastric emptying (gastric residual volume)
- [ ] Malnutrition

when a person wants to travel or participate in sports activities. Explaining one's medical needs to friends and acquaintances may be embarrassing.

Among physical difficulties, people receiving nocturnal feedings often cite disturbed sleep as a major problem. Disruptions may be due to multiple nighttime bathroom visits, noisy infusion pumps, or difficulty finding a comfortable sleeping position when "hooked up." People using parenteral support sometimes prefer infusing solutions during the day to improve their sleeping patterns.

Among social issues, the inability to consume meals with family and friends is often a great concern. Some individuals may be able to eat small amounts of food or may decide to sit at the table to participate in the conversation. Joining friends at restaurants and attending certain types of social events, however, can be problematic.

People who depend on nutrition support face a number of stressful issues that can affect quality of life. Although parenteral and enteral nutrition are life-sustaining therapies, both are associated with serious complications. Many people find that their lifestyles need to be greatly altered to accommodate nutrition therapy and may experience fear, anxiety, and depression. Support groups or counseling resources may help patients cope with ongoing stresses. ■

■ The Oley Foundation is an excellent source of outreach services, emotional support, and current information for people who require home nutrition support (www.oley.org).

Nutrition Assessment Checklist

FOR PEOPLE RECEIVING PARENTERAL NUTRITION

Medical History

Check the medical record for medical conditions that:

- Prevent the use of enteral nutrition
- Indicate the appropriate feeding route (peripheral versus central)
- Suggest the length of time that parenteral nutrition will be required

Monitor the medical record for complications that may influence the parenteral solution formulation or delivery technique, including:

- Acid-base imbalances
- Fluid and electrolyte imbalances
- Hyperglycemia or hypoglycemia
- Hypertriglyceridemia
- Nutrient deficiencies
- Refeeding syndrome

Medications

For medications added to the parenteral solution, determine the:

- Medication's compatibility with the parenteral solution
- Length of time that the medication can remain stable in solution

For medications infused separately, determine:

- Length of time that the feeding may need to be stopped
- Adjustments in solution infusion to compensate for the medication delivery

Dietary Intake

To assess nutritional adequacy, check to see if:

- Patient's nutrient needs were correctly determined
- Solution is administered as prescribed
- Infusion pump is operating correctly

Anthropometric Data

Measure baseline height and weight, and monitor daily weights. If weight is not appropriate:

- Determine whether energy needs have been correctly assessed.
- Check to see if parenteral solution is being delivered as prescribed.
- Check for signs of dehydration or overhydration.

Laboratory Tests

Check serum and urine tests for signs of:

- Fluid, electrolyte, and acid-base imbalances
- Hyperglycemia or hypoglycemia
- Hypertriglyceridemia
- Nutrient deficiencies
- Adequacy of protein intake (serum protein levels)
- Improvement or deterioration of medical condition

Physical Signs

Routinely monitor the following:

- Catheter insertion site for signs of infection or inflammation
- Blood pressure, temperature, pulse, and respiration for signs of fluid, electrolyte, and acid-base imbalances

Look for physical signs of:

- Dehydration or overhydration
- Protein-energy malnutrition
- Malnutrition

■ Candidates for home enteral nutrition services have functional GI tracts but are unable to consume food orally. Parenteral nutrition candidates have illnesses that impair nutrient absorption or cause motility problems.

■ Patients and caregivers should participate in decisions about access sites, formulas, and nutrient delivery methods. Formulas and solutions can be prepared in the home.

■ The use of portable pumps may help individuals lead a normal lifestyle. Nevertheless, lifestyle adjustments to nutrition support may be difficult and stressful.

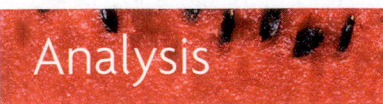

Analysis

CLINICAL APPLICATIONS

1. The administration of tube feedings and parenteral nutrition require attention to many technical details, making it easy to focus on the procedure rather than the patient. Imagine that your brother, sister, or a parent requires a transnasal tube feeding. How might this person react to the need for a tube feeding? How would you explain the benefits and possible problems associated with the procedure? Think about the ways you would want the health practitioner to help.

2. A liter of a TPN solution contains 500 milliliters of 50 percent dextrose solution and 500 milliliters of 5 percent amino acid solution. Determine the daily energy and protein intakes of a person who receives 2 liters of such a solution. Calculate the average daily energy intake if the person also receives 500 milliliters of a 20 percent fat emulsion three times per week.

3. Consider what it might be like to be using home parenteral nutrition, with no foods allowed by mouth. What would be the advantages of living at home instead of in a hospital or other residential facility? Can you think of some disadvantages?

   Think about how you might manage daily feedings: consider the time, cost, and commitment required to maintain the therapy. If not allowed to consume foods, what possible difficulties might you encounter? How would you handle holidays and special occasions that center around food?

Self Check

1. The terms *osmolality* and *osmolarity* refer to:
   a. energy density.
   b. nutrient density.
   c. fiber content.
   d. concentrations of molecules and ionic particles.

2. An important measure that may prevent bacterial contamination in tube feeding formulas is:
   a. nonstop feeding of formula.
   b. using the same feeding bag and tubing each day.
   c. discarding opened containers of formula within 24 hours.
   d. adding formula to the feeding container before it empties completely.

3. Compared to intermittent feedings, continuous feedings:
   a. require an infusion pump.
   b. allow greater freedom of movement.
   c. are more similar to normal patterns of eating.
   d. are associated with more GI side effects.

4. A patient needs 1,800 milliliters of formula per day. If the patient is to receive formula intermittently every 4 hours, how many milliliters of formula will he need at each feeding?
   a. 225
   b. 300
   c. 400
   d. 425

5. The term that describes the volume of formula remaining in the stomach from a previous feeding is:
   a. residue.
   b. osmolar load.
   c. gastric residual.
   d. intermittent feeding.

6. The clinician using the feeding tube to deliver medications recognizes that:
   a. medications given by feeding tube generally do not cause GI complaints.

b. medications can usually be added directly to the feeding container.

c. enteral formulas do not interact with medications in the same way that foods do.

d. thick or sticky liquid medications and crushed tablets can clog feeding tubes.

7. TPN is preferred over PPN for a patient who:
   a. does not have high nutrient requirements.
   b. needs long-term parenteral nutrition support.
   c. has strong peripheral veins and moderate nutrient needs.
   d. needs parenteral feedings as a supplement to tube feedings.

8. For a patient receiving central TPN who also receives intravenous lipid emulsions two or three times per week, the lipid emulsions serve primarily as a source of:
   a. essential fatty acids.
   b. cholesterol.

c. fat-soluble vitamins.

d. concentrated energy.

9. Refeeding syndrome causes dangerous fluctuations in:
   a. electrolytes.
   b. liver enzymes.
   c. triglycerides.
   d. ketone bodies.

10. Patients using home parenteral nutrition:
    a. are unable to use TNA solutions.
    b. are usually given continuous rather than cyclic infusions.
    c. require infusion pumps for use at home.
    d. are generally unable to work out of the home or travel.

Answers to these questions can be found in Appendix H.

Notes

1. C. R. Parrish and S. McCray, Enteral feeding: Dispelling myths, *Practical Gastroenterology* 27 (September 2003): 33–50.

2. L. Gramlich and coauthors, Does enteral nutrition compared to parenteral nutrition result in better outcomes in critically ill adult patients? A systematic review of the literature, *Nutrition* 20 (2004): 843–848.

3. D. A. Neumann and M. H. DeLegge, Gastric versus small-bowel tube feeding in the intensive care unit: A prospective comparison of efficacy, *Critical Care Medicine* 30 (2002): 1436–1438.

4. L. Klein, Is blue dye safe as a method of detection for pulmonary aspiration? *Journal of the American Dietetic Association* 104 (2004): 1651–1652; J. P. Maloney and T. A. Ryan, Detection of aspiration in enterally fed patients: A requiem for bedside monitors of aspiration, *Journal of Parenteral and Enteral Nutrition* 26 (2002): S34–S42.

5. Parrish and McCray, 2003.

6. Parrish and McCray, 2003.

7. J. M. Culebras and coauthors, Practical aspects of peripheral parenteral nutrition, *Current Opinion in Clinical Nutrition and Metabolic Care* 7 (2004): 303–307.

8. L. J. Walshe and coauthors, Complication rates among cancer patients with peripherally inserted central catheters, *Journal of Clinical Oncology* 20 (2002): 3276–3281; C. T. Cowl and coauthors, Complications and cost associated with parenteral nutrition delivered to hospitalized patients through either subclavian or peripherally inserted central catheters, *Clinical Nutrition* 19 (2000): 237–243.

9. M. M. McMahon, Parenteral nutrition, in L. Goldman and D. Ausiello, eds., *Cecil Textbook of Medicine* (Philadelphia: Saunders, 2004), pp. 1322–1326; A.S.P.E.N. Board of Directors and The Clinical Guidelines Task Force, Guidelines for the use of parenteral and enteral nutrition in adult and pediatric patients, *Journal of Parenteral and Enteral Nutrition* 26 (2002): 1SA–138SA.

10. A.S.P.E.N. Board of Directors and The Clinical Guidelines Task Force, 2002.

11. McMahon, 2004.

12. R. O. Brown and G. Minard, Parenteral nutrition, in M. E. Shils and coeditors, *Modern Nutrition in Health and Disease* (Baltimore: Lippincott Williams & Wilkins, 2006), pp. 1567–1597.

CHAPTER 16

Nutrition in Practice

INBORN ERRORS OF METABOLISM

An **inborn error of metabolism** is an inherited trait, caused by a genetic **mutation**, that results in the absence, deficiency, or malfunction of a protein that has a critical metabolic role.[1] The severity of the inborn error is related to the degree of impairment caused by the altered or missing protein. This Nutrition in Practice describes several inborn errors of metabolism and discusses the role of diet in two of these disorders: phenylketonuria and galactosemia. The accompanying Glossary defines the related terms.

What problems can result from inborn errors of metabolism?

The protein affected by an inborn error of metabolism may function as an enzyme, receptor, transport protein, or structural protein. When the body fails to make a protein, body functions that depend on that protein are impaired. For example, when an enzyme is missing or malfunctioning in a metabolic pathway that typically converts compound A to compound B, compound A will accumulate and compound B will not be made. The excess of compound A and the lack of compound B may have harmful effects. Furthermore, the imbalances in one pathway may affect other pathways and ultimately cause a number of metabolic and physiologic disturbances.

What role can diet play in treating inborn errors of metabolism?

Medical nutrition therapy is the primary treatment for many inborn errors that involve nutrient metabolism. Once the biochemical pathway affected by a mutation is identified, a health practitioner may be able to manipulate elements of the diet to compensate for deficiencies and excesses. Dietary intervention generally involves restricting substances that cannot be properly metabolized and supplying substances that cannot be produced. Thus dietary changes may be able to improve outcomes of some inborn errors by:

- Preventing the accumulation of toxic metabolites

- Replacing nutrients that are deficient as a result of a defective metabolic pathway

- Providing a diet that supports normal growth and development and maintains health

Successful treatment of an inborn error of metabolism depends on the ability to screen newborns and diagnose metabolic diseases before irreversible damage can occur. After a genetic disorder is diagnosed, family members undergo **genetic counseling** to evaluate the likelihood that their future offspring may inherit the disorder. During counseling, couples may learn about reproductive options such as artificial insemination, *in vitro* fertilization, or prenatal monitoring after conception.

Are there treatments for inborn errors of metabolism that do not involve dietary changes?

Some nondietary therapies are used to treat inborn errors of metabolism, although the options are somewhat limited. In some cases, the missing protein is infused; this is the primary means of treating **hemophilia**, caused by deficiency of one of the plasma proteins needed for clotting blood. Drug therapy is the main treatment for some inborn errors, including **cystic fibrosis** (discussed in Chapter 19), which is characterized by a defect that prevents normal chloride transport across cell membranes.[2] Future approaches may include **gene therapy**, a treatment that introduces DNA sequences into the chromosomes of affected cells, prompting the cells to express the protein needed to correct the abnormality.

GLOSSARY

cystic fibrosis: an inherited disorder that affects the transport of chloride across epithelial cell membranes; primarily affects the gastrointestinal and respiratory systems.

galactosemia (ga-LACK-toe-SEE-me-ah): an inherited disorder that affects galactose metabolism. Accumulated galactose causes damage to the liver, kidneys, and brain in untreated patients.

gene therapy: treatment for inherited disorders, in which DNA sequences are introduced into the chromosomes of affected cells, prompting the cells to express the protein needed to correct the disease.

genetic counseling: support for families at risk of genetic disorders; involves diagnosis of disease, identification of inheritance patterns within the family, and review of reproductive options.

hemophilia (HE-moh-FEEL-ee-ah): inherited bleeding disorders characterized by deficiency or malfunction of plasma proteins needed for clotting blood.

inborn error of metabolism: an inherited trait (present at birth) that causes the absence, deficiency, or malfunction of a protein that has a critical metabolic role.

metabolites: products of metabolism; the compounds produced by a biochemical pathway.

mutation: an inheritable alteration in the DNA sequence of a gene.

phenylketonuria (FEN-il-KEY-toe-NU-ree-ah) or **PKU:** an inherited disorder that affects the conversion of the essential amino acid phenylalanine to the amino acid tyrosine.

What is an example of an inborn error of metabolism that benefits from dietary treatment?

A classic example is **phenylketonuria (PKU)**, a metabolic disorder that affects amino acid metabolism. PKU occurs in approximately 1 out of every 10,000 births in the United States each year. In PKU, the missing or defective protein is a liver enzyme that converts the essential amino acid phenylalanine to the amino acid tyrosine (see Figure NP16–1). Without this enzyme, phenylalanine and its **metabolites** (metabolic products) accumulate and damage the developing nervous system. The impairment in the metabolic pathway also prevents liver synthesis of tyrosine and tyrosine-derived compounds (such as the neurotransmitter epinephrine). Under these conditions, tyrosine becomes essential; because the body cannot make tyrosine, the diet must supply it.

Although PKU's most debilitating effect is on brain development, other symptoms may manifest if the condition is untreated. Infants with PKU may have poor appetites and grow slowly. They may be irritable or have tremors or seizures. Their bodies and urine may have a musty odor. Their skin coloring may be unusually pale, and they may develop skin rashes.

How is PKU diagnosed?

Although PKU is not evident at birth, it must be diagnosed in the first few days of life so that early treatment can prevent its devastating effects. For this reason, newborns are routinely screened for PKU in all fifty states.[3] A standard blood test for phenylalanine is typically conducted by heel puncture after the infant has consumed several meals containing protein (usually after 24 hours). Abnormal results require further testing. The screening of newborns for PKU is one of the most common genetic tests in the United States and many other countries. Before newborn screening, infants with PKU demonstrated developmental delays (for example, inability to crawl) by six to nine months of age. By the time parents recognized the problem, the damage was irreversible.

What is the treatment for PKU?

The only current treatment for PKU is a diet that restricts phenylalanine and supplies tyrosine so that the blood levels of these amino acids are maintained within safe ranges.

Because phenylalanine is an essential amino acid, the diet cannot exclude it completely. Children with PKU need phenylalanine to grow, but they cannot handle excesses without detrimental effects. Therefore, their diets must provide enough phenylalanine to support growth and health but not so much as to cause harm. The diet must also provide tyrosine, which is an essential nutrient for PKU children. To ensure that blood concentrations of phenylalanine and tyrosine are close to normal, blood tests are performed periodically, and diets are adjusted when necessary. If the dietary treatment is conscientiously followed, it can prevent the symptoms described earlier.

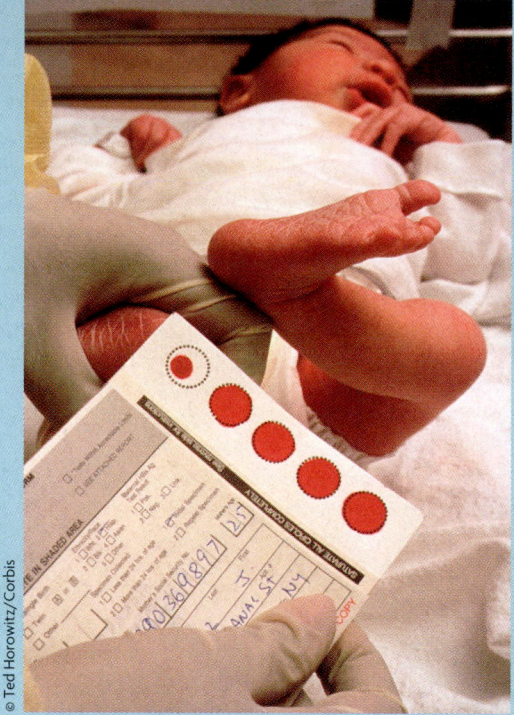

© Ted Horowitz/Corbis

A simple blood test screens newborns for PKU— a common inborn error of metabolism.

How do people affected by PKU manage such a strict diet?

Central to the PKU diet is the use of an enteral formula that is phenylalanine-free yet supplies energy, amino acids, vitamins, and minerals. Some formulas supply small amounts of phenylalanine and are useful for infants who must meet most of their nutrient needs by consuming formula. Formula requirements need to be recalculated periodically to accommodate the growing infant's shifting needs for protein, phenylalanine, tyrosine, and energy.

Once food consumption begins, a phenylalanine-free formula supplies the needed amino acids, and foods that contain phenylalanine are carefully monitored. All proteins contain some phenylalanine; therefore, high-protein foods such as meat, fish, poultry, milk, cheese, legumes, and nuts (including peanut butter) are omitted. Fruits, vegetables, and cereals also contain phenylalanine, so only limited amounts are allowed. Low-protein flours and mixes are available for making low-phenylalanine breads, pasta, cakes, and cookies. Foods that do not contain phenylalanine, such as jams, jellies, and most sweeteners, can be used freely. Growth rates and nutrition status are monitored to ensure that the diet is adequate.

Parents and children may need to develop creative ways to make the diet enjoyable. The formula can be flavored or combined with fruits or juices to make smoothies or frozen juice bars. Sandwiches can include low-phenylalanine breads and fillings such as mashed bananas or avocados, shredded carrots and olives, or tomato slices with mayonnaise. Children often

Normal:

Normally, the amino acid phenylalanine follows two pathways, one in the liver, the other in the kidneys. In the liver, the enzyme phenylalanine hydroxylase adds a hydroxyl group (OH) to produce the amino acid tyrosine. Tyrosine, in turn, produces melanin, the pigmented compound found in skin and brain cells; the neurotransmitters epinephrine and norepinephrine; and the hormone thyroxin. In the kidneys, enzymes convert phenylalanine to by-products that are excreted.

In the liver:

Phenylalanine → (Phenylalanine hydroxylase) → Tyrosine → Melanin, Epinephrine, Norepinephrine, Thyroxin

In the kidneys:

Phenylalanine → Phenylpyruvic acid (a ketone body) → Other phenyl acids (excreted)

In PKU:

Individuals with PKU lack the liver enzyme phenylalanine hydroxylase, impairing conversion of phenylalanine to tyrosine. Phenylalanine accumulates in the liver and blood, reaching the kidneys in abnormally high concentrations. In the kidneys, an aminotransferase enzyme converts phenylalanine to the ketone body phenylpyruvic acid, which spills into the urine—thus the name phenylketonuria.

In the liver:

Phenylalanine (accumulates) → (Phenylalanine hydroxylase (deficient)) → Tyrosine (deficient)

In the kidneys:

Phenylalanine (accumulates) → Phenylpyruvic acid (accumulates) → Other phenyl acids (accumulate)

FIGURE NP16–1

Biochemical Alterations in PKU

enjoy creating special recipes with permitted foods to make their choices more varied and to share meals with friends.

How long should a person with PKU continue the dietary treatment?

Lifelong adherence to a phenylalanine-restricted diet is currently recommended for all individuals with PKU.[4] Elevated phenylalanine levels can adversely affect cognitive function at any age. Case studies have suggested that PKU patients who discontinue dietary management may have problems with attention span, concentration, and memory. It is especially important that women with PKU maintain safe phenylalanine concentrations during pregnancy. Elevated phenylalanine levels, especially during the first trimester, have been associated with birth defects, congenital heart disease, and mental retardation in the offspring of PKU mothers who have discontinued dietary treatment.[5]

What is another example of an inborn error of metabolism that requires dietary changes?

Galactosemia is an example of an inborn error of carbohydrate metabolism. Individuals with galactosemia are deficient in one of the enzymes needed to metabolize galactose, a sugar that is primarily found in milk products (recall that each lactose molecule contains a molecule of galactose). An accumulation of galactose can cause damage in multiple tissues. Infants with galactosemia who are given milk react with severe vomiting and liver jaundice within days of the initial feeding. Serious liver damage can develop and progress to symptomatic cirrhosis. Other complications may include kidney failure, cataracts, and brain damage. Treatment in the first weeks of life can prevent the most detrimental effects of galactose accumulation, but if treatment is delayed, the damage to the brain is irreversible.[6]

What is the dietary treatment for galactosemia?

The diet for galactosemia is much simpler than the diet for PKU. For one thing, galactose is not an essential nutrient. The galactosemia diet essentially eliminates galactose from the diet and does not need to provide a carefully determined amount of any nutrient, like the PKU diet does. In addition, dietary galactose is primarily obtained from lactose (the milk sugar), so the main focus of dietary treatment is the exclusion of milk and milk products. A number of other foods that contain galactose in substantial amounts, such as organ meats and some legumes, fruits, and vegetables, must also be avoided or restricted.

Patients receive food lists that identify the galactose content of common foods.

Infants diagnosed with galactosemia are given lactose-free formulas to meet their nutrient needs. Once a child can consume adequate amounts of regular foods, special formulas are unnecessary. However, care must be taken to ensure that the diet supplies adequate calcium.

How effective is the dietary treatment for galactosemia?

Although the early introduction of a galactose-restricted diet can eliminate the acute toxic effects of galactosemia, complications of the disease may develop despite an individual's compliance with diet therapy. For example, most patients experience delays in speech and language development. Ovarian failure occurs in up to 85 percent of women who have galactosemia.[7] In addition, some evidence suggests that IQ declines as a person with galactosemia ages. The reasons for these long-term complications are not fully understood.

As our scientific understanding of human genetics and biochemistry increases, more inborn errors of metabolism are being recognized. Mainstays of treatment for these diseases include effective diagnosis, early treatment, and control of environmental factors that cause toxicity. In some cases, dietary changes are central to treatment and can prevent serious complications. Not all inborn errors are easily treated, however. Future developments in biotechnology may someday allow medical practitioners to correct genetic errors using gene therapy.

Notes

1. L. J. Elsas II, Inborn errors of metabolism, in L. Goldman and D. Ausiello, eds., *Cecil Textbook of Medicine* (Philadelphia: Saunders, 2004), pp. 185–191.

2. L. M. Bellini and M. A. Grippi, Cystic fibrosis, in A. P. Fishman and coeditors, *Fishman's Manual of Pulmonary Disease and Disorders* (New York: McGraw-Hill, 2002), pp. 176–186.

3. L. J. Elsas II and P. B. Acosta, Inherited metabolic disease: Amino acids, organic acids, and galactose, in M. E. Shils and coeditors, *Modern Nutrition in Health and Disease* (Baltimore: Lippincott Williams & Wilkins, 2006), pp. 909–959.

4. S. D. Cederbaum and C. R. Scriver, Disorders of phenylalanine and tyrosine metabolism, in L. Goldman and D. Ausiello, eds., *Cecil Textbook of Medicine* (Philadelphia: Saunders, 2004), pp. 1282–1285.

5. Cederbaum and Scriver, 2004.

6. L. J. Elsas II, Galactosemia, in L. Goldman and D. Ausiello, eds., *Cecil Textbook of Medicine* (Philadelphia: Saunders, 2004), pp. 1268–1270.

7. L. J. Elsas II, Galactosemia, 2004.

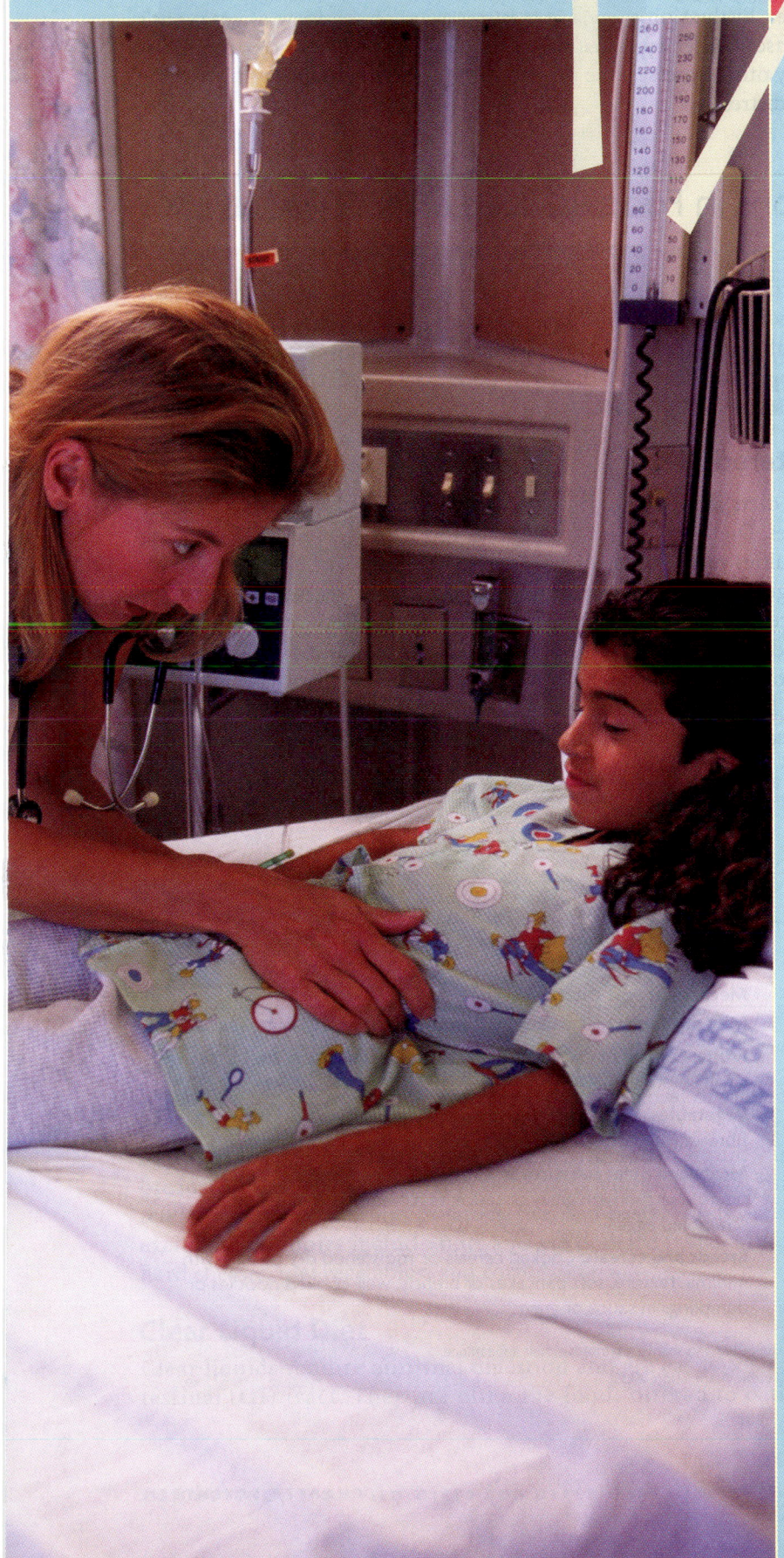

17

Consistency-Modified and Other Diets for Upper Gastrointestinal Disorders

Modifications in Food Texture and Consistency
Mechanically Altered Diets
Clear Liquid Diet

Conditions Affecting the Mouth and Esophagus
Dry Mouth
Dysphagia
Gastroesophageal Reflux Disease

Conditions Affecting the Stomach
Dyspepsia
Nausea and Vomiting
Gastritis
Peptic Ulcer Disease

Gastric Surgery
Gastrectomy
Bariatric Surgery

Nutrition in Practice
Helping People with Feeding Disabilities

© Keith Brofsky/Photodisc

487

feeding problem caused by hypersensitivity to oral stimulation. The health care professional may start by teaching the caregiver to gently and playfully stroke the child's face with a hand, washcloth, or soft toy. Once the child tolerates touch on less sensitive areas of the face, the health care professional may encourage the caregiver to slowly begin to rub the child's lips, gums, palate, and tongue. With time, the child may be better able to tolerate the presence of food in the mouth. Examples of other strategies that can help feeding problems are listed in Table NP17–2.

TABLE NP 17–2

Interventions for Feeding-Related Problems

Inability to suck

- Use squeeze bottles, which do not require sucking, to express liquids into the mouth.
- Place a spoon on the center of the tongue and apply downward pressure to stimulate sucking.
- Apply rhythmic, slow strokes on the tongue to alter tongue position and improve sucking response.

Inability to chew

- Place foods between gums and teeth to promote chewing.
- Improve chewing skills with different textured foods; for example, fruit leathers stimulate jaw movements but dissolve quickly enough to minimize choking.
- Provide soft foods that require minimal chewing or are easily chewed.

Inability to swallow

- Provide thickened liquids, pureed foods, and moist foods that form boluses easily.
- Provide cold formulas, frozen fruit juice bars, and ice; cold substances promote swallowing movements by the tongue and soft palate.
- Make sure the patient's jaw and lips are closed to facilitate swallowing action.
- Correct posture and head position if they interfere with swallowing ability.

Inability to grasp or coordinate movements

- Provide utensils that have modified handles or are smaller or larger as necessary.
- Encourage use of hands for feeding if utensils are difficult to maneuver.
- Provide plates with food guards to prevent spilling.
- Supply clothing protection.

Impaired vision

- Place foods (meats, vegetables) in similar locations on the plate at meals.
- Provide plates with food guards to prevent spilling.

Sources: J. Case-Smith and R. Humphry, Feeding and oral motor skills, in J. Case-Smith, A. S. Allen, and P. N. Pratt, eds., Occupational Therapy for Children (St. Louis, MO: Mosby–Year Book, 1996), pp. 430–460; S. Escott-Stump, Nutrition and Diagnosis-Related Care (Baltimore: Lippincott Williams & Wilkins, 2002), pp. 64–65.

Can special equipment be used to help people with certain feeding difficulties?

Yes. Figure NP17–1 shows a few of the many special feeding devices that are available and describes their uses. These devices can make a remarkable difference in a person's ability to eat independently. Other examples of adaptive equipment include specialized chairs to improve posture, bolsters inserted under arms to improve elbow stability, and raised trays or eating surfaces to simplify hand-to-mouth movements.[5]

Sometimes, despite the best efforts of all involved, a patient is unable to consume enough food by mouth. In these cases, tube feedings can help to improve nutrition status. Tube feedings are also recommended for patients who have severe dysphagia (difficulty swallowing) or aspiration pneumonia.[6]

In what ways can feeding difficulties affect family life?

Mealtimes are a critical time for social interaction, and individuals with feeding problems may encounter emotional and social problems if unable to participate. Children may fail to develop social skills, whereas adults may miss the social stimulation that mealtimes provide. Individuals should be encouraged to sit with family and friends during meals so that they are not deprived of the social and cultural aspects of eating.

The responsibility of caring for a person with a feeding problem can frequently overwhelm the caregiver. Caring for a disabled person requires time and patience—often many new therapies must be learned and administered. The caregiver may spend many hours preparing special foods, monitoring use of adaptive feeding equipment, and helping with feedings. Moreover, a disabled person may need help with other tasks as well, and all may require a considerable amount of time. In many cases, a caregiver may receive little or no assistance. These conditions may lead to strained interactions between caregiver and patient and cause frustration and depression.[7] Psychologists can offer counseling to patients or caregivers to help them adjust, and all members of the health care team can offer emotional support and practical suggestions to ease caregivers' responsibilities and frustrations.

Successful therapy for people with feeding disabilities requires the involvement of many health care professionals and depends on accurate identification of impaired feeding skills and determination of appropriate interventions. Ideally, with training, people with disabilities attain total independence—they are able to prepare, serve, and eat nutritionally adequate food daily without help. In some cases, these goals can be met with the help of caregivers. The combined efforts of the health care team can support both patients and caregivers in enhancing quality of life and in achieving independence to the greatest degree possible.

Utensils

Rocker knife

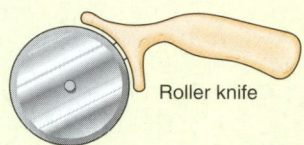

Roller knife

People with only one arm or hand may have difficulty cutting foods and may appreciate using a *rocker knife* or a *roller knife.*

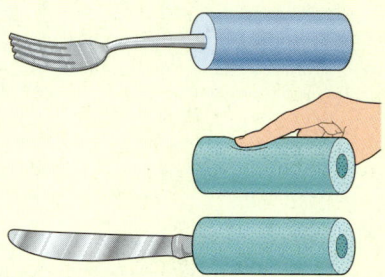

People with a limited range of motion can feed themselves better when they use *flatware with built-up handles.*

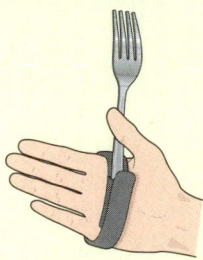

People with extreme muscle weakness may be able to eat with a *utensil holder.*

For people with tremors, spasticity, and uneven jerky movements, *weighted utensils* can aid the feeding process.

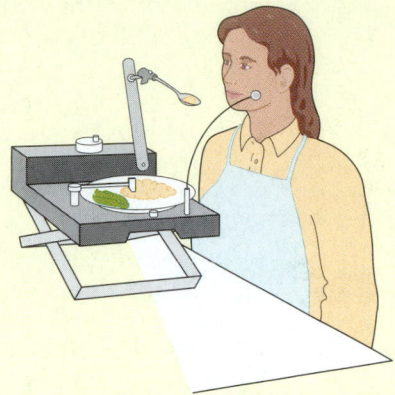

Battery-powered feeding machines enable people with severe limitations to eat with less assistance from others.

Plates

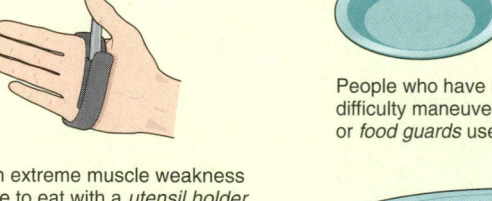

People who have limited dexterity and difficulty maneuvering food find *scoop dishes* or *food guards* useful.

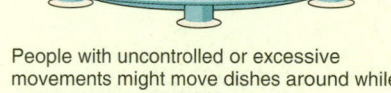

People with uncontrolled or excessive movements might move dishes around while eating and may benefit from using *unbreakable dishes with suction cups.*

Cups

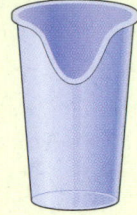

People with limited neck motion can use a *cutout plastic cup.*

Two-handed cups enable people with moderate muscle weakness to lift a cup with two hands.

People with uncontrolled or excessive movements might prefer to drink liquids from a *covered cup* or glass with a *slotted opening* or *spout.*

A soft, flexible long plastic straw may also ease the task of drinking.

FIGURE NP17–1

Examples of Adaptive Feeding Devices

Notes

1. E. B. Fung and coauthors, Feeding dysfunction is associated with poor growth and health status in children with cerebral palsy, *Journal of the American Dietetic Association* 102 (2002): 361–368.

2. Position of the American Dietetic Association: Providing nutrition services for infants, children, and adults with developmental disabilities and special health care needs, *Journal of the American Dietetic Association* 104 (2004): 97–107.

3. Position of the American Dietetic Association, 2004.

4. H. H. Cloud, Expanding roles for dietitians working with persons with developmental disabilities, *Journal of the American Dietetic Association* 97 (1997): 129–130.

5. J. Case-Smith and R. Humphry, Feeding and oral motor skills, in J. Case-Smith, A. S. Allen, and P. N. Pratt, eds., *Occupational Therapy for Children* (St. Louis, MO: Mosby–Year Book, 1996), pp. 430–460.

6. Position of the American Dietetic Association, 2004.

7. Fung and coauthors, 2002.

18

Fiber-Modified Diets for Lower Gastrointestinal Tract Disorders

Modifying Fiber Intake

Disorders of Bowel Function
Constipation
Diarrhea
Irritable Bowel Syndrome

Inflammatory Bowel Diseases

Diverticular Disease of the Colon

Colostomies and Ileostomies

Nutrition in Practice
Probiotics and Intestinal Health

© Bob Daemmrich/The ImageWorks

TABLE 18–1

Indications for Modifying Fiber Intake

Increasing Fiber Intake
Constipation
Diabetes mellitus
Diverticulosis (prevention)
Heart disease
Irritable bowel syndrome
Prediabetes
Weight management
Reducing Fiber Intake
Inflammatory bowel diseases (active episodes)
Postsurgical diet (depending on tolerance)

Unlike the use of consistency-modified diets in treating some conditions of the upper gastrointestinal (GI) tract, lower GI tract disorders can sometimes benefit by modifying dietary fiber. This chapter discusses the potential benefits of adjusting fiber intake when treating some common GI disorders. Chapter 19 describes illnesses that primarily affect nutrient digestion and absorption.

Modifying Fiber Intake

Diets can be modified by either increasing or decreasing dietary fiber. As discussed in Chapter 2, *insoluble fibers* increase fecal weight and speed the passage of wastes through the large intestine. *Soluble, viscous fibers* slow the passage of food through the GI tract, which increases satiety and delays glucose absorption in the small intestine. A diet high in soluble fibers can also lower blood cholesterol levels because the fibers bind and reduce reabsorption of the bile acids. Table 18–1 lists some medical problems that may benefit from modifications in fiber intake.

As Chapter 2 described, health authorities recommend that most people increase their intake of fiber-rich foods by emphasizing the consumption of whole grains, legumes, nuts and seeds, and fruits and vegetables. Figure 2–4 (p. 54) and Appendix A provide information about the fiber content of selected foods. Figure 18–1 shows a one-day menu that includes about 38 grams of fiber, which is the DRI for an adult male. People who need to restrict dietary fiber must eliminate or limit consumption of the high-fiber foods listed.

The fiber DRI for women and men aged 19 to 50 years are 25 grams and 38 grams, respectively.

Increased intestinal gas (**flatulence**) is sometimes an unpleasant side effect of consuming a high-fiber diet; this is because undigested fibers pass into the colon where they are fermented by bacteria, which produce gas as a by-product. Therefore, health practitioners often advise that fiber-containing foods be added gradually at first and portions increased as tolerance improves. Intestinal gas also develops when other carbohydrates are incompletely digested or absorbed; these include fructose, sugar alcohols (sorbitol, mannitol, maltitol), the undigestible carbohydrates in beans (raffinose and stachyose), and some forms of resistant starch, found in grain products and potatoes. Table 18–2 lists foods commonly associated with excessive gas production.[1] Conditions that cause malabsorption (discussed in Chapter 19) also cause flatulence because the undigested nutrients can be metabolized by intestinal bacteria. Swallowed

FIGURE 18–1

Sample One-Day Menu Containing 38 g Fiber

High-Fiber Diet Menu

Breakfast	Lunch	Supper	Snack
1 c multigrain cereal	1 c black bean soup	3 oz baked fish	3 c popcorn
½ c strawberries	3 oz broiled chicken	½ c brown rice	1 c tomato juice
1 c fat-free milk	½ c steamed broccoli	½ c peas	
2 slices whole wheat toast	½ c baked sweet potatoes	1 whole wheat dinner roll	
2 tbs peanut butter	1 fresh pear	2 tsp margarine	
1 c coffee	1 whole wheat dinner roll	1 piece carrot cake	
	1 tsp margarine	1 c fat-free milk	

air also contributes to intestinal gas. Note that people often attribute abdominal bloating or pain to excessive gas, but these symptoms do not correlate well with increased intestinal gas.[2]

IN SUMMARY

- The amount of insoluble fiber in foods affects stool volume and transit time in the large intestine. Soluble, viscous fibers slow the passage of food in the GI tract and help to reduce blood cholesterol levels.

- Fiber intake can be increased by emphasizing the consumption of whole grains, legumes, nuts and seeds, and fruits and vegetables.

- If fiber-containing foods cause excessive intestinal gas, they should be added gradually and portions increased as tolerance improves.

Disorders of Bowel Function

Many factors can alter the transit time and fluid content of stools in the lower GI tract, causing constipation and diarrhea. Consequently, these symptoms are common complaints of people with a wide variety of health problems. Because dietary fiber affects GI transit time as well as stool volume and composition, modifying fiber intake can often serve in the treatment of these problems.

Constipation

A person who complains of constipation is usually concerned about difficulty passing stools or having infrequent bowel movements. However, sometimes people who complain of constipation have a mistaken notion of what constitutes "normal" bowel habits. Constipation may be diagnosed when a person has fewer than three bowel movements per week. Constipation is much more prevalent among women than men and increases somewhat with aging.

Causes of Constipation In Western societies, constipation generally correlates with low-fiber diets, low food intake, and inactivity. All of these factors can extend the time that wastes take to travel through the colon, leading to increased water removal and dry, hard stools that are difficult to pass. Medical conditions often associated with constipation include diabetes mellitus, chronic renal failure, and hypothyroidism. Neurological conditions such as Parkinson's disease, spinal cord lesions, and multiple sclerosis may cause motor problems that lead to constipation. During pregnancy, women may experience constipation because the enlarged uterus presses against the lower portion of the colon. Constipation is also a common side effect of several classes of medications and some dietary supplements including opiate-containing analgesics, tricyclic antidepressants, antihistamines, calcium channel blockers, aluminum-containing antacids, and iron and calcium supplements.

Treatment of Constipation The primary treatment for constipation is a gradual increase in fiber intake. High-fiber diets increase stool weights and promote a more rapid transit of materials through the colon. Foods that increase stool weight the most are wheat bran, fruits, and vegetables.[3] Bran intake can be increased by adding bran cereals and whole wheat bread to the diet or by mixing bran powder with beverages or foods. Because the transition to a high-fiber diet may be difficult for some people because it can increase intestinal gas, high-fiber foods should be added gradually, as tolerated. Fiber supplements like methylcellulose (Citrucel), psyllium (Metamucil, Fiberall), and

TABLE 18-2
Foods That May Increase Intestinal Gas
Apples
Beer
Broccoli
Brussels sprouts
Cabbage
Carbonated beverages
Cauliflower
Corn
Dried beans and peas
Fruit juices
Leeks
Milk products (if lactose intolerant)
Onions
Peanuts
Pears
Potatoes
Turnips

flatulence: the condition of having excessive intestinal gas, which causes abdominal discomfort.

TABLE 18–3

Laxatives and Bulk-Forming Agents

Laxative Type	Active Ingredients	Product Examples	Method of Action	Cautions
Fiber (bulk formers)	Methylcellulose, polycarbophil, psyllium, malt soup extract	Metamucil, Citrucel, FiberLax	Fiber supplements increase stool weight and aid in formation of soft, bulky stools. Similar effects are achieved by adding bran to the diet. For mild constipation. Safe for long-term use.	Some fiber supplements may increase flatulence. Psyllium may cause an allergic reaction.
Emollient (stool softeners)	Docusate sodium	Colace, Surfak	Detergent action promotes the mixing of water with stools. Prevents formation of dry, hard stools.	Does not increase stool weight. Limited effectiveness.
Nonabsorbable sugars (osmotic laxatives)	Lactulose, sorbitol, mannitol	Chronulac, Cephulac	Unabsorbed sugars attract water to large intestine and promote softer stools. Must be used for several days to take effect. Safe for long-term use.	May cause flatulence and cramps. Can lose effectiveness over time.
Saline laxatives (osmotic laxatives)	Magnesium hydroxide, magnesium citrate, sodium sulfate	Milk of magnesia	Unabsorbed salts attract and retain water in large intestine and stimulate contractions.	May cause bloating and watery stools or diarrhea. Should be used with caution. Avoid using in renal patients and children.
Stimulant or irritant laxatives	Senna, bisacodyl, cascara, castor oil, aloe	Ex-Lax, Correctol, Dulcolax	Act as local irritants to colonic tissue; stimulate peristalsis and mucosal secretions. For moderate-to-severe constipation. Long-term use is discouraged.	Usually given only after milder treatments fail. May alter fluid and electrolyte balances. May lead to laxative dependency.

polycarbophil (a synthetic fiber) are also effective (see Table 18–3). Unlike other fibers, methylcellulose and polycarbophil do not increase intestinal gas.

Some other measures may also help constipation. Consuming adequate fluid prevents dehydration, which can draw water from the colon in order to increase hydration in the rest of the body. Adding prunes or prune juice to the diet is often recommended because prunes contain compounds that have a mild laxative effect. Increasing daily exercise can help to stimulate peristalsis.

High-fiber foods promote regular bowel movements.

Laxatives Many laxatives can be purchased without prescription. They work by increasing stool weight, increasing the water content of the stool, or stimulating peristaltic contractions. Table 18–3 includes examples of popular laxatives and describes their modes of action. Enemas and suppositories (chemicals introduced into the rectum) are also used to promote defecation; they work by distending and stimulating the rectum or by lubricating the stool.

Medical Interventions Patients with severe constipation who do not respond to dietary or laxative treatments may require more aggressive therapies. Some prescription drugs (for example, prucalopride) work by targeting the neural receptors involved with colonic peristalsis. Surgical interventions are a last resort and include colonic resections and colostomy operations (see p. 523).

Diarrhea

Diarrhea is characterized by the passage of frequent, watery stools. In most cases, it lasts for only a day or two and subsides without complication. Severe or persistent diarrhea, however, can cause dehydration and electrolyte imbalances. Serious cases of diarrhea are often accompanied by other symptoms, such as fever, cramps, dyspepsia, or bleeding, that help in diagnosing its cause.

Causes of Diarrhea Diarrhea is a complication of various medical problems and may also be caused by infections, medications, or dietary substances; therefore, diarrhea is often classified according to the cause of fluid loss.[4] *Motility disorders* can accelerate the entry of fluids into the colon, where they are inadequately reabsorbed. *Osmotic diarrhea* results from nutrient malabsorption, as when poorly absorbed sugars like sorbitol or fructose attract water to the colon and increase fecal water content. In *secretory diarrhea,* the intestines are stimulated to secrete excessive fluid into the colon. Secretory diarrhea is often due to bacterial food poisoning but can also be caused by conditions that cause severe intestinal inflammation.

Acute cases of diarrhea start abruptly and may persist for several weeks; they are frequently caused by infections or occur as a side effect of medication use. Chronic diarrhea persists for one month or longer and can result from altered GI tract motility, intestinal inflammation, malabsorptive and endocrine disorders, infectious diseases, radiation treatment, and many other conditions.

Treatment of Diarrhea Correcting the underlying medical disorder is the first step in treating diarrhea. Antibiotics are given to treat infections. If a medication causes diarrhea, an alternate drug may be prescribed. If certain foods are responsible, they can be omitted from the diet. Bulk-forming agents such as psyllium (Metamucil) can help to reduce the liquidity of the stool. If chronic diarrhea does not respond to treatment, antidiarrheal drugs may be prescribed to slow GI motility or reduce intestinal secretions. People with severe, **intractable** diarrhea sometimes require total parenteral nutrition.

Nutrition therapy depends on the cause of diarrhea, its severity and duration, and the degree of water loss. Rehydration therapy (see the following section) is necessary if dehydration develops. A low-fiber diet may improve symptoms in some individuals. Fiber restriction is also necessary during episodes of active intestinal inflammation, which can reduce tolerance to fiber. Foods and beverages that contain fructose, sugar alcohols, or lactose may worsen symptoms. Caffeine-containing foods should be avoided because they stimulate GI motility. Table 18–4 (p. 518) lists foods that may either aggravate or alleviate diarrhea, although individual tolerances vary.

Rehydration Therapy Severe diarrhea requires the replacement of lost fluid and electrolytes. Oral rehydration solutions can be purchased or easily mixed using water, salts, and glucose or sucrose (see the recipe in the margin). ■ The addition of carbohydrate to the rehydration solution facilitates sodium and water absorption. Commercial sports drinks are not recommended as oral rehydration solutions because their sodium contents are too low to replace losses that result from severe diarrhea. If diarrhea results in extreme dehydration, intravenous solutions can be used to quickly replenish fluid and electrolytes.

Irritable Bowel Syndrome

People with **irritable bowel syndrome** experience chronic and recurring intestinal symptoms that cannot be explained by specific physical abnormalities. Symptoms include both diarrhea and constipation, as well as flatulence, bloating, and distention. In some patients, symptoms may be mild; in others, the defecation disturbances can

■ An oral rehydration solution can be mixed from the following ingredients:

- ½ tsp sodium chloride (table salt)
- ⅓ tsp potassium chloride (salt substitute)
- ¾ tsp sodium bicarbonate (baking soda)
- 1⅓ tbs sugar
- 1 liter water

intractable: not easily managed or controlled.

irritable bowel syndrome: an intestinal disorder of unknown cause that affects the functioning of the lower bowel; symptoms include abdominal pain, flatulence, diarrhea, and constipation.

TABLE 18–4

Foods That May Affect Diarrhea

Foods That May Worsen Diarrhea	Foods That May Lessen Diarrhea
Apple juice	Applesauce
Caffeine-containing beverages	Bananas
Coffee	Barley
Dates	Cheese
Fried foods	Oat bran
Fructose-sweetened drinks	Oatmeal
Grapes	Peanut butter (smooth)
Honey	Potatoes
Milk and milk products	Rice (boiled)
Pear juice	Soda crackers
Prune juice	Tapioca
Sugar-free candies	Yogurt

Note: Individual tolerances vary.

Sources: American Dietetic Association, *Manual of Clinical Dietetics* (Chicago: American Dietetic Association, 2000), p. 423; M. H. Beers and R. Berkow, eds., *The Merck Manual of Diagnosis and Therapy* (Whitehouse Station, NJ: Merck Research Laboratories, 1999), pp. 275–278.

interfere with work and social activities and dramatically alter the person's lifestyle and sense of well-being. Irritable bowel syndrome accounts for about 12 percent of primary care visits and is more common in women than in men. About 30 percent of people with the disorder eventually become asymptomatic.[5]

Although the causes of irritable bowel syndrome remain elusive, people with the disorder tend to have excessive colonic responses to meals, GI hormones, and stress.[6] Many individuals exhibit hypersensitivity to intestinal distention; for example, they may have an exaggerated response to normal meal transit and intestinal gas. Intestinal motility after meals may be excessive, leading to diarrhea, or reduced, causing constipation. Symptoms often worsen during periods of psychological stress, and in some cases, stressful life events may trigger the illness. Some evidence suggests that an infection may be the cause of the initial GI disturbance and that tissue sensitization persists after the infection has healed.

Treatment of Irritable Bowel Syndrome Medical treatment of irritable bowel syndrome often includes dietary adjustments, stress management, and behavioral therapies. Medications may be prescribed to manage symptoms although they are not always helpful. The drugs prescribed may include antidiarrheal agents, anticholinergics, antidepressants, and laxatives.

Medical Nutrition Therapy for Irritable Bowel Syndrome Although dietary changes may be useful, measures that help one symptom can sometimes make another worse. The usual dietary advice is to increase fiber intake, which helps to reduce constipation and improve stool bulk. To minimize discomfort from intestinal gas, fiber-containing foods should be added gradually. Other foods that produce gas should be avoided unless well tolerated (review Table 18–2). If diarrhea persists, a bulking agent (psyllium) may be effective. Avoidance of milk products may benefit those who are lactose intolerant. Caffeine and alcohol can exacerbate symptoms. The placebo effect has a strong influence on food tolerance, so foods that are perceived to be problematic should be discussed with patients so that the diet is not restricted unnecessarily.

The diet history may reveal dietary behaviors that worsen symptoms. Generally, small, frequent meals are better tolerated than larger ones. Eating quickly should be

Marcy Hudson is a 22-year-old recent college graduate who began her first professional job in a bank one month ago. As a college student, she occasionally experienced abdominal pain and cramping after eating. She also had frequent bouts of diarrhea and felt somewhat better after bowel movements. Once Marcy began her new job, her symptoms occurred more frequently. At first she attributed her symptoms to job stress, but when the symptoms continued for several months, she decided to see her physician. After taking a careful history and conducting tests to rule out other bowel disorders, the physician diagnosed irritable bowel

syndrome. The physician prescribed bulk-forming agents and advised Marcy to keep a record of her food intake and symptoms for one week. Marcy was then referred to a dietitian for a review of her dietary record. The dietitian noticed that Marcy routinely drank several cups of coffee in the morning and had large meals for lunch and dinner. Marcy often ate out in Mexican restaurants and favored highly spiced foods and refried beans. Between meals, she snacked on low-carb foods sweetened with sugar alcohols and drank several cans of soda daily. Her dietary fiber intake, however, totaled only about 15 grams daily.

1. Describe the characteristics of irritable bowel syndrome to Marcy, and indicate the role that stress might play in her illness.

2. Explain how the record of food intake and symptoms might be helpful in devising an appropriate diet plan for Marcy. Are any of the foods in Marcy's diet likely to be aggravating her symptoms?

3. What dietary measures might benefit individuals with irritable bowel syndrome? What problems might the dietary changes cause?

discouraged because it increases the amount of air that is swallowed. Some patients may find a low-fat diet easier to tolerate. Fluid intake should be assessed for adequacy. A careful evaluation of the dietary patterns that exacerbate symptoms may uncover the foods and habits most closely associated with intestinal discomfort. Review the accompanying Case Study to apply your knowledge about irritable bowel syndrome to a clinical situation.

IN SUMMARY

- Constipation and diarrhea can be caused by a wide variety of health problems.

- Constipation is often improved by a high-fiber diet or fiber supplements.

- Diarrhea may result from infection, malabsorption, motility disorders, or dietary substances; treatment depends on its cause, severity, and duration.

- Irritable bowel syndrome is associated with abdominal pain and alternating diarrhea and constipation; a high-fiber diet is often recommended.

Inflammatory Bowel Diseases

Inflammatory bowel diseases are chronic inflammatory disorders involving the GI tract. Both genetic and environmental factors are believed to contribute to the development of these diseases, but the exact triggers are unknown. Table 18–5 (p. 520) compares the two major forms of inflammatory bowel disease, **Crohn's disease** and **ulcerative colitis**. Crohn's disease usually involves the small intestine and may lead to nutrient malabsorption, whereas ulcerative colitis affects the colon, which is past the absorptive areas. Both diseases may cause nutrient losses due to tissue damage, bleeding, and diarrhea.

Complications of Crohn's Disease Crohn's disease may occur in any region of the GI tract, but most cases involve the ileum and/or colon. Lesions may develop in different

Crohn's disease: an inflammatory bowel disease that usually occurs in the lower portion of the small intestine and the colon. Inflammation may pervade the entire intestinal wall.

ulcerative colitis (ko-LYE-tis): an inflammatory bowel disease that involves the colon. Inflammation affects the mucosa and submucosa of the intestinal wall.

TABLE 18–5

Comparison of Crohn's Disease and Ulcerative Colitis

	Crohn's Disease	Ulcerative Colitis
Location of inflammation	Approximately 40 percent of cases involve the ileum and cecum, 30 percent are in the small intestine only, and 20 percent are in the colon.	Inflammation is confined to the rectum and colon; it begins at the rectum and spreads into the colon.
Pattern of inflammation	Discrete areas separated by normal tissue ("skip" lesions)	Continuous inflammation that begins at the rectum and ends abruptly within the colon
Depth of damage	Damage throughout all layers of tissue; causes deep fissures that give intestinal tissue a "cobblestone" appearance	Damage primarily in the mucosa and submucosa
Fistulas	Common	Usually do not occur
Cancer risk	Increased	Greatly increased

areas in the intestine, with normal tissue separating affected regions (called "skip" lesions). The inflammation may extend deeply into intestinal tissue and be accompanied by ulcerations, fissures, and **fistulas** (abnormal passages between tissues). The resultant scar tissue thickens and stiffens the intestine, narrowing the lumen and sometimes causing obstructions. About 60 to 70 percent of patients require surgical resections during the course of illness, although disease often recurs in the remaining intestine. Patients with Crohn's disease are at increased risk of developing intestinal cancers.

Malnutrition may result from nutrient malabsorption, reduced food intake, and surgical resections that shorten the small intestine. If the ileum is affected, bile acids may become depleted, ■ causing malabsorption of fat, fat-soluble vitamins, calcium, magnesium, and zinc (minerals can bind to unabsorbed fatty acids). Because the ileum is the site of vitamin B_{12} absorption, deficiency can develop unless supplements are taken. Anemia may result from bleeding, the inadequate absorption of nutrients involved in blood cell formation, or the metabolic effects of chronic illness (see Nutrition in Practice 19).

■ Most of the bile used during digestion is eventually reabsorbed in the ileum and returned to the liver.

Complications of Ulcerative Colitis Ulcerative colitis always involves the rectum and usually extends into the colon. Inflammation is continuous along the length of intestine affected, ending abruptly at the area where healthy tissue begins. Tissue erosion

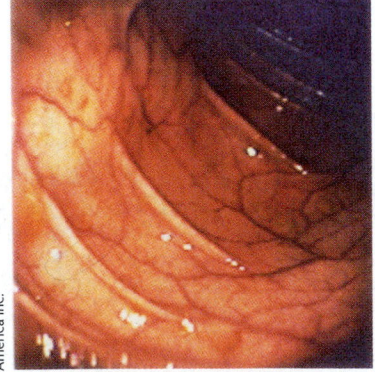

The healthy colon has a smooth surface with a visible pattern of fine blood vessels.

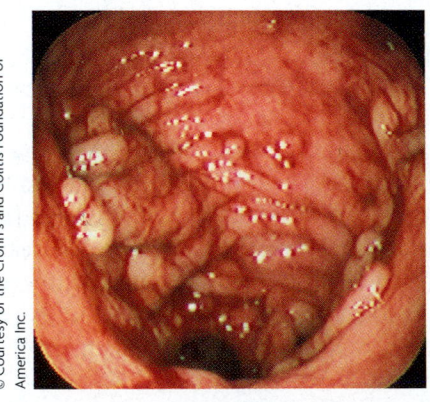

In Crohn's disease, the mucosa has a "cobblestone" appearance due to deep fissuring in the inflamed mucosal tissue.

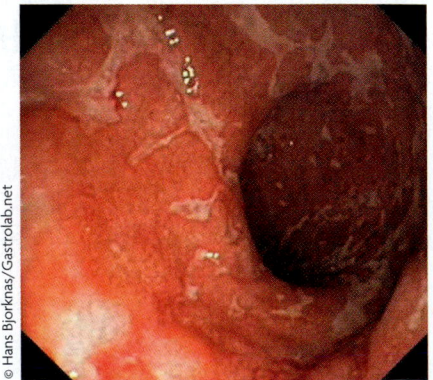

In ulcerative colitis, the colon appears inflamed and reddened, and ulcers are visible.

Check this table for notable nutrition-related effects of the medications discussed in this chapter.

	Gastrointestinal Effects	Interactions with Dietary Substances	Metabolic Effects
Antidiarrheals	Constipation	—	—
Anti-inflammatory drugs (sulfasalazine, corticosteroids)	Nausea, heartburn (sulfasalazine)	Sulfasalazine may decrease folate absorption; supplementation is recommended	Anemia (sulfasalazine); fluid retention, hyperglycemia, hypocalcemia, hypokalemia, hypophosphatemia, increased appetite, protein catabolism (corticosteroids)
Laxatives	Diarrhea	Mineral oil may decrease absorption of fat-soluble vitamins but is not often used	Fluid and electrolyte imbalances, laxative dependency

or ulceration develops primarily in the mucosa and submucosa (the top two layers of intestinal tissue). During active episodes, patients have frequent, urgent bowel movements that are small in volume. Stools are often streaked with blood and contain mucus.

Although mild disease may cause few complications, weight loss, fever, and weakness are common when most of the colon is involved. Severe disease is often associated with anemia (due to blood loss), dehydration, and electrolyte imbalances. Protein losses from the inflamed tissue can be substantial. A **colectomy** (removal of the colon) is performed in 20 to 25 percent of patients and prevents future recurrence.[7] Colon cancer risk is substantially increased in ulcerative colitis patients.

Drug Treatment of Inflammatory Bowel Diseases Medications help to control symptoms, reduce inflammation, and minimize complications. The drugs prescribed include antidiarrheal agents, immunosuppressants, and anti-inflammatory drugs (usually corticosteroids and salicylates). Although these medications help in achieving and maintaining remission, some may cause side effects that are detrimental to nutrition status (check the Diet-Drug Interactions feature above).

Medical Nutrition Therapy for Inflammatory Bowel Diseases Crohn's disease often requires aggressive dietary management because it can lead to protein-energy malnutrition (PEM), nutrient deficiencies, and growth failure in children. Dietary measures depend on the symptoms and complications that develop, so the nutrition care of patients is highly variable (see Table 18–6, p. 522). Some people require high-kcalorie, high-protein diets to prevent or treat malnutrition. Liquid supplements can increase energy intake and improve weight gain. Tube feedings can be used to supplement the diet or may be the sole means of providing nutrients; elemental formulas are most easily tolerated. Some individuals may tolerate meals better if high-fiber and lactose-containing foods are restricted. Vitamin and mineral supplements are generally needed, especially if nutrient malabsorption is present. Table 18–6 includes examples of other adjustments that may be beneficial.

The diet for ulcerative colitis may require few adjustments. During severe illness, dietary goals are to restore fluid and electrolyte balances and correct deficiencies that result from protein and blood losses. A low-fiber diet may reduce irritation by minimizing fecal volume. Sometimes food must be withheld to allow bowel rest. Fluids and

fistulas (FIST-you-luz): abnormal passages between body tissues; may lead from one hollow organ to another or to an organ surface.

colectomy: removal of a portion or all of the colon.

TABLE 18–6

Management of Symptoms and Complications in Crohn's Disease

Symptom or Complication	Possible Dietary Measures
Growth failure/weight loss	High-kcalorie diet Enteral supplements Elemental tube feedings
Anorexia/pain with eating	Small, frequent meals Enteral supplements If long-term (>5 to 7 days): elemental tube feedings
Malabsorption	High-kcalorie diet Nutrient supplementation
Steatorrhea (fat malabsorption)	Fat restriction Medium-chain triglycerides Nutrient supplementation
Diarrhea	Fluid and electrolyte replacement Nutrient supplementation
Lactose intolerance	Avoidance of lactose-containing foods
Nutrient deficiencies	Nutrient-dense diet Nutrient supplementation
Intestinal recovery	High-protein diet Glutamine supplementation
Strictures/fistulas	Low-fiber diet
Severe bowel obstruction/high-output fistulas/severe exacerbations of disease	Total parenteral nutrition

electrolytes may need to be replaced intravenously, and some patients require parenteral nutrition support. As in Crohn's disease, the symptoms and complications that develop can be managed with specific dietary measures (see Table 18–6).

IN SUMMARY

- Crohn's disease causes damage to the intestinal mucosa and impairs its absorptive function. Ulcerative colitis is an inflammatory bowel condition that is confined to the rectum and colon.

- The treatment for inflammatory bowel diseases includes drug therapies, surgical removal of diseased areas, and correction of malnutrition.

- The nutrition therapy for inflammatory bowel diseases varies according to the symptoms and complications that develop. A low-fiber diet can reduce irritation during active episodes of illness.

Diverticular Disease of the Colon

Diverticulosis refers to the presence of pebble-sized outpockets in the intestinal wall, called diverticula (see Figure 18–2). The prevalence of diverticulosis increases with age, occurring in about half of adults over 60 years of age.[8] Most people with diverticulosis are symptom-free and remain unaware of the condition until a complication develops.

Dietary fiber influences the development of diverticulosis. By increasing stool weight and bulk, fiber reduces the workload of the circular muscles that move wastes through the colon. Low-fiber diets require more vigorous muscle contractions, increasing pressure within the segments immediately adjacent to the circular muscles. This increase in pressure induces small areas of intestinal tissue to balloon outward over time.[9]

Diverticulitis Inflammation or infection sometimes develops in the area around a diverticulum. This condition, called **diverticulitis**, is the most common complication of diverticulosis. It is thought to result from hardened fecal matter that abrades the mucosal lining, causing inflammation and possibly a microperforation that leads to subsequent infection. If the infection spreads to adjacent organs, fistulas may develop. More rarely, the infection spreads to the peritoneal cavity, causing life-threatening illness. Symptoms of diverticulitis include persistent abdominal pain, fever, and alternating constipation and diarrhea.

Treatment for Diverticular Disease Treatment for diverticulosis is necessary only if symptoms develop; the treatment focuses on reducing pain and alleviating constipation. Increasing dietary fiber may help to prevent disease progression, so patients are sometimes advised to add wheat bran to meals or use bulk-forming agents such as psyllium. Although dietary advice sometimes includes a recommendation to avoid nuts, popcorn, and foods that contain seeds, there is no evidence that these restrictions can reduce complications.[10]

Patients with diverticulitis may need antibiotics to treat infections and, possibly, pain-control medications. In mild cases, a clear liquid diet may be advised initially, with progression to solid foods as tolerated. In more severe cases, bowel rest is necessary (oral fluids and food are withheld), and fluids are given intravenously. Oral intakes are gradually reintroduced as the condition improves. Surgical interventions are sometimes necessary to treat complications of diverticulitis and may include removal of the affected portion of colon.

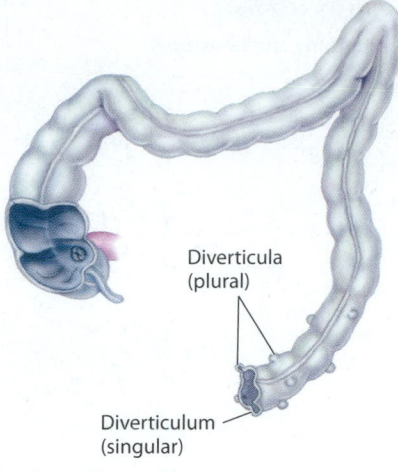

FIGURE 18–2

Diverticula in the Colon
Diverticula are small pouches that develop in weakened areas of the intestinal wall. The condition of having diverticula is known as diverticulosis.

IN SUMMARY

- Diverticulosis is a condition in which outpockets develop in the colon walls; it is often asymptomatic until complications develop.

- Increasing dietary fiber may help to prevent the development and progression of diverticulosis.

- Diverticulitis is an inflammation or infection involving the diverticula; symptoms may include abdominal pain, fever, and alternating constipation and diarrhea.

Colostomies and Ileostomies

An *ostomy* is a surgically created opening (called a **stoma**) in the abdominal wall through which dietary waste can be eliminated. A permanent ostomy is necessary after a partial or total colectomy. A temporary ostomy is sometimes constructed so that part or all of the colon can be bypassed after injury or extensive surgery. To create the stoma, the cut end of the remaining segment of functioning intestine is brought through an opening in the abdominal wall and stitched in place so that it empties to the exterior. The stoma can be formed from a section of the colon (**colostomy**) or ileum (**ileostomy**), as shown in Figure 18–3, p. 524.

To collect waste, a disposable bag is affixed to the skin around the stoma and emptied during the day as needed. Alternatively, an interior pouch can be surgically constructed behind the stoma using intestinal tissue, and it can be emptied with a catheter when convenient. Stool consistency varies according to the length of colon that is

diverticulosis (dye-ver-tic-you-LOH-sis): an intestinal disorder characterized by the presence of small outpockets (called diverticula) in the intestinal wall, which often develop by middle age.

diverticulitis (dye-ver-tic-you-LYE-tis): an inflammation or infection involving diverticula.

stoma (STOE-ma): a surgical opening made in the abdominal wall.

colostomy (co-LAHS-toe-me): a surgical procedure that creates a stoma using a section of the colon.

ileostomy (ill-ee-OS-toe-me): a surgical procedure that creates a stoma using the ileum.

FIGURE 18–3

Colostomy and Ileostomy

Colostomy

Ileostomy

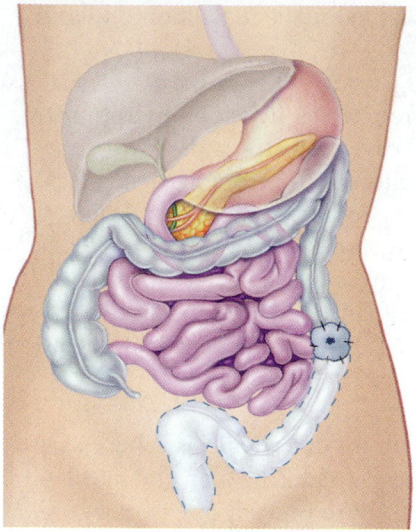
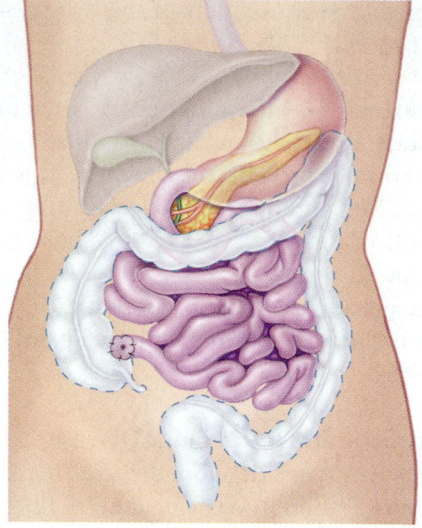

In a colostomy, a portion of the colon is removed or bypassed, and the stoma is formed from the remaining section of functional colon.

In an ileostomy, the entire colon is removed or bypassed, and the stoma is formed from the ileum.

functional. If a small portion of the colon is absent or bypassed, the stools may continue to be semisolid. If the entire colon has been removed or is bypassed, absorption of fluid and electrolytes into the body is reduced substantially, and the output is liquid.

Medical Nutrition Therapy for Ostomies Following surgery, the diet gradually progresses from clear liquids that are low in sugars to a normal meal plan that contains low-fiber foods. Small, frequent meals may be better tolerated at first. Questionable foods should be added back to the diet one at a time and in small amounts to assess their effects. A food that causes problems can be tried again at a later time.

People with ileostomies need to chew thoroughly to ensure that foods can be adequately digested and to prevent obstructions, which are a common complication. Foods high in insoluble fibers are sometimes avoided because they reduce intestinal transit time and may increase output. Because the colon is no longer available for reabsorbing water, the diet should provide at least 8 cups of liquid daily to prevent dehydration.

Dietary concerns after colostomies depend on the length of colon removed. If a large portion is removed, recommendations may be similar to those given to ileostomy patients. In other cases, a high-fiber diet is often recommended to improve stool consistency and promote regularity.

Obstructions As mentioned, foods that are incompletely digested can cause obstructions, a primary concern of ileostomy patients. Although almost any food can be consumed if cut into small pieces and carefully chewed, the following foods may cause difficulty: corn, celery, coconut, dried fruit, grapes, nuts, popcorn, raw cabbage (for example, in coleslaw), and unpeeled apples.[11]

Reducing Gas and Odors Persons with ostomies are often concerned about foods that may increase gas production or cause strong odors. Foods that may cause excessive gas include those listed in Table 18–2 on p. 515; practices that increase gas formation include smoking, gum chewing, tobacco chewing, using drinking straws, and eating quickly. Foods that sometimes produce unpleasant odors include fish, eggs, dried beans and peas, onions, garlic, asparagus, brussels sprouts, and beer.

Foods that may help to reduce odors include buttermilk, cranberry juice, parsley, and yogurt.[12]

Diarrhea Examples of foods that may either aggravate or reduce diarrhea were listed in Table 18–4 on p. 518. What works may differ for each individual, however, and is best determined by trial and error.

IN SUMMARY

- Colostomies and ileostomies are surgically created openings in the abdominal wall using the colon or ileum.

- Fluid and electrolyte requirements are greater after an ostomy because colon function is reduced or absent.

- Foods that are poorly digested may cause obstructions in people with ostomies, although thorough chewing can reduce risk.

- Foods should be avoided if they provoke diarrhea or cause excessive gas or strong odors.

Nutrition Assessment Checklist

FOR PEOPLE WITH LOWER GI TRACT DISORDERS

Medical History

Check the medical record for diseases that:

- Cause chronic GI symptoms, such as irritable bowel syndrome or ulcerative colitis
- Interfere with nutrient absorption, such as Crohn's disease

Check for surgical procedures involving the lower GI tract, such as:

- Intestinal resections
- Ileostomy
- Colostomy

Check for the following symptoms or complications:

- Anemia
- Constipation
- Diarrhea, dehydration
- Fistulas
- Lactose intolerance
- Nutrient deficiencies
- Poor growth, in children

Medications

Check for medications or dietary supplements that may:

- Cause constipation or diarrhea
- Alter appetite or nutrient needs

Dietary Intake

Note the following problems and contact the dietitian if you suspect difficulty with:

- Poor appetite or food intake
- Food intolerances

- Inadequate fiber intake, in those with constipation
- Fluid intake

Anthropometric Data

Measure baseline height and weight. Address weight loss early to prevent malnutrition in patients with:

- Severe or persistent diarrhea
- Nutrient malabsorption

Laboratory Tests

Check laboratory tests for signs of dehydration, electrolyte imbalances, nutrient deficiencies, and anemia in patients with:

- Severe or persistent diarrhea
- Nutrient malabsorption
- Intestinal resections

Physical Signs

Look for physical signs of:

- Dehydration
- Protein-energy malnutrition
- Essential fatty acid and fat-soluble vitamin deficiencies
- Folate and vitamin B_{12} deficiencies
- Mineral deficiencies

Analysis

CLINICAL APPLICATIONS

1. A health practitioner working with a patient with a constipation problem had earlier provided him with detailed information about a high-fiber diet. At a follow-up appointment, he reported no change in symptoms. His food diary for that day showed that he consumed an omelet and toast for breakfast and a sandwich with juice for lunch.

   ■ Considering these two meals only, what additional information would help the health practitioner evaluate the man's compliance with the diet he was given?

   ■ Review the discussion about fiber in Chapter 2 and create a one-day menu that provides the DRI for fiber for an adult male, using the fiber values listed in Appendix A.

2. A number of symptoms and disorders of the GI tract, described in this chapter and the previous one, are associated with aging. Review both chapters and list these symptoms and disorders. Referring back to Chapter 13, describe some effects of aging on the GI tract and explain how these changes might relate to the GI symptoms and disorders you listed.

Self Check

1. Foods that do *NOT* supply fiber include:
   a. cabbage and broccoli.
   b. beef and pork.
   c. oats and corn.
   d. beans and peas.

2. The health practitioner advising an elderly patient with constipation encourages the patient to:
   a. consume a low-fat diet low in sodium.
   b. consume a high-protein diet rich in calcium.
   c. gradually add high-fiber foods to the diet.
   d. eliminate gas-forming foods from the diet.

3. Laxatives help to promote bowel movements by:
   a. stimulating peristalsis.
   b. increasing stool weight.
   c. increasing the water content of stools.
   d. all of the above.

4. Osmotic diarrhea often results from:
   a. excessive motility of fluids within the colon.
   b. excessive fluid secretion by the intestines.
   c. nutrient malabsorption.
   d. viral, bacterial, or protozoal infections.

5. An oral rehydration solution can be mixed at home using:
   a. water, vinegar, honey, and table salt.
   b. water, table salt, and salt substitute.
   c. water, milk, sugar, and table salt.
   d. water, table salt, baking soda, potassium chloride, and sugar.

6. Symptoms of irritable bowel syndrome most often include:
   a. constipation, diarrhea, and flatulence.
   b. weight loss and malnutrition.

   c. strong odors and obstructions.
   d. nauea and vomiting.

7. A patient with Crohn's disease may develop all of the following nutrition problems, *except:*
   a. fat malabsorption.
   b. dumping syndrome.
   c. vitamin B_{12} deficiency.
   d. anemia.

8. Ulcerative colitis can afflict this region of the digestive tract:
   a. ileum, rectum, and colon.
   b. rectum and colon.
   c. stomach and duodenum.
   d. most regions of the GI tract can be affected.

9. Diverticulosis is most often usually associated with:
   a. a low-fiber diet.
   b. inadequate exercise.
   c. intestinal surgery.
   d. a high-fiber diet.

10. After an ileostomy, the most serious concern is that:
    a. the diet is too restrictive to meet nutrient needs.
    b. waste disposal causes frequent daily interruptions.
    c. incompletely digested foods may cause obstructions.
    d. fluid restrictions prevent patients from drinking beverages freely.

Answers to these questions appear in Appendix H.

Notes

1. F. L. Suarez and M. D. Levitt, Intestinal gas, in M. Feldman, L. S. Friedman, and M. H. Sleisenger, eds., *Sleisenger and Fordtran's Gastrointestinal and Liver Disease: Pathophysiology, Diagnosis, Management* (Philadelphia: Saunders, 2002); L. J. Cheskin and D. L. Miller, Nutrition in the prevention and treatment of common gastrointestinal symptoms, in A. M. Coulston, C. L. Rock, and E. R. Monsen, eds., *Nutrition in the Prevention and Treatment of Disease* (San Diego, CA: Academic Press, 2001), pp. 549–562; American Dietetic Association, *Nutrition Care Manual* (Chicago: American Dietetic Association, 2006).

2. S. L. Friedman, K. R. McQuaid, and J. H. Grendell, eds., *Current Diagnosis and Treatment in Gastroenterology* (New York: Lange Medical Books/McGraw-Hill, 2003); M. Feldman, L. S. Friedman, and M. H. Sleisenger, eds., *Sleisenger and Fordtran's Gastrointestinal and Liver Disease: Pathophysiology, Diagnosis, Management* (Philadelphia: Saunders, 2002).

3. Standing Committee on the Scientific Evaluation of Dietary Reference Intakes, Food and Nutrition Board, Institute of Medicine, *Dietary Reference Intakes for Energy, Carbohydrate, Fiber, Fat, Fatty Acids, Cholesterol, Protein, and Amino Acids* (Washington, DC: National Academies Press, 2002).

4. Friedman, McQuaid, and Grendell, 2003; Feldman, Friedman, and Sleisenger, 2002.

5. N. J. Talley, Functional gastrointestinal disorders: Irritable bowel syndrome, nonulcer dyspepsia, and noncardiac chest pain, in L. Goldman and D. Ausiello, eds., *Cecil Textbook of Medicine* (Philadelphia: Saunders, 2004), pp. 806–814.

6. Talley, 2004.

7. W. F. Stenson, Inflammatory bowel disease, in L. Goldman and D. Ausiello, eds., *Cecil Textbook of Medicine* (Philadelphia: Saunders, 2004), pp. 861–868.

8. C. L. Simmang and G. T. Shires, Diverticular disease of the colon, in M. Feldman, L. S. Friedman, and M. H. Sleisenger, eds., *Sleisenger and Fordtran's Gastrointestinal and Liver Disease: Pathophysiology, Diagnosis, Management* (Philadelphia: Saunders, 2002), pp. 2100–2112.

9. B. E. Stabile and T. D. Arnell, Diverticular disease of the colon, in S. L. Friedman, K. R. McQuaid, and J. H. Grendell, eds., *Current Diagnosis and Treatment in Gastroenterology* (New York: Lange Medical Books/McGraw-Hill, 2003), pp. 436–451; Simmang and Shires, 2002.

10. Stabile and Arnell, 2003.

11. American Dietetic Association, 2006.

12. American Dietetic Association, 2006.

PROBIOTICS AND INTESTINAL HEALTH

Soon after birth, the warm, nutrient-rich environment within the gastrointestinal tract is colonized by a wide variety of bacterial species. The bacteria (or **flora**) inhabiting our bodies—amounting to perhaps 10 trillion cells—make up more than 90 percent of the cells in our bodies.[1] Most reside in our colon, which harbors over 500 different bacterial species. Table NP18–1 lists the predominant types of bacteria that colonize the human intestines, and Table NP18–2 shows how the bacterial populations vary within different regions of the GI tract. Although the exact composition of intestinal bacteria varies among individuals, the pattern within an individual tends to remain constant over time, fluctuating somewhat due to illness, antibiotic treatment, and, to some extent, dietary factors. In the past several decades, nutritional scientists and microbiologists have tried to determine whether **probiotics,** foods and supplements that supply live, **nonpathogenic** bacteria in sufficient numbers to possibly benefit our health, can be useful for preventing or treating a wide variety of medical problems. The accompanying glossary defines the relevant terms.

How do intestinal bacteria influence our health?

Intestinal bacteria can benefit our health in a number of different ways. First, the bacteria degrade much of our undigested or unabsorbed dietary carbohydrate, including dietary fibers, starch that is resistant to digestion, and poorly absorbed sugars and sugar alcohols. In turn, the bacteria produce some vitamins as well as short-chain fatty acids that our cells can use as an energy source. Intestinal bacteria also stimulate our immune defenses and may prevent the overgrowth of **pathogenic** bacteria in the gastrointestinal tract. Healthy bacteria may help to prevent invasion of our tissues by pathogenic bacteria by creating a barrier on the intestinal walls.[2]

TABLE NP18–1	
Intestinal Flora	
Predominant Types	**Subdominant Types**
■ *Bacteroides*	■ *Enterobacteria*
■ *Bifidobacteria*	■ *Enterococci*
■ *Clostridia*	■ *Escherichia*
■ *Eubacteria*	■ *Klebsiella*
■ *Peptococci*	■ *Lactobacilli*
■ *Peptostreptococci*	■ *Micrococci*
■ *Ruminococci*	■ *Staphylococci*

Why are certain types of bacteria considered "probiotic"?

For bacteria to be "probiotic," that is, beneficial to health, they must be nonpathogenic when consumed. They must survive their transit through the digestive tract; therefore, they must be resistant to destruction by stomach acid, bile, and other digestive substances. They should be able to alter the intestinal environment in some way that is beneficial to the human host, either by producing antimicrobial substances, altering immune defenses, metabolizing undigested foodstuffs, or adhering to the intestinal walls.[3]

Probiotic bacteria must be consumed in high amounts—between 100 million and 100 billion live bacteria per day—to survive in sufficient numbers to influence the bacterial populations in the large intestine (a serving of yogurt usually provides these amounts). Carefully controlled studies have not found that probiotic bacteria actually *colonize* the intestine, however, as they are no longer recovered from the body once consumption stops.[4] Note that only a few different types of bacteria are used in foods, and the relatively small amounts consumed cannot compete with the huge populations that normally populate our digestive tract.

What specific types of medical problems can probiotics help?

Although results of research studies vary, probiotic bacteria may help to prevent and treat some gastric and intestinal disorders

GLOSSARY

flora: the bacteria that normally reside in a person's body.

nonpathogenic: not capable of causing disease.

pathogenic: capable of causing disease.

prebiotics: nondigestible substances in foods that stimulate the growth of probiotic bacteria within the large intestine.

probiotics: foods and supplements that supply live, nonpathogenic bacteria in sufficient numbers to benefit our health.

TABLE NP18–2	
Bacterial Populations in the Gastrointestinal Tract	
Organ	**Total Bacteria (per mL of contents)**
Stomach	0 to 100
Small intestine: duodenum	0 to 1000
Small intestine: jejunum and ileum	10^5 to 10^8
Colon	10^{10} to 10^{12}

(described in the next section), improve lactose digestion in some people, and improve the availability and digestibility of various nutrients.[5] Certain bacteria may alter susceptibility to food allergens and alleviate some allergy symptoms. There is also some evidence that probiotics may help to prevent or reverse infections in the urethra and vagina.[6] Not all studies show these benefits, however, and beneficial effects of a bacterial strain cannot be extrapolated to other strains of the same species.[7] Thus, individuals who decide to consume probiotic-containing foods and supplements to benefit their health cannot be certain that the substances they use will help their condition. At best, probiotics should be considered an adjunct therapy rather than a primary treatment for an illness.

Which intestinal disorders may be helped by using probiotics?

Much of the research investigating probiotics and intestinal illness has focused on the prevention and treatment of infectious diarrhea. For example, controlled trials have suggested that certain strains of probiotic bacteria may shorten the duration of diarrhea caused by rotavirus infection in infants and children, decrease incidence of traveler's diarrhea in tourists visiting high-risk areas, and prevent recurrence of infectious diarrhea in hospitalized patients. In studies of children and adults using antibiotics, some strains of probiotic bacteria have been shown to reduce the incidence and duration of antibiotic-associated diarrhea.[8] As another example, some studies have suggested that probiotic treatment may be helpful for reducing recurrence of the condition *pouchitis*, an inflammation of the surgical pouch created in patients who have had an ileostomy or colostomy.[9] A problem with extrapolating these data to include the general public, however, is that different studies have used different bacterial strains and there is no clear conclusion about the appropriate probiotics, doses, or durations of treatment for these conditions.[10]

Which foods provide probiotic bacteria?

Probiotics are provided mainly by fermented foods. In the United States, yogurt and acidophilus milk are produced using various species of *Lactobacilli* and *Bifidobacteria*,[11] although the species are chosen due to their ability to produce desirable food products rather than for their potential health benefits.[12] In Europe and Asia, food products containing probiotic bacteria include yogurt, milk, ice cream, oatmeal gruel, and soft drinks.[13] Although *Lactobacilli* are used to produce various other fermented food products, such as sauerkraut, pickles, brined olives, Korean kimchi, and sausages, the foods do not necessarily contain adequate numbers of live bacteria to benefit health.[14]

Are probiotic bacteria available from dietary supplements?

Yes, a number of companies market probiotic supplements, which are available in capsules, tablets, and powders. Because probiotic bacteria are living organisms, storage conditions may affect their viability—heat, moisture, and oxygen can reduce survival time—and therefore the expiration date should be

Various species of *Lactobacilli* are used in the production of fermented food products, such as the foods shown in this photo.

checked before purchasing a product. When a consumer group (ConsumerLab.com) tested 26 probiotic supplements, they found that eight of the products contained less than one percent of the live bacteria claimed, and six products had only one ten-thousandth of the amount listed on the label.[15] Thus, there is no guarantee that a dietary supplement will contain the amount of bacteria expected.

Is there any way to stimulate the growth of our own intestinal bacteria?

Yes. Certain nondigestible substances in food, called **prebiotics**, may stimulate the growth of resident bacteria within the large intestine;[16] these include some of the carbohydrates found in Jerusalem artichokes, chicory root, onions, and other vegetables. Because the intestinal bacteria that degrade these substances produce gas as a by-product, people who consume high amounts of these foods may experience more flatulence than usual.

What safety concerns are associated with the use of probiotics?

Of major concern is the possibility that probiotic bacteria may cause infection in immune-compromised individuals. Various species of probiotic bacteria, including *Lactobacilli*, have been isolated from infection sites of severely ill individuals who were consuming the probiotic orally.[17] The individuals most likely to be susceptible to infectious complications include patients with reduced immunity, such as those with AIDS, cancer, or undergoing organ transplantation. Care should be taken to inquire about probiotic use in these patients.

Other concerns are related to the lack of industry standards for probiotics in foods and supplements, such that the concentrations of probiotic bacteria in foods may vary substantially. Thus, it would be difficult to determine how much of a product to consume in order to achieve the desired effect.

In recent years, the contributions of our intestinal flora to health have been increasingly recognized. Preliminary research suggests that altering our bacterial populations by consuming probiotics or prebiotics may help to improve our defenses against certain illnesses.

Notes

1. J. E. Teitelbaum and W. A. Walker, Nutritional impact of pre- and probiotics as protective gastrointestinal organisms, *Annual Review of Nutrition* 22 (2002): 107–138; C. Dunne and coauthors, In vitro selection criteria for probiotic bacteria of human origin: Correlation with in vivo findings, *American Journal of Clinical Nutrition* 73 (2001): 386S–392S.

2. F. Guarner and J.-R. Malagelada, Gut flora in health and disease, *Lancet* 361 (2003): 512–519.

3. Teitelbaum and Walker, 2002.

4. A. Bezkorovainy, Probiotics: Determinants of survival and growth in the gut, *American Journal of Clinical Nutrition* 73 (2001): 399S–405S.

5. L. Kopp-Hoolihan, Prophylactic and therapeutic uses of probiotics: A review, *Journal of the American Dietetic Association* 101 (2001): 229–238.

6. G. Reid and J. Burton, Use of *Lactobacillus* to prevent infection by pathogenic bacteria, *Microbes and Infection* 4 (2002): 319–324.

7. Guarner and Malagelada, 2003.

8. Guarner and Malagelada, 2003; Kopp-Hoolihan, 2001; P. R. Marteau and coauthors, Protection from gastrointestinal diseases with the use of probiotics, *American Journal of Clinical Nutrition* 73 (2001): 430S–436S.

9. J. J. Jones and A. E. Foxx-Orenstein, Probiotics in inflammatory bowel disease, *Practical Gastroenterology* (March, 2006): 44–50.

10. E. I. Benchimol and D. R. Mack, Safety issues of probiotic ingestion, *Practical Gastroenterology* (November, 2005): 23–34; R. S. Carvalho and M. Oliva-Hemker, Clinical indications for the use of probiotics in the pediatric population, *Practical Gastroenterology* (October, 2005): 51–64.

11. Kopp-Hoolihan, 2001.

12. K. J. Heller, Probiotic bacteria in fermented foods: Product characteristics and starter organisms, *American Journal of Clinical Nutrition* 73 (2001): 374S–379S.

13. C. Stanton and coauthors, Market potential for probiotics, *American Journal of Clinical Nutrition* 73 (2001): 476S–483S; G. Molin, Probiotics in foods not containing milk or milk constituents, with special reference to *Lactobacillus plantarum* 299v, *American Journal of Clinical Nutrition* 73 (2001): 380S–385S.

14. P. Lavermicocca and coauthors, Study of adhesion and survival of Lactobacilli and Bifidobacteria on table olives with the aim of formulating a new probiotic food, *Applied and Environmental Microbiology* 71 (2005): 4233–4240; Heller, 2001.

15. ConsumerLab.com, Product review: Probiotic supplements and foods (Including *Lactobacillus acidophilus* and *Bifidobacterium),* available at www.consumerlab.com (site visited December 3, 2006).

16. J. H. Cummings, G. T. Macfarlane, and H. N. Englyst, Prebiotic digestion and fermentation, *American Journal of Clinical Nutrition* 73 (2001): 415S–420S.

17. Benchimol and Mack, 2005, N. Ishibashi and S. Yamazaki, Probiotics and safety, *American Journal of Clinical Nutrition* 73 (2001): 465S–470S.

19

Carbohydrate- and Fat-Modified Diets for Malabsorption

© Jerzyworks/Masterfile

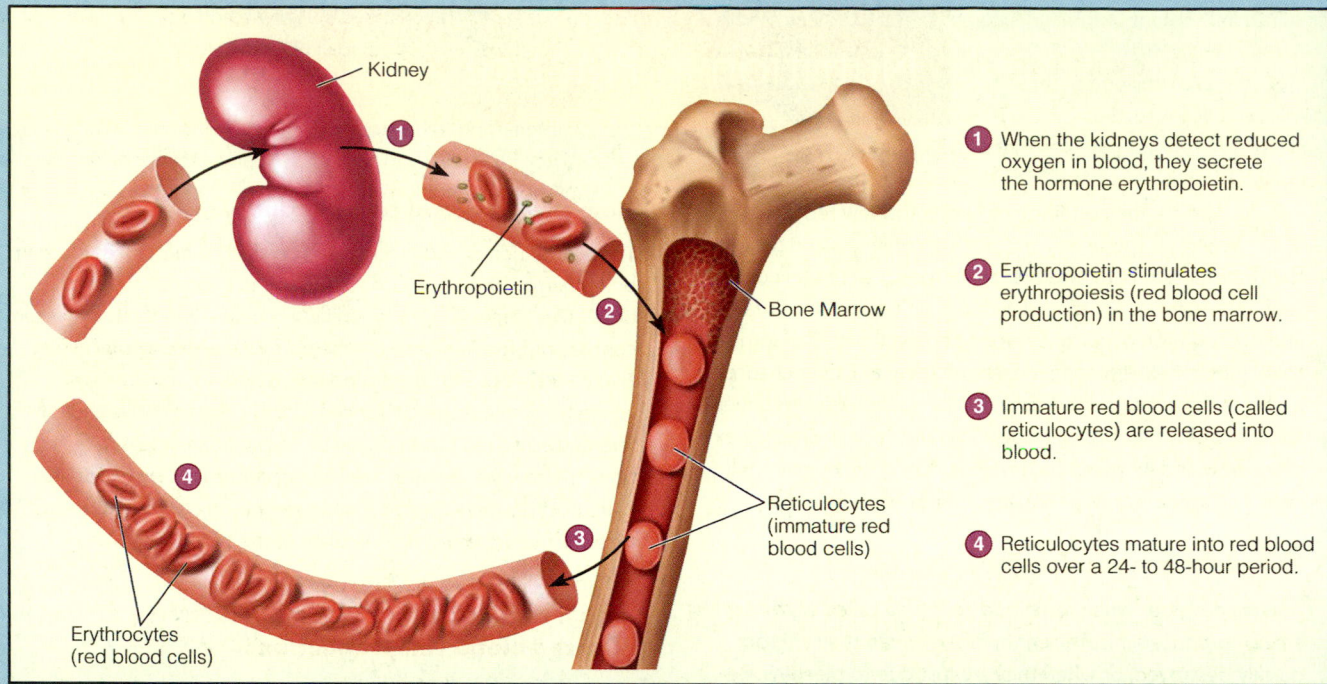

1. When the kidneys detect reduced oxygen in blood, they secrete the hormone erythropoietin.

2. Erythropoietin stimulates erythropoiesis (red blood cell production) in the bone marrow.

3. Immature red blood cells (called reticulocytes) are released into blood.

4. Reticulocytes mature into red blood cells over a 24- to 48-hour period.

FIGURE NP19-1

Erythropoiesis
Source: Reprinted with permission from L. Sherwood, *Human Physiology*, 5th ed., (Brooks/Cole, 2004), Figure 11-4, p. 395.

Which illnesses are associated with blood loss?

Conditions that involve the gastrointestinal tract often cause bleeding; examples include peptic ulcers, inflammatory bowel conditions, and gastrointestinal varices (enlarged veins) that develop in advanced liver disease. Excessive bleeding can also accompany coagulation disorders, which are usually due to liver disease, genetic defects, or vitamin K deficiency. Frequent blood draws or surgical procedures can contribute to blood loss and result in iron deficiency.[2] Unfortunately, slow, chronic bleeding is sometimes difficult to identify before anemia develops.

How do malabsorptive disorders contribute to the development of anemia?

Chapters 18 and 19 explain how nutrient malabsorption often results from disorders that damage the small intestine. For example, Crohn's disease and celiac disease can destroy intestinal mucosa and reduce absorption of all nutrients. Iron is primarily absorbed in the duodenum and upper jejunum, and its absorption is impaired by conditions that reduce hydrochloric acid secretion or result in surgical resection of the upper intestine. Resection of the stomach or ileum can hasten the onset of vitamin B_{12} deficiency because both organs have roles in vitamin B_{12} absorption. Recall from Chapter 8 that the stomach produces a protein called intrinsic factor that is needed for vitamin B_{12} absorption and that the ileum is the site of vitamin B_{12} absorption.

Why are so many chronic illnesses associated with anemia, even when nutrient deficiencies and bleeding disorders are absent?

Chronic disease itself can cause anemia, and anemia is sometimes the initial sign that chronic disease is present.[3] In fact, the **anemia of chronic disease** is the most common type of anemia affecting hospitalized patients and patients with chronic illnesses.[4] This type of anemia usually occurs in individuals who have chronic infections, inflammatory conditions, autoimmune disorders, or cancer. Although often a mild form of anemia, it can progress and become severe enough to require blood transfusions.[5]

The anemia of chronic disease is characterized by alterations both in the distribution of iron among tissues and in the rates of red blood cell production and destruction. As a result of the inflammatory response, macrophages in the liver, spleen, and bone marrow sequester iron, making it unavailable for erythropoiesis and hence slowing the rate of production of new red blood cells. In addition, red blood cells are destroyed more rapidly than usual, and the reduced production of red blood cells cannot keep pace. Finally, iron absorption is impaired, possibly because intestinal cells inhibit iron's release into blood. Eventually, outright iron deficiency may result from inadequate iron absorption.[6]

How is the anemia of chronic disease diagnosed?

Blood tests help to distinguish between the anemia of chronic disease and iron-deficiency anemia (see Table NP19–1). The combination of low serum iron and low total iron-binding capacity

TABLE NP19-1

Laboratory Tests for Evaluating Iron Deficiency and Anemia of Chronic Disease

Laboratory Test	Effect of Iron Deficiency	Effect of Chronic Disease
Red blood cell (RBC) size and number	Microcytic; reduced RBC count	Normocytic or microcytic; reduced RBC count
Serum iron	Low	Low
Serum ferritin	Low	Normal or elevated
Serum transferrin	Elevated	Low
Total iron-binding capacity	High	Low
Bone marrow iron	Low	Normal or elevated

suggests the anemia of chronic disease rather than iron deficiency. In addition, serum ferritin levels are normal or elevated, whereas they are typically low in iron deficiency. Diagnosis is more complicated if both types of anemia are present.[7]

Can medications cause anemia?

Yes, anemia is among the adverse effects that may result from medication use. Medications can alter nutrient metabolism, impair blood coagulation and erythropoiesis, and sometimes lead to increased red blood cell destruction. Because the life span of red blood cells is about 120 days, long-term use of medications is more likely to result in anemia than short-term use.

How do medications alter nutrient metabolism?

There are many different ways in which medications can alter nutrient metabolism; the most common of these are listed in Table 15–4 on p. 446. As an example, a variety of medications are known to influence the absorption or metabolism of folate and lead to macrocytic anemia. Sulfasalazine (for ulcerative colitis) and some anticonvulsant drugs inhibit folate absorption, and methotrexate (an immunosuppressive), triamterene (a diuretic), and pyrimethamine (an antimalarial) interfere with folate metabolism.[8] If a medication is known to result in deficiency, nutrient supplementation is usually recommended as an adjunct therapy.

How do medications impair blood coagulation?

Anticoagulants, which are prescribed specifically to reduce blood clotting, may sometimes lead to excessive bleeding. These medications work by interfering with one of the steps involved in blood clotting, including platelet function, synthesis of clotting proteins, and vitamin K function. Other than anticoagulants, drugs that impair coagulation include aspirin and other nonsteroidal anti-inflammatory drugs (NSAIDs), acetaminophen, cimetidine (Tagamet), ranitidine (Zantac), and thiazide diuretics.[9] The anticoagulant effects may be augmented if several of these drugs are used simultaneously. Sometimes the slow, chronic bleeding that develops may go unnoticed until excessive blood loss has occurred.

Which categories of drugs inhibit erythropoiesis?

The categories of drugs that can inhibit erythropoiesis include anticonvulsants, antibiotics, antidiabetic drugs, diuretics, antithyroid drugs, and anticancer agents.[10] The anemia that occurs when the bone marrow fails to produce adequate numbers of blood cells is called **aplastic anemia.** Aplastic anemia can also be caused by a genetic defect or may result from viral infections or exposure to toxins.

How can medications cause blood cell destruction?

Some patients may develop hemolytic anemia as a result of drug interactions with red blood cells. For example, a drug may alter the red blood cell membrane in such a way that a component of the membrane becomes an antigen and induces an antibody response that destroys the cell.[11] Several types of antibiotics, including penicillin and cephalosporin, may cause this type of response. Withdrawal of the drug can eventually reverse the anemia, and sometimes medications are given to suppress the immune response.

With all the possible ways in which anemia can develop, how is its cause determined?

Identifying the cause of anemia is sometimes quite challenging. In some cases, anemia may be a well-known consequence of disease, as when renal failure impairs the synthesis of the hormone erythropoietin. As mentioned earlier, the results of laboratory tests provide valuable clues, although conditions such as dehydration and inflammation can influence the values. Laboratory results are especially difficult to analyze if several disturbances are present simultaneously. A **peripheral blood smear** (see the photo) is often used to study abnormalities in red blood cell shape and may reveal an underlying cause. Also, an anemia that develops rapidly usually indicates blood loss, whereas a gradual onset is often associated with malnutrition, chronic illness, or slow, chronic

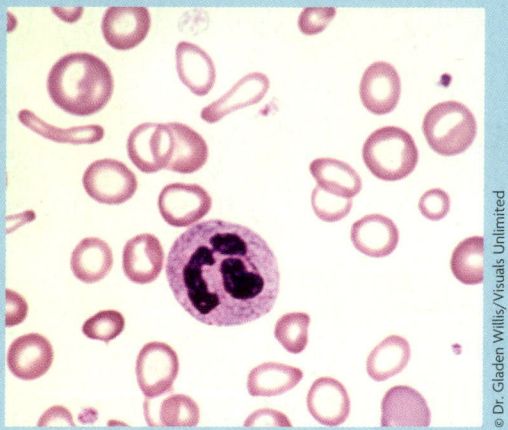

A peripheral blood smear provides information about the number and shape of blood cells.

© Dr. Gladen Willis/Visuals Unlimited

bleeding.

Anemia is a disorder associated with many diseases, and it may also be caused during disease treatment. When it occurs, its causes must be investigated before it leads to complications that worsen prognosis. The medical history, blood tests, and peripheral blood smears may all help to determine the reasons why anemia has developed.

Notes

1. K. S. Zuckerman, Approach to the anemias, in L. Goldman and D. Ausiello, eds., *Cecil Textbook of Medicine* (Philadelphia: Saunders, 2004), pp. 963–971.

2. Zuckerman, 2004; J. E. Ansell, Cardinal manifestations of hematologic disease, anemias, and related conditions, in J. Noble and coeditors, *Textbook of Primary Care Medicine* (St. Louis, MO: Mosby, 2001), pp. 1027–1037.

3. T. P. Duffy, Microcytic and hypochromic anemias, in L. Goldman and D. Ausiello, eds., *Cecil Textbook of Medicine* (Philadelphia: Saunders, 2004), pp. 1003–1008.

4. C. N. Roy, D. A. Weinstein, and N. D. Andrews, 2002 E. Mead Johnson Award for Research in Pediatrics lecture: The molecular biology of the anemia of chronic disease: A hypothesis, *Pediatric Research* 53 (2003): 507–512.

5. Roy, Weinstein, and Andrews, 2003.

6. Duffy, 2004; Roy, Weinstein, and Andrews, 2003; D. A. Weinstein and coauthors, Inappropriate expression of hepcidin is associated with iron refractory anemia: Implications for the anemia of chronic disease, *Blood* 100 (2002): 3776–3781.

7. Roy, Weinstein, and Andrews, 2003.

8. S. P. Stabler and R. H. Allen, Megaloblastic anemias, in L. Goldman and D. Ausiello, eds., *Cecil Textbook of Medicine* (Philadelphia: Saunders, 2004), pp. 1050–1057.

9. M. Shuman, Hemorrhagic disorders: Abnormalities of platelet and vascular function, in L. Goldman and D. Ausiello, eds., *Cecil Textbook of Medicine* (Philadelphia: Saunders, 2004), pp. 1060–1069.

10. H. Castro-Malaspina and R. J. O'Reilly, Aplastic anemia and related disorders, in L. Goldman and D. Ausiello, eds., *Cecil Textbook of Medicine* (Philadelphia: Saunders, 2004), pp. 1044–1050.

11. A. D. Schreiber, Autoimmune and intravascular hemolytic anemias, in L. Goldman and D. Ausiello, eds., *Cecil Textbook of Medicine* (Philadelphia: Saunders, 2004), pp. 1013–1021.

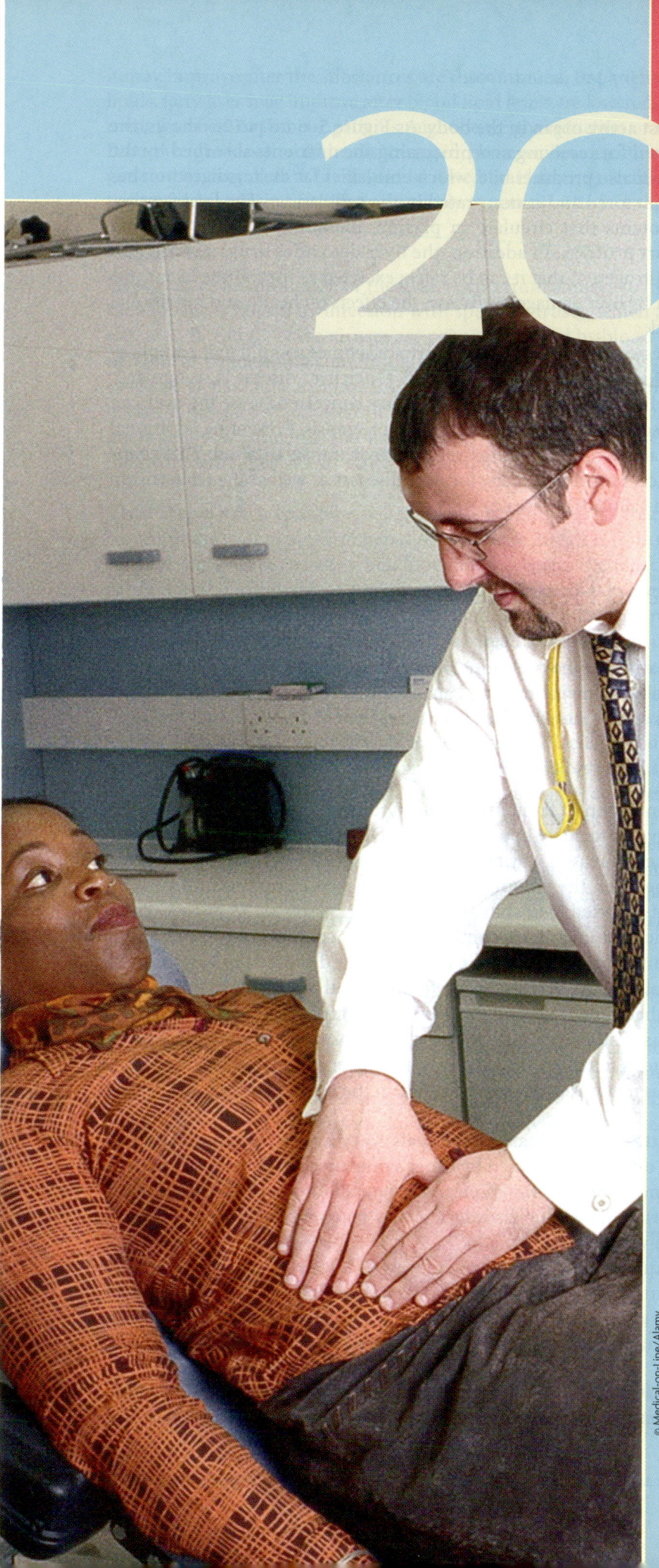

20
Nutrition Therapy for Liver and Gallbladder Diseases

Fatty Liver and Hepatitis
    Fatty Liver
    Hepatitis

Cirrhosis
    Consequences of Cirrhosis
    Treatment of Cirrhosis
    Medical Nutrition Therapy for Cirrhosis

Liver Transplantation
Gallbladder Disease
    Types of Gallstones
    Consequences of Gallstones
    Risk Factors for Gallstones
    Treatment for Gallstones

Nutrition in Practice
    Alcohol in Health and Disease

© Medical-on-Line/Alamy

551

How is alcohol toxic to cells?

Alcohol alters the structure of cell membranes, increases their permeability, and interferes with the actions of cell membrane proteins. Under certain conditions, alcohol exposure can induce cell death. A metabolite of alcohol—**acetaldehyde**—causes numerous adverse effects; it binds to proteins and interferes with their functions, prevents formation of microtubules within cells, alters immune function, inhibits DNA repair, and causes oxidative damage.

How does alcohol affect brain function?

Alcohol acts as a central nervous system depressant; it can cause sedation, slow reaction times, and relieve anxiety. In excess it impairs judgment, reduces inhibitions, and impairs speech and motor function. Extremely high blood alcohol levels can lead to coma, respiratory depression, and death.

Chronic heavy drinking can lead to certain types of neurological damage.[4] Symptoms include tingling in the hands and feet, lack of muscular coordination, and changes in a person's manner of walking. Visual impairments, such as blurred vision and optic nerve degeneration, can also occur.

What are some other long-term consequences of drinking too much alcohol?

Alcoholic liver disease is the most common complication of alcohol abuse, occurring in about 10 to 20 percent of alcoholics.[5] The disease develops in a manner similar to the disease progression described in Chapter 20: fatty liver, hepatitis, and, eventually, cirrhosis and liver failure. In addition to liver damage, alcohol can cause damage to the GI tract, pancreas, and heart (see Figure NP20–1). Damage to GI mucosal tissue increases the

FIGURE NP20–1

Alcohol's Effects on Organ Systems

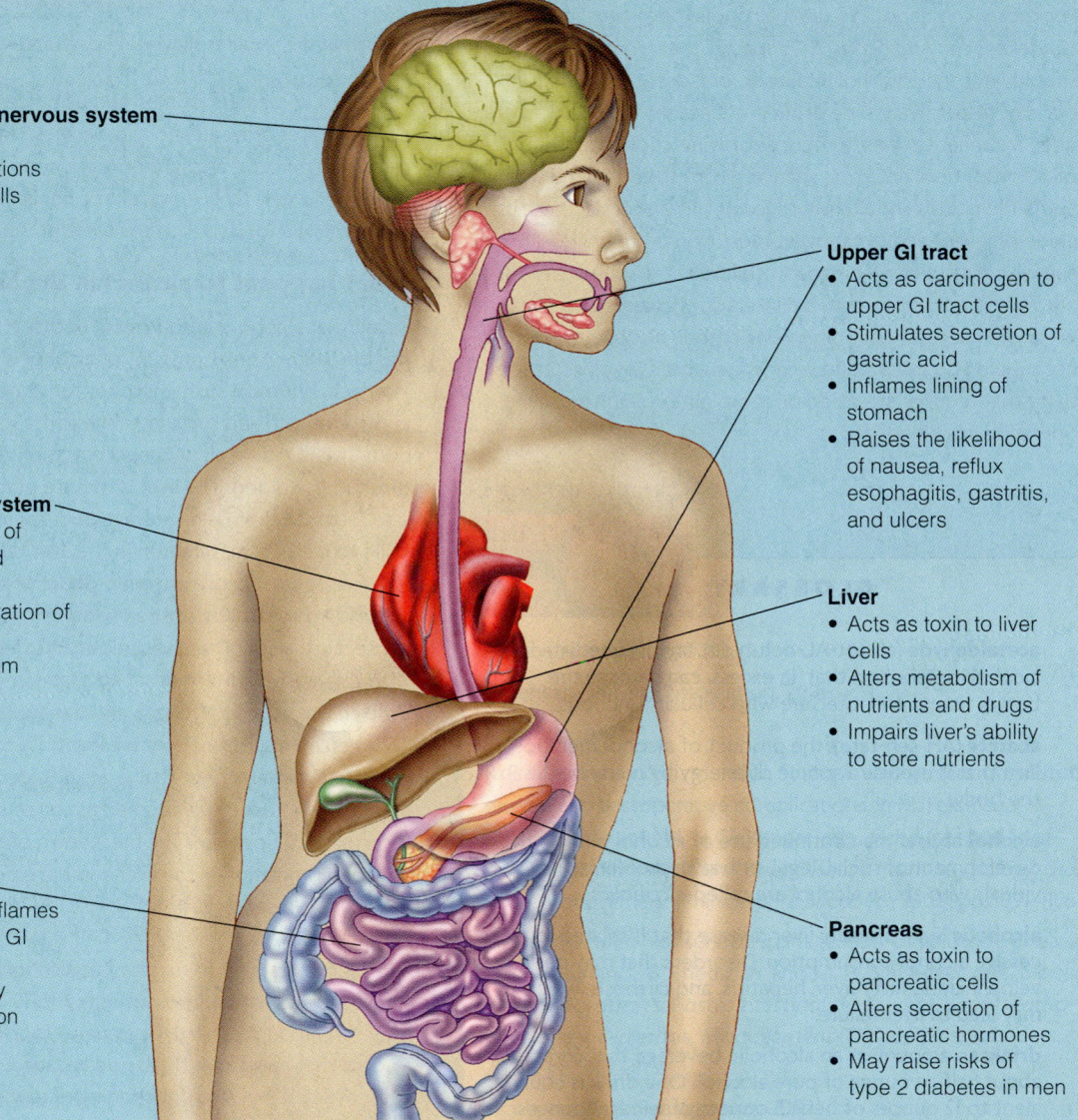

Brain and central nervous system
- Alters judgment
- Depresses inhibitions
- Impairs motor skills
- Slows breathing

Cardiovascular system
- Raises likelihood of hypertension and heart attacks
- Leads to deterioration of heart muscle
- Alters heart rhythm

Lower GI tract
- Damages and inflames cells of the lower GI tract
- Delays GI motility
- Impairs absorption

Upper GI tract
- Acts as carcinogen to upper GI tract cells
- Stimulates secretion of gastric acid
- Inflames lining of stomach
- Raises the likelihood of nausea, reflux esophagitis, gastritis, and ulcers

Liver
- Acts as toxin to liver cells
- Alters metabolism of nutrients and drugs
- Impairs liver's ability to store nutrients

Pancreas
- Acts as toxin to pancreatic cells
- Alters secretion of pancreatic hormones
- May raise risks of type 2 diabetes in men

risks of gastritis, ulcers, and GI cancers and may lead to nutrient malabsorption and diarrhea. Damage to the pancreas can alter pancreatic secretions and eventually result in pancreatitis and diabetes. Excessive alcohol consumption can raise the risk of heart attacks by elevating blood pressure and weakening heart muscle contractility.

What are the effects of excessive alcohol consumption on nutrition status?

An excessive intake of alcohol causes nutrient deficiencies for several reasons. It can upset food intake: because alcohol supplies 7 kcalories per gram, it displaces other energy sources along with the essential nutrients such foods would provide. As mentioned earlier, it can cause widespread malabsorption as a result of direct damage to the GI mucosa. Alcohol also interferes with the way the body processes nutrients. Examples of common deficiencies in persons who abuse alcohol include:[6]

- **Vitamin A.** Alcohol and vitamin A are metabolized by similar enzymes, so an increase in the enzymes that break down alcohol—induced by heavy drinking—can increase degradation of vitamin A. Alcohol also interferes with the synthesis of vitamin A's carrier protein in the blood.

- **Thiamin.** Alcohol abuse is the most frequent cause of thiamin deficiency in the United States. Heavy drinking is associated with low thiamin intakes, and alcohol ingestion dramatically reduces thiamin absorption. Alcohol also interferes with thiamin activation in the body and increases thiamin losses in urine.

- **Folate.** Alcohol reduces the absorption of folate in the small intestine, inhibits the transformation of folate to its coenzyme forms, and increases folate losses in the urine. Because folate has a central role in cell division (and cells of the GI tract turn over rapidly), folate deficiency contributes to malabsorption of other nutrients as well.

Does alcohol also have disruptive effects on the metabolism of medications?

Yes. Some of the liver enzymes that degrade alcohol have the additional role of metabolizing drugs. When alcohol is consumed, the enzymes involved in alcohol degradation cannot metabolize drugs (to prepare them for excretion), so heavy drinking can increase medication potency. Thus it may be dangerous to use alcohol while taking certain medications. It is especially problematic to combine alcohol and drugs if they have similar effects in the body; examples of such drugs include sedatives and blood glucose–lowering medications.[7] Alcohol can also reduce drug absorption because alcohol slows stomach emptying; hence, medications may take longer to exert their effects.[8]

Does alcohol have any beneficial effects on health?

A light-to-moderate alcohol intake can reduce heart disease risk. Alcohol's protective effects are seen mainly in older persons who have one or more classic risk factors for heart disease (discussed in Chapter 22). Although the reasons that alcohol is beneficial are not entirely clear, alcohol increases levels of HDL cholesterol (protecting against further development of atherosclerosis) and also reduces the tendency for blood clotting. These effects occur when any type of alcoholic beverage is consumed and are seen most frequently when alcohol is consumed at least three times per week.[9]

The benefits or harm associated with alcohol consumption depend on a person's health status, age, and the amount consumed. Most people in the United States who consume alcoholic beverages do not drink more alcohol than is recommended; however, health professionals should be alert to those who may be abusing alcohol and potentially endangering their health.

Notes

1. U.S. Department of Agriculture and U.S. Department of Health and Human Services, *Nutrition and Your Health: Dietary Guidelines for Americans 2005*, 6th ed. (Washington, DC: 2005).

2. C. D. Fryar and coauthors, *Smoking and alcohol behaviors reported by adults, United States, 1999–2002, Advance data from vital and health statistics;* no. 378 (Hyattsville, MD: National Center for Health Statistics, 2006).

3. S. B. Masters, The alcohols, in B. G. Katzung, ed., *Basic and Clinical Pharmacology* (New York: McGraw-Hill, 2001), pp. 382–394.

4. Masters, 2001.

5. M. H. Beers and coeditors, *Merck Manual of Diagnosis and Therapy,* (Whitehouse Station, NJ: Merck Research Laboratories, 2006), pp. 211–214.

6. K. E. Light and R. Hakkak, Alcohol and nutrition, in B. J. McCabe, E. H. Frankel, and J. J. Wolfe, eds., *Handbook of Food-Drug Interactions* (Boca Raton, FL: CRC Press, 2003), pp. 167–189.

7. Masters, 2001.

8. Light and Hakkak, 2003.

9. K. J. Mukamal and coauthors, Roles of drinking pattern and type of alcohol consumed in coronary heart disease in men, *New England Journal of Medicine* 348 (2003): 109–118.

21

Carbohydrate-Controlled Diets for Diabetes Mellitus

© Tony Freeman/PhotoEdit

573

■ Reminder: *Insulin* is a pancreatic hormone that regulates blood glucose concentrations. Its actions are countered mainly by the hormone *glucagon*.

■ Reminder: *Osmolarity* refers to the concentration of osmotically active particles in solution. Hyperglycemia causes the body's fluids to become *hyerposmolar,* meaning that they have an abnormally high osmolarity.

■ Normal fasting plasma glucose levels are approximately 75 to 100 mg/dL (published values vary).

diabetes (DYE-ah-BEE-teez) **mellitus:** a group of metabolic disorders characterized by hyperglycemia and disordered insulin metabolism.
   diabetes = siphon (in Greek), referring to the excessive passage of urine that is characteristic of untreated diabetes
   mellitus = honey-sweet

renal threshold: the blood concentration of a substance that exceeds the kidneys' capacity for reabsorption, thus allowing the substance to be passed into the urine.

oral glucose tolerance test: a test that evaluates a person's ability to tolerate a glucose load.

The incidence of **diabetes mellitus** is steadily increasing in the United States and in many other countries. It now affects an estimated 6.6 percent of the U.S. population, or more than 18 million people.[1] About 29 percent of persons with diabetes do not know that they have it,[2] a danger because damage to the body often occurs before symptoms develop. Diabetes ranks sixth among the leading causes of death in the United States. It also contributes to the development of other life-threatening diseases, including heart disease and kidney failure, which are discussed in the two chapters that follow. The Glossary on p. 576 defines diabetes-related symptoms and complications.

Overview of Diabetes Mellitus

The term *diabetes mellitus* refers to metabolic disorders characterized by elevated blood glucose concentrations and disordered insulin metabolism. ■ People with diabetes may have impaired insulin secretion or cells that do not respond to insulin normally. The result is **hyperglycemia,** a marked elevation in blood glucose levels that can ultimately cause damage to blood vessels, nerves, and tissues.

Symptoms of Diabetes

Symptoms of diabetes (see Table 21–1) are usually related to the degree of hyperglycemia present. When the plasma glucose concentration rises above about 200 milligrams per deciliter (mg/dL), it exceeds the **renal threshold**, which is the concentration at which the kidneys begin to pass glucose into the urine (**glycosuria**). The presence of glucose in the urine draws additional water out of the blood, increasing the amount of urine produced. Thus the symptoms that arise in diabetes typically include frequent urination (**polyuria**), dehydration, and increased thirst (**polydipsia**). Some people lose weight and have an increased appetite (**polyphagia**) as a result of the nutrient depletion that occurs when insulin is deficient. Hyperglycemia can also cause blurred vision due to the exposure of eye tissues to hyperosmolar fluids. ■ In some cases, constant fatigue is the only symptom and may be related to altered energy metabolism, dehydration, or other effects of the disease.

Diagnosis of Diabetes

The diagnosis of diabetes is based primarily on plasma glucose levels, which can be measured under fasting conditions or at random times during the day. ■ In some cases, an **oral glucose tolerance test** is given: the patient ingests a 50- or 75-gram glucose load, and plasma glucose is measured at one or more time intervals following glucose ingestion. The following criteria are currently used to diagnose diabetes:

■ The plasma glucose concentration of a blood sample obtained at a random time during the day (without regard to food intake) is 200 mg/dL or greater, and classic symptoms of diabetes (such as polyuria, polydipsia, and unexplained weight loss) are present.

■ The plasma glucose concentration is 126 mg/dL or greater after a fast of at least eight hours.

■ The plasma glucose concentration measured two hours after a 75-gram glucose load is 200 mg/dL or greater.

Overt symptoms of hyperglycemia help to confirm the diagnosis. Otherwise, diagnosis of diabetes is confirmed only if a subsequent test yields similar results.

TABLE 21–2

Features of Type 1 and Type 2 Diabetes

	Type 1	Type 2
Prevalence in diabetic population	5 to 10% of cases	90 to 95% of cases
Age of onset	<30 years	>45 years[a]
Associated conditions	Autoimmune diseases, viral infection, inherited factors	Obesity, aging, inherited factors
Major defect	Destruction of pancreatic beta cells; insulin deficiency	Insulin resistance; insulin deficiency (relative to needs)
Insulin secretion	Little or none	Varies; may be normal, increased, or decreased
Requirement for insulin therapy	Always	Sometimes
Other names	Juvenile-onset diabetes	Adult-onset diabetes
	Insulin-dependent diabetes mellitus (IDDM)	Noninsulin-dependent diabetes mellitus (NIDDM)
	Ketosis-prone diabetes	Ketosis-resistant diabetes

[a] Incidence of type 2 diabetes is increasing in children and adolescents; in over 90 percent of these cases, it is associated with overweight or obesity and a family history of type 2 diabetes.

The term **prediabetes** is used to refer to blood glucose levels between normal and diabetic, that is, between 100 and 125 mg/dL when fasting or between 140 and 200 mg/dL when measured two hours after ingesting a 75-gram glucose load. A person with prediabetes is at risk of developing diabetes and cardiovascular diseases.[3]

Types of Diabetes

Table 21–2 lists the features of the two main types of diabetes, type 1 and type 2 diabetes. Pregnancy can lead to abnormal glucose tolerance and the condition known as *gestational diabetes*, which often resolves after pregnancy but is a risk factor for type 2 diabetes (gestational diabetes is discussed later in this chapter). Diabetes can also be caused by medical conditions that either damage the pancreas or interfere with insulin function.

Type 1 Diabetes **Type 1 diabetes** accounts for about 5 to 10 percent of diabetes cases. It is usually caused by **autoimmune** destruction of the pancreatic beta cells, which produce and secrete insulin. By the time symptoms develop, the damage to the beta cells has usually progressed so far that insulin must be supplied exogenously, most often by injection. Although the trigger for the autoimmune attack is unknown, both inherited and environmental factors are likely to be involved.[4] Individuals who develop type 1 diabetes are at increased risk of developing other autoimmune disorders.

Type 1 diabetes usually develops during childhood or adolescence, and symptoms may appear abruptly in previously healthy children.[5] Classic symptoms are frequent urination, weight loss, and increased thirst. **Ketoacidosis**—acidosis due to excessive production of ketone bodies—is sometimes the first sign of disease.[6] ■ Disease onset tends to be more gradual in individuals who develop type 1 diabetes in later years. Blood tests that detect antibodies to insulin, pancreatic islet cells, and pancreatic enzymes can confirm the diagnosis and help to predict development of the disease in close relatives.

Type 2 Diabetes **Type 2 diabetes** is the most prevalent form of diabetes, accounting for 90 to 95 percent of cases, and is often asymptomatic. The primary defect in type 2 diabetes is **insulin resistance**, a reduced sensitivity to insulin in muscle, adipose, and liver cells. To compensate, the pancreas secretes larger amounts of insulin, and plasma insulin concentrations can rise to abnormally high levels (**hyperinsulinemia**). Over time, the pancreas becomes less able to compensate for the cells' reduced sensitivity to

■ Reminder: *Ketone bodies* are products of fat metabolism that are produced in the liver; they accumulate in tissues when fatty acids are released in abnormally high amounts from adipose tissue.

prediabetes: the condition in which blood glucose levels are higher than normal but not high enough to be diagnosed as diabetes; formerly called **impaired glucose tolerance.**

type 1 diabetes: the type of diabetes that accounts for 5 to 10 percent of diabetes cases and usually results from autoimmune destruction of pancreatic beta cells.

autoimmune: an immune response directed against the body's own tissues.
   auto = self

type 2 diabetes: the type of diabetes that accounts for 90 to 95 percent of diabetes cases and usually results from insulin resistance coupled with insufficient insulin secretion.

insulin resistance: reduced sensitivity to insulin in muscle, adipose, and liver cells.

hyperinsulinemia: abnormally high levels of insulin in the blood.

acetone breath: distinctive fruity odor on the breath of a person with ketosis.

claudication (CLAW-dih-KAY-shun): pain in the legs while walking; usually due to an inadequate supply of blood to muscles.

dawn phenomenon: morning hyperglycemia that is caused by the early morning release of growth hormone, which counteracts insulin's glucose-lowering effects.

diabetic coma: a coma that occurs in uncontrolled diabetes; may be due to diabetic ketoacidosis, the hyperosmolar hyperglycemic state, or excessive doses of insulin or certain antidiabetic drugs.

gangrene: death of tissue due to a deficient blood supply and/or infection.

gastroparesis (GAS-troe-pah-REE-sis): delayed stomach emptying.

glycosuria (GLY-co-SOOR-ee-ah): an abnormal amount of glucose in urine.

hyperglycemia: elevated blood glucose concentrations. Normal fasting blood glucose is less than 100 mg/dL. Fasting blood glucose between 100 and 126 mg/dL suggests prediabetes; values of 126 mg/dL and above suggest diabetes.

hyperosmolar hyperglycemic state: extreme hyperglycemia associated with hyperosmolar blood, dehydration, and altered mental status; formerly called **hyperglycemic hyperosmolar nonketotic coma**.

hypoglycemia: abnormally low concentrations of blood glucose. In diabetes, hypoglycemia is treated when plasma glucose levels fall below 70 mg/dL.

ketoacidosis (KEY-toe-ah-sih-DOE-sis): lowering of pH in the blood and tissues due to the excessive production of ketone bodies.

ketonuria (KEY-toe-NOOR-ee-ah): the presence of ketone bodies in the urine.

ketosis (key-TOE-sis): excessive amounts of ketone bodies in the body's tissues.

macrovascular complications: disorders that affect the large blood vessels, including cardiovascular diseases and arteries of the limbs.

microalbuminuria: the presence of small amounts of albumin (protein) in the urine, a sign of diabetic nephropathy.

microvascular complications: disorders that affect the small blood vessels and capillaries, including those in the retina and kidneys.

nephropathy (neh-FRAH-pah-thee): damage or disease affecting the kidneys.

neuropathy (nur-RAH-pah-thee): disorders affecting the nervous system.

polydipsia (POL-ee-DIP-see-ah): excessive thirst.

polyphagia (POL-ee-FAY-jee-ah): excessive appetite or eating.

polyuria (POL-ee-YOOR-ree-ah): excessive urine secretion.

rebound hyperglycemia: hyperglycemia that results from the release of counterregulatory hormones following nighttime hypoglycemia; also called the **Somogyi phenomenon**.

retinopathy (REH-tih-NAH-pah-thee): damage or disease affecting the retina.

insulin, and hyperglycemia worsens. The high demand for insulin can eventually exhaust the beta cells of the pancreas and lead to impaired insulin secretion and reduced plasma insulin concentrations. Type 2 diabetes is therefore associated both with insulin resistance and with relative insulin deficiency; that is, the amount of insulin is insufficient to compensate for its diminished effect in the cells.

Although the actual causes of type 2 diabetes are unknown, the risk is substantially increased by obesity (especially abdominal obesity), aging, and physical inactivity. An estimated 80 to 90 percent of individuals with type 2 diabetes are obese, and obesity itself can directly cause some degree of insulin resistance.[7] The prevalence of type 2 diabetes increases with age and approaches 18 percent in persons between 65 and 74 years of age; however, many of these cases remain undiagnosed.[8] Inherited factors strongly influence risk, and type 2 diabetes is more common in certain ethnic populations, including Native Americans, Hispanic Americans,

Mexican Americans, African-Americans, Asian-Americans, and Pacific Islanders.

Type 2 Diabetes in Children and Adolescents Although most cases of type 2 diabetes are diagnosed in individuals over 45 years old, children and adolescents who are overweight or have a family history of diabetes are at increased risk. Because type 2 diabetes is frequently asymptomatic, it is generally detected in children only when high-risk groups are screened for the disease. For example, when 167 obese children of different ethnic groups were screened, prediabetes was detected in 25 percent of children between 4 and 10 years of age and in 21 percent of adolescents between 11 and 18 years of age.[9] Among the Pima Indians in Arizona, a population with one of the highest rates of type 2 diabetes in the world, overt type 2 diabetes was reported in 2.2 percent of 10- to 14-year-old children and in 5 percent of 15- to 19-year-old teens.[10] Routine screening and prevention programs that target food intake and activity patterns can be important safeguards for preventing diabetes in children at risk.

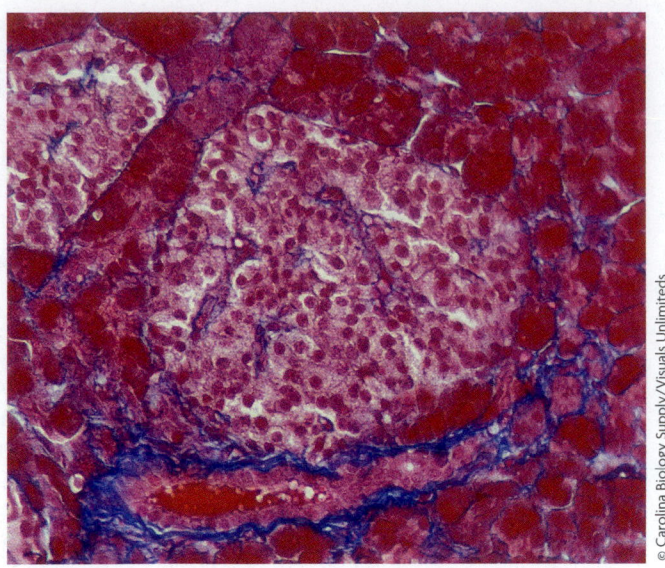

Cross sections of the pancreas reveal distinct areas known as the islets of Langerhans, which contain the beta cells that produce insulin.

Acute Complications of Diabetes

Untreated diabetes may result in life-threatening complications. Insulin deficiency can cause severe disturbances in energy metabolism, and hyperglycemia can lead to fluid and electrolyte imbalances. In treated diabetes, hypoglycemia is a possible complication of inappropriate disease management.

Diabetic Ketoacidosis in Type 1 Diabetes A severe lack of insulin causes diabetic ketoacidosis. The insulin deficiency results in unrestrained breakdown of the triglycerides in adipose tissue and excessive release of fatty acids into the blood. This promotes a substantial increase in the liver's production of ketone bodies (**ketosis**). Ketone bodies are acidic and can reach extremely high levels in the bloodstream (**ketoacidosis**) and spill into the urine (**ketonuria**). Blood pH typically falls below 7.3. ■ Blood glucose concentrations usually exceed 250 mg/dL and may rise above 1,000 mg/dL in severe cases.[11] The main features of diabetic ketoacidosis thus include ketosis, acidosis, and hyperglycemia.

> ■ Normal blood pH ranges from 7.35 to 7.45.

Patients with ketoacidosis may exhibit symptoms of both acidosis and dehydration. Acidosis is partially corrected by exhalation of carbon dioxide, so hyperventilation or deep breathing is characteristic. ■ Ketone accumulation is sometimes evident by a fruity odor on a person's breath (**acetone breath**). The fluid loss (polyuria) that accompanies hyperglycemia lowers blood volume and blood pressure and depletes electrolytes. Mental state may vary from alertness to comatose (**diabetic coma**).[12] Diabetic coma was a frequent cause of death before insulin was routinely used to manage diabetes.

Diabetic ketoacidosis may result from inappropriate treatment (such as missed insulin injections), illness or infection, alcohol abuse, and other physiological stressors.[13] The condition usually develops quickly, within hours or a few days. Although diabetic ketoacidosis can occur in type 2 diabetes as well, it rarely develops because even relatively low insulin concentrations can prevent excessive ketone body production. When ketosis does occur in type 2 diabetes, it is usually associated with metabolic stress due to infection or serious illness.

> ■ Bicarbonate is a buffer in the blood that corrects acidosis. The acid (H^+) and bicarbonate (HCO_3^-) combine to form carbonic acid (H_2CO_3), which breaks down to water (H_2O) and carbon dioxide (CO_2). The carbon dioxide is then exhaled.

Hyperosmolar Hyperglycemic State in Type 2 Diabetes The **hyperosmolar hyperglycemic state** is a condition of severe hyperglycemia that usually develops in the absence of significant ketosis. It often evolves slowly, over several days or weeks, and is most often associated with type 2 diabetes. Blood glucose levels typically exceed 600 mg/dL and may rise above 2,000 mg/dL. The extreme hyperglycemia causes substantial fluid losses, leading to depleted blood volume and electrolyte imbalances. Blood plasma may become so hyperosmolar as to cause neurological abnormalities, such as abnormal reflexes, motor impairments, reduced verbal ability, and seizures; about 10 percent of patients lapse into coma. The hyperosmolar hyperglycemic state often develops because patients are unable to recognize thirst or adequately replace fluid losses due to age, illness, sedation, or incapacity.[14]

Hypoglycemia **Hypoglycemia,** or low blood glucose, arises from the inappropriate management of diabetes rather than from the disease itself. It can result from using excessive amounts of insulin or antidiabetic drugs, prolonged exercise, skipped or delayed meals, inadequate food intake, or consuming alcohol without food. Hypoglycemia most often occurs in type 1 diabetes and accounts for about 3 to 4 percent of deaths in insulin-treated patients.[15]

Symptoms of hypoglycemia include hunger, sweating, shakiness, heart palpitations, slurred speech, and confusion. Mental confusion may prevent a person from recognizing the problem and taking such corrective action as ingesting glucose tablets, juice, or candy. If hypoglycemia occurs during the night, patients may be completely unaware of its presence.

Chronic Complications of Diabetes

Prolonged exposure to high glucose concentrations can destroy cells and tissues. Glucose and glucose fragments react with proteins to form **advanced glycation end products (AGEs)**, compounds that accumulate and cause damage to cells and blood vessels. Chronic complications of diabetes typically affect the large blood vessels (**macrovascular complications**), smaller vessels such as arterioles and capillaries (**microvascular complications**), and the nervous system (**neuropathy**). Increased infections are also common in diabetes, a possible consequence of hyperglycemia, impaired circulation, or depressed immune responses.

Macrovascular Complications The damage caused by diabetes accelerates the development of atherosclerosis in the coronary arteries and the arteries of the limbs. Cardiovascular diseases are the leading cause of death in people with diabetes, accounting for 75 percent of deaths.[16] Type 2 diabetes is often accompanied by multiple risk factors for coronary heart disease, including hypertension, abnormal blood lipids, and obesity. ■ In addition, people with diabetes have increased tendencies for thrombosis (blood clot formation) and abnormal ventricle function, both of which can worsen the clinical course of heart disease.[17]

Impaired blood flow in the arteries of the limbs increases the risk of **claudication** (pain while walking) and contributes to the development of foot ulcers (see the photo). Left untreated, foot ulcers can lead to **gangrene** (tissue death), and some patients require foot amputation, a major cause of disability in diabetes. About 15 to 20 percent of persons with diabetes are hospitalized with foot complications during the course of illness.[18]

Microvascular Complications Long-term diabetes can cause progressive damage to the small blood vessels of the retina (**retinopathy**), resulting in visual impairments and blindness. Damage to the small blood vessels of the kidneys (**nephropathy**) frequently occurs in the later stages of diabetes and may lead to renal failure. Retinopathy and nephropathy progress most rapidly when diabetes is poorly controlled, and intensive management substantially reduces the risks of developing these conditions.[19]

■ People with type 2 diabetes frequently develop the *metabolic syndrome*, a cluster of symptoms associated with insulin resistance (including hyperglycemia, hypertension, and altered blood lipids) that substantially increase heart disease risk (see Nutrition in Practice 22).

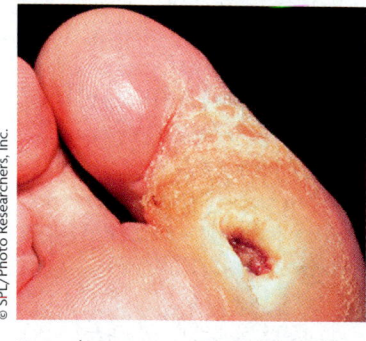

© SPL/Photo Researchers, Inc.

Foot ulcers are a common complication of diabetes because blood circulation is impaired, which slows healing, and nerve damage dampens foot pain, delaying recognition and treatment of cuts and bruises.

Neuropathy Nerve degeneration occurs in about 50 percent of diabetes cases. Symptoms of neuropathy vary and may be experienced as pain or burning, numbness and tingling in the hands and feet, or loss of sensation. Pain and cramping, especially in the legs, are often severe during the night and may interrupt sleep. Neuropathy also contributes to the development of foot ulcers because cuts and bruises may go unnoticed until wounds are severe. Other manifestations of neuropathy include sweating abnormalities, sexual dysfunction, constipation, and delayed stomach emptying (**gastroparesis**).[20]

IN SUMMARY

- In type 1 diabetes, the pancreas secretes little or no insulin, and insulin therapy is necessary for survival. Type 2 diabetes is characterized by insulin resistance coupled with relative insulin deficiency.

- In diabetic ketoacidosis, which is more common in type 1 diabetes, hyperglycemia is accompanied by ketosis and acidosis. The hyperosmolar hyperglycemic state is usually a consequence of type 2 diabetes; it is associated with severe hyperglycemia and depleted blood volume and may lead to mental impairment and coma.

- Chronic complications of diabetes include cardiovascular diseases, impaired blood circulation, retinopathy, nephropathy, and neuropathy.

Treatment of Diabetes Mellitus

Diabetes is a chronic and progressive illness that requires lifelong treatment. Managing blood glucose levels is a delicate balancing act that involves meal planning, proper timing of medications, and physical exercise; frequent adjustments in treatment are often necessary to establish good **glycemic** control. Individuals with type 1 diabetes require insulin therapy for survival. Type 2 diabetes is initially treated with diet therapy and exercise, but most patients are eventually treated with oral medications or insulin. Although the health care team must determine the appropriate therapy, the individual with diabetes ultimately assumes much of the responsibility for treatment and therefore requires education in self-management of the disease.

Treatment Goals

The main goal of diabetes treatment is to maintain blood glucose levels within a desirable range to prevent or reduce the risk of complications. Several multicenter clinical trials have shown that intensive diabetes treatment, which keeps blood glucose levels tightly controlled, can reduce the incidence and severity of nephropathy, retinopathy, and neuropathy.[21] ■ Therefore, maintenance of near-normal glucose levels has become the fundamental objective of all diabetes care plans. Other goals of treatment include maintaining healthy blood lipid concentrations, controlling blood pressure, and managing weight—measures that can help to prevent or delay diabetes complications as well. Table 21–3 (p. 580) provides examples of the major differences between conventional and intensive therapies for type 1 diabetes. Although intensive therapy is also associated with some risks, including an increased risk of hypoglycemia, its benefits outweigh these disadvantages.

Newly diagnosed patients and their families have much to learn about diabetes and its management. Diabetes education provides an individual with the knowledge and skills necessary to implement treatment. The primary instructor is often a **Certified Diabetes Educator (CDE)**, a health care professional (often a nurse or dietitian) who has specialized knowledge about diabetes treatment and the health education process.

■ Studies that evaluated the benefits of intensive diabetes treatment include the *Diabetes Control and Complications Trial* and the *United Kingdom Prospective Diabetes Study*.

advanced glycation end products (AGEs): compounds formed when glucose or glucose fragments combine with proteins. AGEs can damage tissues and lead to diabetic complications.

glycemic (gly-SEE-mic): pertaining to blood glucose.

Certified Diabetes Educator (CDE): a health care professional who specializes in diabetes management education. Certification is obtained from the National Certification Board for Diabetes Educators.

TABLE 21-3

Comparison of Conventional and Intensive Therapies for Type 1 Diabetes

	Conventional Therapy	Intensive Therapy
Blood glucose monitoring	Monitored daily	Monitored at least three times daily
Insulin therapy	One or two daily injections; no daily adjustments	Three or more daily injections or use of external insulin pump; dosage adjusted according to results of glucose monitoring and expected carbohydrate intake
Advantages	Fewer incidences of severe hypoglycemia; less weight gain	Delayed progression of retinopathy, nephropathy, and neuropathy
Disadvantages	More rapid progression of retinopathy, nephropathy, and neuropathy	Two- to threefold increase in severe hypoglycemia; weight gain; increased risk of becoming overweight

To manage diabetes, patients need to learn about appropriate meal planning, medication administration, blood glucose monitoring, weight management, appropriate physical activity, and prevention of complications.

Evaluating Diabetes Treatment

Diabetes treatment is largely evaluated by monitoring glycemic status. Good glycemic control requires frequent home monitoring of blood glucose using a glucose meter, referred to as **self-monitoring of blood glucose**. Glucose testing provides valuable feedback when the patient adjusts food intake, medications, and physical activity and is helpful for preventing hypoglycemia. Ideally, patients with type 1 diabetes should monitor blood glucose three or more times daily—and more frequently when therapy is adjusted. Self-monitoring of blood glucose is also useful in type 2 diabetes, although the recommended frequency depends on the specific needs of individual patients.[22]

Long-Term Glycemic Control Health care providers periodically evaluate long-term glycemic control by measuring **glycated hemoglobin** (abbreviated **HbA$_{1c}$**). The glucose in blood freely enters red blood cells and attaches to hemoglobin molecules in direct proportion to the amount of glucose present. Because the life span of red blood cells averages 120 days, the percentage of HbA$_{1c}$ is a measure of glycemic control during the preceding two to three months. In people without diabetes, HbA$_{1c}$ is typically less than 6 percent of total hemoglobin. The goal of diabetes treatment is a HbA$_{1c}$ value under 7 percent,[23] but it is often markedly higher even in people with diabetes who are maintaining near-normal blood glucose levels.

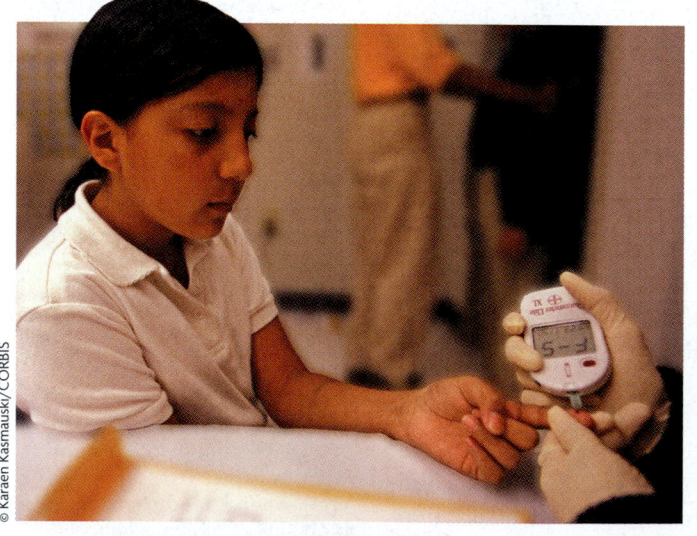

© Karaen Kasmauski/CORBIS

Self-monitoring of blood glucose can help persons with diabetes learn how to maintain blood glucose levels within a desirable range.

Monitoring for Long-Term Complications Individuals with diabetes are routinely monitored for signs of long-term complications. Blood pressure is measured at each checkup. Lipid screening is suggested annually for most adult patients. Routine checks for urinary protein (**microalbuminuria**) help to determine if nephropathy has developed. Physical examinations generally screen for signs of retinopathy, neuropathy, and foot problems.

Ketone Testing Ketone testing checks for the development of ketoacidosis if symptoms are present or if risk has increased due to acute illness, stress, or pregnancy. Both blood and urine tests are available for home use, although the blood tests are currently more reliable. Ketone testing is most useful for patients who have type 1 diabetes or gestational diabetes.

Body Weight Concerns

Whereas individuals with newly diagnosed type 1 diabetes are likely to be thin, most people with type 2 diabetes are overweight or obese. Weight and growth patterns are routinely monitored to evaluate whether energy intakes are appropriate.

Body Weight in Type 1 Diabetes In general, people with type 1 diabetes are less likely to be overweight than those in the general population. However, excessive weight gain is sometimes an unwanted side effect of improved glycemic control, especially in those undergoing intensive insulin therapy. Although the cause of weight gain is unclear, it is possibly related to the insulin treatment, which may increase appetite or reduce metabolic rate.[24] Although efforts should be made to prevent excessive weight gain, concerns about weight should not discourage the use of intensive therapy, which is associated with longer life expectancy and fewer complications than occur with conventional therapy (review Table 21–3). It is also important to ensure that growing children receive sufficient energy for normal growth and development.

Body Weight in Type 2 Diabetes Because excessive body fat can worsen insulin resistance, weight loss is often recommended for those who are overweight or obese. Even moderate weight loss (10 to 20 pounds) can help to improve glycemic control, blood lipid levels, and blood pressure. Weight loss is most beneficial early in the course of diabetes, before insulin secretion has diminished.

Not all persons with type 2 diabetes are overweight or obese. Older adults and those in long-term care facilities are often underweight and may need to gain weight. Low body weight increases risks of morbidity and mortality in these individuals.

Medical Nutrition Therapy: Nutrient Recommendations

Medical nutrition therapy has a considerable influence on diabetes outcome. The appropriate dietary choices can both improve blood glucose levels and slow the progression of diabetic complications. As always, the nutrition care plan must take personal preferences and lifestyle habits into account. Dietary intakes need to be modified to accommodate growth, lifestyle changes, aging, and any complications that develop. Although all members of the diabetes care team should understand the principles of dietary treatment, a registered dietitian is best suited for designing and implementing the medical nutrition therapy provided to diabetes patients. This section presents the nutrient recommendations for diabetes. A later section describes meal planning strategies.

Total Carbohydrate Intake The amount of carbohydrate ingested has the greatest influence on blood glucose levels after meals—the more grams of carbohydrate ingested, the greater the glycemic response. The carbohydrate recommendation is based in part on the person's metabolic needs (that is, the type of diabetes or degree of glucose tolerance) and individual preferences. In addition, the carbohydrate intake must be fairly consistent at meals and snacks to help reduce fluctuations in blood glucose levels between meals. Low-carbohydrate diets, which restrict carbohydrate intake to less than 130 grams per day, are not recommended.[25]

Carbohydrate Sources Different carbohydrate-containing foods have different effects on blood glucose levels; for example, consuming a portion of white rice may cause blood glucose to rise more than would a similar portion of barley. This *glycemic effect* of foods is influenced by a food's fiber content, preparation method, the other foods included in a meal, and individual tolerances. The **glycemic index (GI)**, a ranking of

self-monitoring of blood glucose: home monitoring of blood glucose levels using a glucose meter.

glycated hemoglobin (HbA₁c): hemoglobin molecules to which glucose is attached. The percentage of such molecules is used to evaluate long-term glycemic control; also called **glycosylated hemoglobin.**

glycemic index (GI): a ranking of carbohydrate foods based on their effect on blood glucose levels after ingestion; *low-GI* foods are foods with a lesser glycemic effect; *high-GI* foods are those with a greater glycemic effect.

carbohydrate foods based on their glycemic effect, has been compiled from the scientific literature; using this information when making food choices may help to improve glycemic control in some individuals.[26] However, the GI is not a primary consideration when treating diabetes, because current evidence is insufficient to conclude that low-GI diets improve glycemic control in most people with diabetes.[27] In addition, there is considerable variability in individual responses to specific carbohydrate foods. Nonetheless, high-fiber, whole-grain products, which have more moderate effects on blood glucose than do highly processed starchy foods, are among the foods frequently recommended for persons with diabetes. Nutrition in Practice 21 provides additional information about the glycemic index and its potential use in the prevention and treatment of disease.

Fiber Fiber recommendations for diabetes are similar to those for the general population; ■ thus people with diabetes are encouraged to include fiber-rich foods such as legumes, whole-grain cereals, fruits, and vegetables in their diet. Although some studies have suggested that very high intakes of fiber (50 grams or more per day) may improve glycemic control, the benefits have not been consistent across studies, and many individuals may have difficulty tolerating such large amounts of fiber.[28]

■ The fiber DRI for adult women and men ranges from 21 to 38 g; check the DRI table on the inside front cover of this text for specific values.

Sugars A common misperception is that people with diabetes need to avoid sugar and sugar-containing foods. In reality, table sugar (sucrose), made up of glucose and fructose, has a lower glycemic effect than that of starch. Because moderate consumption of sugar has not been shown to adversely affect glycemic control,[29] sugar recommendations for people with diabetes are similar to those for the general population, which suggest minimizing foods and beverages that contain added sugars. However, sugars and sugary foods must be counted as part of the daily carbohydrate allowance.

Although fructose has a minimal glycemic effect, its use as an added sweetener is not advised because excessive dietary fructose may adversely affect blood lipid levels. (Note that it is not necessary to avoid the naturally occurring fructose in fruits and vegetables.) Sugar alcohols (such as sorbitol and maltitol) have lower glycemic effects than glucose, fructose, or sucrose, but their use has not been found to significantly improve long-term glycemic control. Artificial sweeteners (such as aspartame, saccharin, and sucralose) contain no digestible carbohydrate and can be safely used in place of sugar.

Dietary Fat As mentioned earlier, people with diabetes are at high risk of developing cardiovascular diseases. Guidelines for dietary fat are similar to those for other persons at risk: saturated fat intake should be limited to less than 7 percent of total kcalories, *trans* fat intake should be minimized, and cholesterol intake should be limited to less than 200 milligrams daily.[30] Dietary strategies for cardiovascular disease are discussed further in Chapter 22.

Protein Protein recommendations for people with diabetes are similar to those for the general population, in which protein intake ranges from 15 to 20 percent of total kcalories. Although small, short-term studies have suggested that diets with higher protein intakes may improve glycemic control, increase satiety, and help with weight loss, the long-term effects of such diets on diabetes management and complications are unknown.[31] In addition, high protein intakes are discouraged because they may be detrimental to kidney function in some individuals.

■ Reminder: One drink is equivalent to 12 ounces of beer, 5 ounces of wine, 10 ounces of wine cooler, or 1½ ounces of 80 proof distilled spirits such as gin, rum, vodka, and whiskey.

Alcohol Use in Diabetes Alcohol can be used in moderation by adults with diabetes. Guidelines are similar to those for the general population, which advise a daily limit of one drink for women and two drinks for men. ■ However, those using insulin or medications that promote insulin secretion should consume food when they use alcoholic beverages to avoid hypoglycemia. Alcohol can cause hypoglycemia by interfering with glucose production in the liver. Conversely, excessive alcohol can worsen hyperglycemia, and it can also raise triglyceride levels in susceptible persons. People who should avoid alcohol include pregnant women and individuals with pancreatitis, advanced neuropathy, abnormally high triglyceride levels, or a history of alcohol abuse.[32]

Micronutrients Micronutrient recommendations for people with diabetes are the same as for the general population. Vitamin and mineral supplementation is not recommended unless nutrient deficiencies develop; those at risk include the elderly, pregnant or lactating women, strict vegetarians, and individuals on kcalorie-restricted diets.[33] Although some studies have suggested that supplemental chromium can improve glycemic control in type 2 diabetes, results have not been consistent. At present, chromium supplementation is not recommended for those with type 2 diabetes.[34]

Medical Nutrition Therapy: Meal-Planning Strategies

Dietitians use a number of meal-planning strategies to help people with diabetes maintain glycemic control. Emphasis is given to controlling carbohydrate intake and portion sizes. A regular eating pattern, with carbohydrate intake spaced evenly throughout the day, is typically recommended. Providing sample menus can help to illustrate general principles. People using intensive insulin therapy must coordinate insulin injections with meals and adjust insulin dosages to carbohydrate intake, as discussed in a later section.

Carbohydrate Counting Carbohydrate counting techniques are simpler and more flexible than other menu planning approaches and are widely used for planning diabetes diets. Carbohydrate counting works as follows: after a dietitian determines a person's nutrient and energy needs, the individual is given a daily carbohydrate allowance, often divided into a pattern of meals and snacks according to individual preferences. The carbohydrate allowance can be expressed in grams or as the number of carbohydrate portions allowed per meal (see Table 21–4, p. 584). The user of the plan need only be concerned about meeting carbohydrate goals and can select from any of the carbohydrate-containing food groups when planning meals (see Table 21–5 and Figure 21–1). Although encouraged to make healthy food choices, the individual has the freedom to choose the foods desired at each meal without risking loss of glycemic control. Some people may also need guidance about noncarbohydrate foods to help them choose a healthy diet that improves blood lipids or energy intakes. The "How To" on pp. 584–585 explains how to implement carbohydrate counting in clinical practice.

Carbohydrate counting is taught at different levels of complexity depending on a person's needs and abilities. The basic carbohydrate counting method just described can be helpful for most people, although it requires a consistent carbohydrate intake to match the medication or insulin regimen. Advanced carbohydrate counting allows more flexibility but is best suited for patients using intensive insulin therapy. With this method, a person can determine the specific dosage of insulin needed to cover the amount of carbohydrate consumed at a meal. The person is then free to choose the types and portions of food desired without sacrificing glycemic control. Advanced carbohydrate counting requires some training and should be attempted only after more basic methods are mastered.

Exchange Lists for Meal Planning The exchange list system is still taught by some dietitians, although it is more complex and difficult for patients to learn than carbohydrate counting. The exchange system sorts foods according to their proportions of carbohydrate, fat, and protein so that each item in a food group (or "exchange list") has a similar macronutrient and energy content (see pp. 811 to 812). Thus any food on a list can be exchanged, or traded, for any other food on the same list without

USE CARBOHYDRATE COUNTING IN CLINICAL PRACTICE

1. The first step in basic carbohydrate counting is to determine an appropriate carbohydrate intake and suitable distribution pattern; an example is shown in Table 21–4. A nutrition assessment can help to estimate a person's usual energy and carbohydrate intakes. The carbohydrate level should be acceptable to the person using the plan. Frequent monitoring of blood glucose levels can help determine whether additional carbohydrate restriction would be helpful.

   The example given in Table 21–4 illustrates a meal pattern for a person consuming 2,000 kcalories daily with a carbohydrate allowance of 50 percent of kcalories. This is calculated as follows:

   $$50\% \times 2{,}000 \text{ kcal} = 1{,}000 \text{ kcal of carbohydrate}$$

   $$\frac{1{,}000 \text{ kcal carbohydrate}}{4 \text{ kcal/g carbohydrate}} = 250 \text{ g carbohydrate/day}$$

   $$\frac{250 \text{ g carbohydrate}}{15 \text{ g/1 carbohydrate portion}} = 16.7 \text{ carbohydrate portions/day}$$

2. The distribution of carbohydrates among meals and snacks is based on both individual preferences and metabolic needs. In type 1 diabetes, the insulin regimen must coordinate with the individual's dietary and lifestyle choices. People using conventional insulin therapy must have a consistent carbohydrate intake from day to day to match their particular insulin prescription, whereas those using intensive therapy can alter insulin dosages when carbohydrate intakes change. People with type 2 diabetes are encouraged to develop dietary patterns that suit their lifestyle and medication schedules. For all types of diabetes, the carbohydrate recommendation may need to be altered periodically to improve blood glucose control.

3. Carbohydrate counting can be done in one of two ways:

   - Count the grams of carbohydrate provided by foods.

   - Count carbohydrate portions, expressed in terms of servings that contain approximately 15 grams each.

   Success with carbohydrate counting requires knowledge about the food sources of carbohydrates and an understanding of portion control. As shown in Table 21–5,

---

TABLE 21–4

Sample Carbohydrate Distribution for a 2,000-kCalorie Diet

Meals	Carbohydrate Allowance	
	Grams	Portions[a]
Breakfast	60	4
Lunch	60	4
Afternoon snack	30	2
Dinner	75	5
Evening snack	30	2
Totals	**255 g**	**17**

Note: The carbohydrate allowance in this example is approximately 50 percent of total kcalories.

[a] 1 portion = 15 g carbohydrate = 1 portion of starchy food, milk, or fruit

---

TABLE 21–5

Portion Sizes of Carbohydrate-Containing Foods

Food Groups with Sample Portion Sizes

Bread, cereal, rice, and pasta: 1 portion = 15 g carbohydrate
1 slice of bread or 1 tortilla
1/2 English muffin
3/4 c unsweetened, ready-to-eat cereal
1/2 c cooked oatmeal
1/3 c cooked rice or pasta

Fruit: 1 portion = 15 g carbohydrate
1 medium apple, orange, or peach
1 small banana
3/4 c blueberries or chopped pineapple
1/2 c apple or orange juice

Milk products: 1 portion = 12 g carbohydrate; may be rounded up to 15 g for ease in counting carbohydrate portions
1 c milk (whole, low-fat, or fat-free)
1 c buttermilk
6 oz plain yogurt

Starchy vegetables: 1 portion = 15 g carbohydrate
1 small (3 oz) potato
1/2 c canned or frozen corn
1/3 c baked beans
1 c winter squash, cubed

Sweets and desserts: Considerable variation in carbohydrate content; portions listed contain approximately 15 g
1/2 c ice cream
2 sandwich cookies (with cream filling)
1/2 frosted cupcake
1 granola bar (1 oz)
1 tbs honey

Nonstarchy vegetables: 1 portion = 3 to 6 g carbohydrate; 3 servings are equivalent to 1 carbohydrate portion; can be disregarded if less than 3 servings are consumed
1/2 c cooked cauliflower
1/2 c cooked cabbage, collards, or kale
1/2 c cooked okra
1/2 c diced or raw tomatoes

Note: Unprocessed meats, fish, and poultry contain negligible amounts of carbohydrate.

(continued)

food selections that contain about 15 grams of carbohydrate are interchangeable. The portions of foods that contain 15 grams may vary substantially, however, even among foods in a single food group. Accurate carbohydrate counting often requires instruction and practice in portion control using measuring cups, spoons, and a food scale. Food lists that indicate the carbohydrate contents of common foods are available from the American Diabetes Association and the American Dietetic Association; these are helpful resources for learning carbohydrate counting methods.

When using packaged foods, individuals should check the Nutrition Facts panel of food labels to find the carbohydrate content of a serving. If the fiber content is greater than 5 grams per serving, it should be subtracted from the *Total*

Carbohydrate value, as fiber does not contribute to blood glucose. If the sugar alcohol content is greater than 5 grams per serving, half of the grams of sugar alcohol can be subtracted from the *Total Carbohydrate* value.

4. Once they have learned the basic carbohydrate counting method, individuals can select whatever foods they wish, as long as they do not exceed their carbohydrate goals. Figure 21–1 shows a day's menu that follows the dietary plan shown in Table 21–4. Although carbohydrate counting focuses on a single macronutrient, people using this technique should be encouraged to follow a healthy eating plan that meets other dietary objectives as well.

FIGURE 21–1

Translating Carbohydrate Portions into a Day's Meals

❋ SAMPLE MENU ❋

	Carbohydrate Portions			Carbohydrate Portions
Breakfast:		**Afternoon snack:**		
Carbohydrate goal = 4 portions or 60 g.		**Carbohydrate goal = 2 portions or 30 g.**		
³/₄ c unsweetened, ready-to-eat cereal	1	2 sandwich cookies		1
½ c low-fat milk	½	1 c low-fat millk		1
1 scrambled egg	—			
1 slice whole wheat toast (with margarine or butter)	1	**Dinner:**		
6 oz orange juice	1½	**Carbohydrate goal = 5 portions or 75 g.**		
Coffee (without milk or sugar)	—	4 oz grilled steak		—
		1 small baked potato (with margarine or butter)		1
Lunch:		Corn on cob, 1 large ear		2
Carbohydrate goal = 4 portions or 60 g.		½ c steamed collard greens[a]		1
1 tuna salad sandwich (includes 2 slices whole-grain bread, mayonnaise)	2	1 c sliced, raw tomatoes[a]		
6 oz yogurt (plain) with 3/4 c blueberries and artificial sweetener	2	½ c ice cream		1
Diet cola	—	**Evening snack:**		
		Carbohydrate goal = 2 portions or 30 g.		
		1 medium apple		1
		1 oz granola bar		1

[a] Three servings of nonstarchy vegetables are equivalent to 1 carbohydrate portion.

---

affecting the macronutrient balance in a day's meals. Although the exchange list system may be helpful for individuals who want a structured dietary plan that provides specific percentages of protein, carbohydrate, and fat, it offers no advantages for maintaining glycemic control and is less flexible than carbohydrate counting. This system of meal planning is described further in Appendix C (Appendix B for Canadians).

The exchange lists can be helpful resources for individuals using carbohydrate counting methods because the portions in the exchange lists are interchangeable with the portions used in carbohydrate counting. For example, foods listed in the starch, fruit, and milk exchange lists are equivalent to carbohydrate "portions," as each item contains approximately 15 grams of carbohydrate (see pp. 813 and 814; note that the carbohydrate in milk exchanges can be rounded up to 15 grams). The list labeled Sweets, Desserts, and Other Carbohydrates (p. 815) indicates the number of carbohydrate portions per serving in the far-right column.

TABLE 21–6

Insulin Preparations

Form of Insulin	Common Preparations	Onset of Action	Peak Activity	Duration of Action
Rapid-acting	Lispro Aspart	15 min	30 min to 2 hr	3 to 5 hr
Short-acting	Regular	30 min	2 to 4 hr	5 to 8 hr
Intermediate-acting	Lente NPH	1 to 3 hr	5 to 10 hr	18 to 24 hr
Long-acting	Glargine Ultralente	2 to 4 hr 4 to 6 hr	Steady effects 8 to 12 hr	24 hr Over 30 hr
Insulin mixtures (with sample ratios)	NPH/regular (70:30) NPH/regular (50:50)	Variable; depends on formulation	Variable; depends on formulation	Variable; depends on formulation

Insulin Therapy

Insulin therapy is necessary for people who cannot produce enough insulin to meet their metabolic needs. It is therefore required by individuals with type 1 diabetes and those with type 2 diabetes who cannot maintain glycemic control with oral medications, diet, and exercise. The pancreas normally secretes insulin in relatively low amounts between meals and during the night (called *basal insulin*) and in much higher amounts when meals are ingested. Ideally, the insulin treatment should reproduce the natural pattern of insulin secretion as closely as possible.

Insulin Preparations The forms of insulin available differ by their onset of activity, timing of peak activity, and duration of effects. Table 21–6 and Figure 21–2 show how insulin preparations are classified: they may be rapid-acting (lispro and aspart), short-acting (regular), intermediate-acting (lente and NPH), or long-acting (glargine and ultralente), thereby allowing substantial flexibility in establishing a suitable insulin regimen. Mixtures of several types of insulin can produce greater glycemic control than any one type alone. Several premixed formulations are also available (several examples are listed in Table 21–6).

■ Because insulin is a protein, it would be destroyed by digestive processes if taken orally.

Insulin Delivery Insulin is most often administered by injection, ■ either self-administered or provided by caregivers. A rapid-acting inhalation powder (Exubera)

FIGURE 21–2

Effects of Insulin Preparations

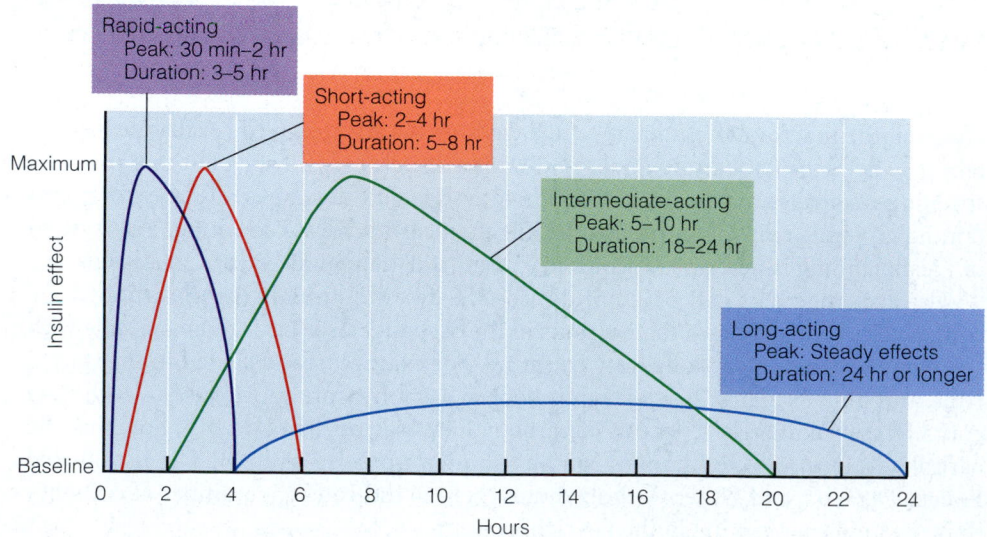

is also available; this form of insulin is taken right before meals and is usually used in combination with longer-acting insulin preparations or various oral medications that treat hyperglycemia (discussed in a later section).

Individual syringes are most commonly used for injecting insulin, although several other options are available. Injection ports, inserted through the skin, can remain in place for several days, eliminating the need for multiple punctures. Another option is to use an insulin pump, a computerized device that can be programmed to deliver basal insulin continuously and bolus doses at mealtimes (see the photo). The pump infuses insulin through thin, flexible tubing that remains in the skin. The pump can be worn under clothes, attached to a belt, or kept in a pocket.

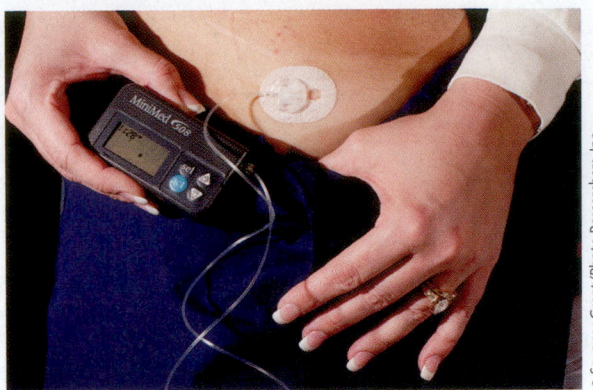

External insulin pumps deliver insulin continuously through thin, flexible tubing inserted into the skin.

Insulin Regimen for Type 1 Diabetes Type 1 diabetes is best managed with intensive insulin therapy, which involves multiple daily injections of several types of insulin or use of an insulin pump (review Table 21–3). Usually, intermediate- or long-acting insulin meets basal insulin needs, and rapid- or short-acting insulin is injected before meals. ■ Three or more daily injections are required for good glycemic control. Simpler regimens involve twice-daily injections of a mixture of intermediate- and short-acting insulin. Regimens that include three or more injections allow for greater flexibility in meal timing. With fewer injections, the timing of both meals and injections must be similar from day to day to avoid periods of insulin deficiency or excess.[35]

A person using intensive therapy must learn to accurately determine the amount of insulin to inject before each meal. The amount required depends on premeal blood glucose levels, the carbohydrate content of the meal, and the person's body weight and sensitivity to insulin. To determine insulin sensitivity, a person keeps careful records of food intake, insulin dosages, and blood glucose levels. Eventually, these records are analyzed to determine the appropriate **carbohydrate-to-insulin ratio** for that individual, which assists in calculating insulin dosages at mealtime.[36]

After insulin therapy is initiated, persons with type 1 diabetes may experience a temporary remission of disease symptoms and a reduced need for insulin, known as the "honeymoon phase." The remission is due to a temporary improvement in pancreatic function and may last for several weeks or months. In all cases, diabetes eventually returns, and full insulin treatment must be reinstated.

Insulin Regimen for Type 2 Diabetes Approximately 30 percent of people diagnosed with type 2 diabetes are treated with insulin therapy.[37] Although initial treatment of type 2 diabetes may involve diet therapy, physical activity, and oral antidiabetic medications, long-term results with these treatments are often disappointing. As the disease progresses, pancreatic function worsens, and many individuals require insulin therapy to maintain glycemic control.

Various regimens can be used to control type 2 diabetes. Some persons may be treated with insulin alone, whereas others may use insulin in combination with oral antidiabetic agents. Often, only one or two daily injections are needed. Some regimens involve a mixture of rapid- and intermediate-acting insulin in the morning and an injection of intermediate- or long-acting insulin at dinner or before bedtime. In other cases, only a single injection of intermediate- or long-acting insulin may be needed at bedtime. Dosages and timing are adjusted according to the results of blood glucose self-monitoring.[38]

Insulin Therapy and Hypoglycemia Hypoglycemia is the most common complication of insulin treatment, although it may also result from the use of some oral antidiabetic

■ Rapid-acting insulin begins working within 15 minutes, so it can be injected right before a meal. Short-acting insulin requires a half hour wait before the meal can begin.

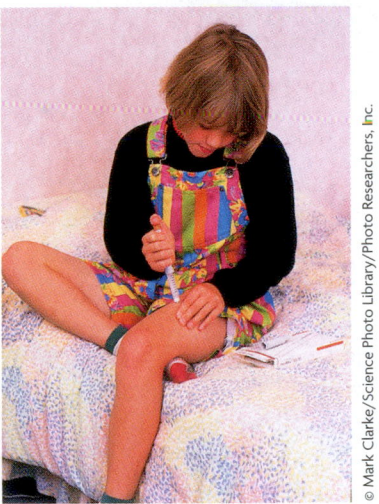

Children often become adept at administering the insulin they require.

carbohydrate-to-insulin ratio: the amount of carbohydrate that can be handled per unit of insulin. On average, every 15 g of carbohydrate requires about 1 unit of rapid- or short-acting insulin.

drugs. It most often results from intensive insulin therapy because the attempt to attain near-normal blood glucose levels increases the risk of overtreatment.

Hypoglycemia can be corrected with the immediate intake of glucose or a carbohydrate-containing food. Usually, 15 to 20 grams of carbohydrate ■ can relieve hypoglycemia in about 15 minutes, although blood glucose levels should be retested after 15 minutes in case additional treatment is necessary.[39] Foods that provide pure glucose yield a better response than foods that contain other sugars, such as sucrose or fructose. People using insulin are usually advised to carry glucose tablets or a source of carbohydrate that can be readily ingested. Those at risk of severe hypoglycemia are often given prescriptions for the hormone glucagon, which can be injected by caregivers in case of unconsciousness.

Fasting Hyperglycemia Insulin therapy must sometimes be adjusted to prevent fasting hyperglycemia, which has three possible causes. The usual cause is a waning of insulin action during the night due to insufficient insulin. A second possibility, known as the **dawn phenomenon**, occurs when blood glucose levels increase in the morning due to the early morning secretion of growth hormone, which counteracts insulin's actions. Less frequently, fasting hyperglycemia develops as a result of nighttime hypoglycemia, which causes hormonal responses that stimulate glucose production; the resulting condition is known as **rebound hyperglycemia**. Whatever the cause, fasting hyperglycemia can be treated by adjusting the dosage or formulation of insulin administered in the evening.

Oral Antidiabetic Agents

Treatment of type 2 diabetes often requires the use of oral medications. These drugs can improve hyperglycemia by four modes of action: they can improve insulin secretion, reduce glucose production in the liver, improve use of glucose by the tissues, or delay carbohydrate absorption. Treatment may involve the use of a single medication (monotherapy) or a combination of several (combination therapy). By utilizing several mechanisms at once, combination therapy achieves more rapid and sustained glycemic control than is possible with monotherapy.[40] Table 21–7 lists examples of oral antidiabetic agents, and the Diet-Drug Interactions feature lists their nutrition-related effects. Because medications cannot replace the benefits offered by dietary modifications and physical activity, persons with diabetes should be advised to continue both.

Physical Activity and Diabetes Management

Regular physical activity can improve glycemic control considerably and is therefore a central feature of disease management. A regular exercise program improves insulin

TABLE 21–7

Oral Antidiabetic Agents

Mode of Action	Drug Category	Common Examples
Stimulate insulin secretion by pancreas	Sulfonylureas	Chlorpropamide Tolbutamide Glyburide Glipizide
	Meglitinides	Repaglinide Nateglinide
Inhibit liver glucose production	Biguanides	Metformin
Increase insulin sensitivity	Thiazolidinediones	Pioglitazone Rosiglitazone
Delay carbohydrate absorption	Alpha-glucosidase inhibitors	Acarbose Miglitol

Diet-Drug Interactions

Check this table for notable nutrition-related effects of the medications discussed in this chapter.

	Gastrointestinal Effects	Interactions with Dietary Substances	Metabolic Effects
Sulfonylureas	Nausea, vomiting, cramps, diarrhea	Avoid using with alcohol due to a toxic reaction that causes flushing, throbbing head and neck pain, shortness of breath, palpitations, and sweating. Avoid using with dietary supplements that contain ginseng, garlic, fenugreek, coriander, and celery; they may increase risk of hypoglycemia.	Hypoglycemia, weight gain, allergic skin reactions
Biguanides (metformin)	Abdominal pain, nausea, vomiting, gas, cramps, diarrhea, metallic taste, anorexia	—	Asymptomatic vitamin B_{12} deficiency
Thiazolidinediones	—	—	Weight gain, fluid retention, edema, anemia
Alpha-glucosidase inhibitors	Abdominal pain, nausea, gas, cramps, diarrhea	—	Elevated liver enzymes, hyperbilirubinemia

sensitivity, which can reduce insulin requirements. Physical activity also benefits other aspects of health, including blood lipid levels, blood pressure, body weight, and cardiovascular functioning. People with diabetes are encouraged to regularly perform both aerobic activity and resistance exercise unless contraindicated. At least 150 minutes of moderate-intensity activity and/or 90 minutes of vigorous activity should be undertaken each week. In addition, a resistance exercise program that targets all major muscle groups should be performed three times weekly.[41]

Physical Activity and Insulin Therapy People with type 1 diabetes must carefully adjust food intake and insulin therapy to prevent hypoglycemia during physical activity. Insulin dosages that precede exercise often need to be reduced substantially. Blood glucose levels should be checked both before and after an activity. If blood glucose is below 100 mg/dL, carbohydrate should be consumed before exercise begins. Additional carbohydrate may be needed during or after prolonged activity or even several hours after the activity is completed. Physical activity should be avoided if blood glucose levels exceed 250 mg/dL with ketosis or 300 mg/dL without ketosis, because hyperglycemia may worsen.[42]

Physical Activity in Type 2 Diabetes People with type 2 diabetes are often overweight and sedentary, and many develop complications during the course of disease. Before an exercise program is planned, a medical evaluation should screen for problems that may be aggravated by certain activities. Complications involving the heart and blood vessels, eyes, kidneys, feet, and nervous system may limit the types of activity recommended.[43]

Only mild or moderate exercise may be prescribed at first. For obese, inactive persons, a short walk at a comfortable pace may be the first activity suggested. Persons with retinopathy should avoid heavy lifting or straining, which may raise blood pressure and damage eye tissue. Peripheral neuropathy precludes repetitive weight-bearing exercises such as jogging and step exercises because these activities may lead to foot ulcerations. To prevent dehydration, which can adversely affect blood glucose levels and heart function, proper hydration should be encouraged before and during exercise.

Sick-Day Management

Illness, infection, or injury can cause hormonal changes that raise blood glucose levels. In people with type 1 diabetes, illness also increases the risk of diabetic ketoacidosis. A patient with type 1 diabetes should be advised to test blood glucose frequently during illness and to test blood or urine for ketones. Some carbohydrate should be ingested regularly to avoid ketosis; a daily intake of 150 to 200 grams of carbohydrate (about 45 to 50 grams every three to four hours) is recommended. If appetite is poor, carbohydrate-sweetened beverages or frozen juice bars may be easier to consume than solid foods. Fluid intake should be monitored to prevent dehydration.[44]

IN SUMMARY

- Diabetes treatment often includes medical nutrition therapy, insulin or oral medications, and appropriate physical activity. Glycemic control is evaluated by monitoring blood glucose and glycated hemoglobin levels.

- Carbohydrate intake is the main factor that influences blood glucose levels after meals. The total amount of carbohydrate ingested is more important than the type of carbohydrate consumed.

- Carbohydrate counting is widely used in menu planning and can be taught at different levels of complexity depending on individual needs.

- Insulin therapy is required for patients who are unable to produce sufficient insulin and may be used in both type 1 and type 2 diabetes. Oral medications can improve insulin secretion and sensitivity, lower glucose production by the liver, and delay carbohydrate absorption.

- Physical activity can improve glycemic control and enhance various aspects of general health. The Case Study provides an opportunity to review the factors that influence treatment for a 12-year-old with type 1 diabetes.

Case Study

CHILD WITH TYPE 1 DIABETES

Nora is a 12-year-old girl who was diagnosed with type 1 diabetes two years ago. She practices intensive therapy and has had the support of her parents and an excellent diabetes management team. With their help, Nora has been able to assume the bulk of the responsibility for her diabetes care and has managed to control her blood glucose remarkably well. In the last few months, however, Nora has been complaining bitterly about the impositions diabetes has placed on her life and her interactions with friends. Sometimes she refuses to monitor her blood glucose levels, and she has skipped insulin injections a few times. Recently, Nora was admitted to the emergency room complaining of fever, nausea, vomiting, and intense thirst. The physician noted that Nora was confused and lethargic. A urine test was positive for ketones, and her blood glucose levels were 400 mg/dL. The diagnosis was diabetic ketoacidosis.

1. Describe the metabolic events that lead to ketoacidosis. Were Nora's symptoms and laboratory tests consistent with the diagnosis?

2. Review Table 21–3 (p. 580) and consider the advantages and disadvantages that intensive therapy might have for Nora.

3. Discuss how Nora's age might influence her ability to cope with and manage her diabetes. Why might she feel that diabetes is disrupting her life? What suggestions may help?

4. Review the list of complications associated with long-term diabetes. How might you explain the importance of glycemic control to a 12-year-old girl?

Diabetes Management in Pregnancy

Women with diabetes face new challenges during pregnancy. Due to hormonal changes, pregnancy increases insulin resistance and the need for insulin, so maintaining glycemic control may be more difficult. In addition, up to 7 percent of nondiabetic women develop gestational diabetes and require treatment during pregnancy.[45] Women with gestational diabetes are at greater risk of developing type 2 diabetes later in life, and their children are at increased risk of developing obesity and type 2 diabetes as they enter into adulthood.

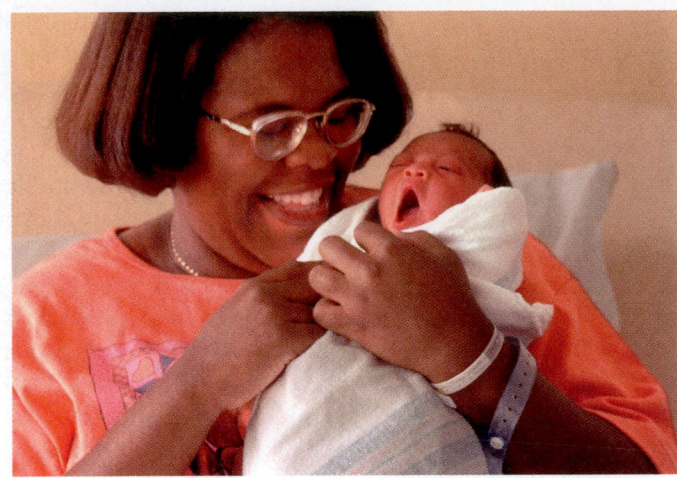

Glycemic control during pregnancy offers the best chance of a safe delivery and a healthy infant.

© Lester Lefkowitz/Corbis

A pregnancy complicated by diabetes increases health risks for both mother and fetus. Uncontrolled diabetes is linked with an increased rate of miscarriages. Incidences of birth defects and fetal deaths are higher than normal. Newborns are more likely to suffer from respiratory distress and to develop metabolic problems such as hypoglycemia, jaundice, and hypocalcemia. Women with type 2 diabetes and gestational diabetes often deliver babies with **macrosomia** (abnormally large bodies), which makes delivery more difficult and can result in birth trauma or a cesarean section.[46]

Pregnancy in Type 1 or Type 2 Diabetes

Women with diabetes who achieve glycemic control at conception and during the first trimester can substantially reduce the risks of birth defects and spontaneous abortion during pregnancy. For this reason, it is recommended that women contemplating pregnancy receive preconception care to avoid the complications associated with poorly controlled diabetes.[47] Maintaining glycemic control during the second and third trimesters can minimize the risks of macrosomia and morbidity in newborn infants.

Nutrient requirements during pregnancy are generally similar for women with and without diabetes. The dietary adjustments suggested for improving glycemic control should be based on a woman's dietary habits and the results of blood glucose monitoring. Regular meals and snacks help to avoid hypoglycemia, which is more likely to occur during pregnancy because glucose is continuously supplied to the fetus. An evening snack is usually required to prevent overnight hypoglycemia and ketosis. Insulin and medication changes are often needed during pregnancy, and the woman may have to adjust her dietary habits further as a result.

Gestational Diabetes

Risk of gestational diabetes is highest in women who have a family history of diabetes; are obese; are members of certain ethnic groups, including Hispanic Americans, Native Americans, Asian-Americans, African-Americans, and Pacific Islanders; or have given birth to infants weighing over 9 pounds. To ensure that problems are dealt with promptly, physicians routinely test all women for gestational diabetes between 24 and 28 weeks of gestation. In high-risk women, screening should begin prior to pregnancy or soon after conception. Even mild hyperglycemia can have adverse effects on a developing fetus and may lead to complications during pregnancy.[48]

macrosomia (MAK-roh-SOH-mee-ah): the condition of having an abnormally large body. In infants, refers to birth weights of 4,000 g (8 lb 13 oz) and above.

Women with gestational diabetes who are overweight or obese may need to adjust their energy intakes during pregnancy. Although adequate energy is needed for fetal development, a modest kcaloric reduction (about 30 percent less than total energy needs) may improve glycemic control without increasing the risk of ketosis. Restricting carbohydrate to 40 to 45 percent of total energy intake can improve blood glucose levels after meals. Because carbohydrate may be poorly tolerated in the morning, reducing carbohydrate at breakfast can be helpful. The remaining carbohydrate intake should be spaced throughout the day into several meals and snacks, including an evening snack to prevent ketosis during the night. Regular aerobic activity is often recommended because it can help to improve glycemic control. Women who fail to achieve glycemic goals by diet and exercise alone may need insulin therapy, which is given to approximately 20 to 25 percent of women with gestational diabetes.[49] Oral antidiabetic drugs generally are not prescribed during pregnancy because they may be toxic to the developing fetus. The Case Study reviews the connections between gestational diabetes and type 2 diabetes.

IN SUMMARY

- Careful management of blood glucose levels before and during pregnancy may reduce complications in mother and infant. Most nutrient requirements during pregnancy are similar for women with and without diabetes.

- Carbohydrate intake should be distributed into several meals and snacks, including an evening snack to prevent overnight ketosis. Carbohydrate restriction may be recommended, especially in women with gestational diabetes.

- Moderate energy restriction may help to improve glycemic control in overweight and obese women.

Case Study

SCHOOL COUNSELOR WITH TYPE 2 DIABETES

Alicia Cordova is a 41-year-old Mexican-American woman recently diagnosed with type 2 diabetes. Mrs. Cordova developed gestational diabetes while she was pregnant with her second child. Her blood glucose levels returned to normal following pregnancy, and she was advised to get regular checkups, maintain a desirable weight, and engage in regular physical activity. Although she reports that she does not overeat and that she exercises regularly, she has been unable to maintain a healthy weight. At 5 feet 3 inches tall, Mrs. Cordova currently weighs 155 pounds. She has decided to lose weight and join a gym because she is concerned about the long-term effects of diabetes and the possibility that she may need insulin injections. She is also concerned about her husband and children because they are overweight and not very active. The physician refers Mrs. Cordova to a dietitian to help her plan a diet.

1. What factors in Mrs. Cordova's medical history increase her risk for diabetes? Are her husband and children also at risk?

2. Describe the general characteristics of a diet and exercise program that would be appropriate for Mrs. Cordova. How might weight loss and physical activity benefit her diabetes?

3. If Mrs. Cordova is unable to control her blood glucose with diet and physical activity, what treatment might be suggested? Can you explain to Mrs. Cordova why she would probably not require insulin at this time?

4. What dietary and lifestyle changes may help to prevent diabetes in Mrs. Cordova's husband and children?

Medical History

Check the medical record to determine:

- ☐ Type of diabetes
- ☐ Duration of diabetes
- ☐ Acute and chronic complications
- ☐ Conditions, including pregnancy, that may alter treatment

Medications

For people with preexisting diabetes who use oral antidiabetic drugs, insulin, or both, note:

- ☐ Type of medication
- ☐ Administration schedule

Check for use of other medications, including:

- ☐ Medications that affect blood glucose levels
- ☐ Cholesterol- and triglyceride-lowering medications
- ☐ Antihypertensive medications

Dietary Intake

To devise an acceptable meal plan and coordinate medications, obtain:

- ☐ An accurate and thorough record of food intake and meal patterns
- ☐ An account of usual physical activities

At medical checkups, reassess the person's ability to:

- ☐ Maintain appropriate carbohydrate intake
- ☐ Maintain appropriate energy intake

- ☐ Monitor blood glucose levels at home
- ☐ Adjust insulin and diet to accommodate sick days
- ☐ Use appropriate foods to treat hypoglycemia

Anthropometric Data

Take accurate baseline height and weight measurements as a basis for:

- ☐ Appropriate energy intake
- ☐ Initial insulin therapy

Periodically reassess height and weight for children and weight for adults and pregnant women to ensure that the meal plan provides an appropriate energy intake.

Laboratory Tests

Monitor the success of diabetes treatment using these tests:

- ☐ Glycated hemoglobin
- ☐ Blood lipid concentrations
- ☐ Blood or urinary ketones
- ☐ Urinary protein (microalbuminuria)

Physical Signs

Look for signs of:

- ☐ Nerve damage
- ☐ Vision problems
- ☐ Foot ulcers
- ☐ Dehydration, especially in older adults

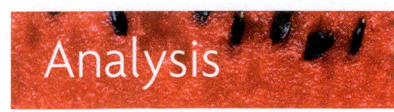

Analysis

CLINICAL APPLICATIONS

1. Using the carbohydrate counting method described in the "How To" on p. 584, determine an appropriate carbohydrate intake (in both grams and portions) for a man with type 2 diabetes who requires approximately 2,600 kcalories daily. Assume he would benefit from a carbohydrate allowance that is 50 percent of his energy intake. Using information from Tables 21–4 and 21–5, develop a one-day sample menu that is likely to meet his carbohydrate goals. Use the exchange lists in Appendix C to find additional examples of foods to include in your menu.

2. Take a trip to a pharmacy or use information from an online drugstore to price these items: blood glucose meter, test strips for the glucose meter selected, lancets, insulin, and syringes. Determine the approximate cost of insulin and syringes for a person who uses a total of 14 units of short-acting insulin (regular) and 26 units of intermediate-acting insulin (lente or NPH) in three injections daily. Also estimate the cost of testing blood glucose four times daily. Approximately how much would these supplies cost per month?

1. Which of the following is characteristic of type 1 diabetes?
   a. It frequently goes undiagnosed.
   b. The pancreas makes little or no insulin.
   c. It is the predominant form of diabetes.
   d. It often arises during pregnancy.

2. Which of the following describes type 2 diabetes?
   a. It is usually an autoimmune disease.
   b. The pancreas makes little or no insulin.
   c. Diabetic ketoacidosis is a common complication.
   d. Chronic complications may develop before it is diagnosed.

3. The chronic complications associated with all types of diabetes result from:
   a. altered kidney function.
   b. infections that deplete nutrient reserves.
   c. weight gain and hypertension.
   d. damage to blood vessels and nerves.

4. Long-term glycemic control is usually evaluated by:
   a. self-monitoring of blood glucose.
   b. testing urinary ketone levels.
   c. measuring glycated hemoglobin.
   d. testing urinary protein levels (microalbuminuria).

5. Regarding dietary carbohydrate, a patient with diabetes should be most concerned about:
   a. consuming the correct quantity of carbohydrate at each meal or snack.
   b. consuming the correct proportion of sugars, starches, and fiber in meals.
   c. avoiding added sugars and kcaloric sweeteners.
   d. choosing meals with ideal proportions of protein, carbohydrate, and fat.

6. Which of the following is true regarding the general use of alcohol in diabetes?
   a. A serving of alcohol is considered part of the carbohydrate allowance.
   b. Alcohol contributes to hyperglycemia and should be avoided completely.
   c. Alcohol can cause hypoglycemia and should therefore be consumed with food if patients use insulin or medications that stimulate insulin secretion.
   d. Patients can use alcohol in unlimited quantities unless they are pregnant.

7. The most ideal meal-planning strategy for people with diabetes is:
   a. carbohydrate counting.
   b. the exchange list system.
   c. following menus and recipes provided by a registered dietitian.
   d. any approach that best helps the patient control blood glucose levels.

8. A patient using intensive insulin therapy is likely to use a regimen that involves:
   a. twice-daily injections that combine short-, intermediate-, and long-acting insulin in each injection.
   b. a mixture of intermediate- and long-acting insulin injected between meals.
   c. multiple daily injections that supply basal insulin and precise insulin dosages for each meal.
   d. use of both insulin and oral antidiabetic agents.

9. Sudden hyperglycemia in a person who has previously maintained good glycemic control can be precipitated by:
   a. infections or illnesses.
   b. chronic alcohol ingestion.
   c. undertreatment of hypoglycemia.
   d. prolonged exercise.

10. Women with pregnancies complicated by diabetes:
    a. generally benefit from larger meals and a snack at bedtime.
    b. often need less carbohydrate at breakfast.
    c. need more carbohydrate than women with diabetes who are not pregnant.
    d. need more kcalories to support the pregnancy than women without diabetes.

Answers to these questions appear in Appendix H.

Notes

1. M. Lethbridge-Çejku and J. Vickerie, Summary health statistics for U.S. adults: National Health Interview Survey, 2003, National Center for Health Statistics, *Vital Health Statistics* 10 (July 2005).

2. Centers for Disease Control and Prevention, Prevalence of diabetes and impaired fasting glucose in adults—United States, 1999–2000, *MMWR Morbidity and Mortality Weekly Reports* 52 (2003): 833–837.

3. American Diabetes Association, Diagnosis and classification of diabetes mellitus, *Diabetes Care* 30 (2007): S42–S47.

4. K. N. Frayn, *Metabolic Regulation: A Human Perspective* (Oxford, UK: Blackwell Science, 2003).

5. R. S. Sherwin, Diabetes mellitus, in L. Goldman and D. Ausiello, eds., *Cecil Textbook of Medicine* (Philadelphia: Saunders, 2004), pp. 1424–1452.

6. A. Peters Harmel and R. Mathur, eds., *Davidson's Diabetes Mellitus: Diagnosis and Treatment* (Philadelphia: Saunders, 2004).

7. American Diabetes Association, 2007; Peters Harmel and Mathur, 2004.

8. Centers for Disease Control and Prevention, 2003.

9. F. R. Kaufman, Diabetes management in children and adolescents, in A. Peters Harmel and R. Mathur, eds., *Davidson's Diabetes Mellitus: Diagnosis and Treatment* (Philadelphia: Saunders, 2004), pp. 299–321.

10. Kaufman, 2004; P. A. Tartaranni and C. Bogardus, Obesity and diabetes mellitus, in D. Porte, Jr., R. S. Sherwin, and A. Baron, eds., *Ellenberg and Rifkin's Diabetes Mellitus* (New York: McGraw-Hill, 2003), pp. 401–413.

11. Sherwin, 2004.

12. Peters Harmel and Mathur, 2004.

13. Sherwin, 2004.

14. R. Matz, Hyperglycemic hyperosmolar syndrome, in D. Porte, Jr., R. S. Sherwin, and A. Baron, eds., *Ellenberg and Rifkin's Diabetes Mellitus* (New York: McGraw-Hill, 2003), pp. 587–599.

15. Sherwin, 2004.

16. D. M. Kendall, Reducing cardiovascular risk in type 2 diabetes and the metabolic syndrome: The emerging role of insulin resistance, in A. Peters Harmel and R. Mathur, eds., *Davidson's Diabetes Mellitus: Diagnosis and Treatment* (Philadelphia: Saunders, 2004), pp. 239–257.

17. L. H. Young and D. A. Chyun, Heart disease in patients with diabetes, in D. Porte, Jr., R. S. Sherwin, and A. Baron, eds., *Ellenberg and Rifkin's Diabetes Mellitus* (New York: McGraw-Hill, 2003), pp. 823–844.

18. R. G. Frykberg, Diabetic foot ulcers: Pathogenesis and management, *American Family Physician* 66 (2002): 1655–1662.

19. Peters Harmel and Mathur, 2004.

20. Peters Harmel and Mathur, 2004.

21. American Diabetes Association, Implications of the United Kingdom Prospective Diabetes Study, *Diabetes Care* 21 (1998): 2180–2184; Diabetes Control and Complications Trial Research Group, The effect of intensive treatment of diabetes on the development and progression of long-term complications in insulin-dependent diabetes mellitus, *New England Journal of Medicine* 329 (1993): 977–986.

22. American Diabetes Association, Standards of medical care in diabetes—2007, *Diabetes Care* 30 (2007): S4–S41.

23. American Diabetes Association, Standards of medical care in diabetes—2007, 2007.

24. K. V. Williams and coauthors, Improved glycemic control reduces the impact of weight gain on cardiovascular risk factors in type 1 diabetes: The Epidemiology of Diabetes Complications Study, *Diabetes Care* 22 (1999): 1084–1091.

25. American Diabetes Association, Nutrition recommendations and interventions for diabetes, *Diabetes Care* 30 (2007): S48-S65.

26. American Diabetes Association, Nutrition recommendations and interventions for diabetes, 2007.

27. American Dietetic Association, *Nutrition Care Manual* (Chicago: American Dietetic Association, 2007); American Diabetes Association, Nutrition recommendations and interventions for diabetes, 2007.

28. American Diabetes Association, Nutrition recommendations and interventions for diabetes, 2007.

29. American Diabetes Association, Nutrition recommendations and interventions for diabetes, 2007; C. A. Beebe, Nutrition and physical activity in diabetes, in A. Peters Harmel and R. Mathur, eds., *Davidson's Diabetes Mellitus: Diagnosis and Treatment* (Philadelphia: Saunders, 2004), pp. 49–69.

30. American Diabetes Association, Nutrition recommendations and interventions for diabetes, 2007.

31. American Diabetes Association, Nutrition recommendations and interventions for diabetes, 2007.

32. American Diabetes Association, Nutrition recommendations and interventions for diabetes, 2007; Beebe, 2004.

33. American Diabetes Association, Nutrition recommendations and interventions for diabetes, 2007.

34. F. Guerrero-Romero and M. Rodriguez-Moran, Complementary therapies for diabetes: The case for chromium, magnesium, and antioxidants, *Archives of Medical Research* 36 (2005): 250–257; G. Y. Yeh and coauthors, Systematic review of herbs and dietary supplements for glycemic control in diabetes, *Diabetes Care* 26 (2003): 1277–1294.

35. Peters Harmel and Mathur, 2004; S. M. Strowig and P. Raskin, Intensive management of type 1 diabetes mellitus, in D. Porte, Jr., R. S. Sherwin, and A. Baron, eds., *Ellenberg and Rifkin's Diabetes Mellitus* (New York: McGraw-Hill, 2003), pp. 501–514.

36. Strowig and Raskin, 2003.

37. D. M. Nathan, Insulin treatment of type 2 diabetes mellitus, in D. Porte, Jr., R. S. Sherwin, and A. Baron, eds., *Ellenberg and Rifkin's Diabetes Mellitus* (New York: McGraw-Hill, 2003), pp. 515–522.

38. Nathan, 2003.

39. American Diabetes Association, Standards of medical care in diabetes—2007, 2007.

40. S. Mudaliar and R. R. Henry, The oral antidiabetic agents, in D. Porte, Jr., R. S. Sherwin, and A. Baron, eds., *Ellenberg and Rifkin's Diabetes Mellitus* (New York: McGraw-Hill, 2003), pp. 531–564.

41. American Diabetes Association, Standards of medical care in diabetes—2007, 2007.

42. American Diabetes Association, Physical activity/exercise and diabetes, *Diabetes Care* 27 (2004): S58–S62.

43. American Diabetes Association, Physical activity/exercise and diabetes, 2004.

44. M. J. Franz and coauthors, Evidence-based nutrition principles and recommendations for the treatment and prevention of diabetes and related complications, *Diabetes Care* 25 (2002): 148–198.

45. American Diabetes Association, Gestational diabetes mellitus, *Diabetes Care* 27 (2004): S88–S90.

46. Peters Harmel and Mathur, 2004.

47. American Diabetes Association, Preconception care of women with diabetes, *Diabetes Care* 27 (2004): S76–S78.

48. Peters Harmel and Mathur, 2004; Sherwin, 2004.

49. American Diabetes Association, Nutrition principles and recommendations in diabetes, *Diabetes Care* 27 (2004): S36–S46.

Nutrition in Practice

THE GLYCEMIC INDEX IN NUTRITION PRACTICE

Chapter 21 introduces the *glycemic index (GI)*, a ranking of carbohydrate foods based on their glycemic effect after ingestion. In diabetes treatment, the total amount of carbohydrate is more important than the type of carbohydrate consumed, and clinical research suggests only minimal improvement in blood glucose levels when mainly low-GI foods are selected. Thus, the use of the glycemic index is rarely recommended for diabetes treatment by dietetics experts. However, in the past decade, a considerable amount of research has been conducted to determine whether low-GI diets may be helpful for improving risk factors for a number of chronic diseases. This Nutrition in Practice will describe the factors that contribute to a food's glycemic effect and the results of research studies that have examined the potential benefits of selecting mainly low-GI foods.

How is the glycemic index measured?

The glycemic index is essentially a measure of how quickly the carbohydrate in a food is digested and absorbed. Although testing methods vary to some degree, the most common protocol is to feed the test food—which contains a measured quantity of digestible carbohydrate—to research subjects and then measure blood glucose levels for two or three hours after the feeding. The increase in blood glucose over the two- or three-hour period is then compared to the blood glucose rise after an identical amount of digestible carbohydrate is ingested from a reference food such as pure glucose or white bread. Figure NP21–1 illustrates the difference in the blood glucose response to a low-GI food and a high-GI food. The blood glucose curve displays the surge in blood glucose above normal

fasting levels after the food is consumed and the subsequent fall over several hours. Table NP21–1 lists the GIs of various carbohydrate-containing foods arranged in order by food group.

The *amount* of carbohydrate consumed also influences the glycemic response, and therefore the GI of a food must be evaluated along with the quantity eaten. A food's total glycemic effect—expressed as the *glycemic load (GL)*—is the product of its GI and the amount of available carbohydrate from the portion consumed.[1] The importance of the GL can be illustrated by considering the GI for pumpkin. Although pumpkin's GI of 75 is high, an 80-gram portion contains only 4 grams of carbohydrate, giving it the relatively low GL of 3 (see Table NP21–1). Because the GL should be considered along with a food's GI, updated GI tables usually contain both values.

What factors influence a food's glycemic effect?

Table NP21–1 shows that starchy foods such as bread and potatoes tend to have high GI values, whereas many fruits and legumes have low GI values. The main factors that influence the GI value of a food include the following:

- **Starch structure.** Starch is present in foods as either a straight chain or branched chain of glucose molecules. Whereas digestion of the branched form tends to release glucose quickly, the straight chain is resistant to digestion. Thus, foods that contain mainly the branched form of starch tend to raise blood glucose levels more quickly and have a high GI value. Due to the subtle differences in starch among foods, different species of the same foods can have substantially different GI values; for example, current GI values for rice range from low

FIGURE NP 21–1

Glycemic Index (GI) of Food
Foods that are digested slowly have a low glycemic index. These types of foods cause a gradual and more moderate response in blood glucose than do foods with a high glycemic index.

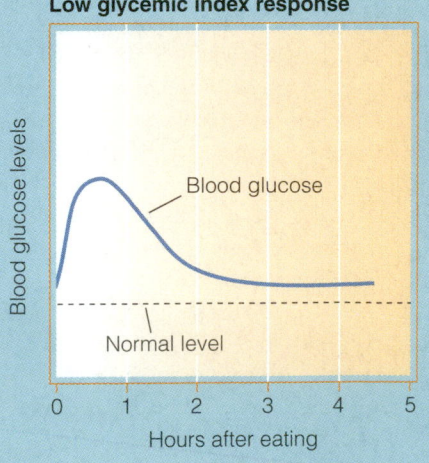

Low glycemic index response

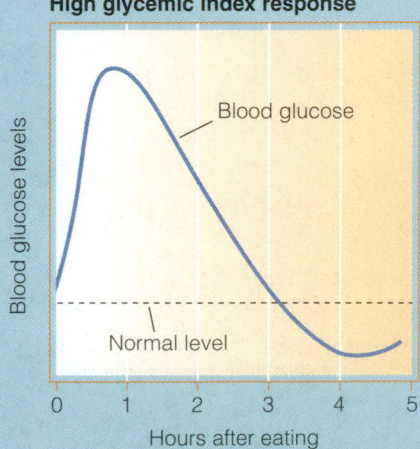

High glycemic index response

Glycemic Index (GI) and Glycemic Load (GL) of Selected Foods

Value	GI	GL
High	≥70	≥20
Medium	56–69	11–19
Low	≤55	≤10

Food Item	Available Carbohydrate per Serving (g)	Glycemic Index[a]	Glycemic Load[b]
Bread and pasta			
White bread, enriched *(30 g)*	14	73	10
Whole wheat bread *(30 g)*	13	71	9
Oat bran bread; 45–50% oat bran *(30 g)*	18	47	9
Spaghetti, boiled 10–15 minutes *(180 g)*	48	44	21
Cereal grains			
White rice *(150 g)*	36	64	23
Brown rice *(150 g)*	33	55	18
Bulgur (cracked wheat) *(150 g)*	26	48	12
Milk products			
Whole milk *(250 g)*	12	27	3
Ice cream *(50 g)*	13	61	8
Legumes			
Garbanzo beans *(150 g)*	30	28	8
Kidney beans *(150 g)*	25	28	7
Vegetables			
Carrots *(80 g)*	6	47	3
Russet potatoes, baked *(150 g)*	30	85	26
Sweet potatoes *(150 g)*	28	61	17
Pumpkin *(80 g)*	4	75	3
Fruits			
Apples *(120 g)*	15	38	6
Bananas *(120 g)*	24	52	12
Oranges *(120 g)*	11	42	5
Sugars			
Fructose *(10 g)*	10	19	2
Honey *(25 g)*	18	55	10
Sucrose *(10 g)*	10	62	7

[a] Reference food: glucose = 100
[b] Glycemic load = glycemic index × available carbohydrate (g) ÷ 100

Source: K. Foster-Powell and coauthors, International table of glycemic index and glycemic load values: 2002, *American Journal of Clinical Nutrition* 76 (2002): 5–56.

(27 for parboiled white Bangladeshi rice) to high (112 for boiled white Kenyan rice).

- **Fiber content.** Certain types of dietary fibers (primarily soluble fibers) increase the viscosity of chyme, slowing the passage of food in the stomach and upper intestine and making it more difficult for enzymes to digest the food. Therefore, foods such as beans, fruits, and vegetables, which contain soluble fibers, tend to have lower GI values.

- **Presence of fat and protein.** The fat in foods tends to slow stomach emptying, thus reducing the rate of digestion and absorption; hence, the presence of fat usually reduces a food's GI value. The protein in foods can also influence the GI because protein promotes insulin secretion, increasing the rate at which glucose is taken up from the blood.[2]

- **Food processing.** The manner in which a food is processed and cooked influences the interactions among starch, protein, and fiber and thus affects the final GI value. For example, both pasta and bread are prepared from wheat flour, but pasta (cooked *al dente*) has a lower GI because the starch granules in pasta are surrounded by a sturdy protein barrier that hampers starch digestion. Cooking the pasta for longer periods can break down its structure and raise the GI value. As another example, the GI values for oatmeal vary according to the size and thickness of the oats used to prepare it: oatmeal prepared from steel-cut oats has a lower GI value than oatmeal prepared from quick oats. This is because the steel-cut oats are solid particles of grain, whereas "quick oats" are small, thin flakes.

Pasta cooked *al dente* has a lower glycemic index than many other starchy foods.

Mixture of foods in a meal. Because foods are rarely consumed in isolation, the GI value of an individual food may be less important than the combination of foods consumed at a meal. For example, in a cheese sandwich, the high GI of the bread is lowered by the addition of fat and protein in the cheese.

Individual glucose tolerance. Cellular responses to insulin vary; thus, individual variability affects the glycemic response to foods. Persons with diabetes or prediabetes exhibit higher blood glucose levels after ingesting carbohydrate foods than do healthy individuals.

What evidence suggests that a low-GI diet may influence chronic disease risk?

Researchers have hypothesized that low-GI diets may improve the risks of developing diabetes, heart disease, and obesity and help individuals achieve weight loss.[3] Although some studies support the benefits of consuming low-GI diets for these problems, others do not. Because the results of research studies are contradictory, heath practitioners do not currently recommend that patients consume low-GI diets to prevent or treat disease. Examples of research include the following:

Diabetes prevention. Some researchers have proposed that a high glycemic load can increase the body's demand for insulin and eventually reduce pancreatic function, resulting in inadequate insulin secretion. Indeed, several population studies have suggested that low-GI diets might prevent or delay the onset of type 2 diabetes in those at risk, and some short-term intervention studies in animals and humans have

indicated that high-GI diets may promote insulin resistance.[4] Conversely, other researchers have been unable to confirm an association between glycemic index or load and diabetes onset, and a number of studies suggest that the benefits of low-GI diets may be primarily due to their dietary fiber content.[5]

Heart disease risk. Although some research studies have indicated that low-GI diets may improve blood lipid levels,[6] an analysis of fifteen controlled trials suggested only weak evidence of a relationship between low-GI diets and lower total cholesterol levels and no significant effect of glycemic index on LDL, HDL, and triglyceride levels. The authors concluded that there was insufficient evidence to prescribe low-GI diets for the purpose of improving heart disease risk.[7]

Appetite and weight loss. Low-GI foods can slow the after-meal rise in blood glucose, supporting the hypothesis that such foods may promote satiety and suppress hunger. Although several studies have confirmed these benefits, a number of trials have found either no difference between low- and high-GI foods or the opposite result.[8] In addition, several controlled trials found no differences between the amounts of weight lost on low-GI and high-GI diets.[9]

Given the mixed results of research studies on chronic disease prevention, are there any benefits associated with consuming low-GI foods?

Yes, if the low-GI foods are nutrient-dense, high-fiber foods. Not all low-GI foods meet these criteria: note that cakes, cookies, and candy bars may have a low GI due to their high fat content. Thus, a food's GI should be considered along with other criteria when assessing the health benefits.

The GI can be a helpful tool for choosing the most healthful food from a food group. For example, low-GI breakfast cereals tend to be high in fiber and low in added sugars, whereas high-GI cereals tend to be those that contain refined flours and significant amounts of added sugars. In other words, low-GI foods are often wholesome foods that have been minimally processed.

In general, should people avoid consuming high-GI foods?

Some people assume that starchy foods such as breads and potatoes should be avoided due to their high-GI values. As mentioned earlier, these foods are rarely consumed in isolation, and their GI values are reduced in a mixed meal. For example, breads often have a GI greater than 70, but adding cheese or peanut butter reduces the GI to 55 and 59, respectively.[10] Another problem with avoiding so-called high-GI foods is that GI values often vary considerably. For example, published values for white potatoes range from 24 to 101, and many samples have values in the mid-50s.[11]

Given the complexity of the GI, what are the current recommendations?

The potential benefits associated with consuming low-GI diets are still under investigation. Additional research is needed to justify the use of these diets for preventing or treating diseases such as diabetes, heart disease, obesity, or other medical problems.

At present, many nutrition scientists advocate consuming a plant-based diet that contains minimally processed grains, legumes, vegetables, and fruits. Such a diet would include abundant fiber and limited amounts of added fats and sugars. Undoubtedly, meals consisting of these foods would tend to have low- or medium-GI values. It is not yet clear whether consideration of the individual foods' GIs offer additional advantages to overall health and disease risk.

Notes

1. K. Foster-Powell, S. H. A. Holt, and J. C. Brand-Miller, International table of glycemic index and glycemic load values: 2002, *American Journal of Clinical Nutrition* 76 (2002): 5–56.

2. A. Flint and coauthors, The use of glycaemic index tables to predict glycaemic index of composite breakfast meals, *British Journal of Nutrition* 91 (2004): 979–989.

3. M. A. Pereira, Weighing in on glycemic index and body weight, *American Journal of Clinical Nutrition* 84 (2006): 677–679; S. Dickinson and J. Brand-Miller, Glycemic index, postprandial glycemia and cardiovascular disease, *Current Opinion in Lipidology* 16 (2005): 69–75; S. Kelly and coauthors, Low glycaemic index diets for coronary heart disease, *Cochrane Database Systematic Reviews* (2004): Oct 18, 2004:CD004467.

4. A. Hodge and coauthors, Glycemic index and dietary fiber and the risk of type 2 diabetes, *Diabetes Care* 27 (2004):2701–2706; A. E. Brynes and coauthors, A randomised four-intervention crossover study investigating the effect of carbohydrates on daytime profiles of insulin, glucose, non-esterified fatty acids and triacylglycerols in middle-aged men, *British Journal of Nutrition* 89 (2003): 207–218; W. Willett, J. Manson, and S. Liu, Glycemic index, glycemic load, and risk of type 2 diabetes, *American Journal of Clinical Nutrition* 76 (2002): 274S–280S.

5. A. D. Liese and coauthors, Dietary glycemic index and glycemic load, carbohydrate and fiber intake, and measures of insulin sensitivity, secretion, and adiposity in the Insulin Resistance Atherosclerosis Study, *Diabetes Care* 28 (2005): 2832–2838.

6. A. D. Liese and coauthors, Carbohydrate nutrition, glycaemic load, and plasma lipids: The Insulin Resistance Atherosclerosis Study, *European Heart Journal* 28 (2006): 80–87; Dickinson and Brand-Miller, 2005.

7. Kelly and coauthors, 2004.

8. A. Flint and coauthors, Glycemic and insulinemic responses as determinants of appetite in humans, *American Journal of Clinical Nutrition* 84 (2006): 1365–1373.

9. S. K. Raatz and coauthors, Reduced glycemic index and glycemic load diets do not increase the effects of energy restriction on weight loss and insulin sensitivity in obese men and women, *Journal of Nutrition* 135 (2005): 2387–2381; B. Sloth and coauthors, No difference in body weight decrease between a low-glycemic-index and a high-glycemic-index diet but reduced LDL cholesterol after 10-wk ad libitum intake of the low-glycemic-index diet, *American Journal of Clinical Nutrition* 80 (2004): 337–347.

10. Foster-Powell, Holt, and Brand-Miller, 2002.

11. Foster-Powell, Holt, and Brand-Miller, 2002.

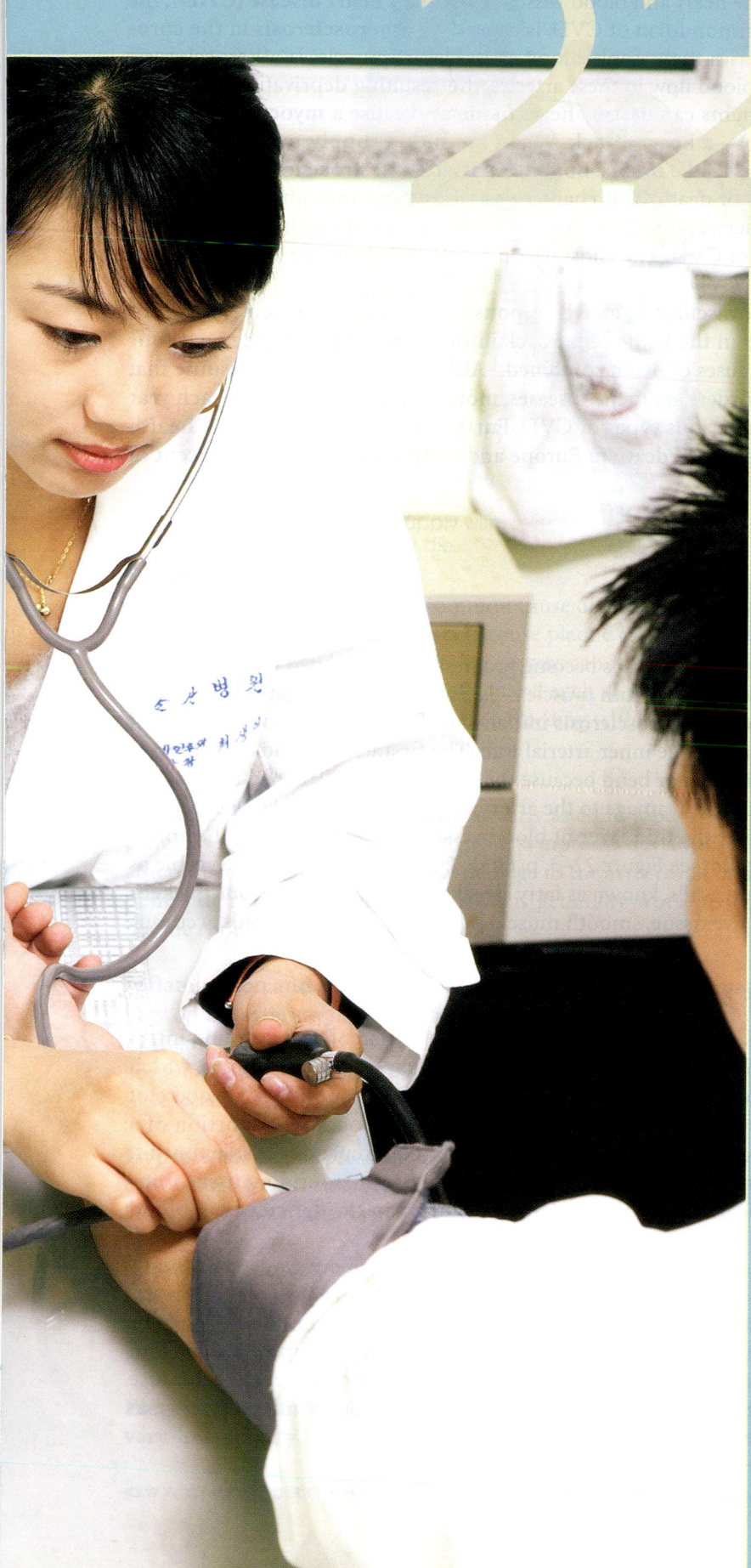

22

Fat-Controlled, Mineral-Modified Diets for Cardiovascular Diseases

Atherosclerosis
Consequences of Atherosclerosis
Causes of Atherosclerosis

Coronary Heart Disease (CHD)
Evaluating Risk for Coronary Heart Disease
Therapeutic Lifestyle Changes for
Lowering CHD Risk
Lifestyle Changes for Hypertriglyceridemia
Vitamin Supplementation and CHD Risk
Drug Therapies for CHD Prevention
Treatment for Heart Attack

Hypertension
Factors That Influence Blood Pressure
Contributing Factors for Hypertension
Treatment of Hypertension

Congestive Heart Failure
Consequences of Congestive Heart Failure
Medical Management of Congestive
Heart Failure

Stroke
Stroke Prevention
Stroke Management

Nutrition in Practice
The Metabolic Syndrome

THE METABOLIC SYNDROME

Chapter 21 describes how insulin resistance—a reduced sensitivity to insulin in muscle, adipose, and liver cells—can lead to hyperglycemia and hyperinsulinemia and, eventually, to type 2 diabetes. Insulin resistance is also a central feature of the **metabolic syndrome**, a group of disorders that substantially increases the risk of developing cardiovascular disease (CVD). The metabolic syndrome is a cluster of at least three of the following: hyperglycemia, obesity, hypertriglyceridemia (elevated blood triglycerides), reduced HDL cholesterol levels, and hypertension (high blood pressure). This Nutrition in Practice describes how the metabolic syndrome is diagnosed, how and why it might develop, its consequences, and current treatment approaches. The accompanying Glossary defines the relevant terms.

How is the metabolic syndrome diagnosed, and how many people in the United States does it affect?

Table NP22–1 lists the laboratory values used to identify metabolic syndrome, which currently affects an estimated 29 percent of the adult population in the United States.[1] As Figure NP22–1 shows, the prevalence of metabolic syndrome increases with age. Risk also varies among ethnic groups: Hispanic Americans have the highest incidence in the United States, with an overall prevalence of 36 percent.[2]

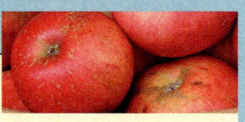

GLOSSARY

adiponectin (AH-dih-poe-NECK-tin): a hormone produced by adipose cells that improves insulin sensitivity.

cytokines (SIGH-toe-kines): proteins produced by white blood cells that regulate immune cell development and immune responses.

fibrinogen (fye-BRIN-oh-jen): a liver protein that promotes blood clot formation.

metabolic syndrome: a cluster of interrelated clinical symptoms, including obesity, insulin resistance, high blood pressure, and abnormal blood lipids, which together increase cardiovascular disease risk two- to three-fold; also called **syndrome X** or **insulin resistance syndrome.**

nitric oxide: a compound produced by blood vessel cells that helps to regulate blood vessel activity, including dilation and constriction.

plasminogen activator inhibitor-1: a protein that promotes blood clotting by inhibiting blood clot degradation within blood vessels.

resistin (re-ZIST-in): a hormone produced by adipose cells that induces insulin resistance.

TABLE NP22–1

Features of the Metabolic Syndrome

Metabolic syndrome is diagnosed when a person has three or more of the following symptoms.

Symptom	Diagnostic Criteria
Hyperglycemia	Fasting plasma glucose ≥100 mg/dL
Abdominal obesity	Waist circumference >40" in men, >35" in women
Hypertriglyceridemia	≥150 mg/dL
Reduced HDL cholesterol	<40 mg/dL in men, <50 mg/dL in women
Hypertension	≥130/85 mm Hg

What causes the metabolic syndrome?

Although the precise cause of the metabolic syndrome is not known, the close relationship between abdominal obesity and insulin resistance suggests that the current obesity crisis in the United States may be partly responsible for its high prevalence. Excessive abdominal fat can induce metabolic changes that lead to insulin resistance, which then leads to hyperglycemia and other abnormalities.

What kinds of metabolic changes are induced by excess abdominal fat?

Adipose (fat) cells that reside in the abdominal area are more metabolically active than adipose cells elsewhere.[3] Hence, triglycerides break down more rapidly, increasing fatty acid levels in the blood. These higher fatty acid concentrations inhibit the actions of insulin receptors, the proteins that recognize and bind insulin at cell surfaces.[4] Unless the pancreas can secrete enough insulin to compensate, glucose uptake from the blood is reduced, contributing to hyperglycemia.

Obesity can also alter production of the hormones and proteins made in adipose cells. For example, it causes reduced secretion of **adiponectin**, a hormone that improves insulin sensitivity.[5] Conversely, **resistin**, a hormone that contributes to insulin resistance, is released in greater amounts. Enlarged adipose cells also boost their production of certain **cytokines** that induce the synthesis of liver proteins that promote inflammation and blood coagulation.[6] People who are obese often have elevated levels of C-reactive protein,[7] a marker of inflammation linked to an increased risk of CVD.

Can obesity cause other problems related to the metabolic syndrome?

Obesity increases the risk of developing high blood pressure, a common component of the metabolic syndrome. Both the insulin resistance and hyperinsulinemia mentioned earlier may be implicated in raising blood pressure.[8] Insulin resistance interferes with the normal relaxation and dilation of blood vessels. Hyperinsulinemia promotes reabsorption of sodium by the kidneys, resulting in fluid retention and increased blood volume. These effects can contribute to increased blood pressure.

Abdominal obesity is often associated with blood lipid abnormalities.[9] Increases in body weight are linked with higher triglyceride and LDL cholesterol levels and lower HDL cholesterol levels. As a result of obesity, adipose cells are less responsive to insulin and release more fatty acids into the bloodstream. At the same time, they are less able to extract and store triglycerides from chylomicrons and VLDL. To keep up with the greater influx of fatty acids, the liver must accelerate its production of VLDL, and hypertriglyceridemia develops.

How does the metabolic syndrome contribute to cardiovascular disease risk?

The disorders that characterize the metabolic syndrome—obesity, lipid abnormalities, and hypertension—are all independent risk factors for CVD. In addition, both insulin resistance and elevated lipoprotein levels can cause damage to blood vessels, accelerating the progression of atherosclerosis.[10] The resulting blood vessel inflammation induces liver secretion of **fibrinogen**, a protein that promotes blood clot formation. C-reactive protein, which is elevated by both inflammation and obesity, inhibits **nitric oxide** production by blood vessel cells, an effect that impairs blood vessel activity and also promotes blood clotting.[11] Another procoagulant factor—**plasminogen activator inhibitor-1**— is overproduced as a consequence of both obesity and hyperinsulinemia. The combined effect of these multiple abnormalities can worsen atherosclerosis and increase the risks of developing heart attack and stroke. Individuals with insulin resistance are also at increased risk of developing diabetes, another major risk factor for CVD.

What is the treatment for the metabolic syndrome?

The metabolic syndrome is primarily treated with dietary and lifestyle changes, with the goal of correcting abnormalities that increase CVD risk.[12] In most individuals, a combination of weight loss and physical activity can improve insulin resistance, blood pressure, and blood lipid levels. Even a small weight loss (10 to 20 pounds) can improve symptoms, although many people find this difficult to achieve. Additional dietary strategies depend on a patient's specific symptoms. If dietary and lifestyle changes are not successful, medications may be prescribed. Because effective treatment requires lifelong commitment, health care providers should work with patients to develop a treatment plan that they are willing to adopt.

What dietary strategies, other than weight loss, are suggested for people with the metabolic syndrome?

In individuals with hypertriglyceridemia, the general recommendation is to reduce intake of added sugars and refined grain

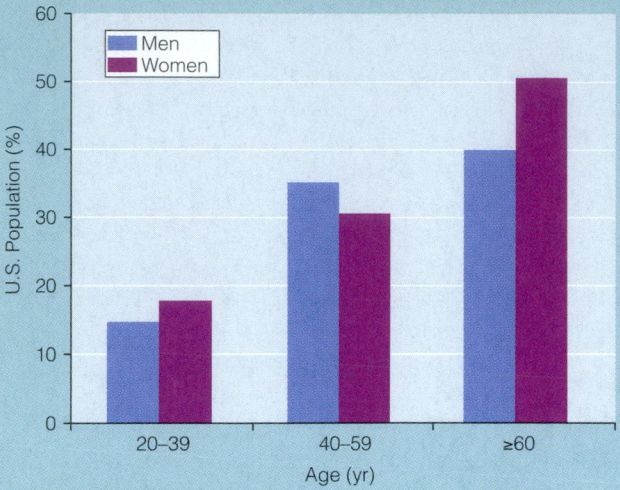

FIGURE NP22–1

Prevalence of Metabolic Syndrome in the U.S. Population

products (soda, juices, white bread, sweetened cereal, desserts) and increase servings of whole grains and foods high in fiber (whole wheat bread, oatmeal, legumes, fruits, vegetables).[13] In some people, carbohydrate restriction may help to reduce blood triglyceride levels and improve hyperglycemia.[14] Including fish in the diet each week may also improve triglyceride levels. Individuals with hypertension are encouraged to reduce sodium intake and increase consumption of fruits and vegetables and low-fat milk products. A diet low in saturated fat, *trans* fats, and cholesterol can help to reduce LDL cholesterol levels. Chapter 22 includes additional information about dietary modifications that can reduce CVD risk.

Regular exercise can reduce the risks of developing the metabolic syndrome, cardiovascular diseases, and type 2 diabetes.

Why is physical activity recommended for people with the metabolic syndrome?

Regular physical activity helps with weight management and may also improve blood lipid concentrations, hypertension, and insulin resistance—all changes that can reduce the risk of developing CVD. A regular exercise program can also prevent or delay the onset of diabetes in persons at risk.[15] A program that includes both aerobic exercise and strength training is best. A minimum of 30 minutes of moderate aerobic activity (brisk walking, jogging, cycling) daily is suggested, although longer periods (one hour daily) are recommended for weight control.[16] A sedentary lifestyle can worsen the progression of metabolic syndrome and should be discouraged.

Are medications used to treat the metabolic syndrome?

If dietary and lifestyle changes are unsuccessful, medications may be prescribed to correct hypertriglyceridemia and hypertension (Chapter 22 provides details). Insulin resistance is not routinely treated with drug therapy in nondiabetic patients due to insufficient evidence that the medications can benefit individuals with the metabolic syndrome.[17]

As explained in this Nutrition in Practice, the metabolic syndrome consists of a cluster of related disorders that increase the risk for developing CVD. Whereas the common features of the metabolic syndrome are independent risk factors for CVD, in combination they may raise risk two- to threefold. Treatment of the metabolic syndrome emphasizes dietary and lifestyle changes.

Notes

1. E. S. Ford, W. H. Giles, and A. H. Mokdad, Increasing prevalence of the metabolic syndrome among U.S. adults, *Diabetes Care* 27 (2004): 2444–2449.

2. Z. T. Bloomgarden, American Association of Clinical Endocrinologists (AACE) consensus conference on the insulin resistance syndrome, *Diabetes Care* 26 (2003): 1297–1303.

3. D. E. Moller and K. D. Kaufman, Metabolic syndrome: A clinical and molecular perspective, *Annual Review of Medicine* 56 (2005): 45–62.

4. G. A. Bray and C. M. Champagne, Obesity and the metabolic syndrome: Implications for dietetics practitioners, *Journal of the American Dietetic Association* 104 (2004): 86–89.

5. Moller and Kaufman, 2005; P. A. Kern and coauthors, Adiponectin expression from human adipose tissue, *Diabetes* 52 (2003): 1779–1785.

6. Bray and Champagne, 2004; S. M. Grundy, Inflammation, hypertension, and the metabolic syndrome, *Journal of the American Medical Association* 290 (2003): 3000–3002.

7. Grundy, 2003.

8. Grundy, 2003.

9. A. Tiengo and A. Avogaro, Cardiovascular disease, in P. Bjorntorp, ed., *International Textbook of Obesity* (West Sussex, UK: John Wiley & Sons, 2001), pp. 365–377.

10. Moller and Kaufman, 2005.

11. S. M. Grundy and coauthors, Clinical management of metabolic syndrome: Report of the American Heart Association/National Heart, Lung, and Blood Institute/American Diabetes Association Conference on Scientific Issues Related to Management, *Circulation* 109 (2004): 551–556.

12. Grundy and coauthors, 2004; D. Deen, Metabolic syndrome: Time for action, *American Family Physician* 69 (2004): 2875–2882.

13. Grundy and coauthors, 2004.

14. D. M. Kendall, Reducing cardiovascular risk in type 2 diabetes and the metabolic syndrome: The emerging role of insulin resistance, in A. Peters Harmel and R. Mathur, eds., *Davidson's Diabetes Mellitus: Diagnosis and Treatment* (Philadelphia: Saunders, 2004), pp. 239–257.

15. D. H. Wasserman, Z.-Q. Shi, and M. Vranic, Metabolic implications of exercise and physical fitness in physiology and diabetes, in D. Porte, Jr., R. S. Sherwin, and A. Baron, eds., *Ellenberg and Rifkin's Diabetes Mellitus* (New York: McGraw-Hill, 2003), pp. 453–480.

16. Grundy and coauthors, 2004; Deen, 2004.

17. Grundy and coauthors, 2004.

23 Protein-, Mineral-, and Fluid-Modified Diets for Kidney Diseases

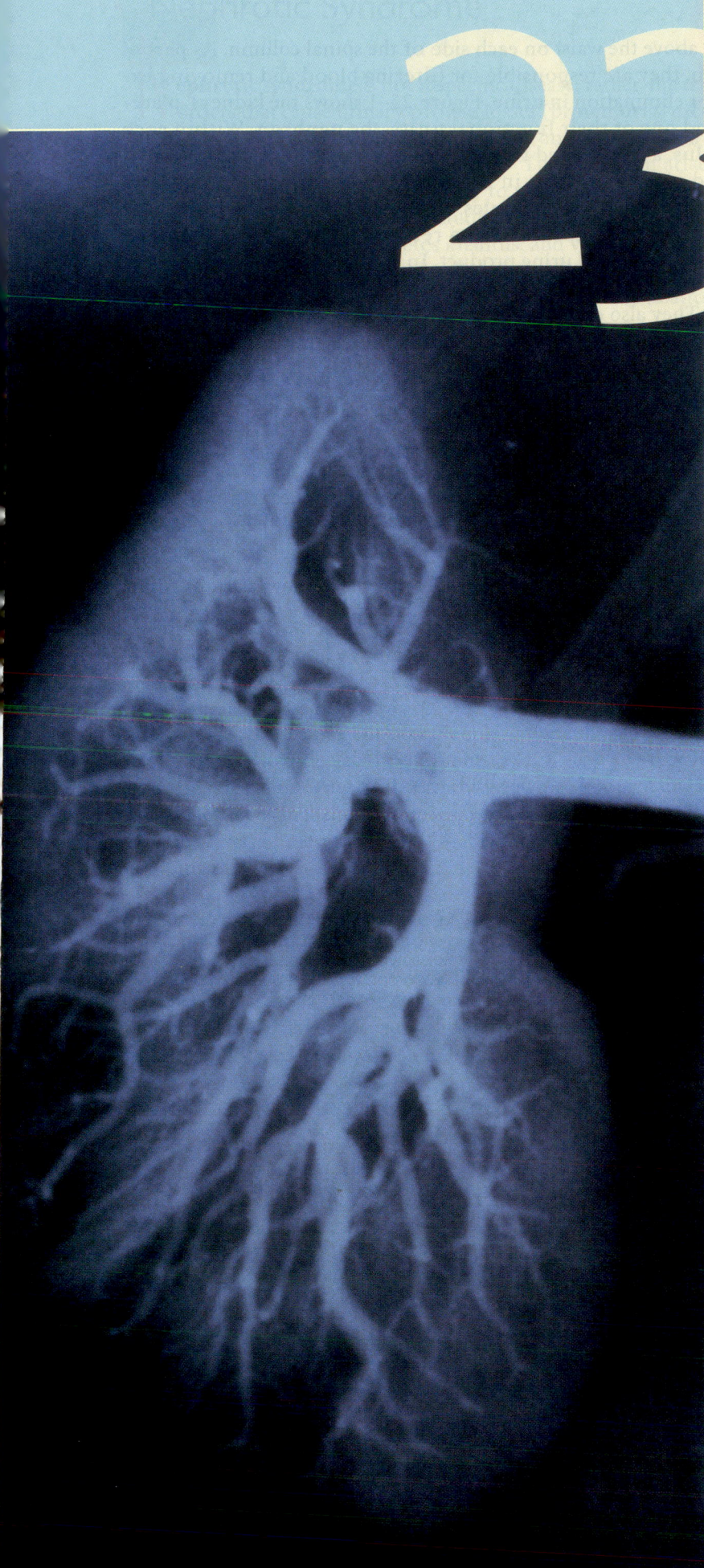

© PHOTOTAKE Inc./Alamy

membrane, will diffuse out of the blood. To maximally remove waste products such as urea from the blood, the dialysate contains no urea. For many other solutes, the dialysate is adjusted so that only excesses will be removed. For example, potassium can be removed from the blood by providing a dialysate with a lower concentration of potassium than is found in the person's blood. The dialysate must contain some potassium, however; otherwise the blood potassium would fall too low.

The dialysate can also be used to add needed components back into the blood. For example, for a person with acidosis, bases such as bicarbonate are added to the dialysate and move by diffusion into the blood to alleviate acidosis.

How is fluid removed from the blood?

Because albumin and other plasma proteins are so adept at retaining fluids in blood, osmosis alone is not an efficient process for removing fluid. In hemodialysis, a **pressure gradient** is created between the blood and the dialysate. Most modern dialyzers produce *positive* pressure in the blood compartment and *negative* pressure in the dialysate compartment,[1] establishing a pressure gradient that "pushes" water (and accompanying solutes) through the pores of the membrane. This process, called ultrafiltration, relies on pumps to establish an appropriate flow rate between the blood and the dialysate.

How does the health practitioner know if the dialysis treatment has been effective?

A number of methods have been devised for gauging the adequacy of dialysis treatment. The most common method is **urea kinetic modeling,** a technique that evaluates the amount of urea cleared from the blood. The formula used most often is Kt/V, where K is the amount of urea cleared, t is the time spent on dialysis, and V is the blood volume. The value obtained indicates whether the patient has undergone sufficient dialysis; the goal is a Kt/V result of approximately 1.2. Because technical data need to be incorporated into the calculation (such as dialyzer clearance data, blood flow rate, and dialysate flow rate), the computation is usually done by computer analysis. Current

treatment guidelines recommend that hemodialysis adequacy be evaluated at least monthly or more often if problems develop or if the patient is noncompliant.[2]

How long does a hemodialysis treatment last?

As described previously, hemodialysis utilizes a dialyzer to cleanse the patient's blood. Although dialyzers vary in efficiency, the treatment usually lasts 3 to 4 hours and is required at least three times weekly. Some studies suggest that patients undergoing daily hemodialysis for briefer periods (2 to 2.5 hours) may tolerate dialysis treatment better and have fewer complications, but this approach has not been widely adopted.[3] Most patients visit dialysis centers to obtain treatment. Home hemodialysis programs are available, but only about 2 percent of patients use them.

Are any complications associated with hemodialysis?

Yes. Although lifesaving, hemodialysis is associated with a substantial number of complications.[4] Problems at the vascular access site include infections and blood clotting. Hypotension can develop while blood is circulated through the dialyzer. Muscle cramping often occurs during the procedure, especially in the hands, legs, and feet. Blood losses can worsen anemia, which is already severe in two-thirds of patients beginning hemodialysis treatment. Patients may also experience headaches, weakness, nausea, vomiting, restlessness, and agitation.[5]

How does peritoneal dialysis work?

In peritoneal dialysis, the peritoneal membrane surrounding the abdominal organs serves as a semipermeable membrane. The dialysate is infused into a catheter that empties into the peritoneal space—the space within the abdomen near the intestines (see Figure NP23–2). In the most common procedure, **continuous ambulatory peritoneal dialysis (CAPD),** the dialysate remains in the peritoneal cavity for 4 to 6 hours, after which it is drained and replaced with fresh dialysate (about 2 to 3 liters in adults). Generally, the dialysate solution is exchanged four times daily and requires only about 30 minutes to drain and replace.

Because a pressure gradient cannot be created in the peritoneal cavity as it can in a dialyzer, the glucose concentration in the dialysate must be high enough to create enough **oncotic pressure** to draw fluid from the blood. As indicated in Chapter 23, a substantial amount of glucose can be absorbed into the patient's blood and may contribute to weight gain over time. The high glucose load may also cause hyperglycemia and hypertriglyceridemia in some patients.

What are the advantages and disadvantages of peritoneal dialysis?

Peritoneal dialysis offers a number of advantages over hemodialysis: vascular access is not required, dietary restrictions are fewer, and the procedure can be scheduled when convenient. The most common complication is infection, which can

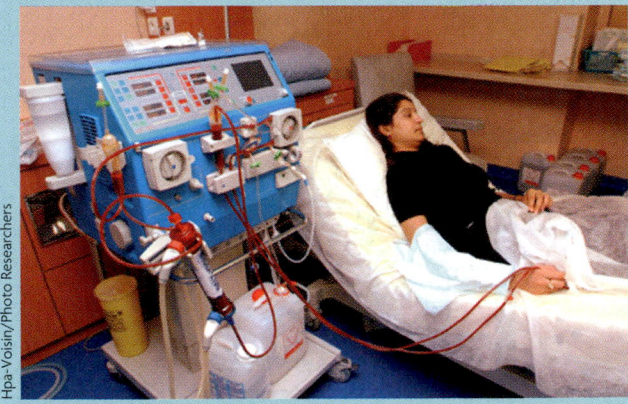

Hpa-Voisin/Photo Researchers

During hemodialysis, blood passes through a dialyzer where wastes are extracted, and the cleansed blood is returned to the body.

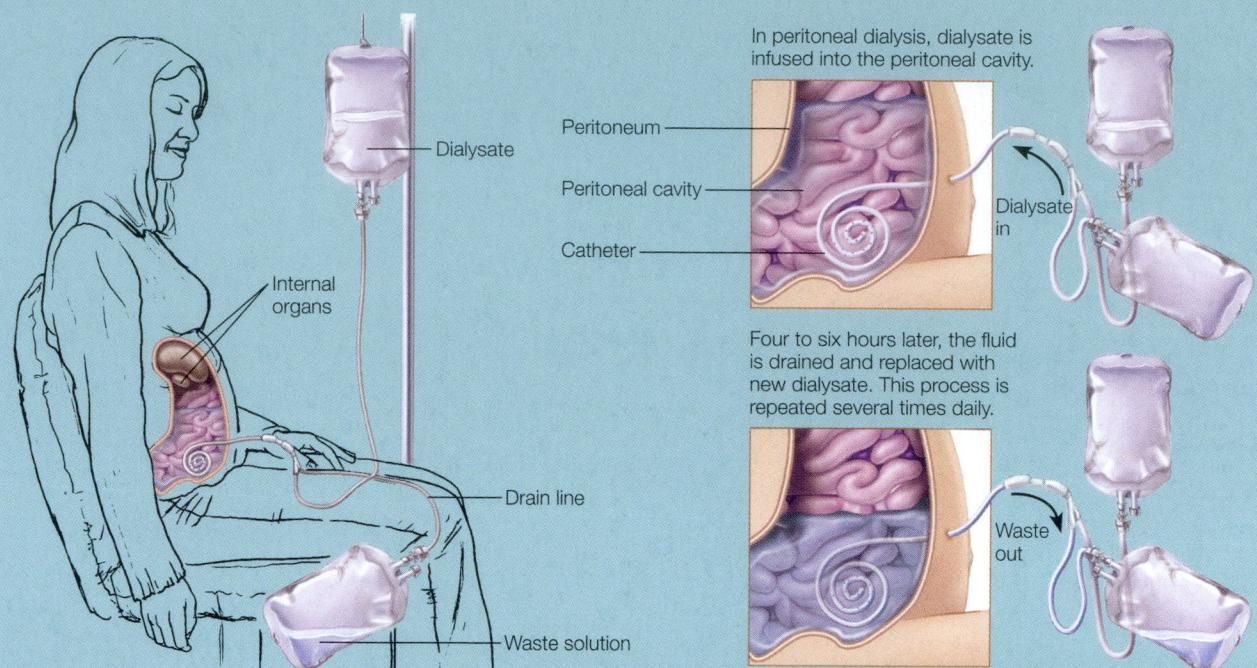

In peritoneal dialysis, dialysate is infused into the peritoneal cavity.

Peritoneum

Peritoneal cavity

Catheter

Dialysate in

Four to six hours later, the fluid is drained and replaced with new dialysate. This process is repeated several times daily.

Waste out

Dialysate

Internal organs

Drain line

Waste solution

FIGURE NP23–2

Peritoneal Dialysis

occur at the catheter site or in the peritoneal cavity (**peritonitis**). Other problems that may arise include blood clotting in the catheter, catheter migration, and abdominal hernia due to the dialysate volume.[6]

What are the features of continuous renal replacement therapy?

In people with acute renal failure, **continuous renal replacement therapy (CRRT)** removes fluids and wastes. CRRT utilizes the process of **hemofiltration**, in which blood is gently pumped across a filtration membrane over a prolonged time period. (This differs from dialysis treatments that rely on the diffusion of wastes across a membrane into the dialysate.) Either a pump or the patient's own blood pressure may move the blood across the membrane. The procedure can be used to remove fluids, solutes, or both. Some patients require fluid replacement during the procedure to maintain adequate blood volume, so hydration status must be closely monitored.

The use of CRRT is advantageous in acute care situations because it corrects imbalances without causing sudden shifts in blood volume, which are poorly tolerated in acute care patients. In addition, replacement fluids can include parenteral feedings without upsetting fluid balance. Complications include clotting problems, damage to arteries, and inadequate blood flow rates in hypotensive patients.

Dialysis and CRRT help to remove the wastes and fluids that are normally removed by healthy kidneys. Although these procedures cannot restore the kidneys' hormonal functions, they provide a lifesaving means for alleviating symptoms of uremia, hypertension, and edema.

Notes

1. C. F. Gutch, Principles of hemodialysis, in C. F. Gutch, M. H. Stoner, and A. L. Corea, eds., *Review of Hemodialysis for Nurses and Dialysis Personnel* (St. Louis, MO: Mosby, 1999), pp. 35–45.

2. National Kidney Foundation, K/DOQI Clinical practice guidelines for hemodialysis adequacy: Update 2006, www.kidney.org/PROFESSIONALS/kdoqi/ guideline_upHD_PD_VA/hd_guide2.htm, visited January 1, 2007.

3. A. Pierratos, New approaches to hemodialysis, *Annual Review of Medicine* 55 (2004): 179–189.

4. N. Tolkoff-Rubin and N. Goes, Treatment of irreversible renal failure, in L. Goldman and D. Ausiello, eds., *Cecil Textbook of Medicine* (Philadelphia: Saunders, 2004), pp. 716–726.

5. Gutch, 1999.

6. Tolkoff-Rubin and Goes, 2004.

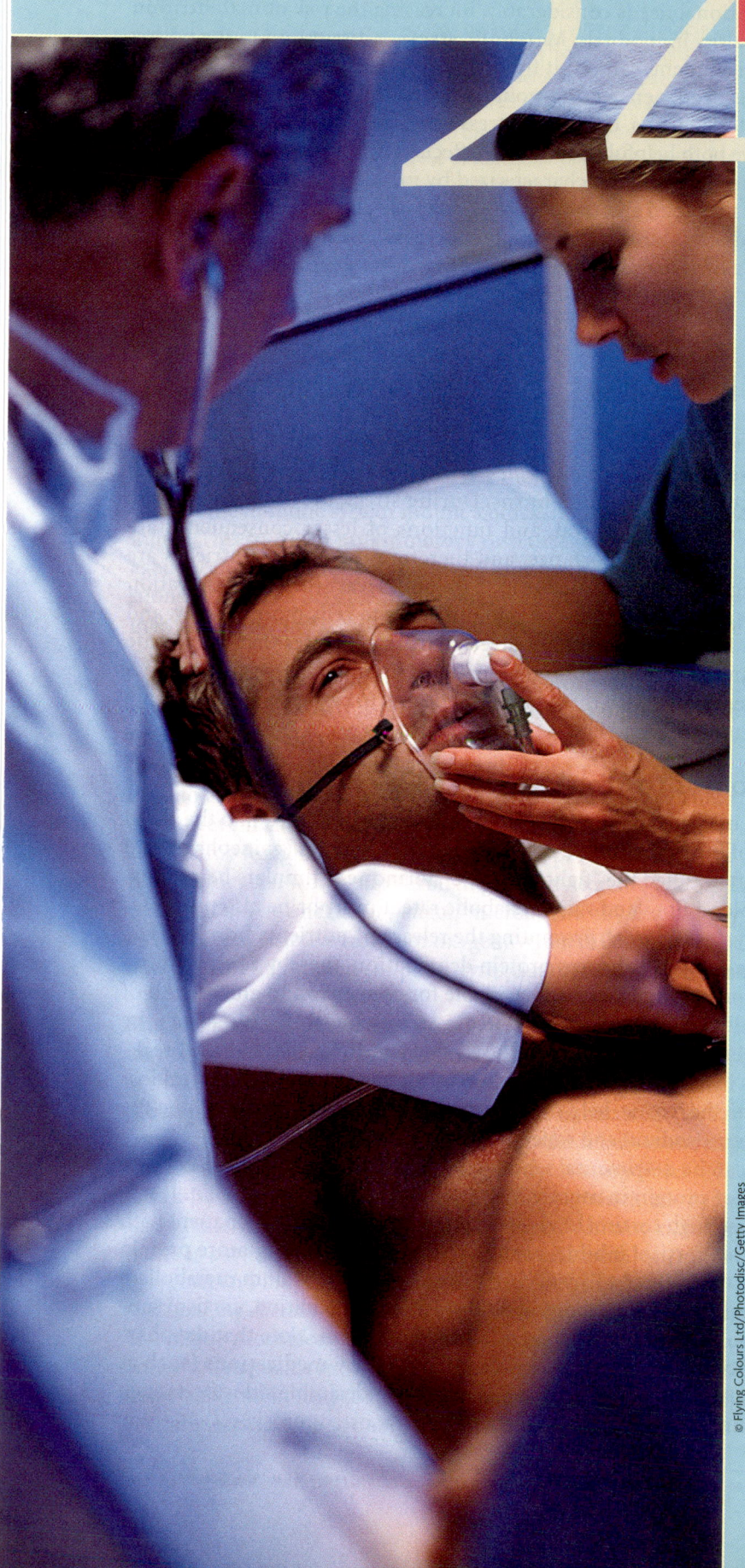

24 Energy- and Protein-Modified Diets for Metabolic and Respiratory Stress

Analysis

CLINICAL APPLICATIONS

1. Adam is a 49-year-old male who is 6 feet 2 inches tall and has a usual body weight of 180 pounds. He was severely injured by an explosion in the chemistry lab where he works and is now in the intensive care unit. Using the method described in the "How To" on p. 426 (Chapter 14), estimate Adam's energy requirement, using the stress factor 1.2. Estimate his protein requirement, using the factor 1.5 gram/kilogram body weight.

2. Turning again to the case described in item 1, assume that Adam requires a tube feeding and can tolerate a standard enteral formula. Check Appendix G to find at least three formulas that the nutrition support team might select for tube feeding. Determine the volume of each formula that would be needed to meet Adam's energy and protein needs. Would this volume also meet the recommendations for vitamins and minerals?

3. Ayla is a 23-year-old law student admitted to the hospital following an automobile accident in which she broke several bones and ruptured part of her small intestine. She has been in the hospital for several weeks and has just begun eating table foods. Her brother, who was driving the vehicle, was also seriously injured and nearly lost his life. Aside from the increased nutritional needs imposed by the stress of the accident, discuss how the following factors might interfere with Ayla's ability to improve nutrition status:

- Ayla's injuries are painful.

- Ayla's medications cause drowsiness.

- Ayla is depressed.

- Ayla is often out of her room for X-rays and other diagnostic tests when the menus and food trays arrive.

- Ayla's food intake is sometimes restricted due to the procedures she is undergoing.

How might these problems be resolved to improve Ayla's food intake?

Self Check

1. Which of the following metabolic changes accompanies acute stress?
   a. reduced plasma concentrations of glucose and fatty acids
   b. reduced blood volume and blood pressure
   c. increased insulin action
   d. catabolism of protein in skeletal muscle and connective tissue

2. Tissue injury is followed by:
   a. fluid accumulation in damaged tissue.
   b. reduced blood flow to injured tissue.
   c. reduced capillary permeability.
   d. decreased body temperature.

3. What is a possible effect of replacing vegetable oils rich in omega-6 fatty acids with oils rich in omega-3 fatty acids?
   a. improvement in blood circulation
   b. suppression of inflammation
   c. protection against sepsis
   d. hypertriglyceridemia

4. The acute-phase response results in increased plasma concentrations of:
   a. albumin.
   b. iron.
   c. C-reactive protein.
   d. zinc.

5. Which of the following statements concerning protein and energy recommendations during acute metabolic stress is true?
   a. Protein and energy recommendations are similar to those for healthy people.
   b. Protein and energy recommendations are reduced because a stressed individual cannot metabolize nutrients normally.
   c. Acutely stressed individuals can benefit from as much protein and energy as can be provided.
   d. Protein and energy recommendations are high in order to minimize muscle tissue losses.

6. The amount of protein recommended for an acutely stressed individual who weighs 150 pounds ranges from about ___ grams of protein per day.
   a. 55 to 85
   b. 68 to 136
   c. 126 to 158
   d. 150 to 300

7. The primary risk factor for COPD is:
   a. alpha-1 antitrypsin deficiency.
   b. occupational exposure to dusts or chemicals.
   c. smoking tobacco.
   d. respiratory infections.

8. A primary feature of emphysema is:
   a. obstruction within the bronchi.
   b. obstruction within the bronchioles.
   c. destruction of the walls separating the alveoli.
   d. excessive lung elasticity.

9. The weight loss and wasting that often occur in COPD can be caused by:
   a. reduced food intake.
   b. increased metabolic rate.
   c. reduced exercise tolerance.
   d. all of the above.

10. Medical nutrition therapy for a person undergoing respiratory failure includes:
    a. careful attention to providing enough, but not too much, energy.
    b. a generous fluid intake to facilitate mucus clearance.
    c. a high-fat intake to prevent weight loss.
    d. a high-carbohydrate intake to limit carbon dioxide production.

Answers to these questions appear in Appendix H.

Notes

1. E. Lopez-Garcia and coauthors, Consumption of (n-3) fatty acids is related to plasma biomarkers of inflammation and endothelial activation in women, *Journal of Nutrition* 134 (2004): 1806–1811; T. Pischon and coauthors, Habitual dietary intake of n-3 and n-6 fatty acids in relation to inflammatory markers among US men and women, *Circulation* 108 (2003): 155–160.

2. C. A. Dinarello and R. Porat, The acute phase response, in L. Goldman and D. Ausiello, eds., *Cecil Textbook of Medicine* (Philadelphia: Saunders, 2004), pp. 1733–1735.

3. A.S.P.E.N. Board of Directors and The Clinical Guidelines Task Force, Guidelines for the use of parenteral and enteral nutrition in adult and pediatric patients, *Journal of Parenteral and Enteral Nutrition* 26 (2002): 1SA–138SA.

4. K. A. Kudsk and G. S. Sacks, Nutrition in the care of the patient with surgery, trauma, and sepsis, in M. E. Shils and coeditors, *Modern Nutrition in Health and Disease* (Baltimore: Lippincott Williams & Wilkins, 2006), pp. 1414–1435.

5. N. Barak, E. Wall-Alonso, and M. D. Sitrin, Evaluation of stress factors and body weight adjustments currently used to estimate energy expenditure in hospitalized patients, *Journal of Parenteral and Enteral Nutrition* 26 (2002): 231–238.

6. Kudsk and Sacks, 2006.

7. F. Novak and coauthors, Glutamine supplementation in serious illness: A systematic review of the evidence, *Critical Care Medicine* 30 (2002): 2022–2029.

8. K. C. McCowen and B. R. Bistrian, Immunonutrition: Problematic or problem solving? *American Journal of Clinical Nutrition* 77 (2003): 764–770.

9. Kudsk and Sacks, 2006.

10. Kudsk and Sacks, 2006.

11. A. Shenkin, Micronutrients and outcome, *Nutrition* 13 (1997): 825–828.

12. A.S.P.E.N. Board of Directors and The Clinical Guidelines Task Force, 2002.

13. N. Anthonisen, Chronic obstructive pulmonary disease, in L. Goldman and D. Ausiello, eds., *Cecil Textbook of Medicine* (Philadelphia: Saunders, 2004), pp. 509–515.

14. A. M. Schols and E. F. Wouters, Nutritional assessment and support of the stable COPD patient, in T. Similowski, W. A. Whitelaw, and J.-P. Derenne, eds., *Clinical Management of Chronic Obstructive Pulmonary Disease* (New York: Marcel Dekker, 2002), pp. 686–687.

15. A. M. Malone, Enteral formula selection: A review of selected product categories, *Practical Gastroenterology* 29 (June 2005): 44–74.

16. M. A. P. Vermeeren and coauthors, Acute effects of different nutritional supplements on symptoms and functional capacity in patients with chronic obstructive pulmonary disease, *American Journal of Clinical Nutrition* 73 (2001): 295–301.

17. Malone, 2005; Schols and Wouters, 2002; Vermeeren and coauthors, 2001.

18. C. F. Donner and A. Patessio, Exercise in stable COPD, in T. Similowski, W. A. Whitelaw, and J.-P. Derenne, eds., *Clinical Management of Chronic Obstructive Pulmonary Disease* (New York: Marcel Dekker, 2002), pp. 731–758; R. M. Senior, Chronic obstructive pulmonary disease: Epidemiology, pathophysiology, pathogenesis, clinical course, management, and rehabilitation, in A. P. Fishman and coeditors, *Fishman's Manual of Pulmonary Diseases and Disorders* (New York: McGraw-Hill, 2002), pp. 118–141.

19. M. C. Steiner and coauthors, Nutritional enhancement of exercise performance in chronic obstructive pulmonary disease: A randomised controlled trial, *Thorax* 58 (2003): 745–751.

20. M. A. Grippi, Acute respiratory failure in the surgical patient, in A. P. Fishman and coeditors, *Fishman's Manual of Pulmonary Diseases and Disorders* (New York: McGraw-Hill, 2002), pp. 1034–1043.

21. L. M. Bellini, Nutrition in acute respiratory failure, in A. P. Fishman and coeditors, *Fishman's Manual of Pulmonary Diseases and Disorders* (New York: McGraw-Hill, 2002), pp. 1082–1089.

22. Barak, Wall-Alonso, and Sitrin, 2002.

MULTIPLE ORGAN FAILURE

Multiple organ failure (also called *multiple organ dysfunction syndrome*) is the cause of death in up to one-half of intensive care patients.[1] Described as a failure of two or more of the body's organ systems, multiple organ failure most frequently involves the lungs, liver, kidneys, and gastrointestinal (GI) tract. Involvement of three or more organ systems is associated with a fatality rate of nearly 100 percent.

Multiple organ failure is not a disease per se, but rather a late stage of severe illness or injury that results from a severe inflammatory response (discussed in Chapter 24). Multiple organ failure can be initiated by a number of very different critical illnesses, including acute respiratory failure, trauma, sepsis, burn injuries, extensive surgery, and pancreatitis. This Nutrition in Practice discusses how multiple organ failure develops, the manner in which it is treated, and the importance of prevention.[2]

How long has multiple organ failure been a major clinical problem?

As a clinical entity, multiple organ failure was first recognized only after World War II. Prior to the mid-twentieth century, patients with severe illnesses or multiple injuries frequently died of shock or circulatory failure. After fluid replacement and blood transfusions became standard treatments, the kidneys became the organs at highest risk, and kidney failure became the most common cause of death. Eventually, physicians learned to better support kidney function by providing appropriate electrolyte solutions and improving urine output. With improved kidney care, the lungs became the most vulnerable organ after severe injury. Improved treatment of respiratory failure eventually led to the current situation: advances in critical care allow patients to survive severe illnesses and injuries, but the body's defenses often overburden organs that were not originally injured.

Why does critical illness lead to multiple organ failure?

As discussed in this chapter, injury and infection cause the release of chemical mediators that have systemic (whole-body) effects. A severe, persistent inflammatory response may cause systemic inflammatory response syndrome (SIRS), which is associated with a constellation of symptoms including fever, raised heart and respiratory rates, and abnormal white blood cell counts. SIRS is a normal adaptive response to a severe insult, but if not reversed quickly enough, it can progress to shock, which is characterized by extremely low blood pressure and an inadequate blood supply for the tissues and organs of the body.

As might be expected from a systemic reduction in blood availability, shock can impair numerous organ systems. The abnormal delivery of oxygen and nutrients to tissues and insufficient removal of wastes result in irreversible injury to cells and tissues. Although each organ system is affected differently, ultimately one or more organs may begin to fail. The failure of one organ may place excessive demands on another, causing the second to fail as well. The progression of SIRS to multiple organ failure reflects the inability of the body's defenses and medical treatments to counter the detrimental effects of a sustained and potent inflammatory response.

The specific pathophysiology of multiple organ failure is poorly understood. Although early reports attempted to link the development of multiple organ failure directly to sepsis, sepsis is not present in all cases. Infection often results from impaired immune function and therefore is a frequent consequence of multiple organ failure, but it may not necessarily be the underlying trigger of organ dysfunction. Recall from this chapter that sepsis gives rise to the identical symptoms seen in SIRS. Figure NP24–1 illustrates the relationships among SIRS, infection, sepsis, and multiple organ failure.

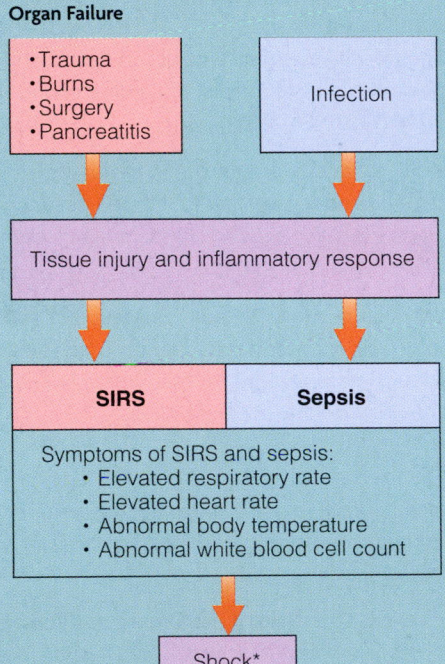

FIGURE NP24–1

Relationships among SIRS, Sepsis, and Multiple Organ Failure

- Trauma
- Burns
- Surgery
- Pancreatitis

Infection

↓

Tissue injury and inflammatory response

↓

| SIRS | Sepsis |

Symptoms of SIRS and sepsis:
- Elevated respiratory rate
- Elevated heart rate
- Abnormal body temperature
- Abnormal white blood cell count

↓

Shock*

↓

Multiple organ failure

*After critical injury, shock may sometimes precede and be the cause of SIRS.

TABLE NP24-1

Physiological Effects of Organ or System Failure

Organ or System	Effects of Failure
Lungs	Inability to maintain gas exchange
Liver	Altered metabolic processes
Kidneys	Inability to regulate blood volume, maintain electrolytes, remove wastes
Heart	Low cardiac output, low blood pressure, inadequate circulation, shock
GI tract	Impaired digestion and absorption, abnormal bleeding, bacterial translocation
Immune system	Infection, sepsis
Coagulation system	Excessive bleeding or coagulation
Central nervous system	Decreased perceptions, brain injury, coma

Do organs fail in a specific pattern?

Although the clinical course differs greatly among patient populations, the sequence of multiple organ failure often follows a similar pattern among patients: first the lungs fail, then the liver, and finally the kidneys, GI tract, or heart.[3] Other organs or systems may also become involved, and each additional failure reduces the likelihood of survival. Table NP24–1 lists the organs and systems most often involved in multiple organ failure and the potential consequences of their failure.

Are there any risk factors for multiple organ failure?

Epidemiological studies have identified a number of factors that increase risk. For example, people who develop multiple organ failure are often older, have multiple or severe injuries, and develop severe infections. Table NP24–2 lists the major risk factors associated with multiple organ failure; some of these are discussed below:

- **Age.** Patients over 55 years old are several times more likely to develop multiple organ failure than are younger patients.

TABLE NP24-2

Factors That Influence Risk of Multiple Organ Failure

Age over 55 years
Prior chronic disease
Persistent SIRS
Major infection
Blood transfusions
Severity of tissue injury
Length of time between injury and arrival at hospital
Malnutrition

In elderly patients, the increased risk may be due to the presence of chronic illnesses that directly affect organ function, such as heart disease, lung disease, diabetes, or liver damage. Aging also decreases the functional reserve of organs, thereby reducing an older patient's ability to deal with the additional stress that arises during critical illness.

- **Severity of SIRS.** The length of time that SIRS persists is related to the development of multiple organ failure. In one study, patients who had SIRS that persisted for more than three days were more likely to develop multiple organ failure than patients who had SIRS for less than two days.[4]

- **Infection.** Prolonged SIRS can suppress immune function and increase the risk of developing an infection. During hospital stays, critically ill patients often contract pneumonia—the principal infection associated with multiple organ failure. The risks of infection and sepsis greatly increase with the use of invasive catheters, which are frequently needed during intensive care to provide oxygen support, intravenous fluid resuscitation, nutrition support, and urine clearance.

- **Blood transfusions.** Blood transfusions are immunosuppressive and may increase a patient's risks of developing infection or sepsis. Blood transfusions frequently have adverse effects that can add further stress; they may cause acute lung injury, allergic reactions, red blood cell hemolysis (breakdown), and other complications.

What is the treatment for multiple organ failure?

Once multiple organ failure has developed, extensive medical support is needed until the inflammatory response has abated. Unfortunately, aggressive treatments can have damaging effects of their own and may cause further injury to organs that are already weakened by illness. Health practitioners must be aware of the adverse effects of aggressive therapies and alert to a patient's responses to treatments. Examples of therapies that are often used to manage organ failure include:

- **Lung support.** Mechanical ventilation is used to assist injured lungs and sustain gas exchange.

- **Fluid resuscitation.** Fluids and electrolytes are supplied to restore blood volume and maintain electrolyte balance.

- **Heart and blood vessel function.** Medications help to sustain or increase cardiac output and maintain adequate blood pressure.

- **Kidney support.** Hemofiltration or dialysis helps to prevent the buildup of toxic metabolites in blood.

- **Protection against infection.** Antibiotic therapy may reverse or prevent infections.

- **Nutrition support.** Enteral and parenteral nutrition support provide nutrients, help to prevent excessive wasting, and promote recovery.

What can be done to reduce the incidence of multiple organ failure?

Because mortality rates for multiple organ failure are so high, prevention must be considered at the earliest stages of injury and treatment before an excessive inflammatory response can cause further damage. Health practitioners have learned to identify the conditions that may increase organ stress whether they are due to a disease process, an inflammatory response, or an aggressive treatment that is intended to provide organ support. Although improvements in care over the past few decades have reduced some of the complications that arise during intensive care, rates of mortality from multiple organ failure have not changed. Thus a focus on prevention is critical until a better understanding of the pathophysiology of multiple organ failure is achieved, which may lead to additional therapeutic options.

Notes

1. D. Johnson and I. Mayers, Multiple organ dysfunction syndrome: A narrative review, *Canadian Journal of Anesthesia* 48 (2001): 502–509.

2. J. Parrillo, Approach to the patient with shock, in L. Goldman and D. Ausiello, eds., *Cecil Textbook of Medicine* (Philadelphia: Saunders, 2004), pp. 608–615; Johnson and Mayers, 2001; A. E. Baue, E. Faist, and D. E. Fry, eds., *Multiple Organ Failure: Pathophysiology, Prevention, and Therapy* (New York: Springer-Verlag, 2000); T. W. Evans and M. Smithies, Organ dysfunction, *British Medical Journal* 318 (1999): 1606–1608.

3. P. J. Offner and E. E. Moore, Risk factors for MOF and pattern of organ failure following severe trauma, in A. E. Baue, E. Faist, and D. E. Fry, eds., *Multiple Organ Failure: Pathophysiology, Prevention, and Therapy* (New York: Springer-Verlag, 2000).

4. Offner and Moore, 2000.

25

Energy- and Protein-Modified Diets for Cancer and HIV Infection

Cancer
 How Cancer Develops
 Nutrition and Cancer Risk
 Consequences of Cancer
 Treatments for Cancer
 Medical Nutrition Therapy for Cancer

HIV Infection
 Consequences of HIV Infection
 Treatments for HIV Infection
 Medical Nutrition Therapy for HIV
  Infection

Nutrition in Practice
 Ethical Issues in Nutrition Care

© Mary Kate Denny/Stone/Getty Images

■ Cancers are classified by the tissues or cells from which they develop:

■ *Adenomas* (ADD-eh-NO-muz) arise from glandular tissues.

■ *Carcinomas* (CAR-sih-NO-muz) arise from epithelial tissues.

■ *Gliomas* (gly-OH-muz) arise from glial cells of the central nervous system.

■ *Leukemias* (loo-KEY-mee-uz) arise from white blood cell precursors.

■ *Lymphomas* (lim-FOE-muz) arise from lymph tissue.

■ *Melanomas* (MEL-ah-NO-muz) arise from pigmented skin cells.

■ *Sarcomas* (sar-KO-muz) arise from connective tissues, such as muscle or bone.

■ An abnormal mass of cells that is noncancerous is called a *benign* tumor.

Although **cancers** and **HIV** (**human immunodeficiency virus**) infections are distinct disorders, from a nutritional standpoint, they share some similarities. Both disorders have debilitating effects that influence nutritional needs, and both can lead to severe wasting in advanced cases. These illnesses require medical nutrition therapy that is highly individualized based on the symptoms manifested and the organ systems involved.

Cancer

Cancer, the growth of **malignant** tissue, ranks just below cardiovascular disease as a cause of death in the United States. Cancer is not a single disorder, however; there are many cancers, that is, many different kinds of malignant growths. The different types of cancer have different characteristics, occur in different locations in the body, take different courses, and require different treatments. ■ Whereas an isolated, nonspreading type of skin cancer may be removed in a physician's office with no effect on nutrition status, advanced cancers, especially those of the gastrointestinal (GI) tract and pancreas, can seriously impair nutrition status.

How Cancer Develops

The development of cancer, called **carcinogenesis**, often proceeds slowly and continues for several decades. A cancer arises from mutations in the genes that control cell division in a single cell. These mutations may promote cellular growth, interfere with growth restraint, or prevent cellular death.[1] The affected cell thereby loses its built-in capacity for halting cell division and produces daughter cells with the same genetic defects. As the abnormal mass of cells, called a **tumor**, grows, ■ blood vessels form to supply the tumor with the nutrients it needs to support its growth. The tumor can disrupt the functioning of the normal tissue around it, and some tumor cells may **metastasize**, spreading to another region in the body. In leukemia (cancer affecting the white blood cells), the cells do not form a tumor but rather accumulate in blood and other tissues. Figure 25–1 illustrates the steps in cancer development.

The reasons that cancers develop are numerous and varied. Vulnerability to cancer is sometimes inherited, as when a person is born with a genetic defect that alters DNA structure, function, or repair. Certain metabolic processes may initiate carcinogenesis, as when phagocytes (immune cells) produce oxidants that cause DNA damage or when chronic inflammation increases the rate of cell division, increasing the risk of a

FIGURE 25–1

Cancer Development

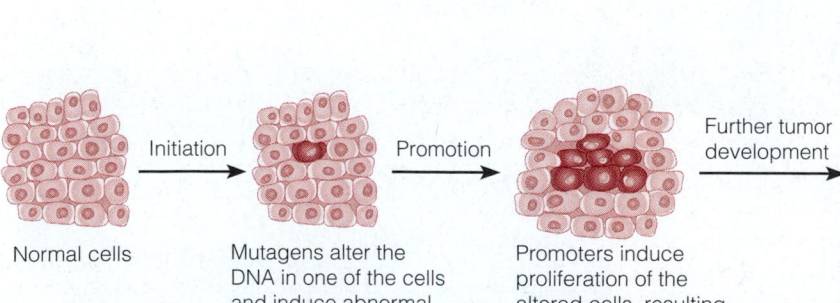

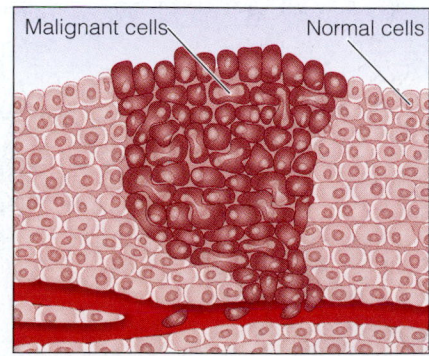

Normal cells

Mutagens alter the DNA in one of the cells and induce abnormal cell division.

Initiation

Promoters induce proliferation of the altered cells, resulting in formation of a tumor.

Promotion

Further tumor development

Malignant cells   Normal cells

The cancerous tumor releases cells into the bloodstream or lymphatic system (metastasis).

TABLE 25-1

Environmental Factors That Increase Cancer Risk

Environmental Factor	Cancer Site(s)
Aflatoxins (toxins in moldy peanuts or grains)	Liver
Alcohol[a]	Oral cavity, pharynx, esophagus, larynx, liver, colon, rectum, breast
Asbestos[b]	Lung, pleura, peritoneum
Chromium (hexavalent) compounds	Lung
Estrogen-progesterone replacement therapy	Breast
Immunosuppressive medications	Lymphoid tissues
Infection with *Helicobacter pylori*	Stomach
Infection with hepatitis B and hepatitis C viruses	Liver
Infection with human papillomavirus (HPV)	Cervix
Radiation	White blood cells (leukemia), breast, thyroid, lung
Sun exposure (ultraviolet radiation)	Skin
Tobacco[a]	Lung, oral cavity, pharynx, esophagus, larynx, renal pelvis, pancreas, bladder, kidney

[a] A combined exposure to alcohol and tobacco multiplies risks of developing cancers of the oral cavity, pharynx, esophagus, and larynx.
[b] Risk is greatly increased in cigarette smokers.

Source: W. J. Blot, Epidemiology of cancer, in L. Goldman and D. Ausiello, eds., *Cecil Textbook of Medicine* (Philadelphia: Saunders, 2004), pp. 1116–1120.

damaging mutation.[2] More often, cancers are caused by interactions between a person's genes and environmental agents. Exposure to cancer-causing substances, or **carcinogens**, may either induce genetic mutations that lead to cancer or promote proliferation of cancerous cells. Table 25–1 provides examples of environmental factors that increase cancer risk.

Nutrition and Cancer Risk

Like other environmental factors, diet and lifestyle strongly influence cancer risk. Certain food components may directly damage DNA, alter the metabolism of carcinogens by liver enzymes, or inhibit the formation of carcinogens in the body.[3] In addition, energy balance and growth rates affect the rate of cell division and consequently influence the rates at which mutations form and are replicated. Table 25–2, p. 676 lists examples of nutrition-related factors that may increase or decrease the risk of developing cancer.

Nutrition and Increased Cancer Risk As shown in Table 25–2, obesity is a risk factor for a number of different cancers, including some relatively common cancers such as colon cancer and postmenopausal breast cancer. Obesity increases cancer risk, in part, by altering levels of hormones that influence cell growth, such as the sex hormones, insulin, and several kinds of growth factors. For example, in the case of breast cancer in postmenopausal women, the hormone estrogen is likely involved: obese women have higher estrogen levels than do lean women because adipose tissue produces estrogen.

Although studies in animals have suggested that high-fat diets can promote tumor growth, studies of humans have not proved that fat's effects are independent from those of energy intake and physical activity.[4] Evidence from population studies is mixed: high-fat diets often, but not always, correlate with high cancer rates. Within single populations, cancer rates do not reliably reflect fat intakes. In addition, the type of fat consumed may be critical: studies of prostate cancer implicate animal fats but not vegetable fat, while consuming fatty fish may be protective.[5]

Food preparation methods are responsible for producing certain types of carcinogens in foods. Cooking meat, poultry, and fish at high temperatures causes

cancers: malignant growths or tumors that result from abnormal and uncontrolled cell division.

HIV (human immunodeficiency virus): the virus that causes acquired immune deficiency syndrome (AIDS). HIV destroys immune cells and progressively impedes the body's ability to fight infections and certain cancers.

malignant (ma-LIG-nent): describes a cancerous cell or tumor, which can injure healthy tissue and spread cancer to other regions of the body.

carcinogenesis (CAR-sin-oh-JEN-eh-sis): the process of cancer development.

tumor: an abnormal tissue mass that has no physiological function; also called a *neoplasm* (NEE-oh-plazm).

metastasize (meh-TAS-tah-size): the spread of cancer cells from one part of the body to another.

carcinogens (CAR-sin-oh-jenz or car-SIN-oh-jenz): substances that can cause cancer (the adjective is *carcinogenic*).

TABLE 25-2

Nutrition-Related Factors That Influence Cancer Risk

	Nutrition-Related Factor[a]	Cancer Site(s)
Factors that increase cancer risk	Obesity	Colon, kidney, pancreas, esophagus, endometrium, gallbladder, breast (in postmenopausal women)
	Total fat[b]	Colon, prostate
	Red meat, processed meats	Prostate, colon, rectum
	Calcium (over 1,500 mg daily)	Prostate
	Salted and salt-preserved foods	Stomach
	Low level of physical activity	Colon, prostate, breast, endometrium
Factors that decrease cancer risk	Fruits and vegetables	Lung, esophagus, stomach, colon, rectum
	Tomato products	Prostate
	Cruciferous vegetables (broccoli, cauliflower, brussels sprouts)	Prostate, bladder, lung
	Allium vegetables (onion, garlic)	Stomach
	Citrus fruits	Lung
	Folate-containing foods and supplements	Colon, esophagus, breast, white blood cells (leukemia)
	Calcium (up to 1,000 mg daily)	Colon, rectum
	High level of physical activity	Colon, prostate, breast, endometrium

[a] Altered cancer risk is associated with high intakes of the dietary substances listed. The risks associated with alcohol are included in Table 25–1.

[b] The effect of fat may be due to its direct association with higher energy intakes (obesity) or higher intakes of red meat.

Sources: W. C. Willett and E. Giovannucci, Epidemiology of diet and cancer risk, in M. E. Shils, M. Shike, A. C. Ross, B. Caballero, and R. Cousins, eds., *Modern Nutrition in Health and Disease* (Philadelphia: Lippincott Williams & Wilkins, 2006), pp. 1267–1279; L. H. Kushi and coauthors, American Cancer Society guidelines on nutrition and physical activity for cancer prevention: Reducing the risk of cancer with healthy food choices and physical activity, *CA: A Cancer Journal for Clinicians* 56 (2006): 254–281.

carcinogens to form on food surfaces.[6] Carcinogens also accompany the smoke that adheres to foods during grilling and are present in the charred surfaces of meat and fish. However, the cancer risk from eating such foods is unclear because the biological actions of these carcinogens are modulated by other dietary components, including compounds in vegetables and other plant foods. In several population studies, consumption of well-cooked meats has been linked to cancers of the colon, breast, and stomach.[7]

Nutrition and Decreased Cancer Risk A considerable number of human studies have found a link between the consumption of fruits and vegetables and reduced incidences of certain cancers (review Table 25–2). Fruits and vegetables contain both nutrients and phytochemicals with antioxidant activity, and these substances may prevent or reduce the oxidative reactions in cells that cause DNA damage. Phytochemicals may also help to inhibit carcinogen production in the body, enhance immune functions that protect against cancer development, and promote enzyme reactions that inactivate carcinogens.[8] In addition, certain fruits and vegetables provide the B vitamin folate, which plays roles in DNA synthesis and repair; thus, inadequate folate intakes may allow DNA damage to accumulate.

Although research reports in the 1970s and 1980s suggested that a fiber-rich diet could protect against colon cancer, recent studies have cast doubt on the earlier analyses.[9] The earlier studies depended on the ability of colon cancer patients to recall the foods they had consumed during the preceding years, whereas more recent studies—considered more

© Polara Studios, Inc.

Cruciferous vegetables, such as cauliflower, broccoli, and brussels sprouts, contain nutrients and phytochemicals that inhibit cancer development.

TABLE 25–3

Recommendations for Reducing Cancer Risk

Maintain a healthy body weight.
- Balance energy intake with appropriate physical activity.
- Avoid excessive weight gain throughout life.
- If overweight or obese, achieve a healthy body weight.

Be physically active.
- For adults: engage in moderate to vigorous activity for 30 minutes on at least 5 days of the week; 45 to 60 minutes is preferable.
- For children and adolescents: engage in moderate to vigorous activity for 60 minutes on at least 5 days of the week.

Choose a healthy diet that emphasizes plant sources.
- Consume five or more servings of a variety of vegetables and fruits daily.
- Choose whole grain products instead of processed (refined) grains.
- Limit consumption of red meats and processed meats.

Limit consumption of alcoholic beverages.
- For women: drink no more than one drink daily.
- For men: drink no more than two drinks daily.

Source: L. H. Kushi and coauthors, American Cancer Society guidelines on nutrition and physical activity for cancer prevention: Reducing the risk of cancer with healthy food choices and physical activity, *CA: A Cancer Journal for Clinicians* 56 (2006): 254–281.

reliable—tracked the subjects' health behaviors and cancer outcomes for extended periods (10 to 20 years). Moreover, some studies that had found fiber to be protective did not analyze factors such as physical activity, smoking, or folate intake, all of which can influence cancer outcome. A fiber-rich diet may be protective, in part, because high-fiber foods usually contain high levels of nutrients and phytochemicals that are protective against cancer. Table 25–3 summarizes the dietary and lifestyle practices that may help to reduce the risk of developing cancer.

Consequences of Cancer

Once cancer develops, its consequences depend on the location of the tumor, its severity, and the treatment. The complications that develop are often due to the tumor's impingement on surrounding tissues. Nonspecific effects of cancer include anorexia, lethargy, weight loss, night sweats, and fever.[10] During the early stages, many cancers produce no symptoms, and the person may be unaware of the threat to health.

Wasting Associated with Cancer Anorexia, muscle wasting, weight loss, and fatigue typify **cancer cachexia**, which occurs in as many as 80 percent of people with cancer.[11] Weight loss is often evident at the time that cancer is diagnosed, and severe malnutrition, often seen in the later stages of cancer, is the ultimate cause of death in many cases. Without adequate energy and nutrients, the body is poorly equipped to maintain organ function, support immune defenses, and mend damaged tissues. An involuntary weight loss of more than 10 percent, which indicates significant malnutrition, is cause for concern.[12]

Many factors play a role in the wasting associated with cancer. Cytokines, released by both tumor cells and immune cells, induce a hypermetabolic, catabolic state. The combined effects of a poor appetite, accelerated and abnormal metabolism, and the diversion of nutrients to support tumor growth result in a lower supply of energy and nutrients at a time when demands are high. Appetite and food

■ Reminder: The immune responses to tissue injury include the release of *cytokines* into the bloodstream (described in Chapter 24).

cancer cachexia (ka-KEK-see-ah): a wasting syndrome, associated with cancer, that is characterized by anorexia, muscle wasting, weight loss, and fatigue.

intake are further disturbed by the effects of treatments and medications prescribed for cancer patients.

Metabolic Changes The metabolic changes that arise in cancer exacerbate the wasting described in the previous section. Cancer patients exhibit an increased rate of protein turnover, but reduced muscle protein synthesis.[13] ■ Muscle contributes amino acids for glucose production, further straining the body's supply of protein. Triglyceride breakdown increases, elevating serum lipids. Many patients develop insulin resistance. These metabolic abnormalities help to explain why people with cancer fail to regain lean tissue or maintain healthy body weights even when they are consuming adequate energy and nutrients.

■ Reminder: *Protein turnover refers to the continuous degradation and synthesis of the body's proteins.*

Anorexia and Reduced Food Intake Anorexia is a major contributor to the wasting associated with cancer. Some factors that contribute to anorexia or otherwise reduce food intake include:

- **Chronic nausea and early satiety.** People with cancer frequently experience nausea and a premature feeling of fullness after eating small amounts of food.

- **Fatigue.** People with cancer often tire easily and lack the energy to prepare and eat meals. Once cancer cachexia develops, these tasks become even more difficult to handle.

- **Pain.** People in pain may have little interest in eating, particularly if eating makes the pain worse.

- **Mental stress.** A cancer diagnosis can cause distress, anxiety, and depression, all of which may reduce appetite. Facing and undergoing cancer treatments causes additional psychological stress.

- **Effects of cancer therapies.** Therapies for cancer (including medications, chemotherapy, radiation therapy, surgery, and bone marrow transplants) can affect food intake by causing nausea, vomiting, altered taste perceptions, food aversions, inflammation of the mouth and esophagus, dry mouth, mouth sores, difficulty swallowing, intestinal cramping, diarrhea, and constipation.

- **Obstructions.** A tumor may partially or completely obstruct a portion of the GI tract, causing complications such as nausea and vomiting, early satiety, delayed gastric emptying, and bacterial overgrowth. Some patients with obstructions are unable to tolerate oral diets.

Treatments for Cancer

The primary medical treatments for cancer—surgery, chemotherapy, radiation therapy, or any combination of the three—aim to remove cancer cells, prevent further tumor growth, and alleviate symptoms. The likelihood of effective treatment is highest with early detection and intervention. Because treatment decisions are difficult and cancer therapies have considerable side effects, patients rely on health care providers to help them make informed decisions.

Surgery Surgery is performed to remove tumors, determine the extent of cancer, and discern whether nearby tissues are involved.[14] Often, surgery must be followed by other cancer treatments to prevent growth of new tumors. The acute metabolic stress caused by surgery raises protein and energy needs and can exacerbate wasting. Surgery also contributes to pain, fatigue, and anorexia, all of which can reduce food intake at a time when nutritional needs are substantial. Blood loss contributes to nutrient losses and further exacerbates malnutrition. Some surgeries can have long-term effects on nutritional status (see Table 25–4).

TABLE 25–4

Nutrition-Related Side Effects of Cancer Surgeries

Head and Neck Surgeries

Difficulty in chewing/swallowing

Inability to chew/swallow

Esophageal Resection

Diarrhea

Fistula formation

Reduced gastric acid secretion

Reduced gastric motility

Steatorrhea (fat malabsorption)

Stenosis (constriction)

Gastric Resection

Dumping syndrome

General malabsorption

Hypoglycemia

Lack of gastric acid

Vitamin B_{12} malabsorption

Intestinal Resection

Blind loop syndrome

Diarrhea

Fluid and electrolyte imbalances

Hyperoxaluria

Malabsorption

Steatorrhea

Pancreatic Resection

Diabetes mellitus

Malabsorption

TABLE 25–5

Nutrition-Related Side Effects of Chemotherapy and Radiation

	Reduced Nutrient Intake	Accelerated Nutrient Losses	Altered Metabolism
Chemotherapy	Abdominal pain Anorexia Mouth ulcers Nausea Taste alterations Vomiting	Diarrhea Intestinal ulcers Malabsorption Vomiting	Fluid and electrolyte imbalances Hyperglycemia Interference with vitamins or other metabolites Negative nitrogen and calcium balances Secondary effects of malnutrition, infection, or tissue damage (inflammation)
Radiation	Anorexia Damage to teeth and jaws Dysphagia Esophagitis Mouth ulcers Nausea Reduced salivary secretions Taste alterations Thick salivary secretions Vomiting	Blood loss from intestine and bladder Diarrhea Fistulas Intestinal obstructions Malabsorption Radiation enteritis Vomiting	Fluid and electrolyte imbalances as a consequence of vomiting, diarrhea, or malabsorption Secondary effects of malnutrition, infection, or tissue damage (inflammation)

Chemotherapy **Chemotherapy** relies on the use of drugs to inhibit tumor growth. Some of these drugs interfere with the process of cell division; ■ others sterilize cells that are in a resting phase and not actively dividing. Ideally, chemotherapy would wipe out cancer cells without destroying healthy ones. Unfortunately, most of these drugs have toxic effects on normal cells as well and are especially damaging to rapidly dividing cells, such as those of the GI tract, skin, and bone marrow. Some of the newer drugs are able to target properties specific to cancer cells and are better tolerated by the body's tissues.[15] Table 25–5 includes a summary of the nutrition-related side effects that may result from chemotherapy.

Radiation Therapy **Radiation therapy** treats cancer by bombarding cancer cells with X-rays, gamma rays, or various atomic particles, which damage DNA and lead to cell death. Newer techniques are able to focus radiation directly at tumors and minimize damage to nearby tissues. An advantage of radiation therapy over surgery is that it can shrink tumors while preserving organ structure and function. Compared with chemotherapy, radiation therapy is better able to target specific regions of the body, rather than involving all body cells. Nonetheless, radiation therapy can damage healthy tissues and sometimes has long-term detrimental effects on nutritional status (see Table 25–5). Radiation to the head and neck area can damage the salivary glands and taste buds, causing inflammation, dry mouth, and a reduced sense of taste; in severe cases, damage may be permanent. Radiation treatment in the lower abdominal area can cause **radiation enteritis**, an inflammatory condition of the small intestine that causes nausea, vomiting, malabsorption, and diarrhea.

Bone Marrow Transplants **Bone marrow transplants** replace bone marrow that has been destroyed by chemotherapy or radiation therapy and are also one of the primary treatments for leukemia. If possible, bone marrow cells are collected from the patient before chemotherapy or radiation treatment begins so that it is not necessary to find a separate donor.[16] If another person's cells are used, the patient must take immunosuppressant drugs to prevent **tissue rejection**.

The procedures used in transplant patients can have a substantial impact on food intake and nutrition status. The intensive chemotherapy, radiation treatments, and

■ One drug that inhibits cell division is *methotrexate,* which closely resembles the B vitamin folate (see Figure 15–1 on p. 448). Folate is required for DNA synthesis. Methotrexate inhibits cell division by blocking activity of the enzyme that converts folate to its active form.

chemotherapy: the use of drugs to arrest or destroy cancer cells. Such drugs are called *antineoplastic agents.*

radiation therapy: the use of X-rays, gamma rays, or atomic particles to destroy cancer cells.

radiation enteritis: inflammation of intestinal tissue caused by exposure to radiation.

bone marrow transplants: procedures that replace bone marrow that has been destroyed by cancer treatments; also used to treat certain types of cancers and blood disorders.

tissue rejection: destruction of donor tissue by the recipient's immune system, which recognizes the donor cells as foreign.

immunosuppressant drugs can all impair immune function substantially, increasing the risk of foodborne illness. Other common complications include anorexia, dry mouth, altered taste sensations, inflamed mucous membranes, malabsorption, nausea, vomiting, and diarrhea. Patients are often unable to consume adequate food and may require nutrition support, as described in a later section.

Medications to Combat Anorexia and Wasting To combat anorexia, medications may be prescribed to stimulate the appetite and promote weight gain. One of the most effective medications, megestrol acetate, is a synthetic compound similar in structure to the hormone progesterone.[17] Dronabinol, which resembles the psychoactive ingredient in marijuana, stimulates the appetite at doses that have minimal mental effects. Under investigation are medications that may help to restore lean tissue, such as anabolic steroids, growth hormone, and insulin-like growth factor.

■ Reminder: The term *complementary and alternative medicine (CAM)* refers to health care practices that have not been proved to be effective and consequently are not included as part of conventional treatment.

Alternative Therapies Many patients turn to *complementary and alternative medicine (CAM)* to assist them with their fight against cancer. ■ Although few abandon conventional medicine, up to 80 percent of cancer patients combine one or more CAM approaches with standard treatment.[18] Few patients discuss their use of CAM with physicians.

Dietary supplements and herbal remedies are among the most frequently used CAM therapies. Although many supplements can be used without risk, some may have adverse effects or interfere with conventional treatments. For example, use of the herbal remedy St. John's wort can reduce the effectiveness of some anticancer drugs.[19] As another example, some studies suggest that antioxidant supplements interfere with chemotherapy and radiation treatments.[20] Clinical trials of several popular supplements are in progress to learn more about their potential effects and interactions with treatments.

Medical Nutrition Therapy for Cancer

The objectives of medical nutrition therapy for cancer patients are to minimize loss of weight and muscle tissue, correct nutrient deficiencies, and provide a diet that can be tolerated and enjoyed despite the complications of illness. Appropriate nutrition care helps patients preserve their strength and improves recovery after stressful cancer treatments. Moreover, malnourished cancer patients develop more complications and have shorter survival times than patients who maintain good nutrition status.[21]

Because there are many forms of cancer and a variety of potential treatments, nutritional needs among cancer patients vary considerably. Furthermore, a person's needs may change at different stages of illness. Patients should be screened for malnutrition when cancer is diagnosed and reassessed during the treatment and recovery periods.

Protein and Energy For patients at risk of weight loss and wasting, protein and energy needs are considerable. Protein recommendations range from 1.0 to 1.2 grams per kilogram body weight for nonstressed patients, 1.2 to 1.5 grams per kilogram for those undergoing treatment, and 1.5 to 2.5 grams per kilogram for patients with cachexia.[22] Energy needs may be 25 to 35 kcalories per kilogram body weight, depending on the patient's current weight, activity level, degree of metabolic stress, and energy needs for weight regain and tissue repair. Health practitioners should regularly monitor patients' weight changes and adjust intake recommendations as necessary. Patients who cannot eat adequate food may be able to meet their needs by supplementing the diet with nutrient-dense formulas. The "How To" provides suggestions that can help to increase the energy and protein content of meals.

Although weight loss is a problem for many cancer patients, breast cancer patients often gain weight. In one survey, 63 percent of women with breast cancer reported weight gains, ranging from 5 to 27 pounds.[23] Weight gain occurs more often in premenopausal women and in those undergoing extensive chemotherapy. By discussing weight maintenance soon after diagnosis and encouraging physical activity, health practitioners can help patients avoid unnecessary weight gain.[24]

To increase the energy content of a meal, try these suggestions:

- *Butter or margarine.* Melt on pasta, potatoes, rice, and cooked vegetables. Add to hot cereals, casseroles, and soups. Spread liberally on bread, crackers, and rolls.

- *Mayonnaise.* Add to pasta, tuna, and potato salads. Use as a dressing for raw or cooked vegetables.

- *Cream cheese.* Spread on raw vegetables, toast, and crackers. Mix into chopped fruit. Use as a spread in sandwiches made with luncheon meats.

- *Half-and-half and cream.* Replace milk or water with half-and-half or cream in soups, sauces, hot chocolate, desserts, mashed potatoes, and cold and cooked cereals.

- *Nuts.* Add chopped nuts to pasta dishes, stir-fried vegetables, fruit salads, and green salads. Use nut meats in baked products.

- *Beverages.* Replace water and nonkcaloric beverages with sweetened drinks, fruit juices, and milk shakes.

These suggestions can help to add protein to a meal:

- *Powdered milk (use full-fat milk powder if available).* Add to recipes that include milk. Dissolve extra milk powder into milk-containing beverages. Stir into hot cereals, potato dishes, casseroles, and sauces. Add to scrambled eggs, hamburger, and meat loaf.

- *Cheese.* Melt on burgers, meat loaf, cooked vegetables, scrambled eggs, casseroles, and potatoes. Add cottage cheese to casseroles, egg dishes, pasta recipes, and salad dressings. Grate hard cheeses and sprinkle on soups, salads, and cooked vegetable dishes.

- *Eggs.* Add raw eggs when preparing casseroles, meatballs, and hamburgers. Add chopped hard-cooked eggs to salads, vegetable dishes, sandwich fillings, and pasta and potato salads.

- *Meats.* Add meat pieces to soups, egg dishes, casseroles, bean dishes, and pasta sauces. Add minced meats to vegetable dishes. Add chunks of cooked chicken or turkey to salads.

Managing Symptoms and Complications A thorough nutrition assessment may uncover specific problems that interfere with food consumption. Table 25–6 lists dietary considerations related to cancers affecting different sites in the body. The "How to" on p. 682 outlines a variety of dietary measures that may improve food intake and alleviate symptoms. Patients' responses to these strategies may vary considerably, and in some cases, a number of adjustments may be necessary.

TABLE 25–6

Dietary Considerations for Specific Cancers

Cancer Sites	Common Complications[a]	Possible Dietary Measures
Brain and nervous system	Chewing and swallowing difficulties, difficulty feeding oneself	Mechanically altered diet, use of adaptive feeding devices (see Nutrition in Practice 17)
Head and neck[b]	Swallowing difficulty, aspiration, inflamed mucosa, dry mouth, altered taste sensation	Tube feeding, mechanically altered diet
Esophagus	Swallowing difficulty, obstruction, acid reflux, inflamed mucosa	Tube feeding, mechanically altered diet
Stomach	Anorexia, delayed stomach emptying, early satiety, dumping syndrome, malabsorption	Tube feeding (for obstruction or unmanageable dumping syndrome), postgastrectomy diet, small frequent meals, limited sugars and insoluble fibers (see Chapter 17)
Intestine	Fluid and electrolyte imbalances, altered bowel function, malabsorption, lactose intolerance, inflamed mucosa, bacterial overgrowth, short bowel syndrome (if resected), obstruction	Tube feeding or total parenteral nutrition for obstruction, enteritis, or short-bowel syndrome; fat- and lactose-restricted diet (see Chapter 19)
Pancreas	Malabsorption, bile insufficiency, hyperglycemia	Fat-restricted diet, enzyme replacement (see Chapter 19), small frequent meals, carbohydrate-controlled diet (Chapter 21)

[a] Actual complications depend on the specific methods used for treating the cancer.
[b] Includes cancers of the pharynx, larynx, salivary glands, and oral and nasal cavities.

HELP PATIENTS HANDLE FOOD-RELATED PROBLEMS

In people with cancer or HIV infections, various complications can interfere with eating. Health practitioners can try to identify the specific problems that patients are having and offer appropriate solutions. Not every suggestion will work for each patient; encourage patients to experiment and find the strategies that work best.

I just don't have an appetite.

- Eat small meals and snacks at regular times each day.
- Eat the largest meal at the time of day when you feel the best.
- Include nutrient-dense foods in meals, and consume them before other foods.
- Indulge in favorite foods throughout the day. Serve foods attractively.
- Avoid drinking large amounts of liquids before or with meals.
- Eat in a pleasant and relaxed environment. Eat with family and friends when possible.
- Listen to your favorite music or enjoy a program on TV while you eat.
- Take a walk before you eat.

I am too tired to fix meals and eat.

- Let family members and friends prepare food for you.
- Obtain foods that are easy to prepare and easy to eat, such as sandwiches, frozen dinners, take-out meals from restaurants, instant breakfast drinks, liquid formulas, and energy bars.

Foods just don't taste right.

- Brush your teeth or use mouthwash before you eat.
- Consume foods chilled or at room temperature.
- Choose eggs, fish, poultry, and milk products instead of meats.
- Experiment with sauces, seasonings, herbs, and spices to improve food flavor.
- Use plastic, rather than metal, eating utensils.
- Save your favorite foods for times when you are not feeling nauseated.

I am nauseated a lot of the time, and sometimes I need to vomit.

- Consume liquids throughout the day to replace fluids.
- If you become nauseated from chemotherapy treatments, avoid eating for at least two hours before treatments.
- Consume smaller meals, and eat slowly.
- Avoid foods and meals that have strong odors or are fatty, greasy, or gas forming.

I am having problems chewing and swallowing food.

- Experiment with food consistencies to find the ones you can manage best. Thin liquids, dry foods, and sticky foods (such as peanut butter) are often difficult to swallow.
- Add sauces and gravies to dry foods.
- Drink fluids with meals to ease chewing and swallowing.
- Try using a straw to drink liquids.
- Tilt your head forward and backward to see if you can swallow more easily when your head is positioned differently.

I have sores in my mouth, and they hurt when I eat.

- Use cold or frozen foods; they are often soothing.
- Try soft foods such as ice cream, milk shakes, bananas, applesauce, mashed potatoes, cottage cheese, and macaroni and cheese.
- Avoid foods that irritate mouth sores such as citrus fruits and juices, tomatoes and tomato-based products, spicy foods, foods that are very salty, foods with seeds (such as poppy seeds and sesame seeds) that can scrape the sore, and coarse foods such as raw vegetables and toast.
- Ask your doctor about using a local anesthetic solution such as lidocaine before eating to reduce pain.
- Use a straw for drinking liquids in order to bypass the sores.

My mouth is really dry.

- Rinse your mouth with warm salt water or mouthwash frequently. Avoid using mouthwash that contains alcohol.
- Drink small amounts of liquid frequently between meals.
- Ask your doctor or pharmacist about medications that can help dry mouth.
- Use sour candy or gum to stimulate the flow of saliva.
- Add broth, sauces, gravies, mayonnaise, butter, or margarine to dry foods.
- Make sure you brush your teeth and floss regularly to prevent cavities and oral infections.

I am having trouble with diarrhea.

- Drink plenty of fluids. Salty broths and soups, diluted fruit juices, and sports drinks are good choices. For severe diarrhea, try oral rehydration formulas that are commercially prepared.
- Avoid foods and beverages that increase gas, such as legumes, onions, vegetables of the cabbage family, foods that contain sorbitol or mannitol, and carbonated beverages.
- Try using lactase enzyme replacements when you use milk products in case you are experiencing lactose intolerance. Yogurt and aged cheeses may be easier to tolerate than milk and fresh cheeses.
- Avoid high-fat foods if you are fat intolerant.
- Avoid caffeine.
- Eat smaller meals, and eat more frequently.
- Check with your doctor about using digestive enzyme replacements if you have had diarrhea for a long time.

I am having trouble with constipation.

- Drink plenty of fluids. Try warm fluids, especially in the morning.
- Eat whole-grain breads and cereals, nuts, fresh fruits and vegetables, prunes, and prune juice. Avoid refined carbohydrate foods such as white bread, white rice, and pasta.
- Engage in physical activity regularly.

Enteral and Parenteral Nutrition Support Nutrition support is used in limited situations during cancer treatment. Generally, tube feedings and parenteral nutrition are provided to patients who have long-term or permanent gastrointestinal impairment or are experiencing complications that interfere with food intake.[25] For example, many patients undergoing radiation treatment for head and neck cancers require long-term tube feeding and may need to continue tube feedings at home. ■ Parenteral nutrition is reserved for patients who have inadequate GI function, such as individuals with chronic radiation enteritis. Whenever possible, enteral nutrition is strongly preferred over parenteral nutrition, to preserve GI function and avoid infection.

■ Radiation to the head and neck regions often causes dysphagia and mouth sores.

Patients who undergo bone marrow transplants may require total parenteral nutrition (TPN) before and after the transplant because the GI tract is often severely damaged by the preparatory procedure (which may include high-dose chemotherapy or radiation treatment). When GI function returns, the patient can begin consuming small amounts of food along with TPN. As oral intake improves, TPN is gradually tapered. Because recipients of bone marrow transplants are severely immunocompromised, they should be instructed to follow safe food-handling practices to minimize the risk of foodborne illness (see Nutrition in Practice 5). In addition, they need to avoid foods that are likely to contain unsafe levels of bacteria, such as fresh fruits and vegetables and undercooked meat, poultry, and eggs.

IN SUMMARY

- Cancer arises from mutations in the genes that control cell division. Some dietary substances promote carcinogenesis, while others may help to prevent cancer.

- Cancer's effects on nutritional status depend on the type of cancer, its severity, and the methods of treatment. Cancer cachexia is a frequent complication of cancer and may be related to anorexia, altered metabolism, and responses to treatment.

- Medical nutrition therapy for cancer patients aims to minimize weight loss and wasting, correct deficiencies, and manage complications that impair food intake.

The accompanying Case Study allows you to apply information about nutrition and cancer to a clinical situation.

Case Study

PUBLIC RELATIONS CONSULTANT WITH CANCER

Jane Woodhouse is a 58-year-old public relations consultant who was recently diagnosed with colon cancer after a routine colonoscopy, a procedure in which the colon is examined using a flexible tube attached to an optical device. Mrs. Woodhouse is scheduled to have surgery to remove the segment of colon that contains the tumor and to determine if the cancer has spread to the surrounding lymph nodes and, possibly, other organs. The health practitioner completing the nutrition assessment finds that Mrs. Woodhouse is 5 feet 5 inches tall and weighs 178 pounds. She spends most of the day sitting and has little time to engage in recreational exercise. Her diet is high in fat and typically includes red meat at both lunch and dinner. She eats two or three servings of fruits and vegetables each day, although she does not like green leafy vegetables very much. She rarely drinks milk or consumes milk products.

1. Review Table 25–2 on p. 676 and describe the factors in Mrs. Woodhouse's diet and lifestyle that may have contributed to the development of colon cancer.

2. What symptoms and complications can arise after colon surgery and impair nutrition status? If the cancer team decides that Mrs. Woodhouse needs follow-up chemotherapy, how might the chemotherapy affect her nutrition status?

3. If Mrs. Woodhouse is unresponsive to treatment and her cancer progresses, she may develop cancer cachexia. Describe this syndrome and its causes. What are the benefits of preventing or correcting the wasting associated with cancer?

4. Provide suggestions that may help Mrs. Woodhouse handle these problems should they develop: poor appetite, fatigue, taste alterations, nausea and vomiting, chewing and swallowing difficulties, mouth sores, dry mouth, diarrhea, and weight loss.

TABLE 25–7

HIV and AIDS Epidemic at a Glance, 2006

	World	North America
Living with HIV or AIDS	39,500,000	1,400,000
Newly infected with HIV	4,300,000	43,000
AIDS deaths	2,900,000	18,000

Source: UNAIDS/WHO, *AIDS epidemic update: December 2006*, http://data.unaids.org/pub/EpiReport/2006/2006_EpiUpdate_en.pdf, site visited January 6, 2007.

TABLE 25–8

Risk Factors for HIV Infection

- History of receiving blood transfusions or clotting factors between 1978 and 1985.
- Infant born to mother with HIV infection.
- Intravenous drug use in which syringes are shared among users.
- Sexual contact with multiple partners.
- Sexual contact with intravenous drug users, prostitutes, or individuals with a history of HIV or other sexually transmitted diseases.
- Unsafe sexual practices.

T cells are lymphocytes that develop in the thymus gland. The other lymphocytes are the *B cells* (which develop in bone marrow) and *natural killer cells*.

acquired immune deficiency syndrome (AIDS): the late stage of illness caused by infection with the human immunodeficiency virus (HIV); characterized by severe damage to immune function.

helper T cells: lymphocytes that have a specific protein called CD4 on their surfaces and therefore are also known as *CD4+ T cells*; the cells most affected in HIV infection.

opportunistic infections: infections from microorganisms that normally do not cause disease in healthy people but are damaging to persons with compromised immune function.

HIV Infection

A diagnosis of HIV infection, which leads to **acquired immune deficiency syndrome (AIDS)**, can be devastating. HIV (human immunodeficiency virus) attacks the immune system and disables a person's defenses against infections and certain cancers. Patients may expect an ever-worsening course of illness and, possibly, death. In recent years, however, treatment options have expanded and patients have benefited by vast improvements in quality of life.

The HIV/AIDS epidemic continues to sweep across countries, especially in sub-Saharan Africa. Table 25–7 shows its impact worldwide and in the United States. For many years, the destructive effects of HIV infection seemed unstoppable, but in the mid-to-late 1990s, the death rate from AIDS began to decline in the United States, and the progression from HIV to AIDS slowed dramatically. The disease still has no cure, but remarkable progress has been made in understanding and treating HIV infection. Without a cure, the best course is prevention. HIV is most often sexually transmitted and can be spread by direct contact with contaminated body fluids, such as blood, semen, vaginal secretions, and breast milk.

Because many people remain symptom-free during the early stages of infection, they may not realize that they can pass the infection to others. To reduce the spread of HIV infection, those at risk (see Table 25–8) are encouraged to undergo testing. A blood test can usually detect HIV antibodies within several months after exposure and, often, after one or two weeks. An estimated 25 percent of persons in the United States who have HIV infection are unaware that they are infected.[26]

Consequences of HIV Infection

HIV infection destroys immune cells that have a protein called CD4 on their surfaces. The cells most affected are the **helper T cells,** also called *CD4+ T cells* because the presence of CD4 is a primary characteristic. Early symptoms of HIV infection are nonspecific and may include fever, sore throat, malaise, skin rashes, nausea, muscle and joint pain, and diarrhea. Afterward, many people remain symptom-free for five to ten years or even longer. However, if the HIV infection is not treated, the depletion of T cells eventually increases the person's susceptibility to **opportunistic infections,** that is, infections caused by microorganisms that normally do not cause disease in healthy individuals.

The term *AIDS* applies to the advanced stages of HIV infection, in which the inability to fight illness allows a number of serious diseases and complications to develop; such **AIDS-defining illnesses** include severe infections, certain cancers, and wasting of lean tissue. Health practitioners evaluate disease progression by measuring the concentrations of helper T cells and circulating virus (called the *viral load*) and by monitoring clinical symptoms. Although current drug therapies can dramatically slow the progression of HIV infection, the drugs' side effects may make it difficult for patients to adhere to treatments, as discussed in the following sections.

Lipodystrophy Many of the drug treatments that suppress HIV infection cause abnormalities in glucose and fat metabolism in an estimated 25 to 50 percent of patients. These complications, collectively known as the **HIV-lipodystrophy syndrome**, include body fat redistribution, abnormal blood lipid levels, and insulin resistance.[27] Patients tend to accumulate abdominal fat and lose fat from the face, arms, and legs; thus, they appear to be thin except for a "pot belly." Also observed are breast enlargement (in both men and women), fat accumulation at the base of the neck (called a **buffalo hump**), and benign growths composed of fat tissue (called **lipomas**). The changes in body composition are often disfiguring and may cause physical discomfort; moreover, patients often develop hypertriglyceridemia, low HDL (high-density lipoprotein) cholesterol levels, glucose intolerance, and hyperinsulinemia. The reasons for the development of lipodystrophy are unknown.

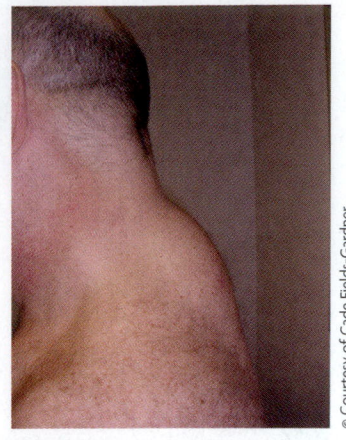

© Courtesy of Cade Fields-Gardner

HIV-lipodystrophy syndrome is sometimes evident by the accumulation of fatty tissue at the base of the neck, referred to as *buffalo hump.*

Weight Loss and Wasting Even with effective treatment of HIV infection, weight loss and wasting are ongoing problems for HIV-infected patients.[28] The Centers for Disease Control defines *AIDS-related wasting syndrome* as a 10 percent weight loss within a six-month period accompanied by diarrhea or fever for more than 30 days without a known cause. The wasting has been linked with accelerated disease progression, reduced strength, and fatigue. In the later stages of AIDS, wasting is severe and increases the risk of death. Much as in cancer, the wasting associated with HIV infection has many causes: anorexia and inadequate food intake, altered metabolism, malabsorption, chronic diarrhea, and diet-drug interactions.

Anorexia and Reduced Food Intake Inadequate food intake is a key factor in the development of wasting. Poor food intake may result from various factors, including the following:

- **Emotional distress, pain, and fatigue.** The physical and social problems that accompany chronic illness may cause fear, anxiety, and depression, which contribute to anorexia. Pain and fatigue, which may be associated with some disease complications, can cause anorexia and difficulty with eating.

- **Oral infections.** The oral infections associated with HIV infection can cause discomfort and interfere with food consumption. Common infections include thrush and herpes simplex virus. **Thrush** can cause mouth pain, dysphagia, and altered taste sensation, and **herpes simplex virus** may cause painful lesions around the lips and in the mouth.

- **Respiratory disorders.** Respiratory infections, including pneumonia and tuberculosis, are common in people with HIV infection. Symptoms often include chest pain, shortness of breath, and cough, which interfere with eating and contribute to anorexia.

- **Cancer.** As described earlier in this chapter, cancer leads to anorexia for numerous reasons. In addition, **Kaposi's sarcoma**, a type of cancer frequently associated with HIV infection, can cause lesions in the mouth and throat that make eating painful.

- **Medications**. The medications given to treat HIV infection, other infections, and cancer often cause anorexia, nausea and vomiting, altered taste sensations, food aversions, and diarrhea.

GI Tract Complications Complications involving the GI tract may result from opportunistic infections, medications, or the HIV infection itself.[29] In addition to the

AIDS-defining illnesses: diseases and complications associated with the later stages of a HIV infection, such as wasting, recurrent bacterial pneumonia, opportunistic infections, and certain cancers.

HIV-lipodystrophy (LIP-oh-DIS-tro-fee) **syndrome:** a collection of abnormalities in fat and glucose metabolism that result from drug treatments for HIV; includes body fat redistribution, abnormal blood lipid levels, and insulin resistance. The accumulation of abdominal fat is sometimes called *protease paunch.*

buffalo hump: the accumulation of fatty tissue at the base of the neck.

lipomas (lih-POE-muz): benign tumors composed of fatty tissue.

thrush: a fungal infection of the mouth and throat, most often caused by *Candida albicans.*

herpes simplex virus: a common virus that can cause blisterlike lesions on the lips and in the mouth.

Kaposi's (cap-OH-seez) **sarcoma:** a common cancer in HIV-infected persons that is characterized by lesions on the skin, in the lungs, and in the GI tract.

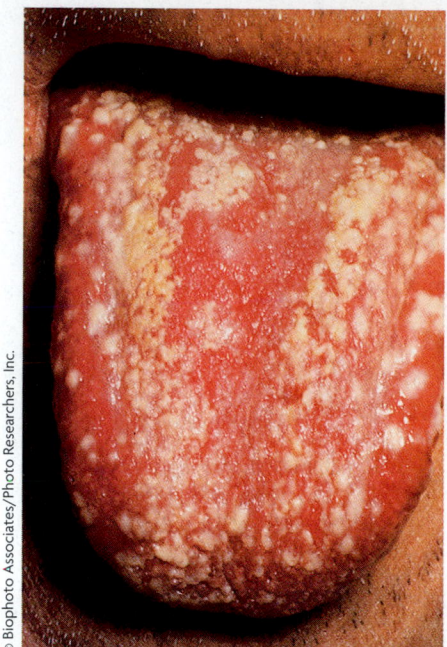

The oral infection thrush is easily identified by the characteristic milky white patches that appear on the tongue.

oral infections described previously, infections commonly develop in the stomach and intestines. Advanced AIDS is often accompanied by characteristic changes in the lining of the small intestine, likely caused by GI infection: the villi appear shortened and flattened, and the absorptive area is substantially reduced. ■ These changes contribute to malabsorption, steatorrhea, and diarrhea.

As described earlier, many patients are unable to tolerate the medications used to suppress HIV and develop nausea, vomiting, and diarrhea. Furthermore, medications that treat GI viral, parasitic, and fungal infections contribute to bacterial overgrowth. Thus HIV-infected patients face an extremely high risk of malnutrition due to the combination of intestinal malabsorption, bacterial overgrowth, and nutrient losses from vomiting and diarrhea.

Treatments for HIV Infection

Although there is no cure for HIV infection, treatments can help to slow its progression, reduce complications, and alleviate pain. The standard treatment for suppressing HIV infection, called *highly active antiretroviral therapy (HAART),* combines three or more antiretroviral drugs.[30] Table 25–9 lists the major drug categories included in antiretroviral therapy and describes the drugs' modes of action. These antiretroviral agents have multiple adverse effects that make their long-term use difficult to tolerate. In addition to the GI effects discussed previously, side effects include skin rashes, headache, anemia, tingling and numbness, hepatitis, pancreatitis, and kidney stones. Thus, although HAART has improved life span and quality of life for many patients, the drug regimens are difficult to adhere to and cause complications that require continual management. The Diet-Drug Interactions feature summarizes the nutrition-related effects of antiretroviral agents and other drugs mentioned in this chapter.

■ The AIDS-related abnormalities in the intestinal mucosa are sometimes referred to as *HIV enteropathy* (EN-ter-OP-ah-thy).

Control of Anorexia and Wasting Appetite stimulants, physical activity, and anabolic hormones have been successful in reversing weight loss and increasing muscle mass in HIV-infected patients.[31] The medications megestrol acetate and dronabinol (described on p. 680) are sometimes prescribed to stimulate appetite and help with weight gain. Testosterone and human growth hormone have demonstrated positive effects on nitrogen balance and lean tissue content, especially in combination with

TABLE 25–9

Antiretroviral Drugs for Treatment of HIV Infection

Category	Example(s)	Mode of Action
Nucleoside reverse transcriptase inhibitors (NRTI)	Zidovudine (AZT) Didanosine Lamivudine	As analogs of the nucleosides needed for DNA synthesis, NRTI impair the ability of HIV's *reverse transcriptase* enzyme to produce usable copies of DNA.
Nonnucleoside reverse transcriptase inhibitors (NNRTI)	Nevirapine Delavirdine Efavirenz	NNRTI bind active sites on HIV's *reverse transcriptase* enzyme, blocking the ability of HIV to produce DNA copies of its genetic material.
Protease inhibitors (PI)	Saquinavir Ritonavir Indinavir	PI inhibit HIV's *protease* enzyme, which cleaves HIV's gene products into usable structural proteins.
Fusion Inhibitors	Enfuvirtide	Fusion inhibitors prevent HIV from entering cells.

Diet-Drug Interactions

Check this table for notable nutrition-related effects of the medications discussed in this chapter.

	Gastrointestinal Effects	Interactions with Dietary Substances	Metabolic Effects
Appetite stimulants (megestrol acetate, dronabinol)	Nausea, vomiting, diarrhea	—	Hyperglycemia (megestrol acetate)
Didanosine	Nausea, vomiting, dry mouth, altered taste perception, anorexia, constipation	Avoid alcohol and aluminum- and magnesium-containing antacids.	Pancreatitis
Enfuvirtide	Nausea, vomiting, anorexia, diarrhea, constipation	—	Pancreatitis, increased blood triglycerides
Methotrexate	Nausea, vomiting, diarrhea, reduced absorption of vitamin B_{12} and calcium	Milk may reduce methotrexate absorption if they are ingested together.	Increased serum uric acid levels, anemia, liver toxicity
Ritonavir	Nausea, vomiting, altered taste perception, anorexia, diarrhea	—	Pancreatitis, diabetes, reduced blood levels of copper and zinc; increased levels of triglycerides, liver enzymes, creatine kinase, and uric acid
Zidovudine (AZT)	Nausea, vomiting, altered taste perception, anorexia, mouth sores, constipation	—	Anemia, reduced blood levels of copper and zinc

Note: Other antiretroviral drugs that treat HIV infection have gastrointestinal and metabolic side effects; only a few are listed here as examples.

resistance training. A regular program of resistance training can improve muscle mass and strength and correct some of the metabolic abnormalities (altered blood lipids and insulin resistance) that are common in HIV-infected patients.

Control of Lipodystrophy Treatment strategies for lipodystrophy are under investigation. Both aerobic activity and resistance training help to reduce abdominal fat, although some patients opt for cosmetic surgery.[32] Patients may be given alternative antiretroviral drugs to alleviate symptoms. Medications may be prescribed to treat abnormal blood lipids and insulin resistance.

Alternative Therapies Like cancer patients, people with HIV infection and AIDS are frequently tempted to try unconventional methods of treatment. Although many alternative therapies are harmless, they can be expensive at a time when financial security is of concern. Monitoring patients' use of dietary supplements is essential to reduce the possibility of nutrient-drug and herb-drug interactions.

Resistance training can help a person with HIV infection maintain muscle mass and strength.

Medical Nutrition Therapy for HIV Infection

The initial nutrition assessment provides baseline data with which to monitor progress throughout the course of the disease. The assessment should include an evaluation of body weight and body composition. Follow-up measurements may indicate the need to adjust dietary recommendations and drug therapies.

Weight Maintenance A primary objective of nutrition therapy is to maintain weight and muscle tissue.[33] Health practitioners should attempt to determine the dietary and lifestyle factors that may interfere with patients' food intake, appetite, and physical activity and provide suggestions that may prevent future weight problems. If food consumption is difficult, small, frequent feedings may be better tolerated than several large meals. The addition of nutrient-dense snacks, protein or energy bars, and oral supplements can improve intakes. Liquid formulas may be especially useful for the person who is too tired to eat or prepare meals. The "How To" on p. 681 provides suggestions for adding energy and protein to the diet.

Vitamins and Minerals Vitamin and mineral needs of people with HIV infections are highly variable, and little information is available concerning specific needs. Because nutrient deficiencies are likely to result from reduced food intake, malabsorption, diet-drug interactions, and nutrient losses, multivitamin-mineral supplements are often recommended.

Metabolic Complications HIV patients using antiretroviral drugs frequently develop insulin resistance and elevated triglyceride and LDL (low-density lipoprotein) cholesterol levels, and dietary adjustments should be attempted before medications are prescribed.[34] Patients should be advised to achieve or maintain desirable weight, replace saturated fats with monounsaturated and polyunsaturated fats, and limit intakes of *trans*-fatty acids and cholesterol. Complex carbohydrates are preferred over foods high in sugars. Regular physical activity is recommended for improving both blood lipid levels and insulin resistance. ■ If problems persist, alternative antiretroviral medications are sometimes attempted.

■ Additional suggestions for managing insulin resistance and hyperlipidemias are available in Chapters 21 and 22, respectively.

Symptom Management As mentioned, antiretroviral medications and opportunistic GI infections can cause problems that interfere with food intake; some of these problems may increase fluid and electrolyte losses as well. The "How To" on p. 682 describes measures that can improve food and fluid intakes and alleviate discomfort in individuals who experience nausea and vomiting, diarrhea, mouth lesions, and altered taste sensitivity.

Food Safety The depressed immunity of people with HIV infections places them at extremely high risk of developing foodborne infections. Patients should be cautioned about their high susceptibility for foodborne illness and given detailed instructions about the safe handling and preparation of foods (see Nutrition in Practice 5). Water can also be a source of foodborne illness and is a common cause of **cryptosporidiosis** in HIV-infected individuals. Because water quality varies throughout the United States, local health departments should be consulted to determine if the local tap water is safe for patients to drink. If not, or to take additional safety measures, water used for drinking and making ice cubes should be boiled for one minute. Some types of filtered and bottled waters are also safe, but not all.

Case Study

FILM PRODUCER WITH HIV INFECTION

Three years ago, Darrell Meckler, a 34-year-old film producer, sought medical help when he began feeling run-down and developed a painful white fungal infection over his mouth and tongue. The presence of thrush, recent weight loss, and anemia alerted Mr. Meckler's physician to the possibility of a HIV infection. When Mr. Meckler tested positive for HIV, he and his family and friends were devastated by the news, but those close to him have remained supportive. During the three years since Mr. Meckler began antiretroviral drug therapy, he has maintained his weight but has also developed lipodystrophy and hypertriglyceridemia. Mr. Meckler is 6 feet tall and currently weighs 185 pounds. He occasionally develops diarrhea and sometimes anorexia.

1. Describe lipodystrophy, and discuss its typical pattern in people who have a HIV infection. What adjustments in treatment and lifestyle may be helpful?

2. Describe an appropriate diet for Mr. Meckler. What strategies may improve his problems with diarrhea and anorexia? Suggest reasons why people with a HIV infection may develop diarrhea and anorexia.

3. Explain why a HIV infection can lead to wasting as the disease progresses to the later stages. How should Mr. Meckler's diet change if wasting becomes a problem for him?

Enteral and Parenteral Nutrition Support In later stages of illness, people with HIV infections may need aggressive nutrition support if they are unable to consume enough food. Tube feedings are preferred whenever the GI tract is functional; they can be given at night to supplement oral diets consumed during the day. Preventing bacterial contamination of the formula is particularly important. Parenteral nutrition is reserved for patients who are unable to tolerate enteral nutrition, such as those with GI obstructions that prevent food intake. For individuals with severe malabsorption, orally administered hydrolyzed formulas containing medium-chain triglycerides may be as effective as parenteral nutrition for reversing weight loss and wasting.

IN SUMMARY

- By attacking immune cells, HIV causes progressive damage to immune function and may eventually lead to AIDS.

- Improved drug therapies have slowed the progression of HIV infection; however, these drugs may promote the HIV-lipodystrophy syndrome, characterized by body fat redistribution, abnormal lipid levels, and insulin resistance.

- HIV infection is often associated with weight loss and wasting, anorexia, and various complications that affect food intake. Dietary adjustments, resistance training, and medications can help patients maintain their weight and prevent wasting.

- People with HIV infections must pay strict attention to food safety guidelines to prevent foodborne infections.

The Case Study provides an opportunity to review the nutrition concerns of a person with HIV infection.

cryptosporidiosis (KRIP-toe-spor-ih-dee-OH-sis): a foodborne illness caused by the parasite *Cryptosporidium parvum*.

FOR PEOPLE WITH CANCER OR HIV INFECTIONS

Medical History

Check the medical record to determine:

- Type and stage of cancer
- Stage of HIV infection

Review the medical record for complications that may alter medical nutrition therapy including:

- Altered organ function
- Altered taste perception
- Anorexia
- Dry mouth and oral infections
- GI symptoms and infections
- Hyperlipidemias
- Insulin resistance
- Malnutrition and wasting

Medications

For patients with cancer or HIV infections:

- Check medications to identify potential diet-drug interactions.
- Recommend using antinauseants at mealtime, if needed.
- Ask about use of dietary supplements, including herbal remedies.

For cancer patients who require chemotherapy:

- Recommend strategies to prevent food aversions.
- Offer suggestions for managing drug-related complications.

For HIV-infected patients using antiretroviral drug therapy:

- Remind patients that some drugs are better absorbed with foods and others must be taken on an empty stomach.
- Help patients work out a medication schedule that suits their lifestyle and is timed appropriately in regard to food intake.
- Offer suggestions for managing drug-related complications.

Dietary Intake

For patients with poor food intakes and weight loss:

- Determine the reasons for reduced food intake.
- Offer appropriate suggestions to improve food intake.
- Provide interventions before weight loss progresses too far.

For patients with HIV infections who experience weight gain, elevated triglyceride or LDL cholesterol levels, or hyperglycemia:

- Assess the diet for energy, total fat, types of fat, carbohydrates, fiber, and sugars.
- For hyperlipidemias, recommend a diet low in saturated fat, trans-fatty acids, and sugars.
- For hyperglycemia, recommend a consistent carbohydrate intake that emphasizes complex carbohydrates.
- Recommend regular physical activity for weight control and for improving blood lipid levels and insulin resistance.

Anthropometric Data

Take baseline height and weight measurements, monitor weight regularly, and suggest dietary adjustments for weight maintenance, if necessary. Remember that body composition may change without affecting body weight.

Perform baseline and periodic body composition measurements in HIV-infected patients who are using antiretroviral drug therapy.

Laboratory Tests

Note that albumin and other serum proteins may be reduced in patients with cancer or HIV infections, especially in those experiencing wasting. Check laboratory tests for indications of:

- Anemia
- Dehydration
- Elevated triglyceride levels
- Elevated LDL cholesterol levels
- Hyperglycemia

For patients with HIV infections, evaluate disease progression by checking:

- Helper T cell counts
- Viral load

Physical Signs

Look for physical signs of:

- Protein-energy malnutrition and wasting
- Dehydration (especially for those with fever, vomiting, or diarrhea)
- Oral infections
- Kaposi's sarcoma

Analysis

CLINICAL APPLICATIONS

1. Consider the nutrition problems that may develop in a 36-year-old woman with a malignant brain tumor that affects her ability to move the right side of her body (including the tongue) and to speak coherently. She is taking a pain medication that makes her nauseated and sleepy. Her expected survival time is only about six months.

   ■ If she is right-handed, how might her impairment interfere with eating? What suggestions do you have for overcoming this problem?

   ■ How might her nutrition status be affected by her inability to communicate effectively? What suggestions may help?

   ■ Describe ways in which the pain medication she is taking can affect her nutrition status.

2. Various types of chronic conditions can lead to weight loss and wasting. For some of these conditions, such as Crohn's disease or celiac disease (Chapters 18 and 19), diet is a cornerstone of treatment. For others, such as cancer and HIV infection, nutrition plays a supportive role. What determines whether nutrition plays a primary role or a supportive role in the treatment of disease?

Self Check

1. Which dietary substances may help to protect against cancer?
   a. alcohol
   b. well-cooked meats, poultry, and fish
   c. animal fats
   d. phytochemicals from fruits and vegetables

2. The metabolic changes that often result from cancer include:
   a. increased fat synthesis.
   b. increased protein turnover.
   c. increased muscle protein synthesis.
   d. reduced serum lipids.

3. An advantage of radiation therapy over chemotherapy is that:
   a. radiation is not damaging to rapidly dividing cells.
   b. radiation's side effects do not include malnutrition.
   c. radiation can be directed toward the regions affected by cancer.
   d. the radiation used is too weak to damage GI tissues.

4. Although many cancer patients lose weight, which type of cancer is often associated with weight gain?
   a. kidney cancer
   b. breast cancer
   c. colon cancer
   d. Kaposi's sarcoma

5. Oral diets after bone marrow transplants may restrict:
   a. fiber.
   b. carbohydrates.
   c. high-protein foods.
   d. raw fruits and vegetables.

6. The immune cells most seriously damaged by HIV are:
   a. B cells.
   b. helper T cells.
   c. natural killer cells.
   d. neutrophils.

7. HIV-lipodystrophy syndrome is characterized by these changes in body composition:
   a. increased central and peripheral fat
   b. decreased central and peripheral fat
   c. increased central and decreased peripheral fat
   d. decreased central and increased peripheral fat

8. Mouth sores in people with HIV infections are most frequently due to:
   a. oral infections.
   b. dehydration.
   c. nutrient deficiency.
   d. foodborne illnesses.

9. The medications megestrol acetate and dronabinol:
   a. are used to promote weight gain.
   b. are protease inhibitors that fight HIV infection.
   c. treat common opportunistic infections that develop in AIDS patients.
   d. treat HIV-lipodystrophy syndrome.

10. To prevent cryptosporidiosis, a person with HIV infection may need to:
    a. cook meat, poultry, and fish to an appropriate internal temperature.
    b. avoid consuming undercooked eggs.
    c. avoid foods prepared by people who are sick or who have skin infections.
    d. boil drinking water for one minute.

Answers to these questions appear in Appendix H.

Notes

1. E. T. Liu, Oncogenes and suppressor genes: Genetic control of cancer, in L. Goldman and D. Ausiello, eds., *Cecil Textbook of Medicine* (Philadelphia: Saunders, 2004), pp. 1108–1116.

2. S. Christen and coauthors, Chronic inflammation, mutation, and cancer, in J. Parsonnet and S. Hornig, eds., *Microbes and Malignancy: Infection as a Cause of Cancer* (New York: Oxford University Press, 1999), pp. 35–88.

3. W. C. Willett and E. Giovannucci, Epidemiology of diet and cancer risk, in M. E. Shils and coeditors, *Modern Nutrition in Health and Disease* (Philadelphia: Lippincott Williams & Wilkins, 2006), pp. 1267–1279.

4. Willett and Giovannucci, 2006.

5. P. D. Terry and coauthors, Intakes of fish and marine fatty acids and the risks of cancers of the breast and prostate and of other hormonerelated cancers: A review of the epidemiologic evidence, *American Journal of Clinical Nutrition* 77 (2003): 532–543.

6. T. Sugimura and coauthors, Heterocyclic amines: Mutagens/carcinogens produced during cooking of meat and fish, *Cancer Science* 95 (2004): 290–299; J. S. Felton and coauthors, Impact of environmental exposures on the mutagenicity/carcinogenicity of heterocyclic amines, *Toxicology* 198 (2004): 135–145; P. Jakszyn and coauthors, Development of a food database of nitrosamines, heterocyclic amines, and polycyclic aromatic hydrocarbons, *Journal of Nutrition* 134 (2004): 2011–2014.

7. G. N. Wogan and coauthors, Environmental and chemical carcinogenesis, *Seminars in Cancer Biology* 14 (2004): 473–486.

8. R. H. Liu, Potential synergy of phytochemicals in cancer prevention: Mechanism of action, *Journal of Nutrition* 134 (2004): 3479S–3485S.

9. Willett and Giovannucci, 2006.

10. H. S. Rugo, Paraneoplastic syndromes and other non-neoplastic effects of cancer, in L. Goldman and D. Ausiello, eds., *Cecil Textbook of Medicine* (Philadelphia: Saunders, 2004), pp. 1124–1131.

11. Rugo, 2004.

12. M. Schattner and M. Shike, Nutrition support of the patient with cancer, in M. E. Shils and coeditors, *Modern Nutrition in Health and Disease* (Philadelphia: Lippincott Williams & Wilkins, 2006), pp. 1290–1313.

13. B. Eldridge and coauthors, Nutrition and the patient with cancer, in A. M. Coulston, C. L. Rock, and E. R. Monsen, eds., *Nutrition in the Prevention and Treatment of Disease* (San Diego: Academic Press, 2001), pp. 397–412.

14. J. R. Bertino and W. Hait, Principles of cancer therapy, in L. Goldman and D. Ausiello, eds., *Cecil Textbook of Medicine* (Philadelphia: Saunders, 2004), pp. 1137–1150.

15. Bertino and Hait, 2004.

16. S. Z. Pavletic and J. M. Vose, Hematopoietic stem cell transplantation, in L. Goldman and D. Ausiello, eds., *Cecil Textbook of Medicine* (Philadelphia: Saunders, 2004), pp. 999–1003.

17. M. Tomiska, Palliative treatment of cancer anorexia with oral suspension of megestrol acetate, *Neoplasma* 50 (2003): 227–233.

18. B. R. Cassileth and G. Deng, Complementary and alternative therapies for cancer, *The Oncologist* 9 (2004): 80–89; M. A. Richardson and coauthors, Complementary/alternative medicine use in a comprehensive cancer center and the implications for oncology, *Journal of Clinical Oncology* 18 (2000): 2505–2514.

19. M. Markman, Safety issues in using complementary and alternative medicine, *Journal of Clinical Oncology* 20 (2002): 39S–41S.

20. B. Bruemmer and coauthors, The association between vitamin C and vitamin E supplement use before hematopoietic stem cell transplant and outcomes to two years, *Journal of the American Dietetic Association* 103 (2003): 982–990; H. E. Seifried and coauthors, The antioxidant conundrum in cancer, *Cancer Research* 63 (2003): 4295–4298.

21. Schattner and Shike, 2006.

22. American Dietetic Association, *Nutrition Care Manual* (Chicago: American Dietetic Association, 2007).

23. J. A. McInnes and M. T. Knobf, Weight gain and quality of life in women treated with adjuvant chemotherapy for early-stage breast cancer, *Oncology Nursing Forum* 28 (2001): 675–684.

24. A. L. Schwartz, Exercise and weight gain in breast cancer patients receiving chemotherapy, *Cancer Practice* 8 (2000): 231–237.

25. Schattner and Shike, 2006.

26. UNAIDS/WHO, *AIDS epidemic update: December 2006*, http://data.unaids.org/pub/EpiReport/2006/2006_EpiUpdate_en.pdf, site visited January 6, 2007.

27. P. Koutkia and S. Grinspoon, HIV-associated lipodystrophy: Pathogenesis, prognosis, treatment, and controversies, *Annual Review of Medicine* 55 (2004): 303–317.

28. S. Grinspoon and K. Mulligan, Weight loss and wasting in patients infected with human immunodeficiency virus, *Clinical Infectious Diseases* 36 (2003): S69–S78.

29. J. G. Bartlett, Gastrointestinal manifestations of AIDS, in L. Gold-man and D. Ausiello, eds., *Cecil Textbook of Medicine* (Philadel-phia: Saunders, 2004), pp. 2168–2170.

30. Panel on Clinical Practices for Treatment of HIV Infection, *Guidelines for the Use of Antiretroviral Agents in HIV-1-Infected Adults and Adolescents,* October 10, 2006, available from http://AIDSinfo.nih.gov (site visited January 8, 2007); S. Safrin, Antiviral agents, in B. G. Katzung, ed., *Basic and Clinical Pharmacology* (New York: Lange Medical Books/McGraw-Hill, 2001), pp. 823–844.

31. Grinspoon and Mulligan, 2003.

32. Koutkia and Grinspoon, 2004.

33. J. Nerad and coauthors, General nutrition management in pa-tients infected with human immunodeficiency virus, *Clinical In-fectious Diseases* 36 (2003): S52–S62.

34. M. Dube and M. Fenton, Lipid abnormalities, *Clinical Infectious Diseases* 36 (2003): S79–S83; M. C. Gelato, Insulin and carbohy-drate dysregulation, *Clinical Infectious Diseases* 36 (2003): S91–S95.

ETHICAL ISSUES IN NUTRITION CARE

As with other medical technologies, the availability of specialized nutrition support forces health care professionals and members of our society to face difficult **ethical** issues. When medical treatments prolong life by merely delaying death, the lifetime that remains may be of extremely low quality. This discussion examines the ethical dilemmas that clinicians must face when dealing with patients in critical care. The accompanying Glossary defines the relevant terms.

If providing nutrition care can do little to promote recovery, is it appropriate to withhold or withdraw nutrition support?

In attempting to answer questions such as this one, health care professionals must consider the following ethical principles:[1]

- A patient has the right to make decisions concerning his or her own well-being (**patient autonomy**), even if refusing treatment could result in death. It is generally accepted that a patient's preferences should take precedence over the desires of others.[2]

- A patient should be fully informed of a treatment's benefits and risks in a fair and honest manner (**disclosure**). A patient's acceptance of a treatment that has been adequately disclosed is considered **informed consent**.

- A patient must have the mental capacity to make appropriate health care decisions (**decision-making capacity**). If a patient is mentally incapable of doing so, a person designated by the patient should serve as a **surrogate** decision-maker.

- The potential benefits (**beneficence**) of any treatment should outweigh its potential harm (**maleficence**).

- Health care providers must determine whether the provision of health care to one patient would unfairly limit the care of other patients (**distributive justice**).

Although these principles may seem simple and obvious, it is often difficult to determine the appropriate action to take during intensive care.[3] When clinicians and families disagree, the courts may be asked to decide.

What kinds of treatments can help to sustain a patient's life?

Nutrition support and hydration are both considered life-sustaining treatments because withholding or withdrawing either can result in death. Other life-sustaining treatments include **cardiopulmonary resuscitation (CPR)**, which supplies oxygen and restores a person's ability to breathe and pump blood; **defibrillation**, in which an electronic device shocks the heart

and reestablishes normal contractions; **mechanical ventilation**, which substitutes for lung function; and **dialysis**, which substitutes for kidney function.

Do patients have a right to life-sustaining treatments?

Although life-sustaining treatments are readily provided to patients who have a reasonable chance of recovering from illness, it is sometimes difficult to determine the best course of action for patients who are dying or who are unlikely to regain consciousness. Under such circumstances, such treatments may be considered **futile** because they are unable to improve the outcome of disease or increase the patient's comfort and well-being. If patients or caregivers demand treatment that health practitioners have determined to be useless, a legal resolution may be required. Conversely, medical personnel may find it objectionable to withdraw life support knowing that the consequence is the death of a patient.

How have the courts resolved conflicts involving nutrition support?

One of the landmark cases involving nutrition support concerned Nancy Cruzan, who suffered permanent and irreversible brain damage after a car crash in 1983, when she was 26 years of age.[4] After she had been in a **persistent vegetative state** for five years,

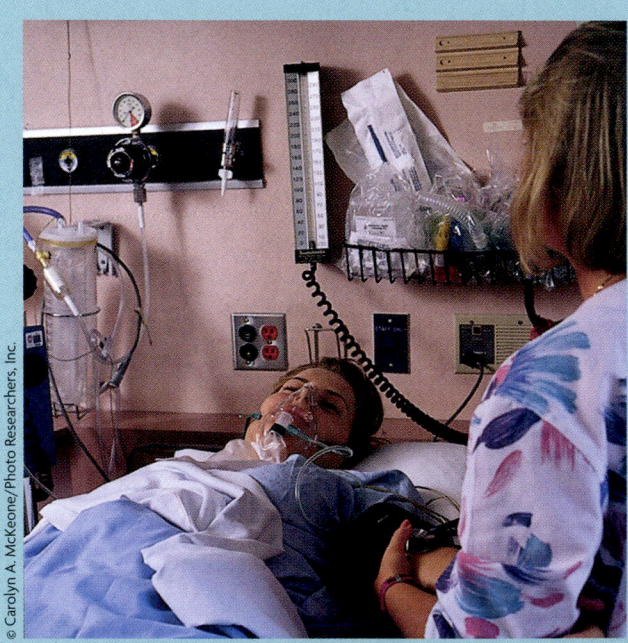

© Carolyn A. McKeone/Photo Researchers, Inc.

Is it ever morally and legally appropriate to withhold or withdraw nutrition support?

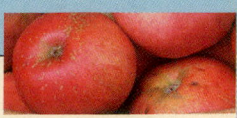

advance directive: a written or oral instruction regarding one's preferences for medical treatment to be used in the event of becoming incapacitated.

beneficence (be-NEF-eh-sense): the act of performing beneficial services rather than harmful ones.

cardiopulmonary resuscitation (CPR): life-sustaining treatment that supplies oxygen and restores a person's ability to breathe and pump blood.

decision-making capacity: the ability to understand pertinent information and make appropriate decisions; known as **decision-making competency** within the legal system.

defibrillation: life-sustaining treatment in which an electronic device is used to shock the heart and reestablish a pattern of normal contractions. Defibrillation is used when the heart has arrhythmias or has experienced cardiac arrest.

dialysis: life-sustaining treatment in which a patient's blood is filtered using selective diffusion through a semipermeable membrane; substitutes for kidney function.

disclosure: the act of revealing pertinent information. For example, clinicians should accurately describe proposed tests and procedures, their benefits and risks, and alternative approaches.

distributive justice: the equitable distribution of resources.

do-not-resuscitate (DNR) order: a request by a patient or surrogate to withhold cardiopulmonary resuscitation.

durable power of attorney: a legal document (sometimes called a **health care proxy**) that gives legal authority to another (a *health care agent*) to make medical decisions in the event of incapacitation.

ethical: in accordance with accepted principles of right and wrong.

futile: medical care that will not improve the medical circumstances of a patient.

health care agent: a person given legal authority to make medical decisions for another in the event of incapacitation.

informed consent: a patient's or caregiver's agreement to undergo a treatment that has been adequately disclosed. Persons must be mentally competent in order to make the decision.

living will: a written statement that specifies the medical procedures desired or not desired in the event that a person is unable to communicate or is incapacitated; also called a **medical directive**.

maleficence (mah-LEF-eh-sense): the act of doing evil or harm.

mechanical ventilation: life-sustaining treatment in which a mechanical ventilator is used to substitute for a patient's failing lungs.

patient autonomy: a principle of self-determination, such that patients (or surrogate decision-makers) are free to choose the medical interventions that are acceptable to them, even if they choose to refuse interventions that may extend their lives.

persistent vegetative state (PVS): a vegetative mental state resulting from brain injury that persists for at least one month. Individuals lose awareness and the ability to think but retain noncognitive brain functions, such as motor reflexes and normal sleep patterns.

surrogate: a substitute; a person who takes the place of another.

---

her parents requested permission to discontinue tube feeding, but hospital staff refused to honor the request, and the matter was taken to court. The Missouri Supreme Court determined that Nancy had never definitively stated her "right to die" wishes and that her parents were unable to make such a request for her. The court also stated that preserving life, no matter what its quality, should take precedence over all other considerations. Nancy's parents appealed the ruling, but in 1990, the U.S. Supreme Court upheld the Missouri Supreme Court in a five-to-four decision. Three witnesses were eventually found who could testify that Nancy would not desire life-sustaining treatment under the circumstances, and the court finally granted permission to remove the feeding tube. This case illustrates the importance of having an **advance directive** (discussed in a later section) that clearly indicates one's preferences for medical treatment in the event of incapacitation.

In a more recent case that received widespread media attention, the spouse and parents of a patient in a persistent vegetative state fought a ten-year legal battle over her medical care. In 1990, at the age of 25, Terri Schiavo suffered a full cardiac arrest.[5] She initially fell into a coma, but her condition evolved into a persistent vegetative state that was considered irreversible. Despite the neurologists' diagnosis and a series of CT and MRI scans showing extensive brain atrophy, her parents maintained that she was minimally conscious and would improve with rigorous treatment. Her husband, who was legally responsible for her care, insisted that she would never have wanted to be kept alive in a vegetative state. Like Nancy Cruzan, Terri had never expressed her wishes in an advance directive.

In 1998, Terri's husband filed a petition to have her feeding tube removed, and a Florida court approved the motion in February 2000. Although Terri's parents appealed, an appeals court affirmed the decision, and the Florida Supreme Court declined to review the case. In April 2001, Terri's physicians removed her feeding tube, but within days, a federal circuit court judge ordered it to be reinserted and reopened the case. Eventually, the motions filed by the parents were dismissed, and Terri's feeding tube was removed for the second time in October 2003. Within days, the Florida legislature passed a bill known as "Terri's Law" that gave the governor the authority to intervene, and Governor Jeb Bush ordered her feeding tube restored. A year later, Florida's Supreme Court declared Terri's Law to be unconstitutional. Although the governor appealed the decision, his appeal was rejected in January 2005. Terri's feeding tube was removed for the third time in March 2005. Despite emergency petitions by her parents and an attempt by the U.S. Congress to have her case reconsidered, the courts refused to grant a restraining order, and Terri died thirteen days after her feeding tube was removed.

How can people ensure that their wishes will be considered in the event that they become incapacitated?

People can declare their preferences about medical treatments in a **living will**, sometimes called a **medical directive**. Living wills can include detailed instructions about life-sustaining procedures that a person does or does not want. Advance directives are incorporated into the medical record and updated when appropriate. They take effect only if a physician determines that a patient lacks the ability to understand and make decisions about available treatments.

Another important directive is a **durable power of attorney** (sometimes called a **health care proxy**) in which another person (a **health care agent**) is appointed to act as decision-maker in the event of incapacitation. The agent should understand one's medical preferences and be absolutely trustworthy. Only one person can be designated, although one or two alternates may also be listed. If an agent is given comprehensive power to supervise care, he or she may make decisions about medical staff, health care facilities, and medical procedures.

Laws regarding advance directives vary from state to state. In some states, nutrition and hydration are not considered life-sustaining treatments, and a person's instructions about them may need to be indicated separately. Some states restrict the use of advance directives to terminal illness or disallow them if a woman is pregnant. Generally, advance directives created in one state are honored in another. If no advance directive is available and a person's preferences are unknown, decisions are based on a patient's best interests as determined by an immediate family member.[6]

How does a "do-not-resuscitate" order differ from other advance directives?

A **do-not-resuscitate (DNR) order** is frequently used to withhold CPR in the event of cardiopulmonary arrest, which occurs too suddenly for deliberate decision-making.[7] A DNR order is written in the medical record as are other directives, but it does not exclude the use of other life-prolonging measures. A DNR order is most often used in patients with serious illnesses or advanced age. Some institutions allow a physician to write a DNR order for a patient who has a poor prognosis, but the physician must inform the patient or surrogate if this is done.

Have advance directives changed the way that medical care is provided?

Not really. Only about 20 percent of people in the United States have completed an advance directive.[8] Furthermore, advance directives are often unavailable when intensive care decisions are made: one study found that only 57.5 percent of patient charts indicating the existence of an advance directive actually contained a copy.[9] In addition, physicians often make treatment decisions without first discussing them with patients or caregivers.[10] In many cases, life-sustaining treatments are begun without prior knowledge of patients or their decision-makers, or treatments continue even if patients want them stopped.

Advance directives are sometimes too general or vague to guide treatment decisions. Patients who are fully aware of treatment options and clearly state their preferences are more likely to be successful at obtaining the care they desire.

Notes

1. M. A. Grippi, Ethics in critical care, in A. P. Fishman and coeditors, *Fishman's Manual of Pulmonary Diseases and Disorders* (New York: McGraw-Hill, 2002), pp. 1111–1114.

2. E. J. Emanuel, Bioethics in the practice of medicine, in L. Goldman and D. Ausiello, eds., *Cecil Textbook of Medicine* (Philadelphia: Saunders, 2004), pp. 5–9.

3. Emanuel, 2004.

4. J. O. Maillet, R. L. Potter, and L. Heller, Position of the American Dietetic Association: Ethical and legal issues in nutrition, hydration, and feeding, *Journal of the American Dietetic Association* 102 (2002): 716–726.

5. R. Cranford, Facts, lies, and videotapes: The permanent vegetative state and the sad case of Terri Schiavo, *Journal of Law, Medicine, and Ethics* 33 (2005): 363–372.

6. American College of Physicians, Ethics manual, *Annals of Internal Medicine* 128 (1998): 576–594.

7. American College of Physicians, 1998.

8. Emanuel, 2004.

9. Institute of Medicine, *Approaching Death: Improving Care at the End of Life* (Washington, DC: National Academy Press, 1997), pp. 202–203.

10. Emanuel, 2004.

A

Table of Food Composition

This edition of the table of food composition includes a wide variety of foods. It is updated with each edition to reflect current nutrient data for foods, to remove outdated foods, and to add foods that are new to the marketplace.[a] The nutrient database for this appendix is compiled from a variety of sources, including the USDA Standard Release database and manufacturers' data. The USDA database provides data for a wider variety of foods and nutrients than other sources. Because laboratory analysis for each nutrient can be quite costly, manufacturers tend to provide data only for those nutrients mandated on food labels. Consequently, data for their foods are often incomplete; any missing information on this table is designated as a dash. Keep in mind that a dash means only that the information is unknown and should not be interpreted as a zero. A zero means that the nutrient is not present in the food.

Whenever using nutrient data, remember that many factors influence the nutrient contents of foods. These factors include the mineral content of the soil, the diet fed to the animal or the fertilizer used on the plant, the season of harvest, the method of processing, the length and method of storage, the method of cooking, the method of analysis, and the moisture content of the sample analyzed. With so many influencing factors, users should view nutrient data as a close approximation of the actual amount.

For updates, corrections, and a list of more than 8,000 foods and codes found in the diet analysis software that accompanies this text, visit www.thomsonedu.com/nutrition and click on Diet Analysis Plus.

- **Fats** Total fats, as well as the breakdown of total fats to saturated, monounsaturated, polyunsaturated, and trans fats, are listed in the table. The fatty acids seldom add up to the total in part due to rounding but also because values are derived from a variety of laboratories.

- **Trans Fats** Trans fat data has been listed in the table. Because food manufacturers have only been required to report trans fats on food labels since January 2006, much of the data is incomplete. Missing trans fat data is designated with a dash. As additional trans fat data becomes available, the table will be updated.

- **Vitamin A and Vitamin E** In keeping with the 2001 RDA for vitamin A, this appendix presents data for vitamin A in micrograms (µg) RAE. Similarly, because the 2000 RDA for vitamin E is based only on the alpha-tocopherol form of vitamin E, this appendix reports vitamin E data in milligrams (mg) alpha-tocopherol, listed on the table as Vit E (mg α).

- **Bioavailability** Keep in mind that the availability of nutrients from foods depends not only on the quantity provided by a food, but also on the amount absorbed and used by the body—the bioavailability. The bioavailability of folate from fortified foods, for example, is greater than from naturally occurring sources. Similarly, the body can make niacin from the amino acid tryptophan, but niacin values in this table (and most databases) report preformed niacin only. Chapter 8 provides conversion factors and additional details.

- **Using the Table** The foods and beverages in this table are organized into several categories, which are listed at the head of each right-hand page. Page numbers are provided, and each group is color-coded to make it easier to find individual foods.

- **Caffeine Sources** Caffeine occurs in several plants, including the familiar coffee bean, the tea leaf, and the cocoa bean from which chocolate is made. Most human societies use caffeine regularly, most often in beverages, for its stimulant effect and flavor. Caffeine contents of beverages vary depending on the plants they are made from, the climates and soils where the plants are grown, the grind or cut size, the method and duration of brewing, and the amounts served. The accompanying table shows that, in general, a cup of coffee contains the most caffeine; a cup of tea, less than half as much; and cocoa or chocolate, less still. As for cola beverages, they are made from kola nuts, which contain caffeine, but most of their caffeine is added, using the

---

[a] This food composition table has been prepared by Wadsworth Publishing Company. The nutritional data are supplied by Axxya Systems.

purified compound obtained from decaffeinated coffee beans. The FDA lists caffeine as a multipurpose GRAS substance that may be added to foods and beverages. Drug manufacturers use caffeine in many products.

Reminder: A GRAS substance is one is "generally recognized as safe."

Caffeine Content of Selected Beverages, Foods, and Medications

Beverages and Foods	Serving Size	Average (mg)
Coffee		
Brewed	8 oz	95
Decaffeinated	8 oz	2
Instant	8 oz	64
Tea		
Brewed, green	8 oz	30
Brewed, herbal	8 oz	0
Brewed, leaf or bag	8 oz	47
Instant	8 oz	26
Lipton Brisk iced tea	12 oz	7
Nestea Cool iced tea	12 oz	12
Snapple iced tea (all flavors)	16 oz	42
Soft drinks		
A & W Creme Soda	12 oz	29
Barq's Root Beer	12 oz	18
Coca-Cola	12 oz	30
Dr. Pepper, Mr. Pibb, Sunkist Orange	12 oz	36
A&W Root Beer, club soda, Fresca, ginger ale, 7-Up, Sierra Mist, Sprite, Squirt, tonic water, caffeine-free soft drinks	12 oz	0
Mello Yello	12 oz	51
Mountain Dew	12 oz	45
Pepsi	12 oz	32
Energy drinks		
Amp	8.4 oz	70
Aqua Blast	.5 L	90
Aqua Java	.5 L	55
E Maxx	8.4 oz	74
Java Water	.5 L	125
KMX	8.4 oz	33
Krank	.5 L	100
Red Bull	8.3 oz	67
Red Devil	8.4 oz	42
Sobe Adrenaline Rush	8.3 oz	77
Sobe No Fear	16 oz	141
Water Joe	.5 L	65

Beverages and Foods	Serving Size	Average (mg)
Other beverages		
Chocolate milk or hot cocoa	8 oz	5
Starbucks Frappuccino Mocha	9.5 oz	72
Starbucks Frappuccino Vanilla	9.5 oz	64
Yoohoo chocolate drink	9 oz	3
Candies		
Baker's chocolate	1 oz	26
Dark chocolate covered coffee beans	1 oz	235
Dark chocolate, semisweet	1 oz	18
Milk chocolate	1 oz	6
Milk chocolate covered coffee beans	1 oz	224
White chocolate	1 oz	0
Foods		
Frozen yogurt, Ben & Jerry's coffee fudge	1 cup	85
Frozen yogurt, Häagen-Dazs coffee	1 cup	40
Ice cream, Starbucks coffee	1 cup	50
Ice cream, Starbucks Frappuccino bar	1 bar	15
Yogurt, Dannon coffee flavored	1 cup	45

Drugs[a]	Serving Size	Average (mg)
Cold remedies		
Coryban-D, Dristan	1 tablet	30
Diuretics		
Aqua-Ban	1 tablet	100
Pre-Mens Forte	1 tablet	100
Pain relievers		
Anacin, BC Fast Pain Reliever	1 tablet	32
Excedrin, Midol, Midol Max Strength	1 tablet	65
Stimulants		
Awake, NoDoz	1 tablet	100
Awake Maximum Strength, Caffedrine, NoDoz Maximum Strength, Stay Awake, Vivarin	1 tablet	200
Weight-control aids		
Dexatrim	1 tablet	200

[a]A pharmacologically active dose of caffeine is defined as 200 milligrams.

NOTE: The FDA su ggests a maximum of 65 milligrams per 12-ounce cola beverage but does not regulate the caffeine contents of other beverages. Because products change, contact the manufacturer for an update on products you use regularly.

SOURCE: Adapted from USDA database Release 18 (**http://www.nal.usda.gov/fnic/foodcomp/Data/**), Caffeine content of foods and drugs, Center for Science and the Public Interest (**www.cspinet.org/new/cafchart.htm**), and R. R. McCusker, B. A. Goldberger, and E. J. Cone, Caffeine content of energy drinks, carbonated sodas, and other beverages, *Journal of Analytical Toxicology* 30 (2006): 112–114.

Table A–1
Food Composition

(DA+ code is for Wadsworth Diet Analysis program)
(For purposes of calculations, use "0" for t, <1, <.1, <.01, etc.)

DA + Code	Food Description	Quantity	Measure	Wt (g)	H2O (g)	Ener (kcal)	Prot (g)	Carb (g)	Fiber (g)	Fat (g)	Sat	Mono	Poly	Trans
	BREADS, BAKED GOODS, CAKES, COOKIES, CRACKERS, CHIPS, PIES													
	Bagels													
8534	Cinnamon & raisin	1	item(s)	71	23	195	7	39	2	1	0.19	0.12	0.48	—
4910	Enriched, all varieties	1	item(s)	71	23	195	7	38	2	1	0.16	0.09	0.49	0
4911	Plain, enriched, toasted	1	item(s)	66	18	195	7	38	2	1	0.16	0.09	0.49	0
8538	Oat bran	1	item(s)	71	23	181	8	38	3	1	0.14	0.18	0.35	—
12079	Whole grain	1	item(s)	85	—	170	9	35	6	2.5	0	—	—	0
	Biscuits													
25008	Biscuits	1	item(s)	41	16	121	3	16	1	5	1.40	1.41	1.82	0
16729	Scone	1	item(s)	42	11	149	4	19	1	6	2.01	2.55	1.26	—
25166	Wheat biscuits	1	item(s)	55	21	162	4	22	1	7	1.90	1.92	2.51	0
	Bread													
325	Boston brown, canned	1	slice(s)	45	21	88	2	19	2	1	0.13	0.09	0.25	—
8716	Bread sticks, plain	4	item(s)	24	1	99	3	16	1	2	0.34	0.86	0.87	—
25176	Cornbread	1	piece(s)	55	26	141	5	18	1	5	2.09	1.44	1.50	0
327	Cracked wheat	1	slice(s)	25	9	65	2	12	1	1	0.23	0.48	0.17	—
9079	Croutons, plain	¼	cup(s)	8	<1	31	1	6	<1	<1	0.11	0.23	0.10	—
8582	Egg	1	slice(s)	40	14	115	4	19	1	2	0.64	0.92	0.44	—
8585	Egg, toasted	1	slice(s)	37	10	117	4	19	1	2	0.60	1.11	0.43	—
329	French	1	slice(s)	25	9	69	2	13	1	1	0.16	0.30	0.17	—
8591	French, toasted	1	slice(s)	23	7	69	2	13	1	1	0.16	0.30	0.17	—
8597	Indian fry	1	item(s)	90	24	296	6	48	2	9	2.08	3.59	2.33	—
332	Italian	1	slice(s)	30	11	81	3	15	1	1	0.26	0.24	0.42	—
1393	Mixed grain	1	slice(s)	26	10	65	3	12	2	1	0.21	0.40	0.24	—
8604	Mixed grain, toasted	1	slice(s)	24	8	65	3	12	2	1	0.21	0.40	0.24	—
8605	Oat bran	1	slice(s)	30	13	71	3	12	1	1	0.21	0.48	0.51	—
8608	Oat bran, toasted	1	slice(s)	27	10	70	3	12	1	1	0.21	0.47	0.50	—
8609	Oatmeal	1	slice(s)	27	10	73	2	13	1	1	0.19	0.43	0.46	—
8613	Oatmeal, toasted	1	slice(s)	25	8	73	2	13	1	1	0.19	0.43	0.46	—
1409	Pita	1	item(s)	60	19	165	5	33	1	1	0.10	0.06	0.32	—
7905	Pita, whole wheat	1	item(s)	64	20	170	6	35	5	2	0.26	0.22	0.68	—
338	Pumpernickel	1	slice(s)	32	12	80	3	15	2	1	0.14	0.30	0.40	—
334	Raisin, enriched	1	slice(s)	26	9	71	2	14	1	1	0.28	0.60	0.18	—
8625	Raisin, toasted	1	slice(s)	24	7	71	2	14	1	1	0.28	0.60	0.18	—
10168	Rice, white	1	slice(s)	42	—	140	1	21	1	6	0.50	—	—	0
8653	Rye	1	slice(s)	32	12	83	3	15	2	1	0.20	0.42	0.26	—
8654	Rye, toasted	1	slice(s)	29	9	82	3	15	2	1	0.20	0.42	0.25	—
336	Rye, light	1	slice(s)	25	9	65	2	12	2	1	0.20	0.30	0.30	—
8588	Sourdough	1	slice(s)	25	9	69	2	13	1	1	0.16	0.30	0.17	—
8592	Sourdough, toasted	1	slice(s)	23	7	69	2	13	1	1	0.16	0.30	0.17	—
491	Submarine or hoagie roll	1	item(s)	135	41	400	11	72	4	8	1.80	3.00	2.20	—
8596	Vienna, toasted	1	slice(s)	23	7	69	2	13	1	1	0.16	0.30	0.17	—
8670	Wheat	1	slice(s)	25	9	65	2	12	1	1	0.22	0.43	0.23	—
8671	Wheat, toasted	1	slice(s)	23	7	65	2	12	1	1	0.22	0.43	0.23	—
340	White	1	slice(s)	25	9	67	2	13	1	1	0.18	0.17	0.34	—
1395	Whole wheat	1	slice(s)	46	15	128	4	24	3	2	0.37	0.53	1.35	—
	Cakes													
386	Angel food, from mix	1	slice(s)	50	16	129	3	29	<1	<1	0.02	0.01	0.06	—
8772	Butter pound, ready to eat, commercially prepared	1	slice(s)	75	18	291	4	37	<1	15	8.67	4.43	0.80	—
8737	Carrot, cream cheese frosting, from mix	1	slice(s)	111	23	484	5	52	1	29	5.43	7.24	15.10	—
4931	Chocolate, chocolate icing, commercially prepared	1	slice(s)	64	15	235	3	35	2	10	3.05	5.61	1.18	—

Chol (mg)	Calc (mg)	Iron (mg)	Magn (mg)	Pota (mg)	Sodi (mg)	Zinc (mg)	Vit A (µg)	Thia (mg)	Vit E (mg α)	Ribo (mg)	Niac (mg)	Vit B_6 (mg)	Fola (µg)	Vit C (mg)	Vit B_{12} (µg)	Sele (µg)
0	13	2.70	20	105	229	0.80	15	0.27	0.22	0.20	2.19	0.04	79	<1	0	22
0	53	2.53	21	72	379	0.62	0	0.38	0.07	0.22	3.24	0.04	75	0	0	23
0	53	2.52	20	72	379	0.62	0	0.31	0.08	0.20	2.91	0.03	64	0	0	23
0	9	2.19	22	82	360	0.64	1	0.24	0.23	0.24	2.10	0.03	70	<1	0	24
0	200	1.08	120	0	200	4.5	0	0.44	—	0.5	8	0.6	—	0	1.79	0
<1	33	1.01	6	37	205	0.27	9	0.13	0.01	0.12	1.08	0.01	26	0	<.1	7
49	80	1.31	7	48	288	0.29	—	0.15	0.43	0.16	1.20	0.03	8	<.1	<1	—
<1	57	1.22	16	81	321	0.42	12	0.16	0.01	0.13	1.49	0.03	29	<.1	<.1	12
<1	32	0.95	28	143	284	0.23	11	0.01	0.14	0.05	0.50	0.04	5	0	<.1	10
0	5	1.03	8	30	158	0.21	0	0.14	0.24	0.13	1.27	0.02	39	0	0	9
21	88	1.01	10	59	209	0.57	38	0.13	0.33	0.16	0.98	0.04	36	2	<1	6
0	11	0.70	13	44	135	0.31	0	0.09	—	0.06	0.92	0.08	15	0	<.1	6
0	6	0.31	2	9	52	0.07	0	0.05	—	0.02	0.41	0.00	10	0	0	3
20	37	1.22	8	46	197	0.32	25	0.18	0.10	0.17	1.94	0.03	42	0	<.1	12
21	38	1.24	8	47	200	0.32	26	0.14	0.11	0.16	1.77	0.02	36	0	<.1	12
0	19	0.63	7	28	152	0.22	0	0.13	0.08	0.08	1.19	0.01	37	0	0	8
0	19	0.63	7	28	152	0.22	0	0.10	0.07	0.07	1.07	0.01	22	0	0	8
0	210	3.24	14	67	626	0.45	0	0.39	—	0.27	3.27	0.02	67	0	0	21
0	23	0.88	8	33	175	0.26	0	0.14	0.09	0.09	1.31	0.01	57	0	0	8
0	24	0.90	14	53	127	0.33	0	0.11	0.09	0.09	1.13	0.09	31	<.1	<.1	8
0	24	0.90	14	53	127	0.33	0	0.08	0.08	0.08	1.02	0.08	28	<.1	<.1	8
0	20	0.94	11	44	122	0.27	1	0.15	0.13	0.10	1.45	0.02	24	0	0	9
0	19	0.93	9	33	121	0.28	1	0.12	0.13	0.09	1.29	0.01	19	0	0	9
0	18	0.73	10	38	162	0.28	1	0.11	0.13	0.06	0.85	0.02	17	0	<.1	7
0	18	0.74	10	39	163	0.28	1	0.09	0.13	0.06	0.77	0.02	13	<.1	<.1	7
0	52	1.57	16	72	322	0.50	0	0.36	0.18	0.20	2.78	0.02	64	0	0	16
0	10	1.96	44	109	340	0.97	0	0.22	0.39	0.05	1.82	0.17	22	0	0	28
0	22	0.92	17	67	215	0.47	0	0.10	0.13	0.10	0.99	0.04	30	0	0	8
0	17	0.75	7	59	101	0.19	0	0.09	0.07	0.10	0.90	0.02	28	<.1	0	5
0	17	0.76	7	59	102	0.19	0	0.07	0.07	0.09	0.81	0.02	24	<.1	0	5
0	40	1.08	—	45	160	—	0	0.23	—	0.14	1.20	—	40	0	—	—
0	23	0.91	13	53	211	0.36	0	0.14	0.11	0.11	1.22	0.02	35	<1	0	10
0	23	0.90	12	53	210	0.36	0	0.11	0.11	0.10	1.09	0.02	30	<.1	0	10
0	20	0.70	4	51	175	0.18	0	0.10	—	0.08	0.80	0.01	5	0	<.1	8
0	19	0.63	7	28	152	0.22	0	0.13	0.08	0.08	1.19	0.01	37	0	0	8
0	19	0.63	7	28	152	0.22	0	0.10	0.07	0.07	1.07	0.01	22	0	0	8
0	100	3.80	—	128	683	—	0	0.54	—	0.33	4.50	0.05	—	0	—	42
0	19	0.63	7	28	152	0.22	0	0.10	0.07	0.07	1.07	0.01	22	0	0	8
0	26	0.83	12	50	133	0.26	0	0.10	0.07	0.07	1.03	0.02	23	0	0	8
0	26	0.83	12	50	132	0.26	0	0.08	0.07	0.06	0.93	0.02	19	0	0	8
0	38	0.94	6	25	170	0.19	0	0.11	0.05	0.08	1.10	0.02	28	0	0	4
0	15	1.43	37	144	159	0.69	0	0.14	0.35	0.10	1.83	0.09	30	0	0	18
0	42	0.12	4	68	255	0.07	0	0.05	0.00	0.10	0.09	0.00	10	0	<.1	8
166	26	1.04	8	89	299	0.35	112	0.10	—	0.98		0.03	31	0	<1	7
60	28	1.39	20	124	273	0.54	—	0.15	—	0.17	1.13	0.08	13	1	<1	—
27	28	1.41	22	128	214	0.44	—	0.02	—	0.09	0.37	0.03	11	<.1	<.1	2

Table A–1
Food Composition

(DA+ code is for Wadsworth Diet Analysis program)
(For purposes of calculations, use "0" for t, <1, <.1, <.01, etc.)

DA + Code	Food Description	Quantity	Measure	Wt (g)	H2O (g)	Ener (kcal)	Prot (g)	Carb (g)	Fiber (g)	Fat (g)	Sat	Mono	Poly	Trans
	BREADS, BAKED GOODS, CAKES, COOKIES, CRACKERS, CHIPS, PIES													
	Cakes–Continued													
8756	Chocolate, from mix	1	slice(s)	95	23	340	5	51	2	14	5.16	5.74	2.62	—
393	Devil's food cupcake, chocolate frosting	1	item(s)	35	8	120	2	20	1	4	1.80	1.60	0.60	—
8757	Fruitcake, ready to eat, commercially prepared	1	piece(s)	43	11	139	1	26	2	4	0.45	1.81	1.43	—
1397	Pineapple upside down, from mix	1	slice(s)	115	37	367	4	58	1	14	3.35	5.97	3.77	—
411	Sponge, from mix	1	slice(s)	63	19	187	5	36	<1	3	0.82	0.99	0.41	—
8817	White, coconut frosting, from mix	1	slice(s)	112	23	399	5	71	1	12	4.36	4.14	2.42	—
8819	Yellow, chocolate frosting, ready to eat, commercially prepared	1	slice(s)	64	14	243	2	35	1	11	2.98	6.14	1.35	—
8822	Yellow, vanilla frosting, ready to eat, commercially prepared	1	slice(s)	64	14	239	2	38	<1	9	1.52	3.91	3.30	—
	Snack cakes													
8791	Chocolate snack cake, creme filled, w/frosting	1	item(s)	50	10	188	2	30	<1	7	1.43	2.85	2.62	—
25010	Cinnamon coffee cake	1	piece(s)	72	23	231	4	36	1	8	2.19	2.65	2.99	0
16777	Funnel cake	1	item(s)	90	37	278	7	29	1	14	2.77	4.46	6.33	—
8794	Sponge snack cake, creme filled	1	item(s)	43	9	155	1	27	<1	5	1.09	1.73	1.40	—
	Snacks, chips, pretzels													
29428	Bagel chips, plain	3	item(s)	29	—	130	3	19	1	5	0.50	—	—	—
29429	Bagel chips, toasted onion	3	item(s)	29	—	130	4	20	1	5	0.50	—	—	—
38192	Chex traditional snack mix	1	cup(s)	46	—	198	3	33	2	6	0.76	—	—	—
654	Potato chips, salted	20	item(s)	28	1	152	2	15	1	10	3.11	2.79	3.46	—
8816	Potato chips, unsalted	20	item(s)	28	1	152	2	15	1	10	3.11	2.79	3.46	—
4641	Tortilla chips, plain	6	item(s)	28	1	142	2	18	2	7	1.43	4.39	1.03	—
5096	Pretzels, plain, hard, twists	5	item(s)	30	1	114	3	24	1	1	0.23	0.41	0.37	—
4632	Pretzels, whole wheat	1	ounce(s)	28	1	103	3	23	2	1	0.16	0.29	0.24	—
	Cookies													
8859	Animal crackers	12	piece(s)	30	0	134	2	22	<1	4	1.03	2.29	0.56	—
8876	Brownie, prepared from mix	1	item(s)	24	3	112	1	12	1	7	1.76	2.60	2.26	—
25207	Chocolate chip cookies	1	item(s)	30	4	140	2	16	1	8	2.09	3.26	2.09	0
8915	Chocolate sandwich cookie, extra creme filling	1	item(s)	13	<1	65	<1	9	<1	3	0.50	1.39	1.22	1.10
14145	Fig Newtons	1	item(s)	16	—	55	1	10	1	1	0.50	0.50	0.00	0.50
8920	Fortune cookie	1	item(s)	8	1	30	<1	7	<1	<1	0.05	0.11	0.04	—
25208	Oatmeal cookies	1	item(s)	69	12	234	6	45	3	4	0.70	1.28	1.85	0
25213	Peanut butter cookies	1	item(s)	35	4	163	4	17	1	9	1.65	4.72	2.43	0
33095	Sugar cookies	1	item(s)	16	4	61	1	7	<1	3	0.63	1.27	0.87	0
9002	Vanilla sandwich cookie, creme filling	1	item(s)	10	<1	48	<1	7	<1	2	0.30	0.84	0.76	—
	Crackers													
9008	Cheese crackers (mini)	30	item(s)	30	1	151	3	17	1	8	2.81	3.63	0.74	—
9010	Cheese crackers (mini), low salt	30	item(s)	30	1	151	3	17	1	8	2.82	2.70	1.44	—
9012	Cheese cracker sandwich w/peanut butter	4	item(s)	28	1	139	3	16	1	7	1.23	3.64	1.43	—
8928	Honey graham crackers	4	item(s)	28	1	118	2	22	1	3	0.43	1.14	1.07	—
9016	Matzo crackers, plain	1	item(s)	28	1	112	3	24	1	<1	0.06	0.04	0.17	—
9024	Melba toast	3	item(s)	15	1	59	2	11	1	<1	0.07	0.12	0.19	—
14189	Ritz crackers	5	item(s)	16	<1	80	1	10	1	4	0.50	1.50	0.00	—
9014	Rye crispbread crackers	1	item(s)	10	1	37	1	8	2	<1	0.01	0.02	0.06	—
9028	Rye melba toast	3	item(s)	15	1	58	2	12	1	1	0.07	0.14	0.20	—
9040	Rye wafer	1	item(s)	11	1	37	1	9	3	<.1	0.01	0.02	0.04	—
432	Saltine crackers	5	item(s)	15	1	65	1	11	<1	2	0.44	0.96	0.25	0.54
9046	Saltine crackers, low salt	5	item(s)	15	1	65	1	11	<1	2	0.44	0.96	0.25	

Chol (mg)	Calc (mg)	Iron (mg)	Magn (mg)	Pota (mg)	Sodi (mg)	Zinc (mg)	Vit A (µg)	Thia (mg)	Vit E (mg α)	Ribo (mg)	Niac (mg)	Vit B_6 (mg)	Fola (µg)	Vit C (mg)	Vit B_{12} (µg)	Sele (µg)
55	57	1.53	30	133	299	0.66	38	0.13	—	0.20	1.08	0.04	26	<1	<1	11
19	21	0.70	—	46	92	—	—	0.04	—	0.05	0.30	—	2	0	—	2
2	14	0.89	7	66	116	0.12	3	0.02	0.39	0.04	0.34	0.02	9	<1	<.1	1
25	138	1.70	15	129	367	0.36	71	0.18	—	0.18	1.37	0.04	30	1	<.1	11
107	26	1.00	6	89	144	0.37	49	0.10	—	0.19	0.76	0.04	25	0	<1	12
1	101	1.30	13	111	318	0.37	13	0.14	0.13	0.21	1.19	0.03	35	<1	<.1	12
35	24	1.33	19	114	216	0.40	21	0.08	—	0.10	0.80	0.02	14	0	<1	2
35	40	0.68	4	34	220	0.16	12	0.06	—	0.04	0.32	0.02	17	0	<.1	4
9	37	1.68	21	61	213	0.26	3	0.11	1.09	0.15	1.21	0.01	20	0	<.1	1
26	50	1.46	10	81	277	0.38	35	0.14	0.23	0.16	1.17	0.02	30	<.1	<1	10
63	128	1.86	18	154	273	0.64	—	0.24	1.55	0.32	1.86	0.05	14	<1	<1	—
7	19	0.55	3	37	155	0.12	2	0.07	0.50	0.06	0.52	0.01	17	<.1	<.1	1
0	0	0.72	—	45	70	—	0	—	—	—	—	—	—	0	0	—
0	0	0.72	—	50	300	—	0	—	—	—	—	—	—	0	0	—
0	0	0.55	0	76	623	0.00	0	0.09	—	0.05	1.22	0.00	12	0	—	—
0	7	0.46	19	362	169	0.31	0	0.05	1.91	0.06	1.09	0.19	13	9	0	2
0	7	0.46	19	362	2	0.31	0	0.05	2.59	0.06	1.09	0.19	13	9	0	2
0	44	0.43	25	56	150	0.43	1	0.02	1	0.05	0.36	0.08	3	0	0	2
0	11	1.30	11	44	515	0.26	0	0.14	—	0.19	1.58	0.03	51	0	0	2
0	8	0.76	9	122	58	0.18	0	0.12	—	0.08	1.86	0.08	15	<1	0	—
0	13	0.82	5	30	1118	0.19	—	0.10	0.04	0.09	1.04	0.00	50	0	<.1	—
18	14	0.44	13	42	82	0.23	42	0.03	—	0.05	0.24	0.02	7	<.1	<.1	3
13	11	0.70	12	62	109	0.24	27	0.07	0.54	0.06	0.82	0.02	16	<.1	<.1	4
0	3	0.37	4	16	64	0.08	0	0.01	0.25	0.02	0.20	0.00	6	0	<.1	<1
0	5	0.36	—	40	60	—	4	0.03	—	0.04	0.22	—	—	<1	—	—
<1	1	0.12	1	3	22	0.01	<.1	0.01	0.00	0.01	0.15	0.00	5	0	<.1	<1
<.1	26	1.94	49	177	311	1.43	48	0.23	0.23	0.12	1.24	0.09	30	<1	<.1	17
13	28	0.67	22	104	157	0.46	51	0.08	0.74	0.09	1.81	0.05	21	<.1	<.1	5
18	5	0.32	2	13	50	0.08	31	0.04	0.28	0.04	0.28	0.01	8	<.1	<.1	3
0	3	0.22	1	9	35	0.04	0	0.03	0.16	0.02	0.27	0.00	5	0	0	<1
4	45	1.43	11	44	299	0.34	9	0.17	0.66	0.13	1.40	0.17	46	0	<1	3
4	45	1.44	11	32	137	0.33	—	0.18	—	0.12	1.41	0.18	8	0	<1	—
0	14	0.76	16	61	199	0.29	0	0.15	0.16	0.08	1.63	0.04	26	0	<.1	2
0	7	1.04	8	38	169	0.23	0	0.06	0.09	0.09	1.15	0.02	13	0	0	3
0	4	0.90	7	32	1	0.19	0	0.11	0.02	0.08	1.11	0.03	5	0	0	10
0	14	0.56	9	30	124	0.30	0	0.06	0.06	0.04	0.62	0.01	19	0	0	5
0	20	0.72	3	10	135	0.23	—	0.07	—	0.04	0.45	0.01	10	1	0	—
0	3	0.24	8	32	26	0.24	0	0.02	0.08	0.01	0.10	0.02	5	0	0	4
0	12	0.55	6	29	135	0.20	0	0.07	—	0.04	0.71	0.01	13	0	0	6
0	4	0.65	13	54	87	0.31	0	0.05	0.09	0.03	0.17	0.03	5	<.1	0	3
0	18	0.81	4	19	195	0.12	0	0.08	0.15	0.07	0.79	0.01	19	0	0	2
0	18	0.81	4	109	95	0.12	0	0.08	0.02	0.07	0.79	0.01	19	0	0	3

Table A–1
Food Composition

(DA+ code is for Wadsworth Diet Analysis program)
(For purposes of calculations, use "0" for t, <1, <.1, <.01, etc.)

DA + Code	Food Description	Quantity	Measure	Wt (g)	H₂O (g)	Ener (kcal)	Prot (g)	Carb (g)	Fiber (g)	Fat (g)	Sat	Mono	Poly	Trans
	BREADS, BAKED GOODS, CAKES, COOKIES, CRACKERS, CHIPS, PIES													
	Crackers–Continued													
9048	Snack crackers, round	10	item(s)	30	1	151	2	18	<1	8	1.13	3.19	2.86	—
9050	Snack crackers, round, low salt	10	item(s)	30	1	151	2	18	<1	8	1.13	3.19	2.86	—
9052	Snack cracker sandwich, cheese filling	4	item(s)	28	1	134	3	17	1	6	1.72	3.15	0.72	—
9054	Snack cracker sandwich, peanut butter filling	4	item(s)	28	1	138	3	16	1	7	1.38	3.86	1.30	—
9044	Soda crackers	5	item(s)	15	1	65	1	11	<1	2	0.44	0.96	0.25	0.54
9055	Wheat crackers	10	item(s)	30	1	142	3	19	1	6	1.55	3.43	0.84	—
9057	Wheat crackers, low salt	10	item(s)	30	1	142	3	19	1	6	1.55	3.43	0.84	—
9059	Wheat cracker sandwich, cheese filling	4	item(s)	28	1	139	3	16	1	7	1.16	2.90	2.57	—
9061	Wheat cracker sandwich, peanut butter filling	4	item(s)	28	1	139	4	15	1	7	1.29	3.29	2.48	—
9022	Whole wheat crackers	7	item(s)	28	1	124	2	19	3	5	0.95	1.65	1.85	—
	Pastry													
16754	Apple fritter	1	item(s)	17	6	62	1	6	<1	4	0.87	1.69	1.13	—
5118	Cinnamon sweet roll w/icing, from refrigerator dough	1	item(s)	30	7	109	2	17	1	4	1.00	2.23	0.52	—
4945	Croissant, butter	1	item(s)	57	13	231	5	26	1	12	6.59	3.15	0.62	—
9096	Danish pastry, nut	1	item(s)	65	13	280	5	30	1	16	3.78	8.90	2.78	—
4947	Doughnut, cake	1	item(s)	47	10	198	2	23	1	11	1.70	4.37	3.70	—
9105	Doughnut, cake, chocolate glazed	1	item(s)	42	7	175	2	24	1	8	2.16	4.74	1.04	—
9115	Doughnut, creme filling	1	item(s)	85	32	307	5	26	1	21	4.62	10.27	2.62	—
437	Doughnut, glazed	1	item(s)	60	15	242	4	27	1	14	3.49	7.72	1.74	—
9117	Doughnut, jelly filling	1	item(s)	85	30	289	5	33	1	16	4.12	8.69	2.02	—
10617	Toaster pastry, brown sugar cinnamon	1	item(s)	50	5	210	3	35	1	6	1.00	4.00	1.00	—
30928	Toaster pastry, cream cheese	1	item(s)	54	—	200	3	23	1	11	3.50	—	—	—
	Muffins													
25015	Blueberry	1	item(s)	63	30	160	3	23	1	6	0.87	1.48	3.25	0
4997	Bran, from mix	1	item(s)	50	18	138	3	23	2	5	1.18	2.34	0.72	—
9189	Corn, ready to eat	1	item(s)	57	19	174	3	29	2	5	0.77	1.20	1.83	—
9121	English muffin, plain, enriched	1	item(s)	57	24	134	4	26	2	1	0.15	0.17	0.51	—
29582	English, toasted	1	item(s)	50	19	128	4	25	1	1	0.14	0.16	0.48	—
9145	English, wheat	1	item(s)	57	24	127	5	26	3	1	0.16	0.16	0.48	—
	Granola bars													
38161	Kudos milk chocolate w/fruit & nuts	1	item(s)	28	—	90	2	15	1	3	1.00	—	—	—
38196	Nature Valley banana nut crunchy	1	item(s)	21	—	95	2	14	1	4	0.50	—	—	—
38187	Nature Valley fruit n nut trail mix	1	item(s)	35	—	140	3	25	2	4	0.50	—	—	—
1383	Plain, hard	1	item(s)	25	1	115	2	16	1	5	0.58	1.07	2.95	—
4606	Plain, soft	1	item(s)	28	2	126	2	19	1	5	2.06	1.08	1.51	—
	Pies													
454	Apple pie, from home recipe	1	slice(s)	155	73	411	4	58	2	19	4.73	8.36	5.17	—
470	Pecan pie, from home recipe	1	slice(s)	122	24	503	6	64	0	27	4.87	13.64	6.97	—
472	Pumpkin pie, from home recipe	1	slice(s)	155	91	316	7	41	0	14	4.92	5.73	2.81	—
9007	Pie crust, frozen, ready to bake, enriched, baked	1	slice(s)	16	2	82	1	8	<1	5	1.69	2.51	0.65	—
5052	Pie crust, prepared w/water, baked	1	slice(s)	20	2	100	1	10	<1	6	1.54	3.46	0.77	—
	Rolls													
8555	Crescent dinner roll	1	item(s)	28	10	80	2	14	1	1	0.34	0.70	0.25	—
489	Hamburger roll or bun, plain	1	item(s)	43	15	120	4	21	1	2	0.47	0.48	0.85	—
490	Hard roll	1	item(s)	57	18	167	6	30	1	2	0.35	0.65	0.98	—

Chol (mg)	Calc (mg)	Iron (mg)	Magn (mg)	Pota (mg)	Sodi (mg)	Zinc (mg)	Vit A (µg)	Thia (mg)	Vit E (mg α)	Ribo (mg)	Niac (mg)	Vit B$_6$ (mg)	Fola (µg)	Vit C (mg)	Vit B$_{12}$ (µg)	Sele (µg)
0	36	1.08	8	40	254	0.20	0	0.12	0.61	0.10	1.21	0.02	27	0	0	2
0	36	1.08	8	107	112	0.20	0	0.12	0.61	0.10	1.21	0.02	27	0	0	2
1	72	0.67	10	120	392	0.17	5	0.12	0.06	0.19	1.05	0.01	28	<.1	<.1	6
0	23	0.78	15	60	201	0.32	0	0.14	0.58	0.08	1.71	0.04	24	0	<.1	3
0	18	0.81	4	19	195	0.12	0	0.08	0.15	0.07	0.79	0.01	19	0	0	2
0	15	1.32	19	55	239	0.48	0	0.15	0.15	0.10	1.49	0.04	35	0	0	2
0	15	1.32	19	61	85	0.48	0	0.15	0.15	0.10	1.49	0.04	15	0	0	10
2	57	0.73	15	86	256	0.24	5	0.10	—	0.12	0.89	0.07	18	<1	<.1	7
0	48	0.75	11	83	226	0.23	0	0.11	—	0.08	1.65	0.04	20	0	0	6
0	14	0.86	28	83	185	0.60	0	0.06	0.24	0.03	1.27	0.05	8	0	0	4
14	9	0.25	2	24	7	0.09	—	0.03	0.07	0.04	0.23	0.01	2	<1	<.1	—
0	10	0.80	4	19	250	0.10	—	0.12	—	0.07	1.09	0.01	14	<.1	<.1	—
38	21	1.16	9	67	424	0.43	101	0.22	—	0.14	1.25	0.03	35	<1	<.1	13
30	61	1.17	21	62	236	0.57	6	0.14	0.53	0.16	1.50	0.07	54	1	<1	9
17	21	0.92	9	60	257	0.26	—	0.10	—	0.11	0.87	0.03	22	<.1	<1	0
24	89	0.95	14	45	143	0.24	5	0.02	0.09	0.03	0.20	0.01	19	<.1	<.1	2
20	21	1.56	17	68	263	0.68	9	0.29	0.25	0.13	1.91	0.06	60	0	<1	9
4	26	0.36	13	65	205	0.46	2	0.53	—	0.04	0.39	0.03	13	<.1	<.1	5
22	21	1.50	17	67	249	0.64	14	0.27	0.37	0.12	1.82	0.09	58	0	<1	11
0	0	1.80	—	70	190	—	—	0.15	—	0.17	2.00	0.20	40	0	0	—
15	0	1.08	—	—	230	—	—	—	—	—	—	—	—	0	—	—
20	50	1.15	7	56	288	0.39	20	0.14	0.76	0.15	1.14	0.03	29	<1	<1	9
34	16	1.27	29	74	234	0.57	—	0.10	—	0.12	1.44	0.09	33	0	<.1	—
15	42	1.60	18	39	297	0.31	30	0.16	0.46	0.19	1.16	0.05	46	0	<.1	9
0	30	1.43	12	75	264	0.40	0	0.25	—	0.16	2.21	0.02	42	0	<.1	—
0	95	1.36	11	72	252	0.38	0	0.19	0.17	0.14	1.90	0.02	15	<.1	<.1	—
0	101	1.64	21	106	218	0.61	0	0.25	0.26	0.17	1.91	0.05	36	0	0	17
0	200	0.36	—	—	60	—	0	—	—	—	—	—	—	0	0	—
0	10	0.54	—	60	80	—	0	—	—	—	—	—	—	0	—	—
0	0	0.00	—	—	95	—	0	—	—	—	—	—	—	0	—	—
0	15	0.72	24	82	72	0.50	2	0.06	—	0.03	0.39	0.02	6	<1	0	4
<1	30	0.73	21	92	79	0.43	0	0.08	—	0.05	0.15	0.03	7	0	<1	5
0	11	1.74	11	122	327	0.29	17	0.23	—	0.17	1.91	0.05	37	3	0	12
106	39	1.81	32	162	320	1.24	100	0.23	—	0.22	1.03	0.07	32	<1	<1	15
65	146	1.97	29	288	349	0.71	660	0.14	—	0.31	1.21	0.07	33	3	<1	11
0	3	0.36	3	18	104	0.05	0	0.04	0.42	0.06	0.39	0.01	9	0	<.1	<1
0	12	0.43	3	12	146	0.08	0	0.06	—	0.04	0.47	0.01	20	0	0	—
0	39	0.89	6	39	157	0.17	0	0.14	0.02	0.09	1.10	0.01	—	0	<.1	—
0	59	1.43	9	40	206	0.28	0	0.17	0.03	0.14	1.79	0.03	48	0	<.1	8
0	54	1.87	15	62	310	0.54	0	0.27	0.24	0.19	2.42	0.02	54	0	0	22

Table A–1
Food Composition
(DA+ code is for Wadsworth Diet Analysis program)
(For purposes of calculations, use "0" for t, <1, <.1, <.01, etc.)

A

DA + Code	Food Description	Quantity	Measure	Wt (g)	H₂O (g)	Ener (kcal)	Prot (g)	Carb (g)	Fiber (g)	Fat (g)	Sat	Mono	Poly	Trans
												Fat Breakdown (g)		
BREADS, BAKED GOODS, CAKES, COOKIES, CRACKERS, CHIPS, PIES														
	Rolls–Continued													
5127	Kaiser roll	1	item(s)	57	18	167	6	30	1	2	0.35	0.65	0.98	—
5130	Whole wheat roll or bun	1	item(s)	28	9	76	2	15	2	1	0.24	0.34	0.62	—
	Sport bars													
37026	Balance original chocolate	1	item(s)	50	—	200	14	22	1	6	3.50	—	—	—
37024	Balance original peanut butter	1	item(s)	50	—	200	14	22	1	6	2.50	—	—	—
36580	Clif Bar chocolate brownie energy bar	1	item(s)	68	—	240	10	41	6	4	1.00	—	—	—
36583	Clif Bar crunchy peanut butter energy bar	1	item(s)	68	—	240	12	39	5	5	0.50	—	—	—
36584	Clif Luna tropical crisp energy bar	1	item(s)	48	—	180	10	24	2	5	3.50	0.00	0.00	—
12005	Powerbar apple cinnamon	1	item(s)	65	—	230	10	45	3	3	0.50	1.50	0.50	—
16078	Powerbar banana	1	item(s)	65	—	230	9	45	3	2	0.50	1.00	0.50	—
16080	Powerbar chocolate	1	item(s)	65	—	230	10	45	3	2	0.50	0.50	1.00	—
16079	Powerbar mocha	1	item(s)	65	—	230	10	45	3	3	1.00	1.00	0.50	—
	Tortillas													
1391	Corn tortillas, soft	1	item(s)	26	11	58	1	12	1	1	0.09	0.17	0.29	—
1669	Flour tortilla	1	item(s)	32	9	104	3	18	1	2	0.56	1.21	0.34	—
1390	Taco shells, hard	1	item(s)	13	1	62	1	8	1	3	0.43	1.19	1.13	—
	Pancakes, waffles													
8926	Pancakes, blueberry, from recipe	3	item(s)	114	61	253	7	33	1	10	2.26	2.64	4.74	—
5037	Pancakes, from mix w/egg & milk	3	item(s)	114	60	249	9	33	2	9	2.33	2.36	3.33	—
9219	Waffle, plain, frozen, toasted	2	item(s)	66	28	174	4	27	2	5	0.95	2.12	1.84	—
500	Waffle, plain, from recipe	1	item(s)	75	32	218	6	25	2	11	2.14	2.64	5.08	—
30311	Waffle, 100% whole grain	1	item(s)	75	32	201	7	25	2	8	2.35	3.38	2.06	—
CEREAL, FLOUR, GRAIN, PASTA, NOODLES, POPCORN														
	Grain													
2861	Amaranth, dry	½	cup(s)	98	10	365	14	65	15	6	1.62	1.40	2.82	—
1953	Barley, pearled, cooked	½	cup(s)	79	54	97	2	22	3	<1	0.07	0.04	0.17	—
1956	Buckwheat groats, cooked, roasted	½	cup(s)	84	64	77	3	17	2	1	0.11	0.16	0.16	—
1957	Bulgur, cooked	½	cup(s)	91	71	76	3	17	4	<1	0.04	0.03	0.09	—
1963	Couscous, cooked	½	cup(s)	79	57	88	3	18	1	<1	0.02	0.02	0.05	—
1967	Millet, cooked	½	cup(s)	120	86	143	4	28	2	1	0.21	0.22	0.61	—
1969	Oat bran, dry	½	cup(s)	47	3	116	8	31	7	3	0.62	1.12	1.30	—
1972	Quinoa, dry	½	cup(s)	85	8	318	11	59	5	5	0.50	1.30	1.99	—
	Rice													
129	Brown, long grain, cooked	½	cup(s)	98	71	108	3	22	2	1	0.18	0.32	0.31	—
2863	Brown, medium grain, cooked	½	cup(s)	98	71	109	2	23	2	<1	0.16	0.29	0.28	—
37488	Jasmine, saffroned, cooked	½	cup(s)	280	—	340	8	78	0	0	0.00	—	—	0
30280	Pilaf, cooked	½	cup(s)	103	74	129	2	22	1	3	0.67	1.61	0.95	—
28066	Spanish, cooked	½	cup(s)	120	3	25	2	1	<1	<1	0.33	0.07	18.31	0
2867	White glutinous, cooked	½	cup(s)	87	67	84	2	18	1	<1	0.03	0.06	0.06	—
482	White, instant long grain, enriched, boiled	½	cup(s)	83	63	81	2	18	<1	<1	0.04	0.04	0.04	—
484	White, long grain, boiled	½	cup(s)	79	54	103	2	22	<1	<1	0.06	0.07	0.06	—
486	White, long grain, enriched, parboiled, cooked	½	cup(s)	88	63	100	2	22	<1	<1	0.06	0.07	0.06	—
1194	Wild cooked	½	cup(s)	82	61	83	3	17	1.47	<1	0.04	0.04	0.17	—
	Flour & grain fractions													
505	All purpose flour, self rising, enriched	½	cup(s)	63	7	221	6	46	2	1	0.10	0.05	0.26	—
503	All purpose flour, white, bleached, enriched	½	cup(s)	63	7	228	6	48	2	1	0.10	0.05	0.26	—

Chol (mg)	Calc (mg)	Iron (mg)	Magn (mg)	Pota (mg)	Sodi (mg)	Zinc (mg)	Vit A (µg)	Thia (mg)	Vit E (mg α)	Ribo (mg)	Niac (mg)	Vit B_6 (mg)	Fola (µg)	Vit C (mg)	Vit B_{12} (µg)	Sele (µg)
0	54	1.87	15	62	310	0.54	0	0.27	—	0.19	2.42	0.02	54	0	0	22
0	30	0.69	24	78	136	0.57	0	0.07	—	0.04	1.05	0.06	9	0	0	14
3	100	4.50	40	160	180	3.75	—	0.38	—	0.43	5.00	0.50	100	60	2	18
3	100	4.50	40	130	230	3.75	—	0.38	—	0.43	5.00	0.50	100	60	2	18
0	250	5.40	120	260	150	3.75	—	0.38	—	0.26	4.00	0.40	80	60	1	18
0	250	5.40	120	300	290	3.75	—	0.38	—	0.34	6.00	0.40	100	60	1	14
0	350	6.30	140	120	135	5.25	—	1.50	—	1.70	20.00	2.00	400	60	6	25
0	300	6.30	140	110	90	5.25	0	1.50	—	1.70	20.00	2.00	400	60	6	—
0	300	6.30	140	200	90	5.25	0	1.50	—	1.70	20.00	2.00	400	60	6	—
0	300	6.30	140	150	90	5.25	0	1.50	—	1.70	20.00	2.00	400	60	6	—
0	300	6.30	140	150	90	5.25	0	1.50	—	1.70	20.00	2.00	400	60	6	—
0	46	0.36	17	40	42	0.24	0	0.03	0.07	0.02	0.39	0.06	26	0	0	1
0	40	1.06	8	42	153	0.23	0	0.17	0.06	0.09	1.14	0.02	33	0	0	7
0	21	0.33	14	24	49	0.19	0	0.03	0.22	0.01	0.18	0.04	17	0	0	2
64	235	1.96	18	157	470	0.62	57	0.22	—	0.31	1.74	0.06	41	3	<1	16
81	245	1.48	25	227	576	0.86	82	0.23	—	0.36	1.40	0.12	105	1	<1	—
16	153	2.95	15	84	519	0.38	253	0.25	0.65	0.31	2.93	0.59	36	0	2	11
52	191	1.73	14	119	383	0.50	49	0.19	—	0.26	1.55	0.04	51	<1	<1	35
71	196	1.56	30	173	374	0.85	—	0.15	0.32	0.25	1.47	0.09	14	<1	<1	—
0	149	7.40	259	357	20	3.10	0	0.08	—	0.20	1.25	0.22	48	4	0	—
0	9	1.04	17	73	2	0.64	0	0.07	0.01	0.05	1.62	0.09	13	0	0	7
0	6	0.67	43	74	3	0.51	0	0.03	0.08	0.03	0.79	0.06	12	0	0	2
0	9	0.87	29	62	5	0.52	0	0.05	0.01	0.03	0.91	0.08	16	0	0	1
0	6	0.30	6	46	4	0.20	0	0.05	0.10	0.02	0.77	0.04	12	0	0	22
0	4	0.76	53	74	2	1.09	0	0.13	0.02	0.10	1.60	0.13	23	0	0	1
0	27	2.54	110	266	2	1.46	0	0.55	0.47	0.10	0.44	0.08	24	0	0	21
0	51	7.86	179	629	18	2.81	0	0.17	—	0.34	2.49	0.19	42	0	0	—
0	10	0.41	42	42	5	0.61	0	0.09	0.03	0.02	1.49	0.14	4	0	0	10
0	9	0.51	43	77	1	0.6	0	0.09	—	0.01	1.29	0.14	3.9	0	0	38
0	—	2.16	—	—	780	—	—	—	—	—	—	—	—	—	—	—
0	13	1.16	9	55	403	0.38	—	0.13	0.28	0.02	1.24	0.06	4	<1	<.1	—
1	47	0.78	48	1	13	0.13	<1	0.03	0.06	0.19	8.71	0.14	<.1	7	<.1	9
0	2	0.12	4	9	4	0.36	0	0.02	0.03	0.01	0.25	0.02	1	0	0	5
0	7	0.52	4	3	2	0.20	0	0.06	0.01	0.04	0.73	0.01	58	0	0	3
0	8	0.95	9	28	1	0.39	0	0.13	0.03	0.01	1.17	0.07	46	0	0	6
0	17	0.99	11	32	3	0.27	0	0.22	0.01	0.02	1.23	0.02	67	0	0	7
0	2	0.49	26	83	2	1.09	0	0.04	—	0.07	1.05	0.11	21	0	0	<1
0	211	2.92	12	78	794	0.39	0	0.42	0.03	0.26	3.65	0.03	123	0	0	22
0	9	2.90	14	67	1	0.44	0	0.49	0.04	0.31	3.69	0.03	114	0	0	21

Table A–1
Food Composition

(DA+ code is for Wadsworth Diet Analysis program)
(For purposes of calculations, use "0" for t, <1, <.1, <.01, etc.)

A

DA + Code	Food Description	Quantity	Measure	Wt (g)	H₂O (g)	Ener (kcal)	Prot (g)	Carb (g)	Fiber (g)	Fat (g)	Sat	Mono	Poly	Trans
												Fat Breakdown (g)		

CEREAL, FLOUR, GRAIN, PASTA, NOODLES, POPCORN

Flour & grain fractions–Continued

DA + Code	Food Description	Quantity	Measure	Wt (g)	H₂O (g)	Ener (kcal)	Prot (g)	Carb (g)	Fiber (g)	Fat (g)	Sat	Mono	Poly	Trans
1643	Barley flour	½	cup(s)	56	6	198	4	45	2	1	0.16	0.10	0.38	—
383	Buckwheat flour, whole groat	½	cup(s)	60	7	201	8	42	6	2	0.41	0.57	0.57	—
504	Cake wheat flour, enriched	½	cup(s)	55	7	197	4	43	1	<1	0.07	0.04	0.21	—
426	Cornmeal, degermed, enriched	½	cup(s)	69	8	253	6	54	5	1	0.16	0.28	0.49	—
424	Cornmeal, yellow whole grain	½	cup(s)	61	6	221	5	47	4	2	0.31	0.58	1.00	—
1644	Masa corn flour, enriched	½	cup(s)	57	5	208	5	43	5	2	0.30	0.57	0.98	—
1976	Rice flour, brown	½	cup(s)	79	9	287	6	60	4	2	0.44	0.80	0.79	—
1645	Rice flour, white	½	cup(s)	79	9	289	5	63	2	1	0.30	0.35	0.30	—
1978	Rye flour, dark	½	cup(s)	64	7	207	9	44	14	2	0.20	0.21	0.77	—
1980	Semolina, enriched	½	cup(s)	84	11	301	11	61	3	1	0.13	0.10	0.36	—
2827	Soy flour, raw	½	cup(s)	43	2	186	15	15	4	9	1.27	1.94	4.96	—
1990	Wheat germ, crude	2	tablespoon(s)	14	2	52	3	7	2	1	0.24	0.20	0.86	—
506	Whole wheat flour	½	cup(s)	60	6	203	8	44	7	1	0.19	0.14	0.47	—

Breakfast bars

DA + Code	Food Description	Quantity	Measure	Wt (g)	H₂O (g)	Ener (kcal)	Prot (g)	Carb (g)	Fiber (g)	Fat (g)	Sat	Mono	Poly	Trans
39230	Atkins Morning Start apple crisp	1	item(s)	37	—	170	11	12	6	9	4.00	—	—	—
10574	Health Valley fat free apple	1	item(s)	38	—	110	2	26	3	0	0.00	0.00	0.00	0
10647	Nutri-Grain blueberry cereal bar	1	item(s)	37	5	140	2	27	1	3	0.50	2.00	0.50	—
10648	Nutri-Grain raspberry cereal bar	1	item(s)	37	5	140	2	27	1	3	0.50	2.00	0.50	—
10649	Nutri-Grain strawberry cereal bar	1	item(s)	37	5	140	2	27	1	3	0.50	2.00	0.50	—

Breakfast cereals, hot

DA + Code	Food Description	Quantity	Measure	Wt (g)	H₂O (g)	Ener (kcal)	Prot (g)	Carb (g)	Fiber (g)	Fat (g)	Sat	Mono	Poly	Trans
363	Corn grits, white, regular & quick, enriched, cooked w/water & salt	½	cup(s)	121	103	71	2	16	<1	<1	0.03	0.06	0.10	—
8636	Corn grits, yellow, regular & quick, enriched, cooked w/salt	½	cup(s)	121	103	71	2	16	<1	<1	0.03	0.06	0.10	—
1260	Cream of Wheat, instant, prepared	½	cup(s)	121	106	61	2	13	<1	<.1	0.01	0.01	0.04	0
365	Farina, enriched, cooked w/water & salt	½	cup(s)	117	102	56	2	12	<1	<.1	0.01	0.01	0.03	—
8657	Oatmeal, cooked w/water	½	cup(s)	117	100	74	3	13	2	1	0.19	0.37	0.44	—
5500	Oatmeal, maple & brown sugar, instant, prepared	1	item(s)	198	150	200	5	40	2	2	0.42	0.74	0.85	—
5510	Oatmeal, ready to serve, packet	1	item(s)	186	158	112	4	20	3	2	0.38	0.66	0.76	—

Breakfast cereals, ready to eat

DA + Code	Food Description	Quantity	Measure	Wt (g)	H₂O (g)	Ener (kcal)	Prot (g)	Carb (g)	Fiber (g)	Fat (g)	Sat	Mono	Poly	Trans
1197	All-Bran	1	cup(s)	62	2	160	8	46	20	2	0.00	0.00	1.00	0
1200	All-Bran Buds	1	cup(s)	91	3	212	6	73	42	3	—	—	—	0
1199	Apple Jacks	1	cup(s)	33	1	130	1	30	1	1	—	—	—	0
13633	Bran Flakes, Post	1	cup(s)	40	1	133	4	32	7	1	0.00	0.00	0.71	—
1204	Cap'n Crunch	1	cup(s)	36	1	144	2	30	1	2	0.53	0.39	0.27	—
1205	Cap'n Crunch Crunchberries w/wildberry colors	1	cup(s)	35	1	139	2	29	1	2	0.49	0.39	0.28	—
1206	Cheerios	1	cup(s)	30	1	110	3	22	3	2	0.00	0.50	0.50	—
3415	Cocoa Puffs	1	cup(s)	30	1	120	1	26	0	1	—	—	—	—
1207	Cocoa Rice Krispies	1	cup(s)	41	1	160	1	36	1	1	0.67	0.00	0.00	—
5522	Complete wheat bran flakes	1	cup(s)	39	1	120	4	31	7	1	—	—	—	0
1211	Corn Flakes	1	cup(s)	28	1	100	2	24	1	0	0.00	0.00	0.00	0
1247	Corn Pops	1	cup(s)	31	1	120	1	28	0	0	0.00	0.00	0.00	0
1937	Cracklin' Oat Bran	1	cup(s)	65	0	266	5	47	7	9	2.70	4.70	1.33	0
1220	Froot Loops	1	cup(s)	32	1	120	1	28	1	1	0.50	0.00	0.50	—
38214	Frosted Cheerios	1	cup(s)	30	—	120	2	25	1	1	0.00	0.00	0.00	—
372	Frosted Flakes	1	cup(s)	41	1	160	1	37	1	0	0.00	0.00	0.00	0
38215	Frosted Mini Chex	1	cup(s)	40	—	146	1	36	0	0	0.00	0.00	0.00	0
10268	Frosted Mini-Wheats	5	item(s)	51	3	180	5	41	5	1	0.00	0.00	0.50	—
38216	Frosted Wheaties	1	cup(s)	40	—	146	1	36	<1	0	0.00	0.00	0.00	—

Chol (mg)	Calc (mg)	Iron (mg)	Magn (mg)	Pota (mg)	Sodi (mg)	Zinc (mg)	Vit A (µg)	Thia (mg)	Vit E (mg α)	Ribo (mg)	Niac (mg)	Vit B_6 (mg)	Fola (µg)	Vit C (mg)	Vit B_{12} (µg)	Sele (µg)
0	16	0.71	45	186	4	1.05	0	0.07	—	0.03	2.57	0.16	13	0	0	2
0	25	2.44	151	346	7	1.87	0	0.25	0.19	0.11	3.69	0.35	32	0	0	3
0	8	3.99	9	57	1	0.34	0	0.49	0.01	0.23	3.70	0.02	101	0	0	3
0	3	2.85	28	112	2	0.50	8	0.49	0.10	0.28	3.47	0.18	161	0	0	5
0	4	2.10	77	175	21	1.11	7	0.23	0.26	0.12	2.22	0.19	15	0	0	9
0	80	4.11	63	170	3	1.01	0	0.81	0.09	0.43	5.61	0.21	133	0	0	9
0	9	1.56	88	228	6	1.94	0	0.35	0.95	0.06	5.01	0.58	13	0	0	—
0	8	0.28	28	60	0	0.63	0	0.11	0.09	0.02	2.05	0.34	3	0	0	12
0	36	4.13	159	467	1	3.60	1	0.20	0.90	0.16	2.73	0.28	38	0	0	23
0	14	3.64	39	155	1	0.88	0	0.68	0.22	0.48	5.00	0.09	153	0	0	75
0	88	2.71	183	1070	6	1.67	3	0.25	0.93	0.49	1.84	0.20	147	0	0	3
0	6	0.90	34	128	2	1.77	0	0.27	—	0.07	0.98	0.19	40	0	0	11
0	20	2.33	83	243	3	1.76	0	0.27	0.49	0.13	3.82	0.20	26	0	0	42
0	200	—	—	90	70	—	—	0.23	—	0.26	3.00	—	—	9	—	—
0	0	0.72	—	160	25	—	—	0.09	—	0.03	0.40	—	—	1	—	—
0	200	1.80	8	75	110	1.50	—	0.38	—	0.43	5.00	0.50	40	0	0	—
0	200	1.80	8	70	110	1.50	—	0.38	—	0.43	5.00	0.50	40	0	0	—
0	200	1.80	8	55	110	1.50	—	0.38	—	0.43	5.00	0.50	40	0	0	—
0	4	0.73	6	25	270	0.08	0	0.10	0.02	0.07	0.87	0.03	40	0	0	4
0	4	0.73	6	25	270	0.08	2	0.10	0.02	0.07	0.87	0.03	40	0	0	3
0	27	8.60	2	17	1	0.10	0	0.07	—	0.04	0.60	0.01	357	0	0	—
0	5	0.58	2	15	383	0.09	0	0.07	0.01	0.05	0.57	0.01	40	0	0	11
0	9	0.80	28	66	1	0.57	0	0.13	0.12	0.02	0.15	0.02	5	0	0	9
0	26	6.84	50	126	404	1.04	0	1.02	—	0.05	1.57	0.31	30	0	0	11
0	21	3.96	45	112	241	0.93	0	0.60	—	0.05	0.78	0.19	19	0	0	4
0	300	9.00	200	700	160	3.00	300	0.75	—	0.85	10.00	4.00	800	12	12	6
0	0	13.64	182	909	606	4.55	455	1.14	—	1.29	15.15	6.06	1212	18	18	26
0	0	4.50	8	35	150	1.50	150	0.38	—	0.43	5.00	0.50	100	15	2	2
0	0	10.77	80	253	293	2.00	—	0.50	—	0.57	6.65	0.67	133	0	2	—
0	5	6.00	20	72	269	4.99	3	0.51	—	0.57	6.66	0.67	133	0	0	7
<.1	7	6.14	19	71	242	5.12	2	0.51	—	0.57	6.66	0.67	133	<.1	0	7
0	100	8.10	40	95	280	3.75	150	0.38	—	0.43	5.00	0.50	200	6	2	11
0	100	4.50	8	50	170	3.75	0	0.38	—	0.43	5.00	0.50	100	6	2	2
0	53	5.99	11	67	253	2.00	200	0.50	—	0.57	6.65	0.67	133	20	2	6
0	0	23.94	53	226	279	19.95	299	2.00	—	2.26	26.60	2.66	532	80	8	4
0	0	8.10	3	25	200	0.17	150	0.38	—	0.43	5.00	0.50	100	6	2	1
0	0	1.80	2	25	120	1.50	150	0.38	—	0.43	5.00	0.50	100	6	2	2
0	27	2.38	80	293	186	2.00	299	0.49	—	0.56	6.65	0.67	218	20	2	14
0	0	4.50	8	35	150	1.50	150	0.38	—	0.43	5.00	0.50	100	15	2	2
0	100	4.50	16	55	210	3.75	—	0.38	—	0.43	5.00	0.50	100	6	2	—
0	0	5.99	4	27	200	0.21	200	0.50	—	0.57	6.65	0.67	133	8	2	2
0	133	11.97	—	33	266	3.99	—	0.50	—	0.57	6.65	0.67	266	8	2	—
0	0	15.30	60	170	5	1.50	0	0.38	—	0.43	5.00	0.50	100	0	2	2
0	133	10.77	0	47	266	9.98	—	1.00	—	1.13	13.30	1.33	532	8	4	—

Table A–1
Food Composition

(DA+ code is for Wadsworth Diet Analysis program)
(For purposes of calculations, use "0" for t, <1, <.1, <.01, etc.)

DA + Code	Food Description	Quantity	Measure	Wt (g)	H₂O (g)	Ener (kcal)	Prot (g)	Carb (g)	Fiber (g)	Fat (g)	Sat	Mono	Poly	Trans
	CEREAL, FLOUR, GRAIN, PASTA, NOODLES, POPCORN													
	Breakfast cereals, ready to eat–Continued													
1223	Granola, prepared	½	cup(s)	61	0	299	9	32	5	15	2.76	4.7	6.53	—
13334	Granola, Quaker 100% natural, oats & honey	½	cup(s)	48	0	219	5	31	3	9	3.83	4.0	1.19	—
13335	Granola, Quaker 100% natural, oats, honey & raisins	½	cup(s)	51	0	225	5	34	3	9	3.57	3.80	1.10	—
2415	Honey Bunches of Oats honey roasted	1	cup(s)	40	1	160	3	33	1	2	0.67	1.20	0.13	—
1227	Honey Nut Cheerios	1	cup(s)	30	1	120	3	24	2	2	0.00	0.50	0.00	—
2424	Honeycomb	1	cup(s)	22	<1	83	2	20	<1	<1	0.00	—	—	—
10286	Kashi puffed	1	cup(s)	25	—	70	3	13	2	1	0.00	—	—	—
1231	Kix	1	cup(s)	23	<1	90	2	20	1	<1	0.00	0.00	0.00	—
30569	Life	1	cup(s)	43	2	160	4	33	3	2	0.35	0.64	0.61	—
1233	Lucky Charms	1	cup(s)	30	1	120	2	25	1	1	0.00	0.00	0.00	—
1201	Multi-Bran Chex	1	cup(s)	58	1	200	4	49	7	2	0.00	0.00	0.00	0
38220	Multi Grain Cheerios	1	cup(s)	30	—	110	3	24	3	1	0.00	0.00	0.00	—
1238	Nutri-Grain golden wheat	1	cup(s)	40	—	133	4	31	5	1	0.00	0.00	0.67	—
1241	Product 19	1	cup(s)	30	1	100	2	25	1	0	0.00	0.00	0.00	0
32432	Puffed rice, fortified	1	cup(s)	14	<1	56	1	13	<1	<.1	0.02	—	—	—
32433	Puffed wheat, fortified	1	cup(s)	12	0	44	2	10	<1	<1	0.02	—	—	—
2420	Raisin Bran	1	cup(s)	59	5	190	4	47	8	1	0.00	0.10	0.36	—
1244	Rice Chex	1	cup(s)	25	1	96	2	22	<1	0	0.00	0.00	0.00	0
1245	Rice Krispies	1	cup(s)	26	1	96	2	23	0	0	0.00	0.00	0.00	0
5593	Shredded Wheat	1	cup(s)	25	1	88	3	20	3	1	0.04	0.01	0.10	—
1248	Smacks	1	cup(s)	36	1	133	3	32	1	1	0.00	0.00	0.00	—
1246	Special K	1	cup(s)	31	1	110	7	22	1	0	0.00	0.00	0.00	0
3428	Total, corn flakes	1	cup(s)	23	1	83	2	18	1	0	0.00	0.00	0.00	0
1253	Total whole grain	1	cup(s)	40	1	146	3	31	4	1	0.00	0.00	0.00	—
1254	Trix	1	cup(s)	30	1	120	1	27	1	1	0.00	0.00	0.00	—
382	Wheat germ, toasted	2	tablespoon(s)	14	0	54	4	7	2	1	0.25	0.21	0.93	—
1257	Wheaties	1	cup(s)	30	1	110	3	24	3	1	0.00	0.00	0.00	—
	Pasta, noodles													
449	Chinese chow mein noodles, cooked	½	cup(s)	23	<1	119	2	13	1	7	0.99	1.73	3.90	—
1995	Corn pasta, cooked	½	cup(s)	70	48	88	2	20	3	1	0.07	0.13	0.23	—
448	Egg noodles, enriched, cooked	½	cup(s)	80	55	106	4	20	1	1	0.25	0.34	0.33	0.02
440	Macaroni, enriched, cooked	½	cup(s)	70	46	99	3	20	1	<1	0.07	0.06	0.19	—
1996	Pasta, plain, fresh-refrigerated, cooked	½	cup(s)	64	44	84	3	16	0	1	0.10	0.08	0.27	—
1725	Ramen noodles, cooked	½	cup(s)	114	95	104	3	15	1	4	0.19	0.22	0.21	—
2878	Soba noodles, cooked	½	cup(s)	95	69	94	5	20	0	<.1	0.02	0.02	0.03	—
2879	Somen noodles, cooked	½	cup(s)	88	60	115	4	24	0	<1	0.02	0.02	0.06	—
493	Spaghetti, al dente, cooked	½	cup(s)	65	42	95	4	20	1	1	0.05	0.05	0.15	—
2884	Spaghetti, whole wheat, cooked	½	cup(s)	70	47	87	4	19	3	<1	0.07	0.05	0.15	—
1563	Spinach egg noodles, enriched, cooked	½	cup(s)	80	55	105	4	19	2	1	0.29	0.39	0.28	—
2000	Tricolor vegetable macaroni, enriched, cooked	½	cup(s)	67	46	86	3	18	3	<.1	0.01	0.01	0.03	—
	Popcorn													
476	Air popped	1	cup(s)	8	<1	31	1	6	1	<1	0.05	0.09	0.15	—
4619	Caramel	1	cup(s)	35	1	152	1	28	2	5	1.27	1.01	1.58	—
4620	Cheese flavored	1	cup(s)	37	1	196	3	19	4	12	2.38	3.61	5.72	—
477	Popped in oil	1	cup(s)	33	1	165	3	19	3	9	1.61	2.70	4.43	—

PAGE KEY: A–2 = Breads/Baked Goods A–6 = Cereal/Rice/Pasta A–10 = Fruit A–14 = Vegetables/Legumes A–24 = Nuts/Seeds A–26 = Vegetarian
A–28 = Dairy A–34 = Eggs A–34 = Seafood A–36 = Meats A–40 = Poultry A–40 = Processed meats A–42 = Beverages A–46 = Fats/Oils A–48 = Sweets
A–50 = Spices/Condiments/Sauces A–52 = Mixed foods/Soups/Sandwiches A–58 = Fast food A–74 = Convenience meals A–76 = Baby foods

A

Chol (mg)	Calc (mg)	Iron (mg)	Magn (mg)	Pota (mg)	Sodi (mg)	Zinc (mg)	Vit A (μg)	Thia (mg)	Vit E (mg α)	Ribo (mg)	Niac (mg)	Vit B6 (mg)	Fola (μg)	Vit C (mg)	Vit B12 (μg)	Sele (μg)
0	48	2.59	107	328	13	2.5	2	0.44	3.59	0.17	1.29	0.18	51	1	0	17
1	61	1.21	51	225	20	1.04	1	0.12	—	0.11	0.81	0.07	17	<1	0.1	8
1	59	1.24	49	250	19	0.99	<1	0.12	—	0.11	0.8	0.07	16	<1	0.1	9
0	0	3.59	21	67	253	0.40	—	0.50	—	0.57	6.65	0.67	133	0	2	—
0	100	4.50	24	95	270	3.75	—	0.38	—	0.43	5.00	0.50	200	6	2	7
0	0	2.03	6	26	165	1.13	—	0.28	—	0.32	3.74	0.37	75	0	1	—
0	0	0.72	—	35	0	—	0	0.03	—	0.03	0.80	0.00	—	0	—	—
0	113	6.08	6	26	203	2.81	113	0.28	—	0.32	3.75	0.38	150	5	1	5
0	124	11.92	41	121	218	5.32	1	0.53	—	0.60	7.10	0.70	142	0	0	11
0	100	4.50	16	60	210	3.75	—	0.38	—	0.43	5.00	0.50	200	6	2	6
0	100	16.20	60	220	390	3.75	158	0.38	—	0.03	5.00	0.50	100	6	2	5
0	100	18.00	24	85	200	15.00	—	1.50	—	1.70	20.00	2.00	400	15	6	—
0	0	1.46	32	146	279	4.99	0	0.50	—	0.57	6.65	0.67	133	20	2	9
0	0	18.00	16	50	210	15.00	225	1.50	—	1.70	20.00	2.00	400	60	6	4
0	1	4.44	4	16	<1	0.14	0	0.36	—	0.25	4.94	0.01	3	0	0	1
0	3	3.8	17	42	<1	0.28	0	0.31	—	0.21	4.23	0.02	4	0	0	15
0	20	10.80	80	340	300	2.25	—	0.53	—	0.60	7.00	0.70	140	0	2	—
0	80	7.20	7	28	232	3.00	—	0.30	—	0.34	4.00	0.40	160	5	1	1
0	0	1.44	13	32	256	0.48	120	0.30	—	0.34	4.80	0.40	80	5	1	4
0	10	1.08	31	92	2	0.70	0	0.07	—	0.06	1.77	0.10	12	0	0	1
0	0	0.48	11	53	67	0.40	200	0.50	—	0.57	6.65	0.67	133	8	2	17
0	0	8.70	16	60	220	0.90	225	0.53	—	0.60	7.00	2.00	400	15	6	7
0	750	13.50	0	23	158	11.25	113	1.13	22.50	1.28	15.00	1.50	300	45	5	1
0	1330	23.94	32	120	253	19.95	200	2.00	31.24	2.26	26.60	2.66	532	80	8	2
0	100	4.50	0	15	190	3.75	150	0.38	—	0.43	5.00	0.50	100	6	2	6
0	6	1.28	45	134	<1	2.35	0	0.23	—	0.11	0.78	0.13	50	<1	0	9
0	0	8.10	32	110	220	7.50	150	0.75	2.26	0.85	10.00	1.00	200	6	3	1
0	5	1.06	12	27	99	0.32	0	0.13	—	0.09	1.34	0.02	20	0	0	10
0	1	0.18	25	22	0	0.44	2	0.04	0.78	0.02	0.39	0.04	4	0	0	2
26	10	1.27	15	22	6	0.49	5	0.15	0.14	0.07	1.19	0.03	51	0	<.1	17
0	5	0.98	13	22	1	0.37	0	0.14	0.04	0.07	1.17	0.02	54	0	0	15
21	4	0.73	12	15	4	0.36	4	0.13	—	0.10	0.63	0.02	41	0	<.1	—
18	9	0.89	9	34	415	0.31	—	0.08	—	0.05	0.71	0.03	4	<.1	<.1	—
0	4	0.45	9	33	57	0.11	0	0.09	—	0.02	0.48	0.04	7	0	0	—
0	7	0.46	2	25	141	0.19	0	0.02	—	0.03	0.09	0.01	2	0	0	—
0	7	1.00	12	52	1	0.35	0	0.12	0.04	0.07	0.90	0.04	8	0	0	40
0	11	0.74	21	31	2	0.57	0	0.08	0.21	0.03	0.49	0.06	4	0	0	18
26	15	0.87	19	30	10	0.50	4	0.20	0.46	0.10	1.18	0.09	51	0	<1	17
0	7	0.33	13	21	4	0.29	3	0.08	0.06	0.04	0.72	0.02	44	0	0	13
0	1	0.22	11	24	<1	0.28	1	0.02	0.02	0.02	0.16	0.02	2	0	0	1
2	15	0.61	12	38	73	0.20	1	0.02	0.42	0.02	0.77	0.01	2	0	<.1	1
4	42	0.83	34	97	331	0.75	14	0.05	—	0.09	0.54	0.09	4	<1	<1	4
0	3	0.92	36	74	292	0.87	3	0.04	—	0.04	0.51	0.07	6	<.1	0	2

Table A–1
Food Composition

(DA+ code is for Wadsworth Diet Analysis program)
(For purposes of calculations, use "0" for t, <1, <.1, <.01, etc.)

A

DA + Code	Food Description	Quantity	Measure	Wt (g)	H₂O (g)	Ener (kcal)	Prot (g)	Carb (g)	Fiber (g)	Fat (g)	Sat	Mono	Poly	Trans
	FRUIT AND FRUIT JUICES													
	Apples													
223	Raw medium, w/peel	1	item(s)	138	118	72	<1	19	3	<1	0.04	0.01	0.07	—
224	Slices	½	cup(s)	55	47	29	<1	8	1	<.1	0.02	0.00	0.03	—
946	Slices w/o skin, boiled	½	cup(s)	85	73	45	<1	12	2	<1	0.05	0.01	0.09	—
948	Dried, sulfured	½	cup(s)	22	7	52	<1	14	2	<.1	0.01	0.00	0.02	—
952	Juice, from frozen concentrate	½	cup(s)	120	105	56	<1	14	<1	<1	0.02	0.00	0.04	—
225	Juice, unsweetened, canned	½	cup(s)	124	109	58	<.1	14	<1	<1	0.02	0.01	0.04	—
226	Applesauce, sweetened, canned	½	cup(s)	128	101	97	<1	25	2	<1	0.04	0.01	0.07	—
227	Applesauce, unsweetened, canned	½	cup(s)	122	108	52	<1	14	1	<.1	0.01	0.00	0.02	—
38492	Crabapples	1	item(s)	35	28	27	<1	7	1	<1	0.02	0.00	0.03	—
	Apricot													
228	Fresh w/o pits	4	item(s)	140	121	67	2	16	3	1	0.04	0.24	0.11	—
230	Halves, dried, sulfured	¼	cup(s)	33	10	79	1	21	2	<1	0.01	0.02	0.02	—
229	Halves w/skin, canned in heavy syrup	½	cup(s)	129	100	107	1	28	2	<1	0.01	0.04	0.02	—
	Avocado													
233	California, whole, w/o skin or pit	1	item(s)	170	123	284	3	15	12	26	3.59	16.61	3.42	—
234	Florida, whole, w/o skin or pit	1	item(s)	304	240	365	7	24	17	31	5.90	16.70	5.00	—
2998	Pureed	⅛	cup(s)	29	21	46	1	2	2	4	0.61	2.82	0.52	—
	Banana													
235	Fresh whole, w/o peel	1	item(s)	118	88	105	1	27	3	<1	0.13	0.04	0.09	—
4580	Dried chips	¼	cup(s)	55	2	287	1	32	4	19	16.00	1.08	0.35	—
	Blackberries													
237	Raw	½	cup(s)	72	63	31	1	7	4	<1	0.01	0.03	0.20	—
958	Unsweetened, frozen	½	cup(s)	76	62	48	1	12	4	<1	0.01	0.03	0.18	—
	Blueberries													
238	Raw	½	cup(s)	72	61	41	1	10	2	<1	0.02	0.03	0.11	—
959	Canned in heavy syrup	½	cup(s)	128	98	113	1	28	2	<1	0.03	0.06	0.18	—
960	Unsweetened, frozen	½	cup(s)	78	67	40	1	10	2	1	0.04	0.07	0.22	—
	Boysenberries													
961	Canned in heavy syrup	½	cup(s)	128	98	113	1	29	3	<1	0.01	0.02	0.09	—
962	Unsweetened, frozen	½	cup(s)	66	57	33	1	8	3	<1	0.01	0.02	0.10	—
35576	**Breadfruit**	1	item(s)	384	271	396	4	104	17	1	0.00	0.00	0.00	—
	Cherries													
3000	Sour red, raw	½	cup(s)	78	67	39	1	9	1	<1	0.05	0.06	0.07	—
967	Sour red, canned in water	½	cup(s)	122	110	44	1	11	1	<1	0.03	0.03	0.04	—
240	Sweet, raw	½	cup(s)	73	60	46	1	12	2	<1	0.03	0.03	0.04	—
3004	Sweet, canned in heavy syrup	½	cup(s)	127	98	105	1	27	2	<1	0.04	0.05	0.06	—
969	Sweet, canned in water	½	cup(s)	124	108	57	1	15	2	<1	0.03	0.04	0.05	—
	Cranberries													
3007	Chopped, raw	½	cup(s)	55	48	25	<1	7	3	<.1	0.01	0.01	0.03	—
1638	Cranberry juice cocktail	½	cup(s)	127	108	72	0	18	<1	<1	0.01	0.02	0.06	—
241	Cranberry juice cocktail, low calorie, w/saccharin	½	cup(s)	127	120	24	<.1	6	0	<.1	0.00	0.00	0.00	—
1717	Cranberry apple juice drink	½	cup(s)	123	100	87	<.1	22	<1	<.1	0.00	0.00	0.00	—
242	Cranberry sauce, sweetened, canned	¼	cup(s)	69	42	105	<1	27	1	<1	0.01	0.01	0.05	—
	Dates													
244	Domestic, chopped	¼	cup(s)	45	0	126	1	33	4	<1	0.01	0.01	0	—
243	Domestic, whole	¼	cup(s)	45	0	126	1	33	4	<1	0.01	0.01	0	—
	Figs													
973	Raw, medium	2	item(s)	101	80	74	1	19	3	<1	0.06	0.07	0.14	—

Chol (mg)	Calc (mg)	Iron (mg)	Magn (mg)	Pota (mg)	Sodi (mg)	Zinc (mg)	Vit A (μg)	Thia (mg)	Vit E (mg α)	Ribo (mg)	Niac (mg)	Vit B_6 (mg)	Fola (μg)	Vit C (mg)	Vit B_{12} (μg)	Sele (μg)
0	8	0.17	7	148	1	0.06	4	0.02	—	0.04	0.13	0.06	4	6	0	0
0	3	0.07	3	59	1	0.02	2	0.01	—	0.01	0.05	0.02	2	3	0	0
0	4	0.16	3	75	1	0.03	2	0.01	0.04	0.01	0.08	0.04	1	<1	0	<1
0	3	0.30	3	97	19	0.04	0	0.00	0.11	0.03	0.20	0.03	0	1	0	<1
0	7	0.31	6	151	8	0.05	0	0.00	0.01	0.02	0.05	0.04	0	1	0	<1
0	9	0.46	4	148	4	0.04	0	0.03	0.01	0.02	0.12	0.04	0	1	0	<1
0	5	0.45	4	78	4	0.05	1	0.02	0.27	0.04	0.24	0.03	1	2	0	<1
0	4	0.15	4	92	2	0.04	1	0.02	0.26	0.03	0.23	0.03	1	1	0	<1
0	6	0.13	2	68	<1	—	0	0.01	—	0.01	0.04	—	2	3	0	—
0	18	0.55	14	363	1	0.28	134	0.04	1.25	0.06	0.84	0.08	13	14	0	<1
0	18	0.88	11	383	3	0.13	59	0.00	1.43	0.02	0.85	0.05	3	<1	0	1
0	12	0.39	9	181	5	0.14	80	0.03	0.77	0.03	0.49	0.07	3	4	0	<1
0	22	1.00	49	861	14	1.12	104	0.12	3.35	0.24	3.24	0.47	105	15	0	1
0	30	0.50	73	1067	6	1.20	185	0.00	0.09	0.10	2.00	0.20	106	53	0	0
0	3	0.16	8	139	2	0.18	2	0.02	0.60	0.04	0.50	0.07	17	3	0	<1
0	6	0.31	32	422	1	0.18	4	0.04	0.12	0.09	0.78	0.43	24	10	0	1
0	10	0.69	42	296	3	0.41	2	0.05	0.13	0.01	0.39	0.14	8	3	0	1
0	21	0.45	14	117	1	0.38	8	0.01	0.84	0.02	0.47	0.02	18	15	0	<1
0	22	0.60	17	106	1	0.19	5	0.02	0.88	0.03	0.91	0.05	26	2	0	<1
0	4	0.20	4	55	1	0.12	2	0.03	0.41	0.03	0.30	0.04	4	7	0	<.1
0	6	0.42	5	51	4	0.09	3	0.04	0.49	0.07	0.14	0.05	3	1	0	<1
0	6	0.14	4	42	1	0.06	2	0.03	0.37	0.03	0.41	0.05	6	2	0	0
0	23	0.55	14	115	4	0.24	3	0.03	—	0.04	0.29	0.05	44	8	0	1
0	18	0.56	11	92	1	0.15	2	0.03	0.57	0.02	0.51	0.04	42	2	0	<1
0	65	2.07	96	1882	8	0.46	8	0.42	—	0.12	3.46	0.00	54	111	0	2
0	12	0.25	7	134	2	0.08	50	0.02	0.05	0.03	0.31	0.03	6	8	0	0
0	13	1.67	7	120	9	0.09	46	0.02	0.28	0.05	0.22	0.05	10	3	0	0
0	9	0.26	8	161	0	0.05	2	0.02	0.05	0.02	0.11	0.04	3	5	0	0
0	11	0.44	11	183	4	0.13	10	0.03	0.29	0.05	0.50	0.04	5	5	0	0
0	14	0.45	11	162	1	0.10	10	0.03	0.29	0.05	0.51	0.04	5	3	0	0
0	4	0.14	3	47	1	0.06	2	0.01	0.66	0.01	0.06	0.03	1	7	0	<.1
0	4	0.19	3	23	3	0.09	0	0.01	0.28	0.01	0.04	0.02	0	45	0	0
0	11	0.05	3	32	4	0.03	0	0.00	0.06	0.00	0.01	0.00	0	41	0	0
0	6	0.15	2	34	9	0.22	0	0.01	0.15	0.02	0.07	0.03	0	39	0	0
0	3	0.15	2	18	20	0.03	1	0.01	0.57	0.01	0.07	0.01	1	1	0	<1
0	17	0.45	19	292	1	0.12	1	0.02	0.02	0.02	0.56	0.07	9	<1	0	1
0	17	0.45	19	292	1	0.12	1	0.02	0.02	0.02	0.56	0.07	9	<1	0	1
0	35	0.37	17	233	1	0.15	7	0.06	0.11	0.05	0.40	0.11	6	2	0	<1

Table A–1
Food Composition

(DA+ code is for Wadsworth Diet Analysis program)
(For purposes of calculations, use "0" for t, <1, <.1, <.01, etc.)

A

DA + Code	Food Description	Quantity	Measure	Wt (g)	H₂O (g)	Ener (kcal)	Prot (g)	Carb (g)	Fiber (g)	Fat (g)	Sat	Mono	Poly	Trans
	FRUIT AND FRUIT JUICES													
	Figs–Continued													
975	Canned in heavy syrup	½	cup(s)	130	99	114	<1	30	3	<1	0.03	0.03	0.06	—
974	Canned in water	½	cup(s)	124	106	66	<1	17	3	<1	0.02	0.03	0.06	—
	Fruit cocktail & salad													
245	Fruit cocktail, canned in heavy syrup	½	cup(s)	124	100	91	<1	23	1	<.1	0.01	0.02	0.04	—
978	Fruit cocktail, canned in juice	½	cup(s)	119	104	55	1	14	1	<.1	0.00	0.00	0.00	—
977	Fruit cocktail, canned in water	½	cup(s)	119	108	38	<1	10	1	<.1	0.01	0.01	0.02	—
979	Fruit salad, canned in water	½	cup(s)	123	112	37	<1	10	1	<.1	0.01	0.02	0.03	—
	Gooseberries													
981	Raw	½	cup(s)	75	66	33	1	8	3	<1	0.03	0.04	0.24	
982	Canned in light syrup	½	cup(s)	126	101	92	1	24	3	<1	0.02	0.02	0.14	
	Grapefruit													
3022	Raw, pink or red	½	cup(s)	115	101	48	1	12	2	<1	0.02	0.02	0.04	—
247	Raw, white	½	item(s)	118	107	39	1	10	1	<1	0.02	0.02	0.03	—
251	Juice, pink, sweetened, canned	½	cup(s)	125	109	58	1	14	<1	<1	0.02	0.02	0.03	—
249	Juice, white	½	cup(s)	124	111	48	1	11	<1	<1	0.02	0.02	0.03	—
248	Sections, canned in light syrup	½	cup(s)	127	106	76	1	20	1	<1	0.02	0.02	0.03	—
983	Sections, canned in water	½	cup(s)	122	110	44	1	11	<1	<1	0.02	0.02	0.03	—
	Grapes													
255	American, slip skin	½	cup(s)	46	37	31	<1	8	<1	<1	0.05	0.01	0.05	—
256	European, red or green, adherent skin	½	cup(s)	80	61	55	1	14	1	<1	0.04	0.01	0.04	—
259	Juice, sweetened, added vitamin C, from frozen concentrate	½	cup(s)	125	109	64	<1	16	<1	<1	0.04	0.01	0.03	—
3159	Juice drink, canned	½	cup(s)	125	109	63	<1	16	0	0	0.00	0.00	0.00	—
3060	Raisins, seeded, packed	¼	cup(s)	41	7	122	1	32	3	<1	0.07	0.01	0.07	—
987	**Guava, raw**	1	item(s)	90	77	46	1	11	5	1	0.15	0.05	0.23	—
35593	**Guava, strawberry**	1	item(s)	6	5	4	<.1	1	<1	<.1	0.01	0.00	0.02	—
3027	**Jackfruit**	½	cup(s)	83	61	78	1	20	1	<1	0.05	0.04	0.07	—
8458	**Kiwi fruit**	1	item(s)	77	63	53	1	11	3	1	0.02	0.03	0.19	—
	Lemon													
992	Raw	1	item(s)	108	94	22	1	12	5	<1	0.04	0.01	0.10	—
262	Juice	1	tablespoon(s)	15	14	4	<.1	1	<.1	0	0.00	0.00	0.00	—
993	Peel	1	teaspoon(s)	2	2	1	<.1	<1	<1	<.1	0.00	0.00	0.00	—
	Lime													
994	Raw	1	item(s)	67	61	15	<1	6	2	<.1	0.01	0.01	0.02	—
269	Juice	1	tablespoon(s)	15	14	4	<.1	1	<.1	<.1	0.00	0.00	0.00	—
995	**Loganberries, frozen**	½	cup(s)	74	62	40	1	10	4	<1	0.01	0.02	0.13	—
	Mandarin orange													
1038	Canned in juice	½	cup(s)	125	111	46	1	12	1	<.1	0.00	0.01	0.01	—
1039	Canned in light syrup	½	cup(s)	126	105	77	1	20	1	<1	0.02	0.02	0.03	—
999	**Mango**	½	item(s)	104	85	67	1	18	2	<1	0.07	0.10	0.05	—
1005	**Nectarine, raw, sliced**	½	cup(s)	69	60	30	1	7	1	<1	0.02	0.06	0.08	—
	Melons													
271	Cantaloupe	½	cup(s)	80	72	27	1	7	1	<1	0.04	0.00	0.07	—
1000	Casaba melon	½	cup(s)	85	78	24	1	6	1	<.1	0.02	0.00	0.03	—
272	Honeydew	½	cup(s)	89	80	32	<1	8	1	<1	0.03	0.00	0.05	—
318	Watermelon	½	cup(s)	77	71	23	<1	6	<1	<1	0.01	0.03	0.04	—
	Orange													
273	Raw	1	item(s)	131	114	62	1	15	3	<1	0.02	0.03	0.03	—
3040	Peel	1	teaspoon(s)	2	1	2	<.1	1	<1	<.1	0.00	0.00	0.00	—
274	Sections	½	cup(s)	90	78	43	1	11	2	<1	0.01	0.02	0.02	—
275	Juice	½	cup(s)	124	109	56	1	13	<1	<1	0.03	0.04	0.05	—

Chol (mg)	Calc (mg)	Iron (mg)	Magn (mg)	Pota (mg)	Sodi (mg)	Zinc (mg)	Vit A (µg)	Thia (mg)	Vit E (mg α)	Ribo (mg)	Niac (mg)	Vit B$_6$ (mg)	Fola (µg)	Vit C (mg)	Vit B$_{12}$ (µg)	Sele (µg)
0	35	0.36	13	128	1	0.14	3	0.03	0.16	0.05	0.55	0.09	3	1	0	<1
0	35	0.36	12	128	1	0.15	2	0.03	0.10	0.05	0.55	0.09	2	1	0	<1
0	7	0.36	6	109	7	0.10	12	0.02	0.50	0.02	0.46	0.06	4	2	0	1
0	9	0.25	8	113	5	0.11	18	0.01	0.47	0.02	0.48	0.06	4	3	0	1
0	6	0.30	8	111	5	0.11	15	0.02	0.47	0.01	0.43	0.06	4	2	0	1
0	9	0.37	6	96	4	0.10	27	0.02	—	0.03	0.46	0.04	4	2	0	1
0	19	0.23	8	149	1	0.09	11	0.03	0.28	0.02	0.23	0.06	5	21	0	<1
0	20	0.42	8	97	3	0.14	9	0.03	—	0.07	0.19	0.02	4	13	0	1
0	25	0.09	10	155	0	0.08	30	0.05	0.15	0.03	0.23	0.06	15	36	0	<1
0	14	0.07	11	175	0	0.08	2	0.04	0.15	0.02	0.32	0.05	12	39	0	2
0	10	0.45	13	203	3	0.08	0	0.05	0.05	0.03	0.40	0.03	13	34	0	<1
0	11	0.25	15	200	1	0.06	2	0.05	0.27	0.02	0.25	0.05	12	47	0	<1
0	18	0.51	13	164	3	0.10	0	0.05	0.11	0.03	0.31	0.03	11	27	0	1
0	18	0.50	12	161	2	0.11	0	0.05	0.11	0.03	0.30	0.02	11	27	0	1
0	6	0.13	2	88	1	0.02	2	0.04	0.09	0.03	0.14	0.05	2	2	0	<.1
0	8	0.29	6	153	2	0.06	6	0.06	0.15	0.06	0.15	0.07	2	9	0	<.1
0	5	0.13	5	26	3	0.05	0	0.02	0.00	0.03	0.16	0.05	1	30	0	<1
0	4	0.13	4	41	1	0.03	0	0.01	0.00	0.02	0.09	0.02	1	20	0	<1
0	12	1.07	12	340	12	0.07	0	0.05	—	0.08	0.46	0.08	1	2	0	<1
0	18	0.28	9	256	3	0.21	28	0.05	0.66	0.05	1.08	0.13	13	165	0	1
0	1	0.01	1	18	2	—	—	0.00	—	0.00	0.04	0.00	—	2	0	—
0	28	0.50	31	251	2	0.35	12	0.02	—	0.09	0.33	0.09	12	6	0	<1
0	30	0.38	14	251	2	0.10	4	—	—	0.02	0.25	0.05	<.1	74	0	—
0	66	0.76	13	157	3	0.11	2	0.05	—	0.04	0.22	0.12	—	83	0	1
0	1	0.00	1	19	<1	0.01	<1	0.00	0.02	0.00	0.02	0.01	2	7	0	<.1
0	3	0.02	<1	3	<1	0.01	<.1	0.00	0.00	0.00	0.01	0.00	<1	3	0	<.1
0	9	0.06	5	78	1	0.05	1	0.02	0.15	0.01	0.10	0.03	7	20	0	<.1
0	1	0.00	1	17	<1	0.01	<1	0.00	0.03	0.00	0.02	0.01	1	5	0	<.1
0	19	0.47	15	107	1	0.25	1	0.04	0.64	0.02	0.62	0.05	19	11	0	<1
0	14	0.34	14	166	6	0.63	54	0.10	0.12	0.04	0.55	0.05	6	43	0	<1
0	9	0.47	10	98	8	0.30	53	0.07	0.13	0.06	0.56	0.05	6	25	0	1
0	10	0.13	9	161	2	0.04	39	0.06	1.16	0.06	0.60	0.14	14	29	0	1
0	4	0.19	6	139	0	0.12	12	0.02	0.53	0.02	0.78	0.02	3	4	0	0
0	7	0.17	10	215	13	0.14	136	0.03	0.04	0.02	0.59	0.06	17	30	0	<1
0	9	0.29	9	155	8	0.06	0	0.01	0.04	0.03	0.20	0.14	7	19	0	<1
0	5	0.15	9	203	16	0.08	3	0.03	0.02	0.01	0.37	0.08	17	16	0	1
0	5	0.19	8	86	1	0.08	22	0.03	0.04	0.02	0.14	0.03	2	6	0	<1
0	52	0.13	13	237	0	0.09	14	0.11	0.24	0.05	0.37	0.08	39	70	0	1
0	3	0.02	<1	4	<.1	0.01	<1	0.00	0.00	0.00	0.02	0.00	1	3	0	<.1
0	36	0.09	9	164	0	0.06	10	0.08	0.16	0.04	0.26	0.05	27	48	0	<1
0	14	0.25	14	248	1	0.06	12	0.11	0.05	0.04	0.50	0.05	37	62	0	<1

Table A–1
Food Composition

(DA+ code is for Wadsworth Diet Analysis program)
(For purposes of calculations, use "0" for t, <1, <.1, <.01, etc.)

DA + Code	Food Description	Quantity	Measure	Wt (g)	H₂O (g)	Ener (kcal)	Prot (g)	Carb (g)	Fiber (g)	Fat (g)	Sat	Mono	Poly	Trans
	FRUIT AND FRUIT JUICES													
	Orange–Continued													
29630	Juice, fresh squeezed	½	cup(s)	124	109	56	1	13	<1	<1	0.03	0.04	0.05	—
14414	Juice w/calcium & extra vitamin C	½	cup(s)	125	109	55	1	13	<1	0	0.00	0.00	0.00	—
278	Juice, unsweetened, from frozen concentrate	½	cup(s)	125	110	56	1	13	<1	<.1	0.01	0.01	0.01	
	Papaya													
282	Raw	½	cup(s)	70	62	27	<1	7	1	<.1	0.03	0.03	0.02	—
16830	Dried, strips	2	item(s)	46	12	119	2	30	5	<1	0.13	0.12	0.09	
35640	**Passion fruit, purple**	1	item(s)	18	13	17	<1	4	3	<1	0.00	0.00	0.00	
	Peach													
283	Raw, medium	1	item(s)	98	87	38	1	9	1	<1	0.02	0.07	0.08	
285	Halves, canned in heavy syrup	½	cup(s)	131	104	97	1	26	2	<1	0.01	0.05	0.06	
286	Halves, canned in water	½	cup(s)	122	114	29	1	7	2	<.1	0.01	0.03	0.03	
290	Slices, sweetened, frozen	½	cup(s)	125	93	118	1	30	2	<1	0.02	0.06	0.08	
	Pear													
291	Raw	1	item(s)	166	139	96	1	26	5	<1	0.01	0.04	0.05	—
8672	Asian	1	item(s)	122	108	51	1	13	4	<1	0.01	0.06	0.07	—
293	Danjou	1	item(s)	200	168	120	1	30	5	1	0.00	0.20	0.20	—
294	Halves, canned in heavy syrup	½	cup(s)	133	107	98	<1	25	2	<1	0.01	0.04	0.04	
1012	Halves, canned in juice	½	cup(s)	124	107	62	<1	16	2	<.1	0.00	0.02	0.02	
1017	**Persimmon**	1	item(s)	25	16	32	<1	8	0	<1	0.01	0.02	0.02	
	Pineapple													
295	Raw, diced	½	cup(s)	78	67	37	<1	10	1	<.1	0.01	0.01	0.03	
3053	Canned in extra heavy syrup	½	cup(s)	130	101	108	<1	28	1	<1	0.01	0.02	0.05	
1019	Canned in juice	½	cup(s)	125	104	75	1	20	1	<.1	0.01	0.01	0.04	
296	Canned in light syrup	½	cup(s)	126	108	66	<1	17	1	<1	0.01	0.02	0.05	
1018	Canned in water	½	cup(s)	123	112	39	1	10	1	<1	0.01	0.01	0.04	
299	Juice, unsweetened, canned	½	cup(s)	125	107	70	<1	17	<1	<1	0.01	0.01	0.04	
1024	**Plantain, cooked**	½	cup(s)	77	52	89	1	24	2	<1	0.05	0.01	0.03	
300	**Plum, raw, large**	1	item(s)	83	72	38	1	9	1	<1	0.01	0.11	0.04	
1027	**Pomegranate**	1	item(s)	154	125	105	1	26	1	<1	0.06	0.07	0.10	
	Prunes													
5644	Dried	2	item(s)	17	5	40	<1	11	1	<.1	0.01	0.06	0.02	
305	Dried, stewed	½	cup(s)	119	86	128	1	33	4	<1	0.00	0.15	0.04	
306	Juice, canned	1	cup(s)	256	208	182	2	45	3	<.1	0.01	0.05	0.02	
	Raisins, *see* grapes													
	Raspberries													
309	Raw	½	cup(s)	62	53	32	1	7	4	<1	0.01	0.04	0.23	—
310	Red, sweetened, frozen	½	cup(s)	125	91	129	1	33	6	<1	0.01	0.02	0.11	—
311	**Rhubarb, cooked with sugar**	½	cup(s)	120	82	140	1	38	3	<.1	0.00	0.00	0.05	
	Strawberries													
313	Raw	½	cup(s)	72	65	23	<1	6	1	<1	0.01	0.03	0.11	—
315	Sweetened, frozen, thawed	½	cup(s)	128	100	99	1	27	2	<1	0.01	0.02	0.09	—
16828	**Tangelo**	1	item(s)	95	82	45	1	11	2	<1	0.01	0.02	0.02	—
	Tangerine													
316	Raw	1	item(s)	84	74	37	1	9	2	<1	0.02	0.03	0.03	
1040	Juice	½	cup(s)	124	110	53	1	12	<1	<1	0.03	0.04	0.05	
	VEGETABLES, LEGUMES													
	Amaranth													
1042	Leaves, raw	1	cup(s)	28	26	6	1	1	0	<.1	0.03	0.02	0.04	
1043	Leaves, boiled, drained	½	cup(s)	66	60	14	1	3	0	<1	0.03	0.03	0.05	
8683	**Arugula leaves, raw**	1	cup(s)	20	18	5	1	1	<1	<1	0.02	0.01	0.06	

A

Chol (mg)	Calc (mg)	Iron (mg)	Magn (mg)	Pota (mg)	Sodi (mg)	Zinc (mg)	Vit A (µg)	Thia (mg)	Vit E (mg α)	Ribo (mg)	Niac (mg)	Vit B_6 (mg)	Fola (µg)	Vit C (mg)	Vit B_{12} (µg)	Sele (µg)
0	14	0.25	14	248	1	0.06	—	0.11	0.05	0.04	0.50	0.05	38	62	0	—
0	176	—	—	226	0	—	5	0.08	—	—	0.40	0.06	30	54	0	—
0	11	0.12	12	237	1	0.06	6	0.10	0.25	0.02	0.25	0.05	55	48	0	<1
0	17	0.07	7	180	2	0.05	39	0.02	0.51	0.02	0.24	0.01	27	43	0	<1
0	73	0.30	30	783	9	0.21	—	0.06	2.22	0.09	0.93	0.05	58	38	0	—
0	2	0.29	5	63	5	—	—	0.00		0.02	0.27	—	3	5	0	<1
0	6	0.25	9	186	0	0.17	16	0.02	0.72	0.03	0.79	0.02	4	6	0	<.1
0	4	0.35	7	121	8	0.12	22	0.01	0.64	0.03	0.80	0.02	4	4	0	<1
0	2	0.39	6	121	4	0.11	33	0.01	0.60	0.02	0.64	0.02	4	4	0	<1
0	4	0.46	6	163	8	0.06	18	0.02	0.77	0.04	0.82	0.02	4	118	0	1
0	15	0.28	12	198	2	0.17	2	0.02	0.20	0.04	0.26	0.05	12	7	0	<1
0	5	0.00	10	148	0	0.02	0	0.01	0.15	0.01	0.27	0.03	10	5	0	<1
0	22	0.50	12	250	0	0.24	—	0.04	1.00	0.08	0.20	0.04	15	8	0	1
0	7	0.29	5	86	7	0.11	0	0.01	0.11	0.03	0.32	0.02	1	1	0	0
0	11	0.36	9	119	5	0.11	0	0.01	0.10	0.01	0.25	0.02	1	2	0	0
0	7	0.63	—	78	<1	—	—	—	—	—	—	—	—	17	0	0
0	10	0.22	9	89	1	0.08	2	0.06	0.02	0.02	0.38	0.09	12	28	0	<.1
0	18	0.49	20	133	1	0.14	1	0.12	—	0.03	0.37	0.10	7	9	0	—
0	17	0.35	17	152	1	0.12	2	0.12	0.01	0.02	0.35	0.09	6	12	0	<1
0	18	0.49	20	132	1	0.15	3	0.11	0.01	0.03	0.37	0.09	6	9	0	1
0	18	0.49	22	156	1	0.15	2	0.11	0.01	0.03	0.37	0.09	6	9	0	<1
0	21	0.33	16	168	1	0.14	0	0.07	0.03	0.03	0.32	0.12	29	13	0	<1
0	2	0.45	25	358	4	0.10	35	0.04	0.10	0.04	0.58	0.18	20	8	0	1
0	5	0.14	6	130	0	0.08	14	0.02	0.21	0.02	0.34	0.02	4	8	0	0
0	5	0.46	5	399	5	0.18	8	0.05	0.92	0.05	0.46	0.16	9	9	0	1
0	9	0.42	8	125	1	0.09	17	0.01	0.00	0.03	0.33	0.04	1	1	0	<1
0	23	0.46	21	383	1	0.19	37	0.00	0.23	0.12	0.85	0.23	0	3	0	<1
0	31	3.02	36	707	10	0.54	0	0.04	0.31	0.18	2.01	0.56	0	10	0	2
0	15	0.42	14	93	1	0.26	1	0.02	0.54	0.02	0.37	0.03	13	16	0	<1
0	19	0.81	16	143	1	0.23	4	0.02	0.90	0.06	0.29	0.04	33	21	0	<1
0	174	0.25	16	115	1	—	—	0.02	—	0.03	0.25	—	—	4	0	—
0	12	0.30	9	110	1	0.10	1	0.02	0.21	0.02	0.28	0.03	17	42	0	<1
0	14	0.60	8	125	1	0.06	1	0.02	0.31	0.10	0.37	0.04	5	50	0	1
0	38	0.10	10	172	0	0.07	—	0.08	0.17	0.04	0.27	0.06	29	51	0	—
0	12	0.08	10	132	1	0.20	29	0.09	0.17	0.02	0.13	0.06	17	26	0	<1
0	22	0.25	10	220	1	0.04	16	0.07	0.16	0.02	0.12	0.05	6	38	0	<1
0	60	0.65	15	171	6	0.25	0	0.01	—	0.04	0.18	0.05	24	12	0	<1
0	138	1.49	36	423	14	0.58	92	0.01	—	0.09	0.37	0.12	38	27	0	1
0	32	0.29	9	74	5	0.09	24	0.01	0.09	0.02	0.06	0.01	19	3	0	<.1

Table A–1
Food Composition

(DA+ code is for Wadsworth Diet Analysis program)
(For purposes of calculations, use "0" for t, <1, <.1, <.01, etc.)

DA + Code	Food Description	Quantity	Measure	Wt (g)	H₂O (g)	Ener (kcal)	Prot (g)	Carb (g)	Fiber (g)	Fat (g)	Sat	Mono	Poly	Trans
	VEGETABLES, LEGUMES–Continued													
	Artichoke													
1044	Boiled, drained	1	item(s)	120	101	60	4	13	6	<1	0.04	0.01	0.08	—
2885	Hearts, boiled, drained	½	cup(s)	84	71	42	3	9	5	<1	0.03	0.00	0.06	—
	Asparagus													
566	Boiled, drained	½	cup(s)	90	83	20	2	4	2	<1	0.06	0	0.12	—
568	Canned, drained	½	cup(s)	121	114	23	3	3	2	1	0.18	0.03	0.34	—
565	Tips, frozen, boiled, drained	½	cup(s)	90	82	25	3	4	1	<1	0.09	0.01	0.17	—
	Bamboo shoots													
1048	Boiled, drained	½	cup(s)	60	58	7	1	1	1	<1	0.03	0.00	0.06	—
1049	Canned, drained	½	cup(s)	65	62	12	1	2	1	<1	0.06	0.01	0.12	—
	Beans													
1801	Adzuki beans, boiled	½	cup(s)	115	76	147	9	28	8	<1	0.04	—	—	—
511	Baked beans w/franks, canned	½	cup(s)	129	89	182	9	20	9	8	3.02	3.64	1.07	—
512	Baked beans w/pork in tomato sauce, canned	½	cup(s)	127	92	124	7	25	6	1	0.50	0.56	0.17	—
513	Baked beans w/pork in sweet sauce, canned	½	cup(s)	127	89	140	7	27	7	2	0.71	0.80	0.24	0
1805	Black beans, boiled	½	cup(s)	86	57	114	8	20	7	<1	0.12	0.04	0.20	—
14597	Chickpeas, garbanzo beans, or bengal gram, boiled	½	cup(s)	82	49	134	7	22	6	2	0.22	0.48	0.95	
569	Fordhook lima beans, frozen, boiled, drained	½	cup(s)	85	62	88	5	16	5	<1	0.07	0.02	0.14	
1806	French beans, boiled	½	cup(s)	89	59	114	6	21	8	1	0.07	0.05	0.40	—
2773	Great northern beans, boiled	½	cup(s)	89	61	104	7	19	6	<1	0.12	0.02	0.17	—
2736	Hyacinth beans, boiled, drained	½	cup(s)	44	38	22	1	4	0	<1	0.05	0.06	0.00	—
515	Lima beans, boiled, drained	½	cup(s)	85	57	105	6	20	5	<1	0.06	0.02	0.13	—
570	Lima beans, baby, frozen, boiled, drained	½	cup(s)	90	65	95	6	18	5	<1	0.06	0.02	0.13	—
579	Mung beans, sprouted, boiled, drained	½	cup(s)	62	62	13	1	3	<1	<.1	0.02	0.00	0.02	—
510	Navy beans, boiled	½	cup(s)	91	57	129	8	24	6	1	0.13	0.05	0.22	0
32816	Pinto beans, boiled, drained, no salt added	½	cup(s)	114	106	25	2	5	0	<1	0.04	0.03	0.21	
1052	Pinto beans, frozen, boiled, drained	½	cup(s)	47	27	76	4	15	4	<1	0.03	0.02	0.13	—
514	Red kidney beans, canned	½	cup(s)	128	99	109	7	20	8	<1	0.06	0.03	0.24	—
1810	Refried beans, canned	½	cup(s)	127	96	119	7	20	7	2	0.60	0.71	0.19	—
1053	Shell beans, canned	½	cup(s)	123	111	37	2	8	4	<1	0.03	0.02	0.13	—
1670	Soybeans, boiled	½	cup(s)	86	54	149	14	9	5	8	1.12	1.70	4.36	—
1108	Soybeans, green, boiled, drained	½	cup(s)	90	62	127	11	10	4	6	0.67	1.09	2.71	—
1807	White beans, small, boiled	½	cup(s)	90	57	127	8	23	9	1	0.15	0.05	0.25	—
574	Green string beans, canned, fat added in cooking	½	cup(s)	93	63	41	1	4	2	3	0.51	1.23	0.75	—
575	Yellow snap, string or wax beans, boiled, drained	½	cup(s)	62	6	22	1	5	2	<1	0.04	0.00	0.09	—
576	Yellow snap, string or wax beans, frozen, boiled, drained	½	cup(s)	68	62	19	1	4	2	<1	0.02	0.00	0.05	—
	Beets													
580	Whole, boiled, drained	2	item(s)	100	87	44	2	10	2	<1	0.03	0.04	0.06	—
581	Sliced, boiled, drained	½	cup(s)	85	74	37	1	8	2	<1	0.02	0.03	0.05	—
583	Sliced, canned, drained	½	cup(s)	85	77	26	1	6	1	<1	0.02	0.02	0.04	—
2730	Pickled, canned with liquid	½	cup(s)	114	93	74	1	18	3	<.1	0.01	0.02	0.03	—
584	Beet greens, boiled, drained	½	cup(s)	72	64	19	2	4	2	<1	0.02	0.03	0.05	—
585	**Cowpeas or black-eyed peas, boiled, drained**	½	cup(s)	83	60	80	3	17	4	0.31	0.07	0.02	0.13	—

Chol (mg)	Calc (mg)	Iron (mg)	Magn (mg)	Pota (mg)	Sodi (mg)	Zinc (mg)	Vit A (µg)	Thia (mg)	Vit E (mg α)	Ribo (mg)	Niac (mg)	Vit B$_6$ (mg)	Fola (µg)	Vit C (mg)	Vit B$_{12}$ (µg)	Sele (µg)
0	54	1.55	72	425	114	0.59	11	0.08	0.23	0.08	1.20	0.13	61	12	0	<1
0	38	1.08	50	297	80	0.41	8	0.05	0.16	0.06	0.84	0.09	43	8	0	<1
0	21	0.81	13	202	13	0.54	49	0.14	1.35	0.12	0.97	0.07	134	7	0	5
0	19	0.73	12	208	347	0.48	50	0.07	0.38	0.12	1.15	0.13	116	22	0	2
0	21	0.58	12	196	4	0.50	—	0.06	1.08	0.09	0.93	0.02	121	22	0	4
0	7	0.14	2	320	2	0.28	0	0.01	—	0.03	0.18	0.06	1	0	0	<1
0	5	0.21	3	52	5	0.43	1	0.02	0.41	0.02	0.09	0.09	2	1	0	<1
0	32	2.30	60	612	9	2.04	0	0.13	—	0.07	0.82	0.11	139	0	0	1
8	62	2.22	36	302	553	2.40	5	0.07	0.59	0.07	1.16	0.06	39	3	0	8
9	71	4.15	44	380	557	7.41	5	0.07	0.13	0.06	0.63	0.09	29	4	0	6
9	77	2.10	43	336	425	1.90	1	0.06	0.04	0.08	0.44	0.11	47	4	0	6
0	23	1.81	60	305	1	0.96	0	0.21	—	0.05	0.43	0.06	128	0	0	1
0	40	2.37	39	239	6	1.25	1	0.10	0.29	0.05	0.43	0.11	141	1	0	3
0	26	1.55	36	258	59	0.63	9	0.06	0.25	0.05	0.91	0.10	18	11	0	1
0	56	0.96	50	327	5	0.57	0	0.12	—	0.05	0.48	0.09	66	1	0	1
0	60	1.89	44	346	2	0.78	0	0.14	—	0.05	0.60	0.10	90	1	0	4
0	18	0.33	18	114	1	0.17	3	0.02	—	0.04	0.21	0.01	20	2	0	1
0	27	2.08	63	485	14	0.67	16	0.12	0.12	0.08	0.88	0.16	22	9	0	2
0	25	1.76	50	370	26	0.50	7	0.06	0.58	0.05	0.69	0.10	14	5	0	2
0	7	0.40	9	63	6	0.29	1	0.03	0.04	0.06	0.50	0.03	18	7	0	<1
0	64	2.26	54	335	1	0.96	0	0.18	0.01	0.06	0.48	0.15	127	1	0	5
0	17	0.75	20	111	58	0.19	0	0.08	—	0.07	0.82	0.06	146	7	0	1
0	24	1.27	25	304	39	0.32	0	0.13	—	0.05	0.30	0.09	16	<1	0	1
0	31	1.61	36	329	436	0.70	0	0.13	0.77	0.11	0.58	0.03	65	1	0	2
10	44	2.10	42	338	378	1.48	0	0.03	0.00	0.02	0.40	0.18	14	8	0	2
0	36	1.21	18	134	409	0.33	13	0.04	0.04	0.07	0.25	0.06	22	4	0	1
0	88	4.42	74	443	1	0.99	0	0.13	0.30	0.25	0.34	0.20	46	1	0	6
0	131	2.25	54	485	13	0.82	7	0.23	—	0.14	1.13	0.05	100	15	0	1
0	65	2.54	61	414	2	0.98	0	0.21	—	0.05	0.24	0.11	123	0	0	1
0	24	0.81	12	100	266	0.26	129	0.01	0.40	0.05	0.18	0.03	—	4	0.00	—
0	29	0.80	16	187	2	0.22	5	0.05	0.28	0.06	0.38	0.03	21	6	0	<1
0	33	0.59	16	85	6	0.32	7	0.02	0.24	0.06	0.26	0.04	16	3	0	<1
0	16	0.79	23	305	77	0.35	2	0.03	0.04	0.04	0.33	0.07	80	4	0	1
0	14	0.67	20	259	65	0.30	2	0.02	0.03	0.03	0.28	0.06	68	3	0	1
0	13	1.55	14	126	165	0.18	1	0.01	0.03	0.03	0.13	0.05	26	3	0	<1
0	12	0.47	17	168	300	0.30	1	0.01	—	0.05	0.28	0.06	31	3	0	1
0	82	1.37	49	654	174	0.36	276	0.08	1.30	0.21	0.36	0.10	10	18	0	1
0	106	0.92	43	345	3	0.84	65	0.08	0.18	0.12	1.15	0.05	105	1.81	0	2

Table A–1
Food Composition

(DA+ code is for Wadsworth Diet Analysis program)
(For purposes of calculations, use "0" for t, <1, <.1, <.01, etc.)

A

DA + Code	Food Description	Quantity	Measure	Wt (g)	H₂O (g)	Ener (kcal)	Prot (g)	Carb (g)	Fiber (g)	Fat (g)	Fat Breakdown (g)			
											Sat	Mono	Poly	Trans
	VEGETABLES, LEGUMES–Continued													
	Broccoli													
587	Raw, chopped	½	cup(s)	44	39	15	1	3	1	<1	0.02	0.00	0.02	—
588	Chopped, boiled, drained	½	cup(s)	78	70	27	2	6	3	<1	0.06	0.03	0.13	—
590	Frozen, chopped, boiled, drained	½	cup(s)	92	83	26	3	5	3	<1	0.02	0.01	0.05	—
16848	**Broccoflower, raw, chopped**	½	cup(s)	32	29	10	1	2	1	<.1	0.01	0.01	0.04	—
	Brussels sprouts													
591	Boiled, drained	½	cup(s)	78	69	28	2	6	2	<1	0.08	0.03	0.20	—
592	Frozen, boiled, drained	½	cup(s)	78	67	33	3	6	3	<1	0.06	0.02	0.16	—
	Cabbage													
594	Raw, shredded	1	cup(s)	70	65	17	1	4	2	<.1	0.01	0.01	0.04	—
595	Boiled, drained, no salt added	1	cup(s)	150	140	33	2	7	3	1	0.08	0.05	0.29	—
35611	Chinese (pak choi or bok choy), boiled w/salt, drained	1	cup(s)	170	162	20	3	3	2	<1	0.04	0.02	0.13	—
16869	Kim chee	1	cup(s)	150	138	31	2	6	2	<1	0.04	0.02	0.15	—
596	Red, shredded, raw	1	cup(s)	70	63	22	1	5	1	<1	0.02	0.01	0.09	—
597	Savoy, shredded, raw	1	cup(s)	70	64	19	1	4	2	<.1	0.01	0.00	0.03	—
11710	**Capers**	1	teaspoon(s)	5	—	0	0	0	0	0	0.00	0.00	0.00	0
	Carrots													
600	Raw	½	cup(s)	61	54	25	1	6	2	<1	0.02	0.01	0.06	0
8691	Raw, baby	8	item(s)	80	72	28	1	7	1	<1	0.02	0.01	0.05	0
601	Grated	½	cup(s)	55	49	23	1	5	2	<1	0.02	0.01	0.06	0
602	Sliced, boiled, drained	½	cup(s)	78	70	27	<1	6	2	0.14	0.02	0	0.08	—
1055	Juice, canned	½	cup(s)	123	109	49	1	11	1	<1	0.03	0.01	0.09	—
32725	**Cassava or manioc**	½	cup(s)	103	61	165	1	39	2	<1	0.08	0.08	0.05	—
	Cauliflower													
605	Raw, chopped,	½	cup(s)	50	46	13	1	3	1	<1	0.02	0.01	0.05	—
606	Boiled, drained	½	cup(s)	62	58	14	1	3	2	<1	0.04	0.02	0.13	—
607	Frozen, boiled, drained	½	cup(s)	90	85	17	1	3	2	<1	0.03	0.01	0.09	—
	Celery													
609	Diced	½	cup(s)	60	58	8	<1	2	1	<1	0.03	0.02	0.05	—
608	Stalk	2	item(s)	80	76	11	1	2	1	<1	0.03	0.03	0.06	—
	Chard													
1056	Swiss chard, raw	1	cup(s)	36	33	7	1	1	1	<.1	0.01	0.01	0.03	—
1057	Swiss chard, boiled, drained	½	cup(s)	88	81	18	2	4	2	<.1	0.01	0.01	0.02	—
	Collard greens													
610	Boiled, drained	½	cup(s)	95	87	25	2	5	3	<1	0.04	0.02	0.16	—
611	Frozen, chopped, boiled, drained	½	cup(s)	85	75	31	3	6	2	<1	0.05	0.02	0.18	—
	Corn													
29614	Yellow corn, fresh, cooked	1	item(s)	100	72	107	3	25	3	1	0.19	0.37	0.59	—
612	Yellow sweet corn, boiled, drained	½	cup(s)	82	57	89	3	21	2	1	0.16	0.31	0.49	—
614	Yellow sweet corn, frozen, boiled, drained	½	cup(s)	82	63	66	2	16	2	1	0.08	0.16	0.26	—
615	Yellow creamed sweet corn, canned	½	cup(s)	128	101	92	2	23	2	1	0.08	0.16	0.25	—
618	**Cucumber**	¼	item(s)	75	72	11	<1	3	<1	<.1	0.03	0.00	0.04	—
16870	**Cucumber, kim chee**	½	cup(s)	75	68	16	1	4	1	<.1	0.02	0.00	0.03	—
	Dandelion greens													
2734	Raw	1	cup(s)	55	47	25	1	5	2	<1	0.09	0.01	0.17	—
620	Chopped, boiled, drained	½	cup(s)	53	47	17	1	3	2	<1	0.08	0.01	0.14	—
1066	**Eggplant, boiled, drained**	½	cup(s)	48	43	17	<1	4	1	<1	0.02	0.01	0.04	—
621	**Endive or escarole, chopped, raw**	1	cup(s)	53	49	9	1	2	2	<1	0.03	0.00	0.05	—
8784	**Jicama or yambean**	½	cup(s)	65	59	25	<1	6	3	<.1	0.01	0.00	0.03	—

PAGE KEY: A–2 = Breads/Baked Goods A–6 = Cereal/Rice/Pasta A–10 = Fruit A–14 = Vegetables/Legumes A–24 = Nuts/Seeds A–26 = Vegetarian
A–28 = Dairy A–34 = Eggs A–34 = Seafood A–36 = Meats A–40 = Poultry A–40 = Processed meats A–42 = Beverages A–46 = Fats/Oils A–48 = Sweets
A–50 = Spices/Condiments/Sauces A–52 = Mixed foods/Soups/Sandwiches A–58 = Fast food A–74 = Convenience meals A–76 = Baby foods

A

Chol (mg)	Calc (mg)	Iron (mg)	Magn (mg)	Pota (mg)	Sodi (mg)	Zinc (mg)	Vit A (µg)	Thia (mg)	Vit E (mg α)	Ribo (mg)	Niac (mg)	Vit B_6 (mg)	Fola (µg)	Vit C (mg)	Vit B_{12} (µg)	Sele (µg)
0	21	0.32	9	139	15	0.18	15	0.03	0.34	0.05	0.28	0.08	28	39	0	1
0	31	0.52	16	229	32	0.35	76	0.05	1.13	0.10	0.43	0.16	84	51	0	1
0	30	0.56	12	131	10	0.26	52	0.05	1.21	0.07	0.42	0.12	52	37	0	1
0	11	0.23	6	96	7	0.20	0	0.03	0.01	0.03	0.23	0.07	18	28	0	—
0	28	0.94	16	247	16	0.26	30	0.08	0.34	0.06	0.47	0.14	47	48	0	1
0	20	0.37	14	225	12	0.19	36	0.08	0.40	0.09	0.42	0.22	78	35	0	<1
0	33	0.41	11	172	13	0.13	6	0.04	0.10	0.03	0.21	0.07	30	23	0	1
0	47	0.26	12	146	12	0.14	11	0.09	0.18	0.08	0.42	0.17	30	30	0	1
0	158	1.77	19	631	459	0.29	360	0.05	0.15	0.11	0.73	0.28	70	44	0	1
0	145	1.28	27	375	995	0.36	—	0.07	0.08	0.10	0.75	0.34	88	80	0	—
0	32	0.56	11	170	19	0.15	39	0.04	0.12	0.05	0.29	0.15	13	40	0	<1
0	25	0.28	20	161	20	0.19	35	0.05	—	0.02	0.21	0.13	56	22	0	1
0	—	—	—	—	105	—	—	—	—	—	—	—	—	—	0	—
0	20	0.18	7	195	42	0.15	367	0.04	0.40	0.04	0.60	0.08	12	4	0	<.1
0	26	0.71	8	190	62	0.14	552	0.02	—	0.03	0.44	0.08	26	7	0	1
0	18	0.17	7	177	38	0.13	333	0.04	0.36	0.03	0.54	0.08	11	3	0	<.1
0	23	0.26	7.8	183	45	0.15	671	0.05	0.80	0.03	0.5	0.11	11	2.8	0	<1
0	30	0.57	17	359	36	0.22	1176	0.11	1.43	0.07	0.47	0.27	5	10	0	1
0	16	0.28	22	279	14	0.35	1	0.09	0.20	0.05	0.88	0.09	28	21	0	1
0	11	0.22	8	152	15	0.14	1	0.03	0.04	0.03	0.26	0.11	29	23	0	<1
0	10	0.20	6	88	9	0.11	1	0.03	0.04	0.03	0.25	0.11	27	27	0	<1
0	15	0.37	8	125	16	0.12	0	0.03	0.05	0.05	0.28	0.08	37	28	0	1
0	24	0.12	7	157	48	0.08	13	0.01	0.16	0.03	0.19	0.04	22	2	0	<1
0	32	0.16	9	208	64	0.10	18	0.02	0.22	0.05	0.26	0.06	29	2	0	<1
0	18	0.65	29	136	77	0.13	110	0.01	0.68	0.03	0.14	0.04	5	11	0	<1
0	51	1.98	75	480	157	0.29	268	0.03	1.65	0.08	0.32	0.07	8	16	0	1
0	133	1.10	19	110	15	0.22	386	0.04	0.84	0.10	0.55	0.12	88	17	0	<1
0	179	0.95	26	213	43	0.23	489	0.04	1.06	0.10	0.54	0.10	65	22	0	1
0	2	0.6	32	248	242	0.47	22	0.21	0.09	0.07	1.6	0.05	—	6	0	—
0	2	0.50	26	204	14	0.39	11	0.18	0.07	0.06	1.32	0.05	38	5	0	<1
0	2	0.39	23	191	1	0.52	8	0.02	0.06	0.05	1.08	0.08	29	3	0	1
0	4	0.49	22	172	365	0.68	5	0.03	0.09	0.07	1.23	0.08	58	6	0	1
0	12	0.21	10	111	2	0.15	4	0.02	0.02	0.02	0.07	0.03	5	2	0	<1
0	7	3.62	6	88	766	0.38	—	0.02	0.36	0.02	0.35	0.08	17	3	0	—
0	103	1.71	20	219	42	0.23	137	0.11	2.65	0.14	0.45	0.14	15	19	0	<1
0	74	0.95	13	122	23	0.15	260	0.07	1.79	0.09	0.27	0.08	7	9	0	<1
0	3	0.12	5	59	<1	0.06	1	0.04	0.20	0.01	0.29	0.04	7	1	0	<.1
0	27	0.44	8	165	12	0.41	57	0.04	0.23	0.04	0.21	0.01	75	3	0	<1
0	8	0.39	8	98	3	0.10	1	0.01	0.30	0.02	0.13	0.03	8	13	0	<1

Table A–1
Food Composition

(DA+ code is for Wadsworth Diet Analysis program)
(For purposes of calculations, use "0" for t, <1, <.1, <.01, etc.)

DA + Code	Food Description	Quantity	Measure	Wt (g)	H₂O (g)	Ener (kcal)	Prot (g)	Carb (g)	Fiber (g)	Fat (g)	Sat	Mono	Poly	Trans
	VEGETABLES, LEGUMES–Continued													
	Kale													
29313	Raw	1	cup(s)	67	57	34	2	7	1	<1	0.06	0.03	0.23	—
623	Frozen, chopped, boiled, drained	½	cup(s)	65	59	20	2	3	1	<1	0.04	0.02	0.15	—
	Kohlrabi													
1071	Raw	1	cup(s)	135	123	36	2	8	5	<1	0.02	0.01	0.06	—
1072	Boiled, drained	½	cup(s)	83	74	24	1	6	1	<.1	0.01	0.01	0.04	—
	Leeks													
1073	Raw	1	cup(s)	89	74	54	1	13	2	<1	0.04	0.00	0.15	—
1074	Boiled, drained	½	cup(s)	52	47	16	<1	4	1	<1	0.01	0.00	0.06	—
	Lentils													
522	Boiled	½	cup(s)	99	69	115	9	20	8	<1	0.05	0.06	0.17	—
1075	Sprouted	1	cup(s)	77	52	82	7	17	0	<1	0.04	0.08	0.17	—
	Lettuce													
624	Butterhead, boston, or bibb	1	cup(s)	55	53	7	1	1	1	<1	0.02	0.00	0.06	—
625	Butterhead leaves	11	piece(s)	83	79	11	1	2	1	<1	0.02	0.01	0.10	—
626	Iceberg	1	cup(s)	55	53	6	<1	1	1	<.1	0.01	0.00	0.03	—
628	Iceberg, chopped	1	cup(s)	55	53	6	<1	1	1	<.1	0.01	0.00	0.03	—
629	Looseleaf	1	cup(s)	56	54	8	1	2	1	<.1	0.01	0.00	0.05	—
1665	Romaine, shredded	1	cup(s)	56	53	10	1	2	1	<1	0.02	0.01	0.09	—
	Mushrooms													
15585	Crimini (about 6)	3	ounce(s)	85	28	4	3	2	0	0	0.00	0.00	0	0
8700	Enoki	30	item(s)	90	80	31	2	6	2	<1	0.04	0.01	0.14	—
630	Mushrooms, raw	½	cup(s)	35	32	8	1	1	<1	<1	0.02	0.00	0.05	—
1079	Mushrooms, boiled, drained	½	cup(s)	78	71	22	2	4	2	<1	0.05	0.01	0.14	—
1080	Mushrooms, canned, drained	½	cup(s)	78	71	20	1	4	2	<1	0.03	0.00	0.09	—
15587	Portobello, grilled	1	item(s)	85	30	3	4	3	0	0	0.00	0.00	0	0
2743	Shiitake, cooked	½	cup(s)	73	61	40	1	10	2	<1	0.04	0.05	0.02	—
	Mustard greens													
29319	Raw	1	cup(s)	56	51	15	2	3	2	<1	0.01	0.05	0.02	—
2744	Frozen, boiled, drained	½	cup(s)	75	70	14	2	2	2	<1	0.01	0.08	0.04	—
	Okra													
632	Sliced, boiled, drained	½	cup(s)	80	74	18	1	4	2	<1	0.04	0.02	0.04	—
32742	Frozen, boiled, drained, no salt added	½	cup(s)	92	84	26	2	5	3	<1	0.07	0.05	0.07	—
16866	Batter coated, fried	11	piece(s)	83	55	160	2	13	2	11	1.50	2.80	6.37	—
	Onions													
633	Raw, chopped	½	cup(s)	80	71	34	1	8	1	<.1	0.02	0.02	0.05	—
635	Chopped, boiled, drained	½	cup(s)	106	93	47	1	11	1	<1	0.03	0.03	0.08	—
2748	Frozen, boiled, drained	½	cup(s)	106	98	30	1	7	2	<1	0.02	0.01	0.04	—
16850	Red onions, sliced, raw	½	cup(s)	58	52	22	1	5	1	<.1	0.02	0.01	0.04	—
636	Scallions, green or spring onions	2	item(s)	30	27	10	1	2	1	<.1	0.01	0.01	0.02	—
1081	Onion rings, breaded & pan fried, frozen, heated	11	item(s)	78	22	318	4	30	1	21	6.70	8.49	3.99	—
16860	**Palm hearts, cooked**	½	cup(s)	73	51	75	2	19	1	<1	0.03	0.00	0.07	—
637	**Parsley, chopped**	1	tablespoon(s)	4	3	1	<1	<1	<1	<.1	0.01	0.01	0.00	—
638	**Parsnips, sliced, boiled, drained**	½	cup(s)	78	63	55	1	13	3	<1	0.04	0.09	0.04	—
	Peas													
639	Green peas, canned, drained	½	cup(s)	85	69	59	4	11	3	<1	0.05	0.03	0.14	—
641	Green peas, frozen, boiled, drained	½	cup(s)	80	64	62	4	11	4	<1	0.04	0.02	0.10	—
35694	Pea pods, boiled w/salt, drained	½	cup(s)	80	71	34	3	6	2	<1	0.04	0.02	0.08	—
1082	Peas & carrots, canned w/liquid	½	cup(s)	128	112	48	3	11	3	<1	0.06	0.03	0.16	—
1083	Peas & carrots, frozen, boiled, drained	½	cup(s)	80	69	38	2	8	2	<1	0.06	0.03	0.16	—

Chol (mg)	Calc (mg)	Iron (mg)	Magn (mg)	Pota (mg)	Sodi (mg)	Zinc (mg)	Vit A (µg)	Thia (mg)	Vit E (mg α)	Ribo (mg)	Niac (mg)	Vit B$_6$ (mg)	Fola (µg)	Vit C (mg)	Vit B$_{12}$ (µg)	Sele (µg)
0	90	1.14	23	299	29	0.29	515	0.07	—	0.09	0.67	0.18	19	80	0	1
0	90	0.61	12	209	10	0.12	478	0.03	0.60	0.07	0.44	0.06	9	16	0	1
0	32	0.54	26	473	27	0.04	3	0.07	0.65	0.03	0.54	0.20	22	84	0	1
0	21	0.33	16	281	17	0.26	2	0.03	0.43	0.02	0.32	0.13	10	45	0	1
0	53	1.87	25	160	18	0.11	74	0.05	0.82	0.03	0.36	0.21	57	11	0	1
0	16	0.57	7	45	5	0.03	1	0.01	—	0.01	0.10	0.06	13	2	0	<1
0	19	3.30	36	365	2	1.26	0	0.17	0.11	0.07	1.05	0.18	179	1	0	3
0	19	2.47	28	248	8	1.16	2	0.18	—	0.10	0.87	0.15	77	13	0	<1
0	19	0.69	7	132	3	0.11	92	0.03	0.10	0.03	0.20	0.05	40	2	0	<1
0	29	1.02	11	196	4	0.17	137	0.05	0.15	0.05	0.29	0.07	60	3	0	<1
0	11	0.19	4	84	5	0.09	9	0.02	0.10	0.01	0.07	0.03	31	2	0	<1
0	11	0.19	4	84	5	0.09	9	0.02	0.10	0.01	0.07	0.03	31	2	0	<1
0	20	0.48	7	109	16	0.10	208	0.04	0.16	0.05	0.21	0.05	21	10	0	<1
0	19	0.55	8	139	5	0.13	163	0.04	0.07	0.04	0.18	0.04	77	14	0	<1
0	1	—	—	33	—	0	—	—	—	—	—	—	—	0	0	—
0	1	0.80	14	343	3	0.51	0	0.08	0.01	0.09	3.28	0.04	27	11	0	14
0	1	0.18	3	110	1	0.18	0	0.03	0.00	0.15	1.35	0.04	6	1	<.1	3
0	5	1.36	9	278	2	0.68	0	0.06	0.01	0.23	3.48	0.07	14	3	0	9
0	9	0.62	12	101	332	0.56	0	0.07	0.01	0.02	1.24	0.05	9	0	0	3
40	<1	—	—	10	—	0	—	—	—	—	—	—	—	0	0	—
0	2	0.32	10	85	3	0.96	0	0.03	0.01	0.12	1.09	0.12	15	<1	0	18
0	58	0.82	18	199	14	0.11	295	0.05	1.13	0.06	0.45	0.10	105	39	0	1
0	76	0.84	10	104	19	0.15	266	0.03	1.01	0.04	0.19	0.08	53	10	0	<1
0	62	0.22	29	108	5	0.34	11	0.11	0.22	0.04	0.70	0.15	37	13	0	<1
0	88	0.62	47	215	3	0.57	16	0.09	0.29	0.11	0.72	0.04	134	11	0	1
2	54	1.13	32	170	110	0.44	—	0.16	1.51	0.13	1.29	0.11	34	9	<.1	—
0	18	0.15	8	115	2	0.13	0	0.04	0.02	0.02	0.07	0.12	15	5	0	<1
0	23	0.26	12	177	3	0.22	0	0.04	0.02	0.02	0.18	0.14	16	6	0	1
0	17	0.32	6	115	13	0.07	0	0.02	0.01	0.03	0.15	0.07	14	3	0	<1
0	11	0.13	6	90	2	0.11	0	0.02	0.01	0.01	0.09	0.07	11	4	0	—
0	22	0.44	6	83	5	0.12	15	0.02	0.17	0.02	0.16	0.02	19	6	0	<1
0	24	1.32	15	101	293	0.33	9	0.22	—	0.11	2.82	0.06	52	1	0	3
0	13	1.23	7	1318	10	2.72	—	0.03	0.37	0.13	0.62	0.53	15	5	0	—
0	5	0.24	2	21	2	0.04	16	0.00	0.03	0.00	0.05	0.00	6	5	0	<.1
0	29	0.45	23	286	8	0.20	0	0.06	0.78	0.04	0.56	0.07	45	10	0	1
0	17	0.81	14	147	214	0.60	23	0.10	0.03	0.07	0.62	0.05	37	8	0	1
0	19	1.22	18	88	58	0.54	84	0.23	0.02	0.08	1.18	0.09	47	8	0	1
0	34	1.58	21	192	192	0.30	43	0.10	0.31	0.06	0.43	0.12	23	38	0	1
0	29	0.96	18	128	332	0.74	368	0.09	—	0.07	0.74	0.11	23	8	0	1
0	18	0.75	13	126	54	0.36	374	0.18	0.42	0.05	0.92	0.07	21	6	0	1

Table A–1
Food Composition

(DA+ code is for Wadsworth Diet Analysis program)
(For purposes of calculations, use "0" for t, <1, <.1, <.01, etc.)

DA + Code	Food Description	Quantity	Measure	Wt (g)	H₂O (g)	Ener (kcal)	Prot (g)	Carb (g)	Fiber (g)	Fat (g)	Sat	Mono	Poly	Trans
	VEGETABLES, LEGUMES													
	Peas–Continued													
640	Snow or sugar peas, raw	½	cup(s)	32	28	13	1	2	1	<.1	0.01	0.01	0.03	—
2750	Snow or sugar peas, frozen, boiled, drained	½	cup(s)	80	69	42	3	7	2	<1	0.06	0.03	0.13	—
29324	Split peas, sprouted	½	cup(s)	60	37	77	5	17	0	<1	0.07	0.04	0.20	—
	Peppers													
643	Green bell or sweet, raw	½	cup(s)	75	70	15	1	3	1	<1	0.04	0.01	0.05	—
644	Green bell or sweet, boiled, drained	½	cup(s)	68	62	19	1	5	1	<1	0.02	0.01	0.07	—
1664	Green hot chili	1	item(s)	45	39	18	1	4	1	<.1	0.01	0.00	0.05	—
1663	Green hot chili, canned w/liquid	½	cup(s)	68	63	14	1	3	1	<.1	0.01	0.00	0.04	—
1086	Jalapeno, canned w/liquid	½	cup(s)	68	60	18	1	3	2	1	0.07	0.04	0.35	—
8703	Yellow bell or sweet	1	item(s)	186	171	50	2	12	2	<1	0.06	0.03	0.21	—
1087	**Poi**	½	cup(s)	122	87	136	<1	33	<1	<1	0.04	0.01	0.07	—
	Potatoes													
5791	Baked, flesh & skin	1	item(s)	202	144	220	5	51	4	<1	0.05	0.00	0.09	—
645	Baked, flesh only	½	cup(s)	61	46	57	1	13	1	<.1	0.02	0.00	0.03	—
1088	Baked, skin only	1	item(s)	58	27	115	2	27	5	<.1	0.02	0.00	0.02	—
5794	Boiled, drained, skin & flesh	1	item(s)	150	116	129	3	30	2	<1	0.04	0.00	0.06	—
647	Boiled, flesh only	½	cup(s)	78	60	67	1	16	1	<.1	0.02	0.00	0.03	—
5795	Boiled in skin, drained, flesh only	1	item(s)	136	105	118	3	27	2	<1	0.04	0.00	0.06	—
2759	Microwaved	1	item(s)	202	146	212	5	49	5	<1	0.05	0.00	0.09	—
5804	Microwaved, skin only	1	item(s)	58	37	77	3	17	4	<.1	0.02	0.00	0.02	—
2760	Microwaved in skin, flesh only	½	cup(s)	78	57	78	2	18	1	<.1	0.02	0.00	0.03	—
1089	Au gratin, prepared w/butter	½	cup(s)	123	91	162	6	14	2	9	5.80	2.63	0.34	—
1090	Au gratin mix, prepared w/water, whole milk, & butter	½	cup(s)	114	90	106	3	15	1	5	2.94	1.34	0.15	—
648	French fried, deep fried, prepared from raw	14	item(s)	70	32	190	3	24	2	10	1.93	4.21	2.97	—
649	French fried, frozen, heated	14	item(s)	70	40	140	2	22	2	5	0.88	3.33	0.55	—
1091	Hashed brown	½	cup(s)	78	37	207	2	27	2	10	1.11	3.13	2.78	—
653	Mashed, from dehydrated granules w/milk, water, & margarine	½	cup(s)	105	80	122	2	17	1	5	1.27	2.05	1.41	—
652	Mashed, w/margarine & whole milk	½	cup(s)	105	79	119	2	18	2	4	1.05	1.83	1.27	—
1097	Potato puffs, frozen, heated	½	cup(s)	64	34	142	2	20	2	7	3.26	2.79	0.51	—
1093	Scalloped, prepared w/butter	½	cup(s)	123	99	105	4	13	2	5	2.76	1.27	0.20	—
1094	Scalloped mix, prepared w/water, whole milk, & butter	½	cup(s)	114	90	106	2	15	1	5	2.99	1.38	0.22	—
	Pumpkin													
1773	Boiled, drained	½	cup(s)	123	115	25	1	6	1	<.1	0.05	0.01	0.00	—
656	Canned	½	cup(s)	123	110	42	1	10	4	<1	0.18	0.05	0.02	—
	Radicchio									<.1				
2498	Raw	1	cup(s)	40	37	9	1	2	<1	<1	0.02	0.00	0.04	—
8731	Raw, leaves	10	item(s)	80	75	18	1	4	1	<1	0.05	0.01	0.09	—
657	**Radishes**	6	item(s)	27	26	4	<1	1	<1	<.1	0.01	0.00	0.01	—
1099	**Rutabaga, boiled, drained**	½	cup(s)	85	76	33	1	7	2	<1	0.02	0.02	0.08	—
658	**Sauerkraut, canned**	½	cup(s)	114	105	22	1	5	3	<1	0.04	0.01	0.07	—
	Seaweed													
1102	Kelp	½	cup(s)	41	33	17	1	4	1	<1	0.10	0.04	0.02	—
1104	Spirulina, dried	½	cup(s)	8	<1	22	4	2	<1	1	0.20	0.05	0.16	—
1106	**Shallots**	3	tablespoon(s)	30	24	22	1	5	0	<.1	0.01	0.00	0.01	—
	Soybeans													
1670	Boiled	½	cup(s)	86	<1	149	14	9	5	8	1.11	1.7	4.35	—
2825	Dry roasted	½	cup(s)	86	1	388	34	28	7	19	2.69	4.11	10.50	—

Chol (mg)	Calc (mg)	Iron (mg)	Magn (mg)	Pota (mg)	Sodi (mg)	Zinc (mg)	Vit A (µg)	Thia (mg)	Vit E (mg α)	Ribo (mg)	Niac (mg)	Vit B6 (mg)	Fola (µg)	Vit C (mg)	Vit B12 (µg)	Sele (µg)
0	14	0.66	8	63	1	0.09	17	0.05	0.12	0.03	0.19	0.05	13	19	0	<1
0	47	1.92	22	174	4	0.39	53	0.05	0.38	0.10	0.45	0.14	28	18	0	1
0	22	1.36	34	229	12	0.63	5	0.14	—	0.09	1.85	0.16	86	6	0	<1
0	7	0.25	7	130	2	0.10	13	0.04	0.28	0.02	0.36	0.17	8	60	0	0
0	6	0.31	7	113	1	0.08	10	0.04	0.36	0.02	0.32	0.16	11	51	0	<1
0	8	0.54	11	153	3	0.14	27	0.04	0.31	0.04	0.43	0.13	10	109	0	<1
0	5	0.34	10	127	798	0.12	24	0.01	0.47	0.03	0.54	0.10	7	46	0	<1
0	16	1.28	10	131	1136	0.23	58	0.03	0.47	0.03	0.27	0.13	10	7	0	<1
0	20	0.86	22	394	4	0.32	19	0.05	—	0.05	1.66	0.31	48	341	0	1
0	19	1.07	29	223	15	0.27	4	0.16	2.80	0.05	1.34	0.33	26	5	0	1
0	20	2.75	55	844	16	0.65	0	0.22	—	0.07	3.32	0.70	22	26	0	2
0	3	0.21	15	239	3	0.18	0	0.06	0.02	0.01	0.85	0.18	5	8	0	<1
0	20	4.08	25	332	12	0.28	1	0.07	0.02	0.06	1.78	0.36	13	8	0	<1
0	13	1.27	34	572	7	0.47	0	0.15	—	0.03	2.13	0.44	15	18	0	—
0	6	0.24	16	256	4	0.21	0	0.08	0.01	0.01	1.02	0.21	7	6	0	<1
0	7	0.42	30	515	5	0.41	0	0.14	—	0.03	1.96	0.41	14	18	0	<1
0	22	2.50	55	903	16	0.73	0	0.24	—	0.06	3.46	0.69	24	31	0	1
0	27	3.45	21	377	9	0.30	0	0.04	—	0.04	1.29	0.29	10	9	0	<1
0	4	0.32	20	321	5	0.26	0	0.10	—	0.02	1.27	0.25	9	12	0	<1
28	146	0.78	25	485	530	0.85	78	0.08	—	0.14	1.22	0.21	13	12	0	3
17	94	0.36	17	249	499	0.27	59	0.02	—	0.09	1.07	0.05	8	4	0	3
0	9	1.02	28	731	8	0.53	0	0.10	0.09	0.05	1.90	0.33	13	21	0	—
0	6	0.87	15	293	21	0.28	0	0.08	0.08	0.02	1.46	0.22	8	7	0	<1
0	11	0.43	27	449	267	0.37	0	0.13	0.01	0.03	1.80	0.37	12	10	0	<1
2	34	0.22	21	163	181	0.25	49	0.09	0.54	0.09	0.91	0.17	8	7	<1	6
1	21	0.27	20	342	350	0.32	43	0.10	0.44	0.05	1.23	0.26	9	11	<.1	1
0	19	1.00	12	243	477	0.19	0	0.13	0.15	0.05	1.38	0.15	11	4	0	<1
15	70	0.70	23	463	410	0.49	39	0.08	—	0.11	1.29	0.22	13	13	0	2
13	41	0.43	16	231	388	0.28	40	0.02	—	0.06	1.17	0.05	11	4	0	2
0	18	0.70	11	282	1	0.28	306	0.04	0.98	0.10	0.51	0.05	11	6	0	<1
0	32	1.70	28	252	6	0.21	953	0.03	1.30	0.07	0.45	0.07	15	5	0	<1
0	8	0.23	5	121	9	0.25	<1	0.01	0.90	0.01	0.10	0.02	24	3	0	<1
0	15	0.46	10	242	18	0.50	1	0.01	1.81	0.02	0.20	0.05	48	6	0	1
0	7	0.09	3	63	11	0.08	0	0.00	0.00	0.01	0.07	0.02	7	4	0	<1
0	41	0.45	20	277	17	0.30	0	0.07	0.27	0.03	0.61	0.09	13	16	0	1
0	34	1.67	15	193	751	0.22	1	0.02	0.11	0.02	0.16	0.15	27	17	0	1
0	68	1.16	49	36	94	0.50	2	0.02	0.35	0.06	0.19	0.00	73	1	0	<1
0	9	2.14	15	102	79	0.15	2	0.18	0.38	0.28	0.96	0.03	7	1	0	1
0	11	0.36	6	100	4	0.12	18	0.02	—	0.01	0.06	0.10	10	2	0	<1
0	88	4.42	74	443	0.86	0.98	0.86	0.13	0.30	0.24	0.34	0.2	46	1	0	6
0	120	3.40	196	1173	2	4.10	0	0.37	—	0.65	0.91	0.19	176	4	0	17

Table A–1
Food Composition

(DA+ code is for Wadsworth Diet Analysis program)
(For purposes of calculations, use "0" for t, <1, <.1, <.01, etc.)

DA + Code	Food Description	Quantity	Measure	Wt (g)	H₂O (g)	Ener (kcal)	Prot (g)	Carb (g)	Fiber (g)	Fat (g)	Sat	Mono	Poly	Trans
	VEGETABLES, LEGUMES													
	Soybeans–Continued													
2824	Roasted, salted	½	cup(s)	86	2	405	30	29	15	22	3.16	4.82	12.33	—
30282	Soup (miso)	1	cup(s)	240	218	85	6	8	2	3	0.59	1.05	1.47	—
8739	Sprouted, stir fried	3	ounce(s)	85	57	106	11	8	1	6	0.84	1.37	3.41	—
	Soy products													
1813	Soy milk	1	cup(s)	240	214	118	9	11	3	5	0.51	0.78	2.00	—
2838	Tofu, dried, frozen (koyadofu)	3	ounce(s)	85	5	408	41	12	6	26	3.73	5.70	14.57	—
13844	Tofu, extra firm	3	ounce(s)	79	—	80	8	2	1	4	0.50	0.87	2.60	—
13843	Tofu, firm	3	ounce(s)	79	—	80	8	2	1	4	0.50	0.87	2.17	—
1816	Tofu, firm, w/calcium sulfate & magnesium chloride (nigari)	3	ounce(s)	85	72	65	7	3	<1	4	0.54	0.83	2.14	—
1817	Tofu, fried	3	ounce(s)	85	43	230	15	9	3	17	2.48	3.79	9.69	—
13841	Tofu, silken	3	ounce(s)	91	—	30	6	0	1	1	0.50	0.51	1.52	—
13842	Tofu, soft	3	ounce(s)	91	—	30	6	1	1	1	0.50	1.00	2.00	—
1671	Tofu, soft, w/calcium sulfate & magnesium chloride (nigari)	3	ounce(s)	85	73	52	6	2	<1	3	0.45	0.69	1.76	—
	Spinach													
659	Raw, chopped	1	cup(s)	30	27	7	1	1	1	<1	0.02	0.00	0.05	—
663	Canned, drained	½	cup(s)	108	100	25	3	4	3	1	0.09	0.02	0.23	—
660	Chopped, boiled, drained	½	cup(s)	90	82	21	3	3	2	<1	0.04	0.01	0.10	—
661	Chopped, frozen, boiled, drained	½	cup(s)	95	84	30	4	5	4	<1	0.09	0.00	0.20	—
662	Leaf, frozen, boiled, drained	½	cup(s)	95	84	30	4	5	4	<1	0.09	0.00	0.20	—
8470	Trimmed leaves	1	cup(s)	32	27	3	1	<.1	3	<.1	—	—	—	—
	Squash													
1662	Acorn, baked	½	cup(s)	103	85	57	1	15	5	<1	0.03	0.01	0.06	—
29702	Acorn, boiled, mashed	½	cup(s)	123	110	42	1	11	3	<.1	0.02	0.01	0.04	—
1661	Butternut, baked	½	cup(s)	103	90	41	1	11	3	<.1	0.02	0.01	0.04	—
29451	Butternut, frozen, boiled	½	cup(s)	132	116	51	2	13	2	<.1	0.02	0.01	0.04	—
32773	Butternut, frozen, boiled, mashed, no salt added	½	cup(s)	122	<1	47	1	12	0	<.1	0.02	0.00	0.03	—
29700	Crookneck & straightneck, boiled, drained	½	cup(s)	90	84	18	<1	4	1	<1	0.05	0.02	0.11	—
29703	Hubbard, baked	v	cup(s)	103	87	51	3	11	0	1	0.13	0.05	0.27	—
1660	Hubbard, boiled, mashed	½	cup(s)	118	107	35	2	8	3	<1	0.09	0.03	0.18	—
29704	Spaghetti, boiled, drained, or baked	½	cup(s)	78	72	21	1	5	1	<1	0.05	0.02	0.10	—
664	Summer, all varieties, sliced, boiled, drained	½	cup(s)	90	84	18	1	4	1	<1	0.06	0.02	0.12	—
665	Winter, all varieties, baked, mashed	½	cup(s)	103	91	38	1	9	3	<1	0.13	0.05	0.27	—
1112	Zucchini, boiled, drained	½	cup(s)	90	85	14	1	4	1	<.1	0.01	0.00	0.02	—
1113	Zucchini, frozen, boiled, drained	½	cup(s)	113	107	19	1	4	1	<1	0.03	0.01	0.06	—
	Sweet potatoes													
666	Baked, peeled	½	cup(s)	100	76	90	2	21	3	<1	0.03	0.00	0.06	—
667	Boiled, mashed	½	cup(s)	166	133	126	2	29	4	<1	0.05	0.00	0.10	—
668	Candied, home recipe	½	cup(s)	84	56	115	1	23	2	3	1.13	0.53	0.12	—
670	Canned, vacuum pack	½	cup(s)	100	76	91	2	21	2	<1	0.04	0.01	0.09	—
2765	Frozen, baked	½	cup(s)	88	65	88	2	21	2	<1	0.02	0.00	0.05	—
1136	Yams, baked or boiled, drained	½	cup(s)	68	48	79	1	19	3	<.1	0.02	0.00	0.04	—
32785	**Taro shoots, cooked, no salt added**	½	cup(s)	70	67	10	1	2	0	<.1	0.01	0.00	0.02	—
	Tomatillo													
8774	Raw	2	item(s)	68	62	22	1	4	1	1	0.09	0.11	0.28	—
8777	Raw, chopped	½	cup(s)	66	60	21	1	4	1	1	0.09	0.10	0.28	—

Chol (mg)	Calc (mg)	Iron (mg)	Magn (mg)	Pota (mg)	Sodi (mg)	Zinc (mg)	Vit A (µg)	Thia (mg)	Vit E (mg α)	Ribo (mg)	Niac (mg)	Vit B$_6$ (mg)	Fola (µg)	Vit C (mg)	Vit B$_{12}$ (µg)	Sele (µg)
0	119	3.35	125	1264	140	2.70	9	0.09	0.78	0.12	1.21	0.18	181	2	0	16
0	64	1.89	37	361	988	0.87	—	0.06	0.96	0.16	2.61	0.17	57	4	<1	—
0	70	0.34	82	482	12	1.79	1	0.36	—	0.16	0.94	0.14	108	10	0	1
0	10	1.39	46	338	29	0.55	5	0.39	3.24	0.17	0.35	0.10	5	0	0	3
0	310	8.28	50	17	5	4.17	22	0.42	—	0.27	1.01	0.24	78	1	0	46
0	60	1.08	78	—	0	—	0	—	0.03	—	—	—	—	0	0	—
0	60	1.08	52	—	0	—	0	—	—	—	—	—	—	0	0	—
0	138	1.23	39	150	7	0.85	1	0.07	—	0.08	0	0.05	28	<1	0	8
0	316	4.14	51	124	14	1.69	1	0.14	0.03	0.04	0.09	0.08	23	0	0	24
0	300	0.73	35	—	65	—	0	—	—	—	—	—	—	0	2	—
0	300	0.72	33	—	65	—	0	—	—	—	—	—	—	0	2	—
0	94	0.94	23	102	7	0.54	1	0.03	0.01	0.03	0.45	0.04	37	<1	0	8
0	30	0.81	24	167	24	0.16	141	0.02	0.61	0.06	0.22	0.06	58	8	0	<1
0	138	2.49	82	375	29	0.50	531	0.02	2.10	0.15	0.42	0.11	106	16	0	2
0	122	3.21	78	419	63	0.68	472	0.09	1.87	0.21	0.44	0.22	131	9	0	1
0	145	1.86	78	287	92	0.47	573	0.07	3.36	0.17	0.42	0.13	115	2	0	5
0	145	1.86	78	287	92	0.47	573	0.07	3.36	0.17	0.42	0.13	115	2	0	5
0	25	2.13	25	134	38	0.18	—	0.03	—	0.06	0.18	0.07	<.1	8	0	—
0	45	0.95	44	448	4	0.17	22	0.17	—	0.01	0.90	0.20	19	11	0	1
0	32	0.69	32	322	4	0.13	50	0.12	—	0.01	0.65	0.14	13	8	0	<1
0	42	0.62	30	291	4	0.13	572	0.07	1.32	0.02	0.99	0.13	19	15	0	1
0	25	0.77	12	176	3	0.16	—	0.07	—	0.05	0.61	0.09	22	5	0	1
0	23	0.70	11	162	2	0.14	406	0.05	—	0.05	0.56	0.08	19	4	0	1
0	18	0.41	18	184	1.73	0.25	29	0.04	—	0.03	0.39	0.09	20	7	0	<1
0	17	0.48	23	367	8	0.15	310	0.08	—	0.05	0.57	0.18	16	10	0	1
0	12	0.33	15	253	6	0.12	236	0.05	0.14	0.03	0.39	0.12	12	8	0	<1
0	16	0.26	9	91	14	0.16	5	0.03	0.09	0.02	0.63	0.08	6	3	0	<1
0	24	0.32	22	173	1	0.35	10	0.04	0.13	0.04	0.46	0.06	18	5	0	<1
0	23	0.45	13	448	1	0.23	268	0.02	0.12	0.07	0.51	0.17	21	10	0	<1
0	12	0.32	20	228	3	0.16	50	0.04	0.11	0.04	0.39	0.07	15	4	0	<1
0	19	0.54	15	219	2	0.23	11	0.05	0.14	0.05	0.44	0.05	9	4	0	<1
0	38	0.69	27	475	36	0.32	961	1.45	0.71	0.11	1.49	0.29	6	20	0	<1
0	45	1.20	30	382	45	0.33	1310	0.09	1.56	0.08	0.89	0.27	10	21	0	<1
7	22	0.95	9	159	59	0.13	176	0.02	—	0.04	0.33	0.03	9	6	0	1
0	22	0.89	22	312	53	0.18	399	0.04	1.00	0.06	0.74	0.19	17	26	0	1
0	31	0.48	18	332	7	0.26	722	0.06	0.68	0.05	0.49	0.16	19	8	0	1
0	10	0.36	12	458	5	0.14	4	0.06	0.26	0.02	0.38	0.16	11	8	0	<1
0	10	0.29	6	241	1	0.38	2	0.03	—	0.04	0.57	0.08	2	13	0	1
0	5	0.42	14	182	1	0.15	4	0.03	0.26	0.02	1.26	0.04	5	8	0	<1
0	5	0.41	13	177	1	0.15	4	0.03	0.25	0.02	1.22	0.04	5	8	0	<1

Table A–1
Food Composition

(DA+ code is for Wadsworth Diet Analysis program)
(For purposes of calculations, use "0" for t, <1, <.1, <.01, etc.)

A

DA + Code	Food Description	Quantity	Measure	Wt (g)	H₂O (g)	Ener (kcal)	Prot (g)	Carb (g)	Fiber (g)	Fat (g)	Sat	Mono	Poly	Trans
	VEGETABLES, LEGUMES–Continued											Fat Breakdown (g)		
	Tomato													
671	Fresh, ripe, red	1	item(s)	123	116	22	1	5	1	<1	0.05	0.06	0.16	—
16846	Fresh, cherry	5	item(s)	85	80	18	1	4	1	<1	0.03	0.04	0.11	—
3952	Diced, red	½	cup(s)	90	85	16	1	4	1	<1	0.04	0.05	0.12	—
1118	Boiled, red	½	cup(s)	120	113	22	1	5	1	<1	0.02	0.02	0.05	—
675	Juice, canned	½	cup(s)	122	115	21	1	5	<1	<.1	0.01	0.01	0.03	—
75	Juice, no salt added	½	cup(s)	122	115	21	1	5	<1	<.1	0.01	0.01	0.03	—
1699	Paste, canned	2	tablespoon(s)	33	24	27	1	6	1	<1	0.04	0.03	0.07	—
1700	Puree, canned	¼	cup(s)	63	55	24	1	6	1	<1	0.02	0.02	0.05	—
1125	Sauce, canned	¼	cup(s)	61	55	20	1	5	1	<1	0.02	0.02	0.06	—
1120	Stewed, canned, red	½	cup(s)	128	117	33	1	8	1	<1	0.03	0.04	0.10	—
8778	Sun dried	½	cup(s)	27	4	70	4	15	3	1	0.12	0.13	0.30	—
8783	Sun dried in oil, drained	¼	cup(s)	28	15	59	1	6	2	4	0.52	2.38	0.57	—
	Turnips													
677	Turnips, cubed, boiled, drained	½	cup(s)	78	73	17	1	4	2	<.1	0.01	0.00	0.03	—
678	Turnip greens, chopped, boiled, drained	½	cup(s)	72	67	14	1	3	3	<1	0.04	0.01	0.07	—
679	Turnip greens, frozen, chopped, boiled, drained	½	cup(s)	82	74	24	3	4	3	<1	0.08	0.02	0.14	—
	Vegetables, mixed													
1132	Canned, drained	½	cup(s)	82	71	40	2	8	2	<1	0.04	0.01	0.10	—
680	Frozen, boiled, drained	½	cup(s)	91	76	59	3	12	4	<1	0.03	0.01	0.07	—
7489	Vegetable juice, V8 100%	½	cup(s)	120	113	25	1	5	1	0	0.00	0.00	0.00	0
7490	Vegetable juice, V8 low sodium	½	cup(s)	120	113	25	0	7	1	0	0.00	0.00	0.00	0
7491	Vegetable juice, V8 spicy hot	½	cup(s)	120	113	25	1	5	1	0	0.00	0.00	0.00	0
	Water chestnuts													
31073	Sliced, drained	½	cup(s)	75	70	20	<1	5	1	0	0.00	0.00	0.00	0
31087	Whole	½	cup(s)	75	70	20	<1	5	1	0	0.00	0.00	0.00	0
1135	**Watercress**	1	cup(s)	34	32	4	1	<1	<1	<.1	0.01	0.00	0.01	—
	NUTS, SEEDS, AND PRODUCTS													
	Almonds													
32886	Blanched	¼	cup(s)	36	2	211	8	7	4	18	1.41	11.70	4.37	—
32887	Dry roasted, no salt added	¼	cup(s)	35	1	206	8	7	4	18	1.40	11.61	4.36	—
29724	Dry roasted, salted	¼	cup(s)	35	1	206	8	7	4	18	1.40	11.61	4.36	—
29725	Oil roasted, salted	¼	cup(s)	39	1	238	8	7	4	22	1.65	13.66	5.31	—
508	Slivered	¼	cup(s)	34	2	195	7	7	4	17	1.31	10.85	4.12	—
1137	Almond butter, no salt added	1	tablespoon(s)	16	<1	101	2	3	1	9	0.90	6.14	1.98	—
32940	Almond butter, salt added	1	tablespoon(s)	16	<1	101	2	3	1	9	0.90	6.14	1.98	—
1138	**Beechnuts, dried**	¼	cup(s)	57	4	327	4	19	5	28	3.25	12.43	11.41	—
517	**Brazil nuts, unblanched, dried**	¼	cup(s)	35	1	230	5	4	3	23	5.30	8.59	7.20	—
1166	**Breadfruit seeds, roasted**	¼	cup(s)	57	28	118	4	23	3	2	0.41	0.20	0.82	—
1139	**Butternuts, dried**	¼	cup(s)	30	1	184	7	4	1	17	0.39	3.13	12.82	—
	Cashews													
1140	Dry roasted	¼	cup(s)	34	1	197	5	11	1	16	3.14	9.36	2.68	—
518	Oil roasted	¼	cup(s)	33	1	189	5	10	1	16	2.76	8.42	2.78	—
32889	Cashew butter, no salt added	1	tablespoon(s)	16	<1	94	3	4	<1	8	1.56	4.66	1.34	—
32931	Cashew butter, salt added	1	tablespoon(s)	16	<1	94	3	4	<1	8	1.56	4.66	1.34	—
	Coconut													
32896	Dried, not sweetened	¼	cup(s)	60	2	393	4	14	10	38	34.06	1.63	0.42	—
1153	Dried, shredded, sweetened	¼	cup(s)	24	3	122	1	12	1	9	7.68	0.37	0.09	—
520	Shredded	¼	cup(s)	21	10	75	1	3	2	7	6.27	0.30	0.08	—

Chol (mg)	Calc (mg)	Iron (mg)	Magn (mg)	Pota (mg)	Sodi (mg)	Zinc (mg)	Vit A (µg)	Thia (mg)	Vit E (mg α)	Ribo (mg)	Niac (mg)	Vit B_6 (mg)	Fola (µg)	Vit C (mg)	Vit B_{12} (µg)	Sele (µg)
0	12	0.33	14	292	6	0.2	76	0.04	0.66	0.02	0.73	0.09	18	16	0	0
0	4	0.37	9	189	8	0.07	53	0.05	0.46	0.03	0.53	0.07	—	16	0	—
0	9	0.24	10	213	5	0.15	38	0.03	0.49	0.02	0.53	0.07	14	11	0	0
0	13	0.82	11	262	13	0.17	29	0.04	0.67	0.03	0.64	0.09	16	27	0	1
0	12	0.52	13	279	328	0.18	28	0.06	0.39	0.04	0.82	0.14	24	22	0	<1
0	12	0.52	13	279	12	0.18	28	0.06	0.39	0.04	0.82	0.14	24	22	0	<1
0	12	0.98	14	333	259	0.21	25	0.02	1.41	0.05	1.01	0.07	4	7	0	2
0	11	1.11	14	274	249	0.23	16	0.02	1.23	0.05	0.92	0.08	7	7	0	3
0	8	0.62	10	203	321	0.12	10	0.01	1.27	0.04	0.60	0.06	6	4	0	<1
0	43	1.70	15	264	282	0.22	11	0.06	1.06	0.04	0.91	0.02	6	10	0	1
0	30	2.45	52	925	566	0.54	12	0.14	0.00	0.13	2.44	0.09	18	11	0	1
0	13	0.74	22	430	73	0.21	18	0.05	—	0.11	1.00	0.09	6	28	0	1
0	26	0.14	7	138	12	0.09	0	0.02	0.02	0.02	0.23	0.05	7	9	0	<1
0	99	0.58	16	146	21	0.10	274	0.03	1.35	0.05	0.30	0.13	85	20	0	1
0	125	1.59	21	184	12	0.34	441	0.04	2.18	0.06	0.38	0.05	32	18	0	1
0	22	0.86	13	237	121	0.33	474	0.04	0.28	0.04	0.47	0.06	20	4	0	<1
0	23	0.75	20	154	32	0.45	195	0.06	0.40	0.11	0.77	0.07	17	3	0	<1
0	20	0.54	13	270	310	0.24	50	0.05	—	0.03	0.87	0.17	—	30	0	—
0	20	0.36	—	420	70	—	63	0.02	—	0.02	0.75	—	—	30	0	—
0	20	0.36	13	255	370	0.24	50	0.05	—	0.03	0.88	0.17	—	18	0	—
0	7	0.23	—	—	6	—	0	—	—	—	—	—	—	2	—	—
0	7	0.23	—	—	6	—	0	—	—	—	—	—	—	2	—	—
0	41	0.07	7	112	14	0.04	80	0.03	0.34	0.04	0.07	0.04	3	15	0	<1
0	78	1.35	100	249	10	1.13	0	0.07	8.96	0.20	1.33	0.04	11	0	0	1
0	92	1.56	99	257	<1	1.22	0	0.03	8.97	0.30	1.33	0.04	11	0	0	1
0	92	1.56	99	257	117	1.22	0	0.03	8.97	0.30	1.33	0.04	11	0	0	1
0	114	1.44	108	274	133	1.20	0	0.04	10.19	0.31	1.44	0.05	11	0	0	1
0	84	1.45	93	246	<1	1.13	0	0.08	8.73	0.27	1.32	0.04	10	0	0	1
0	43	0.59	48	121	2	0.49	0	0.02	—	0.10	0.46	0.01	10	<1	0	—
0	43	0.59	48	121	72	0.49	0	0.02	—	0.10	0.46	0.01	10	<1	0	1
0	1	1.40	0	578	22	0.20	0	0.17	—	0.21	0.50	0.39	64	9	0	4
0	56	0.85	132	231	1	1.42	0	0.22	2.01	0.01	0.10	0.04	8	<1	0	671
0	49	0.51	35	615	16	0.59	9	0.23	—	0.14	4.20	0.24	34	4	0	8
0	16	1.21	71	126	<1	0.94	2	0.11	—	0.04	0.31	0.17	20	1	0	5
0	15	2.06	89	194	5	1.92	0	0.07	0.32	0.07	0.48	0.09	24	0	0	4
0	14	1.97	89	205	4	1.74	0	0.12	0.30	0.07	0.56	0.10	8	<.1	0	7
0	7	0.80	41	87	2	0.83	0	0.05	—	0.03	0.26	0.04	11	0	0	2
0	7	0.80	41	87	98	0.83	0	0.05	0.15	0.03	0.26	0.04	11	0	0	2
0	15	1.98	54	323	22	1.20	0	0.04	0.26	0.06	0.36	0.18	5	1	0	11
0	4	0.47	12	82	64	0.44	0	0.01	0.10	0.00	0.12	0.07	2	<1	0	4
0	3	0.51	7	75	4	0.23	0	0.01	0.05	0.00	0.11	0.01	5	1	0	2

Table A-1
Food Composition

(DA+ code is for Wadsworth Diet Analysis program)
(For purposes of calculations, use "0" for t, <1, <.1, <.01, etc.)

A

DA +Code	Food Description	Quantity	Measure	Wt (g)	H₂O (g)	Ener (kcal)	Prot (g)	Carb (g)	Fiber (g)	Fat (g)	Sat	Mono	Poly	Trans
	NUTS, SEEDS, AND PRODUCTS–Continued													
	Chestnuts													
1152	Chinese, roasted	¼	cup(s)	57	23	136	3	30	0	1	0.10	0.35	0.17	—
32895	European, boiled & steamed	¼	cup(s)	57	39	74	1	16	0	1	0.15	0.27	0.31	—
32911	European, roasted	¼	cup(s)	57	23	139	2	30	3	1	0.23	0.43	0.49	—
32922	Japanese, boiled & steamed	¼	cup(s)	57	49	32	<1	7	0	<1	0.02	0.06	0.03	—
32923	Japanese, roasted	¼	cup(s)	57	28	114	2	26	0	<1	0.07	0.24	0.12	—
4958	**Flaxseeds or linseeds**	¼	cup(s)	57	5	276	11	19	16	19	1.79	3.85	12.54	—
32904	**Ginkgo nuts, dried**	¼	cup(s)	57	7	197	6	41	0	1	0.22	0.42	0.42	—
	Hazelnuts or filberts													
32901	Blanched	¼	cup(s)	57	3	357	8	10	6	35	2.65	27.32	3.15	—
32902	Dry roasted, no salt added	¼	cup(s)	57	1	366	9	10	5	35	2.56	26.43	4.80	—
1156	**Hickorynuts, dried**	¼	cup(s)	30	1	197	4	5	2	19	2.11	9.78	6.57	—
	Macadamias													
1157	Raw	¼	cup(s)	34	<1	241	3	5	3	25	4.04	19.72	0.50	—
32905	Dry roasted, no salt added	¼	cup(s)	34	1	241	3	4	3	25	4.00	19.86	0.50	—
32932	Dry roasted, salt added	¼	cup(s)	34	1	240	3	4	3	25	4.00	19.86	0.50	—
	Mixed nuts													
1159	With peanuts, dry roasted	¼	cup(s)	34	1	203	6	9	3	18	2.36	10.75	3.69	—
32933	With peanuts, dry roasted, salt added	¼	cup(s)	34	1	203	6	9	3	18	2.36	10.75	3.69	—
32906	Without peanuts, oil roasted, no salt added	¼	cup(s)	36	1	221	6	8	2	20	3.27	11.93	4.12	—
	Peanuts													
2807	Dry roasted	¼	cup(s)	37	0	214	9	8	3	18	2.51	8.99	5.72	—
2806	Dry roasted, salted	¼	cup(s)	37	0	214	9	8	3	18	2.51	8.99	5.72	—
1763	Oil roasted, salted	¼	cup(s)	36	0	216	10	5	3	19	3.12	9.33	5.49	—
2804	Raw	¼	cup(s)	37	2	207	9	6	3	18	2.49	8.92	5.68	—
1884	Peanut butter, chunky	1	tablespoon(s)	16	<1	94	4	3	1	8	1.53	3.77	2.27	—
30303	Peanut butter, low sodium	1	tablespoon(s)	16	<1	95	4	3	1	8	1.66	3.88	2.21	—
30305	Peanut butter, reduced fat	1	tablespoon(s)	18	<1	94	5	6	1	6	1.33	2.91	1.85	—
524	Peanut butter, smooth	1	tablespoon(s)	16	<1	96	4	3	1	8	1.60	3.96	2.38	—
	Pecans													
32907	Dry roasted, no salt added	¼	cup(s)	57	1	403	5	8	5	42	3.56	24.92	11.66	—
32936	Dry roasted, salt added	¼	cup(s)	57	1	403	5	8	5	42	3.56	24.92	11.66	—
1162	Halves, oil roasted	¼	cup(s)	28	<1	197	3	4	3	21	1.99	11.27	6.49	—
526	Raw	¼	cup(s)	27	1	187	2	4	3	19	1.67	11.02	5.84	—
12973	**Pine nuts or pignolia, dried**	1	tablespoon(s)	9	<1	58	1	1	<1	6	0.42	1.61	2.93	—
	Pistachios													
1164	Dry roasted	¼	cup(s)	32	1	183	7	9	3	15	1.78	7.75	4.45	—
32938	Dry roasted, salt added	¼	cup(s)	32	1	182	7	9	3	15	1.78	7.75	4.45	—
1167	**Pumpkin or squash seeds, roasted**	¼	cup(s)	57	4	296	19	8	2	24	4.52	7.43	10.90	—
	Sesame													
1169	Sesame seeds, whole, roasted, toasted	3	teaspoon(s)	9	<1	51	2	2	1	4	0.60	1.63	1.89	—
32912	Sesame butter paste	1	tablespoon(s)	16	<1	95	3	4	1	8	1.14	3.07	3.57	—
32941	Tahini or sesame butter	1	tablespoon(s)	15	<1	89	3	3	1	8	1.11	3.00	3.48	—
	Soy nuts													
34173	Deep sea salted	¼	cup(s)	56	—	240	24	18	10	8	2.00	—	—	
34174	Unsalted	¼	cup(s)	56	—	240	24	18	10	8	2.00	—	—	
	Sunflower seeds													
528	Kernels, dried	¼	cup(s)	36	2	205	8	7	4	18	1.87	3.41	11.78	—
29721	Kernels, dry roasted, salted	¼	cup(s)	32	<1	186	6	8	3	16	1.67	3.04	10.52	—

A

Chol (mg)	Calc (mg)	Iron (mg)	Magn (mg)	Pota (mg)	Sodi (mg)	Zinc (mg)	Vit A (µg)	Thia (mg)	Vit E (mg α)	Ribo (mg)	Niac (mg)	Vit B_6 (mg)	Fola (µg)	Vit C (mg)	Vit B_{12} (µg)	Sele (µg)
0	11	0.85	51	271	2	0.53	0	0.09	—	0.05	0.85	0.25	41	22	0	4
0	26	0.98	31	405	15	0.14	1	0.08	—	0.06	0.41	0.13	22	15	0	—
0	16	0.52	19	336	1	0.32	1	0.14	0.28	0.10	0.76	0.28	40	15	0	1
0	6	0.30	10	67	3	0.23	1	0.07	—	0.03	0.31	0.06	10	5	0	—
0	20	1.19	36	242	11	0.81	2	0.26	—	—	0.40	0.24	33	16	0	—
0	111	3.48	203	381	19	2.34	0	0.10	—	0.09	0.78	0.52	156	1	0	3
0	11	0.91	30	566	7	0.38	31	0.24	—	0.10	6.65	0.36	60	17	0	—
0	84	1.87	91	373	0	1.25	1	0.27	9.92	0.06	0.88	0.33	44	1	0	2
0	70	2.48	98	428	0	1.42	2	0.19	8.66	0.07	1.16	0.35	50	2	0	2
0	18	0.64	52	131	<1	1.29	2	0.26	—	0.04	0.27	0.06	12	1	0	2
0	28	1.24	44	123	2	0.44	0	0.40	0.18	0.05	0.83	0.09	4	<1	0	1
0	23	0.89	40	122	1	0.43	0	0.24	0.19	0.03	0.76	0.12	3	<1	0	1
0	23	0.89	40	122	89	0.43	0	0.24	0.19	0.03	0.76	0.12	3	<1	0	4
0	24	1.27	77	204	4	1.30	<1	0.07	—	0.07	1.61	0.10	17	<1	0	1
0	24	1.27	77	204	229	1.30	0	0.07	3.75	0.07	1.61	0.10	17	<1	0	3
0	38	0.93	90	196	4	1.68	<1	0.18	—	0.17	0.71	0.06	20	<1	0	—
0	20	0.82	64	240	2	1.20	0	0.15	2.56	0.03	4.93	0.09	53	0	0	3
0	20	0.82	64	240	297	1.20	0	0.15	2.89	0.03	4.93	0.09	53	0	0	3
0	22	0.54	63	261	115	1.18	0	0.03	2.50	0.03	4.97	0.16	43	<1	0	1
0	34	1.67	61	257	7	1.19	0	0.23	3.04	0.05	4.40	0.13	88	0	0	3
0	8	0.33	31	101	75	0.52	0	0.02	1.01	0.02	2.19	0.07	15	0	0	1
0	6	0.29	25	107	3	0.47	0	0.01	1.23	0.02	2.14	0.07	12	0	0	—
0	6	0.34	31	120	97	0.50	0	0.05	1.20	0.01	2.63	0.06	11	0	0	—
0	8	0.30	28	88	80	0.47	0	0.01	1.44	0.02	2.14	0.07	12	0	0	1
0	41	1.59	75	240	1	2.87	4	0.26	0.74	0.06	0.66	0.11	9	<1	0	2
0	41	1.59	75	240	217	2.87	4	0.26	0.74	0.06	0.66	0.11	9	<1	0	2
0	18	0.68	33	108	<1	1.23	1	0.13	0.70	0.03	0.33	0.05	4	<1	0	2
0	19	0.68	33	111	0	1.22	1	0.18	0.38	0.04	0.32	0.06	6	<1	0	1
0	1	0.48	22	51	<1	0.55	<.1	0.03	0.80	0.02	0.38	0.01	6	<.1	0	<.1
0	35	1.34	38	333	<1	0.74	4	0.27	0.62	0.05	0.46	0.41	16	1	0	3
0	35	1.34	38	333	<1	0.74	4	0.27	0.62	0.05	0.46	0.41	16	1	0	3
0	24	8.48	303	457	10	4.22	11	0.12	0.00	0.18	0.99	0.05	32	1	0	3
0	89	1.33	32	43	1	0.64	0	0.07	—	0.02	0.41	0.07	9	0	0	1
0	154	3.07	58	93	2	1.17	<1	0.04	—	0.03	1.07	0.13	16	0	0	1
0	21	0.66	14	69	5	0.69	<1	0.24	—	0.02	0.85	0.02	15	1	0	<1
0	120	2.16	—	—	300	—	0	—	—	—	—	—	—	0	—	—
0	120	2.16	—	—	20	—	0	—	—	—	—	—	—	0	—	—
0	42	2.44	127	248	1	1.82	1	0.82	12.42	0.09	1.62	0.28	82	1	0	21
0	22	1.22	41	272	250	1.69	<1	0.03	8.35	0.08	2.25	0.26	76	<1	0	25

Table A–1
Food Composition

(DA+ code is for Wadsworth Diet Analysis program)
(For purposes of calculations, use "0" for t, <1, <.1, <.01, etc.)

DA + Code	Food Description	Quantity	Measure	Wt (g)	H₂O (g)	Ener (kcal)	Prot (g)	Carb (g)	Fiber (g)	Fat (g)	Sat	Fat Breakdown (g) Mono	Poly	Trans
	NUTS, SEEDS, AND PRODUCTS													
	Sunflower seeds–Continued													
29723	Kernels, toasted, salted	¼	cup(s)	34	<1	207	6	7	4	19	1.99	3.63	12.56	—
32928	Sunflower seed butter, salt added	1	tablespoon(s)	16	<1	93	3	4	0	8	0.80	1.46	5.04	
	Trail mix													
4646	Trail mix	¼	cup(s)	38	3	173	5	17	2	11	2.08	4.70	3.62	—
4647	Trail mix with chocolate chips	¼	cup(s)	38	2	182	5	17	0	12	2.29	5.08	4.23	—
4648	Tropical trail mix	¼	cup(s)	35	3	142	2	23	0	6	2.97	0.87	1.81	—
	Walnuts													
529	Dried black, chopped	¼	cup(s)	31	1	193	8	3	2	18	1.05	4.69	10.96	—
531	English or persian	¼	cup(s)	30	1	196	5	4	2	20	1.84	2.68	14.15	—
	VEGETARIAN FOODS													
	Prepared													
34222	Brown rice & tofu stir-fry (vegan)	8	ounce(s)	227	183	228	12	13	3	16	1.25	4.03	9.54	0
34368	Cheese enchilada casserole (lacto)	8	ounce(s)	227	86	410	18	41	4	19	10.06	6.54	1.24	0
34247	Five bean casserole (vegan)	8	ounce(s)	228	178	178	6	26	5	6	1.11	2.49	1.96	0
34261	Lentil stew (vegan)	8	ounce(s)	228	152	125	8	24	7	<1	0.08	0.07	0.21	0
34397	Macaroni & cheese (lacto)	8	ounce(s)	226	163	181	8	17	<1	9	4.37	2.88	0.89	0
34238	Steamed rice & vegetables (vegan)	8	ounce(s)	228	100	265	5	40	3	10	1.84	3.91	4.07	0
34308	Tofu rice burgers (ovo-lacto)	1	piece(s)	218	78	435	22	68	6	8	1.69	2.39	3.52	—
34276	Vegan spinach enchiladas (vegan)	1	piece(s)	82	59	93	5	15	2	2	0.34	0.55	1.27	—
34243	Vegetable chow mein (vegan)	8	ounce(s)	227	163	166	6	22	2	6	0.65	2.66	2.47	—
34454	Vegetable lasagna (lacto)	8	ounce(s)	225	154	177	12	25	2	4	1.92	0.93	0.34	—
34339	Vegetable marinara (vegan)	8	ounce(s)	229	182	94	3	15	1	3	0.36	1.32	0.92	—
34356	Vegetable rice casserole (lacto)	8	ounce(s)	227	172	230	9	24	4	12	4.67	3.48	2.96	—
34311	Vegetable strudel (ovo-lacto)	8	ounce(s)	227	100	756	19	51	4	54	18.24	26.38	6.17	—
34371	Vegetable taco (lacto)	1	item(s)	227	147	365	13	43	9	17	6.45	5.81	4.02	—
34282	Vegetarian chili (vegan)	8	ounce(s)	227	196	116	6	21	7	2	0.24	0.29	0.74	0
34367	Vegetarian vegetable soup (vegan)	8	ounce(s)	226	204	92	3	14	2	4	0.77	1.67	1.30	0
	Boca burger													
32067	All American flamed grilled patty	1	item(s)	71	—	110	14	6	4	4	1.00	—	—	0
32070	Bigger chef max's favorite	1	item(s)	99	—	130	18	11	5	4	1.00	1.00	1.50	—
32069	Bigger vegan	1	item(s)	99	—	120	18	11	6	0	0.00	0.00	0.00	0
32074	Boca chik'n nuggets	4	item(s)	87	—	190	16	16	2	7	2.00	—	—	0
32075	Boca meatless ground burger	½	cup(s)	57	—	70	11	7	4	1	0.00	—	—	—
32073	Boca tenders	1	item(s)	85	—	140	20	9	3	3	0.00	2.00	1.00	—
32072	Breakfast links	2	item(s)	45	—	100	10	6	5	4	0.00	—	—	0
32071	Breakfast patties	1	item(s)	38	—	80	8	5	3	4	0.00	—	—	0
32068	Roasted garlic patty	1	item(s)	71	—	100	14	7	5	2	0.50	—	—	0
32066	Vegan original patty	1	item(s)	71	—	90	13	4	0	1	0.00	—	—	0
	Gardenburger													
37810	Bbq chik'n with sauce	1	item(s)	142	—	250	14	30	5	8	1.00	—	—	0
39661	Black bean burger	1	item(s)	71	—	80	8	11	4	2	0.00	—	—	0
39666	Buffalo chick'n wing	3	item(s)	95	—	180	9	8	5	12	1.50	—	—	0
37808	Chik'n grill	1	item(s)	71	—	100	13	5	3	3	0.00	—	—	0
39665	Country fried chicken w/creamy pepper gravy	1	item(s)	142	—	190	9	16	2	9	1.00	—	—	0
37805	Crispy nuggets	6	item(s)	82	—	180	4	22	3	9	1.50	—	—	0
39663	Homestyle classic burger	1	item(s)	71	—	110	12	6	4	5	0.50	—	—	0
37807	Meatless breakfast sausage	1	item(s)	43	—	50	5	2	2	4	0.00	—	—	0
37809	Meatless meatballs	6	item(s)	85	—	110	12	8	4	5	1.00	—	—	0
37806	Meatless riblets w/sauce	1	item(s)	142	—	210	17	11	4	5	0.00	—	—	0
29913	Original	3	ounce(s)	85	—	132	7	19	4	4	1.80	1.80	0.60	0

A

Chol (mg)	Calc (mg)	Iron (mg)	Magn (mg)	Pota (mg)	Sodi (mg)	Zinc (mg)	Vit A (µg)	Thia (mg)	Vit E (mg α)	Ribo (mg)	Niac (mg)	Vit B$_6$ (mg)	Fola (µg)	Vit C (mg)	Vit B$_{12}$ (µg)	Sele (µg)
0	19	2.28	43	164	205	1.78	0	0.11	—	0.10	1.41	0.27	80	<1	0	21
0	20	0.76	59	12	83	0.85	<1	0.05	—	0.05	0.85	0.13	38	<1	0	—
0	29	1.14	59	257	86	1.21	<1	0.17	—	0.07	1.77	0.11	27	1	0	—
2	41	1.27	60	243	45	1.18	1	0.15	—	0.08	1.65	0.10	24	<1	0	—
0	20	0.92	34	248	4	0.41	1	0.16	—	0.04	0.52	0.11	15	3	0	—
0	19	0.98	63	163	1	1.05	1	0.02	0.56	0.04	0.15	0.18	10	1	0	5
0	29	0.87	47	132	1	0.93	<1	0.10	0.21	0.05	0.34	0.16	29	<1	0	1
0	266	4.73	88	375	112	1.51	121	0.14	0.07	0.12	1.08	0.28	32	18	0	11
42	468	2.58	37	204	1219	1.96	107	0.33	0.06	0.38	2.38	0.11	77	22	<1	22
0	48	1.71	42	367	618	0.60	54	0.09	0.53	0.08	0.93	0.11	33	8	<.1	4
0	23	2.35	31	380	289	0.87	18	0.14	0.14	0.10	1.50	0.16	61	13	0	9
22	187	0.77	20	120	768	1.11	82	0.15	0.29	0.24	1.02	0.04	39	<.1	<1	16
0	41	1.43	68	358	1403	0.91	86	0.16	3.05	0.12	2.76	0.30	28	13	<.1	8
51	468	4.78	90	455	2454	2.07	82	0.27	0.12	0.27	3.43	0.30	99	2	<1	43
0	117	1.13	40	168	134	0.68	26	0.07	—	0.07	0.54	0.11	46	1	0	5
0	190	3.65	28	302	371	0.74	8	0.13	0.06	0.12	1.43	0.15	47	7	0	6
10	144	1.91	33	393	637	1.06	31	0.20	0.05	0.27	2.07	0.21	64	15	<1	19
0	15	0.85	17	180	378	0.35	18	0.13	0.50	0.08	1.25	0.11	41	20	0	10
16	176	1.72	28	395	609	1.19	121	0.16	0.35	0.29	1.93	0.18	92	54	<1	6
46	318	3.36	39	299	813	1.98	288	0.45	0.21	0.50	4.52	0.16	123	27	<1	31
21	231	2.58	83	550	893	1.80	81	0.23	0.10	0.18	1.48	0.25	132	12	<1	10
<1	68	2.42	41	532	383	0.78	46	0.13	0.15	0.13	1.26	0.18	58	16	0	5
0	37	1.32	28	443	503	0.44	109	0.11	0.55	0.08	1.54	0.22	38	24	<.1	1
3	150	1.80	—	—	370	—	0	—	—	—	—	—	—	0	—	—
5	150	2.70	—	—	400	—	—	—	—	—	—	—	—	0	—	—
0	60	1.80	—	—	380	—	0	—	—	—	—	—	—	2	—	—
0	80	1.80	—	220	570	—	0	—	—	—	—	—	—	0	—	—
0	80	1.44	—	—	220	—	0	—	—	—	—	—	—	0	—	—
0	80	1.08	—	—	440	—	0	—	—	—	—	—	—	0	—	—
0	60	1.44	—	—	330	—	0	—	—	—	—	—	—	0	—	—
0	60	1.44	—	—	260	—	0	—	—	—	—	—	—	0	—	—
3	100	1.80	—	—	400	—	0	—	—	—	—	—	—	1	—	—
0	80	1.80	—	—	350	—	0	—	—	—	—	—	—	1	—	—
0	150	1.08	—	—	890	—	—	—	—	—	—	—	—	0	—	—
0	40	1.44	—	—	330	—	—	—	—	—	—	—	—	0	—	—
0	40	0.72	—	—	1000	—	—	—	—	—	—	—	—	0	—	—
0	60	3.60	—	—	360	—	—	—	—	—	—	—	—	0	—	—
5	40	1.44	—	—	550	—	—	—	—	—	—	—	—	0	—	—
5	60	0.72	—	—	570	—	—	—	—	—	—	—	—	5	—	—
0	80	1.44	—	—	380	—	—	—	—	—	—	—	—	0	—	—
0	20	0.72	—	—	120	—	—	—	—	—	—	—	—	0	—	—
0	60	1.80	—	—	400	—	—	—	—	—	—	—	—	0	—	—
0	60	1.80	—	—	720	—	—	—	—	—	—	—	—	4	—	—
24	72	0.00	37	232	672	1.07	0	0.12	—	0.18	1.30	0.10	12	0	<1	8

Table A–1
Food Composition

(DA+ code is for Wadsworth Diet Analysis program)
(For purposes of calculations, use "0" for t, <1, <.1, <.01, etc.)

A

DA + Code	Food Description	Quantity	Measure	Wt (g)	H₂O (g)	Ener (kcal)	Prot (g)	Carb (g)	Fiber (g)	Fat (g)	Sat	Mono	Poly	Trans
	NUTS, SEEDS, AND PRODUCTS													
	Gardenburger–Continued													
31707	Santa Fe	3	ounce(s)	85	—	156	—	24	5	3	1.20	—	—	—
29915	Veggie medley	3	ounce(s)	85	—	108	6	22	4	0	0.00	0.00	0.00	0
	Loma Linda													
9311	Big franks	1	item(s)	51	30	110	10	2	2	7	1.00	2.00	4.00	0
9315	Chik'n nuggets	5	item(s)	85	40	240	14	13	4	15	2.00	4.50	8.00	0
9317	Corn dogs	1	item(s)	71	31	150	7	22	3	4	0.50	1.00	2.50	0
9323	Fried chik'n with gravy	2	piece(s)	80	46	150	12	5	2	10	1.50	2.50	5.00	0
9326	Linketts, canned	1	item(s)	35	21	70	7	1	1	5	0.50	1.00	2.50	0
9336	Redi-Burger patties, canned	1	slice(s)	85	50	120	18	7	4	3	0.50	0.50	1.50	0
9354	Tender Rounds meatball substitute, canned in gravy	6	piece(s)	80	54	120	13	6	1	5	0.50	1.00	2.50	0
	Morningstar Farms													
33707	America's Original Veggie Dog links	1	item(s)	57	—	80	11	6	1	1	0.00	0.00	0.00	0
9362	Better n Eggs egg substitute	¼	cup(s)	57	50	20	5	0	0	0	0.00	0.00	0.00	0
9368	Breakfast links	2	item(s)	45	27	80	9	3	2	3	0.50	0.50	2.00	0
9371	Breakfast strips	2	item(s)	16	7	60	2	2	1	5	0.50	1.00	3.00	0
33705	Chik Nuggets	4	piece(s)	86	—	180	13	17	5	6	0.50	1.50	4.00	—
11587	Chik Patties	1	item(s)	71	36	150	9	16	2	6	1.00	1.50	2.50	—
2531	Garden veggie patties	1	item(s)	67	40	100	10	9	4	3	0.50	0.50	1.50	—
9412	Natural Touch low fat vegetarian chili, canned	1	cup(s)	230	173	170	18	21	11	1	—	—	—	0
33702	Spicy black bean veggie burger	1	item(s)	78	47	150	11	16	5	5	0.50	1.50	2.50	0
	Worthington													
9422	Chik Stiks	1	item(s)	47	27	110	10	4	2	6	1.00	1.00	3.00	0
9424	Chili, canned	1	cup(s)	230	167	290	19	21	9	15	2.50	3.50	9.00	0
9432	Crispychik patties	1	item(s)	71	37	150	9	16	2	6	1.00	1.50	3.50	0
9440	Dinner roast, frozen	1	slice(s)	85	53	180	12	5	3	12	1.50	5.00	5.00	0
9442	Fillets, frozen	2	piece(s)	85	48	180	16	8	4	9	1.00	3.50	4.50	0
9478	Meatless smoked beef, sliced	6	slice(s)	57	—	130	11	7	1	7	1.00	2.00	4.00	0
9480	Meatless smoked turkey, sliced	3	slice(s)	57	—	140	10	5	0	9	1.00	2.50	5.00	—
9462	Prosage links	2	item(s)	45	27	80	9	3	2	3	0.50	0.50	2.00	0
9486	Stripples bacon substitute	2	item(s)	16	7	60	2	2	1	5	0.50	1.00	2.50	0
9496	Vegetable Skallops	½	cup(s)	85	65	90	15	3	3	2	0.50	0.50	0.00	0
9434	Vegetarian cutlets	1	slice(s)	61	43	70	11	3	2	1	—	—	—	—
	DAIRY													
	Butter: *see* **Fats & Oils**													
	Cheese													
1433	Blue, crumbled	1	ounce(s)	28	12	100	6	1	0	8	5.29	2.21	0.23	—
884	Brick	1	ounce(s)	28	12	104	7	1	0	8	5.25	2.41	0.22	—
885	Brie	1	ounce(s)	28	14	94	6	<1	0	8	4.87	2.24	0.23	—
34821	Camembert	1	ounce(s)	29	15	87	6	<1	0	7	4.43	2.04	0.21	—
888	Cheddar or colby	1	ounce(s)	28	11	110	7	1	0	9	5.66	2.60	0.27	—
32096	Cheddar or colby, low fat	1	ounce(s)	28	18	49	7	1	0	2	1.23	0.59	0.06	—
5	Cheddar, shredded	¼	cup(s)	28	10	114	7	<1	0	9	5.96	2.65	0.27	—
889	Edam	1	ounce(s)	28	12	100	7	<1	0	8	4.92	2.28	0.19	—
890	Feta	1	ounce(s)	28	15	74	4	1	0	6	4.18	1.29	0.17	—
891	Fontina	1	ounce(s)	28	11	109	7	<1	0	9	5.37	2.43	0.46	—
8527	Goat, soft	1	ounce(s)	28	17	76	5	<1	0	6	4.14	1.37	0.14	—
893	Gouda	1	ounce(s)	28	12	100	7	1	0	8	4.93	2.17	0.18	—
894	Gruyere	1	ounce(s)	28	9	116	8	<1	0	9	5.30	2.81	0.49	—
895	Limburger	1	ounce(s)	28	14	92	6	<1	0	8	4.69	2.41	0.14	—
896	Monterey jack	1	ounce(s)	28	11	104	7	<1	0	8	5.34	2.45	0.25	—

A

Chol (mg)	Calc (mg)	Iron (mg)	Magn (mg)	Pota (mg)	Sodi (mg)	Zinc (mg)	Vit A (µg)	Thia (mg)	Vit E (mg α)	Ribo (mg)	Niac (mg)	Vit B$_6$ (mg)	Fola (µg)	Vit C (mg)	Vit B$_{12}$ (µg)	Sele (µg)
24	96	0.00	—	—	336	—	0	—	—	—	—	—	—	0		0
0	48	0.00	32	218	336	0.55	—	0.08	—	0.10	1.08	0.11	13	0	<.1	5
0	0	0.77	—	50	240	0.89	0	0.23	—	0.43	1.60	0.04	—	0	1	—
0	20	1.44	—	210	410	0.43	0	0.75	—	0.51	6.00	0.90	—	0	3	—
0	0	1.08	—	60	500	0.43	0	0.72	—	0.61	1.47	0.87	—	0	2	—
0	20	1.80	—	70	430	0.34	0	1.05	—	0.34	4.00	0.30	—	0	2	—
0	0	0.36	—	15	160	0.46	0	0.12	—	0.20	0.40	0.20	—	0	1	—
0	0	1.06	—	140	450	1.11	0	0.23	—	0.34	6.00	0.40	—	0	2	—
0	20	1.08	—	80	340	0.66	0	0.75	—	0.17	2.00	0.16	—	0	1	—
0	0	0.72	—	60	580	—	0	—	—	—	—	—	—	0	—	—
0	20	0.63	—	75	90	0.60	75	0.03	—	0.34	0.00	0.08	24	0	1	—
0	0	1.44	—	50	320	0.36	0	1.80	—	0.17	2.00	0.30	—	0	3	—
0	0	0.27	—	15	220	0.05	0	0.75	—	0.04	0.40	0.07	—	0	<1	—
0	40	3.60	—	330	590	—	0	1.20	—	0.26	5.00	0.40	—	0	3	—
0	0	1.80	—	210	540	0.31	0	1.80	—	0.17	2.00	0.20	—	0	1	—
0	40	0.72	—	180	350	0.58	—	6.47	—	0.10	0.00	0.00	—	0	0	—
0	40	1.80	—	480	870	1.36	—	0.60	—	0.21	0.00	0.30	—	0	0	—
0	40	1.80	44	320	470	0.93	0	—	—	0.14	0.00	0.21	—	0	<.1	—
0	20	1.80	—	100	300	0.31	0	0.60	—	0.17	6.00	0.40	—	0	2	—
0	40	3.60	—	420	1130	1.24	0	0.06	—	0.07	2.00	0.70	—	0	2	—
0	0	1.80	—	170	440	0.33	0	1.80	—	0.17	2.00	0.20	—	0	1	—
3	40	0.36	—	55	580	0.64	0	1.80	—	0.26	6.00	0.60	—	0	2	—
0	0	1.80	—	130	750	0.92	0	0.68	—	0.14	0.80	0.40	—	0	3	—
0	20	1.80	—	180	510	0.14	0	1.80	—	0.17	6.00	0.40	—	0	2	—
0	100	2.70	—	60	490	0.23	0	1.80	—	0.17	6.00	0.40	—	0	3	—
0	0	1.44	—	50	320	0.36	0	1.80	—	0.17	2.00	0.30	—	0	3	—
0	0	0.36	—	15	220	0.05	0	0.75	—	0.03	0.40	0.08	—	0	<1	—
0	0	0.72	—	10	410	0.67	0	0.03	—	0.03	0.00	0.01	—	0	0	—
0	0	0.00	—	30	340	0.43	0	0.03	—	0.04	0.00	0.04	—	0	0	—
21	150	0.09	7	73	395	0.75	56	0.01	0.07	0.11	0.29	0.05	10	0	<1	4
26	189	0.12	7	38	157	0.73	82	0.00	0.07	0.10	0.03	0.02	6	0	<1	4
28	52	0.14	6	43	176	0.67	49	0.02	0.07	0.15	0.11	0.07	18	0	<1	4
21	112	0.10	6	54	244	0.69	—	0.01	—	0.14	0.18	0.07	18	0	<1	4
27	192	0.21	7	36	169	0.86	74	0.00	0.08	0.11	0.03	0.02	5	0	<1	4
6	118	0.12	5	19	174	0.52	17	0.00	0.02	0.06	0.01	0.01	3	0	<1	4
30	204	0.19	8	28	175	0.88	75	0.01	0.08	0.11	0.02	0.02	5	0	<1	4
25	205	0.12	8	53	270	1.05	68	0.01	0.07	0.11	0.02	0.02	4	0	<1	4
25	138	0.18	5	17	312	0.81	35	0.04	0.05	0.24	0.28	0.12	9	0	<1	4
32	154	0.06	4	18	224	0.98	73	0.01	0.08	0.06	0.04	0.02	2	0	<1	4
13	40	0.54	5	7	105	0.26	82	0.02	0.05	0.11	0.12	0.07	3	0	<.1	1
32	196	0.07	8	34	229	1.09	46	0.01	0.07	0.09	0.02	0.02	6	0	<1	4
31	283	0.05	10	23	94	1.09	76	0.02	0.08	0.08	0.03	0.02	3	0	<1	4
25	139	0.04	6	36	224	0.59	95	0.02	0.06	0.14	0.04	0.02	16	0	<1	4
25	209	0.20	8	23	150	0.84	55	0.00	0.07	0.11	0.03	0.02	5	0	<1	4

Table A–1
Food Composition

(DA+ code is for Wadsworth Diet Analysis program)
(For purposes of calculations, use "0" for t, <1, <.1, <.01, etc.)

A

DA + Code	Food Description	Quantity	Measure	Wt (g)	H₂O (g)	Ener (kcal)	Prot (g)	Carb (g)	Fiber (g)	Fat (g)	Sat	Fat Breakdown (g) Mono	Poly	Trans
	DAIRY													
	Cheese–Continued													
13	Mozzarella, part skim milk	1	ounce(s)	28	15	71	7	1	0	4	2.83	1.26	0.13	—
12	Mozzarella, whole milk	1	ounce(s)	28	14	84	6	1	0	6	3.68	1.84	0.21	—
897	Muenster	1	ounce(s)	28	12	103	7	<1	0	8	5.35	2.44	0.19	—
898	Neufchatel	1	ounce(s)	28	17	73	3	1	0	7	4.14	1.90	0.18	—
14	Parmesan, grated	1	tablespoon(s)	5	1	22	2	<1	0	1	0.87	0.42	0.06	—
17	Provolone	1	ounce(s)	28	11	98	7	1	0	7	4.78	2.07	0.22	—
19	Ricotta, part skim milk	¼	cup(s)	62	46	85	7	3	0	5	3.03	1.42	0.16	—
18	Ricotta, whole milk	¼	cup(s)	62	44	107	7	2	0	8	5.10	2.23	0.24	—
20	Romano	1	tablespoon(s)	5	2	19	2	<1	0	1	0.86	0.39	0.03	—
900	Roquefort	1	ounce(s)	28	11	103	6	1	0	9	5.39	2.37	0.37	—
21	Swiss	1	ounce(s)	28	10	106	8	2	0	8	4.98	2.04	0.27	—
	Imitation cheese													
7998	Shredded imitation cheddar	¼	cup(s)	28	—	90	5	2	0	7	1.50	—	—	—
8028	Shredded imitation mozzarella	¼	cup(s)	28	—	80	6	1	0	6	1.00	—	—	—
	Cottage Cheese													
9	Low fat, 1% fat	½	cup(s)	113	93	81	14	3	0	1	0.73	0.33	0.04	—
8	Low fat, 2% fat	½	cup(s)	113	90	102	16	4	0	2	1.38	0.62	0.07	—
	Cream cheese													
11	Cream cheese	2	tablespoon(s)	29	16	101	2	1	0	10	6.37	2.85	0.37	—
17366	Fat free cream cheese	2	tablespoon(s)	30	23	29	4	2	0	<1	0.27	0.10	0.02	—
10438	Tofutti Better Than Cream Cheese	2	tablespoon(s)	30	—	80	1	1	0	8	2.00	—	6.00	—
	Processed cheese													
22	American cheese, processed	1	ounce(s)	28	11	106	6	<1	0	9	5.58	2.54	0.28	—
24	American cheese food, processed	1	ounce(s)	28	12	94	5	2	0	7	4.23	2.05	0.31	—
25	American cheese spread, processed	1	ounce(s)	28	14	82	5	2	0	6	3.78	1.77	0.18	—
9110	Kraft deluxe singles pasteurized process American cheese	1	ounce(s)	28	—	110	5	1	0	9	6.00	—	—	—
23	Swiss cheese, processed	1	ounce(s)	28	12	95	7	1	0	7	4.55	2.00	0.18	—
	Soy cheese													
10430	Nu Tofu cheddar flavored cheese alternative	1	ounce(s)	28	—	70	6	1	0	4	0.50	2.50	1.00	—
10435	Nu Tofu mozzarella flavored cheese alternative	1	ounce(s)	28	—	70	6	2	0	4	0.50	2.50	1.00	—
	Cream													
26	Half & half	1	tablespoon(s)	15	12	20	<1	1	0	2	1.07	0.50	0.06	—
28	Light coffee or table, liquid	1	tablespoon(s)	15	11	29	<1	1	0	3	1.80	0.84	0.11	—
30	Light whipping cream, liquid	1	tablespoon(s)	15	10	44	<1	<1	0	5	2.90	1.36	0.13	—
32	Heavy whipping cream, liquid	1	tablespoon(s)	15	9	52	<1	<1	0	6	3.45	1.60	0.21	—
34	Whipped cream topping, pressurized	1	tablespoon(s)	4	2	10	<1	<1	0	1	0.52	0.24	0.03	—
	Sour cream													
36	Sour cream	2	tablespoon(s)	24	17	51	1	1	0	5	3.13	1.45	0.19	—
30556	Fat free sour cream	2	tablespoon(s)	32	26	24	1	5	0	0	0.00	0.00	0.00	0
	Imitation cream													
3659	Coffeemate nondairy creamer, liquid	1	tablespoon(s)	16	—	20	0	2	0	1	0.00	0.50	0.00	—
40	Cream substitute, powder	1	teaspoon(s)	2	<.1	11	<.1	1	0	1	0.65	0.02	0.00	—
35972	Nondairy coffee whitener, liquid, frozen	1	tablespoon(s)	16	12	22	<1	2	0	2	0.31	1.20	0.00	—
35975	Nondairy dessert topping, pressurized	1	tablespoon(s)	5	3	12	<.1	1	0	1	0.88	0.09	0.01	—
35976	Nondairy dessert topping, frozen	1	tablespoon(s)	5	3	16	<.1	1	0	1	1.09	0.08	0.03	—
904	Imitation sour cream	2	tablespoon(s)	24	17	50	1	2	0	5	4.27	0.14	0.01	—

Chol (mg)	Calc (mg)	Iron (mg)	Magn (mg)	Pota (mg)	Sodi (mg)	Zinc (mg)	Vit A (μg)	Thia (mg)	Vit E (mg α)	Ribo (mg)	Niac (mg)	Vit B_6 (mg)	Fola (μg)	Vit C (mg)	Vit B_{12} (μg)	Sele (μg)
18	219	0.06	6	24	173	0.77	36	0.01	0.04	0.08	0.03	0.02	3	0	<1	4
22	141	0.12	6	21	176	0.82	50	0.01	0.05	0.08	0.03	0.01	2	0	1	5
27	201	0.11	8	38	176	0.79	83	0.00	0.07	0.09	0.03	0.02	3	0	<1	4
21	21	0.08	2	32	112	0.15	83	0.00	—	0.05	0.04	0.01	3	0	<.1	1
4	55	0.05	2	6	76	0.19	6	0.00	0.01	0.02	0.01	0.00	1	0	<1	1
19	212	0.15	8	39	245	0.90	66	0.01	0.06	0.09	0.04	0.02	3	0	<1	4
19	167	0.27	9	77	77	0.82	66	0.01	0.04	0.11	0.05	0.01	8	0	<1	10
31	127	0.23	7	65	52	0.71	74	0.01	0.07	0.12	0.06	0.03	7	0	<1	9
5	53	0.04	2	4	60	0.13	5	0.00	0.01	0.02	0.00	0.00	<1	0	<.1	1
25	185	0.16	8	25	507	0.58	82	0.01	—	0.16	0.21	0.03	14	0	<1	4
26	221	0.06	11	22	54	1.22	62	0.02	0.11	0.08	0.03	0.02	2	0	1	5
0	150	0.00	—	—	420	—	—	—	—	—	—	—	—	0	—	—
0	150	0.00	8	—	320	1.20	—	0.00	—	0.26	0.00	0.00	40	0	<1	—
5	69	0.16	6	97	459	0.43	12	0.02	0.01	0.19	0.14	0.08	14	0	1	10
9	78	0.18	7	108	459	0.47	24	0.03	0.02	0.21	0.16	0.09	15	0	1	12
32	23	0.35	2	35	86	0.16	106	0.00	0.09	0.06	0.03	0.01	4	0	<1	1
2	56	0.05	4	49	164	0.26	84	0.02	0.00	0.05	0.05	0.02	11	0	<1	1
0	0	0.00	—	—	135	—	0	—	—	—	—	—	—	0	—	—
27	156	0.05	8	48	422	0.81	72	0.01	0.08	0.10	0.02	0.02	2	0	<1	4
23	162	0.16	9	83	359	0.91	57	0.02	0.06	0.15	0.05	0.02	2	0	<1	5
16	160	0.09	8	69	382	0.74	49	0.01	0.05	0.12	0.04	0.03	2	0	<1	3
25	150	0.00	0	25	450	0.90	84	—	—	0.10	—	—	—	0	<1	—
24	219	0.17	8	61	388	1.02	56	0.00	0.10	0.08	0.01	0.01	2	0	<1	5
0	200	0.36	—	—	190	—	—	—	—	—	—	—	—	0	—	—
0	150	0.36	—	—	190	—	—	—	—	—	—	—	—	0	—	—
6	16	0.01	2	20	6	0.08	15	0.01	0.05	0.02	0.01	0.01	<1	<1	<.1	<1
10	14	0.01	1	18	6	0.04	27	0.00	0.08	0.02	0.01	0.00	<1	<1	<.1	<.1
17	10	0.00	1	15	5	0.04	42	0.00	0.13	0.02	0.01	0.00	1	<.1	<.1	<.1
21	10	0.00	1	11	6	0.03	62	0.00	0.16	0.02	0.01	0.00	1	<.1	<.1	<.1
3	4	0.00	<1	6	5	0.01	7	0.00	0.02	0.00	0.00	0.00	<1	0	<.1	<.1
11	28	0.01	3	35	13	0.06	42	0.01	0.14	0.04	0.02	0.00	3	<1	<.1	1
3	40	0.00	3	41	45	0.16	—	0.01	0.00	0.05	0.02	0.01	4	0	<.1	—
0	0	0.00	—	30	0	—	0	0.02	—	0.02	0.20	—	—	0	—	—
0	<1	0.02	<.1	16	4	0.01	<.1	0.00	0.01	0.00	0.00	0.00	0	0	0	<.1
0	1	0.00	<.1	30	13	0.00	—	0.00	—	0.00	0.00	0.00	0	0	0	<1
0	<1	0.00	<.1	1	3	0.00	—	0.00	—	0.00	0.00	0.00	0	0	0	<.1
0	<1	0.01	<.1	1	1	0.00	—	0.00	—	0.00	0.00	0.00	0	0	0	<1
0	1	0.09	1	39	24	0.28	0	0.00	0.18	0.00	0.00	0.00	0	0	0	1

Table A–1
Food Composition

(DA+ code is for Wadsworth Diet Analysis program)
(For purposes of calculations, use "0" for t, <1, <.1, <.01, etc.)

DA + Code	Food Description	Quantity	Measure	Wt (g)	H₂O (g)	Ener (kcal)	Prot (g)	Carb (g)	Fiber (g)	Fat (g)	Sat	Fat Breakdown (g) Mono	Poly	Trans
	DAIRY–Continued													
	Fluid milk													
57	Fat free, nonfat, or skim	1	cup(s)	245	223	83	8	12	0	<1	0.29	0.12	0.02	—
58	Fat free, nonfat, or skim, w/nonfat milk solids	1	cup(s)	245	221	91	9	12	0	1	0.40	0.16	0.02	—
54	Low fat, 1%	1	cup(s)	244	219	102	8	12	0	2	1.54	0.68	0.09	—
55	Low fat, 1%, w/nonfat milk solids	1	cup(s)	245	220	105	9	12	0	2	1.48	0.69	0.09	—
60	Low fat buttermilk	1	cup(s)	245	221	98	8	12	0	2	1.34	0.62	0.08	—
51	Reduced fat, 2%	1	cup(s)	244	218	122	8	11	0	5	2.35	2.04	0.17	—
52	Reduced fat, 2%, w/nonfat milk solids	1	cup(s)	245	218	125	9	12	0	5	2.93	1.36	0.17	—
50	Whole, 3.3%	1	cup(s)	244	216	146	8	11	0	8	4.55	1.98	0.48	—
	Canned													
61	Whole evaporated	2	tablespoon(s)	32	23	42	2	3	0	2	1.45	0.74	0.08	—
62	Fat free, nonfat, or skim evaporated	2	tablespoon(s)	32	25	25	2	4	0	<.1	0.04	0.02	0.00	—
63	Sweetened condensed	2	tablespoon(s)	38	10	123	3	21	0	3	2.10	0.93	0.13	—
	Dried Milk													
64	Dried buttermilk	¼	cup(s)	30	1	118	10	15	0	2	1.09	0.51	0.07	—
65	Instant nonfat dry milk w/added vitamin A	¼	cup(s)	17	1	63	6	9	0	<1	0.08	0.03	0.00	—
5234	Skim milk powder	¼	cup(s)	18	1	64	6	9	0	<1	0.08	0.03	0.01	—
907	Whole dry milk	¼	cup(s)	32	1	161	9	12	0	9	5.43	2.57	0.22	—
909	**Goat milk**	1	cup(s)	244	212	168	9	11	0	10	6.51	2.71	0.36	—
	Chocolate milk													
69	Low fat	1	cup(s)	250	211	158	8	26	1	3	1.54	0.75	0.09	—
68	Reduced fat	1	cup(s)	250	209	180	8	26	1	5	3.10	1.47	0.18	—
67	Whole milk	1	cup(s)	250	206	208	8	26	2	8	5.26	2.48	0.31	—
33156	Chocolate syrup, fortified, prepared w/milk	1	cup(s)	263	220	197	8	24	<1	8	5.22	2.44	0.31	—
908	Cocoa, hot, prepared w/milk	1	cup(s)	250	206	193	9	27	3	6	3.58	1.69	0.09	0.18
33184	Cocoa mix with aspartame, added sodium & vitamin A, no added calcium or phosphorus, prepared with water	1	cup(s)	192	177	56	2	10	1	<1	0.00	0.15	0.01	—
70	**Eggnog**	1	cup(s)	254	189	343	10	34	0	19	11.29	5.67	0.86	—
	Breakfast drinks													
10093	Carnation Instant Breakfast classic chocolate malt, prepared w/skim milk, no sugar added	1	cup(s)	243	—	142	11	21	<1	1	0.89	—	—	—
10091	Carnation Instant Breakfast strawberry creme, prepared w/skim milk	1	cup(s)	273	—	220	13	39	0	<1	0.40	—	—	—
10094	Carnation Instant Breakfast strawberry creme, prepared w/skim milk, no sugar added	1	cup(s)	243	—	134	12	21	0	<1	0.45	—	—	—
10092	Carnation Instant Breakfast vanilla creme, prepared w/skim milk, no sugar added	1	cup(s)	273	—	220	13	39	0	<1	0.40	—	—	—
1417	Ovaltine rich chocolate flavor, prepared w/skim milk	1	cup(s)	243	—	134	12	21	0	<1	0.45	—	—	—
8539	**Malted milk, chocolate mix, fortified, prepared w/milk**	1	cup(s)	265	216	223	9	29	1	9	4.95	2.17	0.54	—
	Milkshakes													
73	Chocolate	1	cup(s)	227	164	270	7	48	1	6	3.81	1.77	0.23	—
74	Vanilla	1	cup(s)	227	169	254	9	40	0	7	4.28	1.98	0.26	—

Chol (mg)	Calc (mg)	Iron (mg)	Magn (mg)	Pota (mg)	Sodi (mg)	Zinc (mg)	Vit A (µg)	Thia (mg)	Vit E (mg α)	Ribo (mg)	Niac (mg)	Vit B$_6$ (mg)	Fola (µg)	Vit C (mg)	Vit B$_{12}$ (µg)	Sele (µg)
5	223	1.23	22	238	108	2.08	149	0.11	0.02	0.45	0.23	0.09	12	0	1	8
5	316	0.12	37	419	130	1.00	149	0.10	0.00	0.43	0.22	0.11	12	2	1	5
12	264	0.85	27	290	122	2.12	142	0.05	0.02	0.45	0.23	0.09	12	0	1	8
10	314	0.12	34	397	127	0.98	145	0.10	—	0.42	0.22	0.11	12	2	1	6
10	284	0.12	27	370	257	1.03	17	0.08	0.12	0.38	0.14	0.08	12	2	1	5
20	271	0.24	27	342	115	1.17	134	0.10	0.07	0.45	0.22	0.09	12	<1	1	6
20	314	0.12	34	397	127	0.98	137	0.10	—	0.42	0.22	0.11	12	2	1	6
24	246	0.07	24	325	105	0.93	68	0.11	0.15	0.45	0.26	0.09	12	0	1	9
9	82	0.06	8	95	33	0.24	20	0.01	0.04	0.10	0.06	0.02	3	1	<.1	1
1	93	0.09	9	106	37	0.29	38	0.01	0.00	0.10	0.06	0.02	3	<1	<.1	1
13	109	0.07	10	142	49	0.36	28	0.03	0.06	0.16	0.08	0.02	4	1	<1	6
21	360	0.09	33	484	157	1.22	15	0.12	0.03	0.48	0.27	0.10	14	2	1	6
3	215	0.05	20	298	96	0.77	124	0.07	0.00	0.30	0.16	0.06	9	1	1	5
3	222	0.06	21	307	99	0.79	0	0.07	—	0.31	0.16	0.06	9	1	1	5
31	296	0.15	28	431	120	1.08	83	0.09	0.16	0.39	0.21	0.10	12	3	1	5
27	327	0.12	34	498	122	0.73	139	0.12	0.17	0.34	0.68	0.11	2	3	<1	3
8	288	0.60	33	425	153	1.03	145	0.10	0.05	0.42	0.32	0.10	13	2	1	5
18	285	0.60	33	423	150	1.03	138	0.09	0.10	0.41	0.32	0.10	13	2	1	5
30	280	0.60	33	418	150	1.03	65	0.09	0.15	0.41	0.31	0.10	13	2	1	5
34	292	2.68	32	460	147	0.92	—	0.09	—	0.55	6.53	0.11	13	2	1	5
20	263	1.20	58	493	110	1.58	128	0.10	0.08	0.46	0.33	0.10	13	1	1	7
<1	90	0.75	33	405	171	0.52	27	0.04	0.06	0.21	0.16	0.05	2	<1	<1	2
150	330	0.51	48	419	137	1.17	114	0.09	0.51	0.48	0.27	0.13	3	4	1	11
9	445	4.01	89	632	196	3.38	—	0.35	—	0.45	4.45	0.45	4	27	1	8
9	500	4.47	100	638	360	3.75	—	0.38	—	0.51	5.08	0.48	100	30	1	9
9	445	4.01	89	570	187	3.38	—	0.33	—	0.45	4.45	0.45	89	27	1	8
9	500	4.50	100	630	240	3.75	—	0.38	—	0.51	5.00	0.50	100	30	2	9
9	445	4.01	89	570	187	3.38	—	0.33	—	0.45	4.45	0.45	89	27	1	8
27	339	3.76	45	578	231	1.17	904	0.76	0.16	1.32	11.08	1.01	19	32	1	12
25	299	0.70	36	508	252	1.09	41	0.11	0.11	0.50	0.28	0.06	11	0	1	4
27	331	0.23	27	415	215	0.88	57	0.07	0.11	0.44	0.33	0.10	16	0	1	5

Table A–1
Food Composition

(DA+ code is for Wadsworth Diet Analysis program)
(For purposes of calculations, use "0" for t, <1, <.1, <.01, etc.)

DA + Code	Food Description	Quantity	Measure	Wt (g)	H₂O (g)	Ener (kcal)	Prot (g)	Carb (g)	Fiber (g)	Fat (g)	Sat	Mono	Poly	Trans
												Fat Breakdown (g)		
	DAIRY–Continued													
	Ice cream													
4776	Chocolate	½	cup(s)	66	37	143	3	19	1	7	4.49	2.12	0.27	—
16514	Chocolate, soft serve	½	cup(s)	87	50	177	3	24	1	8	5.17	2.43	0.31	—
12137	Chocolate fudge, fat free no sugar added	½	cup(s)	71	—	100	4	22	0	0	0.00	0.00	0.00	0
82	Light vanilla	½	cup(s)	66	42	109	4	18	<1	3	1.71	0.57	0.10	—
78	Light vanilla, soft serve	½	cup(s)	86	60	108	4	19	0	2	1.40	0.65	0.09	—
16523	Sherbet, all flavors	½	cup(s)	97	64	133	1	29	<1	2	1.12	0.51	0.08	—
4778	Strawberry	½	cup(s)	66	40	127	2	18	1	6	3.43	—	—	—
76	Vanilla	½	cup(s)	66	40	133	2	16	<1	7	4.48	1.96	0.30	—
12146	Vanilla chocolate swirl, fat free, no sugar added	½	cup(s)	71	—	100	4	20	0	0	0.00	0.00	0.00	0
	Soy desserts													
10694	Tofutti low fat vanilla fudge nondairy frozen dessert	½	cup(s)	70	—	120	2	24	0	2	1.00	—	—	—
15721	Tofutti premium chocolate supreme nondairy frozen dessert	½	cup(s)	60	—	180	3	18	0	11	2.00	—	—	—
15720	Tofutti premium vanilla nondairy frozen dessert	½	cup(s)	60	—	190	2	20	0	11	2.00	—	—	—
	Ice milk													
16516	Flavored, not chocolate	½	cup(s)	66	45	91	2	15	0	3	1.72	0.81	0.11	—
16517	Chocolate	½	cup(s)	66	43	95	3	17	<1	2	1.29	0.61	0.08	—
	Pudding													
25032	Chocolate	½	cup(s)	144	110	154	5	23	1	5	2.78	1.94	0.23	0
1923	Chocolate, sugar free, prepared w/2% milk	½	cup(s)	133	—	100	5	14	<1	3	1.50	—	—	—
1722	Rice	½	cup(s)	113	73	175	6	26	1	6	1.99	2.14	0.88	—
4747	Tapioca, ready to eat	1	item(s)	142	105	169	3	28	<1	5	0.85	2.24	1.93	—
25031	Vanilla	½	cup(s)	136	110	116	5	17	<.1	3	1.31	1.21	0.16	0
1924	Vanilla, sugar free, prepared w/2% milk	½	cup(s)	133	90	4	12	<1	2	2	10	150	0.00	—
	Frozen yogurt													
4785	Chocolate, soft serve	½	cup(s)	72	46	115	3	18	2	4	2.61	1.26	0.16	—
1747	Fruit varieties	½	cup(s)	113	80	144	3	24	0	4	2.63	1.11	0.11	—
4786	Vanilla, soft serve	½	cup(s)	72	47	117	3	17	0	4	2.46	1.14	0.15	—
	Milk substitutes													
	Lactose free													
16081	Fat free calcium fortified milk	1	cup(s)	240	—	90	9	13	0	0	0.00	—	—	0
36486	Low fat milk	1	cup(s)	240	—	110	8	13	0	3	1.50	—	—	—
36487	Reduced fat milk	1	cup(s)	240	—	130	8	13	0	5	3.00	—	—	—
36488	Whole milk	1	cup(s)	240	—	160	8	12	0	9	5.00	—	—	—
	Rice													
10083	Rice Dream carob rice beverage	1	cup(s)	240	—	150	1	32	0	3	0.00	—	—	—
10087	Rice Dream vanilla enriched rice beverage	1	cup(s)	240	—	130	1	28	0	2	0.00	—	—	—
17089	Rice Dream original rice beverage, enriched	1	cup(s)	240	—	120	1	25	0	2	0.00	—	—	—
	Soy													
34750	Soy Dream chocolate enriched soy beverage	1	cup(s)	240	—	210	7	37	1	4	0.50	—	—	—
34749	Soy Dream vanilla enriched soy beverage	1	cup(s)	240	—	150	7	22	0	4	0.50	—	—	—
13840	Vitasoy light chocolate soymilk	1	cup(s)	237	—	100	4	17	0	2	0.50	0.50	1.00	—
13839	Vitasoy light vanilla soymilk	1	cup(s)	237	—	70	4	10	0	2	0.50	0.50	1.00	—

Chol (mg)	Calc (mg)	Iron (mg)	Magn (mg)	Pota (mg)	Sodi (mg)	Zinc (mg)	Vit A (µg)	Thia (mg)	Vit E (mg α)	Ribo (mg)	Niac (mg)	Vit B6 (mg)	Fola (µg)	Vit C (mg)	Vit B12 (µg)	Sele (µg)
22	72	0.61	19	164	50	0.38	78	0.03	0.20	0.13	0.15	0.04	11	<1	<1	2
22	103	0.33	19	192	44	0.48	—	0.04	0.22	0.13	0.11	0.03	5	1	<1	—
0	80	0.36	—	—	60	—	—	—	—	—	—	—	—	0	—	—
17	77	0.05	9	137	49	0.48	91	0.02	0.08	0.11	0.06	0.02	3	<1	<1	1
10	135	0.05	12	190	60	0.46	25	0.04	0.05	0.17	0.10	0.04	5	1	<1	3
5	52	0.14	8	93	44	0.46	—	0.02	0.03	0.07	0.09	0.03	4	4	<1	—
19	79	0.14	9	124	40	0.22	63	0.03	—	0.17	0.11	0.03	8	5	<1	1
29	84	0.06	9	131	53	0.46	78	0.03	0.20	0.16	0.08	0.03	3	<1	<1	1
0	80	0.00	50	0					—							
0	0	0.00	—	8	90	—	0	—	—	—	—	—	—	0	—	—
0	0	0.00	—	7	180	—	0	—	—	—	—	—	—	0	—	—
0	0	0.00	—	2	210	—	0	—	—	—	—	—	—	0	—	—
9	91	0.07	10	138	56	0.29	—	0.04	0.06	0.17	0.06	0.04	4	1	<1	—
6	94	0.17	13	155	41	0.38	—	0.03	0.05	0.12	0.09	0.03	4	<1	<1	—
35	138	1.04	29	211	135	1.07	73	0.04	0.00	0.25	0.18	0.03	7	<1	<1	5
10	150	0.72	—	330	310	—	—	0.06	—	0.26	—	—	—	0	—	—
71	130	1.21	21	250	253	0.61	—	0.10	0.06	0.26	0.73	0.08	14	1	<1	—
1	119	0.33	11	136	226	0.38	0	0.03	0.43	0.14	0.44	0.03	4	1	<1	2
35	133	0.25	14	173	134	0.63	73	0.03	0.00	0.24	0.11	0.03	6	<1	<1	5
—	190	380	—	—	<1	—	—	0.17	—	—	—	0	—	—		
4	106	0.90	19	188	71	0.35	32	0.03	—	0.15	0.22	0.05	8	<1	<1	2
15	113	0.52	11	176	71	0.32	—	0.05	0.10	0.20	0.08	0.05	5	1	<.1	—
1	103	0.22	10	152	63	0.30	42	0.03	0.08	0.16	0.21	0.06	4	1	<1	2
3	500	0.00	—	—	130	—	100	—	—	—	—	—	—	0	—	—
15	300	0.00	—	—	125	—	100	—	—	—	—	—	—	0	—	—
20	300	0.00	—	—	125	—	98	—	—	—	—	—	—	0	—	—
35	300	0.00	—	—	125	—	58	—	—	—	—	—	—	0	—	—
0	20	0.72	—	—	100	—	—	—	—	—	—	—	—	1	—	—
0	300	0.00	—	—	90	—	—	—	—	—	—	—	—	0	2	—
0	300	0.00	13	60	90	0.24	—	0.07	—	0.00	0.84	0.08	—	0	2	—
0	300	1.80	60	350	160	0.60	33	0.15	—	0.07	0.80	0.12	60	0	3	—
0	300	1.80	40	260	140	0.60	33	0.15	—	0.07	0.80	0.12	60	0	3	—
0	300	0.72	24	200	140	0.90	0	0.09	—	0.34	—	—	24	0	1	—
0	300	0.72	24	200	110	0.90	0	0.09	—	0.34	—	—	24	0	1	—

Table A–1
Food Composition

(DA+ code is for Wadsworth Diet Analysis program)
(For purposes of calculations, use "0" for t, <1, <.1, <.01, etc.)

DA + Code	Food Description	Quantity	Measure	Wt (g)	H₂O (g)	Ener (kcal)	Prot (g)	Carb (g)	Fiber (g)	Fat (g)	Sat	Mono	Poly	Trans
	DAIRY													
	Soy–Continued													
13836	Vitasoy rich chocolate soymilk	1	cup(s)	237	—	160	7	24	1	4	0.50	1.00	2.50	—
13835	Vitasoy vanilla delite soymilk	1	cup(s)	237	—	120	8	13	1	4	0.50	1.00	2.50	—
	Yogurt													
3615	Custard style, fruit flavors	6	ounce(s)	170	127	190	7	32	0	4	2.00	—	—	—
3617	Custard style, vanilla	6	ounce(s)	170	134	190	7	32	0	4	2.00	0.94	0.10	—
32101	Fruit, low fat	1	cup(s)	245	184	243	10	46	0	3	1.82	0.77	0.08	—
29638	Fruit, nonfat, sweetened w/low calorie sweetener	1	cup(s)	241	208	122	11	19	1	<1	0.21	0.10	0.04	—
93	Plain, low fat	1	cup(s)	245	208	154	13	17	0	4	2.45	1.04	0.11	—
94	Plain, nonfat	1	cup(s)	245	209	137	14	19	0	<1	0.28	0.12	0.01	—
32100	Vanilla, low fat	1	cup(s)	245	194	208	12	34	0	3	1.97	0.84	0.09	—
5242	Yogurt beverage	1	cup(s)	245	200	172	6	33	0	2	1.39	0.59	0.06	—
38202	Yogurt smoothie, nonfat, all flavors	1	item(s)	325	—	290	10	60	6	0	0.00	0.00	0.00	0
	Soy yogurt													
10453	White Wave plain silk cultured	8	ounce(s)	227	—	120	5	22	1	3	0.00	—	—	0
34616	Stonyfield Farm Osoy chocolate-vanilla pack organic cultured	1	serving(s)	113	—	90	4	15	3	2	0.00	—	—	
34617	Stonyfield Farm Osoy strawberry-peach pack organic cultured	1	serving(s)	113	—	90	4	15	3	2	0.00	—	—	
	EGGS													
96	Raw, whole	1	item(s)	50	38	74	6	<1	0	5	1.55	1.91	0.68	—
97	Raw, white	1	item(s)	33	29	17	4	<1	0	<.1	0.00	0.00	0.00	—
98	Raw, yolk	1	item(s)	17	9	53	3	1	0	4	1.59	1.95	0.70	—
99	Fried	1	item(s)	46	32	92	6	<1	0	7	1.98	2.92	1.22	—
100	Hard boiled	1	item(s)	50	37	78	6	1	0	5	1.63	2.04	0.71	—
101	Poached	1	item(s)	50	38	74	6	<1	0	5	1.54	1.90	0.68	—
102	Scrambled, prepared w/milk & butter	2	item(s)	122	89	203	14	3	0	15	4.49	5.82	2.62	—
	Egg Substitute													
920	Frozen	¼	cup(s)	60	44	96	7	2	0	7	1.16	1.46	3.74	—
918	Liquid	¼	cup(s)	63	52	53	8	<1	0	2	0.41	0.56	1.01	—
4028	Egg Beaters	¼	cup(s)	61	—	30	6	1	0	0	0.00	0.00	0.00	0
	SEAFOOD													
	Fish													
	Cod													
6040	Atlantic cod or scrod, baked or broiled	3	ounce(s)	44	34	46	10	0	0	<1	0.07	0.05	0.13	—
1573	Atlantic cod, cooked, dry heat	3	ounce(s)	85	65	89	19	0	0	1	0.14	0.11	0.25	—
2905	**Eel, raw**	3	ounce(s)	85	58	156	16	0	0	10	2.01	6.12	0.81	—
	Fish fillets													
25079	Baked	3	ounce(s)	84	80	99	22	0	0	1	0.08	0.07	0.26	—
8615	Batter coated or breaded, fried	3	ounce(s)	85	46	197	12	14	<1	10	2.39	2.19	5.32	—
25082	Broiled fish steaks	3	ounce(s)	86	69	129	24	0	0	3	0.37	0.87	0.84	—
25083	Poached fish steaks	3	ounce(s)	86	68	112	21	0	0	2	0.33	0.76	0.74	—
25084	Steamed fish fillets	3	ounce(s)	86	73	80	17	0	0	1	0.12	0.08	0.22	—
25089	**Flounder, baked**	3	ounce(s)	85	65	114	15	<1	<.1	6	1.15	2.17	1.44	0
1825	**Grouper, cooked, dry heat**	3	ounce(s)	85	62	100	21	0	0	1	0.25	0.23	0.34	—
	Haddock													
6049	Baked or broiled	3	ounce(s)	44	33	50	11	0	0	<1	0.07	0.07	0.14	—
1578	Cooked, dry heat	3	ounce(s)	85	63	95	21	0	0	1	0.14	0.13	0.26	—
1886	**Halibut, Atlantic & Pacific, cooked, dry heat**	3	ounce(s)	85	61	119	23	0	0	2	0.35	0.82	0.80	—
1582	**Herring, Atlantic, pickled**	4	piece(s)	60	33	157	9	6	0	11	1.43	7.17	1.01	—

A

Chol (mg)	Calc (mg)	Iron (mg)	Magn (mg)	Pota (mg)	Sodi (mg)	Zinc (mg)	Vit A (µg)	Thia (mg)	Vit E (mg α)	Ribo (mg)	Niac (mg)	Vit B$_6$ (mg)	Fola (µg)	Vit C (mg)	Vit B$_{12}$ (µg)	Sele (µg)
0	300	1.08	40	320	150	0.90	0	0.15	—	0.34	—	—	60	0	1	—
0	40	0.72	—	320	115	—	0	—	—	—	—	—	—	0	—	—
15	200	0.00	16	310	90	—	0	—	—	0.26	—	—	—	0	—	—
15	200	0.00	16	300	90	—	0	—	—	0.26	—	—	—	0	—	—
12	338	0.15	32	434	130	1.64	27	0.08	0.05	0.40	0.21	0.09	22	1	1	7
3	370	0.62	41	550	139	1.83	0.10	0.17	0.17	0.50	0.11	0.09	26	1		
15	448	0.20	42	573	172	2.18	34	0.11	0.05	0.52	0.28	0.12	27	2	1	8
5	488	0.22	47	625	189	2.38	5	0.12	0.07	0.57	0.30	0.13	29	2	1	9
12	419	0.17	39	537	162	2.03	29	0.10	0.00	0.49	0.26	0.11	27	2	1	12
13	260	0.22	39	399	98	1.10	—	0.11	0.05	0.51	0.30	0.15	29	2	2	—
5	300	2.70	100	580	290	2.25	—	0.38	—	0.43	5.00	0.50	100	15	2	—
0	700	0.90	—	—	30	—	—	—	—	—	—	—	—	0	0	—
0	100	0.72	—	—	20	—	0	—	—	—	—	—	—	0	—	—
0	100	0.72	—	—	20	—	0	—	—	—	—	—	—	0	—	—
212	27	0.92	6	67	70	0.56	70	0.03	0.49	0.24	0.04	0.07	24	0	1	16
0	2	0.03	4	54	55	0.01	0	0.00	0.00	0.15	0.04	0.00	1	0	<.1	7
205	21	0.45	1	18	8	0.38	63	0.03	0.43	0.09	0.00	0.06	24	0	<1	9
210	27	0.91	6	68	94	0.55	91	0.03	0.56	0.24	0.04	0.07	23	0	1	16
212	25	0.60	5	63	62	0.53	85	0.03	0.51	0.26	0.03	0.06	22	0	1	15
211	27	0.92	6	67	147	0.55	70	0.03	0.48	0.24	0.04	0.07	24	0	1	16
429	87	1.46	15	168	342	1.22	174	0.06	1.04	0.53	0.10	0.14	37	<1	1	27
1	44	1.19	9	128	119	0.59	7	0.07	0.95	0.23	0.08	0.08	10	<1	<1	25
1	33	1.32	6	207	111	0.82	11	0.07	0.17	0.19	0.07	0.00	9	0	<1	16
0	20	1.08	4	85	115	0.60	113	0.15	—	0.85	0.20	0.08	60	0	1	—
24	6	0.22	19	108	35	0.26	—	0.04	—	0.03	1.11	0.13	5	<1	<1	17
47	12	0.42	36	207	66	0.49	12	0.07	0.69	0.07	2.14	0.24	7	1	1	32
107	17	0.43	17	231	43	1.38	887	0.13	3.40	0.03	2.98	0.06	13	2	3	6
44	8	0.32	29	489	86	0.49	10	0.03	—	0.05	2.48	0.46	8	3	1	44
29	15	1.79	20	272	452	0.37	10	0.09	—	0.09	1.78	0.08	17	0	1	8
37	55	0.99	98	529	64	0.49	55	0.06	—	0.08	6.88	0.36	13	0	1	43
33	48	0.86	85	460	55	0.43	48	0.06	—	0.08	5.97	0.33	12	0	1	37
42	13	0.30	25	323	42	0.35	12	0.07	—	0.06	1.92	0.22	6	1	1	32
44	19	0.35	47	225	281	0.21	39	0.06	0.41	0.08	2.03	0.19	7	3	2	34
40	18	0.97	31	404	45	0.43	43	0.07	—	0.01	0.32	0.30	9	0	1	40
33	19	0.60	22	177	39	0.21	—	0.02	—	0.02	2.05	0.15	4	0	1	18
63	36	1.15	43	339	74	0.41	16	0.03	—	0.04	3.94	0.29	11	0	1	34
35	51	0.91	91	490	59	0.45	46	0.06	—	0.08	6.05	0.34	12	0	1	40
8	46	0.73	5	41	522	0.32	155	0.02	1.03	0.08	1.98	0.10	1	0	3	35

Table A–1
Food Composition

(DA+ code is for Wadsworth Diet Analysis program)
(For purposes of calculations, use "0" for t, <1, <.1, <.01, etc.)

DA + Code	Food Description	Quantity	Measure	Wt (g)	H₂O (g)	Ener (kcal)	Prot (g)	Carb (g)	Fiber (g)	Fat (g)	Fat Breakdown (g)			
											Sat	Mono	Poly	*Trans*
	SEAFOOD													
	Haddock–Continued													
1587	**Jack mackerel, solids, canned, drained**	2	ounce(s)	57	39	88	13	0	0	4	1.05	1.26	0.94	—
8580	**Octopus, common, cooked, moist heat**	3	ounce(s)	85	51	139	25	4	0	2	0.39	0.28	0.41	—
1831	**Perch, mixed species, cooked, dry heat**	3	ounce(s)	85	62	99	21	0	0	1	0.20	0.17	0.40	—
1592	**Pacific rockfish, cooked, dry heat**	3	ounce(s)	85	62	103	20	0	0	2	0.40	0.38	0.50	—
	Salmon													
29727	Smoked chinook (lox)	2	ounce(s)	57	<.1	66	10	0	0	2	0.52	1.14	0.56	—
1594	Broiled or baked w/butter	3	ounce(s)	85	54	155	23	0	0	6	1.16	2.29	2.33	—
2938	Coho, farmed, raw	3	ounce(s)	85	60	136	18	0	0	7	1.54	2.83	1.58	—
154	**Sardines, Atlantic, with bones, canned in oil**	2	item(s)	24	14	50	6	0	0	3	0.36	0.92	1.23	—
	Scallops													
155	Mixed species, breaded, fried	3	item(s)	47	27	100	8	5	0	5	1.24	2.09	1.32	—
1599	Steamed	3	ounce(s)	85	65	90	14	2	0	3	—	—	—	—
1839	**Snapper, mixed species, cooked, dry heat**	3	ounce(s)	85	60	109	22	0	0	1	0.31	0.27	0.50	—
	Squid													
1868	Mixed species, fried	3	ounce(s)	85	55	149	15	7	0	6	1.60	2.34	1.82	—
16617	Steamed or boiled	3	ounce(s)	85	63	90	15	3	0	1	0.35	0.11	0.51	—
1570	**Striped bass, cooked, dry heat**	3	ounce(s)	85	62	105	19	0	0	3	0.55	0.72	0.85	—
1601	**Sturgeon, steamed**	3	ounce(s)	85	59	111	17	0	0	4	0.97	2.04	0.73	—
1840	**Surimi, formed**	3	ounce(s)	85	65	84	13	6	0	1	0.16	0.13	0.38	—
1842	**Swordfish, cooked, dry heat**	3	ounce(s)	85	58	132	22	0	0	4	1.20	1.68	1.00	—
1846	**Tuna, yellowfin or ahi, raw**	3	ounce(s)	85	60	92	20	0	0	1	0.20	0.13	0.24	—
	Tuna, canned													
159	Light, canned in oil, drained	2	ounce(s)	57	34	113	17	0	0	5	0.87	1.68	1.64	—
355	Light, canned in water, drained	2	ounce(s)	57	42	66	14	0	0	<1	0.13	0.09	0.19	—
33211	Light, no salt, canned in oil, drained	2	ounce(s)	57	34	112	17	0	0	5	0.87	1.67	1.64	—
33212	Light, no salt, canned in water, drained	2	ounce(s)	57	43	66	14	0	0	<1	0.13	0.09	0.19	—
2961	White, canned in oil, drained	2	ounce(s)	57	36	105	15	0	0	5	0.73	1.85	1.69	—
351	White, canned in water, drained	2	ounce(s)	57	41	73	13	0	0	2	0.45	0.44	0.63	—
33213	White, no salt, canned in oil, drained	2	ounce(s)	57	36	105	15	0	0	5	0.94	1.41	1.92	—
33214	White, no salt, canned in water, drained	2	ounce(s)	57	42	73	13	0	0	2	0.45	0.44	0.63	—
	Yellowtail													
2970	Mixed species, raw	2	ounce(s)	57	42	83	13	0	0	3	0.73	1.13	0.81	—
8548	Mixed species, cooked, dry heat	3	ounce(s)	85	57	159	25	0	0	6	1.44	2.21	1.52	—
	Shellfish, meat only													
1857	Abalone, mixed species, fried	3	ounce(s)	85	51	161	17	9	0	6	1.40	2.33	1.42	—
16618	Abalone, steamed or poached	3	ounce(s)	85	41	177	29	10	0	1	0.25	0.18	0.18	—
	Crab													
1851	Blue crab, canned	2	ounce(s)	57	43	56	12	0	0	1	0.14	0.12	0.25	—
1852	Blue crab, cooked, moist heat	3	ounce(s)	85	66	87	17	0	0	2	0.19	0.24	0.58	—
8562	Dungeness crab, cooked, moist heat	3	ounce(s)	85	62	94	19	1	0	1	0.14	0.18	0.35	—
1860	**Clams, cooked, moist heat**	3	ounce(s)	85	54	126	22	4	0	2	0.16	0.15	0.47	—
1853	**Crayfish, farmed, cooked, moist heat**	3	ounce(s)	85	69	74	15	0	0	1	0.18	0.21	0.35	—
	Oysters													
8720	Baked or broiled	3	ounce(s)	85	69	90	6	3	0	6	1.38	2.18	1.88	—
152	Eastern, farmed, raw	3	ounce(s)	85	73	50	4	5	0	1	0.38	0.13	0.50	—

Chol (mg)	Calc (mg)	Iron (mg)	Magn (mg)	Pota (mg)	Sodi (mg)	Zinc (mg)	Vit A (µg)	Thia (mg)	Vit E (mg α)	Ribo (mg)	Niac (mg)	Vit B_6 (mg)	Fola (µg)	Vit C (mg)	Vit B_{12} (µg)	Sele (µg)
45	137	1.16	21	110	215	0.58	74	0.02	0.58	0.12	3.50	0.12	3	1	4	21
82	90	8.11	51	536	391	2.86	77	0.05	1.02	0.06	3.21	0.55	20	7	31	76
98	87	0.99	32	292	67	1.22	9	0.07	—	0.10	1.62	0.12	5	1	2	14
37	10	0.45	29	442	65	0.45	60	0.04	1.33	0.07	3.33	0.23	9	0	1	40
13	6	0.48	10	99	1134	0.17	15	0.01	—	0.05	2.67	0.15	1	0	2	22
40	15	1.02	27	377	99	0.56	—	0.14	1.15	0.05	8.33	0.19	4	2	2	41
43	10	0.29	26	383	40	0.37	48	0.08	—	0.09	5.79	0.56	11	1	2	11
34	108	0.70	9	95	121	0.31	16	0.01	0.49	0.05	1.25	0.04	3	0	2	13
28	20	0.38	27	155	216	0.49	10	0.01	—	0.05	0.69	0.06	23	1	1	13
27	21	0.22	—	238	366	—	—	—	0.16	—	—	—	—	2	—	—
40	34	0.20	31	444	48	0.37	30	0.05	—	0.00	0.29	0.39	5	1	3	42
221	33	0.86	32	237	260	1.48	9	0.05	—	0.39	2.21	0.05	12	4	1	44
227	31	0.63	29	192	356	1.49	—	0.02	1.17	0.32	1.70	0.04	4	3	1	—
88	16	0.92	43	279	75	0.43	26	0.10	—	0.03	2.17	0.29	9	0	4	40
63	11	0.59	30	239	389	0.36	—	0.07	0.53	0.07	8.31	0.19	14	0	2	—
26	8	0.22	37	95	122	0.28	17	0.02	0.54	0.02	0.19	0.03	2	0	1	24
43	5	0.88	29	314	98	1.25	35	0.04	—	0.10	10.02	0.32	2	1	2	52
38	14	0.62	43	378	31	0.44	15	0.37	0.43	0.04	8.33	0.77	2	1	<1	31
10	7	0.79	18	118	202	0.51	13	0.02	0.50	0.07	7.06	0.06	3	0	1	43
17	6	0.87	15	134	192	0.44	10	0.02	0.19	0.04	7.53	0.20	2	0	2	46
10	7	0.79	18	117	28	0.51	13	0.02	—	0.07	7.03	0.06	3	0	1	43
17	6	0.87	15	134	28	0.44	10	0.02	—	0.04	7.53	0.20	2	0	2	46
18	2	0.37	19	189	225	0.27	3	0.01	1.30	0.04	6.63	0.24	3	0	1	34
24	8	0.55	19	134	214	0.27	3	0.00	0.48	0.02	3.29	0.12	1	0	1	37
18	2	0.37	19	189	28	0.27	14	0.01	—	0.04	6.63	0.24	3	0	1	34
24	8	0.55	19	134	28	0.27	3	0.00	—	0.02	3.29	0.12	1	0	1	37
31	13	0.28	17	238	22	0.29	16	0.08	—	0.02	3.86	0.09	2	2	1	21
60	25	0.53	32	457	43	0.56	26	0.14	—	0.04	7.41	0.15	3	2	1	40
80	31	3.23	48	241	502	0.81	2	0.19	—	0.11	1.62	0.13	12	2	1	44
143	50	4.85	69	295	980	1.38	—	0.29	6.74	0.13	1.90	0.22	6	3	1	—
50	57	0.48	22	212	189	2.28	1	0.05	1.04	0.05	0.78	0.09	24	2	<1	18
85	88	0.77	28	275	237	3.59	2	0.09	1.56	0.04	2.81	0.15	43	3	6	34
65	50	0.37	49	347	321	4.65	26	0.05	—	0.17	3.08	0.15	36	3	9	40
57	78	23.77	15	534	95	2.32	145	0.13	—	0.36	2.85	0.09	25	19	84	54
116	43	0.94	28	202	82	1.26	13	0.04	—	0.07	1.42	0.11	9	<1	3	29
42	37	5.30	38	126	418	72.22	60	0.07	0.99	0.06	1.04	0.05	8	3	15	—
21	37	4.91	28	105	151	32.23	7	0.09	—	0.06	1.08	0.05	15	4	14	54

Table A–1
Food Composition

(DA+ code is for Wadsworth Diet Analysis program)
(For purposes of calculations, use "0" for t, <1, <.1, <.01, etc.)

DA + Code	Food Description	Quantity	Measure	Wt (g)	H₂O (g)	Ener (kcal)	Prot (g)	Carb (g)	Fiber (g)	Fat (g)	Sat	Mono	Poly	Trans
	SEAFOOD													
	Oysters–Continued													
8715	Eastern, wild, cooked, moist heat	3	ounce(s)	85	60	116	12	7	0	4	1.31	0.53	1.65	—
8584	Pacific, cooked, moist heat	3	ounce(s)	85	55	139	16	8	0	4	0.87	0.66	1.52	—
1865	Pacific, raw	3	ounce(s)	85	70	69	8	4	0	2	0.43	0.30	0.76	—
1854	**Lobster, northern, cooked, moist heat**	3	ounce(s)	85	65	83	17	1	0	1	0.09	0.14	0.08	—
1862	**Mussels, blue, cooked, moist heat**	3	ounce(s)	85	52	146	20	6	0	4	0.72	0.86	1.03	—
	Shrimp													
1855	Mixed species, cooked, moist heat	3	ounce(s)	85	66	84	18	0	0	1	0.25	0.17	0.37	—
158	Mixed species, breaded, fried	3	ounce(s)	85	45	206	18	10	0.34	10	1.77	3.24	4.32	—
	BEEF, LAMB, PORK													
	Beef													
4450	Breakfast strips, cooked	2	slice(s)	23	0	101	7	<1	0	8	3.24	3.8	0.35	—
174	Corned, canned	3	ounce(s)	85	49	213	23	0	0	13	5.25	5.07	0.54	—
33147	Cured, thin sliced	2	ounce(s)	57	31	87	18	2	0	1	0.54	0.48	0.04	—
4581	Jerky	1	ounce(s)	28	0	116	9	3	1	7	3.08	3.21	0.28	—
	Ground													
4411	Extra lean, broiled, well	3	ounce(s)	85	46	225	24	0	0	13	5.28	5.88	0.50	—
4417	Lean, broiled, medium	3	ounce(s)	85	47	231	21	0	0	16	6.16	6.87	0.59	—
4418	Lean, broiled, well	3	ounce(s)	85	45	238	24	0	0	15	5.89	6.56	0.56	—
4423	Regular, broiled, medium	3	ounce(s)	85	46	246	20	0	0	18	6.91	7.70	0.65	—
	Rib													
4183	Rib, whole, lean & fat, ¼" fat, roasted	3	ounce(s)	85	39	320	19	0	0	27	10.71	11.42	0.94	—
	Roast													
4264	Bottom round, lean & fat, ¼" fat, braised	3	ounce(s)	85	44	241	24	0	0	15	5.71	6.63	0.58	—
169	Bottom round, separable lean, ¼" fat, roasted	3	ounce(s)	85	57	161	24	0	0	6	2.13	2.83	0.24	—
4147	Chuck, arm pot roast, lean & fat, ¼" fat, braised	3	ounce(s)	85	41	282	23	0	0	20	7.97	8.68	0.77	—
4161	Chuck, blade roast, lean & fat, ¼" fat, braised	3	ounce(s)	85	40	293	23	0	0	22	8.70	9.44	0.78	—
5853	Chuck, blade roast, separable lean, ¼" trim, pot roasted	3	ounce(s)	85	39	209	27	0	0	10	3.94	4.37	0.33	—
4295	Eye of round, lean, ¼" fat, roasted	3	ounce(s)	85	55	149	25	0	0	5	1.76	2.06	0.15	—
4285	Eye of round, lean & fat, ¼" fat, roasted	3	ounce(s)	85	51	195	23	0	0	11	4.23	4.66	0.39	—
	Steak													
1757	Rib, small end, lean, ¼" fat, broiled	3	ounce(s)	85	49	188	24	0	0	10	3.84	4.01	0.27	—
4349	Short loin, T-bone steak, lean, ¼" fat, broiled	3	ounce(s)	85	52	174	23	0	0	9	3.05	4.23	0.26	—
4348	Short loin, T-bone steak, lean & fat, ¼" fat, broiled	3	ounce(s)	85	43	274	19	0	0	21	8.29	9.58	0.75	—
4360	Top loin, prime, lean & fat, ¼" fat, broiled	3	ounce(s)	85	43	275	22	0	0	20	8.16	8.61	0.73	—
	Variety													
188	Liver, pan fried	3	ounce(s)	85	53	149	23	4	0	4	1.27	0.56	0.49	0.17
4447	Tongue, simmered	3	ounce(s)	85	49	236	16	0	0	19	6.91	8.59	0.56	0.71
	Lamb													
	Chop													
3275	Loin, domestic, lean & fat, ¼" fat, broiled	3	ounce(s)	85	44	269	21	0	0	20	8.36	8.25	1.43	—
3287	Shoulder, arm, domestic, lean & fat, ¼" fat, braised	3	ounce(s)	85	38	294	26	0	0	20	8.39	8.65	1.45	—

A

Chol (mg)	Calc (mg)	Iron (mg)	Magn (mg)	Pota (mg)	Sodi (mg)	Zinc (mg)	Vit A (µg)	Thia (mg)	Vit E (mg α)	Ribo (mg)	Niac (mg)	Vit B$_6$ (mg)	Fola (µg)	Vit C (mg)	Vit B$_{12}$ (µg)	Sele (µg)
89	77	10.19	81	239	359	154.37	46	0.16	—	0.15	2.11	0.10	12	5	30	61
85	14	7.82	37	257	180	28.25	124	0.11	0.72	0.38	3.08	0.08	13	11	24	131
43	7	4.35	19	143	90	14.14	69	0.06	—	0.20	1.71	0.04	9	7	14	65
61	52	0.33	30	299	323	2.48	22	0.01	0.85	0.06	0.91	0.07	9	0	3	36
28	5.71	31	228	314	2.27	77	0.26	—	0.36	2.55	0.09	65	12	20	76	
166	33	2.63	29	155	190	1.33	58	0.03	1.17	0.03	2.20	0.11	3	2	1	34
150.44	56.95	1.07	34	191.25	292.39	1.17	47.59	0.1	—	0.11	2.6	0.08	20.39	1.27	1.58	35.44
26.89	2.03	0.7	6.1	93.11	509.17	1.43	0	0.02	0.07	0.05	1.46	0.07	1.8	0	0.77	6.05
73	10	1.77	12	116	855	3.03	0	0.02	0.13	0.12	2.07	0.11	8	0	1	36
45	3	1.58	11	140	1582	2.49	0	0.03	0.00	0.12	1.85	0.16	5	0	1	13
13.63	5.67	1.53	14.48	169.54	628.49	2.3	0	0.04	0.14	0.04	0.49	0.05	38.05	0	0.28	3.03
84	8	2.35	21	314	70	5.47	0	0.06	—	0.27	4.97	0.27	9	0	2	19
74	9	1.79	18	256	65	4.56	0	0.04	—	0.18	4.39	0.22	8	0	2	25
86	10	2.08	20	297	76	5.27	0	0.05	—	0.20	5.07	0.26	9	0	2	22
77	9	2.07	17	248	71	4.40	0	0.03	—	0.16	4.90	0.23	8	0	2	16
72	9	1.96	16	252	54	4.45	0	0.06	—	0.14	2.86	0.20	6	0	2	19
82	5	2.65	19	240	43	4.17	0	0.06	0.17	0.20	3.17	0.28	9	0	2	27
66.3	4.25	2.66	23.79	332.35	56.09	3.92	0	0.06	—	0.2	3.45	0.31	10.19	0	2.29	23.29
84	9	2.64	16	209	51	5.81	0	0.06	0.19	0.20	2.70	0.24	8	0	3	21
88	11	2.64	16	196	54	7.07	0	0.06	0.15	0.20	2.06	0.22	4	0	2	21
73.94	11.05	3.12	19.54	223.55	60.34	8.72	0	0.06	—	0.23	0	0.24	—	0	2.09	22.69
59	4	1.66	23	336	53	4.03	0	0.08	—	0.14	3.19	0.32	6	0	2	23
61	5	1.56	20	308	50	3.69	0	0.07	0.15	0.14	2.97	0.30	6	0	2	22
68	11	2.18	23	335	59	5.94	0	0.09	0.12	0.19	4.08	0.34	7	0	3	19
50	5	3.11	22	278	65	4.34	0	0.09	0.12	0.21	3.94	0.33	7	0	2	9
58	7	2.56	18	234	58	3.56	0	0.08	0.19	0.18	3.29	0.28	6	0	2	10
67	8	1.89	20	294	54	3.85	0	0.07	—	0.15	3.96	0.31	6	0	2	19
324	5	5.24	19	298	65	4.45	6582	0.15	0.39	2.91	14.85	0.87	221	1	71	28
112	4	2.22	13	156	55	34.77	0	0.02	0.25	0.25	2.97	0.13	6	1	3	11
85	17	1.54	20	278	65	2.96	0	0.09	0.11	0.21	6.04	0.11	15	0	2	23
102	21	2.03	22	260	61	5.17	0	0.06	0.13	0.21	5.66	0.09	15	0	2	32

Table A–1
Food Composition

(DA+ code is for Wadsworth Diet Analysis program)
(For purposes of calculations, use "0" for t, <1, <.1, <.01, etc.)

DA + Code	Food Description	Quantity	Measure	Wt (g)	H₂O (g)	Ener (kcal)	Prot (g)	Carb (g)	Fiber (g)	Fat (g)	Sat	Fat Breakdown (g) Mono	Poly	Trans
	BEEF, LAMB, PORK													
	Lamb													
	Chop–Continued													
3290	Shoulder, arm, domestic, lean, ¼" fat, braised	3	ounce(s)	85	42	237	30	0	0	12	4.28	5.24	0.78	—
	Leg													
3264	Domestic, lean & fat, ¼" fat, cooked	3	ounce(s)	85	46	250	21	0	0	18	7.51	7.50	1.28	—
	Rib													
183	Domestic, lean, ¼" fat, broiled	3	ounce(s)	85	50	200	24	0	0	11	3.95	4.43	1.00	—
182	Domestic, lean & fat, ¼" fat, broiled	3	ounce(s)	85	40	307	19	0	0	25	10.80	10.30	2.01	—
	Shoulder													
187	Arm & blade, domestic, choice, lean, ¼" fat, roasted	3	ounce(s)	85	54	173	21	0	0	9	3.47	3.71	0.81	—
186	Arm & blade, domestic, choice, lean & fat, ¼" fat, roasted	3	ounce(s)	85	48	235	19	0	0	17	7.17	6.94	1.38	—
	Variety													
3375	Brain, pan fried	3	ounce(s)	85	52	232	14	0	0	19	4.82	3.42	1.94	—
3406	Tongue, braised	3	ounce(s)	85	49	234	18	0	0	17	6.66	8.50	1.06	—
	Pork													
	Cured													
161	Bacon, cured, broiled, pan fried or roasted	2	slice(s)	13	2	68	5	<1	0	5	1.73	2.33	0.57	0
29229	Bacon, Canadian style, cured	2	ounce(s)	57	38	89	12	1	0	4	1.26	1.79	0.36	—
35422	Breakfast strips, cured, cooked	3	slice(s)	34	9	156	10	<1	0	12	4.34	5.58	1.92	—
16561	Ham, smoked or cured, lean, cooked	1	slice(s)	42	28	66	11	0	0	2	0.77	1.06	0.27	—
189	Ham, cured, boneless, 11% fat, roasted	3	ounce(s)	85	55	151	19	0	0	8	2.65	3.77	1.20	—
1316	Ham, cured, extra lean, 5% fat, roasted	3	ounce(s)	85	58	123	18	1	0	5	1.54	2.23	0.46	—
29215	Ham, cured, extra lean, 4% fat, canned	2	ounce(s)	57	42	68	10	0	0	3	0.86	1.25	0.22	—
	Chop													
32671	Loin, blade, lean & fat, pan fried	3	ounce(s)	85	42	291	18	0	0	24	8.65	9.97	2.64	—
32672	Loin, center cut, lean & fat, pan fried	3	ounce(s)	85	45	236	25	0	0	14	5.11	6.00	1.62	—
32682	Loin, center rib, boneless, lean & fat, braised	3	ounce(s)	85	49	217	22	0	0	13	5.21	6.13	1.12	—
32603	Loin, center rib, lean, broiled	3	ounce(s)	85	48	186	26	0	0	8	2.94	3.78	0.53	—
32481	Loin, whole, lean, braised	3	ounce(s)	85	52	174	24	0	0	8	2.87	3.54	0.60	—
32478	Loin, whole, lean & fat, braised	3	ounce(s)	85	50	203	23	0	0	12	4.35	5.15	1.00	—
	Leg or ham													
32471	Rump portion, lean & fat, roasted	3	ounce(s)	85	48	214	25	0	0	12	4.47	5.42	1.17	—
32468	Whole, lean & fat, roasted	3	ounce(s)	85	47	232	23	0	0	15	5.50	6.70	1.43	—
	Ribs													
32696	Loin, country style, lean, roasted	3	ounce(s)	85	49	210	23	0	0	13	4.52	5.49	0.94	—
32693	Loin, country style, lean & fat, roasted	3	ounce(s)	85	43	279	20	0	0	22	7.83	9.36	1.71	—
	Shoulder													
32629	Arm picnic, lean, roasted	3	ounce(s)	85	51	194	23	0	0	11	3.66	5.09	1.02	—
32626	Arm picnic, lean & fat, roasted	3	ounce(s)	85	44	270	20	0	0	20	7.47	9.12	2.00	—
	Rabbit													
3366	Domesticated, roasted	3	ounce(s)	85	52	167	25	0	0	7	2.04	1.84	1.33	—
3367	Domesticated, stewed	3	ounce(s)	85	50	175	26	0	0	7	2.13	1.93	1.39	—
	Veal													
3391	Liver, braised	3	ounce(s)	85	51	163	24	3	0	5	1.69	0.97	0.88	0.26
3319	Rib, lean only, roasted	3	ounce(s)	85	55	150	22	0	0	6	1.77	2.26	0.57	—
1732	**Deer or venison, roasted**	3	ounce(s)	85	55	134	26	0	0	3	1.06	0.75	0.53	—

Chol (mg)	Calc (mg)	Iron (mg)	Magn (mg)	Pota (mg)	Sodi (mg)	Zinc (mg)	Vit A (µg)	Thia (mg)	Vit E (mg α)	Ribo (mg)	Niac (mg)	Vit B$_6$ (mg)	Fola (µg)	Vit C (mg)	Vit B$_{12}$ (µg)	Sele (µg)
103	22	2.30	25	287	65	6.21	0	0.06	0.15	0.23	5.38	0.11	19	0	2	32
82	14	1.60	20	264	61	3.79	0	0.09	0.12	0.21	5.66	0.11	15	0	2	22
77	14	1.88	25	266	72	4.48	0	0.09	0.15	0.21	5.57	0.13	18	0	2	26
84	16	1.60	20	230	65	3.40	0	0.08	0.10	0.19	5.95	0.09	12	0	2	20
74	16	1.81	21	225	58	5.13	0	0.08	0.15	0.22	4.90	0.13	21	0	2	24
78	17	1.67	20	213	56	4.45	0	0.08	0.12	0.20	5.23	0.11	18	0	2	22
2128	18	1.73	19	304	133	1.70	0	0.14	—	0.31	3.87	0.20	6	20	20	10
161	9	2.24	14	134	57	2.54	0	0.07	—	0.36	3.14	0.14	3	6	5	24
14	1	0.18	4	71	291	0.44	1	0.05	0.04	0.03	1.40	0.04	<1	0	<1	8
28	5	0.39	10	195	799	0.79	0	0.43	0.12	0.10	3.53	0.22	2	0	<1	14
36	5	0.67	9	158	714	1.25	0	0.25	0.09	0.13	2.58	0.12	1	0	1	8
23	3	0.40	9	133	557	1.08	0	0.29	0.11	0.11	2.11	0.20	2	0	<1	—
50	7	1.14	19	348	1275	2.10	0	0.62	0.26	0.28	5.23	0.26	3	0	1	17
45	7	1.26	12	244	1023	2.45	0	0.64	0.21	0.17	3.42	0.34	3	0	1	17
22	3	0.53	10	206	712	1.09	0	0.47	0.10	0.13	3.01	0.26	3	0	<1	8
72	26	0.75	18	282	57	2.71	3	0.53	0.17	0.25	3.36	0.29	3	1	1	30
78	23	0.77	25	361	68	1.96	2	0.97	0.21	0.26	4.76	0.40	5	1	1	33
62	4	0.78	14	329	34	1.76	2	0.45	0.21	0.21	3.67	0.26	3	<1	<1	28
69	26	0.70	24	357	55	2.02	2	0.95	0.25	0.28	5.25	0.40	3	<1	1	40
67	15	0.96	17	329	43	2.11	2	0.56	0.18	0.23	3.90	0.33	3	1	<1	41
68	18	0.91	16	318	41	2.02	2	0.54	0.20	0.22	3.76	0.31	3	1	<1	39
82	10	0.89	23	318	53	2.40	3	0.64	0.19	0.28	3.96	0.27	3	<1	1	40
80	12	0.86	19	299	51	2.52	3	0.54	0.19	0.27	3.89	0.34	9	<1	1	39
79	25	1.10	20	297	25	3.24	2	0.49	—	0.29	3.97	0.37	4	<1	1	36
78	21	0.90	20	293	44	2.01	3	0.76	—	0.29	3.67	0.38	4	<1	1	32
81	8	1.21	17	299	68	3.46	2	0.49	—	0.30	3.67	0.35	4	<1	1	33
80	16	1.00	14	276	60	2.93	2	0.44	—	0.26	3.33	0.30	3	<1	1	29
70	16	1.93	18	326	40	1.93	0	0.08	—	0.18	7.17	0.40	9	0	7	33
73	17	2.01	17	255	31	2.01	0	0.05	0.37	0.14	6.09	0.29	8	0	6	33
434	5	4.34	17	280	66	9.55	17973	0.15	0.58	2.43	11.18	0.78	281	1	72	16
98	10	0.82	20	264	82	3.82	0	0.05	0.31	0.25	6.38	0.23	12	0	1	9
95	6	3.80	20	285	46	2.34	0	0.15	—	0.51	5.70	—	—	0	—	11

Table A–1
Food Composition

(DA+ code is for Wadsworth Diet Analysis program)
(For purposes of calculations, use "0" for t, <1, <.1, <.01, etc.)

DA + Code	Food Description	Quantity	Measure	Wt (g)	H₂O (g)	Ener (kcal)	Prot (g)	Carb (g)	Fiber (g)	Fat (g)	Fat Breakdown (g)			
											Sat	Mono	Poly	*Trans*
	POULTRY													
	Chicken													
29562	Flaked, canned	2	ounce(s)	57	37	97	10	0.05	0	6	1.62	2.32	1.29	—
	Fried													
29632	Breast, meat only, breaded, baked or fried	3	ounce(s)	85	44	193	25	7	<1	7	1.62	2.66	1.73	—
35327	Broiler breast, meat only, fried	3	ounce(s)	85	51	159	28	<1	0	4	1.10	1.46	0.91	
36413	Broiler breast, meat & skin, flour coated, fried	3	ounce(s)	85	48	189	27	1	<.1	8	2.08	2.98	1.67	—
35389	Broiler drumstick, meat only, fried	3	ounce(s)	85	53	166	24	0	0	7	1.81	2.50	1.68	
36414	Broiler drumstick, meat & skin, flour coated, fried	3	ounce(s)	85	48	208	23	1	<.1	12	3.11	4.61	2.75	—
35406	Broiler leg, meat only, fried	3	ounce(s)	85	52	177	24	1	0	8	2.12	2.92	1.89	—
35484	Broiler wing, meat only, fried	3	ounce(s)	85	51	179	26	0	0	8	2.13	2.62	1.76	—
29580	Patty, fillet, or tenders, breaded, cooked	3	ounce(s)	85	42	241	14	13	<1	15	4.62	7.25	1.87	—
	Roasted, meat only													
35409	Broiler chicken leg	3	ounce(s)	85	55	162	23	0	0	7	1.95	2.59	1.68	
35486	Broiler chicken wing	3	ounce(s)	85	53	173	26	0	0	7	1.92	2.22	1.51	
35138	Roasting chicken, dark meat	3	ounce(s)	85	57	151	20	0	0	7	2.07	2.82	1.70	
35136	Roasting chicken, light meat	3	ounce(s)	85	58	130	23	0	0	3	0.92	1.29	0.79	
35132	Roasting chicken	3	ounce(s)	85	57	142	21	0	0	6	1.54	2.13	1.28	
	Stewed													
3174	Meat only, stewed	3	ounce(s)	85	48	150	23	0	0	6	1.56	2.03	1.3	—
1268	Gizzard, simmered	3	ounce(s)	85	58	124	26	0	0	2	0.57	0.45	0.30	0.11
1270	Liver, simmered	3	ounce(s)	85	57	142	21	1	0	6	1.75	1.20	1.08	0.08
	Duck													
1286	Domesticated, meat & skin, roasted	3	ounce(s)	85	44	286	16	0	0	24	8.22	10.97	3.10	—
1287	Domesticated, meat only, roasted	3	ounce(s)	85	55	171	20	0	0	10	3.54	3.15	1.22	—
	Goose													
35507	Domesticated, meat & skin, roasted	3	ounce(s)	85	44	259	21	0	0	19	5.84	8.72	2.14	—
35524	Domesticated, meat only, roasted	3	ounce(s)	85	49	202	25	0	0	11	3.88	3.69	1.31	—
1297	Liver pâté, smoked, canned	4	tablespoon(s)	52	19	240	6	2	0	23	7.51	13.32	0.44	—
	Turkey													
3256	Ground turkey, cooked	3	ounce(s)	85	51	200	23	0	0	11	2.88	4.16	2.75	
222	Roasted, fryer roaster breast, meat only	3	ounce(s)	85	58	115	26	0	0	1	0.20	0.11	0.17	
219	Roasted, dark meat, meat only	3	ounce(s)	85	54	159	24	0	0	6	2.06	1.39	1.84	
220	Roasted, light meat, meat only	3	ounce(s)	85	56	133	25	0	0	3	0.88	0.48	0.73	
3263	Patty, batter coated, breaded, fried	1	item(s)	94	47	266	13	15	<1	17	4.41	7.02	4.43	
1302	Turkey roll, light meat	2	slice(s)	57	41	83	11	<1	0	4	1.15	1.42	0.99	
1303	Turkey roll, light & dark meat	2	slice(s)	57	40	84	10	1	0	4	1.16	1.30	1.01	
	PROCESSED MEATS													
	Beef													
1331	Corned beef loaf, jellied, sliced	2	slice(s)	57	39	87	13	0	0	3	1.47	1.52	0.18	—
	Bologna													
13458	Made w/chicken, pork, & beef	1	slice(s)	28	15	90	3	1	0	8	3.00	4.05	1.10	—
13461	Light, made w/pork, chicken, & beef	1	slice(s)	28	18	60	3	2	0	4	1.50	2.04	0.43	—
13459	Beef	1	slice(s)	28	15	90	3	1	0	8	3.50	4.26	0.31	—
13565	Turkey	1	slice(s)	28	19	50	3	1	0	4	1.00	1.09	0.98	—
	Chicken													
13562	Oven roasted white chicken	1	slice(s)	28	20	40	4	1	0	3	0.50	—	—	

A

Chol (mg)	Calc (mg)	Iron (mg)	Magn (mg)	Pota (mg)	Sodi (mg)	Zinc (mg)	Vit A (µg)	Thia (mg)	Vit E (mg α)	Ribo (mg)	Niac (mg)	Vit B_6 (mg)	Fola (µg)	Vit C (mg)	Vit B_{12} (µg)	Sele (µg)
35	8	0.90	7	148	410	0.8	19.37	0	—	0.07	3.60	0.19	—	0	<1	—
67	19	1.05	25	223	450	0.84	—	0.08	—	0.10	10.98	0.47	4	0	<1	—
77	14	0.97	26	235	67	0.92	—	0.07	—	0.11	12.57	0.54	3	0	<1	22
76	14	1.01	26	220	65	0.94	—	0.07	—	0.11	11.69	0.49	5	0	<1	20
80	10	1.12	20	212	82	2.74	—	0.07	—	0.20	5.23	0.33	8	0	<1	17
77	10	1.14	20	195	76	2.46	—	0.07	—	0.19	5.13	0.30	9	0	<1	16
84	11	1.19	21	216	82	2.53	—	0.07	—	0.21	5.69	0.33	8	0	<1	16
71	13	0.97	18	177	77	1.80	—	0.04	—	0.11	6.16	0.50	3	0	<1	22
51	14	1.06	17	209	452	0.88	—	0.08	—	0.12	5.71	0.26	9	<1	<1	—
80	10	1.11	20	206	77	2.43	—	0.06	—	0.20	5.37	0.32	7	0	<1	19
72	14	0.99	18	179	78	1.82	—	0.04	—	0.11	6.22	0.50	3	0	<1	21
64	9	1.13	17	191	81	1.81	14	0.05	—	0.16	4.88	0.26	6	0	<1	17
64	11	0.92	20	201	43	0.66	7	0.05	0.23	0.08	8.90	0.46	3	0	<1	22
64	10	1.03	18	195	64	1.29	10	0.05	—	0.13	6.70	0.35	4	0	<1	21
71	12	0.99	18	153	60	1.69	13	0.04	0.23	0.13	5.19	0.22	5	0	<1	18
315	14	2.71	3	152	48	3.76	0	0.02	0.17	0.18	2.65	0.06	4	0	1	35
479	9	9.89	21	224	65	3.38	3384	0.25	0.70	1.69	9.39	0.64	491	24	14	70
71	9	2.30	14	173	50	1.58	54	0.15	0.59	0.23	4.10	0.15	5	0	<1	17
76	10	2.30	17	214	55	2.21	20	0.22	0.59	0.40	4.34	0.21	9	0	<1	19
77	11	2.41	19	280	60	2.23	18	0.07	—	0.28	3.55	0.32	2	0	<1	19
82	12	2.44	21	330	65	2.70	10	0.08	—	0.33	3.47	0.40	10	0	<1	22
78	36	2.86	7	72	362	0.48	521	0.05	—	0.16	1.31	0.03	31	0	5	23
87	21	1.64	20	230	91	2.43	0	0.05	0.29	0.14	4.10	0.33	6	0	<1	32
71	10	1.30	25	248	44	1.48	0	0.04	0.08	0.11	6.37	0.48	5	0	<1	27
72	27	1.98	20	247	67	3.79	0	0.05	0.54	0.21	3.10	0.31	8	0	<1	35
59	16	1.15	24	259	54	1.73	0	0.05	0.08	0.11	5.81	0.46	5	0	<1	27
58	13	2.07	14	259	752	1.35	10	0.09	1.18	0.18	2.16	0.19	26	0	<1	19
24	23	0.73	9	142	277	0.88	0	0.05	0.07	0.13	3.97	0.18	2	0	<1	13
31	18	0.77	10	153	332	1.13	0	0.05	0.19	0.16	2.72	0.15	3	0	<1	17
27	6	1.16	6	57	540	2.32	0	0.00	—	0.06	1.00	0.07	5	0	1	10
30	0	0.36	6	43	290	0.40	0	—	—	—	—	—	—	0	—	—
15	0	0.36	6	46	310	0.45	0	—	—	—	—	—	—	0	—	—
20	0	0.36	4	47	310	0.57	0	0.01	—	0.03	0.68	0.05	4	0	<1	—
20	40	0.36	6	43	270	0.52	0	—	—	—	—	—	—	0	—	—
15	0	0.36	7	85	350	0.32	0	—	—	—	—	—	—	0	—	—

Table A–1
Food Composition

(DA+ code is for Wadsworth Diet Analysis program)
(For purposes of calculations, use "0" for t, <1, <.1, <.01, etc.)

DA + Code	Food Description	Quantity	Measure	Wt (g)	H₂O (g)	Ener (kcal)	Prot (g)	Carb (g)	Fiber (g)	Fat (g)	Sat	Mono	Poly	Trans
	PROCESSED MEATS—Continued													
	Ham													
13581	Honey glazed, traditional carved	2	slice(s)	45	—	50	8	1	0	2	0.50	0.68	0.18	—
13777	Deli sliced cooked	1	slice(s)	28	—	30	5	1	0	1	0.50	0.39	0.11	—
13778	Deli sliced honey	1	slice(s)	28	—	35	5	1	0	1	0.50	0.39	0.11	—
8614	**Pork & beef mortadella, sliced**	2	slice(s)	46	24	143	8	1	0	12	4.37	5.23	1.44	—
1323	**Pork olive loaf**	2	slice(s)	57	33	133	7	5	0	9	3.32	4.47	1.10	—
1324	**Pork pickle & pimento loaf**	2	slice(s)	57	32	149	7	3	0	12	4.45	5.45	1.47	—
	Sausages & frankfurters													
37296	Beerwurst beef beer salami (bierwurst)	1	slice(s)	29	17	74	4	1	0	6	2.50	2.69	0.21	—
37257	Beerwurst pork beer salami	1	slice(s)	21	13	50	3	<1	0	4	1.32	1.89	0.50	—
35338	Berliner, pork & beef	1	ounce(s)	28	17	65	4	1	0	5	1.72	2.27	0.45	—
37299	Braunschweiger pork liver sausage	1	slice(s)	15	0	51.34	1.97	0.34	0	4.48	1.52	2.08	0.52	—
37298	Bratwurst pork, cooked	1	piece(s)	74	42	181	10	2	0	14	5.15	6.73	1.51	—
1329	Cheesefurter or cheese smokie, beef & pork	1	item(s)	43	23	141	6	1	0	12	4.52	5.89	1.30	—
1330	Chorizo, beef & pork	2	ounce(s)	57	18	258	14	1	0	22	8.15	10.43	1.96	—
8600	Frankfurter, beef	1	item(s)	45	23	149	5	2	0	13	5.26	6.44	0.53	—
202	Frankfurter, beef & pork	1	item(s)	57	32	174	7	1	1	16	6.14	7.79	1.56	—
1293	Frankfurter, chicken	1	item(s)	45	26	116	6	3	0	9	2.49	3.82	1.82	—
3261	Frankfurter, turkey	1	item(s)	45	28	102	6	1	0	8	2.65	2.51	2.25	—
37275	Italian sausage, pork, cooked	1	item(s)	68	34	220	14	1	0	17	6.14	8.13	2.23	—
37307	Kielbasa, kolbassa, pork & beef	2⅛	ounce(s)	61	37	135	10	2	0	9	3.40	4.44	1.06	—
1333	Knockwurst or knackwurst, beef & pork	2	ounce(s)	57	31	174	6	2	0	16	5.79	7.26	1.66	—
37285	Pepperoni, beef & pork	1	slice(s)	11	3	55	2	<1	0	5	1.77	2.32	0.48	—
37313	Polish sausage, pork	2	slice(s)	57	31	163	8	2	—	14	4.91	6.42	1.46	—
206	Salami, beef, cooked, sliced	2	slice(s)	46	28	119	6	1	0	10	4.54	4.90	0.48	—
37272	Salami, pork, dry or hard	1	slice(s)	13	5	52	3	<1	0	4	1.52	2.05	0.48	—
3262	Salami, turkey	2	slice(s)	57	31	125	8	11	<.1	5	1.98	1.80	1.43	0
7162	Sausage, breakfast, turkey	2½	ounce(s)	100	67	190	17	<1	0	13	3.90	6.23	3.33	0
8620	Smoked sausage, beef & pork	2	ounce(s)	57	31	181	7	1	0	16	5.54	6.94	2.23	0
8619	Smoked, sausage, pork	2	ounce(s)	57	22	221	13	1	0	18	6.42	8.30	2.13	—
37273	Smoked, sausage, pork link	1	piece(s)	76	30	295	17	2	—	24	8.58	11.09	2.85	—
1336	Summer sausage, thuringer, or cervelat, beef & pork	2	ounce(s)	57	29	190	9	<1	0	17	6.82	7.35	0.68	—
37294	Vienna sausage, cocktail, beef & pork, canned	1	piece(s)	16	10	45	2	<1	0	4	1.49	2.01	0.27	—
	Spreads													
32419	Pork & beef sandwich spread	4	tablespoon(s)	60	36	141	5	7	<1	10	3.59	4.57	1.54	—
1318	Ham salad spread	¼	cup(s)	60	38	130	5	6	0	9	3.04	4.32	1.62	—
	Turkey													
16049	Breast, hickory smoked, slices	1	slice(s)	56	—	50	11	1	0	0	0.00	0.00	0.00	0
13606	Breast, hickory smoked fat free	1	slice(s)	28	—	25	4	1	0	0	0.00	0.00	0.00	0
16047	Breast, honey roasted, slices	1	slice(s)	56	—	60	11	2	0	0	0.00	0.00	0.00	0
16048	Breast, oven roasted, slices	1	slice(s)	56	—	50	11	1	0	0	0.00	0.00	0.00	0
13583	Breast, traditional carved	2	slice(s)	45	—	40	9	0	0	1	0.00	0.07	0.14	—
13604	Breast, oven roasted, fat free	1	slice(s)	28	—	25	4	1	0	0	0.00	0.00	0.00	0
13567	Turkey ham, 10% water added	1	slice(s)	28	20	35	5	0	0	1	0.00	0.22	0.31	—
13596	Turkey pastrami	2	ounce(s)	56	—	70	11	1	0	2	1.00	—	—	—
13597	Turkey salami	2	ounce(s)	56	—	120	8	1	0	9	2.50	2.92	2.30	—

PAGE KEY: **A–2 = Breads/Baked Goods**   A–6 = Cereal/Rice/Pasta   A–10 = Fruit   A–14 = Vegetables/Legumes   A–24 = Nuts/Seeds   A–26 = Vegetarian
A–28 = Dairy   A–34 = Eggs   A–34 = Seafood   A–36 = Meats   A–40 = Poultry   A–40 = Processed meats   A–42 = Beverages   A–46 = Fats/Oils   A–48 = Sweets
A–50 = Spices/Condiments/Sauces   A–52 = Mixed foods/Soups/Sandwiches   A–58 = Fast food   A–74 = Convenience meals   A–76 = Baby foods

A

Chol (mg)	Calc (mg)	Iron (mg)	Magn (mg)	Pota (mg)	Sodi (mg)	Zinc (mg)	Vit A (µg)	Thia (mg)	Vit E (mg α)	Ribo (mg)	Niac (mg)	Vit B$_6$ (mg)	Fola (µg)	Vit C (mg)	Vit B$_{12}$ (µg)	Sele (µg)
25	0	0.72	—	—	560	—	0	—	—	—	—	—	—	0	—	—
15	0	0.00	—	—	240	—	0	—	—	—	—	—	—	0	—	—
15	0	0.00	—	—	240	—	0	—	—	—	—	—	—	0	—	—
26	8	0.64	5	75	573	0.97	0	0.05	0.10	0.07	1.23	0.06	1	0	1	10
22	62	0.31	11	169	843	0.78	34	0.17	0.14	0.15	1.04	0.13	1	0	1	9
21	54	0.58	10	193	789	0.80	12	0.17	0.24	0.14	1.17	0.11	3	0	1	8
18	3	0.44	4	67	265	0.71	0	0.02	—	0.04	0.99	0.05	1	0	1	5
12	2	0.16	3	53	261	0.36	0	0.12	—	0.04	0.69	0.07	1	0	<1	—
13	3	0.33	4	80	368	0.70	0	0.11	—	0.06	0.88	0.06	1	0	1	4
23.69	1.36	1.42	1.67	27.49	131.54	0.42	641.01	0.03	—	0.23	1.27	0.05	—	0	3.05	8.81
44	33	0.96	11	157	412	1.70	0	0.37	—	0.14	2.37	0.16	1	1	1	16
29	25	0.46	6	89	465	0.97	20	0.11	0.00	0.07	1.25	0.06	1	0	1	7
50	5	0.90	10	226	700	1.93	0	0.36	0.12	0.17	2.91	0.30	1	0	1	12
24	6	0.68	6	70	513	1.11	0	0.02	0.09	0.07	1.07	0.04	2	0	1	4
29	6	0.66	6	95	638	1.05	10	0.11	0.14	0.07	1.50	0.07	2	0	1	8
45	43	0.90	5	38	617	0.47	18	0.03	0.10	0.05	1.39	0.14	2	0	<1	8
48	48	0.83	6	81	642	1.40	0	0.02	0.28	0.08	1.86	0.10	4	0	<1	7
53	16	1.02	12	207	627	1.62	0	0.42	—	0.16	2.83	0.22	3	1	1	15
41	27	0.88	10	169	566	1.23	0	0.14	—	0.13	1.75	0.11	3	0	1	11
34	6	0.37	6	113	527	0.94	0	0.19	—	0.08	1.55	0.10	1	0	1	8
9	1	0.15	2	38	224	0.28	0	0.04	—	0.03	0.55	0.03	<1	0	<1	—
40	7	0.82	8	102	546	1.10	0	0.29	—	0.08	1.96	0.11	1	1	1	10
33	3	1.01	6	86	524	0.81	0	0.05	0.09	0.09	1.49	0.08	1	0	1	7
10	2	0.17	3	48	289	0.54	0	0.12	—	0.04	0.72	0.07	<1	0	<1	3
45	42	0.87	15	225	616	1.76	1	0.24	0.14	0.17	2.26	0.24	6	12	1	11
92	57	2.20	18	188	665	2.07	0	0.04	0.00	0.12	3.55	0.29	5	1	<1	—
33	7	0.43	7	101	517	0.71	7	0.11	0.07	0.06	1.67	0.09	1	0	<1	0
39	17	0.66	11	191	851	1.60	0	0.40	0.14	0.15	2.57	0.20	3	1	1	12
52	23	0.88	14	255	1137	2.14	0	0.53	—	0.20	3.43	0.27	4	0	1	16
43	7	1.44	8	154	704	1.45	0	0.09	0.12	0.19	2.44	0.15	1	0	3	12
8	2	0.14	1	16	152	0.26	0	0.01	—	0.02	0.26	0.02	1	0	<1	3
23	7	0.47	5	66	608	0.61	16	0.10	1.04	0.08	1.04	0.07	1	0	1	6
22	5	0.35	6	90	547	0.66	0	0.26	1.04	0.07	1.26	0.09	1	0	<1	11
25	0	0.72	—	—	730	—	0	—	—	—	—	—	—	0	—	—
10	0	0.00	—	—	300	—	0	—	—	—	—	—	—	0	—	—
20	0	0.72	—	—	640	—	0	—	—	—	—	—	—	0	—	—
20	0	0.72	—	—	620	—	0	—	—	—	—	—	—	0	—	—
20	0	0.72	—	—	540	—	0	—	—	—	—	—	—	0	—	—
10	0	0.00	—	—	330	—	0	—	—	—	—	—	—	0	—	—
20	0	0.36	6	81	310	0.73	0	—	—	—	—	—	—	0	—	—
40	0	0.72	—	—	590	—	0	—	—	—	—	—	—	0	—	—
50	40	0.72	—	—	500	—	0	—	—	—	—	—	—	0	—	—

Table A–1
Food Composition

(DA+ code is for Wadsworth Diet Analysis program)
(For purposes of calculations, use "0" for t, <1, <.1, <.01, etc.)

DA+ Code	Food Description	Quantity	Measure	Wt (g)	H₂O (g)	Ener (kcal)	Prot (g)	Carb (g)	Fiber (g)	Fat (g)	Sat	Mono	Poly	Trans
	BEVERAGES													
	Alcoholic													
	Beer													
866	Ale, mild	12	fluid ounce(s)	360	332	148	1	13	1	0	0.00	0.00	0.00	—
686	Beer	12	fluid ounce(s)	356	336	118	1	6	<1	<1	0.00	0.00	0.00	0
869	Beer, light	12	fluid ounce(s)	354	337	99	1	5	0	0	0.00	0.00	0.00	0
16886	Beer, nonalcoholic	12	fluid ounce(s)	360	353	32	1	5	0	0	0.00	0.00	0.00	0
31608	Budweiser beer	12	fluid ounce(s)	355	328	143	1	11	0	0	0.00	0.00	0.00	0
31609	Bud Light beer	12	fluid ounce(s)	355	335	110	1	7	0	0	0.00	0.00	0.00	0
31613	Michelob Beer	12	fluid ounce(s)	355	323	155	1	13	0	0	0.00	0.00	0.00	0
31614	Michelob Light beer	12	fluid ounce(s)	355	330	134	1	12	0	0	0.00	0.00	0.00	0
	Gin, rum, vodka, whiskey													
687	Distilled alcohol, 80 proof	1	fluid ounce(s)	28	19	64	0	0	0	0	0.00	0.00	0.00	0
688	Distilled alcohol, 86 proof	1	fluid ounce(s)	28	18	70	0	<.1	0	0	0.00	0.00	0.00	0
689	Distilled alcohol, 90 proof	1	fluid ounce(s)	28	17	73	0	0	0	0	0.00	0.00	0.00	0
856	Distilled alcohol, 94 proof	1	fluid ounce(s)	28	17	76	0	0	0	0	0.00	0.00	0.00	0
857	Distilled alcohol, 100 proof	1	fluid ounce(s)	28	16	82	0	0	0	0	0.00	0.00	0.00	0
	Liqueurs													
3142	Coffee liqueur, 63 proof	1	fluid ounce(s)	35	14	107	<.1	11	0	<1	0.04	0.01	0.04	—
33187	Coffee liqueur, 53 proof	1	fluid ounce(s)	35	11	117	<.1	16	0	<1	0.04	0.01	0.04	—
736	Cordials, 54 proof	1	fluid ounce(s)	30	9	106	<.1	13	0	<.1	0.02	0.01	0.04	—
	Wine													
858	Champagne, domestic	5	fluid ounce(s)	150	—	105	<1	4	0	0	0.00	0.00	0.00	0
861	Red wine, California	5	fluid ounce(s)	150	133	125	<1	4	0	0	0.00	0.00	0.00	0
690	Sweet dessert wine	5	fluid ounce(s)	150	106	240	<1	21	0	0	0.00	0.00	0.00	0
1481	White wine	5	fluid ounce(s)	148	132	100	<1	1	0	0	0.00	0.00	0.00	0
1811	Wine cooler	10	fluid ounce(s)	300	270	150	<1	18	<.1	<.1	0.01	0.00	0.02	—
	Carbonated													
692	Club soda	12	fluid ounce(s)	355	355	0	0	0	0	0	0.00	0.00	0.00	0
12010	Coca-Cola Classic cola soda	12	fluid ounce(s)	360	—	146	0	41	0	0	0.00	0.00	0.00	0
12031	Coke diet cola soda	12	fluid ounce(s)	360	—	2	0	<1	0	0	0.00	0.00	0.00	0
693	Cola	12	fluid ounce(s)	426	380	179	<1	46	0	0	0.00	0.00	0.00	0
9522	Cola soda, decaffeinated	12	fluid ounce(s)	372	331	156	<1	40	0	0	0.00	0.00	0.00	0
1415	Cola, low calorie w/aspartame	12	fluid ounce(s)	355	354	4	<1	<1	0	0	0.00	0.00	0.00	0
9524	Cola, decaffeinated, low calorie w/aspartame	12	fluid ounce(s)	355	354	4	<1	<1	0	0	0.00	0.00	0.00	0
1412	Cream soda	12	fluid ounce(s)	371	321	189	0	49	0	0	0.00	0.00	0.00	0
31899	Diet 7 Up	12	fluid ounce(s)	360	—	0	0	0	0	0	0.00	0.00	0.00	0
695	Ginger ale	12	fluid ounce(s)	366	334	124	0	32	0	0	0.00	0.00	0.00	0
694	Grape soda	12	fluid ounce(s)	372	330	160	0	42	0	0	0.00	0.00	0.00	0
1876	Lemon lime soda	12	fluid ounce(s)	368	330	147	0	38	0	0	0.00	0.00	0.00	—
29392	Mountain Dew diet soda	12	fluid ounce(s)	360	—	0	0	0	0	0	0.00	0.00	0.00	0
29391	Mountain Dew soda	12	fluid ounce(s)	360	—	170	0	46	0	0	0.00	0.00	0.00	0
3145	Orange soda	12	fluid ounce(s)	372	326	179	0	46	0	<1	0.00	0.00	0.00	0
1414	Pepper-type soda	12	fluid ounce(s)	368	329	151	0	38	0	<1	0.26	0.00	0.00	—
2391	Pepper-type or cola soda, low calorie w/saccharin	12	fluid ounce(s)	355	354	0	0	<1	0	0	0.00	0.00	0.00	0
29389	Pepsi diet cola soda	12	fluid ounce(s)	360	—	0	0	0	0	0	0.00	0.00	0.00	0
29388	Pepsi regular cola soda	12	fluid ounce(s)	360	—	150	0	41	0	0	0.00	0.00	0.00	0
696	Root beer	12	fluid ounce(s)	370	330	152	0	39	0	0	0.00	0.00	0.00	0
31898	7 Up	12	fluid ounce(s)	360	—	240	0	59	0	0	0.00	0.00	0.00	0
12034	Sprite diet soda	12	fluid ounce(s)	360	—	4	0	0	0	0	0.00	0.00	0.00	0
12044	Sprite soda	12	fluid ounce(s)	360	—	144	0	39	0	0	0.00	0.00	0.00	0

PAGE KEY: A–2 = Breads/Baked Goods   A–6 = Cereal/Rice/Pasta   A–10 = Fruit   A–14 = Vegetables/Legumes   A–24 = Nuts/Seeds   A–26 = Vegetarian
A–28 = Dairy   A–34 = Eggs   A–34 = Seafood   A–36 = Meats   A–40 = Poultry   A–40 = Processed meats   A–42 = Beverages   A–46 = Fats/Oils   A–48 = Sweets
A–50 = Spices/Condiments/Sauces   A–52 = Mixed foods/Soups/Sandwiches   A–58 = Fast food   A–74 = Convenience meals   A–76 = Baby foods

A

Chol (mg)	Calc (mg)	Iron (mg)	Magn (mg)	Pota (mg)	Sodi (mg)	Zinc (mg)	Vit A (µg)	Thia (mg)	Vit E (mg α)	Ribo (mg)	Niac (mg)	Vit B$_6$ (mg)	Fola (µg)	Vit C (mg)	Vit B$_{12}$ (µg)	Sele (µg)
0	18	0.11	—	—	18	—	0	0.02	0.00	0.10	1.63	—	—	0	<.1	—
0	18	0.07	21	89	14	0.04	0	0.02	0.00	0.09	1.61	0.18	21	0	<.1	2
0	18	0.14	18	64	11	0.11	0	0.03	0.00	0.11	1.39	0.12	14	0	<.1	2
0	25	0.04	32	90	18	0.04	—	0.02	0.00	0.10	1.63	0.18	22	0	<.1	—
0	18	0.11	21	89	9	0.07	0	0.02	0.00	0.09	1.61	0.18	21	0	<.1	4
0	18	0.14	18	64	9	0.11	0	0.03	0.00	0.11	1.39	0.12	15	0	<.1	4
0	18	0.11	21	89	9	0.07	0	0.02	0.00	0.09	1.61	0.18	21	0	<.1	4
0	18	0.14	18	64	9	0.11	0	0.03	0.00	0.11	1.39	0.12	15	0	<.1	4
0	0	0.01	0	1	<1	0.01	0	0.00	0.00	0.00	0.00	0.00	0	0	0	0
0	0	0.01	0	1	<1	0.01	0	0.00	0.00	0.00	0.00	0.00	0	0	0	0
0	0	0.01	0	1	<1	0.01	0	0.00	0.00	0.00	0.00	0.00	0	0	0	0
0	0	0.01	0	1	<1	0.01	0	0.00	0.00	0.00	0.00	0.00	0	0	0	0
0	0	0.01	0	1	<1	0.01	0	0.00	0.00	0.00	0.00	0.00	0	0	0	0
0	<1	0.02	1	10	3	0.01	0	0.00	—	0.00	0.05	0.00	0	0	0	<1
0	<1	0.02	1	10	3	0.01	0	0.00	0.00	0.00	0.05	0.00	0	0	0	<1
0	<1	0.02	<1	5	2	0.01	0	0.00	0.00	0.00	0.02	0.00	0	0	0	—
0	—	—	—	—	—	—	—	—	—	—	0.00	—	—	—	0	—
0	12	1.43	16	171	15	0.15	0	0.02	0.00	0.04	0.12	0.05	1	0	<.1	—
0	12	0.36	14	138	14	0.11	0	0.03	0.00	0.03	0.32	0.00	0	0	0	1
0	13	0.47	15	118	7	0.10	0	0.01	—	0.01	0.10	0.02	0	0	0	<1
0	17	0.81	16	135	25	0.17		0.01	0.03	0.02	0.13	0.04	4	5	<.1	—
0	18	0.04	4	7	75	0.36	0	0.00	0.00	0.00	0.00	0.00	0	0	0	0
0	—	—	—	0	50	—	0	—	—	—	—	—	—	0	—	—
0	—	—	—	18	42	—	0	—	—	—	—	—	—	0	—	—
0	13	0.09	4	4	17	0.04	0	0.00	0.00	0.00	0.00	0.00	0	0	0	<1
0	11	0.07	4	4	15	0.04	0	0.00	0.00	0.00	0.00	0.00	0	0	0	<1
0	11	0.11	4	21	18	0.00	0	0.02	0.00	0.08	0.00	0.00	0	0	0	0
0	14	0.11	4	0	21	0.28	0	0.02	0.00	0.08	0.00	0.00	0	0	0	<1
0	19	0.19	4	4	44	0.26	0	0.00	0.00	0.00	0.00	0.00	0	0	0	0
0	—	—	—	116	53	—	—	—	0.00	—	—	—	—	—	—	—
0	11	0.66	4	4	26	0.18	0	0.00	0.00	0.00	0.00	0.00	0	0	0	<1
0	11	0.30	4	4	56	0.26	0	0.00	0.00	0.00	0.00	0.00	0	0	0	0
0	7	0.26	4	4	41	0.18	0	0.00	0.00	0.00	0.00	0.06	0	0	0	0
0	—	—	—	70	35	—	—	—	—	—	—	—	—	—	—	—
0	—	—	—	0	70	—	—	—	—	—	—	—	—	—	—	—
0	19	0.22	4	7	45	0.37	0	0.00	—	0.00	0.00	0.00	0	0	0	0
0	11	0.15	0	4	37	0.15	0	0.00	—	0.00	0.00	0	0	0	0	<1
0	14	0.07	4	14	57	0.11	0	0.00	0.00	0.00	0.00	0.00	0	0	0	<1
0	—	—	—	30	35	—	—	—	—	—	—	—	—	—	—	—
0	—	—	—	0	35	—	—	—	—	—	—	—	—	—	—	—
0	18	0.18	4	4	48	0.26	0	0.00	0.00	0.00	0.00	0.00	0	0	0	<1
0	—	—	—	0	113	—	—	—	—	—	—	—	—	—	—	—
0	—	—	—	110	36	—	0	—	—	—	—	—	—	0	—	—
0	—	—	—	0	71	—	0	—	—	—	—	—	—	0	—	—

Table A–1
Food Composition

(DA+ code is for Wadsworth Diet Analysis program)
(For purposes of calculations, use "0" for t, <1, <.1, <.01, etc.)

DA + Code	Food Description	Quantity	Measure	Wt (g)	H₂O (g)	Ener (kcal)	Prot (g)	Carb (g)	Fiber (g)	Fat (g)	Sat	Mono	Poly	Trans
	BEVERAGES–Continued													
	Coffee													
731	Brewed	8	fluid ounce(s)	237	236	9	<1	0	0	0	0.00	0.00	0.00	0
9520	Brewed, decaffeinated	8	fluid ounce(s)	237	235	5	<1	1	0	0	0.00	0.00	0.00	0
16882	Cappuccino	8	fluid ounce(s)	240	224	78	4	6	<1	4	2.53	1.18	0.15	—
16883	Cappuccino, decaffeinated	8	fluid ounce(s)	240	224	78	4	6	<1	4	2.53	1.18	0.15	—
16880	Espresso	8	fluid ounce(s)	237	235	5	<1	1	0	0	0.00	0.00	0.00	0
16881	Espresso, decaffeinated	8	fluid ounce(s)	237	235	5	<1	1	0	0	0.00	0.00	0.00	0
732	Instant, prepared	8	fluid ounce(s)	239	237	5	<1	1	0	0	0.00	0.00	0.00	0
	Fruit drinks													
29357	Crystal Light low calorie lemonade drink	8	fluid ounce(s)	240	—	5	0	0	0	0	0.00	0.00	0.00	0
6012	Fruit punch drink w/added vitamin C, canned	8	fluid ounce(s)	276	242	129	0	33	<1	<.1	0.01	0.01	0.01	0
260	Grape drink, canned	8	fluid ounce(s)	250	221	113	<.1	29	0	0	0.00	0.00	0.00	0
266	Lemonade, from frozen concentrate	8	fluid ounce(s)	248	213	131	<1	34	<1	<1	0.02	0.00	0.04	—
268	Limeade, from frozen concentrate	8	fluid ounce(s)	247	220	104	<.1	26	0	<.1	0.00	0.00	0.00	—
31143	Gatorade Thirst Quencher, all flavors	8	fluid ounce(s)	240	—	50	0	14	0	0	0.00	0.00	0.00	0
17372	Kool-Aid (lemonade/punch/ fruit drink)	8	fluid ounce(s)	248	220	108	<1	28	<1	<.1	0.01	0.01	0.02	0
17225	Kool-Aid sugar free, low calorie tropical punch mix, prepared	8	fluid ounce(s)	240	—	5	0	0	0	0	0.00	0.00	0.00	0
14266	Odwalla strawberry 'c' monster fruit drink	8	fluid ounce(s)	240	—	150	2	34	1	1	0.00	—	—	0
10080	Odwalla strawberry lemonade quencher	8	fluid ounce(s)	240	—	120	1	28	1	0	0.00	0.00	0.00	0
10099	Snapple fruit punch	8	fluid ounce(s)	240	—	110	0	29	0	0	0.00	0.00	0.00	0
10096	Snapple kiwi strawberry	8	fluid ounce(s)	240	211	110	0	28	0	0	0.00	0.00	0.00	0
	Slim Fast ready to drink shake													
16056	Dark chocolate fudge	11	fluid ounce(s)	325	—	220	10	42	5	3	1.00	1.50	0.50	—
16054	French vanilla	11	fluid ounce(s)	325	—	220	10	40	5	3	0.50	1.50	0.50	—
16055	Strawberries n cream	11	fluid ounce(s)	325	—	220	10	40	5	3	0.50	1.50	0.50	—
	Tea													
733	Tea, prepared	8	fluid ounce(s)	237	236	2	0	1	0	0	0.00	0.00	0.01	0
33179	Decaffeinated, prepared	8	fluid ounce(s)	237	236	2	0	1	0	0	0.00	0.00	0.01	0
1877	Herbal, prepared	8	fluid ounce(s)	237	236	2	0	<1	0	0	0.00	0.00	0.01	0
734	Instant tea mix, unsweetened, prepared	8	fluid ounce(s)	237	236	2	<.1	<1	0	0	0.00	0.00	0.00	0
735	Instant lemon flavored tea mix w/sugar, prepared	8	fluid ounce(s)	259	236	88	<1	22	0	<.1	0.01	0.00	0.02	0
	Water													
1413	Mineral water, carbonated	8	fluid ounce(s)	237	237	0	0	0	0	0	0.00	0.00	0.00	0
33183	Poland spring water, bottled	8	fluid ounce(s)	237	237	0	0	0	0	0	0.00	0.00	0.00	0
1	Tap water	8	fluid ounce(s)	237	237	0	0	0	0	0	0.00	0.00	0.00	0
1879	Tonic water	8	fluid ounce(s)	244	222	83	0	21	0	0	0.00	0.00	0.00	0
	FATS AND OILS													
	Butter													
104	Butter	1	tablespoon(s)	15	2	108	<1	<.1	0	12	6.13	5.00	0.43	—
921	Unsalted	1	tablespoon(s)	15	3	108	<1	<.1	0	12	7.71	3.15	0.46	—
107	Whipped	1	tablespoon(s)	11	2	82	<.1	<.1	0	9	5.76	2.67	0.34	—
944	Whipped, unsalted	1	tablespoon(s)	11	2	82	<.1	<.1	0	9	5.76	2.67	0.34	—
2522	Butter Buds, dry butter substitute	1	teaspoon(s)	2	—	8	0	2	0	0	0.00	0.00	0.00	0
	Fats, cooking													
2671	Beef tallow, semisolid	1	tablespoon(s)	13	0	115	0	0	0	13	6.37	5.35	0.51	—
922	Chicken fat	1	tablespoon(s)	13	<.1	115	0	0	0	13	3.81	5.72	2.68	—

Chol (mg)	Calc (mg)	Iron (mg)	Magn (mg)	Pota (mg)	Sodi (mg)	Zinc (mg)	Vit A (µg)	Thia (mg)	Vit E (mg α)	Ribo (mg)	Niac (mg)	Vit B_6 (mg)	Fola (µg)	Vit C (mg)	Vit B_{12} (µg)	Sele (µg)
0	2	0.02	5	114	2	0.02	0	0.00	0.02	0.12	0.00	0.00	5	0	0	0
0	5	0.12	12	128	5	0.05	0	0.00	0.00	0.00	0.53	0.00	<1	0	0	0
17	152	0.26	22	250	62	0.50	—	0.04	0.10	0.20	0.37	0.05	5	1	<1	—
17	152	0.26	22	250	62	0.50	—	0.04	0.10	0.20	0.37	0.05	5	1	<1	—
0	5	0.12	12	128	5	0.05	0	0.00	0.05	0.00	0.53	0.00	<1	0	0	—
0	5	0.12	12	128	5	0.05	0	0.00	0.05	0.00	0.53	0.00	<1	0	0	—
0	10	0.10	7	72	5	0.02	0	0.00	0.00	0.00	0.56	0.00	0	0	0	<1
0	0	0.00	—	160	20	—	0	—	—	—	—	—	—	0	—	—
0	22	0.58	6	69	61	0.33	—	0.06	0.00	0.06	0.06	0.00	4	99	0	0
0	5	0.45	3	30	15	0.30	0	0.00	0.00	0.01	0.03	0.01	0	85	0	<1
0	10	0.52	5	50	7	0.07	0	0.02	0.02	0.07	0.05	0.02	2	13	0	<1
0	7	0.02	2	22	5	0.02	0	0.00	0.00	0.01	0.02	0.01	2	6	0	<1
0	10	0.18	—	30	110	—	—	—	—	—	—	—	—	1	—	—
0	14	0.46	5	50	31	0.20	—	0.04	—	0.05	0.05	0.01	4	42	0	1
0	0	0.00	—	10	10	—	0	—	—	—	—	—	—	6	—	—
0	20	1.44	—	330	40	—	—	—	—	—	—	—	—	600	0	—
0	20	0.00	—	70	30	—	0	—	—	—	—	—	—	60	0	—
0	0	0.00	—	20	10	—	0	—	—	—	—	—	—	0	0	—
0	0	0.00	—	40	10	—	0	—	—	—	—	—	—	0	0	—
5	400	2.70	140	600	220	2.25	—	0.53	—	0.60	7.00	0.70	120	60	2	18
5	400	2.70	140	600	220	2.25	—	0.53	—	0.60	7.00	0.70	120	60	2	18
5	400	2.70	140	600	220	2.25	—	0.53	—	0.60	7.00	0.70	120	60	2	18
0	0	0.05	7	88	7	0.05	0	0.00	0.00	0.03	0.00	0.00	12	0	0	0
0	0	0.05	7	88	7	0.05	0	0.00	0.00	0.03	0.00	0.00	12	0	0	0
0	5	0.19	2	21	2	0.09	0	0.02	0.00	0.01	0.00	0.00	2	0	0	0
0	7	0.05	5	47	7	0.02	0	0.00	0.00	0.00	0.09	0.00	0	0	0	0
0	5	0.05	5	49	8	0.03	0	0.00	0.00	0.04	0.09	0.01	0	<1	0	<1
0	33	0.00	0	0	2	0.00	0	0.00	—	0.00	0.00	0.00	0	0	0	0
0	2	0.02	2	0	2	0.00	0	0.00	—	0.00	0.00	0.00	0	0	0	0
0	4.74	0.00	2.37	0	4.74	0	0	0	0.57	0	0	0	0	0	0	0
0	2	0.02	0	0	10	0.24	0	0.00	0.00	0.00	0.00	0.00	0	0	0	0
32	4	0.00	<1	4	86	0.01	103	0.00	0.35	0.01	0.01	0.00	<1	0	<.1	<1
32	4	0.00	<1	4	2	0.01	103	0.00	0.35	0.01	0.01	0.00	<1	0	<.1	<1
25	3	0.02	<1	3	94	0.01	78	0.00	0.26	0.00	0.00	0.00	<1	0	<.1	<1
25	3	0.02	<1	3	1	0.01	—	0.00	0.26	0.00	0.01	0.00	<1	0	<.1	—
0	0	0.00	0	2	70	0.00	0	0.00	0.00	0.00	0.00	0.00	<1	0	0	—
14	0	0.00	0	0	0	0.00	0	0.00	0.35	0.00	0.00	0.00	0	0	0	<.1
11	0	0.00	0	0	0	0.00	0	0.00	0.35	0.00	0.00	0.00	0	0	0	<.1

Table A–1
Food Composition

(DA+ code is for Wadsworth Diet Analysis program)
(For purposes of calculations, use "0" for t, <1, <.1, <.01, etc.)

DA + Code	Food Description	Quantity	Measure	Wt (g)	H₂O (g)	Ener (kcal)	Prot (g)	Carb (g)	Fiber (g)	Fat (g)	Sat	Mono	Poly	Trans
	FATS AND OILS													
	Fats, cooking–Continued													
5454	Household shortening w/vegetable oil	1	tablespoon(s)	13	0	115	0	0	0	13	3.39	5.56	2.75	2.20
111	Lard	1	tablespoon(s)	13	0	114	0	0	0	13	4.94	5.68	1.41	—
	Margarine													
114	Margarine	1	tablespoon(s)	14	2	101	<1	<1	0	11	2.23	5.05	3.58	—
116	Soft	1	tablespoon(s)	14	2	101	<1	<.1	0	11	1.95	4.02	4.88	—
117	Soft, unsalted	1	tablespoon(s)	14	3	101	<1	<1	0	11	1.95	5.26	3.62	—
928	Unsalted	1	tablespoon(s)	14	3	101	<.1	<.1	0	11	2.12	5.17	3.53	—
119	Whipped	1	tablespoon(s)	9	1	64	<.1	<.1	0	7	1.17	3.25	2.51	—
	Spreads													
16164	I Can't Believe It's Not Butter! whipped spread	1	tablespoon(s)	14	4	60	0	0	0	7	1.50	1.50	2.50	—
16157	Promise vegetable oil spread, stick	1	tablespoon(s)	14	4	90	0	0	0	10	2.50	2.00	4.00	—
	Oils													
2681	Canola	1	tablespoon(s)	14	0	120	0	0	0	14	0.97	8.01	4.03	—
120	Corn	1	tablespoon(s)	14	0	120	0	0	0	14	1.73	3.29	7.98	0.04
122	Olive	1	tablespoon(s)	14	0	119	0	0	0	14	1.82	9.98	1.35	—
124	Peanut	1	tablespoon(s)	14	0	119	0	0	0	14	2.28	6.24	4.32	—
2693	Safflower	1	tablespoon(s)	14	0	120	0	0	0	14	0.84	10.15	1.95	—
923	Sesame	1	tablespoon(s)	14	0	120	0	0	0	14	1.93	5.40	5.67	—
130	Soybean w/cottonseed oil	1	tablespoon(s)	14	0	120	0	0	0	14	2.45	4.01	6.54	—
128	Soybean, hydrogenated	1	tablespoon(s)	14	0	120	0	0	0	14	2.03	5.85	5.11	—
2700	Sunflower	1	tablespoon(s)	14	0	120	0	0	0	14	1.77	6.28	4.95	—
357	**Pam original no stick cooking spray**	1	serving(s)	0	—	0	0	0	0	0	0.00	0.00	0.00	
	Salad dressing													
132	Blue cheese	2	tablespoon(s)	31	10	154	1	2	0	16	3.03	3.76	8.51	—
133	Blue cheese, low calorie	2	tablespoon(s)	32	25	32	2	1	0	2	0.82	0.57	0.78	—
1764	Caesar	2	tablespoon(s)	30	10	158	<1	1	<.1	17	2.64	4.05	9.86	—
29654	Creamy, reduced calorie, fat free, cholesterol free, sour cream and/or buttermilk & oil	2	tablespoon(s)	32	24	34	<1	6	0	1	0.16	0.21	0.46	—
29617	Creamy, reduced calorie, sour cream and/or buttermilk & oil	2	tablespoon(s)	30	22	48	<1	2	0	4	0.63	0.98	2.40	—
134	French	2	tablespoon(s)	31	11	143	<1	5	0	14	1.76	2.63	6.56	—
135	French, low fat	2	tablespoon(s)	33	18	76	<1	10	<1	4	0.36	1.92	1.64	—
136	Italian	2	tablespoon(s)	29	17	86	<1	3	0	8	1.32	1.86	3.80	—
137	Italian, diet	2	tablespoon(s)	30	25	23	<1	1	0	2	0.14	0.66	0.51	—
139	Mayonnaise type	2	tablespoon(s)	29	12	115	<1	7	0	10	1.44	2.65	5.29	—
942	Oil & vinegar	2	tablespoon(s)	31	15	140	0	1	0	16	2.84	4.62	7.52	—
1765	Ranch	2	tablespoon(s)	30	12	146	<1	2	<.1	16	2.32	3.85	8.92	—
3666	Ranch, reduced calorie	2	tablespoon(s)	30	21	62	<1	2	<.1	6	1.13	1.79	2.89	—
940	Russian	2	tablespoon(s)	31	11	151	<1	3	0	16	2.23	3.61	9.00	—
939	Russian, low calorie	2	tablespoon(s)	33	21	46	<1	9	<.1	1	0.20	0.29	0.75	—
941	Sesame seed	2	tablespoon(s)	31	12	136	1	3	<1	14	1.90	3.64	7.68	—
142	Thousand island	2	tablespoon(s)	31	15	115	<1	5	<1	11	1.59	2.46	5.68	—
143	Thousand island, low calorie	2	tablespoon(s)	31	19	62	<1	7	<1	4	0.23	1.98	0.82	—
	Sandwich spreads													
138	Mayonnaise w/soybean oil	1	tablespoon(s)	14	2	99	<1	1	0	11	1.64	2.70	5.89	0.04
2708	Mayonnaise w/soybean & safflower oils	1	tablespoon(s)	14	0	98.94	0.15	0.37	0	10.95	1.18	1.79	7.59	—
140	Mayonnaise, low calorie	1	tablespoon(s)	16	10	37	<.1	3	0	3	0.53	0.72	1.70	—
141	Tartar sauce	2	tablespoon(s)	28	9	144	<1	4	<.1	14	2.14	4.13	7.57	—

Chol (mg)	Calc (mg)	Iron (mg)	Magn (mg)	Pota (mg)	Sodi (mg)	Zinc (mg)	Vit A (µg)	Thia (mg)	Vit E (mg α)	Ribo (mg)	Niac (mg)	Vit B6 (mg)	Fola (µg)	Vit C (mg)	Vit B12 (µg)	Sele (µg)
0	0	0.00	0	0	0	0.00	0	0.00	—	0.00	0.00	0.00	0	0	0	—
12	0	0.00	0	0	0	0.01	0	0.00	0.08	0.00	0.00	0.00	0	0	0	<.1
0	4	0.01	<1	6	133	0.00	115	0.00	1.27	0.01	0.00	0.00	<1	<.1	<.1	0
0	4	0.00	<1	5	152	0.00	103	0.00	0.99	0.00	0.00	0.00	<1	<.1	<.1	0
0	4	0.00	<1	5	4	0.00	103	0.00	1.23	0.00	0.00	0.00	<1	<.1	<.1	0
0	2	0.00	<1	4	<1	0.00	115	0.00	1.80	0.00	0.00	0.00	<1	<.1	<.1	0
0	2	0.00	<1	3	97	0.00	—	0.00	0.45	0.00	0.00	0.00	<.1	<.1	<.1	—
0	10	0.18	—	4	70	—	—	1.65	0.00	0.00	0.00	—	—	1	—	—
0	10	0.18	—	9	90	—	—	0.00	—	0.00	0.00	—	—	1	—	—
0	0	0.00	0	0	0	0.00	0	0.00	2.33	0.00	0.00	0.00	0	0	0	0
0	0	0.00	0	0	0	0.00	0	0.00	1.94	0.00	0.00	0.00	0	0	0	0
0	<1	0.09	0	<1	<1	0.00	0	0.00	1.94	0.00	0.00	0.00	0	0	0	0
0	0	0.00	0	0	0	0.00	0	0.00	2.12	0.00	0.00	0.00	0	0	0	0
0	0	0.00	0	0	0	0.00	0	0.00	4.64	0.00	0.00	0.00	0	0	0	0
0	0	0.00	0	0	0	0.00	0	0.00	0.19	0.00	0.00	0.00	0	0	0	0
0	0	0.00	0	0	0	0.00	0	0.00	1.65	0.00	0.00	0.00	0	0	0	0
0	0	0.00	0	0	0	0.00	0	0.00	1.10	0.00	0.00	0.00	0	0	0	0
0	0	0.00	0	0	0	0.00	0	0.00	—	0.00	0.00	0.00	0	0	0	0
0	0	0.00	—	0	0	—	0	—	0.00	—	—	—	0	0	—	
5	25	0.06	0	11	335	0.08	21	0.00	1.84	0.03	0.03	0.01	9	1	<.1	<1
<1	28	0.16	2	2	384	0.08	—	0.01	0.08	0.03	0.02	0.01	1	<.1	<.1	—
1	7	0.05	1	9	323	0.03	—	0.00	1.57	0.00	0.01	0.00	1	0	<.1	—
0	12	0.08	2	43	320	0.06	0	0.00	0.21	0.02	0.01	0.01	1	0	0	
0	2	0.04	1	11	307	0.01	—	0.00	0.72	0.00	0.01	0.01	4	<1	<.1	—
0	7	0.25	2	21	261	0.09	7	0.01	1.56	0.02	0.06	0.00	0	0	<.1	0
0	4	0.28	3	35	262	0.07	9	0.01	0.10	0.02	0.15	0.02	1	0	0	1
0	2	0.19	1	14	486	0.04	1	0.00	1.47	0.01	0.00	0.02	0	0	0	1
2	3	0.20	1	26	410	0.06	<1	0.00	0.06	0.00	0.00	0.02	0	0	0	2
8	4	0.06	1	3	209	0.05	19	0.00	0.61	0.01	0.00	0.00	2	0	<.1	<1
0	0	0.00	0	2	<1	0.00	0	0.00	1.44	0.00	0.00	0.00	0	0	0	0
1	4	0.03	1	8	354	0.01	—	0.00	1.85	0.00	0.00	0.00	<1	<.1	<.1	—
<1	5	0.01	1	8	414	0.02	—	0.00	0.73	0.01	0.01	0.00	<1	<1	<.1	—
6	6	0.18	1	48	266	0.13	5	0.02	1.02	0.02	0.18	0.01	3	2	<.1	<1
2	6	0.20	0	51	283	0.03	1	0.00	0.13	0.00	0.00	0.00	1	2	<.1	1
0	6	0.18	0	48	306	0.03	1	0.00	1.53	0.00	0.00	0.00	0	0	0	<1
8	5	0.37	2	33	269	0.08	3	0.45	1.25	0.02	0.13	0.00	0	0	0	<1
<1	5	0.28	2	62	254	0.06	5	0.01	0.31	0.01	0.13	0.00	0	0	0	0
5	2	0.07	<1	5	78	0.02	12	0.00	0.72	0.00	0.00	0.08	1	0	<.1	<1
8.14	2.48	0.06	0.13	4.69	78.38	0.01	11.59	0	3.04	0	0	0.07	1.1	0	0.03	0.22
4	<.1	0.00	<.1	2	80	0.02	0	0.00	0.32	0.00	0.00	0.00	0	0	0	—
11	6	0.21	1	10	200	0.05	—	0.00	0.97	0.00	0.01	0.08	2	<1	<.1	—

Table A–1
Food Composition

(DA+ code is for Wadsworth Diet Analysis program)
(For purposes of calculations, use "0" for t, <1, <.1, <.01, etc.)

DA + Code	Food Description	Quantity	Measure	Wt (g)	H₂O (g)	Ener (kcal)	Prot (g)	Carb (g)	Fiber (g)	Fat (g)	Sat	Mono	Poly	Trans
	SWEETS													
4799	**Butterscotch or caramel topping**	2	tablespoon(s)	41	13	103	1	27	<1	<.1	0.05	0.01	0.00	—
	Candy													
1786	Almond Joy candy bar	1	item(s)	49	5	240	2	29	2	13	9.00	3.63	0.74	0
1785	Bit-o-Honey candy	6	item(s)	40	2	170	1	34	0	3	2.00	0	20	—
33375	Butterscotch candy	2	piece(s)	12	1	47	<.1	11	0	<1	0.25	0.10	0.01	—
1701	Chewing gum, stick	1	item(s)	3	<.1	7	0	2	<.1	<.1	0.00	0.00	0.00	—
33378	Chocolate fudge w/nuts, prepared	2	piece(s)	38	3	175	2	26	1	7	2.29	1.41	2.81	—
1787	Jelly beans	15	item(s)	43	3	159	0	40	<.1	<.1	0.00	0.00	0.00	—
1784	Kit Kat wafer bar	1	item(s)	42	1	220	3	27	1	11	7.00	3.53	0.34	0
4674	Krackel candy bar	1	item(s)	41	1	220	3	26	1	11	6.00	3.94	0.37	0
4934	Licorice	4	piece(s)	44	7	147	1	34	1	1	0.18	0.07	0.00	—
1780	Life Savers candy	1	item(s)	2	—	8	0	2	0	<.1	0.00	—	—	0
1790	Lollipop	1	item(s)	28	—	108	0	28	0	0	0.00	0.00	0.00	0
4679	M & Ms peanut chocolate candy, small bag	1	item(s)	49	1	250	5	30	2	13	5.00	5.42	2.07	—
1781	M & Ms plain chocolate candy, small bag	1	item(s)	48	1	240	2	34	1	10	6.00	3.30	0.30	—
4673	Milk chocolate bar	1	item(s)	91	1	483	8	53	2	28	16.69	7.20	0.63	—
1783	Milky Way bar	1	item(s)	58	4	270	2	41	1	10	5.00	3.50	0.35	—
1788	Peanut brittle	1½	ounce(s)	43	<1	206	3	30	1	8	1.76	3.43	1.94	—
1789	Reese's peanut butter cups	2	piece(s)	45	1	250	5	25	1	14	5.00	6.17	2.34	0
4689	Reese's pieces candy, small bag	1	item(s)	46	1	230	6	26	1	11	7.00	0.97	0.46	0
33399	Semisweet chocolate candy, made w/butter	½	ounce(s)	14	<.1	68	1	9	1	4	2.49	1.41	0.13	—
1782	Snickers bar	1	item(s)	59	3	280	4	35	1	14	5.00	6.13	2.89	—
4694	Special Dark chocolate bar	1	item(s)	41	<1	220	2	24	3	13	8.00	4.59	0.41	0
4695	Starburst fruit chews, original fruits	1	package	59	4	240	0	48	0	5	1.00	2.10	1.83	—
4698	Taffy	3	piece(s)	45	2	169	<.1	41	0	1	0.92	0.43	0.05	—
4699	Three Musketeers bar	1	item(s)	60	4	260	2	46	1	8	4.50	2.59	0.27	—
4702	Twix caramel cookie bars	2	item(s)	58	2	280	3	37	1	14	5.00	7.75	0.49	—
4705	York peppermint pattie	1	item(s)	42	4	170	1	34	1	3	2.00	1.32	0.12	—
	Frosting, icing													
4760	Chocolate frosting, ready to eat	2	tablespoon(s)	28	5	112	<1	18	<1	5	1.55	2.54	0.60	—
4771	Creamy vanilla frosting, ready to eat	2	tablespoon(s)	28	4	118	0	19	<.1	5	0.84	1.37	2.24	0
17291	Dec-a-Cake variety pack candy decoration	1	teaspoon(s)	4	—	15	0	3	0	1	0.00	—	—	—
536	White icing	2	tablespoon(s)	40	3	163	<1	32	0	4	0.86	2.07	1.19	—
	Gelatin													
13697	Gelatin snack, all flavors	1	item(s)	99	97	70	1	17	0	0	0.00	0.00	0.00	0
2616	Mixed fruit gelatin mix, sugar free, low calorie, prepared	½	cup(s)	121	—	10	1	0	0	0	0.00	0.00	0.00	0
548	**Honey**	1	tablespoon(s)	21	4	64	<.1	17	<.1	0	0.00	0.00	0.00	0
	Jams, Jellies													
23054	Jams, jellies, preserves, all flavors	1	tablespoon(s)	20	<.1	56	<.1	14	<1	<.1	0.00	0.01	0.00	—
23278	Jams, jellies, preserves, all flavors, low sugar	1	tablespoon(s)	18	<.1	25	<.1	6	<1	<.1	0.00	0.01	0.02	—
545	**Marshmallows**	4	item(s)	29	5	92	1	23	<.1	<.1	0.02	0.02	0.01	—
4800	**Marshmallow cream topping**	2	tablespoon(s)	28	6	91	<1	22	<.1	<.1	0.02	0.02	0.01	—
555	**Molasses**	1	tablespoon(s)	20	4	58	0	15	0	<.1	0.00	0.01	0.01	—
4780	**Popsicle or ice pop**	1	item(s)	59	47	42	0	11	0	0	0.00	0.00	0.00	—
	Sugar													
559	Brown, packed	1	teaspoon(s)	5	<.1	17	0	4	0	0	0.00	0.00	0.00	0

PAGE KEY: A–2 = Breads/Baked Goods A–6 = Cereal/Rice/Pasta A–10 = Fruit A–14 = Vegetables/Legumes A–24 = Nuts/Seeds A–26 = Vegetarian A–28 = Dairy A–34 = Eggs A–34 = Seafood A–36 = Meats A–40 = Poultry A–40 = Processed meats A–42 = Beverages A–46 = Fats/Oils A–48 = Sweets A–50 = Spices/Condiments/Sauces A–52 = Mixed foods/Soups/Sandwiches A–58 = Fast food A–74 = Convenience meals A–76 = Baby foods

A

Chol (mg)	Calc (mg)	Iron (mg)	Magn (mg)	Pota (mg)	Sodi (mg)	Zinc (mg)	Vit A (µg)	Thia (mg)	Vit E (mg α)	Ribo (mg)	Niac (mg)	Vit B$_6$ (mg)	Fola (µg)	Vit C (mg)	Vit B$_{12}$ (µg)	Sele (µg)
<1	22	0.08	3	34	143	0.08	11	0.00	—	0.04	0.02	0.01	1	<1	<.1	0
3	20	0.36	33	138	70	0.40	0	0.02	—	0.08	0.24	—	—	0	—	—
0.00	—	—	85	—	0	—	—	—	—	—	0	—	—	—	—	
1	<1	0.00	<1	<1	47	0.00	3	0.00	0.01	0.00	0.00	0.00	0	0	0	<.1
0	0	0.00	0	<.1	<.1	0.00	0	0.00	0.00	0.00	0.00	0.00	0	0	0	<.1
5	21	0.75	21	68	16	0.54	14	0.03	0.10	0.04	0.12	0.03	6	<.1	<.1	1
0	1	0.06	1	16	21	0.02	0	0.00	0.00	0.00	0.00	0.00	0	0	0	<1
3	40	0.36	16	126	25	0.52	8	0.07	—	0.23	1.07	0.05	60	0	<.1	2
3	60	0.37	—	169	80	—	0	—	—	—	—	—	—	0	—	—
0	3	0.13	3	28	109	0.07	0	0.01	0.08	0.02	0.04	0.00	0	0	0	—
0	<1	0.04	—	0	1	—	0	0.00	—	0.00	0.00	—	—	0	—	0
0	0	0.00	—	—	11	—	0	0.00	—	0.00	0.00	—	—	0	—	1
5	40	0.36	36	171	25	1.13	15	0.03	—	0.07	1.60	0.04	17	1	<.1	2
5	40	0.36	20	127	30	0.46	15	0.03	—	0.07	0.11	0.01	3	1	<1	1
22	228	0.83	61	399	92	1.00	20	0.06	—	0.26	0.15	0.10	11	2	<1	—
5	60	0.18	20	140	95	0.41	15	0.02	—	0.07	0.20	0.03	6	1	<1	3
5	11	0.52	18	71	189	0.37	17	0.06	1.09	0.02	1.13	0.03	20	0	<.1	1
3	20	0.36	40	233	140	0.82	7	0.11	—	0.08	2.08	0.07	25	0	<.1	2
0	40	0.00	20	182	90	0.35	25	0.04	—	0.07	1.31	0.03	13	0	<.1	1
3	5	0.44	16	52	2	0.23	<1	0.01	—	0.01	0.06	0.01	<1	0	0	<1
5	40	0.36	42	—	140	1.38	15	0.03	—	0.07	1.60	0.05	23	1	<.1	3
3	0	0.72	46	136	0	0.60	0	0.01	—	0.03	0.16	0.01	1	0	0	1
0	10	0.18	1	1	0	0.00	—	0.00	—	0.00	0.00	0.00	0	30	0	<1
4	1	0.03	<1	2	40	0.02	—	0.00	—	0.01	0.01	0.00	0	0	<.1	—
5	20	0.36	18	80	110	0.33	14	0.02	—	0.03	0.20	0.01	0	1	<1	2
5	40	0.36	18	117	115	0.45	15	0.09	—	0.13	0.69	0.02	14	1	<1	1
0	0	0.36	25	71	10	0.31	0	0.01	—	0.04	0.34	0.01	2	0	<.1	—
0	2	0.40	6	55	51	0.08	0	0.00	0.44	0.00	0.03	0.00	<1	0	0	<1
0	1	0.04	<1	10	52	0.02	0	0.00	0.43	0.08	0.06	0.00	2	0	0	<.1
0	0	0.00	—	—	15	—	0	—	—	—	—	—	—	0	—	—
<1	5	0.02	—	7	92	—	—	0.00	0.33	0.01	0.00	—	—	<.1	—	—
0	0	0.00	—	0	40	—	0	—	—	—	—	—	—	0	—	—
0	0	0.00	0	0	50	0.00	0	0.00	0.00	0.00	0.00	0.00	0	0	0	—
0	1	0.09	<1	11	1	0.05	0	0.00	0.00	0.01	0.03	0.01	<1	<1	0	<1
0	4	0.10	1	15	6	0.01	0.00	0.00	0.00	0.02	0.01	0.00	2.20	1.76	0.00	—
0	2	0.05	1	19	<1	0.02	0.76	0.00	0.01	0.01	0.03	0.01	—	4.93	0.00	—
0	1	0.07	1	1	23	0.01	0	0.00	0.00	0.00	0.02	0.00	<1	0	0	<1
0	1	0.06	1	1	23	0.01	0	0.00	0.00	0.00	0.02	0.00	<1	0	0	1
0	41	0.94	48	293	7	0.06	0	0.01	0.00	0.00	0.19	0.13	0	0	0	4
0	0	0.00	1	2	7	0.01	0	0.00	0.00	0.00	0.00	0.00	0	0	0	0
0	4	0.09	1	16	2	0.01	0	0.00	0.00	0.00	0.00	0.00	<.1	0	0	<.1

Table A–1
Food Composition

(DA+ code is for Wadsworth Diet Analysis program)
(For purposes of calculations, use "0" for t, <1, <.1, <.01, etc.)

DA+ Code	Food Description	Quantity	Measure	Wt (g)	H₂O (g)	Ener (kcal)	Prot (g)	Carb (g)	Fiber (g)	Fat (g)	Sat	Mono	Poly	Trans
	SWEETS													
	Sugar–Continued													
563	Powdered, sifted	⅓	cup(s)	33	<.1	130	0	33	0	<.1	0.01	0.01	0.02	—
561	White granulated	1	teaspoon(s)	4	<.1	15	0	4	0	0	0.00	0.00	0.00	
	Sugar Substitute													
1760	Equal sweetener, packet	1	item(s)	1	<.1	4	<.1	1	0	0	0.00	0.00	0.00	0
13029	Splenda granular no calorie sweetener	1	teaspoon(s)	1	—	2	0	1	0	0	0.00	0.00	0.00	0
1759	Sweet n Low sugar substitute, packet	1	item(s)	1	<.1	4	0	1	0	0	0.00	0.00	0.00	0
	Syrup													
3148	Chocolate	2	tablespoon(s)	38	12	105	1	24	1	<1	0.19	0.11	0.01	—
29676	Maple	¼	cup(s)	80	26	209	0	54	0	<1	0.03	0.05	0.08	—
4795	Pancake	¼	cup(s)	80	30	187	0	49	1	0	0.00	0.00	0.00	0
	SPICES, CONDIMENTS, SAUCES													
	Spices													
807	Allspice, ground	1	teaspoon(s)	2	<1	5	<1	1	<1	<1	0.05	0.01	0.04	—
1171	Anise seeds	1	teaspoon(s)	2	<1	7	<1	1	<1	<1	0.01	0.21	0.07	—
729	Baker's yeast active	1	teaspoon(s)	4	<1	12	2	2	1	<1	0.02	0.10	0.00	—
683	Baking powder, double acting, w/phosphate	1	teaspoon(s)	5	<1	2	<.1	1	<.1	0	0.00	0.00	0.00	0
1611	Baking soda	1	teaspoon(s)	5	<.1	0	0	0	0	0	0.00	0.00	0.00	
8552	Basil	1	teaspoon(s)	1	1	<1	<.1	<.1	<.1	<.1	0.00	0.00	0.00	—
34959	Basil, fresh	1	piece(s)	1	<1	<1	<.1	<.1	<.1	<.1	0.00	0.00	0.00	—
808	Basil, ground	1	teaspoon(s)	1	<.1	4	<1	1	1	<1	0.00	0.01	0.03	—
809	Bay leaf	1	teaspoon(s)	1	<.1	2	<.1	<1	<1	<.1	0.01	0.01	0.01	—
11720	Betel leaves	1	ounce(s)	28	—	17	2	2	0	<.1	—	—	—	—
818	Black pepper	1	teaspoon(s)	2	<1	5	<1	1	1	<.1	0.02	0.02	0.02	—
730	Brewer's yeast	1	teaspoon(s)	3	<1	8	1	1	1	0	0.00	0.00	0.00	0
35417	Capers	1	teaspoon(s)	4	—	2	0	0	0	0	0.00	0.00	0.00	—
1172	Caraway seeds	1	teaspoon(s)	2	<1	7	<1	1	1	<1	0.01	0.15	0.07	—
819	Cayenne pepper	1	teaspoon(s)	2	<1	6	<1	1	<1	<1	0.06	0.05	0.15	—
1173	Celery seeds	1	teaspoon(s)	2	<1	8	<1	1	<1	1	0.04	0.32	0.07	—
1174	Chervil, dried	1	teaspoon(s)	1	<.1	1	<1	<1	<.1	<.1	0.00	0.01	0.01	—
810	Chili powder	1	teaspoon(s)	3	<1	8	<1	1	1	<1	0.08	0.09	0.19	—
8553	Chives, chopped	1	teaspoon(s)	1	1	<1	<.1	<.1	<.1	<.1	0.00	0.00	0.00	—
8556	Cilantro	1	teaspoon(s)	2	1	<1	<.1	<.1	<.1	<.1	0.00	0.00	0.00	—
811	Cinnamon, ground	1	teaspoon(s)	2	<1	6	<.1	2	1	<.1	0.01	0.01	0.01	—
812	Cloves, ground	1	teaspoon(s)	2	<1	7	<1	1	1	<1	0.11	0.03	0.15	—
1175	Coriander leaf, dried	1	teaspoon(s)	1	<.1	2	<1	<1	<.1	<.1	0.00	0.01	0.00	—
1176	Coriander seeds	1	teaspoon(s)	2	<1	5	<1	1	1	<1	0.02	0.24	0.03	—
1706	Cornstarch	1	tablespoon(s)	8	1	30	<.1	7	<.1	<.1	0.00	0.00	0.00	—
11729	Cumin, ground	1	teaspoon(s)	5	—	11	<1	1	1	<1	—	—	—	—
1177	Cumin seeds	1	teaspoon(s)	2	<1	8	<1	1	<1	<1	0.03	0.29	0.07	—
1178	Curry powder	1	teaspoon(s)	2	<1	7	<1	1	1	<1	0.04	0.11	0.05	—
1179	Dill seeds	1	teaspoon(s)	2	<1	6	<1	1	<1	<1	0.02	0.20	0.02	—
1180	Dill weed, dried	1	teaspoon(s)	1	<.1	3	<1	1	<1	<.1	0.00	0.01	0.00	—
34949	Dill weed, fresh	5	piece(s)	1	1	<1	<.1	<.1	<.1	<.1	0.00	0.01	0.00	—
4949	Fennel leaves, fresh	1	teaspoon(s)	1	1	<1	<.1	<.1	0	<.1	0.00	0.00	0.00	—
1181	Fennel seeds	1	teaspoon(s)	2	<1	7	<1	1	1	<1	0.01	0.20	0.03	—
1182	Fenugreek seeds	1	teaspoon(s)	4	<1	12	1	2	1	<1	0.05	—	—	—
11733	Garam masala, powder	1	ounce(s)	28	—	107	4	13	0	4	—	—	—	—
1067	Garlic clove	1	item(s)	3	2	4	<1	1	<.1	<.1	0.00	0.00	0.01	—
813	Garlic powder	1	teaspoon(s)	3	<1	9	<1	2	<1	<.1	0.00	0.00	0.01	—

Chol (mg)	Calc (mg)	Iron (mg)	Magn (mg)	Pota (mg)	Sodi (mg)	Zinc (mg)	Vit A (µg)	Thia (mg)	Vit E (mg α)	Ribo (mg)	Niac (mg)	Vit B_6 (mg)	Fola (µg)	Vit C (mg)	Vit B_{12} (µg)	Sele (µg)
0	<1	0.01	0	1	<1	0.00	0	0.00	0.00	0.01	0.00	0.00	0	0	0	<1
0	<.1	0.00	0	<.1	0	0.00	0	0.00	0.00	0.00	0.00	0.00	0	0	0	<.1
0	0	0.00	0	0	0	0.00	0	0.00	0.00	0.00	0.00	0.00	0	0	0	0
0	10	0.18	—	—	<1	—	—	0.02	0.00	0.02	0.20	—	—	1	0	—
0	0	0.00	0	—	0	0.00	0	0.00	0.00	0.00	0.00	0.00	0	0	0	0
0	5	0.79	24	84	27	0.27	0	0.00	0.00	0.02	0.12	0.00	1	<.1	0	1
0	54	0.96	11	163	7	3.33	0	0.00	0.00	0.01	0.02	0.00	0	0	0	<1
0	2	0.02	2	12	66	0.06	0	0.00	0.00	0.01	0.01	0.00	0	0	0	0
0	13	0.13	3	20	1	0.02	1	0.00	—	0.00	0.05	0.00	1	1	0	<.1
0	14	0.78	4	30	<1	0.11	<1	0.01	—	0.01	0.06	0.01	<1	<1	0	<1
0	3	0.66	4	80	2	0.26	0	0.09	0.00	0.22	1.59	0.06	94	<.1	<.1	1
0	339	0.52	2	<1	363	0.00	0	0.00	0.00	0.00	0.00	0.00	0	0	0	<.1
0	0	0.00	0	0	1259	0.00	0	0.00	0.00	0.00	0.00	0.00	0	0	0	<.1
0	1	0.03	1	4	<.1	0.01	2	0.00	—	0.00	0.01	0.00	1	<1	0	<.1
0	1	—	<1	2	<.1	0.00	—	0.00	—	0.00	0.01	0.00	<1	—	0	<.1
0	30	0.59	6	48	<1	0.08	7	0.00	0.10	0.00	0.10	0.03	4	1	0	<.1
0	5	0.26	1	3	<1	0.02	2	0.00	—	0.00	0.01	0.01	1	<1	0	<.1
0	110	2.29	—	156	2	—	—	0.04	—	0.07	0.20	—	—	1	0	—
0	9	0.61	4	26	1	0.03	<1	0.00	0.02	0.01	0.02	0.01	<1	<1	0	<.1
0	6	0.47	6	51	3	0.21	0	0.42	—	0.11	1.00	0.07	104	0	0	0
0	0	0.00	—	—	140	—	0	—	—	—	—	—	—	0	—	—
0	14	0.34	5	28	<1	0.12	<1	0.01	0.05	0.01	0.08	0.01	<1	<1	0	<1
0	3	0.14	3	36	1	0.04	37	0.01	0.54	0.02	0.16	0.04	2	1	0	<1
0	35	0.90	9	28	3	0.14	<.1	0.01	0.02	0.01	0.06	0.02	<1	<1	0	<1
0	8	0.19	1	28	<1	0.05	2	0.00	—	0.00	0.03	0.01	2	<1	0	<1
0	7	0.37	4	50	26	0.07	39	0.01	—	0.02	0.21	0.10	3	2	0	<1
0	1	0.02	<1	3	<.1	0.01	2	0.00	0.76	0.00	0.00	0.00	1	1	0	<.1
0	1	0.03	<1	8	1	0.00	—	0.00	—	0.00	0.02	0.00	1	1	0	<.1
0	28	0.88	1	12	1	0.05	<1	0.00	0.02	0.00	0.03	0.01	1	1	0	<.1
0	14	0.18	6	23	5	0.02	1	0.00	0.18	0.01	0.03	0.01	2	2	0	<1
0	7	0.25	4	27	1	0.03	2	0.01	—	0.01	0.06	0.00	2	3	0	<1
0	13	0.29	6	23	1	0.08	0	0.00	—	0.01	0.04	—	0	<1	0	<1
0	<1	0.04	<1	<1	1	0.00	0	0.00	0.00	0.00	0.00	0.00	0	0	0	<1
0	20	—	—	44	5	—	—	—	—	—	—	—	—	—	—	—
0	20	1.39	8	38	4	0.10	1	0.01	0.07	0.01	0.10	0.01	<1	<1	0	<1
0	10	0.59	5	31	1	0.08	1	0.01	0.44	0.01	0.07	0.02	3	<1	0	<1
0	32	0.34	5	25	<1	0.11	<.1	0.01	—	0.01	0.06	0.01	<1	<1	0	<1
0	18	0.49	5	33	2	0.03	3	0.00	—	0.00	0.03	0.02	2	1	0	0
0	2	—	1	7	1	0.01	—	0.00	—	0.00	0.02	0.00	2	—	—	—
0	1	0.03	—	4	<.1	—	—	0.00	—	0.00	0.01	0.00	—	<1	0	—
0	24	0.37	8	34	2	0.07	<1	0.01	—	0.01	0.12	0.01	—	<1	0	0
0	7	1.24	7	28	2	0.09	<1	0.01	—	0.01	0.06	0.02	2	<1	0	<1
0	215	9.25	94	411	28	1.07	—	0.10	—	0.09	0.71	—	0	0	0	—
0	5	0.05	1	12	1	0.03	0	0.01	0.00	0.00	0.02	0.04	<.1	1	0	<1
0	2	0.08	2	31	1	0.07	0	0.01	0.02	0.00	0.02	0.08	<.1	1	0	1

A

Table A–1
Food Composition

(DA+ code is for Wadsworth Diet Analysis program)
(For purposes of calculations, use "0" for t, <1, <.1, <.01, etc.)

DA + Code	Food Description	Quantity	Measure	Wt (g)	H₂O (g)	Ener (kcal)	Prot (g)	Carb (g)	Fiber (g)	Fat (g)	Sat	Mono	Poly	Trans
	SPICES, CONDIMENTS, SAUCES													
	Spices–Continued													
1183	Ginger, ground	1	teaspoon(s)	2	<1	6	<1	1	<1	<1	0.03	0.02	0.02	—
1068	Ginger root	2	teaspoon(s)	4	3	3	<.1	1	<.1	<.1	0.01	0.01	0.01	—
35497	Leeks, bulb & lower leaf, freeze-dried	¼	cup(s)	1	<.1	3	<1	1	<.1	<.1	0.00	0.00	0.01	
1184	Mace, ground	1	teaspoon(s)	2	<1	8	<1	1	<1	1	0.16	0.19	0.07	
1185	Marjoram, dried	1	teaspoon(s)	1	<.1	2	<.1	<1	<1	<.1	0.00	0.01	0.03	
1186	Mustard seeds, yellow	1	teaspoon(s)	3	<1	15	1	1	<1	1	0.05	0.65	0.18	
814	Nutmeg, ground	1	teaspoon(s)	2	<1	12	<1	1	<1	1	0.57	0.07	0.01	
2747	Onion flakes, dehydrated	1	teaspoon(s)	2	<.1	6	<1	1	<1	<.1	0.00	0.00	0.00	
1187	Onion powder	1	teaspoon(s)	2	<1	7	<1	2	<1	<.1	0.00	0.00	0.01	
815	Oregano, ground	1	teaspoon(s)	2	<1	5	<1	1	1	<1	0.04	0.01	0.08	
816	Paprika	1	teaspoon(s)	2	<1	6	<1	1	1	<1	0.04	0.03	0.17	
817	Parsley, dried	1	teaspoon(s)	0	<.1	1	<.1	<1	<.1	<.1	0.00	0.01	0.00	
1189	Poppy seeds	1	teaspoon(s)	3	<1	15	1	1	<1	1	0.14	0.18	0.86	
1190	Poultry seasoning	1	teaspoon(s)	2	<1	5	<1	1	<1	<1	0.05	0.02	0.03	
1191	Pumpkin pie spice, powder	1	teaspoon(s)	2	<1	6	<.1	1	<1	<1	0.11	0.02	0.01	
1192	Rosemary, dried	1	teaspoon(s)	1	<1	4	<.1	1	1	<1	0.09	0.04	0.03	
11723	Rosemary, fresh	1	teaspoon(s)	1	<1	1	<.1	<1	<.1	<.1	0.02	0.01	0.01	
2722	Saffron powder	1	teaspoon(s)	1	<.1	2	<.1	<1	<.1	<.1	0.01	0.00	0.01	
11724	Sage	1	ounce(s)	28	—	34	1	4	0	1	—	—	—	—
1193	Sage, ground	1	teaspoon(s)	1	<.1	2	<.1	<1	<1	<.1	0.05	0.01	0.01	
822	Salt, table	¼	teaspoon(s)	2	<.1	0	0	0	0	0	0.00	0.00	0.00	0
30189	Salt substitute	¼	teaspoon(s)	1	—	<.1	0	<.1	0	0	0.00	0.00	0.00	0
30190	Salt substitute, seasoned	¼	teaspoon(s)	1	—	1	<.1	<1	0	<.1	0.00	—	—	
1194	Savory, ground	1	teaspoon(s)	1	<1	4	<.1	1	1	<.1	0.05			
820	Sesame seed kernels, toasted	1	teaspoon(s)	3	<1	15	<1	1	<1	1	0.18	0.49	0.57	
11725	Sorrel	1	tablespoon(s)	9	—	2	<1	<1	<.1	<.1	0.00	—	—	
11721	Spearmint	1	teaspoon(s)	2	2	1	<.1	<1	<1	<.1	0.00	0.00	0.01	
35498	Sweet green peppers, freeze-dried	¼	cup(s)	2	<.1	5	<1	1	<1	<.1	0.01	0.00	0.03	
11726	Tamarind leaves	1	ounce(s)	28	—	33	2	5	0	1	—	—	—	—
11727	Tarragon	1	ounce(s)	28	—	14	1	2	0	<1	—	—	—	—
1195	Tarragon, ground	1	teaspoon(s)	2	<1	5	<1	1	<1	<1	0.03	0.01	0.06	
11728	Thyme, fresh	1	teaspoon(s)	1	1	1	<.1	<1	<1	<.1	0.00	0.00	0.00	
821	Thyme, ground	1	teaspoon(s)	1	<1	4	<1	1	1	<1	0.04	0.01	0.02	
1196	Turmeric, ground	1	teaspoon(s)	2	<1	8	<1	1	<1	<1	0.07	0.04	0.05	
11995	Wasabi	1	tablespoon(s)	14	11	11	1	2	<1	<.1	—	—	—	—
1188	White pepper	1	teaspoon(s)	2	<1	7	<1	2	1	<.1	0.02	0.02	0.01	
	Condiments													
674	Catsup or ketchup	1	tablespoon(s)	15	11	14	<1	4	<1	<.1	0.01	0.01	0.04	
703	Dill pickle	1	ounce(s)	28	26	5	<1	1	<1	<.1	0.01	0.00	0.02	
1641	Horseradish sauce, prepared	1	teaspoon(s)	5	3	10	<1	<1	<.1	1	0.59	0.28	0.04	
140	Mayonnaise, low calorie	1	tablespoon(s)	16	10	37	<.1	3	0	3	0.53	0.72	1.70	—
138	Mayonnaise w/soybean oil	1	tablespoon(s)	14	2	99	<1	1	0	11	1.64	2.70	5.89	0.04
1682	Mustard, brown	1	teaspoon(s)	5	4	5	<1	<1	<.1	<1	—	—	—	—
700	Mustard, yellow	1	teaspoon(s)	5	4	3	<1	<1	<1	<1	0.01	0.11	0.03	
706	Sweet pickle relish	1	tablespoon(s)	15	9	20	<.1	5	<1	<.1	0.01	0.03	0.02	
141	Tartar sauce	2	tablespoon(s)	28	9	144	<1	4	<.1	14	2.14	4.13	7.57	
	Sauces													
685	Barbecue sauce	2	tablespoon(s)	31	25	23	1	4	<1	<1	0.08	0.24	0.21	
834	Cheese sauce	¼	cup(s)	70	49	121	5	5	<1	9	4.19	2.67	1.81	—
32123	Chili enchilada sauce, green	2	tablespoon(s)	57	53	15	1	3	1	<1	0.04	0.04	0.13	0
32122	Chili enchilada sauce, red	2	tablespoon(s)	32	24	27	1	5	2	1	0.08	0.05	0.43	0

Chol (mg)	Calc (mg)	Iron (mg)	Magn (mg)	Pota (mg)	Sodi (mg)	Zinc (mg)	Vit A (µg)	Thia (mg)	Vit E (mg α)	Ribo (mg)	Niac (mg)	Vit B6 (mg)	Fola (µg)	Vit C (mg)	Vit B12 (µg)	Sele (µg)
0	2	0.21	3	24	1	0.08	<1	0.00	0.32	0.00	0.09	0.02	1	<1	0	1
0	1	0.02	2	17	1	0.01	0	0.00	0.01	0.00	0.03	0.01	<1	<1	0	<.1
0	3	0.06	1	19	<1	0.01	<1	0.01	—	0.00	0.03	0.01	3	1	0	<.1
0	4	0.24	3	8	1	0.04	1	0.01	—	0.01	0.02	0.00	1	<1	0	<.1
0	12	0.50	2	9	<1	0.02	2	0.00	0.01	0.00	0.02	0.01	2	<1	0	<.1
0	17	0.33	10	23	<1	0.19	<.1	0.02	0.10	0.01	0.26	0.01	3	<.1	0	4
0	4	0.07	4	8	<1	0.05	<1	0.01	0.00	0.00	0.03	0.00	2	<.1	0	<.1
0	4	0.03	2	27	<1	0.03	<.1	0.01	0.00	0.00	0.02	0.03	3	1	0	<.1
0	8	0.05	3	20	1	0.05	0	0.01	0.01	0.00	0.01	0.03	3	<1	0	<.1
0	24	0.66	4	25	<1	0.07	5	0.01	0.28	0.00	0.09	0.02	4	1	0	<.1
0	4	0.50	4	49	1	0.09	55	0.01	0.63	0.04	0.32	0.08	2	1	0	<.1
0	4	0.29	1	11	1	0.01	2	0.00	0.02	0.00	0.02	0.00	1	<1	0	<.1
0	41	0.26	9	20	1	0.29	0	0.02	0.03	0.00	0.03	0.01	2	<.1	0	<.1
0	15	0.53	3	10	<1	0.05	2	0.00	0.03	0.00	0.04	0.02	2	<1	0	<1
0	12	0.34	2	11	1	0.04	<1	0.00	0.02	0.00	0.04	0.01	1	<1	0	<1
0	15	0.35	3	11	1	0.04	2	0.01	—	0.01	0.01	0.02	4	1	0	<.1
0	2	0.05	1	5	<1	0.01	1	0.00	—	0.00	0.01	0.00	1	<1	0	—
0	1	0.08	2	12	1	0.01	<1	0.00	—	0.00	0.01	0.01	1	1	0	<.1
0	170	—	45	110	1	0.48	—	0.03	—	—	—	—	—	—	0	—
0	12	0.20	3	7	<.1	0.03	2	0.01	0.05	0.00	0.04	0.02	2	<1	0	<.1
0	<1	0.00	<.1	<1	581	0.00	0	0.00	0.00	0.00	0.00	0.00	0	0	0	<.1
0	7	0.00	<.1	604	<.1	—	0	—	—	—	—	—	—	0	—	—
0	0	0	476	<1	—	0	—	—	—	—	—	—	0	—	—	
0	30	0.53	5	15	<1	0.06	4	0.01	—	—	0.06	0.03	—	1	0	<.1
0	4	0.21	9	11	1	0.28	<.1	0.03	0.01	0.01	0.15	0.00	3	0	0	<.1
0	—	—	—	—	<1	—	—	—	—	—	—	—	—	—	—	—
0	4	0.23	1	9	1	0.02	4	0.00	—	0.00	0.02	0.00	2	<1	0	—
0	2	0.17	3	51	3	0.04	3	0.02	0.06	0.02	0.12	0.04	4	30	0	<.1
0	85	1.48	20	—	—	—	—	0.07	—	0.03	1.16	—	—	1	0	—
0	48	—	14	128	3	0.17	—	0.04	—	—	—	—	—	1	0	—
0	18	0.52	6	48	1	0.06	3	0.00	0.10	0.02	0.14	0.04	4	1	0	<.1
0	3	0.14	1	5	<.1	0.01	2	0.00	—	0.00	0.01	0.00	<1	1	0	—
0	26	1.73	3	11	1	0.09	3	0.01	—	0.01	0.07	0.01	4	1	0	<.1
0	4	0.91	4	56	1	0.10	0	0.00	—	0.01	0.11	0.04	1	1	0	<.1
0	13	0.11	—	—	—	—	—	0.02	—	0.01	0.07	—	—	11	0	—
0	6	0.34	2	2	<1	0.03	0	0.00	0.10	0.00	0.01	0.00	<1	1	0	<.1
0	3	0.08	3	57	167	0.04	7	0.00	0.22	0.07	0.23	0.02	2	2	0	<.1
0	3	0.15	3	33	363	0.04	3	0.00	0.03	0.01	0.02	0.00	<1	1	0	0
2	5	0.00	1	7	15	0.01	—	0.00	0.03	0.01	0.00	0.00	1	<.1	<.1	—
4	<.1	0.00	<.1	2	80	0.02	0	0.00	0.32	0.00	0.00	0.00	0	0	0	—
5	2	0.07	<1	5	78	0.02	12	0.00	0.72	0.00	0.00	0.08	1	0	<.1	<1
0	6	0.09	1	7	68	0.02	0	0.00	0.09	0.00	0.01	0.00	<1	<.1	0	—
0	4	0.09	2	8	56	0.03	<1	0.00	0.01	0.00	0.02	0.00	<1	<1	0	2
0	<1	0.13	1	4	122	0.02	1	0.00	0.06	0.00	0.03	0.00	<1	<1	0	0
11	6	0.21	1	10	200	0.05	—	0.00	0.97	0.00	0.01	0.08	2	<1	<.1	—
0	6	0.28	6	54	255	0.06	<1	0.01	0.01	0.01	0.28	0.02	1	2	0	<1
20	128	0.15	6	21	578	0.68	56	0.00	—	0.08	0.02	0.01	3	<1	<.1	2
0	5	0.36	9	126	62	0.11	—	0.03	0.00	0.02	0.63	0.06	6	44	0	0
0	7	1.05	11	231	114	0.15	—	0.02	0.00	0.22	0.61	0.34	7	<1	0	<1

Table A–1
Food Composition

(DA+ code is for Wadsworth Diet Analysis program)
(For purposes of calculations, use "0" for t, <1, <.1, <.01, etc.)

DA + Code	Food Description	Quantity	Measure	Wt (g)	H₂O (g)	Ener (kcal)	Prot (g)	Carb (g)	Fiber (g)	Fat (g)	Sat	Mono	Poly	Trans
	SPICES, CONDIMENTS, SAUCES													
	Sauces–Continued													
29688	Hoisin sauce	1	tablespoon(s)	16	7	35	1	7	<1	1	0.09	0.15	0.27	—
16670	Mole poblano sauce	½	cup(s)	133	103	155	5	11	2	11	2.67	5.15	2.91	—
29689	Oyster sauce	1	tablespoon(s)	16	13	8	<1	2	<.1	<.1	0.01	0.01	0.01	—
1655	Pepper sauce or tabasco	1	teaspoon(s)	5	5	1	<.1	<.1	<.1	<.1	0.01	0.00	0.02	—
347	Salsa	2	tablespoon(s)	16	14	4	<1	1	<1	<.1	0.00	0.00	0.02	—
841	Soy sauce	1	tablespoon(s)	18	13	10	1	2	0	<.1	0.00	0.00	0.01	—
839	Sweet & sour sauce	2	tablespoon(s)	39	30	37	<.1	9	<.1	<.1	0.00	0.00	0.00	—
1613	Teriyaki sauce	1	tablespoon(s)	18	12	15	1	3	<.1	0	0.00	0.00	0.00	0
25294	Tomato sauce	½	cup(s)	112	100	46	2	8	2	1	0.18	0.29	0.72	0
728	White sauce, medium	¼	cup(s)	63	47	92	2	6	<1	7	1.78	2.78	1.79	—
1654	Worcestershire sauce	1	teaspoon(s)	6	4	4	0	1	0	0	0.00	0.00	0.00	0
	Vinegar													
30853	Balsamic	1	tablespoon(s)	15	—	10	0	2	0	0	0.00	0.00	0.00	—
727	Cider	1	tablespoon(s)	15	14	2	0	1	0	0	0.00	0.00	0.00	—
1673	Distilled	1	tablespoon(s)	15	14	2	0	1	0	0	0.00	0.00	0.00	—
15439	Tarragon	1	tablespoon(s)	16	—	0	0	0	0	0	0.00	0.00	0.00	0
	MIXED FOODS, SOUPS, SANDWICHES													
	Mixed Dishes													
16652	Almond chicken	1	cup(s)	242	186	280	22	16	3	15	1.91	6.07	5.62	—
25224	Barbecued chicken	2	piece(s)	177	100	325	27	15	<1	17	4.63	6.78	3.71	0
25227	Bean burrito	1	item(s)	149	82	327	17	33	6	15	8.30	4.73	0.85	—
9516	Beef & vegetable fajita	1	item(s)	223	144	397	23	35	3	18	5.50	7.53	3.45	—
16796	Beef or pork egg roll	2	item(s)	128	85	227	10	19	1	12	2.88	5.96	2.64	—
177	Beef stew w/vegetables, prepared	1	cup(s)	245	201	220	16	15	3	11	4.40	4.50	0.50	—
30233	Beef stroganoff w/noodles	1	cup(s)	256	190	343	20	23	2	19	7.37	5.62	4.47	—
16651	Cashew chicken	1	cup(s)	242	131	644	43	17	3	46	7.75	20.83	14.47	—
475	Cheese pizza	2	slice(s)	126	60	281	15	41	0	6	3.08	1.98	0.98	—
30330	Cheese quesadilla	1	item(s)	54	19	183	6	18	1	10	3.49	3.42	2.16	—
215	Chicken & noodles, prepared	1	cup(s)	240	170	365	22	26	1	18	5.10	7.10	3.90	—
30239	Chicken & vegetables w/broccoli, onion, bamboo shoots in soy based sauce	1	cup(s)	162	112	287	22	6	1	19	5.13	7.65	4.68	—
25093	Chicken cacciatore	1	cup(s)	230	166	266	28	5	1	14	3.98	5.78	3.11	0
28020	Chicken fried turkey steak	3	ounce(s)	85	48	122	13	12	1	2	0.59	0.37	0.78	—
218	Chicken pot pie	1	cup(s)	252	154	542	23	42	3	31	9.79	12.52	7.03	—
30240	Chicken teriyaki	1	cup(s)	244	163	339	51	13	1	7	1.78	2.03	1.71	—
25119	Chicken waldorf salad	½	cup(s)	100	68	178	14	6	1	11	1.76	3.18	5.05	0
25099	Chili con carne	¾	cup(s)	215	175	197	14	21	7	7	2.55	2.83	0.54	0
1062	Coleslaw	¾	cup(s)	90	73	62	1	11	1	2	0.35	0.64	1.22	—
1896	Combination pizza, w/meat & vegetables	2	slice(s)	158	75	368	26	43	5	11	3.07	5.09	1.83	—
1574	Crab cakes, from blue crab	1	item(s)	60	43	93	12	<1	0	5	0.89	1.69	1.36	—
32144	Enchiladas w/green chili sauce (enchiladas verdes)	1	item(s)	144	104	207	9	18	3	12	6.35	3.65	0.96	0
2793	Falafel patty	3	item(s)	51	18	170	7	16	0	9	1.22	5.19	2.12	—
28546	Fettuccine alfredo	1	cup(s)	222	81	247	11	42	1	3	1.61	0.79	0.43	0
32146	Flautas	3	item(s)	162	78	438	25	36	4	22	8.22	8.80	2.29	—
29629	Fried rice w/meat or poultry	1	cup(s)	198	129	329	12	41	1	12	2.27	3.53	5.69	—
16649	General tso chicken	1	cup(s)	146	91	293	19	16	1	17	3.98	6.27	5.27	—
1826	Green salad	¾	cup(s)	104	99	17	1	3	2	<.1	0.01	0.00	0.04	—
1814	Hummus	½	cup(s)	123	80	218	6	25	5	11	1.38	6.04	2.56	—
16650	Kung pao chicken	1	cup(s)	162	88	431	29	11	2	31	5.19	13.95	9.69	—

Chol (mg)	Calc (mg)	Iron (mg)	Magn (mg)	Pota (mg)	Sodi (mg)	Zinc (mg)	Vit A (µg)	Thia (mg)	Vit E (mg α)	Ribo (mg)	Niac (mg)	Vit B_6 (mg)	Fola (µg)	Vit C (mg)	Vit B_{12} (µg)	Sele (µg)
<1	5	0.16	4	19	258	0.05	0	0.00	0.04	0.03	0.19	0.01	4	<.1	0	<1
1	37	1.51	57	283	305	0.95	—	0.07	1.72	0.09	1.82	0.09	14	5	<.1	—
0	5	0.03	1	9	437	0.01	0	0.00	0.00	0.02	0.24	0.00	2	<.1	<.1	1
0	1	0.06	1	6	32	0.01	4	0.00	—	0.00	0.01	0.01	<.1	<1	0	—
0	5	0.16	2	34	69	0.04	5	0.01	0.19	0.01	0.13	0.02	3	2	0	<.1
0	3	0.36	6	32	1029	0.07	0	0.01	0.00	0.02	0.61	0.03	3	0	0	—
0	5	0.20	1	8	98	0.01	0	0.00	—	0.01	0.12	0.04	<1	0	0	—
0	5	0.31	11	41	690	0.02	0	0.01	0.00	0.01	0.23	0.02	4	0	0	<1
0	21	1.08	19	431	199	0.30	48	0.05	0.39	0.05	1.18	0.13	15	15	0	1
4	74	0.21	9	98	221	0.26	—	0.04	—	0.12	0.25	0.03	3	1	<1	—
0	6	0.30	1	45	56	0.01	0	0.00	0.00	0.01	0.04	0.00	0	1	0	—
0	0	0.00	—	—	0	—	0	—	—	—	—	—	—	—	0	—
0	1	0.09	3	15	<1	0.00	0	0.00	0.00	0.00	0.00	0.00	0	0	0	<.1
0	1	0.09	0	2	<1	0.00	0	0.00	0.00	0.00	0.00	0.00	0	0	0	5
0	0	0.00	—	0	0	—	0	—	—	—	—	—	—	0	0	—
40	69	1.97	60	549	526	1.62	—	0.09	4.11	0.20	9.48	0.44	26	7	<1	—
120	26	1.64	31	387	477	2.69	69	0.07	0.01	0.37	6.92	0.39	15	5	<1	19
38	331	2.95	45	384	514	1.92	119	0.24	0.01	0.29	1.82	0.15	115	4	<1	18
45	84	3.74	37	476	757	3.51	—	0.39	0.80	0.30	5.37	0.38	23	27	2	—
74	30	1.66	20	248	547	0.91	—	0.32	1.28	0.25	2.55	0.19	20	4	<1	—
71	29	2.90	—	613	292	—	—	0.15	0.51	0.17	4.70	—	—	17	<.1	15
74	70	3.26	37	393	818	3.66	—	0.21	1.25	0.31	3.80	0.21	17	1	2	—
96	74	2.92	94	640	1355	2.24	—	0.23	4.11	0.22	19.76	0.88	64	11	<1	—
19	233	1.16	32	219	672	1.63	147	0.37	—	0.33	4.96	0.09	69	3	1	27
13	132	1.21	13	77	230	0.64	—	0.13	0.43	0.14	1.09	0.04	6	15	<.1	—
103	26	2.20	—	149	600	—	—	0.05	—	0.17	4.30	—	—	0	—	29
84	22	1.38	29	344	962	1.70	—	0.08	1.12	0.17	7.90	0.32	13	8	<1	—
103	45	2.21	37	444	451	2.01	53	0.10	0.00	0.21	9.20	0.54	15	8	<1	22
27	69	1.34	19	197	139	1.08	5	0.15	0.00	0.18	3.46	0.22	21	<1	<1	16
69	64	3.38	38	393	651	1.93	607	0.40	1.06	0.40	7.24	0.24	31	11	<1	—
157	52	3.27	67	589	3209	3.75	—	0.15	0.59	0.37	16.69	0.89	23	6	1	—
42	20	0.78	24	197	246	1.13	21	0.04	0.62	0.10	4.05	0.25	15	2	<1	11
27	43	3.16	50	646	865	2.44	25	0.13	0.02	0.23	3.01	0.18	56	10	1	10
7	41	0.53	9	163	21	0.18	48	0.06	—	0.06	0.24	0.11	24	29	0	1
41	202	3.07	36	357	765	2.23	117	0.43	—	0.35	3.92	0.19	65	3	1	22
90	63	0.65	20	194	198	2.45	34	0.05	—	0.05	1.74	0.10	32	2	4	24
27	266	1.08	38	251	276	1.27	—	0.07	0.03	0.16	1.28	0.18	45	59	<1	6
0	28	1.74	42	298	150	0.77	1	0.07	—	0.08	0.53	0.06	47	1	0	1
9	153	1.88	32	123	386	1.48	51	0.35	0.00	0.34	2.60	0.06	103	1	<1	35
73	146	2.66	61	223	886	3.44	0	0.10	0.10	0.17	3.00	0.27	96	0	1	37
102	36	2.66	31	182	821	1.42	—	0.30	1.60	0.19	3.51	0.24	24	3	<1	—
65	27	1.49	24	250	906	1.40	—	0.10	1.62	0.19	6.28	0.28	17	12	<1	—
0	13	0.65	11	178	27	0.22	59	0.03	—	0.05	0.57	0.08	38	24	<1	—
0	60	1.93	36	213	298	1.34	0	0.11	0.92	0.06	0.49	0.49	73	10	0	3
64	49	1.96	63	428	907	1.50	—	0.15	4.32	0.15	13.23	0.59	43	8	<1	—

A

Table A–1
Food Composition

(DA+ code is for Wadsworth Diet Analysis program)
(For purposes of calculations, use "0" for t, <1, <.1, <.01, etc.)

DA + Code	Food Description	Quantity	Measure	Wt (g)	H₂O (g)	Ener (kcal)	Prot (g)	Carb (g)	Fiber (g)	Fat (g)	Sat	Mono	Poly	Trans
	MIXED FOODS, SOUPS, SANDWICHES													
	Mixed Dishes–Continued													
16622	Lamb curry	1	cup(s)	236	188	256	28	3	1	14	3.93	4.92	3.35	—
25253	Lasagna w/ground beef	1	cup(s)	237	157	288	18	22	2	15	7.47	4.84	0.84	0
442	Macaroni & cheese	1	cup(s)	200	122	393	15	40	1	19	8.18	6.72	2.66	—
25105	Meat loaf	1	slice(s)	115	85	244	17	7	<1	16	6.15	6.89	0.83	0
16646	Moo shi pork	1	cup(s)	151	77	512	19	5	1	46	6.84	14.80	22.07	—
16788	Nachos w/beef, beans, cheese, tomatoes, & onions	7	item(s)	551	284	1496	40	119	19	99	22.34	40.19	30.69	—
1668	Pepperoni pizza	2	slice(s)	142	66	362	20	40	1	14	4.47	6.28	2.33	—
655	Potato salad	½	cup(s)	125	95	179	3	14	2	10	1.79	3.10	4.67	
29637	Ravioli, meat filled, w/tomato or meat sauce, canned	1	cup(s)	251	196	220	9	38	2	4	1.58	1.49	0.41	
25109	Salisbury steaks w/mushroom sauce	1	serving(s)	135	102	251	17	9	1	15	5.98	6.67	0.76	0
16637	Shrimp creole w/rice	1	cup(s)	243	176	311	27	28	1	9	1.83	3.79	2.88	
497	Spaghetti & meat balls w/tomato sauce, prepared	1	cup(s)	248	174	330	19	39	3	12	3.90	4.40	2.20	
28585	Spicy thai noodles (pad thai)	8	ounce(s)	231	74	222	9	36	3	6	0.83	3.33	1.83	0
33073	Stir fried pork & vegetables w/rice	1	cup(s)	235	173	349	15	34	2	16	5.55	6.87	2.62	0
28588	Stuffed shells	2½	item(s)	299	189	292	18	33	3	10	3.81	3.57	1.62	0
16821	Sushi w/egg in seaweed	6	piece(s)	156	117	190	9	20	<1	8	2.09	3.02	1.55	
16819	Sushi w/vegetables & fish	6	piece(s)	156	102	217	8	44	2	1	0.16	0.14	0.20	
16820	Sushi w/vegetables in seaweed	6	piece(s)	156	110	182	3	41	1	<1	0.10	0.11	0.11	
25266	Sweet & sour pork	¾	cup(s)	249	206	264	29	17	1	8	2.59	3.51	1.48	0
16824	Tabouli, tabbouleh, or tabuli	1	cup(s)	160	124	199	3	16	4	15	2.04	10.83	1.37	—
25276	Three bean salad	½	cup(s)	99	82	95	2	10	3	6	0.76	1.41	3.48	0
160	Tuna salad	½	cup(s)	103	65	192	16	10	0	9	1.58	2.96	4.23	0
25241	Turkey & noodles	1	cup(s)	319	228	271	24	21	1	9	2.39	3.48	2.27	0
16794	Vegetable egg roll	2	item(s)	128	90	202	5	20	2	12	2.46	5.71	2.65	—
16818	Vegetable sushi, no fish	6	piece(s)	156	99	225	5	50	2	<1	0.11	0.10	0.14	—
	Sandwiches													
1744	Bacon, lettuce & tomato w/mayonnaise	1	item(s)	164	97	349	11	34	2	19	4.54	7.22	6.07	—
30287	Bologna & cheese w/margarine	1	item(s)	111	46	350	13	28	1	20	8.55	8.40	2.28	—
30286	Bologna w/margarine	1	item(s)	83	34	256	7	26	1	13	4.08	6.31	2.07	—
16546	Cheese	1	item(s)	83	31	262	10	27	1	13	5.59	4.77	1.67	—
8789	Cheeseburger, large, plain	1	item(s)	185	72	609	30	47	0	33	14.84	12.74	2.44	—
8624	Cheeseburger, large, w/bacon, vegetables, & condiments	1	item(s)	195	85	608	32	37	2	37	16.24	14.49	2.71	—
1745	Club w/bacon, chicken, tomato, lettuce, & mayonnaise	1	item(s)	246	137	555	31	48	3	26	5.94	—	—	—
1908	Cold cut submarine w/cheese & vegetables	1	item(s)	228	132	456	22	51	2	19	6.81	8.23	2.28	—
30247	Corned beef	1	item(s)	130	75	268	19	25	2	10	3.75	3.96	0.80	—
25283	Egg salad	1	item(s)	126	72	278	10	29	1	13	2.96	3.97	4.79	—
16686	Fried egg	1	item(s)	96	50	226	10	26	1	9	2.29	3.51	1.64	—
16547	Grilled cheese	1	item(s)	83	27	292	10	27	1	16	6.22	6.29	2.54	—
16659	Gyro w/onion & tomato	1	item(s)	105	67	170	12	21	1	4	1.53	1.41	0.43	—
1906	Ham & cheese	1	item(s)	146	74	352	21	33	2	15	6.44	6.74	1.38	—
31890	Ham w/mayonnaise	1	item(s)	112	55	282	14	27	1	13	3.06	5.04	3.79	—
756	Hamburger, double patty, large, w/condiments & vegetables	1	item(s)	226	121	540	34	40	0	27	10.52	10.33	2.80	—
8793	Hamburger, large, plain	1	item(s)	137	58	426	23	32	2	23	8.38	9.88	2.14	—
8795	Hamburger, large, w/vegetables & condiments	1	item(s)	218	121	512	26	40	3	27	10.42	11.42	2.20	—

Chol (mg)	Calc (mg)	Iron (mg)	Magn (mg)	Pota (mg)	Sodi (mg)	Zinc (mg)	Vit A (µg)	Thia (mg)	Vit E (mg α)	Ribo (mg)	Niac (mg)	Vit B$_6$ (mg)	Fola (µg)	Vit C (mg)	Vit B$_{12}$ (µg)	Sele (µg)
89	36	2.97	40	495	495	6.62	—	0.09	1.30	0.28	8.05	0.20	27	1	3	—
68	222	2.33	40	437	493	2.81	108	0.19	0.22	0.29	3.02	0.20	50	10	1	22
30	323	2.26	42	263	800	1.95	327	0.25	0.72	0.40	2.18	0.10	12	<1	<1	—
85	54	2.09	21	278	423	3.55	27	0.08	0.00	0.29	3.77	0.13	20	<1	2	17
172	30	1.45	26	330	1078	1.83	—	0.50	5.39	0.38	2.90	0.31	22	8	1	—
82	699	6.71	205	1067	1611	7.55	—	0.31	7.71	0.50	5.62	0.85	59	14	1	—
28	129	1.87	17	305	534	1.04	105	0.27	—	0.47	6.09	0.11	74	3	<1	26
85	24	0.81	19	318	661	0.39	40	0.10	—	0.08	1.11	0.18	9	13	0	5
17	28	2.04	23	337	1354	1.19	—	0.22	0.70	0.20	2.88	0.14	17	22	<1	—
60	64	2.21	23	282	370	3.66	27	0.11	0.00	0.30	4.00	0.13	22	<1	2	17
181	101	4.44	64	439	381	1.73	—	0.29	2.07	0.10	4.77	0.22	12	18	1	—
89	124	3.70	—	665	1009	—	82	0.25	—	0.30	4.00	—	—	22	—	22
37	32	1.58	50	187	598	1.08	38	0.18	0.36	0.13	1.88	0.17	44	22	<.1	3
46	39	2.65	32	394	574	2.07	80	0.51	0.38	0.20	5.07	0.30	102	18	<1	23
35	241	3.18	63	462	543	1.68	280	0.32	0.00	0.36	4.64	0.30	109	15	<1	36
217	42	1.63	18	128	527	0.98	—	0.12	0.67	0.29	1.33	0.13	29	2	<1	—
11	24	2.18	25	204	340	0.79	—	0.26	0.25	0.07	2.77	0.15	14	4	<1	—
0	20	1.54	20	99	153	0.70	—	0.20	0.12	0.04	1.86	0.14	10	2	0	—
74	41	1.78	35	622	624	2.53	64	0.80	0.20	0.37	6.69	0.65	14	10	1	50
0	29	1.25	36	246	799	0.48	—	0.08	2.43	0.05	1.14	0.11	31	29	0	—
0	26	0.96	15	144	224	0.31	12	0.04	0.89	0.06	0.26	0.06	31	9	0	3
13	17	1.03	19	182	412	0.57	25	0.03	0.00	0.07	6.87	0.08	8	2	1	42
77	60	2.69	33	379	576	2.64	108	0.23	0.29	0.32	6.40	0.30	60	1	1	34
60	29	1.61	18	193	548	0.51	—	0.16	1.28	0.21	1.59	0.10	27	6	<1	—
0	23	2.40	23	158	369	0.84	—	0.28	0.16	0.06	2.44	0.13	15	4	0	—
20	76	2.54	27	328	837	0.98	—	0.39	1.16	0.27	3.81	0.20	31	15	<1	—
35	221	2.18	24	185	940	1.68	—	0.30	0.56	0.33	2.77	0.12	21	<.1	1	—
16	60	1.96	15	112	598	0.85	—	0.29	0.50	0.21	2.73	0.08	19	<.1	<1	—
19	216	1.75	20	135	655	1.14	—	0.25	0.47	0.29	2.04	0.07	19	<.1	<1	—
96	91	5.46	39	644	1589	5.55	185	0.48	—	0.57	11.17	0.28	74	0	3	39
111	162	4.74	45	332	1043	6.83	82	0.31	—	0.41	6.63	0.31	86	2	2	33
72	116	4.05	47	463	855	1.65	—	0.61	1.53	0.44	11.92	0.59	48	9	1	—
36	189	2.51	68	394	1651	2.58	71	1.00	—	0.80	5.49	0.14	87	12	1	31
46	67	2.67	20	187	1177	2.24	—	0.24	0.21	0.25	3.23	0.10	22	2	1	—
217	107	2.60	18	147	494	0.94	94	0.26	0.13	0.43	2.27	0.16	82	1	1	24
207	80	2.25	17	120	433	0.85	—	0.27	0.66	0.41	2.06	0.10	34	0	<1	—
19	219	1.76	21	137	696	1.15	—	0.19	0.72	0.28	1.86	0.06	13	<.1	<1	—
34	46	1.85	21	209	272	2.30	—	0.24	0.26	0.21	3.14	0.13	18	4	1	—
58	130	3.24	16	291	771	1.37	96	0.31	0.29	0.48	2.69	0.20	76	3	1	23
36	59	2.10	23	245	1033	1.50	—	0.71	0.50	0.31	4.89	0.26	19	0	<1	—
122	102	5.85	50	570	791	5.67	5	0.36	—	0.38	7.57	0.54	77	1	4	26
71	74	3.58	27	267	474	4.11	0	0.29	—	0.29	6.25	0.23	60	0	2	27
87	96	4.93	44	480	824	4.88	24	0.41	—	0.37	7.28	0.33	83	3	2	34

Table A–1
Food Composition

(DA+ code is for Wadsworth Diet Analysis program)
(For purposes of calculations, use "0" for t, <1, <.1, <.01, etc.)

DA + Code	Food Description	Quantity	Measure	Wt (g)	H₂O (g)	Ener (kcal)	Prot (g)	Carb (g)	Fiber (g)	Fat (g)	Fat Breakdown (g)			
											Sat	Mono	Poly	Trans
	MIXED FOODS, SOUPS, SANDWICHES													
	Sandwiches–Continued													
25134	Hot chicken salad	1	item(s)	98	49	239	16	23	1	9	2.83	2.61	2.76	0
1411	Hot dog w/bun, plain	1	item(s)	98	53	242	10	18	2	15	5.11	6.85	1.71	—
25133	Hot turkey salad	1	item(s)	98	50	221	16	23	1	7	2.23	1.76	2.28	0
30249	Pastrami	1	item(s)	134	71	331	14	27	2	18	6.18	8.74	1.02	—
16701	Peanut butter	1	item(s)	93	24	344	13	37	3	17	3.55	8.16	4.58	—
30306	Peanut butter & jelly	1	item(s)	93	24	330	11	42	3	15	3.00	6.87	3.82	—
1910	Roast beef, plain	1	item(s)	139	68	346	22	33	1	14	3.61	6.80	1.71	—
1909	Roast beef submarine w/mayonnaise & vegetables	1	item(s)	216	127	410	29	44	—	13	7.09	1.84	2.61	—
1907	Steak w/mayonnaise & vegetables	1	item(s)	204	104	459	30	52	2	14	3.81	5.34	3.35	—
25288	Tuna salad	1	item(s)	179	102	414	24	29	2	22	3.61	5.46	11.43	—
31891	Turkey w/mayonnaise	1	item(s)	143	75	330	29	26	1	11	2.61	3.25	4.40	—
30283	Turkey submarine w/cheese, lettuce, tomato, & mayonnaise	1	item(s)	277	156	583	37	51	3	25	7.15	8.03	7.81	—
	Soups													
25296	Bean	1	cup(s)	301	253	191	14	29	6	2	0.67	0.83	0.53	0
711	Bean with pork, condensed, prepared w/water	1	cup(s)	265	223	180	8	24	9	6	1.59	2.28	1.91	—
713	Beef noodle, condensed, prepared w/water	1	cup(s)	244	224	83	5	9	1	3	1.15	1.24	0.49	—
825	Cheese, condensed, prepared w/milk	1	cup(s)	251	207	231	9	16	1	15	9.11	4.09	0.45	—
826	Chicken broth, condensed, prepared w/water	1	cup(s)	244	234	39	5	1	0	1	0.39	0.59	0.27	—
25297	Chicken noodle	1	cup(s)	286	258	117	11	11	1	3	0.78	1.10	0.66	—
827	Chicken noodle, condensed, prepared w/water	1	cup(s)	241	222	75	4	9	1	2	0.65	1.11	0.55	—
724	Chicken noodle, dehydrated, prepared w/water	1	cup(s)	252	237	58	2	9	<1	1	0.31	0.52	0.39	—
823	Cream of asparagus, condensed, prepared w/milk	1	cup(s)	248	213	161	6	16	1	8	3.32	2.08	2.23	—
824	Cream of celery, condensed, prepared w/milk	1	cup(s)	248	214	164	6	15	1	10	3.94	2.46	2.65	—
708	Cream of chicken, condensed, prepared w/milk	1	cup(s)	248	210	191	7	15	<1	11	4.64	4.46	1.64	—
715	Cream of chicken, condensed, prepared w/water	1	cup(s)	244	221	117	3	9	<1	7	2.07	3.27	1.49	—
709	Cream of mushroom, condensed, prepared w/milk	1	cup(s)	248	210	203	6	15	<1	14	5.13	2.98	4.61	—
716	Cream of mushroom, condensed, prepared w/water	1	cup(s)	244	220	129	2	9	<1	9	2.44	1.71	4.22	—
25298	Cream of vegetable	1	cup(s)	285	251	165	7	15	2	9	1.56	4.62	1.92	—
16689	Egg drop	1	cup(s)	244	229	73	8	1	0	4	1.15	1.52	0.59	—
25138	Golden squash	1	cup(s)	258	224	144	8	21	2	4	0.84	2.18	0.88	0
16663	Hot & sour	1	cup(s)	244	210	161	15	5	1	8	2.72	3.40	1.20	—
28054	Lentil chowder	1	cup(s)	229	188	150	11	27	12	<1	0.09	0.08	0.22	0
28560	Macaroni & bean	1	cup(s)	229	129	136	6	21	5	3	0.48	2.06	0.59	0
714	Manhattan clam chowder, condensed, prepared w/water	1	cup(s)	244	224	78	2	12	1	2	0.38	0.38	1.29	—
28561	Minestrone	1	cup(s)	230	177	99	4	16	5	2	0.32	1.30	0.43	0
717	Minestrone, condensed, prepared w/water	1	cup(s)	241	220	82	4	11	1	3	0.55	0.70	1.11	—
28038	Mushroom & wild rice	1	cup(s)	230	188	81	4	12	2	<1	0.05	0.02	0.15	0
828	New England clam chowder, condensed, prepared w/milk	1	cup(s)	248	211	164	9	17	1	7	2.95	2.26	1.09	—

Chol (mg)	Calc (mg)	Iron (mg)	Magn (mg)	Pota (mg)	Sodi (mg)	Zinc (mg)	Vit A (µg)	Thia (mg)	Vit E (mg α)	Ribo (mg)	Niac (mg)	Vit B$_6$ (mg)	Fola (µg)	Vit C (mg)	Vit B$_{12}$ (µg)	Sele (µg)
39	114	1.93	20	150	470	1.22	28	0.20	0.28	0.23	4.93	0.20	54	<1	<1	17
44	24	2.31	13	143	670	1.98	0	0.24	—	0.27	3.65	0.05	48	<.1	1	26
37	113	2.04	22	167	459	1.09	23	0.19	0.29	0.21	4.36	0.23	54	<1	<1	20
51	68	2.64	23	243	1335	2.69	—	0.29	0.27	0.27	4.77	0.13	21	2	1	—
1	80	2.47	62	272	479	1.25	0	0.33	2.39	0.25	6.46	0.17	43	0	<.1	—
1	68	2.11	53	239	409	1.06	—	0.27	2.02	0.21	5.45	0.15	37	<1	<.1	—
51	54	4.23	31	316	792	3.39	11	0.38	—	0.31	5.87	0.26	57	2	1	29
73	41	2.81	67	330	845	4.38	30	0.41	—	0.41	5.96	0.32	71	6	2	26
73	92	5.16	49	524	798	4.53	20	0.41	—	0.37	7.30	0.37	90	6	2	42
53	100	3.29	35	302	795	1.08	46	0.26	0.35	0.26	12.29	0.48	70	1	2	71
69	78	3.10	34	315	490	2.94	—	0.30	0.74	0.33	6.64	0.46	24	0	<1	—
70	324	3.88	51	552	2408	2.66	—	0.53	1.19	0.49	12.50	0.54	46	5	2	—
5	80	3.08	61	590	690	1.41	26	0.27	0.03	0.15	3.61	0.23	139	3	<1	8
3	85	2.15	48	421	996	1.09	48	0.09	0.80	0.03	0.59	0.04	34	2	<.1	8
5	15	1.10	5	100	952	1.54	7	0.07	0.68	0.06	1.07	0.04	20	<1	<1	7
48	289	0.80	20	341	1019	0.68	359	0.06	—	0.33	0.50	0.08	10	1	<1	7
0	10	0.51	2	210	776	0.24	0	0.01	0.05	0.07	3.35	0.02	5	0	<1	0
24	26	1.34	16	335	776	0.77	49	0.15	0.02	0.16	5.57	0.13	40	1	<1	10
7	17	0.77	5	55	1106	0.39	36	0.05	0.10	0.06	1.39	0.03	22	<1	<1	6
10	5	0.50	8	33	577	0.20	3	0.20	0.13	0.08	1.09	0.03	18	0	<.1	10
22	174	0.87	20	360	1042	0.92	62	0.10	—	0.28	0.88	0.06	30	4	<1	8
32	186	0.69	22	310	1009	0.20	114	0.07	—	0.25	0.44	0.06	7	1	<1	5
27	181	0.67	17	273	1047	0.67	179	0.07	—	0.26	0.92	0.07	7	1	1	8
10	34	0.61	2	88	986	0.63	163	0.03	—	0.06	0.82	0.02	2	<1	<.1	7
20	179	0.60	20	270	918	0.64	35	0.08	1.24	0.28	0.91	0.06	10	2	<1	4
2	46	0.51	5	100	881	0.59	15	0.05	0.95	0.09	0.72	0.01	5	1	<.1	1
1	68	1.38	17	312	784	0.74	100	0.12	1.06	0.20	3.27	0.12	37	10	<1	5
103	21	0.75	5	220	729	0.48	—	0.02	0.29	0.19	3.03	0.05	15	0	<1	—
4	203	1.63	39	412	500	1.72	454	0.17	0.53	0.38	1.15	0.15	32	10	1	8
34	29	1.89	29	382	1561	1.51	—	0.27	0.12	0.25	4.97	0.20	13	1	<1	—
<1	47	4.07	55	590	26	1.44	163	0.21	0.06	0.12	1.69	0.30	164	13	0	3
<1	64	1.86	32	254	489	0.46	174	0.15	0.35	0.13	1.36	0.09	59	7	0	9
2	27	1.63	12	188	578	0.98	56	0.03	0.34	0.04	0.82	0.10	10	4	4	9
0	68	1.76	31	273	423	0.38	138	0.10	0.23	0.10	0.69	0.07	47	12	0	4
2	34	0.92	7	313	911	0.75	118	0.05	—	0.04	0.94	0.10	36	1	0	8
0	27	1.08	26	332	267	0.87	4	0.06	0.07	0.21	2.97	0.14	18	4	<.1	4
22	186	1.49	22	300	992	0.79	57	0.07	0.45	0.24	1.03	0.13	10	3	10	13

Table A–1
Food Composition

(DA+ code is for Wadsworth Diet Analysis program)
(For purposes of calculations, use "0" for t, <1, <.1, <.01, etc.)

DA + Code	Food Description	Quantity	Measure	Wt (g)	H₂O (g)	Ener (kcal)	Prot (g)	Carb (g)	Fiber (g)	Fat (g)	Sat	Fat Breakdown (g) Mono	Poly	Trans
	MIXED FOODS, SOUPS, SANDWICHE													
	Soups–Continued													
28036	New England style clam chowder	1	cup(s)	229	207	83	3	15	2	<1	0.08	0.03	0.05	0
28566	Old country pasta	1	cup(s)	228	164	135	6	20	3	3	1.17	1.60	0.63	0
725	Onion, dehydrated, prepared w/water	1	cup(s)	246	237	27	1	5	1	1	0.12	0.32	0.07	—
16667	Shrimp gumbo	1	cup(s)	244	206	171	10	19	3	7	1.34	3.02	2.05	—
28037	Southwestern corn chowder	1	cup(s)	229	202	102	5	18	2	<1	0.12	0.12	0.20	0
25140	Split pea	1	cup(s)	165	117	85	4	19	2	<1	0.07	0.03	0.18	0
718	Split pea with ham, condensed, prepared w/water	1	cup(s)	253	207	190	10	28	2	4	1.77	1.80	0.63	—
710	Tomato, condensed, prepared w/milk	1	cup(s)	248	210	161	6	22	3	6	2.90	1.61	1.12	—
719	Tomato, condensed, prepared w/water	1	cup(s)	244	220	85	2	17	<1	2	0.37	0.44	0.95	—
726	Tomato vegetable, dehydrated, prepared w/water	1	cup(s)	253	237	56	2	10	1	1	0.38	0.30	0.08	—
28595	Turkey noodle	1	cup(s)	228	203	106	8	14	2	2	0.27	1.06	0.67	0
28051	Turkey vegetable	1	cup(s)	227	203	98	11	8	2	1	0.32	0.17	0.30	0
720	Vegetable beef, condensed, prepared w/water	1	cup(s)	244	224	78	6	10	<1	2	0.85	0.81	0.12	—
28598	Vegetable gumbo	1	cup(s)	229	168	153	4	26	3	4	0.61	2.93	0.56	0
25141	Vegetable	1	cup(s)	252	225	96	5	20	4	—	0.06	0.04	0.16	0
721	Vegetarian vegetable, condensed, prepared w/water	1	cup(s)	241	223	72	2	12	—	2	0.29	0.82	0.72	
	FAST FOOD													
	Arby's													
36094	Au jus sauce	1	serving(s)	85	—	5	<1	1	<.1	<.1	0.02	—	—	—
751	Beef 'n cheddar sandwich	1	item(s)	198	—	480	23	43	2	24	8.00	—	—	—
9279	Cheddar curly fries	1	serving(s)	170	—	460	6	54	4	24	6.00	—	—	—
36131	Chocolate shake	1	serving(s)	397	—	480	10	84	0	16	8.00	—	—	—
36045	Curly fries, large	1	serving(s)	198	—	620	8	78	7	30	7.00	—	—	—
36044	Curly fries, medium	1	serving(s)	128	—	400	5	50	4	20	5.00	—	—	—
9265	Fish fillet sandwich	1	item(s)	220	—	529	23	50	2	27	7.00	9.20	10.60	—
752	Ham 'n cheese sandwich	1	item(s)	170	—	340	23	35	1	13	4.50	—	—	—
36048	Homestyle fries, large	1	serving(s)	213	—	560	6	79	6	24	6.00	—	—	—
36047	Homestyle fries, medium	1	serving(s)	142	—	370	4	53	4	16	4.00	—	—	—
33465	Homestyle fries, small	1	serving(s)	113	—	300	3	42	3	13	3.50	—	—	—
9267	Italian sub sandwich	1	item(s)	312	—	780	29	49	3	53	15.00	—	—	—
36041	Market Fresh grilled chicken caesar salad w/o dressing	1	serving(s)	338	—	230	33	8	3	8	3.50	—	—	—
9291	Roast beef deluxe sandwich, light	1	item(s)	182	—	296	18	33	6	10	3.00	5.00	2.00	—
9251	Roast beef sandwich, giant	1	item(s)	228	—	480	32	41	3	23	10.00	—	—	—
9249	Roast beef sandwich, junior	1	item(s)	129	—	310	16	34	2	13	4.50	—	—	—
750	Roast beef sandwich, regular	1	item(s)	157	—	350	21	34	2	16	6.00	—	—	—
2009	Roast beef sandwich, super	1	item(s)	245	—	470	22	47	3	23	7.00	—	—	—
9269	Roast beef sub sandwich	1	item(s)	334	—	760	35	47	3	48	16.00	—	—	—
9295	Roast chicken deluxe sandwich, light	1	item(s)	194	—	260	23	33	3	5	1.00	—	—	—
9293	Roast turkey deluxe sandwich, light	1	item(s)	194	—	260	23	33	3	5	0.50	—	—	—
36132	Strawberry shake	1	serving(s)	397	—	500	11	87	0	13	8.00	—	—	—
9273	Turkey sub sandwich	1	item(s)	306	—	630	26	51	2	37	9.00	—	—	—
36130	Vanilla shake	1	serving(s)	397	—	470	10	83	0	15	7.00	—	—	—
	Auntie Anne's													
35371	Cheese dipping sauce	1	serving(s)	35	—	100	3	4	0	8	4.00	—	—	—
35353	Cinnamon sugar soft pretzel	1	item(s)	120	—	350	9	74	2	2	0.00	—	—	—
35354	Cinnamon sugar soft pretzel w/butter	1	item(s)	120	—	450	8	83	3	9	5.00	—	—	—

A

Chol (mg)	Calc (mg)	Iron (mg)	Magn (mg)	Pota (mg)	Sodi (mg)	Zinc (mg)	Vit A (µg)	Thia (mg)	Vit E (mg α)	Ribo (mg)	Niac (mg)	Vit B₆ (mg)	Fola (µg)	Vit C (mg)	Vit B₁₂ (µg)	Sele (µg)
2	69	1.29	26	430	236	0.66	34	0.07	0.02	0.12	1.02	0.20	17	12	3	4
6	51	2.32	47	434	319	0.69	114	0.20	0.01	0.15	2.42	0.23	65	17	<.1	9
0	12	0.15	5	64	849	0.05	0	0.03	0.00	0.06	0.48	0.00	2	<1	0	2
51	99	2.34	51	515	515	0.93	—	0.19	1.90	0.10	2.54	0.19	59	26	<1	—
1	65	1.10	24	374	200	0.73	46	0.08	0.09	0.14	1.65	0.22	27	37	<1	2
0	30	1.25	33	352	608	0.57	112	0.12	0.00	0.09	1.67	0.21	61	9	0	<1
8	23	2.28	48	400	1007	1.32	23	0.15	—	0.08	1.47	0.07	3	2	<1	8
17	159	1.81	22	449	744	0.30	64	0.13	1.24	0.25	1.52	0.16	17	68	<1	2
0	12	1.76	7	264	695	0.24	29	0.09	2.32	0.05	1.42	0.11	15	66	0	<1
0	8	0.63	20	104	1146	0.18	10	0.06	0.35	0.05	0.79	0.05	10	6	0	5
24	27	1.40	22	200	372	0.67	81	0.20	0.02	0.11	2.68	0.15	45	5	<1	13
20	36	1.30	22	383	328	0.90	110	0.08	0.01	0.09	3.33	0.27	21	10	<1	9
5	17	1.12	5	173	791	1.54	95	0.04	0.37	0.05	1.03	0.08	10	2	<1	4
0	52	1.90	35	313	471	0.56	15	0.17	0.58	0.07	1.59	0.16	51	18	0	4
0	41	2.45	38	688	674	0.78	118	0.12	0.00	0.13	2.37	0.27	33	23	0	5
0	22	1.08	7	210	822	0.46	116	0.05	—	0.05	0.92	0.06	10	1	0	4
0	0	0.00	—	—	386	—	0	—	—	—	—	—	—	0	—	—
90	100	3.60	—	—	1240	—	0	—	—	—	—	—	—	1	—	—
5	60	1.80	—	—	1290	—	0	—	—	—	—	—	—	15	—	—
45	500	0.72	—	—	370	—	38	—	—	—	—	—	—	2	—	—
0	0	2.70	—	—	1540	—	0	—	—	—	—	—	—	21	—	—
0	0	1.80	—	—	990	—	0	—	—	—	—	—	—	15	—	—
43	90	3.78	—	450	864	—	10	0.35	—	0.31	5.60	—	—	1	—	—
90	150	2.70	—	—	1450	—	20	—	—	—	—	—	—	1	—	—
0	0	1.80	—	—	1070	—	0	—	—	—	—	—	—	30	—	—
0	0	1.08	—	—	710	—	0	—	—	—	—	—	—	21	—	—
0	0	0.72	—	—	570	—	0	—	—	—	—	—	—	15	—	—
120	250	2.70	—	—	2440	—	—	—	—	—	—	—	—	2	—	—
80	200	1.80	—	—	920	—	—	—	—	—	—	—	—	42	—	—
42	130	4.50	—	392	826	—	40	0.27	—	0.49	8.40	—	—	8	—	—
110	60	5.40	—	—	1440	—	0	—	—	—	—	—	—	0	—	—
70	60	2.70	—	—	740	—	0	—	—	—	—	—	—	0	—	—
85	60	3.60	—	—	950	—	0	—	—	—	—	—	—	0	—	—
85	80	3.60	—	—	1130	—	40	—	—	—	—	—	—	1	—	—
130	300	4.50	—	—	2230	—	40	—	—	—	—	—	—	4	—	—
40	100	2.70	—	—	1010	—	—	—	—	—	—	—	—	2	—	—
40	80	1.80	—	—	980	—	—	—	—	—	—	—	—	1	—	—
15	350	0.36	—	—	340	—	36	—	—	—	—	—	—	1	—	—
100	200	0.36	—	—	2170	—	—	—	—	—	—	—	—	2	—	—
45	500	1.08	—	—	360	—	39	—	—	—	—	—	—	2	—	—
10	100	0.00	—	—	510	—	—	—	—	—	—	—	—	0	—	—
0	20	1.98	—	—	410	—	0	—	—	—	—	—	—	0	—	—
25	30	2.34	—	—	430	—	—	—	—	—	—	—	—	0	—	—

Table A–1
Food Composition

(DA+ code is for Wadsworth Diet Analysis program)
(For purposes of calculations, use "0" for t, <1, <.1, <.01, etc.)

A

DA + Code	Food Description	Quantity	Measure	Wt (g)	H₂O (g)	Ener (kcal)	Prot (g)	Carb (g)	Fiber (g)	Fat (g)	Sat	Fat Breakdown (g)		
												Mono	Poly	Trans
	FAST FOOD													
	Auntie Annee–Continued													
35372	Marinara dipping sauce	1	serving(s)	35	—	10	0	4	0	0	0.00	0.00	0.00	0
35357	Original soft pretzel	1	item(s)	120	—	340	10	72	3	1	0.00	—	—	—
35358	Original soft pretzel w/butter	1	item(s)	120	—	370	10	72	3	4	2.00	—	—	—
35359	Parmesan herb soft pretzel	1	item(s)	120	—	390	11	74	4	5	2.50	—	—	—
35360	Parmesan herb soft pretzel w/butter	1	item(s)	120	—	440	10	72	9	13	7.00	—	—	—
35361	Sesame soft pretzel	1	item(s)	120	—	350	11	63	3	6	1.00	—	—	—
35362	Sesame soft pretzel w/butter	1	item(s)	120	—	410	12	64	7	12	4.00	—	—	—
35364	Sour cream & onion soft pretzel	1	item(s)	120	—	310	9	66	2	1	0.00	—	—	—
35366	Sour cream & onion soft pretzel w/butter	1	item(s)	120	—	340	9	66	2	5	3.00	—	—	—
35373	Sweet mustard dipping sauce	1	serving(s)	35	—	60	1	8	0	2	1.00	—	—	—
35367	Whole wheat soft pretzel	1	item(s)	120	—	350	11	72	7	2	0.00	—	—	—
35368	Whole wheat soft pretzel w/butter	1	item(s)	120	—	370	11	72	7	5	1.50	—	—	—
	Boston Market													
34975	Bbq baked beans	¾	cup(s)	201	—	270	8	48	12	5	2.00	—	—	—
34976	Black beans & rice	1	cup(s)	227	—	300	8	45	5	10	1.50	—	—	—
34978	Butternut squash	¾	cup(s)	193	—	150	2	25	6	6	4.00	—	—	—
35006	Caesar side salad	1	serving(s)	119	—	300	5	13	1	26	4.50	—	—	—
34979	Chicken gravy	1	ounce(s)	28	—	15	0	2	0	1	0.00	—	—	—
34973	Chicken pot pie	1	item(s)	425	—	750	26	57	2	46	14.00	—	—	—
35007	Cole slaw	¾	cup(s)	184	—	300	2	30	3	19	3.00	—	—	—
35057	Cornbread	1	item(s)	68	—	200	3	33	1	6	1.50	—	—	—
35008	Cranberry walnut relish	¾	cup(s)	210	—	350	3	75	3	5	0.00	—	—	—
34980	Creamed spinach	¾	cup(s)	181	—	260	9	11	2	20	13.00	—	—	—
34981	Glazed carrots	¾	cup(s)	153	—	280	1	35	4	15	3.00	—	—	—
34983	Green bean casserole	¾	cup(s)	170	—	80	1	9	2	5	1.50	—	—	—
34982	Green beans	¾	cup(s)	85	—	70	1	6	2	4	0.50	—	—	—
34967	Half chicken, w/skin	1	item(s)	277	—	590	70	4	0	33	10.00	—	—	—
34984	Homestyle mashed potatoes	¾	cup(s)	173	—	210	4	30	2	9	5.00	—	—	—
34985	Homestyle mashed potatoes & gravy	1	cup(s)	201	—	230	4	32	3	9	5.00	—	—	—
34969	Honey glazed ham	5	ounce(s)	142	—	210	24	10	0	8	3.00	—	—	—
34988	Hot cinnamon apples	¾	cup(s)	181	—	250	0	56	3	5	0.50	—	—	—
34989	Macaroni & cheese	¾	cup(s)	192	—	280	13	33	1	11	6.00	—	—	—
34970	Meatloaf	5	ounce(s)	142	—	282	20	15	1	17	7.28	—	—	—
35012	Old-fashioned potato salad	¾	cup(s)	150	—	200	3	22	2	12	2.00	—	—	—
34965	Quarter chicken, dark meat, no skin	1	item(s)	95	—	190	22	1	0	10	3.00	—	—	—
34966	Quarter chicken, dark meat, w/skin	1	item(s)	125	—	320	30	2	0	21	6.00	—	—	—
34963	Quarter chicken, white meat, no skin or wing	1	item(s)	140	—	170	33	2	0	4	1.00	—	—	—
34964	Quarter chicken, white meat, w/skin & wing	1	item(s)	152	—	280	40	2	0	12	3.50	—	—	—
34993	Rice pilaf	1	cup(s)	137	—	140	2	24	1	4	0.50	—	—	—
34968	Rotisserie turkey breast, skinless	5	ounce(s)	142	—	170	36	3	0	1	0.00	—	—	—
34998	Savory stuffing	1	cup(s)	132	—	190	4	27	2	8	1.50	—	—	—
34999	Squash casserole	¾	cup(s)	187	—	330	7	20	3	24	13.00	—	—	—
35003	Steamed vegetables	1	cup(s)	102	—	30	2	6	2	0	0.00	—	—	0
35004	Sweet potato casserole	¾	cup(s)	181	—	280	3	39	2	13	4.50	—	—	—
35005	Whole kernel corn	¾	cup(s)	146	—	180	5	30	2	4	0.50	—	—	—
	Burger King													
29731	Biscuit with sausage, egg, & cheese	1	item(s)	189	—	650	20	38	1	46	14.00	—	—	1
3739	BK Broiler chicken sandwich	1	item(s)	258	—	550	30	52	3	25	5.00	—	—	—
14249	Cheeseburger	1	item(s)	133	—	360	19	31	2	17	8.00	—	—	0.50

Chol (mg)	Calc (mg)	Iron (mg)	Magn (mg)	Pota (mg)	Sodi (mg)	Zinc (mg)	Vit A (µg)	Thia (mg)	Vit E (mg α)	Ribo (mg)	Niac (mg)	Vit B$_6$ (mg)	Fola (µg)	Vit C (mg)	Vit B$_{12}$ (µg)	Sele (µg)
0	0	0.00	—	—	180	—	0	—	—	—	—	—	—	0	—	—
0	30	2.34	—	—	900	—	0	—	—	—	—	—	—	0	—	—
10	30	2.16	—	—	930	—	—	—	—	—	—	—	—	0	—	—
10	80	1.80	—	—	780	—	—	—	—	—	—	—	—	1	—	—
30	60	1.80	—	—	660	—	—	—	—	—	—	—	—	1	—	—
0	20	2.88	—	—	840	—	0	—	—	—	—	—	—	0	—	—
15	20	2.70	—	—	860	—	—	—	—	—	—	—	—	0	—	—
0	30	1.98	—	—	920	—	—	—	—	—	—	—	—	0	—	—
10	40	2.16	—	—	930	—	—	—	—	—	—	—	—	0	—	—
40	0	0.00	—	—	120	—	0	—	—	—	—	—	—	0	—	—
0	30	1.98	—	—	1100	—	0	—	—	—	—	—	—	0	—	—
10	30	2.34	—	—	1120	—	—	—	—	—	—	—	—	0	—	—
0	100	3.60	—	—	540	—	42	—	—	—	—	—	—	6	—	—
0	40	1.80	—	—	1050	—	0	—	—	—	—	—	—	4	—	—
20	80	1.08	—	—	560	—	1150	—	—	—	—	—	—	30	—	—
15	100	0.72	—	—	690	—	—	—	—	—	—	—	—	9	—	—
0	0	0.00	—	—	180	—	0	—	—	—	—	—	—	0	—	—
110	40	4.50	—	—	1530	—	—	—	—	—	—	—	—	1	—	—
20	60	0.72	—	—	540	—	108	—	—	—	—	—	—	36	—	—
25	0	1.08	—	—	390	—	0	—	—	—	—	—	—	0	—	—
0	0	5.40	—	—	0	—	0	—	—	—	—	—	—	0	—	—
55	250	2.70	—	—	740	—	—	—	—	—	—	—	—	9	—	—
0	40	1.08	—	—	80	—	1000	—	—	—	—	—	—	1	—	—
5	20	0.72	—	—	670	—	—	—	—	—	—	—	—	2	—	—
0	40	0.36	—	—	250	—	30	—	—	—	—	—	—	5	—	—
290	0	2.70	—	—	1010	—	0	—	—	—	—	—	—	0	—	—
25	40	0.36	—	—	590	—	53	—	—	—	—	—	—	15	—	—
25	60	0.36	—	—	780	—	—	—	—	—	—	—	—	15	—	—
75	0	1.08	—	—	1460	—	0	—	—	—	—	—	—	0	—	—
0	20	0.36	—	—	45	—	—	—	—	—	—	—	—	0	—	—
30	300	1.44	—	—	890	—	—	—	—	—	—	—	—	0	—	—
68	91	2.46	—	—	592	—	—	—	—	—	—	—	—	1	—	—
15	60	1.08	—	—	450	—	0	—	—	—	—	—	—	6	—	—
115	0	1.08	—	—	440	—	0	—	—	—	—	—	—	0	—	—
155	0	1.80	—	—	500	—	0	—	—	—	—	—	—	0	—	—
85	0	0.72	—	—	480	—	0	—	—	—	—	—	—	0	—	—
135	0	1.08	—	—	510	—	0	—	—	—	—	—	—	0	—	—
0	20	1.08	—	—	520	—	—	—	—	—	—	—	—	4	—	—
100	20	1.80	—	—	850	—	0	—	—	—	—	—	—	0	—	—
5	40	1.44	—	—	620	—	—	—	—	—	—	—	—	2	—	—
70	200	0.72	—	—	1110	—	—	—	—	—	—	—	—	5	—	—
0	40	0.35	—	—	135	—	389	—	—	—	—	—	—	18	—	—
10	40	1.08	—	—	190	—	—	—	—	—	—	—	—	9	—	—
0	0	0.36	—	—	170	—	20	—	—	—	—	—	—	5	—	—
190	150	2.70	—	—	1600	—	90	—	—	—	—	—	—	0	—	—
105	60	3.60	—	—	1110	—	—	0.46	—	0.23	10.50	—	—	6	—	—
50	150	3.60	—	—	790	—	63	0.25	—	0.32	4.18	—	—	1	—	—

Table A–1
Food Composition

(DA+ code is for Wadsworth Diet Analysis program)
(For purposes of calculations, use "0" for t, <1, <.1, <.01, etc.)

DA+ Code	Food Description	Quantity	Measure	Wt (g)	H₂O (g)	Ener (kcal)	Prot (g)	Carb (g)	Fiber (g)	Fat (g)	Sat	Mono	Poly	Trans
	FAST FOOD													
	Burger King–Continued													
14251	Chicken sandwich	1	item(s)	224	—	660	25	53	3	39	8.00	—	—	2.20
3808	Chicken Tenders, 8 pieces	1	serving(s)	123	—	340	22	20	1	19	5.00	—	—	3.50
14259	Chocolate shake, small	1	item(s)	333	—	620	12	72	2	32	21.00	—	—	0
29732	Croissanwich w/sausage & cheese	1	item(s)	107	—	420	14	23	1	31	11.00	—	—	2
14261	Croissanwich w/sausage, egg, & cheese	1	item(s)	157	—	520	19	24	1	39	14.00	—	—	1.93
3809	Double cheeseburger	1	item(s)	189	—	540	32	32	2	31	15.00	—	—	1.50
14244	Double Whopper	1	item(s)	374	—	980	52	52	4	62	22.00	—	—	2
14245	Double Whopper w/cheese	1	item(s)	399	—	1070	57	53	4	70	27.00	—	—	2.50
14250	Fish Fillet sandwich	1	item(s)	185	—	520	18	44	2	30	8.00	—	—	1.12
14255	French fries, medium, salted	1	item(s)	117	—	360	4	46	4	18	5.00	—	—	4.50
14262	French toast sticks	1	serving(s)	112	—	390	6	46	2	20	4.50	—	—	4.50
14248	Hamburger	1	item(s)	121	—	310	17	31	2	13	5.00	—	—	0.50
14263	Hash brown rounds, small	1	serving(s)	75	—	230	2	23	2	15	4.00	—	—	5.0
14256	Onion rings, medium	1	serving(s)	91	—	320	4	40	3	16	4.00	—	—	3.50
39000	Tendercrisp chicken sandwich	1	item(s)	310	—	810	28	72	6	47	8.00	—	—	4.28
14258	Vanilla shake, small	1	item(s)	305	—	560	11	56	1	32	21.00	—	—	0
1736	Whopper	1	item(s)	291	—	710	31	52	4	43	13.00	—	—	1
14243	Whopper w/cheese	1	item(s)	316	—	800	36	53	4	50	18.00	—	—	2
	Carl's Jr													
10801	Carl's Catch fish sandwich	1	item(s)	201	—	530	18	55	2	28	7.00	—	1.89	—
10862	Carl's Famous Star hamburger	1	item(s)	254	—	590	24	50	3	32	9.00	—	—	—
10866	Charboiled chicken salad-to-go	1	item(s)	350	—	200	25	12	4	7	3.00	—	1.02	—
10855	Charboiled Sante Fe chicken sandwich	1	item(s)	220	—	540	28	37	2	31	8.00	—	—	—
10790	Chicken stars (6 pieces)	6	item(s)	90	—	260	13	14	1	16	4.50	—	1.71	—
34864	Chocolate shake, small	1	item(s)	595	—	530	14	96	0	10	7.00	—	—	—
10797	Crisscut fries	1	serving(s)	139	—	410	5	43	4	24	5.00	—	—	—
10799	Double western bacon cheeseburger	1	item(s)	308	—	920	51	65	3	50	21.00	—	6.55	—
34855	Famous bacon cheeseburger	1	item(s)	279	—	700	31	51	3	41	13.00	—	—	—
14238	French fries, small	1	serving(s)	92	—	290	5	37	3	14	3.00	—	—	—
10798	French toast dips w/o syrup	1	serving(s)	105	—	370	6	42	1	20	2.50	—	1.35	—
34856	Hamburger	1	item(s)	119	—	280	14	36	1	9	3.50	—	—	—
10802	Onion rings	1	serving(s)	127	—	430	7	53	3	22	5.00	—	0.84	—
38925	Six Dollar burger	1	item(s)	539	—	1000	39	72	6	82	25.00	—	—	—
34858	Spicy chicken sandwich	1	item(s)	198	—	480	14	47	2	26	5.00	—	—	—
34867	Strawberry shake, small	1	item(s)	595	—	510	14	91	0	10	7.00	—	—	—
10865	Super Star hamburger	1	item(s)	345	—	790	41	51	3	47	15.00	—	—	—
10818	Vanilla shake, small	1	item(s)	595	—	470	15	78	0	11	7.00	—	—	—
10770	Western bacon cheeseburger	1	item(s)	225	—	660	31	64	3	30	12.00	—	4.85	—
	Chick Fil-A													
38746	Biscuit w/bacon, egg, & cheese	1	item(s)	155	—	430	16	38	1	24	9.00	—	—	2.85
38747	Biscuit w/egg	1	item(s)	135	—	340	11	38	1	16	4.50	—	—	3
38748	Biscuit w/egg & cheese	1	item(s)	148	—	390	13	38	1	21	7.00	—	—	2.98
38753	Biscuit w/gravy	1	item(s)	191	—	310	5	44	1	13	3.50	—	—	3.98
38752	Biscuit w/sausage, egg, & cheese	1	item(s)	189	—	540	18	43	1	33	13.00	—	—	2.67
38741	Biscuit, plain	1	item(s)	78	—	260	4	38	1	11	2.50	—	—	2.97
38771	Carrot & raisin salad	1	item(s)	91	—	130	1	22	2	5	1.00	—	—	0
38761	Chargrilled chicken cool wrap	1	item(s)	245	—	380	29	54	3	6	3.00	—	—	0
38766	Chargrilled chicken garden salad	1	item(s)	275	—	180	22	9	3	6	3.00	—	—	0
38758	Chargrilled chicken sandwich	1	item(s)	157	—	280	26	30	1	7	1.50	—	—	0
38759	Chargrilled deluxe chicken sandwich	1	item(s)	195	—	290	27	31	2	7	1.50	—	—	0

Chol (mg)	Calc (mg)	Iron (mg)	Magn (mg)	Pota (mg)	Sodi (mg)	Zinc (mg)	Vit A (µg)	Thia (mg)	Vit E (mg α)	Ribo (mg)	Niac (mg)	Vit B$_6$ (mg)	Fola (µg)	Vit C (mg)	Vit B$_{12}$ (µg)	Sele (µg)
70	80	2.70	—	—	1330	—	—	0.47	—	0.30	9.59	—	—	0	—	—
50	20	0.72	—	—	840	—	—	0.14	—	0.12	10.93	—	—	0	—	—
95	350	1.08	—	—	310	—	42	0.11	—	0.56	0.24	—	—	0	—	—
45	100	3.60	—	—	840	—	—	—	—	—	—	—	—	0	—	—
210	300	4.50	—	—	1090	—	140	0.36	—	0.42	4.35	—	—	0	—	—
100	250	4.50	—	—	1050	—	100	0.26	—	0.45	6.37	—	—	1	—	—
160	150	9.00	—	—	1070	—	—	0.40	—	0.60	11.08	—	—	9	—	—
185	300	9.00	—	—	1500	—	—	0.40	—	0.67	11.07	—	—	9	—	—
55	150	2.70	—	—	840	—	14	—	—	—	—	—	—	1	—	—
0	20	0.72	—	—	640	—	0	0.16	—	0.48	2.32	—	—	9	—	—
0	60	1.80	—	—	440	—	0	0.19	—	0.22	2.86	—	—	0	—	—
40	76	3.60	—	—	580	—	9	0.25	—	0.29	4.26	—	—	1	—	—
0	0	0.36	—	—	450	—	0	0.11	—	0.07	2.11	—	—	1	—	—
0	97	0.00	—	—	460	—	0	0.14	—	0.09	2.33	—	—	0	—	—
60	80	4.50	—	—	1800	—	—	—	—	—	—	—	—	9	—	—
95	300	0.36	—	—	220	—	39	0.11	—	0.64	0.22	—	—	0	—	—
85	150	6.30	—	—	980	—	52	0.39	—	0.44	7.33	—	—	9	—	—
110	250	6.30	—	—	1420	—	157	0.39	—	0.51	7.31	—	—	9	—	—
80	150	1.80	—	—	1030	—	60	—	—	—	—	—	—	2	—	—
70	100	4.50	—	—	910	—	—	—	—	—	—	—	—	6	—	—
75	150	1.80	—	—	440	—	—	—	—	—	—	—	—	5	—	—
95	200	2.70	—	—	1210	—	—	—	—	—	—	—	—	6	—	—
40	20	1.08	—	—	480	—	0	—	—	—	—	—	—	0	—	—
45	600	1.08	—	—	350	—	0	—	—	—	—	—	—	0	—	—
0	20	1.80	—	—	950	—	0	—	—	—	—	—	—	12	—	—
155	300	7.20	—	—	1770	—	—	—	—	—	—	—	—	1	—	—
95	200	5.40	—	—	1310	—	102	—	—	—	—	—	—	6	—	—
0	0	1.08	—	—	180	—	0	—	—	—	—	—	—	21	—	—
0	40	1.08	—	—	430	—	0	0.26	—	0.24	2.00	—	—	0	—	—
35	80	2.70	—	—	480	—	0	—	—	—	—	—	—	1	—	—
0	20	0.72	—	—	700	—	0	—	—	—	—	—	—	4	—	—
135	350	5.40	—	—	1690	—	—	—	—	—	—	—	—	21	—	—
40	100	2.70	—	—	1220	—	—	—	—	—	—	—	—	6	—	—
45	600	0.00	—	—	330	—	0	—	—	—	—	—	—	0	—	—
130	100	7.20	—	—	980	—	—	—	—	—	—	—	—	9	—	—
50	600	0.00	—	—	350	—	0	—	—	—	—	—	—	0	—	—
85	200	5.40	—	—	1410	—	40	—	—	—	—	—	—	1	—	—
265	150	3.60	—	—	1070	—	—	—	—	—	—	—	—	0	—	—
245	80	2.70	—	—	740	—	—	—	—	—	—	—	—	0	—	—
260	150	2.70	—	—	960	—	—	—	—	—	—	—	—	0	—	—
5	60	1.80	—	—	930	—	0	—	—	—	—	—	—	0	—	—
280	150	3.60	—	—	1030	—	—	—	—	—	—	—	—	0	—	—
0	60	1.80	—	—	670	—	0	—	—	—	—	—	—	0	—	—
0	20	0.36	—	—	90	—	—	—	—	—	—	—	—	4	—	—
70	200	2.70	—	—	1060	—	—	—	—	—	—	—	—	6	—	—
70	150	0.72	—	—	660	—	—	—	—	—	—	—	—	30	—	—
70	80	1.80	—	—	980	—	0	—	—	—	—	—	—	2	—	—
70	80	1.80	—	—	990	—	—	—	—	—	—	—	—	5	—	—

Table A–1
Food Composition

(DA+ code is for Wadsworth Diet Analysis program)
(For purposes of calculations, use "0" for t, <1, <.1, <.01, etc.)

A

DA+ Code	Food Description	Quantity	Measure	Wt (g)	H₂O (g)	Ener (kcal)	Prot (g)	Carb (g)	Fiber (g)	Fat (g)	Sat	Fat Breakdown (g) Mono	Poly	Trans
	FAST FOOD													
	Chick Fil-A–Continued													
38742	Chicken biscuit	1	item(s)	137	—	400	16	43	2	18	4.50	—	—	2.83
38743	Chicken biscuit w/cheese	1	item(s)	151	—	450	19	43	2	23	7.00	—	—	2.85
38762	Chicken caesar wrap	1	item(s)	227	—	460	36	52	2	10	6.00	—	—	0
38757	Chicken deluxe sandwich	1	item(s)	208	—	420	28	39	2	16	3.50	—	—	0
38764	Chicken salad sandwich	1	item(s)	153	—	350	20	32	5	15	3.00	—	—	0
38756	Chicken sandwich	1	item(s)	170	—	410	28	38	1	15	3.50	—	—	0
38768	Chick-n-Strip salad	1	item(s)	331	—	390	34	22	4	18	5.00	—	—	0
38763	Chick-n-Strips	4	item(s)	127	—	290	29	14	1	13	2.50	—	—	0
38770	Coleslaw	1	item(s)	105	—	210	1	14	2	17	2.50	—	—	0
38755	Hash browns	1	serving(s)	84	—	170	2	20	2	9	4.50	—	—	1
38765	Hearty breast of soup	1	cup(s)	241	—	140	8	18	1	4	1.00	—	—	0
38778	Icedream, small cone	1	item(s)	135	—	160	4	28	0	4	2.00	—	—	0
38774	Icedream, small cup	1	serving(s)	213	—	230	5	38	0	6	3.50	—	—	0
38775	Lemonade	1	cup(s)	255	—	170	0	41	0	1	0.00	—	—	0
38776	Lemonade, diet	1	cup(s)	255	—	25	0	5	0	0	0.00	0.00	0.00	0
38777	Nuggets	8	item(s)	113	—	260	26	12	1	12	2.50	—	—	0
38769	Side salad	1	item(s)	108	—	60	3	4	2	3	1.50	—	—	0
38767	Southwest chargrilled salad	1	item(s)	303	—	240	22	17	5	8	3.50	—	—	0
38772	Waffle potato fries, small, salted	1	serving(s)	85	—	280	3	37	5	14	5.00	—	—	1.50
	Cinnabon													
39569	Caramel Pecanbon	1	item(s)	272	—	1100	16	141	8	56	10.00	—	—	5
39572	Caramellata Chill w/whipped cream	16	fluid ounce(s)	480	—	406	10	61	0	14	8.00	—	—	—
39571	Cinnapoppers	1	serving(s)	74	—	368	4	41	2	21	11.00	—	—	1
39567	Classic roll	1	item(s)	221	—	813	15	117	4	32	8.00	—	—	5
39568	Minibon	1	item(s)	92	—	339	6	49	2	13	3.00	—	—	2
39573	Mochalatta chill w/whipped cream	16	fluid ounce(s)	480	—	362	9	55	0	13	8.00	—	—	—
39570	Stix	5	item(s)	85	—	379	6	41	1	21	6.00	—	—	4
	Dairy Queen													
1466	Banana split	1	item(s)	369	—	510	8	96	3	12	8.00	3.00	0.50	0
38552	Brownie Earthquake	1	serving(s)	304	—	740	10	112	0	27	16.00	—	—	3
38561	Chocolate chip cookie dough blizzard, small	1	item(s)	319	—	720	12	105	0	28	14.00	—	—	2.50
1464	Chocolate malt, small	1	item(s)	418	—	650	15	111	0	16	10.00	—	—	0.50
38541	Chocolate shake, small	1	item(s)	397	—	560	13	93	1	15	10.00	—	—	0.50
17257	Chocolate soft serve	½	cup(s)	94	—	150	4	22	0	5	3.50	—	—	0
1463	Chocolate sundae, small	1	item(s)	163	—	280	5	49	0	7	4.50	1.00	1.00	0
1462	Dipped cone, small	1	item(s)	156	—	340	6	42	1	17	9.00	4.00	3.00	1
38555	Oreo cookies blizzard, small	1	item(s)	283	—	570	11	83	1	21	10.00	—	—	2.50
38547	Royal Treats Peanut Buster parfait	1	item(s)	305	—	730	16	99	2	31	17.00	—	—	0
17256	Vanilla soft serve	½	cup(s)	94	—	140	3	22	0	5	3.00	—	—	0
	Domino's													
31606	Barbeque wings	1	item(s)	25	—	50	6	2	<1	2	0.65	—	—	
31604	Breadsticks	1	item(s)	37	—	116	3	18	1	4	0.79	—	—	
37551	Buffalo chicken kickers	1	item(s)	24	14	47	4	3	<1	2	0.39	—	—	
37548	Cinnastix	1	item(s)	32	8	122	2	15	1	6	1.15	—	—	
	Classic hand tossed pizza													
31573	America's favorite feast, 12"	2	slice(s)	205	99	508	22	57	4	22	9.20	—	—	
31574	America's favorite feast, 14"	2	slice(s)	283	138	697	30	79	5	30	12.70	—	—	
37543	Bacon cheeseburger feast, 12"	2	slice(s)	198	60	549	25	55	3	26	11.62	—	—	
37545	Bacon cheeseburger feast, 14"	2	slice(s)	275	121	762	35	75	4	36	16.10	—	—	
37546	Barbeque feast, 12"	2	slice(s)	192	85	506	22	62	3	20	9.08	—	—	
37547	Barbeque feast, 14"	2	slice(s)	262	115	691	30	85	4	27	12.24	—	—	

Chol (mg)	Calc (mg)	Iron (mg)	Magn (mg)	Pota (mg)	Sodi (mg)	Zinc (mg)	Vit A (µg)	Thia (mg)	Vit E (mg α)	Ribo (mg)	Niac (mg)	Vit B_6 (mg)	Fola (µg)	Vit C (mg)	Vit B_{12} (µg)	Sele (µg)
30	60	2.70	—	—	1200	—	0	—	—	—	—	—	—	0	—	—
45	150	2.70	—	—	1430	—	—	—	—	—	—	—	—	0	—	—
80	500	2.70	—	—	1390	—	—	—	—	—	—	—	—	1	—	—
60	100	2.70	—	—	1300	—	—	—	—	—	—	—	—	2	—	—
65	150	1.80	—	—	880	—	—	—	—	—	—	—	—	0	—	—
60	100	2.70	—	—	1300	—	—	—	—	—	—	—	—	0	—	—
80	200	0.36	—	—	860	—	—	—	—	—	—	—	—	30	—	—
65	20	0.36	—	—	730	—	—	—	—	—	—	—	—	1	—	—
20	40	0.36	—	—	180	—	—	—	—	—	—	—	—	27	—	—
10	0	0.72	—	—	350	—	—	—	—	—	—	—	—	0	—	—
25	40	1.08	—	—	900	—	—	—	—	—	—	—	—	0	—	—
15	100	0.36	—	—	80	—	—	—	—	—	—	—	—	0	—	—
25	150	0.00	—	—	100	—	—	—	—	—	—	—	—	0	—	—
0	0	0.36	—	—	10	—	0	—	—	—	—	—	—	15	—	—
0	0	0.36	—	—	5	—	0	—	—	—	—	—	—	15	—	—
70	40	1.08	—	—	1090	—	0	—	—	—	—	—	—	0	—	—
10	100	0.00	—	—	75	—	—	—	—	—	—	—	—	15	—	—
60	200	1.08	—	—	770	—	—	—	—	—	—	—	—	24	—	—
15	20	0.00	—	—	105	—	0	—	—	—	—	—	—	21	—	—
63	—	—	—	—	600	—	—	—	—	—	—	—	—	—	—	—
46	—	—	—	—	187	—	—	—	—	—	—	—	—	—	—	—
62	—	—	—	—	104	—	—	—	—	—	—	—	—	—	—	—
67	—	—	—	—	801	—	—	—	—	—	—	—	—	—	—	—
27	—	—	—	—	337	—	—	—	—	—	—	—	—	—	—	—
46	100	0.00	—	—	252	—	—	—	—	—	—	—	—	0	—	—
16	—	—	—	—	413	—	—	—	—	—	—	—	—	—	—	—
30	250	1.80	—	860	180	—	—	0.15	—	0.60	0.20	—	—	15	—	—
50	250	1.80	—	—	350	—	—	—	—	—	—	—	—	0	—	—
50	350	2.70	—	—	370	—	—	—	—	—	—	—	—	1	—	—
55	450	1.80	—	—	370	—	—	—	—	—	—	—	—	2	—	—
50	450	1.44	—	—	280	—	—	0.12	—	—	—	—	—	2	—	—
15	100	0.72	—	—	75	—	—	—	—	—	—	—	—	0	—	—
20	200	1.08	—	278	140	—	—	0.06	—	0.24	0.20	—	—	0	—	—
20	200	1.08	—	290	130	—	—	0.06	—	0.26	0.20	—	—	1	—	—
40	350	2.70	—	—	430	—	—	—	—	—	—	—	—	1	—	—
35	300	1.80	—	—	400	—	—	—	—	—	—	—	—	1	—	—
15	150	0.72	—	—	70	—	150	—	—	—	—	—	—	0	—	—
26	6	0.32	—	—	175	—	—	—	—	—	—	—	—	<.1	—	—
0	<.1	0.87	—	—	152	—	—	—	—	—	—	—	—	6	—	—
9	3	0.00	—	—	163	—	—	—	—	—	—	—	—	0	—	—
0	6	0.70	—	—	110	—	—	—	—	—	—	—	—	<.1	—	—
49	202	3.70	—	—	1221	—	—	—	—	—	—	—	—	1	—	—
68	281	5.10	—	—	1685	—	—	—	—	—	—	—	—	1	—	—
60	293	3.56	—	—	1274	—	—	—	—	—	—	—	—	0	—	—
84	395	4.96	—	—	1809	—	—	—	—	—	—	—	—	0	—	—
46	—	—	—	—	1206	—	—	—	—	—	—	—	—	—	—	—
63	393	4.42	—	—	1672	—	—	—	—	—	—	—	—	2	—	—

Table A–1
Food Composition

(DA+ code is for Wadsworth Diet Analysis program)
(For purposes of calculations, use "0" for t, <1, <.1, <.01, etc.)

DA + Code	Food Description	Quantity	Measure	Wt (g)	H₂O (g)	Ener (kcal)	Prot (g)	Carb (g)	Fiber (g)	Fat (g)	Sat	Mono	Poly	Trans
	FAST FOOD													
	Classic hand tossed pizza–Continued													
31569	Cheese, 12"	2	slice(s)	159	—	375	15	55	3	11	4.81	—	—	—
31570	Cheese, 14"	2	slice(s)	219	—	516	21	75	4	15	6.72	—	—	—
37538	Deluxe feast, 12"	2	slice(s)	201	102	465	20	57	3	18	7.66	—	—	—
37540	Deluxe feast, 14"	2	slice(s)	273	138	627	26	78	5	24	10.20	—	—	—
31685	Deluxe, 12"	2	slice(s)	213	—	465	20	57	3	18	7.65	—	—	—
31694	Deluxe, 14"	2	slice(s)	273	—	627	26	78	5	24	10.20	—	—	—
31686	Extravaganzza, 12"	2	slice(s)	245	127	576	27	59	4	27	11.56	—	—	—
31695	Extravaganzza, 14"	2	slice(s)	329	171	773	36	88	5	36	15.42	—	—	—
31575	Hawaiian feast, 12"	2	slice(s)	204	105	450	21	58	3	16	7.20	—	—	—
31576	Hawaiian feast, 14"	2	slice(s)	283	147	623	29	80	5	22	10.09	—	—	—
31687	Meatzza, 12"	2	slice(s)	213	—	560	26	57	3	26	11.40	—	—	—
31696	Meatzza, 14"	2	slice(s)	293	139	753	35	78	5	34	15.24	—	—	—
31571	Pepperoni feast, extra pepperoni & cheese, 12"	2	slice(s)	196	87	534	24	56	3	25	10.92	—	—	—
31572	Pepperoni feast, extra pepperoni & cheese, 14"	2	slice(s)	270	121	732	33	77	4	34	15.00	—	—	—
31577	Vegi feast, 12"	2	slice(s)	203	107	439	19	57	4	16	7.09	—	—	—
31578	Vegi feast, 14"	2	slice(s)	278	147	304	27	78	5	22	9.89	—	—	—
37549	Dot cinnamon	1	item(s)	28	8	99	2	15	1	4	0.68	—	—	—
31605	Double cheesy bread	1	item(s)	35	11	123	4	13	1	6	2.06	—	—	—
31607	Hot wings	1	item(s)	25	—	45	5	1	<1	2	0.65	—	—	—
	Thin crust pizza													
31583	America's favorite, 12"	¼	item(s)	159	—	408	19	34	2	23	9.77	—	—	—
31584	America's favorite, 14"	¼	item(s)	202	—	557	26	47	3	31	13.19	—	—	—
31579	Cheese, 12"	¼	item(s)	106	—	273	12	31	2	12	9.37	—	—	—
31580	Cheese, 14"	¼	item(s)	148	—	382	17	43	2	17	6.72	—	—	—
31688	Deluxe, 12"	¼	item(s)	159	—	363	16	34	2	19	7.64	—	—	—
31697	Deluxe, 14"	¼	item(s)	202	—	494	22	47	3	25	10.20	—	—	—
31689	Extravaganzza, 12"	¼	item(s)	159	—	425	20	34	3	24	9.41	—	—	—
31698	Extravaganzza, 14"	¼	item(s)	202	—	571	27	48	4	31	12.44	—	—	—
31585	Hawaiian, 12"	¼	item(s)	159	—	349	18	35	2	16	7.20	—	—	—
31586	Hawaiian, 14"	¼	item(s)	202	—	489	25	48	3	23	10.09	—	—	—
31690	Meatzza, 12"	¼	item(s)	159	—	458	23	33	2	27	11.39	—	—	—
31699	Meatzza, 14"	¼	item(s)	202	—	619	31	46	3	36	15.24	—	—	—
31581	Pepperoni, extra pepperoni & cheese 12"	¼	item(s)	159	—	420	20	32	2	24	10.46	—	—	—
31582	Pepperoni, extra pepperoni & cheese 14"	¼	item(s)	202	—	586	28	45	3	34	14.55	—	—	—
31587	Vegi, 12"	¼	item(s)	159	—	338	16	34	3	17	7.08	—	—	—
31588	Vegi, 14"	¼	item(s)	202	—	471	22	47	3	23	9.89	—	—	—
	Ultimate deep dish pizza													
31596	America's favorite, 12"	2	slice(s)	235	—	617	26	59	4	33	12.88	—	—	—
31702	America's favorite, 14"	2	slice(s)	311	—	851	36	84	5	44	17.35	—	—	—
31590	Cheese, 12"	2	slice(s)	181	—	482	19	56	3	22	7.91	—	—	—
31591	Cheese, 14"	2	slice(s)	257	—	677	26	80	5	30	10.88	—	—	—
31589	Cheese, 6"	1	item(s)	215	—	598	23	68	4	28	9.94	—	—	—
31691	Deluxe, 12"	2	slice(s)	235	—	527	23	59	4	29	10.75	—	—	—
31700	Deluxe, 14"	2	slice(s)	311	—	788	31	84	5	38	14.36	—	—	—
31692	Extravaganzza, 12"	2	slice(s)	235	—	635	27	59	4	34	12.52	—	—	—
31701	Extravaganzza, 14"	2	slice(s)	311	—	866	36	85	6	45	16.60	—	—	—
31599	Hawaiian, 12"	2	slice(s)	235	—	558	24	60	4	26	10.31	—	—	—
31600	Hawaiian, 14"	2	slice(s)	311	—	784	35	85	5	36	14.25	—	—	—

Chol (mg)	Calc (mg)	Iron (mg)	Magn (mg)	Pota (mg)	Sodi (mg)	Zinc (mg)	Vit A (µg)	Thia (mg)	Vit E (mg α)	Ribo (mg)	Niac (mg)	Vit B$_6$ (mg)	Fola (µg)	Vit C (mg)	Vit B$_{12}$ (µg)	Sele (µg)
23	187	2.99	—	—	776	—	131	—	—	—	—	—	—	0	—	—
32	261	4.13	—	—	1080	—	184	—	—	—	—	—	—	0	—	—
40	199	3.56	—	—	1063	—	—	—	—	—	—	—	—	1	—	—
53	276	4.84	—	—	1432	—	—	—	—	—	—	—	—	2	—	—
40	199	3.56	—	—	1063	—	—	—	—	—	—	—	—	1	—	—
53	276	4.85	—	—	1432	—	—	—	—	—	—	—	—	2	—	—
60	290	4.08	—	—	1348	—	—	—	—	—	—	—	—	1	—	—
89	403	5.48	—	—	1780	—	—	—	—	—	—	—	—	2	—	—
41	274	3.30	—	—	1102	—	—	—	—	—	—	—	—	2	—	—
57	384	4.57	—	—	1544	—	—	—	—	—	—	—	—	3	—	—
344	282	3.71	—	—	1463	—	—	—	—	—	—	—	—	<1	—	—
85	393	5.04	—	—	1947	—	—	—	—	—	—	—	—	<1	—	—
57	279	3.36	—	—	1349	—	155	—	—	—	—	—	—	<1	—	—
78	390	4.66	—	—	1855	—	233	—	—	—	—	—	—	<1	—	—
34	279	3.44	—	—	987	—	—	—	—	—	—	—	—	1	—	—
47	389	4.71	—	—	1369	—	—	—	—	—	—	—	—	2	—	—
0	6	0.59	—	—	86	—	—	—	—	—	—	—	—	<.1	—	—
6	47	0.66	—	—	164	—	—	—	—	—	—	—	—	<1	—	—
26	5	0.30	—	—	354	—	—	—	—	—	—	—	—	1	—	—
51	318	1.52	—	—	1285	—	—	—	—	—	—	—	—	<1	—	—
69	444	2.07	—	—	1751	—	—	—	—	—	—	—	—	1	—	—
23	225	0.97	—	—	835	—	125	—	—	—	—	—	—	0	—	—
32	315	1.36	—	—	1172	—	175	—	—	—	—	—	—	0	—	—
40	237	1.54	—	—	1123	—	—	—	—	—	—	—	—	1	—	—
53	330	2.08	—	—	1523	—	—	—	—	—	—	—	—	2	—	—
53	245	1.95	—	—	1408	—	—	—	—	—	—	—	—	1	—	—
69	340	2.59	—	—	1871	—	—	—	—	—	—	—	—	2	—	—
41	312	1.28	—	—	1162	—	—	—	—	—	—	—	—	2	—	—
57	437	1.80	—	—	1635	—	—	—	—	—	—	—	—	3	—	—
64	320	1.69	—	—	1523	—	—	—	—	—	—	—	—	<1	—	—
454	446	2.27	—	—	2039	—	—	—	—	—	—	—	—	<1	—	—
54	316	1.34	—	—	1362	—	162	—	—	—	—	—	—	<1	—	—
76	442	1.87	—	—	1900	—	227	—	—	—	—	—	—	<1	—	—
34	317	1.42	—	—	1047	—	—	—	—	—	—	—	—	1	—	—
47	442	1.94	—	—	1460	—	—	—	—	—	—	—	—	2	—	—
58	334	4.43	—	—	1573	—	—	—	—	—	—	—	—	1	—	—
78	464	6.24	—	—	2155	—	—	—	—	—	—	—	—	1	—	—
30	241	3.88	—	—	1123	—	151	—	—	—	—	—	—	<1	—	—
41	335	5.53	—	—	1575	—	210	—	—	—	—	—	—	1	—	—
36	295	4.67	—	—	1341	—	174	—	—	—	—	—	—	1	—	—
47	253	4.45	—	—	1410	—	—	—	—	—	—	—	—	2	—	—
62	349	6.25	—	—	1927	—	—	—	—	—	—	—	—	2	—	—
60	261	4.86	—	—	1696	—	—	—	—	—	—	—	—	2	—	—
78	359	6.76	—	—	2275	—	—	—	—	—	—	—	—	2	—	—
48	328	4.19	—	—	1449	—	—	—	—	—	—	—	—	2	—	—
67	457	5.97	—	—	2039	—	—	—	—	—	—	—	—	3	—	—

A

Table A–1
Food Composition

(DA+ code is for Wadsworth Diet Analysis program)
(For purposes of calculations, use "0" for t, <1, <.1, <.01, etc.)

A

DA +Code	Food Description	Quantity	Measure	Wt (g)	H₂O (g)	Ener (kcal)	Prot (g)	Carb (g)	Fiber (g)	Fat (g)	Sat	Fat Breakdown (g) Mono	Poly	Trans
	FAST FOOD													
	Ultimate deep dish pizza–Continued													
31693	Meatzza, 12"	2	slice(s)	235	—	667	30	58	4	37	14.50	—	—	—
31703	Meatzza, 14"	2	slice(s)	311	—	914	40	83	5	49	19.40	—	—	—
31593	Pepperoni, extra pepperoni & cheese 12"	2	slice(s)	235	—	629	26	57	4	34	13.57	—	—	—
31594	Pepperoni, extra pepperoni & cheese 14"	2	slice(s)	311	—	880	37	82	5	47	18.71	—	—	—
31602	Vegi, 12"	2	slice(s)	235	—	547	22	59	4	26	10.19	—	—	—
31603	Vegi, 14"	2	slice(s)	311	—	765	32	84	6	36	14.05	—	—	—
31598	With ham & pineapple tidbits, 6"	1	item(s)	430	—	619	25	70	4	28	10.19	—	—	—
31595	With Italian sausage, 6"	1	item(s)	430	—	642	25	70	4	31	11.33	—	—	—
31592	With pepperoni, 6"	1	item(s)	430	—	647	25	69	4	32	11.70	—	—	—
31601	With vegetables, 6"	1	item(s)	430	—	619	23	71	5	29	10.11	—	—	—
	In-n-Out Burger													
34374	Cheeseburger	1	item(s)	268	—	480	22	39	3	27	10.00	—	—	—
34391	Cheesburger w/mustard & ketchup	1	item(s)	268	—	400	22	41	3	18	9.00	—	—	—
34390	Cheeseburger, lettuce leaves instead of buns	1	item(s)	300	—	330	18	11	2	25	9.00	—	—	—
34377	Chocolate shake	1	item(s)	425	—	690	9	83	0	36	24.00	—	—	—
34375	Double-Double cheeseburger	1	item(s)	328	—	670	37	40	3	41	18.00	—	—	—
34393	Double-Double cheeseburger w/mustard & ketchup	1	item(s)	328	—	590	37	42	3	32	17.00	—	—	—
34392	Double-Double cheeseburger, lettuce leaves instead of buns	1	item(s)	361	—	520	33	11	3	39	17.00	—	—	—
34376	French fries	1	item(s)	125	—	400	7	54	2	18	5.00	—	—	—
34373	Hamburger	1	item(s)	243	—	390	16	39	3	19	5.00	—	—	—
34389	Hamburger w/mustard & ketchup	1	item(s)	243	—	310	16	41	3	10	4.00	—	—	—
34388	Hamburger, lettuce leaves instead of buns	1	item(s)	275	—	240	12	10	2	17	4.50	—	—	—
34379	Strawberry shake	1	item(s)	425	—	690	8	91	2	33	22.00	—	—	—
34378	Vanilla shake	1	item(s)	425	—	680	9	78	2	37	25.00	—	—	—
	Jack in the Box													
30392	Bacon ultimate cheeseburger	1	item(s)	353	—	1120	52	59	2	55	28.00	—	—	3.13
1740	Breakfast Jack	1	item(s)	133	—	310	14	34	1	14	5.00	—	—	0
14074	Cheeseburger	1	item(s)	116	—	300	14	31	2	13	6.00	—	—	0.89
14106	Chicken breast pieces	5	piece(s)	150	—	360	27	24	1	17	3.00	—	—	4.48
37241	Chicken club salad	1	item(s)	535	—	310	28	15	5	16	6.00	—	—	0
14111	Chocolate ice cream shake	1	item(s)	315	—	660	11	89	1	29	18.00	—	—	1
14075	Double cheeseburger	1	item(s)	155	—	410	20	32	1	22	11.00	—	—	—
14098	French fries, jumbo	1	serving(s)	142	—	410	4	55	4	20	4.50	—	—	5.34
14099	French fries, super scoop	1	serving(s)	198	—	580	6	77	6	28	6.00	—	—	7.07
14073	Hamburger	1	item(s)	104	—	250	12	30	2	9	3.50	—	—	0.88
14090	Hash browns	1	serving(s)	57	—	150	1	13	2	10	2.50	—	—	3
14072	Jack's Spicy Chicken sandwich	1	item(s)	253	—	580	24	53	3	31	6.00	—	—	2.81
1468	Jumbo Jack hamburger	1	item(s)	269	—	600	22	58	3	31	11.00	—	—	1.55
1469	Jumbo Jack hamburger w/cheese	1	item(s)	294	—	690	26	60	3	38	16.00	—	—	1.55
1470	Onion rings	1	serving(s)	119	—	500	6	51	3	30	5.00	—	—	10
33141	Sausage, egg, & cheese biscuit	1	item(s)	223	—	760	25	33	2	60	20.00	—	—	5.72
14095	Seasoned curly fries	1	serving(s)	125	—	400	6	45	5	23	5.00	—	—	7
14077	Sourdough Jack	1	item(s)	244	—	700	30	36	3	49	16.00	—	—	2.98
37249	Southwest chicken salad	1	serving(s)	598	—	340	28	31	9	13	6.00	—	—	0
14112	Strawberry ice cream shake	1	item(s)	313	—	640	10	84	0	28	18.00	—	—	1
14078	Ultimate cheeseburger	1	item(s)	328	—	990	41	59	2	66	28.00	—	—	3.05
14110	Vanilla ice cream shake	1	item(s)	285	—	570	12	65	0	29	18.00	—	—	1

PAGE KEY: **A–2 = Breads/Baked Goods**   A–6 = Cereal/Rice/Pasta   A–10 = Fruit   A–14 = Vegetables/Legumes   A–24 = Nuts/Seeds   A–26 = Vegetarian
A–28 = Dairy   A–34 = Eggs   A–34 = Seafood   A–36 = Meats   A–40 = Poultry   A–40 = Processed meats   A–42 = Beverages   A–46 = Fats/Oils   A–48 = Sweets
A–50 = Spices/Condiments/Sauces   A–52 = Mixed foods/Soups/Sandwiches   A–58 = Fast food   A–74 = Convenience meals   A–76 = Baby foods

A

Chol (mg)	Calc (mg)	Iron (mg)	Magn (mg)	Pota (mg)	Sodi (mg)	Zinc (mg)	Vit A (µg)	Thia (mg)	Vit E (mg α)	Ribo (mg)	Niac (mg)	Vit B_6 (mg)	Fola (µg)	Vit C (mg)	Vit B_{12} (µg)	Sele (µg)
379	336	4.60	—	—	1810	—	—	—	—	—	—	—	—	1	—	—
501	466	6.44	—	—	2443	—	—	—	—	—	—	—	—	1	—	—
61	332	4.25	—	—	1650	—	187	—	—	—	—	—	—	1	—	—
85	462	6.04	—	—	2304	—	260	—	—	—	—	—	—	1	—	—
41	333	4.33	—	—	1334	—	—	—	—	—	—	—	—	2	—	—
57	462	6.11	—	—	1864	—	—	—	—	—	—	—	—	2	—	—
43	298	4.84	—	—	1498	—	—	—	—	—	—	—	—	1	—	—
45	302	4.89	—	—	1478	—	—	—	—	—	—	—	—	1	—	—
47	299	4.81	—	—	1524	—	168	—	—	—	—	—	—	1	—	—
36	307	5.10	—	—	1472	—	—	—	—	—	—	—	—	5	—	—
60	200	3.60	—	—	1000	—	188	—	—	—	—	—	—	15	—	—
55	200	3.60	—	—	1080	—	182	—	—	—	—	—	—	15	—	—
60	200	1.08	—	—	720	—	—	—	—	—	—	—	—	18	—	—
95	300	0.72	—	—	350	—	143	—	—	—	—	—	—	0	—	—
120	350	5.40	—	—	1430	—	184	—	—	—	—	—	—	15	—	—
115	350	5.40	—	—	1510	—	229	—	—	—	—	—	—	15	—	—
120	350	1.08	—	—	1160	—	275	—	—	—	—	—	—	18	—	—
0	20	1.80	—	—	245	—	0	—	—	—	—	—	—	0	—	—
40	40	3.60	—	—	640	—	50	—	—	—	—	—	—	15	—	—
35	40	3.60	—	—	720	—	75	—	—	—	—	—	—	15	—	—
40	40	1.08	—	—	370	—	—	—	—	—	—	—	—	18	—	—
85	250	0.00	—	—	280	—	134	—	—	—	—	—	—	0	—	—
90	300	0.00	—	—	390	—	145	—	—	—	—	—	—	0	—	—
160	300	7.20	—	600	2260	—	—	—	—	—	—	—	—	1	—	—
210	150	3.60	—	210	770	—	—	—	—	—	—	—	—	4	—	—
40	150	3.60	—	180	840	—	40	—	—	—	—	—	—	0	—	—
80	20	1.80	—	430	970	—	—	—	—	—	—	—	—	1	—	—
65	300	3.60	—	1010	890	—	—	—	—	—	—	—	—	54	—	—
110	350	0.36	—	720	270	—	215	—	—	—	—	—	—	0	—	—
70	250	4.50	—	280	920	—	—	—	—	—	—	—	—	1	—	—
0	20	1.08	—	550	690	—	0	—	—	—	—	—	—	6	—	—
0	20	1.44	—	770	960	—	0	—	—	—	—	—	—	9	—	—
30	100	3.60	—	155	610	—	0	—	—	—	—	—	—	0	—	—
0	10	0.18	—	190	230	—	0	—	—	—	—	—	—	0	—	—
60	150	1.80	—	470	950	—	—	—	—	—	—	—	—	9	—	—
45	164	4.92	—	390	980	—	—	—	—	—	—	—	—	10	—	—
75	250	4.50	—	420	1360	—	—	—	—	—	—	—	—	9	—	—
0	40	2.70	—	140	420	—	40	—	—	—	—	—	—	18	—	—
280	100	2.70	—	240	1390	—	—	—	—	—	—	—	—	0	—	—
0	40	1.80	—	580	890	—	—	—	—	—	—	—	—	0	—	—
80	200	4.50	—	450	1220	—	—	—	—	—	—	—	—	9	—	—
60	300	4.50	—	1020	920	—	—	—	—	—	—	—	—	48	—	—
110	350	0.00	—	610	220	—	202	—	—	—	—	—	—	0	—	—
130	300	7.20	—	480	1670	—	—	—	—	—	—	—	—	1	—	—
115	400	0.00	—	630	220	—	218	—	—	—	—	—	—	0	—	—

Table A–1
Food Composition

(DA+ code is for Wadsworth Diet Analysis program)
(For purposes of calculations, use "0" for t, <1, <.1, <.01, etc.)

DA + Code	Food Description	Quantity	Measure	Wt (g)	H₂O (g)	Ener (kcal)	Prot (g)	Carb (g)	Fiber (g)	Fat (g)	Sat	Mono	Poly	Trans
	FAST FOOD–Continued													
	Jamba Juice													
31646	Banana berry smoothie	24	fluid ounce(s)	719	—	470	5	112	5	2	0.50	—	—	—
31647	Caribbean passion smoothie	24	fluid ounce(s)	730	—	440	4	102	4	2	1.00	—	—	—
38422	Carrot juice	16	fluid ounce(s)	472	—	100	3	23	0	1	0.00	—	—	—
31648	Chocolate mood smoothie	24	fluid ounce(s)	612	—	690	16	142	2	8	4.50	—	—	—
31649	Citrus squeeze smoothie	24	fluid ounce(s)	729	—	450	4	105	5	2	1.00	—	—	—
31650	Coffee mood smoothie	24	fluid ounce(s)	560	—	596	13	121	1	6	4.00	—	—	—
31651	Coldbuster smoothie	24	fluid ounce(s)	724	—	430	5	100	5	3	1.00	—	—	—
31652	Cranberry craze smoothie	24	fluid ounce(s)	731	—	420	6	97	4	2	1.00	—	—	—
31654	Jamba powerboost smoothie	24	fluid ounce(s)	730	—	440	6	103	7	2	0.00	—	—	—
38423	Lemonade	16	fluid ounce(s)	483	—	300	1	75	0	0	0.00	0.00	0.00	0
31656	Lime sublime smoothie	24	fluid ounce(s)	721	—	450	3	104	6	2	1.00	—	—	—
31657	Mango-a-go-go smoothie	24	fluid ounce(s)	739	—	500	4	117	4	2	1.00	—	—	—
38424	Orange juice, freshly squeezed	16	fluid ounce(s)	496	—	220	3	52	1	1	0.00	—	—	—
38426	Orange/carrot juice	16	fluid ounce(s)	484	—	160	3	37	0	1	0.00	—	—	—
31660	Orange-a-peel smoothie	24	fluid ounce(s)	726	—	440	9	102	5	1	0.00	—	—	—
31665	Protein berry pizzaz smoothie	24	fluid ounce(s)	710	—	440	20	92	6	2	0.00	—	—	—
31667	Raspberry refresher smoothie	24	fluid ounce(s)	636	—	442	3	101	8	3	0.90	—	—	—
31668	Razzmatazz smoothie	24	fluid ounce(s)	730	—	480	3	112	4	2	1.00	—	—	—
31669	Strawberries wild smoothie	24	fluid ounce(s)	725	—	450	6	105	4	0	0.00	—	—	—
38421	Strawberry tsunami smoothie	24	fluid ounce(s)	740	—	530	4	128	4	2	1.00	—	—	—
38427	Vibrant C juice	16	fluid ounce(s)	448	—	210	2	50	1	0	0.00	0.00	0.00	0
38428	Wheatgrass juice, freshly squeezed	1	ounce(s)	32	—	5	1	1	0	0	0.00	0.00	0.00	0
	Kentucky Fried Chicken (KFC)													
31850	BBQ baked beans	1	serving(s)	156	—	190	6	33	6	3	1.00	—	—	0.29
31853	Biscuit	1	item(s)	56	—	180	4	20	1	10	2.50	—	—	3.44
31851	Coleslaw	1	serving(s)	142	—	232	2	26	3	14	2.00	—	—	0.27
31842	Colonel's Crispy Strips	3	item(s)	150	—	340	28	20	0	16	4.50	—	—	4.47
31849	Corn on the cob	1	item(s)	162	—	150	5	35	2	2	0.00	—	—	0
3761	Extra Crispy chicken, breast	1	item(s)	162	—	470	34	19	0	28	8.00	—	—	4.50
3762	Extra Crispy chicken, drumstick	1	item(s)	60	—	160	12	5	0	10	2.50	—	—	1.50
3763	Extra Crispy chicken, thigh	1	item(s)	114	—	370	21	12	0	26	7.00	—	—	3
3764	Extra Crispy chicken, whole wing	1	item(s)	52	—	190	10	10	0	12	3.50	—	—	2
31833	Honey BBQ wing pieces	6	item(s)	189	—	607	33	33	1	38	10.00	—	—	5.42
10810	Hot & spicy chicken, breast	1	item(s)	179	—	450	33	20	0	27	8.00	—	—	0
10813	Hot & spicy chicken, drumstick	1	item(s)	60	—	140	13	4	0	9	2.50	—	—	0
10811	Hot & spicy chicken, thigh	1	item(s)	128	—	390	22	14	0	28	8.00	—	—	0
10812	Hot & spicy chicken, whole wing	1	item(s)	55	—	180	11	9	0	11	3.00	—	—	0
10859	Hot wings pieces	6	piece(s)	135	—	471	27	18	2	33	8.00	—	—	4.03
31848	Macaroni & cheese	1	serving(s)	153	—	180	7	21	2	8	3.00	—	—	2.81
31847	Mashed potatoes with gravy	1	serving(s)	136	—	120	1	17	2	6	1.00	—	—	0.50
10825	Original Recipe chicken, breast	1	item(s)	161	—	370	40	11	0	19	6.00	—	—	2.50
10826	Original Recipe chicken, drumstick	1	item(s)	59	—	140	14	4	0	8	2.00	—	—	1
10827	Original Recipe chicken, thigh	1	item(s)	126	—	360	22	12	0	25	7.00	—	—	1.50
10828	Original Recipe chicken, whole wing	1	item(s)	47	—	145	11	5	0	9	2.50	—	—	1
3760	Original Recipe chicken sandwich w/sauce	1	item(s)	200	—	450	29	33	2	22	5.00	—	—	—
31834	Original Recipe chicken sandwich w/o sauce	1	item(s)	187	—	360	29	21	1	13	3.50	—	—	—
31852	Potato salad	1	serving(s)	160	—	230	4	23	3	14	2.00	—	—	0.31
10845	Potato wedges	1	serving(s)	156	—	376	6	53	5	15	4.20	—	—	6.12
10853	Rotisserie Gold chicken, breast & wing w/skin	4	ounce(s)	114	—	218	26	1	0	12	3.51	—	—	—

Chol (mg)	Calc (mg)	Iron (mg)	Magn (mg)	Pota (mg)	Sodi (mg)	Zinc (mg)	Vit A (µg)	Thia (mg)	Vit E (mg α)	Ribo (mg)	Niac (mg)	Vit B_6 (mg)	Fola (µg)	Vit C (mg)	Vit B_{12} (µg)	Sele (µg)
5	200	1.08	32	1000	85	0.30	—	0.06	0.32	0.26	1.20	0.40	33	15	0	0
5	100	1.80	24	810	60	0.30	—	0.09	0.64	0.26	5.00	0.50	100	78	0	1
0	150	2.70	80	1030	250	0.90	0	0.53	—	0.26	5.00	0.70	80	18	0	6
25	500	1.08	32	760	280	0.60	0	0.09	0.00	0.85	0.40	0.08	9	6	1	4
5	150	1.80	60	1150	50	0.30	—	0.30	0.40	0.26	1.90	0.40	100	168	0	1
28	455	0.30	49	634	429	1.50	—	0.10	0.16	0.60	0.30	0.10	18	7	1	3
5	100	1.08	60	1240	35	15.00	—	0.38	17.71	0.34	3.00	0.40	122	1302	0	1
5	250	1.44	16	500	90	0.30	—	0.03	0.64	0.26	5.00	0.50	100	54	0	1
0	1100	1.44	480	1110	40	15.00	—	5.25	17.71	5.78	66.00	6.80	640	294	10	70
0	20	0.00	8	200	10	0.00	0	0.03	0.00	0.17	14.00	1.80	320	36	0	0
5	150	1.80	32	660	75	0.60	—	0.12	0.32	0.26	7.00	0.80	160	66	<1	1
5	100	1.08	24	800	60	0.30	—	0.15	1.61	0.26	5.00	0.70	120	72	0	1
0	60	1.08	60	990	0	0.30	0	0.45	—	0.14	2.00	0.20	160	246	0	0
0	100	1.80	60	1010	125	0.60	0	0.45	—	0.26	3.00	0.50	120	132	0	3
0	250	1.80	60	1350	100	0.30	—	0.38	0.64	0.43	3.00	0.40	140	240	0	1
0	1100	2.62	39	650	240	0.58	—	0.09	0.31	0.10	1.55	0.40	58	60	0	4
3	104	2.20	56	806	47	0.80	—	0.10	0.40	0.30	1.60	0.40	43	35	<1	1
5	150	1.80	32	790	70	0.60	—	0.09	0.32	0.26	6.00	0.90	160	60	0	1
0	250	1.80	32	1020	115	0.30	—	0.03	0.32	0.34	1.20	0.20	32	60	0	1
5	100	1.08	24	480	10	0.30	0	0.06	—	0.34	14.00	1.80	320	90	0	1
0	20	1.08	40	720	0	0.30	0	0.30	—	0.10	1.60	0.40	80	678	0	0
0	0	1.80	8	80	0	0.00	0	0.03	—	0.03	0.40	0.04	16	4	0	3
5	80	1.80	—	—	760	—	—	—	—	—	—	—	—	1	—	—
0	20	1.08	—	—	560	—	—	—	—	—	—	—	—	1	—	—
8	30	0.18	—	—	284	—	65	—	—	—	—	—	—	34	—	—
70	10	0.72	—	—	1140	—	—	—	—	—	—	—	—	1	—	—
0	10	0.18	—	—	20	—	10	—	—	—	—	—	—	4	—	—
135	19	1.44	—	—	1230	—	—	—	—	—	—	—	—	1	—	—
70	9	0.65	—	—	415	—	—	—	—	—	—	—	—	1	—	—
120	19	1.04	—	—	710	—	—	—	—	—	—	—	—	1	—	—
55	9	0.34	—	—	390	—	—	—	—	—	—	—	—	1	—	—
193	40	1.44	—	—	1145	—	—	—	—	—	—	—	—	5	—	—
130	10	1.07	—	—	1450	—	—	—	—	—	—	—	—	1	—	—
65	20	0.68	—	—	380	—	—	—	—	—	—	—	—	1	—	—
125	10	1.44	—	—	1240	—	—	—	—	—	—	—	—	1	—	—
60	10	0.72	—	—	420	—	—	—	—	—	—	—	—	1	—	—
150	40	1.44	—	—	1230	—	—	—	—	—	—	—	—	1	—	—
10	150	0.18	—	—	860	—	350	—	—	—	—	—	—	1	—	—
1	10	0.36	—	—	440	—	—	—	—	—	—	—	—	1	—	—
145	20	1.14	—	—	1145	—	—	—	—	—	—	—	—	1	—	—
75	10	0.70	—	—	440	—	—	—	—	—	—	—	—	1	—	—
165	10	1.00	—	—	1060	—	—	—	—	—	—	—	—	1	—	—
60	10	0.36	—	—	370	—	—	—	—	—	—	—	—	1	—	—
70	40	1.80	—	—	940	—	—	—	—	—	—	—	—	1	—	—
60	40	1.80	—	—	890	—	—	—	—	—	—	—	—	1	—	—
15	20	2.70	—	—	540	—	100	—	—	—	—	—	—	1	—	—
4	36	1.55	—	—	1323	—	—	—	—	—	—	—	—	8	—	—
102	7	0.12	—	—	718	—	—	—	—	—	—	—	—	1	—	—

A

Table A–1
Food Composition

(DA+ code is for Wadsworth Diet Analysis program)
(For purposes of calculations, use "0" for t, <1, <.1, <.01, etc.)

DA + Code	Food Description	Quantity	Measure	Wt (g)	H₂O (g)	Ener (kcal)	Prot (g)	Carb (g)	Fiber (g)	Fat (g)	Sat	Mono	Poly	Trans
	FAST FOOD													
	Kentucky Fried Chicken (KFC)–Continued													
10851	Rotisserie Gold chicken, thigh & leg w/skin	4	ounce(s)	114	—	260	23	1	0	18	5.15	—	—	—
10852	Rotisserie Gold chicken, thigh & leg w/o skin	4	ounce(s)	117	—	217	27	0	0	12	3.50	—	—	—
31843	Spicy Crispy Strips	3	item(s)	115	—	335	25	23	1	15	4.00	—	—	—
10854	Tender Roast chicken, breast w/o skin	1	item(s)	118	—	169	31	1	0	4	1.20	—	—	—
	Long John Silver													
39392	Baked cod	1	serving(s)	101	—	120	22	1	0	5	1.00	—	—	—
3777	Batter dipped fish sandwich	1	item(s)	177	—	440	17	48	3	20	5.00	—	—	—
37568	Battered fish	1	item(s)	92	—	230	11	16	0	13	4.00	—	—	—
37569	Breaded clams	1	serving(s)	85	—	240	8	22	1	13	2.00	—	—	—
39404	Clam chowder	1	item(s)	227	—	220	9	23	0	10	4.00	—	—	—
39398	Cocktail sauce	1	ounce(s)	28	—	25	0	6	0	0	0.00	0.00	0.00	0
3770	Coleslaw	1	serving(s)	113	—	200	1	15	4	15	2.50	1.76	4.10	
39394	Crunchy shrimp basket	21	item(s)	114	—	340	12	32	2	19	5.00	—	—	—
39400	French fries, large	1	item(s)	142	—	390	4	56	5	17	4.00	—	—	—
3774	Fries regular	1	serving(s)	85	—	230	3	34	3	10	2.50	7.40	5.10	
3779	Hushpuppy	1	piece(s)	23	—	60	1	9	1	3	0.50	—	—	—
3781	Shrimp batter-dipped	1	piece(s)	14	—	45	2	3	0	3	1.00	—	—	—
39399	Tartar sauce	1	ounce(s)	28	—	100	0	4	0	9	1.50	—	—	—
39395	Ultimate fish sandwich	1	item(s)	199	—	500	20	48	3	25	8.00	—	—	—
	McDonald's													
2247	Barbecue sauce	1	serving(s)	28	—	45	0	10	0	0	0.00	0.00	0.00	0
737	Big Mac hamburger	1	item(s)	216	—	590	24	47	3	34	11.00	—	—	1.48
738	Cheeseburger	1	item(s)	121	—	330	15	36	2	14	6.00	—	—	1.02
29775	Chicken McGrill sandwich	1	item(s)	213	—	400	25	37	2	17	3.00	—	—	0
3792	Chicken McNuggets	4	item(s)	72	—	210	10	12	1	13	2.50	—	—	1.13
1873	Chicken McNuggets	6	item(s)	108	—	310	15	18	2	20	4.00	—	—	1.69
73	Chocolate milkshake	8	fluid ounce(s)	227	164	270	7	48	1	6	3.81	1.77	0.23	—
29774	Crispy chicken sandwich	1	item(s)	219	—	500	22	46	2	26	4.50	—	—	1.50
743	Egg McMuffin	1	item(s)	138	—	300	18	29	2	12	4.50	—	—	0.42
742	Filet-o-fish sandwich	1	item(s)	156	—	470	15	45	1	26	5.00	—	—	1.11
2257	French fries, large	1	serving(s)	176	—	540	8	68	6	26	4.50	—	—	6.18
1872	French fries, small	1	serving(s)	68	—	210	3	26	2	10	1.50	—	—	2.30
2244	French fries, super size	1	serving(s)	198	—	610	9	77	7	29	5.00	—	—	—
33822	Fruit n' yogurt parfait	1	item(s)	338	—	380	10	76	2	5	2.00	—	—	0.18
2251	Garden salad	1	item(s)	177	—	35	2	7	3	0	0.00	0.00	0.00	0
739	Hamburger	1	item(s)	107	—	280	12	35	2	10	4.00	—	—	0.51
2003	Hash browns	1	item(s)	53	—	130	1	14	1	8	1.50	—	—	2
2249	Honey sauce	1	item(s)	14	—	45	0	12	0	0	0.00	0.00	0.00	—
33816	McSalad Shaker chef salad	1	item(s)	206	—	150	17	5	2	8	3.50	—	—	—
33817	McSalad Shaker garden salad	1	item(s)	149	—	100	7	4	2	6	3.00	—	—	—
33818	McSalad Shaker grilled chicken caesar salad	1	item(s)	163	—	100	17	3	2	3	1.50	—	—	—
38396	Newman's Own cobb salad dressing	1	item(s)	59	—	120	1	9	0	9	1.50	—	—	0.01
38397	Newman's Own creamy caesar salad dressing	1	item(s)	59	—	190	2	4	0	18	3.50	—	—	0.29
38398	Newman's Own low fat balsamic vinaigrette salad dressing	1	item(s)	44	—	40	0	4	0	3	0.00	—	—	0.01
38399	Newman's Own ranch salad dressing	1	item(s)	59	—	290	1	4	0	30	4.50	—	—	0.22
1874	Plain hotcakes w/syrup & margarine	3	item(s)	228	—	600	9	104	0	17	3.00	—	—	4
740	Quarter Pounder hamburger	1	item(s)	172	—	430	23	37	2	21	8.00	—	—	1.01

Chol (mg)	Calc (mg)	Iron (mg)	Magn (mg)	Pota (mg)	Sodi (mg)	Zinc (mg)	Vit A (µg)	Thia (mg)	Vit E (mg α)	Ribo (mg)	Niac (mg)	Vit B$_6$ (mg)	Fola (µg)	Vit C (mg)	Vit B$_{12}$ (µg)	Sele (µg)
127	8	0.14	—	—	764	—	—	—	—	—	—	—	—	1	—	—
128	10	0.18	—	—	772	—	—	—	—	—	—	—	—	1	—	—
70	20	0.90	—	—	1140	—	—	—	—	—	—	—	—	1	—	—
112	10	0.18	—	—	797	—	—	—	—	—	—	—	—	1	—	—
90	20	0.72	—	—	240	—	—	—	—	—	—	—	—	0	—	—
35	60	3.60	—	—	1120	—	—	—	—	—	—	—	—	9	—	—
30	20	1.80	—	—	700	—	—	—	—	—	—	—	—	5	—	—
10	20	1.08	—	—	1110	—	—	—	—	—	—	—	—	0	—	—
25	150	0.72	—	—	810	—	—	—	—	—	—	—	—	0	—	—
0	0	0.00	—	—	250	—	—	—	—	—	—	—	—	0	—	—
20	40	0.36	—	223	340	0.70	34	0.07	—	0.08	2.35	—	—	18	—	—
105	500	1.80	—	—	720	—	—	—	—	—	—	—	—	1	—	—
0	0	0.00	—	—	580	—	—	—	—	—	—	—	—	24	—	—
0	0	0.00	—	370	350	0.30	—	0.09	—	0.02	1.60	—	—	15	—	—
0	20	0.36	—	—	200	—	—	—	—	—	—	—	—	0	—	—
15	0	0.00	—	—	125	—	—	—	—	—	—	—	—	1	—	—
15	0	0.00	—	—	250	—	—	—	—	—	—	—	—	0	—	—
50	150	3.60	—	—	1310	—	—	—	—	—	—	—	—	9	—	—
0	10	0.18	—	45	250	—	3	—	—	—	—	—	—	4	—	—
85	300	4.50	—	430	1090	—	60	—	—	—	—	—	—	4	—	—
45	250	2.70	—	250	830	—	60	—	—	—	—	—	—	2	—	—
60	200	2.70	—	440	890	—	—	—	—	—	—	—	—	6	—	—
35	20	0.72	—	180	460	—	—	—	—	—	—	—	—	1	—	—
50	20	0.72	—	260	680	—	—	—	—	—	—	—	—	1	—	—
25	299	0.70	36	508	252	1.09	41	0.11	0.11	0.50	0.28	0.06	11	0	1	4
50	200	2.70	—	400	1100	—	—	—	—	—	—	—	—	6	—	—
235	300	2.70	—	210	830	—	—	—	0.72	—	—	—	—	1	—	—
50	200	1.80	—	280	890	—	40	—	—	—	—	—	—	1	—	—
0	20	1.44	—	1210	350	—	—	—	—	—	—	—	—	21	—	—
0	10	0.36	—	470	135	—	—	—	—	—	—	—	—	9	—	—
0	20	1.44	—	1370	390	—	—	—	—	—	—	—	—	24	—	—
15	300	1.80	—	550	240	—	—	—	—	—	—	—	—	24	—	—
0	40	1.09	—	410	20	—	—	—	—	—	—	—	—	24	—	—
30	200	2.70	—	230	590	—	5	—	—	—	—	—	—	2	—	—
0	10	0.36	—	210	330	—	—	—	—	—	—	—	—	2	—	—
0	10	0.18	—	7	0	—	—	—	—	—	—	—	—	1	—	—
95	150	1.44	—	360	740	—	323	—	—	—	—	—	—	15	—	—
75	150	1.08	—	290	120	—	273	—	—	—	—	—	—	15	—	—
40	100	1.08	—	420	240	—	—	—	—	—	—	—	—	12	—	—
10	40	0.18	—	13	440	—	—	—	0.00	—	—	—	—	1	—	—
20	60	0.18	—	16	500	—	—	—	15.40	—	—	—	—	1	—	—
0	10	0.18	—	9	730	—	—	—	0.00	—	—	—	—	2	—	—
20	40	0.18	—	64	530	—	—	—	—	—	—	—	—	1	—	—
20	100	4.50	—	280	770	—	—	—	—	—	—	—	—	1	—	—
70	200	4.50	—	370	840	—	10	—	—	—	—	—	—	2	—	—

Table A–1
Food Composition

(DA+ code is for Wadsworth Diet Analysis program)
(For purposes of calculations, use "0" for t, <1, <.1, <.01, etc.)

A

DA + Code	Food Description	Quantity	Measure	Wt (g)	H₂O (g)	Ener (kcal)	Prot (g)	Carb (g)	Fiber (g)	Fat (g)	Sat	Mono	Poly	Trans
												Fat Breakdown (g)		
	FAST FOOD													
	McDonald's–Continued													
741	Quarter Pounder hamburger w/cheese	1	item(s)	200	—	530	28	38	2	30	13.00	—	—	1.51
2005	Sausage McMuffin w/egg	1	item(s)	164	—	450	20	29	2	28	10.00	—	—	0.59
3163	Strawberry milkshake	8	fluid ounce(s)	226	168	256	8	43	1	6	3.93	—	—	—
74	Vanilla milkshake	8	fluid ounce(s)	227	169	254	9	40	0	7	4.28	1.98	0.26	—
	Pizza Hut													
39009	Hot chicken wings	2	item(s)	57	—	110	11	1	0	6	2.00	—	—	0.25
14025	Meat Lovers hand tossed pizza	1	slice(s)	125	—	320	16	30	2	15	7.00	—	—	0.53
14026	Meat Lovers pan pizza	1	slice(s)	130	—	360	16	29	2	20	7.00	—	—	0.53
31009	Meat Lovers stuffed crust pizza	1	slice(s)	188	—	500	25	44	3	25	11.00	—	—	1.11
14024	Meat Lovers thin 'n crispy pizza	1	slice(s)	112	—	310	15	22	2	18	8.00	—	—	0.57
14031	Pepperoni Lovers hand tossed pizza	1	slice(s)	114	—	300	15	30	2	14	7.00	—	—	0.50
14032	Pepperoni Lovers pan pizza	1	slice(s)	119	—	350	15	29	2	19	8.00	—	—	0.50
31011	Pepperoni Lovers stuffed crust pizza	1	slice(s)	171	—	480	23	44	3	24	11.00	—	—	1.05
14030	Pepperoni Lovers thin 'n crispy pizza	1	slice(s)	94	—	270	13	22	2	14	7.00	—	—	0.51
10834	Personal Pan pepperoni pizza	1	slice(s)	59	—	150	7	18	—	6	2.50	—	—	0.97
10842	Personal Pan supreme pizza	1	slice(s)	73	—	170	8	19	1	7	3.00	—	—	0.95
39013	Personal Pan Veggie Lovers pizza	1	slice(s)	69	—	150	6	19	1	6	2.00	—	—	0.50
14028	Veggie Lovers hand tossed pizza	1	slice(s)	120	—	220	10	31	2	6	3.00	—	—	0.25
14029	Veggie Lovers pan pizza	1	slice(s)	125	—	260	10	31	2	12	4.00	—	—	0.26
31010	Veggie Lovers stuffed crust pizza	1	slice(s)	181	—	370	17	45	3	14	7.00	—	—	0.53
14027	Veggie Lovers thin 'n crispy pizza	1	slice(s)	110	—	190	8	23	2	7	3.00	—	—	0.54
39012	Wing blue cheese dipping sauce	1	item(s)	43	—	230	2	2	0	24	5.00	—	—	1
39011	Wing ranch dipping sauce	1	item(s)	43	—	210	1	4	0	22	3.50	—	—	0.50
	Starbucks													
38042	Apple cider, tall steamed	12	fluid ounce(s)	360	—	180	0	45	0	0	0.00	0.00	0.00	0
38052	Cappuccino, tall	12	fluid ounce(s)	360	—	120	7	10	0	6	4.00	—	—	—
38053	Cappuccino, tall nonfat	12	fluid ounce(s)	360	—	80	7	11	0	0	0.00	0.00	0.00	0
38054	Cappuccino, tall soy milk	12	fluid ounce(s)	360	—	100	5	13	1	3	0.00	—	—	—
38059	Cinnamon spice mocha, tall nonfat w/o whipped cream	12	fluid ounce(s)	360	—	170	11	32	0	0	0.50	0.00	0.00	0
38057	Cinnamon spice mocha, tall w/whipped cream	12	fluid ounce(s)	360	—	320	10	31	0	17	11.00	—	—	—
38051	Espresso, single shot	1	fluid ounce(s)	30	—	5	0	1	0	0	0.00	0.00	0.00	—
38088	Flavored syrup, 1 pump	1	serving(s)	10	—	20	0	5	0	0	0.00	0.00	0.00	0
32562	Frappuccino coffee drink, lite mocha	9½	fluid ounce(s)	281	—	100	7	12	3	3	2.00	—	—	0
38079	Frappuccino, grande chocolate malt	16	fluid ounce(s)	480	—	470	15	87	2	10	3.50	—	—	—
38075	Frappuccino, grande mocha malt	12	fluid ounce(s)	360	—	430	14	91	1	7	4.00	—	—	—
32561	Frappuccino low fat coffee drink, all flavors	9½	fluid ounce(s)	281	—	190	6	39	0	3	2.00	—	—	—
38067	Frappuccino, tall caramel	12	fluid ounce(s)	360	—	210	4	43	0	3	1.50	—	—	—
38078	Frappuccino, tall chocolate	12	fluid ounce(s)	360	—	290	13	52	1	5	1.00	—	—	—
38069	Frappuccino, tall chocolate brownie	12	fluid ounce(s)	360	—	270	5	51	1	7	4.50	—	—	—
38070	Frappuccino, tall coffee	12	fluid ounce(s)	360	—	190	4	38	0	3	1.50	—	—	—
38071	Frappuccino, tall espresso	12	fluid ounce(s)	360	—	160	4	33	0	2	1.50	—	—	—
38073	Frappuccino, mocha	12	fluid ounce(s)	360	—	220	5	44	0	3	1.50	—	—	—
38072	Frappuccino, tall mocha coconut	12	fluid ounce(s)	360	—	300	5	58	2	7	5.00	—	—	—
38080	Frappuccino, tall vanilla	12	fluid ounce(s)	360	—	260	11	47	0	4	1.00	—	—	—
38074	Frappuccino, tall white chocolate	12	fluid ounce(s)	360	—	240	5	48	0	4	2.50	—	—	—
33111	Latte, tall w/nonfat milk	12	fluid ounce(s)	360	335	123	12	17	0	1	0.40	0.16	0.02	0
33112	Latte, tall w/whole milk	12	fluid ounce(s)	360	325	212	11	17	0	11	6.90	3.24	0.42	—
33109	Macchiato, tall caramel w/nonfat milk	12	fluid ounce(s)	360	—	140	7	27	0	1	0.40	—	—	—
33110	Macchiato, tall caramel w/whole milk	12	fluid ounce(s)	360	—	190	6	27	0	7	4.00	—	—	—

Chol (mg)	Calc (mg)	Iron (mg)	Magn (mg)	Pota (mg)	Sodi (mg)	Zinc (mg)	Vit A (µg)	Thia (mg)	Vit E (mg α)	Ribo (mg)	Niac (mg)	Vit B$_6$ (mg)	Fola (µg)	Vit C (mg)	Vit B$_{12}$ (µg)	Sele (µg)
95	350	4.50	—	420	1310	—	100	—	—	—	—	—	—	2	—	—
255	300	2.70	—	260	930	—	115	—	0.72	—	—	—	—	1	—	—
25	256	0.25	29	412	188	0.82	59	0.10	—	0.44	0.40	0.10	7	2	1	5
27	331	0.23	27	415	215	0.88	57	0.07	0.11	0.44	0.33	0.10	16	0	1	5
70	0	0.36	—	—	450	—	—	—	—	—	—	—	—	0	—	—
40	150	1.80	—	—	830	—	—	—	—	—	—	—	—	6	—	—
40	150	2.70	—	—	810	—	—	—	—	—	—	—	—	6	—	—
65	250	2.70	—	—	1450	—	—	—	—	—	—	—	—	9	—	—
45	150	1.80	—	—	880	—	—	—	—	—	—	—	—	9	—	—
40	200	1.80	—	—	730	—	58	—	—	—	—	—	—	2	—	—
40	200	2.70	—	—	710	—	58	—	—	—	—	—	—	2	—	—
65	300	2.70	—	—	1300	—	—	—	—	—	—	—	—	4	—	—
40	200	1.44	—	—	700	—	58	—	—	—	—	—	—	2	—	—
15	80	1.44	—	—	340	—	38	—	—	—	—	—	—	1	—	—
15	80	1.86	—	—	400	—	—	—	—	—	—	—	—	4	—	—
10	80	1.80	—	—	280	—	—	—	—	—	—	—	—	4	—	—
15	150	1.80	—	—	490	—	—	—	—	—	—	—	—	9	—	—
15	150	2.70	—	—	470	—	—	—	—	—	—	—	—	9	—	—
35	250	2.70	—	—	980	—	—	—	—	—	—	—	—	12	—	—
15	150	1.44	—	—	480	—	—	—	—	—	—	—	—	12	—	—
25	20	0.00	—	—	550	—	0	—	—	—	—	—	—	0	—	—
10	0	0.00	—	—	340	—	0	—	—	—	—	—	—	0	—	—
0	0	1.08	—	—	15	—	0	—	—	—	—	—	—	0	0	—
25	250	0.00	—	—	95	—	0	—	—	—	—	—	—	1	0	—
3	200	0.00	—	—	100	—	0	—	—	—	—	—	—	0	0	—
0	250	0.72	—	—	75	—	0	—	—	—	—	—	—	0	0	—
5	300	0.72	—	—	150	—	0	—	—	—	—	—	—	0	0	—
70	350	1.08	—	—	140	—	0	—	—	—	—	—	—	2	0	—
0	0	0.00	—	—	0	—	0	—	—	—	—	—	—	0	0	—
0	0	0.00	—	—	0	—	0	—	—	—	—	—	—	0	0	—
13	200	1.08	—	—	80	—	—	—	—	—	—	—	—	0	—	—
15	250	2.70	—	—	420	—	0	—	—	—	—	—	—	12	—	—
20	250	1.08	—	—	390	—	0	—	—	—	—	—	—	0	—	—
12	220	0.00	—	—	110	—	—	—	—	—	—	—	—	0	—	—
10	150	0.00	—	—	180	—	0	—	—	—	—	—	—	0	0	—
3	400	1.80	—	—	300	—	0	—	—	—	—	—	—	5	0	—
10	150	1.44	—	—	220	—	0	—	—	—	—	—	—	0	0	—
10	150	0.00	—	—	180	—	0	—	—	—	—	—	—	0	0	—
10	100	0.00	—	—	160	—	0	—	—	—	—	—	—	0	0	—
10	150	0.72	—	—	180	—	0	—	—	—	—	—	—	0	0	—
10	150	1.08	—	—	220	—	0	—	—	—	—	—	—	0	0	—
3	400	0.00	—	—	280	—	0	—	—	—	—	—	—	4	0	—
10	150	0.00	—	—	210	—	0	—	—	—	—	—	—	0	0	—
6	420	0.18	40	—	174	1.35	—	0.12	—	0.47	0.36	0.14	18	4	1	—
46	400	0.18	47	254	165	1.28	—	0.13	—	0.54	0.35	0.14	17	3	1	—
25	250	0.36	—	—	110	—	—	—	—	—	—	—	—	2	—	—
25	200	0.36	—	—	105	—	—	—	—	—	—	—	—	1	—	—

Table A–1
Food Composition

(DA+ code is for Wadsworth Diet Analysis program)
(For purposes of calculations, use "0" for t, <1, <.1, <.01, etc.)

A

DA + Code	Food Description	Quantity	Measure	Wt (g)	H₂O (g)	Ener (kcal)	Prot (g)	Carb (g)	Fiber (g)	Fat (g)	Sat	Fat Breakdown (g) Mono	Poly	Trans
	FAST FOOD													
	Starbucks–Continued													
33107	Mocha coffee drink, tall nonfat, w/o whipped cream	12	fluid ounce(s)	360	—	180	12	33	1	2	1.50	0.68	0.08	—
38089	Mocha syrup	1	serving(s)	17	—	25	1	6	0	1	0.00	—	—	—
33108	Mocha, tall w/whole milk	12	fluid ounce(s)	360	—	340	12	33	1	20	12.00	3.48	0.44	—
38084	Tazo chai black tea, tall	12	fluid ounce(s)	360	—	210	6	36	0	5	3.50	—	—	—
38083	Tazo chai black tea, tall nonfat	12	fluid ounce(s)	360	—	170	6	37	0	0	0.00	0.00	0.00	0
38087	Tazo chai black tea, tall soy milk	12	fluid ounce(s)	360	—	190	4	39	1	2	0.00	—	—	—
38063	Tazo chai creme frappuccino, tall	12	fluid ounce(s)	360	—	280	11	51	0	4	1.00	—	—	—
38076	Tazo iced tea, tall	12	fluid ounce(s)	360	—	60	0	16	0	0	0.00	0.00	0.00	0
38077	Tazo tea, grande lemonade	16	fluid ounce(s)	480	—	120	0	31	0	0	0.00	0.00	0.00	0
38065	Tazoberry creme frappuccino, tall	12	fluid ounce(s)	360	—	240	4	54	1	1	0.00	—	—	—
38066	Tazoberry frappuccino, tall	12	fluid ounce(s)	360	—	140	1	36	1	0	0.00	0.00	0.00	0
38045	Vanilla creme steamed nonfat milk, tall w/whipped cream	12	fluid ounce(s)	360	—	180	12	32	0	0	0.00	0.00	0.00	—
38046	Vanilla creme steamed soy milk, tall w/whipped cream	12	fluid ounce(s)	360	—	300	8	37	1	12	6.00	—	—	—
38044	Vanilla creme steamed whole milk, tall w/whipped cream	12	fluid ounce(s)	360	—	340	10	31	0	18	12.00	—	—	—
38090	Whipped cream	1	serving(s)	27	—	100	0	2	0	9	6.00	—	—	—
38062	White chocolate mocha, tall nonfat w/o whipped cream	12	fluid ounce(s)	360	—	260	12	45	0	4	3.00	—	—	—
38061	White chocolate mocha, tall w/whipped cream	12	fluid ounce(s)	360	—	410	11	44	0	20	13.00	—	—	—
38048	White hot chocolate, tall w/o whipped cream	12	fluid ounce(s)	360	—	300	15	51	0	5	3.50	—	—	—
38047	White hot chocolate, tall w/whipped cream	12	fluid ounce(s)	360	—	460	13	50	0	22	15.00	—	—	—
38050	White hot chocolate soy milk, tall w/whipped cream	12	fluid ounce(s)	360	—	420	11	56	1	16	9.00	—	—	—
	Subway													
34023	Asiago caesar chicken wrap	1	item(s)	244	—	413	22	47	2	15	3.00	—	—	0
38622	Atkins-friendly chicken bacon ranch wrap	1	item(s)	213	—	480	40	19	11	27	9.00	—	—	0
38623	Atkins-friendly turkey bacon melt wrap	1	item(s)	199	—	430	32	22	12	25	9.00	—	—	0
34029	Bacon & egg breakfast sandwich	1	item(s)	127	—	302	14	29	1	15	4.00	—	—	0
32045	Chocolate chip cookie	1	item(s)	48	—	209	3	29	1	10	3.50	—	—	1.07
32048	Chocolate chip M&M cookie	1	item(s)	48	—	210	2	29	1	10	3.00	—	—	2.67
32049	Chocolate chunk cookie	1	item(s)	48	—	210	2	30	1	10	3.00	—	—	2.67
4024	Classic Italian B.M.T. sandwich, 6", white bread	1	item(s)	250	—	453	21	40	3	24	8.00	—	—	0
16397	Club salad	1	item(s)	323	—	145	17	12	3	4	1.00	—	—	0
3422	Club sandwich, 6", white bread	1	item(s)	253	—	294	22	40	3	5	1.50	—	—	0
4030	Cold cut trio sandwich, 6", white bread	1	item(s)	254	—	415	19	40	3	20	7.00	—	—	0
34030	Ham & egg breakfast sandwich	1	item(s)	147	—	291	15	30	1	12	3.00	—	—	0
3885	Ham sandwich, 6", white bread	1	item(s)	219	—	261	17	39	3	5	1.50	—	—	0
34026	Honey mustard melt sandwich, 6", Italian bread	1	item(s)	258	—	373	23	47	3	11	5.00	—	—	—
34027	Horseradish roast beef sandwich, 6", Italian bread	1	item(s)	230	—	401	18	42	3	17	3.00	—	—	—
4651	Meatball sandwich, 6", white bread	1	item(s)	284	—	501	23	46	4	25	10.00	—	—	0.75
15839	Melt sandwich, 6", white bread	1	item(s)	256	—	380	23	41	3	15	5.00	—	—	—
32046	Oatmeal raisin cookie	1	item(s)	48	—	197	3	29	1	8	2.00	—	—	2.67

Chol (mg)	Calc (mg)	Iron (mg)	Magn (mg)	Pota (mg)	Sodi (mg)	Zinc (mg)	Vit A (µg)	Thia (mg)	Vit E (mg α)	Ribo (mg)	Niac (mg)	Vit B$_6$ (mg)	Fola (µg)	Vit C (mg)	Vit B$_{12}$ (µg)	Sele (µg)
5	350	2.70	—	—	150	—	—	—	—	—	—	—	—	2	—	—
0	0	0.72	—	—	0	—	0	—	—	—	—	—	—	0	0	—
47	300	0.18	—	—	169	—	—	—	—	—	—	—	—	2	—	—
20	200	0.36	—	—	85	—	0	—	—	—	—	—	—	1	0	—
5	200	0.36	—	—	95	—	0	—	—	—	—	—	—	0	0	—
0	200	0.72	—	—	70	—	0	—	—	—	—	—	—	0	0	—
3	400	0.00	—	—	280	—	0	—	—	—	—	—	—	4	0	—
0	0	0.00	—	—	0	—	0	—	—	—	—	—	—	0	0	—
0	0	0.00	—	—	15	—	0	—	—	—	—	—	—	5	0	—
0	150	0.00	—	—	125	—	0	—	—	—	—	—	—	1	0	—
0	0	0.00	—	—	30	—	0	—	—	—	—	—	—	0	0	—
5	350	0.00	—	—	170	—	0	—	—	—	—	—	—	0	0	—
30	400	1.44	—	—	130	—	0	—	—	—	—	—	—	0	0	—
75	40	0.00	—	—	160	—	0	—	—	—	—	—	—	2	0	—
40	0	0.00	—	—	10	—	0	—	—	—	—	—	—	0	0	—
5	400	0.00	—	—	210	—	0	—	—	—	—	—	—	0	0	—
70	400	0.00	—	—	210	—	0	—	—	—	—	—	—	2	0	—
10	450	0.00	—	—	250	—	0	—	—	—	—	—	—	0	0	—
75	500	0.00	—	—	250	—	0	—	—	—	—	—	—	4	0	—
35	500	1.44	—	—	210	—	0	—	—	—	—	—	—	0	0	—
46	40	2.70	—	—	1320	—	—	—	—	—	—	—	—	15	—	—
90	350	2.70	—	—	1340	—	—	—	—	—	—	—	—	7	—	—
65	300	2.70	—	—	1650	—	—	—	—	—	—	—	—	5	—	—
185	60	1.80	—	—	480	—	—	—	—	—	—	—	—	15	—	—
12	0	1.00	—	—	135	—	0	—	—	—	—	—	—	0	—	—
13	0	1.00	—	—	135	—	0	—	—	—	—	—	—	0	—	—
12	0	1.00	—	—	150	—	0	—	—	—	—	—	—	0	—	—
56	100	2.70	—	—	1740	—	—	—	—	—	—	—	—	24	—	—
30	40	1.80	—	—	1070	—	—	—	—	—	—	—	—	30	—	—
30	40	3.60	—	—	1250	—	60	—	—	—	—	—	—	24	—	—
57	150	3.60	—	—	1670	—	100	—	—	—	—	—	—	24	—	—
189	60	2.70	—	—	700	—	67	—	—	—	—	—	—	15	—	—
25	40	2.70	—	—	1260	—	—	—	—	—	—	—	—	24	—	—
41	100	2.70	—	—	1570	—	—	—	—	—	—	—	—	24	—	—
27	40	3.60	—	—	880	—	—	—	—	—	—	—	—	24	—	—
56	100	3.60	—	—	1350	—	—	—	—	—	—	—	—	24	—	—
41	100	2.70	—	—	1690	—	—	—	—	—	—	—	—	24	—	—
14	0	1.00	—	—	180	—	0	—	—	—	—	—	—	0	—	—

A

Table A–1
Food Composition

(DA+ code is for Wadsworth Diet Analysis program)
(For purposes of calculations, use "0" for t, <1, <.1, <.01, etc.)

DA + Code	Food Description	Quantity	Measure	Wt (g)	H₂O (g)	Ener (kcal)	Prot (g)	Carb (g)	Fiber (g)	Fat (g)	Fat Breakdown (g)			
											Sat	Mono	Poly	*Trans*
	FAST FOOD													
	Subway–Continued													
32047	Peanut butter cookie	1	item(s)	48	—	220	3	26	1	12	3.00	—	—	1.07
3957	Roast beef sandwich, 6", white bread	1	item(s)	220	—	264	18	39	3	5	1.00	—	—	0
16403	Roasted chicken breast salad	1	item(s)	304	—	137	16	12	3	3	0.50	—	—	—
16378	Roasted chicken breast sandwich, 6", white bread	1	item(s)	234	—	311	25	40	3	6	1.50	—	—	0
34028	Southwest steak & cheese sandwich, 6", Italian bread	1	item(s)	255	—	412	23	42	4	18	6.00	—	—	—
4032	Spicy italian sandwich, 6", white bread	1	item(s)	213	—	458	19	42	2	24	9.00	—	—	0
4031	Steak & cheese sandwich, 6", white bread	1	item(s)	253	—	362	23	41	4	13	4.50	—	—	0
34024	Steak & cheese wrap	1	item(s)	245	—	353	22	46	3	9	4.00	—	—	—
32050	Sugar cookie	1	item(s)	48	—	222	2	28	1	12	3.00	—	—	3.73
16402	Tuna salad	1	item(s)	314	—	238	13	11	3	16	4.00	—	—	—
15844	Tuna sandwich, 6", white bread	1	item(s)	252	—	419	18	39	3	21	5.00	—	—	—
15834	Turkey breast & ham sandwich, 6", white bread	1	item(s)	229	—	267	18	40	3	5	1.00	—	—	0
34025	Turkey breast & bacon wrap	1	item(s)	228	—	318	19	45	2	7	2.50	—	—	—
16376	Turkey breast sandwich, 6", white bread	1	item(s)	220	—	254	16	39	3	4	1.00	—	—	0
16375	Veggie delite, 6", white bread	1	item(s)	163	—	200	7	37	3	3	0.50	—	—	0
32051	White macadamia nut cookie	1	item(s)	48	—	221	2	27	1	12	3.00	—	—	1.07
	Taco Bell													
29906	7-layer burrito	1	item(s)	283	—	530	18	67	10	22	8.00	—	—	3
744	Bean burrito	1	item(s)	198	—	370	14	55	8	10	3.50	—	—	2
749	Beef burrito supreme	1	item(s)	248	—	440	18	51	7	18	8.00	—	—	2
33417	Beef chalupa supreme	1	item(s)	153	—	390	14	31	3	24	10.00	—	—	3
29910	Beef gordita supreme	1	item(s)	153	—	310	14	30	3	16	7.00	—	—	0.50
2014	Beef soft taco	1	item(s)	99	—	210	10	21	2	10	4.50	—	—	1
10860	Beef soft taco supreme	1	item(s)	134	—	260	11	22	3	14	7.00	—	—	1
2018	Big beef burrito supreme	1	item(s)	291	—	510	23	52	11	23	9.00	6.55	1.61	—
14467	Big chicken burrito supreme	1	item(s)	255	—	460	27	50	3	17	6.00	—	—	—
34472	Chicken burrito supreme	1	item(s)	248	—	410	21	50	5	14	6.00	—	—	2
33418	Chicken chalupa supreme	1	item(s)	153	—	370	17	30	1	20	8.00	—	—	3
29900	Chicken fajita wrap supreme	1	item(s)	255	—	510	20	53	3	24	7.76	—	—	—
29895	Choco taco ice cream dessert	1	item(s)	113	—	310	3	37	1	17	10.00	—	—	—
10794	Cinnamon twists	1	serving(s)	35	—	160	0	28	0	5	1.00	—	—	1.50
14465	Grilled chicken burrito	1	item(s)	198	—	390	19	49	3	13	4.00	—	—	—
29911	Grilled chicken gordita supreme	1	item(s)	153	—	290	17	28	2	12	5.00	—	—	0
14463	Grilled chicken soft taco	1	item(s)	99	—	190	14	19	0	6	2.50	—	—	—
29912	Grilled steak gordita supreme	1	item(s)	153	—	290	16	28	2	13	6.00	—	—	0.50
29904	Grilled steak soft taco	1	item(s)	127	—	280	12	21	1	17	4.50	—	—	1
29905	Grilled steak soft taco supreme	1	item(s)	135	—	240	15	20	2	11	5.00	—	—	—
2021	Mexican pizza	1	serving(s)	216	—	550	21	46	7	31	11.00	—	—	5
2011	Nachos	1	serving(s)	99	—	320	5	33	2	19	4.50	—	—	5
2012	Nachos bellgrande	1	serving(s)	308	—	780	20	80	12	43	13.00	—	—	10
34473	Steak burrito supreme	1	item(s)	248	—	420	19	50	6	16	7.00	—	—	2
33419	Steak chalupa supreme	1	item(s)	153	—	370	15	29	2	22	8.00	—	—	3
29899	Steak fajita wrap supreme	1	item(s)	255	—	510	21	52	3	25	8.00	—	—	—
747	Taco	1	item(s)	78	—	170	8	13	3	10	4.00	—	—	0.50
2015	Taco salad w/salsa, with shell	1	serving(s)	533	—	790	31	73	13	42	15.00	—	—	8.75
14459	Taco supreme	1	item(s)	113	—	220	9	14	3	14	7.00	—	—	1
748	Tostada	1	item(s)	170	—	250	11	29	7	10	4.00	—	—	1.50
29901	Veggie fajita wrap supreme	1	item(s)	255	—	470	11	55	3	22	7.00	—	—	—

Chol (mg)	Calc (mg)	Iron (mg)	Magn (mg)	Pota (mg)	Sodi (mg)	Zinc (mg)	Vit A (µg)	Thia (mg)	Vit E (mg α)	Ribo (mg)	Niac (mg)	Vit B6 (mg)	Fola (µg)	Vit C (mg)	Vit B12 (µg)	Sele (µg)
0	0	1.00	—	—	200	—	0	—	—	—	—	—	—	0	—	—
20	40	3.60	—	—	840	—	60	—	—	—	—	—	—	24	—	—
36	40	1.08	—	—	730	—	—	—	—	—	—	—	—	30	—	—
48	60	3.60	—	—	880	—	—	—	—	—	—	—	—	24	—	—
44	100	6.30	—	—	1120	—	—	—	—	—	—	—	—	24	—	—
57	30	3.00	—	—	1498	—	—	—	—	—	—	—	—	13	—	—
37	100	6.30	—	—	1200	—	—	—	—	—	—	—	—	24	—	—
37	150	7.20	—	—	1400	—	—	—	—	—	—	—	—	15	—	—
18	0	1.00	—	—	170	—	0	—	—	—	—	—	—	0	—	—
42	100	1.08	—	—	880	—	177	—	—	—	—	—	—	30	—	—
42	100	2.70	—	—	1180	—	100	—	—	—	—	—	—	24	—	—
23	40	2.70	—	—	1210	—	—	—	—	—	—	—	—	24	—	—
24	60	2.70	—	—	1490	—	—	—	—	—	—	—	—	15	—	—
15	40	2.70	—	—	1000	—	—	—	—	—	—	—	—	24	—	—
0	40	1.80	—	—	500	—	—	—	—	—	—	—	—	24	—	—
13	0	1.00	—	—	140	—	0	—	—	—	—	—	—	0	—	—
25	300	3.59	—	—	1360	—	—	—	—	—	—	—	—	5	—	—
10	200	2.69	—	—	1200	—	53	—	—	—	—	—	—	5	—	—
40	200	2.70	—	—	1330	—	351	—	—	—	—	—	—	9	—	—
40	150	1.80	—	—	600	—	—	—	—	—	—	—	—	5	—	—
35	150	2.70	—	—	590	—	—	—	—	—	—	—	—	5	—	—
25	100	1.80	—	—	620	—	44	—	—	—	—	—	—	2	—	—
40	150	1.80	—	—	630	—	73	—	—	—	—	—	—	5	—	—
60	150	2.70	—	493	1500	—	877	—	—	0.07	—	—	—	5	—	—
70	101	1.46	—	—	1200	—	—	—	—	—	—	—	—	2	—	—
45	200	2.70	—	—	1270	—	—	—	—	—	—	—	—	9	—	—
45	100	1.08	—	—	530	—	—	—	—	—	—	—	—	5	—	—
57	165	1.52	—	—	1182	—	—	—	—	—	—	—	—	7	—	—
20	60	0.72	—	—	100	—	—	—	—	—	—	—	—	0	—	—
0	0	0.37	—	—	150	—	0	—	—	—	—	—	—	0	—	—
40	151	1.44	—	—	1240	—	—	—	—	—	—	—	—	2	—	—
45	100	1.80	—	—	530	—	—	—	—	—	—	—	—	5	—	—
30	100	1.08	—	—	550	—	15	—	—	—	—	—	—	1	—	—
35	100	2.70	—	—	520	—	—	—	—	—	—	—	—	4	—	—
30	100	1.44	—	—	650	—	29	—	—	—	—	—	—	4	—	—
35	100	1.08	—	—	510	—	29	—	—	—	—	—	—	4	—	—
45	350	3.60	—	—	1030	—	—	—	—	—	—	—	—	6	—	—
4	80	0.72	—	—	530	—	0	—	—	—	—	—	—	0	—	—
35	200	2.70	—	—	1300	—	162	—	—	—	—	—	—	6	—	—
35	200	2.70	—	—	1260	—	789	—	—	—	—	—	—	9	—	—
35	100	1.44	—	—	520	—	—	—	—	—	—	—	—	4	—	—
50	150	1.80	—	—	1200	—	—	—	—	—	—	—	—	6	—	—
25	60	1.08	—	—	350	—	44	—	—	—	—	—	—	2	—	—
65	400	6.23	—	—	1670	—	—	—	—	—	—	—	—	21	—	—
40	80	1.44	—	—	360	—	73	—	—	—	—	—	—	5	—	—
15	150	1.44	—	—	710	—	281	—	—	—	—	—	—	5	—	—
30	150	1.44	—	—	990	—	—	—	—	—	—	—	—	6	—	—

Table A-1
Food Composition

(DA+ code is for Wadsworth Diet Analysis program)
(For purposes of calculations, use "0" for t, <1, <.1, <.01, etc.)

A

DA + Code	Food Description	Quantity	Measure	Wt (g)	H2O (g)	Ener (kcal)	Prot (g)	Carb (g)	Fiber (g)	Fat (g)	Sat	Mono	Poly	Trans
	CONVENIENCE MEALS													
	Banquet													
29961	Barbeque chicken meal	1	item(s)	281	—	330	16	37	2	13	3.00	—	—	—
14788	Boneless white fried chicken meal	1	item(s)	234	—	490	14	49	2	27	7.00	—	—	—
29960	Fish sticks meal	1	item(s)	187	—	270	13	31	3	10	3.00	—	—	—
29957	Lasagna with meat sauce meal	1	item(s)	312	—	320	15	46	7	9	4.00	—	—	—
14777	Macaroni & cheese meal	1	item(s)	340	—	420	15	57	5	14	8.00	—	—	—
1741	Meatloaf meal	1	item(s)	269	—	240	14	20	4	11	4.00	—	—	—
39418	Pepperoni pizza meal	1	item(s)	191	—	480	11	56	5	23	8.00	—	—	—
33759	Roasted white turkey meal	1	item(s)	255	—	230	14	30	5	6	2.00	—	—	—
1743	Salisbury steak meal	1	item(s)	269	197	380	12	28	3	24	12.00	—	—	—
	Budget Gourmet													
1914	Cheese manicotti w/meat sauce	1	item(s)	284	194	420	18	38	4	22	11.00	6.00	1.34	
1915	Chicken w/fettucini	1	item(s)	284	—	380	20	33	3	19	10.00	—	—	
3986	Light beef stroganoff	1	item(s)	248	177	290	20	32	3	7	4.00	—	—	
3996	Light sirloin of beef in herb sauce	1	item(s)	269	214	260	19	30	5	7	4.00	2.30	0.31	
3987	Light vegetable lasagna	1	item(s)	298	227	290	15	36	5	9	1.79	0.89	0.60	
	Healthy Choice													
36979	Bowls chicken teriyaki with rice	1	item(s)	298	—	330	19	50	5	6	2.00	2.00	2.00	
9425	Cheese French bread pizza	1	item(s)	170	—	360	20	57	5	5	1.50	—	—	
9306	Chicken enchilada suprema meal	1	item(s)	320	252	360	13	59	8	7	3.00	2.00	2.00	
9316	Lemon pepper fish meal	1	item(s)	303	—	280	11	49	5	5	2.00	1.00	2.00	
9322	Traditional salisbury steak meal	1	item(s)	354	250	360	23	45	5	9	3.50	4.00	1.00	
9359	Traditional turkey breasts meal	1	item(s)	298	—	330	21	50	4	5	2.00	1.50	1.50	
9451	Zucchini lasagna	1	item(s)	383	—	280	13	47	5	4	2.50	—	—	
	Stouffers													
2363	Cheese enchiladas with mexican rice	1	serving(s)	276	—	370	12	48	5	14	5.00	—	—	
2313	Cheese French bread pizza	1	serving(s)	294	—	370	14	43	3	16	6.00	—	—	
11138	Cheese manicotti w/tomato sauce	1	item(s)	255	—	330	17	35	3	13	8.00	—	—	
2366	Chicken pot pie	1	item(s)	284	—	740	23	56	4	47	18.00	12.41	10.48	
11116	Homestyle baked chicken breast w/mashed potatoes & gravy	1	item(s)	252	—	260	19	21	1	11	3.00	—	—	
11146	Homestyle beef pot roast & potatoes	1	item(s)	252	—	270	16	25	3	12	4.50	—	—	
11152	Homestyle roast turkey breast w/stuffing & mashed potatoes	1	item(s)	273	—	300	16	34	2	11	3.00	—	—	
11043	Lean Cuisine Cafe Classics baked chicken & whipped potatoes w/stuffing	1	item(s)	227	—	240	17	33	3	5	1.50	1.50	1.00	0
11046	Lean Cuisine Cafe Classics honey mustard chicken	1	item(s)	213	—	260	18	37	1	4	1.50	1.00	1.00	0
360	Lean Cuisine Everyday Favorites chicken chow mein w/rice	1	item(s)	255	—	210	12	33	2	3	1.00	1.00	0.50	0
9467	Lean Cuisine Everyday Favorites fettucini alfredo	1	item(s)	262	—	280	13	40	2	7	3.50	2.00	1.00	0
11055	Lean Cuisine Everyday Favorites lasagna w/meat sauce	1	item(s)	291	—	300	19	41	3	8	4.00	2.00	0.50	0
9479	Lean Cuisine French bread deluxe pizza	1	item(s)	174	—	330	18	44	3	9	3.50	1.50	1.00	0
	Weight Watchers													
11164	Smart Ones chicken enchiladas suiza entree	1	serving(s)	255	—	270	15	33	2	9	3.50	—	—	
11155	Smart Ones garden lasagna entree	1	item(s)	312	—	270	14	36	5	7	3.50	—	—	
11187	Smart Ones pepperoni pizza	1	item(s)	158	—	390	23	46	4	12	4.00	—	—	

A

Chol (mg)	Calc (mg)	Iron (mg)	Magn (mg)	Pota (mg)	Sodi (mg)	Zinc (mg)	Vit A (µg)	Thia (mg)	Vit E (mg α)	Ribo (mg)	Niac (mg)	Vit B$_6$ (mg)	Fola (µg)	Vit C (mg)	Vit B$_{12}$ (µg)	Sele (µg)
50	40	1.08	—	—	1210	—	0	—	—	—	—	—	—	5	—	—
65	60	1.08	—	—	1150	—	—	—	—	—	—	—	—	0	—	—
30	60	1.44	—	—	690	—	—	—	—	—	—	—	—	2	—	—
20	100	2.70	—	—	1170	—	—	—	—	—	—	—	—	0	—	—
20	150	1.44	—	—	1330	—	0	—	—	—	—	—	—	0	—	—
30	0	1.80	—	—	1040	—	0	—	—	—	—	—	—	0	—	—
35	150	1.80	—	—	870	—	0	—	—	—	—	—	—	0	—	—
25	60	1.80	—	—	1070	—	—	—	—	—	—	—	—	4	—	—
60	40	1.44	—	—	1140	—	0	—	—	—	—	—	—	0	—	—
85	300	2.70	45	484	810	2.29	—	0.45	—	0.51	4.00	0.23	31	0	1	—
85	100	2.70	—	—	810	—	—	0.15	—	0.43	6.00	—	—	0	—	—
35	40	1.80	39	280	580	4.71	—	0.17	—	0.37	4.28	0.27	19	2	3	—
30	40	1.80	58	540	850	4.81	—	0.16	—	0.29	5.53	0.37	38	6	2	—
15	283	3.03	79	420	780	1.39	—	0.22	—	0.45	3.13	0.32	75	59	<1	—
40	20	0.72	—	—	600	—	—	—	—	—	—	—	—	15	—	—
10	350	3.60	—	—	600	—	—	—	—	—	—	—	—	12	—	—
30	40	1.44	—	—	580	—	—	—	—	—	—	—	—	4	—	—
30	40	0.36	—	—	580	—	—	—	—	—	—	—	—	30	—	—
45	80	2.70	—	—	580	—	—	—	—	—	—	—	—	21	—	—
35	40	1.44	—	—	600	—	—	—	—	—	—	—	—	0	—	—
10	200	1.80	—	—	310	—	—	—	—	—	—	—	—	0	—	—
25	200	1.44	—	360	890	—	—	—	—	—	—	—	—	12	—	—
15	200	1.80	—	240	880	—	—	—	—	—	—	—	—	0	—	—
40	350	1.08	—	430	810	—	—	—	—	—	—	—	—	1	—	—
65	150	2.70	—	—	1170	—	—	—	—	—	—	—	—	2	—	—
50	20	0.72	—	500	760	—	0	—	—	—	—	—	—	0	—	—
35	20	1.80	—	790	820	—	—	—	—	—	—	—	—	6	—	—
35	40	0.72	—	450	1190	—	0	—	—	—	—	—	—	0	—	—
30	80	0.72	—	480	690	—	—	—	—	—	—	—	—	0	—	—
35	60	0.36	—	370	640	—	—	—	—	—	—	—	—	0	—	—
30	20	0.36	—	310	620	—	—	—	—	—	—	—	—	0	—	—
20	200	0.36	—	260	670	—	0	—	—	—	—	—	—	0	—	—
30	200	1.08	—	590	650	—	—	—	—	—	—	—	—	5	—	—
20	100	1.80	—	390	630	—	—	—	—	—	—	—	—	9	—	—
50	250	1.08	—	—	660	—	—	—	—	—	—	—	—	4	—	—
30	350	1.80	—	—	610	—	—	—	—	—	—	—	—	6	—	—
45	450	1.80	—	320	650	—	55	—	—	—	—	—	—	5	—	—

Table A–1
Food Composition

(DA+ code is for Wadsworth Diet Analysis program)
(For purposes of calculations, use "0" for t, <1, <.1, <.01, etc.)

DA + Code	Food Description	Quantity	Measure	Wt (g)	H₂O (g)	Ener (kcal)	Prot (g)	Carb (g)	Fiber (g)	Fat (g)	Fat Breakdown (g)			
											Sat	Mono	Poly	Trans
	CONVENIENCE MEALS													
	Weight Watchers–Continued													
31514	Smart Ones spicy penne pasta & ricotta	1	item(s)	289	—	280	11	45	4	6	2.00	—	—	—
31512	Smart Ones spicy szechuan style vegetables & chicken	1	item(s)	255	—	220	11	39	3	2	0.50	—	—	—
	BABY FOODS													
787	Apple juice	4	fluid ounce(s)	127	112	60	0	15	<1	<1	0.02	0.00	0.04	—
778	Applesauce, strained	4	tablespoon(s)	64	55	31	<1	8	1	<1	0.02	0.01	0.04	—
779	Bananas w/tapioca, strained	4	tablespoon(s)	60	50	34	<1	9	1	<.1	0.02	0.01	0.01	—
604	Carrots, strained	4	tablespoon(s)	56	52	15	<1	3	1	<.1	0.01	0.00	0.03	—
770	Chicken noodle dinner, strained	4	tablespoon(s)	64	55	42	2	6	1	1	0.38	0.55	0.30	—
801	Green beans, strained	4	tablespoon(s)	60	0.05	15	0.77	3.53	1.13	0.05	0.01	0	0.03	—
910	Human milk, mature	2	fluid ounce(s)	62	54	43	1	4	0	3	1.24	1.02	0.31	—
760	Mixed cereal, prepared w/whole milk	4	ounce(s)	114	85	128	5	18	1	4	2.19	1.25	0.43	—
772	Mixed vegetable dinner, strained	2	ounce(s)	57	50	23	1	5	1	<.1	0.00	0.00	0.06	—
762	Rice cereal, prepared w/whole milk	4	ounce(s)	114	85	131	4	19	<1	4	2.64	1.02	0.16	—
758	Teething biscuits	1	item(s)	11	1	43	1	8	<1	<1	0.17	0.16	0.09	—

Chol (mg)	Calc (mg)	Iron (mg)	Magn (mg)	Pota (mg)	Sodi (mg)	Zinc (mg)	Vit A (µg)	Thia (mg)	Vit E (mg α)	Ribo (mg)	Niac (mg)	Vit B$_6$ (mg)	Fola (µg)	Vit C (mg)	Vit B$_{12}$ (µg)	Sele (µg)
5	150	2.70	—	250	400	—	—	—	—	—	—	—	—	6	—	—
10	150	1.80	—	—	730	—	—	—	—	—	—	—	—	2	—	—
0	5	0.72	4	115	4	0.04	1	0.01	0.76	0.02	0.11	0.04	0	73	0	<1
0	3	0.14	2	45	1	0.01	1	0.01	0.38	0.02	0.04	0.02	1	25	0	<1
0	3	0.12	6	53	5	0.04	1	0.01	0.36	0.02	0.11	0.07	4	10	0	<1
0	12	0.21	5	110	21	0.08	321	0.01	0.29	0.02	0.26	0.04	8	3	0	<1
10	17	0.41	9	89	15	0.35	70	0.03	0.13	0.04	0.46	0.04	7	<.1	<.1	2
0	23.39	0.44	14.39	94.8	1.2	0.12	27	0.01	0.31	0.05	0.2	0.02	21	3.11	0	0.18
9	20	0.02	2	31	10	0.10	38	0.01	0.05	0.02	0.11	0.01	3	3	<.1	1
12	250	11.85	31	226	53	0.81	28	0.49	—	0.66	6.56	0.07	12	1	<.1	—
0	12	0.19	6	69	5	0.09	77	0.01	—	0.02	0.29	0.04	5	2	0	<1
12	272	13.85	51	216	52	0.73	25	0.53	—	0.57	5.91	0.13	9	1	<1	4
0	29	0.39	4	36	40	0.10	3	0.03	0.03	0.06	0.48	0.01	5	1	<.1	3

WHO: Nutrition Recommendations
Canada: Guidelines and Meal Planning

This appendix presents nutrition recommendations from the World Health Organization (WHO) and details for Canadians on the *Eating Well with Canada's Food Guide* and the *Beyond the Basics* meal planning system.

B

Nutrition Recommendations from WHO

The World Health Organization (WHO) has assessed the relationships between diet and the development of chronic diseases. Its recommendations include:

- Energy: sufficient to support growth, physical activity, and a healthy body weight (BMI between 18.5 and 24.9) and to avoid weight gain greater than 11 pounds (5 kilograms) during adult life

- Total fat: 15 to 30 percent of total energy

- Saturated fatty acids: <10 percent of total energy

- Polyunsaturated fatty acids: 6 to 10 percent of total energy

- Omega-6 polyunsaturated fatty acids: 5 to 8 percent of total energy

- Omega-3 polyunsaturated fatty acids: 1 to 2 percent of total energy

- *Trans* fatty acids: <1 percent of total energy

- Total carbohydrate: 55 to 75 percent of total energy

- Sugars: <10 percent of total energy

- Protein: 10 to 15 percent of total energy

- Cholesterol: <300 mg per day

- Salt (sodium): <5 g salt per day (<2 g sodium per day), appropriately iodized

- Fruits and vegetables: ≥400 g per day (about 1 pound)

- Total dietary fiber: >25 g per day from foods

- Physical activity: one hour of moderate-intensity activity, such as walking, on most days of the week

Eating Well with Canada's Food Guide

Figure B–1 presents the 2007 *Eating Well with Canada's Food Guide,* which interprets Canada's *Guidelines for Healthy Eating* (see Table 1-5 on p. 17) for consumers and recommends a range of servings to consume daily from each of the four food groups. Additional publications, which are available from Health Canada ■ through its website, provide many more details.

- Search for "*Canada's* food guide" at Health Canada: **www.hc-sc.gc.ca**

FIGURE B-1

Eating Well with Canada's Food Guide

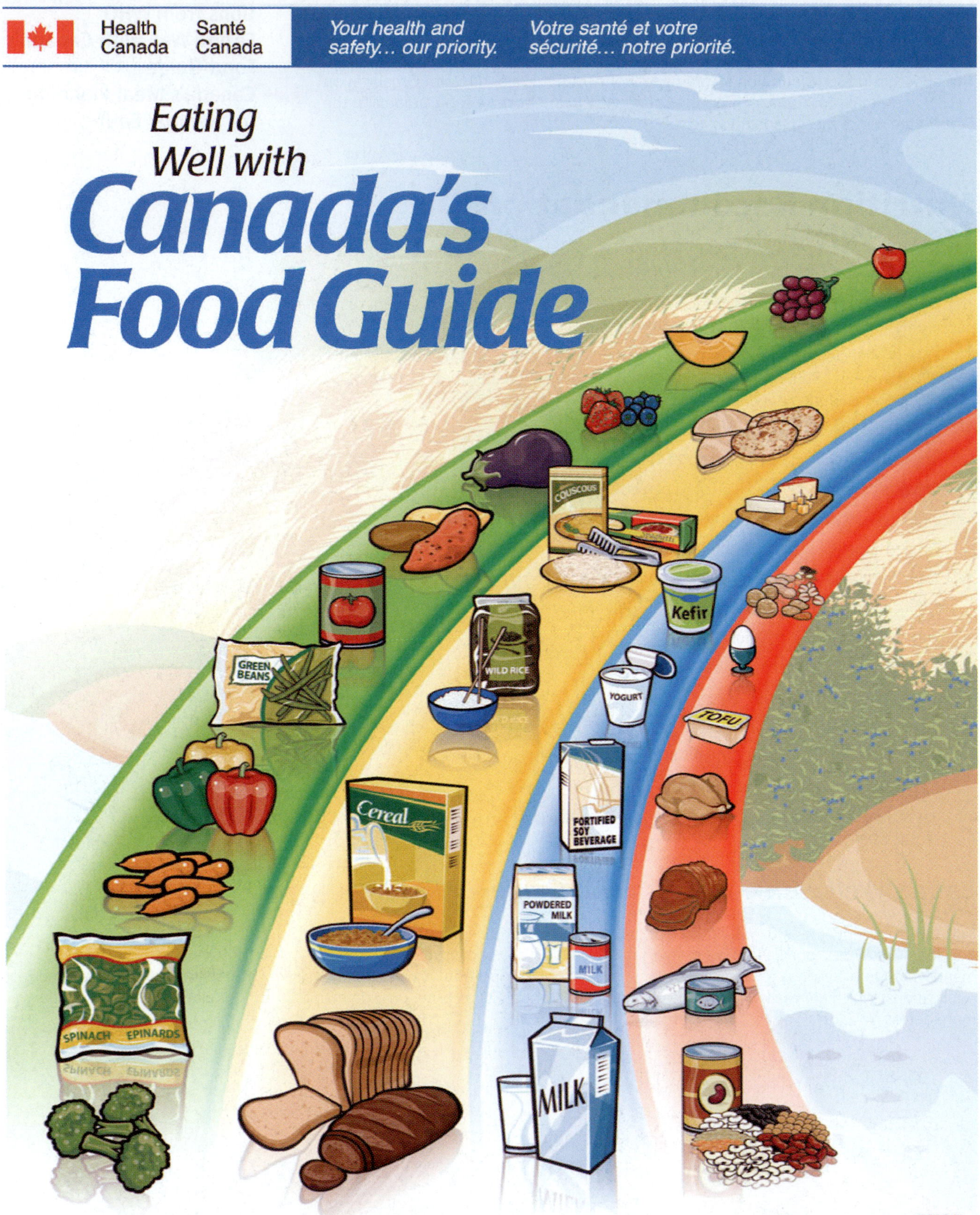

Health Canada Santé Canada
Your health and safety... our priority.
Votre santé et votre sécurité... notre priorité.
Eating Well with Canada's Food Guide
COUSCOUS
Kefir
GREEN BEANS
WILD RICE
YOGURT
TOFU
Cereal
FORTIFIED SOY BEVERAGE
POWDERED MILK
MILK
SPINACH EPINARDS
MILK
Canada

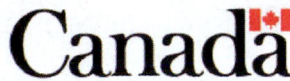

Recommended Number of *Food Guide Servings* per Day

	Children			Teens		Adults			
Age in Years	2-3	4-8	9-13	14-18		19-50		51+	
Sex	Girls and Boys			Females	Males	Females	Males	Females	Males
Vegetables and Fruit	4	5	6	7	8	7-8	8-10	7	7
Grain Products	3	4	6	6	7	6-7	8	6	7
Milk and Alternatives	2	2	3-4	3-4	3-4	2	2	3	3
Meat and Alternatives	1	1	1-2	2	3	2	3	2	3

The chart above shows how many Food Guide Servings you need from each of the four food groups every day.

Having the amount and type of food recommended and following the tips in *Canada's Food Guide* will help:

- Meet your needs for vitamins, minerals and other nutrients.
- Reduce your risk of obesity, type 2 diabetes, heart disease, certain types of cancer and osteoporosis.
- Contribute to your overall health and vitality.

FIGURE B-1

Eating Well with Canada's Food Guide—continued

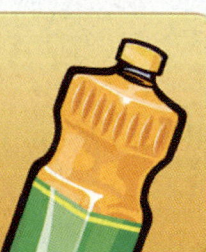

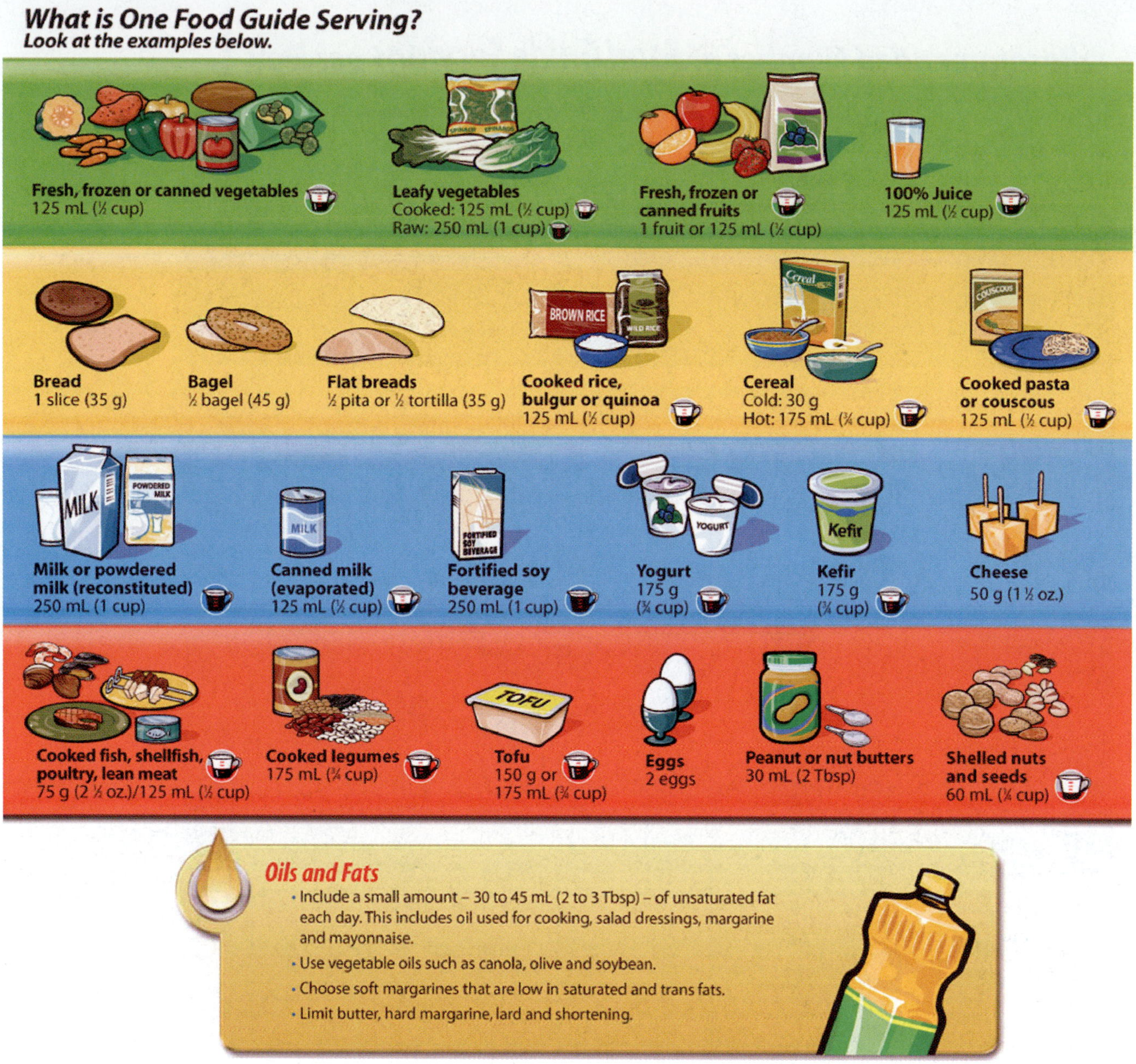
What is One Food Guide Serving?
Look at the examples below.
Fresh, frozen or canned vegetables
125 mL (½ cup)
Leafy vegetables
Cooked: 125 mL (½ cup)
Raw: 250 mL (1 cup)
Fresh, frozen or canned fruits
1 fruit or 125 mL (½ cup)
100% Juice
125 mL (½ cup)
Bread
1 slice (35 g)
Bagel
½ bagel (45 g)
Flat breads
½ pita or ½ tortilla (35 g)
Cooked rice, bulgur or quinoa
125 mL (½ cup)
Cereal
Cold: 30 g
Hot: 175 mL (¾ cup)
Cooked pasta or couscous
125 mL (½ cup)
Milk or powdered milk (reconstituted)
250 mL (1 cup)
Canned milk (evaporated)
125 mL (½ cup)
Fortified soy beverage
250 mL (1 cup)
Yogurt
175 g
(¾ cup)
Kefir
175 g
(¾ cup)
Cheese
50 g (1 ½ oz.)
Cooked fish, shellfish, poultry, lean meat
75 g (2 ½ oz.)/125 mL (½ cup)
Cooked legumes
175 mL (¾ cup)
Tofu
150 g or
175 mL (¾ cup)
Eggs
2 eggs
Peanut or nut butters
30 mL (2 Tbsp)
Shelled nuts and seeds
60 mL (¼ cup)
Oils and Fats
• Include a small amount – 30 to 45 mL (2 to 3 Tbsp) – of unsaturated fat each day. This includes oil used for cooking, salad dressings, margarine and mayonnaise.
• Use vegetable oils such as canola, olive and soybean.
• Choose soft margarines that are low in saturated and trans fats.
• Limit butter, hard margarine, lard and shortening.

Make each Food Guide Serving count...
wherever you are – at home, at school, at work or when eating out!

▸ **Eat at least one dark green and one orange vegetable each day.**
- Go for dark green vegetables such as broccoli, romaine lettuce and spinach.
- Go for orange vegetables such as carrots, sweet potatoes and winter squash.

▸ **Choose vegetables and fruit prepared with little or no added fat, sugar or salt.**
- Enjoy vegetables steamed, baked or stir-fried instead of deep-fried.

▸ **Have vegetables and fruit more often than juice.**

▸ **Make at least half of your grain products whole grain each day.**
- Eat a variety of whole grains such as barley, brown rice, oats, quinoa and wild rice.
- Enjoy whole grain breads, oatmeal or whole wheat pasta.

▸ **Choose grain products that are lower in fat, sugar or salt.**
- Compare the Nutrition Facts table on labels to make wise choices.
- Enjoy the true taste of grain products. When adding sauces or spreads, use small amounts.

▸ **Drink skim, 1%, or 2% milk each day.**
- Have 500 mL (2 cups) of milk every day for adequate vitamin D.
- Drink fortified soy beverages if you do not drink milk.

▸ **Select lower fat milk alternatives.**
- Compare the Nutrition Facts table on yogurts or cheeses to make wise choices.

▸ **Have meat alternatives such as beans, lentils and tofu often.**

▸ **Eat at least two Food Guide Servings of fish each week.***
- Choose fish such as char, herring, mackerel, salmon, sardines and trout.

▸ **Select lean meat and alternatives prepared with little or no added fat or salt.**
- Trim the visible fat from meats. Remove the skin on poultry.
- Use cooking methods such as roasting, baking or poaching that require little or no added fat.
- If you eat luncheon meats, sausages or prepackaged meats, choose those lower in salt (sodium) and fat.

Enjoy a variety of foods from the four food groups.

Satisfy your thirst with water!

Drink water regularly. It's a calorie-free way to quench your thirst. Drink more water in hot weather or when you are very active.

* Health Canada provides advice for limiting exposure to mercury from certain types of fish. Refer to www.healthcanada.gc.ca for the latest information.

B

Advice for different ages and stages...

Children

Following *Canada's Food Guide* helps children grow and thrive.

Young children have small appetites and need calories for growth and development.

- Serve small nutritious meals and snacks each day.

- Do not restrict nutritious foods because of their fat content. Offer a variety of foods from the four food groups.

- Most of all... be a good role model.

Women of childbearing age

All women who could become pregnant and those who are pregnant or breastfeeding need a multivitamin containing **folic acid** every day. Pregnant women need to ensure that their multivitamin also contains **iron**. A health care professional can help you find the multivitamin that's right for you.

Pregnant and breastfeeding women need more calories. Include an extra 2 to 3 Food Guide Servings each day.

Here are two examples:
- Have fruit and yogurt for a snack, or
- Have an extra slice of toast at breakfast and an extra glass of milk at supper.

Men and women over 50

The need for **vitamin D** increases after the age of 50.

In addition to following *Canada's Food Guide*, everyone over the age of 50 should take a daily vitamin D supplement of 10 µg (400 IU).

How do I count Food Guide Servings in a meal?

Here is an example:

Vegetable and beef stir-fry with rice, a glass of milk and an apple for dessert		
250 mL (1 cup) mixed broccoli, carrot and sweet red pepper	=	2 **Vegetables and Fruit** Food Guide Servings
75 g (2 ½ oz.) lean beef	=	1 **Meat and Alternatives** Food Guide Serving
250 mL (1 cup) brown rice	=	2 **Grain Products** Food Guide Servings
5 mL (1 tsp) canola oil	=	part of your **Oils and Fats** intake for the day
250 mL (1 cup) 1% milk	=	1 **Milk and Alternatives** Food Guide Serving
1 apple	=	1 **Vegetables and Fruit** Food Guide Serving

Eat well and be active today and every day!

The benefits of eating well and being active include:

- Better overall health.
- Lower risk of disease.
- A healthy body weight.
- Feeling and looking better.
- More energy.
- Stronger muscles and bones.

Be active

To be active every day is a step towards better health and a healthy body weight.

Canada's Physical Activity Guide recommends building 30 to 60 minutes of moderate physical activity into daily life for adults and at least 90 minutes a day for children and youth. You don't have to do it all at once. Add it up in periods of at least 10 minutes at a time for adults and five minutes at a time for children and youth.

Start slowly and build up.

Eat well

Another important step towards better health and a healthy body weight is to follow *Canada's Food Guide* by:

- Eating the recommended amount and type of food each day.
- Limiting foods and beverages high in calories, fat, sugar or salt (sodium) such as cakes and pastries, chocolate and candies, cookies and granola bars, doughnuts and muffins, ice cream and frozen desserts, french fries, potato chips, nachos and other salty snacks, alcohol, fruit flavoured drinks, soft drinks, sports and energy drinks, and sweetened hot or cold drinks.

Read the label

- Compare the Nutrition Facts table on food labels to choose products that contain less fat, saturated fat, trans fat, sugar and sodium.
- Keep in mind that the calories and nutrients listed are for the amount of food found at the top of the Nutrition Facts table.

Limit trans fat

When a Nutrition Facts table is not available, ask for nutrition information to choose foods lower in trans and saturated fats.

Nutrition Facts
Per 0 mL (0 g)

Amount	% Daily Value
Calories 0	
Fat 0 g	0 %
Saturates 0 g	0 %
+ Trans 0 g	
Cholesterol 0 mg	
Sodium 0 mg	0 %
Carbohydrate 0 g	0 %
Fibre 0 g	0 %
Sugars 0 g	
Protein 0 g	

Vitamin A	0 %	Vitamin C	0 %
Calcium	0 %	Iron	0 %

Take a step today...

✓ Have breakfast every day. It may help control your hunger later in the day.

✓ Walk wherever you can – get off the bus early, use the stairs.

✓ Benefit from eating vegetables and fruit at all meals and as snacks.

✓ Spend less time being inactive such as watching TV or playing computer games.

✓ Request nutrition information about menu items when eating out to help you make healthier choices.

✓ Enjoy eating with family and friends!

✓ Take time to eat and savour every bite!

For more information, interactive tools, or additional copies visit Canada's Food Guide on-line at:
www.healthcanada.gc.ca/foodguide

or contact:
Publications
Health Canada
Ottawa, Ontario K1A 0K9
E-Mail: publications@hc-sc.gc.ca
Tel.: 1-866-225-0709
Fax: (613) 941-5366
TTY: 1-800-267-1245

Également disponible en français sous le titre :
Bien manger avec le Guide alimentaire canadien

This publication can be made available on request on diskette, large print, audio-cassette and braille.

Canada's *Meal Planning for Healthy Eating*

Beyond the Basics: Meal Planning for Healthy Eating, Diabetes Prevention and Management is Canada's system of meal planning.[1] Similar to the U.S. exchange system, *Beyond the Basics* sorts foods into groups and defines portion sizes to help people manage their blood glucose and maintain a healthy weight. Because foods that contain carbohydrate raise blood glucose, the food groups are organized into two sections—those that contain carbohydrate (presented in Table B–1) and those that contain little or no carbohydrate (shown in Table B–2). One portion from any of the food groups listed in Table B–1 provides about 15 grams of available carbohydrate (total carbohydrate minus fiber) and counts as one carbohydrate choice. Within each group, foods are identified as those to "choose more often" (generally higher in vitamins, minerals, and fiber) and those to "choose less often" (generally higher in sugar, saturated fat, or *trans* fat).

[1]The tables for the Canadian meal planning system are adapted from *Beyond the Basics: Meal Planning for Healthy Eating, Diabetes Prevention and Management*, copyright 2005, with permission of the Canadian Diabetes Association. Additional information is available from **www.diabetes.ca**.

TABLE B–1
Food Groups that Contain Carbohydrate

1 serving = 15 g carbohydrate or 1 carbohydrate choice

Key:
- ■ Choose more often
- ■ Choose less often

Food	Measure
GRAINS AND STARCHES: 15 G CARBOHYDRATE, 2 G PROTEIN, 0 G FAT, 286 KJ (68 KCAL)	
■ Bagel, large	¼
■ Bagel, small	½
■ Bannock, fried	1.5″ × 2.5″
■ Bannock, whole grain baked	1.5″ × 2.5″
■ Barley, cooked	125 mL (½ c)
■ Bread, white	30 g (1 oz)
■ Bread, whole grain	30 g (1 oz)
■ Bulgur, cooked	125 mL (½ c)
■ Bun, hamburger or hotdog	½
■ Cereal, flaked unsweetened	125 mL (½ c)
■ Cereal, hot	¾ c
■ Chapati, whole wheat (6″)	1
■ Corn	125 mL (½ c)
■ Couscous, cooked	125 mL (½ c)
■ Crackers, soda type	7
■ Croutons	⅔ c
■ English muffin, whole grain	½
■ French fries	10
■ Millet, cooked	⅓ c
■ Naan bread (6″)	¼
■ Pancake (4″)	1
■ Pasta, cooked	125 mL (½ c)
■ Pita bread, white (6″)	1
■ Pita bread, whole wheat (6″)	1
■ Pizza crust (12″)	¹⁄₁₂
■ Plantain, mashed	⅓ c
■ Potatoes, boiled or baked	½ medium
■ Rice, cooked	⅓ c
■ Roti, whole wheat (6″)	1
■ Soup, thick type	250 mL (1 c)
■ Sweet potato, mashed	⅓ c
■ Taco shells (5″)	2
■ Tortilla, whole wheat (6″)	1
■ Waffle (4″)	1

TABLE B–1
Food Groups that Contain Carbohydrate

1 serving = 15 g carbohydrate or 1 carbohydrate choice

Food	Measure
FRUITS:15 G CARBOHYDRATE, 1 G PROTEIN, 0 G FAT, 269 KJ (64 KCAL)	
■ Apple	1 medium
■ Apple sauce, unsweetened	125 mL (½ c)
■ Banana	1 small
■ Blackberries	500 mL (2 c)
■ Cherries	15
■ Fruit, canned in juice	125 mL (½ c)
■ Fruit, dried	50 mL (¼ c)
■ Grapefruit	1 small
■ Grapes	15
■ Kiwi	2 medium
■ Juice	125 mL (½ c)
■ Mango	½ medium
■ Melon	250 mL (1 c)
■ Orange	1 medium
■ Other berries	250 mL (1 c)
■ Pear	1 medium
■ Pineapple	¾ c
■ Plum	2 medium
■ Raspberries	500 mL (2 c)
■ Strawberries	500 mL (2 c)
MILK AND ALTERNATIVES: 15 G CARBOHYDRATE, 8 G PROTEIN, VARIABLE FAT, 386–651 KJ (92–155 KCAL)	
■ Chocolate milk, 1%	125 mL (½ c)
■ Evaporated milk, canned	125 mL (½ c)
■ Milk, fluid	250 mL (1 c)
■ Milk powder, skim	30 mL (2 tbs)
■ Soy beverage, flavored	125 mL (½ c)
■ Soy beverage, plain	250 mL (1 c)
■ Soy yogurt, flavored	⅓ c
■ Yogurt, nonfat, plain	¾ c
■ Yogurt, skim, artificially sweetened	250 mL (1 c)
OTHER CHOICES (SWEET FOODS AND SNACKS): 15 G CARBOHYDRATE, VARIABLE PROTEIN AND FAT	
■ Brownies, unfrosted	2" × 2"
■ Cake, unfrosted	2" × 2"
■ Cookies, arrowroot or gingersnap	3–4
■ Jam, jelly, marmalade	15 mL (1 tbs)
■ Milk pudding, skim, no sugar added	125 mL (½ c)
■ Muffin	1 small (2")
■ Oatmeal granola bar	1 (28 g)
■ Popcorn, low fat	750 mL (3 c)
■ Pretzels, low fat, large	7
■ Pretzels, low fat, sticks	30
■ Sugar, white	15 mL (3 tsp or packets)

B

Food	Measure
VEGETABLES: TO ENCOURAGE CONSUMPTION, MOST VEGETABLES ARE CONSIDERED "FREE"	
■ Asparagus	
■ Beans, yellow or green	
■ Bean sprouts	
■ Beets	
■ Broccoli	
■ Cabbage	
■ Carrots	
■ Cauliflower	
■ Celery	
■ Cucumber	
■ Eggplant	
■ Greens	
■ Leeks	
■ Mushrooms	
■ Okra	
■ Parsnips[a]	
■ Peas[a]	
■ Peppers	
■ Rutabagas (turnips)[a]	
■ Salad vegetables	
■ Snow peas	
■ Squash, winter[a]	
■ Tomatoes	
MEAT AND ALTERNATIVES: 0 G CARBOHYDRATE, 7 G PROTEIN, 3–5 G FAT, 307 KJ (73 KCAL)	
■ Cheese, skim (<7% milk fat)	30 g (1 oz)
■ Cheese, light (<17% milk fat)	30 g (1 oz)
■ Cheese, regular (17–33% milk fat)	30 g (1 oz)
■ Cottage cheese (1–2% milk fat)	50 mL (¼ c)
■ Egg	1 large
■ Fish, canned in oil	50 mL (¼ c)
■ Fish, canned in water	50 mL (¼ c)
■ Fish, fresh, cooked	30 g (1 oz)
■ Hummus[b]	⅓ c
■ Legumes, cooked[b]	125 mL (½ c)
■ Meat, game, cooked	30 g (1 oz)
MEAT AND ALTERNATIVES: 0 G CARBOHYDRATE, 7 G PROTEIN, 3–5 G FAT, 307 KJ (73 KCAL)	
■ Meat, ground, lean, cooked	30 g (1 oz)
■ Meat, ground, medium-regular, cooked	30 g (1 oz)
■ Meat, lean, cooked	30 g (1 oz)
■ Meat, organ or tripe, cooked	30 g (1 oz)
■ Meat, prepared, low fat	30 g (1 oz)
■ Meat, prepared, regular fat	30 g (1 oz)
■ Meat, regular, cooked	30 g (1 oz)
■ Peameal/back bacon, cooked	30 g (1 oz)
■ Poultry, ground, lean, cooked	30 g (1 oz)
■ Poultry, skinless, cooked	30 g (1 oz)
■ Poultry/wings, skin on, cooked	30 g (1 oz)
■ Shellfish, cooked	30 g (1 oz)
■ Tofu (soybean)	½ block (100 g)
■ Vegetarian meat alternatives	30 g (1 oz)

TABLE B–2
Food Groups that Contain
Little or No Carbohydrate

Food	Measure
FATS: 0 G CARBOHYDRATE, 0 G PROTEIN, 5 G FAT, 189 KJ (45 KCAL)	
■ Avocado	⅙
■ Bacon	30 g (1 oz)
■ Butter	5 mL (1 tsp)
■ Cheese, spreadable	15 mL (1 tbs)
■ Margarine, non-hydrogenated	5 mL (1 tsp)
■ Mayonnaise, light	30 mL (2 tbs)
■ Nuts	15 mL (1 tbs)
■ Oil, canola or olive	5 mL (1 tsp)
■ Salad dressing, regular	15 mL (1 tbs)
■ Seeds	15 mL (1 tbs)
■ Tahini	7.5 mL (½ tbs)
EXTRAS: <5 G CARBOHYDRATE, 84 KJ (20 KCAL)	
Broth	
Coffee	
Herbs and spices	
Ketchup	
Mustard	
Sugar-free soft drinks	
Sugar-free gelatin	
Tea	

[a] These vegetables provide significant carbohydrate when more than 125 mL (½ c) is eaten.
[b] Legumes contain 15 g carbohydrate in a 125 mL (½ c) serving.

United States: Exchange Lists

Chapter 21 introduced the exchange system, and this appendix provides details from the 2003 edition. Appendix B presents Canada's meal planning system.

The Exchange Groups and Lists

The exchange system sorts foods into three main groups by their proportions of carbohydrate, fat, and protein. These three groups—the carbohydrate group, the fat group, and the meat and meat substitutes group (protein)—organize foods into several exchange lists (see Table C–1). Then any food on a list can be "exchanged" for any other on that same list. The carbohydrate group covers these exchange lists:

- Starch (cereals, grains, pasta, breads, crackers, snacks, starchy vegetables, and dried beans, peas, and lentils).

- Fruit.

- Milk (fat-free, reduced fat, and whole).

- Other carbohydrates (desserts and snacks with added sugars and fats).

- Vegetables.

TABLE C–1
The Exchange Groups and Lists

Group/Lists	Typical Item/Portion Size	Carbohydrate (g)	Protein (g)	Fat (g)	Energy[a] (kcal)
CARBOHYDRATE GROUP					
Starch[b]	1 slice bread	15	3	0–1	80
Fruit	1 small apple	15	—	—	60
Milk					
Fat-free, low-fat	1 c fat-free milk	12	8	0–3	90
Reduced-fat	1 c reduced-fat milk	12	8	5	120
Whole	1 c whole milk	12	8	8	150
Other carbohydrates[c]	2 small cookies	15	varies	varies	varies
Vegetable (nonstarchy)	1/2 c cooked carrots	5	2	—	25
MEAT AND MEAT SUBSTITUTE GROUP[d]					
Meat					
Very lean	1 oz chicken (white meat, no skin)	—	7	0–1	35
Lean	1 oz lean beef	—	7	3	55
Medium-fat	1 oz ground beef	—	7	5	75
High-fat	1 oz pork sausage	—	7	8	100
FAT GROUP					
Fat	1 tsp butter	—	—	5	45

[a]The energy value for each exchange list represents an approximate average for the group and does not reflect the precise number of grams of carbohydrate, protein, and fat. For example, a slice of bread contains 15 grams of carbohydrate (that's 60 kcalories), 3 grams protein (that's another 12 kcalories), and a little fat—rounded to 80 kcalories for ease in calculating. A half-cup of vegetables (not including starchy vegetables) contains 5 grams carbohydrate (20 kcalories) and 2 grams protein (8 more), which has been rounded down to 25 kcalories.

[b]The starch list includes cereals, grains, breads, crackers, snacks, starchy vegetables (such as corn, peas, and potatoes), and legumes (dried beans, peas, and lentils).

[c]The other carbohydrates list includes foods that contain added sugars and fats such as cakes, cookies, doughnuts, ice cream, potato chips, pudding, syrup, and frozen yogurt.

[d]The meat and meat substitutes list includes legumes, cheeses, and peanut butter.

The fat group covers this exchange list:

- Fats.

The meat and meat substitutes group (protein) covers these exchange lists:

- Meat and meat substitutes (very lean, lean, medium-fat, and high-fat).

See Tables C–2 through C–10 for the exchange lists.

Portion Sizes The exchange system helps people control their energy intakes by paying close attention to portion sizes. The portion sizes have been carefully adjusted and defined so that a portion of any food on a given list provides roughly the same amount of carbohydrate, fat, and protein and, therefore, total kcalories. Any food on a list can then be exchanged, or traded, for any other food on that same list without significantly affecting the diet's balance or total kcalories. For example, a person may select either 17 small grapes or 1/2 large grapefruit as one fruit exchange, and either choice would provide roughly 60 kcalories. A whole grapefruit, however, would count as 2 exchanges.

Both the exchange system and the USDA Food Guide list meats in single ounces; that is, one *portion* (or *exchange*) of meat is 1 ounce. Calculating meat by the ounce encourages a person to keep close track of the exact amounts eaten. This in turn helps control energy and fat intakes. Be aware, though, that most people do not serve foods in carefully measured portions, nor do the amounts reflect the exchange system or USDA Food Guide amounts.

To apply the system successfully, users must become familiar with portion sizes. A convenient way to remember the portion sizes and energy values is to keep in mind a typical item from each list (review Table C–1).

The Foods on the Lists Foods do not always appear on the exchange list where you might first expect to find them. They are grouped according to their energy-nutrient contents rather than by their source (such as milks), their outward appearance, or their vitamin and mineral contents. Notice, for example, that cheeses are grouped with meats (not milk) because, like meats, cheeses contribute energy from protein and fat but provide negligible carbohydrate. Similarly, starchy vegetables such as potatoes are found on the starch list with breads and cereals, not with the vegetables, and bacon is with the fats and oils, not with the meats.

Users of the exchange lists learn to view mixtures of foods, such as casseroles and soups, as combinations of foods from different exchange lists. They also learn to interpret food labels with the exchange system in mind (see Figure C–1).

Controlling Energy and Fat By assigning items like bacon to the fat list, the exchange system alerts consumers to foods that are unexpectedly high in fat. Even the starch list specifies which grain products contain added fat (such as biscuits, muffins, and waffles). In addition, the exchange system encourages users to think of fat-free milk as milk and of whole milk as milk with added fat, and to think of very lean meats as meats and of lean, medium-fat, and high-fat meats as meats with added fat. To that end, foods on the milk and meat lists are separated into categories based on their fat contents. The milk group is classed as fat-free, reduced-fat, and whole; the meat group as very lean, lean, medium-fat, and high-fat.

Control of food energy and fat intake can be highly successful with the exchange system. Exchange plans do not, however, guarantee adequate intakes of vitamins and minerals. Food group plans work better from that standpoint because the food groupings are based on similarities in vitamin-mineral content. In the exchange system, for example, meats are grouped with cheeses, yet the meats are iron-rich and calcium-poor, whereas the cheeses are iron-poor and calcium-rich. To take advantage of the strengths of both food group plans and exchange patterns, and to compensate for their weaknesses, diet planners often combine these two diet-planning tools.

HOME▲TASTE
Lasagna Dinner
WITH MEAT SAUCE

Nutrition Facts
Serving size 10 1/2 oz (298 g)
Servings per Package 1

Amount per serving

Calories 361	Calories from Fat 117

	% Daily Value
Total Fat 13 g	**20%**
Saturated Fat 8 g	**40%**
Cholesterol 87 mg	**29%**
Sodium 860 mg	**36%**
Total Carbohydrate 37 g	**12%**
Dietary fiber 0 g	
Sugars 8 g	
Protein 26 g	

Can you "see" these exchanges in the label above?

Exchange	Carbohydrate	Protein	Fat
2 starches	30 g	6 g	—
1 vegetable	5 g	2 g	—
3 medium-fat meats	—	21 g	15 g
Exchange totals	**35**	**29**	**15**
Label totals	**37**	**26**	**13**

FIGURE C–1

Seeing Exchanges on a Food Label
Knowing that foods on the starch list provide 15 grams of carbohydrate and those on the vegetable list provide 5, you can count a lasagna dinner that provides 37 grams of carbohydrate as "2 starches and 1 vegetable"; knowing that foods on the meat list provide 7 grams of protein, you might count it as "3 meats"; the grams of fat suggest that the meat (and cheese) is probably medium-fat.

Table C–2
U.S. Exchange System: Starch List

1 starch exchange = 15 g carbohydrate, 3 g protein, 0–1 g fat, and 80 kcal
NOTE: In general, one starch exchange is $1/2$ c cooked cereal, grain, or starchy vegetable; $1/2$ c cooked rice or pasta; 1 oz of bread; $3/4$ to 1 oz snack food.

Serving Size	Food
BREAD	
$1/4$ (1 oz)	Bagel, 4 oz
2 slices ($1\frac{1}{2}$ oz)	Bread, reduced-kcalorie
1 slice (1 oz)	Bread, white (including French and Italian), whole-wheat, pumpernickel, rye
4 ($2/3$ oz)	Bread sticks, crisp, 4″ × $1/2$″
$1/2$	English muffin
$1/2$ (1 oz)	Hot dog or hamburger bun
$1/4$	Naan, 8″ × 2″
1	Pancake, 4″ across, $1/4$″ thick
$1/2$	Pita, 6″ across
1 (1 oz)	Plain roll, small
1 slice (1 oz)	Raisin bread, unfrosted
1	Tortilla, corn, 6″ across
1	Tortilla, flour, 6″ across
$1/3$	Tortilla, flour, 10″ across
1	Waffle, 4″ square or across, reduced-fat
CEREALS AND GRAINS	
$1/2$ c	Bran cereals
$1/2$ c	Bulgur, cooked
$1/2$ c	Cereals, cooked
$3/4$ c	Cereals, unsweetened, ready-to-eat
3 tbs	Cornmeal (dry)
$1/3$ c	Couscous
3 tbs	Flour (dry)
$1/4$ c	Granola, low-fat
$1/4$ c	Grape nuts
$1/2$ c	Grits, cooked
$1/2$ c	Kasha
$1/3$ c	Millet
$1/4$ c	Muesli
$1/2$ c	Oats
$1/3$ c	Pasta, cooked
$1\frac{1}{2}$ c	Puffed cereals
$1/3$ c	Rice, white or brown, cooked
$1/2$ c	Shredded wheat
$1/2$ c	Sugar-frosted cereal
3 tbs	Wheat germ
STARCHY VEGETABLES	
$1/3$ c	Baked beans
$1/2$ c	Corn
$1/2$ cob (5 oz)	Corn on cob, large
1 c	Mixed vegetables with corn, peas, or pasta
$1/2$ c	Peas, green

Serving Size	Food
$1/2$ c	Plantains
$1/2$ medium (3 oz) or $1/2$ c	Potato, boiled
$1/4$ large (3 oz)	Potato, baked with skin
$1/2$ c	Potatoes, mashed
1 c	Squash, winter (acorn, butternut, pumpkin)
$1/2$ c	Yams, sweet potatoes, plain
CRACKERS AND SNACKS	
8	Animal crackers
3	Graham crackers, $2\frac{1}{2}$″ square
$3/4$ oz	Matzoh
4 slices	Melba toast
24	Oyster crackers
3 c	Popcorn (popped, no fat added or low-fat microwave)
$3/4$ oz	Pretzels
2	Rice cakes, 4″ across
6	Saltine-type crackers
15–20 ($3/4$ oz)	Snack chips, fat-free or baked (tortilla, potato)
2–5 ($3/4$ oz)	Whole-wheat crackers, no fat added
BEANS, PEAS, AND LENTILS (COUNT AS 1 STARCH + 1 VERY LEAN MEAT)	
$1/2$ c	Beans and peas, cooked (garbanzo, lentils, pinto, kidney, white, split, black-eyed)
$2/3$ c	Lima beans
3 tbs	Miso 🖊
STARCHY FOODS PREPARED WITH FAT (COUNT AS 1 STARCH + 1 FAT)	
1	Biscuit, $2\frac{1}{2}$″ across
$1/2$ c	Chow mein noodles
1 (2 oz)	Cornbread, 2″ cube
6	Crackers, round butter type
1 c	Croutons
1 c (2 oz)	French-fried potatoes (oven baked)
$1/4$ c	Granola
$1/3$ c	Hummus
$1/5$ (1 oz)	Muffin, 5 oz
3 c	Popcorn, microwave
3	Sandwich crackers, cheese or peanut butter filling
9–13 ($3/4$ oz)	Snack chips (potato, tortilla)
$1/3$ c	Stuffing, bread (prepared)
2	Taco shells, 6″ across
1	Waffle, $4\frac{1}{2}$″ square or across
4–6 (1 oz)	Whole-wheat crackers, fat added

🖊 = 400 mg or more of sodium per serving.

Table C–3
U.S. Exchange System: Fruit List

1 fruit exchange = 15 g carbohydrate and 60 kcal
NOTE: In general, one fruit exchange is 1 small fresh fruit; 1/2 c canned or fresh fruit or unsweetened fruit juice; 1/4 c dried fruit.

Serving Size	Food	Serving Size	Food
1 (4 oz)	Apple, unpeeled, small	1 (4 oz)	Peach, medium, fresh
1/2 c	Applesauce, unsweetened	1/2 c	Peaches, canned
4 rings	Apples, dried	1/2 (4 oz)	Pear, large, fresh
4 whole (5 1/2 oz)	Apricots, fresh	1/2 c	Pears, canned
8 halves	Apricots, dried	3/4 c	Pineapple, fresh
1/2 c	Apricots, canned	1/2 c	Pineapple, canned
1 (4 oz)	Banana, small	2 (5 oz)	Plums, small
3/4 c	Blackberries	1/2 c	Plums, canned
3/4 c	Blueberries	3	Plums, dried (prunes)
1/3 melon (11 oz) or 1 c cubes	Cantaloupe, small	2 tbs	Raisins
12 (3 oz)	Cherries, sweet, fresh	1 c	Raspberries
1/2 c	Cherries, sweet, canned	1 1/4 c whole berries	Strawberries
3	Dates	2 (8 oz)	Tangerines, small
1 1/2 large or 2 medium (3 1/2 oz)	Figs, fresh	1 slice (13 1/2 oz) or 1 1/4 c cubes	Watermelon
1 1/2	Figs, dried	**FRUIT JUICE, UNSWEETENED**	
1/2 c	Fruit cocktail	1/2 c	Apple juice/cider
1/2 (11 oz)	Grapefruit, large	1/3 c	Cranberry juice cocktail
3/4 c	Grapefruit sections, canned	1 c	Cranberry juice cocktail, reduced-kcalorie
17 (3 oz)	Grapes, small	1/3 c	Fruit juice blends, 100% juice
1 slice (10 oz) or 1 c cubes	Honeydew melon	1/3 c	Grape juice
1 (3 1/2 oz)	Kiwi	1/2 c	Grapefruit juice
3/4 c	Mandarin oranges, canned	1/2 c	Orange juice
1/2 (5 1/2 oz) or 1/2 c	Mango, small	1/2 c	Pineapple juice
1 (5 oz)	Nectarine, small	1/3 c	Prune juice
1 (6 1/2 oz)	Orange, small		
1/2 (8 oz) or 1 c cubes	Papaya		

Table C–4
U.S. Exchange System: Milk List

NOTE: In general, one milk exchange is 1 c milk or yogurt.

Serving Size	Food	Serving Size	Food
FAT-FREE AND LOW-FAT MILK		**REDUCED-FAT MILK**	
1 fat-free/low-fat milk exchange = 12 g carbohydrate, 8 g protein, 0–3 g fat, 90 kcal		1 reduced-fat milk exchange = 12 g carbohydrate, 8 g protein, 5 g fat, 120 kcal	
1 c	Fat-free milk	1 c	2% milk
1 c	1/2% milk	1 c	Soy milk
1 c	1% milk	1 c	Sweet acidophilus milk
1 c	Fat-free or low-fat buttermilk	3/4 c	Yogurt, plain low-fat
1/2 c	Evaporated fat-free milk	**WHOLE MILK**	
1/3 c dry	Fat-free dry milk	1 whole milk exchange = 12 g carbohydrate, 8 g protein, 8 g fat, 150 kcal	
1 c	Soy milk, low-fat or fat-free	1 c	Whole milk
2/3 c (6 oz)	Yogurt, fat-free or low-fat, flavored, sweetened with nonnutritive sweetener and fructose	1/2 c	Evaporated whole milk
		1 c	Goat's milk
2/3 c (6 oz)	Yogurt, plain fat-free	1 c	Kefir
		3/4 c	Yogurt, plain (made from whole milk)

1 other carbohydrate exchange 5 15 g carbohydrate, or 1 starch, or 1 fruit, or 1 milk exchange

Food	Serving Size	Exchanges per Serving
Angel food cake, unfrosted	$1/12$ cake (2 oz)	2 carbohydrates
Brownies, small, unfrosted	2" square (1 oz)	1 carbohydrate, 1 fat
Cake, unfrosted	2" square (1 oz)	1 carbohydrate, 1 fat
Cake, frosted	2" square (2 oz)	2 carbohydrates, 1 fat
Cookies or sandwich cookies with creme filling	2 small ($2/3$ oz)	1 carbohydrate, 1 fat
Cookies, sugar-free	3 small or 1 large ($3/4$–1 oz)	1 carbohydrate, 1–2 fats
Cranberry sauce, jellied	$1/4$ c	$1 1/2$ carbohydrates
Cupcake, frosted	1 small (2 oz)	2 carbohydrates, 1 fat
Doughnut, plain cake	1 medium ($1 1/2$ oz)	$1 1/2$ carbohydrates, 2 fats
Doughnut, glazed	$3 3/4$" across (2 oz)	2 carbohydrates, 2 fats
Energy, sport, or breakfast bar	1 bar ($1 1/3$ oz)	$1 1/2$ carbohydrates, 0–1 fat
Energy, sport, or breakfast bar	1 bar (2 oz)	2 carbohydrates, 1 fat
Fruit cobbler	$1/2$ c ($3 1/2$ oz)	3 carbohydrates, 1 fat
Fruit juice bar, frozen, 100% juice	1 bar (3 oz)	1 carbohydrate
Fruit snacks, chewy (pureed fruit concentrate)	1 roll ($3/4$ oz)	1 carbohydrate
Fruit spreads, 100% fruit	$1 1/2$ tbs	1 carbohydrate
Gelatin, regular	$1/2$ c	1 carbohydrate
Gingersnaps	3	1 carbohydrate
Granola or snack bar, regular or low-fat	1 bar (1 oz)	$1 1/2$ carbohydrates
Honey	1 tbs	1 carbohydrate
Ice cream	$1/2$ c	1 carbohydrate, 2 fats
Ice cream, light	$1/2$ c	1 carbohydrate, 1 fat
Ice cream, low-fat	$1/2$ c	$1 1/2$ carbohydrates
Ice cream, fat-free, no sugar added	$1/2$ c	1 carbohydrate
Jam or jelly, regular	1 tbs	1 carbohydrate
Milk, chocolate, whole	1 c	2 carbohydrates, 1 fat
Pie, fruit, 2 crusts	$1/6$ of 8" commercially prepared pie	3 carbohydrates, 2 fats
Pie, pumpkin or custard	$1/8$ of 8" commercially prepared pie	2 carbohydrates, 2 fats
Pudding, regular (made with reduced-fat milk)	$1/2$ c	2 carbohydrates
Pudding, sugar-free (made with fat-free milk)	$1/2$ c	1 carbohydrate
Reduced-calorie meal replacement (shake)	1 can (10–11 oz)	$1 1/2$ carbohydrates, 0–1 fats
Rice milk, low-fat or fat-free, plain	1 c	1 carbohydrate
Rice milk, low-fat, flavored	1 c	$1 1/2$ carbohydrates
Salad dressing, fat-free 🖋	$1/4$ c	1 carbohydrate
Sherbet, sorbet	$1/2$ c	2 carbohydrates
Spaghetti or pasta sauce, canned 🖋	$1/2$ c	1 carbohydrate, 1 fat
Sports drinks	8 oz (1 c)	1 carbohydrate
Sugar	1 tbs	1 carbohydrate
Sweet roll or danish	1 ($2 1/2$ oz)	$2 1/2$ carbohydrates, 2 fats
Syrup, light	2 tbs	1 carbohydrate
Syrup, regular	1 tbs	1 carbohydrate
Syrup, regular	$1/4$ c	4 carbohydrates
Vanilla wafers	5	1 carbohydrate, 1 fat
Yogurt, frozen	$1/2$ c	1 carbohydrate, 0–1 fat
Yogurt, frozen, fat-free	$1/3$ c	1 carbohydrate
Yogurt, low-fat with fruit	1 c	3 carbohydrates, 0–1 fat

🖋 = 400 mg or more of sodium per serving.

Table C–6
U.S. Exchange System: Nonstarchy Vegetable List

1 vegetable exchange = 5 g carbohydrate, 2 g protein, 0 g fat, and 25 kcal
NOTE: In general, one vegetable exchange is $1/2$ c cooked vegetables or vegetable juice; 1 c raw vegetables. Starchy vegetables such as corn, peas, and potatoes are on the starch list (Table C–2).

Artichokes	Mushrooms
Artichoke hearts	Okra
Asparagus	Onions
Beans (green, wax, Italian)	Pea pods
Bean sprouts	Peppers (all varieties)
Beets	Radishes
Broccoli	Salad greens (endive, escarole, lettuce, romaine, spinach)
Brussels sprouts	Sauerkraut 🖋
Cabbage	Spinach
Carrots	Summer squash (crookneck)
Cauliflower	Tomatoes
Celery	Tomatoes, canned
Cucumbers	Tomato sauce 🖋
Eggplant	Tomato/vegetable juice 🖋
Green onions or scallions	Turnips
Greens (collard, kale, mustard, turnip)	Water chestnuts
Kohlrabi	Watercress
Leeks	Zucchini
Mixed vegetables (without corn, peas, or pasta)	

🖋 = 400 mg or more of sodium per serving.

Table C–7
U.S. Exchange System: Meat and Meat Substitutes List

NOTE: In general, a meat exchange is 1 oz meat, poultry, or cheese; $\frac{1}{2}$ c dried beans (weigh meat and poultry and measure beans after cooking).

Serving Size	Food
VERY LEAN MEAT AND SUBSTITUTES	
1 very lean meat exchange = 7 g protein, 0–1 g fat, 35 kcal	
1 oz	Poultry: Chicken or turkey (white meat, no skin), Cornish hen (no skin)
1 oz	Fish: Fresh or frozen cod, flounder, haddock, halibut, trout, lox (smoked salmon ✎); tuna, fresh or canned in water
1 oz	Shellfish: Clams, crab, lobster, scallops, shrimp, imitation shellfish
1 oz	Game: Duck or pheasant (no skin), venison, buffalo, ostrich
	Cheese with ≤1g fat/oz:
$\frac{1}{4}$ c	Fat-free or low-fat cottage cheese
1 oz	Fat-free cheese
1 oz	Processed sandwich meats with ≤1 g fat/oz (such as deli thin, shaved meats, chipped beef ✎, turkey ham)
2	Egg whites
$\frac{1}{4}$ c	Egg substitutes, plain
1 oz	Hot dogs with ≤1 g fat/oz ✎
1 oz	Kidney (high in cholesterol)
1 oz	Sausage with ≤1 g fat/oz
Count as 1 very lean meat + 1 starch exchange:	
$\frac{1}{2}$ c	Beans, peas, lentils (cooked)
LEAN MEAT AND SUBSTITUTES	
1 lean meat exchange = 7 g protein, 3 g fat, 55 kcal	
1 oz	Beef: USDA Select or Choice grades of lean beef trimmed of fat (round, sirloin, and flank steak); tenderloin; roast (rib, chuck, rump); steak (T-bone, porterhouse, cubed), ground round
1 oz	Pork: Lean pork (fresh ham); canned, cured, or boiled ham; Canadian bacon ✎; tenderloin, center loin chop
1 oz	Lamb: Roast, chop, leg
1 oz	Veal: Lean chop, roast
1 oz	Poultry: Chicken, turkey (dark meat, no skin), chicken (white meat, with skin), domestic duck or goose (well drained of fat, no skin)
	Fish:
1 oz	Herring (uncreamed or smoked)
6 medium	Oysters
1 oz	Salmon (fresh or canned), catfish
2 medium	Sardines (canned)
1 oz	Tuna (canned in oil, drained)
1 oz	Game: Goose (no skin), rabbit

Serving Size	Food
	Cheese:
$\frac{1}{4}$ c	4.5%-fat cottage cheese
2 tbs	Grated Parmesan
1 oz	Cheeses with ≤3 g fat/oz
1$\frac{1}{2}$ oz	Hot dogs with ≤3 g fat/oz ✎
1 oz	Processed sandwich meat with ≤3 g fat/oz (turkey pastrami or kielbasa)
1 oz	Liver, heart (high in cholesterol)
MEDIUM-FAT MEAT AND SUBSTITUTES	
1 medium-fat meat exchange = 7 g protein, 5 g fat, and 75 kcal	
1 oz	Beef: Most beef products (ground beef, meat loaf, corned beef, short ribs, prime grades of meat trimmed of fat, such as prime rib)
1 oz	Pork: Top loin, chop, Boston butt, cutlet
1 oz	Lamb: Rib roast, ground
1 oz	Veal: Cutlet (ground or cubed, unbreaded)
1 oz	Poultry: Chicken (dark meat, with skin), ground turkey or ground chicken, fried chicken (with skin)
1 oz	Fish: Any fried fish product
	Cheese with ≤5 g fat/oz:
1 oz	Feta
1 oz	Mozzarella
$\frac{1}{4}$ c (2 oz)	Ricotta
1	Egg (high in cholesterol, limit to 3/week)
1 oz	Sausage with ≤5 g fat/oz
1 c	Soy milk
$\frac{1}{4}$ c	Tempeh
4 oz or $\frac{1}{2}$ c	Tofu
HIGH-FAT MEAT AND SUBSTITUTES	
1 high-fat meat exchange = 7 g protein, 8 g fat, 100 kcal	
1 oz	Pork: Spareribs, ground pork, pork sausage
1 oz	Cheese: All regular cheeses (American ✎, cheddar, Monterey Jack, swiss)
1 oz	Processed sandwich meats with ≤8 g fat/oz (bologna, pimento loaf, salami)
1 oz	Sausage (bratwurst, Italian, knockwurst, Polish, smoked)
1 (10/lb)	Hot dog (turkey or chicken) ✎
3 slices (20 slices/lb)	Bacon
1 tbs	Peanut butter (contains unsaturated fat)
Count as 1 high-fat meat + 1 fat exchange:	
1 (10/lb)	Hot dog (beef, pork, or combination) ✎

✎ = 400 mg or more of sodium per serving.

Table C–8
U.S. Exchange System: Fat List

1 fat exchange = 5 g fat and 45 kcal

NOTE: In general, one fat exchange is 1 tsp regular butter, margarine, or vegetable oil; 1 tbs regular salad dressing. Many fat-free and reduced-fat foods are on the Free Foods List (Table C–9).

Serving Size	Food
MONOUNSATURATED FATS	
2 tbs (1 oz)	Avocado
1 tsp	Oil (canola, olive, peanut)
8 large	Olives, ripe (black)
10 large	Olives, green, stuffed 🖊
6 nuts	Almonds, cashews
6 nuts	Mixed nuts (50% peanuts)
10 nuts	Peanuts
4 halves	Pecans
½ tbs	Peanut butter, smooth or crunchy
1 tbs	Sesame seeds
2 tsp	Tahini or sesame paste
POLYUNSATURATED FATS	
4 halves	English walnuts
1 tsp	Margarine, stick, tub, or squeeze
1 tbs	Margarine, lower-fat spread (30% to 50% vegetable oil)
1 tsp	Mayonnaise, regular
1 tbs	Mayonnaise, reduced-fat
1 tsp	Oil (corn, safflower, soybean)
1 tbs	Salad dressing, regular 🖊
2 tbs	Salad dressing, reduced-fat
2 tsp	Mayonnaise type salad dressing, regular
1 tbs	Mayonnaise type salad dressing, reduced-fat
1 tbs	Seeds (pumpkin, sunflower)
SATURATED FATS*	
1 slice (20 slices/lb)	Bacon, cooked
1 tsp	Bacon, grease
1 tsp	Butter, stick
2 tsp	Butter, whipped
1 tbs	Butter, reduced-fat
2 tbs (½ oz)	Chitterlings, boiled
2 tbs	Coconut, sweetened, shredded
1 tbs	Coconut milk
2 tbs	Cream, half and half
1 tbs (½ oz)	Cream cheese, regular
1½ tbs (¾ oz)	Cream cheese, reduced-fat
	Fatback or salt pork† 🖊
1 tsp	Shortening or lard
2 tbs	Sour cream, regular
3 tbs	Sour cream, reduced-fat

🖊 = 400 mg or more of sodium per serving.

*Saturated fats can raise blood cholesterol levels.

†Use a piece 1″ × 1″ × ¼″ if you plan to eat the fatback cooked with vegetables. Use a piece 2″ × 1″ × ½″ when eating only the vegetables with the fatback removed.

Table C–9
U.S. Exchange System: Free Foods List

NOTE: A serving of free food contains less than 20 kcalories or no more than 5 grams of carbohydrate; those with serving sizes should be limited to 3 servings a day whereas those without serving sizes can be eaten freely.

Serving Size	Food
FAT-FREE OR REDUCED-FAT FOODS	
1 tbs (½ oz)	Cream cheese, fat-free
1 tbs	Creamers, nondairy, liquid
2 tsp	Creamers, nondairy, powdered
4 tbs	Margarine spread, fat-free
1 tsp	Margarine spread, reduced-fat
1 tbs	Mayonnaise, fat-free
1 tsp	Mayonnaise, reduced-fat
1 tbs	Mayonnaise type salad dressing, fat-free
1 tsp	Mayonnaise type salad dressing, reduced-fat
	Nonstick cooking spray
1 tbs	Salad dressing, fat-free or low-fat
2 tbs	Salad dressing, fat-free, Italian
1 tbs	Sour cream, fat-free, reduced-fat
1 tbs	Whipped topping, regular
2 tbs	Whipped topping, light or fat-free
SUGAR-FREE FOODS	
1 piece	Candy, hard, sugar-free
	Gelatin dessert, sugar-free
	Gelatin, unflavored
	Gum, sugar-free
2 tsp	Jam or jelly, light
	Sugar substitutes
2 tbs	Syrup, sugar-free
DRINKS	
	Bouillon, broth, consommé 🖊)
	Bouillon or broth, low-sodium
	Carbonated or mineral water
	Club soda

Serving Size	Food
1 tbs	Cocoa powder, unsweetened
	Coffee
	Diet soft drinks, sugar-free
	Drink mixes, sugar-free
	Tea
	Tonic water, sugar-free
CONDIMENTS	
1 tbs	Catsup
	Horseradish
	Lemon juice
	Lime juice
	Mustard
1 tbs	Pickle relish
1½ medium	Pickles, dill 🖊)
2 slices	Pickles, sweet (bread and butter)
¾ oz	Pickles, sweet (gherkin)
¼ c	Salsa
1 tbs	Soy sauce, regular or light 🖊)
1 tbs	Taco sauce
	Vinegar
2 tbs	Yogurt
SEASONINGS	
	Flavoring extracts
	Garlic
	Herbs, fresh or dried
	Hot pepper sauces
	Pimento
	Spices
	Wine, used in cooking

🖊 = 400 mg or more of sodium per serving.

Table C–10
U.S. Exchange System: Combination Foods List

C

Entrées

Tuna noodle casserole, lasagna, spaghetti with meatballs, chili with beans, macaroni and cheese ✦	1 c (8 oz)	2 carbohydrates, 2 medium-fat meats
Chow mein (without noodles or rice)	2 c (16 oz)	1 carbohydrate, 2 lean meats
Tuna or chicken salad	$1/2$ c ($3^1/2$ oz)	$1/2$ carbohydrate, 2 lean meats, 1 fat

Frozen Entrées and Meals

Dinner-type meal ✦	Generally 14–17 oz	3 carbohydrates, 3 medium-fat meats, 3 fats
Entrée or meal with <340 kcal ✦	About 8–11 oz	2–3 carbohydrates, 1–2 lean meats
Meatless burger, soy based	3 oz	$1/2$ carbohydrate, 2 lean meats
Meatless burger, vegetable and starch based	3 oz	1 carbohydrate, 1 lean meat
Pizza, cheese, thin crust ✦	$1/4$ of 12" (6 oz)	2 carbohydrates, 2 medium-fat meats, 1 fat
Pizza, meat topping, thin crust ✦	$1/4$ of 12" (6 oz)	2 carbohydrates, 2 medium-fat meats, 2 fats
Pot pie ✦	1 (7 oz)	$2^1/2$ carbohydrates, 1 medium-fat meat, 3 fats

Soups

Bean ✦	1 c	1 carbohydrate, 1 very lean meat
Cream (made with water) ✦	1 c (8 oz)	1 carbohydrate, 1 fat
Instant ✦	6 oz prepared	1 carbohydrate
Instant with beans/lentils ✦	8 oz prepared	$2^1/2$ carbohydrates, 1 very lean meat
Split pea (made with water) ✦	$1/2$ c (4 oz)	1 carbohydrate
Tomato (made with water)	1 c (8 oz)	1 carbohydrate
Vegetable beef, chicken noodle, or other broth-type ✦	1 c (8 oz)	1 carbohydrate

Fast Foods

Burrito with beef ✦	1 (5–7 oz)	3 carbohydrates, 1 medium-fat meat, 1 fat
Chicken nuggets ✦	6	1 carbohydrate, 2 medium-fat meats, 1 fat
Chicken breast and wing, breaded and fried ✦	1 each	1 carbohydrate, 4 medium-fat meats, 2 fats
Chicken sandwich, grilled ✦	1	2 carbohydrates, 3 very lean meats
Chicken wings, hot	6 (5 oz)	1 carbohydrate, 3 medium-fat meats, 4 fats
Fish sandwich/tartar sauce ✦	1	3 carbohydrates, 1 medium-fat meat, 3 fats
French fries ✦	1 medium serving (5 oz)	4 carbohydrates, 4 fats
Hamburger, regular	1	2 carbohydrates, 2 medium-fat meats
Hamburger, large ✦	1	2 carbohydrates, 3 medium-fat meats, 1 fat
Hot dog with bun ✦	1	1 carbohydrate, 1 high-fat meat, 1 fat
Individual pan pizza ✦	1	5 carbohydrates, 3 medium-fat meats, 3 fats
Pizza, cheese, thin crust ✦	$1/4$ of 12" (about 6 oz)	$2^1/2$ carbohydrates, 2 medium-fat meats
Pizza, meat, thin crust ✦	$1/4$ of 12" (about 6 oz)	$2^1/2$ carbohydrates, 2 medium-fat meats, 1 fat
Soft serve cone	1 small (5 oz)	$2^1/2$ carbohydrates, 1 fat
Submarine sandwich ✦	1 sub (6")	3 carbohydrates, 1 vegetable, 2 medium-fat meats, 1 fat
Submarine sandwich (<6 g fat) ✦	1 sub (6")	$2^1/2$ carbohydrates, 2 lean meats
Taco, hard or soft shell	1 (3–$3^1/2$ oz)	1 carbohydrate, 1 medium-fat meat, 1 fat

✦ = 400 mg or more of sodium per serving.

Physical Activity and Energy Requirements

D

Chapter 6 described how to calculate estimated energy requirements (EER) for adults by using an equation that accounts for gender, age, weight, height, and physical activity level. Table D–1 presents additional equations to determine the EER for infants, children, adolescents, and pregnant and lactating women.

This appendix helps you determine the correct physical activity (PA) factor to use in the equations, either by calculating the physical activity level or by estimating it. For those who prefer to bypass these steps, the appendix presents tables that provide a shortcut to estimating total energy expenditure.*

Calculating Physical Activity Level

To calculate your physical activity level, record all of your activities for a typical 24-hour day, noting the type of activity, the level of intensity, and the duration. Then, using a copy of Table D–2, find your activity in the first column (or an activity that is reasonably similar) and multiply the number of minutes spent on that activity by the factor in the third column. Put your answer in the last column and total the accumulated values for the day. Now add the subtotal of the last column to 1.1 (to account for basal energy and the thermic effect of food) as shown. This score indicates your physical activity level. Using Table D–3, find the PA factor for your age and gender that correlates with your physical activity level and use it in the energy equations presented in Table D–1.

Estimating Physical Activity Level

As an alternative to recording your activities for a day, you can use the third column of Table D–3 to decide if your daily activity is sedentary, low active, active, or very active. Find the PA factor for your age and gender that correlates with your typical physical activity level and use it in the energy equations presented in Table D–1.

Using a Shortcut to Estimate Total Energy Expenditure

The DRI Committee has developed estimates of total energy expenditure based on the equations for adults presented in Table D–1. These estimates are presented in Table D–4 for women and Table D–5 for men. You can use these tables to estimate your energy requirement—that is, the number of kcalories needed to maintain your current body weight. On the table appropriate for your gender, find your height in meters (or inches) in the left-hand column. Then follow the row across to find your weight in kilograms (or pounds). (If you can't find your exact height and weight, choose a value between the two closest ones.) Look down the column to find the number of kcalories that corresponds to your activity level.

Importantly, the values given in the tables are for 30-year-old people. Women 19 to 29 should add 7 kcalories per day for each year below age 30; older women should subtract 7 kcalories per day for each year above age 30. Similarly, men 19 to 29 should add 10 kcalories per day for each year below age 30; older men should subtract 10 kcalories per day for each year above age 30.

---

*This appendix, including the tables, is adapted from Committee on Dietary Reference Intakes, *Dietary Reference Intakes for Energy, Carbohydrate, Fiber, Fat, Fatty Acids, Cholesterol, Protein, and Amino Acids* (Washington, DC: National Academies Press, 2002/2005).

Table D–1
Equations to Determine Estimated Energy Requirement (EER)

INFANTS

0–3 months	EER = (89 × weight − 100) + 175
4–6 months	EER = (89 × weight − 100) + 56
7–12 months	EER = (89 × weight − 100) + 22
13–15 months	EER = (89 × weight − 100) + 20

CHILDREN AND ADOLESCENTS

Boys

3–8 years	EER = 88.5 − (61.9 × age + PA × [(26.7 × weight) + (903 × height)] + 20
9–18 years	EER = 88.5 − (61.9 × age + PA × [(26.7 × weight) + (903 × height)] + 25

Girls

3–8 years	EER = 135.3 − (30.8 × age + PA × [(10.0 × weight) + (934 × height)] + 20
9–18 years	EER = 135.3 − (30.8 × age + PA × [(10.0 × weight) + (934 × height)] + 25

Adults

Men	EER = 662 − (9.53 × age + PA × [(15.91 × weight) + (539.6 × height)]
Women	EER = 354 − (6.91 × age + PA × [(9.36 × weight) + (726 × height)]

Pregnancy

1st trimester	EER = nonpregnant EER + 0
2nd trimester	EER = nonpregnant EER + 340
3rd trimester	EER = nonpregnant EER + 452

Lactation

0–6 months postpartum	EER = nonpregnant EER + 500 − 170
7–12 months postpartum	EER = nonpregnant EER + 400 − 0

NOTE: Select the appropriate equation for gender and age and insert weight in kilograms, height in meters, and age in years. See the text and Table D–3 to determine PA.

Table D–2
Physical Activities and Their Scores

If your activity was equivalent to this …	Then list the number of minutes here and …	Multiply by this factor …	Add this column to get your physical activity level score:
ACTIVITIES OF DAILY LIVING			
Gardening (no lifting)		0.0032	
Household tasks (moderate effort)		0.0024	
Lifting items continuously		0.0029	
Loading/unloading car		0.0019	
Lying quietly		0.0000	
Mopping		0.0024	
Mowing lawn (power mower)		0.0033	
Raking lawn		0.0029	
Riding in a vehicle		0.0000	
Sitting (idle)		0.0000	
Sitting (doing light activity)		0.0005	
Taking out trash		0.0019	
Vacuuming		0.0024	
Walking the dog		0.0019	
Walking from house to car or bus		0.0014	
Watering plants		0.0014	
ADDITIONAL ACTIVITIES			
Billiards		0.0013	
Calisthenics (no weight)		0.0029	

Table D–1
Equations to Determine Estimated Energy Requirement (EER) (continued)

If your activity was equivalent to this …	Then list the number of minutes here and …	Multiply by this factor …	Add this column to get your physical activity level score:
ADDITIONAL ACTIVITIES—Continued			
Canoeing (leisurely)		0.0014	
Chopping wood		0.0037	
Climbing hills (carrying 11 lb load)		0.0061	
Climbing hills (no load)		0.0056	
Cycling (leisurely)		0.0024	
Cycling (moderately)		0.0045	
Dancing (aerobic or ballet)		0.0048	
Dancing (ballroom, leisurely)		0.0018	
Dancing (fast ballroom or square)		0.0043	
Golf (with cart)		0.0014	
Golf (without cart)		0.0032	
Horseback riding (walking)		0.0012	
Horseback riding (trotting)		0.0053	
Jogging (6 mph)		0.0088	
Music (playing accordion)		0.0008	
Music (playing cello)		0.0012	
Music (playing flute)		0.0010	
Music (playing piano)		0.0012	
Music (playing violin)		0.0014	
Rope skipping		0.0105	
Skating (ice)		0.0043	
Skating (roller)		0.0052	
Skiing (water or downhill)		0.0055	
Squash		0.0106	
Surfing		0.0048	
Swimming (slow)		0.0033	
Swimming (fast)		0.0057	
Tennis (doubles)		0.0038	
Tennis (singles)		0.0057	
Volleyball (noncompetitive)		0.0018	
Walking (2 mph)		0.0014	
Walking (3 mph)		0.0022	
Walking (4 mph)		0.0033	
Walking (5 mph)		0.0067	
Subtotal			
Factor for basal energy and the thermic effect of food			**1.1**
Your physical activity level score			

Table D–3
Physical Activity Equivalents and Their PA Factors

Physical Activity Level	Description	Physical Activity Equivalents	Men, 19+ yr PA Factor	Women, 19+ yr PA FActor	Boys, 3–18 yr PA Factor	Girls, 3–18 yr PA Factor
1.0 to 1.39	Sedentary	Only those physical activities required for typical daily living	1.0	1.0	1.0	1.0
1.4 to 1.59	Low active	Daily living + 30–60 min moderate activity[a]	1.11	1.12	1.13	1.16
1.6 to 1.89	Active	Daily living + ≥ 60 min moderate activity	1.25	1.27	1.26	1.31
1.9 and above	Very active	Daily living + ≥ 60 min moderate activity and ≥ 60 min vigorous activity or ≥ 120 min moderate activity	1.48	1.45	1.42	1.56

[a]Moderate activity is equivalent to walking at a pace of 3 to 4½ mph.

Table D–4
Total Energy Expenditure (TEE in kcalories per Day) for Women 30 Years of Age[a] at Various Levels of Activity and Various Heights and Weights

Heights m (in)	Physical Activity Level			Weight[b] kg (lb)			
1.45 (57)		38.9 (86)	45.2 (100)	52.6 (116)	63.1 (139)	73.6 (162)	84.1 (185)
				kcalories			
	Sedentary	1564	1623	1698	1813	1927	2042
	Low active	1734	1800	1912	2043	2174	2304
	Active	1946	2021	2112	2257	2403	2548
	Very active	2201	2287	2387	2553	2719	2886
1.50 (59)		41.6 (92)	48.4 (107)	56.3 (124)	67.5 (149)	78.8 (174)	90.0 (198)
				kcalories			
	Sedentary	1625	1689	1771	1894	2017	2139
	Low active	1803	1874	1996	2136	2276	2415
	Active	2025	2105	2205	2360	2516	2672
	Very active	2291	2382	2493	2671	2849	3027
1.55 (61)		44.4 (98)	51.7 (114)	60.1 (132)	72.1 (159)	84.1 (185)	96.1 (212)
				kcalories			
	Sedentary	1688	1756	1846	1977	2108	2239
	Low active	1873	1949	2081	2230	2380	2529
	Active	2104	2190	2299	2466	2632	2798
	Very active	2382	2480	2601	2791	2981	3171
1.60 (63)		47.4 (104)	55.0 (121)	64.0 (141)	76.8 (169)	89.6 (197)	102.4 (226)
				kcalories			
	Sedentary	1752	1824	1922	2061	2201	2340
	Low active	1944	2025	2168	2327	2486	2645
	Active	2185	2276	2396	2573	2750	2927
	Very active	2474	2578	2712	2914	3116	3318
1.65 (65)		50.4 (111)	58.5 (129)	68.1 (150)	81.7 (180)	95.3 (210)	108.9 (240)
				kcalories			
	Sedentary	1816	1893	1999	2148	2296	2444
	Low active	2016	2102	2556	2425	2594	2763
	Active	2267	2364	2494	2682	2871	3059
	Very active	2567	2678	2824	3039	3254	3469

[a]For each year below 30, add 7 kcalories/day to TEE. For each year above 30, subtract 7 kcalories/day from TEE.

[b]These columns represent a BMI of 18.5, 22.5, 25, 30, 35, and 40, respectively.

Table D–4
Total Energy Expenditure (TEE in kcalories per Day) for Women 30 Years of Age[a] at Various Levels of Activity and Various Heights and Weights   (continued)

Heights m (in)	Physical Activity Level	Weight[b] kg (lb)					
1.70 (67)		53.5 (118)	62.1 (137)	72.3 (159)	86.7 (191)	101.2 (223)	115.6 (255)
		kcalories					
	Sedentary	1881	1963	2078	2235	2393	2550
	Low active	2090	2180	2345	2525	2705	2884
	Active	2350	2453	2594	2794	2994	3194
	Very active	2662	2780	2938	3166	3395	3623
1.75 (69)		56.7 (125)	65.8 (145)	76.6 (169)	91.9 (202)	107.2 (236)	122.5 (270)
		kcalories					
	Sedentary	1948	2034	2158	2325	2492	2659
	Low active	2164	2260	2437	2627	2817	3007
	Active	2434	2543	2695	2907	3119	3331
	Very active	2758	2883	3054	3296	3538	3780
1.80 (71)		59.9 (132)	69.7 (154)	81.0 (178)	97.2 (214)	113.4 (250)	129.6 (285)
		kcalories					
	Sedentary	2015	2106	2239	2416	2593	2769
	Low active	2239	2341	2529	2731	2932	3133
	Active	2519	2634	2799	3023	3247	3472
	Very active	2855	2987	3172	3428	3684	3940
1.85 (73)		63.3 (139)	73.6 (162)	85.6 (189)	102.7 (226)	119.8 (264)	136.9 (302)
		kcalories					
	Sedentary	2083	2179	2322	2509	2695	2882
	Low active	2315	2422	2624	2836	3049	3262
	Active	2605	2727	2904	3141	3378	3615
	Very active	2954	3093	3292	3562	3833	4103
1.90 (75)		66.8 (147)	77.6 (171)	90.3 (199)	108.3 (239)	126.4 (278)	144.4 (318)
		kcalories					
	Sedentary	2151	2253	2406	2603	2800	2996
	Low active	2392	2505	2720	2944	3168	3393
	Active	2693	2821	3011	3261	3511	3760
	Very active	3053	3200	3414	3699	3984	4270
1.95 (77)		70.3 (155)	81.8 (180)	95.1 (209)	114.1 (251)	133.1 (293)	152.1 (335)
		kcalories					
	Sedentary	2221	2328	2492	2699	2906	3113
	Low active	2470	2589	2817	3053	3290	3526
	Active	2781	2917	3119	3383	3646	3909
	Very active	3154	3309	3538	3838	4139	4439

Table D–5
Total Energy Expenditure (TEE in kcalories per Day) for men 30 Years of Age[a] at Various Levels of Activity and Various Heights and Weights

Heights m (in)	Physical Activity Level	Weight[b] kg (lb)					
1.45 (57)		38.9 (86)	47.3 (100)	52.6 (116)	63.1 (139)	73.6 (163)	84.1 (185)
		kcalories					
	Sedentary	1777	1911	2048	2198	2347	2496
	Low active	1931	2080	2225	2393	2560	2727
	Active	2127	2295	2447	2636	2826	3015
	Very active	2450	2648	2845	3075	3305	3535

[a]For each year below 30, add 10 kcalories/day to TEE. For each year above 30, subtract 10 kcalories/day from TEE.

[b]These columns represent a BMI of 18.5, 22.5, 25, 30, 35, and 40, respectively.

Table D–5
Total Energy Expenditure (TEE in kcalories per Day) for men 30 Years of Age[a] at Various Levels of Activity and Various Heights and Weights   (continued)

Heights m (in)	Physical Activity Level	Weight[b] kg (lb)					
1.50 (59)		41.6 (92)	50.6 (107)	56.3 (124)	67.5 (149)	78.8 (174)	90.0 (198)
				kcalories			
	Sedentary	1848	1991	2126	2286	2445	2605
	Low active	2009	2168	2312	2491	2670	2849
	Active	2215	2394	2545	2748	2951	3154
	Very active	2554	2766	2965	3211	3457	3703
1.55 (61)		44.4 (98)	54.1 (114)	60.1 (132)	72.1 (159)	84.1 (185)	96.1 (212)
				kcalories			
	Sedentary	1919	2072	2205	2376	2546	2717
	Low active	2089	2259	2401	2592	2783	2974
	Active	2305	2496	2646	2862	3079	3296
	Very active	2660	2887	3087	3349	3612	3875
1.60 (63)		47.4 (104)	57.6 (121)	64.0 (141)	76.8 (169)	89.6 (197)	102.4 (226)
				kcalories			
	Sedentary	1993	2156	2286	2468	2650	2831
	Low active	2171	2351	2492	2695	2899	3102
	Active	2397	2601	2749	2980	3210	3441
	Very active	2769	3010	3211	3491	3771	4051
1.65 (65)		50.4 (111)	61.3 (129)	68.1 (150)	81.7 (180)	95.3 (210)	108.9 (240)
				kcalories			
	Sedentary	2068	2241	2369	2562	2756	2949
	Low active	2254	2446	2585	2801	3017	3234
	Active	2490	2707	2854	3099	3345	3590
	Very active	2880	3136	3339	3637	3934	4232
1.70 (67)		53.5 (118)	65.0 (137)	72.3 (159)	86.7 (191)	101.2 (223)	115.6 (255)
				kcalories			
	Sedentary	2144	2328	2454	2659	2864	3069
	Low active	2338	2542	2679	2909	3139	3369
	Active	2586	2816	2961	3222	3483	3743
	Very active	2992	3265	3469	3785	4101	4417
1.75 (69)		56.7 (125)	68.9 (145)	76.6 (169)	91.9 (202)	107.2 (236)	122.5 (270)
				kcalories			
	Sedentary	2222	2416	2540	2757	2975	3192
	Low active	2425	2641	2776	3020	3263	3507
	Active	2683	2927	3071	3347	3623	3900
	Very active	3108	3396	3602	3937	4272	4607
1.80 (71)		59.9 (132)	72.9 (154)	81.0 (178)	97.2 (214)	113.4 (250)	129.6 (285)
				kcalories			
	Sedentary	2301	2507	2628	2858	3088	3318
	Low active	2513	2741	2875	3132	3390	3648
	Active	2782	3040	3183	3475	3767	4060
	Very active	3225	3530	3738	4092	4447	4801
1.85 (73)		63.3 (139)	77.0 (162)	85.6 (189)	102.7 (226)	119.8 (264)	136.9 (302)
				kcalories			
	Sedentary	2382	2599	2718	2961	3204	3447
	Low active	2602	2844	2976	3248	3520	3792
	Active	2883	3155	3297	3606	3915	4223
	Very active	3344	3667	3877	4251	4625	4999

[a] For each year below 30, add 10 kcalories/day to TEE. For each year above 30, subtract 10 kcalories/day from TEE.

[b] These columns represent a BMI of 18.5, 22.5, 25, 30, 35, and 40, respectively.

D

1.90 (75)		66.8 (147)	81.2 (171)	90.3 (199)	108.3 (239)	126.4 (278)	144.4 (318)
				kcalories			
	Sedentary	2464	2693	2810	3066	3322	3579
	Low active	2693	2948	3078	3365	3652	3939
	Active	2986	3273	3414	3739	4065	4390
	Very active	3466	3806	4018	4413	4807	5202
1.95 (77)		70.3 (155)	85.6 (180)	95.1 (209)	114.1 (251)	133.1 (293)	152.1 (335)
				kcalories			
	Sedentary	2547	2789	2903	3173	3443	3713
	Low active	2786	3055	3183	3485	3788	4090
	Active	3090	3393	3533	3875	4218	4561
	Very active	3590	3948	4162	4578	4993	5409

[a] For each year below 30, add 10 kcalories/day to TEE. For each year above 30, subtract 10 kcalories/day from TEE.

[b] These columns represent a BMI of 18.5, 22.5, 25, 30, 35, and 40, respectively.

Nutrition Assessment: Supplemental Information

E

Several chapters in this book described data from nutrition assessments that help health professionals evaluate patients' nutrition status and nutrient needs. This appendix provides additional information that may be useful for complete assessments.

Weight Gain during Pregnancy

Chapter 11 described desirable weight-gain patterns during pregnancy. Figure E–1 shows prenatal weight-gain grids, used to plot the rate of weight gain during pregnancy.

Growth Charts

Health professionals generally evaluate physical development by monitoring the growth rate of a child and comparing this rate with standard charts. Standard charts compare length or height to age, weight to age, and body mass index to age; ideally, height and weight are in roughly the same percentile. Although individual growth patterns may vary, a child's growth curve will generally stay at about the same percentile throughout childhood. In children whose growth has been retarded, nutrition rehabilitation will ideally induce height and weight to increase to higher percentiles. In overweight children, the goal is for weight to remain stable as height increases, until weight becomes appropriate for height.

To evaluate growth in infants, an assessor uses charts such as those in Figures E–2 (A and B) through E–5 (A and B). The assessor follows these steps to plot a weight measurement on a percentile graph:

- Select the appropriate chart based on age and gender.

- Locate the child's age along the horizontal axis on the bottom or top of the chart.

- Locate the child's weight in pounds or kilograms along the vertical axis.

- Mark the chart where the age and weight lines intersect, and read off the percentile.

Reminder: The body mass index (BMI) is an index of a person's weight in relation to height, determined by dividing the weight in kilograms by the square of the height in meters:

$$BMI = \frac{Weight\ (kg)}{Height\ (m)^2}.$$

Additional growth charts are available at www.cdc.gov/growthcharts.

FIGURE E–1

Recommended Prenatal Weight Gain Based on Prepregnancy Weight

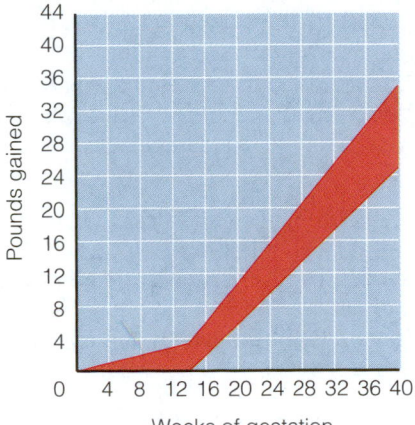

Normal-weight women should gain about 3½ pounds in the first trimester and just under 1 pound/week thereafter, achieving a total gain of 25 to 35 pounds by term.

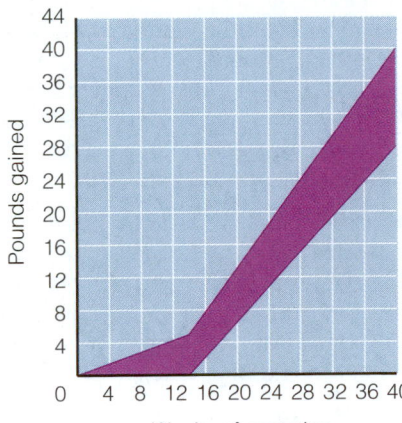

Underweight women should gain about 5 pounds in the first trimester and just over 1 pound/week thereafter, achieving a total gain of 28 to 40 pounds by term.

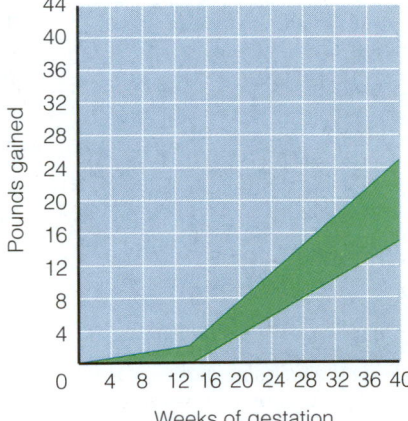

Overweight women should gain about 2 pounds in the first trimester and ⅔ pound/week thereafter, achieving a total gain of 15 to 25 pounds.

To assess length, height, or head circumference, the assessor follows the same procedure, using the appropriate chart. (When length is measured, use the chart for birth to 36 months; when height is measured, use the chart for 2 to 20 years.) Head circumference percentile should be similar to the child's height and weight percentiles. With height, weight, and head circumference measures plotted on growth percentile charts, a skilled clinician can begin to interpret the data.

Percentile charts divide the measures of a population into 100 equal divisions. Thus half of the population falls above the 50th percentile, and half falls below. The use of percentile measures allows for comparisons among people of the same age and gender. For example, a six-month-old female infant whose weight is at the 75th percentile weighs more than 75 percent of the female infants her age.

Head circumference is generally measured in children under two years of age. Since the brain grows rapidly before birth and during early infancy, extreme and chronic malnutrition during these times can impair brain development, curtailing the number of brain cells and the size of head circumference. Nonnutritional factors, such as certain disorders and genetic variation, can also influence head circumference.

Measures of Body Fat and Lean Tissue

Significant weight changes in both children and adults can reflect overnutrition or undernutrition with respect to energy and protein. To estimate the degree to which fat stores or lean tissues are affected by overnutrition or malnutrition, several anthropometric measurements are useful.

Skinfold Measures Skinfold measures provide a good estimate of total body fat and a fair assessment of the fat's location. Approximately half the fat in the body lies directly beneath the skin, and the thickness of this subcutaneous fat reflects total body fat. In some parts of the body, such as the back and the back of the arm over the triceps muscle, this fat is loosely attached; a person can pull it up between the thumb and forefinger to obtain a measure of skinfold thickness. To measure the skinfold, a skilled assessor follows a standard procedure using reliable calipers (illustrated in Figure E–6 on p. 835) and then compares the measurement with standards. Triceps skinfold measures greater than 15 millimeters in men or 25 millimeters in women suggest excessive body fat.

Skinfold measurements correlate directly with the risk of heart disease. They assess central obesity and its associated risks better than do weight measures alone. If a person gains body fat, the skinfold increases proportionately; if the person loses fat, it decreases. Measurements taken from central-body sites (around the abdomen) better reflect changes in fatness than those taken from upper sites (arm and back). A major limitation of the skinfold test is that fat may be thicker under the skin in one area than in another. A pinch at the side of the waistline may not yield the same measurement as a pinch on the back of the arm. This limitation can be overcome by taking skinfold measurements at several (often three) different places on the body (including upper-, central-, and lower-body sites) and comparing each measurement with standards for that site. Multiple measures are not always practical in clinical settings, however, and most often, the triceps skinfold measurement alone is used because it is easily accessible.

Waist Circumference Chapter 6 described how fat distribution correlates with health risks and mentioned that the waist circumference is a valuable indicator of fat distribution. To measure waist circumference, the assessor places a nonstretchable tape around the person's body, crossing just above the upper hip bones and making sure that the tape remains on a level horizontal plane on all sides (see Figure E–7 on p. 835). The tape is tightened slightly, but without compressing the skin.

Waist-to-Hip Ratio Alternatively, some clinicians measure both the waist and the hips. The waist-to-hip ratio also assesses abdominal obesity, but provides no more information than using the waist circumference alone. In general, women with a waist-to-hip ratio of 0.80 or greater and men with a waist-to-hip ratio of 0.90 or greater have a high risk of health problems.

Common sites for skinfold measures:
- Triceps.
- Biceps.
- Subscapular (below shoulder blade).
- Suprailiac (above hip bone).
- Abdomen.
- Upper thigh.

To calculate the waist-to-hip ratio, divide the waistline measurement by the hip measurement. For example, a woman with a 28-inch waist and 38-inch hips would have a ratio of 28 ÷ 38 = 0.74.

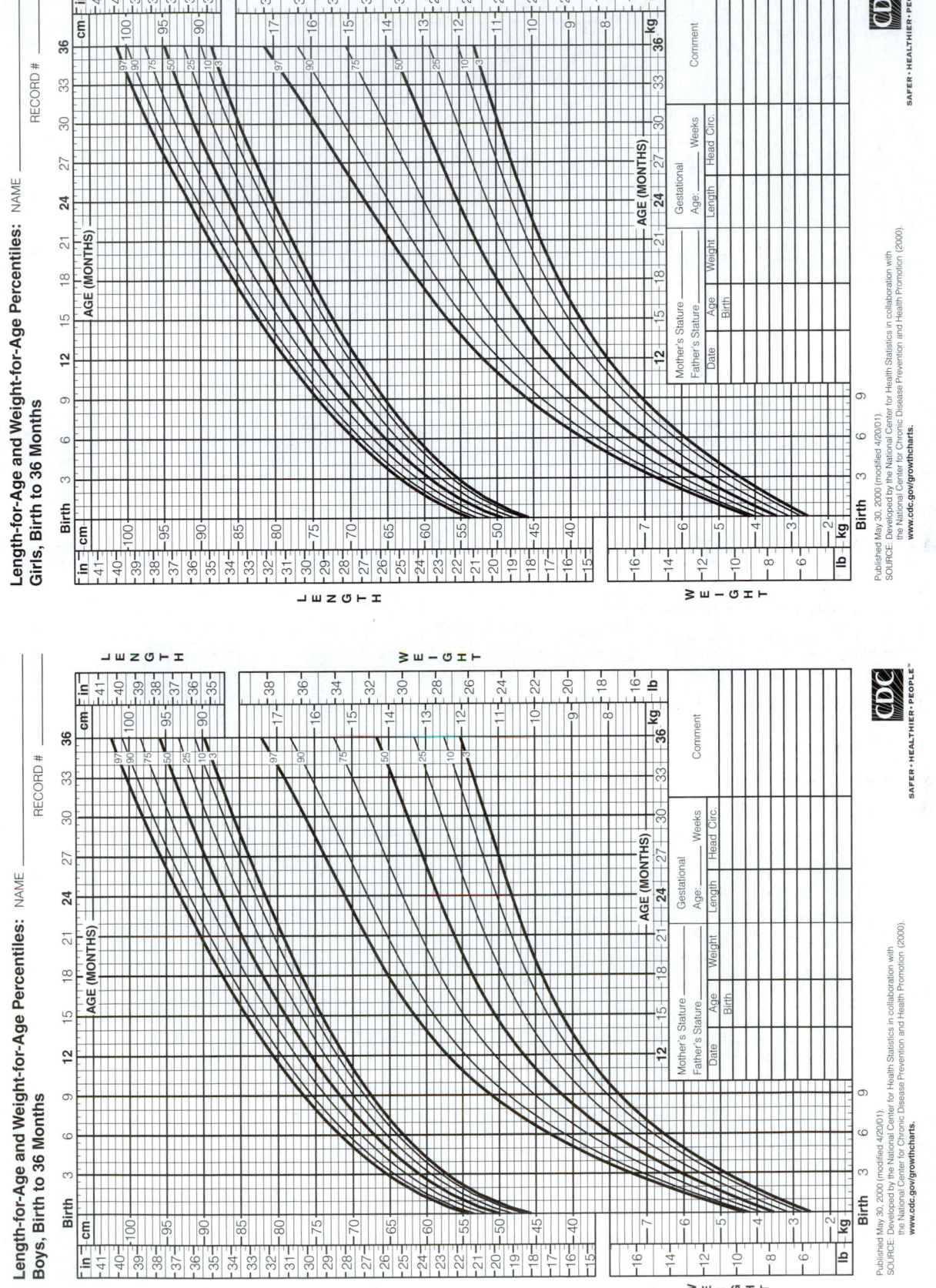

Length-for-Age and Weight-for-Age Percentiles:
Girls, Birth to 36 Months

Length-for-Age and Weight-for-Age Percentiles:
Boys, Birth to 36 Months

Published May 30, 2000 (modified 4/20/01).
SOURCE: Developed by the National Center for Health Statistics in collaboration with
the National Center for Chronic Disease Prevention and Health Promotion (2000).
www.cdc.gov/growthcharts.

FIGURE E–2 B

Length-for-Age and Weight-for-Age Percentiles: Girls, Birth to 36 Months

http://www.cdc.gov/nchs/data/nhanes/growthcharts/set1clinical/cj41l018.pdf

Published May 30, 2000 (modified 4/20/01).
SOURCE: Developed by the National Center for Health Statistics in collaboration with
the National Center for Chronic Disease Prevention and Health Promotion (2000).
www.cdc.gov/growthcharts.

FIGURE E–2 A

Length-for-Age and Weight-for-Age Percentiles: Boys, Birth to 36 Months

http://www.cdc.gov/nchs/data/nhanes/growthcharts/set1clinical/cj41l017.pdf

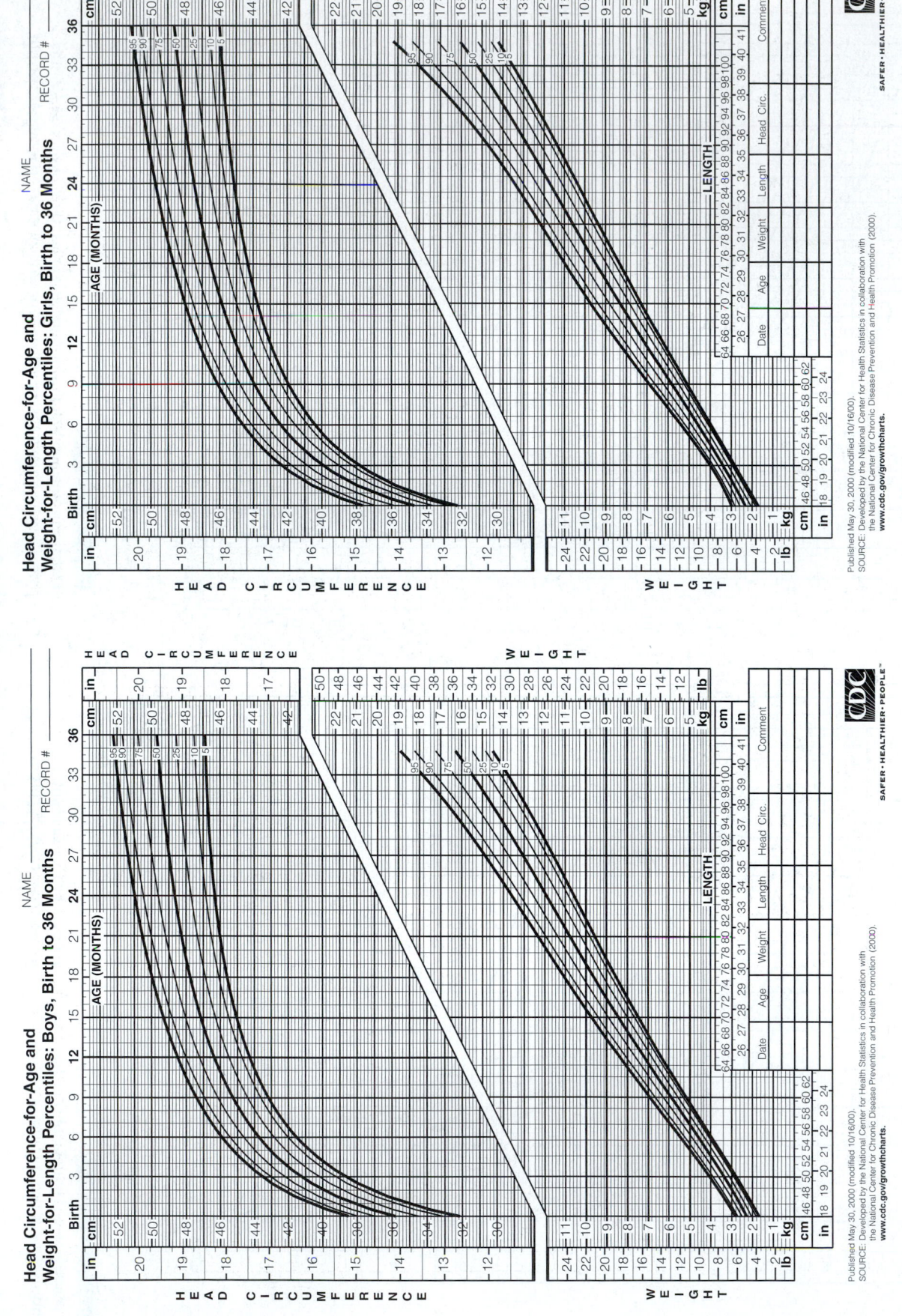

Head Circumference-for-Age and
Weight-for-Length Percentiles: Boys, Birth to 36 Months

Published May 30, 2000 (modified 10/16/00).
SOURCE: Developed by the National Center for Health Statistics in collaboration with
the National Center for Chronic Disease Prevention and Health Promotion (2000).
www.cdc.gov/growthcharts.

FIGURE E-3A

Head Circumference-for-Age and Weight-for-Length Percentiles: Boys, Birth to 36 Months
http://www.cdc.gov/nchs/data/nhanes/growthcharts/set1clinical/cj41l019.pdf

Head Circumference-for-Age and
Weight-for-Length Percentiles: Girls, Birth to 36 Months

Published May 30, 2000 (modified 10/16/00).
SOURCE: Developed by the National Center for Health Statistics in collaboration with
the National Center for Chronic Disease Prevention and Health Promotion (2000).
www.cdc.gov/growthcharts.

FIGURE E-3B

Head Circumference-for-Age and Weight-for-Length Percentiles: Girls, Birth to 36 Months
http://www.cdc.gov/nchs/data/nhanes/growthcharts/set1clinical/cj41l020.pdf

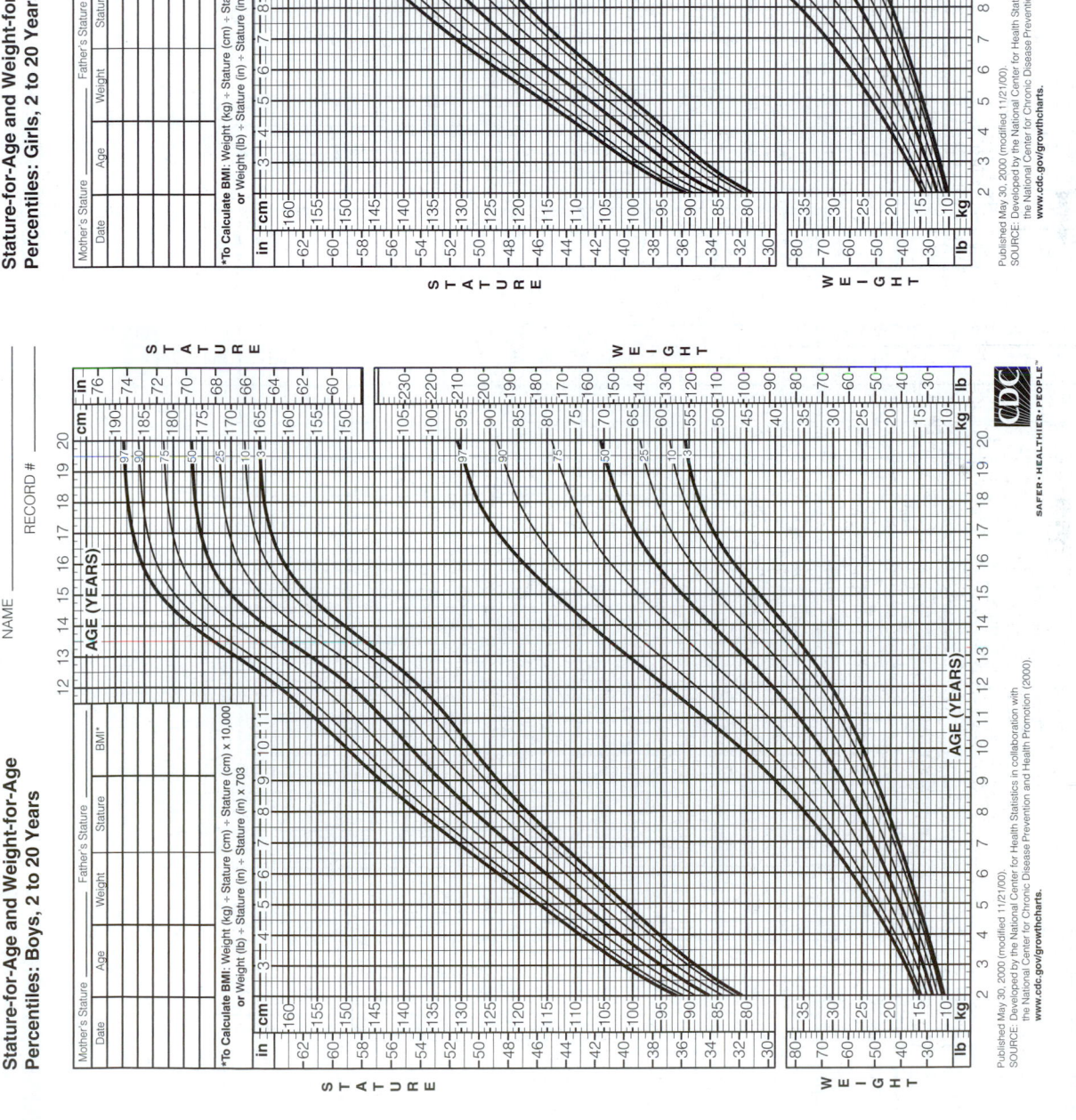

Stature-for-Age and Weight-for-Age Percentiles: Boys, 2 to 20 Years

Published May 30, 2000 (modified 11/21/00).
SOURCE: Developed by the National Center for Health Statistics in collaboration with the National Center for Chronic Disease Prevention and Health Promotion (2000).
www.cdc.gov/growthcharts.

FIGURE E–4A

Stature-for-Age and Weight-for-Age Percentiles: Boys, 2 to 20 Years

http://www.cdc.gov/nchs/data/nhanes/growthcharts/set1clinical/cj41l021.pdf

Stature-for-Age and Weight-for-Age Percentiles: Girls, 2 to 20 Years

Published May 30, 2000 (modified 11/21/00).
SOURCE: Developed by the National Center for Health Statistics in collaboration with the National Center for Chronic Disease Prevention and Health Promotion (2000).
www.cdc.gov/growthcharts.

FIGURE E–4B

Stature-for-Age and Weight-for-Age Percentiles: Girls, 2 to 20 Years

http://www.cdc.gov/nchs/data/nhanes/growthcharts/set1clinical/cj41l022.pdf

E

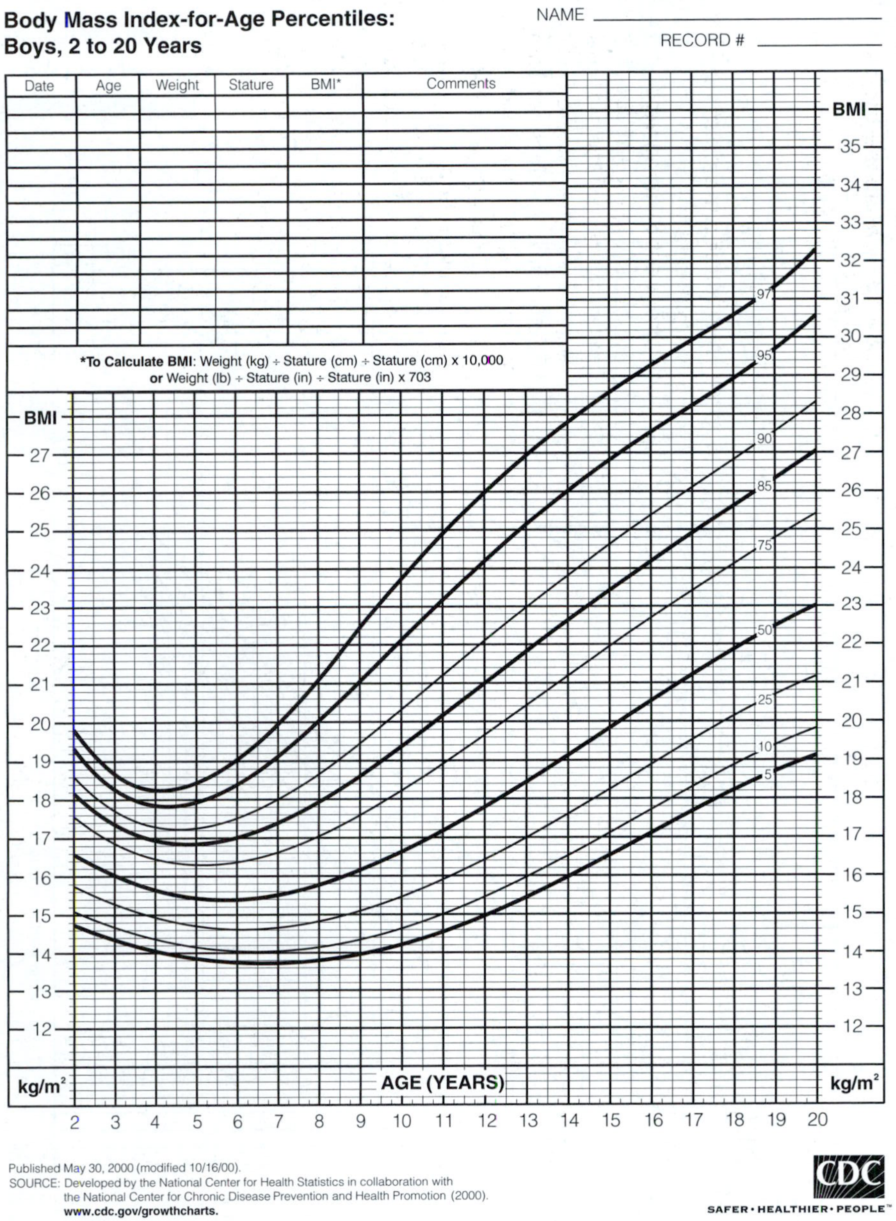
Body Mass Index-for-Age Percentiles:
Boys, 2 to 20 Years
NAME
RECORD #
Date
Age
Weight
Stature
BMI*
Comments
*To Calculate BMI: Weight (kg) ÷ Stature (cm) ÷ Stature (cm) x 10,000
or Weight (lb) ÷ Stature (in) ÷ Stature (in) x 703
BMI
kg/m²
AGE (YEARS)
97
95
90
85
75
50
25
10
5
Published May 30, 2000 (modified 10/16/00).
SOURCE: Developed by the National Center for Health Statistics in collaboration with the National Center for Chronic Disease Prevention and Health Promotion (2000).
www.cdc.gov/growthcharts.
CDC
SAFER · HEALTHIER · PEOPLE™

FIGURE E-5A

Body Mass Index-for-Age Percentiles: Boys, 2 to 20 Years
http://www.cdc.gov/nchs/data/nhanes/growthcharts/set1clinical/cj41l023.pdf

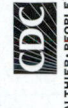

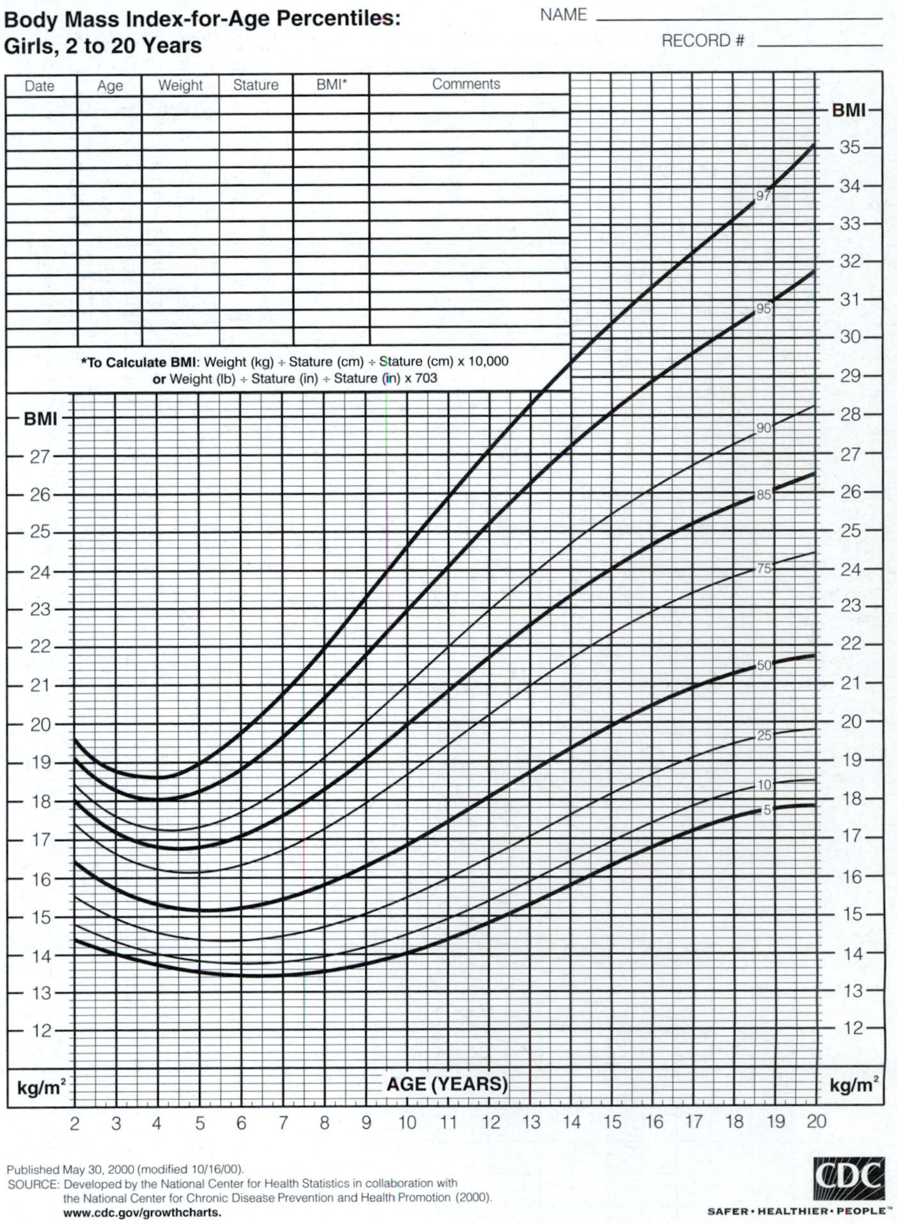
Body Mass Index-for-Age Percentiles:
Girls, 2 to 20 Years
NAME
RECORD #
Date
Age
Weight
Stature
BMI*
Comments
*To Calculate BMI: Weight (kg) ÷ Stature (cm) ÷ Stature (cm) x 10,000
or Weight (lb) ÷ Stature (in) ÷ Stature (in) x 703
BMI
kg/m²
AGE (YEARS)
97
95
90
85
75
50
25
10
5
Published May 30, 2000 (modified 10/16/00).
SOURCE: Developed by the National Center for Health Statistics in collaboration with the National Center for Chronic Disease Prevention and Health Promotion (2000).
www.cdc.gov/growthcharts.
CDC
SAFER · HEALTHIER · PEOPLE™

FIGURE E-5B

Body Mass Index-for-Age Percentiles: Girls, 2 to 20 Years
http://www.cdc.gov/nchs/data/nhanes/growthcharts/set1clinical/cj41l024.pdf

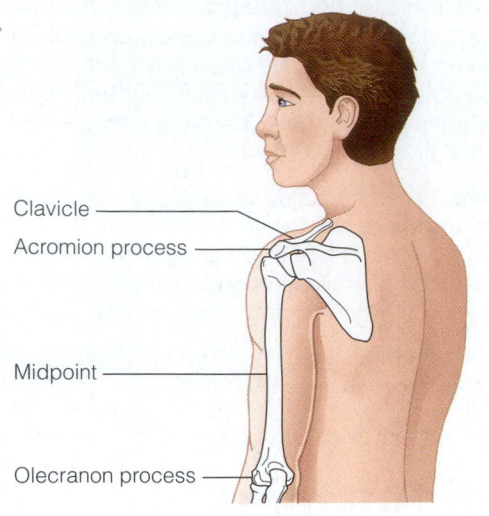
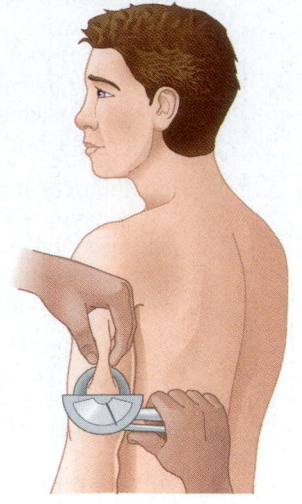

Clavicle

Acromion process

Midpoint

Olecranon process

A. Find the midpoint of the arm:
   1. Ask the subject to bend his or her arm at the elbow and lay the hand across the stomach. (If he or she is right-handed, measure the left arm, and vice versa.)
   2. Feel the shoulder to locate the acromion process. It helps to slide your fingers along the clavicle to find the acromion process. The olecranon process is the tip of the elbow.
   3. Place a measuring tape from the acromion process to the tip of the elbow.

Divide this measurement by 2 and mark the midpoint of the arm with a pen.
B. Measure the skinfold:
   1. Ask the subject to let his or her arm hang loosely to the side.
   2. Grasp a fold of skin and subcutaneous fat between the thumb and forefinger slightly above the midpoint mark. Gently pull the skin away from the underlying muscle. (This step takes a lot of practice. If you want to be sure you don't have muscle as well as fat, ask the subject to con-

tract and relax the muscle. You should be able to feel if you are pinching muscle.)
   3. Place the calipers over the skinfold at the midpoint mark, and read the measurement to the nearest 1.0 millimeter in two to three seconds. (If using plastic calipers, align pressure lines, and read the measurement to the nearest 1.0 millimeter in two to three seconds.)
   4. Repeat steps 2 and 3 twice more. Add the three readings, and then divide by 3 to find the average.

FIGURE E–6

How to Measure the Triceps Skinfold

FIGURE E–7

How to Measure Waist Circumference
Place the measuring tape around the waist just above the bony crest of the hip. The tape runs parallel to the floor and is snug (but does not compress the skin). The measurement is taken at the end of normal expiration.
Source: National Institutes of Health Obesity Education Initiative, *Clinical Guidelines on the Identification, Evaluation, and Treatment of Overweight and Obesity in Adults* (Washington, DC: U.S. Department of Health and Human Services, 1998), p. 59.

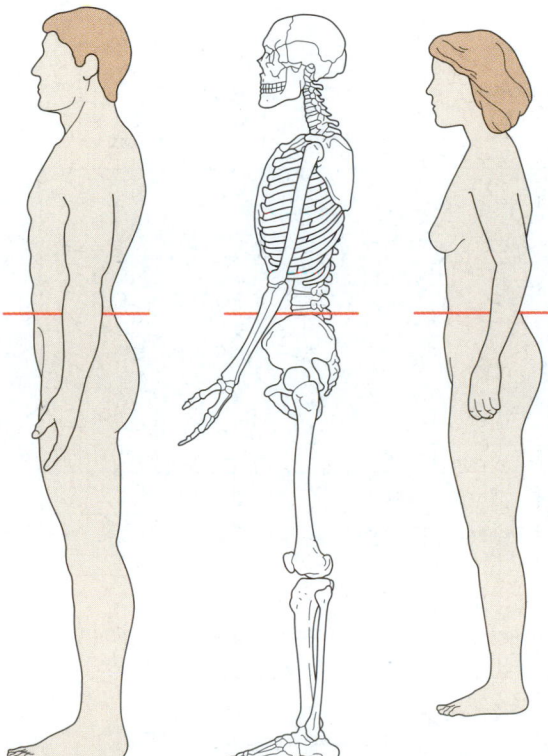
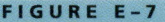

Hydrodensitometry To estimate body density using hydrodensitometry, the person is weighed twice—first on land and then again when submerged under water. Underwater weighing usually generates a good estimate of body fat and is useful in research, although the technique has drawbacks: it requires bulky, expensive, and nonportable equipment. Furthermore, submerging some people (especially those who are very young, very old, ill, or fearful) under water is not always practical.

Bioelectrical Impedance To measure body fat using the bioelectrical impedance technique, a very-low-intensity electrical current is briefly sent through the body by way of electrodes placed on the wrist and ankle. As is true of other anthropometric techniques, bioelectrical impedance requires standardized procedures and calibrated instruments to provide reliable results. Recent food intake and hydration status, for example, influence results.

Clinicians use many other methods to estimate body fat and its distribution. Each has its advantages and disadvantages, as Table E–1 summarizes.

Nutritional Anemias

Anemia, a symptom of a wide variety of nutrition- and nonnutrition-related disorders, is characterized by a reduced number of red blood cells. Iron, folate, and vitamin B12 deficiencies caused by inadequate intake, poor absorption, or abnormal metabolism of these nutrients are the most common nutritional anemias. Some nonnutrition-related causes of anemia include massive blood loss, infections, hereditary blood disorders such as sickle-cell anemia, and chronic liver or kidney disease.

Assessment of Iron-Deficiency Anemia

Iron deficiency, a common mineral deficiency, develops in stages. Chapter 9 describes iron deficiency in detail. This section describes tests used to uncover iron deficiency as it progresses. Table E–2 shows which laboratory tests detect various nutrition-related anemias, and Table E–3 (p. 838) provides values used for assessing iron status. Although other tests are more specific in detecting early deficiencies, hemoglobin and hematocrit are the commonly available tests.

Hemoglobin Iron forms an integral part of the hemoglobin molecule that transports oxygen to the cells. In iron deficiency, the body cannot synthesize hemoglobin. Low hemoglobin values signal depleted iron stores. Table E–3 (p. 838) provides hemoglobin values

TAble E–1
Methods of Estimating Body Fat and Its Distribution

Method	cost	Ease of Use	Accuracy	Measures Fat Distribution
Height and weight	Low	Easy	High	No
Skinfolds	Low	Easy	Low	Yes
Circumferences	Low	Easy	Moderate	Yes
Ultrasound	Moderate	Moderate	Moderate	Yes
Hydrodensitometry	Low	Moderate	High	No
Heavy water tritiated	Moderate	Moderate	High	No
Deuterium oxide, or heavy oxygen	High	Moderate	High	No
Potassium isotope (^{40}K)	Very high	Difficult	High	No
Total body electrical conductivity (TOBEC)	High	Moderate	High	No
Bioelectrical impedance (BIA)	Moderate	Easy	High	No
Dual energy X-ray absorptiometry (DEXA)	High	Easy	High	No
Computed tomography (CT)	Very high	Difficult	High	Yes
Magnetic resonance imaging (MRI)	Very high	Difficult	High	Yes

SOURCE: Adapted with permission from G. A. Bray, a handout presented at the North American Association for the Study of Obesity and Emory University School of Medicine Conference on Obesity Update: Pathophysiology, Clinical Consequences, and Therapeutic Options, Atlanta, Georgia, August 31–September 2, 1992.

used in nutrition assessment. Hemoglobin's usefulness in evaluating iron status is limited, however, because hemoglobin concentrations drop fairly late in the development of iron deficiency, and other nutrient deficiencies and medical conditions can also alter hemoglobin concentrations.

Hematocrit Hematocrit is commonly used to diagnose iron deficiency, even though it is an inconclusive measure of iron status. To measure the hematocrit, a clinician spins a volume of blood in a centrifuge to separate the red blood cells from the plasma. The hematocrit is the percentage of red blood cells in the total blood volume. Table E–3 includes values used to assess hematocrit status. Low values indicate incomplete hemoglobin formation, which is manifested by microcytic (abnormally small-celled), hypochromic (abnormally lacking in color) red blood cells.

Low hemoglobin and hematocrit values alert the assessor to the possibility of iron deficiency. However, many nutrients and other conditions can affect hemoglobin and hematocrit. The other tests of iron status help pinpoint true iron deficiency.

Serum Ferritin In the first stage of iron deficiency, iron stores diminish. Serum ferritin measures provide a noninvasive estimate of iron stores. Such information is most valuable to iron assessment. Table E–3 shows serum ferritin cutoff values that indicate iron store depletion in children and adults. Serum ferritin is not reliable for diagnosing iron deficiency in infants because normal serum ferritin values are often present in conjunction with iron-responsive anemia.

A decrease in transport iron characterizes the second stage of iron deficiency. This is revealed by an increase in the iron-binding capacity of the protein transferrin and a decrease in serum iron. These changes are reflected by the transferrin saturation, which is calculated from the

Stages of iron deficiency:

1. Iron stores diminish.
2. Transport iron decreases.
3. Hemoglobin production falls.

E

TAble E–2
Laboratory Tests Useful in Evaluating Nutrition-Related Anemias

Test or Test Result	What It Reflects
FOR ANEMIA (GENERAL)	
Hemoglobin (Hg)	Total amount of hemoglobin in the red blood cells (RBC)
Hematocrit (Hct)	Percentage of RBC in the total blood volume
Red blood cell (RBC) count	Number of RBC
Mean corpuscular volume (MCV)	RBC size; helps to determine if anemia is microcytic (iron deficiency) or macrocytic (folate or vitamin B_{12} deficiency)
Mean corpuscular hemoglobin concentration (MCHC)	Hemoglobin concentration within the average RBC; helps to determine if anemia is hypochromic (iron deficiency) or normochromic (folate or vitamin B_{12} deficiency)
Bone marrow aspiration	The manufacture of blood cells in different developmental states
FOR IRON-DEFICIENCY ANEMIA	
↓ Serum ferritin	Early deficiency state with depleted iron stores
↓ Transferrin saturation	Progressing deficiency state with diminished transport iron
↑ Erythrocyte protoporphyrin	Later deficiency state with limited hemoglobin production
FOR FOLATE-DEFICIENCY ANEMIA	
↓ Serum folate	Progressing deficiency state
↓ RBC folate	Later deficiency state
FOR VITAMIN B12-DEFICIENCY ANEMIA	
↓ Serum vitamin B_{12}	Progressing deficiency state
Schilling test	Absorption of vitamin B_{12}

ratio of the other two values as described in the following paragraphs.

Total Iron-Binding Capacity (TIBC) Iron travels through the blood bound to the protein transferrin. TIBC is a measure of the total amount of iron that transferrin can carry. Lab technicians measure iron-binding capacity directly. Table E–3 includes the value cutoff for TIBC.

Serum Iron Lab technicians can also measure serum iron directly. Elevated values indicate iron overload; reduced values indicate iron deficiency. Table E–3 shows the deficient value for serum iron.

Transferrin Saturation The percentage of transferrin that is saturated with iron is an indirect measure that is derived from the serum iron and total iron-binding capacity measures as follows:

$$\% \text{Transferrin} = \frac{\text{serum iron}}{\text{total iron-binding capacity}} \times 100$$

Table E–3 shows deficient transferrin saturation values for various age groups.

The third stage of iron deficiency occurs when the supply of transport iron diminishes to the point that it limits hemoglobin production. It is characterized by increases in erythrocyte protoporphyrin, a decrease in mean corpuscular volume, and decreased hemoglobin and hematocrit.

Erythrocyte Protoporphyrin The iron-containing portion of the hemoglobin molecule is heme. Heme is a combination of iron and protoporphyrin. Protoporphyrin accumulates in the blood when iron supplies are inadequate for the formation of heme. Lab technicians can measure erythrocyte protoporphyrin directly in a blood sample. The cutoffs for abnormal values of erythrocyte protoporphyrin are shown in Table E–3.

Mean Corpuscular Volume (MCV) A direct or calculated measure of the mean corpuscular volume (MCV) determines the average size of a red blood cell. Such a measure helps to classify the type of nutrient anemia. In iron deficiency, the red blood cells are smaller than average.

TAble E–3
Criteria for Assessing Iron Status

Test	Age (yr)	Gender	Deficiency Value
Hemoglobin (g/dL)	0.5–10	M–F	<11
	11–15	M	<12
		F	<11.5
	>15	M	<13
		F	<12
	Pregnancy		<11
Hematocrit (%)	0.5–4	M–F	<32
	5–10	M–F	<33
	11–15	M	<35
		F	<34
	>15	M	<40
		F	<36
Serum ferritin (μg/L)	0.5–15	M–F	<10
	>15	M–F	<12
Total iron-binding capacity (μg/dL)	>15	M–F	<400
Serum iron (μg/dL)	>15	M–F	<60
Transferrin saturation (%)	0.5–4	M–F	<12
	5–10	M–F	<14
	>10	M–F	<16
Erythrocyte protoporphyrin (μg/dL RBC)	0.5–4	M–F	>80
	>4	M–F	>70

Assessment of Folate and Vitamin B$_{12}$ Anemias

Folate deficiency and vitamin B$_{12}$ deficiency present a similar clinical picture—an anemia characterized by abnormally large red blood cell precursors (megaloblasts) in the bone marrow and abnormally large, mature red blood cells (macrocytic cells) in the blood. Distinguishing between these two deficiencies is particularly important because their treatments differ. Giving folate to a person with vitamin B$_{12}$ deficiency improves many of the lab test results indicative of vitamin B$_{12}$ deficiency, but this is a dangerous error because vitamin B$_{12}$ deficiency causes nerve damage that folate cannot correct. Thus inappropriate folate administration masks vitamin B$_{12}$-deficiency anemia, and nerve damage worsens. For this reason, it is critical to determine whether the anemia results from a folate deficiency or from a vitamin B$_{12}$ deficiency. The following biochemical assessment techniques help to make this distinction.

Mean Corpuscular Volume (MCV) As previously mentioned, the MCV is a measure of red blood cell size. In folate and vitamin B$_{12}$ deficiencies, the red blood cells are larger than average (macrocytic). Additional tests must be performed to differentiate folate from vitamin B$_{12}$ deficiency.

Folate Levels Serum folate levels fluctuate with changes in folate intake and metabolism. Thus serum folate concentrations reflect current status, but provide little information about folate stores. As folate deficiency progresses and low serum levels persist, folate stores decline, resulting in folate depletion. Folate depletion is characterized by a fall in the folate concentrations of red blood cells (erythrocytes). As erythrocyte folate levels diminish, folate-deficiency anemia develops. Because low erythrocyte folate concentrations also occur with vitamin B$_{12}$ deficiency, serum vitamin B$_{12}$ concentrations must also be measured. Table E–4 (p. E-12) shows standards for folate assessment.

Vitamin B$_{12}$ Levels Serum and urinary methylmalonic acid are elevated in vitamin B$_{12}$ deficiency, but not in folate deficiency. Thus this measure is useful in distinguishing between the two. Vitamin B$_{12}$ deficiency usually arises from malabsorption. To determine whether malabsorption is the cause, a small oral dose of vitamin B$_{12}$ is given, and urinary excretion is measured. This procedure measures vitamin B$_{12}$ absorption and is called a Schilling test.

Early stages of vitamin B$_{12}$ deficiency can be detected by a low percentage saturation of its transport protein, a measure similar to iron's transferrin saturation. As the deficiency progresses, serum vitamin B$_{12}$ concentrations fall. Table E–4 shows standards for vitamin B$_{12}$ assessment.

TAble E–4
Criteria for Assessing Folate and Vitamin B$_{12}$

	Deficient	Borderline	Acceptable
Serum folate (ng/mL)[a]	<3.0	3.0–5.9	>6.0
Erythrocyte folate (ng/mL)[a]	<140	140–159	>160
Serum vitamin B$_{12}$ (pg/mL)	<150	150–200	$201
Serum methylmalonic acid (nmol/L)	<376	—	—

NOTE: A nanogram (ng) is one-billionth of a gram; a picogram (pg) is one-trillionth of a gram.
[a]To convert folate values (ng/mL) to international standard units (nmol/L), multiply by 2.266.

Aids to Calculation

Many mathematical problems have been worked out in the "How to" sections of the text. These pages provide additional help and examples.

Conversion Factors

A *conversion factor* is a fraction that converts a measurement expressed in one unit to another unit; for example, the fraction may be used to convert pounds to kilograms or feet to inches. To create a conversion factor, an equality (such as *1 kilogram = 2.2 pounds*) is expressed as a fraction:

$$\frac{1 \text{ kg}}{2.2 \text{ lb}} \quad \text{and} \quad \frac{2.2 \text{ lb}}{1 \text{ kg}}$$

Because a conversion factor has a value of *1* (the value in the numerator is equal to the value in the denominator), it can be used as a multiplier to change the *unit* of measure without changing the *value* of the measurement. To convert the units of a measurement, the fraction must have the desired unit in the numerator, as the unit in the denominator will cancel out the original unit.

Example 1 Convert the weight of 130 pounds to kilograms.

- Multiply 130 pounds by a conversion factor (fraction) that includes both pounds and kilograms and is arranged so that the desired unit (kilograms) is in the numerator:

$$130 \text{ lb} \times \frac{1 \text{ kg}}{2.2 \text{ lb}} = \frac{130 \text{ kg}}{2.2} = 59 \text{ kg.}$$

As this example shows, the unit for *pounds* cancels out and the unit for *kilograms* remains in the solution.

Example 2 The food label on a bottle of apple juice shows the contents in both fluid ounces and liters. How many liters are contained in a bottle that holds 64 fluid ounces?

- Multiply 64 ounces by a conversion factor that includes both fluid ounces and liters, with the desired unit (liters) in the numerator:

$$64 \text{ fl oz} \times \frac{1 \text{ L}}{33.8 \text{ fl oz}} = \frac{64 \text{ L}}{33.8} = 1.89 \text{ L.}$$

Percentages

A percentage expresses a fraction that has 100 in the denominator; for example, the term *50 percent* is equivalent to the fraction 50/100. Similar to other fractions, percentages are used to express a *proportion* of the *whole;* therefore, the units in the numerator and denominator must be similar. Any fraction can be expressed in hundredths and converted to a percentage by dividing the numerator by the denominator and multiplying by 100:

$$^1/_4 = 0.25.$$
$$0.25 \times 100 = 25\%.$$

Example 3 Suppose your energy intake for the day is 2000 kcalories (kcal) and your recommended energy intake is 2400 kcalories. What percent of the recommended energy intake did you consume?

- Divide your intake by the recommended intake:

$$2000 \text{ kcal (your intake)} \div$$
$$2400 \text{ kcal (recommended intake)} = 0.83.$$

- Multiply by 100 to express the decimal as a percentage:

$$0.83 \times 100 = 83\%.$$

Example 4 A percentage can also be more than 100. Suppose your intake of vitamin C is 120 milligrams and your RDA (male) is 90 milligrams. What percent of the RDA for vitamin C did you consume?

120 mg (your intake) ÷ 90 mg (RDA) = 1.33.

$$1.33 \times 100 = 133\%.$$

Weights and Measures

Length

1 meter (m) = 39 in.
1 centimeter (cm) = 0.4 in.
1 inch (in) = 2.5 cm.
1 foot (ft) = 30 cm.

Temperature

Boling point	— 100°C	212°F —	Boiling point
Body temperature	— 37°C	98.6°F —	Body temperature
Freezing	— 0°C	32°F —	Freezing
	Celsius*	Fahrenheit	

■ To find degrees Fahrenheit (°F) when you know degrees Celsius (°C), multiply by 9/5 and then add 32.

■ To find degrees Celsius (°C) when you know degrees Fahrenheit (°F), subtract 32 and then multiply by 5/9.

Volume

1 liter (L) = 1000 mL, 0.26 gal, 1.06 qt, 2.1 pt, or 33.8 fl. oz.
1 milliliter (mL) = 1/1000 L or 0.03 fluid oz.
1 gallon (gal) = 128 oz, 8 c, or 3.8 L.
1 quart (qt) = 32 oz, 4 c, or 0.95 L.
1 pint (pt) = 16 oz, 2 c, or 0.47 L.
1 cup (c) = 8 oz, 16 tbs, about 250 mL, or 0.25 L.
1 ounce (oz) = 30 mL.
1 tablespoon (tbs) = 3 tsp or 15 mL.
1 teaspoon (tsp) = 5 mL.

Weight

1 kilogram (kg) = 1000 g or 2.2 lb.
1 gram (g) = 1/1000 kg, 1000 mg, or 0.035 oz.
1 milligram (mg) = 1/1000 g or 1000 μg.
1 microgram (μg) = 1/1000 mg.
1 pound (lb) = 16 oz, 454 g, or 0.45 kg.
1 ounce (oz) = about 28 g.

Energy

1 kilojoule (kJ) = 0.24 kcal.
1 millijoule (mJ) = 240 kcal.
1 kcalorie (kcal) = 4.2 kJ.
1 g carbohydrate = 4 kcal = 17 kJ.
1 g fat = 9 kcal = 37 kJ.
1 g protein = 4 kcal = 17 kJ.
1 g alcohol = 7 kcal = 29 kJ.

*Also known as centigrade.

Enteral Formulas

The large number of enteral formulas available allows health care professionals to meet a variety of their patients' medical needs, but also complicates the process of selecting an appropriate formula. The first step in narrowing the choice of formulas is to determine the patient's ability to digest and absorb nutrients. Table G–1 on pp. 844 to 845 lists examples of standard formulas for patients who can adequately digest and absorb nutrients, and Table G–2 on p. 846 provides examples of hydrolyzed formulas for patients with limited ability to digest or absorb nutrients. Products promoted to the general public and intended primarily as oral supplements, such as Carnation Instant Breakfast® (Nestlé), Boost® (Novartis), and Ensure (Ross), are not included as examples. Each formula is listed only once, although a formula may have more than one use. A high-protein formula, for example, may also be a fiber-containing formula. Tables G–3 through G–5 on p. 847 list modular formulas.

The information listed in this appendix reflects the literature provided by manufacturers and does not suggest endorsement by the authors. Manufacturers frequently add new formulas, discontinue old ones, and change formula composition. Consult the manufacturers' literature and websites for updates and additional examples of enteral formulas. The following products are listed in this appendix:

- Mead Johnson Nutritionals:[a]
  Kindercal® TF

- Nestlé Nutrition:[b]
  Crucial®
  Nutren® 1.0
  Nutren® 1.0 Fiber
  Nutren® 1.5
  Nutren® 2.0
  Nutren® Glytrol
  Nutren® Junior
  Nutren® Probalance
  Nutren® Pulmonary
  Nutren® Renal
  Nutren® Replete
  NutriHep®
  Peptamen®
  Peptamen Junior®

- Novartis Medical Nutrition:[c]
  Compleat® Pediatric
  Impact®
  Impact® 1.5
  Impact® Glutamine
  Isosource®
  Isosource® HN
  Isosource® VHN
  MCT Oil
  Microlipid®
  Novasource® Pulmonary
  Novasource® Renal
  Resource® Beneprotein® Instant Protein Powder
  Resource® Diabetic TF
  Vivonex® Pediatric
  Vivonex® Plus
  Vivonex® T.E.N.

- Ross Medical Nutritionals:[d]
  Glucerna®
  Jevity® 1 Cal
  Nepro® with Carb Steady™
  Optimental®
  Osmolite®
  Oxepa®
  Perative®
  Polycose Liquid®
  Polycose Powder®
  Promote®
  Promote® with Fiber
  Pulmocare®

---

[a]Mead Johnson Nutritionals, www.meadjohnson.com, visited May 18, 2007.

[b]Nestlé Nutrition, www.nestleclinicalnutrition.com, visited May 18, 2007.

[c]Novartis Medical Nutrition, www.novartis.com, visited May 18, 2007.

[d]Ross Medical Nutritionals, www.ross.com, visited May 18, 2007.

Table G–1
Standard Formulas

Product[a]	Volume to Meet 100% RDI[b] (mL)	Energy (kcal/mL)	Protein or Amino Acids (g/L)	Carbohydrate (g/L)	Fat (g/L)	Osmolality[c] (mOsm/kg)	Notes
LACTOSE-FREE, STANDARD FORMULAS							
Isosource®	1165	1.20	43	170	39	490	50% fat from MCT
Nutren® 1.0	1500	1.00	40	127	38	315	25% fat from MCT
Osmolite®	1887	1.06	37	151	35	300	20% fat from MCT
LACTOSE-FREE, FIBER-CONTAINING FORMULAS							
Jevity® 1 Cal	1321	1.06	44	155	35	300	14 g fiber/L
Nutren® 1.0 Fiber	1500	1.00	40	127	38	330	14 g fiber/L
Nutren® ProBalance	1000	1.20	54	156	41	350	10 g fiber/L
Promote® with Fiber	1000	1.00	63	138	28	380	14 g fiber/L
LACTOSE-FREE, HIGH-KCALORIE FORMULAS							
Nutren® 1.5	1000	1.50	60	169	68	430	50% fat from MCT
Nutren® 2.0	750	2.00	80	196	104	745	75% fat from MCT
LACTOSE-FREE, HIGH-PROTEIN FORMULAS							
Isosource® HN	1165	1.20	53	160	39	490	Low residue
Promote®	1000	1.00	63	130	26	340	20% fat from MCT, low residue
SPECIAL-USE FORMULAS: PEDIATRIC (1 TO 10 YEARS)							
Compleat® Pediatric	Varies[d]	1.00	38	130	39	380	Blenderized formula, 6.8 g fiber/L
Kindercal® TF	Varies[d]	1.06	30	135	44	345	12% fat from MCT
Nutren® Junior	Varies[d]	1.00	30	110	50	350	21% fat from MCT
SPECIAL-USE FORMULAS: GLUCOSE INTOLERANCE							
Glucerna®	1420	1.00	42	96	54	355	14 g fiber/L
Nutren® Glytrol	1400	1.00	45	100	48	280	15 g fiber/L; 20% fat from MCT
Resource® Diabetic TF	1180	1.06	58	85	50	400	15 g fiber/L
SPECIAL-USE FORMULAS: IMMUNE SYSTEM SUPPORT							
Impact®	1500	1.00	56	130	28	375	Enriched with arginine, nucleic acids, and omega-3 fatty acids
Impact® 1.5	1250	1.50	84	140	69	550	Same as above
Impact® Glutamine	1000	1.30	78	150	43	630	Same as above and enriched with glutamine; 10 g fiber/L

NOTE: MCT = Medium-chain triglycerides.

[a]Formulas come in ready-to-use (liquid) form unless specified under "Notes."

[b]RDI = Reference Daily Intakes, which are labeling standards for vitamins, minerals, and protein. Consuming 100 percent of the RDI will meet the nutrient needs of most people using the product.

[c]Osmolality may vary, depending on the flavorings added to a product.

[d]Depends on age of child.

Table G–1
Standard Formulas—continued

Product[a]	Volume to Meet 100% RDI[b] (mL)	Energy (kcal/mL)	Protein or Amino Acids (g/L)	Carbohydrate (g/L)	Fat (g/L)	Osmolality[c] (mOsm/kg)	Notes
SPECIAL-USE FORMULAS: RENAL FAILURE							
Nepro® with Carb Steady™	948	1.80	81	167	96	600	Low-potassium, low-phosphorus; intended for use once dialysis has been instituted
Novasource® Renal	1000	2.00	74	200	100	700	Low in electrolytes; intended for use once dialysis has been instituted
Nutren® Renal	750	2.00	70	205	104	650	50% fat from MCT; enriched with vitamins C and B$_6$, folate, zinc, and selenium; intended for use once dialysis has been instituted
SPECIAL-USE FORMULAS: RESPIRATORY INSUFFICIENCY							
Novasource® Pulmonary	933	1.50	75	150	68	650	8 g fiber/L
Nutren® Pulmonary	1000	1.50	68	100	95	330	55% kcal from fat, 40% fat from MCT
Oxepa	947	1.50	63	106	94	493	55% kcal from fat, enriched with antioxidant nutrients
Pulmocare®	947	1.50	63	106	93	475	55% kcal from fat, 20% fat from MCT, enriched with antioxidant nutrients
SPECIAL-USE FORMULAS: WOUND HEALING							
Isosource® VHN	1250	1.00	62	130	29	300	10 g fiber/L, enriched with vitamins A and C and zinc
Nutren® Replete	1000	1.00	62	113	34	300	Enriched with vitamins A and C and zinc; 25% fat from MCT

NOTE: MCT = Medium-chain triglycerides.

[a]Formulas come in ready-to-use (liquid) form unless specified under "Notes."

[b]RDI = Reference Daily Intakes, which are labeling standards for vitamins, minerals, and protein. Consuming 100 percent of the RDI will meet the nutrient needs of most people using the product.

[c]Osmolality may vary, depending on the flavorings added to a product.

[d]Depends on age of child.

TABLE G–2
Hydrolyzed Protein Formulas

Product	Volume to Meet 100% RDI[a] (mL)	Energy (kcal/mL)	Protein or Amino Acids (g/L)	Carbohydrate (g/L)	Fat (g/L)	Osmolality[b] (mOsm/kg)	Notes
SPECIAL-USE HYDROLYZED FORMULAS: HEPATIC INSUFFICIENCY							
NutriHep®	1000	1.50	40	290	21	790	Free amino acids, high in branched-chain amino acids, low in aromatic amino acids
SPECIAL-USE HYDROLYZED FORMULAS: IMMUNE SYSTEM SUPPORT							
Crucial®	1000	1.50	94	134	68	490	Enriched with arginine, antioxidant nutrients, and zinc
Perative®	1155	1.30	67	180	37	460	Enriched with arginine and beta-carotene
Vivonex® Plus	1800	1.00	45	190	7	650	Powder form; 100% free amino acids, enriched with glutamine, arginine, and branched-chain amino acids
SPECIAL-USE HYDROLYZED FORMULAS: MALABSORPTION							
Optimental®	1422	1.00	51	139	28	540	Contains MCT and arginine; enriched with vitamins C and E and beta-carotene
Peptamen®	1500	1.00	40	127	39	270	70% fat from MCT
Vivonex® T.E.N.	2000	1.00	38	210	3	630	Powder form; 100% free amino acids, enriched with glutamine
SPECIAL-USE HYDROLYZED FORMULAS: PEDIATRIC (1 TO 10 YEARS)							
Peptamen Junior®	Varies[c]	1.0	30	138	39	260	60% fat from MCT
Vivonex® Pediatric	Varies[c]	0.8	24	130	24	360	Powder form; 100% free amino acids

[a]RDI = Reference Daily Intakes, which are labeling standards for vitamins, minerals, and protein. Consuming 100 percent of the RDI will meet the nutrient needs of most people using the product.

[b]Osmolality may vary depending on the flavorings added to a product.

[c]Depends on age of child.

TABLE G–3
Protein Modules

Product	Form	Major Protein Source	Energy (kcal/g)	Protein (g/100 g)
Resource® Beneprotein® Instant Protein Powder	Powder	Whey protein	3.6	86

TABLE G–4
Carbohydrate Modules

Product	Form	Major Carbohydrate Source	Energy (kcal/mL or g)
Polycose Liquid®	Liquid	Hydrolyzed cornstarch	2.0 kcal/mL
Polycose Powder®	Powder	Hydrolyzed cornstarch	3.8 kcal/g

TABLE G–5
Fat Modules

Product	Form	Major Fat Source	Energy (kcal/mL)	Fat (g/100 mL)
MCT Oil	Liquid	Coconut oil	7.7	93
Microlipid®	Liquid	Safflower oil	4.5	50

Answers to Self Check Questions

Chapter 1
1. d, 2. d, 3. c, 4. d, 5. c, 6. b, 7. c, 8. b, 9. c, 10. c

Chapter 2
1. b, 2. b, 3. c, 4. a, 5. a, 6. a, 7. b, 8. a, 9. d, 10. c

Chapter 3
1. b, 2. c, 3. a, 4. d, 5. a, 6. a, 7. d, 8. d, 9. b, 10. a

Chapter 4
1. c, 2. b, 3. d, 4. d, 5. a, 6. d, 7. c, 8. b, 9. c, 10. d

Chapter 5
1. b, 2. a, 3. d, 4. a, 5. c, 6. a, 7. d, 8. c, 9. d, 10. d

Chapter 6
1. d, 2. b, 3. c, 4. d, 5. d, 6. b, 7. a, 8. a, 9. b, 10. c

Chapter 7
1. b, 2. a, 3. a, 4. c, 5. b, 6. c, 7. d, 8. b, 9. c, 10. b

Chapter 8
1. d, 2. d, 3. c, 4. a, 5. d, 6. d, 7. b, 8. c, 9. c, 10. a

Chapter 9
1. d, 2. b, 3. a, 4. d, 5. c, 6. d, 7. a, 8. d, 9. d, 10. b

Chapter 10
1. b, 2. a, 3. b, 4. c, 5. d, 6. c, 7. d, 8. b, 9. a, 10. d

Chapter 11
1. a, 2. d, 3. c, 4. c, 5. b, 6. d, 7. b, 8. d, 9. b, 10. c

Chapter 12
1. c, 2. b, 3. a, 4. b, 5. d, 6. c, 7. d, 8. c, 9. b, 10. c

Chapter 13
1. d, 2. c, 3. d, 4. d, 5. d, 6. d, 7. b, 8. b, 9. a, 10. b

Chapter 14
1. b, 2. d, 3. c, 4. d, 5. a, 6. d, 7. a, 8. c, 9. c, 10. b

Chapter 15
1. d, 2. d, 3. a, 4. c, 5. d, 6. b, 7. d, 8. c, 9. c, 10. b

Chapter 16
1. d, 2. c, 3. a, 4. b, 5. c, 6. d, 7. b, 8. a 9. a, 10. c

Chapter 17
1. a, 2. c, 3. b, 4. d, 5. a, 6. c, 7. d, 8. b, 9. b, 10. c

Chapter 18
1. b, 2. c, 3. d, 4. c, 5. d, 6. a, 7. b, 8. b, 9. a, 10. c

Chapter 19
1. d, 2. c, 3. b, 4. b, 5. a, 6. c, 7. a, 8. c, 9. a, 10. d

Chapter 20
1. d, 2. a, 3. b, 4. a, 5. b, 6. d, 7. d, 8. a, 9. b, 10. c

Chapter 21
1. b, 2. d, 3. d, 4. c, 5. a, 6. c, 7. d, 8. c, 9. a, 10. b

Chapter 22
1. a, 2. d, 3. c, 4. a, 5. d, 6. c, 7. d, 8. b, 9. c, 10. d

Chapter 23
1. a, 2. c, 3. b, 4. d, 5. c, 6. a, 7. b, 8. b, 9. c, 10. d

Chapter 24
1. d, 2. a, 3. b, 4. c, 5. d, 6. b, 7. c, 8. c, 9. d, 10. a

Chapter 25
1. d, 2. b, 3. c, 4. b, 5. d, 6. b, 7. c, 8. a, 9. a, 10. d

Glossary

2-in-1 solution: a parenteral solution that contains dextrose and amino acids but excludes lipids.

24-hour recall: a record of foods consumed in the previous 24 hours; sometimes modified to include foods consumed in a typical day.

abscesses (AB-sess-es): accumulated pus that is surrounded by inflamed tissue.

Acceptable Daily Intake (ADI): the amount of an artificial sweetener that individuals can safely consume each day over the course of a lifetime without adverse effect. It includes a 100-fold safety factor.

Acceptable Macronutrient Distribution Ranges (AMDR): ranges of intakes for the energy-yielding nutrients that provide adequate energy and nutrients and reduce the risk of chronic disease.

acetyl CoA (ASS-eh-teel, or ah-SEET-il, coh-AY): a 2-carbon compound (acetate, or acetic acid) with a molecule of CoA attached to it.

achalasia (ack-ah-LAY-zhah): an esophageal disorder characterized by weakened peristalsis and impaired relaxation of the lower esophageal sphincter.

achlorhydria (AY-clor-HIGH-dree-ah): absence of gastric acid secretion.

acid-base balance: the balance maintained between acid and base concentrations in the blood and body fluids.

acidosis: acid accumulation in the blood and body fluids; depresses the central nervous system and can lead to disorientation and, eventually, coma.

acids: compounds that release hydrogen ions in a solution.

acquired immune deficiency syndrome (AIDS): the late stage of illness caused by infection with the human immunodeficiency virus (HIV); characterized by severe damage to immune function.

acute erosive gastritis: a condition in which the gastric mucosa is acutely injured, often by the toxic effects of chemical substances or radiation treatment. Damage may include hemorrhaging, tissue erosion, and ulcers.

acute-phase response: changes in body chemistry resulting from infection, inflammation, or injury; characterized by alterations in plasma proteins.

acute renal failure: abrupt loss of kidney function over a period of hours or days.

acute respiratory distress syndrome (ARDS): respiratory failure triggered by acute lung injury; a medical emergency that causes dyspnea and pulmonary edema and usually requires assisted (mechanical) ventilation.

added sugars: sugars and syrups added to a food for any purpose, such as to add sweetness or bulk or to aid in browning (baked foods). Also called *carbohydrate sweeteners,* they include glucose, fructose, corn syrup, concentrated fruit juice, and other sweet carbohydrates.

adequacy: the characteristic of a diet that provides all the essential nutrients, fiber, and energy necessary to maintain health and body weight.

Adequate Intakes (AI): a set of values that are used as guides for nutrient intakes when scientific evidence is insufficient to determine a RDA.

adipose tissue: the body's fat, which consists of masses of fat-storing cells called adipose cells.

adolescence: the period of growth from the beginning of puberty until full maturity. Timing of adolescence varies from person to person.

advanced glycation end products (AGEs): compounds formed when glucose or glucose fragments combine with proteins. AGEs can damage tissues and lead to diabetic complications.

aerobic (air-ROH-bic): requiring oxygen. Aerobic activity strengthens the heart and lungs by requiring them to work harder than normal to deliver oxygen to the tissues.

AIDS-defining illnesses: diseases and complications associated with the later stages of a HIV infection, such as wasting, recurrent bacterial pneumonia, opportunistic infections, and certain cancers.

alcohol-related birth defects (ARBD): a condition caused by prenatal alcohol exposure. ARBD is diagnosed when there is a history of substantial, regular maternal alcohol intake or heavy episodic drinking and birth defects known to be associated with alcohol exposure.

alcohol-related neurodevelopmental disorder (ARND): a condition caused by prenatal alcohol exposure. ARND is diagnosed when there is a confirmed history of substantial, regular maternal alcohol intake or heavy episodic drinking and behavioral, cognitive, or central nervous system abnormalities known to be associated with alcohol exposure.

aldosterone (al-DOS-ter-own): a hormone secreted by the adrenal glands that stimulates the reabsorption of sodium by the kidneys; also regulates chloride and potassium concentrations.

alkalosis: excessive base in the blood and body fluids.

alpha-lactalbumin (lackt-AL-byoo-min): the chief protein in human breast milk, as *casein* (CAY-seen) is the chief protein in cow's milk.

alveoli (al-VEE-oh-lie): air sacs in the lungs. One sac is an *alveolus.*

Alzheimer's disease: a progressive, degenerative disease that attacks the brain and impairs thinking, behavior, and memory.

amino (a-MEEN-oh) **acids:** building blocks of protein. Each has a hydrogen atom, an amino group, and an acid group attached to a central carbon, which also carries a distinctive side chain.

amniotic (am-nee-OTT-ic) **sac:** the "bag of waters" in the uterus in which the fetus floats.

anabolism (an-ABB-o-lism): reactions in which small molecules are put together to build larger ones. Anabolic reactions require energy.

anaerobic (AN-air-ROH-bic): not requiring oxygen. Anaerobic activity may require strength but does not work the heart and lungs very hard for a sustained period.

anaphylactic (AN-ah-feh-LAC-tic) **shock:** a life-threatening whole-body allergic reaction to an offending substance.

anencephaly (AN-en-SEF-a-lee): an uncommon and always fatal type of neural tube defect; characterized by the absence of a brain.

anthropometric (AN-throw-poe-MEH-trik): related to physical measurements of the human body, such as height, weight, body circumferences, and percentage of body fat.

antibodies: large proteins of the blood and body fluids, produced in response to invasion of the body by unfamiliar molecules (mostly proteins) called *antigens.* Antibodies inactivate the invaders and so protect the body.

antidiuretic hormone (ADH): a hormone released by the pituitary gland in response to high salt concentrations in the blood. The kidneys respond by reabsorbing water.

antioxidant (anti-OX-ih-dant): a compound that protects other compounds from oxygen by itself reacting with oxygen. *Oxidation* is a potentially damaging effect of normal cell chemistry involving oxygen.

appetite: the psychological desire to eat; a learned motivation that is experienced as a pleasant sensation that accompanies the sight, smell, or thought of appealing foods.

artery: a vessel that carries blood away from the heart.

arthritis: inflammation of a joint, usually accompanied by pain, swelling, and structural changes.

artificial fats: zero-energy fat replacers that are chemically synthesized to mimic the sensory and cooking qualities of naturally occurring fats but are totally or partially resistant to digestion.

artificial sweeteners: noncarbohydrate, nonkcaloric synthetic sweetening agents; sometimes called *nonnutritive sweeteners.*

ascites (ah-SIGH-teez): the abnormal accumulation of fluid in the abdominal cavity.

ascorbic acid: one of the two active forms of vitamin C. Many people refer to vitamin C by this name.

aspiration: drawing in by suction or breathing; a common complication of enteral feedings in which foreign material enters the lungs, often from reflux of stomach contents.

atherosclerosis (ath-er-oh-scler-OH-sis): a type of artery disease characterized by accumulations of lipid-containing material on the inner walls of the arteries.

ATP or **adenosine** (ah-DEN-oh-seen) **triphosphate** (tri-FOS-fate): a common high-energy compound that contains three phosphate groups. The bonds between the phosphate groups are often described as "high-energy" because of their readiness to release their energy.

atrophic gastritis (a-TRO-fik gas-TRI-tis): a condition characterized by chronic inflammation of the stomach accompanied by a diminished size and functioning of the mucosa and glands.

atrophy (AT-tro-fee): a decrease in size (for example, of a muscle) because of disuse.

autoimmune: an immune response directed against the body's own tissues.

balance: the dietary characteristic of providing foods of a number of types in proportion to each other, such that foods rich in some nutrients do not crowd out foods that are rich in other nutrients.

bacterial overgrowth: excessive bacterial colonization of the stomach and small intestine; may be caused by low gastric acidity, altered gastrointestinal motility, mucosal damage, or contamination.

bariatric (BAH-ree-AH-trik) **surgery:** surgery that treats severe obesity.

Barrett's esophagus: a condition in which esophageal cells damaged by chronic exposure to stomach acid are replaced by cells that resemble those in the stomach or small intestine, sometimes becoming cancerous.

basal metabolic rate (BMR): the rate of energy use for metabolism under specified conditions: after a 12-hour fast and restful sleep, without any physical activity or emotional excitement, and in a comfortable setting. It is usually expressed as kcalories per kilogram of body weight per hour.

basal metabolism: the energy needed to maintain life when a person is at complete digestive, physical, and emotional rest. Basal metabolism is normally the largest part of a person's daily energy expenditure.

bases: compounds that accept hydrogen ions in a solution.

behavior modification: the changing of behavior by the manipulation of *antecedents* (cues or environmental factors that trigger behavior), the behavior itself, and *consequences* (the penalties or rewards attached to behavior).

beriberi: the thiamin-deficiency disease; characterized by loss of sensation in the hands and feet, muscular weakness, advancing paralysis, and abnormal heart action.

beta-carotene: a vitamin A precursor made by plants and stored in human fat tissue; an orange pigment.

BHA, BHT: food preservatives commonly used to slow the development of "off" flavors, odors, and color changes caused by oxidation.

bifidus (BIFF-id-us, by-FEED-us) **factors:** factors in colostrum and breast milk that favor the growth of the "friendly" bacterium *Lactobacillus* (lack-toe-ba-SILL-us) *bifidus* in the infant's intestinal tract. These bacteria prevent other, less desirable intestinal inhabitants from flourishing.

bile: a compound that prepares fats and oils for digestion; made by the liver, stored in the gallbladder, and released into the small intestine.

biliary system: the gallbladder and ducts that deliver bile from the liver and gallbladder to the small intestine.

binders: chemical compounds in foods that combine with nutrients (especially minerals) to form complexes the body cannot absorb. Examples include *phytates* and *oxalates*.

bioavailability: the rate and extent to which a nutrient is absorbed and used.

blenderized formulas: enteral formulas that are made by blenderizing whole foods.

body mass index (BMI): an index of a person's weight in relation to height; determined by dividing the weight (in kilograms) by the square of the height (in meters).

bolus (BOH-lus): the portion of food swallowed at one time.

bolus (BOH-lus) **feeding:** delivery of about 250 to 500 mL of formula in less than 15 minutes.

bone marrow transplants: procedures that replace bone marrow that has been destroyed by cancer treatments; also used to treat certain types of cancers and blood disorders.

bronchi (BRON-key), **bronchioles:** the main airways of the lungs. The singular form of bronchi is *bronchus.*

buffalo hump: the accumulation of fatty tissue at the base of the neck.

buffers: compounds that can reversibly combine with hydrogen ions to help keep a solution's acidity or alkalinity constant.

calcium rigor: hardness or stiffness of the muscles caused by high blood calcium.

calcium tetany: intermittent spasms of the extremities due to nervous and muscular excitability caused by low blood calcium.

calories: units by which energy is measured. Food energy is measured in *kilocalories* (1,000 calories equal 1 kilocalorie), abbreviated *kcalories* or kcal. One kcalorie is the amount of heat necessary to raise the temperature of 1 kilogram (kg) of water 1°C. The scientific use of the term kcalorie is the same as the popular use of the term calorie.

cancers: malignant growths or tumors that result from abnormal and uncontrolled cell division.

capillaries: small vessels that branch from an artery. Capillaries connect arteries to veins. Oxygen, nutrients, and waste materials are exchanged across capillary walls.

carbohydrate loading: a regimen of moderate exercise, followed by eating a high-carbohydrate diet, which enables muscles to temporarily store glycogen beyond their normal capacity; also called *glycogen loading* or *glycogen supercompensation.*

carbohydrate-to-insulin ratio: the amount of carbohydrate that can be handled per unit of insulin. On average, every 15 g of carbohydrate requires about 1 unit of rapid- or short-acting insulin.

carbohydrates: energy nutrients composed of monosaccharides.

carcinogenesis (CAR-sin-oh-JEN-eh-sis): the process of cancer development.

carcinogens (CAR-sin-oh-jenz or car-SIN-oh-jenz): substances that can cause cancer (the adjective is *carcinogenic*).

cardiac cachexia: the severe muscle wasting and weight loss that accompany congestive heart failure.

cancer cachexia (ka-KEK-see-ah): a wasting syndrome, associated with cancer, that is characterized by anorexia, muscle wasting, weight loss, and fatigue.

cardiac output: the volume of blood discharged by the heart each minute.

cardiorespiratory endurance: the ability to perform large-muscle dynamic exercise of moderate-to-high intensity for prolonged periods.

cardiovascular disease (CVD): a general term for all diseases of the heart and blood vessels.

catabolism (ca-TAB-o-lism): reactions in which large molecules are broken down to smaller ones. Catabolic reactions release energy.

cataracts: thickenings of the eye lenses that impair vision and can lead to blindness.

cathartics (ca-THART-ics): strong laxatives.

catheter: a thin tube placed within a narrow lumen (such as a blood vessel) or body cavity; can be used to infuse or withdraw fluids or to keep a passage open.

celiac (SEE-lee-ack) **disease:** a condition characterized by an abnormal immune reaction to wheat gluten that causes severe intestinal damage and nutrient malabsorption; also called *gluten-sensitive enteropathy* or *celiac sprue.*

cellulite (SELL-you-light or SELL-you-leet): supposedly, a lumpy form of fat; actually, a fraud. Fatty areas of the body may appear lumpy when the strands of connective tissue that attach the skin to underlying muscles pull tight where the fat is thick. The fat itself is the same as fat anywhere else in the body. If the fat in these areas is lost, the lumpy appearance disappears.

central obesity: excess fat around the trunk of the body; also called *abdominal fat* or *upper-body fat.*

central veins: large-diameter veins located close to the heart.

cerebral cortex: the outer surface of the cerebrum, which is the largest part of the brain.

Certified Diabetes Educator (CDE): a health care professional who specializes in diabetes management education. Certification is obtained from the National Certification Board for Diabetes Educators.

chemotherapy: the use of drugs to arrest or destroy cancer cells. Such drugs are called *antineoplastic agents.*

cholecystectomy (KOH-leh-sis-TEK-toe-mee): surgical removal of the gallbladder.

cholecystitis (KOH-leh-sih-STY-tis): inflammation of the gallbladder, usually caused by obstruction of the cystic duct by gallstones.

cholelithiasis (KOH-leh-lih-THIGH-ah-sis): formation of gallstones.

choline (KOH-leen): a nitrogen-containing compound found in foods and made in the body from an amino acid.

chronic bronchitis (bron-KYE-tis): persistent inflammation of the mucous membranes lining the main airways of the lungs. Chronic inflammation leads to narrower airways and difficulty with breathing.

chronic diseases: degenerative diseases characterized by deterioration of the body organs. Examples include heart disease, cancer, and diabetes.

chronic obstructive pulmonary disease (COPD): a group of lung diseases characterized by persistent obstructed airflow through the lungs and airways; includes chronic bronchitis and emphysema.

chronological age: a person's age in years from his or her date of birth.

chylomicrons (kye-lo-MY-crons): the lipoproteins that transport lipids from the intestinal cells into the body. The cells of the body remove the lipids they need from the chylomicrons, leaving chylomicron remnants to be picked up by the liver cells.

chyme (KIME): the semiliquid mass of partly digested food expelled by the stomach into the duodenum (the top portion of the small intestine).

cirrhosis (sih-ROE-sis): an advanced stage of liver disease in which extensive scarring replaces healthy liver tissue, causing impaired liver function and liver failure.

clear liquid diet: a diet that consists of foods that are liquid at room temperature and leave almost no residue (undigested material) in the intestines after digestion and absorption.

clinically severe obesity: a BMI of 40 or greater or a BMI of 35 or greater and one or more serious conditions such as hypertension. Another term used to describe the same condition is *morbid obesity*.

closed feeding systems: delivery systems in which formula comes prepackaged in containers that are ready to be attached to feeding tubes for administration.

CoA (coh-AY): coenzyme A; the coenzyme derived from the B vitamin pantothenic acid and central to energy metabolism.

coenzyme (co-EN-zime): a small molecule that works with an enzyme to promote the enzyme's activity. Many coenzymes have B vitamins as part of their structure.

cofactor: a mineral element that, like a coenzyme, works with an enzyme to facilitate a chemical reaction.

colectomy: removal of a portion or all of the colon.

collagen: the characteristic protein of connective tissue.

collaterals: blood vessels that enlarge to allow an alternative pathway for diverted blood.

colostomy (co-LAHS-toe-me): a surgical procedure that creates a stoma using a section of the colon.

colostrum (co-LAHS-trum): a milklike secretion from the breasts that is rich in protective factors. Colostrum is present during the first day or so after delivery, before milk appears.

complement: a group of plasma proteins that assist the activities of antibodies.

complementary proteins: two or more proteins whose amino acid assortments complement each other in such a way that the essential amino acids missing from one are supplied by the other.

conditionally essential amino acid: an amino acid that is normally nonessential but must be supplied by the diet in special circumstances when the need for it becomes greater than the body's ability to produce it.

congestive heart failure (CHF): a condition in which the heart is unable to pump adequate blood, resulting in fluid congestion in tissues and in the veins leading to the heart.

conjugated linoleic acid: a collective term for several fatty acids that have the same chemical formulas as linoleic acid but with different configurations.

continuous feedings: slow delivery of formula at a constant rate over an 8- to 24-hour period.

continuous parenteral nutrition: continuous administration of parenteral solutions over a 24-hour period.

cornea (KOR-nee-uh): the hard, transparent membrane covering the outside of the eye.

C-reactive protein: an acute-phase protein released from the liver during acute inflammation or stress.

cretinism (CREE-tin-ism): an iodine-deficiency disease characterized by mental and physical retardation.

critical pathways: coordinated programs of treatment that merge the care plans of different health practitioners; also called *clinical pathways*.

critical period: a finite period during development in which certain events occur that will have irreversible effects on later developmental stages; usually a period of rapid cell division.

Crohn's disease: an inflammatory bowel disease that usually occurs in the lower portion of the small intestine and the colon. Inflammation may pervade the entire intestinal wall.

cryptosporidiosis (KRIP-toe-spor-ih-dee-OH-sis): a foodborne illness caused by the parasite *Cryptosporidium parvum*.

cyanosis (sigh-ah-NOH-sis): a bluish cast in the skin due to the color of deoxygenated hemoglobin. Cyanosis is most evident in individuals with lighter, thinner skin; it is mostly seen on lips, cheeks, and ears and under the nails.

cyclic parenteral nutrition: administration of a parenteral solution over a 10- to 16-hour period.

cystic fibrosis: an inherited disease characterized by the production of abnormally viscous exocrine secretions; often leads to respiratory illness and pancreatic insufficiency.

cystinuria (SIS-tin-NOO-ree-ah): an inherited disorder characterized by elevated urinary excretion of several amino acids, including cystine.

cytokines (SIGH-toe-kynes): proteins produced by white blood cells that regulate immune cell development and immune responses.

Daily Values: reference values developed by the FDA specifically for use on food labels.

deamination: removal of the amino group (NH_2) from a compound such as an amino acid.

debridement: the surgical removal of dead, damaged, or contaminated tissue resulting from burns or wounds; helps to prevent infection and hasten healing.

deficient: in regard to nutrient intake, the amount below which almost all healthy people can be expected, over time, to experience deficiency symptoms.

dehydration: the loss of water from the body that occurs when water output exceeds water input. The symptoms progress rapidly from thirst to weakness, to exhaustion and delirium, and end in death if not corrected.

denaturation (dee-nay-cher-AY-shun): the change in a protein's shape brought about by heat, acid, or other agents. Past a certain point, denaturation is irreversible.

dental caries: the gradual decay and disintegration of a tooth.

dermatitis herpetiformis (DERM-ah-TYE-tis HER-peh-tih-FOR-mis): a gluten-sensitive disorder characterized by a severe skin rash. Gastrointestinal symptoms may be mild or absent.

diabetes (DYE-ah-BEE-teez) **mellitus:** a group of metabolic disorders characterized by hyperglycemia and disordered insulin metabolism.

dialysate (dye-AL-ih-sate): the solution used in dialysis to draw wastes and fluids from the blood.

dialysis (dye-AH-lih-sis): a procedure for removing wastes and excess fluid from the blood after the kidneys have stopped functioning. The most common types are *hemodialysis* and *peritoneal dialysis*.

dialyzer (DYE-ah-LYE-zer): a machine used for hemodialysis; also called an *artificial kidney*.

diet history: a comprehensive record of a person's food intake and dietary practices.

diet manual: a resource that specifies the foods allowed and restricted in modified diets and provides sample menus.

diet orders: specific instructions concerning dietary management; also called *diet prescriptions*.

diet progression: a change in diet as a patient's tolerances permit.

dietary fibers: a general term denoting in plant foods the polysaccharides cellulose, hemicellulose, pectins, gums, and mucilages, as well as the nonpolysaccharide lignins, that are not digested by human digestive enzymes, although some are digested by GI tract bacteria.

dietary folate equivalents (DFE): the amount of folate available to the body from naturally occurring sources, fortified foods, and supplements,

accounting for differences in bioavailability from each source.

Dietary Reference Intakes (DRI): a set of values for the dietary nutrient intakes of healthy people in the United States and Canada. These values are used for planning and assessing diets.

differentiation: the development of specific functions different from those of the original.

digestion: the process by which complex food particles are broken down to smaller absorbable particles.

digestive system: all the organs and glands associated with the ingestion and digestion of food.

dipeptide: two amino acids bonded together.

disaccharides (dye-SACK-uh-rides): a pair of sugar units bonded together.

discretionary kcalorie allowance: the kcalories remaining in a person's energy allowance after consuming enough nutrient-dense foods to meet all nutrient needs for one day.

disease-specific formulas: enteral formulas designed to meet the nutrient needs of patients with specific illnesses.

diuresis (DYE-uh-REE-sis): increased urine production.

diuretics (dye-yoo-RET-ics): medications that promote the excretion of water through the kidneys. Not all diuretics increase the urinary loss of potassium. Some, called potassium-sparing diuretics, are less likely to result in a potassium deficiency.

diverticulitis (dye-ver-tic-you-LYE-tis): an inflammation or infection involving diverticula.

diverticulosis (dye-ver-tic-you-LOH-sis): an intestinal disorder characterized by the presence of small outpockets (called diverticula) in the intestinal wall, which often develop by middle age.

dumping syndrome: symptoms that result from the rapid emptying of an osmotic load from the stomach into the small intestine. Early symptoms include nausea, abdominal cramps, weakness, and diarrhea; later symptoms are those of hypoglycemia.

duodenal ulcers: peptic ulcers that develop in the duodenum.

dysentery (DIS-en-terry): an infection of the gastrointestinal tract caused by an amoeba or bacterium that gives rise to severe diarrhea.

dyspepsia: a feeling of pain, bloating, or discomfort in the upper abdominal area, often called "indigestion"; a symptom of illness rather than a disease itself.

dysphagia (dis-FAY-ja): difficulty swallowing.

dyspnea (DISP-nee-ah): shortness of breath.

eclampsia: a severe complication during pregnancy in which convulsions occur.

edema (eh-DEEM-uh): the swelling of body tissue caused by leakage of fluid from the blood vessels and accumulation of the fluid in the interstitial spaces.

eicosanoids (eye-KO-sa-noids): 20-carbon molecules derived from dietary fatty acids that help to regulate blood pressure, blood clotting, and other body functions.

electrolyte: a salt that dissolves in water and dissociates into charged particles called ions.

electrolyte solutions: solutions that can conduct electricity.

electron transport chain: the final pathway in energy metabolism that transports electrons from hydrogen to oxygen and captures the energy released in the bonds of a high-energy compound, ATP.

embryo (EM-bree-oh): the developing infant from two to eight weeks after conception.

emphysema (EM-fih-ZEE-mah): lung disease characterized by progressive damage to the alveoli (air sacs) in the lungs; causes difficulty with breathing.

emulsifiers: substances that mix with both fat and water and that disperse the fat in the water, forming an emulsion.

end-stage renal disease (ESRD): an advanced stage of chronic kidney disease in which dialysis or a kidney transplant is necessary to sustain life.

energy density: a measure of the energy a food provides relative to the amount of food (kcalories per gram).

energy metabolism: all the reactions by which the body obtains and expends the energy from food.

energy-yielding nutrients: the nutrients that break down to yield energy the body can use. The three energy-yielding nutrients are carbohydrate, protein, and fat.

enrichment: the addition to a food of nutrients to meet a specified standard. In the case of refined bread or cereal, five nutrients have been added: thiamin, riboflavin, niacin, and folate in amounts approximately equivalent to, or higher than, those originally present and iron in amounts to alleviate the prevalence of iron-deficiency anemia.

enteral (EN-ter-al) **nutrition:** the provision of nutrients using the GI tract, including the use of tube feedings and oral diets.

enteric-coated: refers to medications or enzyme preparations that can withstand gastric acidity and dissolve only at a higher pH.

environmental tobacco smoke (ETS): the combination of exhaled smoke (mainstream smoke) and smoke from lighted cigarettes, pipes, or cigars (sidestream smoke) that enters the air and may be inhaled by other people.

enzymes: protein catalysts. A catalyst is a compound that facilitates chemical reactions without itself being changed in the process..

EPA, DHA: omega-3 fatty acids made from linolenic acid. The full name for EPA is *eicosapentaenoic* (EYE-cosa-PENTA-ee-NO-ick) *acid*. The full name for DHA is *docosahexaenoic* (DOE-cosa-HEXA-ee-NO-ick) *acid*.

epinephrine: one of the stress hormones secreted whenever emergency action is needed; prescribed therapeutically to relax the bronchioles during allergy or asthma attacks.

epithelial (ep-i-THEE-lee-ul) **cells:** cells on the surface of the skin and mucous membranes.

epithelial tissue: tissue composing the layers of the body that serve as selective barriers between the body's interior and the environment (examples are the cornea, skin, respiratory lining, and lining of the digestive tract).

erythrocyte (er-RITH-ro-cite) **hemolysis** (he-MOLL- uh-sis): rupture of the red blood cells, caused by vitamin E deficiency.

erythrocyte protoporphyrin (PRO-toh-PORE-fe-rin): a precursor to hemoglobin.

erythropoietin (eh-RITH-ro-POY-eh-tin): a hormone made by the kidneys that stimulates red blood cell production.

esophageal dysphagia: an inability to move food through the esophagus; usually caused by an obstruction or a motility disorder.

essential amino acids: amino acids that the body cannot synthesize in amounts sufficient to meet physiological need; also called *indispensable* amino acids. Nine amino acids are known to be essential for human adults.

essential fatty acids: fatty acids that the body requires but cannot make in amounts sufficient to meet its physiological needs.

essential nutrients: nutrients a person must obtain from food because the body cannot make them for itself in sufficient quantities to meet physiological needs.

Estimated Average Requirements (EAR): the average daily nutrient intake levels estimated to meet the requirements of half of the healthy individuals in a given age and gender group; used in nutrition research and policymaking and the basis on which RDA values are set.

Estimated Energy Requirement (EER): the average dietary energy intake that maintains energy balance in a healthy person of a defined age, gender, and weight, and a level of physical activity that is consistent with good health.

ethnic diets: foodways and cuisines typical of national origins, races, cultural heritages, or geographic locations.

exercise: planned, structured, and repetitive bodily movement that promotes or maintains physical fitness.

extracellular fluid: fluid residing outside the cells; includes the fluid between the cells (*interstitial fluid*), plasma, and the water of structures such as the skin and bones. Extracellular fluid accounts for about one-third of the body's water.

fat replacers: ingredients that replace some or all of the functions of fat in foods and may or may not provide energy.

fats: lipids that are solid at room temperature (70° F or 25° C).

fatty acids: organic compounds composed of a chain of carbon atoms with hydrogens attached and an acid group at one end.

fatty liver: an accumulation of fat in the liver.

fertility: the capacity of a woman to produce a normal ovum periodically and of a man to produce normal sperm; the ability to reproduce.

fetal alcohol spectrum disorders (FASD): a spectrum of physical, behavioral, and cognitive disabilities caused by prenatal alcohol exposure.

fetal alcohol syndrome (FAS): the cluster of symptoms seen in an infant or child whose mother consumed excessive alcohol during her pregnancy. FAS includes, but is not limited to, brain damage, growth retardation, mental retardation, and facial abnormalities.

fetus (FEET-us): the developing infant from eight weeks after conception until its birth.

fistulas (FIST-you-luz): abnormal passages between body tissues; may lead from one hollow organ to another or to an organ surface.

fitness: the characteristics that enable the body to perform physical activity; more broadly, the ability to meet routine physical demands with enough reserve energy to rise to a physical challenge; the body's ability to withstand stress of all kinds.

flatulence: the condition of having excessive intestinal gas, which causes abdominal discomfort.

flexibility: the capacity of the joints to move through a full range of motion; the ability to bend and recover without injury.

fluid and electrolyte balance: maintenance of the necessary amounts and types of fluid and minerals in each compartment of the body fluids.

fluorapatite (floor-APP-uh-tite): the stabilized form of bone and tooth crystal, in which fluoride has replaced the hydroxy portion of hydroxyapatite.

fluorosis (floor-OH-sis): mottling of the tooth enamel from ingestion of too much fluoride during tooth development.

follicle (FOLL-i-cul): a group of cells in the skin from which a hair grows.

food allergies: adverse reactions to foods that involve an immune response; also called *food-hypersensitivity reactions*.

food aversions: strong desires to avoid particular foods.

food cravings: deep longings for particular foods.

food frequency questionnaire: a survey of foods routinely consumed. Some questionnaires ask about the types of food eaten and yield only qualitative information, whereas others include questions about portions consumed and yield semiquantitative data as well.

food group plan: a diet-planning tool that sorts foods into groups based on nutrient content and then specifies that people should eat certain amounts of food from each group.

food insecurity: limited or uncertain access to foods of sufficient quality or quantity to sustain a healthy and active life.

food intolerance: an adverse response to a food or food additive that does not involve the immune system.

food record: a detailed log of food eaten during a specified time period, usually several days. A food record may also include information regarding disease symptoms, physical activity, and medication use; also called a *food diary*.

fortification: the addition to a food of nutrients that were either not originally present or present in insignificant amounts. Fortification can be used to correct or prevent a widespread nutrient deficiency, to balance the total nutrient profile of a food, or to restore nutrients lost in processing.

free radicals: highly reactive chemical forms that can cause destructive changes in nearby compounds, sometimes setting up a chain reaction.

French units: units of measure used to indicate the size of a feeding tube's outer diameter. One French unit equals 1/3 millimeter.

fructose: a monosaccharide; sometimes known as *fruit sugar*. It is abundant in fruits, honey, and saps.

full liquid diet: a liquid diet that includes milk and other opaque liquids.

functional foods: foods that may provide health benefits beyond their nutrient contributions. Functional foods may include whole foods, fortified foods, and modified foods.

galactose: a monosaccharide; part of the disaccharide lactose.

gallstones: stones that form in the gallbladder from crystalline deposits of cholesterol or bilirubin.

gastrectomy (gah-STREK-ta-mee): the surgical removal of part of the stomach (partial gastrectomy) or the entire stomach (total gastrectomy).

gastric banding: a surgical means of producing weight loss by restricting stomach size with a constricting band; used in people whose severe obesity brings extreme health risks.

gastric bypass: surgery that restricts stomach size and reroutes food from the stomach to the lower part of the small intestine; creates a chronic, lifelong state of malabsorption by preventing normal digestion and absorption of nutrients.

gastric decompression: the removal of the stomach contents (including swallowed saliva, stomach secretions, and gas) of patients who have motility disorders or obstructions that prevent stomach emptying.

gastric residual volume: the volume of formula remaining in the stomach from a previous feeding.

gastric ulcers: peptic ulcers that develop in stomach tissue.

gastritis: inflammation of stomach tissue.

gastrointestinal motility: spontaneous motion in the digestive tract accomplished by involuntary muscular contractions.

gastrointestinal (GI) tract: the digestive tract. The principal organs are the stomach and intestines.

gatekeeper: with respect to nutrition, a key person who controls other people's access to foods and thereby exerts a profound impact on their nutrition. Examples are the spouse who buys and cooks the food, the parent who feeds the children, and the caregiver in a day care center.

gestation: the period of about 40 weeks (three trimesters) from conception to birth; the term of a pregnancy.

gestational diabetes: the presence of abnormal glucose tolerance during pregnancy.

ghrelin (GRELL-in): a hormone produced primarily by the stomach cells. It signals the hypothalamus of the brain to stimulate appetite and food intake.

glands: single cells or groups of cells that secrete materials for special uses in the body. Glands may be *exocrine glands*, secreting their materials "out" (into the digestive tract or onto the surface of the skin), or *endocrine glands*, secreting their materials "in" (into the blood).

glomerular filtration rate (GFR): the rate at which filtrate is formed within the kidneys, normally approximately 125 mL/min.

glomerulus (gloh-MEHR-yoo-lus): a tuft of capillaries within the nephron that filters water and solutes from blood as urine production begins (plural: *glomeruli*).

glucagon (GLOO-ka-gon): a hormone that is secreted by special cells in the pancreas in response to low blood glucose concentration and elicits release of glucose from storage.

gluconeogenesis (gloo-co-nee-oh-GEN-ih-sis): the making of glucose from a noncarbohydrate source.

glucose: a monosaccharide, the sugar common to all disaccharides and polysaccharides; also called *blood sugar* or *dextrose*.

glucose polymers: compounds that supply glucose, not as single molecules, but linked in chains somewhat like starch. The objective is to attract less water from the body into the digestive tract.

gluten: a family of water-insoluble proteins in wheat. The protein in wheat gluten that has toxic effects in celiac disease is called *gliadin* (GLY-ah-din).

glycated hemoglobin (HbA$_{1c}$): hemoglobin molecules to which glucose is attached. The percentage of such molecules is used to evaluate long-term glycemic control; also called *glycosylated hemoglobin.*

glycemic (gly-SEE-mic): pertaining to blood glucose.

glycemic (gligh-SEEM-ic) **effect:** the extent to which a food raises the blood glucose concentration and elicits an insulin response.

glycemic index (GI): a ranking of carbohydrate foods based on their effect on blood glucose levels after ingestion; *low-GI* foods are foods with a lesser glycemic effect; *high-GI* foods are those with a greater glycemic effect.

glycerol (GLISS-er-ol): an organic compound, three carbons long, that can form the backbone of triglycerides and phospholipids.

glycogen (GLY-co-gen): a polysaccharide composed of glucose, made and stored by liver and muscle tissues of human beings and animals as a storage form of glucose. Glycogen is not a significant food source of carbohydrate and is not counted as one of the polysaccharides in foods.

glycolysis (gligh-COLL-ih-sis): the metabolic breakdown of glucose to pyruvate.

goiter (GOY-ter): an enlargement of the thyroid gland due to an iodine deficiency, malfunction of the gland, or overconsumption of a thyroid antagonist. Goiter caused by iodine deficiency is *simple goiter.*

goitrogen (GOY-troh-jen): a substance that enlarges the thyroid gland and causes *toxic goiter.* Goitrogens occur naturally in such foods as cabbage, kale, brussels sprouts, cauliflower, broccoli, and kohlrabi.

gout (GOWT): a metabolic disease in which crystals of uric acid precipitate in the joints, causing acute joint inflammation.

health: a range of states with physical, mental, emotional, spiritual, and social components. At a minimum, health means freedom from physical disease, mental disturbances, emotional distress, spiritual discontent, social maladjustment, and other negative states. At a maximum, health means "wellness."

health claims: statements that characterize the relationship between a nutrient or other substance in food and a disease or health-related condition.

heat stroke: an acute and life-threatening reaction to heat buildup in the body.

Helicobacter pylori: a type of bacterium that colonizes gastric mucosa; a major cause of gastritis and peptic ulcer disease.

helper T cells: lymphocytes that have a specific protein called CD4 on their surfaces and therefore are also known as *CD4+ T cells;* the cells most affected in HIV infection.

hematocrit (hee-MAT-oh-crit): measurement of the volume of the red blood cells packed by centrifuge in a given volume of blood.

hematuria (HE-mah-TOO-ree-ah): blood in the urine.

heme (HEEM): the iron-holding part of the hemoglobin and myoglobin proteins. About 40% of the iron in meat, fish, and poultry is bound into heme; the other 60% is *nonheme* iron.

hemochromatosis (heem-oh-crome-a-TOH-sis): iron overload characterized by deposits of iron-containing pigment in many tissues, with tissue damage. Hemochromatosis is usually caused by a hereditary defect in iron absorption.

hemodialysis (HEE-moh-dye-AL-ih-sis): removal of fluids and wastes from blood by passing the blood through a dialyzer.

hemorrhage: severe bleeding; a copious flow of blood from blood vessels.

hemorrhagic strokes: strokes that result from bleeding within the brain, which destroys or compresses brain tissue.

hemoglobin: the oxygen-carrying protein of the red blood cells.

hemorrhagic (hem-oh-RAJ-ik) **disease:** the vitamin K–deficiency disease in which blood fails to clot.

hepatic coma: loss of consciousness resulting from severe liver disease.

hepatic encephalopathy (en-sef-ah-LOP-ah-thie): a condition in advanced liver disease characterized by altered neurological functioning, including personality changes, reduced mental abilities, and disturbances in motor function.

hepatitis (hep-ah-TYE-tis): inflammation of the liver.

hepatomegaly (HEP-ah-toe-MEG-ah-lee): enlargement of the liver.

herpes simplex virus: a common virus that can cause blisterlike lesions on the lips and in the mouth.

hiatal hernia: a condition in which the upper portion of the stomach protrudes above the diaphragm; most cases are asymptomatic.

high-carbohydrate energy drinks: fruit-flavored commercial beverages used to restore muscle glycogen after exercise or as a pregame beverage.

high-density lipoproteins (HDL): the type of lipoproteins that transport cholesterol back to the liver from peripheral cells; composed primarily of protein.

high-quality proteins: dietary proteins containing all the essential amino acids in relatively the same amounts that human beings require. They may also contain nonessential amino acids.

histamine-2 receptor blockers: a class of drugs that suppresses acid secretion by inhibiting receptors on acid-producing cells (commonly called H2 blockers). Examples include cimetidine (Tagamet), ranitidine (Zantac), and famotidine (Pepcid).

HIV (human immunodeficiency virus): the virus that causes acquired immune deficiency syndrome (AIDS). HIV destroys immune cells and progressively impedes the body's ability to fight infections and certain cancers.

HIV-lipodystrophy (LIP-oh-DIS-tro-fee) **syndrome:** a collection of abnormalities in fat and glucose metabolism that result from drug treatments for HIV; includes body fat redistribution, abnormal blood lipid levels, and insulin resistance. The accumulation of abdominal fat is sometimes called *protease paunch.*

homeostasis (HOME-ee-oh-STAY-sis): the maintenance of constant internal conditions (such as chemistry, temperature, and blood pressure) by the body's control system.

hormones: chemical messengers. Hormones are secreted by a variety of glands in the body in response to altered conditions. Each travels to one or more target tissues or organs and elicits specific responses to restore normal conditions.

hourly sweat rate: the amount of weight lost plus fluid consumed during exercise per hour.

hunger: the physiological need to eat, experienced as a drive to obtain food; an unpleasant sensation that demands relief.

hydrogenation (high-dro-gen-AY-shun): a chemical process by which hydrogens are added to monounsaturated or polyunsaturated fats to reduce the number of double bonds, making the fats more saturated (solid) and more resistant to oxidation (protecting against rancidity). Hydrogenation produces *trans*-fatty acids.

hydrolyzed formulas: enteral formulas that contain macronutrients that have been partially or fully hydrolyzed; also called *monomeric* or *elemental formulas.*

hyperactivity: inattentive and impulsive behavior that is more frequent and severe than is typical of others a similar age; professionally called *attention deficit/hyperactivity disorder (ADHD)*.

hypercalcemia (HIGH-per-kal-SEE-me-ah): elevated serum calcium levels.

hypercalciuria (HIGH-per-kal-see-YOO-ee-ah): elevated urinary calcium levels.

hypercapnia (high-per-CAP-nee-ah): excessive carbon dioxide in the blood.

hyperinsulinemia: abnormally high levels of insulin in the blood.

hyperkalemia (HIGH-per-ka-LEE-me-ah): elevated serum potassium levels.

hypermetabolism: a higher-than-normal metabolic rate.

hyperoxaluria (HIGH-per-ox-ah-LOO-ree-ah): elevated urinary oxalate levels.

hyperphosphatemia (HIGH-per-fos-fa-TEE-me-ah): elevated serum phosphate levels.

hypertension: high blood pressure.

hyperthermia: an above-normal body temperature.

hypertonic formula: a formula with an osmolality greater than that of blood serum.

hypertrophy (high-PURR-tro-fee): an increase in size (for example, of a muscle) in response to use.

hypoallergenic formulas: clinically tested infant formulas that do not provoke reactions in 90 percent of infants or children with confirmed cow's milk allergy. Like all infant formulas, hypoallergenic formulas must demonstrate nutritional suitability to support infant growth and development. Extensively hydrolyzed and free amino acid-based formulas are examples.

hypochlorhydria (HIGH-poe-clor-HIGH-dree-ah): a reduction in gastric acid secretion.

hypokalemia (HIGH-po-ka-LEE-me-ah): low serum potassium levels.

hyponatremia (HIGH-poh-na-TREE-mee-ah): a decreased concentration of sodium in the blood.

hypothalamus (high-poh-THALL-uh-mus): a part of the brain that helps regulate many body balances, including fluid balance.

hypothermia: a below-normal body temperature.

hypoxemia (high-pock-SEE-me-ah): a low level of oxygen in the blood.

hypoxia (high-POCK-see-ah): a low amount of oxygen in body tissues.

ileostomy (ill-ee-OS-toe-me): a surgical procedure that creates a stoma using the ileum.

implantation: the stage of development in which the zygote embeds itself in the wall of the uterus and begins to develop; occurs during the first two weeks after conception.

indirect calorimetry (kal-uh-RIM-eh-tree): a determination of resting energy expenditure by measuring a person's oxygen consumption and carbon dioxide elimination during a period of rest.

inflammatory response: the metabolic responses of the immune system to infection or injury.

inorganic: not containing carbon or pertaining to living things.

insoluble fibers: the tough, fibrous structures of fruits, vegetables, and grains; indigestible food components that do not dissolve in water.

insulin: a hormone secreted by the pancreas in response to high blood glucose. It promotes cellular glucose uptake for use or storage.

insulin resistance: reduced sensitivity to insulin in muscle, adipose, and liver cells.

intermittent feedings: delivery of about 250 to 400 mL of formula over 20 to 40 minutes.

intestinal adaptation: after resection, the process of intestinal recovery that leads to improved absorptive capacity.

intestinal flora: the bacterial inhabitants of the GI tract.

intra-abdominal fat: fat stored within the abdominal cavity in association with the internal abdominal organs, as opposed to fat stored directly under the skin (subcutaneous fat); also called *visceral fat*.

intradialytic parenteral nutrition: the infusion of nutrients during hemodialysis, often providing amino acids, dextrose, lipids, and some trace minerals.

intractable: not easily managed or controlled.

intractable vomiting: vomiting that is not easily managed or controlled.

intravenous feedings: the provision of nutrients through a vein, bypassing the intestine. The intravenous provision of nutrients is called *parenteral nutrition.*

intrinsic factor: inside the system. Anemia that reflects a vitamin B_{12} deficiency caused by lack of intrinsic factor is known as *pernicious anemia.*

irritable bowel syndrome: an intestinal disorder of unknown cause that affects the functioning of the lower bowel; symptoms include abdominal pain, flatulence, diarrhea, and constipation.

iron deficiency: having depleted iron stores.

iron-deficiency anemia: a blood iron deficiency characterized by small, pale red blood cells; also called microcytic hypochromic anemia.

iron overload: toxicity from excess iron.

ischemic strokes: strokes that result from the obstruction of blood flow to brain tissue.

isotonic formula: a formula with an osmolality similar to that of blood serum (300 mOsm/kg).

jaundice (JAWN-dis): yellow discoloration of the skin and eyes due to an accumulation of bilirubin, a breakdown product of hemoglobin that normally exits the body via bile secretions.

Joint Commission on Accreditation of Healthcare Organizations (JCAHO): a nonprofit organization that sets standards for health care performance and safety and confers accreditation to health care organizations that meet those standards.

Kaposi's (cap-OH-seez) **sarcoma:** a common cancer in HIV-infected persons that is characterized by lesions on the skin, in the lungs, and in the GI tract.

kcalorie control: management of food energy intake.

kcalorie counts: the determination of food energy (and often, protein) consumed by patients for one or more days.

keratin (KERR-uh-tin): a water-insoluble protein; the normal protein of hair and nails. Keratin-producing cells may replace mucus-producing cells in vitamin A deficiency.

ketone (KEY-tones) **bodies:** acidic, fat-related compounds formed from the incomplete breakdown of fat when carbohydrate is not available.

kidney stones: crystalline masses that form in the urinary tract; also called *renal calculi* and *nephrolithiasis.*

kwashiorkor (kwash-ee-OR-core or kwash-ee-or-CORE): a severe form of PEM that occurs more frequently after 18 months of age. Kwashiorkor is characterized by failure to grow and

develop, changes in the pigmentation of the hair and skin, edema, and fatty liver. Kwashiorkor is associated with inadequate protein intake and infections.

lactadherin (lack-tad-HAIR-in): a protein in breast milk that attacks diarrhea-causing viruses.

lactate: a compound produced during the breakdown of glucose in anaerobic metabolism.

lactation: production and secretion of breast milk for the purpose of nourishing an infant.

lactoferrin (lack-toe-FERR-in): protein in breast milk that binds iron and keeps it from supporting the growth of the infant's intestinal bacteria.

lactose: a disaccharide composed of glucose and galactose; commonly known as *milk sugar.*

lactose intolerance: a condition that results from inability to digest the milk sugar lactose; characterized by bloating, gas, abdominal discomfort, and diarrhea. Lactose intolerance differs from milk allergy, which is caused by an immune reaction to the protein in milk.

laparoscopic: pertaining to procedures that use a laparoscope for internal examination or surgery. A laparoscope is a narrow surgical telescope that is inserted into the abdominal cavity through a small incision. A video camera is usually attached so that the procedure can be viewed on a television monitor.

laparoscopic weight-loss surgery: a procedure in which surgeons gain access to the abdomen via several small incisions. A tiny video camera is inserted through one of the incisions and surgical instruments through the others. The surgeons watch their work on a large-screen monitor.

latent: the period in the course of a disease when the condition is present but the symptoms have not begun to appear.

lecithins: one type of phospholipid.

legumes (lay-GYOOMS, LEG-yooms): plants of the bean and pea family, with seeds that are rich in protein compared with other plant-derived foods.

leptin: a hormone produced by fat cells under the direction of the (*ob*) gene. It decreases appetite and increases energy expenditure.

life expectancy: the average number of years lived by people in a given society.

life span: the maximum number of years of life attainable by a member of a species.

limiting amino acid: an essential amino acid that is present in dietary protein in the shortest supply relative to the amount needed for protein synthesis in the body.

linoleic acid, linolenic acid: polyunsaturated fatty acids that are essential for human beings.

lipids: a family of compounds that includes triglycerides (fats and oils), phospholipids, and sterols.

lipomas (lih-POE-muz): benign tumors composed of fatty tissue.

lipoproteins: clusters of lipids associated with proteins that serve as transport vehicles for lipids in the lymph and blood.

lipoprotein lipase (LPL): an enzyme that hydrolyzes triglycerides in the blood into fatty acids and glycerol for absorption into the cells. There they are metabolized or reassembled for storage.

listeriosis: a serious foodborne infection that can cause severe brain infection or death in a fetus or newborn; caused by the bacterium *Listeria monocytogenes*, which is found in soil and water.

longevity: long duration of life.

low birthweight (LBW): a birth weight less than 5 lb (2,500 g); indicates probable poor health in the newborn and poor nutrition status of the mother during pregnancy. Normal birthweight for a full-term infant is 6 to 9 lb (about 3,000 to 4,000 g). Low-birthweight infants are of two different types. Some are *premature*; they are born early and are of a weight *appropriate for gestational age (AGA)*. Others have suffered growth failure in the uterus; they may or may not be born early, but they are *small for gestational age (SGA)*.

low-density lipoproteins (LDL): the type of lipoproteins that carry cholesterol and triglycerides from the liver to the cells of the body and are composed primarily of cholesterol.

lymph (LIMF): the body fluid found in lymphatic vessels. Lymph consists of all the constituents of blood except red blood cells

lymphatic system: a loosely organized system of vessels and ducts that conveys the products of digestion toward the heart.

macrosomia (MAK-roh-SOH-mee-ah): the condition of having an abnormally large body. In infants, refers to birth weights of 4,000 g (8 lb 13 oz) and above.

macular (MACK-you-lar) **degeneration:** deterioration of the macular area of the eye that can lead to loss of central vision and eventual blindness. The *macula* is a small, oval, yellowish region in the center of the retina that provides the sharp, straight-ahead vision so critical to reading and driving.

malignant (ma-LIG-nent): describes a cancerous cell or tumor, which can injure healthy tissue and spread cancer to other regions of the body.

malnutrition: any condition caused by deficient or excess energy or nutrient intake or by an imbalance of nutrients.

maltose: a disaccharide composed of two glucose units; sometimes known as *malt sugar*.

marasmus (ma-RAZZ-mus): the most common form of severe PEM before one year of age. Marasmus is characterized by generalized muscle wasting associated with extreme deprivation, or impaired absorption, of energy, protein, vitamins, and minerals.

mast cells: cells within connective tissue (close to blood vessels) that produce and release histamine.

medical nutrition therapy: nutrition care provided by a registered dietitian; includes diagnosing nutrition problems, prescribing diet plans, and providing dietary counseling.

medium-chain triglycerides (MCT): triglycerides that contain fatty acids that are 8 to 10 carbons in length. MCT do not require digestion and can be absorbed in the absence of lipase or bile.

metabolic stress: a disruption in the body's chemical environment due to the effects of disease or injury. Metabolic stress is characterized by changes in metabolic rate, heart rate, blood pressure, hormonal status, and nutrient metabolism.

metabolism: the sum total of all the chemical reactions that go on in living cells.

metastasize (meh-TAS-tah-size): the spread of cancer cells from one part of the body to another.

microvilli (MY-cro-VILL-ee or MY-cro-VILL-eye): tiny, hairlike projections on each cell of every villus that can trap nutrient particles and transport them into the cells. The singular form is *microvillus*.

moderation: providing enough, but not too much, of a substance.

modified diet: a diet that is adjusted in consistency, energy or nutrient content, or by the inclusion or elimination of certain foods; sometimes called a *therapeutic diet*.

modular formulas: enteral formulas that contain only one or two macronutrients; used to enhance other formulas, meet specific nutrient needs, or create individualized formulas for people with unique needs.

molybdenum (mo-LIB-duh-num): a trace element.

monosaccharides (mon-oh-SACK-uh-rides): a single sugar unit.

monounsaturated fatty acid (MUFA): a fatty acid that has one point of unsaturation; for example, the oleic acid found in olive oil.

mucous membrane: membrane composed of mucus-secreting cells that lines the surfaces of body tissues. (Reminder: *Mucus* is the smooth, slippery substance secreted by these cells.)

muscle endurance: the ability of a muscle to contract repeatedly within a given time without becoming exhausted.

muscle strength: the ability of muscles to work against resistance.

muscular dystrophy (DIS-tro-fee): a hereditary disease in which the muscles gradually weaken, with the most debilitating effects occurring in the lungs. This disease should not be confused with *nutritional* muscular dystrophy, a vitamin E-deficiency disease of animals characterized by gradual paralysis of the muscles.

myoglobin: the muscles' iron-containing protein that stores and releases oxygen in response to the muscles' energy needs.

naturally occurring sugars: sugars that are not added to a food but are present as its original constituents, such as the sugars of fruit or milk.

nephron (NEF-ron): the functional unit of the kidneys, consisting of a glomerulus and tubules.

nephrotic (neh-FROT-ik) **syndrome:** a kidney disorder characterized by urinary protein losses exceeding 3.5 g per day. Accompanying symptoms include low serum albumin, elevated blood lipids, and edema.

neural tube: the embryonic tissue that later forms the brain and spinal cord.

neural tube defects (NTD): malformations of the brain, spinal cord, or both during embryonic development.

neurons: nerve cells; the structural and functional units of the nervous system. Neurons initiate and conduct nerve transmissions.

niacin equivalents (NE): the amount of niacin present in food, including the niacin that can theoretically be made from tryptophan, its precursor, present in the food.

night blindness: the slow recovery of vision after exposure to flashes of bright light at night; an early symptom of vitamin A deficiency.

nitrogen balance: the amount of nitrogen consumed (N in) as compared with the amount of nitrogen excreted (N out) in a given period of time. The laboratory scientist can estimate the protein in a sample of food, body tissue, or excreta by measuring the nitrogen in it.

nonessential amino acids: amino acids that the body can synthesize; also called *dispensable amino acids*.

nursing bottle tooth decay: extensive tooth decay due to prolonged tooth contact with formula, milk, fruit juice, or other carbohydrate-rich liquid offered to an infant in a bottle.

nutrient claims: statements that characterize the quantity of a nutrient in a food.

nutrient density: a measure of the nutrients a food provides relative to the energy it provides; the more nutrients and the fewer kcalories, the higher the nutrient density.

nutrients: substances obtained from food and used in the body to provide energy and structural materials and to serve as regulating agents to promote growth, maintenance, and repair. Nutrients may also reduce the risks of some diseases.

nutrition: the science of foods and the nutrients and other substances they contain and of their ingestion, digestion, absorption, transport, metabolism, interaction, storage, and excretion. A broader definition includes the study of the environment and of human behavior as it relates to these processes.

nutrition care plans: strategies for meeting an individual's nutritional needs.

nutrition care process: a problem-solving method that dietetics professionals use to evaluate and treat nutrition-related problems.

nutrition screening: an assessment tool that quickly identifies patients at risk of developing malnutrition.

nutrition support: the delivery of formulated nutrients by feeding tube or intravenous infusion.

nutrition support teams: health care professionals responsible for the provision of nutrients by tube feeding or intravenous infusion.

nutritive sweeteners: sweeteners that yield energy, including both the sugars and the sugar alcohols.

obesity: overfatness with adverse health effects, as determined by reliable measures and interpreted with good medical judgment. Obesity is officially defined as a body mass index (BMI) of 30 or higher.

oils: lipids that are liquid at room temperature (70° F or 25° C).

olestra: a synthetic fat made from sucrose and fatty acids that provides zero kcalories per gram; also known as *sucrose polyester*.

oliguria (OL-lih-GOO-ree-ah): an abnormally low amount of urine, often less than 400 mL per day.

omega-3 fatty acids: polyunsaturated fatty acids in which the endmost double bond is three carbons back from the end of the carbon chain; relatively newly recognized as important in nutrition. *Linolenic acid* is an example.

omega-6 fatty acid: a polyunsaturated fatty acid with its endmost double bond six carbons back from the end of its carbon chain; long recognized as important in nutrition. *Linoleic acid* is an example.

open feeding systems: delivery systems that require formula to be transferred from the original packaging to feeding containers before being administered through feeding tubes.

opportunistic infections: infections from microorganisms that normally do not cause disease in healthy people but are damaging to persons with compromised immune function.

oral glucose tolerance test: a test that evaluates a person's ability to tolerate a glucose load.

organic: carbon containing. The four organic nutrients are carbohydrate, fat, protein, and vitamins.

oropharyngeal dysphagia (OR-oh-fah-ren-JEE-al diss-FAY-jee-ah): an inability to transfer food from the mouth and pharynx to the esophagus; usually caused by a neurological or muscular disorder.

osmolality (OZ-moe-LAL-ih-tee): the osmotic property of a solution, based on its concentrations of molecules and ionic particles. Osmolality is expressed as milliosmoles (mOsm) per kilogram.

osmolarity: the concentration of osmotically active particles in a solution, expressed as milliosmoles per liter (mOsm/L).

osteoarthritis: a painful, chronic disease of the joints that occurs when the cushioning cartilage in a joint breaks down; joint structure is usually altered, with loss of function; also called *degenerative arthritis.*

osteomalacia (os-tee-oh-mal-AY-shuh): a bone disease characterized by softening of the bones. Symptoms include bending of the spine and bowing of the legs. The disease occurs most often in adults with renal failure or malabsorption disorders.

osteoporosis: (os-tee-oh-pore-OH-sis): literally, porous bones; reduced density of the bones, also known as *adult bone loss.*

overload: an extra physical demand placed on the body; an increase in the frequency, duration, or intensity of an activity. A principle of training is that for a body system to improve, it must be worked at frequencies, durations, or intensities that increase by increments.

overnutrition: overconsumption of food energy or nutrients sufficient to cause disease or increased susceptibility to disease; a form of malnutrition.

overt: out in the open, full-blown.

overweight: overfatness of a moderate degree; defined as a body mass index (BMI) of 25.0 through 29.9.

ovum (OH-vum): the female reproductive cell, capable of developing into a new organism upon fertilization; commonly referred to as an egg.

oxidation (OKS-ee-day-shun): the process of a substance combining with oxygen.

paracentesis (pah-rah-sen-TEE-sis): a surgical puncture of a body cavity with an aspirator to draw out excess fluid.

parenteral (par-EN-ter-al) **nutrition:** the intravenous provision of nutrients that bypasses the GI tract.

pellagra (pell-AY-gra): the niacin-deficiency disease. Symptoms include the "4 Ds": diarrhea, dermatitis, dementia, and, ultimately, death.

peptic ulcers: ulcers in the gastrointestinal mucosa resulting from exposure to gastric secretions; may develop in the esophagus, stomach, or duodenum.

peripheral parenteral nutrition (PPN): a type of nutrition support in which intravenous feedings are delivered into peripheral veins.

peripheral resistance: the resistance to pumped blood in the small arterial branches (arterioles) that carry blood to tissues.

peripheral veins: small-diameter veins that carry blood from the arms and legs.

peristalsis (peri-STALL-sis): successive waves of involuntary muscular contractions passing along the walls of the GI tract that push the contents along.

peritoneal (PEH-rih-toe-NEE-al) **dialysis:** removal of fluids and wastes by using the peritoneal membrane to filter blood.

peritoneovenous (PEH-rih-toe-NEE-oh-VEE-nus) **shunt:** a surgical passage created between the peritoneum and the jugular vein to divert fluid and relieve ascites. The peritoneum is the membrane that surrounds the abdominal cavity.

peritonitis: inflammation of the peritoneal membrane, which lines the abdominal cavity.

pH: the concentration of hydrogen ions. The lower the pH, the stronger the acid. Thus pH 2 is a strong acid; pH 6 is a weak acid; pH 7 is neutral; and a pH above 7 is alkaline.

phagocytes (FAG-oh-sites): white blood cells (neutrophils and macrophages) that have the ability to engulf and destroy antigens.

phospholipids: one of the three main classes of lipids; compounds similar to triglycerides but have choline (or another compound) and a phosphorus-containing acid in place of one of the fatty acids.

physical activity: bodily movement produced by muscle contractions that substantially increase energy expenditure.

physiological age: a person's age as estimated from his or her body's health and probable life expectancy.

phytates: nonnutrient components of grains, legumes, and seeds. Phytates can bind minerals such as iron, zinc, calcium, and magnesium in insoluble complexes in the intestine, and the body excretes them unused.

phytochemicals: nonnutrient compounds in plant-derived foods that have biological activity in the body.

pica (PIE-ka): a craving for nonfood substances; also known as *geophagia* (jee-oh-FAY-jee-uh) when referring to clay-eating behavior.

pituitary (pit-TOO-ih-tary) **gland:** in the brain, the "king gland" that regulates the operation of many other glands.

placenta (pla-SEN-tuh): an organ that develops inside the uterus early in pregnancy, in which maternal and fetal blood circulate in close proximity and exchange materials. The fetus receives nutrients and oxygen across the placenta; the mother's blood picks up carbon dioxide and other waste materials to be excreted via her lungs and kidneys.

polypeptide: ten or more amino acids bonded together. An intermediate strand of between four and ten amino acids is an *oligopeptide.*

polysaccharides: long chains of monosaccharide units arranged as starch, glycogen, or fiber.

polyunsaturated fatty acids (PUFA): fatty acids with two or more points of unsaturation. For example, linoleic acid has two such points, and linolenic acid has three. Thus polyunsaturated *fat* is composed of triglycerides containing a high percentage of PUFA.

portal hypertension: elevated blood pressure in the portal vein due to obstructed blood flow through the liver.

potassium iodide: a medication approved by the FDA as safe and effective for the prevention of thyroid cancer caused by radioactive iodine known to be released during radiation emergencies.

precursors: compounds that can be converted into other compounds; with regard to vitamins, compounds that can be converted into active vitamins; also known as *provitamins.*

prediabetes: the condition in which blood glucose levels are higher than normal but not high enough to be diagnosed as diabetes; formerly called *impaired glucose tolerance.*

preeclampsia: a condition characterized by hypertension, fluid retention, and protein in the urine.

preformed vitamin A: vitamin A in its active form.

pregame meal: a meal eaten three to four hours before athletic competition.

pressure sores: regions of damaged skin and tissue due to prolonged pressure on the affected area by an external object, such as a bed, wheelchair, or cast; vulnerable areas of the body include buttocks, hips, and heels. Also called *decubitus* (deh-KYU-bih-tus) *ulcers.*

pressure ulcers: damage to the skin and underlying tissues as a result of compression and poor circulation; commonly seen in people who are bedridden or chairbound.

protein digestibility: a measure of the amount of amino acids absorbed from a given protein intake.

protein-energy malnutrition (PEM): a deficiency of protein and food energy; the world's most widespread malnutrition problem, including both marasmus and kwashiorkor. **proteins:** compounds composed of carbon, hydrogen, oxygen, and nitrogen atoms made from strands of amino acids. Some amino acids also contain sulfur atoms.

protein turnover: the continuous breakdown and synthesis of body proteins involving the recycling of amino acids.

proteinuria (PRO-teen-NOO-ree-ah): loss of protein, especially albumin, in the urine; also known as *albuminuria.*

proton-pump inhibitors: a class of drugs that inhibits the enzyme that pumps hydrogen ions (protons) into the stomach. Examples include omeprazole (Prilosec) and lansoprazole (Prevacid).

puberty: the period in life in which a person becomes physically capable of reproduction.

purines (PYOO-reenz) products of nucleotide metabolism that degrade to uric acid.

pyloroplasty (pye-LORE-oh-PLAS-tee): surgery that enlarges the pyloric sphincter.

pyruvate (PIE-roo-vate): a 3-carbon compound that plays a key role in energy metabolism.

quality of life: a person's perceived physical and mental well-being.

radiation enteritis: inflammation of intestinal tissue caused by exposure to radiation.

radiation therapy: the use of X-rays, gamma rays, or atomic particles to destroy cancer cells.

rancid: the term used to describe fats when they have deteriorated, usually by oxidation. Rancid fats often have an "off" odor.

Recommended Dietary Allowances (RDA): a set of values reflecting the average daily amounts of nutrients considered adequate to meet the known nutrient needs of practically all healthy people in a particular life stage and gender group; a goal for dietary intake by individuals.

refeeding syndrome: a condition that sometimes develops when a severely malnourished person is aggressively fed; characterized by electrolyte and fluid imbalances and hyperglycemia.

refined grain: a product from which the bran, germ, and husk have been removed, leaving only the endosperm.

reflux esophagitis: inflammation in the esophagus related to the reflux of acidic stomach contents.

renal colic: the intense pain that occurs when a kidney stone passes through the ureter.

renal osteodystrophy: a bone disorder that develops in patients with chronic kidney disease as a consequence of increased secretion of parathyroid hormone (which stimulates calcium release from bone), reduced serum calcium, acidosis, and impaired vitamin D activation by the kidneys.

renal threshold: the blood concentration of a substance that exceeds the kidneys' capacity for reabsorption, thus allowing the substance to be passed into the urine.

requirement: the lowest continuing intake of a nutrient that will maintain a specified criterion of adequacy.

resection: the surgical removal of part of an organ or body structure.

residue: material left in the intestine after digestion; includes mostly dietary fiber and undigested starches and proteins.

resistant starches: starches that escape digestion and absorption in the small intestine of healthy people.

respiratory stress: inadequate gas exchange between the air and blood, resulting in lower-than-normal oxygen levels and higher-than-normal carbon dioxide levels.

resting metabolic rate (RMR): a measure of the energy use of a person at rest in a comfortable setting; similar to the BMR but with less stringent criteria for recent food intake and physical activity. Consequently, the RMR is slightly higher than the BMR.

retina (RET-in-uh): the layer of light-sensitive nerve cells lining the back of the inside of the eye; consists of rods and cones.

retinol activity equivalents (RAE): a measure of vitamin A activity; the amount of retinol that the body will derive from a food containing preformed retinol or its precursor beta-carotene.

retinol-binding protein (RBP): the specific protein responsible for transporting retinol.

rheumatoid arthritis: a disease of the immune system involving painful inflammation of the joints and related structures.

rickets: the vitamin D-deficiency disease in children.

salt-sensitive: an individual who responds to a high salt intake with an increase in blood pressure or to a low salt intake with a decrease in blood pressure.

salts: compounds composed of charged particles (ions). An example of a salt is potassium chloride (K^+Cl^-).

sarcopenia (SAR-koh-PEE-nee-ah): loss of skeletal muscle mass, strength, and quality.

saturated fatty acid: a fatty acid carrying the maximum possible number of hydrogen atoms (having no points of unsaturation).

scurvy: the vitamin C-deficiency disease.

sedentary: physically inactive (literally "sitting down a lot").

segmentation: a periodic squeezing or partitioning of the intestine by its circular muscles that both mixes and slowly pushes the contents along.

self-monitoring of blood glucose: home monitoring of blood glucose levels using a glucose meter.

senile dementia: the loss of brain function beyond the normal loss of physical adeptness and memory that occurs with aging.

sepsis: an acute inflammatory response caused by infection; characterized by symptoms similar to those of SIRS.

set-point theory: the theory that proposes that the body tends to maintain a certain weight by means of its own internal controls.

shock: a dangerous physiological response to injury, bleeding, or infection resulting from an insufficient blood supply; associated with reduced blood pressure, raised heart and respiratory rates, and muscle weakness.

shock-wave lithotripsy: a non surgical procedure that uses high-amplitude sound waves to fragment gallstones.

short-bowel syndrome: a malabsorption syndrome that follows resection of the small intestine, causing insufficient absorptive capacity in the remaining intestine.

simple sugars: the monosaccharides (glucose, fructose, and galactose) and the disaccharides (sucrose, lactose, and maltose).

sinusoids: the small capillary-like passages that carry blood through liver tissue.

Sjögren's (SHOW-grenz) **syndrome:** an autoimmune disease characterized by the destruction of secretory glands, resulting in dry mouth and dry eyes.

skinfold measure: a clinical estimate of total body fatness in which the thickness of a fold of skin on the back of the arm (over the triceps muscle), below the shoulder blade (subscapular), or in other places is measured with a caliper.

sludge: literally, a semisolid mass. Biliary sludge is made up of mucus, cholesterol crystals, and bilirubin granules.

soaps: chemical compounds that form between fatty acids and positively charged minerals.

soluble fibers: indigestible food components that readily dissolve in water and often impart gummy or gel-like characteristics to foods. An example is pectin from fruit, which is used to thicken jellies.

Special Supplemental Food Program for Women, Infants, and Children (WIC): a high quality, cost-effective health care and nutrition services program administered by the U.S. Department of Agriculture for low-income women, infants, and children who are nutritionally at risk. WIC provides supplemental foods, nutrition education, and referrals to health care and other social services.

sphincter (SFINK-ter): a circular muscle surrounding and able to close a body opening.

spina (SPY-nah) **bifida** (BIFF-ih-dah): one of the most common types of neural tube defects; characterized by the incomplete closure of the spinal cord and its bony encasement.

standard diet: a diet that includes all foods and meets the nutrient needs of healthy people; sometimes called a *regular diet.*

standard formulas: general-purpose enteral formulas that contain mostly intact proteins and polysaccharides; also called *polymeric formulas.*

starch: a plant polysaccharide composed of glucose and digestible by human beings.

steatorrhea (stee-AH-tor-REE-ah): excessive fat in the stools resulting from fat malabsorption; characterized by stools that are loose, frothy, and foul smelling due to a high fat content.

sterile: free of microorganisms, such as bacteria.

steroids (STARE-oids): medications used to reduce tissue inflammation, to suppress the immune response, or to replace certain steroid hormones in people who cannot synthesize them.

sterol esters: compounds, derived from plants, that belong to the sterol family of lipids and have been shown experimentally to reduce blood cholesterol when consumed in place of other fats in a low-saturated fat diet.

sterols: one of the main classes of lipids; includes cholesterol, vitamin D, and the sex hormones (such as testosterone).

stoma (STOE-ma): a surgical opening made in the abdominal wall.

stress response: the chemical and physical changes that occur within the body during stress.

stricture: abnormal narrowing of a passageway due to inflammation, scarring, or other structural changes.

stroke volume: the amount of oxygenated blood ejected from the heart toward body tissues at each beat.

structure-function claims: statements that describe how a product may affect a structure or function of the body; for example, "calcium builds strong bones." Structure-function claims do not require FDA authorization.

struvite (STROO-vite): crystals of magnesium ammonium phosphate.

sucrose: a disaccharide composed of glucose and fructose; commonly known as *table sugar, beet sugar,* or *cane sugar.*

sugar alcohols: sugarlike compounds. Like sugars, they are sweet to taste but yield 2 to 3 kcal per gram, slightly less than sucrose. Examples are maltitol, mannitol, sorbitol, isomalt, lactitol, and xylitol.

systemic (sih-STEM-ic): relating to the entire body.

systemic inflammatory response syndrome (SIRS): a whole-body response to acute inflammation; characterized by raised heart and respiratory rates, abnormal white blood cell counts, and altered body temperature.

tannins: compounds in tea (especially black tea) and coffee that bind iron.

TCA cycle or **tricarboxylic** (try-car-box-ILL-ick) **acid cycle:** a series of metabolic reactions that break down molecules of acetyl CoA to carbon dioxide and hydrogen atoms; also called the *Kreb's cycle* after the biochemist who elucidated its reactions.

teratogenic (ter-AT-oh-jen-ik): causing abnormal fetal development and birth defects

thermic effect of food: an estimation of the energy required to process food (digest, absorb, transport, metabolize, and store ingested nutrients).

thrush: a fungal infection of the mouth and throat, most often caused by *Candida albicans.*

tissue rejection: destruction of donor tissue by the recipient's immune system, which recognizes the donor cells as foreign.

tocopherol (tuh-KOFF-er-ol): a general term for several chemically related compounds, one of which has vitamin E activity.

Tolerable Upper Intake Levels (UL): a set of values reflecting the highest average daily nutrient intake levels that are likely to pose no risk of toxicity to almost all healthy individuals in a particular life stage and gender group. As intake increases above the UL, the potential risk of adverse health effects increases.

total nutrient admixture (TNA): a parenteral solution that contains dextrose, amino acids, and lipids; also called a *3-in-1* or an *all-in-one* solution.

total parenteral nutrition (TPN): a type of nutrition support in which intravenous feedings are delivered into a central vein.

training: regular practice of an activity, which leads to physical adaptations of the body with improvement in flexibility, strength, or endurance.

*trans***-fatty acids:** fatty acids in which the hydrogens next to the double bond are on opposite sides of the carbon chain.

transferrin (trans-FERR-in): the body's iron-carrying protein.

transient ischemic attacks: brief ischemic episodes that cause short-term neurological symptoms, such as blurred vision, slurred speech, numbness, paralysis, or difficulty speaking.

triglycerides (try-GLISS-er-rides): one of the main classes of lipids; the chief form of fat in foods and the major storage form of fat in the body; composed of glycerol with three fatty acids attached.

trimester: a period representing one-third of the term of gestation. A trimester is about 13 to 14 weeks.

tripeptide: three amino acids bonded together.

tube feedings: liquid formulas delivered through a tube placed in the stomach or intestine.

tubules: tubelike structures of the nephron that process filtrate during urine production. The tubules are surrounded by capillaries that reabsorb the substances retained by tubule cells.

tumor: an abnormal tissue mass that has no physiological function; also called a *neoplasm* (NEE-oh-plazm).

type 1 diabetes: the type of diabetes that accounts for 5 to 10 percent of diabetes cases and usually results from autoimmune destruction of pancreatic beta cells.

type 2 diabetes: the type of diabetes that accounts for 90 to 95% of diabetes cases and usually results from insulin resistance coupled with insufficient insulin secretion.

ulcerative colitis (ko-LYE-tis): an inflammatory bowel disease that involves the colon. Inflammation affects the mucosa and submucosa of the intestinal wall.

umbilical (um-BIL-ih-cul) **cord:** the ropelike structure through which the fetus's veins and arteries reach the placenta; the route of nourishment and oxygen into the fetus and the route of waste disposal from the fetus.

undernutrition: underconsumption of food energy or nutrients severe enough to cause disease or increased susceptibility to disease; a form of malnutrition.

unsaturated fatty acid: a fatty acid with one or more points of unsaturation where hydrogens are missing (includes monounsaturated and polyunsaturated fatty acids).

urea (yoo-REE-uh): the principal nitrogen-excretion product of protein metabolism.

USDA Food Guide: the USDA's food group plan for ensuring dietary adequacy that assigns foods to five major food groups.

uremia (you-REE-me-ah): abnormal accumulation of nitrogen-containing substances, especially urea, in the blood; also called *azotemia* (AZE-oh-TEE-me-ah).

uremic syndrome: the cluster of symptoms associated with a GFR below 15 mL/min, including uremia, anemia, bone disease, hormonal imbalances, bleeding impairment, increased cardiovascular disease risk, and reduced immunity.

uterus (YOO-ter-us): the womb, the muscular organ within which the infant develops before birth.

vagotomy (vay-GOT-oh-mee): surgery that severs the vagus nerve in order to suppress gastric acid secretion. This surgery may require a follow-up *pyloroplasty* procedure to allow stomach drainage.

vagus nerve: the cranial nerve that regulates hydrochloric acid secretion and peristalsis. Effects elsewhere in the body include regulation of heart rate and bronchiole constriction.

variety: eating a wide selection of foods within and among the major food groups (the opposite of monotony).

varices (VAH-rih-seez): abnormally dilated blood vessels (singular: *varix*).

vein: a vessel that carries blood back to the heart.

very-low-density lipoproteins (VLDL): the type of lipoproteins made primarily by liver cells to transport lipids to various tissues in the body; composed primarily of triglycerides.

villi (VILL-ee or VILL-eye): fingerlike projections from the folds of the small intestine. The singular form is *villus*.

viscous: a gel-like consistency.

vitamin A: a fat-soluble vitamin. Its three chemical forms are *retinol* (the alcohol form), *retinal* (the aldehyde form), and *retinoic acid* (the acid form).

vitamins: essential, nonkcaloric, organic nutrients needed in tiny amounts in the diet.

voluntary activities: the component of a person's daily energy expenditure that involves conscious and deliberate muscular work—walking, lifting, climbing, and other physical activities. Voluntary activities normally require less energy per day than basal metabolism does.

VO₂ max: the maximum rate of oxygen consumption by an individual (measured at sea level).

waist circumference: a measurement used to assess a person's abdominal fat.

wasting: the breakdown of muscle tissue that results from disease or malnutrition.

water balance: the balance between water intake and water excretion that keeps the body's water content constant.

water intoxication: the rare condition in which body water contents are too high. The symptoms may include confusion, convulsion, coma, and even death in extreme cases.

weight training: the use of free weights or weight machines to provide resistance for developing muscle strength and endurance. A person's own body weight may also be used to provide resistance, as when a person does push-ups, pull-ups, or sit-ups; also called *resistance training*.

wellness: maximum well-being; the top range of health states; the goal of the person who strives toward realizing his or her full potential physically, mentally, emotionally, spiritually, and socially.

whey protein: a by-product of cheese production; falsely promoted as increasing muscle mass. Whey is the watery part of milk that separates from the curds.

xerostomia (ZEE-roh-STOE-me-ah): dry mouth caused by reduced salivary flow.

zygote (ZY-goat): the product of the union of ovum and sperm; so-called for the first two weeks after fertilization.

Index

E

Early childhood, 347–57
  energy/nutrient needs, 347–50
Eating disorder, **163**
  combatting, 165
Eating patterns, establishment, 168
Eating plan, 181–3
Eclampsia, 320, **321**
Edema, 95, **97**, 631
Eggs
  carbohydrate content, 56
  fat content, 76
Eicosanoids, **657**
Eicosapentaenoic acid (EPA), 68, **69**
Electrolyte, 242, **243**
  losses/replacement, 291
  solutions, 242, **243**
Electron transport chain, **144**
Embolism, **602**, 604
Embolus, **602**, 604
Embryo, **309**
Embryonic development, stages, 309f
Emergency shelters, **402**
Emetic, **163**
Emotional distress, 685
Emotional state, eating, 5
Emotional stress, impact, 499
Emphysema, **662**
  case study, 665
Emulsifiers, 68, **69**, 122, **123**
Enamel, **60**
End-stage renal disease (ESRD),
  637
Energy
  balance, 149–53
  deficit, 147
  expenditure, 184
    components, **150**
    ranking, 151t
  imbalance, 146–9
  intake, 149
  measure, 7–8
  metabolism, 142–4, **143**
    pathways, 145f
  output, 149–52
  providing, 97
  recommendations, setting, 11
  requirements
    determination, 425–7
    estimation, 152–3
Energy balance
Energy-dense foods, consumption,
  188–9

Energy density, 8, **9**
  comparison, 183f
  reduction, 182–183
Energy-yielding nutrients, **7**
Engorgement, **324**
Enrichment, **216**
Enteral formulas, 458–60,
    Appendix G
  delivery, 466–7
  disease-specific formulas, 459
  hydrolized formulas, 459
  modular formulas, 459
  nutrient/energy densities, 459
  oral use, 460
  selection, 463–4
    flowchart, 464f
  standard formulas, 459
  tube feedings, 460–1
  types, 459
  volume/strength, 467
Enteral nutrition, **458**
  support, 458–71, 560–1
    case study, 469
Enteric-coated, 538, **539**
Enterostomy, 462, **463**
Environmental stimuli, 175
Environmental tobacco smoke
    (ETS // secondhand smoke), **321**
Enzymes, 94–5, **95**, **121**. *See also*
    Coenzymes; Digestive enzymes
  action, 94f
Epiglottis, 114, **115**
Epinephrine, **281**, 356, **357**
Epithelial cells, **204**
Epithelial tissue, **204**
Ergogenic, **301**
Ergogenic aides, substances
    (glossary), 303
Erythrocyte hemolysis, **211**
Erythrocyte protoporphyrin, 256,
    257
Erythropoiesis, **547**
  process, 548f
Erythropoietin, **547**, 630, **631**
Esophageal dysphagia, **491**
Esophageal reflux
    conditions/substances, associa-
      tion, 495t
Esophageal sphincter, **114**. *See also*
    Lower esophageal sphincter
Esophageal varices, appearance,
    556f
Esophagus, 114, **115**
  conditions, 490–7

Essential amino acids, 93–4, **95**. *See
    also* Conditionally essential
    amino acid
Essential fatty acids, **67**
Essential nutrients, **7**
Estimated Average Requirements
    (EAR), **10**
Estimated Energy Requirements
    (EER), **11**
  physical activity (PA) factors, 153
Ethical, **694**
Ethnic cuisines, 4t
Ethnic diets, **3**
Ethnic food choices, 23
Ethnic heritage/tradition, 3
Ethnic meals, 5f
Exercise, 274, **275**
Extracellular fluid, 244, **245**

F

Fad diets
  claims/truths, 194t–195t
  identification guidelines, 199t
Faith healing, **454**, 455
Family gatherings, 5f
Fashion, criterion, 154
Fasting, 147–9
  hazards, 148–9
  schematic, 148f
Fasting hypoglycemia, 588
Fat cells
  development, 174, 174f
  enlargement, 146f
  illustration, 64f
Fat replacers, **79**, 79–80
Fat-restricted diet, 534t
  following, 535
Fats, 7, 64, **65**. *See also* Artificial fats;
    Hard fat; Soft fat
  balancing, 182
  emulsification, bile (usage), 122f
  excess, 146
  health effects, 70–4
  heart healthy choices, 78
  intake, reduction, 77–80
  intakes, recommendations, 70–4
  kCalories, relationship, 77
  malabsorption, 504, 533–5
    consequences, 533f
    dietary adjustments, 533–5
  presence, 598
  recommendations, 73–4

U.S. population
    aging, 380f
    metabolic syndrome, prevalence, 626f
Usual body weight, percentage (%UBW) estimation/evaluation, 417
Uterus, **307**

V

Vagotomy, **501**
Vagus nerve, **501**
Valine, **95**
Values, impact, 3–4
Varices (varix), **555**
Variety, **15**
Vascular system, 125
Vegans, **109**
    vitamin B$_{12}$ deficiency, 222
Vegetables
    carbohydrate content, 55
    fat content, 76–7
    heart healthy choices, 78
    phytochemical content, 235f
    subgroups, weekly amounts (recommendation), 21t
Vegetarians. *See* Lacto-ovo vegetarians; Lacto-vegetarians; Semivegetarians
    food guide, 23
    food pyramid, example, 110f
    protein, obtaining, 105f
Vegetarian terms, glossary, 109
Vein, **125**
Venous access, 471–4
Very-low-density lipoprotein (VLDL), **126**, 126–7, 552
    elevation, 605
Vibrio, **131t**
Villi, **123**. *See also* Microvilli; Small intestine
    appearance, 540f
Viral hepatitis, 553
Viscous, **41**
Vision, vitamin A (role), 204, 204f
Vitamin A, 203–8
    beta-carotene, relationship, 207f
    deficiency, 205–6
    toxicity, 206
Vitamin B$_6$, 218–9
    deficiency, 218
    metabolic roles, 218

recommendations, 219
    toxicity, 218–9
Vitamin B$_{12}$, 220–2
    absorption, 221
    cell division, relationship, 220
    nervous system, relationship, 220–1
Vitamin C, 222–5
    antihistamine, possibility, 223
    antioxidant role, 223
    deficiency, 223
    intakes, recommendations, 224
    iron absorption, relationship, 224
    limits, 224
    metabolic roles, 222
    sources, 225f
    special needs, 224
    toxicity, 223–4
Vitamin D, 208–10
    actions, 208
    deficiency, 209
    metabolic conversions, 208
    sun derivation, 209–10
    toxicity, 209
Vitamin E, 210–2, 288–9
    antioxidant effect, 210–1
    deficiency, 211
    myth, 211
    toxicity, 211–2
Vitamin K, 212–3
    deficiency, 213
    food sources, 213f
    intestinal synthesis, 213
    toxicity, 213
Vitamins, 7, **203**. *See also* Fat-soluble vitamins; Water-soluble vitamins
    activity support, 288–9
    names, 202t
    overview, 202–3
    supplements, 288
Voluntary activities, **149**
Vomiting, 498. *See also* Intractable vomiting
    dietary interventions, 498
    treatment, 498

W

Waist circumference, **157**, 418
    measurement, 157f
Wasting, 422, **423**, **656**
    control, 686–7
    medications, 680
Water, 7, 291
    adequacy, 183

balance, 240–2, **241**
    ranking, 241t
    body fluids, relationship, 240–3
    excretion regulation, 240–1
    intake regulation, 240
    intoxication, 240, **241**
    maximum, 241
    needs, meeting, 468
    recommendations/sources, 242
Water-soluble vitamins, 214–26
    summary, 225t–226t
Web sites, reliability (evaluation), 35t
Weight
    cycling, **163**
    gain, strategies, 188–9
    maintenance, 186
    measurement, 417
    reduction, 610
Weight loss
    diet, recommendations, 181t
    gimmicks, 178
    scams, identification guidelines, 199t
    strategies, 180–7
Weight management, 171-199
    carbohydrates, impact, 52
    strategies, 197
Weight training, **189**, **278**, 278–9
Wellness, 2, **3**
Wheat gluten, 540, **541**
Whey protein, **102**, **103**
Women, thighs (cross sections), 388f

X

Xerostomia, 490, **491**

Y

Yersiniosis, **132t**
Yogurt
    carbohydrate content, 55–6
    fat content, 76

Z

Zinc, 259–61
    deficiency, 259–60
    food sources, 260–1
    recommendations, 260–1
    sources, 260f
    toxicity, 260
Zygote, **308**

Photo and Art Credits

Credits

This page constitutes an extension of the copyright page. We have made every effort to trace the ownership of all copyrighted material and to secure permission from copyright holders. In the event of any question arising as to the use of any material, we will be pleased to make the necessary corrections in future printings. Thanks are due to the following authors, publishers, and agents for permission to use the material indicated.

Chapter 1. 13: Source: Health People 2010, www.healthypeople.gov; **14:** Source: National Center for Health Statistics, 2006; **16:** Source: The Dietary Guidelines for Americans 2005, available at www.healthierus.gov/dietaryguidelines; **17:** These guidelines derive from Action Towards Healthy Eating - Canada's Guidelines for Healthy Eating and Recommended Strategies for Implementation; **18:** Source: USDA Food Guide; **24:** USDA, 2005; **34:** Adapted from Position of the American Dietetic Association: Food and nutrition misinformation, Journal of the American Dietetic Association, 106 (2006): 601–607.

Chapter 4. 110: Copyright © GC Nutrition Council, 2006, adapted from USDA 2005 Dietary Guidelines and www.mypyramid.gov. Copies can be ordered from 301-680-6717.

Chapter 5. 135: Courtesy of the USDA.

Chapter 6. 157: Source: National Heart, Lung, and Blood Institute, National Institutes of Health, The Practical Guide: Identification, Evaluation, and Treatment of Overweight and Obesity in Adults, NIH publication no. 00-4084 (Washington, D.C.: Government Printing Office, 2000), p. 9; **158:** Source: www.cdc.nccdphp/dnpa/obesity/trend/maps/index; **160:** Source: National Heart, Lung, and Blood Institute, National Institutes of Health, The Practical Guide: Identification, Evaluation, and Treatment of Overweight and Obesity in Adults, NIH publication 00-4084 (Washington, D.C.: Government Printing Office, 2000); **168:** Reprinted with permission from the Diagnostic and Statistical Manual of Mental Disorders, 4th ed., Text Revision (Copyright 2000). American Psychiatric Association.

Chapter 7. 181: Source: National Heart, Lung, and Blood Institutes of Health, "The Practical Guide: Identification, Evaluation, and Treatment of Overweight and Obesity in Adults", NIH publication no. 00-4084 (Washington, DC: Government Printing Office, 2000) p. 27.

Chapter 8. 235: Source: Data from P.M. Kris-Etherton and coauthors. Bioactive compounds in nutrition and health—research methodologies for establishing biological function: the antioxidant and anti-inflammatory effects of flavonoids on atherosclerosis, 'Annual Review of Nutrition' 24 (2004): 511-538; **236:** Position of the American Dietetic Association: Functional foods, published in Journal of the American Dietetic Association 104 (2004): 814–826, Copyright Elsevier 2004.

Chapter 10. 277: Adapted from American College of Sports Medicine, General principles of exercise prescription, in ASCM's Guidelines for Exercise Testing and Prescription, 7th ed. (Philadelphia, PA: Lippincott Williams & Wilkins, 2006), pp. 133–173; **291:** Adapted from R. Murray, 'Fluid, electrolytes, and exercise' in "Sports Nutrition: A Practice Manual for Professionals", 4th ed., edited by M. Dunford (Chicago: The American Dietetic Association, 2005) pp. 94–115. Copyright © American Dietetic Association. Adapted with permission.

Chapter 11. 307: From Donna L. Hoyert, T.J. Mathews, Fay Menacker, Donna M. Strobino, and Bernard Guyer, "Annual Summary of Vital Statistics: 2004", in 'Pediatrics' (Jan. 2006); 117: pp. 168–183; **333:** Sources: U.S. Department of Health and Human Services, Office on Women's Health, "Breastfeeding: HHS Blueprint for Action on Breastfeeding" (Washington, DC: U.S. Department of Health and Human Services, 2000); United Nations Children's Fund and World Health Organization, "The UNICEF/Baby-Friendly Hospital Initiative: Ten Steps to Successful Breastfeeding" (New York: UNICEF, 1992).

Chapter 12. 361: Source: U.S. Department of Agriculture, National School Lunch Program Regulations, revised January 1, 1998; **365:** From S.A. French, B.H. Lin, and J.F. Guthrie, "National trends in soft drink consumption among adolescents age 6–17 years; Prevalence, amounts, and sources, 1977/1978 to 1994/1998" in 'Journal of the American Dietetic Association' 103 (2003): 1326–1331.

Chapter 13. 382: Source: Exercise: A Guide from the National Institute on Aging, www.nia.nih.gov, accessed May 2005. **401:** Source: United States Department of Agriculture, Household Food Security in the United States, 2005, available at www.ers.usda.gov/publications/err29; **402:** Source: Economic Research Service. U. S. Department of Agriculture, www.ers.usda.gov/publications/fanrr29, posted November 2006 and visited on November 27, 2006.

Chapter 14. 419: Reprinted from L. Goldman and D. Ausiello, eds., "Cecil Textbook of Medicine", Copyright © 2004, with permission from Philadelphia: Saunders; **424:** Adapted from "Nutrition Care Manual" (Chicago: American Dietetic Association, 2006) and "Manual of Clinical Dietetics", 6th ed. (Chicago: American Dietetic Association, 2000). Copyright © American Dietetic Association. Adapted with permission; **433:** Source: Adapted from "A Primer: From DNA to Life," Human Genome Project, U. S. Department of Energy Office of Science, www.ornl.gove/sci/techresources/Human_Genome/primer_pic.shtml.

Chapter 15. 442: From M. Rotblatt and I. Ziment, "Evidence-Based Herbal Medicine" (Philadelphia: Hanley & Belfus, Inc., 2002); **449:** From D.G. Bailey, M.O. Arnold, and J.D. Spence, "Inhibitors in the diet: Grapefruit juice-drug interactions" in R.H. Levy and coeditors, "Metabolic Drug Interactions" (Philadelphia: Lippincott Williams & Wilkins, 2000) pp. 661–669.

Chapter 18. 518: Adapted from "Manual of Clinical Dietetics", 6th ed. (Chicago: American Dietetic Association, 2000) p. 423. Copyright © American Dietetic Association. Adapted with permission.

Chapter 19. 534: Adapted from "Manual of Clinical Dietetics", 6th ed. (Chicago: American Dietetic Association, 2000), Chapter 58. Copyright © American Dietetic Association. Adapted with permission; **548:** From "Human Physiology, From Cells to Systems" (with CD-ROM and InfoTrac) 5th edition by SHERWOOD, 2004. Reprinted with permission of Brooks/Cole, a division of Thomson Learning: www.thomsonrights.com. Fax 800-730-2215.

Chapter 21. 598: Source: K. Foster-Powell and coauthors, "International table of Glycemic index and glycemic load values" in 'American Journal of Clinical Nutrition' 76 (2002): 5–56.

Chapter 22. 602: Source: Based on preliminary data for 2003 from the Centers for Disease Control/National Center for Health Statistics and the National Heart, Lung, and Blood Institute; **606:** Source: Expert Panel on Detection, Evaluation, and Treatment of High Blood Cholesterol in Adults (Adult Treatment Panel III), Third Report of the National Cholesterol Education Program (NCEP), NIH publication no. 02-5215 (Bethesda, MD: National Heart, Lung, and Blood Institute, 2002), pp. II-15 to II-20; **608:** Source: Adapted from Expert Panel on Detection, Evaluation, and Treatment of High Blood Cholesterol in Adults (Adult Treatment Panel III), Third Report of the National Cholesterol Education Program (NCEP), NIH publicaiton no. 02-5215 (Bethesda, MD: National Heart, Lung, and Blood Institute, 2002), section III; **617:** Source: Adapted from "Reference Card from the Seventh Report of the Joint National Committee on Prevention, Detection, Evaluation, and Treatment of High Blood Pressure" (JNC 7), NIH publication no. 03-5231 (Bethesda, MD: National Institutes of Health, National Heart, Lung, and Blood Institute, and National High Blood Pressure Education Program, May 2003).

Chapter 23. 637: From A.S. Levey and coauthors, National Kidney Foundation practice guidelines for chronic kidney disease: Evaluation, classification, and stratification, 'Annals of Internal Medicine' 139 (2003): 137–147; **639:** Published the Journal of the American Dietetic Association 104 (2004): 404-409, J. A. Beto and V. K. Bansal, Medical nutrition therapy in chronic kidney failure: Integrating clinical practice guidelines. © 2004, with permission of the American Dietetic Association and Elsevier; **644:** Published the Journal of the American Dietetic Association 104 (2004): 404–409, J. A. Beto and V. K. Bansal, Medical nutrition therapy in chronic kidney failure: Integrating clinical practice guidelines. © 2004, with permission of the American Dietetic Association and Elsevier; **646:** From J.A.T. Pennington, "Bowes and Church's Food Values of Portions Commonly Used" (Philadelphia: J.B. Lippincott, 1994) p. 387; **651:** Reprinted with permission of Oklahoma Medical Research Foundation. (www.omrf.org.)

Chapter 24. 663, center: Source: "Medical Encyclopedia: Bronchitis and Normal Condition in Tertiary Bronchus" in 'Medicine Plus', 2005, A.D.A.M., Inc. (http://www.nim_nih.gov/medlineplus/ency/imagepages/19357.htm); **663, center right:** Adapted from "Causes and Risk Factors for Emphysema", Emphysema-Symptoms.com. Copyright morefocus.com. All rights reserved. http://www.emphysema-symptoms.com/html/emphysema-causes.php3; **663, center top:** Adapted from "Medical Encyclopedia: Bronchitis and Normal Condition in Tertiary Bronchus" in 'Medicine Plus', 2005, A.D.A.M., Inc. (http://www.nim_nih.gov/medlineplus/ency/imagepages/19357.htm); **663, top:** Based on a drawing in Carol Mattson Porth, "Pathophysiology", 5th ed. (Lippincott Williams and Wilkins, 1998).

Chapter 25. 675: Reprinted from W. J. Blot, 'Epidemiology of Cancer' in L. Goldman and D. Ausiello, eds., "Cecil Textbook of Medicine", pp. 1116-1120, Copyright © 2004, with permission from Philadelphia: Saunders; **677:** Source: Table 1, Page 256, L. H. Kushi and coauthors, American Cancer Society guidelines on nutrition and physical activity for cancer prevention: Reducing the risk of cancer with healthy food choices and physical activity, CA: A Cancer Journal for Clinicians 56 (2006): 254–281; **684:** Source: UNAIDS/WHO, AIDS epidemic update: December 2006, http://data.unaids.org/pub/EpiReport/2006/2006_Epi_en.pdf. site visited January 6, 2007.

Daily Values for Food Labels

The Daily Values are standard values developed by the Food and Drug Administration (FDA) for use on food labels. Daily Values for protein, vitamins, and minerals reflect average allowances based on the RDA. Daily Values for nutrients and food components, such as fat and fiber, that do not have an established RDA but do have important relationships with health are based on recommended calculation factors as noted.

Nutrient	Amount
Protein[a]	50 g
Thiamin	1.5 mg
Riboflavin	1.7 mg
Niacin	20 mg NE
Biotin	300 µg
Pantothenic acid	10 mg
Vitamin B_6	2 mg
Folate	400 µg
Vitamin B_{12}	6 µg
Vitamin C	60 mg
Vitamin A	5000 IU
Vitamin D	400 IU
Vitamin E	30 IU
Vitamin K	80 µg
Calcium	1000 mg
Iron	18 mg
Zinc	15 mg
Iodine	150 µg
Copper	2 mg
Chromium	120 µg
Selenium	70 µg
Molybdenum	75 µg
Manganese	2 mg
Chloride	3400 mg
Magnesium	400 mg
Phosphorus	1000 mg

Food Component	Amount	Calculation Factors
Fat	65 g	30% of kcalories
Saturated fat	20 g	10% of kcalories
Cholesterol	300 mg	Same regardless of kcalories
Carbohydrate (total)	300 g	60% of kcalories
Fiber	25 g	11.5 g per 1000 kcalories
Protein	50 g	10% of kcalories
Sodium	2400 mg	Same regardless of kcalories
Potassium	3500 mg	Same regardless of kcalories

Note: Daily Values were established for adults and children over 4 years old. The values for energy-yielding nutrients are based on 2000 kcalories a day.

Glossary of Nutrient Measures

kcal: kcalories; a unit by which energy is measured (Chapter 1 provides more details).

g: grams; a unit of weight equivalent to about 0.03 ounces.

mg: milligrams; one-thousandth of a gram.

µg: micrograms; one-millionth of a gram.

IU: international units; an old measure of vitamin activity determined by biological methods (as opposed to new measures that are determined by direct chemical analyses). Many fortified foods and supplements use IU on their labels.

- For vitamin A, 1 IU = 0.3 µg retinol, 3.6 µg β-carotene, or 7.2 µg other vitamin A carotenoids.
- For vitamin D, 1 IU = 0.025 µg cholecalciferol.
- For vitamin E, 1 IU = 0.67 natural α-tocopherol (other conversion factors are used for different forms of vitamin E).

mg NE: milligrams niacin equivalents; a measure of niacin activity (Chapter 8 provides more details).

- 1 NE = 1 mg niacin.
  - = 60 mg tryptophan (an amino acid).

µg DFE: micrograms dietary folate equivalents; a measure of folate activity (Chapter 8 provides more details).

- 1 µg DFE = 1 µg food folate.
  - = 0.6 µg fortified food or supplement folate.
  - = 0.5 µg supplement folate taken on an empty stomach.

µg RAE: micrograms retinol activity equivalents; a measure of vitamin A activity (Chapter 8 provides more details).

- 1 µg RE = 1 µg retinol.
  - = 12 µg β-carotene.
  - = 24 µg other vitamin A carotenoids.

[a]The Daily Values for protein vary for different groups of people: pregnant women, 60 g; nursing mothers, 65 g; infants under 1 year, 14 g; children 1 to 4 years, 16 g.

Body Mass Index (BMI)

Height	18	19	20	21	22	23	24	25	26	27	28	29	30	31	32	33	34	35	36	37	38	39	40
											Body Weight (pounds)												
4'10"	86	91	96	100	105	110	115	119	124	129	134	138	143	148	153	158	162	167	172	177	181	186	191
4'11"	89	94	99	104	109	114	119	124	128	133	138	143	148	153	158	163	168	173	178	183	188	193	198
5'0"	92	97	102	107	112	118	123	128	133	138	143	148	153	158	163	168	174	179	184	189	194	199	204
5'1"	95	100	106	111	116	122	127	132	137	143	148	153	158	164	169	174	180	185	190	195	201	206	211
5'2"	98	104	109	115	120	126	131	136	142	147	153	158	164	169	175	180	186	191	196	202	207	213	218
5'3"	102	107	113	118	124	130	135	141	146	152	158	163	169	175	180	186	191	197	203	208	214	220	225
5'4"	105	110	116	122	128	134	140	145	151	157	163	169	174	180	186	192	197	204	209	215	221	227	232
5'5"	108	114	120	126	132	138	144	150	156	162	168	174	180	186	192	198	204	210	216	222	228	234	240
5'6"	112	118	124	130	136	142	148	155	161	167	173	179	186	192	198	204	210	216	223	229	235	241	247
5'7"	115	121	127	134	140	146	153	159	166	172	178	185	191	198	204	211	217	223	230	236	242	249	255
5'8"	118	125	131	138	144	151	158	164	171	177	184	190	197	203	210	216	223	230	236	243	249	256	262
5'9"	122	128	135	142	149	155	162	169	176	182	189	196	203	209	216	223	230	236	243	250	257	263	270
5'10"	126	132	139	146	153	160	167	174	181	188	195	202	209	216	222	229	236	243	250	257	264	271	278
5'11"	129	136	143	150	157	165	172	179	186	193	200	208	215	222	229	236	243	250	257	265	272	279	286
6'0"	132	140	147	154	162	169	177	184	191	199	206	213	221	228	235	242	250	258	265	272	279	287	294
6'1"	136	144	151	159	166	174	182	189	197	204	212	219	227	235	242	250	257	265	272	280	288	295	302
6'2"	141	148	155	163	171	179	186	194	202	210	218	225	233	241	249	256	264	272	280	287	295	303	311
6'3"	144	152	160	168	176	184	192	200	208	216	224	232	240	248	256	264	272	279	287	295	303	311	319
6'4"	148	156	164	172	180	189	197	205	213	221	230	238	246	254	263	271	279	287	295	304	312	320	328
6'5"	151	160	168	176	185	193	202	210	218	227	235	244	252	261	269	277	286	294	303	311	319	328	336
6'6"	155	164	172	181	190	198	207	216	224	233	241	250	259	267	276	284	293	302	310	319	328	336	345

Under-weight (<18.5)   **Healthy Weight** (18.5–24.9)   **Overweight** (25–29.9)   **Obese** (≥30)

Find your height along the left-hand column and look across the row until you find the number that is closest to your weight. The number at the top of that column identifies your BMI. Chapter 6 describes how BMI correlates with disease risks and defines obesity. The area shaded in green represents healthy weight ranges.